中国探月工程科学探测成果系列丛书

绕月探测工程月球科学与探测技术研究

欧阳自远　李春来　主编

科学出版社
北京

内 容 简 介

本文集主要收录了自绕月探测工程立项以来月球探测有效载荷的设计与定标、数据处理与管理方法所获得的研究成果,共收录论文88篇。研究成果涵盖了综述、CCD立体相机、激光高度计、干涉成像光谱仪、伽马与X射线谱仪、微波探测仪和其他研究成果,是我国首次月球探测有效载荷、探测技术与数据接收处理研究成果的系统总结。

本文集所收集论文源自多种学术期刊,各源刊格式标准可能不统一,本着尊重历史、忠于原著的精神,所用物理量单位、符号、图例、参考文献等尽量保留了原文风貌。

本文集可供月球科学、深空探测等相关专业的大专院校教师、研究生与高年级本科生,科研院所相关科研与技术人员参考。

图书在版编目(CIP)数据

绕月探测工程月球科学与探测技术研究/欧阳自远,李春来主编. —北京:科学出版社,2015.2
(中国探月工程科学探测成果系列丛书)
ISBN 978-7-03-043011-3

I. ①绕⋯ II. ①欧⋯ ②李⋯ III. ①月球探索–文集 IV. ①V1-53

中国版本图书馆 CIP 数据核字 (2015) 第 009065 号

责任编辑:韩　鹏　宋云华　张井飞/责任校对:张小霞　赵桂芬
责任印制:赵　博/封面设计:耕者设计工作室

科学出版社　出版
北京东黄城根北街16号
邮政编码: 100717
http://www.sciencep.com

三河市春园印刷有限公司印刷
科学出版社发行　各地新华书店经销
*

2015年2月第 一 版　　开本:889×1194　1/16
2025年3月第二次印刷　印张:42 1/4　插页:10
字数:1 287 000

定价: 389.00元
(如有印装质量问题,我社负责调换)

《中国探月工程科学探测成果系列丛书》出版委员会

顾问委员会

名誉主任	路甬祥　韩启德　万　钢
主　　任	陈求发
副 主 任	江绵恒　曹健林　刘东奎　牛红光
	阴和俊　马兴瑞　熊群力
委　　员	宋　健　徐冠华　张建启　栾恩杰
	孙家栋　欧阳自远　陈炳忠　姜景山
	龙乐豪　吴伟仁

出版编辑委员会

名誉主任	栾恩杰　孙家栋
主　　任	欧阳自远
副 主 任	严　俊　吴志坚　李春来
委　　员	胡　浩　张荣桥　廖小罕　董永初
	刘晓群　艾国祥　刘先林　王任享
	高　俊　李德仁　陈俊勇　王家耀
	童庆禧　于登云　孙辉先　钱卫平
	裴照宇　吴　季　洪晓瑜　王建宇
	王焕玉　赵葆常　常　进　邹永廖
	刘建忠

《绕月探测工程科学目标专题研究》
编辑委员会

主　　编　欧阳自远　李春来

副 主 编　邹永廖　左　维

编　　委　郑永春　徐　琳　王晓倩　耿　良
　　　　　付　强　肖　媛　邢丽萍

《中国探月工程科学探测成果系列丛书》序言

月球是地球唯一的天然卫星。40 多亿年以来,月球是地球的忠实伴侣,伴随着地球共同经历荒古的演化过程,抵御小天体对地球的撞击,掀起汹涌澎湃的海洋潮汐。月球将圣洁的光辉洒向大地,自古以来,激起人们无限的遐想和憧憬,萌发出各种神话传说、宗教信仰、哲学思想、文学艺术和风俗传统,并为古代的历法编制、农耕时令和社会发展发挥过重要作用。

1609 年,伽利略将刚发明的望远镜对准月球进行观测,标志着现代天文观测的开始。1957 年,苏联在人造地球卫星发射成功后,即将探索外层空间的雄心瞄准了月球,于 1959 年成功地发射了第一颗月球探测器,开创了人类探测太阳系的先河。1969 年 7 月,美国阿波罗 11 号飞船成功登陆月球,实现了人类的登月梦想。

飞出地球,探索月球,也是中华民族的千年夙愿。2007 年 10 月 24 日,嫦娥一号满载着中华儿女的梦想奔向月球,11 月 5 日成功实现了绕月飞行;11 月 20—26 日,嫦娥一号成功获取并发布了第一幅月面图像,标志着我国首次月球探测工程取得了圆满成功。2007 年 12 月 12 日,胡锦涛总书记在庆祝我国首次月球探测工程圆满成功大会上发表的重要讲话指出:"实施月球探测工程,是党中央、国务院、中央军委着眼我国社会主义现代化建设全局,把握世界科技发展大势,为推动我国航天事业发展、促进我国科技进步和创新、提高我国综合国力作出的一项重大战略决策","我国首次月球探测工程的成功,是继人造地球卫星、载人航天飞行取得成功之后我国航天事业发展的又一座里程碑,实现了中华民族的千年奔月梦想,开启了中国人走向深空探索宇宙奥秘的时代,标志着我国已经进入世界具有深空探测能力的国家行列。这是我国推进自主创新、建设创新型国家取得的又一标志性成果,是中华民族在攀登世界科技高峰征程上实现的又一历史性跨越,是中华民族为人类和平开发利用外层空间作出的又一重大贡献。"

我国首次月球探测工程的成功实施,不仅突破了一大批具有自主知识产权的核心技术和关键技术,取得了一系列重大科技创新成果,也带动了我国基础前沿研究和应用研究若干领域的深入发展,推动了信息技术和工业技术进步,促进了众多学科的交叉和融合。对未知领域的探索,是人类社会发展进步的不懈追求。深空探测是航天高技术进步的重要推动力,更是人类探索太阳系、认知宇宙的主要途径,被赋予了明确的科学目标和艰巨的科学探索重任。绕月探测工程是我国开展深空探测的第一步,嫦娥一号是我国的第一个飞出地球的探测器,携带了 8 套科学探测仪器,经过一年四个月的在轨探测,获得了海量科学探测数据,取得了一系列的科学研究成果。同时,在国家中长期科技发展规划中,月球探测的后续工程已经明确,我国将向月球发射更多的轨道器、着陆器、月球车,并实现自动采集月球样品返回地球,持续获取月球探测数据甚至月球样品,必将持续产生大量、系统的科学研究成果。将我国月球探测的研究成果以系列丛书为平台集中体现，非常必要，也很有意义。

无限的未知世界，深邃的太空，是科学家遐想和探索的天地。我们相信，《中国探月工程科学探测成果系列丛书》能够承载广大航天科技工作者空间探索的累累硕果，推动空间科学、行星科学、月球科学与地球科学的交融和蓬勃发展，丰富我们对客观世界的科学认知，促进科学技术更好地服务国家、服务人民、服务人类。

2010 年 1 月 22 日

目　　录

《中国探月工程科学探测成果系列丛书》序言

第一部分　综　　述

嫦娥一号月球探测卫星研制综述 ··· 3
信息化技术在嫦娥一号卫星研制中的应用 ·· 10
嫦娥一号月球探测卫星技术特点分析 ··· 16
Policy Making in China's Space Program: A History and Analysis of the Chang'E Lunar Orbiter Project ········ 22
Introduction to the Payloads and the Initial Observation Results of Chang'E-1 ······················· 35
Advances in Lunar Exploration Detectors ·· 49
Scientific Objectives and Payloads of Chang'E-1 Lunar Satellite ··· 57

第二部分　CCD 立体相机

绕月探测工程 CCD 立体相机的实验室辐射定标 ·· 67
嫦娥一号卫星 CCD 立体相机的设计与在轨运行 ··· 72
月球卫星 CE-1 三线阵影像数据的解算试验 ·· 81
三线阵 CCD 摄影测量理论在月球探测中的应用 ··· 85
月球卫星三线阵 CCD 影像 EFP 光束法空中三角测量 ·· 89

第三部分　激光高度计

嫦娥一号卫星载激光高度计 ·· 97
激光测距技术在空间的应用 ··· 106
先进激光雷达探测技术研究进展 ·· 112

第四部分　干涉成像光谱仪

实体 Sagnac 干涉仪的设计 ··· 123
嫦娥一号卫星成像光谱仪光学系统设计与在轨评估 ·· 129
嫦娥一号卫星干涉成像光谱仪现场性能检测实验 ··· 135
嫦娥一号卫星干涉成像光谱仪电子学设计 ··· 142
嫦娥一号卫星干涉成像光谱仪时序设计 ·· 151
嫦娥一号 IIM 数据应用处理流程分析 ··· 157
干涉成像光谱仪中宽谱段傅氏光学系统设计 ·· 164

嫦娥一号干涉成像光谱仪的定标	171
嫦娥一号 IIM 数据绝对定标与初步应用	180
嫦娥一号 IIM 数据定标的改进方法	187
NASVD 方法在 CE1-GRS 谱线分析中的应用研究	197
中红外光谱在月球探测中的应用	203
月球紫外-可见-近红外反射光谱的基本特征及解析方法	209
A Preliminary Experience in the Use of Chang'E-1 IIM Data	216
Absolute Calibration of the Chang'E-1 IIM Camera and its Preliminary Application	228

第五部分　伽马与 X 射线谱仪

嫦娥一号卫星 X 射线谱仪的性能模拟	239
嫦娥一号卫星 X 射线谱仪磁屏蔽设计与模拟计算	245
空间 X 射线成像谱仪系统及其软件研制	249
Gamma-ray Detector on Board Lunar Mission Chang'E-1	254
Time Series Data Correction for the Chang'E-1 Gamma-ray Spectrometer	259
Automatically Smoothing the Spectroscopic Data by Cubic B-Spline Basis Functions	272

第六部分　微波探测仪

嫦娥 1 号卫星微波探月技术机理和应用研究	279
利用微波辐射计对月壤厚度进行研究	287
月表温度剖面对于"嫦娥一号"卫星微波探测仪探测亮温影响的模拟研究	293
地基雷达技术及其在太阳系天体探测中的应用	307
天然掺杂铁氧体的电磁参数调控机制分析及其在吸波材料中的应用	315
空间微波环境对"嫦娥一号"微波探测仪在轨定标影响分析	323
基于 SVM 和"嫦娥一号"数据的月球表面亮温分布	332
Microwave Brightness Temperature Imaging and Dielectric Properties of Lunar Soil	341
The Analysis of Affections to the Cold Space Calibration Source of Chang'E-1 Payload Microwave Detector	348

第七部分　其　他

嫦娥一号卫星的制导、导航与控制	359
嫦娥一号卫星热设计及计算分析	365
"嫦娥一号"月球探测卫星真空热试验的初步思路	372
嫦娥一号卫星定向天线动力学仿真分析	379
绕月探测工程地面接收站通用解调处理机性能测试	385
基于 SAN 的绕月探测工程数据存储系统架构的设计与实现	394
基于复用的软件构架评估方法及其在嫦娥工程中的应用	400

月球探测计划中影像数据的格式 406
绕月探测工程卫星数据的存储与管理 413
单频干扰下BPSK接收性能恶化分析及应用 420
嫦娥一号卫星星地时差校正量计算方法研究 427
嫦娥一号卫星热控系统及其特点 432
嫦娥一号绕月探测器轨道投入过程实时监测判定的原理与技术实现 438
月球重力场对"嫦娥一号"近月轨道的影响 446
嫦娥一号绕月卫星对月球重力场模型的优化 455
嫦娥一号绕月探测卫星精密定轨实现 464
深空探测用数字开环多普勒技术初步研制及其在嫦娥一号探测任务中的应用 471
嫦娥一号卫星双轴天线轨迹规划 482
"嫦娥一号"卫星轨控标定方法研究与实现 490
嫦娥-1卫星绕月捕获分析与快速判断 496
嫦娥一号卫星的地月转移变轨控制 507
瞬时状态归算用于嫦娥一号卫星关键轨道段监测 515
环月卫星可见时段的计算和分析 524
"嫦娥"卫星绕月飞行的星载激光定轨法 530
"嫦娥一号"卫星的调相轨道设计 535
基于大倾角卫星轨道跟踪数据的月球重力场模型仿真解算 542
射电望远镜指向误差的广义延拓插值修正方法 550
"嫦娥"卫星绕月飞行轨道的激光测定法 555
嫦娥一号卫星热控设计中热管的应用及验证 559
嫦娥一号月球卫星缩短阴影时间的分析与实现 567
嫦娥一号卫星数据高可靠性保护设计 575
"嫦娥一号"任务全球地形实时仿真技术研究 579
星载毫米波辐射计地面定标实验 586
月球软着陆点的选择与几个预选点的初步对比分析 591
Advances in the Study of Lunar Opposition Effect 599
The Application of the Instantaneous States Reduction to the Orbital Monitoring of Pivotal Arcs of the Chang'E-1 Satellite 606
A Digital Open-loop Doppler Processing Prototype for Deep-space Navigation 617
Chang'E-1 Orbiter Discovers a Lunar Nearside Volcano: YUTU Mountain 628
Space Operation System for Chang'E Program and its Capability Evaluation 632
Preliminary Evaluation of Radio Data Orbit Determination Capabilities of China's First Lunar Orbiter 637
LUT: A Lunar-based Ultraviolet Telescope 644
Measurements of Electronic Properties of the Miyun 50 m Radio Telescope 651
Design and Implementation of Space Dust Database 661

图版

第一部分 综 述

嫦娥一号月球探测卫星研制综述

叶培建 孙泽洲 饶 炜

（中国空间技术研究院，北京 100094）

摘　要　嫦娥一号卫星是我国第一个月球探测卫星，将实现对月球的环绕探测。嫦娥一号卫星的研制和发射是我国深空探测活动的开端，在我国航天史上将成为继人造卫星和载人航天后的第三个里程碑。与近地卫星相比，嫦娥一号卫星面临更复杂的控制过程和环境，因此，嫦娥一号卫星必须突破一系列的关键技术，实现既定的任务目标。文章介绍了嫦娥一号卫星的任务目标、主要技术方案和研制过程；概要性地说明了嫦娥一号卫星的研制过程。
关键词　嫦娥一号卫星　月球探测　研制

1　前言

空间探测是空间科学和技术创新的重要环节，它将在新的世纪成为国际竞争的重要焦点之一，可极大地开拓人类的视野和科学实验领域，提高人类认识自然和利用自然的能力。月球是距离地球最近的天体，是地球唯一的天然卫星，它蕴藏着丰富的自然资源和能源，一直是人类开展空间探测的首选目标。

在综合分析国际月球探测的发展历程以及近年来主要航天国家和组织提出的月球探测战略目标和实施计划的基础上，结合国家整体发展战略、科学技术水平和综合国力，我国提出了如下月球探测工程发展思路：在2020年前，我国月球探测工程分为三个发展阶段，概括为"绕、落、回"。一期工程的主要目标是实现绕月探测；二期工程是实现月球软着陆探测和自动巡视勘测；三期工程是实现自动采样返回。

一期工程的嫦娥一号(CE-1)卫星的研制和发射是我国深空探测活动的开端，在我国航天史上将成为继人造卫星和载人航天后的第三个里程碑。

2　任务及技术要求

根据我国月球探测工程的总体规划，月球探测一期工程的主要目标是研制环月探测卫星，突破地月转移和环月飞行的关键技术，对月球进行环绕遥感探测，初步建立月球探测工程系统。根据这一总目标，嫦娥一号卫星需完成以下工程目标和科学目标。

2.1　工程目标

绕月探测工程要实现以下基本目标：
(1) 研制和发射我国第一颗月球探测卫星；
(2) 初步掌握绕月探测基本技术；
(3) 首次开展月球科学探测；
(4) 初步构建月球探测航天系统；

(5) 为月球探测后继工程积累经验。

2.2 科学探测目标

1) 获取月球表面三维影像

利用面阵 CCD 相机结合激光高度计获取月球表面三维影像，精细划分月球表面的基本地貌和构造单元，初步编制月球三维地形图、地质图和构造纲要图，划分月球断裂和环形影像纲要图，并为月面软着陆提供参考依据。

2) 分析月球表面有用元素含量和物质类型的分布特点

通过 γ/X 射线谱仪和干涉程序光谱仪分析月球表面有用元素含量和不同物质类型的分布特点，为研究月球形成和演化历史提供直接和有效的证据，为未来开发和利用月球资源提供依据。

3) 探测月壤厚度与特性

利用微波辐射计探测月壤厚度及其分布，分析月壤成熟度与表面年龄的关系，概略估算月球表面 ^{3}He 资源的蕴藏量。

4) 探测地-月空间环境

利用太阳高能粒子探测器、太阳风低能离子探测器探测太阳宇宙线高能带电粒子和太阳风等离子体，研究太阳风和月球的相互作用，深入认识空间物理现象对地球空间以及对月球空间的影响。

3 国内外月球探测卫星的比较

1958 年 10 月美国发射了第一个月球探测卫星——先驱者-1，标志着人类开始利用航天器进行月球探测。在近 50 年的时间里，人类共发射了 100 多个月球探测器对月球进行了深入的探测和了解。

表 1 列出了嫦娥一号月球探测卫星与现阶段国外几颗典型绕月探测器的比较。

表 1 近期月球探测器比较表

	嫦娥一号 CE-1 （中国，2007）	智能-1 Smart-1 （欧洲航天局，2004）	"月球探测者" Luna Prospector （美国，1998）	"克莱门汀" Clementine （美国，1994）	月球初航-1 Chandrayaan-1 （印度，2008）
任务目标	对月三维成像、多光谱成像； 月面元素探测(14种)； 月壤厚度探测及 ^{3}He 资源量评估； 地月空间环境探测	利用电推进飞向月球，验证先进技术。包括： 电推进； Ka 频段通信； 载荷技术等。 科学目标： (1)地形学和表面构造研究； (2)月表元素分布； (3)探寻极区的水冰	勘测月面的资源包括： (1)矿物、水冰和某些气体； (2)绘制月球重力场及磁场图； (3)进一步了解月核的尺寸和组分	由国防部和 NASA 联合，借探测月球、小行星 1620 及空间环境来验证敏感器及航天器技术。 对月观测包括不同谱段的月面成像（可见光、紫外、红外）； 激光测高； 高能粒子探测等	三维全月面成像，空间和高度分辨 5m； 高分辨率的 Fe, Ti, Al, Mg 等矿物成分成像； Mg, Al, Si, Ca, Fe, Ti 等元素的分布与地层学关系； ^{222}Rn(氡)，^{210}Pb(铅) 及其他放射性元素等地球稀有金属的探测
工作轨道	200km 极月圆轨道	近月点 300～1 000km 远月点 10 000km	100km 圆形极轨道	周期 5h 的椭圆轨道，近月点 400km 在南纬 28°，一个月后近月点转到北纬 29°，覆盖南北纬 60°之间	100km 圆形极轨道

续表

	嫦娥一号 CE-1（中国，2007）	智能-1 Smart-1（欧洲航天局，2004）	"月球探测者" Luna Prospector（美国，1998）	"克莱门汀" Clementine（美国，1994）	月球初航-1 Chandrayaan-1（印度，2008）
载荷种类	(1)CCD立体相机；(2)成像光谱仪；(3)激光高度计：11kg，25W；(4)γ/X线谱仪：35kg，15W；(5)微波探测仪：30kg，42W；(6)太阳高能粒子探测器：2.4kg，3W；(7)低能离子探测器：7kg，7W	(1)电推进诊断包：2.4W，18W；(2)潜在电子和尘埃试验：0.8kg，1.8W；(3)Ka频段试验：6.2kg，2W；(4)小型化X射线成像光谱仪和X射线监测仪：5.2kg，18W；(5)红外光谱仪：2.3kg，4W；(6)先进的月球微成像仪(包括激光链路和自主导航设备)：2.3kg，9W	(1)中子谱仪：探测氢元素，以此来推断是否有水存在，3.9kg，2.5W；(2)γ射线谱仪(GRS)：探测10种关键元素的存在，以研究月球的起源与演化，8.6kg，3W；(3)磁力计和电子反射计：探测月面及轨道高度上的磁感应强度，以了解月球的磁场异常：共5kg，4.5W；(4)S频段多普勒重力试验：通过测量航天器速度和位置的变化绘制月球重力场图，用于未来的任务；(5)α粒子探测器：探测挥发气体中的氡(当前的挥发物)和钋(近50a的挥发物)：4kg，7W	(1)充电粒子探测仪：探测能量质子和电子的通量和光谱：0.21kg；(2)高分辨率相机：1.12kg，9.5W；(3)激光测高系统：2.37kg，6.8W；(4)长波红外相机：2.1kg，13W；(5)近红外CCD相机：1.92kg，11W；(6)收发雷达：用于测南极是否有水冰：13.6kg；(7)S频段转发器多普勒重力场试验：8.27kg；(8)星相机：0.58kg，4.5W；(9)可见光/紫外CCD相机：0.41kg，4.5W，环形坑统计	(1)成像光谱仪：8kg，15W；(2)立体相机：7kg，20W；(3)激光高度计：3.5～5kg，8W；(4)低能X射线谱仪：5～8kg，15～25W；(5)低能伽马射线谱仪：15kg，20W；(6)太阳X射线监测仪：3kg，3W；(7)征集载荷：10kg，10W
发射质量/kg	2 350	370	296	424	1 050
干重/kg	1 150	286	158	227	523
载荷质量/kg	140	19	21.5	30.58	55
载荷与干重比/%	12.2	6.7	13.6	13.5	11.1
总功率/W	1 574	1 850	202	360	670
发射时间	1a	6个月	1a+0.5a降轨至30km	1994年1月25日发射对月球工作2个月，飞越小行星失败	2a

4 主要技术方案

嫦娥一号月球探测卫星的构型如图1所示，卫星的发射质量约2 350kg，干重为1150kg，推进剂1200kg，携带140kg的载荷在距月面200km的圆形极轨道上对月球及月球空间进行科学探测。卫星的寿命为1年。

卫星共有9个分系统，可分为服务系统和载荷两个部分，服务系统包括：结构、热控、制导、导航与控制(GNC)、推进、供配电、数据管理、测控数传、定向天线等分系统。

4.1 结构分系统

CE-1卫星结构是由中心承力筒和28块蜂窝夹层板组成的一个长方体箱形结构。从区域划分的角度，卫星结构分为上舱和下舱。从结构装配的角度，卫星结构是由主构架和12块外侧板组成(图2)。主构架是由中心承力筒和 16 块与其有直接连接关系的结构板组成。结构分系统的质量为 217kg,承载能力为 2 350kg。

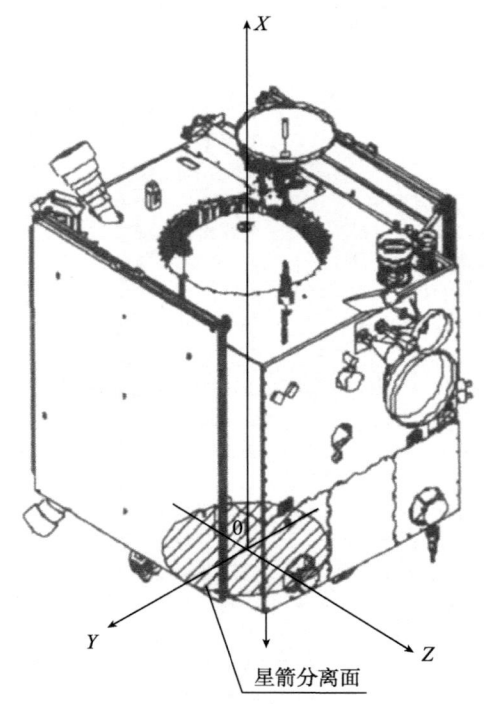

图 1 嫦娥一号卫星总体构型

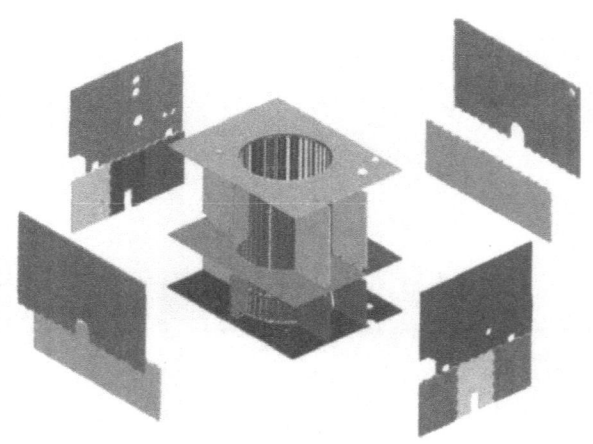

图 2 结构分系统分解图

4.2 热控分系统

热控分系统采用主动和被动热控技术保证寿命期内卫星有效载荷系统及其他各分系统的仪器设备温度要求,其组成主要包括散热面涂层、多层隔热材料、加热器、温度传感器、热管、加热控制器等(图 3)。鉴于热设计边界条件复杂,系统较多地采用了主动控温设计。加热器的通断控制由数管分系统完成。

4.3 GNC 分系统

制导、导航与控制(GNC)分系统的主要任务是完成卫星奔月过程所需的多种姿态的变换和控制,实现卫星对月定向的三轴稳定姿态、太阳翼对日定向跟踪、定向天线对地定向。

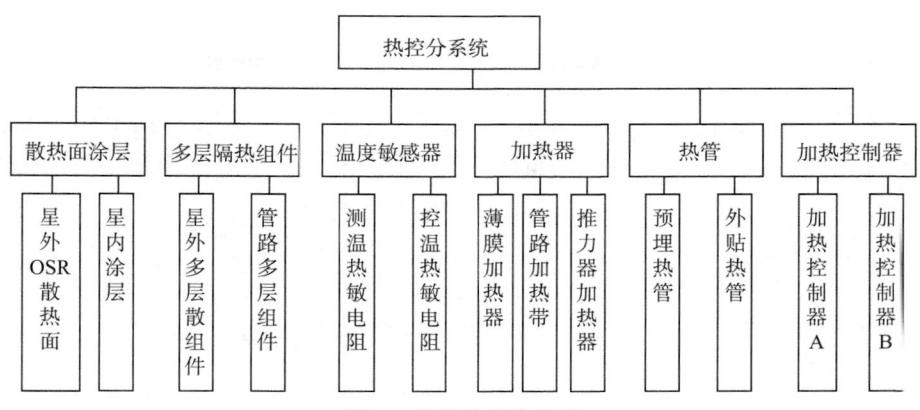

图 3 热控分系统组成

GNC 分系统由敏感器、执行机构和控制器组成。敏感器包括紫外月球敏感器、星敏感器、太阳敏感器、陀螺、加速度计。执行机构包括 490N 变轨发动机和 10N 推力器(属推进分系统)、动量轮、太阳翼驱动机构、天线驱动机构。控制器包括控制计算机、应急线路、GNC 配电器和 GNC 二次电源。

4.4 推进分系统

推进分系统采用双组元统一推进系统,主要任务是与 GNC 分系统配合,在从星箭分离开始到卫星寿命终了的时间内,向卫星提供慢旋、速率阻尼、目标捕获、各种姿态的建立与保持、轨道控制和修正等所需动力。

推进分系统配置了 1 台 490N 变轨发动机,12 台 10N 推力器分成 A、B 两分支,每分支 6 台互为备份。

4.5 供配电分系统

供配电分系统包括一次电源、二次电源和总体电路。一次电源采用有背场的单晶硅太阳电池阵(最大输出功率大于 1 450W)和镍氢蓄电池组(末期容量大于 48Ah)联合电源,为卫星产生、贮存和调节电能,以满足卫星在整个飞行过程中的供电需求。二次电源采用分散供电方式,将卫星的一次母线电压变换成星上各分系统及设备所需要的电压;总体电路实现星上一次电源分配和控制,以及火工品的管理和控制。

4.6 数管分系统

数管分系统是二级分布式容错计算机系统,由中央单元、远置单元和遥控单元,以及一套双冗余的串行数据总线和数管分系统软件组成。用以实现卫星遥测、遥控、程控、自主控制、校时等整星控制和管理功能。

4.7 测控数传分系统

测控数传分系统由测控、数传和甚长基线干涉(VLBI)等部分构成。

测控子系统由 S 频段测控全向天线、低放单元、功放合成单元和测控应答机等设备组成。主动段、调相轨道段、地月转移轨道段和环月轨道段的初期及应急阶段由 S 频段全向天线、测控应答机、低放合成单元、功放合成单元 A 和 B 等完成测控任务。在环月轨道期间,当定向天线对地跟踪正常工作后,测控应答机的发射机输出信号转送到功放合成单元 C,通过定向天线发送下行射频信号。

数据传输子系统由统一频率源、S 频段 BPSK 调制器(S-BPSK)、X 频段 BPSK 调制器(X-BPSK)、S 频段固态放大器(S-SSPA)、X 频段固态放大器(X-SSPA)、X 频段全向天线和大功率微波开关组成,数传能力为3Mbit/s；VLBI设备由X-BPSK、X-SSPA、信标天线等设备组成。需要VLBI测轨时，X-BPSK和X-SSPA工作，将伪码信号通过信标天线传输到VLBI接收站。

4.8 定向天线分系统

定向天线采用双自由度机构实现半空间覆盖，为数传下行信道和遥测下行信道提供满足任务要求的天线增益。定向天线主要由射频部分、双轴驱动机构、压紧释放机构、展开机构和一些结构件组成。射频部分保证±5°范围净增益大于+18dBi，双轴驱动机构的运动范围是±90°。

4.9 有效载荷分系统

为实现月球探测科学目标，卫星的有效载荷配置了五类科学探测仪器和一套有效载荷数据管理系统，具体配置及探测目标如表2。

表2 有效载荷探测仪器

有效载荷		功能简介
光学成像探测系统		获取月球表面立体图像和多光谱图像，绘制月球三维图像和地质学专题图
激光高度计		获取卫星星下点月表地形高度数据
γ/X射线谱仪		测量月表物质的γ射线和荧光X射线谱，探测有用元素的含量和分布
微波探测仪		对不同深度月壤微波辐射亮温进行测量，给出月壤厚度的信息
空间环境探测	高能粒子探测器	探测高能带电粒子的成分、能谱、通量和随时间的变化特征
	太阳风离子探测器	探测原始太阳风等离子体的能谱，包括太阳风的体速度、离子温度以及数密度
有效载荷数据管理系统		采集、存储、处理和传输有效载荷探测数据，有效载荷的在轨管理有效载荷的供配电管理

5 轨道及飞行过程

嫦娥一号卫星整个飞行任务可划分为相对独立的7个阶段：射前准备阶段、主动段、调相轨道阶段、地月转移轨道段、月球捕获阶段、环月工作状态建立阶段和环月运行阶段。嫦娥一号卫星飞行程序示意如图4。卫星由运载火箭送入近地点200km、远地点51 000km、倾角31°的地球椭圆轨道；先进行一次远地点变轨，使近地点变为600km，然后进行三次近地点变轨，进入地-月转移轨道(近地点600km，远地点

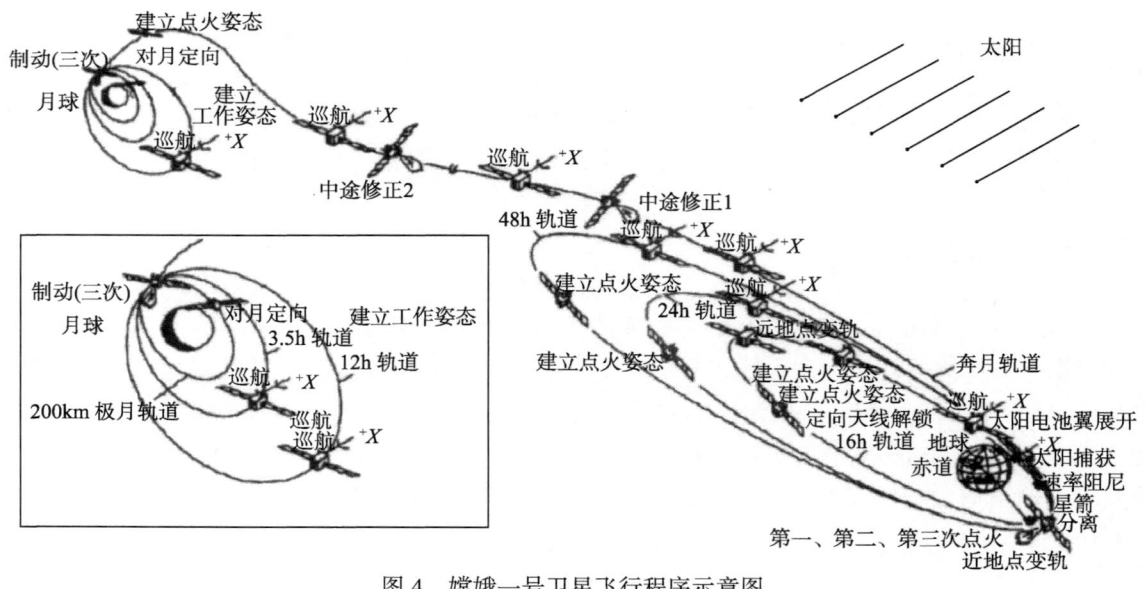

图4 嫦娥一号卫星飞行程序示意图

380 000km)；卫星在此轨道上要飞行5天，在此过程中卫星要进行2~3次中途修正；当卫星接近月球时，将通过第一次近月点制动，使卫星进入围绕月球运行的周期为12h的椭圆轨道，然后再进行第二次近月点制动，把轨道周期调整为3.5h，再进行第三次近月点制动，把轨道变为高度为200km，周期为127min的圆轨道。在这一轨道上，一个月内(27.3d)CCD立体相机可对全月球(极区除外)覆盖一遍，微波遥感覆盖两遍；干涉成像光谱仪经过两个月可以对全月球(极区除外)覆盖一遍。

6 研制过程概述

我国月球探测的预先研究早在20世纪90年代初就已开始进行。经过10多年的研究积累，2001年，国防科工委组织中国科学院、航天科技集团、总装备部等单位正式启动月球探测工程的相关论证工作，经过近一年的工程综合论证，完成了《月球探测工程综合立项论证报告》。其后，嫦娥一号卫星进入预发展阶段，即方案阶段。

在方案设计阶段，总体和各分系统完成了方案的细化和指标的分解工作。2003年底卫星初样初步设计工作完成。2004年4月国防科工委下发了《绕月探测工程研制总要求》，各分系统转入详细设计阶段。

2004年4月30日嫦娥一号卫星转入初样研制阶段。

在历时20个月的初样研制过程中，嫦娥一号研制队伍的主要任务是进行系统、分系统及设备的详细设计、制造、组装、总装、测试和试验，以验证各级设计的合理性，验证是否符合规范所规定的性能指标及接口技术要求。先后完成了初样结构/热控星和电性星的投产和验收、结构星力学试验、热控星热平衡试验、电性星电性能综合测试、卫星与地面测控系统及地面应用系统对接试验，同时还完成了专项试验、关键技术攻关、软件研制、正样技术状态协调确定，以及鉴定件研制和试验等工作。在整个初样研制阶段，总体和各分系统进行了大量的研制、试验、分析、验证和改进工作；着眼于卫星所担负的使命，从元器件、原材料做起，攻克了一个个难关；完成了卫星初样阶段的研制任务，明确了正样产品的技术状态。

2005年12月嫦娥一号卫星转入正样研制阶段。在历时13个月的正样研制过程中，各分系统进行了产品研制和试验、正样设计复核复算；整星进行了卫星的部装、总装、电性能综合测试、环境试验和可靠性增长测试工作；并分别完成了与地面测控系统、地面应用系统和运载火箭系统对接试验；开展了与北京指控中心的飞控协调。

2007年2月，上级机关要求将CE-1卫星的发射窗口调整到2007年10月。由于卫星发射窗口的变化，卫星研制队伍组织完成了新窗口条件下的综合论证，并重点在卫星可靠性增长、飞控准备、质量复查方面等开展工作。

7 小结与展望

经过3年的研制，嫦娥一号卫星已经完成全部的研制和发射准备工作。与此同时，嫦娥一号卫星备份星的研制工作也在紧锣密鼓的展开，利用嫦娥一号平台进行火星探测的可行性研究工作也正在进行。

作为第一颗月球探测卫星，嫦娥一号卫星的研制过程攻克了很多近地卫星不曾遇到的难题，开辟了我国航天的新领域，为我国执行未来的深空探测任务积累了丰富的经验。

作为我国深空探测任务的开端之作，嫦娥卫星不仅具备重要的科学意义，更承载着党和人民的希望和重托。嫦娥一号卫星探测任务的顺利进行，将推动我国航天水平的进步，谱写我国航天历史的新篇章。

信息化技术在嫦娥一号卫星研制中的应用

叶培建[1] 于 萍[2]

（1. 中国空间技术研究院，北京 100081；2. 北京控制工程研究所，北京 100080）

摘 要 嫦娥一号卫星首发成功，是我国航天事业发展的第三个里程碑。信息化技术在嫦娥一号卫星的管理、设计、加工及测试等各个环节都得到了充分应用，为确保高效、高质量地完成嫦娥一号卫星的研制任务起到了至关重要的作用。报告对嫦娥一号卫星的主要特点进行了归纳；全面回顾了在嫦娥一号卫星研制过程中信息化技术的应用情况及其成效；简要介绍了信息化技术在星上产品中的应用情况，详细阐述了信息化技术在制导、导航与控制(GNC)系统中的应用情况；并对信息化技术在飞控任务中的应用作了介绍。

关键词 信息化 计算机 嫦娥一号卫星 GNC系统

1 引言

嫦娥一号卫星的研制、发射和成功环月是我国深空探测活动的开端，开创了我国航天史上继人造卫星和载人航天后的第三个里程碑。作为我国第一颗月球探测卫星，嫦娥一号的研制过程攻克了很多近地卫星不曾遇到的难题，开辟了我国航天的新领域，为我国执行未来的深空探测任务积累了较丰富的经验。信息化技术在嫦娥一号卫星自身及其研制过程中的管理、设计、加工及测试等各个环节得到了充分的应用，为确保准时、高效、高质量的完成嫦娥一号卫星的研制任务起到了至关重要的作用。

本文对嫦娥一号卫星的主要特点进行了归纳；全面回顾了在嫦娥一号卫星研制过程中，信息化技术的应用情况及其成效；简要介绍了信息化技术在星上产品中的应用情况，并重点地详细阐述了信息化技术在GNC系统研制中的应用；同时对信息化技术在飞控任务中的应用作了介绍。

2 嫦娥一号卫星简介

根据我国月球探测工程的总体规划，月球探测一期工程的主要目标是研制环月探测卫星，突破地月转移和环月飞行的关键技术，对月球进行环绕遥感探测，初步建立月球探测工程系统。嫦娥一号卫星的工程目标是研制和发射我国第一颗月球探测卫星，初步掌握绕月球探测基本技术，首次开展月球科学探测，初步构建月球探测航天系统，为月球探测后继工程积累经验；嫦娥一号卫星的科学探测目标是获取月球表面三维影像，分析月球表面有用元素含量和物质类型的分布特点，探测月壤厚度与特性，探测地-月空间环境等。

嫦娥一号卫星共有11个分系统组成：即总体，测试两个综合分系统；平台部分的结构，热控，制导、导航与控制、能源、推进、数据管理(OBDH)、测控数传、定向天线八个分系统及有效载荷分系统。各自相对独立又互相密不可分，构成一个完整的整体，共同实现嫦娥一号卫星的各项功能。

嫦娥一号卫星整个飞行任务可划分为相对独立的7个阶段：射前准备阶段、主动段、调相轨道阶段、地月转移轨道段、月球捕获阶段、环月工作状态建立阶段和环月运行阶段。

本文原载于《空间控制技术与应用》，2008，Vol.34，No.1，9~13，50。

2007年10月24日嫦娥一号卫星由运载火箭送入近地点200km、远地点51000km、倾角31°的地球椭圆轨道；2007年10月31日进入地-月转移轨道(近地点600km，远地点380000km)，期间进行了1次远地点变轨、3次近地点变轨；11月5日卫星飞抵月球完成了第一次近月制动，进入围绕月球运行的周期为12h的椭圆轨道；11月6日进行了第二次近月点制动，把轨道周期调整为3.5h；11月7日完成第三次近月点制动，把轨道变为高度为200km，周期127min的使命轨道；11月20日传回第一幅图像。

嫦娥一号卫星的成功研制、发射和绕月飞行，表明我国在空间技术领域突破并掌握了探月轨道设计、制导、导航与控制，远距离测控与通信，探月卫星热控等一大批核心技术和关键技术。

3 嫦娥一号卫星研制与信息化技术

嫦娥一号卫星是我国深空探测的第一个飞行器，其研制仅用了三年多的时间，且一次成功。整个工程在技术与管理上有多项创新点，得益于多方面的贡献，信息技术的应用是其中重要因素。

在嫦娥一号卫星的研制过程中，始终坚持以信息技术为核心，综合运用信息技术、建模与仿真技术、管理技术、设计与生产技术、系统工程技术及与产品有关的专业技术，重点解决了四个矛盾：第一，任务与能力之间的矛盾；第二，产品性能与新的要求之间的矛盾；第三，多学科交叉与专业分工的矛盾；第四，系统高度集成性与研制单位地理分散性的矛盾。这四个矛盾的解决依赖于两个方面，一是在产品设计、仿真、制造及管理中充分应用了信息化技术，二是星上产品自身中含有大量先进的数字化技术与成果。下面从仿真设计、项目管理两个方面，说明信息化技术应用的成效。

3.1 信息化技术在设计、仿真及制造中的应用

嫦娥一号卫星在研制过程中，广泛采用数字化手段，高可靠的验证了新设计与新方案，减少了设计差错，提高了效率，缩短了研制时间。数字化设计/制造促进了流程再造，实现了质量、进度双重提升。具体说来，有以下几点：

(1) 嫦娥一号卫星在立项之初就充分利用仿真手段确立了合理的方案、优化的参数，特别是最优的轨道设计为顺利完成任务奠定了坚实的基础；

(2) 建立了复杂而逼真的数字环境模型，为设计正确性验证和各项试验(力学、热、电磁等)奠定了基础；在产品设计中大量采用了改造引进及自研的各种机械、电子、高频、无线等软件，保证了设计一次到位和质量；

(3) 充分利用历史数据，通过数据重用和分析仿真减少了部分试验，节省时间、经费；采用数字化手段完成卫星的总体布局、总装设计和结构设计，取消了传统的实物模装星，缩短了设计时间，提高了设计更改的快速反应能力；

(4) 充分应用CAD/CAM一体化技术，减少了三维管路设计误差，减少或取消了取样工作，协调性由设计直接保证。管路加工计划安排由与卫星生产流程主线串行变为与卫星生产流程主线并行；

(5) 数字化设计改变了管路设计流程，传统流程需要20天，而新流程只需1至2天；

(6) 卫星结构的研制中，广泛应用CAD/CAM/CAE技术，在各种工装与模具的设计中全面采用了三维设计，三/二维CAD出图率达到100%；

(7) 通过对重要、关键件的加工仿真，取消了零件的试切削，不仅缩短了加工周期，创造了经济效益，而且提高了加工质量，减少了差错。

3.2 信息化技术在项目管理中的应用

嫦娥一号卫星项目采用以总指挥与总设计师为主要责任人的"两总"管理体制，强调整体资源平衡下

的综合协调，采用自上至下的任务分解与自下而上的产品汇总的管理模式。在卫星研制过程中，全面推行项目管理的信息化，提升了科研生产综合计划管理能力。具体说来，有以下几点：

(1) 利用网络技术，建立了统一的桌面办公和近、远距视频交流环境，加速了研制部门、发射场、飞控中心间的信息流通。网络跨越六省一市，实现中国空间技术研究院所有下属单位与发射基地的互联互通，京区骨干带宽达到了千兆，构建了稳定、畅通的中国空间技术研究院科研网络。网络信息点8700个，覆盖率达100%；

(2) 依靠计算机网络实现了工程文档的电子管理。用航天PDM软件-AVIDM进行工程电子文档的管理，包括报告、图纸、图表的流转、签署、分发和版本控制，形成了统一规范、保密性好、技术状态统一受控的管理模式；

(3) 基于计算机网络平台AVIDM实现了电子仓库及文档管理、产品结构及配置管理、工作流程管理、项目计划管理、建模向导工具、协同工作环境支持及应用集成等多项功能；

(4) 依托AVIDM系统，建立了贯穿决策层、职能管理层、项目经理、计划经理和项目参与人员的全员数字化管理环境，采用统一的数字化平台对工程技术和管理两个层面进行管理；

(5) 依托AVIDM系统，将任务逐级分解到每个研制人员的桌面，实现项目的精细化管理。规范计划管理主要业务流程的执行、控制过程；实现了项目计划执行情况的实时反馈、查询与统计。任务计划与进展和完成信息进行关联，实现了多级计划编制、下发、执行和控制跟踪，从而达到了项目闭环控制的目的；

(6) 通过项目计划与财务、物资等信息关联，实现了对各型号、部门的计划执行，资源占用、经费开支等多要素运行的监控，为型号综合管理提供及时、准确、量化的决策信息。实现了以项目计划为纽带的多系统信息集成。

综上，正是成功、充分的应用信息化技术，改造了工作流程、改进了研制手段、提高了管理水平和研制能力，才确保了嫦娥一号卫星高性能、短周期交付和首发成功。

3.3 信息化技术在星上产品中的应用

与其他卫星一样，嫦娥一号卫星具备需要完成任务的各种设备，但这些设备的功能和性能需满足更高、更新的要求。在资源约束下，要在有限的时间内，用较少的经费，按规定的重量、功耗、体积完成任务，向信息技术要出路是最有效的。从嫦娥一号卫星星上产品来看，其信息化技术的应用具有以下两个特点，本节简要的介绍这方面的情况。

3.3.1 信息技术在星上产品应用中的多学科性

所谓信息技术，其本质是以计算机硬、软件为基础的，卫星产品也不例外。在嫦娥一号卫星上，涉及的信息技术分支颇多。由于星上应用了大量计算机，因此，它在环境上关联计算机体系结构、计算机操作系统、网络与总线、多媒体技术等等；当然，也离不开最基本的科学计算技术；软件方面关联软件开发工具、软件开发方法、软件工程规范、软件工程管理等技术；星上星敏感器、紫外敏感器、有效载荷的大量信息需处理，离不开图像处理、模式识别及人工智能等技术的支持；考虑到信息的安全性、可靠性，尤其是在空间、深空这一特定环境，有关的信息保护、防空间负效应技术是必须采取的；测试是卫星产品保证质量的重要手段，信息技术在测试中也扮演着重要的角色等。

3.3.2 信息技术在星上产品应用中的广泛性

前面谈到，星上除两个综合分系统外，整星由九个物理上的分系统组成。信息技术几乎遍及每个分系统。其中最为典型的是GNC分系统，这在3.4节单独叙述。此外，还有数据管理分系统，它负责对全星进行星务管理，并在一定程度上与GNC相互支持，保存重要数据。具有CTU(中央单元)和多个RTU(远置单元)。确保在深空探测信噪比低的情况下实现低误码率,数管系统采用了卷积编码、多码速率多格式的遥测信息流处理技术,且具有良好的功能与自主性;有效载荷由8台设备组成,运行工作模式多,信息多样。对这些设备进行管理和数据处理的核心是智能化。为克服嫦娥一号卫星在飞行过程中我国国内测控站有观测盲区的实际情况,星上热控、能源系统的控制也与以往的飞行器大不相同,具备了很好的智能,可以按照设定的控制软件进行温度、电量式电压的精确控制;为解决38万公里的遥控遥测,星上产品采用了一系列先进的信息化技术手段,加之其他方面的贡献,从而克服了无大天线支持下的38万公里遥测遥控难题。

3.4 信息化技术在GNC产品中的应用

在制导、导航与控制系统方面,主要挑战表现在复杂的轨道与姿态控制,以及对多个目标天体的指向控制。信息化技术在GNC系统中的应用,为解决上述控制问题起了重要的作用。本节以GNC系统为例,详细介绍信息化技术在星上产品中的应用情况。

嫦娥一号卫星在GNC系统控制下要完成的任务如下:首先,在调相轨道,GNC系统要执行一系列姿态机动和轨道控制,以使飞行器不断提升轨道,在适当时间进入奔月轨道;然后,在奔月轨道,GNC分系统要保证卫星向太阳惯性定向,并根据测轨结果执行轨道中途修正,以使卫星可在预期点捕获月球;接下来,在月球轨道捕获阶段,GNC分系统要执行三次轨控发动机点火,以使飞行器捕获月球轨道并进入标称环月轨道;最后,在环月轨道,GNC分系统要完成"三体定向"的控制功能,即探测器本体对月球定向、太阳帆板对太阳定向、数传天线对地球定向。

嫦娥一号卫星在GNC系统控制下及时、准确、高可靠的完成了上述任务,轨道控制与姿态控制精度都优于设计指标,这与GNC系统研制过程中成功应用了信息化技术有着密切的关系。

3.4.1 GNC系统的信息化体系结构

GNC系统采用了分布式采集、集中控制的系统结构,基于信息流的层次结构实现了系统模式的正常调度和故障情况下的系统重构。

1) GNC系统的信息流结构

嫦娥一号卫星GNC系统层次结构底层为各敏感器和执行机构的局部终端处理单元(LTU);层次结构顶层为控制计算机和应急控制器。在信息流的层次结构支持下,各LTU完成分布式信息采集功能,计算机和应急控制器完成集中控制的功能。系统正常模式调度下,根据模式的不同启用不同的软件模块和设备;当发生故障时,根据故障的具体情况启用备份设备或进行系统降级,自主更改信息流方向,进行系统重构,保证卫星安全。

2) GNC系统的信息交换

控制计算机通过系统的内部串行总线与惯性测量单元、各光学敏感器、帆板和天线驱动线路进行通信,通过并行或模拟量接口控制推进系统和动量轮。

应急线路通过模拟接口采集陀螺和太阳敏感器的信息,通过并行接口控制推力器电磁阀,实现卫星对日定向。

控制计算机通过数管系统1553B总线实现接收地面上行注入指令;同时也通过数管系统1553B总线将数字量遥测发送到数管分系统的CTU,进而由CTU发送地面。

3) GNC系统的信息流程

嫦娥一号卫星GNC系统的信息流程可划分为两类:一般信息流程和应急信息流程,分别以控制计算机和应急线路为核心。系统运行的一般情况下，控制计算机采集各敏感器信息，根据控制目标集中计算各执行机构的控制量，通过控制推进系统和动量轮，实现轨道控制和姿态控制，通过向帆板和天线驱动线路发出控制指令，实现对帆板和天线的控制。应急信息流程仅用于卫星飞行时应急状态的控制。在控制计算机控制功能正常时，应急线路不工作。当控制计算机不能完成控制功能时，应急线路加电，接替工作。此时，系统功能降级，不能完成轨道控制和多种姿态控制要求，仅保证卫星处于太阳定向姿态，避免控制计算机功能不正常而带来的灾难性的故障。

4) GNC 系统的信息检测功能

嫦娥一号卫星 GNC 系统具有自主信息监测功能，可实现星上自主故障诊断和故障处理能力，在星上发生一次故障的情况下，可进行系统重构以确保地月转移关键性变轨如期完成，确保卫星安全。

星上自主故障诊断分为部件级和系统级两个层次进行。部件级故障诊断由部件 LTU 智能单元完成，可进行部件的工况自检及进行数据有效性判别。系统级诊断由控制计算机实现，包含两方面：一是根据系统多路测量信息的一致性检测部件故障；二是发现和确认卫星姿态异常。

星上自主重构手段有更换敏感器组合、启动备份部件，系统工作模式切换等。同时利用地面注入手段可实现控制参数和控制软件的在轨修改，以实现系统设计的在轨完善。基于自主故障诊断与系统重构技术，嫦娥一号卫星在国内首次设计实现了轨控故障自主恢复功能。

3.4.2 多任务实时操作系统在 GNC 中的应用

嫦娥一号卫星 GNC 系统控制需求多样，任务复杂，要求很高的实时性。采用传统的单任务机制很难适应嫦娥一号卫星 GNC 系统复杂的控制任务。在详细分析 GNC 任务的基础上，嫦娥一号卫星在 GNC 控制计算机(GNCC)中采用了多任务实时操作系统，构建了规则清楚、接口明确、时序顺畅的复杂多任务实时处理系统。

根据飞行阶段和飞行任务的不同，嫦娥一号卫星 GNC 分系统设计了 10 个模式，其中多数模式可采用较慢的控制周期来完成，也就是说从测量信息的采集到执行机构按照计算机指令执行控制命令的时间相对可以较长。当发动机点火时，扰动剧烈，为了保证轨控精度就必须采用快周期，以便控制系统对扰动做出快速反应，修正姿态偏差。这就对 GNC 系统提出变周期控制需求，要求系统具有高实时性。

基于实时操作系统时间片调度方式，可有效的解决变周期控制问题。具体地说，是将 GNC 系统每个模式细分为多个功能模块，各功能模块具体划分为优先级不同的多项任务。根据各任务的实时性需求，同时考虑到可能发生的突发事件，由多任务实时操作系统按时间片处理周期性任务，并采用中断方式处理突发性任务。控制计算机采用固定优先级的调度策略，使优先级高的任务被实时处理。同时根据固定时间片，分配任务执行时间，从而实现了变周期的 GNC 控制系统，解决了轨控模式快周期控制与其它模式慢周期控制的冲突。

多任务实时操作系统在嫦娥一号卫星上的应用，在 GNC 控制计算机体系结构层面确保了复杂任务的实时控制顺利完成。

3.4.3 GNC 测试系统的信息化技术

GNC 系统地面测试的目的在于利用地面测试设备与星上产品一起构成模拟飞行系统，由此验证 GNC 系统星上产品的硬件和软件的正确性。嫦娥一号卫星是我国第一颗深空探测卫星，地面测试系统的难点在于要模拟其深空运行环境，以便对 GNC 系统姿轨控功能和性能进行测试。

首先，地面测试系统要模拟卫星深空运行的力学与光学环境。为此基于嵌入式多任务操作系统技术和工业组态软件技术开发了姿态和轨道动力学仿真软件，嵌入了高模态柔性动力学仿真模型、姿轨控耦合动力学模型，实现了卫星绕地、奔月及环月的全过程多体摄动轨道和姿态的连贯仿真，实现了卫星光照条件动态仿真，为验证星上产品性能提供了高精度仿真测试环境。其次，地面测试系统以数据库技术为依托，基于实时与非实时隔离的多层网络拓扑结构，利用先进的数据显示技术，实现了测试数据的大量存储和动态显示。数据库存储功能为在测试结束后对测试数据进行深层次处理提供了便利条件。星上数据与地面仿真数据实时比对，实现了数据判读的可视化。最后，地面测试系统基于工业组态软件技术实现了跳时测试、加速测试等测试功能，提高了测试效率。

嫦娥一号卫星GNC地面测试系统运用了信息技术发展的新成果，丰富了测试手段，提高了测试效率，提升了测试能力，在较短的时间内做到了GNC系统功能和性能的充分测试，为确保GNC系统的高质量提供了有力保障。

4 信息化技术在飞控任务中的应用

为了确保嫦娥一号卫星首发成功，卫星系统还研制了两个飞控辅助系统，一个是卫星飞控模拟器，另一个是飞控仿真与支持系统，为我国首颗探月卫星飞控任务顺利实施提供了保障。

卫星飞控模拟器主要用于卫星发射前，满足测控系统进行飞行控制演练的需求。飞控模拟器的主要组成部分就是GNC和OBDH模拟器，它采用基于网络的分布式结构，由多台商用计算机最大限度地模拟了真实卫星的状态，为实战前的飞控任务演练提供了保障条件。

嫦娥一号卫星的飞控仿真与支持系统是我国为探月卫星首次研制。依托该系统，可对轨控任务的实施、意外/突发事件处理、在轨故障的快速分析、控制参数和措施的正确性进行仿真验证，从而为制定卫星在飞行中的应对策略提供依据，确保嫦娥一号卫星奔月、环月的可靠性。

飞控仿真与支持系统以GNC系统数字和半实物仿真系统为主要支撑，应用了数字仿真手段、综合了MATLAB、STK等工具软件，可全面真实模拟卫星系统的硬件状态和软件状态，具有成本低、手段全、效率高的特点。

此外，飞控决策支持系统还具有仿真动画视频显示功能，在实际飞行数据驱动下，可逼真地展示卫星飞行全过程，方便全国人民观察、了解卫星飞行过程。

5 结束语

嫦娥一号卫星作为我国首颗探月卫星，首发成功，在我国航天技术领域取得了新突破，是我国航天事业发展的里程碑事件。嫦娥一号卫星的成功是航天领域多学科技术进步的共同结晶。在研制周期短、任务重、技术难度大的情况下，充分应用信息化技术为确保嫦娥一号卫星圆满成功做出了重要贡献。因而获得了2007年全国工厂大会颁发的信息化建设项目成就奖。

实践表明，信息化技术是当今最为活跃的一个领域，它在其他技术的需求刺激下迅猛发展，而它的发展又极大的改变了其它科学技术的现状，完全是一个相辅相成、共赢共发展的关系，我国航天器40年的发展也验证了这一点。随着航天器，特别是深空探测器向更高水平进军，由于其飞行时间长、距离非常遥远，对智能化、自主性要求很高，进而对信息技术有着更为强烈的需求。今后的研制工作中，信息化技术效益发挥的水准如何，将会对航天器的技术水平、研制周期、经费等起着更大的制约作用。

致谢 本文在形成过程中，得到了张洪华研究员、刘霞研究员及袁利高级工程师等的关心帮助，在此一并表示感谢！

参 考 文 献

[1] 叶培建. 信息化在嫦娥一号卫星上的应用[N]. 计算机世界报，2007-12-10(47—A7)
[2] 徐福祥. 卫星工程概论[M]. 北京：宇航出版社，2003

嫦娥一号月球探测卫星技术特点分析

叶培建 饶 炜 孙泽洲 张 伍

(中国空间技术研究院,北京 100094)

摘 要 嫦娥一号卫星是我国的第一个月球探测卫星,将飞行至距地球 380 000km 的月球,实现环绕月球对其遥感探测。由于任务目标不同,嫦娥一号卫星将遇到比近地轨道卫星更复杂的空间环境和飞行控制过程,所以必须解决面临的所有新技术问题。文章介绍了嫦娥一号卫星在轨道设计、月食、热设计、制导导航、测控、数传等方面的技术特点及研制验证方法。

关键词 嫦娥一号 月球探测卫星 技术特点

1 前言

嫦娥一号(CE-1)卫星是我国的第一个月球探测卫星,将飞行至距地球 380 000km 的月球,实现环绕月球对其遥感探测。目前只有美国、原苏联/俄罗斯、欧洲航天局和日本实现了月球环绕探测。

由于任务目标不同,嫦娥一号卫星将面临比近地轨道卫星更复杂的空间环境和飞行控制过程。为了满足任务目标要求,适应其面临的特殊环境,嫦娥一号卫星必须采用特殊的设计和技术克服轨道设计、月食、热设计、制导导航、测控数传等所遇到的一系列困难,从而完成探测任务。因此,嫦娥一号卫星也就具备了与近地轨道卫星不同的特点。

2 嫦娥一号卫星技术特点[1-3]

2.1 轨道设计

嫦娥卫星的轨道分析与近地卫星有很大不同,要飞过 380 000km 并实现月球捕获,最终成为环月卫星,是一个复杂三体的分析求解问题。而卫星进入环月轨道后,由于月球引力场的异常复杂性,卫星轨道的变化与地球卫星的情况有很大的不同,需要进行大量的分析工作。此外,嫦娥一号卫星的轨道设计还面临着地月相对位置、测控要求、运载发射条件、燃料携带量、月影分布、月食时机等一系列条件的限制。因此在轨道设计过程中,需综合考虑上述各约束条件,并对其中的一些不利的结果加以甄别和排除。

针对月球轨道特点,经过详细设计,形成的轨道方案是卫星在近地点进行三次近地点机动,逐步增加近地点的速度使远地点高度逐步提高,最终进入地月转移轨道;嫦娥一号卫星采用的是最小能量的转移轨道,在转移轨道进行 2~3 次轨道修正,在近月时卫星进行制动,被月球捕获,进入环月轨道,然后经过三次减速制动和轨道调整,变为 200km×200km 的极月圆轨道。为了确保轨道设计的正确性,组织全国范围内的同行专家,多次对轨道设计结果进行复核复审。

2.2 月食问题

嫦娥一号卫星的设计寿命为 1 年,在寿命期内,卫星不可避免地需经历两次月食,每次月食的有效阴

本文原载于《航天器工程》,2008,Vol.17,No.1,7~11。

影时间均在 3h 左右,在这段时间里,卫星无法获得光照能源,同时卫星温度将会迅速降低,这对星上电源系统的供电能力、热控系统的温度维持能力、星上状态设置的准确性和最小功耗模式的稳定性提出了更高的要求。蓄电池组低温放电能力、蓄电池组的温度维持能力、各设备的低温耐受能力成为保证卫星能够顺利度过月食的关键环节。

在卫星的设计过程中,针对月食问题进行了攻关论证,全面分析各种不确定因素,对月食有效阴影时间、供配电分系统供电能力、热控分系统温度维持能力和测控条件进行了充分的分析。对整星月食期间的工作模式、热控和供配电分系统设计进行了完善。在研制过程中,还针对关键环节进行了专项试验验证。包括太阳翼弱光发电试验、蓄电池单体低温放电试验、设备月食低温耐受试验,并在电池舱段热平衡试验、整星热平衡试验以及整星电性能综合测试中安排了月食工况项目试验及测试验证。另一方面,为确保月食问题的解决,对卫星舱段的热设计进行了修改,以提高蓄电池组在月食期间的温度,保障整星能源的供给。同时,针对月食特点,对相关软件进行了适应性修改,以确保月食任务的可靠完成。

为了考核卫星度过月食的能力,在正样整星级热平衡试验过程中,安排了两个与月食有关的试验工况,通过月食工况和月食校核工况,验证了月食阶段设备工作的协调性、整星温度维持能力和整星能量平衡。试验证明了卫星具有度过两次月食的能力。

2.3 卫星热设计

月球的外热流条件非常恶劣,卫星的热设计面临以下难点:

(1) 卫星在发射、调相、奔月、环月飞行过程中受到太阳、月球、月球阴影、地球阴影(月食)的影响,外热流条件非常复杂,星体各个表面外热流变化很大;

(2) 月面的太阳反照、红外辐照外热流是第一次遇到,目前无法进行地面验证,无飞行经验可以借鉴,月球红外热流值波动很大,从每平方米超过 1 000W 到几十瓦的范围内波动,加大了对月观测有效载荷设备温度保证的难度;

(3) 月食造成卫星要经历月球阴影时间长达3h以上,给热控分系统带来很大影响;

(4) 特殊热控要求的设备较多,关键设备如氢镍蓄电池组、激光高度计、CCD 相机、定向天线等热控要求高,设计余量小,难度大;

(5) 整星没有相对稳定的散热面,散热面布局困难。

由于这些特点和要求,给热控分系统方案的确立和设计带来很大困难。为了解决以上问题,初样阶段进行了专项的蓄电池单体发热量测试、定向天线热平衡试验、载荷舱段热平衡试验、蓄电池舱段热平衡试验等,对特殊热控要求的设备的热设计进行了充分的验证;此外,尽可能地收集了国外各种相关工程和观测数据,经过分析和比对建立了月球环境外热流模型,同时,随整星技术状态的逐步明确和冻结,及时修正热分析模型进行预示,并相应安排热设计复核。

在正样阶段,整星级热平衡试验结果也表明所有设备的平衡温度均在其工作温度范围内,且设备的高温、低温都留有足够的余量,证明嫦娥一号卫星的热设计是合理的。为了确保整星热设计的正确性,热控分系统出厂前再次组织了对月表热环境的复核,确认了嫦娥一号卫星热设计采用的月表热模型是合理的。

2.4 制导、导航与控制

月球探测卫星的制导、导航与控制(GNC)设计要求和常规卫星相比,有以下几个特点:

(1) 从环绕地球飞行到准确进入环绕月球的飞行轨道,卫星需经历多次复杂的轨道和姿态机动,控制任务复杂,可靠性要求高,实时性强;

(2) 在环月运行期间,卫星本体要对月定向,太阳翼要对太阳定向,定向天线要对地球定向,需要高可靠性的 GNC 分系统;

(3) 环月期间的卫星对月姿态确定与一般卫星不同，不能采用红外敏感器，需采用紫外敏感器、星敏感器、陀螺等多种手段，并结合轨道外推算法进行多种模式设计；

(4) GNC 各项成熟技术对新的任务特点要进行适应性分析，同时考虑与新技术的集成和融合。

为了顺利完成任务，GNC 分系统要采取以下主要解决措施：

(1) 控制器的冗余设计、重要敏感器(陀螺、星敏感器)的冗余设计、高性能控制算法的开发。它们能够保证 GNC 分系统的高实时性，并具有故障诊断和在轨重构能力，从而实现高可靠性。

(2) 三体组合控制(星体对月指向控制、数传定向天线对地指向控制、太阳电池阵对日指向控制)技术的开发，使得三体控制满足任务需求。

(3) 紫外月球敏感器和双轴天线驱动控制技术属于攻关项目。它们能够保证三体指向控制的高可靠性。

(4) 通过对任务需求分析、软件开发与测试、分系统及整星各环节测试严格把关，保证系统可靠性。

正样阶段，GNC 分系统通过仿真试验和整星综合测试，验证了 GNC 分系统设计的功能和指标均满足要求。

2.5 紫外月球敏感器

紫外敏感器与陀螺联合定姿是环月阶段 GNC 分系统姿态确定的第一方案，紫外月球敏感器对我国来说是一个全新的设备。在设计与研制中，紫外月球敏感器需完成月球光谱特性研究、光学系统设计与实现、图像处理算法、测试和标定设备研制等关键技术攻关工作。

为顺利完成紫外月球敏感器的研制，保证嫦娥一号卫星的 GNC 功能，主要采取以下措施：

(1) 确定月球光谱特性：月球光谱特性是敏感器设计的前提。通过委托国家天文台和紫金山天文台，完成了月球光谱特性的研究。并以此研究成果，结合现有技术，确定了紫外敏感器工作波段。之后又依托自有技术力量，对月球辐射能量进行了分析，确定了紫外敏感器光学系统设计参数、光积分时间、月球模拟器亮度等重要指标。

(2) 紫外敏感器的光学系统设计：光学设计有两大难点，一是视场达到了 70°，像差校正困难；二是紫外波段可选光学材料非常有限。在综合比较了多个方案之后，确定了"反远距广角物镜"结构形式，并通过优选光学材料，解决了大视场紫外光学系统设计这一具有国际水平的难题。

(3) CCD 工作电路设计：由于不同月相下月面亮度差别很大，就需要紫外敏感器能在比较大的范围调整光积分时间。受光学系统参数限制，要求 CCD 最小光积分时间必须达到几毫秒，经过不懈努力，终于攻克了毫秒级光积分时间控制问题，实现了大范围曝光时间调整。

(4) 月球图像处理及姿态确定算法：采用中心截线法实现了对月球边缘可靠地提取；采用矢量法进行对月姿态确定，解决了月球图像轮廓不为圆弧时的姿态确定问题。

2.6 测控数传分系统设计

由于地月距离约为 40 万千米，射频信道的自由空间损耗与地球同步轨道卫星相比，高出约 20dB，与近地轨道卫星相比，则高出约 46dB。同时，CE-1 卫星较地球同步轨道卫星和中低轨道卫星的入轨过程复杂，测定轨精度要求高，增加了卫星测控与数传的难度，给测控数传分系统的设计提出了很高的要求。

在我国目前没有建立深空站的条件下，嫦娥一号卫星测控数传分系统的设计必须立足于国内现有的地面测控系统和地面应用系统的资源，根据国内这些地面站的能力，嫦娥一号卫星测控数传分系统必须保证射频信道的 EIRP 和 G/T 值，同时适当地降低码速率，以保证星地链路指标满足要求。

为了克服空间衰减大、信道余量较小的困难，系统设计时主要考虑了以下几种措施：

(1) 提高全向天线的增益和全空间覆盖范围；

(2) 取消功率分配环节，以减少信号的功率损失；
(3) 提供高低两种遥测码速率，并增加遥测信道卷积编码，以降低遥控和遥测信道解调所需的信噪比；
(4) 降低数传信道解调所需的信噪比；
(5) 通过灵活的多种工作模式来弥补信道余量的不足；
(6) 降低环月轨道的功耗，增大卫星在环月轨道时的 EIRP 值；
(7) 地面测控系统为减小信道余量偏小而带来的工程风险，新增地面站天线。

经过初样卫星与地面测控系统和卫星与地面应用系统的对接试验，以及正样阶段安排的测试验证和星地测控对接试验、星地应用对接试验，验证了星地之间接口的匹配性和正确性。

2.7 定向天线

根据嫦娥一号卫星的特点，卫星在实现环月飞行后，必须保证星体对月球，太阳翼对太阳，定向天线对地球区域的定向。为了解决三体定向的矛盾，要求嫦娥一号卫星的天线必须具备两维大角度转动的能力。

为满足设计要求，嫦娥一号卫星的定向天线成为国内卫星第一次采用的大角度机械扫描天线，在空间应用方面国内没有成熟的经验。天线的设计和研制涉及多学科综合和多单位协同工作，单个部件和整个分系统的加工、装配、精度标定、工装、调试和试验等过程都需要从头摸索，给研制工作带来很大困难。其面临的难点在于以下四方面：

1) 一体化集成系统的构型与布局

定向天线是一个复杂的机、电、热、控多学科组合系统，机构运动规律复杂，天线布局时，不但要考虑尽量避免星体和太阳翼对电性能的影响，还要保证系统具有较好的刚度和受力环境，避免和星上其他设备谐振；三自由度、多间隙、刚度离散性以及较小的可用空间，给压紧点的布局和配置造成很大困难；在系统高、低频电缆的走向和布局上，既要保证不干涉系统转动，使对系统运动的阻力最小，又要保证展开过程中电缆的安全；在双轴的驱动机构的布置中不仅要考虑运动功能，还要考虑旋转关节的安装以及热控措施的实施。

2) 热控分析、设计与实施

定向天线在整个飞行过程中要保持对地定向，而本身又随着卫星整体绕月球运行，受照面和受照强度随着太阳、地球、月球间的关系不断变化，同时还受到卫星本体和天线面的遮挡，所以定向天线在整个飞行过程中所受外热流变化剧烈，工作环境较为恶劣，对热控分析和设计提出了更高的要求。此外，定向天线结构复杂，活动部件较多，活动范围大，热控措施可操作空间小，热试验方案和工装复杂，给热控实施和热试验带来很大的困难。

3) 系统接口

定向天线内部零部件较多、部件之间接口复杂，使系统设计加工和装配的难度较大；系统与卫星的接口关系复杂，依次涉及总体、GNC、星上数据处理(OBDH)、电源、结构、热控、测控、数传、总体电路等多个分系统，几乎与卫星的每一个分系统都有联系。

4) 系统装调、测试、试验

定向天线是一个二维运动的三自由度系统，运动范围大，在地面装调时要求处在模拟零重力的环境下，目前国内的零重力模拟设备还都是一维的，必须创造一个能够近似模拟二维三自由度运动的系统来进行装调。

根据以上特点，定向天线的热控设计、实施与验证，系统装调、测试与试验验证成为设计和研制的关键。为保证定向天线满足系统要求，设计和研制过程中，主要采取以下解决措施：

(1) 对于定向天线热设计，通过了双轴热平衡试验和天线整机热平衡试验验证，试验结果满足热设计要求。

(2) 对于定向天线的活动部件——双轴驱动机构，为保证其长寿命、高可靠要求，采取了以下措施：对于活动部件轴承，完成了加速寿命试验；线束经过了一万次弯曲寿命试验考核，结果正常；电机、轴承、零位传感器等部件都经过在轨飞行考验，最长在轨正常工作时间已达4年，以上措施保证了长寿命；双轴结构设计简单，所有技术均经过验证；电机绕组、零位传感器采用备份设计，保证了高可靠性要求。双轴驱动机构还顺利通过了热真空、力学、线束寿命和阻力矩等试验考核。

(3) 为在过程控制中保证产品质量，压紧展开机构和双轴驱动机构设置了强制检验点，加强质量管理和过程控制，确保产品质量。

(4) 正样定向天线在研制过程中进行了充分的试验验证，充分验证了定向天线的功能符合总体要求。

2.8 数管分系统设计

数管分系统方案虽然主要继承已有的近地卫星数管分系统的成熟技术，但为适应探月任务的使用背景和限制条件，采取了一些不同于一般卫星的设计，主要体现在以下方面：

1) 信道编码

由于地月间距离遥远，测控信号的空间衰减增大。这是我们以往的卫星型号都未遇到过的情况，因此为了降低误码率，提高卫星下行遥测信道的抗干扰能力，在嫦娥一号卫星上首次采用了卷积码作为遥测信道编码，并对遥控、遥测的码速率进行了调整。

2) 遥测源包

由于卫星遥测下行通道数量有限，数管分系统将重要的遥测数据，通过1553总线送到大容量存储器，经数传实时或延时下发，使得遥测信息量增加并更新及时。

3) 采用FPGA

为了实现设备小型化，降低产品的体积、重量和功耗，提高可靠性，采用现场可编程门阵列(FPGA)进行集成化设计。

4) 软件设计

为有效解决"测控不可见弧段长"的问题，数管CTU软件作为整星的控制管理核心，采取以下措施：
(1) 增加了自主管理功能，提高卫星的自主管理能力；
(2) 设计了多种遥测格式、多种存储模式、多种压缩速率等手段，以满足卫星飞行任务的需求；
(3) 设计了"定向天线展开监控与应急控制"、"整星安全应急控制"等模式，以增强卫星应急能力。

正样阶段，通过分系统测试、系统间联试和整星综合测试，验证了数管分系统设计的功能和指标，均满足要求。

3 小结

嫦娥一号卫星的研制和发射是我国深空探测活动的开端，作为我国航天史上继人造卫星和载人航天后的第三个里程碑项目，在面临时间紧、任务新等不利条件下，总体和分系统着眼于卫星所担负的使命，攻克了很多近地卫星不曾遇到的难题，为月球探测二三期工程的立项和研制奠定了良好的技术基础，也为我国执行未来的深空探测任务积累了丰富的经验。

参 考 文 献

[1] 欧阳自远，主编. 月球科学概论[M]. 北京：中国宇航出版社，2005
[2] 欧阳自远.嫦娥工程月球手册[M]. 北京：宇航出版社，2006
[3] NASA. Lunar surface models[R]. NASA SP-8023, 1969

Policy Making in China's Space Program: A History and Analysis of the Chang'E Lunar Orbiter Project

Patrick Besha

(Space Policy Institute, George Washington University, Washington, DC 20052, USA)

Abstract China's space program is one of the most advanced, rapidly improving and opaque in the world. Insight into the program's policy-making process could help eliminate misunderstandings, make intentions more clear and promote stability in US-China relations. This case study of China's first lunar orbiting probe traces the project from initial policy proposal through agenda-setting, policy approval and final policy implementation. It reveals a highly rational decision-making process that is ruled by incrementalism, consensus building, scientific judgment and the use of leading small groups to coordinate among ministries. This research was guided by several relevant theories, including the "fragmented authoritarian" framework of power, the theory of the "policy entrepreneur" and the recently developed "inside access model". The paper is one of the first published accounts in the English language to detail, from policy proposal to policy implementation, China's first mission to the Moon.

1 Introduction

China's space program is one of the most advanced, rapidly expanding and opaque in the world. The mystery surrounding it heightens fears and fuels speculation. Much of our knowledge of the space program is revealed only in highly publicized glimpses, such as when the nation achieves noble triumphs like placing a human into orbit or successfully orbiting the Moon. Military observers tend to focus on the less public accomplishments of China's rocket program, like the recent ballistic missile intercept test and the 2007 anti-satellite missile test. Much of the national space program, a vast conglomeration of military and civilian organizations, is kept intentionally opaque. However, in recent years information on certain projects of limited military utility has been released to the public. The first lunar orbiter mission, Chang'E-1, falls into this category of project. It is unique in this respect because, while it was a project of national prestige that involved coordination at the highest levels of government, military and Party, there is also an abundance of information available on its development. Insight into this policy-making process could provide much-needed transparency into the decision making that guides the world's newest scientific and technological superpower.

China's space program provides several key benefits to the nation. First, it expands the high-technology sector and helps to integrate key military and technology industries. The joint development of "dual-use" technologies is an increasingly important goal as China seeks to make state defense industries more efficient, while developing indigenous technological capabilities. Second, the space program is an important driver of science and technology innovation in business and of technical education in schools. Third, the space program is a universally recognized symbol of national prestige. In China it is the crown jewel of a burgeoning technology and industrial empire. Fourth, the program is a useful propaganda tool for the ruling Party. Major missions stoke nationalism and provide popular legitimacy.

The national lunar program consists of three major project phases: the successful launch and orbit of a spacecraft to map the Moon's surface, landing an unmanned lunar rover on the Moon, and a lunar sample return mission to carry lunar material back to Earth. The first phase and the subject of this case study, Chang'E-1, was successfully completed in March 2009.[1] Chinese space officials recently announced that there are studying the possibility of a manned Moon mission around 2025.

This article seeks to explain decision making in the Chinese space program bureaucracy. Focusing on Lieberthal and Oksenberg's insight that the nature of the Chinese political system is "fragmented authoritarian", the specific strategies used by policy entrepreneurs to propose policy, set the agenda and implement policy will be explored. Additional theories of policy-making will also be considered in order to develop a model of policy-making in China's lunar space program.

2 Theories of policy making

"Fragmented authoritarianism" continues to be the most robust framework available to understand China's political system. The term was first coined by Kenneth Lieberthal and Michel Oksenberg to characterize Chinese policy-making in the modern reform era. The system it describes is authoritarian in the discipline it exerts over subordinates, and relatively closed to public policy input from citizens; it is fragmented in that territorial and functional lines of authority below the center are often disjointed and policy making incremental.[2] Decision making is characterized by inter-bureaucracy negotiations, bargaining and consensus building. The fragmented authoritarian framework is institutionalist and pluralist in that bureaucratic structure itself helps to guide policy decisions, but factions may compete for influence.

The process is protracted, with most policies shaped over a long period and acquiring a considerable history that is well known to many of the participants; it is disjointed, with key decisions made in a number of different and only loosely coordinated agencies and inter-agency decision bodies; and, it is incremental, with policy in reality usually changing gradually.[3]

During the initial policy proposal stage, John Kingdon's theory of "policy entrepreneurs" is particularly useful. He defined policy entrepreneurs as "advocates for proposals or for the prominence of an idea". They are defined by "their willingness to invest their resources-time, energy, reputation, and sometimes money-in the hope of a future return… in the form of policies of which they approve."[4] This theory seeks to account for policy proposals offered by citizens not firmly situated in the ranks of power. Elite scientists, for example, do not necessarily have the political power to make policy decisions yet they may propose major policies.

The method by which their policies are placed on the agenda is detailed by Wang Shaoguang's "inside access model", which accurately describes the policy subsystem involving policy entrepreneurs. Policy solutions are proposed by advisors and members of the scientific elite within government. These advisors do not necessarily seek public approval, and interaction is primarily between policy advisors and policy makers.[5] This model is premised on the notion that China has a closed authoritarian political system, and most decisions are made within government with little or no input from citizen groups.

Policy making in the lunar program is remarkably similar to other areas of policy making in China, with the major caveat that the program frequently does not consider any input from citizens or the media and most decisions are made behind closed doors.[6] This case study utilizes original Chinese source material and interviews

---

1 A near-duplicate of the Chang'E-1 orbiter, known as Chang'E-2, is slated to launch in 2010-11. This mission will feature enhanced instruments and will mark the actual conclusion of the first phase of the lunar program.
2 Kenneth Lieberthal, Governing China: From Revolution through Reform 2nd ed.(W.W. Norton & Co., 2002).
3 Ibid.
4 John W. Kingdon, Agendas, Alternatives, and Public Policies, 2nd ed. (New York:HarperCollins, 1995), pp. 122-23.
5 Shaoguang Wang, "Changing Models of China's Policy Agenda Setting," *Modern China* 34, no. 1 (2004).
6 Recent news reports suggest this may be changing for the second phase of the Chang'E program, which will place a robotic lander on the lunar surface. The State Administration for Science and Technology Industry for National Defense (SASTIND, the successor to COSTIND - see below in text) has reported that dozens of universities are competing to develop key parts of the lander.

to explore the actual processes and methods utilized to propose policies, set agendas and implement policy.[7] There are several key strategies for proposing policy ideas, setting agendas, making policy and implementing policy in the Chinese political system:

- advocacy groups of elite policy entrepreneurs (e.g. Chinese Academy of Sciences scientists);
- *Guanxi* (interpersonal relationship networks);
- the use of Leading Small Groups (*lingdaoxiaozu*- LSGs) tocoordinate various organizations.

3 New policy idea

The idea of a Chinese voyage to the Moon is as old as the ancient story of the Moon goddess Chang'E, who is said to have floated away to the celestial body after an argument with her lover. Contemporary policy proposals for a lunar program have existed since the1960s, and they gained renewed traction in the late 1980s, after Deng Xiaoping introduced the multi-billion dollar "863" high-technology funding program.[8] This program would provide seed money to thousands of projects that would provide the technology and industrial foundation for China's current growth.

The genesis of the policy proposal that would later become the successful Chang'E-1 lunar probe project can be found in a confluence of events in the early 1990s. The Chinese human spaceflight program, a product of the "863" plan, would begin operations in 1992 after years of planning and development.[9] Also, the launch of the Japanese lunar probe *Hiten* in 1990 had a catalyzing effect on space scientists in China. One of the most outspoken of these scientists was Ouyang Ziyuan, who in 1990 was a prominent geologist in Guizhou province with a deep knowledge of lunar geology. A year later, he would transfer to the most prestigious academic institution in China, the Chinese Academy of Sciences (CAS) in Beijing. He assumed the leadership of the Institute for Geochemistry and quickly set out to engage the scientific community in a new policy idea: the first-ever Chinese mission to the Moon.

Scientists at the Chinese Academy of Sciences occupy a special position within the government and scientific community. They are regarded as the nation's pre-eminent scientists and are highly respected in society.[10] While political elites may desire CAS scientists to be both "red and expert", party membership is not a prerequisite for entry into CAS.[11] However, top scientists are regularly offered membership in the Chinese Communist Party(CPC), and those with particular influence may be awarded seats in the National People's Congress, Chinese People's Political Consultative Conference and CPC Central Committee[12].

After the reform and opening up period, the concept of "scientific decision making" became paramount and signaled a new reliance upon think-tanks and educated advisors.[13] The elite scientists at institutions such as the

---

7 *Note on source*—Information about the Chinese space program is very closely monitored and carefully scrubbed by authorities. However, there are pockets of reliable information on the first phase of the lunar program available in the published and open-source literature. Much of the first phase of the program (Chang'E-1) has been completed and is no longer operational, as the orbiter crashed into the moon in March 2009. The first drafts of history have recently been published and program scientists and officials can more freely discuss project details (and embellish their historical roles), providing useful primary source information. Nearly all information on the program is written in Chinese. This study relied extensively on official Chinese government sources, including documents published by the State Council Information Office and state-owned science and technology, and defense industry publishing houses. Additionally, newspaper reports from the *People's Daily* and Xinhua were utilized. Books written by members of the lunarteam (such as Ouyang Ziyuan) and outside journalists illuminated the social and political context of decision-making. Finally, personal interviews with China space program officials supported and verified published accounts.
8 "863" refers to the date of its inception: March 1986.
9 The human spaceflight program is known as codename Project 921 or China Manned Space Engineering.
10 Cao Cong, *China's Scientific Elite* (London: Routledge Curzon, 2004).
11 Ibid.
12 Examples of space scientists who also hold high political office include Qi Faren, Chief Designer of Shenzhou 1-4 (CPPCC) and Luan Enjie, General Commander of Lunar Mission (CPC Central Committee).
13 See for example Xinhua, "Zhongnanhai Qingting Kexue Sixiangku Jingyan (Chinese Leaders Listen to a Scientific Think Tank," in *Xinhuanet.com* (November 9, 2004).

CAS can be influential policy advisors.[14] Similarly, Hu Jintao's "theory of scientific development" champions the utility of science to solve the nation's most pressing social and economic problems. As the scholars Zhao Quansheng and Cao Cong have noted, think tanks and academies are assuming powerful roles as advisors to top policy makers.[15]

Occasionally, CAS scientists are also policy entrepreneurs. For example, the State High Technology Research and Development Program (commonly known as the "863" program) was implemented in 1986 as a way for China to quickly and broadly accelerate its science and technology activities. The program provided billions of dollars to the military, government ministries and academic institutions for new projects, including space. The program itself was advocated and intensively pushed by a small group of like-minded scientists in the early 1980s. Wang Daheng, Wang Ganchang, Yang Jiachi and Chen Fanyun were influential pioneers in China's nuclear weapons program. They drafted a short letter to paramount leader Deng Xiaoping, suggesting that China must develop a science and technology program to develop indigenous capabilities. The proposal was partially in response to the recently announced US Strategic Defense Initiative. The scientists sought out personal connections close to Deng, and effectively jumped several levels of the bureaucracy to have their letter heard. They were successful and the "863" program was launched.[16]

Similarly, the decision to create the *Beidou*, or "Compass" national satellite navigation system resulted from the efforts of a core group of dedicated scientists tirelessly pushing their policy proposal upon decision makers.[17] Further study is required before a definitive model of this type of policy advocacy can be formulated, but in addition to "863" and Beidou, this study suggests that the Chang'E-1 lunar project is another example of this type of policy advocacy.

When Ouyang Ziyuan first floated the idea of a lunar exploration program, he drew criticism from some fellow scholars. His idea was primarily driven by political motivations, because he suggested that the handover of Hong Kong to China in 1997 should coincide with the launching of a rocket to the moon. The boost to patriotism and nationalist sentiment would be tremendous, he argued. His proposal caught the attention of many scientists at CAS, including high-ranking Academician Min Guirong. With a small group of supporters, he convinced the central government to consider the proposal, but it was swiftly rejected on the grounds that it was primarily politically motivated and lacked solid scientific value.[18]

In the Mao era numerous national defense programs were developed to address key military and strategic issues, in addition to serving political goals. One of the most famous is the "Two Bombs, One Satellite" (*liang dan, yi xing*) program. Mao Zedong sought to develop an atomic bomb, a hydrogen bomb and to launch a satellite as part of a massive national security push with broadly distributed social and economic development benefits. The program was inspired and quickly co-opted by members of the ideological and political elite, who proclaimed that the nation's science and technology industries would be revamped and a new communist-patriot spirit would pervade all work. Mao's intense political motivation and ideological excesses had a seriously detrimental effect on S&T development and nearly derailed the nascent rocket and space program.[19]

Academician Min Guirong, the director of an "863"-sponsored group on aerospace at CAS and a supporter

---

14 Science, "Science Interview: China's Leader Commits to Basic Research, Global Science," in *Science* (2000), Jiang Zemin, "Science in China," in *Science* (2000).
15 The Chinese Academy of Social Sciences is just one example. Zhao Quansheng has written extensively on the role of think-tanks in formulating Chinese foreign policy.
16 Gregory Kulacki, and Lewis, Jeffrey G., "A Place for One's Mat: China's Space Program, 1956-2003," (American Academy of Arts and Sciences, 2009).
17 Interview with Chinese space official 1.
18 Shao Gai He, "Chang'E Benyue Zhilu (The Path of Chang'E to the Moon)," in *GuofangKeji Gongye* (National Defense Science & Technology Industry) (2007).
19 The story of QianXuesen, a pioneer of China's space program, is particularly noteworthy. As biographer Iris Chang recounted in *Thread of The Silkworm* (1996), upon his expulsion from the USA and return to China, he adopted the Communist ideology and became a senior science advisor to Mao on a broad array of policies. Noting the tremendous harm done by unscientific or ill-advised science policies during the Great Leap Forward, it would appear that commitment to ideology often trumped sound scientific rationale.

of the lunar program, suggested the formation of a task force for a lunar mission. The task force would bring like-minded scientists and policy entrepreneurs from CAS together. Most importantly, Ouyang Ziyuan and the team could receive initial funding from the "863" program to continue working on their draft report. For Ouyang Ziyuan, it now became clear that China had the money and it nearly had the technology for lunar exploration.[20]

In 1993, the newly created China National Space Administration(NSA), led by Administrator Liu Jiyuan and Deputy Administrator Luan Enjie, conducted a feasibility study that suggested the Long March 3A rocket could reach the moon and deliver a satellite into orbit. The study was presented to central government policy makers, who again concluded that there were no sound scientific objectives and the excessive costs did not warrant a mission.[21] According to key scientists, the CNSA effort was independent of the CAS effort.[22]

The CAS scientists would later recall that they were beginning to worry whether the project would ever be successful. After the successful launch of NASA's Clementine lunar probe in 1994, Ouyang Ziyuan lamented that, in the face of global competition, if China were indifferent, the country would fall behind internationally and lose its voice.[23] He was not alone, as key advocates have also defined the program as needing to keep pace with the international community of lunar exploration.[24] Since the beginning a key rationale of the program was to maintain China's international competitiveness and to match the contemporary Japanese and US lunar efforts.

In 1995 the "863"-funded lunar task force led by Min Guirong at CAS finally finished its report, "The Necessity and Feasibility of China's Lunar Exploration Program". Ouyang Ziyuan and Chu Guibo were the principal authors.[25] The report was circulated among colleagues, rather than higher-level decision makers. A wildly optimistic and controversial feature of the report was the prediction that the moon was a likely source of the element Helium-3 (theoretically, an ideal element for nuclear fusion) and could one day support China's energy needs. The report is generally acknowledged to be the first formal investigation of a potential lunar mission by official CAS experts.[26]

The strategy of linking up interested scientists and policy advocates was paying off. Nevertheless, the coalition was largely academic and lacked prominent support from the central government and even initial support from the military or the Party. Given the type of project proposed—an expensive, high-profile scientific lunar mission to be launched on military rockets—high levels of support from all three would be necessary to achieve policy approval.

In 1995, the rocket development group at CASC announced that a new *Long March* 2F rocket would loft the first ShenZhou-1 capsule into orbit in 1997, heralding the beginning of the manned space program. The group suggested that the LM-2F could be used for lunar exploration, specifically, as a way to welcome the return of Hong Kong in 1997. This plan was also dismissed. According to the official journal of the national defense industry, clear-cut scientific objectives or an engineering plan had still not materialized.[27] Proposals for a lunar mission had been offered by CAS academics, CNSA, and a key institute of the military-industrial complex. None was successful and all were cited for a lack of specificity, scientific rigor and/or a satisfactory engineering plan.

The CAS group led by Min Guirong and Ouyang Ziyuan was composed of "policy entrepreneurs", or a group of advocates that pushed for a specific policy proposal. In Kingdon's formulation, small groups of elites may initiate and develop policy initiatives. The scientists held a curious position as government officials, but not

---

20 Yang Zheng, "Xin You Yue Hui Zhao Zhao Yao Tu," in *Jing jiribao* (Economic Daily) (2009).
21 Shao Gai He, "Chang'E Benyue Zhilu (The Path of Chang'E to the Moon)," *Guofang Keji Gongye*, (National Defense Science & Technology Industry)2007) pp. 76-81.
22 Yang Zheng, "Xin You Yue Hui Zhao Zhao Yao Tu," *Jing jiribao* (Economic Daily) (2009).
23 Translated from Chinese by author. This is also commonly referred to as 'losing one's seat at the table', or "yi xi zhi di (a place for one's mat)" Source: http://www.amacad.org/publications/spaceChina.aspx.
24 People's Daily, "Why Does China Start a Lunar Exploration Program?," in *The People's Daily* (2003).
25 Ouyang Ziyuan and Chu Guibo. "Wo guo kaizhan yueqiu tance de biyaoxing yu kexingxing (Necessity and Feasibility of Developing a Chinese Lunar Probe)" (1994).
26 He, "Chang'E BenyueZhilu (The Path of Chang'E to the Moon)."
27 Ibid.

government policy makers, which placed them in the position of acting as both advisors and advocates for a specific space policy.

4 Setting the agenda

While Ouyang Ziyuan, Min Guirong and other colleagues at the CAS were influential in their academic fields and in some political circles, their policy proposals were met with silence or dismissed by policy makers, who insisted upon having a clear scientific rationale for any lunar mission. Similarly, Luan Enjie of CNSA was an early supporter of the lunar exploration program and provided early engineering studies to strengthen the case for a lunar program. But, in his position at CNSA (a small inter-agency organization primarily tasked with coordinating space cooperation among various organizations domestic and international), he was too far from the levers of power to effect change.

While the CAS group and Luan Enjie worked toward a common goal, there is no evidence that they worked closely together to develop proposals; indeed, the opposite appears to be the case. Ouyang recalled in a later interview that he, Luan Enjie and Sun Jiadong (Chief Designer) would really only know each other after the lunar program was approved.[28] Without high-level support ora coordinated effort between the various policy advocates, the multiple lunar program proposals could not gain traction. This distribution of power and stove-piping in the government is a key feature of "fragmented authoritarianism". Multiple institutions may push similar policy proposals unknown to each other. Only at the highest levels of decision making can these parallel efforts be detected. These high-level organs include the National Development and Reform Commission (NDRC), which often provides initia approval or the State Council, which approves nearly all projects before submission to the National People's Congress.

A crucial breakthrough came in 1998, when the State Council reorganized the Commission for Science and Technology Industryfor National Defense (COSTIND).[29] Founded in 1982, it had Ministry-level authority and reported directly to the State Council. It originally had a large portfolio that included defense industry procurement, development and production. Subsequent reforms weakened the institution, and in 1998, it was civilianized and military work was shifted to the General Armaments Department of the PLA. In its new role, the organization would provide independent oversight of the development of defense products, promote efficiency through civil-military integration and eliminate corruption. In its role as the key institution responsible for civil-military integration, it occupied a unique position straddling the government/military divide. The military functions very much like a "state within the state" and its leading body, the Central Military Commission, has equivalent authority to the government's State Council.[30] Thus, coordination between military and government can be difficult, and COSTIND was created and re-organization in part to facilitate coordination. The CNSA was placed under COSTIND administration. Luan Enjie was promoted from his position as Deputy Administrator of CNSA to Administrator of CNSA and also Deputy Director of the superior COSTIND. He wielded his power quickly and the organization immediately set about developing a long-term space program that would include a lunar exploration project.

By early 2000, the CAS lunar team had broadened its effort and now consisted of representatives from other CAS centers such as the National Observatories of China, the Center for Space Science and Applied Research, Xi'an Institute of Optics and Precision Mechanics and the Shanghai Astronomical Observatory. In 2000,they submitted a final report, titled "The Scientific Objectives of a Chinese Lunar Resources Orbiter".[31] This was their ultimate response to official requests to explain the lunar program's scientific rationale. The report

---

28 Yang Zheng, "Xin You Yue Hui Zhao Zhao Yao Tu," *Jing ji ri bao* (*Economic Daily*)(2009).
29 According to Chinese law, the leader of COSTIND is chosen by the Standing Committee of the National People's Congress and confirmed by the President. COSTIND was known by the full Chinese name Guofang Kexue Jishu Gongye Weiyuanhui, or its shortened form Guofang Kegong Wei, which is often used in documents.
30 Government and military are reconciled at the apex, by the Party General Secretary (currently President Hu Jintao).
31 "Wo guo yueqiu ziyuan tance weixing kexue mubiao". Cited in Shao Gai He, "*Chang'E* BenyueZhilu (*Chang'E* Lunar Exploration Road)," in Guofang Keji Gongye (National Defense Science & Technology Industry) (2007).

highlighted several important points. First, it suggested that China should seek to discover and recover Helium-3 on the Moon. Such an element has been theorized to bean ideal starter fuel for fusion. Second, the report proposed that China embark on a "three phase" lunar program to include a probe, a lunar rover and a lunar sample return mission.[32]

Important pieces of the puzzle were now addressed. In Luan Enjie, a strong supporter of the project had ascended to a government position with influential power. Using his "inside access" and interpersonal connections, he was able to advocate the policy to top policy makers and contributed to its placement on the agenda. As Wang Shaoguang noted in his "inside access model", this method of agenda-setting typically occurs when the policy initiators are advisors to those in power (as opposed to decision makers or ordinary citizens) and when the degree of public participation is low. A high-level government insider is able to utilize both the power of his office and personal connections to push a policy onto the agenda. However, simply getting an item on the agenda is no guarantee of success. In policy making, the jump from an agenda item for discussion to an action item for implementation can be fraught with obstacles.

In the context of the Chinese lunar program, these obstacles included the approval of scientific studies and engineering plans, as well as intense bureaucratic negotiations over funding. In late 2000, the scientific rationale for the mission won approval from central government policy makers in the former State Planning Commission and the State Council. This approval was sufficient to get initial project funding for further studies. Full funding would be approved after the State Council and the military reviewed the final feasibility study and engineering plan. Additionally, the fragmented authoritarian framework suggests that a complex process of negotiating, bargaining and consensus building would occur before final funding approval.

5 Funding the project

In November 2000, the State Council published the *White Paper on China's Space Activities*, a policy outline which included lunar explorationas part of the country's national development plan.[33] The White Paper indicated that a pre-study for a lunar project had been approved. Planning for the mission would continue along two main lines, focusing on the scientific and engineering plans. Luan Enjie continued to serve as an influential policy advocate for the project.

In January 2001, COSTIND held a conference in the frigid northern capital of Harbin to discuss the way forward, focusing on the various engineering aspects associated with lunar exploration. Shortly thereafter, CAS held its own conference to discuss the science behind lunar exploration. In late 2001, Sun Jiadong, aneminent aerospace engineer and member of the Chinese Academyof Engineering, was designated as the new Chief Designer and began developing an engineering work plan.

In October 2002, Premier Zhu Rongji was briefed on the progress of lunar exploration program studies and he expressed his supportfor continuing work on the feasibility reports. A month later, Sun Jiadong organized COSTIND, the CAS, the General Armaments Department of the PLA, China Aerospace Science and Technology Corporation and over 200 scientists from universities around the country to begin the final demonstration phase of the project. Three months later, in March 2003, the project published its final report, "Feasibility Report on the First Phase of the Lunar Orbiter Project",[34] along with seven other related feasibility studies. This marked a significant step in the policy-approval process, and was a key requirement of the National Development and Reform Commission, State Council and military. The final engineering plan would be completed shortly thereafter.

---

32 This three-phase plan was adopted and is popularly known as "Rao, La, Hui(Orbit, Land, Return).

33 The report states: "To develop space science and explore outer space by developing a scientific research and technological experiment satellite group of the next generation, strengthening studies of space micro-gravity, space material science, space life science, space environment and space astronomy, and carrying out pre-study for outer space exploration centering on the exploration of the moon."

34 The report: "Yue qiu tance yi qi gongcheng zonghe lixiang lunzheng baogao".

Such feasibility studies are an essential part of policy-making in the space program, and in policy-making in general. Major state project proposals in China are usually forwarded by ministries (such as COSTIND) to the National Development and Reform Commission (NDRC, previously the State Planning Commission, which controlled the centrally planned economy). The NDRC requires feasibility studies for major projects, and it can either accept the work done by the ministry or insist upon an independent feasibility study.[35] The back-and-forth over the feasibility study between ministry and NDRC illuminates the process of consensus building essential to the fragmented authoritarianism of the political system. Major players contribute to the feasibility study, as a way to submit inputs, voice opinions and protect departmental resources. The NDRC may also reject the study and request resubmission, as happened several times previously with lunar probe studies.

The NDRC considers a long list of project proposals from various ministries and then approves them before sending to the State Council for inclusion in the Five-Year Plan. The Five-Year Plan outlines the nation's primary development goals from one major meeting of the Communist Party of China Central Committee meeting to the next. The State Council is usually the final arbiter of which projects will be included in the Five-Year Plan, which is ultimately ratified by the National People's Congress. Projects of sufficient scale and expense may also require review by the Ministry of Finance. Approval for some projects is made at the State Council level and even higher in certain leadership groups associated with the Central Committee.[36]

In January 2004 Premier Wen Jiabao and other members of the Party Central Committee approved the general plan to conduct long-term lunar exploration in three phases (orbit, land, return), and formally approved the budget for phase one.[37] The lunar orbiter project was given a name, Chang'E-1, which recalled the ancient Chinese Moon goddess myth. The Committee also approved four scientific objectives and five engineering objectives for the mission.[38] It seems likely that this elevated decision making was the result of the huge number of ministries and organizations involved in the activity, rather than the cost of the project, which was relatively modest for a major project (reportedly about US$190 million). The process whereby specific funding is granted is still highly opaque, and is probably dependent upon the complex social, political and economic environments in which the decision is made. Interestingly, funding for the project was reportedly increased after President Hu reviewed it and insisted a second back-up orbiter be manufactured, as a precaution against failure.[39]

6 Policy implementation and leading small groups

In order to implement policy across the various stove-piped ministries in China, a high-level coordinating body is necessary. In the case of the lunar project, both civilian and military ministries would need to work together, necessitating the need for a leading small group (*lingdaoxiaozu*- LSG) of substantial authority.

In China, there are a variety of LSGs that coordinate government policy on influential issues. Examples of leading small groups includethe Foreign Affairs, Finance and Economic Affairs, Agricultural Affairs Taiwan Affairs and others. These groups are formed by the State Council, the top government organ. The military also creates its own leading groups to coordinate policy, such as the PLA Leading Group on Earthquake Rescue and

---

35 Kenneth Lieberthal, and Michel Oksenberg, *Policy-Making in China: Leaders, Structures and Processes* (Princeton: Princeton University Press, 1988).
36 Kenneth Lieberthal, and Michel Oksenberg, *Policy-Making in China: Leaders, Structures and Processes* (Princeton: Princeton University Press, 1988).
37 Xinhua News, "Wo guo raoyue tance gongcheng dashiji" (The Record of Major Events in Our National Lunar Orbiter Exploration Program) http://news.xinhuanet.com/newscenter/2007-10/24/content_6933859.htm; "Chang'E Tanyue Jihua Shi Zenme Chulu De (the Genesis of the Chang'E Lunar Program)," in *DangzhengLuntan* (*GanbuWenzhai*) (*Party and Government Forum*) (2008).
38 Science objectives: obtain three-dimensional image of lunar surface; analysis of lunar surface characteristics; analysis of lunar soil; measurements of the lunar environment. Engineering objectives: develop and launch China's first lunar exploration satellite; develop lunar orbiting technology; develop scientific instruments to take lunar measurements; continue to develop lunar engineering expertise.
39 Xinhua, "Kexue Juece Zhu Hui Huang - Dang Zhong Yang Guanxin Yueqiu Tance GongchengJishi (a Brilliant Scientific Plan-a History of Center-Lunar Exploration Program Relations)," *Xinhuanet.com* December 12, 2007.

Relief Work and others.[40] Such groups are often formed by the Central Military Commission, the military's equivalent to the State Council. In this context, the unusual mix of government and military bureaucracies exhibited by the lunar program would require a hybrid leading small group.

The LSG for the Lunar Orbiter Project was created by the Central Committee of the Communist Party of China and the State Councilon 19 February 2004.[41] It is also believed that a "Lunar Probe Project Office" under the Party Central Committee was formed. This office is similar to the "Human Spaceflight Project Office" under the Party Central Committee.[42] Other space offices under the Committee relate to the five primary budget lines for the national space program: human spaceflight, lunar projects, navigation satellite constellation, high-resolution Earth observation and next-generation carrier rockets.[43] The Central Committee is one of the highest organs of power in China and exists just under the Politburo and the supremely important Politburo Standing Committee. Among its 300 or so members are the top leaders of the Party, government and military.[44] While Premier Wen Jiabao apparently chairs many key central committee meetings, including those associated with major space programs, the Central Committee itself is currently chaired by General Secretary Hu Jintao.

The new LSG was led by COSTIND Director and Central Committee member Zhang Yunchuan, and included leaders from COSTIND, the National Development and Reform Commission, the Ministry of Science and Technology, Ministry of Finance, General Armaments Department of the PLA, Chinese Academy of Sciences and China Aerospace Science and Technology Corporation (CASC).[45]

The leading group nominated key leadership positions within the program: Luan Enjie would be General Commander, Sun Jiadong was named Chief Designer and Ouyang Ziyuan was named Chief Scientist. Chief Designer Sun Jiadong would oversee the overall engineering plan, research and construction.[46] This "Chang'E Iron Triangle" (Chang'E *tie sanjiao*) formed the core programmatic leadership of the project.

Sun Laiyan (CNSA), Jiang Mianheng (CAS), Wang Wenbao (COSTIND) and Lei Fanpei (CASC) were selected as Deputy Commanders.[47] Chen Bingzhong, Jiang Jingshan and Long Lehao were named Deputy Engineers.[48]

The ministries represented in the initial LSG indicate that the lunar probe project was an unusually broad effort that spanned several centers of national power. The NDRC and the Ministry of Finance are state organs with supra-ministerial capacity for coordination. These organizations operate at the highest levels and serve to integrate and align the fragmented efforts of multiple ministries. In the pre-reform era, ideology played an important role in aligning efforts. However, to the extent that ideology now plays a coordinating role, it is limited to the bland pronouncements issued by top leaders to 'strive hard', follow the 'model of scientific development' and 'always remember the pioneering spirit of "two bombs, one satellite"'. While important to creating a sense of shared mission, such appeals to nationalism rarely serve to settle inter-agency disputes or solve budget problems. Supra-ministerial organs and leading small groups are now the most important national coordinating bodies and are centers of cross-ministry negotiation and consultation.

As is common with other major space engineering projects in China, the Leading Small Group established a "two-line command structure"[49]: The Lunar Orbiter Leading Small Group Office led by the General Commander

---

40 Alice Miller, "The CCP Central Committee's Leading Small Groups," in *China Leadership Monitor* (Palo Alto: Stanford University, 2008).
41 Ibid.
42 Kulacki, "A Place for One's Mat: China's Space Program, 1956-2003".
43 "China's Space Activities in 2006" and interview with Chinese space official.
44 Lieberthal, Governing China: From Revolution through Reform.
45 He, "Chang'E Benyue Zhilu (The Path of Chang'E to the Moon)".
46 "Chang'E Tanyue Jihua Shi Zenme Chulu De (the Genesis of the Chang'E Lunar Program)."
47 Notes of interest: Jiang Mianheng (Vice President of CAS) is former President Jiang Zemin's son. Wang Wenbao would later assume the role of Director of the China Manned Space Engineering Office in 2007. Additional biographical information: http://www.chinavitae.com/ and http://baike.baidu.com/.
48 He, "Chang'E Benyue Zhilu (The Path of Chang'E to the Moon)."
49 The human spaceflight program Shenzhou uses the same structure of authority, according to CMSEO Director General Wang Wenbao in an interview published on the China Manned Space Engineering website: http://www.cmse.gov.cn/.

and the COSTIND Lunar Exploration Engineering Center led by the Chief Designer.[50] These two lines managed the various subsystems of the project, as indicated in Fig.1. At the very top, and overseeing the two command lines was the Leading Small Group and its Director, Zhang Yunchuan of COSTIND. The only related office superior to this is believed to be the Lunar Probe Project Office of the CPC Central Committee. ZhangYunchuan was both the chair of the LSG and a member of the CPC Central Committee during the lunar probe project.[51] Note that the position of Chief Scientist does not carry the authority of the Commander or Designer positions.

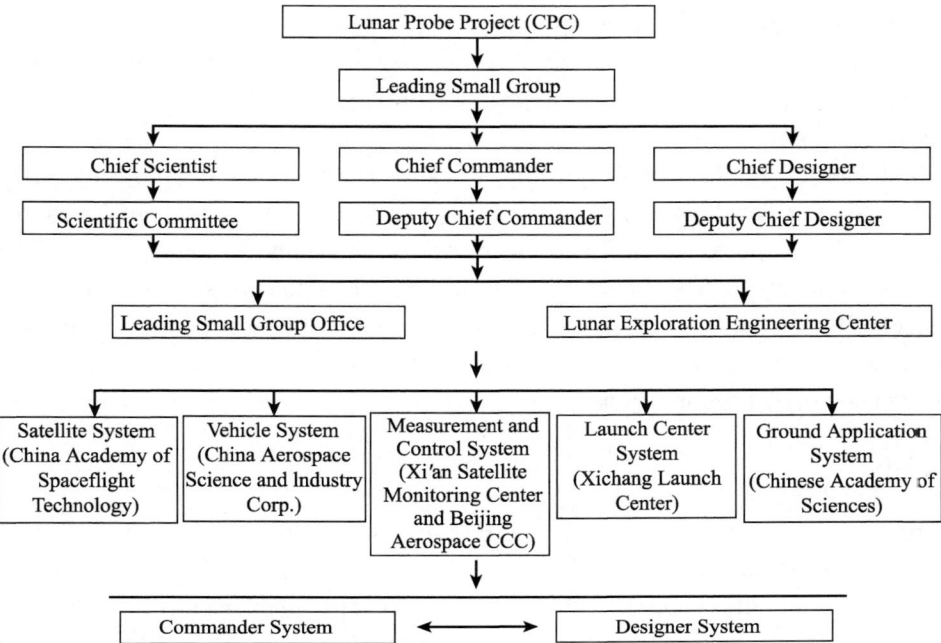

Fig. 1 China's Lunar Program Organization Source: Adapted in part from: WuWeiren, *Ben Xiang YueQiu(To the Moon)* (Beijing: Zhong guo yu hang chu ban she, 2007). Schematic has been translated and expanded with information from other sources

The program made rapid progress in 2004, completing work on several major subsystems. China's news agency reported that, on 4 February 2005, General Secretary Hu Jintao chaired a Politburo Standing Committee meeting to hear the progress of the lunar exploration project report. Participants regarded the "one satellite, one shot" approach as quite risky and declared that two satellites should be manufactured, so a back-up would exist.[52] Naturally, given the perceived authority of the General Secretary, this advice was followed to the letter and a duplicate satellite was made.[53]

The project entered into cooperative partnerships with Europe (ESA) and Russia (Roscosmos) for tracking support, data sharing and other activities. At the end of 2005, the leading group held its third meeting to review progress. In February 2006 Premier Wen Jiabao again inspected the project. In late January 2007, he would also visit the development team to inspect the work.

Anticipating the deluge of international and domestic attention that China's first lunar shot would command, COSTIND set up an additional Leading Small Group on News Propaganda to coordinate the media message in

---

50 Raoyue tance gongcheng lingdao xiaozu ban gong shi (Lunar Orbiter Project Leading Small Group Office) and Guofang Kegong Wei Yueqiu Tance Gongcheng Zhongxin (National Defense Working Group Lunar Exploration Engineering Center). Source:http://www.miit.gov. cn/n 11293472/n11293877/n12079125/n12079351/12082808.html.
51 China Vitae website: http://www.chinavitae.com/biography/Zhang_Yunchuan%7C323.
52 Xinhua, "Kexue Juece Zhu Hui Huang - Dang Zhong Yang Guanxin Yueqiu Tance Gongcheng Jishi (a Brilliant Scientific Plan - a History of Center-Lunar Exploration Program Relations)," *Xinhuanet.com* December 12, 2007.
53 According to Chinese lunar officials interviewed for this article, the back-up satellite, Chang'E 2, has been outfitted with new instruments and is scheduled tolaunch in 2011.

June 2007.[54] The LSG on News Propaganda Office was located in the COSTIND Propaganda Office. At the time of its inception, the group consisted of the powerful COSTIND Vice Director Chen Qiufa, and lower-level members of PLA-GAD and military departments.[55] This LSG was a deputy-level group chartered by COSTIND to coordinate on a specific policy subsystem within the larger program.

In October 2007 the LSG arrived at Xichang Satellite Launch Center in Taiyuan, Shanxi Province to review the mission. The group now consisted of a new leader, COSTIND Director Zhang Qingwei, and his deputies COSTIND Deputy Director and Chief Designer Sun Jiaodong, General Armament Department Deputy Minister Zhang Jianqi, Chang'E General Commander Luan Enjie and CASC Director Ma Xingrui.[56]

On 24 October 2007 the Chang'E-1 mission was launched and, just over a week later, China's first lunar satellite entered its orbit around the Moon. Newspapers around the country heralded the success of "China's two-thousand year old dream".

The Chang'E-1 mission can be regarded as a significant success. While the project experienced numerous setbacks and delays during development, from public appearances, the orbiter performed quite well during operations with only a few minor glitches. The Director of SASTIND and Administrator of CNSA held a news conference in late November 2008 to announce the key results of the mission, which included a complete high-resolution map of the moon, analysis of fundamental lunar surface elements and detailed observations of the space environment. According to the public data archive plan, data from the mission will be stored at several government centers and universities in China.

7 Conclusion

A careful review and analysis of the Chang'E lunar probe project, from policy proposal to policy implementation, reveals insights into the decision and policy-making processes of the Chinese space program.

First, the project was initially proposed and pushed by a small group of scientists at the Chinese Academy of Sciences. The *guanxi* relations between Ouyang Ziyuan and other CAS scientists enabled him to get initial "863" funding and begin drafting studies for consideration. Furthermore, the unique position of and status of CAS enabled them to garner attention at fairly high levels within government. However, this "inside access" was not enough to push a policy onto the agenda. Ouyang Ziyuan, Min Guirong and others submitted several feasibility studies and scientific reports, but all were initially rejected. A second explanation for the initial proposal rejection was that the initial rationale for a lunar probe project was based on political considerations, rather than scientific objectives. The authors were repeatedly instructed to re-submit with a more solid scientific plan.

Second, gaining a real supporter and advocate of the program at a high level of decision-making power was necessary to advance the project. The ascension of Luan Enjie to Deputy Director of COSTIND in 1998 greatly contributed to program success. He was able to advocate the project to top policy makers, who placed it on the agenda. As a result, the project was granted initial pre-study funding and included in a State Council-issued *2000 White Paper on Space Activities*. In this instance, CAS scientists did not initially have much *guanxi* with COSTIND officials. However, they did share the common goal of developing China's first lunar mission. *Guanxi* can often be used as a nebulous catch-all for processes not well understood by researchers. It is the mysterious force behind the scenes. While it undoubtedly plays a key role in policy making, its importance should not be overestimated. Similarly, the highly personalistic style that characterizes some decision making in the government is often in tension with the formal bureaucracy, which seeks to remove personal discretion by

---

54 http://english.gov.cn/wszb/zhibo114/content_712268.htm. Accessed on November 6, 2009.
55 Chen Qiufa would later become Vice Minister of the Ministry of Industry and Information Technology (MIIT), and also the Director of the newly reorganized State Administration for Science and Technology Industry for National Defense Bureau (SASTIND). In July 2010, he was appointed the new Administrator of China National Space Administration (CNSA), succeeding Sun Laiyan.
56 Two years after the mission launched, with the major task of coordination completed, the leading small group met again in Beijing. The membership included SASTIND (formerly COSTIND) Director Chen Qiufa, Minister of Education Chen Xi, GAD Deputy Minister Maj. General NiuHongguang, CAS President Yin Hejun, Chinese Academy of Engineering Vice President Pan Yunhe and CASC Director Ma Xingrui.

bolstering the power of office, position and process. In the lunar program example, this tension is readily apparent, with some decisions the result of personal connections and others of bureaucratic process.

Third, getting the project on the agenda involved both the support of an influential supporter and the complex inter-agency process of bargaining and consensus making. The consideration and re-consideration of feasibility studies by the NDRC and State Council is a way to build group consensus regarding a policy. Not only did the core group of CAS supporters and authors grow each year a proposal was submitted, but the policy makers in central government changed, enabling more people to become familiar with the policy idea. The process also forced the authors to continually refine their plans, moving away from political rationales and toward scientific rationales. To create the final, accepted feasibility studies COSTIND coordinated hundreds of representatives' inputs from dozens of organizations across the nation. Only once the report was approved, could the State Council and others approve initial funding for the project.

Fourth, the implementation of the lunar probe project was a vast undertaking coordinated at the highest levels of government. Because of the fragmented authoritarian nature of power, the participation of the CPC Central Committee in key decisions was necessary because the equally powerful policy-making organs of the government and military were involved. Several reports indicate that Premier Wen Jiabao made key program decisions and President Hu Jintao approved some decisions as well (such as the creation of a second back-up satellite).

Fifth, an assiduously incrementalist approach to policy making was revealed. Major decisions were carefully reviewed by numerous organizations before initial approval was given. Initial approval was often simply an instruction to conduct a pre-study feasibility report, to be completed before the next stage of approvals could be granted. Policy making is slow, methodical, highly rational and incrementalist.

Finally, membership in the LSG is not dependent upon personality, but upon position and organization. Throughout the project's lifetime, the head of the LSG has always been the Director of COSTIND (in 2003 COSTIND was restructured to become the State Administration for Science and Technology Industries for National Defense (SASTIND)).[57] Other positions have been filled as members have retired or taken up new jobs elsewhere. Some positions on the LSG have been eliminated as the function of the LSG changed (e.g., once the project completed the budgeting phase and major engineering phases, it appears that NDRC and Ministry of Finance left the group). This reveals a rigidly hierarchical and impersonal bureaucracy, much like the Max Weber's ideal bureaucracy. Weber insisted that duties in the bureaucracy were the function of an office or position, and not the individual. In the Chinese context, this is significantly different from previous eras in which the personality of the office holder was of fundamental importance.

The use of policy-making frameworks and models such as "fragmented authoritarianism" and the "inside access model" is helpful. The fragmented authoritarian nature of power in China explains the endless sessions of consensus building and incrementalism that exist in the policy-making system. This theory also explains the utility and need for leading small groups, which are able to coordinate across bureaucratic organs more likely to fight with each other over scarce resources than cooperate together on a national project. The inside access model is a much more sharply limited and defined model of agenda-setting. With regard to the lunar probe project, elite government scientists and policy entrepreneurs proposed and pushed the project onto the agenda. Clearly, this model describes a key agenda-setting process in contemporary China.

This study represents an early, detailed examination of the Chinese lunar probe project written in English and based on original Chinese sources. The mission only recently ended in March 2009, so the first draft of history has only just been sketched.[58] While this case study hopefully illuminates some of the rigorous push-and-pull between elite policy entrepreneurs and top decision makers, much more study is necessary in order

---

57 COSTIND was downgraded to a bureau, the State Administration for Science and Technology for National Defense Bureau (SASTIND), as a result of NPC legislation on 18 March 2008.
58 The Chang'E mission was terminated with a planned crash into the lunar surface on 1 March 2009.

to expand this model. Additional studies, especially with the cooperation of Chinese space experts, can begin to open up the "black box" of policy making in China's space program. The human spaceflight program is considerably more opaque but my research leads me to believe that there are a large number of similarities between policy-making in the two programs. For example, they share many of the same leadership personnel, with the exception that the highest levels of human spaceflight program leadership command greater authority and are more closely affiliated with the Central Military Commission and the General Armaments Department of the PLA.

Insight into China's space program could serve a number of purposes, and perhaps the most important would be a better understanding of how the nation operates and implements major projects. This may yield two results. First, it provides information about China's space strategy and its position and relationship to the country's development and modernization strategy. Understanding that important parts of China's space program are guided by highly rational, incrementalist policies that are concurred upon at the highest levels of government allows analysts to discern the policy-making processes and, potentially, the intentions of top leaders. Second, such insight can help to guide cooperation between the two nations. Transparency is one of the most important conditions that must be achieved before two countries can meaningfully cooperate on high-risk and large-scale projects. Indeed, before China can more fully cooperate with the U.S. space program on endeavors such as the International Space Station, civil space officials believe three essential principles must be demonstrated: transparency, reciprocity and mutual benefit.

Introduction to the Payloads and the Initial Observation Results of Chang'E-1

Sun Huixian[1]   Wu Ji[1]   Dai Shuwu[1]   Zhao Baochang[2]
Shu Rong[3]   Chang Jin[4]   Wang Huanyu[5]   Zhang Xiaohui[1]
Ren Qiongying[1]   Chen Xiaomin[1]   Ouyang Ziyuan[6]   Zou Yongliao[6]

(1. Center for Space Science and Applied Research, Chinese Academy of Sciences, Beijing 100190;
2. Xi'an Institute of Optics and Precision Mechanics, Chinese Academy of Sciences;
3. Shanghai Institute of Technical Physics, Chinese Academy of Sciences;
4. Purple Mountain Observatory, Chinese Academy of Sciences;
5. Institute of High Energy Physics, Chinese Academy of Sciences;
6. National Astronomy Observatory, Chinese Academy of Sciences)

Abstract Chang'E-1, the orbiter circling the moon 200 km above the moon surface, is the first Chinese Lunar exploration satellite. The satellite was successfully launched on 24th October 2007. There are 8 kinds of scientific payloads onboard, including the stereo camera, the laser altimeter, the Sagnac-based interferometer image spectrometer, the Gamma ray spectrometer, the X-ray spectrometer, the microwave radiometer, the high energy particle detector, the solar wind plasma detector and a supporting payload data management system. Chang'E-1 opened her eyes to look at the moon and took the first batch of lunar pictures after her stereo camera was switched on in 20th November 2007. Henceforth all the instruments are successfully switched on one by one. After a period of parameter adjustment and initial check out, all scientific instruments are now in their normal operating phase. In this paper, the payloads and the initial observation results are introduced.

Key words   Payloads   Observation results of Chang'E-1

1 Preface

Chang'E-1 was successfully launched form Xichang satellite launching site at 18:05 LT on 24th Oct. 2007. After 12 days long journey Chang'E-1 enters the observation orbit cycling the moon. At 16:49 LT on 20th Nov. 2007 stereo camera was turned on first time, Chang'E-1 opened her eyes to look the moon. The first moon picture taken by Chang'E-1 was published on 26th Nov. 2007; this is the symbol to indicate Chang'E-1 mission to get the success. All other payloads was switched on one by one from 26th Nov. 2007. After several days commissioning Chang'E-1 started the normal observation phase. So far Chang'E-1 payloads have been working for about 5 months, a lot of raw data have been collected, and data processing and analysis have been conducted. Initial results show that all payloads are working well and the exploration results agree with anticipation, the data can be used for science analysis. In the following paragraphs the payloads and the initial observation results are introduced.

2 Mission Objectives

The mission objectives of Chang'E-1[1] are:

(1) Obtaining the lunar surface three-dimensional stereo image. Three-dimensional stereo image is helpful for scientists and the public to study the Moon. A stereo camera and a laser altimeter are designed for this purpose.

(2) Analyzing the distribution of the useful elements and estimating their abundance. The interested elements include K, Th, U, O, Si, Mg, Al, Ca, Te, Ti, Na, Mn, Cr, La, etc. A spectrometer imager and a Gamma and X-ray spectrometer are designed for this purpose.

(3) Surveying the thickness of lunar soil. A microwave radiometer is designed to survey the brightness temperature of lunar surface and to estimate the corresponding thickness of lunar soil.

(4) Exploring the environment of the Moon. One high-energy particle detector and two solar wind ion detectors are designed for this purpose.

3 Payloads and the Initial Observation Results

3.1 Overview

To achieve the above mission goal, eight scientific instruments are chosen as the payloads of the Chang'E-1. These include stereo camera and laser altimeter for obtaining the lunar surface three-dimensional stereo image; sagnac-based interferometer spectrometer imager, Gamma ray spectrometer and X-ray spectrometer for analyzing the distribution of the useful elements of the moon surface and estimating their abundance; microwave radiometer for surveying the thickness of lunar soil; high-energy particle detector and solar wind ion detectors for exploring the environment of the Moon. In order to collect, process, store and transmit the scientific data of the payloads a special Payload Data Management System (PDMS) is also included.

PDMS is a distributed system based on the 1553B data bus, consisting of Bus Controller (BC), Solid State Recorder (SSR), High Rate Multiplexer (HRM), Remote Terminal (RT) and Power Distributor (PD). Most of the payloads access the system via 1553B data bus. Laser altimeter, high-energy particle detector and solar wind ion detectors are connected to the RT. PDMS acquires the science and housekeeping data of the payloads through 1553B data bus and store the data in the SSR or in the payload memories. When the spacecraft passes the ground station access range, the stored data and the real time data will be interpolated and encapsulated by the HRM to form a serial of Transfer Frame according the CCSDS standard[2] and be transmitted to the Earth by S-band transmitters. The data rate is 3 Mbit/s. The PDMS is flexible and effective, if any payload re-treats from the exploration the others will share its store and transmission resources. The capacity of the SSR is 48 Gbits. An image data compression board is included in the SSR, compression radio$\geq$2, which depends on the complicity of the original image.

The block diagram of the PDMS is shown in Fig.1.

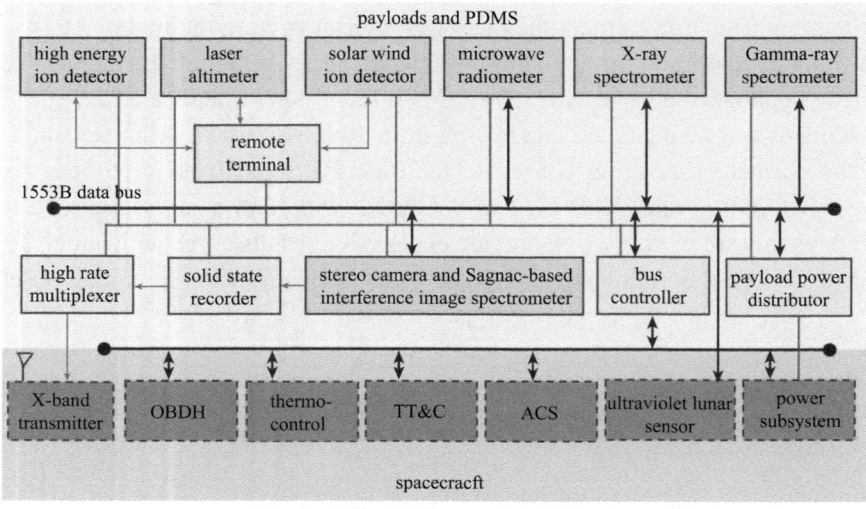

Fig. 1 Block diagram of the PDMS

3.2 Stereo Camera

The stereo camera can get the nadir, forward, and backward view of the moon. As the spacecraft moves, three two dimension lunar surface maps will be acquired. After data processing, a stereo image of the lunar surface could be obtained, as illustrated by Fig.2.

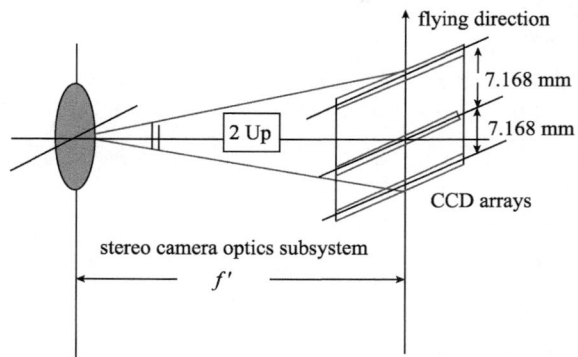

Fig.2 Schematic diagram of stereo camera

The stereo camera consists of an optics subsystem, a framework to support optics lens, the plane CCD array and corresponding signal processing subsystem. The three parallel rows of the plane CCD arrays can get the nadir, forward (17°), backward (−17°) view of the moon as the spacecraft moves forward.

The main specifications of the stereo camera are as the following.
- Spatial resolution is 120 m.
- Swath width L =60 km.
- Spectral range 0.5~0.75 μm.
- Optical channel 1.
- Data quantization 8 bit.
- MTF ≥ 0.2 at Nyquist frequency.
- S/N ($\rho = 0.2, \theta = 60°$) ≥ 100.
- B/H ≥ 0.6.
- Optical parameters:
focal length 23.3 mm;
FOV 40°;
f/number F/5;
pixel exposure time 3.2 ms, 7 ms, 20 ms, 84 ms.

Stereo camera was turned on first time at 16:49 LT on 20th Nov. 2007. The operation condition was chosen in the way to make solar elevation angle suitable for observation. Satellite was in TTC and data receiving ground station access range. Real time image transmission model was used, so the push broom picture could be displayed on the large screen in real time. This way image quality evaluation and parameters adjustment become easy. Actually the preset parameter for stereo camera was proper since the first line of the picture was pretty good and used for the first moon picture of Chang'E-1.Till now the full coverage of moon surface between 75° S to 75° N is almost finished, and the polar area picture will be scanned in the next period. Fig.1 and Fig.3 at page 363 and 364 of this book shows the first picture of the moon surface taken by Chang'E-1, and the one strip picture of the South Pole.

3.3 Laser Altimeter

Laser altimeter is designed to measure the distance between the spacecraft and the nadir point of the lunar surface. Laser altimeter consists of a laser transmitter and a laser receiver. The laser transmitter utilizes a laser diode pumped Q-switched Nd: YAG laser. The output beam divergence is improved to 0.6 mrad by Galileo refractor-type collimator. The return pulses are captured by Cassegrain-type reflector whose aperture is 120 mm and detected by Si-APD detector. The travel time of a pulse gives the information of the distance between satellite and the lunar surface.

The main specifications are as the following.
- Operating altitude:200(±25) km.
- Spot size (foot print):< Φ120 m.
- Transmitter wavelength: 1064 nm.
- Laser energy/pulse: 150 mJ.
- Pulse width:< 7 ns.
- Telescope:>120 mm.
- Pulse Repetition Frequency: 1 Hz.
- Range resolution: ≤ 1 m.
- Range Uncertainty: ≤ 5 m.

The layout of the laser altimeter is illustrated in Fig.3.

Fig.3   Layout of the laser altimeter

On 28th Nov. 2007 the laser altimeter was powered on. After several days adjustment and test in orbit, it was changed into normal operating status. The Laser Altimeter can obtain the elevation data of the whole lunar surface. The data will be used to produce the DEM map of the whole lunar surface. The first lunar surface DEM map based on the data taken by Chang'E-1 has been produced successfully. During all the operating time, the Laser Altimeter works normally, the detective rate of the raw data remains 97% and 76% after cullingpseudo-elevation. Fig.4 shows the one orbit raw data taken by the laser altimeter of Chang'E-1.

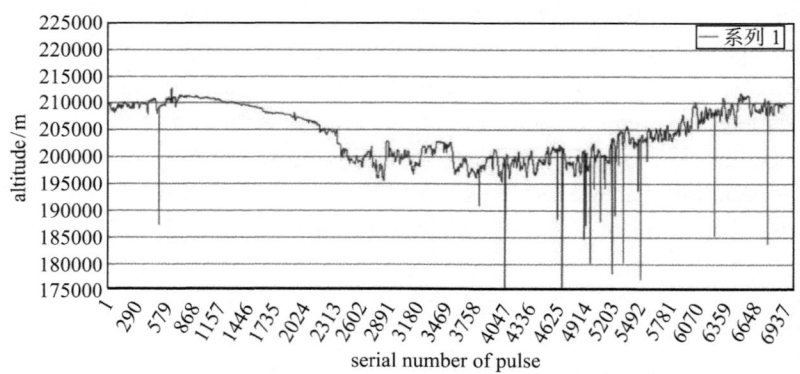

Fig.4   One orbit raw data taken by LA start from 11:25:00 LT 14th Dec. 2007

Sagnac-based Interferometer Image Spectrometer

The Sagnac-based interferometer spectrometer imager is designed to get the multispectral image of lunar surface. It contains three major optical subsystems, a Sagnac interferometer which produces the spatially modulated interferogram, a Fourier transform lens which frees the spectral properties of dependence on aperture geometry and enables the wide field of view, and a cylindrical lens which re-images one axis of the input aperture onto the CCD arrays. Fig.5 is the schematic diagram of the Sagnac-based interferometer spectrometer.

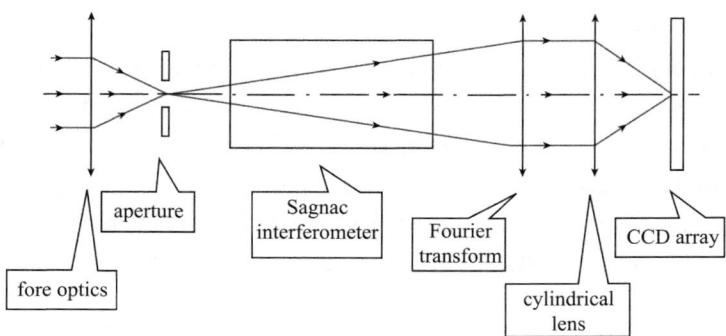

Fig.5 Schematic diagram of the spectrometer imager

The main specifications are as the following.
- Swath width 25.6 km.
- Spatial resolution 200 m.
- Spectral band: $\lambda = 0.48 \sim 0.96$ μm.
- Bandwidth: 32bands (9.6 nm at 0.5435 μm, 13.1 nm at 0.6328 μm, 20 nm at 0.7838μm, 22.5 nm at 0.8312 μm).
- Data quantization: 12 bit.
- Detector type: 256×256 pixels (2×2 pixels combined).
- MTF $\geq$ 0.2 at Nyquist frequency.
- $S/N(\rho \geq 0.2, \theta = 60°) \geq 100$.
- Optical parameters:
    focal length 34 mm;
    FOV 7.34°;
    f/number F/2.4;
    pixel exposure time 140 ms, 70 ms.

The stereo camera and the Sagnac-based interferometer spectrometer imager are integrated together. The layout of the instrument is shown in Fig.6.

Fig.6 Stereo camera and spectrometer imager

The image spectrometer were powered on first time at 00:04 LT on 27 Nov. 2007. Based on the reference of parameter adjustment results of stereo camera the preset parameter for image spectrometer was quite well. The Pixel exposure time and the dynamic range are satisfied for the requirements. The obtained pictures are clear and sharp, S/N satisfies the requirements for spectrum calculation. Fig.5 at page 365 of this book shows the calculated 3 bands picture, the synthesized pseudo-color picture using the 3 bands, the interference pattern and the recovered spectrogram.

3.4 Gamma Ray Spectrometer

Gamma-Ray Spectrometer (GRS) has been proven to be a kind of powerful instrument for remote measuring the abundance of chemical elements, like C, O, Mg, Al, Si, K, Ca, Fe, Th and U on the planetary surface. The main detector of the GRD is a12 cm diameter×7.6 cm long CsI (Tl) crystal. It is surrounded on the sides and back by a single CsI crystal shield, approximately 3 cm thick. This CsI reduces the gamma rays coming from the spacecraft, shield as a charged particle shield, and reduces the Compton background. Two gamma-ray spectra are collected simultaneously, the raw CsI shield spectrum and the CsI (Tl) spectrum in anticoincidence with the CsI shield. The main specifications are as the following.

- Effective area of the sensor. Main CsI crystal φ118 mm. Anticoincidence CsI crystal 30 mm thick.
- Energy resolution ⩽ 9% at 662 keV.
- Energy range: 300 keV~9 MeV.
- Total energy channels: 512.
- Anticoincidence efficiency⩾ 90%.

The layout of the GRS is shown in Fig.7.

Fig.7 Layout of Gamma ray spectrometer

GRS started the normal observation since 28 Nov. 2007. Sample spectrums from the GRS that illustrate the excellent quality of the data are shown in Fig.8. It can be seen that typical gamma-ray lines from lunar surface are clearly identified in the spectrum. The data analysis shows that the main performance parameters of GRS met all the requirements of design.

3.5 X-ray Spectrometer

X-Ray Spectrometer (XRS) consists of lunar X-ray detector and solar X-ray monitor. The main goal of XRS is to detect the fluorescent X-ray from the lunar surface and provide abundance distribution of three major rock-forming elements Mg, Al, Si on the Moon. When solar X-rays or Cosmic-rays bombard the lunar surface, some elements will emit fluorescent X-rays. The species of these elements can be identified using the characteristic lines and the abundances can be determined from the intensity of the emitted X-rays. Elemental composition is crucial in learning not only the geochemical nature of terrains on the Moon surface, but also the history of impact activity and volcano-tectonic on the Moon. XRS is based on Si-PIN diode, a kind of semicon-

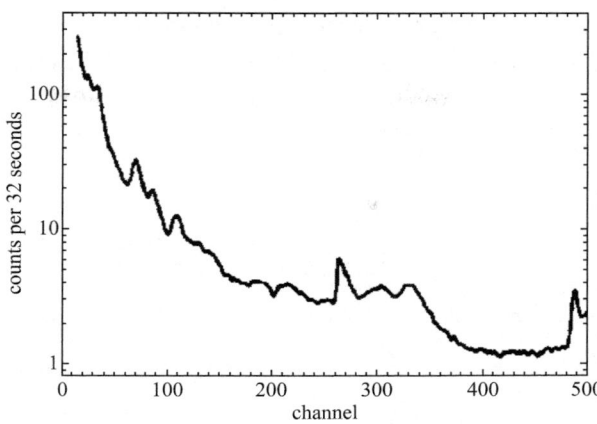

Fig.8 Pulse height spectrum measured during the first 8 h of 28 Nov. 2007 by the GRS

doctor detector, having a better energy resolution and less weight than that of proportional counters which were ever used in Apollo15 and 16 and the NEAR mission to the asteroid Eros. Two types of Si-PIN sensors are used in XRS. One is for lunar Soft X-ray (1~10 keV) Detector (SXD) and the other is for Hard X-ray (10~60 keV) Detector (HXD).The main specifications are as the follows:

- Detection area: 1 cm^2 (SXD: 4 units), 0.75 mm^2 (solar X-ray monitor: 1 unit), 16 cm^2 (HXD: 16 units).
- Energy range: SXD and solar X-ray monitor 1~10 keV, HXD 10~60 keV).
- Energy resolution: $\leq$ 10% at 59.5 keV (HXD), $\leq$ 600 eV at 5.95 keV (SXD and solar X-ray monitor).
- Spatial resolution: 170 km × 170 km at 200 km altitude.

The layout of the XRS is shown in Fig.9.

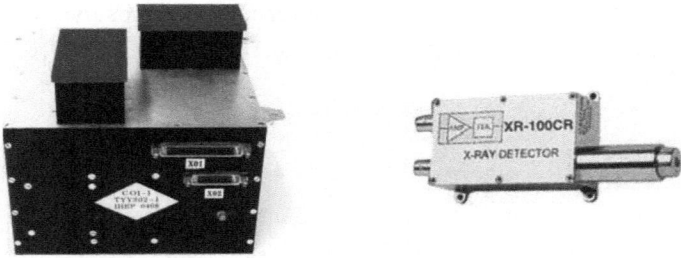

Fig.9 Layout of X-ray spectrometer (right: solar X-ray monitor)

At the end of Nov. 2007, XRS began commissioning. Unfortunately, 2007 is the solar minimum year in a solar cycle. Data from the GOES-XRS which measured the solar X-ray flux in a soft (1~8 Å) and hard (0.5~3 Å)energy band also showed that during the quiescent period solar X-ray had the lowest flux around A0.3 level. As the result, no significant elemental characteristic line was found even in the coadding spectrum from all the SXD in several hours' integration. This situation lasted no longer than 10 days. On December 5th, 2007, a sunspot appeared on the east limb of the sun. Then the solar X-ray flux increased gradually and reached the B1-level in a few days. Until December 20th, 2007, solar X-ray flux started to fall as the sunspot rotated around to the far side of the Sun. During 15 days of solar flare period, spectra obtained by XRS indicated that low energy lines (Mg:1.25 keV; Al:1.49 keV and Si: 1.74 keV) are observed and the Calcium K$_\alpha$ line (3.69 keV) is also unambiguous. At the end of year 2007, a big long duration C-class solar flare began at 00:30 UTC December 31st, 2007, as shown in Fig.10 (Plate I). The flare lasted for more than 2 hours and nearly 50 minutes corresponded to XRS observations. When the flare reached its peak C8.7 level, Chang'E-1 satellite was just flying over the south pole and started to travel northward on the far side of the Moon.

XRS's ground track is along the 93° W longitude during the big X-ray flare and the footprints of the 4 SXD units included areas of only highland lithologies expect the Mare Orientale. Fig.1 at page 363 of this book shows the rough spectra of 1 SXD unit summed for the interval 00:57 UTC to 01:25 UTC.

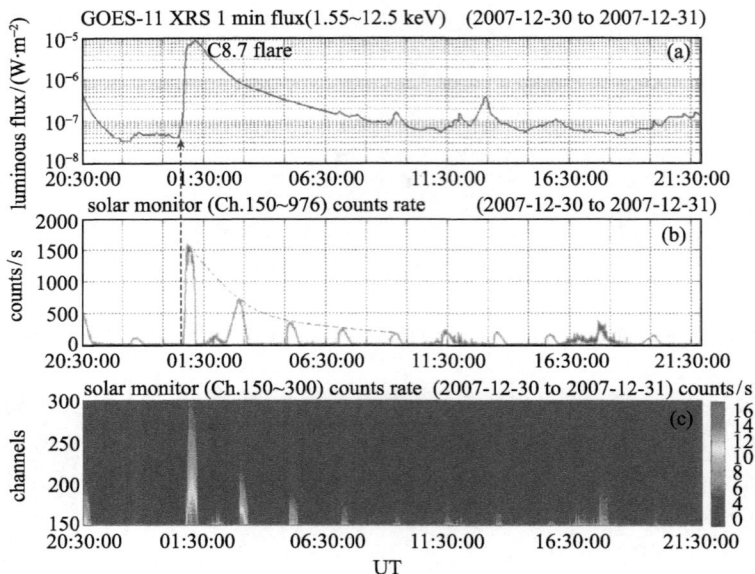

Fig.10 Counts rate (1.5~10 keV) and instantaneous spectra (1.5~3 keV) from solar X-ray monitor (b) and (c). Data from GOES are shown (in units of W/m^2) for comparison (a)

The spectra shown in Fig.11 indicates that a merged peak of low energy lines (Mg:1.25 keV, Al:1.49 keV, Si:1.74 keV) is detected and the Ca K$_\alpha$ - line (3.69 keV)and the Fe K$_\alpha$(6.40 keV)and K$_\beta$ (7.06 keV) lines are also prominent. The quite strong Ca peak implies that the area of lunar surface observed should be enriched in calcium, which coincides with the highland components (eg: anorthosite).

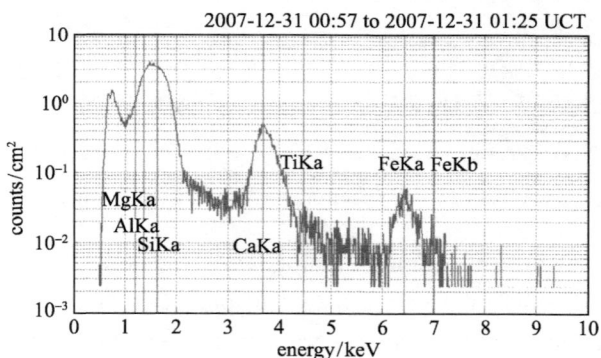

Fig.11 Rough spectra from 1 SXD unit represent a merged peak of magnesium, aluminum and silicon and the unambiguous characteristic lines from calcium and iron

3.6 Microwave Radiometer

Microwave radiometer System (MWS) is designed to survey the thickness of lunar soil. Microwave radiometer uses 4 different microwave bands to get the radiation of the lunar surface. The receivers can survey the radiate brightness temperature of lunar surface, and the temperature resolution is about 0.5 K. The brightness temperature radiation received by receivers consists of two parts, one part is radiated by lunar soil and other part is radiated by lunar rock, as illustrated by Fig.12.

The radiation intensity is the function of dielectric constant, temperature, wave frequency, soil thickness, surveying angle. By surveying the radiate brightness temperature, we can get the thickness of lunar surface soil. 4

frequencies are chosen for microwave radiometer receivers. Besides the receivers, four calibrating antenna are used to compensate the cosmic radiation. The main specifications are showed in Table 1.

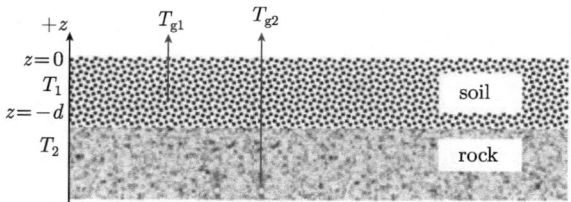

Fig.12 Brightness temperature radiation of lunar surface

Table 1 Main specifications of microwave radiometer

Frequency/GHz	3.0	7.8	19.35	37
Bandwidth/MHz	100	200	500	500
Time/ms	200	200	200	200
NEDT/K	0.5	0.5	0.5	0.5
Linearity	0.99	0.99	0.99	0.99
Resolution/km	50	35	35	35

The configuration of antennas for sounding and calibration are shown in Fig.13.

Fig.13 MWS antennas configuration

In Table 1, the left facet is installed with four calibration antennas which point to cold space at $+x$-axis direction, and the right facet is installed with sounding antennas which point to lunar surface at $-z$-axis direction in satellite local coordinate system. The satellite is flying toward $+x$-axis direction.

After commissioning since November 2007, MWS has been in their normal operating phase except in the period of parameter adjustment and initial check out. Till now, some raw data have been decoded and pre-processed for a series of quality check, algorithm-based analysis, two-point calibration, non-linearity correction, and brightness temperature reversion. These data are going through a detailed calibration and validation using pre-launch experiment results and microwave data from all possible sources of measurements and theoretical simulations. Here we give some initial results of measurements of MWS during its first month operation. Fig.14 shows lunar night brightness temperature at 3 GHz from Dec. 4 to Dec.30 of 2007. We may find that the trend of brightness temperatures changing with latitudes is very clear. At polar area brightness temperatures are lower than 100 K. It should be noted that the data shown in Fig.14 (Plate I) have not been validated with cold space brightness temperature 2.7 K. Since the cosmos background is complicated and changeable, the temperature of lower point of two-point-calibration may be contaminated and should be examined carefully before final data published. Fig.15 (Plate II) gives similar range of data of 37 GHz at lunar day. The obvious influences heated by the Sun exit in Fig.15 because the temperatures are much higher in the equator than them in the polar. The brightness temperatures generally increase with frequencies at daytime, while at night; the phenomena are not the same as at day because of variable temperature profiles and depths[3].

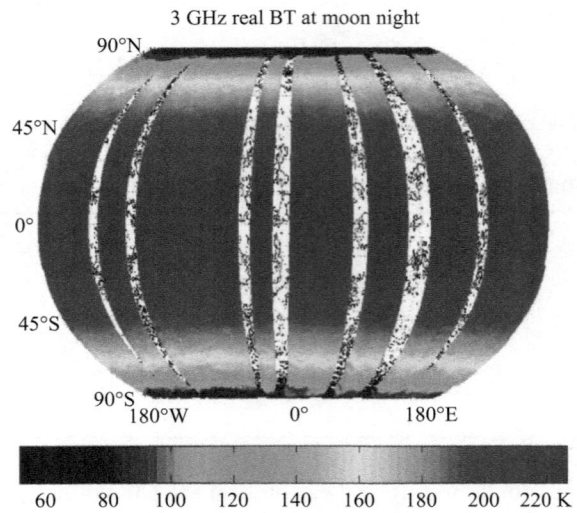

Fig.14 lunar night Brightness Temperature (BT) at 3 GHz from Dec. 4 to Dec. 30 of 2007

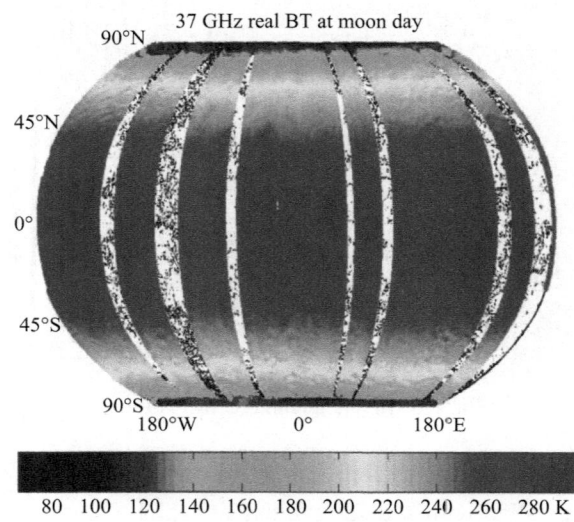

Fig.15 lunar day Brightness Temperature (BT) at 37 GHz from Dec. 4 to Dec. 30 of 2007

3.7 High Energy Particle Detector

The high energy particle detector is designed to analyze the heavy ions and protons on the space around the Moon. The protons with energy between 4 Mev and 400 MeV can be detected. The heavy ions, such as He, Li and C, will also be analyzed. Three slices of semiconductor detector construct telescope sensor system. When comic particles go through the semiconductor detectors, their deposit energy could form electric pulse that is amplified for count. Analyzing the height of pulse can identify different particles. The main specifications are as the following.

High-energy particle detector:

• Electron: $E_1 \geqslant 0.095$ MeV, $E_2 \geqslant 2.2$ MeV.

- Proton.

P1: 4~8 MeV.
P2: 8~15 MeV.
P3: 15~32 MeV.
P4: 32~70 MeV.
P5: 70~160 MeV.
P6: 160~400 MeV.

- Heavy ions.

He: 13~130 MeV.
Li: 34~260 MeV.
C: 117~730 MeV.

The layout of the instrument is shown in Fig.16.

Fig.16 High energy particle detector

High-energy particle detector was powered on at 21:55 LT Oct. 26, 2007. Fig.17 shows the house keeping parameters response while satellite cross the radiation belt from 2007-10-27 to 2007-10-29. When the satellite enters the moon orbit, the science data shows the background noise only and no high energy particle flux was measured due to the solar quiet year. This result is consistent with the anticipation.

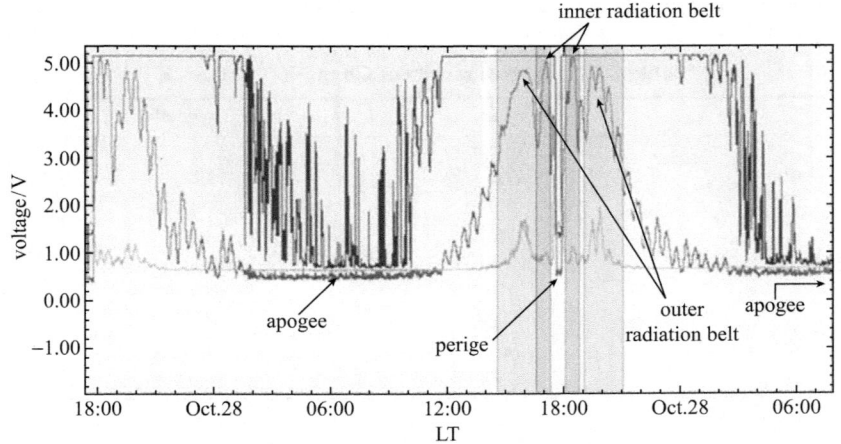

Fig.17 House keeping parameters show satellite cross the radiation belt

3.8 Solar Wind Ion Detector

The two Solar Wind Ion Detectors (SWID)are designed to analyze the ions with low energy on the same space with high-energy particle detector. The two detectors are vertical to each other. The solar wind ion detector

consists of a collimator, an ion analyzer and a MCP amplifier. The main specifications are as the following.
- Detect particles: Solar wind ions.
- Energy range: 0.05~20 keV.
- Energy sweep step: 48.
- Field of View: $6.7° \times 180°$.
- Angular resolution: $6.7° \times 15°$.
- Geometric factor: $8.1 \sim 10^{-6} E$ $(cm^2 \cdot sr \cdot keV)$.

The layout of the instrument is shown in Fig.18.

Fig.18 Solar wind detector

At 22:01 LT Nov. 26, 2007 SWID A was powered on, at 22.21 SWID B was powered on. The observation results show that the data detected by SWID changed periodically, 127 minutes per cycle, and this is identical with the satellite orbit period. Due to the two detectors are installed on craft in different directions, SWID A detected the solar wind in certain polar angles and the counts increase then decrease. However, SWID B detected the solar wind in all polar angles with the similar diversification of SWID A. All the results are consistent with expect and show the instrument is normal and the detected data are available.

Dec. 8 to 9, 2007, the moon was in line between the sun and the earth, the instruments detected the solar wind plasma. Table 2 shows the comparative measurement's result of Chang'E-1 with the similar instruments on ACE satellite. The results are close, show that the detected results of the two instruments are identical.

Table 2 Compares results of Chang'E-1 and ACE

Time	unit	ACE satellite	Chang'E-1
2007-12-08 16:30 LT	N_p /cm^3	1.7	2.0
	V_p /(km·s^{-1})	297	275
	T /K	1.81×10^4	3.83×10^4
2007-12-08 18:20 LT	N_p /cm^3	1.8	1.6
	V_p /(km·s^{-1})	299	271
	T /K	1.58×10^4	3.36×10^4
2007-12-09 18:00 LT	N_p /cm^3	1.2	1.2
	V_p /(km·s^{-1})	330	297
	T /K	4.89×10^4	6.53×10^4

Fig.19 shows the periodic changes of counts with accumulator of 48 energy steps together at 9th polar angle while a detector in solar wind.

3.9 SEL During the Operation

Generally speaking, the payloads of Chang'E-1 are working pretty well, no fault was found in the instruments. But SEL did take place in some payload instruments caused by the space environment. According to the observed results the SEL rates are much higher than that in the earth orbit. After investigation and tests two reasons were found.

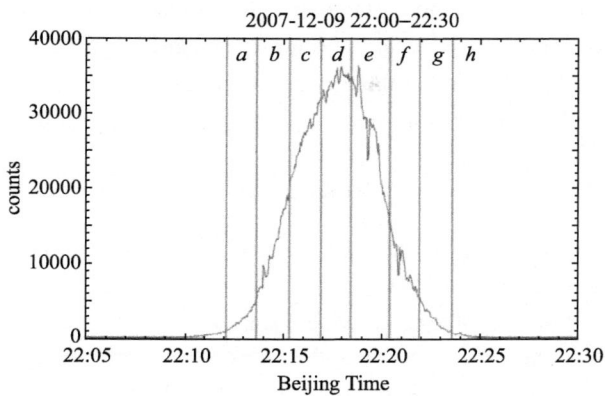

Fig.19　Accumulator of 48 energy steps at 9th polar angle

First of all at earth adjacent orbit quite a lot cosmic ray were blocked due to the shield of magnetic field of earth. Especially the cosmic ray with lower energy is difficulty to pass through the magnetic layer and to enter the earth space. For example, the fluxes of a-particles and cosmic Carbon are quite different at GEO and lunar orbit (see Fig.20, Plate II)[3].

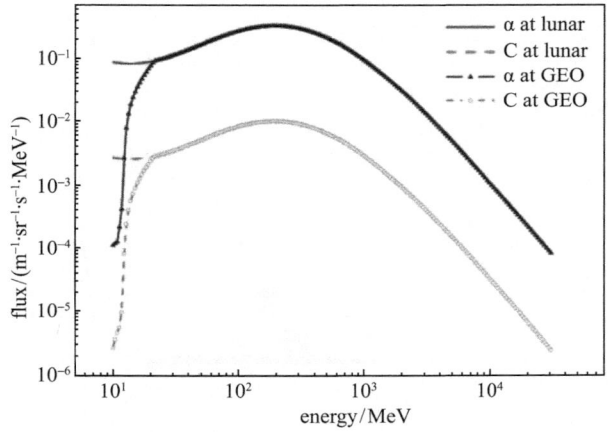

Fig.20　Differential fluxes of cosmic Carbon and a-particles at GEO and lunar orbit

Secondly, all the instruments that SEL occurred used a kind of extreme latchup susceptible SRAM. The ground tested devices have demonstrated a very low Linear-Energy-Transfer (LET) threshold to latchup —— as low as 1 MeV·cm^2/mg. These devices have been used in satellites at earth orbit, but no obvious problems were found. Based on the above analysis the reason is that very a few heavy ions can enter the earth adjacent space due to the shield of magnetic field. However the situation for Chang'E-1 is different, the higher fluxes of heavy ions with lower energy are easy to cause the susceptible SRAM to latch up.

Fortunately the current value are not very high when the device latchup, so it will not lead to the hard or permanent failure. After turn-off the instrument and turn-on again, the device can recover to normal condition. Some instrument for example the soiled state data recorder will loss the data when it turned off. So it effected the acquisition of the whole moon's surface map. The loosed data could be reacquired during second or even third lunar coverage cycle.

4 Conclusions

All eight payloads of Chang'E-1 are normal during the commissioning and preliminary observation phase. The sounding results agree with the expected; the data can be used for science analysis. Latchup of some instruments are well investigated and have a series of countermeasures, the negative effects could be reduced to as low as possible. The predetermined science objectives could be reached with prolonged life time. So far the health of the satellite is quite well and the remaining fuel is enough to support extended operation period.

References

[1] Sun Huixian, Dai Shuwu, Yang Jianfeng, Wu Ji, Jiang Jingshan. Scientific objectives and payloads of Chang'E-1 lunar satellite. *J. Earth Sys. Sci.*, 2005, 114(6):789—794

[2] CCSDS. AOS Space Data Link Protocol. Blue Book, Washington, DC, USA, 732.0-B-2, 2006

[3] Wang Zhenzhan *et al*. Methods of remote sensing Lunar surface by Microwave sounder on CE-1.Submitted to 2008 International Conference on Microwave and Millimeter Wave Technology, Nanjing, 2008

[4] Allan J *et al*. CREME96: a revision of the cosmic ray effects on micro-electronics code. *IEEE Trans. Nuc. Sci.*, 1997, 44(6):2150—2160

Advances in Lunar Exploration Detectors

Xu Tao[1,3]   Ouyang Ziyuan[1]   Li Chunlai[2]   Xu Lin[1,2,3]

(1. Geochemistry Institute, Chinese Academy of Sciences, Guiyang 550002, China;
2. National Astronomical Observatories, Chinese Academy of Sciences, Beijing 100012, China;
3. Graduate School, Chinese Academy of Sciences, Beijing 100039, China)

Abstract   Due to the rapid development of modern science and technology, many advanced sensors have been put into use to explore our solar system, including the Moon. With the help of those detectors, we can retrieve more information to about the Moon's composition and evolution. The Clementine(January, 1994), Lunar Prospector (January, 1998) and especially Smart-1(September, 2003) launched successively have demonstrated the next-generation planet exploration techniques. Now China has decided to send a probe to the Moon. So it is necessary to overview the development of detectors used for the scientific observation of the Moon. In this paper, some main instruments used to acquire geochemistry information are described, which include UV-VIS-NIR CCD imaging spectroscope, neutronray, gamma-ray, and X-ray spectrometers. Moreover, the payloads of China's first lunar satellite are introduced briefly.
Key words   Advance   Detection instrument   Lunar exploration

1 Introduction

Samples returned from the Moon can provide much information about the composition and evolution of the Moon. However, the regions studied were very limited (Craw ford, 2001). The earth-based observation can be effected by using a telescope to get reflectance data from the Moon (Geosciences Node, 2004). But its spatial resolution is very poor, only the near side of the Moon can be observed. The remote-sensing method can overcome the above shortcoming.

The remote sensing of the Moon was first tested during the Apollo program, such as Gamma-ray experiment (Lunar and Planetary Institute, 2004). It is suitable for continuous and long-term observations. It can make it possible the global observation of the Moon, depending on the designation of the orbit. Detectors are held in probes. The sensors can collect the geochemistry and geophysics data about the Moon. The quality of the detectors has been improved with the rapid development of modern computer and electronics techniques, ushering in the lunar remote sensing into a new era, as is marked by the successive launching of Clementine(McEwen and Robinson, 1997), Lunar Prospector (Hubbard et al., 2002) and Smart-1 (Racca et al., 2002). All of them are characterized by lightweight, high efficiency and low manufacturing cost.

China has decided to launch its own lunar exploration program and send a lunar probe in three years (Space Today On line, 2003). So it is necessary to overview the development of detectors. Only sensors, which can bring us geochemistry information, are discussed below.

2 Development of optic sensors

2.1 Introduction

Optic sensors detect and record the intensity of light from the Moon. The light can be divided into three

本文原载于 CHINESE JOURNAL OF GEOCHEMISTRY, 2005, Vol.24, No.1, 95~100。

spectral regions according to its wave length or frequency. The first is ultraviolet ray (UV, 0.003-0.380μm), the second is visible light (VIS, 0.38-0.76μm), and the third is infrared ray (IR, 0.76-1000μm). There are two energy sources from the Moon, which can be detected by optic sensors. The first is the incoming and reflected sunlight. Its energy peak is located in the UV-VIS region. The second is thermal emission. It has a continuous energy distribution. But its maximum emission is related to the surface temperature of the Moon, typically in the mid-infrared region.

Up to now, only the UV-VIS-NIR region has been applied to remote-sensing of the Moon. The photo in the UV-VIS region can excite some transition elements, resulting in absorption features in the reflected sunlight. The photo in the NIR region can also excite some transition elements. The mid-infrared region has never been applied to remote-sensing to the Moon. But the energy in the mid-infrared region can cause lattice vibration of silicates. The fundamental vibration absorption feature of silicates can be used to investigate the distribution of minerals and rocks on the lunar surface.

Unlike the Mars, the Moon's atmosphere is very thin (Heiken et al., 1991). And the absorption effect of the atmosphere can be ignored. So the remote sensing in the UV-VIS-NIR range is very effective, which can be used to investigate the mineralogy of the lunar surface. And water molecules have a strong capability of vibration absorption in the NIR region, which can be used to confirm the existence of water ice in the lunar poles.

In the past, the UV-VIS-NIR sensors often used several spectral channels to receive light. The channels are also called the bands. Only some necessary spectral channels are chosen according to the spectral characteristics of lunar samples, the general practice is to choose the bands characteristic of typical lunar minerals (mainly pyroxene, olivine and plagioclase). The main absorption feature of these minerals is due to the 1-μm and 2-μm Fe^{2+} electron transfer absorption feature. So more filters should be distributed around those features.

Taking the USA Clementine for example, its US-VIS sensor has five channels: 450, 750, 900, 950 and 1000 nm. And the NIR camera observes the Moon through other six band-passes: 1100, 1250, 1500, 2000, 2600, and 2780 nm. You can find there are four different filters (900, 950, 1000, and 1100 nm) around the 1-μm feature.

2.2 CCD detecting system

Early optic sensors used light-sensitive film to measure and record the radiation intensity of the lunar surface, just like the camera we use in our daily life. But the spectral information is difficult to derive. Often topography information and images are gained. The spectral radiance intensity cannot be gained reliably until the use of the charged-coupled device (CCD) detecting system. Not only can CCD cameras get the image of the lunar surface, but also they can record the intensity of light from the Moon at different wavelengths if used with spectrometer techniques. The equipment is called imaging spectroscope. Imaging spectroscopy refers to the application of reflectance/emittance spectroscopy to every pixel in a spatial image (USGS Spectroscopy Lab., 2002).

The CCD camera includes a multi-cell array in two dimensions. When lunar probes circle around the Moon, every cell receives light from a fixed area on the lunar surface; the area is determined by the instantaneous field-of-view of the cell. The light-sensitive semiconductive material determines the sensitivity of the CCD system. When the material encounters light photon, it can generate free electrons from its surface according to the photoelectric effect of Einstein. The ability of a detector to turn incoming photos into useful output is termed quantum efficiency (QE). Generally it is defined as the ratio of incoming photos to those actually detected and stored in the device. The increase of QE can improve the quality of optic sensors, such as SNR. The QE is determined by the CCD material and the manufacturing skill (Howell, 2000).

The charge collected within each pixel is measured as a voltage and converted to an output digital number, which is often referred to as counts or ADUs (analog-to-digital units). The amount of voltage needed (i.e., the number of collected electrons or received photons) to produce 1ADU is termed the gain of the device. The process of conversion of the output voltage signal to a DN is performed within a device called an analog-to-digital converter (A/D or ADC). The counts per unit time correspond to the light intensity if the gain is fixed (Howell, 2000). Commonly, the relationship between the income light intensity and the output counts intensity is linear. So the counts intensity can be used for spectral quantitative analysis if the quantitative model is valid (Hapke, 1993).

The spectral data of different wavelengths for the same area can tell us about its possible geochemical composition. But before spectral analysis, the raw count number must be calibrated and converted to radiance or other planetophysical property. The calibration step is necessary to the final utilization of the raw digital data. Take the calibration of Clementine UV-VIS-NIR raw data for example. From raw data DN (j,i) to final image data, it will involve many steps, such as readout errors or dark current reduction, spectral calibration, and photometric correction. The spectral calibration needs a spectral standard to gain the calibration factor of each filter. The spectral standard of Clementine is an Apollo 16 soil sample numbered 62231. After spectral calibration, the data are normalized to a set of standard observation geometries with the incidence angle $i=30°$ and the emergence angle, $e=0°$ (Planetary Geosciences Group of Brown University, 1995).

Once the calibration of the raw data is finished, the output image data can be used to infer the composition of the lunar surface. For example, the 750/415nm ratios are sensitive to lunar soil maturity and can be represented by the red-channel in Remote Sensing software. The 750/950 ratios are sensitive to the presence of mafic minerals and are displayed as the green channel. Areas that are blue (415/750) in the image are the result of fresh soils or a concentration of the mineral ilmenite ($FeTiO_3$) in mare (USGS Astrogeology Research Program, 2004). If to study mineralogical composition, we can plot those spectral values to get discrete spectra (Fig.1). Then we can estimate the possible mineral composition based on the knowledge of lunar reflectance spectra features. The mineral composition within the scene of a pixel is not pure, so we need to choose suitable end-members and use linear unmixing or unlinear unmixing to deconvolve the calibrated spectral data (Hapke, 1993). But the number of end-members cannot exceed the channels of the UV-VIS camera.

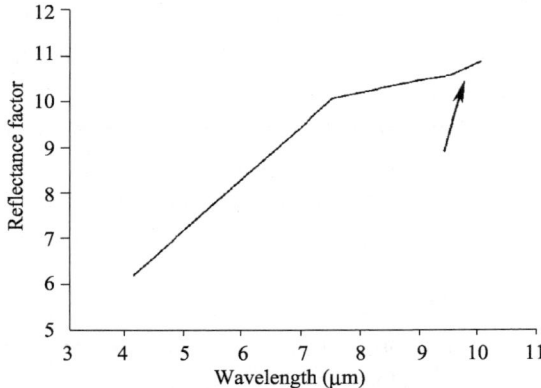

Fig.1 Illustration of the five-color spectra of Clementine multi-spectra UV-VIS data, the arrow points to absorption feature

The linear relationship between the spectral response of CCD and the received radiance is limited. When the light intensity is too strong, the CCD device maybe saturated. Then the linear relationship would be destroyed. The data gained would not be reliable. The saturation can be controlled by the gain state of the CCD camera. When the reflected light is very strong, we can adjust the gain value and the bit number of the A/D converter to avoid saturation, allowing images of scenes with both shadow and bright light to be recorded without saturation. Another way is to choose the CCD material with a wide dynamic range.

2.3 Advanced optic instruments of Smart-1

The Clementine launched in 1994 has many advanced features. It weights only 410g. The dynamic range of its UV/VIS camera was 15000 (The Center for Computational Science of NRL, 2002). But the Smart-1 of European Space Agency (ESA) launched recently are more advanced than Clementine, including the most advanced optic sensors (Marini et al., 2002).

The brightness of any illuminated surface depends on the illumination/observation geometry. This relationship is called the photometric properties. The photometric properties of the Moon's surface bear information about its structure (Kreslavsky and Shkuratov, 2003). The Smart-1 carried camera AMIE (Marini et al., 2002; Kreslavsky and Shkuratov, 2003; Muinonen et al., 2002). It can look at the selected regions from different angles under different lighting conditions, and then provide new clues to how the lunar surface has evolved.

On the other hand, the infrared spectrometer SIR (VIS/NIR) on Smart-1 has used powerful hyper-spectral techniques (Marini et al., 2002). The hyper-spectral technique means high spectral resolution. The number of bands is much higher than that of multi-spectra technique. So the information is much richer. SIR has 256 wave length bands, from 0.9 to 2.4 μm.

The UV-VIS-NIR optic sensors have the best spatial resolution as compared to other detection instruments. Up to now, the resolution has reached 30-100m (Clementine). The spatial resolution of hyper-spectral optic sensors is a little lower than that of common optic sensors. The more the spectral channels are, the weaker the collected energy with each channel will be. So the spatial resolution must be lower to maintain the necessary SNR. The detection system of NIR and thermal infrared is different from that of the UV-VIS CCD system. The background noise increases significantly in the IR region, and also increases with increasing wave length. The cooling system must be used to reduce the background. And special light-sensitive material must be used, such as HgCdTe (1-3 μm) and InSb (3-5 μm) in the NIR region.

An advanced optic detector is the key to the accomplishment of scientific objectives of China's lunar exploration, which is to investigate the mineralogical composition of the Moon's surface. China plans to use a medium-resolution CCD imaging spectroscope, with 32 channels.

3 Development of energy detectors

Energy detectors refer to those high-energy detection instruments, such as gamma ray, neutron, and X-ray fluorescence.

The Sun emits numerous high-energy solar particles. Other mysterious celestial bodies also produce all kinds of high-energy particles. When they hit the planet surface, nuclear reaction will occur, and gamma-rays and neutrons will be emitted. Meanwhile, X-ray fluorescence can be excited. Measuring the intensity and energy-spectra of those high-energy particles can reveal the elemental composition of the lunar surface.

3.1 Gamma-ray spectrometer

The gamma ray is a very energetic photon, more energetic than a visible light ray or an X-ray. The gamma rays on the Moon come from two sources. One is "natural" gamma-rays emitted spontaneously by radioactive elements like thorium and uranium. The other is induced gamma-rays emitted by elements like iron, silicon, and oxygen on the Moon's surface when they are bombarded by cosmic rays (Reedy, 1978).

The energy of a gamma ray serves as a distinctive signature of the atom that emits it. So it can be used for element mapping, especially suitable for some elements, such as Th, K, U, Fe, O, Si, Al, Ca, Mg, and Ti (Reedy, 1978). It is especially sensitive to the heavy, radioactive element thorium and the light element potassium. Thus, gamma-ray detectors are able to determine the global distribution of KREEP (K-potassium, REE rare-earth elements, and P-phosphorous), a chemical "tracer" of sorts which helps to tell the story of the Moon's volcanic and impact history (Zou Yongliao et al., 2004).

The gamma-ray spectrometer uses energy-sensitive crystal as detecting material. The crystal atoms will give off a flash of light when the radiation hits them; the more intense the gamma ray is, the brighter the flash will be. The energy of a gamma-ray, in turn, tells researchers exactly about which kind of atom emitted it. The early detectors used NaI crystal. It was ever used in the Apollo 15, 16 gamma-ray experiments. Its linear response region is 0.2-10 MeV. Its energy resolution is poor.

The gamma-ray detector on the Lunar Prospector used bismuth germinates (Feldman et al., 1999). Its density is higher than that of NaI. So it can stop the gamma-ray more efficiently. And the sensitivity can be improved. But its efficiency is still poorer than that of a pure germanium detector (Kobayashi et al., 2002).

The gamma-ray spectrometer of Apollo 15, 16 couldn't detect uranium, magnesium, aluminum and calcium. Now, it can be mapped using bismuth germanate as detecting material (Feldman et al., 1999). The gamma-ray detectors on Apollo and Lunar Prospector were all omnidirectional detectors. So only the altitude of orbits determines their surface resolution. The resolution is about 150 km (FWHM) at high altitude of 100km, and 45 km at low orbit (10-30 km).

Long-term integration is needed to give scientists enough information to determine the concentrations of the radioactive elements. In addition, the stable elements do not emit gamma rays as readily as the naturally-occurring radioactive ones, so the gamma-ray detector will take one year or more to collect enough data to estimate their concentrations.

China's lunar probe plans to use gamma-ray detector to map some major elements (Ti, Fe, U, REE, Mg, P, Si, Na, K, Ni, Cr, Mn) on the lunar surface. It is an inevitable choice to use gamma-ray spectrometer to detect useful elements on the lunar surface, because it can detect many species of elements at the same time. If successful, we can get more information about the elemental composition of the lunar surface than the Lunar Prospector.

3.2 Neutron spectrometer

The neutrons are also produced by interaction between planet surface and high-energy particles. When cosmic rays collide with atoms in the crust, they will violently dislodge neutrons and other subatomic particles (such as gamma-rays). Some of the neutrons escape directly to space, as hot or "fast" neutrons. Other neutrons shoot off into the crust, where they collide with other atoms. If they only run into heavy atoms, they do not lose very much energy in the collision, and are still traveling at close to their original speed when they finally bounce off into outer space. They will still be warm (or "epithermal") when they reach the neutron detector (NASA, 2001).

The only effective way to slow down a rapid neutron is to let it collide with something of its own size. There is only one atom of the same size as a neutron-hydrogen. If the Moon's crust contains a lot of hydrogen in some locations, any neutron that bounces around in the crust before heading out to space will cool down rapidly. When the neutron detector flies over such a crater, the NS will detect a surge in the number of cool ("thermal") neutrons, and a decrease in the number of warm ("epithermal") neutrons. So the neutron spectrometer can be used to find possible deposits of water on the lunar surface.

The neutron spectrometer (NS) was first used on Lunar Prospector (Feldman et al., 1999). It was designed to search for water. But the scientific objective design is not perfect; it cannot distinguish possible water deposits from hydrogen atoms implanted by solar winds. Maybe, there is no better alternative method except sampling by robots in the future.

The NS is also a kind of unidirectional detector. So its spatial resolution is at the same level of the gamma-ray detector. The detection material of NS is ^{3}He (NASA, 2001).

China doesn't plan to use NS spectrometer this time. It can be used in future missions as a powerful tool in search of hydrogen-rich region and as a powerful supplement to the gamma-ray detector in element mapping.

3.3 X-ray fluorescence spectrometer

When X-rays from the Sun hit the lunar surface, they can cause some elements to emit fluorescence. The elements that emit the X-rays can be identified based on the energy of the X-rays that are emitted. The abundances of these elements can be determined from the intensity of the emitted X-rays (Grande et al., 2003; Dunkin, 2003).

The X-ray fluorescence spectrometer was first used on Apollo-15, -16 spacecrafts (Lunar and Planetary Institute, 2004; Adler et al., 1973). It is used to measure the abundances of the elements magnesium, aluminum, and silicon. Only 9% of the Moon's surface was studied during this experiment. But it has never been used to investigate the Moon within the next twenty years until the launch of Smart-1.

The X-rays from the Moon are very weak, so it is necessary to accumulate a statistically significant number of photons within a given mission duration. The ideal detector should have a large collecting area and angular acceptance. The newly designed D-CIXS Compact X-ray Spectrometer on Smart-1 mission can satisfy the above requirements. The D-CIXS works using a different concept to traditional X-ray telescopes, and is conceived as a modern version of "X–ray detecting paper" (Grande et al., 2003). The energy range is 0.5-10 keV, sufficient to resolve the main fluorescence lines of interest.

D-CIXS also carries an X-ray solar monitor (XSM) (Huovelin et al., 2002) to record the incident solar X-ray flux at the Moon, which is needed to derive absolute elemental surface abundances. Under normal solar conditions, D-CIXS will be able to detect the elements Fe, Mg, Al and Si on the lunar surface and derive absolute elemental abundances rather than the ratios produced by the Apollo missions. The high quality global maps of Mg, Al, and Si can help us to understand the Moon more completely.

During solar flare events, it will be possible to detect other elements such as Ca, Ti, V, Cr, Mn, Co, K, P and Na (Grande et al., 2003; Adler et al., 1973; Kamata et al., 1998). Compared to the X-ray detector on Apollo-15, -16, the species of elements that can be detected efficiently have increased.

The surface resolution of the X-ray detector is better than that of NS and gamma-ray detector (Adler et al., 1972). D-CIXS can derive 45 km spatial resolution images of the entire lunar surface from a 300 km orbiting spacecraft with a spectral resolution of 180 eV.

The Apollo-15, -16 X-ray detectors used gas proportional counter (90% Ar, 9.5% CO_2, 0.5% He). The detector used by D-CIXS is a kind of CCD detector. It is called the swept charge device (SCD). It has the virtue of providing superior X-ray detection and spectroscopic measurement capabilities, so the detection efficiency will be improved greatly (Grande et al., 2003).

4 New detectors China plans to develop and use

China plans to carry the stereo CCDs in its first journey to the Moon. Two CCD cameras are arranged to provide stereo imaging of the same area (51 km×51 km). The camera angles are each offset by plus or minus 25 degrees.

Also, China plans to develop a microwave radiometer to measure the thickness of lunar soil. If we know the thickness of the lunar soil, the total amount of ^{3}He may be estimated. The ^{3}He is commonly believed to be the most prospective replacement of petroleum and coal in the future. It is rare on the Earth, but it is abundant in the lunar regolith. Four different bands would be used, at 3, 8, 19, and 37 GHz. The antenna aperture is 50 cm.

5 Conclusions

It has been proved that the development of planetary sciences cannot continue without the innovation of techniques and manufacturing skills. It can be concluded from the above that the development of detectors can help us to throw insight into the composition and evolution of the Moon.

The above discussion only covers those instruments that can be employed for remote sensing. Some other instruments also can reveal the composition of the Moon, such as Raman spectrometer, but it only suits for the *in-situ* observation.

It can be predicted that the hyper-spectral technique would be used widely in planetary exploration. At the same time, those detectors in use tend to be miniature and compact in size, such as the D-CIXS of Smart-1. China should keep pace with the tendency.

But it still needs improvement to measure and record the light signals and high-energy particles more efficiently and more accurately, relying on the development of material science. And the spatial resolution needs improvement, such as that of neutron spectrometer.

Finally, we think planetary scientists should work together with instrument manufactures to solve more challenging science issues, for example, whether there exists water ice on lunar poles or not.

References

[1] Adler I., Gerard J., Trombka J. I., Schmadebeck R. L., Lowman P.D., Blodget H. W., Yin L. I., Eller E. L., Lamothe R. E., Gorenstein P., Bjorkholm P., Harris B., and Gursky H. (1972) The Apollo X-ray fluorescence experiment [J]. *Geochim. et Cosmochim. Acta (suppl. 3)*. 3, 2157—2178

[2] Adler I., Schmadebeck J. I., Lowman R. L., Blodget P. D. et al. (1973) Results of Apollo 15 and 16 X-ray experiment [J]. *Geochim. et Cosmochim. Acta (suppl. 4)*. 1, 2783—2791

[3] Crawford I. A. (2001) The scientific case for human space exploration [J]. *Space Policy*. 17, 155—159

[4] Dunkin S. K. (2003) Scientific rationale for the D-CIXS X-ray spectrometer on board ESA's Smart-1 mission to the Moon [J]. *Planetary and Space Science*. 51, 35—442

[5] Feldman W. C., Barraclough B. L., Fuller K. R., Lawrence D. J., Maurice S., Mille r M. C., Prettyman T. H., and Binder A. B. (1999) The Lunar Prospector gamma-ray and neutron spectrometers[J]. *Nuclear Instruments and Methods in Physics Research Section A : Accelerators, Spectrometers, Detectors and Associated Equipment*. 422, 562—566

[6] Geosciences Node (2004) Lunar spectroscopy data. http://pds-geosciences.wustl.edu/missions/lunarspec/

[7] Grande M., Browning R., Waltham N., Parker D., Dunkin S. K., Kent B., Kellett B., Perry C. H., Swinyard B., Perry A. et al. (2003) The D-CIXS X-ray mapping spectrometer on Smart-1[J]. *Planetary and Space Science*. 51, 427—433

[8] Hapke B. (1993) *Theory of Reflectance and Emittance Spectroscopy*[M]. Published by the Pres of Syndicate of the University of Cambridge, New York

[9] Heiken G., Vaniman D., and French B.M. (1991) *Lunar Sourcebook: A Users' Guide to the Moon*[M]. pp. 736. Cambridge University Press, New York

[10] Howell S. B. (2000) *Handbook of CCD Astronomy*[M]. pp. 164. Cambridge University Press, New York

[11] Hubbard G. S., Feldman W., Cox S. A., Smith M. A., and Lisa ChuThielbar (2002) Lunar prospector: First results and lessons learned[J]. *Acta Astronautica*. 50, 39—47

[12] Huovelin J., Alha L., Andersson H., Andersson T., Browning R., Drummond D., Foing B., Grande M., Hämäläinen K., Laukkanen J. et al. (2002) The Smart-1 X-ray solar monitor (XSM): Calibration for D-CIXS and independent coronal science[J]. *Planetary and Space Science*. 50, 1345—1354

[13] Kamata Y., Takeshima T., Okada T., and Terada K. (1998) Detection of X-ray fluorescence line feature from the lunar surface[J]. *Advances in Space Research*. 23, 1829—1832

[14] Kobayashi M. N., Hasebe N., Miyachi T., Doke T., Kikuchi J., Okada H., OkaA., Okudaira O., Souri H., Yamashita N. et al. (2002) High-purity germanium gamma-ray spectrometer with stirling cycle cryocooler[J]. *Advances in Space Research*. 30, 1927—1931

[15] Kreslavsky M. A. and Shkuratov Y. G. (2003) Photometric anomalies of the lunar surface: Results from Clementine data[J]. *Journal of Geophysical Research*. 108(E3), 5015, doi: 10.1029/2002JE001937

[16] Lunar and Planetary Institute (2004) Apollo 16 gamma-ray spectrometer experiment. http://www.lp.iusra.edu/expmoon/Apollo16/A16_Orbital_gamma.html

[17] Lunar and Planetary Institute (2004) Apollo16 X-ray fluorescence spectrometer experiment. http://www.lp.iusra.edu/expmoon/Apollo16/A16_Orbital_xray.html

[18] Marini A. E., Racca G. D., and Foing B. H. (2002) Smart-1 technology preparation for future planetary missions[J]. *Advances in Space Research*. 30, 1895—1900

[19] McEwen A. S. and RobinsonM. S. (1997) Mapping of the Moon by Clementine [J]. *Advances in Space Research*. 19, 1523—1533

[20] Muinonen K., Shkuratov Yu., Ovcharenk oA., Piironen J., Stankevich D., Miloslavskaya O., Kaasalainen S., and Josset J. L. (2002) The Smart-1 AMIE experiment: Implication to the lunar opposition effect[J]. *Planetary and Space Science*. 50, 1339—1344

[21] National Aeronautics and Space Administration (NASA) (2001) An introduction to the neutron spectrometer. http://lunar.arc.nasa.gov/results/neutron.htm

[22] Planetary Geosciences Group of Brown University (1995) Clementine UV-VIS data calibration and processing. http://www.planetary.brown.edu/clementine/calibration.html

[23] Racca G. D., Marini A., Stagnaro L., Dooren Van J., Napolidi L., Foing B. H., Lumb R., Volp J., Brinkmann J., Grünagel R. et al. (2002) Smart-1 mission description and development status[J]. *Planetary and Space Science*. 50, 1323—1338

[24] Reedy R. C. (1978) Planetary *Gammar-Ray Spectrcopy*[C]. pp. 2963-2984. Proc. Lunar. Sci. Con. f 9th, Houston

[25] Space Today Online (2003) China's Moon flights. http://www.spacetoday.org/China/ChinaMoonflight.html
[26] The Center for Computational Science of NRL (2002) Clementine DSPSE. http://www.cm.fnr.lnavy.mil/clementine/clementine.html
[27] USGS Astrogeology Research Program (2004) Clementine-USGS IMAG-ES: Clementine-A return to the Moon. http://astrogeology.usgs.gov/Projects/Clementine/index.html
[28] USGS Spectroscopy Lab (2002) About imaging spectroscopy: http://speclab.cr.usgs.gov
[29] ZOU Yong liao, XU Lin, and OU YANG Ziyuan (2004) KREEP rocks[J]. *Chinese Journal of Geochemistry*. 23, 65—70

Scientific Objectives and Payloads of Chang'E-1 Lunar Satellite

Sun Huixian[1]   Dai Shuwu[1]   Yang Jianfeng[2]   Wu Ji[1]   Jiang Jingshan[1]

(1. Center for Space Science and Applied Research (CSSAR), Post Box 8701, Beijing 100080, China;
2. Xi'an Institute of Optics and Precision Mechanics, Chinese Academy of Sciences (CAS), Xi'an 710068, China)

Abstract   China plans to implement its first lunar exploration mission Chang'E-1 by 2007. The mission objectives are
- to obtain a three-dimensional stereo image of the lunar surface,
- to determine distribution of some useful elements and to estimate their abundance,
- to survey the thickness of lunar soil and to evaluate resource of ^{3}He and
- to explore the environment between the Moon and Earth.

To achieve the above mission goals, five types of scientific instruments are selected as payloads of the lunar craft. These include stereo camera and spectrometer imager, laser altimeter, microwave radiometer, gamma and X-ray spectrometers and space environment monitor system. In order to collect, process, store and transmit the scientific data of various payloads a special payload data management system is also included. In this paper the goals of Chang'E-1 and its payloads are described.

Key words   Chang'E-1   Lunar exploration   Microwave radiometer   X-ray spectrometry   Gamma-ray spectrometry

1 Introduction

China plans to initiate its first unmanned lunar exploration by 2007. The project is known as the Chang'E Program after a Chinese legend about a young fairy that flew to the Moon in ancient times. In the first phase, China will send a spacecraft Chang'E-1 to circle the Moon. We describe here the objectives of this mission and its payloads.

2 Mission objectives

The Chinese first lunar exploration includes four objectives:
- Obtaining a three-dimensional stereo image of the lunar surface (except the polar areas). Three-dimensional stereo image is useful to study the surface features of the Moon. A stereo camera and a laser altimeter are designed for this experiment.
- Analyzing the distribution of some useful elements and estimating their abundance. The elements of interest include K, Th, U, O, Si, Mg, Al, Ca, Te, Ti, Na, Mn, Cr, La, etc. A spectrometer imager and a gamma and X-ray spectrometer are designed for this experiment.
- Surveying the thickness of lunar soil. A microwave radiometer is designed to survey the brightness temperature of lunar surface and to estimate the corresponding thickness of lunar soil.
- Exploring the environment between the Moon and Earth. One high-energy particle detector and two solar wind detectors are designed for this purpose.

3 Payloads

3.1 Overview

To achieve the above mission goals, five sets of scientific instruments are chosen as payloads of the lunar satellite. These include stereo camera and Sagnac-based interferometer spectrometer imager, laser altimeter, microwave radiometer, gamma and X-ray spectrometer system and space environment monitor system. In order to collect, process, store and transmit the payload data a special Payload Data Management System (PDMS) is also included.

The stereo camera and laser altimeter are used to obtain the lunar surface three-dimensional image and to achieve the first mission objective. The Sagnac-based interferometer spectrometer imager, the gamma and X-ray spectrometer systems are designed to implement the second objective, i.e., analyzing the composition of lunar surface for useful elements and the distribution of various materials. The microwave radiometer surveys the thickness of lunar soil. The space environment monitor system will survey the Earth–Moon space environment.

3.2 The configuration and composition of the PDMS

PDMS is a distributed system based on the 1553B data bus consisting of Bus Controller (BC), Solid State Recorder (SSR), High Rate Multiplexer (HRM), Remote Terminal (RT) and Power Distributor (PD). Most of the payloads access the system via 1553B data bus. Laser altimeter and space environment monitor system are connected to the RT. PDMS acquires the science and housekeeping data of the payloads through 1553B data bus and stores the data in the SSR or in the memory systems of various payloads. When the spacecraft passes the ground station access range, the stored data and the real time data will be interpolated and capsulated by the HRM to form a serial of Coded Virtual Channel Data Units (CVCDU) according to the CCSDS standard and will be transmitted to the Earth by S-band transmitters. The data rate is 3 Mbps. The PDMS is designed to be flexible and effective; if any payload discontinues its operation, the others will share its storage and transmission resources. The capacity of the SSR is 48 Gbits. An image data compression board is included in the SSR, with compression ratio ≥ 2, which depends on the complexity of the original image.

The block diagram of the PDMS is shown in Fig. 1.

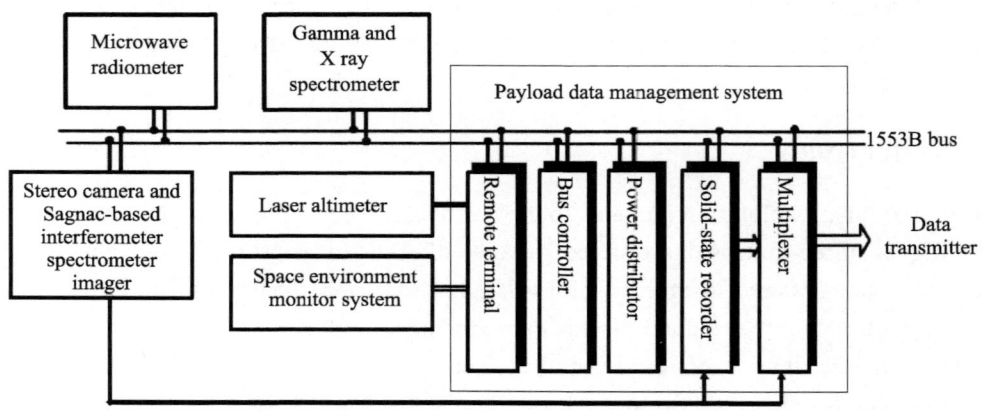

Fig. 1 Payloads and data management scheme for the Chang'E-1 mission

3.3 Stereo camera and spectrometer imager

Stereo camera and interferometer spectrometer imager are the principal payloads of the satellite. The stereo camera can get the nadir, forward, and backward view of the moon. As the spacecraft moves, three two-dimension lunar surface maps will be acquired. After data processing, a stereo image of the lunar surface could be obtained, as illustrated by Fig. 2.

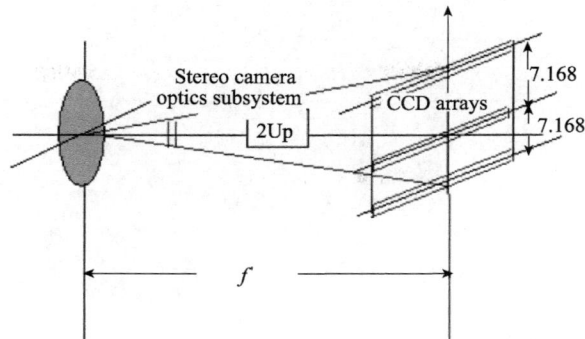

Fig. 2　Schematic diagram of Chang'E-1 stereo camera

The stereo camera consists of an optics subsystem, a framework to support optics lens, the plane CCD array and corresponding signal processing subsystem. The three parallel rows of the plane CCD arrays can get the nadir, forward (17°), backward (−17°) view of the Moon as the spacecraft moves forward.

The image spatial resolution of the stereo camera is about 120 m and the swath width is about 60 km. The Sagnac-based interferometer spectrometer imager is designed to get the multispectral image of the lunar surface. It contains three major optical subsystems: a Sagnac interferometer, which produces spatially modulated interferogram; a Fourier transform lens, which gives the spectral properties depending on aperture geometry and enables us to obtain a wide field of view; and a cylindrical lens, which re-images one axis of the input aperture onto the CCD arrays. Fig. 3 shows the schematic of the interferometer spectrometer.

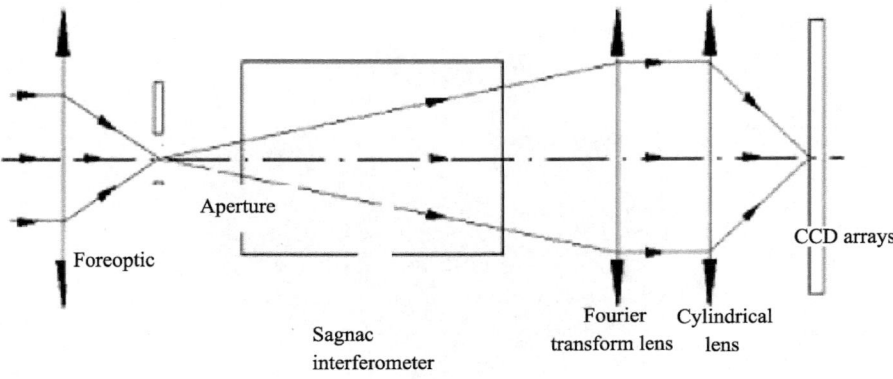

Fig. 3　Schematic of the spectrometer imager

The main specifications are as follows:
1. Stereo camera:
- Resolution: 120 m
- Swath: 60 km

2. Spectrometer imager:
- Resolution: 200 m
- Band: $\lambda = 0.48\sim0.96$ μm
- Swath: 25.6 km

The stereo camera and the interferometer spectrometer imager are integrated together. The layout of the instrument is shown in Fig. 4.

Fig. 4   Stereo camera and spectrometer imager

3.4   Laser altimeter system

Laser altimeter system is designed to measure the altitude of the spacecraft from the nadir point of the lunar surface. The laser altimeter system consists of a laser transmitter and a receiver. The transmitter emits laser pulses towards the lunar surface, and the receiver, which includes a telescope, receives the reflected pulse. The travel time of a pulse gives the information of the distance between the satellite and the lunar surface.

The main specifications are:
- Wavelength: 1064 nm
- Laser emerged: 150 mJ
- Resolution: 1 m

The laser altimeter is illustrated in Fig. 5.

Fig. 5   Laser altimeter

3.5   Gamma and X-ray spectrometer system

Gamma and X-ray spectrometer systems are designed to map the elemental abundances of the lunar crust.

The lunar surface material radiates gamma rays when they are irradiated by cosmic rays and other elements with natural radioactivity such as K, Th and U (and their series nuclides) emit gamma rays by radioactive decay. Therefore, by measuring the flux and energy of these gamma rays, the elemental abundances of the lunar crust can be deduced.

Fig. 6 illustrates the gamma-ray spectrometer, which consists of a main scintillator and anticoincidence scintillator, photomultiplier tube, signal amplifier and data acquisition system. Gamma-ray spectrometer is mounted inside the spacecraft. The spacecraft will also radiate gamma rays due to cosmic ray interactions. Therefore, an anticoincidence scintillator is required to measure the flux and energy of the gamma rays emitted by the Moon.

Fig. 6　Chang'E's gamma-ray spectrometer

The main specifications are:
- Energy range: 300 keV to 9 MeV
- Energy resolution: 9% for ^{137}Cs@662 keV

X-ray spectrometer is similar to gamma-ray spectrometer. It can measure the flux and energy of X-rays emitted from lunar surface material. The X-rays with energy between 0.5 keV and 60 keV can be detected. So gamma-ray spectrometer and X-ray spectrometer are complementary to each other.

Fig. 7 is the illustration of X-ray spectrometer, which consists of sensors, collimator, signal amplifier, scientific data collector and controller. The collimator can absorb the stray cosmic rays reducing the background.

Fig. 7　Chang'E's X-ray spectrometer

The main specifications are:
- Energy range: 0.5 to 60 keV
- Resolution: 600 eV@5.95 keV

3.6　Microwave radiometer

Microwave radiometer system is designed to survey the thickness of lunar soil. It uses four different microwave bands to get the radiation emanating from the lunar surface. The receivers can survey the radiation brightness temperature of the lunar surface, and the temperature resolution is about 0.5 K.

The brightness temperature of the lunar surface is due to radiation emitted by the lunar soil as well as by lunar rocks, as illustrated in Fig. 8.

The radiation intensity is a function of dielectric constant, temperature, wave frequency, soil thickness and surveying angle. By surveying the radiation brightness temperature, we can get the thickness of the lunar regolith. Four frequencies are chosen for the microwave radiometer receivers. Besides the receivers, four calibrating antennae are used to compensate for the cosmic radiation.

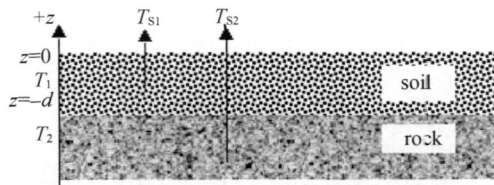

Fig. 8 Schematic showing the emission of the microwave brightness temperature radiation from the lunar surface

The main specifications are:
- Frequencies: 3 GHz, 7.8 GHz, 19.35 GHz, and 37 GHz
- Expected penetration thickness: 30 m, 20 m, 10 m and 1 m
- Temperature resolution: 0.5 K

The instrument is shown in Fig. 9.

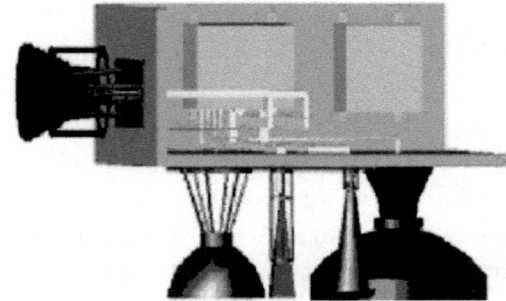

Fig. 9 Microwave radiometer

3.7 Space environment monitor system

The space environment monitor system includes one high-energy particle detector and two solar wind detectors.

The high-energy particle detector is designed to analyze the heavy ions and protons during the journey of Chang'E-1 from the Earth to the Moon and around the Moon. The protons with energy between 4 MeV and 400 MeV can be detected by the system. The flux of heavy ions, such as He, Li and C will also be measured. The high-energy particle detector consists of 6 sensors and a signal processing subsystem.

The two solar wind detectors are designed to measure the flux of solar wind ions with energy up to 20 keV. The solar wind detector consists of a collimator, an ion analyzer and an MCP amplifier.

The specifications of the detector system are:

High-energy particle detector:
- Electrons

E1: $\geqslant 0.095$ MeV

E2: $\geqslant 2.2$ MeV

- Protons

P1: 4 MeV ~ 8 MeV

P2: 8 MeV ~ 15 MeV

P3: 15 MeV ~ 32 MeV

P4: 32 MeV ~ 70 MeV

P5: 70 MeV ~ 160 MeV

P6: 160 MeV ~ 400 MeV
- Heavy ions

He: 13 MeV ~130 MeV
Li: 34 MeV ~260 MeV
C: 117 MeV ~730 MeV
Solar wind detector:
- Energy range: 0.5~20 keV

The two instruments are shown in Fig. 10 and 11 respectively.

Fig. 10   High-energy particle detector

Fig. 11   Solar wind detector

第二部分　CCD 立体相机

绕月探测工程 CCD 立体相机的实验室辐射定标

王珏[1]　李春来[1]　赵葆常[2]

(1. 中国科学院国家天文台，北京　100012；
2. 中国科学院西安光学精密机械研究所，西安　710068)

摘　要　探月光学有效载荷系统含 CCD 立体相机与干涉成像光谱仪两台光学遥感器，CCD 立体相机完成的科学目标主要是与激光高度计配合获取月球表面三维立体图像。文章主要叙述了三线阵立体相机的工作原理，定标内容、目的以及 CCD 裸片像元检测和整机的相对定标和绝对定标过程。

关键词　线阵　CCD 立体相机　相对定标　平场　绝对定标　三线阵

我国首次绕月探测计划中，用于月球表面测绘的主要有效载荷是搭载在 CE-1 号卫星上的 CCD 立体相机，空间分辨率为 120m，光谱范围 0.5~0.75μm，视场角为 40°的一次成像系统，图像探测器采用法国 Thomson 公司的线阵 CCD 芯片。在波长 0.5~0.75μm 范围内，平均量子效率为 15%，水平 MTF 均值为 0.70，满阱容量均值为 320ke-，光谱响应详见图 1。该相机采用三线阵工作原理，获取月球表面三维影像。三线阵 CCD 立体相机主要由窗口、光学镜头、光机结构、CCD 焦平面组件、视频处理器、高速 A/D 转换器、实时数据压缩单元、图像存储器和相机控制器构成[1]。仪器在上星之前毫无例外先要进行定标。定标的目的是为了把光学遥感器电路输出数据（一般为电压）与被探测的物理量，如目标的辐射亮度、反射光谱强度等在一定的精度范围内直接联系起来，使其真实地反映被探测目标的状况，精度愈高，反映的真实性愈好。定标分为实验室定标和星上定标两种。因为近月轨道的空间辐照条件优于地球卫星，且光学遥感器采取了充分的防护措施，而且月球上没有大气，所以此次探月卫星光学遥感器不作星上定标，只做实验室定标。

1　立体相机工作原理

立体相机采用三线阵 CCD 推扫原理，三线阵 CCD 相机的光电扫描成像部分是由光学系统焦面上的 3 个线阵 CCD 传感器组成的。这 3 个 CCD 阵列（A,B,C）相互平行排列并与航天飞行器飞行方向垂直。当航天器飞行时，每个 CCD 阵列以一个同步的周期 N 连续扫描月球表面并产生 3 条相互重叠的航带图像 As，Bs，Cs。这 3 个 CCD 阵列的成像角度不同，如图 2 所示，B 为正视传感器，A 为前视传感器，C 为后视传感器。推扫所获得的航带图像 As，Bs，Cs 的视角也各不相同，从而可以构成立体影像[2]。

定标内容包括 CCD 裸片的检测和相机的整机定标。裸片检测包括暗电流检测和像元响应均匀性检测。整机定标包括相对定标和绝对定标。

裸片 CCD 暗电流的检测目的是把暗电流从信号强度中减去，增加信号"纯度"，并对进口 CCD 芯片的暗电流指标进行复核。

裸片像元响应均匀性检测的目的是复检进口 CCD 芯片的像元响应均匀性，找出盲元、过热及过冷的像元，为随后进行的整机相对定标提供对照。

本文原载于《天文研究与技术》，2007，Vol.4，No.1，30~35。

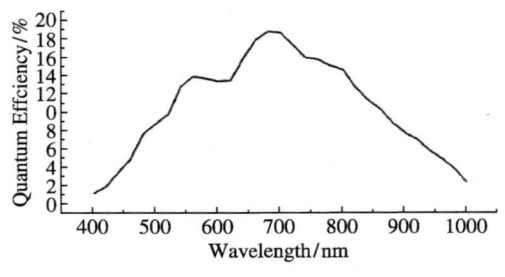
图 1　CCD 光谱响应曲线

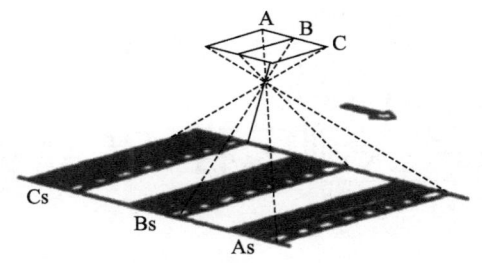
图 2　三线阵立体相机工作原理

相对定标是确定场景中各像元之间测得的辐射度量的相对值,用于消除由于探测器响应不一致引起的图像中的条纹[3],包括光、机、电及 CCD 像元响应的不一致性。

绝对定标是通过各种标准辐射源,在不同光谱段建立遥感探测器入瞳处的光谱辐射度量与遥感探测器输出量之间的定量关系[4],即把月球表面反射光谱在 500~750nm 间的辐亮度（W/m²·str）与 CCD 立体相机的电路输出（V 或 mV）在一定的精度范围内建立定量的关系。

2　定标原理、方法及设备

2.1　相对定标

当 CCD 被一个完全均匀的光场照明时,每个 CCD 像元的输出理论上应该完全相同。但事实上它们的输出会有差异,特别是 CCD 像元中的瞎像元,对光没有光电转换功能,此外还有若干响应偏高或偏低的像元（即热像元或冷像元）。

相对定标的原理是,建立一个发光均匀的面光源作为 CCD 相机的目标。检查并记录每个像元的输出。均值归化为 1,把冷、热像元的输出按均值归化。譬如说有 4 个像元,如图 3 所示,在相同的照度、曝光时间等条件下,它们的输出分别为：495 mV、480 mV、465 mV、450 mV。经相对定标后它们的修正因子分别为：0.97、1、1.03、1.07。国内称为相对定标,国外称为平场（Flat-field）改正。

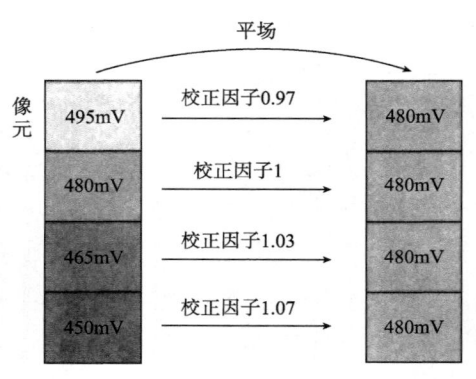
图 3　平场示意图

相对定标所采用的方法是,利用多波段标准辐亮度计检测积分球出射口横截面的均匀性,再用这台标定过的积分球对 CCD 相机响应均匀性进行检测,同时以光谱辐射度计监控积分球在不同辐亮度输出时的光谱一致性。

2.2　绝对定标

绝对定标是对平场改正后的 CCD 像元在 500~750nm 间光谱响应的电路输出（V、mV）与目标同谱段的辐亮度间建立定量的关系,即获得：

$$V_{输出}(\text{mV}) = f(B, \Delta t, k)$$

上式中 B 为目标辐亮度 W/(m²·str),Δt 为曝光时间,k 为电路增益。

绝对定标所采用的方法是用一台经过计量标定的光谱辐射度计对同一目标源进行测量,与 CCD 立体相机进行对照,测量时两者测量的光谱范围 $\Delta \lambda$ 及曝光时间 Δt 应保持相同,根据光谱辐亮度计的输出给 CCD 立体相机的纵坐标（横坐标为波长或波段）赋值,得到绝对定标系数 K,即以 CCD 输出电压 V（mV）除以输入辐亮度 L（W/(m²·str)）,$K=V/L$。

2.3 主要定标设备

主要定标设备详见表1。

表1 主要定标设备

序号	设备名称	用途
1	光谱辐射度计	通过对积分球和太阳模拟器的标定，实施绝对定标。
2	大积分球系统	提供均匀面光源。通过光谱辐射度计对它的标定，实施对CCD立体相机的绝对定标。
3	多波段标准辐亮度计	用于对太阳模拟器及积分球辐亮度均匀性及稳定性监测并进行绝对定标复核验证。
4	光谱辐照度标准灯	与灯板一起实施光谱辐射度计标定。
5	漫反射参考标准板	与标准灯一起实施对光谱辐亮度计标定。
6	漫反射参考标准板	用途同上。
7	辐照度探测器	裸片像元响应均匀性检测时，检测辐照度均匀性。
8	成像单色仪/光谱仪	用作单色光源。

3 CCD定标实验步骤

3.1 CCD裸片检测

CCD裸片检测包括暗电流检测和像元响应均匀性检测两项。

CCD裸片暗电流检测在暗室内进行。"焦平面组件"安放在黑色防静电箱内，打开电源使焦平面组件处于正常工作状态，对3×500个像元，检测暗电流25次以上，得到25张暗电流输出表格，如表2所示。

表2 CCD裸片暗电流检测表(3行×512列)片断

列数	第0行均值	第0行方差	第1行均值	第1行方差	第2行均值	第2行方差
1	5.00000	0.00000	5.00000	0.00000	5.00000	0.00000
2	4.98000	0.14000	5.00000	0.00000	5.00000	0.00000
3	4.95000	0.21794	5.00000	0.00000	5.00000	0.00000
4	4.99000	0.09950	5.00000	0.00000	5.00000	0.00000
5	4.97000	0.17059	5.00000	0.00000	5.00000	0.00000
6	4.97000	0.17059	5.00000	0.00000	5.00000	0.00000
7	4.96000	0.19596	5.00000	0.00000	5.00000	0.00000
⋮	⋮	⋮	⋮	⋮	⋮	⋮
509	5.99000	0.09550	6.00000	0.00000	6.00000	0.00000
510	6.00000	0.00000	6.00000	0.00000	6.00000	0.00000
511	6.00000	0.00000	6.00000	0.00000	6.00000	0.00000
512	6.00000	0.00000	6.00000	0.00000	6.00000	0.00000

温度=25℃，曝光时间=3.2ms，电子学增益=3.0，数值为0~255之间的DN值

CCD裸片像元响应均匀性检测也在夜间进行，并应在工作台上的黑箱内进行，以防止杂散光干扰检测，装置如图4所示。

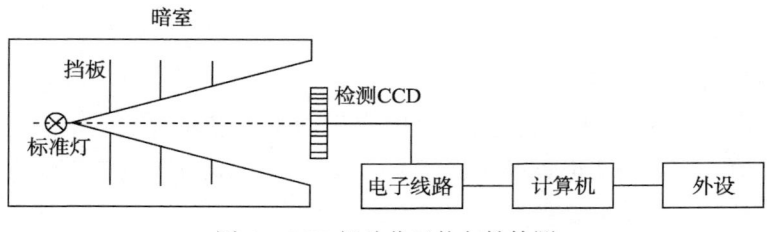

图4 CCD裸片像元均匀性检测

3.2 CCD立体相机整机相对定标(平场改正)

用黑绒布遮挡周边杂散光源,装置如图5所示。

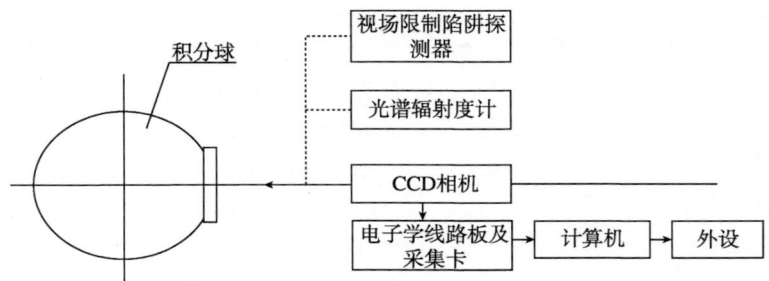

图5 CCD立体相机整机相对定标(平场改正)装置

实验步骤:

(1) 积分球输出辐亮度模拟月表光谱在500~750nm间的辐亮度。

(2) 检测积分球出射面的均匀性。

(3) 以标准辐亮度计的输出变化作为输入相机能量变化的依据,对整机响应的线性性进行检测,同时以光谱辐射度计监控积分球在不同辐亮度输出时的光谱一致性。

相对定标系数矩阵详见表3。

表3 CCD立体相机整体相对定标(平场改正,3行×512列)片段

列数	第0行修正值	第1行修正值	第2行修正值
1	1.731991	1.059605	0.935899
2	1.740145	1.059944	0.936947
3	1.740837	1.065329	0.936449
4	1.723229	1.055659	0.932579
5	1.720243	1.056208	0.931885
6	1.723790	1.059771	0.935148
7	1.734012	1.059101	0.932730
⋮	⋮	⋮	⋮
509	1.720564	1.021624	0.894204
510	1.711674	1.022264	0.896198
511	1.727293	1.023033	0.895974
512	1.725397	1.023520	0.897094
温度=25℃,曝光时间=3.2ms,电子学增益=3.0			

3.3 CCD立体相机绝对定标

用黑绒布遮挡周边杂散光,装置如图6所示。

实验步骤:

(1) 调节目标辐亮度值为希望值。

(2) 记录光谱辐射度计在500~750nm间的辐亮度值及CCD相机的输出。

(3) 对不同辐亮度值、不同曝光时间及不同增益,核对CCD立体相机整机的线性性。

(4) 以中心视场为基准,做绝对定标。

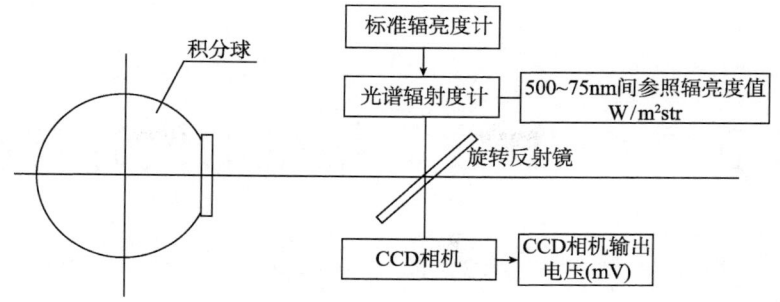

图 6 CCD 立体相机绝对定标

4 辐射定标数据的处理

辐射定标校正数据处理流程包括暗电流校正，相对定标，曝光时间校正，增益校正，绝对定标和噪声消除。

暗电流校正算法如下：

$$\mathrm{Img\_2\_Step1}(i,j) = \mathrm{Img\_1}(i,j) - DC(i,j)$$

其中：$\mathrm{Img\_2\_Step1}(i,j)$ 是剔除暗电流后的中间数据；$\mathrm{Img\_1}(i,j)$ 是 CCD 立体相机 1 级数据；$DC(i,j)$ 是测量的 CCD 探测单元的暗电流矩阵；$i=1,2,3$；$j=1,2,3,\cdots,500$。

相对定标算法如下：

$$\mathrm{Img\_2\_Step2}(i,j)=\mathrm{Img\_2\_Step}(i,j)\times \mathrm{Cali}(i,j)$$

其中：$\mathrm{Img\_2\_Step2}(i,j)$ 是相对定标后的中间数据；Cali 是 3×500 的相对定标矩阵；$i=1,2,3$；$j=1,2,3,\cdots,500$。

绝对定标算法如下：

$$\mathrm{Img\_2\_Step6}(i,j)=\mathrm{Img\_2\_Step5}(i,j)\times K$$

其中：$\mathrm{Img\_2\_Step6}(i,j)$ 是经过绝对定标后的数据；$\mathrm{Img\_2\_Step5}(i,j)$ 是 $\mathrm{Img\_2\_Step2}(i,j)$ 经过曝光校正和增益校正后的中间数据；$i=1,2,3$；$j=1,2,3,\cdots,500$；K 是绝对定标系数，由 CCD 输出电压 V 除以输入辐亮度 L 得来，即 $K=V/L$。

5 结束语

采用了 4 台光谱辐射度计，两种面光源，积分球和太阳模拟器，两种漫反射参考标准板。它们之间都可以互为校对，相互间符合较好，说明过程控制是有效的。CCD 立体相机的实验室定标工作量大，有一定技术难度，克服了这些困难后，形成了最终的定标算法，精度也达到了预期 8%的要求。

参 考 文 献

[1] 金光，南寿松. 立体测绘小卫星有效载荷——传输型三线阵 CCD 摄影测量相机[J]. 遥感技术与应用，1999,14(3):34—37
[2] 刘金国，李杰，郝志航. 三线阵 CCD 相机亚像元精度几何标定方法研究[J]. 光电工程，2004,31(1):36—39
[3] 陈世平. 空间相机设计与试验[M]. 北京：宇航出版社，2003.12
[4] 顾名澧. 多光谱扫描仪的星上辐射定标系统[J]. 航天返回与遥感，1998, 19(3):21—25

嫦娥一号卫星 CCD 立体相机的设计与在轨运行

赵葆常　杨建峰　汶德胜　高伟　阮萍　贺应红

(中国科学院西安光学精密机械研究所，西安　710119)

摘　要　介绍了立体成像原理、参数选择、相机方案及设计结果。发射前的实验室检测表明：整机 MTF 达 0.48，S/N 为 235。由嫦娥一号卫星 CCD 立体相机获取的图像清晰，层次丰富。用户根据在轨运行前期探测数据对 CCD 立体相机的科学探测数据质量进行了初步评估，认为 CCD 立体相机工作正常，数据格式正确，科学数据有效，图像质量优良，能够进行稳定的数据观测、采集与下传。最后介绍了相机在研制过程中采用的一些新技术。

关键词　嫦娥一号卫星　有效载荷　CCD 立体相机　在轨检测　月面图像

1　前言

嫦娥一号卫星有效载荷的配置是从实现科学目标出发制定的，在8个有效载荷中有3个是光学遥感器。它们是 CCD 立体相机、激光高度计及成像光谱仪。CCD 立体相机与激光高度计是为了获取月表的三维立体影像，以实现我国首次探月的第一个科学目标[1-2]。

目前的月球三维地图尚有不少空白，精度不高，特别是纬度在 70°~90°间的南北极区[3]。CCD 立体相机提供三个视角的二维平面图像，以此重构三维立体影像。科学家据此进行月球的地形地貌、表面年龄、地质构造和演化历史的研究[3-4]。

2007 年 10 月 24 日嫦娥一号卫星零窗口发射，11 月 5 日被月球捕获，11 月 20 日 16 时 49 分 CCD 立体相机第一轨三视角图像下传。通过下传工程参数，设备功能正常，下传科学数据表明图像清晰，层次丰富。用户根据在轨运行前期的探测数据对嫦娥一号卫星有效载荷 CCD 立体相机科学探测数据质量进行了初步评估，认为 CCD 立体相机工作正常、数据格式正确、科学数据有效、图像质量优良，能够进行稳定的数据观测、采集与下传。

2　嫦娥一号卫星中 CCD 立体相机的方案

在很多摄影测量学的著作中，都有立体测绘原理的详细论述[5-6]，本文仅就嫦娥一号卫星 CCD 立体相机的方案作一简要说明[7]。

在嫦娥一号卫星的 CCD 立体相机中，配置了三条线阵 CCD，它们分别用于获取正视、前视、后视的航迹图像，其构像原理见图 1。其中正视是垂直向下看的图像，它也是航空摄影中通常获取的图像形式，但是一幅正视图像无法获取月表的高程数据，为此在嫦娥一号卫星的 CCD 立体相机中，还有前视与后视，前、后视与正视间的夹角 ϕ_0 绝对值相同、符号相反，即 $\phi_0=\pm17°$。

夹角 ϕ_0 又称交会角，它是立体测量中的重要参数，它决定了理论的高程精确度，在嫦娥一号卫星 CCD 立体相机的情况下，基高比 $B/H=2\tan\phi_0=0.6115$，所以其理论高程精度为 $R_g/(B/H)\approx200$ m，式中，R_g 为地面像元分辨率，在嫦娥一号卫星 CCD 立体相机中为 120 m。

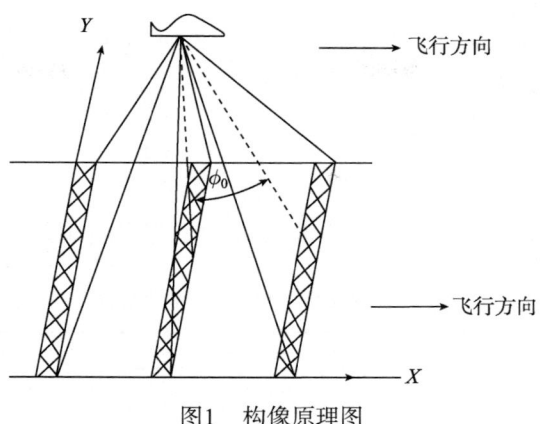

图 1　构像原理图

仅就获取立体图像的要求来说，只要有两个视角的图像即可计算月表任一点的三个坐标值 X、Y、Z。日本的"月女神"探测器就是采用两线阵方案[8]。立体相机的视角可以根据用户要求任意选取，前、后视与星下点正视光轴的夹角既可相同，也可不同。但两视角的夹角一般希望尽可能大，这样可以获得较高的高程精度。

在嫦娥一号卫星 CCD 立体相机的考虑中，用正视图像获得正射照片，前后视图像用于重构立体影像，其构像矩阵为[9]

$$\begin{bmatrix} \cos\phi_0 & 0 & -\sin\phi_0 \\ 0 & 1 & 0 \\ \sin\phi_0 & 0 & \cos\phi_0 \end{bmatrix} \begin{bmatrix} 0 \\ y \\ -f \end{bmatrix} = \begin{bmatrix} f\sin\phi_0 \\ y \\ -f\cos\phi_0 \end{bmatrix} = \lambda R^{\mathrm{T}} \begin{bmatrix} X - X_\mathrm{S} \\ Y - Y_\mathrm{S} \\ Z - Z_\mathrm{S} \end{bmatrix} \tag{1}$$

式中 ϕ_0 为前、后视与指向月心的光轴间夹角，y 为像点坐标；f 为光学系统焦距；X、Y、Z 为月表物点坐标；X_S、Y_S、Z_S 为摄站坐标，它们均在月心坐标系中；λ 为摄影比例常数；R^{T} 为旋转矩阵，它由卫星三个轴的三个方向余弦组成，是一个 3×3 的矩阵[5-6]。

它的共线方程为

$$\begin{aligned} (x) &= f\tan\phi_0 = -f \times \frac{a_1(X-X_\mathrm{S}) + b_1(Y-Y_\mathrm{S}) + c_1(Z-Z_\mathrm{S})}{a_3(X-X_\mathrm{S}) + b_3(Y-Y_\mathrm{S}) + c_3(Z-Z_\mathrm{S})} \\ (y) &= y/\cos\phi_0 = -f \times \frac{a_2(X-X_\mathrm{S}) + b_2(Y-Y_\mathrm{S}) + c_2(Z-Z_\mathrm{S})}{a_3(X-X_\mathrm{S}) + b_3(Y-Y_\mathrm{S}) + c_3(Z-Z_\mathrm{S})} \end{aligned} \tag{2}$$

它的反演公式为

$$\begin{aligned} X &= \left[\frac{a_1 x_b + a_2 y_b - a_3 f}{c_1 x + c_2 y_b - c_3 f}\right](z - z_\mathrm{S}) + X_\mathrm{S} \\ Y &= \left[\frac{b_1 x_b + b_2 y_b - b_3 f}{c_1 x_b + c_2 y_b - c_3 f}\right](z - z_\mathrm{S}) + Y_\mathrm{S} \end{aligned} \tag{3}$$

从(3)式可以看出一个视角 ϕ_0 的图像是不能求解月表物点的坐标 X、Y、Z 的，因为两个方程不能解析出 X、Y、Z 三个未知数，因此还需要有另一个 ϕ_0 的像方数据，这样又可以得到类似于公式(3)的两个方程，四个方程用最小二乘法求解 X、Y、Z，从而获得月表物体的坐标。

公式中摄站坐标 X_S、Y_S、Z_S 即卫星在月心坐标系中的位置，旋转矩阵 R^{T} 的元素是由卫星三根轴的方向余弦得到，亦即卫星的姿态。它们被称为外方位元素，是重构立体影像的参数，由轨测与姿测得到。因此嫦娥一号卫星 CCD 立体相机在传输科学图像数据的同时还必须下传外方位元素。

从上面的讨论可以看出，CCD 立体相机自身就有高程信息，而在嫦娥一号卫星中还配有激光高度计，它同样也是为了获取月表的高程信息,通常激光高度计的测量精度要比CCD立体相机高,但它的采样点很稀,用于两者的相互核对与数据融合。

根据嫦娥一号卫星CCD立体相机的地面像元分辨率R_g=120 m,理论上满足1:150万比例尺的月表地图制作,考虑到各种误差因素,经分析可实现1:250万的月表地图制作。

3 CCD立体相机的设计指标

当卫星轨道高度为200km时,CCD立体相机达到如表1所示的技术指标。

表1 CCD立体相机总体技术指标

指标	数值
月表空间分辨率/ m	120
月表成像宽度/ km	60
光谱范围/ nm	500~750
光谱通道数	1
量化等级/ bit	8
S/N	≥100
MTF	≥0.2
光照条件	太阳高度角≥15°
基高比	≥0.6

4 CCD立体相机方案选择、光学参数计算及设计

4.1 总体方案

在前期的方案论证阶段,曾考虑如下三种实施方案。

(1) 三台独立的相机。三个完全相同的光学系统,在它们各自的焦平面上各配置一条线阵CCD。该方案的优点是光学系统十分简单,可以有大的基高比,但结构松散,体积质量较大,航天环境一致性较差,调整配准较复杂。

(2) 一个广角光学系统,在焦平面上平行地配置三片线阵CCD。由于空间分辨率要求较低,所以光学系统的焦距短(估计在20mm左右),即使视场角做得很大,线视场仍较小,在空间上难以在同一焦平面上配置三条线阵CCD。

(3) 一个广角光学系统,在焦平面上配置一个1k×1k面阵CCD,如图2所示。

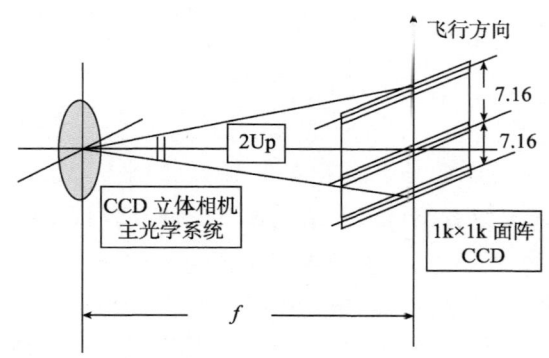

图2 广角物镜加面阵CCD方案图示

此方案的优点是结构紧凑、体积质量较小、航天环境适应性强，装配简单、配准容易且配置精确度高，其缺点是由于帧转移面阵CCD在转移过程中仍在曝光，会影响定标精度。

经分析、计算，认为在面阵CCD前增加一个带有三条狭缝的金属面罩，可以使面阵CCD在转移过程中仍在曝光的影响控制在定标不确定度的范围内。三条缝的宽度可以控制在0.15～0.2mm，约占10～15个像元，三条平行狭缝对准三行CCD，从而把转移过程中的曝光时间减小到帧转移时间的1%～1.5%，因为帧转移时间为2ms，所以其附加曝光时间不会超过0.03ms。同时还在定标中作了进一步的修正。

4.2 探测器参数

CCD立体相机探测器主要参数见表2。

表2　CCD参数

参数	值
CCD 像元数	1024×1024
CCD 像元尺寸	14 μm×14 μm
像元饱和输出电压/ V	1.9
响应度（灵敏度）/V·μJ·cm^{-2}	6.5
动态范围	10 000
MTF	0.7 (λ=500nm)
帧频/ (f/ s)	30

4.3 光学系统参数计算及设计结果

根据轨道高度 $H = 200$ km，地面像元分辨率 $R_g = 120$ m 及像元尺寸 $\delta_x = 0.014$ mm，即可求出焦距 f：

$$f = \frac{H \cdot \delta_x}{Dx} = \frac{2 \cdot 10^5 \cdot 14 \cdot 10^{-6}}{120} = 23.33 \text{mm}$$

根据月表成像幅宽以及基高比的要求可以计算得到最大视场角：

$2\theta = 38°$，光学系统按 $2\theta = 40°$ 设计。

根据衍射MTF及能量，确定系统相对孔径为F/5。

由于采用同一面阵CCD，三条线阵的曝光时间不能分别单独控制，而且系统中有正视CCD，所以视场角从0°到20°间连续使用。

从像面照度的均匀性出发，要求系统为像方远心，从测绘要求出发，要求畸变小于0.5%。

4.4 光学系统设计

采用六组九片远心消畸变光组结构，达到上述所有设计要求，见图3。

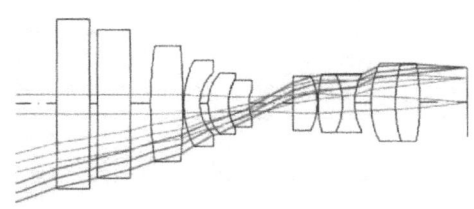

图3　CCD立体相机光学系统图

图4为CCD立体相机白光MTF曲线，表3为CCD立体相机光学镜头MTF值。MTF均值为0.82，最小值与均值差0.047。系统轴上点已达衍射极限，20°边视场下降0.079。

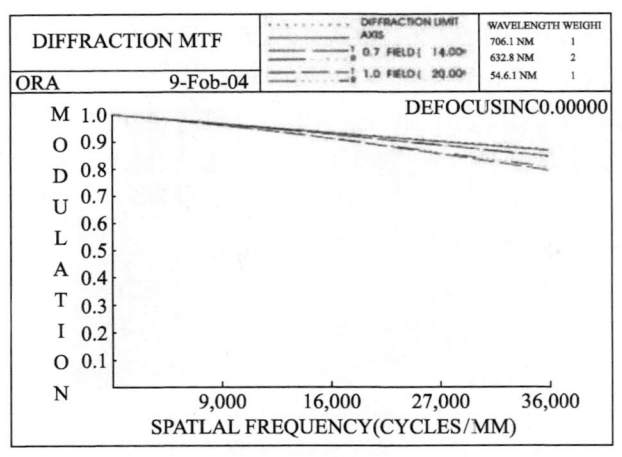

图4 CCD 立体相机白光 MTF 曲线

表3 CCD 立体相机光学镜头MTF 值(奈奎斯脱频率为36lp/mm)

视场	MTF	
	子午	弧矢
0 视场	0.852	0.852
0.7 视场	0.826	0.829
边视场	0.786	0.773

设计完成后，还就真空、温度变化以及离焦对 MTF 的影响作了分析，采取相应的措施。检查了系统像差分布的合理性，没有发现有敏感结构参数（N、ΔN、d），并按照本所所能达到的加工及装配精度，预估了加工、装配 MTF 值，事后检测表明，预估值与检测结果相符合。

5 检测及在轨科学数据质量评估

5.1 检测

CCD 立体相机于 2006 年 2 月上旬，在完成各项环试及实验后以最终状态由所检测中心进行了实验室静态检测，检测结果见表4。

表4 CCD 立体相机整机检测结果

检测项目	设计值	检测值
焦距/mm	23.33	23.36
相对孔径	5	5.13
视场角/(°)	38	≥40
基高比	≥0.6	0.6
光谱通道数	1	1
MTF	≥0.2	0.48
S/N	≥100	235

设备装星后，又在装星状态下，进行了三次复检，其变化都在公差范围内。

5.2 在轨评估

2008 年 2 月用户对开机来一个多月的图像数据进行了初步评估[11]。

由于在轨状态下无法检测地面像元分辨率及成像幅宽,用户通过检测国际上已公布撞击坑直径的特定月坑作了对比检测。

图 5 是对哈密顿(Hamilton)撞击坑的测量情况。

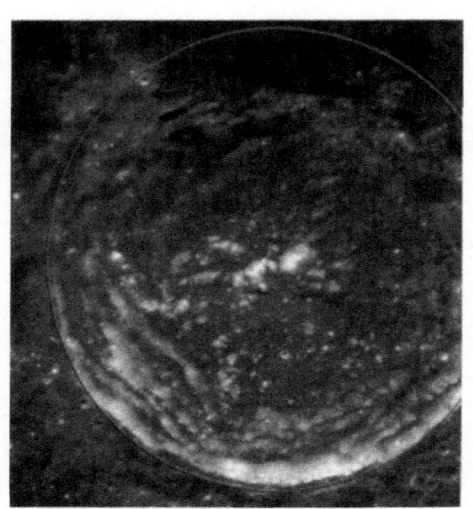

图5 CCD 立体相机获取的哈密顿(Hamilton)撞击坑

国际上公布哈密顿撞击坑的直径为 57km,CCD 立体相机按同时下传的轨道高度计算得到的检测值为 58.4km,两者相差 2.4%。因为国际上公布数值也并非"真值",人工判读及地形地貌均会引入误差,所以认为 2.4%的相对误差是合理的,从中可以换算 CCD 立体相机的地元分辨率及成像宽度。

从图 6 还可以说明 CCD 立体相机可以清晰地显示出 3×3 个像元、4×4 个像元的圆形目标,相当于 ϕ360m～ϕ480m 的小坑。由于小点很清晰,说明相机识别细节的能力较强。

 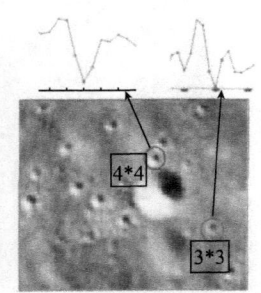

A 原始1:1图像　　　　B 局部放大4倍的图像

图 6 放大后的小坑

对地球卫星,可以通过在地面人工铺设鉴别率板或灰板检测相机的在轨 MTF 以及黑白层次,而月球卫星目前尚不具备上述条件。但是 MTF、S/N 等表征图像质量的参数最终都反映在图像的清晰度、层次上,从嫦娥一号卫星 CCD 立体相机所获得的图像质量来分析,该相机的设计、研制都是比较成功的,达到了预期的目标。

图 7 是经过多轨拼接处理后的月表图像。图 8 是未经加工处理的极区原始图像,虽然图像偏暗,说明信号动态范围较低,但仍然十分清晰。图 9(图版 II)是立体伪彩色图。图 10 是嫦娥一号卫星的图像与美国克林门汀月球探测器的图像比较,拍摄的是月表同一地区(但光照方向不同),两者成像比例尺接近,左边为嫦娥一号的图像,右边为克林门汀探测器的图像[10-11]。所有图像均由用户中科院国家天文台提供。

按设计任务书要求,光照角大于 15°相机工作。而实际上,CCD 立体相机采用 20ms 的曝光时间已获得很多南北纬 75°以上的极区照片(见图 8)。相机具有 3.2ms、7ms、20ms 及 84ms 四档曝光时间及 1×、1.5×、3×、3.5×四档电子学增益，共 16 种组合。通过地面注入指令，可以实现同轨不同纬度分区的不同曝光时间与电子学增益，从而使在不同光照条件下，都能获得较为合适的信号强度。

图 7　CCD 立体相机拍摄的月面图像

图 8　极区原生图像

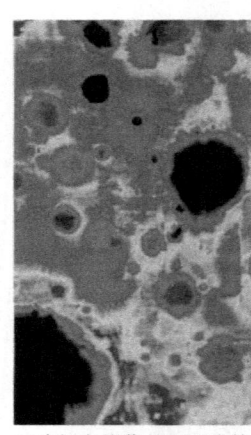  

三个视角影像处理形成的　　正视影像与数字高程模型　　正视影像与数字高程模型
　　数字高程模型图　　　　　处理形成的正射影像图　　　处理形成的数字高程
　　　　　　　　　　　　　　　　　　　　　　　　　　　　色彩编码地形图

图 9　月面图像局域区域形貌图

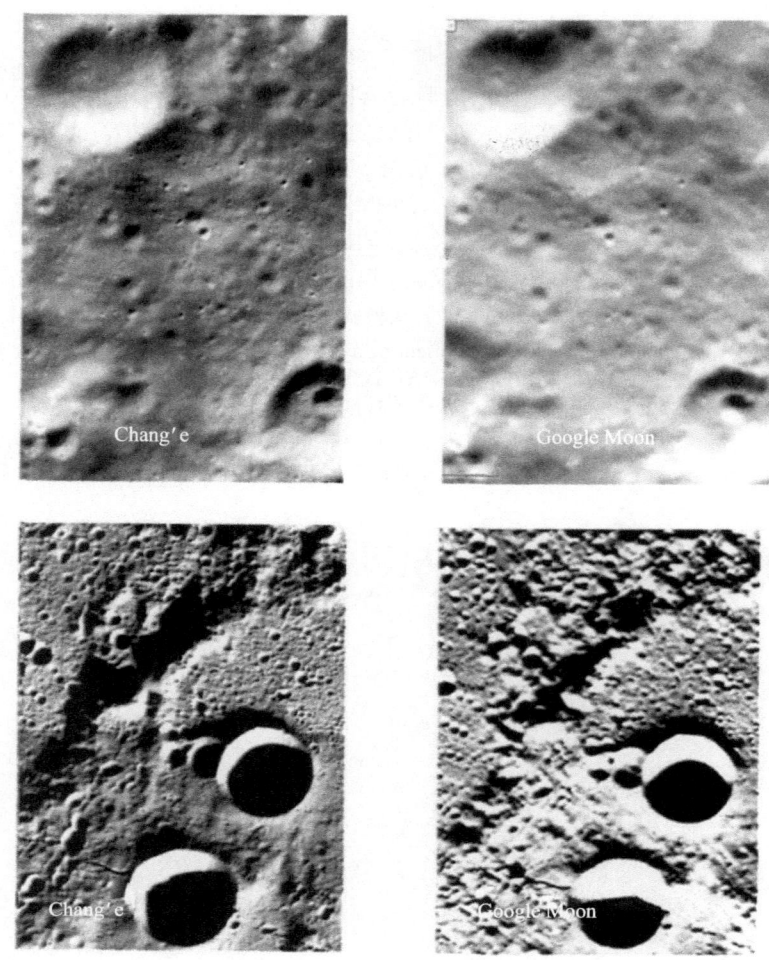

图 10　嫦娥一号卫星 CCD 立体相机图像(左)与美国克林门汀月球探测器图像(右)清晰度比较

6　总结

采用了三线阵 CCD 推扫成像原理。设计研制的嫦娥一号卫星 CCD 立体相机获取的月面照片图像清晰、层次丰富，重构了立体影像，圆满完成了预期的科学应用目标。在研制过程中，碰到了若干技术上的难点，针对这些问题，采取了对策，在轨运行表明措施有效。主要内容有：

(1) 采用一个广角远心消畸变光学系统和一块 1k×1k 的面阵 CCD，在 CCD 前加入一个带有三条细缝的金属面罩，分别对准前视、正视和后视的三行 CCD，实现三线阵立体成像。在轨运行的下传原生图像表明：方案正确，措施有效。

(2) 工作波长范围为 500～700nm 时，根据调研、分析，确定月表最大反照率 $\rho_{max}=0.26$，经计算按 33W/m^2·Sr 的辐亮度，在最短曝光时间为 3.2ms 时，使 CCD 原级输出接近饱和进行标定，经检查在轨下传原生图像的灰度值，仅有个别饱和点，说明对月表漫反射系数的估计及对月球的辐亮度计算基本正确。

(3) 采用了 4 档曝光时间与 4 档电子学增益，共 16 种组合，可以通过地面指令注入，实现同轨按纬度分区的曝光时间控制，从而在同一轨中在不同纬度都获得较为合适的曝光量。

(4) 超过原科学目标及设计指标，获取了极区的清晰图像。预期在第二个正飞阶段结束时，可以获得全月的立体影像。它为建立具有自主知识产权的月表图像数据库提供了科学数据。

参 考 文 献

[1] 欧阳自远, 李春来, 邹永廖, 等. 我国月球探测一期工程的科学目标[J]. 航天器工程, 2005, 14(1):1—5
[2] 叶培健, 孙泽洲, 饶伟. 嫦娥一号卫星月球探测卫星研制综述[J]. 航天器工程, 2007, 16(6):9—15
[3] 欧阳自远. 月球探测对推动科学技术发展的作用[J]. 航天器工程, 2007, 16(6):5—8
[4] 欧阳自远. 月球科学概论[M]. 北京:中国宇航出版社, 2005
[5] 王任享. 三线阵 CCD 影像卫星摄影测量原理[M]. 北京：测绘出版社, 2006
[6] 张剑清, 潘励, 王树根. 摄影测量学[M]. 武汉：武汉大学出版社, 2003
[7] 林宗坚, 殷礼明, 葛之江. 环月遥感立体影像构建的几何模式比较[J]. 遥感器工程, 2007, 16(1)：69—73
[8] Matsunaga T, Ohtake M, Hirahara Y, et al. Development of a Visible and near infrared spectrometer for selenological and engineering explore (SELENE) [C]. Proc. SPIE Vol. 4151, 2001, 32—39
[9] 刘少创, 贾阳, 陈建新. 月面巡视探测器立体相机共线方程的建立[J]. 航天器工程, 2007, 16(3)：17—20
[10] Kordas J F, Lewis I T, Priest R E, et al. U V/ visible camera for the clementine mission[J]. SPIE Vol 2478, 1995, 175—186
[11] Google Corporation. A mosaic of landing and a tour of the Apollo landings [DB/OL]. [2008-12-25]. http ://www. google. com/moon

月球卫星 CE-1 三线阵影像数据的解算试验

王 涛[1,2] 项 琳[1,2] 曹 锋[2]

（1. 中国测绘科学研究院，北京 100039；2. 北京四维远见信息技术有限公司，北京 100070）

摘 要 利用"嫦娥一号"卫星上搭载的 CCD 立体相机获取的月面三线阵影像数据，再配合激光高度计（LAM）数据，经过摄影测量解算可以制作月面数字高程模型（DEM）和数字正射影像图（DOM）。本文采用两种方法对月球卫星 CE-1 三线阵影像数据进行了月面点坐标的解算试验，方法一是根据测控给出的外方位元素直接解算，方法二是对影像上提取的像点进行区域网平差解算，根据两种方法的解算结果，比较了各方法的精度，达到了预期目的。

关键词 "嫦娥一号"卫星 月球 三线阵影像 区域网平差

1 引言

"嫦娥一号"(Chang'E-1)是我国自主研制并发射的首个月球探测器。它获得的三线阵 CCD 影像数据，为人类研究月球提供了丰富的信息源。利用"嫦娥一号"卫星上搭载的 CCD 立体相机获取的月面三视角影像数据，配合激光高度计（LAM）数据，经过摄影测量解算，便能获得月面数字高程模型（Digital Elevation Model，DEM）和数字正射影像图（Digital Orthophoto Model，DOM）。这两种数据产品反映了月球的地形和地貌特征。要想实现这个目标，思路有两个：一是要求卫星摄影中外方位元素测定值精度足够高，用前方交会确定月面点坐标；二是在外方位元素测定值精度不理想的情况下，采用区域网平差方法，削弱误差影响，得到符合精度要求的月面点坐标[2~3]。本文根据这两个思路分别对月球卫星 CE-1 三线阵影像的两轨数据（A 轨、B 轨）进行了月面点坐标的解算试验，并对解算结果进行了分析比较。

2 数据准备

月球卫星 CE-1 三线阵影像特征同名点的提取采用基于局部不变量描述符[6]（Local Invariant Descriptor）的匹配与传统特征点匹配相结合方法：利用局部不变量描述符匹配方法快速、高可靠性地寻找影像间的同名点建立两者之间的仿射变换关系，然后根据仿射变换关系式对重叠区域进行特征点（Forstner，Harris）的高精度匹配。提取结果如图 1 所示，提取的同名点像点坐标结果如表 1、表 2 所示。

表 1 A 轨道的像点坐标   （单位：像素）

	A 轨前视像点坐标			A 轨正视像点坐标			A 轨后视像点坐标	
编号	x	y	编号	x	y	编号	x	y
000	1009.8	251.3	000	516.3	252.0	000	21.7	252.5
001	1009.5	313.7	001	515.3	313.6	001	18.7	314.2
002	1008.3	382.7	002	513.5	383.2	002	17.2	382.7
003	1005.9	462.8	003	516.4	464.6	003	24.7	463.3
…	…	…	…	…	…	…	…	…

本文原载于《遥感信息》，2010，No.3，18~20。

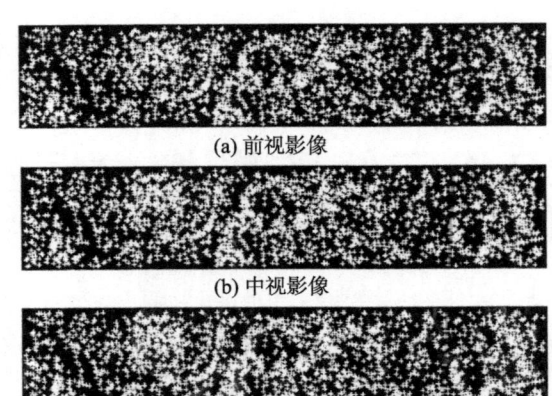

(a) 前视影像

(b) 中视影像

(c) 中视影像

图 1　A 轨影像提取的同名像点

表 2　B 轨道的像点坐标　　　　　　　　　　　　　　　　　　　　　　　（单位：像素）

B 轨前视像点坐标			B 轨正视像点坐标			B 轨后视像点坐标		
编号	x	y	编号	X	y	编号	x	y
000	1010.3	231.3	000	521.5	231.5	000	30.9	232.5
004	994.1	235.8	004	505.4	236.1	004	14.7	237.5
006	1151.2	62.7	006	656.9	62.1	006	161.2	64.9
010	1270.8	77.1	010	777.9	76.4	010	283.3	79.3
…	…	…	…	…	…	…	…	…

3　解算方法

3.1　月球卫星 CE-1 三线阵影像直接解算

月球卫星 CE-1 三线阵影像直接解算方法利用共线方程的严格解法，分别解算出相邻 A、B 轨道的坐标，并对同名点进行比较，得到一组实验结果数据。解算代入的数据有轨道姿态参数、相机参数、像点坐标。轨道姿态参数用 Lagrange 多项式进行拟合。解算过程如下：

对像点（单位：像素）坐标通过内定向参数转换成像平面（单位：mm）坐标系。月球三线阵测量是基于数字影像与摄影测量的基本原理，通过三个视角的像平面坐标用前方交会来计算特征点的月面坐标。空间前方交会是在内、外方位元素已知，地面点为未知待求，因此可以对每一片像点写出公式（1）[1]：

$$v_x = \frac{\partial x}{\partial X}\Delta X + \frac{\partial x}{\partial Y}\Delta Y + \frac{\partial x}{\partial Z}\Delta Z + x° - x$$
$$v_y = \frac{\partial y}{\partial X}\Delta X + \frac{\partial y}{\partial Y}\Delta Y + \frac{\partial y}{\partial Z}\Delta Z + y° - y$$
(1)

外方位元素用 Lagrange 多项式进行拟合。

用 CE-1 的姿态数据、内定向参数、像平面坐标，进行空间前方交会解算。分别解求 A、B 轨像点物方坐标，然后对两轨同名点进行统计，得到中误差。解算结果如表 3 和表 4 所示。

表 3　A 和 B 轨道点的坐标　　　　　　　　　　　　　　　　　　　　　　　（单位：m）

	A 轨道点的切面直角坐标			A、B 轨道同名点坐标较差		
编号	X	Y	Z	dX	dY	dZ
646	−74172.82	2923.38	−198423.95	157.33	78.51	−46.16
000	−75044.77	27368.95	−199733.31	−51.04	−62.27	191.52

续表

编号	A 轨道点的切面直角坐标			A、B 轨道同名点坐标较差		
	X	Y	Z	dX	dY	dZ
001	−75015.54	20106.81	−200188.60	65.36	27.71	34.03
002	−75041.88	11869.65	−200000.30	114.2	4.21	242.07
003	−74763.76	2560.67	−198565.63	97.04	52.67	−225.17
...	...	...	...	...	...	...

表 4　A、B 轨道坐标较差精度　　　　　　　　　　　　　　　　(单位：m)

点数	最大误差			最小误差			中误差		
	Xmax	Ymax	Zmax	Xmin	Ymin	Zmin	M_x	M_y	M_z
3802	258.617	259.908	722.712	0.2039	0	46.166	108.78	68.73	213.99

3.2　区域网平差解算

由于月面点坐标直接解算完全取决于测控外方位元素的精度，在测控给出的外方位元素不理想的情况下，直接解算出的月面点数据无论在卫星飞行方向或轨道连接上都可能出现问题，尤其在轨道连接方面的问题有时会很严重。因此进行区域网平差处理是非常必要的。所谓三线阵影像区域网平差，就是将线阵影像按一定的规则等效成小的面阵影像，然后按面中心投影的方式进行数据平差处理。由于区域网平差不但考虑了航带内模型的连接关系，而且同时考虑相邻航带间的连接关系，因此能够更好地解决整个区域内的月面数据拼接问题。本次试验用独立模型法区域网平差程序对实验数据进行了平差处理，解算出每个像点的地面点中误差及整体统计中误差，结果如表 5 和表 6 所示：

表 5　A 和 B 轨道点的坐标　　　　　　　　　　　　　　　　(单位：m)

编号	A 轨道点的坐标			A、B 轨道同名点坐标较差		
	X	Y	Z	dX	dY	dZ
646	−74146.12	2930.50	−198411.42	29.30	77.94	125.17
000	−75046.50	27859.65	−199783.06	35.45	70.15	151.17
001	−75070.16	20435.37	−200328.49	43.06	36.80	204.35
002	−75091.71	12069.61	−200260.52	63.62	27.04	48.46
003	−74705.24	2577.43	−198400.76	32.53	72.94	219.43
...	...	...	...	...	...	...

表 6　A、B 轨道坐标较差精度　　　　　　　　　　　　　　　　(单位：m)

点数	最大误差			最小误差			中误差		
	Xmax	Ymax	Zmax	Xmin	Ymin	Zmin	M_x	M_y	M_z
3802	110.6519	92.1329	229.3493	2.3496	2.4760	0.7486	53.79	32.82	105.21

3.3　试验结果比较

对于以上两种方法的解算结果，通过统计的中误差分析比较，可以看出区域网平差解算精度要比直接解算精度高出一倍。通过对比生成的 DEM 局部放大影像图（图 2、图 3），可以看出采用直接解算结果生成的图像纹理比较粗糙，显的地形比较破碎，说明 DEM 数据在高程方向的一致性较差，主要是由于外方位元素误差引起的。而采用区域网平差解算结果的影像纹理纹理过度更加自然、细腻，表现出来的地形较光滑，有效地消除了外方位元素的误差影像。

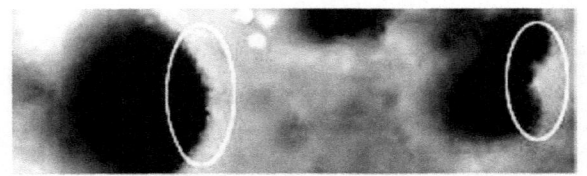

图 2　直接解算结果生成的 DEM 影像图(局部)(1:1)

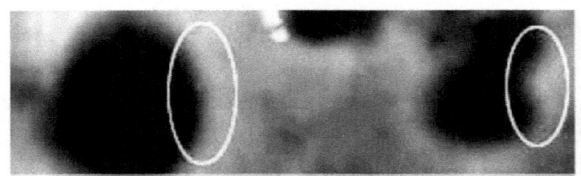

图 3　区域网平差解算结果生成的 DEM 影像图(局部)(1:1)

4　结束语

本文试验采用的数据为月球卫星 CE-1 三线阵影像中的两轨（一小段）数据（A 轨、B 轨）。从试验结果来看，由于范围较小，测控外方位元素数据的精度还比较好，基本能够满足 1:250 万的制图要求。但如果区域较大，测控外方位元素条件不理想，就可能无法满足 1:250 万的制图要求。通过区域网平差试验可以看出，平差后精度可以提高一倍。由于有连接条件的约束，通过平差可以有效地消除测控外方位元素的误差，确保月面点数据的相对精度能够满足制图需求。当然本文的试验仅是月球一块区域数据，不能代表整个月球影像数据情况。对于全月的测绘应用生产还要作进一步的分析、研究，以求得更高精度的解决方案。

参 考 文 献

[1]　王之卓. 摄影测量原理[M]. 武汉大学出版社，2007
[2]　王任享, 胡莘. 无地面控制点卫星摄影测量的技术难点[J]. 测绘科学，2004,29(3)
[3]　王任享. 月球卫星三线阵 CCD 影像 EFP 光束法空中三角测量[J]. 测绘科学，2008, 33(4)
[4]　王任享. 卫星摄影三线阵 CCD 影像的 EFP 法空中三角测量[J]. 测绘科学，2001, 26(4)
[5]　王任享. 卫星三线阵 CCD 影像光束法平差研究[J]. 武汉大学学报（信息科学版），2003, 28(24)
[6]　Distinctive image features from scale-invariant keypoints[EB/OL]. http://www.cs.ubc.ca/~lowe/pubs.html

三线阵CCD摄影测量理论在月球探测中的应用

王建荣　王新义　李晶　赵斐

(工西安测绘研究所，西安 710054)

摘要 月球是目前世界各国进行深空探测的首选星球，月球探测卫星属于传输型卫星。本文利用三线阵CCD摄影测量理论对月球摄影测量处理进行了可行性分析，指出其关键技术并提出解决方案，最后通过模拟实验进行验证。研究结果表明：三线阵CCD摄影测量理论可以较好地解决我国月球探测中的某些难题，完全可以应用于月球摄影测量中。

关键词 三线阵CCD影像　空中三角测量　影像匹配

1 引言

月球是离我们最近的星球，也是地球唯一的天然卫星，是我们人类走出地球，进行深空探测的首选目标。月球上蕴藏着丰富的矿产资源，月球上的超高真空和失重环境也是制造各种精密仪器的天然场所[1]。自20世纪50年代末开始，世界各国就开始对月球进行研究，1969年，美国"阿波罗11"首次实现了人类登上月球的伟大壮举，并把人类的脚印深深地印在了月球表面上；前苏联从1959年开始对月球的探测研究，先后发射了47个无人月球探测器，其中"月神"系列完成了月球全景影像的拍摄、月球资源的取样等任务。目前，欧空局、日本及印度等也在从事对月球的探测研究，并已制定了相应的计划。我国也于20世纪90年代进行了探月活动的可行性研究，目前已制定了到2020年的目标[2]，即"绕、落、回"三期工程。第一期绕月工程计划发射探月卫星"嫦娥一号"，对月球表面进行三维立体成像，获取月球三维影像图，实现对月球表面环境、地形、地貌及地质构造等进行探测。

月球探测卫星属于传输型卫星，月球光学成像系统采用的是CCD相机，可以利用三线阵构像方式获取大量的月球表面影像数据。因此，摄影测量数据处理的理论、技术和方法可用三线阵CCD影像为基础展开研究。

2 月球摄影测量中的关键技术

在利用线阵CCD月球影像进行摄影测量处理中，由于线阵CCD影像成像的特殊性，使得许多传统的技术方法、技术工艺不能满足处理要求，其主要技术难点在于空中三角测量、DEM的采集和编辑及正射影像的纠正等方面。

2.1 月球摄影测量中空中三角测量

针对月球摄影测量的特点，依从月球获取的辅助摄影测量数据和三线阵CCD影像等资料情况，可采用自由外方位平差方法或等效框幅式光束法平差方法(简称EFP法)[3~4]，计算对应影像的外方位元素和同名点三维坐标,完成空间模型的恢复。

本文原载于《测绘科学》，2008，Vol.33，No.6，19~20。

自由外方位元素平差方法基本思想是在无月面控制点参与的条件下，依靠三线阵CCD影像的像点量测值，通过光束法平差计算，建立一个无视差立体模型和相应的自由外方位元素列，类似于框幅像片的相对定向。如果有外方位元素情况下，可直接参与平差建立绝对定向的无视差月面立体模型。

EFP光束法平差方法基本思想是首先利用三线阵影像，建立"等效框幅像片"(EFP是Equivalent Frame Photo的缩写)。所谓"等效框幅像片"，从理论上讲每一取样时刻都有一组独立的外方位元素(简称EO)值，三线阵CCD相机的一个取样时刻内都有前、正、后三条影像，但受外方位元素变化及地形起伏等影响，满足经典框幅像片空中三角测量定向点(含有定向和三角锁本身模型连接作用)影像不可能都落在此三条影像上，如果利用落在此三条影像周围的影像取作定向点参与计算，则每一像点也只能提供两个观测方程，但代入了摄取此定向时刻的额外待解的6个EO值，及其待解的地面坐标，因而理论上无法解算出每一个取样时刻的EO值。当进行卫星摄影时，平台比较平稳，EO变化率不大，如果采用适当大间距的时刻，称作EFP时刻(通常在一条基线内取十个EFP时刻)，将航线模型进一步离散化，近似地表达航线模型和外方位元素，从而有可能采用CCD影像自身解算离散取样时刻的外方位元素，即EFP时刻的外方位元素，这种离散时刻取样的、能够构建三角锁的立体的影像，就称之为等效框幅像片。利用等效框幅像片，采用光束法平差方法，计算EFP时刻的外方位元素和加密点坐标。

2.2 月球摄影测量中DEM自动提取

利用三线阵CCD影像进行DEM自动提取过程中，影像自动匹配技术是其关键技术之一[5]。在影像匹配过程中，精度、速度及可靠性一直是人们所关注的，许多学者也研究出各具特色的影像匹配理论和算法，如松弛法、动态规划法、最小二乘算法及遗传算法等；匹配的方式和策略也多种多样。利用影像匹配的方法，从三线阵CCD数字影像中提取DEM，原则上既可以按像方匹配，也可以按物方匹配。像方匹配时，如果要实现一维搜索，需要对影像作核线排列再取样，像、地坐标计算不需要迭代，但要得到栅格DEM，采集的高程点密度要高，以便于内插栅格DEM。影像匹配有时也称为"病态"问题，为了解决该问题，通常引入核线纠正条件加以限制，对待匹配影像首先进行核线重采样，不仅可以实现把二维影像的搜索和匹配转变成一维的搜索和匹配，提高匹配效率，还可消除左右影像之间因姿态差异而引起的几何变形，提高匹配结果的可靠性。对于线阵CCD影像，由于其摄影原理和复杂的几何特性，无法用经典的核线重采样原理对待匹配影像进行处理，目前对线阵CCD影像尚无成熟的核线理论和技术，广泛采用的都是似核线理论，如基于投影轨迹法的核线理论与模型、基于简化传感器模型的核线理论等。结合针对三线阵CCD影像匹配的策略，我们也采用了一种似核线纠正的方法，对前视、正视及后视影像进行似核线纠正。卫星摄影通常左右摄站高差很小，可根据正射纠正的方法将左右影像生成准正射影像，此时y方向视差可以忽略不计，可看作"似核线纠正影像"，便于影像匹配在一维方向进行。本系统采用将月球影像纠正为正射影像(隐含一维搜索)的物方匹配，利用断面引导匹配原理[6]，计算获取月面栅格位置高程。

从影像匹配的策略上看有以下考虑与特点：

①采用纠正为正射影像进行匹配，使匹配在物方进行，有利于减少影像透视变形对影像匹配精度的影响；②采用由粗到精策略，将匹配影像生成金字塔影像，按分频道匹配，断面引导逼近(PGA)原理贯穿金字塔影像匹配的各层；③匹配中采用相关系数法、多点组匹配及视差平滑的方法或多点最小二乘法相结合的方法；④以正视影像为目标窗口，分别对前视、后视影像进行搜索，匹配计算前、后视同名影像的月面坐标；⑤金字塔第0层的DEM各点高程近似值采用该地区高程概略平均值，每一层匹配完成后，对DEM进行平滑并按栅格间距缩小一倍进行DEM内插，再作为下一层的DEM近似值。

利用影像匹配提取DEM的过程中，对于影像纹理贫乏地区、地形起伏较大地区，即使运用精度较高的匹配算法(如多点最小二乘匹配)，也无法达到预期的匹配精度。对这一问题还没有好的解决办法，目前通常的做法是通过人机交互的方式对错误匹配区域和点进行编辑[7]。

2.3 正射影像生成

正射影像的生成是利用已提取的 DEM 及三线阵 CCD 数字影像(实际工作中最好采用正视相机摄取的影像)，按共线方程，逐点计算月面格网点的像元位置，按双线性内插提取灰度信息，最终完成正射影像的纠正工作。

3 模拟实验

利用美国的阿波罗卫星影像生成的正射影像及其 DEM 等数据，根据线阵影像成像特点，分别模拟三线阵 CCD 相机的前视影像、正视影像以及后视影像，如图 1、图 2 及图 3 所示。并利用生成前视和后视影像进行影像匹配生成实验的 DEM，如图 4、图 5 及图 6 所示。

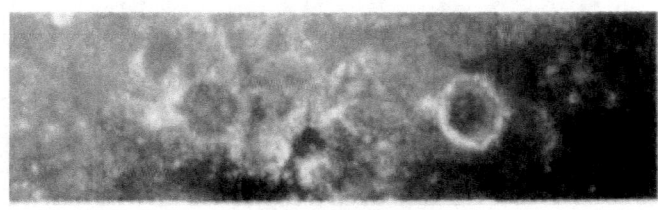

图 1 前视影像

图 2 正视影像

图 3 后视影像

4 结束语

将影像纠正为正射影像进行匹配，使匹配在物方进行，有利于减少影像透视变形对影像匹配精度的影响，同时结合断面引导匹配直接获得栅格 DEM。利用模拟生成的外方位元素，按 480m 格网间距，全航线含 6 条基线共采集约 10 万个月面点高程生成 DEM 后，与真 DEM 相比较后，中误差为 40 m，当像元分辨率为 120 m 时，中误差相当于 0.3 像元。而且从由 DEM 生成的等高线来看(图 4 和图 5)，地形走势起伏与真 DEM 生成的等高线基本一致，验证了本文中所提出的三线阵影像摄影测量理论的可行性，并可以用于"嫦娥一号"工程中的月球影像的初期处理。

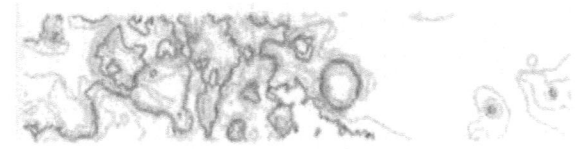

图 4　由真 DEM 生成的等高线

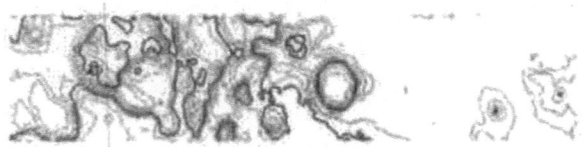

图 5　由模拟影像自动提取 DEM 生成的等高线

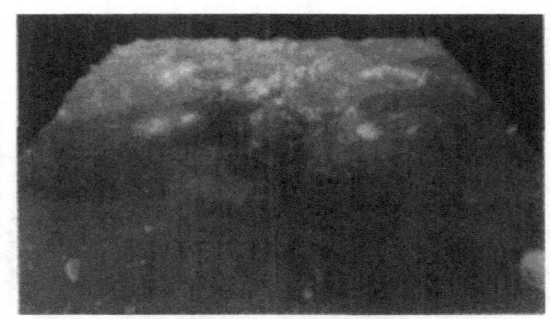

图 6　三维景观模型

参 考 文 献

[1]　卢波. 月球探测的意义及发展态势[J]. 国际太空, 1998, (4): 1—4
[2]　欧阳自远. 月球探测的进展与前景[J]. 学会, 2003, (11): 49—54
[3]　王任享. 卫星三线阵 CCD 影像光束法平差研究[J]. 武汉大学学报. 信息科学版, 2003, 28(4)
[4]　王任享. 利用卫星三线阵 CCD 影像进行光束法平差的数学模拟实验研究[J]. 武汉测绘科技大学学报, 1998, 23(4)
[5]　王建荣. 三线阵 CCD 影像 DEM 自动生成技术的研究与实践[D]. 西安：长安大学, 2006—06
[6]　王任享, 李晶. 物方多点匹配在"断面引导逼近(PGA)"原理的应用[J]. 测绘科技, 1997, (9)
[7]　马东洋, 邱振戈, 钱曾波. 数字高程模型编辑系统的设计与实现[J]. 测绘通报, 2003, (10)
[8]　王任享, 等. 将卫星三线阵 CCD 影响变换为正直影像进行立体测绘[J]. 测绘科学, 2007, 32(3)

月球卫星三线阵 CCD 影像 EFP 光束法空中三角测量

王任享

(西安测绘研究所，西安 710054)

摘 要 本文对月球卫星摄影三线阵 CCD 影像的 EFP (等效框幅相片)光束法空中三角测量作两种处理：一是与现行摄影测量常用的将平差转到切面坐标系进行，二是在摄影测量坐标系内，长航线自由网 EFP 光束法平差利用三线 CCD 推扫特点，在 EFP 平差中增加对前、后视影像的相机主距的附加改正项，用以补偿由于球面曲率产生的前、后视影像比例尺的差异，平差得到的是平面基准的地面坐标及外方位元素的平差值。前者计算，数学上严格，但长航线要适当分段为切面处理；后者计算数学上有近似性，可方便地用于估算卫星姿态变化率，或作地面模型的几何反演等实验研究。利用嫦娥一号获取的第一条航线，并给出相应的结果。

关键词 光束法平差　三线阵 CCD 影像　长航线空中三角测量　嫦娥一号卫星影像

1 引言

如果卫星摄影中外方位元素(摄站坐标 X_s，Y_s，Z_s 及角元素 φ，ω，κ，以下简称 EO)测定值精度足够高，那么三线阵 CCD 影像的摄影测量将非常简单，只要用共线方程就可以计算得精确的地面坐标。但至今卫星摄影中的测定的 EO 值都还达不到理想的程度，因此，很多情况下希望采用光束法平差以便削弱偶然误差和剔除粗差。笔者在文献[2]中比较详细地研究了 EFP 光束法空中三角测量原理，但所有研究均没有考虑摄影地区是球面问题。而且所有的实验工作均是计算机模拟。我国探月工程嫦娥一号取得的可贵的三线 CCD 影像数据，为深入探讨提供了良好条件，本文将依此对 EFP 光束法平差做一些实验研究。

2 在切面坐标系的 EFP 光束法空中三角测量

不将外方位元素用多项式表达是 EFP 光束法平差的重要特点，因而平差航线的长度可以不考虑受多项式阶数的制约。但在球面摄影影像处理时，航线长度依然要考虑随着航线加长，摄站 X_s 和 Z_s 以及倾角(在数值甚至符号上均有大变化，因而要将航线按适当长度分段，并逐段按切面坐标系进行平差计算。这是摄影测量常用的方法[1]。

作为对 EFP 光束法平差的探讨,实验用的 EO 初值可以应用探月工程嫦娥一号卫星测定的 EO 观测值，由于该工程至今未能提供这些数据，所以笔者只好采用卫星摄影参数按推扫式原理加以推算 EO 初值。

按探月工程额定参数：

相机：

焦距 f= 23.33mm，其焦面 CCD 面阵为 1024× 1024 像元，取左、中、右三线构成前视、正视和后视三个线阵，前、后视线阵对正视线阵夹角为 17.5°，截取的线阵数为 512 像元，像元尺寸为 14μm。

轨道：圆形近极轨道，额定对地高度 200km ±25km，标准摄影基线长 60.3km。

月球大地参数：平均半径为 1738km。

选择与月心坐标系的关系最简单的切面坐标系，并分别按实验选用的航线包含不同的基线数，按以上参考数据推算在切面坐标系内的 EO 初值，并参与平差，其中摄站坐标 Zs 要以权重较低的带权观测值处理。利用嫦娥一号第一条影像，按不同基线数平差，残余视差统计列于表 1，从实验结果看，切面坐标系内平差，航线的基线数小于 8 为佳。

表 1　视差统计

基线数	初始视差（像元）		收敛视差	
	mpx	mpy	mpx	mpy
2	0.10	0.38	0.02	0.25
6	0.63	0.47	0.31	0.22
8	0.79	0.51	0.33	0.23
12	1.85	1.01	0.37	0.32
15	2.64	1.33	0.48	0.56

三线交会一点是光束法平差成功的基本标志。在无地面控制点可资考核平差精度情况下，上下视差，左右视差的中误差不失为衡量平差内部精度的重要判据。实验中用于平差的定向点，连接点同名像点坐标均按相关像数法匹配，理论精度为 0.25 像元，表 1 数据表明，平差后视差中误差与影像匹配误差相当，因而嫦娥工程预估中采用影像匹配误差为 0.3 像元是适宜的。

现将基线数为 8 的切面坐标系平差的 EO 值参与自动采集 DEM(含月球曲率高差约为 17km)，及其生成的等高线示于图 1。图 1 等高线中可以感觉到月球曲率高差的存在。利用 EO 值对正视和后视影像作变形纠正于图 2。

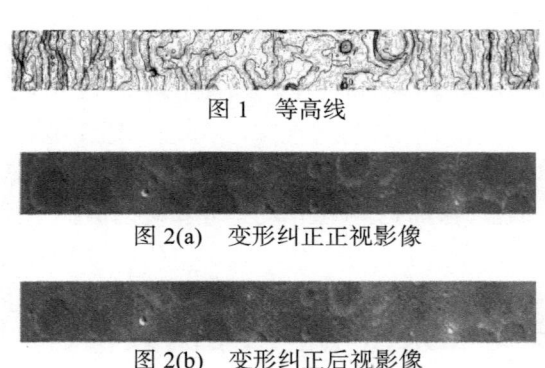

图 1　等高线

图 2(a)　变形纠正正视影像

图 2(b)　变形纠正后视影像

图 1 的等高线是 DEM 含有月球曲率的数据生成的等高线，图 2 的影像是变形纠正后，目视立体能看得出立体模型带有月球曲率的弯曲。

3　摄影测量坐标系长航线自由网 EFP 光束法平差

3.1　三线阵 CCD 影像几何特点

三线阵 CCD 相机推扫摄影的每一个取样周期可获得各自三条线阵影像，这些影像按时序排列成三个图像，由于推扫时倾角不同，使得获取的图像的同名点存在左右视差，立体目视时可以看出地形起伏。不管是在平面基准，还是球面基准的摄影，立体观察到的起伏地面都是如同在平面上的起伏，而无球面感觉。虽然球面基准摄影时，垂直于飞行方向(y 方向)各线阵属于中心投影，恢复的地面模型存在球面曲率，但卫星摄影航线宽度很小，目视立体影像也没有曲率感觉。在飞行方向，即使航线很长，由于推扫摄影属于

"正射"采样,大跨度地面曲率也不会记录在影像中,尽管就以某一时刻的三线CCD影像之间也具有框幅相片的性质,甚至采用笔者的等效框幅相片(EFP)思想,自由网空中三角测量也构不成框幅相片那样,一个单模型像对可以经过相对定向构成立体模型,而且相片三度重叠带有坚强的几何连接成与实际地面相似的航线模型(不计偶然误差系统累积)。也就是说三线阵CCD影像空中三角测量恢复与地面相似的立体模型,必须像上一节关于在切面坐标计算。那么在自由网空中三角测量会有什么特点呢?在球面基准推扫摄影时,在两条基线范围内存在球面曲率的缘故,前、后视影像的摄影高度比正视影像大ΔH(参见图3)。

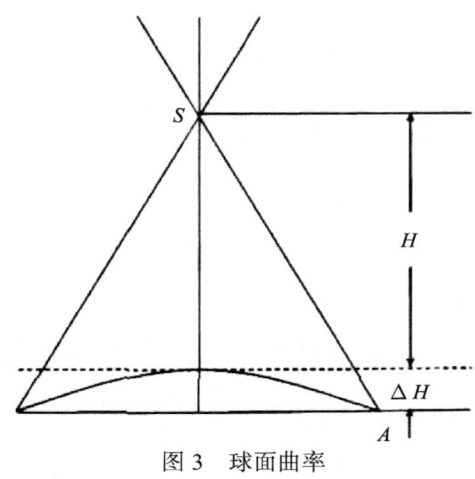

图3 球面曲率

在月球上当基线为60km时,两条基线范围内$\Delta H \approx 1.05$km,其值在整个航线中几乎为常数。因而球面基准摄影时,任意点的正视影像比例尺总是略大于前后视影像,同名点的正视影像与前后视影像之坐标差为:

$$y_v = \frac{y_A}{H}$$
$$y_{lr} = \frac{Y_A}{H + \Delta H}$$

令 $P_y = y_v - y_{lr}$,则 $P_y \cdot \frac{y_A}{H} \cdot \frac{f}{H} \cdot \Delta H$

上式:y_v=点A的正视影像坐标

y_{lr}=点A的前后视影像坐标

P_y=上下视差

在切面坐标系计算中,由于恢复了地面的曲率,上下视差P_y在平差迭代中自然被消除,但在自由网空中三角测量中,EO的初值中各时刻的Z_S为相同值,角元素均为零,所以平差迭代中上述的上下视差P_y无法被外方位元素的6个未知改正数所消除,为此可以引进一个新的未知改正数Δf_{FA},即$P_y = \frac{y_A}{H} \cdot \frac{f}{H} \cdot \Delta H = \frac{y_A}{H} \cdot \Delta f_{FA}$,此外,$\Delta f_{FA} = \frac{\Delta H}{H} \cdot f$

如果在自由网EFP空中三角平差方程系中增加解算Δf_{FA}项,并对前后视影像的相机主距f_l, f_r加以改正,迭代中便可消除上下视差,并且自动地对前后视影像x坐标也作调整。Δf_{FA}在方程式求解的数学形式与求解相机焦距检校值相当,只不过在列方程时仅对前后视影像的误差方程带有Δf_{FA}项,经过这样改化以后的EFP程序就可以进行超长航线的自由网空中三角测量,平差的结果将是长航线沿飞行方向属于平面基准的地面立体模型。

虽然这种计算在理论上有近似性,但由于计算可以在很长的航线中不间断地连续进行,在某些应用中有其方便的地方。

3.2 计算实例

同样采用嫦娥一号获取的第一条影像，基线数47，航线长约2840km，约占月球平均周长的1/4。

平差设置：基线长＝60.3km，地面像元分辨率120m，卫星对月面高200km，EO角元素取值为零，三线CCD影像主距23.33mm。

按照附加 Δf_{FA} 项的 EFP 程序计算，平差结果统计列于表2。

表2 长航线自由网平差统计

基线数	初始视差	（像元）	收敛视差	（像元）	调整主距
	mpx	mpy	mpx	mpy	fl = fr (mm)
8	0.90	1.37	0.38	0.25	23.169
15	1.25	1.55	0.39	0.24	23.177
47	1.18	1.58	0.45	0.28	23.167

从表2的视差看，长航线平差是成功的。

3.2.1 利用长航线平差的 EO 值估算卫星飞行的姿态变化率

将 EO 角元素平差值，图解显示于图4、图5、图6，并按影像地面分辨率120m，卫星地速1.5km/s，计算各轴变化值列于表3。

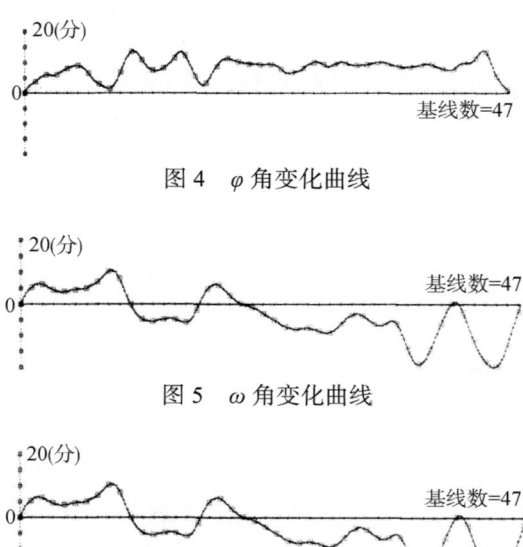

图4 φ 角变化曲线

图5 ω 角变化曲线

图6 κ 角变化曲线

表3 姿态变化值

基线数	$\varphi(10^{-3}°/s)$	$\omega(10^{-3}°/s)$	$\kappa(10^{-3}°/s)$	三轴总和$(10^{-3}°/s)$
8	0.7	0.8	0.6	1.3
15	1.4	1.5	1.2	2.4
47	1.2	1.9	1.8	2.9

如果姿态平均变化率取为 $2.5E^{-3}°/s$，则由此引起的相邻像元的混叠约0.005像元，影像质量对影像匹配精度影响不大。但对一般卫星而言，这样的变化率不甚理想，也可能这是第一条摄影航线，兴许更长时

间飞行后，会有改善。可惜笔者没有更多的嫦娥影像可资应用。本文计算的结果只是从摄影测量角度，对卫星姿态变化的估算，只作参考。

此外，利用长航线自由网 EO 平差值进行 DEM 的自动采集也很成功，这些数据对月面几何反演的应用尤为方便。

笔者根据在探月工程地面应用系统工作期间的观测数据进行平差整理，由于时间短促及应用资料有限，所作结果难免不妥之处，由笔者负责。工作期间地面应用系统在资料、工作条件等方面给予支持，特表感谢。

参 考 文 献

[1]   王之卓. 摄影测量原理[M]. 北京：测绘出版社，1979
[2]   王任享. 三线阵 CCD 影像卫星摄影测量原理[M]. 北京：测绘出版社，2006
[3]   王任享. 等, 将卫星三线阵 CCD 影像变换为正直影像进行立体测绘[J]. 测绘科学，2007，32 (3)

第三部分　激光高度计

嫦娥一号卫星载激光高度计

王建宇[1]　舒嵘[1]　陈卫标[2]　贾建军[1]　黄庚华[1]　王斌永[1]　侯霞[2]

(1. 中国科学院上海技术物理研究所，上海 200083；
2. 中国科学院上海精密光学机械研究所，上海 201800)

摘　要　介绍了嫦娥一号卫星激光高度计的设计和运行情况，包括系统的总体设计、空间激光器设计、激光发射系统设计、接收系统设计；分析了激光高度计的测量精度；阐述了激光高度计的地面测试情况；最后介绍了激光高度计在轨运行和在轨测试的情况。该激光高度计是我国第一次自行研制的空间应用的激光主动遥感仪器，自 2007 年 11 月 28 日在环月轨道上开机后，获取了共计 912 万点有效月球高层数据，圆满地完成了探测任务。
关键词　嫦娥一号卫星　激光高度计　系统设计

激光高度计是我国于 2007 年发射的第一颗月球探测卫星嫦娥一号的主要有效载荷之一。实现了卫星星下点月表地形高度数据的获取，为月球表面三维影像的获取提供服务。通过星上激光高度计测量卫星到星下点月球表面的距离，为光学成像探测系统的立体成图提供修正参数；并通过地面应用系统将距离数据与卫星轨道参数、地月坐标关系进行综合数据处理，获得卫星星下点月表地形高度数据[1]。激光高度计技术是从激光测距中演化来的，激光测距在我国的应用已经有好多年的历史，但是在此前仅限于在地面或机载的情况，嫦娥一号激光高度计是我国第一个星载激光高度计，设备由探头和电路箱两部分组成，如图 1 所示，其主要性能参数如表 1 所示。本文将详细介绍激光高度计的系统设计、核心元部件的研制、地面模拟测试、标定和仪器上天后的工作状况。

图 1　嫦娥一号激光高度计探头和电路箱

本文原载于《中国科学：物理学　力学　天文学》，2010，Vol.40，No.8，1063~1070。

表 1 嫦娥一号卫星激光高度计主要性能参数

作用距离(km)	月面激光足印大小(m)	激光波长(nm)	激光能量(mJ)	激光脉宽(ns)
200±25	<Φ200	1064	150±10	<7
接收望远镜有效口径(mm)	测距频率(Hz)	测距分辨率(m)	测距不确定度(m)	沿卫星飞行方向上月面足印点距离(km)
>120	1	1	5(3δ)	~1.4

1 激光高度计设计

1.1 激光高度计信号链路分析和系统设计

嫦娥一号卫星激光高度计任务的核心是激光测距。该激光高度计中的测量结果是月表目标到卫星的距离。图 2 显示了激光高度计的测量原理和工作流程,激光器向目标发射一束功率为 P,脉宽为 τ 的脉冲激光,目标表面返回的散射光被光学系统接收,光电探测器件将发射脉冲的一小部分及探测到的激光回波信号转变为电信号,分别触发测距计数器开始和结束计时,由此获得光脉冲飞行时间,经数据计算得到距离值 $z = c\Delta T/2$,其中 c 表示真空中的光速,ΔT 表示激光往返时间。

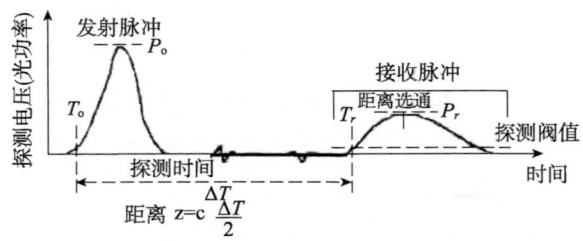

图 2 激光高度计工作原理

激光高度计在其内置的计算机产生的控制时序下同步工作。在总同步脉冲的触发下,激光器每一秒钟发射一次激光,与此同时一个激光发射的主波启动信号使得时间测量电路开始计数。激光回波被探测器接收后,经放大、阈值检波,输出激光回波脉冲,时间测量电路停止计数,由星上计算机读出计数值,并计算出高度计到月表的距离值。同时,计算机还将接收载荷数据管理系统提供数据注入,完成激光高度计主要点状态参数监测和控制任务。激光高度计系统的总体框图如图 3 所示。

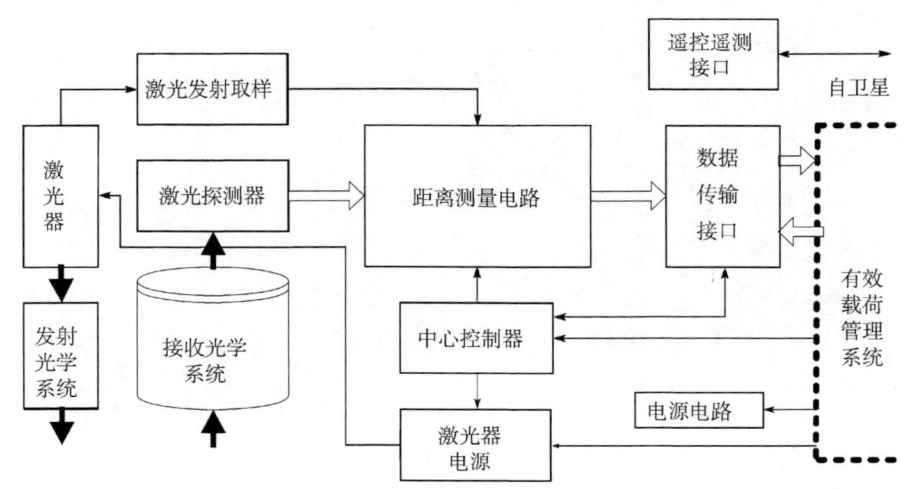

图 3 激光高度计总体框图

根据激光对漫反射面目标的探测方程,激光高度计接收到的回波信号功率 P_r 由功率测距方程给出为[2]

$$P_r = \frac{P_t \cdot S_r \cdot T_r \cdot \rho_1 \cdot \cos^2\theta}{\pi \cdot R^2} \tag{1}$$

式中 P_t 为激光发射功率,S_r 为有效接收口径面积,T_r 为接收系统光学透过率,ρ 为月面反射率,θ 为激光束与月面之间的夹角,R 为激光高度计与月面距离。

激光高度计的发射功率为

$$P_r = \frac{W_t}{\tau} \tag{2}$$

其中 W_t 为激光发射能量,τ 为激光发射脉冲的半脉宽。

激光高度计的工作轨道高度距离月面约为(200 ± 25) km,激光束与月面之间的夹角 θ 接近 90°,月面最低的反射率 ρ 为 3%,从表 1 的激光高度计的其他参数,根据(1)式可以计算得出在距离远、月面反射率低的时候,激光高度计接收系统接收到的回波功率约为几十纳瓦,接近激光高度计的设计探测灵敏度。考虑回波展宽变形以及系统最大测程的系统设计余量,因此,系统设计时需对激光光源输出功率稳定性、激光收发光轴同轴稳定度、背景光噪声抑制技术等方面展开研究。

1.2 激光高度计的发射系统设计

激光高度计的发射系统负责向目标发射高能量密度的激光脉冲光束,主要由激光器和激光扩束器组成。考虑到星载激光高度计的作用距离很远,要求脉冲具有较高的瞬时功率,同时体积、重量又有严格的控制,因此,系统采用了 Nd:YAG 主动电光调 Q 激光器作为发射激光器。为减少发射光束的发散角,系统采用了激光光束扩束准直技术。为解决发射激光能量过强,激光扩束镜采用伽利略式的球面透镜系统,使得激光在镜筒内具有没有会聚点,可以有效防止激光损伤;系统无中心挡光,不损失发射的激光能量;光路短,且结构简洁等特点。扩束镜的扩束倍数为 5 倍,扩束后的激光发散角为 0.6 mrad。图 4 为扩束镜和激光器对接的光路示意图。

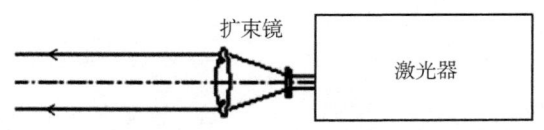

图 4 嫦娥一号激光高度计发射系统光路示意图

1.3 激光高度计的接收系统设计

激光高度计的接收系统由接收望远镜和激光回波接收电路组成。接收望远镜用于将目标反射回来的激光能量高效率地会聚到探测器光敏面上;激光探测电路由雪崩二极管探测器、信号处理电路、脉冲形成电路、峰值检测电路和偏压调整电路组成,将收集到的光信号转换为电信号,并进行处理后得到测量距离数据。

1.3.1 接收光学系统设计

由于月球探测距离较远,为了收集到更多的回波能量,需要接收系统口径较大,为了减小体积和重量,并兼顾到系统焦距设计要求,接收望远镜采用了结构紧凑的双非球面反射式卡式系统。在系统中设置有中继镜组,中继镜组产生平行光路。中继镜组的平行光路中设置有前截止滤光片和窄带滤光片,用于滤除外界的非信号光。光学系统组成如图 5 所示。设计结果如表 2 所示。

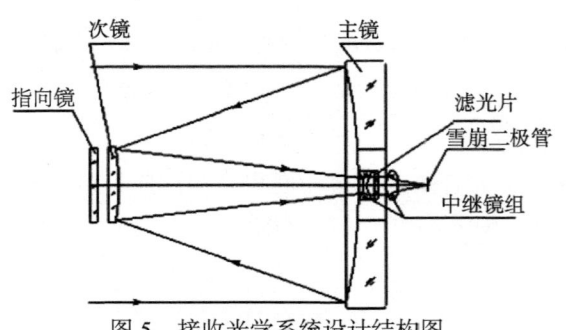

图 5 接收光学系统设计结构图

表 2 光学系统参数

光学参数	接收望远镜有效口径(mm)	系统焦距(mm)	瞬时视场(mrad)	光学效率	弥散斑直径(μm)
设计值	128(通光口径134)	533.333	1.5	0.82	<10

1.3.2 激光回波接收电路设计

激光高度计的回波接收电路工作原理如图 6 所示,雪崩二极管将回波脉冲光信号转化为电信号,经匹配放大器进行放大后,由阈值检波比较器检出回波 TTL 信号,触发时间计数电路停止计时。

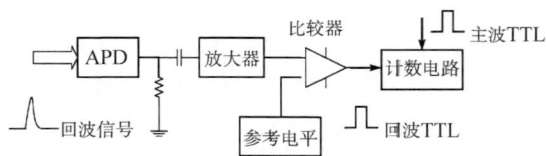

图 6 激光高度计回波接收电路原理图

回波接收单元采用的雪崩二极管(APD)为国产 SPD-0522 近红外增强型探测器,主要指标如表 3。

表 3 SPD-0522 雪崩二极管参数

工作波长(μm)	最佳工作电压(V)	电压响应度(1.06 μm)	等效噪声功率(1.06 μm)	脉冲响应时间(ns)	工作温度(℃)
0.4~1.1	250~450	>0.7×10^5 V/W	<0.35pW/Hz$^{1/2}$	<10	−20~+50

为了达到最佳信噪比,设计匹配放大器的带宽与雪崩二极管带宽相当。当激光回波信号为 $1×10^{-8}$ W 时,雪崩二极管产生的激光回波脉冲信号峰值为 $1.4×10^{-3}$ V。要能达到阈值检波电路的工作范围,匹配放大器设计增益需控制在 60 dB 左右。

激光高度计在对月球的探测过程中,由于月面背景光照条件噪声和雪崩光电二极管工作温度的变化,在放大器的输出端的噪声电平就会不断变化,从而导致在恒定阈值条件下的虚警率变化。由于地面测控系统无法实时对激光高度计的比较阈值进行注入调整,为了保证对应一个阈值门限下的恒定虚警率,需要在不同的工作环境下实现噪声电平的基本不变,以保证不变的虚警率。雪崩光电二极管经匹配滤波器输出的噪声可以近似为高斯分布,如果对该噪声作检波处理,则噪声的包络将是瑞利分布的,其概率密度为[3]

$$P_b(x) = \frac{x}{\sigma_0^2} \exp\left(\frac{-x^2}{2\sigma_0^2}\right) \tag{3}$$

σ_0 为噪声方差。噪声包络的统计平均值是

$$M(x) = \int_0^\infty x P_b(x) \mathrm{d}x = \sqrt{\pi/2} \cdot \sigma_0 \tag{4}$$

由此可见,噪声包络的统计平均值与总输出噪声的方差成正比,从而在理论上来说,可以先求得噪声

的方差值,然后由它去对噪声进行归一化,就可以使噪声恒定,从而达到一定阈值电平下的恒定虚警的目的[4]。

信号处理框图如图7。噪声检波后的平滑结果将通过控制器改变探测器的工作状态,从而实现类似的归一化操作,使得电路总输出的噪声电平基本保持恒定。

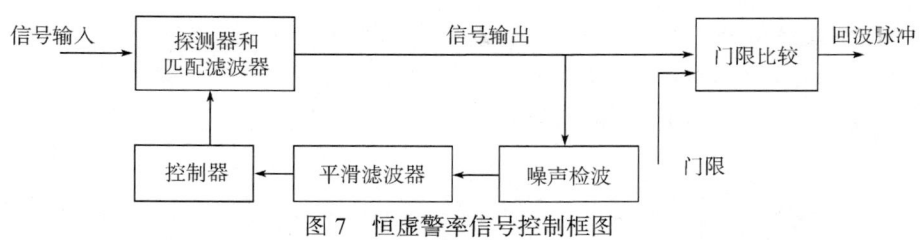

图7 恒虚警率信号控制框图

激光高度计上还设计了回波峰值采样电路,可对月面返回的回波信号进行峰值采样,在一定程度上可以表征月面对1.06 μm波段激光能量的反射强度。

1.4 激光高度计的空间适应性设计

嫦娥一号卫星的发射和空间运行环境比较复杂,空间的太阳辐照、紫外辐照、真空、极端温度、等离子体、带电粒子辐射都有可能给系统带来很大的影响,所以在系统设计时要充分考虑到空间环境给系统结构、光学和电路带来的影响,并进行相应的空间适应性设计。

嫦娥一号卫星激光高度计的研制过程中在材料选型、结构厚度、防辐照加固等方面进行了辐照分析与设计,并进行了紫外辐照、Co_{60}总剂量辐照、单粒子效应试验。

1.4.1 结构的空间适应性设计

激光高度计是一种主动光学系统,对结构的稳定性要求很高,它需要保证发射和接收光学系统保持很稳定的同轴关系,这给结构的强度和抗震性、热变形特性提出了非常高的要求,尤其是激光器部分,其光学部件之间的位置关系误差要在秒级范围之内。所以,在机械结构设计时,采取了高强度、高温度稳定性设计,周密设计外形结构和布局。为了满足力学环境,激光高度计的所有光学元件均采用修磨基座来调整光路、安装螺钉直接固定的安装。并按照力学环境鉴定级实验要求,进行了环境模拟试验,以确保结构满足空间应用的需要。

1.4.2 光学的环境适应性设计

激光高度计为单波长工作系统,为了增加探测灵敏度,探测器采用灵敏度极高的雪崩管。雪崩管响应波长为0.4~1.1μm,极容易受到外部太阳光线的影响。在光学系统中设计了窄带滤光片和中继光学镜组,以便减少太阳光直射和辐照对探测器的影响;设计了遮光罩以抑制空间杂散光进入探测器;光学系统选用了JGS1作镜体材料,该材料对1.064μm波长的光有较好的透过率,并且在空间性能稳定,能够经受空间恶劣环境(如高真空、温度交变、高能粒子轰击、电子束照射等)。

1.4.3 环境适应性热设计

激光二极管是个温度敏感元件,中心波长随温度的变化率为:0.3~0.4 nm/℃。为了保证激光输出能量的稳定性,必须对激光器进行温度控制。激光高度计整机在舱内,舱内的工作温度是-10~45℃。激光发射器的安装板与卫星隔热安装,通过温度控制,确保激光二极管工作区域的温度范围达到(18±3)℃。结构上光学元件的夹具选择与光学元件接近的材料,发射部件与导热良好的金属连接在一起,确保良好的导热。激光器的底部与安装底板之间采用填钢膜,增加传热效果。

2 空间激光器设计

嫦娥一号激光高度计中的激光器是我国第一个应用到空间中的激光器,在综合分析国外星载激光器及国内研制水平和加工工艺的前提下,其采用的技术路线为[5]: (i) 激光二极管泵浦的 Nd:YAG; (ii) 采用直角棱镜和平面镜输出的谐振腔形式和直线结构; (iii) 主动电光调 Q。采用激光二极管泵浦、电光调 Q 的固体 Nd:YAG 激光器,利用 Porro 棱镜改善激光器的失调灵敏度。

激光器的光学设计主要是指激光谐振腔的设计,激光器输出后接扩束望远镜,腔内偏振片输出的一部分激光衰减后,由 PIN 管接收,作为能量监测和主波取样。整个光学系统的组成如图 8 所示。

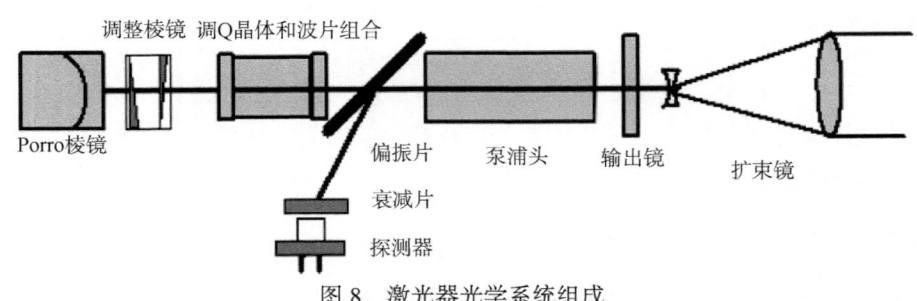

图 8 激光器光学系统组成

激光器的结构设计的原则是根据光学设计要求,满足所有光学元件的安装要求,确保激光输出达到性能指标。光学元件的固定结构,严格根据光学元件的加工净度,公差设计机械尺寸,选用与光学元件的热膨胀系数相近的材料,如紧固各个熔石英材料光学件的各个机械部件均使用钛合金,避免由热膨胀造成的光学件破坏或者光轴的偏移。各个光学元件的固定部位留有一定部位用来封胶,以达到机械件与光学件之间的"软接触"。

激光器对外界的污染非常敏感,因此其结构必须实现密封,使激光器内部光学组件与外界隔绝。同时在整机存储温度范围内,要求密封结构的漏气率必须控制在一定的范围内,避免激光器内的高压电源出现低气压放电情况,损坏器件。其密封方式采用橡胶密封方法,经实物检漏,在常温下漏率优于 2×10^{-8} Pa·m^3/s,适用温度范围$-15\sim 30$℃。满足激光器的漏率要求。激光发射系统的技术指标如表 4 所示。

表 4 激光发射系统的性能参数

工作波长(nm)	单脉冲能量(mJ)	脉冲宽度(ns)	脉冲重复频率(Hz)	激光发散角(mrad)	光斑直径(mm)	波束指向稳定度(mrad)	激光模式
1064±1	150,能量起伏不高于 10%	5~7	1	<0.6(光束扩展后)	<30	0.05	低阶模,>90%

3 激光高度计的地面标定

激光高度计的主要功能是在距离月表 200 km 的轨道上实现测距,在地面无法对其工作能力进行实际测量,所以需要在地面通过间接的办法对其测距精度和最大测程进行标定,以验证其能否具备月表探测的能力。

3.1 地面测距不确定度的标定

测距不确定度综合反映了计数时间分辨率引起的量化测距误差 δ_1、计时晶振的不稳定度引起的测距误差 δ_2、测距回波信号变形和上升沿抖动引起的测距误差 δ_3 以及定标后引起的系统测距残差 δ_4,在这四项误差中的前三项为独立随机事件,而第四项是固定的系统误差[6],所以激光高度计的总不确定度 δ 如(5)式所示。这三项分别在地面进行了测试和标定。嫦娥一号卫星激光高度计的测距不确定度指标要求为<5 m。

$$3\delta = 3\sqrt{\delta_1^2 + \delta_2^2 + \delta_3^2} + \delta_4 \tag{5}$$

计数时间分辨率引起的量化测距误差 δ_1 主要由计数电路时间分辨率引起的量化误差，包括主波起始时间量化误差和回波停止时间量化误差，系统的计数晶振频率 $f=155.2$ MHz，经计算由于计数时间分辨率引起的量化误差引起的测距误差 δ_1 为 0.422 m。

计时晶振的不稳定度引起的测距误差 δ_2，为激光高度计测距电路的计时不确定度，$\delta_{22} = R \times \Delta f_{Tref}$。$R$ 为工作距离，Δf_{Tref} 为时基稳定度，激光高度计采用温补型晶振，时基稳定度是 1.5×10^{-6}，在 200 km 距离上对应的计数不确定度 $\delta_2 = 200 \times 10^3 \times 1.5 \times 10^{-6} = 0.3$ m。

测距回波信号变形和上升沿抖动引起的测距误差 δ_3 主要由回波幅度变化所引起的时延误差、回波脉冲宽度变化所引起的时延误差和噪声所引起的时延误差，经分析和测量，该项误差引起的测距误差不超过 0.66 m；

$$3 \times \sqrt{\delta_1^2 + \delta_2^2 + \delta_3^2} = 2.52 \tag{6}$$

定标后引起的系统测距残差 δ_4 是激光高度计为校正测距中的系统误差而进行的标定过程中留下的残余误差，经测量该项误差不大于 1 m。

综上所述，系统测距不确定度 $3\delta \leq 3.52$，满足 ≤ 5 m 的总体指标要求。

3.2 地面最大测程模拟

嫦娥一号激光高度计的工作环境是月球外层空间，为真空环境；探测目标是月球表面。根据激光高度计要求的发散角和轨道高度，探测时月球目标可以认为是朗伯体大目标，反射率为 3%~7%。最大测程如 (7)式所示：

$$R_{max}^2 = P_t \tau_0 \tau_a^2 \rho A_r / \pi P_{rmin} \tag{7}$$

式中，R_{max} 为最大测程，P_t 为激光发射功率，τ_0 为光学系统效率，τ_a^2 为双程大气透过率 ρ 为目标反射率，A_r 为系统接收孔径，P_{rmin} 为系统最小探测功率。

由(3)式可以看出，只要求出系统最小探测功率，也即系统探测灵敏度，再结合探测目标和环境，就可以对激光高度计的最大测程进行标定。对于嫦娥一号激光高度计，采用了外场消光比模拟测试和目标回波模拟器进行直接探测的办法对其最大测程进行标定，两种标定方法互为补充。

3.2.1 消光系数法测最大测程

消光系数法是指根据激光高度计对环境条件、距离和目标特性均已知的模拟目标进行测距时，接收到的目标回波等于系统探测灵敏度时所需的消光比 M，推算得到激光高度计在规定条件下的最大测程。激光高度计对一个已知距离、已知反射率的面目标进行探测，在激光接收路径上放置光学衰减片，不断增加衰减量，直至探测概率降低到95%时，以此时激光接收路径上的光学衰减量作为消光比(Extinction Ratio, ER)。当激光光路上的光学衰减量衰减到临界值时，激光测距仪接收到的光功率即为系统最小探测功率，如下式所示[7]：

$$P_{rmax} = \frac{P_t \rho A_r \tau_a^2 \tau_0}{\pi R_b^2} 10^{-\frac{ER}{10}} \tag{8}$$

式中 R_b 为地面测试时模拟目标的距离。测试框图如图9所示。

激光高度计研制完成以后，在西安某基地对其进行了最大测程消光比测试。靶面目标距离为1194.34 m，目标漫反射率为0.85，尺寸为 1.6 m×1.6 m。测试得出激光高度计的消光比为60 dB，根据(7)和(8)式得出在月表反射率为0.03时，其最大测程为279.773 km。

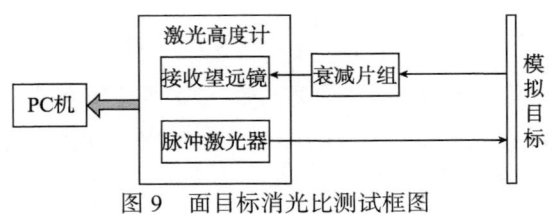

图 9 面目标消光比测试框图

3.2.2 目标回波模拟器测最大测程

直接模拟测量法是指当激光高度计的所有其他参数都已经确定后,通过某种设备直接测定系统的探测灵敏度,再根据测距公式计算得到系统的最大测程。嫦娥一号激光高度计专门研制了目标回波模拟器,原理框图如图10所示。

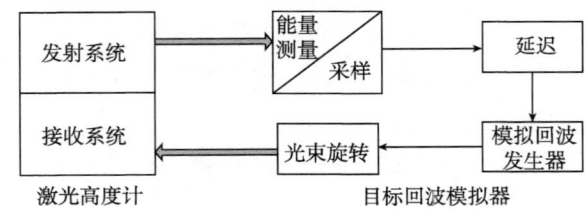

图 10 目标回波模拟器原理框图

该目标回波模拟器可用于产生有一定延时的模拟回波,该模拟回波发散角为 1.5 mrad(与激光高度计接收系统视场相同),脉宽为 25 ns,回波功率最大可达到 $2.6×10^2$ W,最小可小于 $1.4×10^{-15}$ W,与主波的延迟在 0~1s 内可调,调整幅度最小为 1 ps。目标回波模拟器可用于直接测量激光高度计系统探测灵敏度得到系统的最大测程,精度达到±2.24%;还可用于测量发射激光能量、距离范围等参数。该模拟器具有不受外界环境条件影响、适用性强、精度高等特点。

在激光高度计发射之前利用此模拟器对其最小探测灵敏度进行了测试,夜间的最小探测灵敏度为 $1.6×10^{-8}$ W、正午的最小探测灵敏度为 $3.2×10^{-8}$ W。综合激光高度计的系统参数,计算得出激光高度计在假设月面反射率为 0.03 时,结合其他月表环境,根据(2)式得到其正午最大测程为 308.09 km。

4 激光高度计的在轨运行

2007 年 10 月 24 日 18 时 05 分,嫦娥一号卫星发射升空,进入奔月轨道;11 月 5 日成功实施月球捕获;11 月 7 日进入月球 200 km 高的极轨工作轨道;随后开始卫星检查,平台状态调整,进入科学探测阶段。

2007 年 11 月 28 日 1 时 32 分,激光高度计开机,3 时 49 分激光高度计在月球 200 km 轨道向月球表面发出了第一束激光;至 2008 年 10 月 24 日,激光高度计在轨累计开机 3309h,获得了 1369 轨探测数据,有效测距点 912 万个,对月球表面实现了全覆盖。

在激光高度计获取科学探测数据期间,对每次开机后下行的工程、遥测参数进行实时的监视,入轨后主要的控制均在正常工作范围内,仪器工作正常。开机后的第一阶段为激光高度计在轨测试和状态调整期间,该期间激光高度计的每轨探测率为 82.3%,两极探测率可以达到 90%。经工作参数调整,激光高度计探测率总体在 97% 左右,最高时达到了 99%。自 2007 年 12 月 1 日起,激光高度计进入长期管理状态。

激光高度计获取的高程数据结合卫星轨道、姿态、仪器几何参数以及精密星历,通过高程模型解算,制作了空间分辨率为 3 km 全月球 DEM(Digital Elevation Model)图(图11,图版 III)。

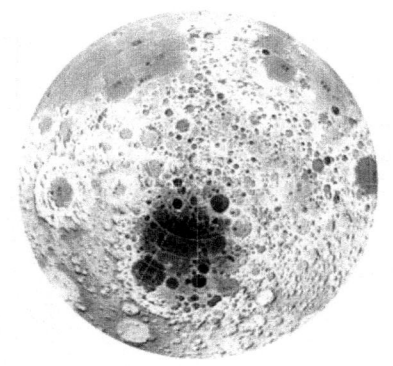 

图 11　由激光高度计数据绘制的全月球 DEM 图

国家天文台月球与深空探测应用中心对激光高度计获取的科学数据进行了处理，并且将之与 Clementine、DEM、Apollo DEM 数据的月表定位结果、高程测量结果、撞击坑大小测量结果、地形剖面线总体趋势等比对，证明激光高度计的定位结果正确、高程测量结果正确、地形剖面线反映的月表地形地貌特征正确。此结果也验证了激光高度计系统设计的合理性和可靠性。

参 考 文 献

[1]　欧阳自远. 我国月球探测的总体科学目标与发展战略. 地球科学进展, 2004, 19(3): 355—357
[2]　陈育伟, 张立, 胡以华, 等. 对地观测激光成像的回波阵列探测技术. 红外与毫米波学报, 2004, 23(3): 170
[3]　左群声, 徐国良, 马林, 等. 雷达系统导论. 第 3 版. 北京:电子工业出版社, 2006. 30—31
[4]　霍联正. 雪崩光电二极管对空激光测距接收机. 激光技术, 1993, 17(3): 137—141
[5]　侯霞, 陆雨田, 胡企铨, 等. 折返式激光二极管侧抽运 Nd:YAG 激光器. 光学学报, 2004, 10: 11
[6]　黄庚华, 王斌永, 舒嵘, 等. 月球探测卫星激光高度计地面定标与性能验证技术. 红外与激光工程, 2006, 35: 375—377
[7]　国防科学技术工业委员会. 脉冲激光测距仪性能试验方法. GJB2241.2—2006

激光测距技术在空间的应用

齐炜胤　尤　政　张高飞　孙　剑　张　弛

随着空间技术和航天工业的发展，空间距离测量已成为空间领域的重要研究内容。传统雷达测距在太空中极易受到高能粒子和电磁波的干扰，测量精度低，无法满足高精度测量的要求。宇宙空间空气稀薄、温度变化剧烈，无法进行超声波测距。因此，测量空间距离需要一种适合空间环境、抗干扰能力强和测量精度高的测距方法。

激光测距技术是一种自动非接触测量方法，对电磁干扰不敏感，抗干扰能力强，测量精度高。与一般光学测距技术相比，它具有操作方便、系统简单及白天和夜晚都可以工作的优点。与雷达测距相比，激光测距具有良好的抗干扰性和很高的精度。

在重复测距的同时，以细激光束对空间扫描，同时获得目标的距离、角度和速度等信息，这就是激光雷达。激光雷达能实现很多传统雷达达不到的性能要求。激光的发散角小、能量集中，能够实现极高的探测灵敏度和分辨率；其极短的波长使得天线和系统尺寸可以很小，这些都是传统雷达所不可比拟的。与微波雷达相比，激光测距仪方向性好、体积小、重量轻，非常适用于搭载在航天器上进行空间目标距离测量。

激光测距技术综合了激光器技术、光子探测技术、信号处理技术等多项技术，测距精度高，测程大，可靠性高，能够满足空间目标高精度、大测程测距的要求，在空间测量领域获得了广泛应用。

1　激光测距技术的基本原理

激光测距技术按照测程可以分为绝对距离测量法和微位移测量法。按照测距方法细分，绝对距离测距法主要有脉冲式激光测距和相位式激光测距，微位移测量法主要有三角法激光测距和干涉法激光测距。

脉冲激光测距的原理是：由脉冲激光器发出一持续时间极短的脉冲激光(主波)，经过待测距离L后射到被测目标，有一部分能量会被反射回来，被反射回来的脉冲激光称为回波。回波返回测距仪，由光电探测器接收。根据主波信号和回波信号之间的间隔，即激光脉冲从激光器到被测目标之间的往返时间t，就可以算出待测目标的距离。

$$D = \frac{1}{2}ct$$

式中c为光速。脉冲法精度一般在米量级。

相位激光测距的原理是：对发射的激光进行光强调制，利用激光空间传播时调制信号的相位变化量，根据调制波的波长，计算出该相位延迟所代表的距离。即用相位延迟测量的间接方法代替直接测量激光往返所需的时间，实现距离的测量。这种方法精度可达到毫米级。

三角法激光测距是由激光器发出的光线，经过会聚透镜聚焦后入射到被测物体表面上，接收透镜接收来自入射光点处的散射光，并将其成像在光电位置探测器敏感面上。当物体移动时，通过光点在成像面上的位移来计算出物体移动的相对距离。三角法激光测距的分辨率很高，可以达到微米数量级。

干涉法激光测距是通过移动被测目标并对相干光进行测量，经计数完成距离增量的测量，因此干涉法测量的灵敏度非常高，可以达到纳米级。

固体激光器和半导体激光器技术的发展以及高功率、高亮度、高效率半导体激光二极管的出现，使得激光测距装置具有结构紧凑、质量轻、寿命长、效率高等特点，非常适合空间环境的应用。从20世纪80年代后期开始，除了美国之外，欧洲和日本也开始研究开发空间用激光测距装置。激光测距装置在空间任务中的运用越来越广泛。

2 激光测距在空间技术中的应用简况

2.1 空间碎片探测

空间碎片俗称太空垃圾，是指宇宙空间中除正常工作的飞行器外的所有人造物体，大到废弃的卫星、运载火箭末级，小到固体火箭发动机燃烧后的三氧化二铝小颗粒或从航天器上剥落下来的漆片。

空间碎片的存在严重威胁着在轨运行航天器的安全。空间碎片的不断产生对有限的轨道资源也构成了严重威胁，尤其是当某一轨道高度的空间碎片密度达到一个临界密度时，碎片之间的链式碰撞过程将会造成轨道资源的永久破坏。

为了安全、持续地开发和利用空间资源，必须不断提高对空间碎片的跟踪监视技术，增强对空间碎片环境的分析预测能力，同时寻求控制空间碎片的有效措施。

空间碎片监测可以通过地基监测和天基监测两种方式。一般来说，大尺度空间碎片主要依靠地基手段；中小尺度空间碎片探测可以依靠天基手段。而基于激光测距技术的激光雷达探测系统在空间碎片探测方面具有独特的优点。它采用主动探测方式，不受光照条件限制，波束窄，探测距离远，空间分辨率高，测量精度高，并且可以同时进行测距和测速。

如毛伊岛光学站基于激光雷达的美国空军地基光电深空监视系统就采用了激光测距技术。该系统由光学分系统(AMOS)和跟踪识别分系统(MOTIF)组成，前者包括一台1.58m卡塞格林望远镜、一台激光发射器和一台AMOS获取设备，主要用于测量、跟踪、红外目标识别和补偿成像；后者由两台并联安装的1.22m卡塞格林望远镜组成，主要用于测量轨道高度在4800km以下的卫星的反射特性、热辐射特性并对其成像。

美国弹道导弹防御局在20世纪90年代初开始研制"快速光束操纵系统"(ROBS)。它是一种基于激光雷达的天基探测系统。ROBS在结构上包括目标识别捕获分系统、跟踪成像分系统和激光雷达分系统三部分，其中激光雷达分系统用来测量目标距空间站的距离和目标的多普勒频移，进而确定目标的运动速度和轨迹。另外，Visdyne公司和菲利浦斯实验室还联合提出了一种用于监测尺寸小于10cm的空间碎片的监测系统。该系统由成像分系统、信号处理分系统和激光雷达分系统三部分组成。

2.2 空间交会对接

航天器空间交会对接技术是发展空间技术的关键途径。它包括两部分相互衔接的空间操作：空间交会和空间对接。所谓交会是指航天器之间在轨道上按预定要求相互接近的过程，即两个或两个以上航天器通过轨道参数的协调在同一时间到达同一空间位置的过程。而对接则是在交会的基础上，通过专门的对接装置将其在结构上连成一个整体（表1）。

表1 交会对接过程与导航方式

阶段	地面导引	自动寻的	最终逼近	对接锁撞
距离范围	大于100km	100~1km	1km~100m	100~10m
导航方式	地面引导，GPS	微波雷达，激光雷达，GPS	激光雷达	激光雷达，光学敏感器

由上表可以看出，基于激光测距技术的激光雷达在整个交会对接过程中起着很关键的作用，特别是在几十公里到几米这一范围内起着主要导航作用。这是由交会对接的实际要求和激光雷达的性能所决定的。因为在这个阶段，交会对接的精度要求很高，很短的距离对于微波雷达来说是测量盲区，而且其精度也远远不能满足要求。激光雷达由于自身的优点，如动态范围很宽、精度极高等，最适合于交会对接。

由于在太空中不存在大气的影响，加上激光雷达自身的巨大优势，使得激光雷达在空间交会对接中获得了广泛的应用。表2为在各国空间交会对接中激光雷达的使用情况。

表2 各国的空间交会激光雷达

系统名称	消息来源	作用距离	工作方式
激光对接系统	美国约翰逊空间中心，1986年	远距离：22~100km 近距离：100~0m	连续波半导体激光器光源；光电二极管接收，检流计式扫描，位置敏感控测器和沃拉斯顿棱镜姿态测量
多目标和单目标定向敏感器	美国航空航天局(NASA)，1986年	多目标：100~6m 单目标：6~0m	析像管接收，相位式测距，远距离为析像管子测角
空间交会对接用扫描激光雷达	日本宇宙航空研究所，1987年	远距离：20km~200m 近距离：200~0m	连续波-镓铝砷激光二极管，雪崩光电二级管(APD)接收，相位法测距，检流计式扫描
自主交会对接光电敏感器	德国MBB公司，1983年	20km接近	连续波-镓铝砷激光二极管光源，APD接收，检流计式扫描，姿态测量用CCD
交会对接跟踪激光雷达	日本电气、三菱电机公司，1989年	30km~0.2m	砷化镓激光二极管光源，四象限检测和CCD成像，意频测距
交会对接光学敏感器系统	日本宇宙开发事业团(NASDA)，1995年	600~0.3m	半导体激光连续测距CCD成像
有源传感器用于空间交会对接	NASA，1997年	110~0m 仰角±8° 方位角±10.5°	850nm脉冲光源，安装角反射器，CCD成像检测

目前美、俄所实现的空间交会对接都需要宇航员手动介入，而在未来的许多太空任务如卫星服务计划、空间站自动补给、深空探索、无人飞船等，则需要无人自主交会对接。因此，美、俄、日及欧洲空间局等都在发展自主自动交会对接测量系统，特别是复合式激光雷达测量系统。

2.3 对地观测及深空探测

利用卫星或航天飞机等航天器搭载激光测距装置在空间轨道上对地球或其他星球表面进行观测，这种激光测距装置通常称激光高度计。它测量航天器到表面的距离，再根据航天器的位置和飞行姿态，计算出表面点的坐标。与地面及机载激光测距设备相比，星载激光器具有不少优势：首先，可在卫星上采集和处理数据，具有观察整个天体的能力，有助于制作天体的综合地形图，所以月球和火星等探测计划都包含了激光高度计；其次，在北极等不能用飞机执行观测任务的地方，可用星载激光高度计观察北极地区冰层和海洋冰川的变化。因此，星载激光高度计在天体特征研究、陆地表面冰川海平面高度变化和植被分布状况研究、云层和气溶胶的垂直分布和光学密度研究以及特殊气候现象监测等方面可发挥重要作用。

早在20世纪70年代，激光测距装置就在阿波罗登月工程中得到应用。1971~1972年间发射的阿波罗15、16和17号飞船上，均搭载了闪光灯泵浦的红宝石激光高度计。不过，闪光灯泵浦器件的寿命和效率问题极大地限制了它在空间环境中的应用。据报道，阿波罗上的激光高度计寿命仅为10^4个脉冲。

20世纪90年代美国航空航天局(NASA)先后发射了装有火星轨道器激光高度计(MOLA和MOLA-2)的探测器对火星进行探测；于1996年和1997年分别发射了返回式激光测高卫星SLA-01和SLA-02，用于观测地表植被和其他自然特性。后来，载有"地学激光测高系统"(GLAS)的"冰卫星"(ICESat)于2003年1月13日发射升空，其使命是监测冰川，观测云层中悬浮微粒的垂直分布密度和光学密度，并测量植被分布和地面地形。

1996年美国的"近地小行星交会探测器"(NEAR)发射升空，于2000年成功进入爱神小行星的运行轨道，进行为期1年的近距离小行星观测计划。激光测距仪作为其装载的5套精密观测系统之一，用以观测计算爱神星的体积大小和了解其密度。

NEAR系统中的激光测距仪主要由5个部分组成：带光纤延迟的激光发射和激光电源部分、光学接收部分、带探测器件的模拟电路及处理器、数字处理电路和低电压供电电源（图1）。这是世界上第一个进入小行星轨道的激光测距仪，在绕小行星轨道工作的1年时间中一直持续工作。

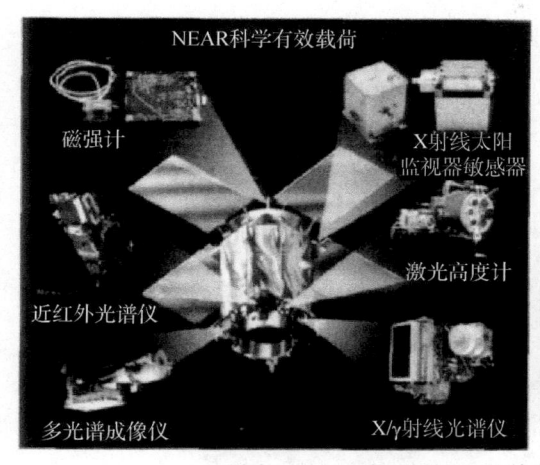

图1　NEAR探测器(左)及其激光测距仪(右)

我国最近发射的嫦娥一号探月卫星的重要有效载荷——月球轨道激光高度计是我国第一套进入太空的激光应用系统。通过激光高度计与CCD立体相机相结合，可以获取月球表面的三维影像和地形高度数据（图2）。

嫦娥一号上的激光高度计开机两个多月来，至今已随星围绕月球转了720圈以上，差不多把月球覆盖了两遍（包括南北极）。它几乎每隔一秒就向月面发射一束激光并接收反射光，使得月球上间隔十几公里就可以"有"一个点，而且分辨率精确到5米。目前"激光足印"的密度已达每平方公里0.87个点，收发之间的成功采集率达99%左右。

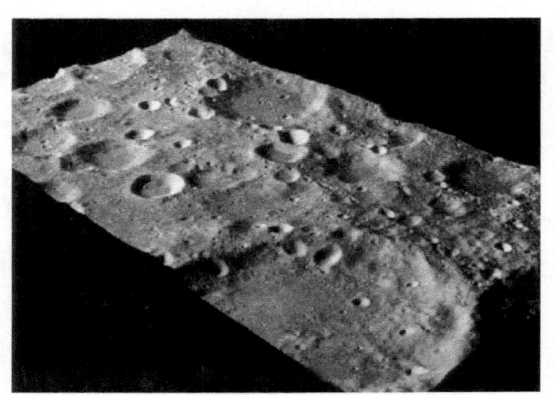

图2　根据嫦娥一号数据绘制的月面图像局部三维景观图

2.4　卫星星座与编队飞行

卫星编队飞行是近年来国内外航天技术研究的重点问题之一。其目的是采用多颗小卫星编队飞行组成星座来实现传统单颗大卫星所不具备的强大功能。随着微型航天器技术的不断发展，由几颗甚至十几颗低成本微型航天器构成的编队星座受到广泛重视。它能在同一时刻对同一目标实现立体探测，并能提供大孔径和长测量基线，在通信、遥感、导航、电子侦察、立体成像、精确定位以及大气、天文和地球物理观测等领域都有着非常重要的意义。星间实时、高精度的距离自主测量是编队飞行卫星进行队形保持、协同控制的重要保障。利用激光相干性好和方向性强的特点，可以通过激光相位实现高精度的星间测量，并通过激光的干涉来实现距离相对固定的两星间距离变化测量。美国和日本都在这方面进行了很多研究和实验。

1) "激光干涉仪空间天线"(LISA)计划

由欧空局和NASA共同实施、预计在2011年左右发射的LISA计划用于探测空间由双星系统产生的重力波，对拥有强大能量的黑洞进行研究以验证爱因斯坦的广义相对论，以及对早期宇宙进行探测等。LISA任务由3个航天器组成。它们运行在以太阳为中心的轨道上，每个航天器之间的相对距离为500万公里。LISA相当于一个天基迈克尔逊干涉仪，通过激光干涉技术来测量相对距离的变化（图3）。

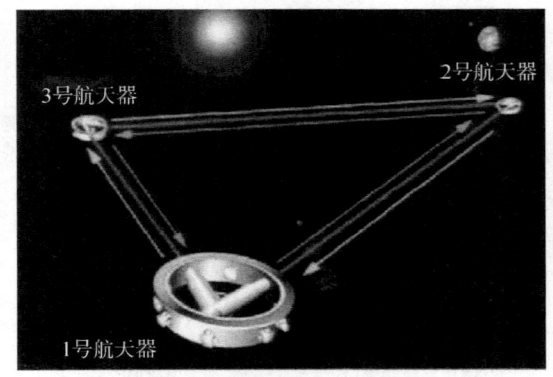

图3　LISA飞行示意图

2) "微扫描激光测距仪"(MS-LRF)

MS-LRF是日本宇宙科学研究本部(ISAS)计划用于微卫星编队中星间状态测量的激光测量系统。由于激光波束窄，采用传统的激光雷达无法同时观测多颗卫星，因此对于编队卫星而言，需要采用扫描型的激光探测设备，通过扫描装置的旋转实现多个目标的搜索与测量。

MS-LRF是基于微机械技术的双轴扫描测量系统(图4)。该系统主要由光学扫描装置、微透镜、分光镜、激光管、雪崩式光电二极管、用于驱动扫描器的压电传动装置、驱动电路和信号处理环路组成。MS-LRF的测量原理为：激光二极管产生的激光分别经分光镜和微扫描平面镜的反射后射向目标。被目标接收到的激光一部分通过角反射器发射，沿原路径返回，并最终被敏感雪崩式光电二极管探测到（图5）。应用于编队卫星的MS-LRF要求有非常高的测量精度和帧频，所以采用的是连续波激光相位式测距。对于0.1~10km的星间距离，半导体激光器分别采用5kHz和1MHz两种频率进行调制，测量精度为1m。

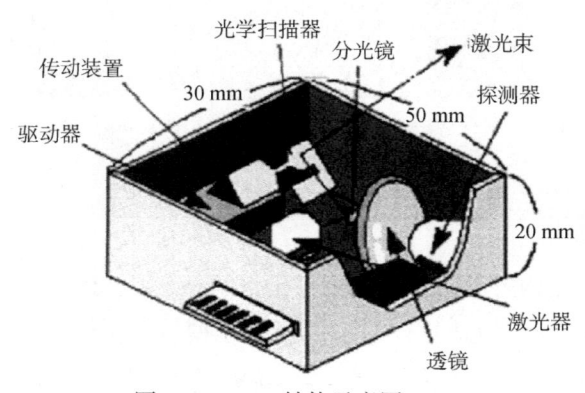

图4　MS-LRF结构示意图

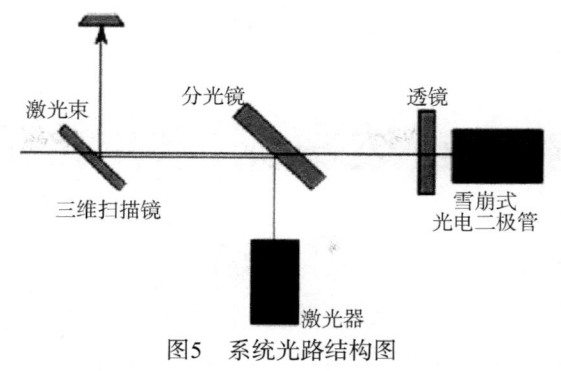

图5 系统光路结构图

3 结束语

综上所述,激光测距技术在空间的应用日益广泛。由于其全天候、高精度、抗干扰、小型化等得天独厚的优势,激光测距越来越受到关注,已成为空间探测领域一个重要技术手段,在军事和工业方面有着极高的应用价值。

由于工程应用特别是航天应用的需要,激光测距技术也在不断发展进步。激光测距装置目前正向着小(体积小、质量小、功耗低)、精(精度高)、远(测程远)、快(测量时间短、重复频率高)等方向发展。激光测距技术极易和其他探测技术相融合,从而得到功能更强大的复合式探测系统。

近年来,我国对激光测距技术的研究非常重视,国内许多大学和研究所都在进行这方面的研究及工程应用工作。但空间激光测距是一项很复杂的技术,仍有许多关键技术需要解决,因此针对空间探测领域的应用研究并不多,在航天领域的真正实际应用则更少。嫦娥一号探月卫星所使用的激光高度计是我国第一套也是目前唯一一套进入太空的激光测距设备。作为空间探测领域的一项重要技术,加快开展激光测距技术在空间领域的应用研究工作是非常必要和急需的。

此外,由于雪崩式光电二极管、激光高频电流调制技术等激光测距关键器件和技术与国际先进水平还有差距,国产激光测距设备在测距精度、可靠性等方面与国际先进水平还存在一定的差距。因此,在对激光测距方法进行研究与应用的同时,还要在某些关键器件、关键技术方面开展重点攻关,这对于推动国内激光测距技术的进步及在空间领域的可靠应用是十分必要的。可以肯定地说,在不久的将来,激光测距技术会在军事和国民经济中发挥越来越重要的作用。

先进激光雷达探测技术研究进展

华灯鑫[1]　宋小全[2]

(1. 西安理工大学 机械与精密仪器工程学院，西安 710048; 2. 中国海洋大学 海洋遥感研究所，青岛 266100)

摘　要　激光雷达作为近年快速发展的新型光波主动式遥感技术，由于具有高精度及高时空分辨率的遥测特性，已经在大气及海洋环境探测等领域得到广泛的应用。主要介绍了激光雷达探测技术的基本原理，重点分析大气环境监测激光雷达，气象观测激光雷达及空间激光雷达的测量原理、关键技术及其应用前景，介绍国内外相关激光雷达的系统特色及其最新进展。

关键词　激光雷达　大气环境　气象参数　遥感探测

0　引言

激光雷达[1]作为一种主动遥感探测技术和工具已有近 50 年的历史，目前广泛用于地球科学和气象学、物理学和天文学、生物学与生态保持、军事等领域。其中，传统意义上的激光雷达主要用于陆地植被监测、激光大气传输、精细气象探测、全球气候预测、海洋环境监测等。随着激光器技术、精细分光技术、光电检测技术和计算机控制技术的飞速发展，激光雷达在遥感探测的高度、空间分辨率、时间上的连续监测和测量精度等方面具有独到的优势。尤其在大气探测方面取得显著发展，对各种参数的测量空间覆盖高度已经可以实现从地面到 120 km 的高度[2]，其应用前景得到普遍的关注。相对于微波、电磁波雷达，激光雷达采用的光波波长较短，与大气中存在的分子和气溶胶及浮尘的相互作用所产生的散射效果复杂、散射形式多样，适用于陆地、大气及海洋环境监测，大气光学及物理特性、气象/气候参数的高时空分辨率的精细探测。激光雷达可以实现地基、车载、机载及星载探测，也是现代雷达探测技术从厘米波、毫米波向光波探测技术的延伸，实现了遥感探测技术向高时空分辨率、高精度领域的发展。

1　激光雷达探测技术

1.1　激光雷达探测原理

光波的物理量可由强度、波长（频率）、相位、偏振态及指向性等来表示。光与物质相互作用主要表现为吸收及散射现象，按作用机理可以分为气溶胶等颗粒物引起的米氏散射，大气分子及原子等引起的瑞利散射、拉曼散射、荧光及共振散射和吸收等现象。通过对各种散射机理及效果进行分析，可以探测物质的物理及化学信息。

大气探测激光雷达工作原理与微波雷达相似，其基本系统构成如图 1 所示。一般采用脉冲激光器作为发射源，向大气中发射一束具有高指向性、高能量的窄脉冲宽度的激光束，通过望远镜收集大气中物质产生的后向散射光，并对散射光进行光谱分析，剔除杂散光信号，经光电转换后获得电信号，由计算机进行数据采集、信号分析及数据反演即可得到所需大气参数或信息[1]。

本文原载于《红外与激光工程》，2008，Vol.37，Supplement.，21~27。

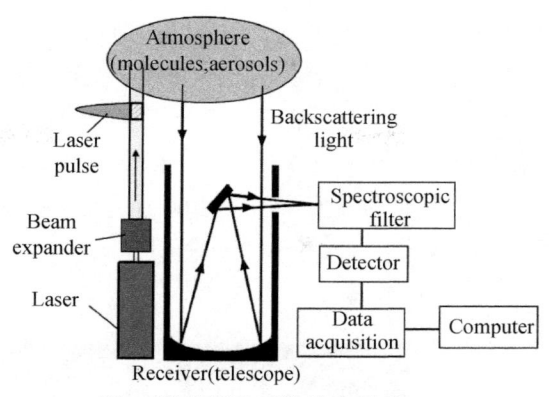

图1 激光雷达系统基本构成

1.2 激光波长及对眼安全特性

随着近年激光技术的快速发展，高效大功率脉冲激光器已经实现商品化生产，并广泛应用于激光雷达及各种光加工领域。因此，激光的安全性问题得到越来越多的关注。激光安全性一般用肉眼及皮肤对激光能量密度的最大允许曝光量（Maximum Permissible Exposure, MPE）来定义，而 MPE 值取决于激光波长，选用合适的激光波长对激光雷达的设计可以取得事半功倍的效果。国际上较为常用的 MPE 值与激光波长的关系包括日本工业标准（JISC6802-1991）、美国国家标准研究院 ANSI 的 Z136 系列激光标准（ANSI Z136）、国际电工协会 IEC 60825 系列标准等。图 2 为根据日本工业标准得到 MPE 值与激光波长的关系。可以看出，在相同激光能量密度下，紫外及中红外波长的安全性能是可见波长的几千倍，因此这类波长常被称作对眼睛安全激光波长。

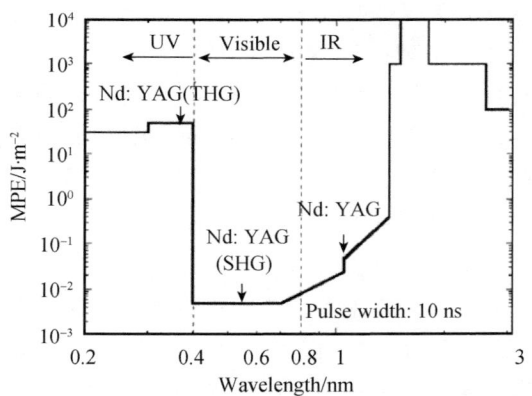

图2 眼睛的最大允许曝光量与激光波长关系

2 大气气溶胶的探测

大气气溶胶是指悬浮在大气中直径约为 100 μm 以下的液体或固体微粒体系，地面扬尘、沙尘暴、林火烟灰、花粉、空气中的气态污染物等，都是对流层气溶胶的自然源。气溶胶对气候变化、云的形成、能见度的改变、大气微量成分的循环及人类健康有着重要影响。

2.1 米散射激光雷达

气溶胶探测的激光雷达主要是以单波长或多波长(如 Nd:YAG 激光器 1064、532、355 nm)米散射激光雷达为主，技术比较成熟。欧美及日本等国已实现商品化，国内的中科院大气所、安徽光机所、上海光机所、中国海洋大学及西安理工大学等单位也先后开展该项技术的研发。

单波长米散射激光雷达可以用探测大气气溶胶的光学特性，主要有散射系数，消光系数，雷达比。而多波长激光雷达除了可以得到上述参数以外，还常用于反演气溶胶的粒谱分布及不同波长的气溶胶消光系数，为研究激光在大气中的传输特性，大气湍流等提供科学依据。图3为利用西安理工大学研发的小型米散射激光雷达，观察到西安上空气溶胶及卷云的THI（Time-Height-Indications）时空分布。该系统采用半导体泵浦的微脉冲激光器的532 nm波长，具有对眼安全特性及三维扫描功能，可用于城市气溶胶及大气质量监测、大气能见度及边界层高度观测等。

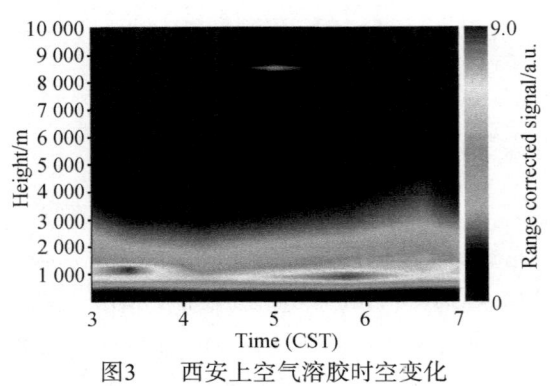

图3　西安上空气溶胶时空变化

2.2　高光谱分辨率激光雷达

米散射激光雷达在反演气溶胶参数如消光系数时，必须对当时的大气状态等做一些假设，因而限制了其探测及数据反演精度，不利于大气的精准探测。高光谱分辨率激光雷达（HSRL）是在米散射激光雷达的基础上发展而来的一种高精度气溶胶探测技术，也是目前公认的与气溶胶拉曼探测激光雷达并列的两种可不需假定、直接探测气溶胶消光参数的技术之一。

HSRL的探测原理是利用大气分子引起的瑞利散射光谱宽度依存大气温度，其谱线宽度一般为GHz级，而气溶胶散射谱宽约等于激励激光谱宽，一般为100 MHz级，通过使用单频率脉冲激光器，高光谱分辨率分光器，如干涉仪、原子吸收滤光器或分子吸收滤光器，从大气散射中分离米散射和瑞利散射光谱。在数据反演中借助于同时获得的瑞利散射信号，可以不需要假设大气粒子消光/散射参数，直接导出消光系数，从而实现高精度的气溶胶探测，提高了参量反演的准确性。

HSRL技术是目前气溶胶的光学及物理特性参数精细激光探测的热门研究课题。美国的NASA，欧空局（ESA），日本的国立环境研究所以及国内外很多高校也在开展该技术的研究。图4、图5为美国Wisconsin大学的HSRL系统构成及探测结果，其激光为波长532 nm，使用碘分子吸收滤光器分光[3]。西安理工大学的研究小组成员，正在采用Fabry-Perot（FP）标准具分光，开展355 nm波段上的HSRL的技术研发。

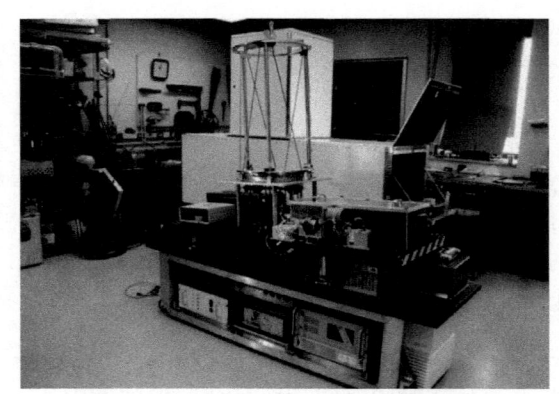

图4　美国Wisconsin大学高光谱分辨率激光雷达系统

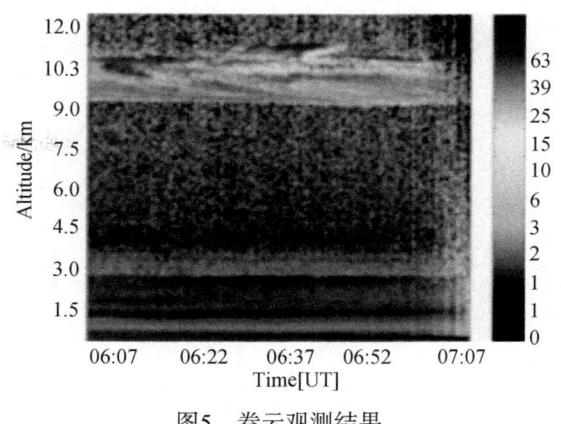

图5 卷云观测结果

近年来，气溶胶米散射激光雷达的发展也具备了更多的功能，如通过增加偏振检测器[4]、多视场角光阑[5]、多波长发射[4-7]等手段，获取气溶胶、沙尘暴、云等的粒径等微物理信息，从而进一步应用于大气辐射、云物理及辐射、气候影响等。欧美等国为了应对日益增加的反恐需要，也加快了可探测生物气溶胶荧光的激光雷达，通过检测附着在气溶胶生物物质被激光激发出的荧光实现气溶胶生物成分的鉴别，从而对生物武器的袭击实现预警[8-9]。此外，气溶胶激光雷达的载体平台也从最初的地基，发展为车载移动、机载飞行、星载全球观测等多种形式。

3 大气温湿度探测

激光雷达探测大气温度分布的主要方法有瑞利散射法密度法[10]，高光谱分辨率瑞利散射法[11-12]，转动拉曼散射法[13]和差分吸收法[14]等。瑞利散射法密度法主要利用激光雷达探测大气分子密度变化，利用大气方程反演温度，所以主要用于气溶胶影响较小的对流层顶部及平流层的大气温度探测。而底层对流层范围内的大气温度探测，由于受温度的遥感灵敏度较低及易受地表产生的高密度气溶胶和白天太阳背景光的影响，底层大气高精度测温技术的研究一直是国际上激光雷达研究的前沿课题。目前对流层内的大气温度探测主要是高光谱分辨率瑞利散射法和转动拉曼散射法。

激光雷达探测水汽的主要方法有振动拉曼散射激光雷达，即利用水汽分子和氮气分子所产生的振动拉曼散射谱线的强度进行水汽密度探测。差分吸收激光雷达，即通过发射2个激光波长，其中一个波长与水汽分子的某一吸收谱线重叠，利用2个波长的回波信号的强度差进行水汽密度探测。相对湿度需要利用温度，所以温湿度是一对相关性很强的大气参数。

3.1 高光谱分辨率瑞利-拉曼散射激光雷达

高光谱分辨率瑞利散射激光雷达是一种利用大气中的原子和分子的瑞利散射机制而工作的激光雷达，通过高光谱分辨率滤光器，对大气分子瑞利散射的光谱宽度进行分析而实现温度测量。由于瑞利谱宽较窄，在常温及355 nm激光波长激励下，其宽度一般为3 GHz，要在如此窄带光谱内，除去中心谱重叠的米散射信号，用于分光的滤光器需要具有MHz的光谱分辨率。目前利用该技术探测大气温度的主要单位有美国Colorado大学[11]及日本EKO公司[12]。采用的高光谱分辨率分光器主要有FP干涉仪、原子吸收滤光器或分子吸收滤光器等。

图6为华灯鑫等利用在日本EKO公司研发的高光谱分辨率瑞利-拉曼散射激光雷达系统对日本东京城区上空的温、湿度廓线进行白天观测，首次实现对流层低层白天大气温，湿度激光雷达高精度探测[12,15-16]。

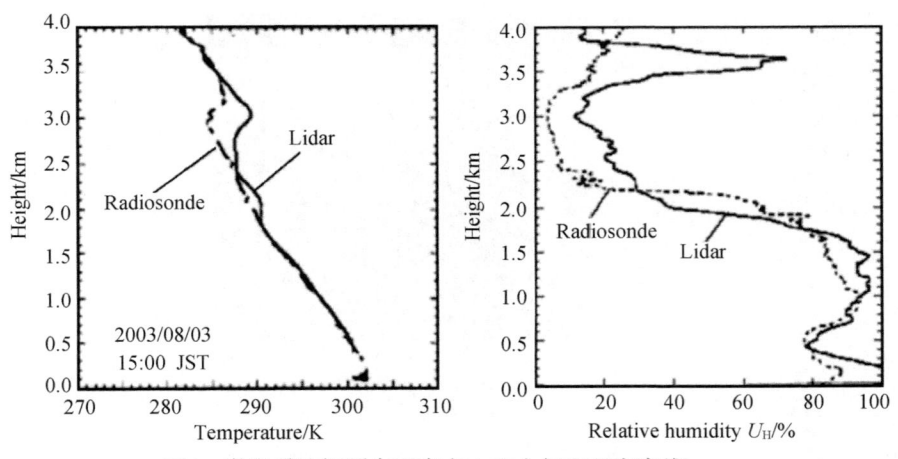

图6 激光雷达探测白天东京上空大气温湿度廓线

3.2 回转拉曼散射激光雷达

回转拉曼散射激光雷达是利用大气分子引起的非弹性散射（拉曼散射）的转动拉曼谱线的强度随温度变化的特性来探测大气温度廓线。相对于瑞利散射信号，拉曼散射信号强度要弱4~5个数量级，主要关键技术是强背景噪声下的微弱信号提取技术。目前普遍采用高光谱分辨率光栅及窄带通滤波器分离2个具有反向温度灵敏度系数的回转拉曼波长，结合光子计数技术来实现强背景噪音下的微弱信号提取。虽然该技术已经实现夜间低层对流层到平流层内30 km高度范围内的大气温度探测，但白天高精度温度探测还需要进一步的研究。目前德国的GKSS[13]，日本的京都大学[17]及国内的西安理工大学，北京理工大学，武汉大学及中科院安徽光机所等单位也正在开展从低层对流层到平流层内大气温度探测的研究。图7为华灯鑫等采用高分辨率光栅结合窄带滤光片分光，采用模拟检测技术探测到大气边界层内昼夜大气温度廓线[18]。

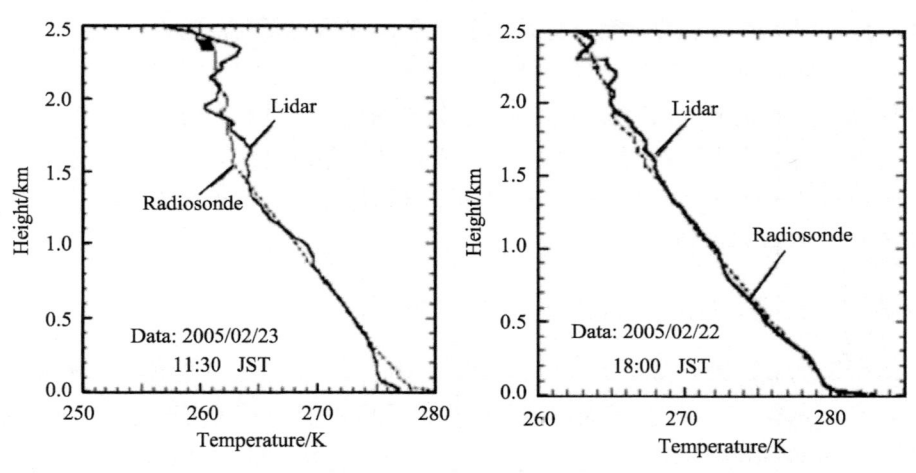

图7 拉曼激光雷达探测昼夜大气温度廓线

4 大气风廓线测量

测风激光雷达通过测量大气中自然出现的气溶胶颗粒或分子运动（风速引起）产生的具有多普勒频移的后向散射信号，利用对回波信号频率进行鉴频或相干，测量出后向散射信号的多普勒频移 Δv，利用 Δv 与风速的关系就可反演得到径向风速数值，通过扫描激光光束得到不同方向上径向速度，矢量合成即可得到风速、风向，实现检测风速、探测紊流、实时测量风廓线风场等。

目前，激光雷达探测风速风向的主要技术有相干激光雷达技术及非相干技术。相干激光雷达主要适用于气溶胶密度较大的对流层，探测范围最大可达 10 km 左右，精度可达 0.1 m/s。而非相干激光雷达主要利用气溶胶及大气分子测速，属于能量检测，其适用范围较广，适合对流层到平流层的风廓线探测，测速精度可到 1m/s 以内。相干测风激光雷达主要的厂家有美国相干公司及日本三菱重工等单位。香港国际机场的测风激光雷达的购置与使用，成为测风激光雷达的市场化应用典范。通过配合原有的多普勒微波雷达共同探测飞机起降前后的大气风场信息、湍流强度等信息，提高了飞机起降的气象灾害预警与安全。

非相干激光雷达主要有采用 4 个不同方向设置的接收望远镜探测 4 个不同径向风速，反演风速风向的法国 CNRS 的地基系统[19]及美国 NASA 的车载系统[20]。图 8 为美国 NASA 研发的车载测风激光雷达系统及其风廓线探测结果，该系统可以探测平流层内的风速风向。图 9 为该研究小组正在研究的机载系统[21]，其主要特点是采用光束扫描(右上角)，多通道 FP 鉴频及脉冲激光锁频（左上角），系统结构紧凑，便于车载、机载。我国的中国海洋大学[22]，中科院安徽光机所[23]，上海光机所[24]等单位也在进行非相干、相干测风激光雷达的研制，其中中国海洋大学研制的车载非相干测风激光雷达系统已经为 2008 年北京奥运会帆船比赛提供气象服务。

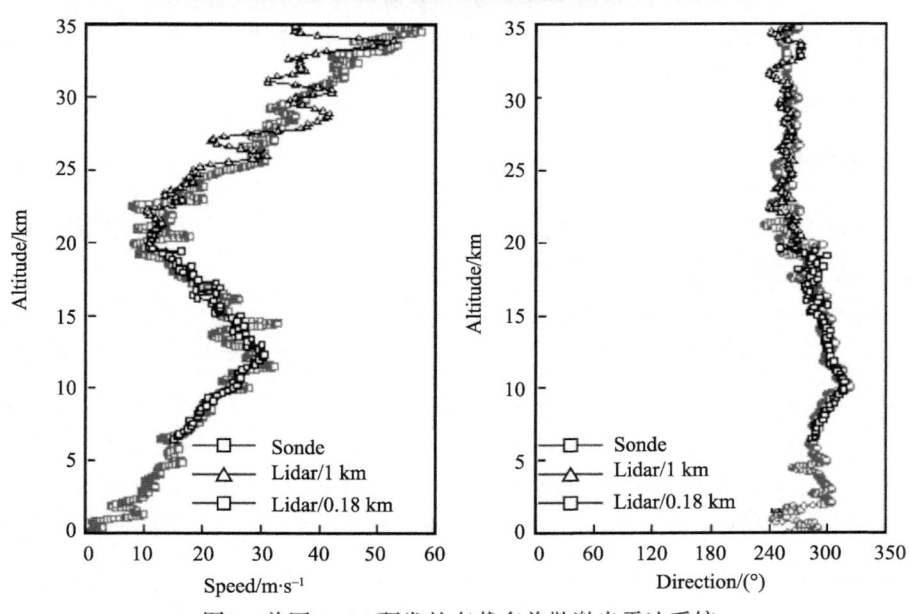

图8 美国NASA研发的车载多普勒激光雷达系统

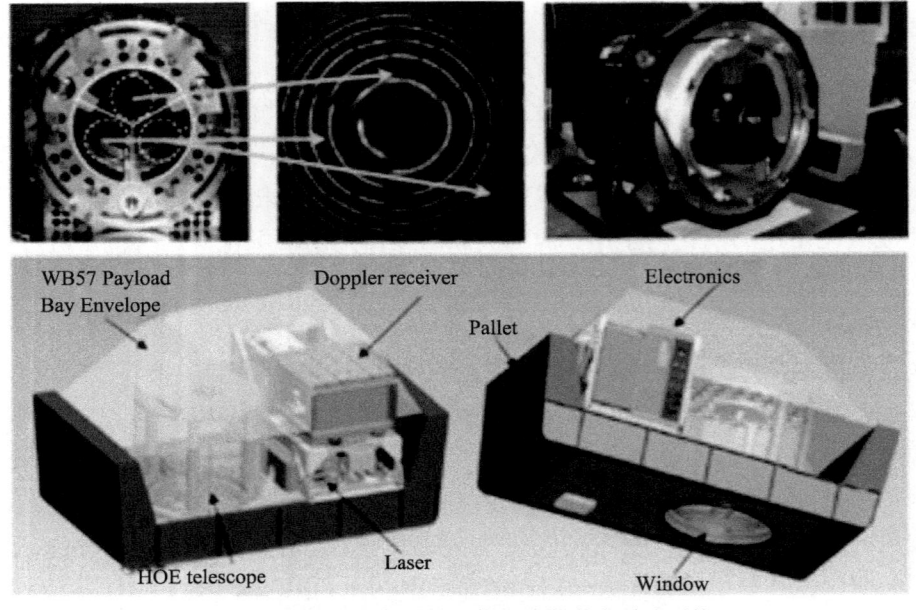

图9 美国NASA在研的机载多普勒激光雷达系统

5 大气污染及生态环境探测技术

大气污染物主要指大气中存在的微量污染气体，如 $NO_x, SO_2, O_3, CO, CH_4$ 等，利用污染气体的吸收谱线特性而构成的差分吸收激光雷达可以有效地对这些微量污染物进行实时高精度探测及监测，研发波长可调谐的脉冲激光器是差分吸收激光雷达的关键技术。近年随着LD泵浦固体激光技术及光参量振荡技术（OPO技术）的发展，波长 $2\sim 10\ \mu m$ 的红外波段可调谐激光器技术得到很快的进步，为激光雷达大气污染监测的应用提供了光源保障。

生态环境中存在的有机物质（植物及有机颗粒物），一般都含有叶绿素及蛋白质等有机物质，脉冲激光束照射这些物质时诱发荧光谱，有机物浓度含量（主要是叶绿素）与荧光谱强度成线性关系，荧光激光雷达主要是通过对特定波长的荧光光谱进行分光检测，可以用于判断植物、水体藻类等的生长情况及监测水体有机污染物及气体中的花粉等有机颗粒物。

6 星载激光雷达

将激光雷达安装于卫星上进行遥感探测的优越性已经得到世界各国专家的肯定，美国 NASA、欧空局 ESA、日本 NASDA 等航天部门都在积极发展星载遥感设备。

原有正在运行的地球卫星观测系统没能提供的一种重要的参数就是气溶胶的剖面信息，针对这一问题美国 NASA 于 1994 年 9 月成功地发射了载有激光雷达的航天飞机，进行了空间激光雷达探测全球云和气溶胶的首次实验(Lidar In-space Technology Experiment，LITE)，取得了令人鼓舞的探测结果，证明了空间激光雷达在测量大气气溶胶和云方面的潜力。此后，美国于 2003 年 1 月发射了地学激光高度计系统（GLAS）、于 2006 年 4 月发射了气溶胶-云激光雷达红外卫星观测系统 CALIPSO 等，用于提供云和气溶胶的垂直分布信息，云的水／冰相态，气溶胶粒径的定性分类等，图 10 为卫星观测系统 CALIPSO 的示意图。为获得对流层的高精度风场数据、提高目前天气预报水平，欧洲近期将要发射星载测风激光雷达系统 ADM-Aeolus，获取全球范围的视线风速。可以预计在不远的将来，星载激光雷达与现有的微波雷达系统将构建一个全天候、高精细的全球气象及大气参数的探测平台。

图10 NASA星载气溶胶（卷云）探测激光雷达（CALIPSO）

未来星载激光雷达系统的准确度和精度可以通过利用更高输出功率的光源、更大口径的望远镜、高增益低噪音的光电检测技术等得到提高，而这些技术的改进将有望在未来几年内实现。美国 NASA 已经资助了几项研究用于研制大口径（直径 20~30 m）反射镜原型样机。而且，高功率、有效的激光器对很多不同类型的未来激光雷达系统而言都是关键技术，满足星载要求的 100~400 W 功率的激光雷达用激光器的可行性目前也在研究中。

国内激光和激光雷达技术在航天领域也得到了广泛的应用，随着 2007 年 10 月 24 日首颗月球探测卫星"嫦娥一号"的发射成功，我国也将自己的地面、机载激光系统装置开始投入太空应用。"嫦娥一号"的有效载荷之一，月球轨道激光高度计由中国科学院上海技术物理所负责总体研制、上海光学精密机械研究所承担其中激光发射器[25]的研究，这成为我国第一套进入太空的固体激光应用系统，为我国开展激光雷达星载系统揭开了新的一页。

7 激光雷达的主要研究热点及其发展

激光雷达技术发展至今，已有近半个世纪的经历，随着现代激光技术，光电检测技术及半导体集成技术的飞速发展，激光雷达技术日趋成熟，已在大气环境监测，气象观测及军事领域得到广泛的应用。目前地球大气环境恶化，如全球变暖，气候异常，大型自然灾害多发，全球性的高时空分辨率，高精度的大气环境监测，气象灾害的早期预警预测等技术的研究已经成为当前热门话题，也为激光雷达技术的应用及发展提供了机遇。

从早期不定期举行到近年来每两年一届的国际激光雷达会议已经经过了整整 40 年的历程，把激光雷达技术应用于揭示地球、了解大气、观测生物圈和海洋方面，成为国际上级别最高、影响范围最广的激光雷达领域国际会议，成为全世界激光遥感和激光雷达领域科技人员共同关注的焦点。

在近些年的 ILRC 会议上，大气物理要素如气溶胶光学性质、大气风场、温度与湿度等的方面的进展始终占相当比重。全球各区域的学者将各自的定点系统组成地面观测网络，如基本以拉曼激光雷达系统为主的欧洲激光雷达观测网 EARLINET，以全自动运行的米散射激光雷达为主的 MPLNET，东亚地区以沙尘暴观测为主的 AD-net，世界气象组织正在讨论建立的 GALION 等，均可对同一过程或事件进行不同时间和地点的综合观测，得到很多新的现象和结果。除此，针对同一对象不同应用目的的需要，激光雷达向多平台载体（地基、车载、机载、星载等）方面迅速发展，美国、欧洲以及我国等均发展了各自的星载激光雷达系统。

参 考 文 献

[1] RAYMOND M. Measures. Laser Remote Sensing: Fundamentals and Applications [M]. Malabar: Krieger Publishing Company, 1992: 205
[2] Diettrich J C, Espy P J, Nott G J, X. Chu, et al, Atmospheric temperature structure (0-120km) at Rothera (Antarctica. 67_S, 68_W): Seasonal Variation, Geophysical Research Abstracts, 2005, 7: 07611
[3] Razenkov et al. Improvement of the arctic high spectral resolution lidar [C]//22nd ILRC, ESA-561, S2P-6, 2004
[4] Stefanutti et al. A four-wavelength depolarization backscattering lidar for polar stratospheric cloud monitoring [J]. Appl Phys

B, 1992, 55: 13—17

[5] Boichenko et al. Four-wavelengths lidar sensing of atmospheric aerosol [J]. Atmos Opt, 1989, 2: 66—72

[6] Flesia et al. Remote measurement of the aerosols size distribution by lidar [J]. J Aerosol Sci. 1989, 20: 1213—1216

[7] G. S. Kent, G. M. Hansen. Multiwavelength lidar observations of the decay phase of the stratospheric aerosol layer produced by the eruption of Mount Pinatubo in June 1991 [J]. Appl. Opt. 1998, 37, 3861

[8] CH IR ISTESEN SD, MERROW C N, DESHA M S. UV fluorescence lidar detection of bio-aerosols [C]. Proc SPIE, 1994, 2222: 2282237

[9] PRIMMERMAN C. A. Detection of Biological Agents. LINCOLN Laboratory Journal, 2000, 12, NUMBER 1, 18

[10] HAUCHECORNE A, CHANIN M L. Density and temperature profiles obtained by lidar between 35 and 70 km [J]. Geophys Res Lett, 1980, 7: 565—568

[11] HAIR J W, CALDWELL L M, KRUEGER D A, et al. High-spectralresolution lidar with iodine- vapor filters: measurement of atmospheric-state and aerosol profiles [J]. Appl Opt, 2001, 40: 5280—5294

[12] HUA D, UCHIDA M, KOBAYASHI T. UV high- spectral resolution Rayleigh-Mie lidar with a dual-pass Fabry-Perot etalon for measuring atmospheric temperature profiles of the troposphere [J]. Opt Lett, 2004, 29 (10): 1063—1065

[13] BEHRENDT A, REICHARD J. Atmospheric temperature profiling in the presence of clouds with a pure rotational Raman lidar by use of an interference-filter-based polychromator [J]. Appl Opt, 2000, 39 (9): 1372—1378

[14] BÖENBERG J. Ground-based differential absorption lidar for water-vaper and temperature profiling: methodology [J]. Appl Opt, 1998, 37 (18): 3845—3860

[15] HUA D, KOBAYASHI T. UV Rayleigh-Mie lidar by multicavity Fabry-Perot Filter for accurate temperature profiling of the troposphere [J]. Appl Opt, 2005, 44 (3): 6474—6478

[16] HUA D, KOBAYASHI T. UV Rayleigh-Mie Raman lidar for simultaneous measurement of atmospheric temperature and relative humidity profiles in the troposphere [J]. Jpn J Appl Phys, 2005, 44 (3):1287—1291

[17] BEHRENDT A, NAKAMURA T. ONISHI M, et al. Combined Raman lidar for the measurement of atmospheric temperature, water vapor, particle extinction coefficient, and particle backscatter coefficient [J]. Appl Opt, 2002, 41 (36): 7657—7666

[18] HUA D, LIU J, UCHIDA K, et al. Daytime temperature profiling of planetary boundary layer with ultraviolet rotational Raman lidar [J]. Jpn J Appl Phys, 2007, 46 (9A): 5849—5852

[19] SOUPRAYEN C, GARNIER A, HERTZOG A, et al. Rayleigh–Mie Doppler wind lidar for atmospheric measurements. I. Instrumental setup, validation, and first climatological results [J]. Appl Opt, 1999, 38 (12): 2410—2421

[20] GENTRY B M, CHEN H, LI S X. Wind measurements with 355 nm molecular Doppler lidar [J]. Opt Lett, 2000, 25 (17): 1231—1233

[21] GENTRY et al. New technologies for direct detection Doppler lidar: Status of the TWiLiTE airborne molecular Doppler lidar project [C]//Proc of 24nd ILRC, 2008, S02P-03, 239—243

[22] LIU Z, LIU B, WU S, et al. High spatial and temporal resolution mobile incoherent Doppler lidar for sea surface wind measurements [J], OPTICS LETTERS. 2008, 33 (13), 1485—1487

[23] XIA H, SUN D, YANG Y, et al. Fabry-Perot interferometer based Mie Doppler lidar for low tropospheric wind observation [J]. Appl Opt, 2007; 46 (29): 7120—7131

[24] 陈卫标，周军，刘继桥，等. 多普勒激光雷达及其单纵模全固态激光器[J]。红外与激光工程，2008，37 (1)，57—60

[25] 施翔春，陈卫标，侯霞. 全固态激光技术在航天领域的应用[J]。红外与激光工程，2005，34 (2)；127—131

第四部分　干涉成像光谱仪

实体 Sagnac 干涉仪的设计

赵葆常[1]　扬建峰[1]　薛彬[1]　马小龙[1]　常凌疑[1,2]　陈立武[1]

(1. 中国科学院西安光学精密机械研究所，西安 710119; 2. 中国科学院研究生院，北京 100049)

摘　要　以嫦娥一号卫星干涉成像光谱仪为例，介绍了 Sagnac 干涉仪的设计思想，共光路及横向剪切原理、剪切量的计算方法以及干涉仪分束面上分束区与透射区的设计。该 Sagnac 横向剪切干涉仪作为嫦娥一号卫星干涉成像光谱仪的核心部件，经检测与在轨运行，表明设计正确，经受了包括月食低温在内的各种恶劣的航天环境的考验，获得了大量清晰的月表多光谱图像。

关键词　成像光谱仪　Sagnac 干涉仪　嫦娥一号

0　引言

成像光谱仪是集照相机与光谱仪两者功能为一体的新型光学遥感探测仪器，它采用面阵探测器作为图像传感器时可以同时获取二维空间信息与一维光谱信息。二维面阵探测器中的一维探测器(行)提供横向空间信息，另一维提供干涉强度信息 $I(x)$，干涉强度分布 $I(x)$ 经傅氏变换后，即获得光谱强度 $B(\sigma)$。所以它相当于色散型成像光谱仪中的光谱信息，另一维的空间信息通过空间平台与探测目标(地球、月球、火星等)的相对运动得到。

同时获得的二维空间信息与一维光谱信息，称为数据立方体。目前能够实现该功能的技术手段很多，主要可分为：色散型成像光谱仪、干涉型成像光谱仪及滤光片型成像光谱仪、声光调制型(AOTF)成像光谱仪等，它们各有利弊，应该根据应用目标，本着扬长避短的原则来选择。

国际上第一台上星的干涉型成像光谱仪是美国的强力小卫星所搭载的干涉成像光谱仪[1-2]，在诸多类型的成像光谱仪中选择了 Sagnac 空间调制型干涉成像光谱仪的主要原因是它具有很强的航天环境适应性，这是它的共光路原理所致。出于同一原因，在嫦娥一号卫星中也选择了 Sagnac 型空间调制干涉成像光谱仪，另一个原因是：因为嫦娥一号卫星是我国的首次探月，科学家对月球的情况尚有很多未知数，所以希望仪器能够获取任一波长的光谱强度信息，仅干涉型成像光谱仪具有这一功能。

1　Sagnac 干涉仪横向剪切原理与性质

Sagnac 干涉仪具有横向剪切性质首先被吉原邦夫教授提出[3]，它可把一个实体光源横向剪切成两个孪生虚光源，实体光源可以位于有限距离也可以位于无穷远。

Sagnac 干涉仪用于空间调制型干涉成像光谱仪时，光源应位于有限距离，本文所要讨论的就是这种形式。当光源位于无穷远时，就出现另一类无狭缝的高通量干涉成像光谱仪，这时 Sagnac 干涉仪同样可以把它横向剪切成两个位于无穷远的虚光源。两者的区别是：在用于空间对地侦察时，前者光学系统必须是

本文原载于《光子学报》，2009，Vol.，38，No.3，474~478。

一个二次成像系统，而后者可以是一次成像系统(当然为了减小杂散光或为了减小干涉仪的尺寸，也可以是二次成像系统)；此外前者的光谱(干涉强度分布)是同一时刻得到的，故称为空间调制型，而后者是通过卫星平台与被观察目标的相对运动(卫星推扫)，通过视场角调制光程差[4-5]，因此它的干涉强度分布不是同一时刻得到的，国外学者仍把它归入时间调制型一类，而国内有学者把它称为时空混合调制，前者的优点是可以探测光谱快速变化的目标，后者的优点是具有高通量(灵敏度)，由于后者对运动平台(飞机、卫星等)有极高要求，所以目前进入实际应用的均为光源位于近距离的空间调制型干涉成像光谱仪。图1为嫦娥一号卫星干涉成像光谱仪的光学系统配置。

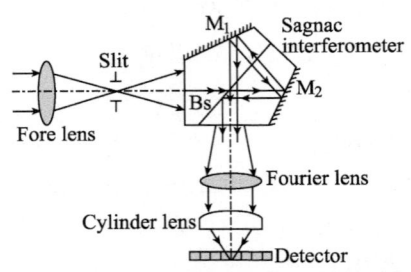

图1　空间调制型干涉成像光谱仪光学系统

在航天对地侦察中，目标(光源)总是位于无穷远，所以在Sagnac干涉仪前必须要有一个前置光学系统，它把目标的像成在它的后焦面上，同时在此处加入一个狭缝，狭缝是一个视场光阑，狭缝的宽度与长度对应于地元分辨率与成像幅宽。

这样Sagnac干涉仪所剪切的是一个近距离的光源(目标)，实际上为狭缝，Sagnac横向剪切干涉仪可以是一个玻璃实体，也可以是分离式的[6]，其剪切原理相同。可以看出它是二块反射镜M₁，M₂与一个分束膜BS所组成(见图2)，把三者做成单独光学部件，就变成了分离式的Sagnac干涉仪，由于实体Sagnac干涉仪具有更好的航天环境适应性，所以目前实际上星的都采用实体形式。下图是近距离的物点S(狭缝)被横向剪切成二个虚像S'S″的光束结构图。用平面反射镜系统来理解对物点S的成像可以得到相同的结果。

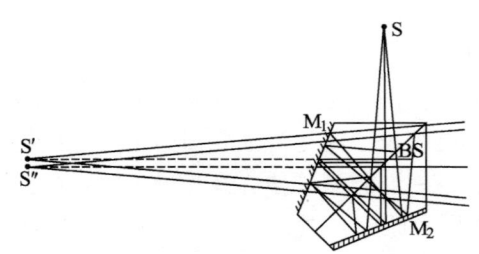

图2　Sagnac干涉仪横向剪切光束结构原理

该类Sagnac横向剪切干涉仪具有如下特征：

(1) 以分束膜为基准，两边必须是非对称的，可以有很多方法实现两臂的非对称，在分离形式的Sagnac干涉仪中，常用分束板的厚度与折射率来实现[4]，在上述实体Sagnac干涉仪中，用两块完全相同的半五角棱镜沿分束面的上下移动来实现，两臂完全对称的干涉仪无剪切作用，因此不产生干涉条纹。

(2) 由于前置光学系统的存在，进入Sagnac干涉仪的光束不是自由光束，而是受限光束。

(3) 实体狭缝与被剪切形成的两个虚狭缝位于垂直于光轴同一平面内，而且两个孪生虚狭缝均为正立像。

(4) 在分束面上，分为两个区域，一个区域镀有分束膜，而另一区域为增透膜，采用胶合时，可以不镀膜。

(5) 为了使各视场在干涉仪中产生相同的光程差，并且在分束面上入射光束具有相同的结构，以便使各视场的色漂移一致,进入干涉仪的光束结构必须是远心光束,即要求前置光学系统为像方远心系统,傅氏光学系统为物方远心系统。

(6) 进入 Sagnac 干涉仪的光能量有一半沿原路返回光源,变成杂散源,所以在干涉仪与狭缝间需要采取有效的防杂散光措施。

(7) 被分束膜分成的两束光线,其一束两次通过分束膜,而另一束两次被分束膜反射,因此与迈氏干涉仪相比,对分束膜有很高的透反比要求与偏振要求。

(8) 工程上,两块半五角棱镜由优质 K9 玻璃磨制而成,且尽量用同一块玻璃切割而成。

2 Sagnac 干涉仪的共光路原理

图 3(a)是由完全相同的两块半五角棱镜构成的"Sagnac 干涉仪",虽然两者间镀了分束膜 BS,但入射光轴经分束膜后,出射时完成重合,不能起到横向剪切的作用,这时它相当于在光路中加入一块平板玻璃。而图 3(b)是把完全相同的两块半五角棱镜沿分束面上下移动了 Δ,这时,同一入射光轴,经 Sagnac 干涉仪后被剪切成两条出射光轴,所产生的两个虚狭缝的中心位于它们各自光线的反向延长线上,两个虚光源是相干的。属于振幅分割的一种。

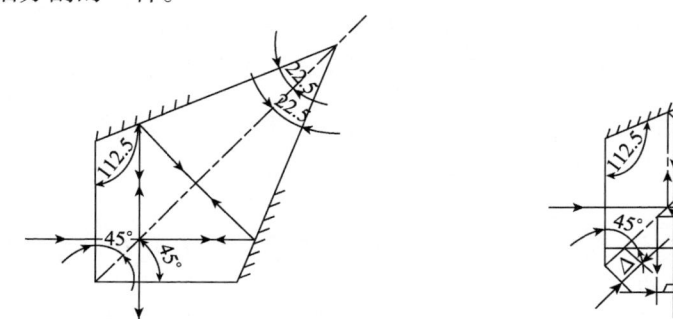

(a) Symmetry between up-prism and down-prism　　(b) Asymetry between up-prism and down-prism

图3　实体Sagnac干涉仪的共光路剪切原理

从图 3(b)可以看出被分束膜分成的两束光线都沿相同的三角形途径传播,但一束沿顺时针方向,另一束沿逆时针方向。通过后面的计算,可以看出,两束光间非常接近,可以把它看作是一种共光路干涉仪,因此它对环境条件不敏感,这也是它被首选作为上星成像光谱仪的主要原因。

3 Sagnac 干涉仪横向剪切量的计算

把两块完全相同的半五角棱镜,沿分束面上下移动一个 Δ 来实现横向剪切,Δ 值是根据仪器的性能指标:光谱分辨率 $\delta\sigma$,最大波数 σ_{max} 以及面阵探测器的像元尺寸 b 及单边像元数 N 以及傅氏透镜的焦距 f' 所确定。

成像光谱仪的光谱分辨率有两种表达形式,一种是色散型的,它基本上以波长等间隔,用 $\delta\lambda$ 表示,而干涉型的和声光调制型的成像光谱仪都是以波数 $\delta\sigma$ 来表示,波数是指以厘米为单位的波长的倒数,如对 λ=500nm,则波数 σ=20 000cm^{-1},因此最大波数 σ_{max} 即为仪器工作波段中最小波长之倒数(同样波长以 cm 计),波长与波数的关系式是

$$\sigma = \lambda^{-1} \tag{1}$$

两边微分,即可得到

$$\frac{\delta\sigma}{\sigma} = \frac{\delta\lambda}{\lambda} \tag{2}$$

它也可以写成

$$\frac{\sigma}{\delta\sigma} = \frac{\lambda}{\delta\lambda} = R \qquad (3)$$

因此若用光谱分辨本领 R 来表示，则两者相同。

如图 4，由于狭缝位于傅氏透镜前焦面上，所以经傅氏透镜后的输出为平面波，两个虚光源 S_1' 与 S'' 的出射平面波在傅氏透镜后焦平面上相交，并形成干涉，两平面波 S_1'，S_1'' 间的距离即为光程差，它是 y 的函数。

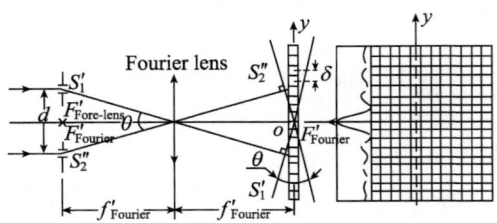

图4　Sagnac干涉成像光谱仪中相干原理

众所周知，干涉成像光谱技术的基础理论与干涉光谱技术是完全相同的，所谓干涉成像光谱技术是指通过测量干涉强度分布 $I(x)$ 来复原获得产生该干涉的光源的光谱强度分布 $B(\sigma)$ 的一种技术，其中的双光束干涉光谱技术，由于 $I(x)$ 与 $B(\sigma)$ 间的关系是傅氏变换与反傅氏变换的关系[7-8]，即

$$I(x) = \int_{-\infty}^{\infty} B(\sigma) \cdot e^{i2\pi\sigma \cdot x} \cdot d\sigma \qquad (4)$$

$$B(\sigma) = \int_{-\infty}^{\infty} I(x) \cdot e^{-i2\pi\sigma \cdot x} \cdot dx \qquad (5)$$

因此双光束的干涉成像光谱仪更确切的名称是"傅里叶变换成像光谱仪"。对式(4)和(5)的运算通常变成余弦傅氏变换并采取离散数值积分方法。

由于它是一个周期性函数，所以当用 $I(x)$ 复原 $B(\sigma)$ 时，会周期性地出现 $B(\sigma)$，为了避免相互间的谱混叠，根据采样定律，它必须满足式(6)

$$X \leq (2\sigma_{max})^{-1} \qquad (6)$$

X 的定义为干涉图的采样步长，从图 4 可以发现，当采用面阵探测器作为图像传感器时

$$X \approx b \cdot \theta \qquad (7)$$

式中 b 为探测器像元的中心距，θ 称为剪切角，它是二出射平面波间的夹角，也是二个虚狭缝对傅氏透镜光心的张角，因为 θ 通常都非常小，所以工程上可以按式(7)以弧度值计算。

在嫦娥一号卫星干涉成像光谱仪的情况中，$b=0.034$ mm，$\lambda_{min}=480$ nm，即 $\sigma_{max}=20833.3$ cm^{-1}，所以根据式(6)，有

$$X \leq \frac{1}{2 \times 20\,833.3\,\text{cm}^{-1}} = 2.4 \times 10^{-5}\,\text{cm} = 0.24\,\mu\text{m}$$

根据式(7)有

$$\theta \leq X/b = 0.0070588\,\text{弧度}$$

再回到图 4，有

$$d/f_{Fourier} = \theta \qquad (8)$$

在嫦娥一号卫星干涉成像光谱仪中，$f'_{Fourier} = 80$ mm。所以有

$$d = f'_{Fourier} \cdot \theta = 80\,\text{mm} * 0.0070588\,\text{弧度} = 0.5647\,\text{mm}$$

d 是嫦娥一号卫星干涉成像光谱仪中 Sagnac 干涉仪要求的横向剪切量，在所采用的 Sagnac 干涉仪的角度关系情况下，从图 3(b)可以推出

$$d = \sqrt{2} \cdot \Delta \cdot \tan(22.5°) = 0.58578 \cdot \Delta \tag{9}$$

得 $\Delta=0.964$mm, Δ 是嫦娥一号卫星干涉仪中两个半五角棱镜沿分束面上下错位量。需要指出的是，在嫦娥一号卫星干涉成像光谱仪中的 Sagnac 干涉仪正好把光轴折转 90°。其实可以根据需要折转为任一角度，其原理相同。根据干涉光谱技术，光谱分辨率只取决于最大光程差 $\text{OPD}_{\max}^{(7)}$，即

$$\delta\sigma = (2 \cdot \text{OPD}_{\max})^{-1} \tag{10}$$

在嫦娥一号卫星干涉成像光谱仪的情况下，指标要求。$\delta\sigma=325.25\text{cm}^{-1}$，所以根据式(10)有

$$\text{OPD}_{\max} = (2 \cdot \delta\sigma)^{-1} = 15.37\mu m$$

由图 4，得到

$$\text{OPD}_{\max} = X \cdot N \tag{11}$$

N 为采集单边干涉图的探测器像元数，对嫦娥一号卫星干涉成像光谱仪有

$$N = \frac{15.37\mu m}{0.24\mu m} = 64 \text{pixel}$$

即只需要 128pixel，就可以得到两边采样的全干涉图。

4 分束面上的镀膜分区

在 Sagnac 干涉仪的分束面 AD 上，需要分成三个区，其中 AB 区镀分束膜，BC 区不镀膜，CD 区与 AB 区同时镀分束膜。CD 区镀膜的目的是胶合时，有一个平的基准面。

图 5 只有一根光轴，其实进入 Sagnac 干涉仪的是一发散光束(见图 2)，对于圆孔光学系统而言，光束在分束面上的投影是一个椭圆。

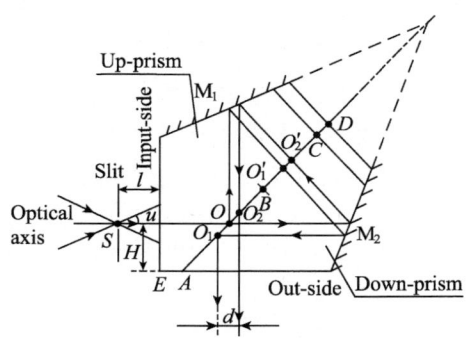

图5 Sagnac干涉仪分束区分布

为了使分束区 AB 与透射区 BC 分离，系统光轴对 Sagnac 干涉仪入射端是向下偏置的。首先选择光轴在入射端面上的入射点高度 H，其原则是从前置光学系统来的光束在入射端面上的投影都处于入射端面的有效口径之内，它很容易由前置光学系统像方孔径角 u 及狭缝距 Sagnac 干涉仪入射端面的距离 l 计算得到，锥光束进入 Sagnac 干涉仪时有一次折射，使发散角 u 减小，尔后光束投射到分束面上的分束区 AB，它是一个椭圆形光斑，其中透过分束面的一束光经 M_2、M_1 反射后再次入射到分束区 AB 上，它同样是一个椭圆形光斑，但中心从 O 移到了 O_2，并且光斑尺寸扩大，当它通过透射区 BC 时也同样有一个椭圆形光(在图 5，光轴折射 90°时，为圆)。对于进入 Sagnac 干涉仪后在分束面上被反射的一束光，情况相同，它经 M_1、M_2 反射后，也再次入射到分束区 AB 上，而椭圆的中心 O 移到 O_1，从图 5 可以看出在分束面上与透射区最近的光斑是透射光束第二次入射到分束面上以 O_2 为中心的大椭圆，而在透射面上离开分束区最近的光斑是反射光束在分束面上以 O'_1 为中心的椭圆，设计中必须使以 O_2 为中心的椭圆与以 O'_1 为中心的椭圆(或圆)二者分离，按经验，二者间隔应大于等于 3mm，同时还要保证透射光在干涉仪出射端面上的光斑位

于 A 点的右侧区域内，且与 A 点应该有一定的距离。上述计算可以列出一套公式，但计算十分烦琐，而用计算机作图，可以非常直观地看出各类光斑在分束面上的位置与大小，并判别是否满足设计要求。

当发现不满足设计要求时，可以调节的环节有，减小 l，改变 H，当仍然不能满足要求时，必须增大 Sagnac 干涉仪的尺寸。采用高折射率的玻璃常常可以缩小干涉仪的尺寸，但不容易获得高品质的材料。

图 6 中左半部分是嫦娥一号卫星干涉成像光谱仪中的 Sagnac 干涉仪分束面上的光斑位置与尺寸的示意图，右图半部分为相应的下棱镜分束面上的镀膜分区。实际光斑为与圆的上下顶点相切的水平状态的椭圆，图中以圆表示，不影响设计。

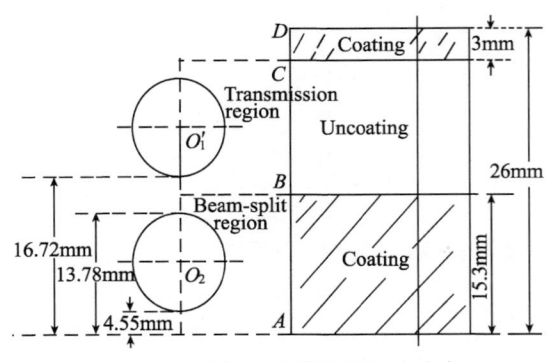

图6 Sagnac干涉仪下棱镜镀膜区域分布

5 结论

在空间调制型的干涉成像光谱仪中，Sagnac 横向剪切干涉仪是核心部件，它的横向剪切量 d 与傅氏光学系统的焦距以及 CCD 像元尺寸、像元数决定了最大光程差，从而决定了空间调制型干涉成像光谱仪的光谱分辨率，本文介绍了 Sagnac 干涉仪的设计思想、共光路及横向剪切原理、剪切量的计算方法以及干涉仪分束面上分束区与透射区的设计。经检测与在轨运行，表明设计正确，经受了包括月食低温在内的各种恶劣的航天环境的考验，获得了大量清晰的月表多光谱图像。

参 考 文 献

[1] OTTEN L J, MEIGS A D, JONES B A, et al. The engineering model for the MightySatII.1 Hyperspectral imager [C]. *SPIE*, 1997, **3221**:412-420
[2] RAFERT J B, OTTEN L J, BUTLER W, et al. Meigs. Satellite sends Hyperspectral images from space [J]. *Laser Focus World*, 2001, **37**(5):181-193
[3] YOSHIHARA K, KITED E. Holographic spectra using a triangle path Interferometer [J]. *J ph J Appl Phys*, 1967 (6):116-116.
[4] TIMRA D C, FRIEDMAN Z, et al. United States patent, 5539517.1995-2-21
[5] BENNETT C L, CARTER M, Fields D, et al. Imaging Fourier Transform Spectromter [C]. *SPIE*. 1993, **1937**:191-200
[6] YANG Jian feng, RUAN Ping, et al. Large aperture static imaging spectroscopy (LASIS) [C]. *Proc SPIE*, 2003, **4897**:318-325
[7] NISHI M. Fourier Spectroscopy [M]. Cirulars of the Electrotechnical Laboratory, 1970:6-20
[8] BELL R J. Introductory Fourier Transform Spectroscopy [M]. New York: Academic Press. 1972:33-50

嫦娥一号卫星成像光谱仪光学系统设计与在轨评估

赵葆常[1]　杨建峰[1]　常凌颖[1,2]　陈立武[1]　贺应红[1]　薛　彬[1]

(1. 中国科学院西安光学精密机械研究所，西安 710119；2. 中国科学院研究生院，北京 100049)

摘　要　介绍了嫦娥一号卫星干涉成像光谱仪的科学目标、总体设计思想、方案选型、工作原理、光学实施方案、光学系统设计及评价、空间环境适应性考虑、发射前光图像质量检测、在轨运行情况及在轨性能评测。
关键词　月球探测　嫦娥一号　CCD立体相机　干涉成像光谱仪　在轨图像质量

0　引言

我国首颗月球探测卫星嫦娥一号于2007年10月24日成功发射，11月5日被月球捕获，在完成卫星调轨、调姿后从11月20日起各种探测设备相继开机开展各项科学探测，干涉成像光谱仪是继CCD立体相机之后第二台开机工作的科学仪器，到2008年12月工作时间已超过设计寿命一年，仪器在轨工作一直正常。已获得了大量中高纬度清晰的多光谱图像。它是国际上首次采用干涉成像光谱技术，实现对月球的可见光/近红外连续宽谱段的多光谱探测，它与X/γ射线谱仪共同完成了分析月球表面有用元素成分及物质类型的含量与分布的科学目标。

1　科学目标

根据嫦娥一号卫星地面应用系统制订的我国首次绕月探测的科学目标，嫦娥一号卫星共配置了八种科学探测设备。

从表1可以看出，在8种探测设备中，有三台属于精密光学仪器，即CCD立体相机、激光高度计及成像光谱仪。其中CCD立体相机与激光高度计完成第一科学目标：获取月表三维立体影像；成像光谱仪与X射线谱仪、γ射线谱仪完成第二科学目标：分析月表有用元素成分与物质类型的含量与分布[1]。

表1　科学目标及相应的探测设备

科学目标	探测设备
获取月表的三维立体影像	CCD立体相机、激光高度计
分析月表有用元素成分及物质类型的含量与分布	成像光谱仪、X射线谱仪、γ射线谱仪
探测月壤厚度	微波探测器
4万~40万公里地月空间环境探测	高能粒子探测器、太阳风探测器

在分析月表有用元素成分与物质类型上，三台设备还是有各自的侧重点，X射线谱仪与γ射线谱仪侧重于元素的探测，而成像光谱仪则侧重于物质类型的分类。其实光谱单通道的照相机也具有区分物质类型的功能，如从形态上看水泥马路、沥青路面与土路是十分相似的，但在相同光照条件下，水泥马路会比沥

本文原载于《光子学报》，2009，Vol.38，No.3，479~483。

青路面显得更亮些,这是因为水泥马路路面比沥青路面具有较高的反射系数所致,但是它尚不具备光谱分辨能力,对一台黑白照相机来说,若在草地上停有一辆颜色与背景不同的汽车,但在照相机的工作波段内探测器具有相同信号输出强度时,则无法发现该汽车,但是由于两者的颜色不同,采用多光谱图像,则可以十分容易地进行识别。月球自身并不发光,它是依靠太阳光照的反射而发光,称为被照发光体,不同类型的物质对不同的波长具有不同的反射系数,因此在由成像光谱仪获得的多光谱序列图像(即相当于波段很窄的准单色图像)中可以把具有不同光谱反射特性的物质类型区分得非常清楚。

2 技术指标

根据科学目标与任务要求,嫦娥一号卫星干涉成像光谱仪当轨道高度 $H=200$ km 时达到的技术指标为:1) 月表地元分辨率 GSD=200 m;2) 月表成像宽度 L=25.6 km;3) 光谱范围 λ=0.48～0.96 μm;4) 光谱通道数 N=32 个谱段;5) 光谱分辨率 δ_σ=325 cm^{-1};6) 量化等级为 12 bit;7) MTF$\geq$0.2;8) $S/N\geq$100;9) 太阳高度角 $\theta\geq$15°。即要求当太阳高度角 $\theta\geq$15° 时,仪器可以获得有用数据图像。

3 成像光谱仪选型

从任务的角度出发,该成像光谱仪是为了实现对月表可见光至近红外的连续反射光谱的测量,提供系统科学家分析月表物质类型的分布。因为它是一台星载设备,所以首先必须考虑对航天环境的适应性。月球距离地球为 35.64×10^4(近地点)~40.67×10^4 km(远地点),卫星需要飞行十天左右的时间。首先在发射场把卫星发射到近地点 200km,远地点 5.1×10^4 km 的椭圆轨道,然后三次扩大椭圆再进入奔月轨道,到达月球并被月球引力场捕获,最后进入 200 km 绕月圆轨道。由此可知,探月卫星要经历多次加速,当它到达月球时,需要激烈减速,因此它的力学环境条件非常严峻,所以成像光谱仪必须要有经受多次加速、减速的能力。同时温度环境条件也非常恶劣,月表受光照时温度最高超过 130℃,背光时最低降至–180℃,而且还要经历月蚀低温的考验,因此何种类型的成像光谱仪具有最强的力学与温度环境适应能力是总体方案选择的重点。

经分析比较各种形式的成像光谱仪后,决定采用 Sagnac 型的干涉成像光谱仪的方案,这是因为它的共光路原理所致,而且它是一个实体干涉仪。

Sagnac 横向剪切干涉仪最早由吉原邦夫教授提出[2],他将一个实体光源(目标)横向剪切成两个相干虚光源。从图 1 中可以看出,在分束面上,光束被分成透、反的两束光。它们在干涉仪中的途径是相同的,但传播方向相反,一束为顺时针方向传播,另一束为逆时针传播。在多数情况下,它们错开得很小,研究中出射端的最大值只有 0.5647 mm,称为剪切量 d,而在光线两次入射到分束时的间隔仅为 $d/2$。由于它是一种共光路干涉仪,所以两路相干光经受外界环境条件,不论是力学条件还是温度环境都是相同的,因此它具有非常强的航天环境适应能力,这是选取它的最主要原因。它同样也是美国于 2000 年发射的强力小卫星搭载的成像光谱仪选用的类型,而且取得了完满成功[3-5]。另一个原因是,本次为首次对月球的探测,尚有很多不清楚之处,所以应用系统科学家希望能计算任一波长上的光谱强度,只有干涉型的成像光谱仪具有该功能。

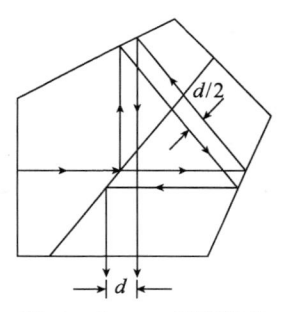

图 1 Sagnac 干涉仪共光路横向剪切原理

4 Sagnac 空间调制型傅氏变换光谱仪原理

Sagnac 空间调制型傅氏变换光谱仪的原理[6-7]与色散型成像光谱仪相同。它必须是一个二次成像光学系统，在一次像面上需插入一个狭缝，狭缝宽度经前置物镜投影到月表上的尺寸即为月表的空间地元分辨率。其长度与月表成像宽度相对应。

由于目标距离为 200 km，所以狭缝应位于前置光学系统的后焦面上，同时它又必须位于傅氏透镜的前焦面上，为使轴上点与轴外点等光程，前置物镜为像方远心，傅氏透镜为物方远心。而且由于狭缝位于傅氏透镜的前焦面上，所以从傅氏透镜出射的光束为平行光，柱面透镜的作用是在一个方向上把由傅氏透镜出射的相干平面波压缩为一条线，它对应面阵 CCD 的一列，傅氏透镜与柱面透镜像方共焦，面阵 CCD 位于两者的共同焦平面上，狭缝的长度方向与柱面镜母线相垂直，面阵 CCD 垂直于飞行方向的称为行，对应于空间方向，沿飞行方向的称为列，对应于干涉图方向，也就是说空间方向上每一个像元都带了一条干涉强度分布曲线 $I(x)$，经滤波、去基线(bias)及位相修正后经余弦傅氏变换，即可得到该地元的光谱强度分布 $B(\sigma)$，即

$$B(\sigma) = \int_{-\infty}^{\infty} I(x) \cdot \cos 2\pi\sigma x \cdot dx \tag{1}$$

实际采用离散数据求和的方法即

$$\overline{B(\sigma)} = \sum_{i=0}^{N} I(xi) \cdot \cos 2\pi(ix) \cdot \sigma \cdot dx \tag{2}$$

图 2 是抽取干涉仪后的干涉原理。图 2 中 S'_1 与 S''_1 是由同一实体狭缝被 Sagnac 干涉仪横向剪切作用而产生的一对孪生相干虚光源，它们对傅氏透镜光心的张角 θ 称为剪切角，由于 S'_1 与 S''_1 位于傅氏透镜的前焦面上，所以经傅氏透镜后出射的光束结构为两个平面波 S'_1 与 S''_1，两个平面波在傅氏透镜的后焦点上相交，夹角也为 θ，并且上下对称，从图 2 可以看出在轴上点，两个波面的光程差理论上为零(当 CCD 像元尺寸不计时)，而在 Y 方向上，随着距离 O 点距离的增加两个波面间的光程增大，如果 CCD 探测器的相邻像元间中心距为 δ，则相邻两个像元的光程差增量 $x \approx \delta \cdot \theta$，最大光程差为干涉图单边像元数 N 乘以 x 即

$$OPD_{max} = N \cdot x \tag{3}$$

根据采样定律，x 必须满足

$$x \leq (1/2)\sigma_{max} \tag{4}$$

当 $x \leq (1/2)\sigma_{max}$ 时，在光谱复原时，会产生谱重叠。而光谱分辨率为

$$\delta\sigma = (1/2)OPD_{max} \tag{5}$$

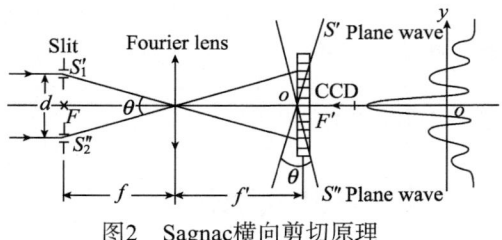

图 2 Sagnac 横向剪切原理

与色散型成像光谱仪不同，干涉型成像光谱仪的光谱分辨率是波数等间隔，它与声光调制型(AOTF)成像光谱仪相同，波数的定义是以厘米为单位的波长之倒数。σ_{max} 称为最大波数，它是以厘米为单位的最小波长之倒数。

5 光学系统设计

它是一个二次成像光学系统，可以把它看作是前置照相物镜加上一个由傅氏透镜与柱面镜组成的投影物镜，设系统总焦距为 f'_G，前置物镜焦距为 f'_F，由傅氏透镜与柱面镜组成的投影倍率为 β，则有

$$f'_G = f'_F \times \beta \tag{6}$$

由于前置物镜处于对月观察窗口附近,温度环境条件最差,所以在总体方案考虑上必须使其具有宽松的公差以及对空间环境条件引起的变化不敏感,由此决定它们的参数分配。

CCD 像元尺寸为 $17 \times 17 \ \mu m^2$,采用了 2×2 像元合并,以消除盲元并提高像元间响应的均匀性,同时它还增大了信号强度。从表 2 的分配可以看出由傅氏光学系统与柱面光学系统组成的投影物镜为缩小 3.08×,沿轴方向缩小近 10×,因此由于空间环境条件造成前置物镜的变化,对最终图像质量灵敏度大大降低,而且这样的系统容易达到好的像质,2×2 像元合并降低了奈奎斯特空间频率。实际的空间频率为 15 lp/mm。图 3 是嫦娥一号卫星干涉成像光谱仪光学系统,图 4 是其全系统白光 MTF 曲线,表 3 为全系统各视场在奈奎斯特空间频率上(15 lp/mm)的白光 MTF 值。

表 2 组成干涉成像光谱仪各分系统参数分配表备

分系统\参量	焦距/mm	F 数	视场/(°)	特殊要求
前置光学系统	104.6	7.34	7.34°	像方远心
傅氏光学系统	80	7.34	9.56°	物方远心,傅氏透镜的出瞳及后焦点与柱面镜后焦面重合
柱面光学系统	26	2.4	9.56°	
全系统	34	2.4	7.34°	

表 3 全系统各视场的白光 MTF 值

视场	白光 MTF	
	子午	弧矢
0 视场	0.916	0.916
0.7 视场	0.929	0.916
边视场	0.906	0.897

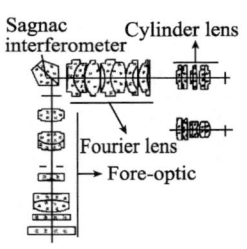

图3 嫦娥一号卫星中干涉成像光谱仪光学系统

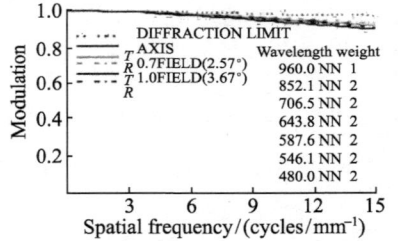

图4 干涉成像光谱仪全系统白光MTF曲线

MTF 均值为 0.9133,最大差值为 0.0163(与均值)。

6 发射前检测与在轨图像质量及性能评测

光机系统完成装配、检测后,与焦平面组件进行了对接,主要是找到最佳焦面位置,使焦平面与光轴垂直,旋转焦平面使干涉图与 CCD 的列重合。在此基础上对整机(光、机、CCD 器件及电子线路)进行了检测,检测结果为整机 MTF=0.51,S/N=375,都达到了很高的指标。其中 MTF 完全是由于 CCD 器件的 MTF 所致。成像光谱仪于 2006 年 2 月 15 日移交总体后,经过了有效载荷总体验收与联试,3 月 15 日交付卫星总体,两次验收均为零缺陷产品。设备装星后,在装星状态下进行了三次复检,检测结果与出所时检测结果相符,说明仪器性能稳定。

干涉成像光谱仪于 2007 年 11 月 27 日凌晨开机,在检查工程参数正常的基础上,直传在轨的黑白快视图及干涉条纹数据图像,复原重构后的多光谱序列图像清晰,到目前为止已获得了大量中、高纬度月表的多光谱图像。

图 5 中(a)~(d)分别是第 4 通道(λ=504 nm)、第 17 通道(λ=645 nm)、第 30 通道(λ=891 nm)的三张光谱图像以及由它们合成的伪彩色图像[8]。

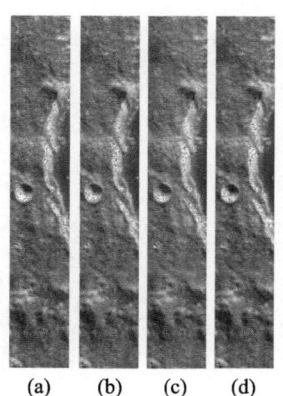

图5　干涉成像光谱仪获得的多光谱序列图像

用该干涉成像光谱仪测量了赫歇尔(Herschel)、维诺格拉多夫(Vinogradov)及欧几里得C(Euclides C)三个位于不同纬度的撞击坑，测量结果表明，它们与国际发表的撞击坑直径基本相符[9]。图 6 是由干涉成像光谱仪获得的撞击坑图像，表 4 是检查结果。

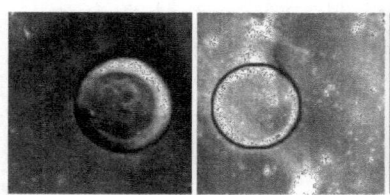

(a) Herschel crater　　(b) Vinogradov crater　　(c) Euclides C crater

图 6　干涉成像光谱仪获取的Herschel、Vinogradov和Euclides C撞击坑图像

表4　干涉成像光谱仪月表空间分辨率与成像幅宽检测结果

试验区	检测目标	轨道高度/km	月表空间分辨率/m	成像幅宽/km	国际发表直径/km	干涉成像光谱仪检测结果/km	相对偏差 百分比/%	单边像元数
测试区 1	Herschel 撞击坑	205.0	205	26.2	13	13.9	~6	2.25
测试区 2	Vinogradov 撞击坑	203.3	203	26.0	11	12.0	~9	2.5
测试区 3	Euclides C 撞击坑	198.4	198	25.4	11	10.5	~5	1.25

进行了在轨辐射场定标试验，来评估干涉成像光谱仪的在轨性能。选择了一块 $3.2 \times 3.2 \ km^2$ 的月表区域，从宏观上看该区域光谱辐亮度较为均一，取代实验室定标中的积分球，进行了相对辐亮度在轨复核试验。由于设备的地元分辨率为 200 m，所以 $3.2 \times 3.2 \ km^2$ 的区域共有 256 个像素点(pixel)，把它们视为均一。试验表明相对不确定度在 2.5%~9.5%，优于设计任务书 15%的指标。由于所选的月表区域并不真正具备朗伯源的要求，所以仪器的实际指标应该优于上述测量值。图 7 中用方框表示了 $3.2 \times 3.2 \ km^2$ 区域的被选辐射场。图 8 为 256 个像元的相对光谱辐射度偏差。

7　结论

由中科院西安光机所研制的嫦娥一号卫星有效载荷干涉成像光谱仪在可见光/近红外连续宽谱段范围内，圆满实现了对月表太阳反射光谱的多光谱探测，为国际首次。发射前的实验室检测及在轨图像质量都表明了该设备的卓越性能,复原光谱 S/N 接近光子噪声极限。获得的数据为科学家分析物质类型的分布具有重要价值。

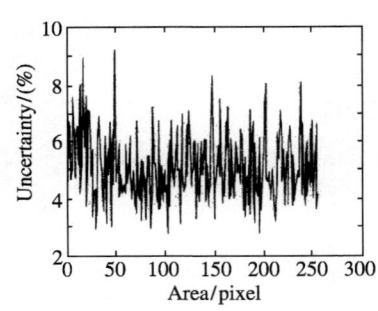

图7　干涉成像光谱仪获得的多光谱序列图像　　　　图8　各地元点的相对偏差

致谢　感谢工业与信息化部航天局探月工程中心对该项工作的支持,以及中国科学院、西安光机所各级领导有力的组织协调,在设备研制过程中还得到了中科院空间中心以及其他兄弟单位诸多帮助。特别需要指出的是用户单位"中国科学院国家天文台月球与深空探测科学应用中心"在科学数据处理中所做出的杰出工作,本文中在轨图像均由他们提供,在此一并致谢。

参 考 文 献

[1] OU YANG Zi-yuan, LI Chun-la, ZOU Yong-liao, et al. The scientific object of the first phase project of Chinese lunar exploration[J]. *Spacecraft Engineering*, 2005, 14(1):1—5
　　欧阳自远,李春来,邹永廖,等. 我国月球探测一期工程的科学目标[J]. 航天器工程,2005,14(1): 1—5
[2] YOSHIHARA K, KITADE A. Holographic spectra using a triangle path interferometer[J]. *Jpn J Appl Phys*, 1967, 6 : 116
[3] OTTEN L J, MEIGS A D, PORTIGAL F P, et al. MightySat II. 1 : an optical design and performance update [C]. *SPIE*, 2005, 2957 : 390—398
[4] YARBROUGH S, CAUDILL T R, KOUBA E T, et al. MightySat II. 1: hyperspectral imager: summary of on-orbit performance[C]. *SPIE*, 2005, 4480 : 186—197
[5] OTTEN L J, MEIGS A D, PORTIGAL F P, et al. The engineering model for the Mightysat II. 1 : hyper spectral imager[C]. *SPIE*, 2005, 3221 : 412—420
[6] SELLAR G R, RAFERT B J. Effects of aberration on spatially modulated Fourier transform spectrometer[J]. *Opt Eng*, 1994, 33(9): 3087—3092
[7] RAFERT B J, SELLAR G R, BLATT H J, et al. Monolithic Fourier-transform imaging spectrometer [J]. *Appl Opt*, 1995, 34(31): 7228—7230
[8] 嫦娥一号拍摄的首幅月球极区图像发布(组图)[EB/OL]. (2008-01-31) [2008-12-15]. http://news.xinhuanet.com/photo/2008-01/31/content_7535200_9.htm
[9] 嫦娥一号卫星科学探测数据质量初步评估报告(内部). 中国科学院国家天文台月球与深空探测科学应用中心

嫦娥一号卫星干涉成像光谱仪现场性能检测实验

邱跃洪 赵葆常 赵建科 薛 彬 乔卫东 汶德胜

(中国科学院西安光学精密机械研究所,西安 710119)

摘 要 介绍了嫦娥一号卫星有效载荷—干涉成像光谱仪现场性能检测实验,包括焦距检测实验、MTF检测实验、谱线位置和光谱分辨率定标检测实验、光谱辐射度定标检测实验。分析了检测的必要性,确定了检测判据,制订了检测方案,规定了实验条件,给出了检测结果。通过实验,保证了干涉成像光谱仪以确定的和良好的技术状态发射。

关键词 嫦娥一号卫星 干涉成像光谱仪 焦距检测 MTF检测 谱线位置定标 光谱辐射度定标 光谱分辨率

0 引言

探月工程是继载人航天工程之后我国航天领域又一重大项目,我国《国家中长期科学和技术发展规划纲要(2006—2020年)》和《中国的航天》白皮书也明确提出开展以月球探测为主的深空探测研究。今年我国首颗月球探测卫星——嫦娥一号卫星圆满完成绕月探测任务,为我国深空探测领域迈出了里程碑的第一步。嫦娥一号卫星载荷之———干涉成像光谱仪主要任务是获取月球矿物的光谱信息并进而获得全球分布信息,从而与X/γ谱仪一起完成首次月球探测的四大科学和应用目标之———分析月球表面有用元素及物质类型的含量与分布[1-2]。

嫦娥一号卫星有效载荷光学成像探测系统由实现两个科学目标的仪器——CCD立体相机和干涉成像光谱仪组成,两台仪器除共用二次电源和下位机之外,其他部分均相互独立。光学成像探测系统正样产品于2006年2月15日交付给中科院探月工程应用系统总体部后到发射间有一年多的时间间隔,为了保证干涉成像光谱仪功能及性能指标的可靠性,为了确保绕月探测成功,在进行室内摸底实验的基础上,在5院卫星联调大厅进行两次光学系统参数及主要指标的复检。本文主要介绍干涉成像光谱仪卫星现场性能检测实验。

1 检测内容、必要性及判据

从最能反映干涉成像光谱仪状态的参数出发,考虑到现场检验条件及环境可行性,对干涉成像光谱仪进行焦距检测、MTF检测、光谱定标检测和光谱辐射度定标检测。

1.1 焦距检测

1.1.1 检测的必要性

一个光学系统最能反映其状态变化的参数是焦距,若焦距正常,可以判知该光学系统光机部件基本正常。

本文原载于《光子学报》,2009,Vol.38,No.3,484~488。

1.1.2 检测合格性判据

以研制任务书为依据，合格性判据为

$$f' = 34\text{mm} \tag{1}$$
$$\Delta f' \leqslant \pm 1\% \times f' \tag{2}$$

1.2 MTF 检测

1.2.1 检测的必要性

干涉成像光谱仪干涉图像质量的最重要检测指标之一是整机 MTF，通过对 MTF 的检测还可以检查干涉成像光谱仪焦面位置和光学系统焦面的配准情况。

1.2.2 检测合格性判据

以研制任务书为依据，合格性判据为

$$\text{MTF(系统)} \geqslant 0.2 \tag{3}$$

1.3 光谱定标检测

1.3.1 检测的必要性

对成像光谱仪而言，除图像质量外，还有一项主要指标———光谱分辨率及谱线位置，通过对它的检测可以判断干涉仪的状态是否正常以及剪切量的变化。

1.3.2 检测合格性判据

采用 $\lambda = 632.8$ nm 及 $\lambda = 543.5$ nm 两种波长的气体激光器检测光谱分辨率及谱线位置。依据研制任务书要求，以允差 10% 作为合格性判据。

1.4 光谱辐射定标的检测

1.4.1 检测的必要性

通过最大饱和曝光辐亮度的检测可以判断光学系统透过率的衰减情况，通过标准灯加白板的定标检测可以检查焦面能量变化情况。

1.4.2 检测合格性判据

考虑到现场检测的条件、环境及时间的限制，采用了一种简化的定标检测方案：标准灯加白板，对现场两次检测数据作比对，以不确定度不超过 15% 为判据。

2 检测方案及实施步骤

2.1 焦距和 MTF 检测

焦距和 MTF 检测实验装置如图 1[3-5]，主要由积分球、平行光管以及其他辅助装置组成[3]。

2.1.1 焦距检测步骤

根据图 1 的检测装置进行配置，检测步骤为：
(1) 把平行光管的光轴与现场被检设备干涉成像光谱仪光轴调平。

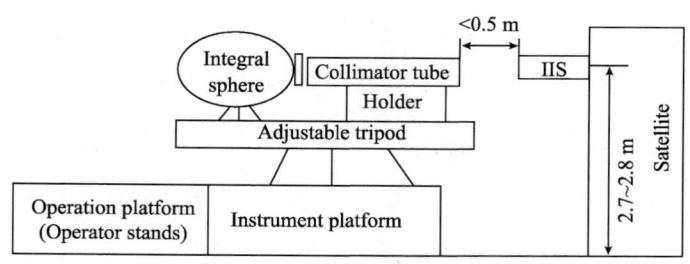

图 1　焦距和 MTF 检测实验

(2) 调节小积分球出口处的亮度，使被测设备的输出合适，并照明均匀。
(3) 由采集被测设备的双狭缝图像数据。
(4) 按式(4)计算被测设备光学系统焦距

$$f' = (n*X/y)f'_0 \tag{4}$$

式中 f' 为被测设备焦距，f'_0 为平行光管焦距，y 为双狭缝板物高，n 为双狭缝间隔像元数，X 为像元尺寸。

平行光管和双狭缝板事先进行标定。

2.1.2　MTF 检测步骤

把焦距检测中的双狭缝板换成多条纹分划板。检测步骤为：
(1) 把平行光管的光轴与现场被检设备干涉成像光谱仪光轴调平；
(2) 调节小积分球出口处的亮度，使被测设备的输出合适，并照明均匀；
(3) 调节多条纹分划板的相位与被测设备 CCD 配准；
(4) 采集被检设备的黑白条纹图像；
(5) 按式(5)计算 MTF

$$\text{CTE} = (\overline{U_\text{W}} - \overline{U_\text{D}})/(\overline{U_\text{W}} + \overline{U_\text{D}}) \tag{5}$$

$$\text{MTF} = (\pi/4)\cdot\text{CTF} \tag{6}$$

式中 $\overline{U_\text{W}}$ 为透光条带对应的平均输出值；$\overline{U_\text{D}}$ 为不透光条带对应的平均输出值。

2.1.3　谱线位置和光谱分辨率定标检测

如图 2，谱线位置和光谱分辨率定标检测实验装置主要由激光器、激光扩束器以及其他辅助装置组成。

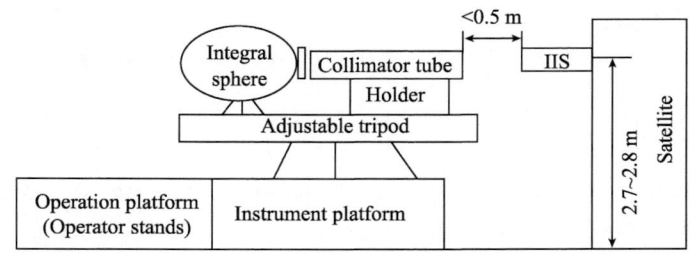

图 2　谱线位置和光谱分辨率定标检测实验

检测步骤为：
(1) 把被测设备、激光扩束镜和激光器三者光轴调平；
(2) 检查激光扩束镜输出光斑尺寸是否充满被测设备物镜入瞳；
(3) 检查激光扩束镜输出光斑尺寸及质量；

(4) 对得到的激光干涉条纹作数学处理，复原仪器函数，计算 FWHM 及谱线位置。

按检测步骤依次用 $\lambda=632.8$ nm 及 $\lambda=543.5$ nm 两种波长的气体激光器检测光谱分辨率及谱线位置。

2.2 光谱辐射定标检测实验

如图 3，光谱辐射度定标检测实验装置主要由标准灯、白板及其他辅助装置组成。实验步骤及注意事项：

(1) 标准灯/白板事先做成一体(见图 4)，并在西安光机所实验室用鉴定级产品做试验，以验证设计参数；

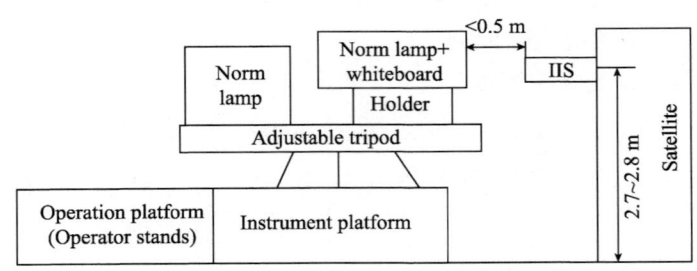

图 3 白板十标准灯定标检验实验

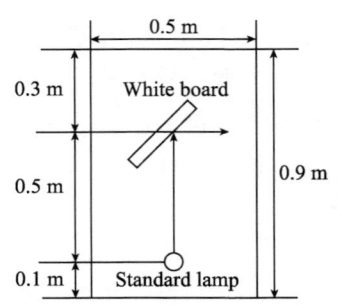

图 4 白板十标准灯实验装置

(2) 标准灯/白板固定在可调节三角架上，调平三角架平台，并调整与被测设备的角度和距离；
(3) 干涉成像光谱仪图像采集(干涉图)；
(4) 光谱辐射度计在相同的状态下同时采集光谱数据；
(5) 对光谱辐射度计采集的光谱数据作二次傅氏变换[6]，使其可以与干涉成像光谱仪获得的数据比对；
(6) 对干涉成像光谱仪采集的图像数据进行位相修正和傅氏变换后，得到光谱数据立方体；
(7) 对两者数据进行比对和处理，计算干涉成像光谱仪光谱辐射度定标的不确定度。

2.3 检测设备及辅助装置

由于在整星测试状态下，光学成像探测系统离地面高度为 2.7~2.8 m，这给检测带来了诸多困难，为了实施上述检测，除检测设备外还要自制若干平台（表 1）。

2.4 实验条件

温度：20℃±5℃；湿度：30%~50%；气压：当地大气压+10 Pa；洁净度：10 万级；静电：局部防静电、防静电鞋、帽、衣服及手套等；辐射：无要求；光照：尽量创造一个局部暗室条件；动力电源：需要 2 路 10A/220V 电源；在每次实验时，为了实时采集、存储和显示图像数据，系统设置为实时非压缩模式。

表1 检测设备及辅助装置一览表

名称	规格型号	性能及用途
仪器平台	自制,尺寸为 0.9 m(长)×0.9 m(宽)×1.3 m(高)	为达到 2.7~2.8m 的检测高度,在该平台上再架设可调节三脚架平台
操作平台	自制,尺寸为 0.5 m(长)×0.5 m(宽)×1.3 m(高)	检测人员站在该平台上实施检测
可调节三脚架平台	高度调节范围 1.1~1.9 m	该三脚架平台放在仪器平台上,再在它上面配置测量仪器,该三脚架平台具有高低、水平调节等功能
平行光管	f=550 mm,上海光学仪器厂产品	检测焦距、MTF 用,配有特定的专用双狭缝板及鉴别率板
激光扩束镜	自制,倍数 20X	用于光谱定标
小积分球	外径	作为焦距、MTF 检测时的照明光源
漫反射参考标准板(标准白板)	Labsphere 公司 STR-PP-240 尺寸为 617 mm×617 mm	定标检测用
光谱辐射度计	Fieldspec R ProVV/VNIR 光谱范围 350~1050 nm,采样间隔 1.4 nm,光谱分辨率 3 nm,量化等级 16 bit	定标检测用
气体激光器	进口气体激光:05-LGR-193-381 激光波长,精度 0.124 nm	光谱定标用
气体激光器	进口气体激光器:1322412 激光波长,精度 0.156 nm	光谱定标用
夹具等	卓立汉光	实验附件

3 结果

在对实验方案和装置进行充分室内摸底实验的基础上,在总体单位的领导和协调下,先后于 2006 年 8 月和 2006 年 12 月在 5 院卫星联调大厅进行了检测试验。

实验由西安光机所在卫星总体和有效载荷总体配合下在卫星测试大厅进行。卫星总体负责环境保障和整星控制,有效载荷总体负责图像快视和存储,西安光机所负责准备实验器具、配置实验装置、进行现场测试和数据结果分析处理。

实验时,卫星配置为有线数传方式,图像数据从高速复接器送到分配器,然后分别送到有效载荷地面综合电测设备和测试现场的快视设备;有效载荷数管配置为实时非压缩模式。

所得结果表明,干涉成像光谱仪性能参数与出所检测相比没有发生变化,从而保证了干涉成像光谱仪以确定的和良好的技术状态发射升空。

图 5 为光学成像成像探测系统的正样产品实物。其中仪器头部左边的是成像光谱仪。

(a) Head of the spectrometer　　(b) Electric control box

图 5 干涉成像光谱仪正样产品实物图

图 6 和图 7[7]为在轨获得的干涉图像和某一地元的干涉曲线。可以看出,干涉成像光谱仪经受住主动段力学环境和在轨空间环境的考验,展示良好的可靠性。

图 8[7]是根据 11 月 27 日 04:54:27 至 04:56:10 获取的月表干涉成像光谱仪数据(星下点经度为 11°W,纬度范围 50~55°N,区域宽约 25.6 km,长约 150 km)复原后的光谱图和合成假彩色图。图中包括第 4 谱段光谱图(500~509 nm)、第 17 谱段光谱图(638~651 nm)、第 30 谱段光谱图(878~904 nm)以及第 4、

17、30 谱段合成图。

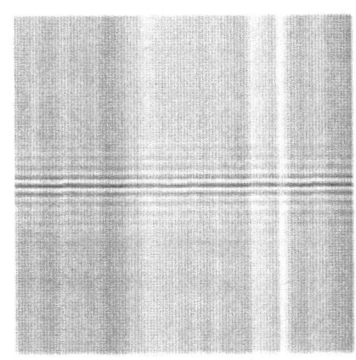

图 6　干涉图像

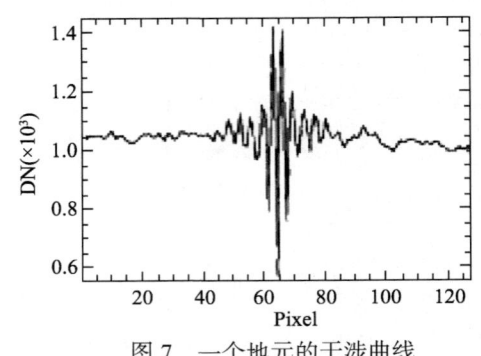

图 7　一个地元的干涉曲线

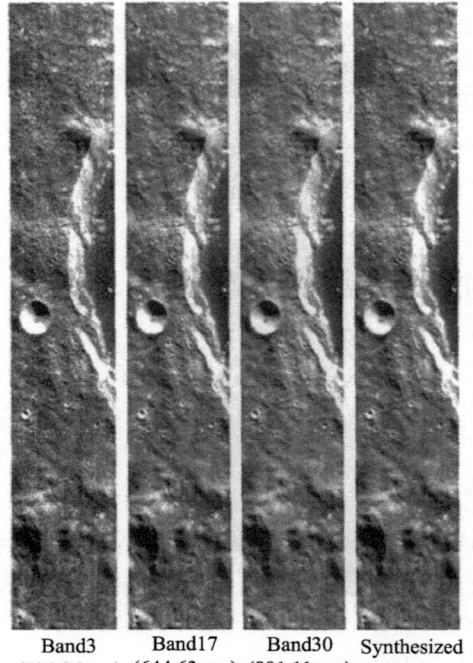

Band3　　Band17　　Band30　Synthesized
(504.96 nm)　(644.63 nm)　(891.11 nm)　image

图 8　月球光谱复原单色图[3]

4　结论

本文对嫦娥一号卫星有效载荷——干涉成像光谱仪现场实验方案进行了简要的介绍。干涉成像光谱仪于 2007 年 10 月 24 日以确定的和良好的技术状态发射升空，并于 2007 年 11 月 26 日首次在轨开机成功。在一年的在轨运行期间，干涉成像光谱仪工作稳定正常，至今已圆满完成预定的目标，这表明本文论述的现场实验方案设计合理。

参 考 文 献

[1] OU YANG Zi-yuan. Scientific Objectives of Chinese Lunar Exploration Project and Development Strategy [J]. *Advance in Earth Sciences*, 2004, 19 (3): 252-259
 欧阳自远. 我国月球探测的总体科学目标与发展战略[J]. 地球科学进展，2004，19 (3)：351—358
[2] OU YANG Zi-yuan, ZOU Yong-liao, LI Chun-lai, *et al*. Prospect of Exploration and Utilization of SomeLunar Resources [J]. *Earth Science-Journal of China University of Geosciences*, 2002, 27 (5): 498—503

欧阳自远，邹永廖，李春来，等. 月球某些资源的开发利用前景[J]. 地球科学——中国地质大学，2002，27 (5)：498—503

[3] Handbook of Optical Instruments [M]. Beijing: National Defense Industry Press, 1971: 538-546
光学仪器手册[M]. 北京：国防工业出版社，1971：538—546

[4] GJB2705—96, Generic Speifications for Space-borne CCD Camera [S]. GJB2705—96, 星载CCD相机通用规范[S]

[5] YANG Hua, ZHU Yong-hong, JIAO Wen-chun, LIU Ying. Output Response of CCD and MTF Test Of CCD Camera [J]. *Optical Technique*, 2001, 27 (5): 444—446
杨桦，朱永红，焦文春，等. CCD的输出响应与相机MTF测试[J]. 2001，27 (5)：444—446

[6] ZHAO Bao-chang, XUE Bin, YANG Jian-feng. Campare Method for Different Spectrometer, Invention Patent: China, ZL 200710018P74.2 [P]. 2009
赵葆常，薛彬，杨建峰，等. 不同类型光谱仪对比方法[P].中国，ZL 200710018P74.2.2009

[7] http://www.clep.org.cn

嫦娥一号卫星干涉成像光谱仪电子学设计

邱跃洪　汶德胜　赵葆常　陈　智　乔卫东

(中国科学院西安光学精密机械研究所，西安 710119)

摘　要　介绍了嫦娥一号卫星用于获取矿物光谱信息的干涉成像光谱仪电子学系统设计。概述了电子学系统的技术指标、系统组成、工作模式、外部接口和设计原则，详细描述了焦平面组件设计、视频处理和时序控制组件设计和 EMC 设计，给出部分正样产品实物照片以及地面实验和在轨飞行图像。各项地面试验和一年的在轨飞行结果表明该设计达到了预定的技术要求。

关键词　嫦娥一号卫星　干涉成像光谱仪　电子学设计　焦平面组件　视频处理和时序控制组件

0　引言

探月工程是继载人航天工程之后我国航天领域又一重大项目。我国《国家中长期科学和技术发展规划纲要(2006~2020 年)》和《中国的航天》白皮书也明确提出开展以月球探测为主的深空探测研究。今年我国首颗月球探测卫星——嫦娥一号卫星圆满完成绕月探测任务，为我国深空探测领域迈出了里程碑的第一步。嫦娥一号卫星载荷之一——干涉成像光谱仪(Interference Imaging Spectrometer，IIS)主要任务是获取月球矿物的光谱信息并进而反演出全球分布信息，从而与 X/γ 谱仪一起完成首次月球探测的四大科学和应用目标之一——分析月球表面有用元素及物质类型的含量与分布[1-2]。本文主要论述干涉成像光谱仪电子学系统的主要技术指标、结构组成及功能、各部分设计和一些实验结果。

1　电子学系统概述

1.1　主要技术指标

根据干涉成像光谱仪的总体技术指标要求，确定电子学系统的主要技术指标为：1)帧频：7.134fps；2)像元数：128×128(2×2 像元合并后)；3)量化等级：12 bits；4)增益设置：3 档(×1、×1.5、×2)；5)曝光控制：2 档(140 ms 和 70 ms)；6)信噪比：>200；7)工作模式：隔点采样和截短采样；8)环境温度：−10~45℃；9)峰值数据率<2 Mbps。

1.2　结构组成

干涉成像光谱仪电子学系统的主要任务是将 CCD 上每列(与飞行方向平行)所成的干涉图转换为数字信号，通过数据输出接口传送到大容量存储器和高速复接器。干涉成像光谱仪的工作参数和模式由下位机通过底板总线进行控制。

如图 1，依据功能电子学系统可以分为：焦平面组件和视频处理控制组件两部分。焦平面组件位于光谱仪主体焦平面盒中，主要功能是电平转换、电压调节、视频缓冲和放大。视频处理组件在光学成像探测系统电控箱中，主要实现时序控制和低噪声视频信号处理功能。下位机与二次电源模块为光谱仪和立体相机共用模块。

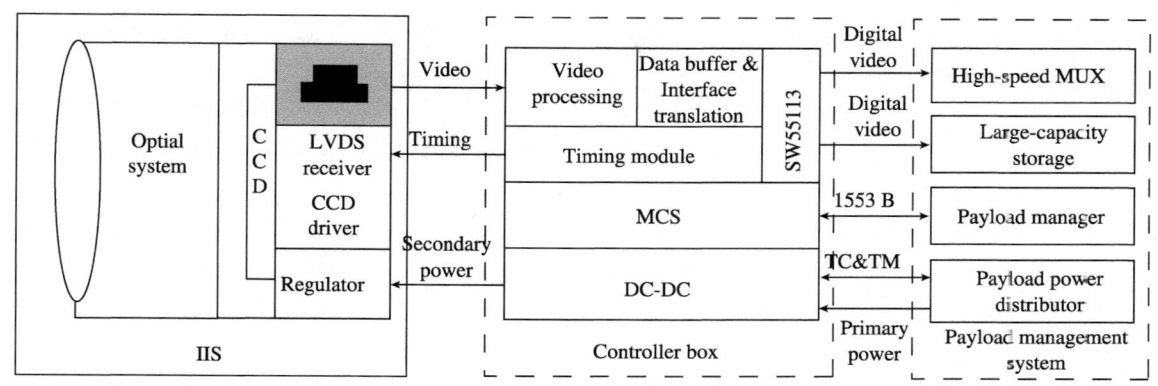

图 1 干涉成像光谱仪电子学系统结构

1.3 工作模式

为了提高动态范围，干涉成像光谱仪设置两档曝光控制 140 ms(缺省模式)和 70 ms，如果 140 ms 曝光时 CCD 饱和，则将帧频加倍，每两帧图像取一帧，从而将曝光时间调整为 70 ms 而输出数据率不变。为了提高信噪比，在行、列方向进行 2×2 像元合并，得到 256×256 图像，其中，在空间方向上只使用中间的 128 像元，在干涉图方向上使用 256 像元。为了减小输出数据率，干涉图采样采取两种工作模式：一是截短采样模式(缺省模式)，即干涉图只取中间一半，这时对应于 32 个光谱通道；一是隔点采样模式，即取一点，丢一点，它对应的光谱通道数为 64 个。

下位机根据所接收的数据注入指令通过底板总线设置曝光时间和工作模式，由视频处理组件内部的时序模块实现。

1.4 外部接口

干涉成像光谱仪输出的串行数字图像数据同时送往高速复接器和大容量存储器。向高速多路复接器输出间断串行时钟(SCL K1)和串行数据(DATA)；向大容量存储器输出连续串行时钟(SCL K2)、串行数据、并行时钟(PCL K)、时钟选通(VAL ID)和帧同步(VSYN)。图像帧头信息由下位机通过地板总线在帧逆程期间注入。

1.5 设计原则

根据航天工程各项规范要求，在设计过程中，始终遵循设计原则：

(1) 模块化：进行模块化设计、方便技术继承和调试；

(2) 轻量化：充分考虑星上资源如功耗、体积、重量等约束条件,采用轻量化材料、轻量化设计及 FPGA 等技术；

(3) 高可靠性：空间环境下具有高的可靠性是航天设备基本要求，从元器件、原材料选择和设计、工艺等方面严格把关，在电路设计上考虑冗余、备份等技术，提高仪器的可靠性；

(4) 工程实施可行性：在设计时既考虑原理的先进性，更考虑工程实施的可行性及安全可靠性，设计留有足够的工程裕量。

2 各部分设计

2.1 焦平面组件设计

焦平面组件由 CCD、电源调整电路、LVDS 接收电路、CCD 驱动电路、CCD 视频缓冲电路和放大电路等部分组成[3-4]，见图 2。

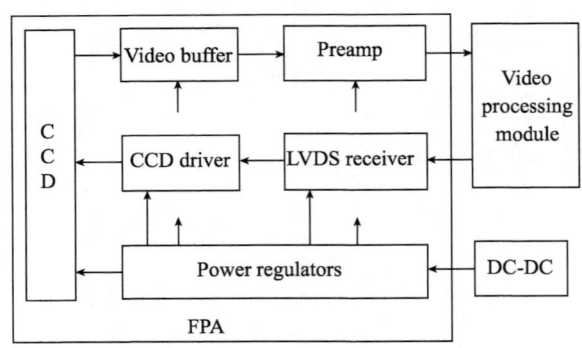

图 2 焦平面组件电路结构

焦平面组件接收视频处理组件通过屏蔽双绞线送来的 RS644 电平差分 CCD 驱动信号，经电平转换后驱动 CCD，CCD 输出的视频信号经缓冲和放大后通过 50Ω 同轴电缆送到视频处理组件。此外，电源调整电路将来自电控箱的二次电源模块输出调整后，产生 CCD 和其他各部分电路所需的低噪声电源。

CCD 为仪器的核心器件，主要技术指标见表 1。

表 1 CCD 主要技术指标

结构	512×512 帧转移
像元尺寸	7 μm×17 μm
填充因子	100%
光谱范围	400~1100 nm
暗电流	5pA/cm^2(@25°C)
暗信号	<1 mV
暗信号非均匀性	0.9 mV
光电响应非均匀性	1%
转换因子	4 μV/e$^-$
动态范围	12 bit
输出放大器功耗	70 mW
帧转移频率	<600 KHz
像元读出频率	<7 MHz
饱和输出电压	1 V
输出放大器热噪声	180 μV
电荷转移效率(每级)	0.99999

图 3 为 CCD 的结构。由三个区组成：感光区、存储区和读出区。感光区和存储区各有 526 行，包括 10 个光屏蔽行、512 个有用行和 4 个隔离行。水平读出移位寄存器由 548 级组成：16 个前扫像元、20 个隔离像元和 512 个有用像元。

图 4 为 CCD 的引脚图。CCD 驱动脉冲有：感光区转移脉冲 $\Phi P_i(i = 1, 2, 3, 4)$、存储区转移脉冲 $\Phi M_i(i=1, 2, 3, 4)$、存储区到水平读出移位寄存器转移脉冲 ΦM、水平读出移位寄存器转移脉冲 $\Phi L_i(i=1, 2)$ 和读出放大器复位脉冲 ΦR。

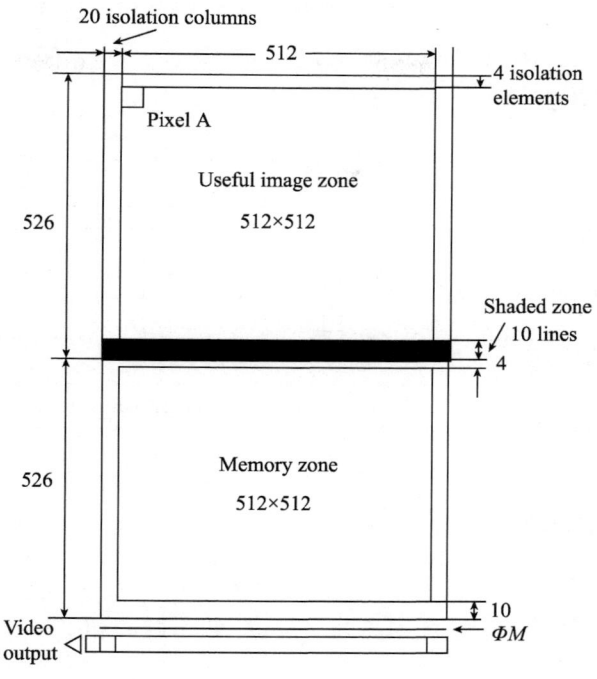

图 3 感光区和存储区配置

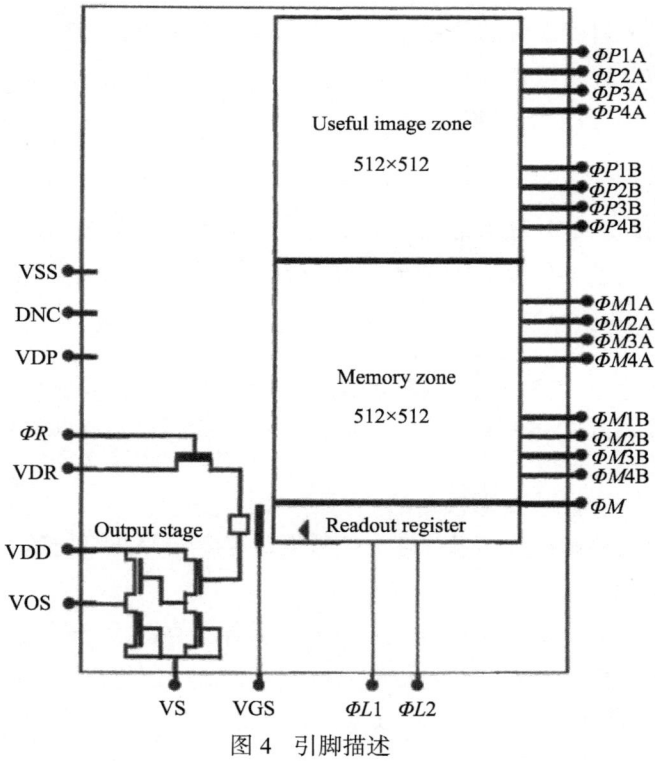

图 4 引脚描述

2.2 视频处理和时序控制组件设计

视频处理和时序控制组件(图 5)主要功能是：1)视频信号处理：对来自焦平面组件的视频信号进行后放、相关双采样(Correlated Double Sampling，CDS)、可编程增益放大(Programmable Gain Amplification，PGA)、低通滤波、A/D 转换、数据缓存和格式化，然后通过高速输出接口分别是送到高速复接器和大容量存储器；

2)时序控制：接收下位机通过底板总线写入输出数据帧帧头信息和设定视频处理组件曝光时间、工作模式和增益设置等工作参数，并进而产生各种控制时序(CCD驱动、CDS、ADC、缓存、并传串转和输出接口等部分的时序)和设置PGA增益。

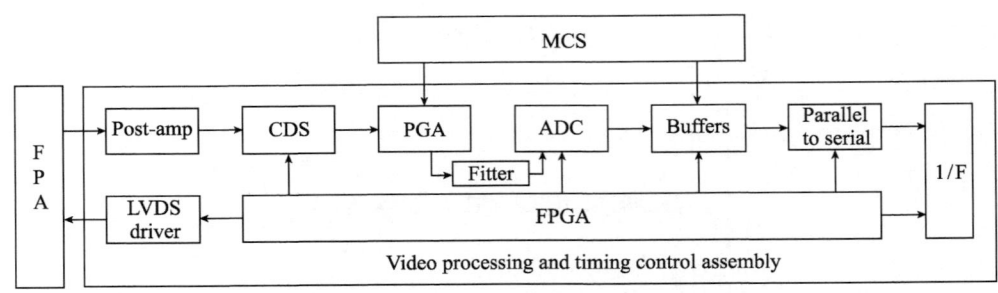

图 5 视频处理组件电路结构

2.2.1 相关双采样(CDS)

为了减小 CCD 输出信号的复位噪声，需要对 CCD 输出的信号进行相关双采样。CDS 采用 Clamp-Sample 结构形式，由低噪声运放和模拟开关构建，箝位脉冲和采样脉冲由时序电路产生，整个 CDS 的增益为 1。

2.2.2 可编程增益放大(PGA)

为了提高低照度情况下目标图像的信噪比，信号链路上设置 PGA 模块，采用模块开关切换运放反馈通道电阻的结构形式。增益分为三档：2、3、4，分别对应于增益设置的 ×1、×1.5、×2。增益设置由下位机通过底板总线的两个信号线来实现。

2.2.3 有源滤波

为了限制整个模拟前端的等效噪声带宽，加入二阶有源低通滤波器。滤波器由一片运放组成，采用阶跃响应无超调的切比雪夫滤波器结构，滤波器的截止频率为 2 MHz，直流增益为–1。

2.2.4 A/D 转换

A/D 转换电路由 12 bit、1.25 MHz 的 A/D 转换器 AD1671 和精密基准源 AD780 构成，140 ms 曝光时间设置时，A/D 转换速率为 0.5265 MHz，70 ms 曝光时间设置时，A/D 转换速率为 1.125 MHz。

2.2.5 数据缓存

数据缓存包括帧数据缓存和帧头信息缓存。帧数据缓存存储 A/D 转换后的数字视频数据，帧头信息缓存存储由下位机写入的帧头信息。

ADC 每帧输出数据为 256×256×12 bits。在写入帧数据缓存时，256 行数据只写其中的 128 行，每行(空间方向)只写入中间的 128 个像元数据，行的选取方式(干涉图方向)有两种：隔点采样和截短采样。这样，每帧存入帧数据缓存的数据为 128×128×12 bits，因此选用两片 16 K×9 bit 的 FIFO ID T7206 做帧数据缓存。帧数据缓存的平均写入速率为 1.4 Mbps，输出接口的读出速率为 1.8 Mbps，输出帧逆程时间为 30.88 ms。当帧频变为 14.258 fps 时，每两帧数据只存一帧，每帧的存入方式与帧频为 7.134 fps 时相同，帧数据缓存的平均写入速率和读出速率不变。为了防止 FIFO 的读写时序受干扰而紊乱，在每帧逆程期间对其复位。

下位机收到时序电路产生的帧同步信号后，在帧逆程期间将帧头信息打入帧头缓存，在每帧读帧数据缓存之前读出。帧头缓存由两片 4 K×9 bit 的 FIFO ID T7203 组成。

2.2.6 串并转换和输出接口

数据缓存输出的并行数据经首先由 54HCT574 锁存，然后经 54HCT165 进行并/串转换，最后将 TTL 电平信号由 SNJ55113 转变为差分信号分别送到高速复接器和大容量存储器。

2.2.7 时序电路

时序电路用来产生 CCD 驱动时序、视频处理时序和输出接口时序，由一片 ACTEL 公司的 A1020B 实现，时钟频率为 18MHz。

2.3 EMC 设计

按照《CE-1 卫星设计和建造规范》和《CE-1 卫星电磁兼容性要求》进行干涉成像光谱仪电磁兼容性设计，主要包括接地、搭接、屏蔽、滤波、接口、电缆和布局设计等，主要采取电磁兼容设计方法。

2.3.1 屏蔽

(1) 光谱仪主体与电控箱之间的驱动信号连接电缆均采用屏蔽双绞线，视频信号电缆采用屏蔽同轴电缆；

(2) 采用多层板 PCB 版工艺；

(3) 电路板在机箱中安装时，要考虑相互间的影响。尽量按数字、模拟、电源分类安装，必要时加装屏蔽隔离板；所有线路板的加固金属框架要与线路板的地线隔离；

(4) 线路板与机箱插座的连接线要分类布置，尽量使用信号/回线双绞连接；

(5) 机箱上的所有缝隙和通孔都会导致对外电磁辐射和外界的电磁进入机箱，机箱结构设计时会充分考虑这些因素，保证机箱结构设计的电连续性；

(6) 二次电源独立屏蔽，减少 EMI 干扰和电磁泄露。

2.3.2 滤波

(1) 为减小对公用电源的干扰，在一次电源输入端接 EMI 滤波器，二次电源输出端采取滤波措施；

(2) 电路设计中核心芯片电源输入端都采用两并两串 RC 滤波电路，部件减少 EMI，还预防 CMOS 电路的闭锁；

(3) 数字信号布线区域中，在 PCB 板电源入口端和最远端各放置一组滤波电路，以防电源尖峰脉冲引发的噪声干扰；

(4) 电源及临界信号走线使用宽线，在 IC 的电源/地间配置去耦电容；清除地线环路，以防电流回馈影响电源；所有地线走线尽量宽，所有 IC 电源/地间的电容走线尽量短。

2.3.3 接地

(1) 每个功能模块电源应分离，每个功能模块的电源/地只能在电源/地的源点相连，地线通过一点相连；

(2) 继电器电路在供电段处单点接地、低频信号屏蔽层单点接地，视频信号单点接地；

(3) 机械结构件电性连接，直接接触的表面部位不发黑、紧密接触，用螺纹紧固件连接。

2.3.4 电路设计、布局和布线

(1) 正确布置元件位置，使敏感元件远离干扰源；正确布置元件方向，使元件之间的互电容、互电感最小；印刷版布局合理，减小电磁干扰；对电源转换模块进行屏蔽，减少对电路的干扰；

(2) 模拟电路和集成电路电源入口加去耦电容，CMOS 器件的电源入口加限制电阻；

(3) 所有地线走线尽量宽，所有 IC 电源/地间的电容走线尽量短；每个功能模块电源应分离，每个功能模块的电源/地只能在电源/地的源点相连，地线通过一点相连；

(4) 晶振电路尽量靠近其驱动器件，所有信号走线远离晶振电路；所有连到晶振输入及输出端的走线尽量短，以减少噪声干扰及分布电容对晶振的影响；晶振电容地线使用尽量宽。

(5) 尽量采用 SMT 器件，减少引线电磁辐射和电磁干扰；

(6) 所有集成电路不留空脚。

2.3.5 电缆

(1) 精心设计电缆组合和走线，一次电源、二次电源、模拟信号、数字信号、加热信号等电缆、电连接器尽量分开，内部布线相互绝缘，避免交叉干扰；

(2) 各种电缆中信号线与其回线一般使用双绞线和屏蔽，并与连接器触点一一对应；

(3) 导线、电缆、电连接器按规范选用和布局，关键信号线采用双点双线、多点多线备份。

2.3.6 隔离与保护

(1) 电源的输入输出隔离，信号地、电源地与机壳地绝缘，绝缘电阻大于 50MΩ；

(2) 模拟电路、数字电路等用二次电源供电，电源模块各自分开；

(3) 一次电源地与二次电源地隔离，一次电源输入端有短路保护；

(4) 单机输入电路、输出电路、遥测接口电路等按要求设计，符合回流原则，具有过压和短路保护。

3 实物照片和实验图像

本设计经过电性星产品、初样鉴定级产品和正样产品的验证，性能满足技术指标要求，工作稳定可靠，取得满意的结果。

3.1 实物照片

图 6 为光谱仪 CCD 驱动板实物，图 7 为光谱仪视频处理和时发控制组件实物。图 8 为光学成像成像探测系统的正样产品实物，其中仪器头部左边的是成像光谱仪，仪器电控箱包括立体相机视频处理组件、干涉成像光谱仪视频处理组件和两台仪器共用的下位机和二次电源模块。

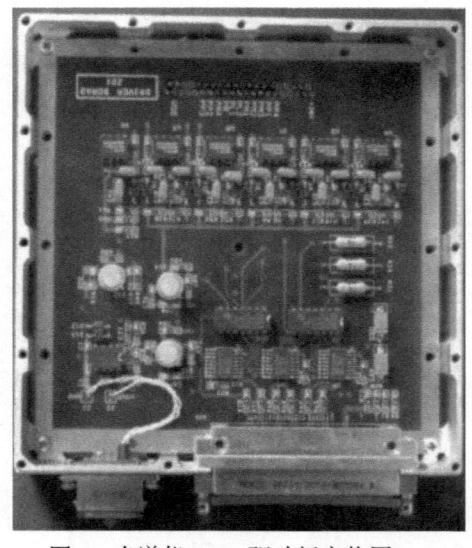

图 6 光谱仪 CCD 驱动板实物图

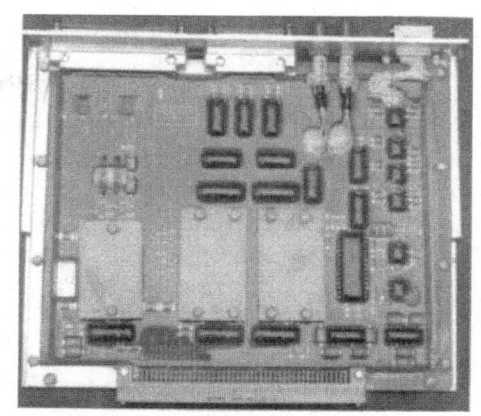

图 7 光谱仪视频处理和时序控制板实物图

(a) Head of the spectrometer

(b) Electric control box

图 8 干涉成像光谱仪正样产品

3.2 图像

图 9 为干涉成像光谱仪初样鉴定级产品地面推扫实验获得的干涉条纹图和经光谱复原后获得的多光谱序列图。

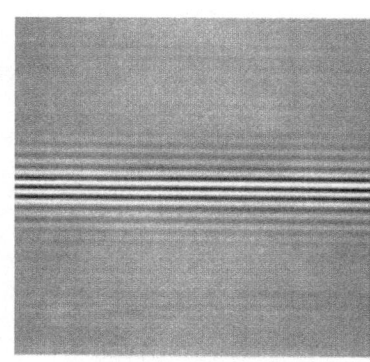
(a) Interference pattern

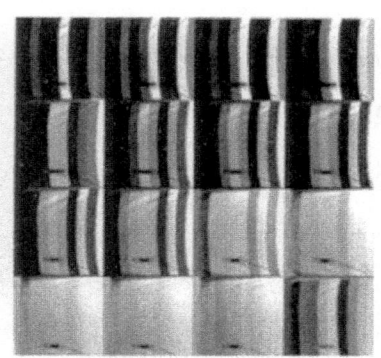
(b) Spectral image of 16 bands

图 9 干涉成像光谱仪地面推扫实验图像

4 结论

本文对嫦娥一号卫星有效载荷——干涉成像光谱仪电子学系统设计作了简要介绍。经过仔细的优化设计、深入的理论分析、充分的计算机仿真验证和严格的环境实验考核，干涉成像光谱仪以良好的技术状态发射升空并于 2007 年 11 月 26 日首次在轨开机成功。在一年的在轨运行期间，干涉成像光谱仪工作稳定正常，至今已圆满完成预定的目标，这表明本文论述的电子学系统设计合理可靠。

参 考 文 献

[1] OU YANG Zi-yuan. Scientific objectives of Chinese lunar exploration project and development strategy [J]. *Advance in Earth Sciences*, 2004, 19 (3): 252—259
欧阳自远. 我国月球探测的总体科学目标与发展战略[J]. 地球科学进展，2004，19 (3)：351—358

[2] OU YANG Zi-yuan, ZOU Yong-liao, LI Chun-lai, et al. Prospect of exploration and utilization of some lunar resources [J]. *Earth Science-Journal of China University of Geosciences*, 2002, 27 (5): 498—503
欧阳自远，邹永廖，李春来，等. 月球某些资源的开发利用前景[J]. 地球科学——中国地质大学，2002，27 (5): 498—503

[3] WANG Qing-you. Image sensor application technology [M]. Beijing: Publishing House of Electronics Industry, 2003.王庆有. 图像传感器应用技术[M]. 北京：电子工业出版社，2003

[4] JANESICK J R. Scientific charge-couple devices [M]. SPIE Press, 2001

[5] http://www.clep.org.cn

嫦娥一号卫星干涉成像光谱仪时序设计

邱跃洪

(中国科学院西安光学精密机械研究所，西安 710119)

摘 要 介绍了基于 Actel 反熔丝 FPGA 的嫦娥一号卫星干涉成像光谱仪时序控制模块设计。概述了时序控制模块的功能要求，详细描述了时序控制模块外部时序接口和内部时序接口，给出了时序控制模块设计的参量配置、逻辑结构和简要结果。各项地面试验和一年的在轨飞行结果表明该设计合理可靠。

关键词 嫦娥一号卫星　干涉成像光谱仪　FPGA　VHDL　时序

0 引言

探月工程是继载人航天工程之后我国航天领域又一重大项目。2008 年我国首颗月球探测卫星—嫦娥一号卫星圆满完成了绕月探测任务，为我国深空探测领域迈出了里程碑的第一步。干涉成像光谱仪(Interference Imaging Spectrometer，IIS)是嫦娥一号卫星的有效载荷之一，用于获取月球矿物的光谱信息进而反演出全球分布信息[1-2]。时序控制模块是整个 IIS 电子学系统的核心。本文主要论述时序控制模块的技术要求、接口关系、设计实现和实验结果。

1 概述

时序控制模块主要功能是根据下位机(MCS)传送的 IIS 工作模式工作和参量数设置指令，产生相应的 CCD 驱动时序、视频处理时序和输出接口时序，具体要求为：

①实现 CCD 驱动时序，具有两档曝光控制(140 ms 和 70 ms)和 2 pixel×2 pixel 合并功能[3-4]；②实现视频处理时序，具有隔点采样和截短采样两种工作模式；③根据接口协议实现与大容量存储器和高速复接器的串行接口时序；④具有足够的时序裕量和可靠性。

2 接口关系描述

时序控制模块与其他部分接口关系如图 1，分为外部时序接口(大容量存储器接口时序和高速复接器接口时序)和内部时序接口(CCD 驱动时序、视频处理时序、缓存控制时序)。

2.1 外部时序接口

与时序控制模块有关的外部时序接口有高速复接器时序接口和大容量存储器时序接口(图 2，图 3)。向高速多路复接器(High-speed MUX)输出间断串行时钟(SCLK1)和串行数据(SDATA)；向大容量存储器(Large-capacity Storage)输出连续串行时钟(SCLK2)、并行时钟(PCLK)、时钟选通(VALID)、帧同步(VSYN)

和串行数据(SDATA)。

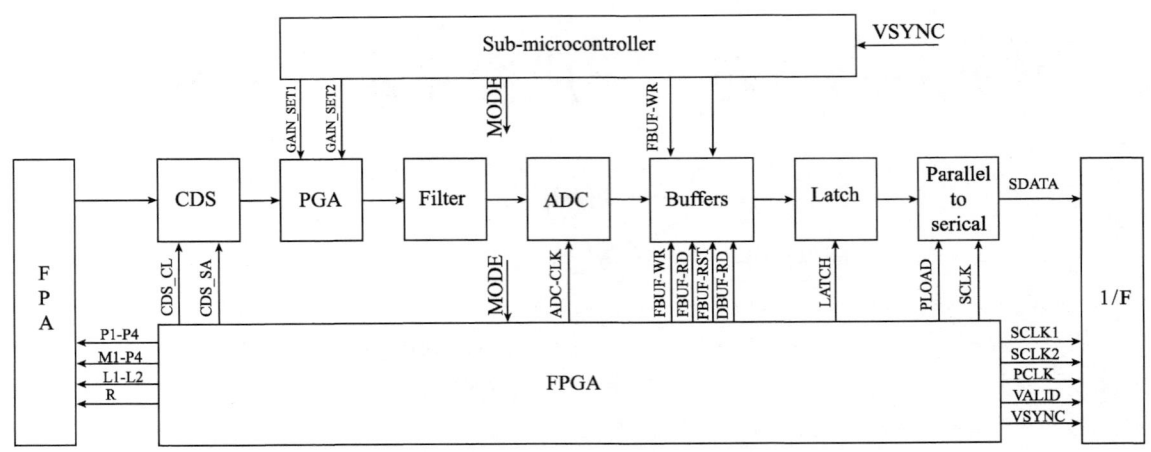

图 1　时序控制模块接口关系

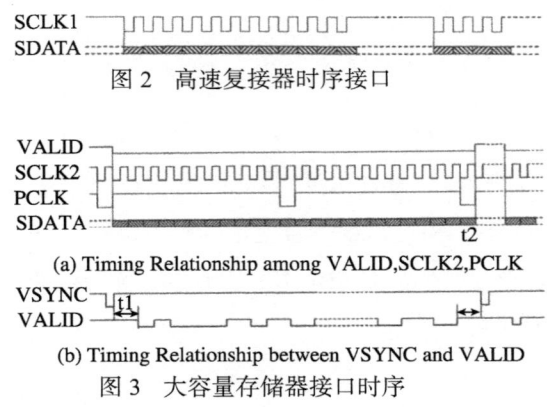

(a) Timing Relationship among VALID,SCLK2,PCLK

(b) Timing Relationship between VSYNC and VALID

图 3　大容量存储器接口时序

　　间断串行时钟是发送给高速多路复接器的串行数据的驱动时钟，瞬时频率小于 2 MHz，在 1.7 ms 内传输的数据位不超过 4 096 bit。时钟下降沿与数据跳变沿对齐(高速多路复接器以时钟上跳沿作为数据的采样时刻)。时钟信号在没有有效数据的期间保持高电平，而数据信号在无效期间不做特别要求。

　　发送给大容量存储器的连续串行时钟是串行数据的驱动时钟，瞬时频率小于 2 MHz，时钟下降沿与数据跳变沿对齐。由于并不是在所有时刻都能输出有效数据，所以需要一个选通信号来判断当前数据的有效性。选通信号约定为低有效，有效宽度完整包括每个 12 位有效数据字所对应的 12 个串行时钟周期，选通的跳变沿与串行时钟的下降沿对齐。并行时钟是串行时钟的 12 分频率，宽度为 1 个串行时钟周期的负脉冲信号，其脉冲位于每第 12 个串行时钟周期处，跳变沿与串行时钟的下降沿对齐。

2.2　内部时序接口

　　内部时序主要包括 CCD 驱动时序、视频处理时序、缓存控制时序和下位机接口时序。

　　CCD 为本仪器的核心器件(512×512 帧转移结构)。感光区和存储区各有 526 行，包括 10 个光屏蔽行、512 个有用行和 4 个隔离行。水平读出移位寄存器由 548 级组成：16 个前扫像元、20 个隔离像元和 512 个有用像元。CCD 驱动脉冲有：感光区转移脉冲 P1、P2、P3、P4，存储区转移脉冲 M1、M2、M3 和 M4、存储区到水平读出移位寄存器转移脉冲 M、水平读出移位寄存器转移脉冲 L1 和 L2、读出放大器复位脉冲 R。图 4~7 为 CCD 各驱动脉冲的时序关系。

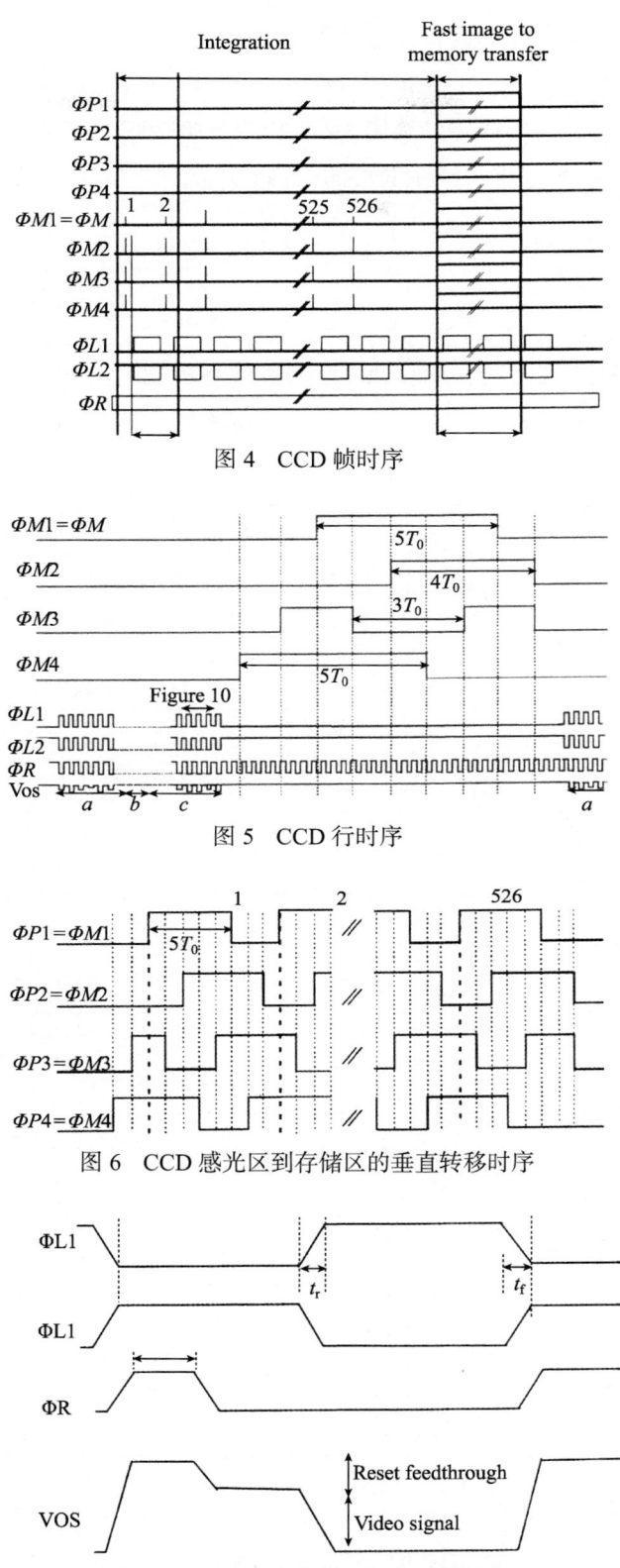

图 4 CCD 帧时序

图 5 CCD 行时序

图 6 CCD 感光区到存储区的垂直转移时序

图 7 CCD 读出寄存器和复位时钟时序

视频处理时序主要包括 CDS 的箝位脉冲 CDS_SA 和采样脉冲 CDS_CL、ADC 的控制时钟 ADC_CLK。缓存控制时序主要由帧数据缓存写脉冲 FBUF_WR、帧数据缓存读脉冲 FBUF_RD、帧数据缓存复位脉冲 FBUF_RST、帧头数据缓存读脉冲 DBUF_RD 和串行移位寄存器并行加载脉冲 PLOAD。缓存控制时序根据工作模式(截短采样和间隔采样)设置情况而不同。

与下位机的时序接口主要关系是：1)在帧逆程期间发送帧同步信号 VSYNC，下位机将帧头信息写入帧头缓存，时序模块在随后一帧开始时将其读出；2)下位机根据接收的卫星注入数据来设置模式控制信号 Mode[0：1]，时序模块根据 Mode 设置值产生对应的控制时序从而设置干涉成像光谱仪的曝光时间(140 ms 和 70 ms)和工作模式(截短采样和间隔采样)。

3 设计与实现

时序电路用来产生 CCD 驱动时序、视频处理时序、缓存控制时序和输出接口时序，由一片 ACTEL 公司的 A1020B[5]实现，时钟频率为 18MHz。表 1 为整个系统的时序参量表。

表 1 时序参量表

序号	名称	描述	频率/MHz	去向
1	P1	CCD 感光区转移脉冲 P1	0.28125/0.5625	焦平面组件
2	P2	CCD 感光区转移脉冲 P2	0.28125/0.5625	焦平面组件
3	P3	CCD 感光区转移脉冲 P3	0.28125/0.5625	焦平面组件
4	P4	CCD 感光区转移脉冲 P4	0.28125/0.5625	焦平面组件
5	M1	CCD 存储区转移脉冲 M1	0.28125/0.5625	焦平面组件
6	M2	CCD 存储区转移脉冲 M2	0.28125/0.5625	焦平面组件
7	M3	CCD 存储区转移脉冲 M3	0.28125/0.5625	焦平面组件
8	M4	CCD 存储区转移脉冲 M4	0.28125/0.5625	焦平面组件
9	M	CCD 存储区到水平读出移位寄存器转移脉冲 M	0.28125/0.5625	焦平面组件
10	L1	CCD 水平读出移位寄存器转移脉冲 L1	1.125/2.25	焦平面组件
11	L2	CCD 水平读出移位寄存器转移脉冲 L2	1.125/2.25	焦平面组件
12	R	读出放大器复位脉冲 R	0.5625/1.125	焦平面组件
13	CDS_SA	CDS 采样脉冲	0.5625/1.125	CDS
14	CDS_CL	CDS 箝位脉冲	0.5625/1.125	CDS
15	ADC_CLK	ADC 转换时钟	0.5625/1.125	ADC
16	FBUF_WR	帧数据缓存写脉冲	0.5625/1.125	帧数据缓存
17	FBUF_RD	帧数据缓存读脉冲	0.15	帧数据缓存
18	FBUF_RST	帧数据缓存复位	7.134Hz	帧数据缓存
19	DBUF_RD	帧头数据缓存读脉冲	0.15	帧头缓存
20	LATCH	锁存器锁存信号	0.15	锁存器
21	PLOAD	并串转换器并行加载信号	0.15	并串转换器
22	SCLK	并串转换器串行移位时钟	1.8	并串转换器
23	SCLK1	间断串行时钟	1.8	输出接口
24	SCLK2	连续串行时钟	1.8	输出接口
25	PCLK	并行时钟	0.15	输出接口
26	VALID	时钟选通	7.134Hz	输出接口
27	VSYN	帧同步	7.134Hz	输出接口和下位机

干涉成像光谱仪具有两档曝光控制：140 ms(缺省模式)和 70 ms，140 ms 曝光设置时帧频为 7.134 fps，70 ms 曝光设置时将帧频加倍，每两帧图像取一帧，从而将曝光时间调整为 70 ms 而输出数据率不变。当曝光时间为 140 ms 时，帧转移时钟频率 P1~P4 和 M1~M4 为 0.28125 MHz，水平读出时钟 L1、L2 频率为 1.125 MHz，读出放大器复位时钟 R、CDS 时钟 CDS-CL 和 CDS-SA、ADC 时钟 ADC-CLK 频率为 0.5625 MHz。当曝光时间为 70 ms 时，帧转移时钟频率 P1-P4 和 M1-M4 为 0.5652 MHz，水平读出时钟 L1、L2 频率为 2.25 MHz，读出放大器复位时钟 R、CDS 时钟 CDS-CL 和 CDS-SA、ADC 时钟 ADC-CLK 频率为

1.125 MHz。

图8为时序模块的逻辑结构。整个结构围绕计数器组实现,其长度由帧周期长度决定。计数器组受下位机送来的IIS工作模式设置信号MODE[1：0]控制。计数器组输出送到CCD垂直时钟发生器、CCD读出和视频处理时钟发生器、缓存时钟发生器和接口时钟发生器,由它们产生相应的输出时序信号。

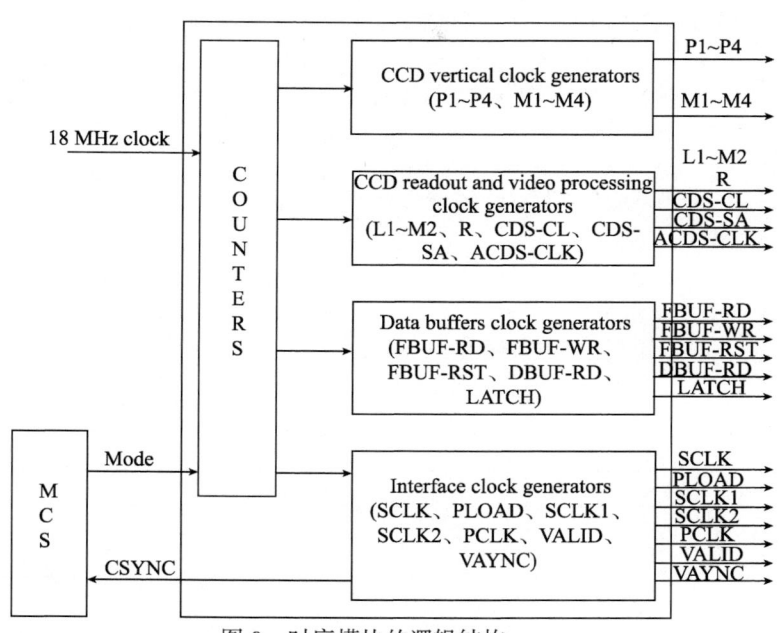

图8 时序模块的逻辑结构

本设计由VHDL语言实现并在Libero集成开发环境下进行编写、综合、前仿真、布局布线、后仿真和烧写文件生成[6]。由于A1020B是一次烧写型的反熔丝结构,充分的仿真优化既可以检查设计的正确性和可靠性,又可以大大减少写片次数。值得指出的是,Libero的仿真结果比较精确,因而,本设计在充分仿真后,只需几次烧写后就彻底定型。

4 结论

本文对嫦娥一号卫星有效载荷——干涉成像光谱仪时序控制模块设计作了简要介绍。经过仔细的需求分析和优化设计、充分的计算机仿真验证和全面的系统调试,该设计逐步完善。从鉴定级产品到正样产品直至发射,经历各种严格的环境实验考核,未作任何修改,也没发现任何问题,从而保证干涉成像光谱仪以良好的技术状态于2007年10月24日发射升空并于2007年11月26日首次在轨开机成功。在随后的一年在轨运行期间,干涉成像光谱仪工作稳定正常,至今已圆满完成预定的目标,这表明本文论述的时序设计合理可靠。

参 考 文 献

[1] OU-YANG Zi-yuan. Scientific objectives of chinese lunar exploration project and development strategy [J]. *Advance in Earth Sciences*, 2004, **19** (3): 252—259
欧阳自远. 我国月球探测的总体科学目标与发展战略[J]. 地球科学进展,2004,**19** (3):351—358

[2] OU-YANG Zi-yuan, ZOU Yong-liao, LI Chun lai, et al. Prospect of Exploration and Utilization of Some Lunar Resources [J]. *Earth Science-Journal of China University of Geosciences*, 2002, **27** (5): 498—503
欧阳自远,邹永廖,李春来,等. 月球某些资源的开发利用前景[J]. 地球科学——中国地质大学,2002,**27** (5):498—

[3] WANG Qing-you. Image sensor application technology [M]. Beijing: Publishing House of Electronics Industry, 2003
王庆有. 图像传感器应用技术[M]. 北京：电子工业出版社，2003
[4] JANESICK J R. Scientific charge-couple devices [M]. SPIE Press, 2001
[5] Actel Inc. A1020B Datasheet [Z], http://www.actel.com
[6] HOU Bo-heng, GU Xin. Hardware discription language VHDL and digital logical circuit design [M]. Xi'an: Xidian University Press, 1999
侯伯亨，顾新. 硬件描述语言VHDL与数字逻辑电路设计[M]. 西安：西安电子科学技术大学出版社，1999

嫦娥一号 IIM 数据应用处理流程分析

吴昀昭[1,2]　唐泽圣[1]

(1. 澳门科技大学月球与行星探测科学应用研究联合实验室，澳门；2. 南京大学地理与海洋科学学院，南京 210093)

摘　要　对嫦娥一号干涉成像光谱仪(IIM)数据的特点进行了分析，并就存在的一些问题提出了解决方案，制定了 IIM 数据应用处理流程，为该数据的正确使用提供方法参考。研究结果表明，在空间域传感器左侧响应偏低，右侧响应偏高；在波谱域长波段响应存在较大偏差。经过绝对定标和辐射畸变校正后的反射率与地基望远镜光谱匹配良好，可以用于应用研究。利用校正后的数据对 Aristarchus 地区岩石类型开展初步研究的结果表明，该地区在纵向和横向上都存在岩性的多样性。校正后的图像不仅提高了分类精度，还被识别出撞击坑可能存在的滑坡。嫦娥一号 IIM 能够在全球、区域和局部尺度上以较高的空间分辨率和光谱分辨率获取月表元素和矿物成分信息，有助于深化对月球形成和演化的认识。

关键词　嫦娥一号　IIM　数据处理流程　月球

0　引言

21世纪，人类掀起了新一轮的月球探测热潮，美、欧、俄、日、印及巴西等国都在组织实施月球探测计划。与20世纪70年代冷战背景下的探测推动力不同，此次探月热潮的总目标主要是科学研究、开发和利用月球的资源和能源，为人类社会的可持续发展服务。2007年10月24日，我国首颗探月卫星"嫦娥一号(CE-1)"成功发射，至2009年3月1日受控撞击丰富海，圆满完成了绕月探测使命，收集了海量科学数据。因此，极有必要尽快对这些数据进行分析和解译，促进我国月球科学研究。

CE-1携带的干涉成像光谱仪(IIM)首次将干涉光谱成像技术用于深空探测，它具有能量利用率高、采样超连续、数据量低、空间稳定性高及光谱不受卫星姿态影响等优势[1]。IIM对月表进行成像的同时获得每个像元的光谱信息，具有图谱合一的特点，是完成我国首次探月4大科学任务的必备仪器[1]。目前绕月探测工程地面应用系统提供的最高级别数据是2C级，该数据经过了暗电流纠正、平场纠正、辐射亮度转换及光学归一化等预处理。然而，由于绝对定标和传感器辐射畸变等问题制约了数据的直接使用，目前尚没有文献介绍对2C数据的进一步处理分析方法。

针对IIM数据的特点，本文从实际应用的角度介绍了2C级IIM数据绝对定标和辐射畸变校正方法，并制定了相应的数据处理流程，生成可以直接使用的反射率产品，为今后研究者使用该数据提供方法参考。

1　IIM 反射率产品处理流程

CE-1卫星与地球的距离远大于地球卫星与地球的距离，遥控、数传难度远超出了地球卫星的难度。加之CE-1携带的科学载荷较多，卫星设计要求仪器轻便，这些因素都影响了数据质量。本文列出了使用数据时需要注意的一些问题，并基于这些问题建立了数据应用处理流程。

本文原载于《国土资源遥感》，2009，No.4，25~30。

首先，数据的刈幅较窄(仅25.6 km)，不便于分析和寻找目标物。数据是PDS格式，经纬度信息存放于头文件，可将头文件中的经纬度信息读出并赋予图像，即几何粗校正。在此基础上，可参考具有精确几何信息的数据(如绕月探测工程地面应用系统发布的CCD 3A级产品或克莱门汀全月基本图像)和地形数据进行几何精校正，并对研究区进行图像镶嵌。

其次，2C级数据是辐射亮度，没有转化为反射率，无法直接使用该数据进行地质解译。Apollo 16着陆点附近地区由于地物成分比较均一，而且月壤非常成熟，常被用来作为卫星遥感或地基光谱观测的定标点。国际上常选择North Ray和South Ray撞击坑组成的等边三角形第3个顶点作为定标点[2,3]。IM没有覆盖该地区。本文经过多次试验以及对比高分辨率克莱门汀图像，在2 225轨选取第11 151~11 167行、69~73列的5像元×17像元作为定标区域[4]。

第三，图像的行向响应不一致。图1(a)和(c)分别是玄武质澄海地区(MS2)和辉长岩质高地撞击坑(Tycho)IIM真彩色合成图像。由于2个地区月表成分、遭受撞击程度和地理位置明显不同，它们从左至右图像色调都不均一。作为对比，图1(b)和(d)是相应地区的克莱门汀图像，在克莱门汀图像上这2个地区没有行向上的色调差异，这一方面说明色调的变化不是月表成分引起的，另一方面也说明这种行向响应的不均一与地区无关，这为对其校正提供了基础(图版III)。

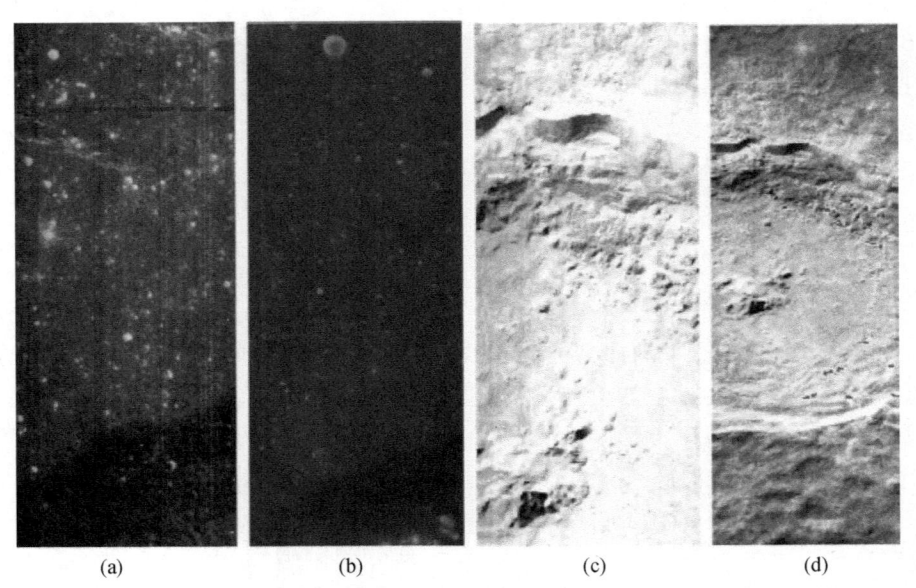

图 1 MS2(a)和Tycho撞击坑(c)IIM真彩色合成图像及相应的克莱门汀图像(b)和(d)

图2是澄海地区(MS2)光谱曲线，图2(左)为MS2左侧和中间(图1(a)上部红色方框)反射率和地基望远镜(简称"地基")光谱；图2(右)为918 nm、891 nm和865 nm 3个波段在一行(图1(a)上部红色虚线)的反射率。如图1(b)所示，该地区地物成分和结构较为均一。由图2(左)可见，尽管行向上物质成分近乎一致(均为月海玄武岩)，然而光谱却不同，左侧像元的反射率低于右侧像元的反射率，而在长波段(波长>800 nm)这种差异尤其明显，而且都偏离地基光谱。这种辐射畸变导致无法直接利用该数据进行基于地物光谱特征的定性矿物识别或定量矿物反演。由图2(右)可见，3个波段从左至右反射率都逐渐增大，波长最长的918 nm波段行向响应不均一性最为明显。

目前已开发了多种图像辐射畸变校正方法，例如直方图匹配、地面同步光谱测量、基于统计理论的校正方法等。针对探月数据的特点和定性与定量分析的需求，笔者提出了基于光谱特征的反射率归一化校正方法[4]。该方法以前述定标区域作为基准，将各列相对于750 nm波段的反射率都归一化到与标准光谱一致的相对反射率，通过二阶多项式拟合获得校正因子。基于该校正，既消除了边缘辐射畸变效应，又确保了校正后的光谱信息具有物理意义，即月表反射率。

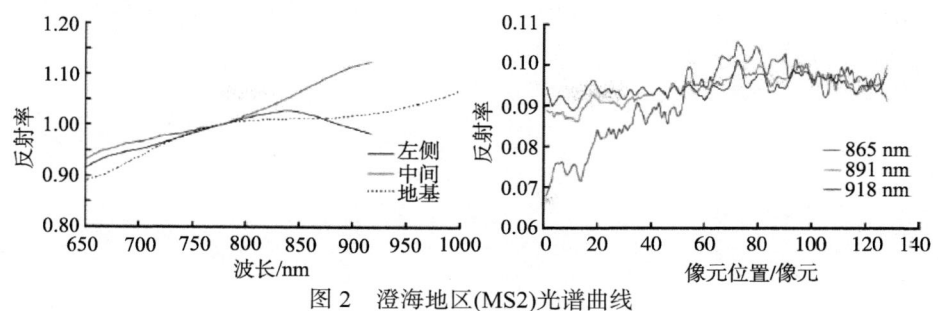

图 2 澄海地区(MS2)光谱曲线

基于上述对IIM 2C数据的分析和处理，本文制定了IIM数据应用处理流程(图3)，使用基于此流程生成的反射率产品来开展研究。

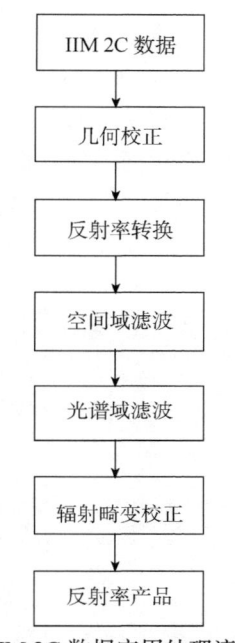

图 3 IIM 2C 数据应用处理流程

2 结果与应用

图4是基于上述校正获得的IIM行向响应不均一性校正因子。可以看出，在空间域，校正因子在图像左侧最高，向右逐渐降低，表明传感器左侧响应较低；在波谱域，波长越长校正因子越大，表明传感器长波段响应不如短波段精确。

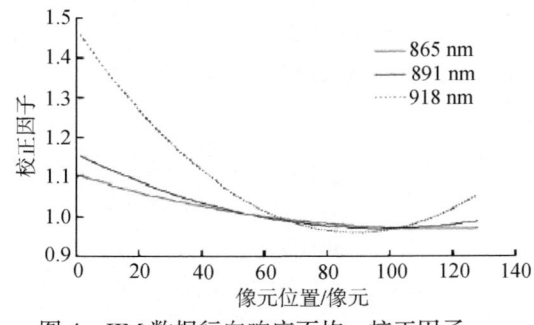

图 4 IIM 数据行向响应不均一校正因子

图5是利用图3所示数据处理流程所得的MS2地区光谱数据。对比图5和图2(左)可见，经过本处理流程操作后数据都为反射率形式，行向上的响应趋于均一化，相对反射率都与地基光谱匹配，可以使用此数据开展研究。

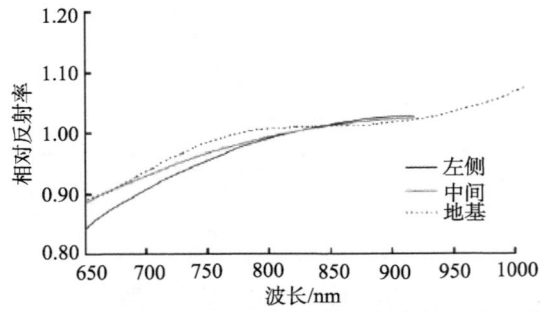

图 5　MS2左侧和中间校正后的相对反射率和地基光谱

为检验本校正因子在其他地区的适应性，对月球正面几个大型撞击坑Copernicus、Petavius、Tycho和Arzachel等进行了校正，都取得了不错的效果。以Arzachel撞击坑(2858轨)为例，图6(左)、(右)分别为Arzachel撞击坑校正前、后的918 nm(R)、776 nm(G)、688 nm(B)波段彩色合成图像。由图6(图版Ⅲ)可见，校正前行向上色调的不均一性在校正后得到消除。校正前图像右侧色调偏红，与月球正常色调不一致，这是由于右侧红波段918 nm响应偏高所致；校正后色调行向上变得均一，而且与月表真实色调一致(通过与克莱门汀彩色合成图像对比)。

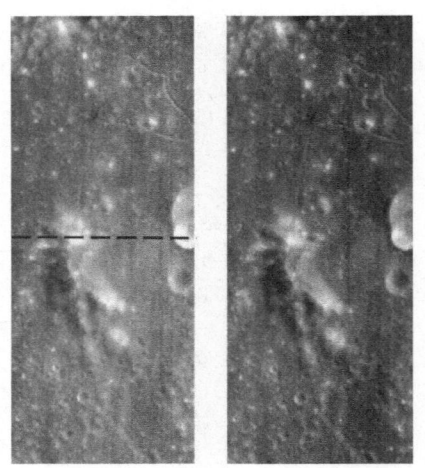

图 6　Arzachel撞击坑校正前(左)、后(右)彩色合成图像

图7是图6(左)中红色虚线所示位置918 nm、776 nm和688 nm波段行向光谱剖面，可以看出，3个近红外波段反射率在横向上的响应也变得均一。以上说明，可以利用本文应用处理流程所得到的数据开展研究。值得注意的是，在本数据处理流程中没有进行新的光度纠正，即仍使用2C数据所应用的光度纠正模型。比值方法可以增强吸收峰，而且有助于消除光度效应。因此，在比值图像上光度效应所引起的光谱差异可以忽略。

图8(图版Ⅳ)是Aristarchus高地B_{757}/B_{644}(R)、B_{757}/B_{865}(G)和B_{644}/B_{757}(B)校正前的比值(左)和校正后的比值(右)合成图像(2 897、2 898和2 899轨镶嵌)。Aristarchus高地表层被大量火山碎屑物质覆盖，并且具有地球物理、地球化学的多种异常，受到许多月球研究者的关注[5~7]。在校正前的比值合成图像上，Aristarchus撞击坑上部火山碎屑物质在横向上色调不一致，中间的偏蓝，被分为两类；经过校正后，色调均一，分为一类。可见，经过光度校正，提高了岩性分类的精度。

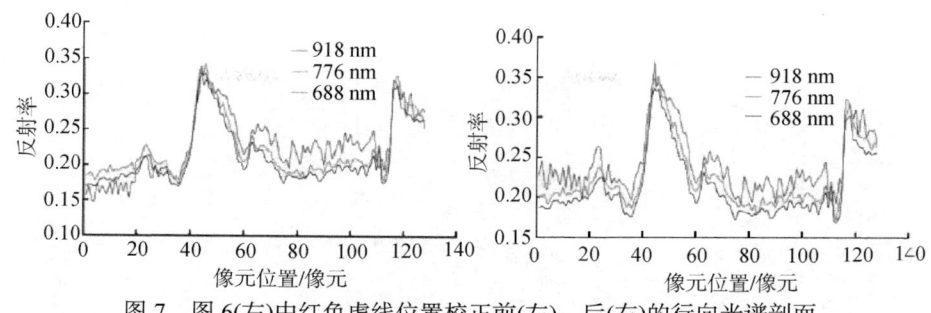

图7 图6(左)中红色虚线位置校正前(左)、后(右)的行向光谱剖面

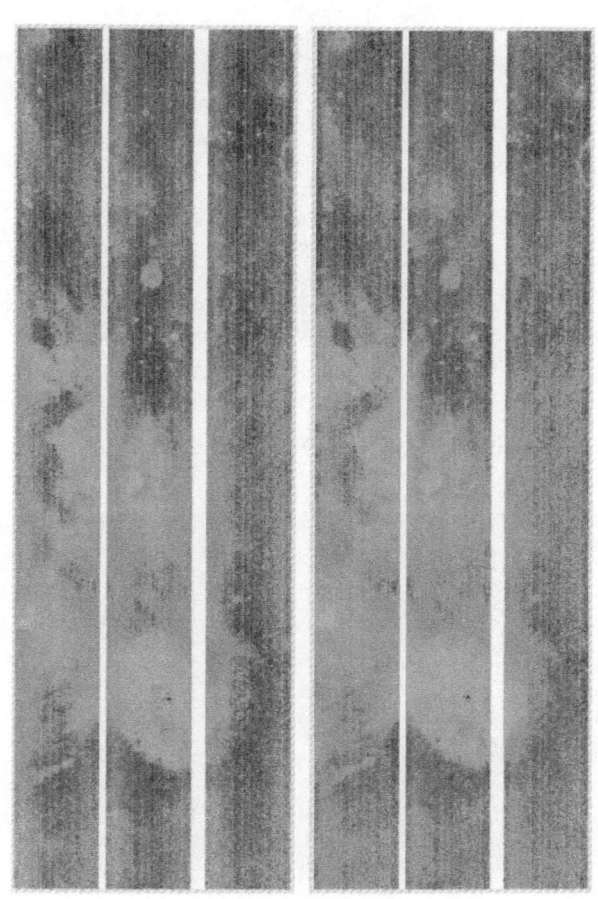

图8 Aristarchus 高地 B_{757}/B_{644}(R)、B_{757}/B_{865}(G)和 B_{644}/B_{757}(B)校正前的比值(左)和校正后的比值(右)合成图像(2 897、2 898 和 2 899 轨镶嵌)

  遥感只能获取月球表层成分,撞击坑是开启月球内部物质奥秘大门的一把钥匙。通过对撞击坑的研究,可以了解月壳在纵向和横向上的成分变异,从而研究月球的起源和演化。由图8可见,Aristarchus撞击坑坑底与坑壁色调不一致,这表明在纵向上,该地区存在成分的多样性。就撞击坑坑壁自身来说,西南坑壁、西北坑壁和东部坑壁展示了不同的色调,这表明在横向上撞击坑存在成分的多样性。西南坑壁、坑底的色调为深蓝,西北坑壁和东北坑壁颜色为浅蓝色,这说明西南坑壁和坑底在短波段的反射率比其他地方高。

  图9(图版Ⅳ)是Aristarchus撞击坑B_{918}(R)、B_{757}(G)和B_{618}(B)波段合成图像。由于传感器行向响应不均一,校正前撞击坑西部坑壁和东部坑壁都呈现蓝色调,这说明918 nm波段反射率较低,可能存在铁镁质矿物吸收。校正后,各轨左侧蓝色纵向条带得到消除,只有一些蓝色像元零星散布于西北坑壁。这一方面表明校正提高了岩性识别精度,另一方面这些蓝色像元较高的反射率和918 nm处的吸收也说明新鲜的表面可能被暴露。从其新鲜程度以及零散的分布,我们推断Aristarchus撞击坑西北坑壁可能存在滑坡,致使新鲜物质

得到暴露。Lunar Orbiter高清晰照片支持了这一推断。

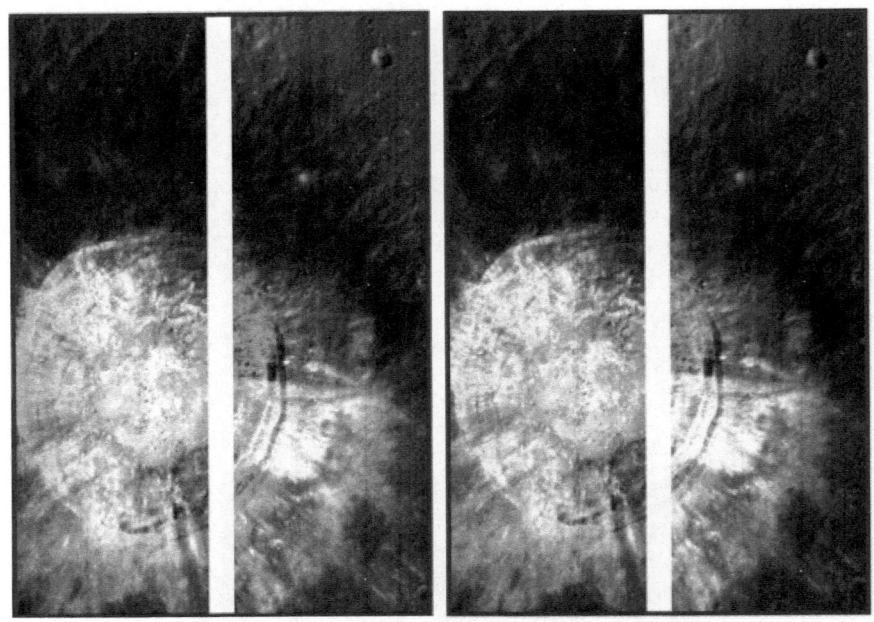

图 9　Aristarchus 撞击坑校正前(左)和校正后(右)B_{918}(R)、B_{757}(G)和 B_{618}(B)合成图像

图10为Aristarchus撞击坑Lunar Orbiter高清晰照片和DEM。

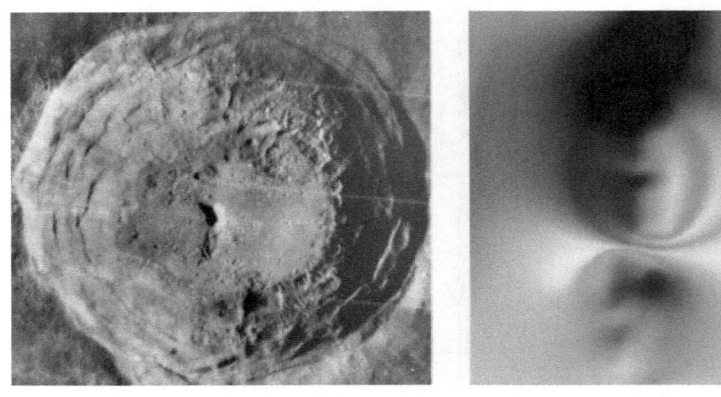

图 10　Aristarchus 撞击坑 Lunar Orbiter 高清晰照片(左)和 DEM(右)

由图10(左)可见(图版V), 东部坑底较为光滑, 而西部坑底较为粗糙, 存在类似滑坡物质。此外, 由嫦娥一号激光高度计获得的DEM可见, 西北坑壁要比东部坑壁陡峭(图10(右)), 增大了西部坑壁存在滑坡的可能性。这些只是推测, 有待进一步研究。我国未来发射的嫦娥二号月球卫星可以获取10 m分辨率数据, 将有可能对该推测给出答案。

3　结论

本文对现有IIM 2C数据进行了分析, 并就应用过程中存在的问题提出了解决方案, 制定了生成可用数据的处理流程。基于本流程处理后的数据, 利用比值方法对Aristarchus高地进行了岩性分类, 结果表明校正后分类精度提高。Aristarchus撞击坑在纵向和横向上都存在成分变异。根据918 nm波段高吸收物质零星分布及其较高的反射率, 推断Aristarchus撞击坑西北坑壁可能存在滑坡。

本文没有考虑光谱信噪比和光度纠正。下一步工作将是对图像光谱信噪比和光度纠正进行研究，进一步提高数据精度。嫦娥一号IIM能够以较高的空间分辨率和光谱分辨率获取月表元素和矿物成分信息，基于本数据应用处理流程，利用我国自主研制的IIM数据开展研究有助于深化对月球形成和演化的认识。

致谢 本文所用 IIM 2C 数据由月球探测工程中心授权、绕月探测工程地面应用系统提供。DEM 数据由中国科学院上海天文台平劲松研究员和黄倩博士提供。感谢国家天文台欧阳自远院士、李春来研究员、张文喜博士等多位专家指导。

参 考 文 献

[1] 薛彬. CE-1 干涉成像光谱仪信息处理及应用研究[D]. 西安：中国科学院西安光学精密机械研究所，2006
[2] Pieters C M. Compositional Diversity and Stratigraphy of the Lunar Crust Derived from Reflectance Spectroscopy [A]. Pieters C, Englert P. Remote Geochemical Analysis: Elemental and Mineralogical Composition [M]. Cambridge: Cambridge University Press, 1993
[3] Blewett D T, Lucey P G, Hawke B R, et al. Clementine Images of the Lunar Sample-return Stations: Refinement of FeO and TiO_2 Mapping Techniques[J]. J. Geophys. Res., 1997, 102(E7): 16319—16325
[4] 吴昀昭，徐夕生，谢志东，等. 嫦娥一号 IIM 数据绝对定标与初步应用[J]. 中国科学(G 辑)，2009, 39(10): 1387—1392
[5] Lucey P G, Hawke B R, Pieters C M, et al. A Compositional Study of the Aristarchus Region of the Moon Using Near Infrared Reflectance Spectroscopy [J]. J. Geophys. Res., 1986, 91: 344—354
[6] McEwen M S, Robinson E M, Eliason P G, et al. Clementine Observations of the Aristarchus Region of the Moon [J]. Science, 1994, 266: 1858—1862
[7] Chevrel S D, Pinet P C, Daydou Y, et al. The Aristarchus Plateau on the Moon: Mineralogical and Structural Study from Integrated Clementine UVVis NIR Spectral Data[J]. Icarus, 2009, 199: 9—24

干涉成像光谱仪中宽谱段傅氏光学系统设计

赵葆常　杨建峰　陈立武　常凌颖　贺应红　薛彬

(中国科学院西安光学精密机械研究所，西安 710119)

摘　要　设计了一种宽谱段前置光孔远心光学系统，其孔径光阑位于透镜组的焦面上，该设计集宽谱段自动复消色差、长工作距、前置光孔及远心等特点于一体。讨论了这样一种特殊光组的设计。选用国产牌号的光学玻璃配对，价格低廉且在校正色差后，自动校正宽谱段二级光谱，并实现了透镜组两侧同时具有长工作距。设计结果表明，光学系统的残余二级光谱很容易控制在千分之一焦距以内，波面平行差小于 0.000 22 弧度。

关键词　光学设计　宽谱段　傅氏光学系统　干涉光谱

0　引言

在嫦娥一号卫星上搭载了一台成像光谱仪，它与 X/γ 相配合，用以分析月球表面有用元素成分与物质类型的含量与分布[1]。成像光谱仪采用了实体 Sagnac 空间调制型干涉成像光谱仪[2]，其原因是：由于对月球探测的不可预见性，科学家希望获得任意波长上的光谱强度数据，仅干涉型成像光谱仪具有这种功能，其次是 Sagnac 型干涉仪由于它的共光路原理，使它具有非常强的航天环境适应能力，嫦娥一号卫星的干涉成像光谱仪在轨运行已超过设计寿命一年，从未出现故障，验证了该优点。本文设计了一种用于该干涉仪成像光谱仪的宽谱段前置光孔远心光学系统，该系统可以自动复消色差，能将残余二级光谱很容易控制在千分之一焦距以内。

1　傅氏光学系统设计要求

与色散型成像光谱仪一样，Sagnac 空间调制型干涉成像光谱仪也必须是一个二次成像光学系统，它由前置光学系统、狭缝、Sagnac 干涉仪、傅氏光学系统、柱面光学系统及二维焦平面阵列组成。在前置光学系统后焦面上设置的狭缝，其宽度与要求和地面分辨率 GSD 相匹配，该狭缝同时又位于傅氏透镜的物方焦平面上，系统结构见图 1。

该光学系统可以被看作是前置光学系统(前置照相物镜)后加入一组投影光学系统(接续光学系统 relay lens)。为了获取高品质的设计结果与增强航天环境的适应性能力，在本设计中，由傅氏光学系统与柱面光学系统构成的投影物镜是一个缩小倍率为 –3.1× 的大比尺缩小系统。这样使位于窗口附近，温度环境条件最为恶劣的前置光学系统具有强的环境适应能力，因为它造成的误差被后续投影物镜大大地缩小了，使它对环境条件变得不敏感。

设计中，三组光学系统的参量分配见表 1。

前置光学系统、傅氏光学系统与柱面光学系统三个光组对系统的空间分辨率与光谱分辨率的影响是不相同的。其中对傅氏光学系统的设计要求最为苛刻，它以从狭缝作为物点，追迹出射光束的角像差来衡量。本文研究的是嫦娥一号卫星干涉成像光谱仪中傅氏透镜的光学设计[3]。

本文原载于《光子学报》，2009，Vol.38，No.3，468~475。

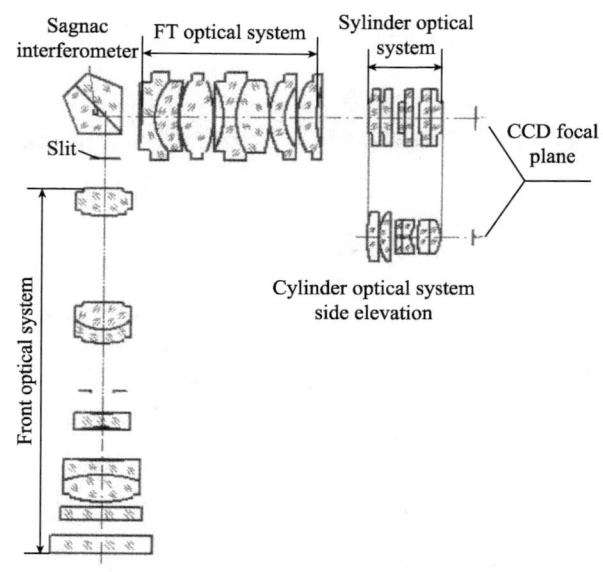

图 1　嫦娥一号卫星中的干涉成像光谱仪光学系统

表 1　干涉成像光谱仪分系统参量分配

分系统名称	焦距/mm	F 数	视场角	特殊要求
前置光学系统	104.6	7.34	7.34°	像方远心
傅氏光学系统	80	7.34	9.56°	物方远心，傅氏镜出瞳与柱面系统焦面共面
柱面光学系统	26	2.4	9.56°	
全系统	34	2.4	7.34°	

2　傅氏光学系统设计要求

根据表 1 的光焦度分配，傅氏光学系统的设计参量为：1)f'=80 mm；2)$2\omega_p$=9.56°；3)F 数=7.34；4)工作波段=480~960 nm；其他约束条件为：

(1) 在狭缝与傅氏透镜间需插入一块厚度为 60 mm 的平板，它是 Sagnac 干涉仪展开为平板的形式。

(2) 在 CCD 与傅氏透镜之间，需插入一个柱面光学系统，在傅氏透镜设计中柱面零件按平板计算。

(3) 傅氏透镜与柱面光学系统共焦，在焦平面上放置面阵 CCD，全系统出射光孔与 CCD 焦平面重合，亦即傅氏透镜焦平面、柱面透镜焦平面、系统出射光孔与 CCD 焦平面四者共面。

(4) 根据要求(3)，傅氏透镜为物方远心系统。

(5) 傅氏透镜在 480~960 nm 宽谱段范围内，需同时校正色差与二级光谱。

(6) 厚度为 60 mm 的平板玻璃，会产生可观的负球差与负色差，所以应与傅氏透镜共同校正像差。

(7) 在狭缝、干涉仪、傅氏透镜、柱面透镜及 CCD 焦平面间均应有不小于 10 mm 的间隙，以满足结构设计的要求。

(8) 以狭缝作为物体，追迹光线，出射平行光线的角像差 $\delta\theta$ 应满足[4-5]

$$\delta\theta \leqslant \theta/N \tag{1}$$

式中 θ 为剪切角，它是二虚狭缝中心对傅氏透镜光心之夹角，也等于在焦平面处二波面之夹角，N 为光谱通道数。

3　玻璃选配

玻璃选配的目的是为了使系统校正色差后，二级光谱达到自动校正之目的。

3.1 二级光谱的概念及基本理论

二级光谱是高级色差的一种，它是由于光学系统工作波段变大所致，它与相对孔径的平方和焦距的一次方成正比，其几何度量是对两端色光校正色差后与主色光之间的位置差。

对于宽谱段的光学系统在满足光焦度要求的前提下校正位置色差的条件为[6]

$$C_1 = \sum_{i=1}^{n} h_i^2 (\varphi_i / v_i) = 0 \tag{2}$$

同时校正二级光谱的条件为

$$\sum_{i}^{n} h_i^2 (\varphi_i / v_i) P_i = 0 \tag{3}$$

为了同时校正色差与二级光谱，通常的办法是选择三种玻璃，它们在部分色散系数 P_i 与 v 的图上所围的三角形具有尽可能大的面积。实际上是其中的两种玻璃组合成一种新的虚拟玻璃，它与第三种玻璃具有相同或接近的部分色散系数，而同时又有一定的阿贝数之差。因为普通玻璃在 P_i-v 图上基本上都在一条直线上，所以它围成的三角形面积总是非常小，所以通常在三种玻璃中需要有一种玻璃它在 P_i-v 图上离开正常色散线。以期扩大所围三角形的面积，用该方法时，还需要通过合适的光焦度分配才能达到同时校正位置色差与二级光谱之目的[7]。偏离正常色散线的玻璃，常称为特种玻璃，其中效果最好的是 CaF_2，但因为氟化钙价格昂贵，尺寸做不大，所以主要用于复消色差显微镜中。为同时校正位置色差与二级光谱的复消色差照相物镜设计中常常采用中国牌号的特种火石玻璃 TF 系列，它相当于德国牌号的 KZFSN 系列玻璃。

另一种做法是只选取两种玻璃，它们在光学系统的工作波段内具有很接近的部分色散系数 P_i，同时又具有一定的阿贝数之差，这种方法的优点是其二级光谱的校正与系统中光焦度分配无关，解除了光学系统自动优化设计中二级光谱对光焦度的约束条件，亦即它满足消位置色差条件的式(1)后，自动满足式(2)，容易设计出高品质的光组。其缺点是通常其 Δv 值都比较小，因而造成光组复杂化，据本文作者的经验两种玻璃的 Δv 只要大于 5 就可以有实用价值的设计。本傅氏透镜从设计要求来看，它的约束条件太多(详见第二部分)，所以采用了选择适合的玻璃配对，以达到在位置消色差的情况下，自动校正二级光谱的目的，并取得了预期的效果。

3.2 傅氏光学系统玻璃选配分析

傅氏光学系统的工作波段范围为 480~960 nm，谱段宽度 Δλ=480 nm。与一般目视光学仪器的工作波段相比，它大大地向长波扩展了。在傅氏光学系统设计中，选择了国产牌号的 Lak2 与 TF3 分别作为光组中正、负透镜之玻璃材料。这样系统中的光焦度分配只需满足消色差的约束条件。色差校正后，二级光谱一定很小。三种玻璃在各波长处的折射率和消色差时的阿贝数见表 2 和表 3。

表 2 所选玻璃的折射率

波长 玻璃	0.48	0.5	0.55	0.60	0.65	0.7	0.75	0.8	0.85	0.90	0.96
Lak2	1.70169	1.69942	1.69482	1.69132	1.68858	1.68636	1.68452	1.68297	1.68162	1.68043	1.67916
TF3	1.62295	1.62044	1.61538	1.61156	1.60858	1.60617	1.60417	1.60248	1.60100	1.59969	1.59829
K9	1.52240	1.52099	1.51810	1.51587	1.51410	1.51264	1.51141	1.51035	1.50941	1.50856	1.50764

表 3 所选玻璃配对对 480~960 nm 消色差时各波长的阿贝数

波长 玻璃	0.5	0.55	0.60	0.65	0.7	0.75	0.8	0.85	0.90	均值
Lak2	31.05	30.85	30.69	30.57	30.47	30.39	30.32	30.26	30.21	30.54
TF3	25.16	24.96	24.80	24.68	24.58	24.50	24.43	24.37	24.32	24.64
K9	35.29	35.09	34.94	34.82	34.72	34.64	34.57	34.50	34.45	34.78

从表3可以得到：

(1) 与可见光波段对 F、C 消色差时，d 线的阿贝数 v_d 相比，由于波段拓宽使阿贝数的值大大减小，如 K9 玻璃 v_d=64.1 而在 480~960 nm 宽谱段时，减小为 34.78；

(2) 对各条谱线的阿贝数 v_λ 变化不大；

(3) 两种玻璃阿贝数的差值 Δv 与可见光消色差相比，大大减小，如 TF3~K9 之间阿贝数之差 Δv 从 20 减小为 10.14，TF3~Lak2 之间阿贝数之差 Δv 从 10 减小为 5.9。

为了分析采用 Lak2 配 TF3 在消色差后，二级光谱很小的原因，以 K9~TF3 的配对作为比对值。

表 4 为在 480~960 nm 的宽谱段内，Lak2 与 TF3 的部分色散系数之差值，同表给出了 K9 与 TF3 的部分色散系数之差值，作为比对。

表 4　两种配对玻璃的部分色散系数之差值×10^{-3}

玻璃配对＼波长	0.5	0.55	0.60	0.65	0.7	0.75	0.8	0.85	0.90
Lak2/TF3	−1.17	−2.01	−1.53	−0.72	0.016	0.53	0.79	0.79	−0.65
K9/TF3	−5.91	−15	−19	−20	−19	−16.8	−13.64	−9.87	−6.83

部分色散系数的计算公式为

$$p(\lambda) = \frac{n480 - n\lambda}{n480 - n960} \quad (4)$$

或者

$$p(\lambda) = \frac{n\lambda - n960}{n480 - n960} \quad (5)$$

由于 480 nm 与 960 nm 二谱线已校正色差，所以式(4)与(5)两者计算结果相同。

从表 4 可以看出采用 Lak2 配对 TF3，在各个波长上二者的部分色散系数都非常接近，所以一个光学系统采用 Lak2 与 TF3 两种玻璃，在 480~960 nm 二条谱线校正色差后，二级光谱一定非常小，从而把消二级光谱对光焦度的约束条件予以释放，使光组获得更好的像差校正，但是 Lak2 与 TF3 的 Δv 值远比 K9 与 TF3 的 Δv 值小，所以光组必须适当复杂化。

4　光组结构选型及最优化设计

4.1　结构选型

根据设计要求，傅氏光学系统两边分别要插入 Sagnac 干涉仪展开平板与柱面光学系统，所以希望物方焦点与像方焦点都远离光组，也就是希望物方主面向前推，像方主面向后推，因此采用"负-正-负"的特殊光组结构型式。

实际的设计结果为前焦点 F 离光组距离为 $0.78f'$，后焦点 F' 离光组的距离为 $0.71f'$ 满足插入 Saganc 干涉仪与柱面镜的要求。

4.2　最优化设计

为了全面评价系统的像质，采用了自右向左倒追光线，在这种模式下，孔径光阑位于焦面 F，以达到远心，保证光线在 Sagnac 干涉仪中，各视场的 OPD 相同，并且各视场在分束面上光束结构相同，使各视场的色漂移相同（图 2）。

图 3 为系统的色球差曲线，可以看出 480 nm 与 960 nm 的二条谱线基本上校正了色差，但交点偏低了一点，光线追迹发现这时离开最远的谱线为 λ=750 nm，其间隔为 0.06 mm 即 $0.00075f'$。如果不作玻璃配对选择，采用普通玻璃，则对双胶合而言，其值应为 $0.0026f'$[5]光组复杂化还会使系数 0.0026 进一步增

大。针对性的玻璃选配使二级光谱值至少减小 3.5×以上。

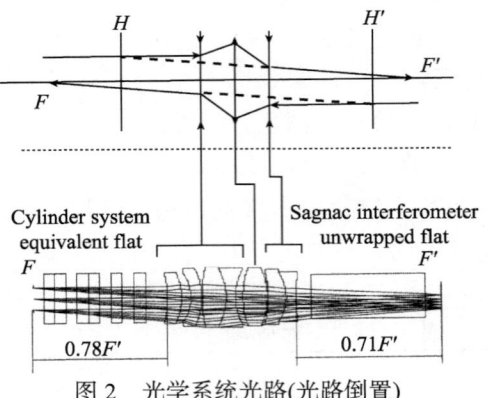

图 2　光学系统光路(光路倒置)

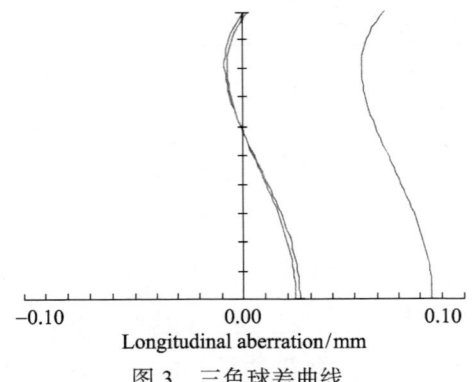

图 3　三色球差曲线

系统的像散和畸变曲线见图 4。虽然该系统是一个密集型结构，但由于航天力学环境条件的要求透镜都被大大加厚了，因此负-正-负的光焦度分布形式，仍然起到了正负光焦度分离的作用，使其具有小的 S_{IV}，同时光组中的二个弯月透镜与胶合面都弯向孔径光阑，减小 S_{III} 的值，因此子午与弧矢场曲都较小。同时由于远心要求造成系统严重失对称，但仍然很好的校正了的畸变 S_V，系统的最大畸变小于 −0.3%。

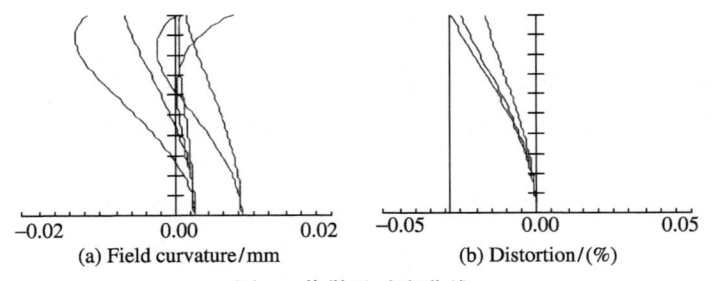

图 4　像散及畸变曲线

系统的光学传递函数见图 5。传递函数的计算结果表明，中心视场 MTF 达到了衍射极限，边缘视场 MTF 仅下降 0.08，当空间频率等于 50 lp/mm 时，复色光各视场 MTF 均大于 0.6。

从弥散斑尺寸(图 6)看出：中心视场最大半径值仅为：0.765 μm，均方差为 0.48 μm；最大视场最大弥散斑半径为 5.8 μm，均方差为 2.6 μm。对 10 μm² 的 CCD 像元，最大弥散基本上控制在一个像元之内。

由于傅氏透镜与干涉仪相配，它输出的是宽谱段的相干平面波，所以必须计算各视场各色光的波像差(见表 5)，至少优于瑞利判据。设计上把宽谱段的波像差量级控制在 λ/40 左右(RMS)。

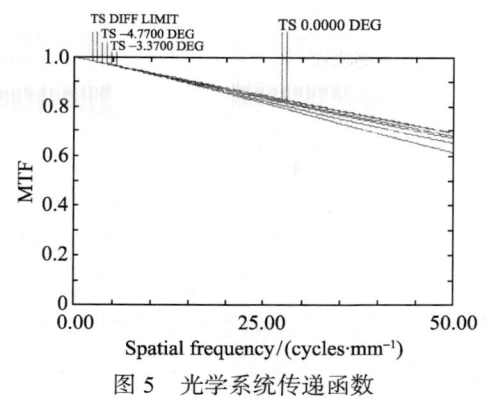

图 5 光学系统传递函数

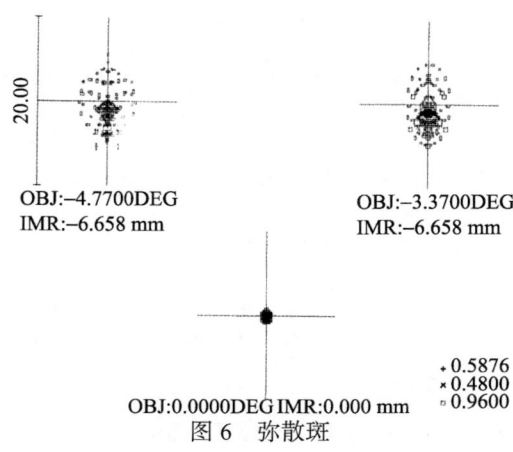

图 6 弥散斑

表 5 波像差(RMS)

波长 \ 场	4.77°	4.055°	3.548°	2.534°	1.52°	0°	综合
0.48000	0.0711	0.0470	0.0332	0.0255	0.0205	0.0167	
0.51200	0.0634	0.0409	0.0288	0.0211	0.0150	0.0103	
0.54400	0.0572	0.0375	0.0280	0.0209	0.0140	0.0087	
0.57600	0.0522	0.0324	0.0227	0.0151	0.0079	0.0042	
0.60800	0.0489	0.0268	0.0148	0.0066	0.0065	0.0115	
0.64000	0.0486	0.0248	0.0113	0.0093	0.0166	0.0226	
0.67200	0.0503	0.0267	0.0156	0.0180	0.0260	0.0322	
0.70400	0.0520	0.0294	0.0206	0.0245	0.0326	0.0389	
0.73600	0.0523	0.0307	0.0233	0.0278	0.0360	0.0422	
0.76800	0.0506	0.0298	0.0231	0.0281	0.0363	0.0425	
0.80000	0.0470	0.0267	0.0203	0.0256	0.0338	0.0399	
0.83200	0.0419	0.0219	0.0154	0.0208	0.0290	0.0351	
0.86400	0.0364	0.0166	0.0091	0.0142	0.0223	0.0283	
0.89600	0.0315	0.0137	0.0057	0.0070	0.0142	0.0201	
0.92800	0.0294	0.0170	0.0124	0.0064	0.0058	0.0109	
0.96000	0.0316	0.0252	0.0224	0.0154	0.0070	0.0026	

为了检查出射平面波的角弥散，追迹了从狭缝上的物点发出的各色光经傅氏透镜后出射光线的角弥散，根据前面的判据，它应该小于 0.000 22 弧度。实际计算表明，在最大孔径处，角弥散值小于 0.000 22 弧度，当孔径略有缩小时，其角弥散(即平面波间的平行误差)均小于 0.000 1 弧度。

5 结论

傅氏光学系统，通过选择合适的玻璃配对，在480~960 nm的宽光谱波段范围内校正色差后，自动校正二级光谱，其残余二级光谱大致与可见光中对F、C校正色差后，它们对d线的二级光谱相当，该方法同样适用于其他的可见光/近红外谱段。如450~950 nm的二级光谱校正。采用玻璃选配的方法，释放了二级光谱对光焦度的约束条件。同时为了获得物方与像方有大的工作距，采用了"负-正-负"的特殊光组结构形式，在上述设计思想的指导下，完成了一个用于嫦娥一号卫星干涉成像光谱仪中的傅氏光学系统的设计，得到了很好的优化结果。该傅氏光学系统已成功应用于嫦娥一号卫星有效载荷干涉成像光谱仪中，经受了航天环境的考验，并获得了清晰的光谱图像。

参 考 文 献

[1] OUYANG Zi-yuan, LI Chun-lai. ZOU Yong-liao. The scientific object of the first phase project of Chinese lunar exploration [J]. *Spacecraft Engineering*, 2005, **14**(1): 1—5
欧阳自远，李春来，邹永廖，等. 我国月球探测一期工程的科学目标[J]. 航天器工程，2005，**14**(1)：1—5

[2] BRUCE R J, GLENN S R, BLATT J H. Monolithic Fourier-transform imaging spectrometer [J]. *Appl Opt*, 1995, **34**(31): 7228—7230

[3] ZHAO bao-chang, YANG Jian-feng, CHEN Li-wu. A kind of wide waveband front aperture and collimated optical system [P]. Patent(right applying), 2008
赵葆常，杨建峰，陈立武，等. 一种宽谱段前置光孔远心光学系统[P]. 发明专利(受理中)，2008

[4] GLENN S R, BRUCE R. J. Effects of aberration on spatially modulated Fourier transform spectrometers [J]. *Opt Eng*, 1994, **33**(9): 3087—3092

[5] CHEN Li-wu. The theoretical study of the optical technology of the engineerization of the interference imaging spectrometer of the Chang'e-1 satellite doctoral dissertation [D]. Xi'an: Xi'an Institute of Optics and Precision Mechnics, CAS, 2006: 59—61
陈立武，Chang'e-1 成像光谱仪工程化光学技术理论研究[D]. 西安：中国科学院西安光学精密机械与物理研究所，2006：59—61

[6] WANG Zhi-jiang. The basic theory of lens design [M]. Beijing: Science Press, 1965: 178—191
王之江. 光学设计理论基础[M]. 北京：科学出版社，1965：178—191

[7] YU Tao. Applying optics [M]. Beijing: Science Press, 1966: 291—296
喻焘. 应用光学[M]. 北京：科学出版社，1966：291—296

嫦娥一号干涉成像光谱仪的定标

赵葆常　杨建峰　薛　彬　乔卫东　邱跃洪

(中国科学院西安光学精密机械研究所，西安 710119)

摘　要　介绍了我国首次探月卫星的有效载荷之一 Sagnac 空间调制型干涉成像光谱仪的定标。提出了"行"平场原理以及不同类型的光谱仪对比方法，分别用于相对定标与光谱辐射度绝对定标，取得了好的实验结果。检测了谱线位置不确定度、光谱分辨率及在轨光谱辐射度的相对不确定度。给出了探月卫星干涉成像光谱仪的定标及检测结果。首次采用干涉型成像光谱仪实现了对月的可见光/近红外宽谱段连续光谱探测。

关键词　干涉成像光谱仪　定标　在轨评测　平场　不同类型光谱仪对比方法

0　引言

根据我国首次探月的科学目标[1]，嫦娥一号卫星需配置一台可见光/近红外的成像光谱仪，它的科学目标定位是：与 X/γ 谱仪相配合，共同完成"分析月球表面有用元素成分与物质类型的含量与分布"。X/γ 谱仪侧重于元素成分的探测，成像光谱仪侧重于物质类型的分析。

在方案论证阶段，通过比较各种类型成像光谱仪之优缺点，选取 Sagnac 空间调制型成像光谱技术方案，其理由是：1)从空间环境条件出发。由于 Sagnac 型干涉成像光谱仪具有非常强的力学和温度环境适应能力，除无运动部件外，其干涉仪是一个实体，而且两束相干光在干涉仪中的路径几乎相同，但方向相反，所以对振动、温度等环境的影响不敏感，这一点也是美国于 2000 年 7 月发射的强力小卫星 MightSat II 中选择 Sagnac 空间调制型成像光谱技术方案的重要原因[2~4]；2)由于我国是首次探测月球，尚有很多未知性，所以探月科学家提出希望能获得任一波长的谱强度信息，这正好是干涉型成像光谱仪技术的特点[5]，它不同于其他类型的成像光谱仪，直接测量得到干涉图，所有谱线的信息都含在干涉强度分布中，可以通过傅里叶变换获取任一波长的真实谱强度信息，因此决定在嫦娥一号选用 Sagnac 空间调制型成像光谱技术方案。

嫦娥一号卫星干涉成像光谱仪(Interference Imaging Spectrometer，IIS)于 2007 年 11 月 27 日开机以来获得了大量月面的多光谱图像，图像清晰、层次纹理丰富，受到用户广泛好评，目前已基本完成了辉石的矿物填图。

嫦娥一号卫星干涉成像光谱仪研制过程中碰到的最大困难之一是平场和光谱辐亮度的绝对定标，国内外几乎找不到有价值的文献资料，在个别文献资料中，虽也提到定标，但描述都十分简单[6-7]。然而定标必不可少，否则干涉成像光谱仪无法达到识别物质类型的目标。"行"平场原理及方法[8]可以很好地解决相对定标的困难，不同类型光谱仪的对比方法[9]可以解决绝对辐射定标难题。

1　嫦娥一号卫星干涉成像光谱仪光学方案

图 1 和表 1 分别给出了嫦娥一号卫星干涉成像光谱仪的光学系统结构和参量分配[10]。

本文原载于《光子学报》，2010，Vol.39，No.5，769~775。

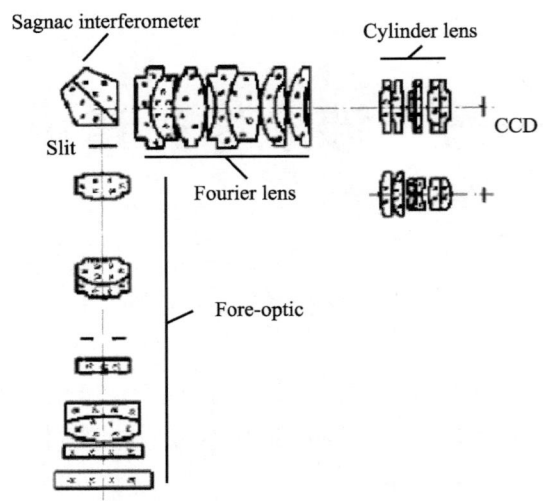

图 1　嫦娥一号卫星中干涉成像光谱仪光学系统图

表 1　组成干涉成像光谱仪各分系统参量分配表

ame\parameter	Focus/mm	F/#	FOV
Fore-Optic	104.6	F/7.34	7.34°
Fourier Lens	80	F/7.34	9.56°
Cylinder Lens	26	F/2.4	9.56°
system	34	F/2.4	7.34°

从图 1 可以看出，它是一个二次成像系统。在前置光学系统(Fore-optic)的像平面上有一个 0.06 mm 宽的狭缝(Slit)，从而形成视场，狭缝又位于傅氏光学系统(Fourier Lens)的前焦平面上，所以由傅氏光学系统出射的是平行光，光场为平面波，柱面光学系统(Cylinder Lens)的作用是在一个方向上把平面波压缩为一条线，再成像在沿飞行方向的一列 CCD 上，成为一组离散的干涉图数据 $I(x)$，经傅氏变换后即可得到复原光谱[11~13]。

$$B(\sigma) = \Sigma I(x)\cos(2\pi\sigma_x)\mathrm{d}x \tag{1}$$

定标是对整机实施的，它包含了光、机、电、CCD 各方面的因素。

2　定标的内容

嫦娥一号探月卫星的干涉成像光谱仪定标工作主要包括：①标准具的选择与校验；②暗电流的检测；③最大饱和辐亮度的配准；④整机相对定标(平场)；⑤整机绝对光谱辐射度定标；⑥输出 Vs 输入线性检测；⑦光谱分辨率检测；⑧谱线位置检测。

2.1　标准具的选择与校验

定标中采用设备(标准)见表 2。

表 2　定标中采用设备

No.	Name	Model	Purpose
1	Spectrometer	FieldSpace Pro UV/VNIR	For spectral radiometric calibration
2	Spectrometer	FieldSpace Pro UV/VNFR	For spectral radiometric calibration
3	Spectrometer	VF921	For spectral radiometric calibration
4	Intergrating sphere	LabsPhere VSS-6500	uniform area lamp-house
5	Solar simulation	LabsPhere XTH-2000	uniform area lamp-house

续表

No.	Name	Model	Purpose
6	Trap detector	Made by AIOPM	Detect change of intergrating sphere
7	Spectral Irradiance standard Lamp	Made by NIM, F07	For the calibration of spectrometer
8	Standard white plate	LabsPhere STR-99-240	For the calibration of spectrometer
9	Laser	05-L GR-193-381、1322412、56ICS006、56ICS008	detect the resolution and wavelength position
10	Laser beam expander	Made by XIOPM	20 rate focus

标准源的传递见图2。

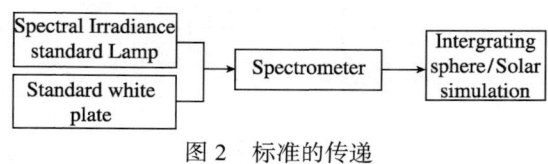

图2 标准的传递

其中光谱辐射度标准灯与标准白板由中国计量研究院校验，并出具校验报告，它们是实验室定标的一级标准源，然后把标准传递到光谱辐射度计，再传递到积分球与太阳模拟器。

国外学者报导标准直接由灯/板系统传递到积分球，本文用三台光谱辐射度计，由中国计量研究院与安徽光机所两个渠道分别传递，最后核对两者数据，以防止出错。把三台光谱辐射度计对准同一积分球，分别读数，经检测相互间差异小于2%。另外，采用光谱辐射度计[14]作为标准还有利于对积分球的输出辐亮度作实时监视与检测。

积分球校验时，主要检测积分球出口处辐亮度的面均匀性及角均匀性，见图3。这部分工作由安徽光机所完成并出具校验报告。

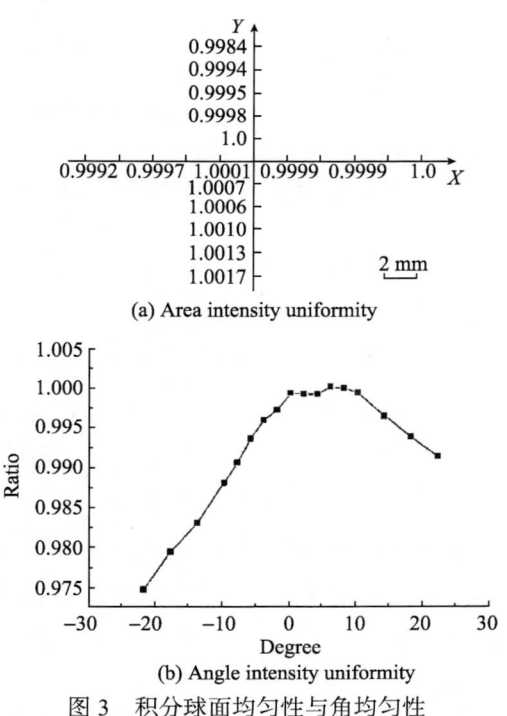

图3 积分球面均匀性与角均匀性

2.2 暗电流的检测

任何一台光电仪器，当没有光输入时仍有电信号输出，这就是暗电流，又称本底噪声。

检测暗电流的方法十分简单，即把镜头盖盖上，并在暗室中或夜间用黑绒布把干涉成像光谱仪包裹起来，让设备电子学部分在设定的温度条件下开机工作，读出电信号的输出 DN。

因为它是一种随机噪声，所以要检测 25 次以上取均值，暗电流噪声还与工作温度、曝光时间、电子学增益有关。如有必要需检测不同情况下，不同的暗电流噪声，嫦娥一号卫星干涉成像光谱仪按二档曝光时间和三档电子学增益共 6 种组合，每种组合含 128 元*128 元的暗电流值。

仪器在轨运行时，CCD 像元的输出应减去相应的暗电流值，以提高光谱辐亮度值的真实性。表 3 为暗电流检测值，单位为 DN，量化等级为 12bit，DN_{max}=4096，检测温度 t=25℃±5℃。

表 3 暗电流检测值

Time of Exposure \ Gain	K=1		K=1.5		K=2	
	DC-bias	AC-bias	DC-bias	AC-bias	DC-bias	AC-bias
t=140 ms	241.554	0.899	357.646	1.331	457.702	1.645
t=70 ms	240.438	0.971	356.163	1.198	455.623	1.453

2.3 最大饱和辐亮度的配准

最大饱和辐亮度的配准是指对预估的目标最大辐亮度值，当干涉成像光谱仪采用最小一档曝光时间与电子学增益时，CCD 像元输出不出现饱和[15]。

对嫦娥一号卫星干涉成像光谱仪，其工作波段是 480 nm～960 nm，在此波段内太阳大气外的辐照度为 630 W/m^2，月表最大漫反射系数经调研为 ρ_{max}=0.26，故目标的最大辐亮度约为 52 W/m^2·sr(480~960 nm)，采用积分球为光源，通过调节积分球内开灯数目，把积分球出口处在 480 nm~960 nm 间的辐亮度调节到预估的最大辐亮度 52 W/m^{2}sr(480~960 nm)，检查主极大的输出是否饱和，在嫦娥一号卫星的干涉成像光谱仪中，由于分束膜产生了 π 位相突变，信号极大成为负值，见图 4，这时可检查次极大是否饱和。

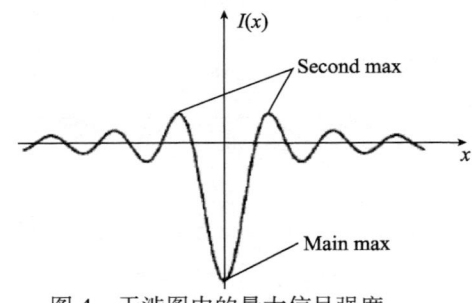

图 4 干涉图中的最大信号强度

还要注意一点，就是积分球输出的光谱辐亮度分布与月表对太阳光谱的反射光辐亮度分布是不相同的，而探测器对不同波长的响应又是不相同的，所以需要建立一个模型予以校正。在嫦娥一号卫星干涉成像光谱仪中是通过改变入射狭缝的宽度来配准最大饱和辐亮度的，理论入射狭缝的宽度为 0.1 mm，对最大辐亮度配准后实际装配的狭缝宽度为 0.06 mm，实际狭缝的装配宽度小于理论值不但无害，而且还可以减小由于推扫造成的 MTF 下降，而当狭缝宽度大于理论值时，则会降低地元分辨率。

能做到这一点的前提条件是在做总体方案设计时，使设计光能量大于预估值，在嫦娥一号卫星中相对孔径为 F/2.4，就是出于这样的考虑。

在最短曝光时间下，使最大目标辐亮度接近饱和，可以有效利用探测器的动态范围，这是一条重要的设计原则。

在轨运行表明，这样的考虑起到了很好的作用，在月表很大的辐亮度变化范围内，都得到了清晰的多光谱序列图像。

2.4 相对定标

就一台照相机来说，如果输入的面光源是均匀的，那么焦平面上的每个CCD像元的输出也应该相同，然而一台实际的照相机不是这样。每个CCD像元的实际输出都不相同，通常中心视场处的CCD输出大，边缘就逐渐降低，它是由于像面照度与视场角余弦的四次方成正比、系统有渐晕、不同视场角光线通过系统时损失不同、以及CCD器件像元间响应的不一致性以及电路性能的不一致等原因所致。

但是它是系统性的缺陷，可以通过平场[8]把它降至最低。平场就是在均匀面光源输入条件下，把响应输出低的CCD像元乘以一个略大于1的系数，而把响应输出高的CCD像元乘以一个略小于1的系数。

照相机的平场技术国内外都十分成熟，做法也大致相同。但它不适用于干涉成像光谱仪，因为对干涉成像光谱仪而言，一个均匀的光场的输入，其输出始终是干涉条纹，为此提出了所谓的"行"平场原理[8]。

图5表示干涉成像光谱仪所用的面阵CCD。其中i方向定义为空间方向，称为行方向，共有128元，j方向与飞行方向相同，称为列方向，每一列都是一个地元的一张干涉图，所以共有128列。其中$j=0$的一行是所有像元处于干涉图中零光程差的行，对一个均匀的输入$j=0$这一行的128个像元的输出应该相同，进行平场，同理对于$j=+1$及$j=-1$的256个像元的输出也应该相同，同样进行平场，为此类推直到$j=+64$及$j=-64$为止。

如果需要检查另一个方向的平场情况，可以把CCD芯片绕中心点旋转90°，重复上面的工作。在平场中，采用积分球作为均匀输入光源，积分球具有很好的平面及角度均匀性，而且光谱辐亮度稳定性高。把被检干涉成像光谱仪直接对准积分球中心即可。

对二档曝光时间与三档电子学增益共六种组合，分别给出了$R(128，128)$平场系数矩阵，相对定标不确定度为1.9%。

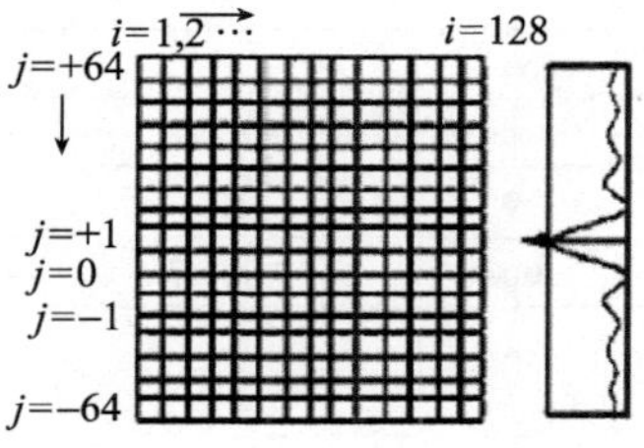

图5 "行"平场原理图

2.5 光谱绝对定标

光谱的绝对定标可以施加于干涉图也可以施加于复原光谱图，实验证明后者直观，简单、可靠。但是作为光谱绝对定标的标准源目前国际上都是色散型的光谱仪，虽然它也可以直接通过比对来给复原光谱定标，但由于两者的原理不同，光谱分辨率的表征方法不同。色散型基本都是波长等间隔，而干涉型的是波数等间隔。

实验室采用美国ASD Fieldspace Pro UV/VNIR色散型光谱仪[14]，经计量校验后作为标准源，它的光谱分辨率为2.5 nm，而探月的干涉成像光谱仪在工作波段最小波长上折算为波长的分辨率为7 nm(波数分辨率为325.5 cm^{-1})。试验时，两台设备同时对积分球采集数据，把由干涉成像光谱仪采集的数据做后处理，复原得到32个通道的未经标定的曲线，而把作为标准的色散型光谱仪测量得到的光谱图做一次傅氏变换，并使最大光程差等于探月干涉成像光谱仪的最大光程差，然后再做一次反傅氏变换，这时就得到一张以波数等间隔，光谱分辨率与探月干涉仪相同的光谱图，以后者对前者赋值[9]。

光谱绝对定标中的标准源是一台色散型光谱仪，没有空间分辨率，但由于干涉成像光谱仪在绝对光谱辐射度定标前已完成相对定标平场，所以在空间上可以视为同一。

在实验室定标中，见图6。不论是被定标的干涉成像光谱仪还是作为标准源的光谱辐射度计都要同时对准积分球采集25次以上的数据，采用均值处理，并做数据的离散性分析。此外还要对不同曝光时间与电子学增益的6中组合，分别给出(1×32)的绝对定标矩阵，光谱辐射度绝对定标的最大不确定度为6.39%，均值为3.56%。

在最大辐亮度配准、平场、光谱辐亮度绝对定标中，采用同一套检测装置，而且一次就采集所有数据。

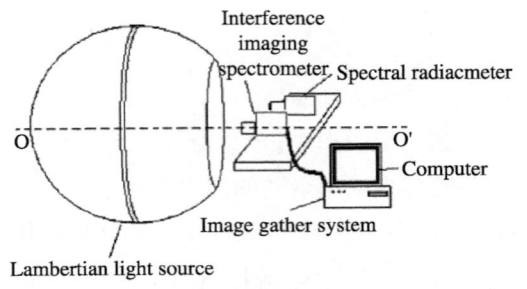

图 6　定标装置配置

2.6　输出 VS 输入的线性检测

积分球内共有 20 个灯,可以逐一开启,以改变积分球出口处的光谱辐亮度值,并由光谱辐射度计测量辐亮度,干涉成像光谱仪输出 DN,得到输出 VS 输入的线性度[15],试验中共采用了 17 档不同的辐亮度,表 4 是线性的均值,图 7 为 17 档不同辐亮度的线性关系。

表 4　响应线性度检测

No.	1	2	3	4	5	6	7	8
linearity	0.025 2	0.006 3	0.001 5	0.000 6	0.000 3	0.000 3	0.000 8	0.001 6
9	10	11	12	13	14	15	16	17
0.002 1	0.001 7	0.001 5	0.001 3	0.000 8	0.000 3	0.000 5	0.001 2	0.002 0

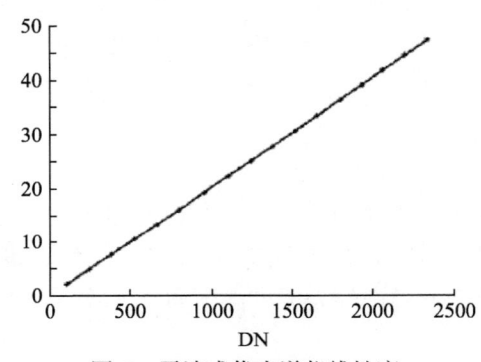

图 7　干涉成像光谱仪线性度

从表 3 可以看出在 17 档不同的辐亮度中除最小档的线性度为 2.5% 之外,其余都小于 1%。

2.7　光谱分辨率与谱线位置检测

光谱分辨率及谱线位置的检测[16]是采用高稳定的激光来进行的,检测装置见图 8。共用了 4 种激光器,其中 2 种为气体激光器,2 种为半导体激光器。激光具有单色性好的特点,它的半高宽 FWHM 一般都在埃的量级,同时谱线位置稳定,然而在近红外波段就没有合适的气体激光器,而只能用半导体激光器,与气体激光器相比,它的谱线位置会产生温漂,所以气体激光器的检测数据较为可靠。

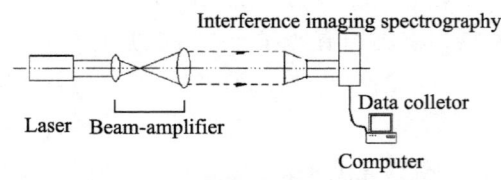

图 8　谱线位置与光谱分辨率检测装置

由于激光束的光束直径很小，通常小于 1 mm，所以在检测时加了一个 20 倍激光扩束镜，使光束口径充满干涉成像光谱仪的入瞳。

数据采集卡得到的激光干涉图经软件处理后，可获得复原光谱，它的半高宽值表示光谱分辨率，其峰值点对应的波长位置为谱线位置，表 5 给出了四种激光波长的检测结果。

表 5 光谱分辨率与谱线位置的检测结果

laser wavelength/nm	spectral resolution			Wavelength location			Comment
	theory	test	error	theory	test	error	
543.5	9.6 nm	10.4 nm	7.7%	543.5 nm	543.46 nm	0.04 nm	Gas laser
632.8	13.1 nm	13.8 nm	5.3%	632.8 nm	633.11 nm	0.31 nm	Gas laser
783.8	20 nm	21.8 nm	9%	783.8 nm	784.56 nm	0.76 nm	semiconductor laser
831.2	22.5 nm	24.2 nm	7.5%	831.2 nm	828.72 nm	2.48 nm	semiconductor laser

表 5 中采用了激光波长而不是波数，所以处理时，按式(2)进行换算。

$$\frac{\delta\sigma}{\sigma} = \frac{\delta\lambda}{\lambda} \tag{2}$$

式中 λ 表示波长，$\delta\lambda$ 表示波长分辨率，σ 表示相应的波数，$\delta\sigma$ 为波数分辨率，用式(2)可以方便地把测量得到的 $\delta\sigma$ 转换为 $\delta\lambda$ 值。

3 在轨光谱辐射度的相对检测

3.1 在轨数据获取情况

嫦娥一号成像光谱仪是国内首台上天的干涉型成像光谱仪，同时又是国际首台对月球进行多光谱探测的干涉成像光谱仪，自 2007 年 11 月 27 日开机以来，设备工作一直正常，到 2009 年 3 月 1 日卫星受控撞击月球为止，已获得全月面 78%区域的多光谱图像，图 9 是部分典型的多光谱图像。

3.2 在轨光谱辐射度的相对检测

在实验室内，可以用标准的光谱辐射度计、积分球定量地进行定标。对地球观察而言，在轨可以通过特定地点的辐射场定标，但是对于月球就十分困难，因为它不具备这种测量条件。检测时，首先在月表面选择一块 3.2 km×3.2 km 的区域，它的光谱辐射度具有较好的均一性，但没有可供比对的绝对值，因此这种检测只是相对的。而且有一个重要前提：即被选的 3.2 km×3.2 km 的区域确实仅有较小的光谱辐亮度差异，但目前无法证实。

干涉成像光谱仪的地元分辨率 GSD=200 m，3.2 km×3.2 km 区域占 16*16 个地元，可以获得 16×16=256 个可分辨地元的复原光谱图，在每一地元的复原光谱图中有 32 个采样点上的波长值以及按地面实验室定标得到的相应辐亮度值，对 256 个地元中同一波长的辐亮度值，求均值作为"真值"对每个地元的值与"均值"求标准差，然后对 32 个波长的标准差再求一次均值，最后得到 256 个地元间的相对光谱辐射度不确定度，见图 10。检测结果为 2.5% ~ 9.5%[17]。

4 结论

根据嫦娥一号探月卫星干涉成像光谱仪的任务要求及科学目标，完成了实验室的光谱辐射度定标，根据干涉型成像光谱仪的特点，在定标过程中，提出了"行"平场、不同类型光谱仪的对比方法，都得到了好的应用成果。文中某些内容，如最大饱和辐亮度的配准等严格的意义上并不属于定标的内容，但它们与定标有一定关联，另有一些项目属质量检测，如光谱分辨率与谱线位置，在文中一并做了介绍。

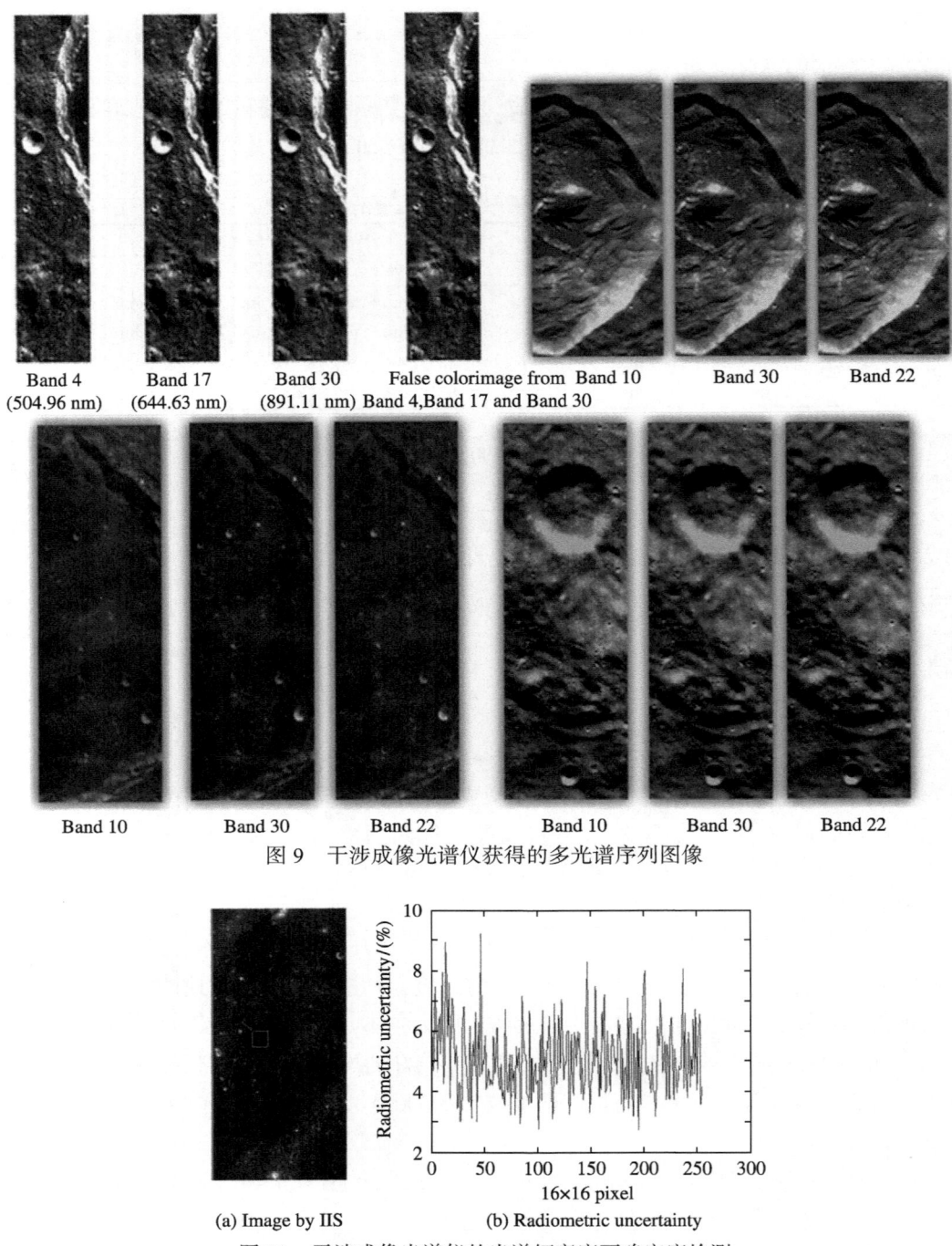

图9 干涉成像光谱仪获得的多光谱序列图像

图10 干涉成像光谱仪的光谱辐亮度不确定度检测

致谢 首先感谢中科院安徽光机所定标研究室的技术人员,他们为定标设备进行了校验,并参与了部分定标工作,同时还要感谢中科院国家天文台,他们完成了某些在轨检测工作。

参 考 文 献

[1] OU YANG Zi-yuan, LI Chun-lai, ZOU Yong-liao, et al. The scientific object of first phase project of chinese lunar exploration [J]. *Spacecraft Engineering*, 2005, 14(1):1—5

[2] LEONARD J O, EUGENE W B, et al. Design of an airborne Fourier transform visible hyperspectral imaging system for light aircraft environmental remote sensing [C]. *SPIE*, 1995, 2480:418—424

[3] SUMMER Y, THOMAS R C, et al. MightySat II. 1 hyperspectral imager: summary of on-orbit performance[C]. *SPIE*, 2002,

4480:186—197

[4] LEONARD J O, ANDREWDM, et al. Payload qualification and optical performance test results for the MightySat II. 1 hyperspectral imager [C]. *SPIE*, 1998, 3498: 231—238
[5] RAFFART B J, SELLAR G R, BLATT H J, et al. Monolithic Fourier-transform imaging spectrometer [J]. *Appl Opt*, 1995, 34:7228—7230
[6] ZHANG C M, ZHAO B C, et al. Analysis of signal-to-noise ratio of ultra-compact static polarization interference imaging spectrometer [J]. J Opt A: *Pure Appl Opt*, 2009, 11: 085401
[7] HAMMER P D, DAVID L P. Imaging Interferometry for terrestrial remote sensing [C]. *SPIE*, 1995, 2480: 153—164
[8] 干涉成像光谱仪的平场方法. 发明专利，200710018975.7
[9] 不同类型光谱仪的对比方法. 发明专利，ZL20071008974.2
[10] ZHAO Bao-chang, YANG Jian-feng, et al. Optical design and on-orbit performance evaluation of the imaging spectrometer for chang'e-1 lunar satellite [J]. *Acta Photonica Sinica*, 2009, 38(3): 4792483
[11] 西师毅，Fourier Spectroscopy. 电子技术综合研究所调查报告，第 169 号
[12] ZHANG Chun-min, YAN Xin-ge, ZHAO Bao-chang. A novel model for obtaining interferogram and spectrum based on the temporarily and spatially mixed modulated polarization interference imaging spectrometer [J]. *Optics Communications*, 2008, 281(8): 2050-2056
[13] ZHANG Chun-min, JIAN Xiao-hua. Wide-spectrum reconstruction method for a birefringence interference imaging spectrometer [J]. *Optics Letters*, 2010, 35(3): 366—368
[14] Analytical spectral devices. FieldSpace Pro FR User's Guide. Boulder: Colo, 2002
[15] GU Y, ANDERSON J M, MONKJ G C. An approach to the spectral and radiometric calibration of the VIFIS system [J]. *International Journal of Remote Sensing*, 1999, 20(3): 535—548
[16] THOMAS G C, ROBERT O G, et al. Accuracy of the spectral and radiometric laboratory calibration of the airborne visible/infrared imaging spectrometer [C]. *SPIE*, 1990, 1298: 37—49
[17] 嫦娥一号卫星科学探测数据质量初步评估报告(内部). 中国科学院国家天文台月球与深空探测科学应用中心

嫦娥一号 IIM 数据绝对定标与初步应用

吴昀昭[1,2]　徐夕生[3]　谢志东[3]　唐泽圣[1]

(1. 澳门科技大学月球与行星探测科学应用研究联合实验室，澳门；2. 南京大学地理与海洋科学学院，南京 210093；3. 南京大学地球科学与工程学院，南京 210093)

摘要 我国首颗探月卫星嫦娥一号携带的干涉成像光谱仪 IIM 能够以较高的空间分辨率和光谱分辨率获取月表元素和矿物信息，是实现我国首次探月四大任务的必备仪器。现有数据尚未转换为反射率且传感器响应不均一，尚无法直接使用。本文主要目的是对该数据进行绝对定标和辐射畸变校正，为数据的正确使用提供方便。研究结果表明，在空间域传感器左侧响应偏低；在波谱域长波段辐射畸变比较明显。针对探月数据特点和定量分析的需求，提出了基于光谱特征的反射率归一化校正方法，对多个地区的验证表明该方法具有实用性和通用性。经过校正后的反射率与地基望远镜光谱匹配良好，可以开展月球研究。最后，利用校正后的数据对两个直径>35 km 的撞击坑开展了初步的应用，发现在千米尺度上月壳存在成分差异。IIM 数据能够在全球、区域和局部尺度上获取月球表面的化学组成，有助于深化对月球形成和演化的认识。

关键词 嫦娥一号　IIM　定标　辐射校正　月球

　　月球岩石的元素和矿物组成反映了其形成时的物质来源、温度、压力、冷却速度、氧逸度等物理化学条件。了解月球元素和矿物组成对于研究月球的起源和演化有重要意义。月球的地质演化在31亿年前基本结束，保持了月球早期的特征[1]。地球的物质调整现今仍很活跃，早期的岩石和演化痕迹已被抹杀殆尽。通过对月球物质组成的研究将有助于了解地球、类地行星乃至太阳系的起源和演化[2]。美国6次 Apollo 和前苏联3次 Luna 登月所采集的样品为我们了解月球的元素和矿物组成提供了宝贵的信息。然而这9个着陆点集中于月球正面低纬度地区，收集的月球样品无论从数量、地域分布以及岩石化学组成的代表性来说，都有很大的局限性。目前对于月球表面元素和矿物空间分布的认知还很不完全，遥感技术仍是当前获取月球化学组成的主要手段。

　　可见光近红外反射光谱(visible and near infrared region, VNIR)遥感已被广泛用于获取月表元素和矿物成分[3~8]。日本的"SELENE"、我国的"嫦娥一号(CE-1)"以及印度的"Chandrayaan-1"等新近发射的3颗月球卫星都携带有成像光谱仪，成像光谱(又称高光谱)遥感已成为月球光学遥感的趋势。成像光谱仪能够在二维空间获取月表连续光谱，具有图谱合一的功能，是完成我国首次探月四大任务的必选仪器[9]。我国自主研制的干涉成像光谱仪(IIM)首次将干涉光谱成像技术用于深空探测。目前嫦娥一号已经圆满完成绕月探测任务，IIM 数据也已整理结束并提交用户，极有必要尽快对数据进行分析和解译，利用我们自己的第一手月球科学探测数据来认识月球。

　　目前所发布的最高级别 IIM 数据为2C级，该数据已经过实验室定标等一系列预处理，然而仍需进一步处理方可使用。目前的主要问题包括：(i)数据单位是辐亮度而非反射率；(ii)传感器响应不均一，以致相同的月表成分在行方向上展示不同的光谱特征(详见下文)，这就给基于光谱特征的月球地质解译带来多解性。针对上述两个问题，本文的主要目的是建立2C级IIM数据绝对定标和辐射畸变校正方法，为

本文原载于《中国科学 G 辑：物理学 力学 天文学》，2009，Vol.39，No.10，1387~1392。

今后研究者使用该数据提供参考。为此，主要工作包括将辐亮度数据转换为反射率数据，并对128个像元线阵光谱响应的差异进行校正，并利用校正的数据开展初步月球地质应用。

1 嫦娥一号 IIM 数据定标

本文所用数据是经过了暗电流纠正，平场纠正，辐亮度转换，光学归一化($i = 30°$，$e = 0°$)等校正的2C级数据。表1是IIM数据主要技术和性能指标。关于2C数据的详细介绍读者可参阅《绕月探测工程科学应用专家委员会工作参考材料》。

表1　嫦娥一号IIM的主要技术和性能指标表

成像宽度/km	像元分辨率/m	成像区域/(°)
25.6	200(星下点)	N75~S75(太阳高度角大于15°时)
光谱范围/nm	光谱波段数-分辨率/nm	量化等级/bit
480~960	32(9.6nm@543.5, 13.1nm@632.8, 20nm@783.8, 22.5nm@831.2)	12
像元数	MTF	S/N
128像元	≥0.2(黑白快视图)	≥100(月表反射率$r=0.2$，太阳入射角30°，黑白快视图)

研究表明，使用月表作为定标标准比采用恒星定标精确[10]。目前月壤样品的光谱已经能够在实验室内精细测量，极大方便了月球卫星的定标。利用月壤样品进行定标的关键是样品成熟度高、有代表性，图像定标点比较均一、未被其他撞击溅射物质混染。Apollo 16样品62231代表了着陆区附近典型的高成熟的斜长岩物质。Pieters等在实验室测试了其双向反射光谱，并用它对地基望远镜和克莱门汀数据进行定标[11]。IIM 2C数据已经校正到与该双向反射率相同的观测几何，本文利用该双向反射率对IIM 2C数据进行定标。

国际上常选择South Ray、North Ray两个撞击坑所构成等边三角形的第三个顶点作为图像定标点[12,13]。该地区恰位于IIM 2224和2225轨之间，没有IIM数据。我们通过将这两轨IIM数据与高分辨率克莱门汀全色数据(100 m)和多光谱数据(500 m)进行对比，在2225轨选择与上述定标点距离近、光谱特征类似、地表均一、没有其他撞击溅射物质混染的一块区域作为IIM数据定标点(图1)。具体步骤如下：

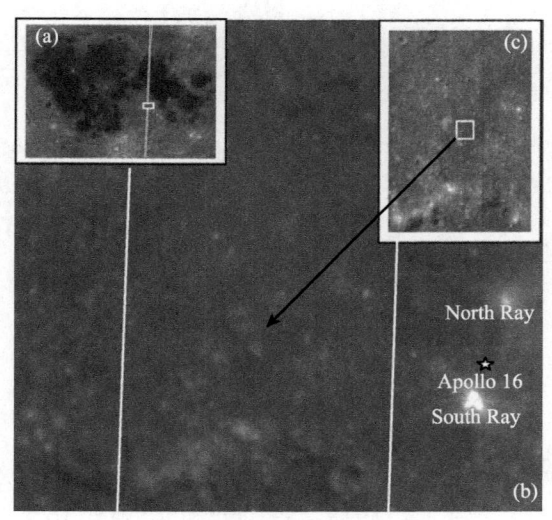

图 1　IIM 定标点位置示意图

(a) 中条带是 2225 轨覆盖月面位置；(b) 100 m 分辨率克莱门汀数据；(c) IIM 数据，方框显示定标点位置

(i) 仔细研究定标点地区色调和结构特征，选取第 11151 至 11167 行，69 至 73 列的 17×5 像素作为定

标区域, 对其光谱进行平均。

(ii) 采用 Gaussian 曲线作为光谱响应函数, 将月壤样品 62231 双向反射率转换为 IIM 反射率。光谱记为 R_{standard}。

(iii) 获得定标因子, 将图像转换为反射率为

$$R = DN_{\text{IIM}} / DN_{\text{Ap16}} \times R_{\text{standard}}$$

图2是定标前、后的光谱曲线。由于最两端两个波段信噪比很低, 本文研究排除这两个波段。

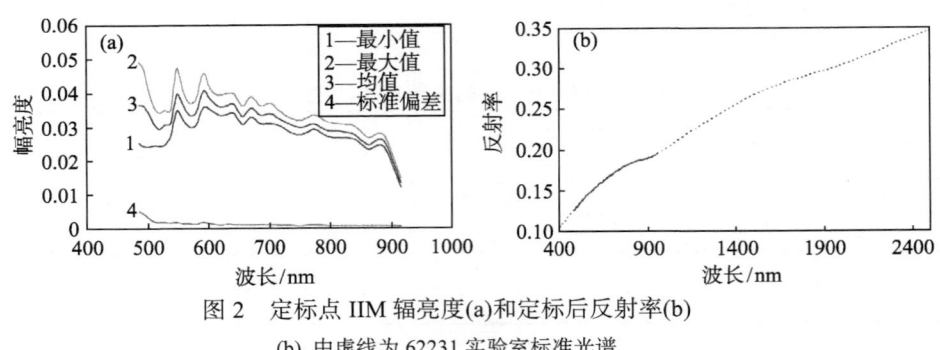

图 2　定标点 IIM 辐亮度(a)和定标后反射率(b)

(b) 中虚线为 62231 实验室标准光谱

2　IIM 数据行向响应不均一性校正

图3(a)和(c)是玄武质月海地区(MS2)和较大的撞击坑(Petavius)IIM真彩色合成影像。尽管两个地区成分、遭受撞击程度和地理位置截然不同, 它们有一种共性, 即图像从左至右色调不均一。这种辐射畸变现象在比值图像上更为明显。图3(b)和(d)分别是上述两个地区对应的891/918 nm(R), 918/891 nm(G)和 757/918 nm(B)比值合成影像。在图3(b)和(d)中, 左侧1/5呈粉红色, 右侧图像偏绿色, 而撞击坑、月溪等真实的月表特征却无法显示。图3(图版V)一方面说明这种色调的变化不是月表成分引起的, 另一方面也说明这种行向响应的不均一不受地区影响, 这为对其校正提供了基础。

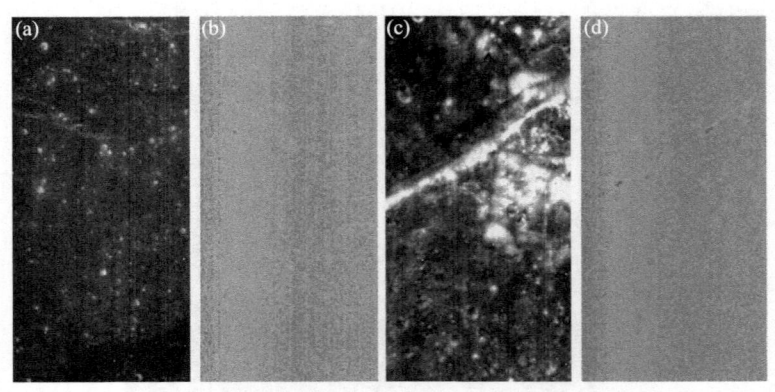

图 3　MS2(a)和 Petavius 撞击坑(c)真彩色合成影像以及两个地区相应的 891/918 nm(R), 918/891 nm(G), 757/918 nm(B)比值合成影像(b)和(d)

为进一步研究IIM各波段横向响应不均一性, 我们在成分和结构上都较为均一的澄海地区(MS2)提取了光谱曲线(图3(a)中两个红色方框位置)以及918, 891 nm和865 nm 3个波段在同一行的反射率(图3(a)中红线位置)。由图4(a)可见, 尽管横向上物质成分和地表结构近乎一致, 然而光谱却不同, 长波段差异尤其明显, 而且都偏离地基望远镜光谱。因此, 在开展地质应用研究之前如不对此进行校正将会产生错误的解译。由图4(b)可见, 3个波段从左至右反射率都逐渐增大, 波长最长的918 nm行向响应不均一性最为

明显。

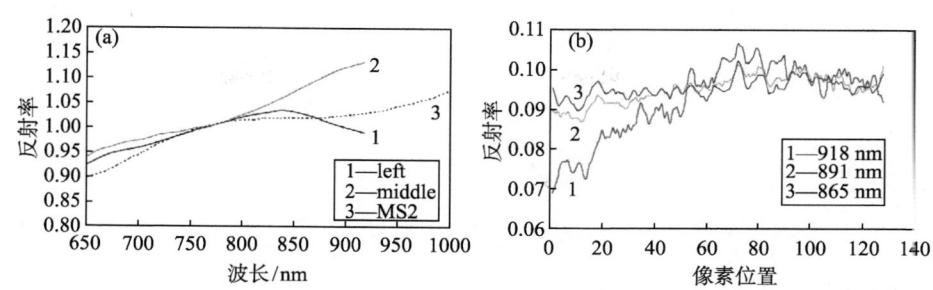

图 4 (a)MS2 处左侧和中间(图 3(a)上部红色方框)反射率和地基望远镜光谱;(b)MS2 处从左至右 128 个像元在 918,891 和 865nm 反射率(图 3(a)上部红色虚线)

地面同步光谱测量是较为精确的遥感图像辐射畸变校正方法,然而该方法对于月球遥感数据难以适用。针对探月数据特点和定量分析的需求,本文提出了基于光谱特征的反射率归一化校正方法。我们选择前述定标点地区校正横向响应不均一性。选择该地区,一是由于此处具有实测的月壤光谱,IIM反射率都以此为基准所产生。二是该地区月表成分比较均一、地形平坦、撞击坑较少,这样行向光谱的差异可以认为主要是传感器响应不均一所致,地表成分差异的影响可以忽略。具体校正方法如下:

(i) 以地基望远镜光谱为标准,求取 IIM 光谱各波段相对 750 nm 反射率。
(ii) 将从左至右每个像元在每个波段的反射率都归一化到(i)所得相对反射率。
(iii) 分别对各个波段利用二阶多项式拟合(ii)所得值,获得校正因子,利用该校正因子对图像进行校正。

该纠正不仅仅是色调上的调整,两个因素确保校正后的像元光谱都具有物理意义,即地物反射率。首先获取校正因子所选择的地区为IIM反射率转换的地区,即Apollo 16着陆点附近。其次,校正因子的获得是将每一列光谱相对750 nm的反射率都校正到与标准光谱一致的相对反射率。

3 校正结果

我们利用前述方法获得了IIM数据行向响应不均一性校正因子。由图5可见在空间域校正因子在左侧最高,向右逐渐降低,表明传感器左侧响应较低。在波谱域,波长越长校正因子越大,表明传感器长波段辐射畸变比较明显。通过对比校正前(图4(a))后(图6)反射率,可见经过校正后,行向上的响应趋于均一化,基于此校正可开展应用研究。为检验本校正因子在其他地区的适应性,我们选择月球高纬度、斜长岩成分地区进行了校正。由图7(a)和(b)可见(图版V),校正前横向上色调的不均一性在校正后得到消除。3个近红外波段反射率在横向上的响应也变得均一(图7(c))。这说明本文提出的辐射畸变校正方法具有实用性和通用性。

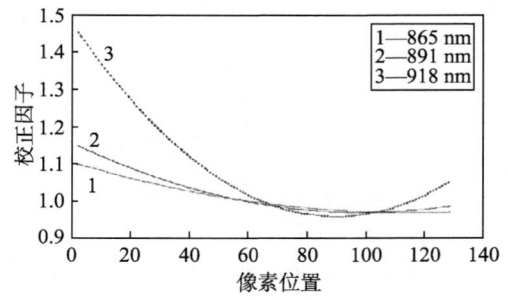

图 5 IIM 数据横向响应不均一校正因子

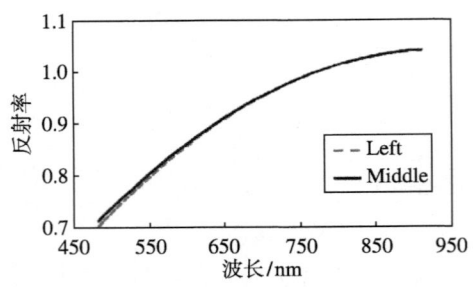

图 6　MS2 处左侧和中间(图 3(a)上部红色方框)校正后的反射率

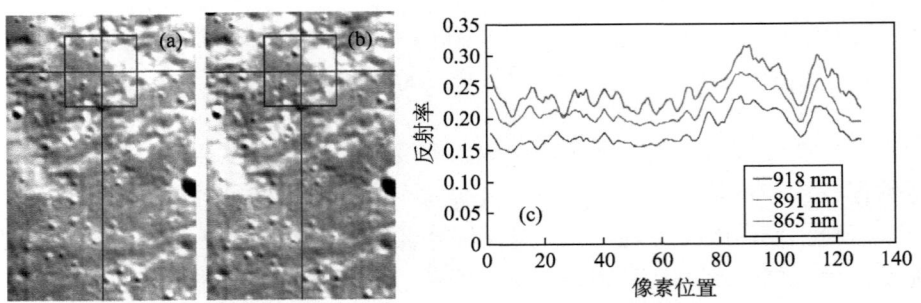

图 7　高纬度、斜长岩地区校正前(a)后(b)影像和校正后的横向光谱剖面(c)

4　嫦娥一号 IIM 数据应用

当前较为流行的月球形成理论是岩浆洋假说，即月球形成早期存在全球范围的熔融岩浆，密度较轻的斜长岩浮在月球表面，形成斜长岩月壳。为验证该模型，除了需要知道全球范围的元素和矿物组成，重要的一点是确定下部月壳的组成。月球上直径>35 km 的撞击坑通常含有中央峰。组成中央峰的物质认为是该撞击坑最深的物质，它们可能暴露了下月壳或者上月幔成分。Pieters利用地基望远镜光谱研究了许多撞击坑从周围到撞击坑中央峰的成分，显示了月壳在纵向上存在成分的变化[14]。然而由于地基望远镜分辨率较低，制约了月壳成分横向变化的研究。

我们利用校正后的IIM光谱研究了Petavius撞击坑(2495轨)中央峰的成分。过去地基望远镜光谱只发现该地区存在斜长岩，没有铁镁质矿物吸收[13]。然而校正后的IIM光谱显示，该撞击坑除了含有斜长岩中央峰之外，还有苏长岩性质的中央峰(图8)。这两个中央峰相距约为6 km，这表明除了纵向上月壳成分存在变化，横向上在很小的距离范围内月壳成分也存在巨大差异。IIM的光谱范围是480~960 nm，其岩性识别能力不如波段范围更宽的地基望远镜光谱。然而地基望远镜光谱空间分辨率较低，视场内含量较少的岩石光谱特征无法显示，本文对Petavius撞击坑的研究也体现了IIM高空间分辨率的优势。

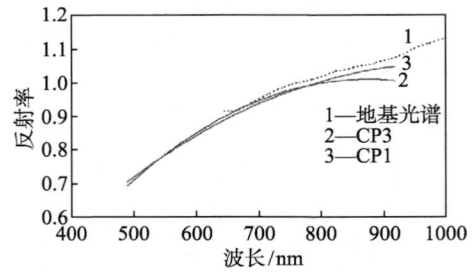

图 8　Petavius 撞击坑两个中央峰校正后反射光谱

IIM有32个谱段，波段之间高度相关，数据存在大量冗余。主成分分析(PCA)有助于对数据降维并减弱噪声增强有用信息。我们利用PCA方法对Aristarchus撞击坑(2897和2898轨)进行了岩性分类研究。图9

(图版VI)是前3个主成分(PC1，PC2，PC3)假彩色合成图。选择月表成分均一地区对比校正前后分类结果，有助于排除成分变化的干扰而充分体现校正效果。结合克莱门汀影像和地基望远镜观测，图9中虚线位置成分比较均一，为黑色火山碎屑沉积物，在校正前图9(a)显示两种色调，被分为两类。校正后图9(b)色调变为一致，分为一类，可见校正提高了分类精度。Aristarchus撞击坑西南壁、西北壁和东部壁展示了不同的色调，这表明这个40 km直径的撞击坑在横向上存在成分的多样性。由于IIM波段范围较窄，没有完全覆盖铁镁质矿物吸收峰，制约了月表岩石矿物的鉴定。我们下一步研究将是开发针对IIM数据特征的岩石矿物识别方法，以促进IIM数据的进一步应用。

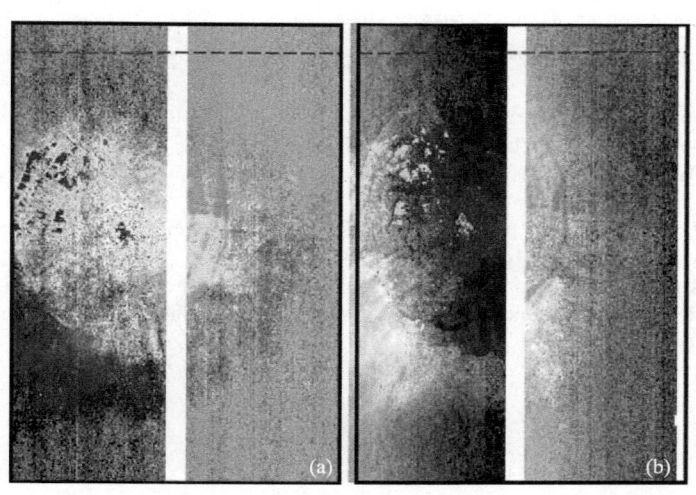

图9 Aristarchus撞击坑校正前(a)和校正后(b)PC1(R)，PC2(G)，PC3(B)合成图

人类对月球形成与演化的认识水平依赖于其化学成分数据的完善程度。月球演化模型都必须合理解释月壳不同尺度成分的不均匀性。此次探月热潮的主要目标之一是在全球、区域和局部尺度上详细了解月球的化学组成，探索月球的形成和演化。IIM数据具有高的光谱分辨率和高的空间分辨率，为实现该目标提供了数据支持。

5 结论

IIM能够以较高的空间分辨率和光谱分辨率获取月表元素和矿物成分信息。但现有数据尚未转换为反射率，而且传感器响应不均一，这些数据尚不能直接使用。本文对该数据进行了绝对定标和辐射畸变校正方法的研究。结果表明在空间域上传感器左侧响应较低；在波谱域，传感器长波段辐射畸变比较明显。本文提出的基于光谱特征的反射率归一化校正方法具有实用性和通用性，校正后的光谱曲线与地基望远镜光谱十分吻合，校正后的数据可以用于月球研究。通过对Petavius和Aristarchus两个直径皆>35 km的撞击坑进行研究，发现在千米尺度上月壳成分存在显著的不均匀性。IIM以其较高的光谱和空间分辨率在全球、区域和局部尺度上获取月球的化学组成，有助于深化对月球形成和演化的认识。

致谢 本文所用IIM 2C数据由月球探测工程中心授权、绕月探测工程地面应用系统提供。感谢国家天文台欧阳自远院士、李春来研究员、张文喜老师等多位专家指导。由于获得数据较晚、时间仓促以及作者水平有限，文中难免有不足之处，敬请原谅。

参 考 文 献

[1] 欧阳自远. 月球科学概论. 北京：中国宇航出版社，2005
[2] 肖龙，Ronald G，曾佐勋，等. 比较行星地质学的研究方法、现状和展望. 地质科技情报，2008，27(3): 1—13

[3] Lucey P G, Taylor G J, Malaret E. Abundance and distribution of iron on the Moon. Science，1995, 268: 1150—1153

[4] Pieters C M, Mustard J F, Sunshine J M. Quantitative mineral analyses of planetary surfaces using reflectance spectroscopy. In: Dyar M D, McCammon C, Schaefer M W, eds. Mineral Spectroscopy: A Tribute to Roger G. Burns. Houston: Geochemical Society, 1996.307—325

[5] 胥涛, 李春来. 月球表面元素含量的定量分析方法. 空间科学学报，2001，21(4): 332—340

[6] 胥涛, David T B, 李春来, 等. 月球紫外-可见-近红外反射光谱的基本特征及解析方法. 地球与环境，2004，32(3-4)：27—33

[7] 薛彬, 杨建峰, 赵葆常. 月球表面主要矿物反射光谱特性研究. 地球物理学进展，2004，19(3): 717—720

[8] Chevrel S D, Pinet P C, Daydou Y, et al. The Aristarchus Plateau on the Moon: Mineralogical and structural study from integrated Clementine UV–VIS–NIR spectral data. Icarus, 2009，199: 9—24

[9] 薛彬. CE-1 干涉成像光谱仪信息处理及应用研究. 博士学位论文. 西安：中国科学院西安光学精密机械研究所，2006

[10] Pieters C M, Pratt S. Earth-based near-infrared collection of spectra for the Moon: A new PDS data set. Lunar Planet Sci XXXI, 2000, 2059

[11] Pieters C M. The moon as a calibration standard enabled by lunar samples.In: New Views of the Moon II: Understanding the Moon Through the Integration of Diverse Datasets, Flagstaff. 1999

[12] Pieters C M, Head J W, Isaacson P, et al. Lunar international science coordination/calibration targets (L-ISCT). Adv Space Res, 2008, 42: 248—258

[13] Blewett D T, Lucey P G, Hawke B R, et al. Clementine images of the lunar sample-return stations: refinement of FeO and TiO_2 mapping techniques. J Geophys Res, 1997, 102: 16319—16326

[14] Pieters C M. Compositional diversity and stratigraphy of the Lunar crust derived from reflectance spectroscopy. In: Pieters C, Englert P, eds. Remote Geochemical Analysis: Elemental and Mineralogical Composition. Cambridge: Cambridge University Press, 1993. 309—339

嫦娥一号 IIM 数据定标的改进方法

胡 森　林杨挺

(中国科学院地质与地球物理研究所地球深部研究重点实验室，北京 100029)

摘　要　嫦娥一号月球探测卫星的 IIM 数据在应用之前，必须进行反射率定标。本文在反射率定标之前，对原始 2C 级 IIM 数据进行了坏线和坏点修复，以及条纹去除。由于条纹去除后的 IIM 影像不存在平场效应，有效地抑制了坏线、坏点和平场效应对反射率转换的影响，获得较已有 IIM 数据定标更好的结果。通过对北月球典型地区的 IIM 光谱与地基望远镜光谱，计算得到两者之间的增益和漂移因子。将上述增益和漂移因子应用到月球的其他区域，包括 Apollo 16 和 14 登陆点，Mare Serenitatis 和 Copernicus 撞击坑等区域，均得到理想的校正结果，表明上述增益和漂移因子可有效地校正月表的光谱。

关键词　嫦娥一号　IIM　反射率　定标

自嫦娥探月一期工程结束至今，围绕嫦娥一号获得的月球数据进行科学研究的工作正在全面展开，并取得了一系列的科研成果[1,2]，如全月球的高分辨率影像[3]，高分辨率高程图[4]，全月球的 U，Th 和 K 的分布图[1,2]，月表的微波亮温反演[5]，月球的 FeO 和 TiO_2 含量分布[6]。矿物的光谱特征与其成分和结构存在对应关系，因此，干涉成像光谱仪(IIM)是月球等深空探测的重要有效载荷之一。嫦娥一号绕月卫星搭载的干涉成像光谱仪共获得了 170 G 的 2C 级数据。这些海量的数据蕴含了月表的矿物组成、粒度，以及矿物化学成分等信息，是深入研究月球形成和演化历史不可或缺的重要内容。

从 IIM 光谱数据解译月表矿物的含量以及化学组成等信息，首先需要对所有 IIM 光谱数据进行反射率定标，将辐亮度转换成反射率。由于我国开展深空探测还处于起步阶段，有关 IIM 光谱数据的反射率定标工作报道很少，其中 Wu 等人[7]的分析表明嫦娥一号的 IIM 数据可以反演月表的矿物成分和含量，Chen 等人[8]对虹湾地区进行了 FeO 和 TiO_2 的分布研究，Liu 等人[9]基于 IIM 的光谱吸收特征来获取月表的 TiO_2 含量，以及月表 FeO 含量的分析[6]。

2C 级 IIM 数据虽然完成了一系列的预处理[10]，但给出的是辐射亮度，需要将其转换成反射率后，才能用于矿物成分和含量的反演。在所有的 IIM 数据应用中都必须先对 IIM 数据进行定标[7,9,11]，因而这一过程对后续的深入研究影响至关重要。2C 级 IIM 数据存在非常明显的行向响应不一致(即平场效应)[12]，所以在以往的反射率定标过程中，定标区域的列向像元远多于行向像元(如 2225 轨的 69~73 列，11151~11167 行)，尽量降低平场效应对反射率定标的影响。从文献[12]对平场效应的校正结果可以看出，918 nm 波段的平场效应在 85 列最低，891 和 865 nm 波段的平场效应单调递减，那么选择 69~73 列作为绝对定标区域势必会引入一定的定标风险。本文主要针对 IIM 数据的定标问题，对原来的定标方法进行了改进，本文先对 2C 级 IIM 数据进行坏线、坏点修复、条纹去除，之后再将辐射亮度转换成反射率。本文条纹去除后的 IIM 影像不存在平场效应，有效地消除了坏线、坏点和平场效应对定标的影响。

1 IIM 数据说明

IIM 的成像原理与主要优势在文献[11]中有详细说明，故本文不再赘述，此处着重描述 IIM 的数据存储格式及相关内容。

本次工作使用的是 2C 级 IIM 数据，这些数据已经完成了一系列校正并转换成辐亮度。2C 级 IIM 数据以 PDS 格式存储，头文件和影像数据均存储在同一文件。头文件定义数据的存储格式、大小、经纬度等必要的信息，是读取影像数据和进行几何校正的基础。每一个 IIM 文件的列数(经度)是固定的，均为 128 个像元，空间分辨率为 200 m，每幅影像的空间跨度约 25.6 km；行(纬度)向跨度比较大，从北纬 70°至南纬 70°均有覆盖，每幅影像的纬度覆盖略有差异。

2 IIM 数据处理流程

以往处理 2C 级 IIM 数据的流程为：先将 IIM 辐亮度转换成反射率，再修复坏线和坏点，去除条纹，进行平场校正、几何校正和反射率交叉定标，最终得到校正后的反射率数据[7,11]。从该处理流程可以看出，坏线修复、坏点修复、条纹去除等步骤在辐射亮度转换成反射率之后。由于存在坏线，所以在选择绝对定标区域的时候，需要避开坏线，以免引入误差。另外，以往 2C 级 IIM 数据的处理结果表明，由于条纹去除后 IIM 影像行向响应的不均一性(平场效应)非常明显，所以在条纹去除后，还需要对平场效应进行校正[7,11]。平场效应的存在，必然会影响反射率定标区域的像元值，最终会影响反射率转换的精度。以往选定的反射率定标区域为第 2225 轨 IIM 影像的 11151~11167 行、69~73 列所覆盖的范围[7]。上述定标区域落在平场校正曲线最小值的左侧[7]，最终会导致影像转换成反射率后数值偏低。

所以本文在 IIM 数据处理流程上进行一些调整：先修复坏线和坏点，之后去除条纹，再转换成反射率，并对反射率进行交叉定标，最后获得反射率数据。该处理流程不仅获得了好结果，同时也可以省略平场效应校正的处理步骤。

3 坏线、坏点修复及条纹去除

有关 IIM 数据中坏线和坏点的处理方式参考前人的方法，详细内容可参考相关文献[7,11]。

在修复坏线过程中，本次工作发现，计算像元灰度斜率的公式中有一项除数，导致计算行向相邻像元值的差可能为 0，产生被 0 除的情况，故本次工作把像元灰度斜率(S)的公式修改为

$$S_{i,j,k} = \frac{(DN_{i-1,j,k} + DN_{i+1,j,k} - 2 \times DN_{i,j,k})}{|DN_{i-1,j,k} - DN_{i+1,j,k}| + \text{Delta}} \tag{1}$$

其中 DN 为像元值，i, j, k 分别代表影像的列号、行号和波段号。当 $|DN_{i-1,j,k} - DN_{i+1,j,k}|$ 不等于 0 时，Delta 赋值为 0；当 $|DN_{i-1,j,k} - DN_{i+1,j,k}|$ 等于 0 时，Delta 赋值为 0.00001 进行像元灰度斜率计算。

去除条纹时采用"全局去条纹"方法[13]。设传感器的增益为 $a_{i,k}$，偏移量为 $b_{i,k}$，则图像中第 k 波段的 i 列、j 行的辐射值 $I_{i,j,k}$ 应修正为，

$$I'_{i,j,k} = a_{i,k} \times I_{i,j,k} + b_{i,k}, \tag{2}$$

其中 $I_{i,j,k}$ 为像素原始反射率，$I'_{i,j,k}$ 为修正值，$a_{i,k} = d_{\text{all}}/d_{i,k}$，$b_{i,k} = m_{\text{all}} - m_{i,k} \times d_{\text{all}}/d_{i,k}$ 式中 d_{all} 和 $d_{i,k}$ 分别是某波段整个图像以及像元所在列的反射率标准差。m_{all} 和 $m_{i,k}$ 分别是某波段整个图像以及像元所在列的反射率的平均值。从"全局去条纹"的原理可以看出，每列的增益和偏移量表明传感器在每列的响应不均一，即存在平场效应，那么去除条纹后可以校正影像行向响应不均一的差异。

以 2225 轨影像为例,坏线修复前后的图像如图 1(a)和(b)所示,坏线修复后的影像修复了像元异常多的列。坏点修复后的影像如图 1(c)所示,消除了原图像孤零异常的像元,降低了图像的标准偏差,该步骤是做好条纹去除的必要保证。条纹去除后的影像如图 1(d)所示,对比图 1(a)~(d)的多波段(RGB,R 为 918 nm,G 为 757 nm,B 为 658 nm)彩色合成影像容易发现,条纹去除后的影像消除了行向响应不均一的现象,如原始图像(图 1(a))、坏线修复后的图像(图 1(b))和坏点修复后的图像(图 1(c))的 RGB 影像左侧偏蓝,中央呈深红,右侧稍微暗红,而条纹去除后的影像(图 1(d))色调均匀(图版 VI)。

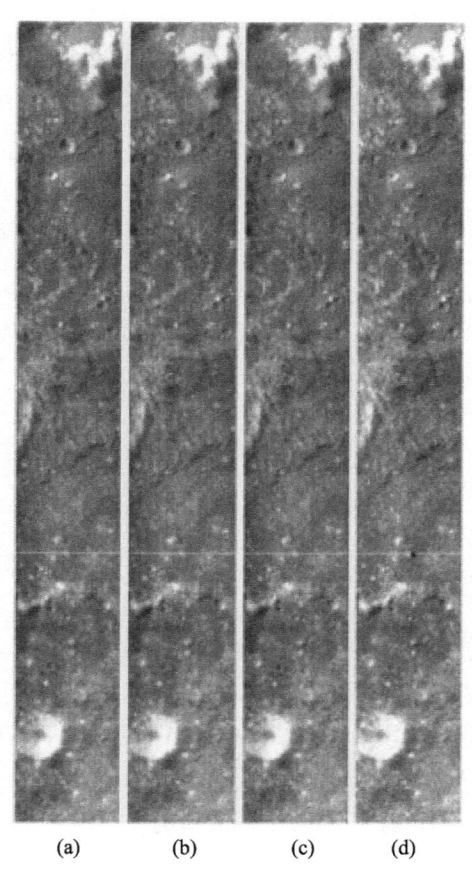

图 1 原始2225轨(a)、坏线修复(b)、坏点修复(c)以及条纹去除后(d)的RGB
(R: 918 nm, G: 757 nm, B: 658 nm)彩色合成影像

(a)~(c)中均可见明显的色调不均匀,左侧偏蓝,中央呈深红,而右侧呈暗红,表明从左到右的响应不一致。(d)是条纹去除后的 RGB 合成影像,色调均一,表明消除了平场效应。(d)中蓝色方框所示范围为定标区域。(a)~(d)中白色横线为图 2 的切面位置

条纹去除后,影像的行向响应比较均一,与以往的处理结果存在较大差异[7,11]。以 2225 轨影像的 11160 行向切面为例,图 2(a)~(d)分别为原始、坏线修复、坏点修复和条纹去除后的影像切面图。图 2 中 3 条曲线分别为图 1 中彩色影像的合成波段(R 为 918 nm, G 为 757 nm, B 为 658 nm)。原始影像(图 2(a))、坏线修复后的影像(图 2(b))和坏点修复后的影像(图 2(c))像元在左侧最低,往右呈升高趋势,在 60 列附近达到较高值后,像元值呈水平波动,至最后数列稍有下降。与 757 和 658 nm 波段相比,918 nm 波段的行向变化最明显(图 2(a)~ (c))。与原始影像(图 2(a))、坏线修复(图 2(b))和坏点修复(图 2(c))影像的三波段切面图相比,条纹去除后的切面图(图 2d)从左至右没有明显的辐亮度单调递增或递减的变化趋势,仅在水平方向波动。从三波段切面图说明条纹去除过程可以消除各波段的平场效应,这也是图 1(d)中影像色调均匀的原因。

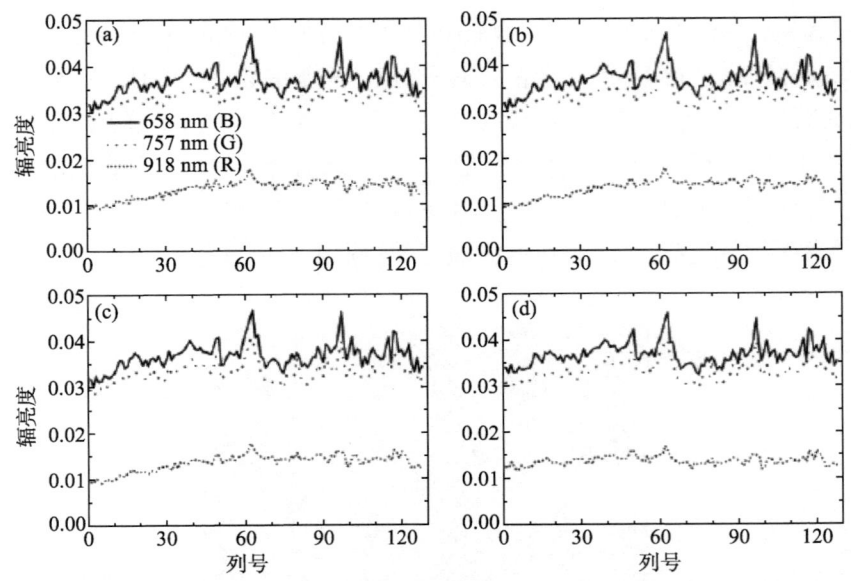

图2 2225轨IIM原始(a)、坏线修复(b)、坏点修复(c)和条纹去除后(d)的11160行切面图

(a)~(c)中可见918 nm，757 nm和658 nm波段均有明显的平场效应，像元值(DN)从左侧往右呈现递增趋势，918 nm波段最明显。从条纹去除后的11160行切面(d)。可见，条纹去除过程有效地消除了行向响应不一致的现象，918 nm，757 nm和658 nm从左至右的辐亮度值切面表现平坦

4 反射率转换

由于本文先修复了坏线和坏点，并采用"全局条纹去除法"[13]消除了影像异常的明亮变化和平场效应(图1和2)。最终本文结合以往对嫦娥一号IIM数据的校正方法[11]和月亮女神MI数据的校正方法[14]，选取2225轨的66~75列、11157~11166行的范围作为定标的区域，如图1(d)中蓝色方框所示区域。该区域的反射率采用Apollo 16月壤(62231)的实验室双向反射率作为定标的光谱(图3(a))。计算该区域的辐亮度如图3(b)。与文献[7]的IIM定标方法相比，除前两个波段外，3~25波段的辐亮度的平均值均低于文献[7]，26~32波段的辐射亮度的平均值均高于文献[7]，表明平场效应对定标工作有一定的影响；与文献[15]的定标方法相比，较多波段的辐亮度有一些差异，可能与选择不同的定标区域有关。从前人两种定标方法相比，本文定标场辐亮度的标准偏差最小(图3(b))。

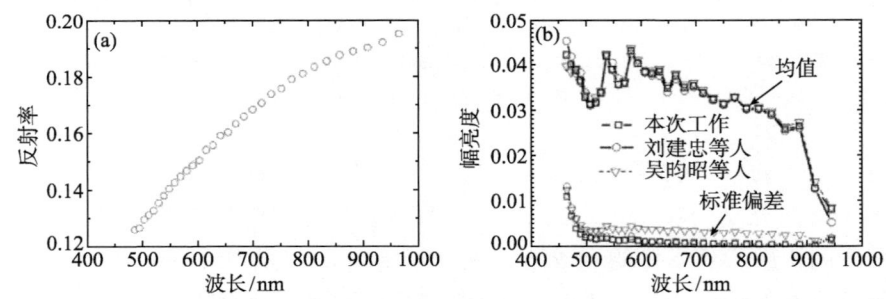

图3 Apollo 16的实验室双向反射率光谱(a)和本次工作定标区域的辐亮度统计图与刘等人[15]和Wu等人工作的比较(b)

5 反射率交叉校正

5.1 交叉定标区域选择

地基望远镜对月球的典型撞击坑、高地、月海等区域进行了光谱观测，获得了月球局部的望远镜光谱，

这些光谱是月表IIM数据的良好参考，可以显而易见地评估反射率转换的好坏。本次工作使用的望远镜数据均从布朗大学Relab实验室的数据库中下载(http：//www.planetary.brown.edu/relab/)。参考Lucey (http：//astrogeology.usgs.gov/Projects/ClementineNIR/files/Comparison_clem_usgs_Lucey.pdf)对Clementine数据的校正方法，本文工作主要选取Aristarchus撞击坑的高地、撞击坑中央隆起、撞击坑西南壁和撞击坑南侧底部作为反射率交叉定标的区域。Aristarchus撞击坑中央隆起位于2898轨IIM影像中，对应80~99列、7590~7609行范围，共400(20×20)个像素点，空间跨度约4×4 km²，略小于望远镜空间覆盖的5×5 km²，Aristarchus撞击坑中央隆起的望远镜数据选择HA1034(图4)；同理，可确定其他五个区域对应的位置和望远镜数据，具体参数详述于表1。在像元选择上，由于2902轨影像没有完全覆盖Aristarchus高地1，所以仅选择了184(8×23)个像元进行计算。

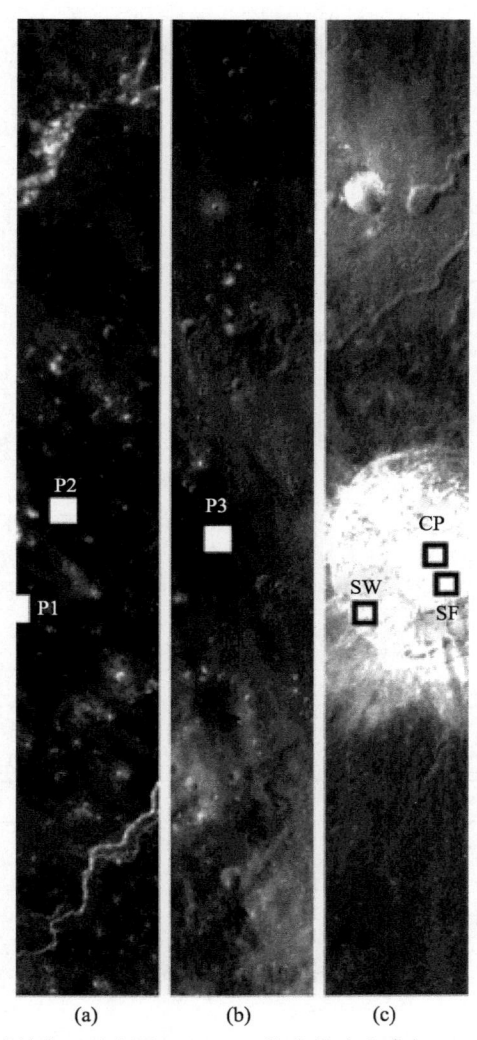

图4 六个反射率交叉定标区域的位置示意图Aristarchus撞击坑中央隆起(CP)、南侧底部(SF)、西南壁(SW)和3个高地(P1，P2和P3)

5.2 交叉定标结果

在选定的交叉定标区域，将IIM数据和望远镜数据在757 nm处归一化(图5)。利用迭代算法计算$(IIM_\lambda \times gain+offset)/IIM_{757}$与$Telescopic_\lambda/Telescopic_{757}$之间的增益(gain)和漂移因子(offset)，其中λ为波长。计算得到的增益与漂移因子如图6所示，详细参数列于表2。从增益的分布看(图6)，短波段的反射率略有抑制，长波段的增益几乎均为1。前两个波段和最后三个波段的漂移不为0，918 nm处的漂移可达0.006。

表1　六个交叉定标区域的影像位置和相对应的望远镜光谱

位置	望远镜光谱	IIM 轨道号	列号	行号	像元
Aristarchus 撞击坑中央峰(CP)	HA1034	2898	80~99	7590~7609	20×20
Aristarchus 撞击坑西南壁(SW)	HA0973	2898	28~47	7639~7658	20×20
Aristarchus 撞击坑南侧底部(SF)	HA0993	2898	90~109	7615~7634	20×20
Aristarchus 高地1(P1)	HA0979	2902	1~8	7014~7036	8×23
Aristarchus 高地2(P2)	HA0130	2902	33~52	6946~6965	20×20
Aristarchus 高地3(P3)	HA1032	2899	33~52	6974~6993	20×20

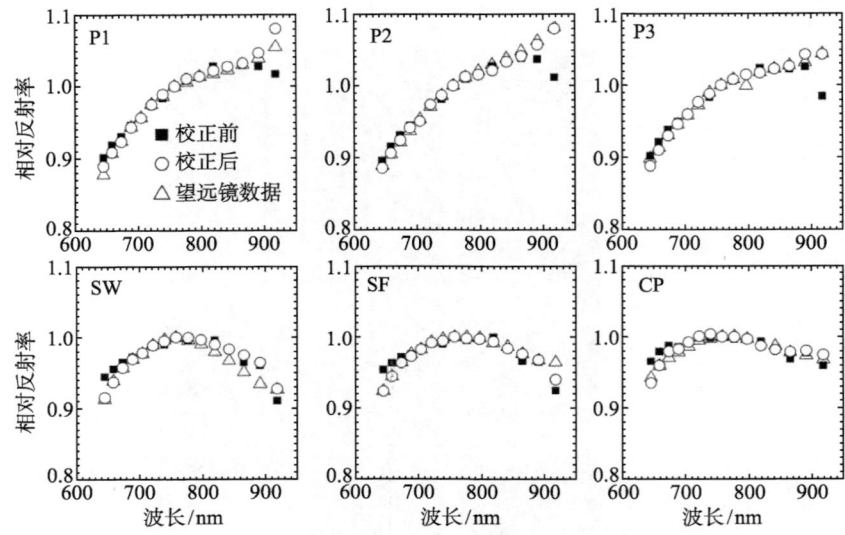

图5　交叉校正前后的IIM光谱与望远镜光谱曲线

在757nm处归一化。经过交叉校正后，Aristarchus Plateau 2(P2)和Plateau 3(P3)、撞击坑中央峰(CP)和撞击坑西南壁(SW)的IIM光谱与望远镜数据几乎完全一样；Aristarchus Plateau 1(P1)和撞击坑南侧底部(SF)在918 nm处的反射率校正效果不是特别理想

表2　嫦娥一号 IIM 光谱在不同波段的增益和漂移因子

波长/nm	644.6	658.6	673.3	688.6	704.6	721.4	739.0	757.4	776.9	797.3	818.9	841.6	865.6	891.1	918.1
增益	0.962	0.977	0.992	0.997	0.999	1.002	1.005	1	1.003	1.001	0.994	1.001	1.014	0.997	0.997
漂移因	0.002	0.001	0	0	0	0	0	0	0	0	0	0	−0.001	0.002	0.006

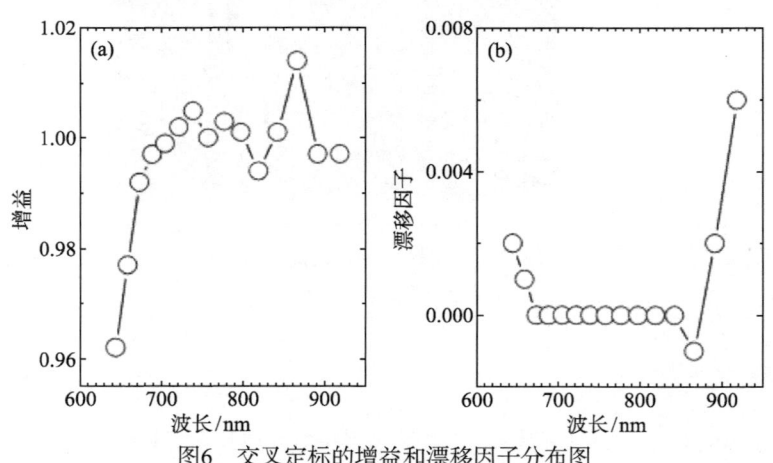

图6　交叉定标的增益和漂移因子分布图

根据计算得到的增益和漂移因子对IIM反射率进行交叉校正，获得交叉校正后的反射率曲线，如图5

所示。从图5中可以发现,反射率校正前后的光谱存在较大变化,特别是光谱曲线两端差异最明显。与撞击坑相比,Aristarchus的3个高地(P1,P2和P3)在短波段的差异不明显,而最后两个波段(891 nm和918 nm)的差异最大(图5),如这3个高地在918 nm处的相对反射率变化分别为0.065,0.068和0.059,而撞击坑内3个交叉定标区域(SW,SF和CP)的相对反射率变化均小于0.017(图7)。Aristarchus撞击坑内的3个定标区域(SW,SF和CP)的相对反射率在短波段略有降低,在长波段略有升高(图5)。从交叉定标前后的结果看,交叉定标过程消除了高地的相对反射率在891和918nm处突然降低的现象,交叉校正后的光谱与典型的高地光谱(如Apollo 16)比较接近(图3)。

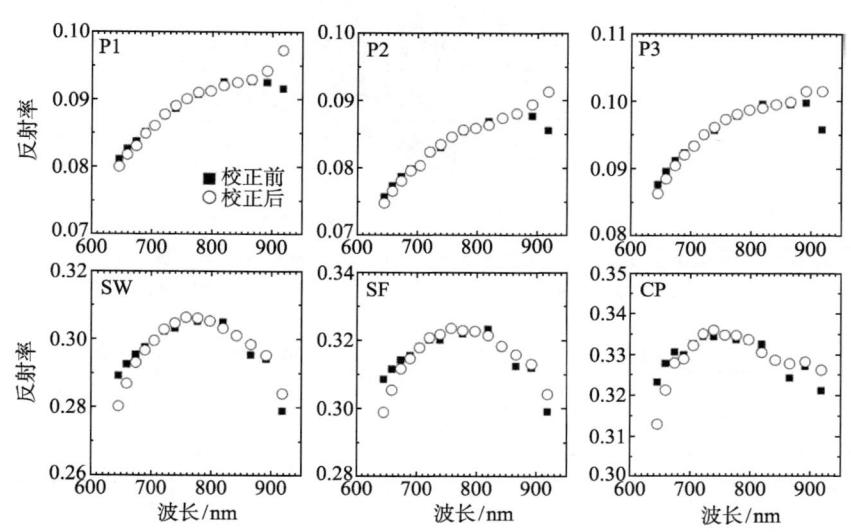

图7 六个交叉校正区域在反射率交叉校正前后的绝对反射率曲线

P1,P2和P3分别为Aristarchus Plateau1,2和3;SW为Aristarchus撞击坑西南壁;SF为撞击坑南侧底部;CP为撞击坑中央峰

图7是交叉定标前后的反射率曲线,可以发现,在交叉定标前,高地的反射率在891和918 nm处突然降低,指示该地区可能富含铁镁质矿物,与该区域的矿物组成和成熟度都不相符。经过交叉定标校正后,高地的光谱在891和918 nm处反射率得到修正,与典型的月表高地光谱(如Apollo 16)相似(图3)。与高地(P1,P2和P3)相比,Aristarchus撞击坑内的三个定标区域的反射率明显偏高(图7)。比较交叉校正前后的光谱,Aristarchus撞击坑内的相对反射率变化也小于高地的相对反射率变化(图5和7)。从高地和撞击坑内的交叉定标结果看,交叉定标主要消除了高地光谱长波段的反射率异常,对撞击坑中富铁镁质区域的光谱特征不会有太大的影响。

5.3 交叉定标检验

为了检验本文计算的增益和漂移因子对月球其他区域的有效性,我们选定Apollo 16(A16),Mare Serenitatis(MS2),Copernicus撞击坑壁(CW)和Apollo 14(A14)作为检验交叉定标好坏的区域。这四个区域涵盖了月球高地、月海和撞击坑,代表了典型的月球地质单元。图8(a)为Apollo 16(A16)在2225轨影像的位置,选取绝对定标的100(10×10)个像元,对应的望远镜光谱选择HA0028;同理可确定与Apollo 14(A14),Mare Serenitatis (MS2)和Copernicus撞击坑壁(CW)望远镜光谱相对应的IIM光谱,分别示意于图8(b)~(d),详细参数列于表3。图9为经过增益和漂移校正后的相对反射率。为了方便比较,将IIM数据与望远镜数据在757 nm处归一化。从图9可见,在反射率交叉校正前,除Apollo 16外,其他3个区域的IIM光谱在891和918 nm处均低于相对应的望远镜光谱;在反射率交叉校正后,高地(Apollo 14)、月海(MS2)和撞击坑(Coper-nicus)的IIM光谱与望远镜光谱几乎没有差异,表明反射率交叉校正的结果非常好。交叉校正后,Apollo 16(A16)在918nm处的反射率略高于望远镜的反射率。从整体上看,反射率交叉校正后的光谱质量得到改善。

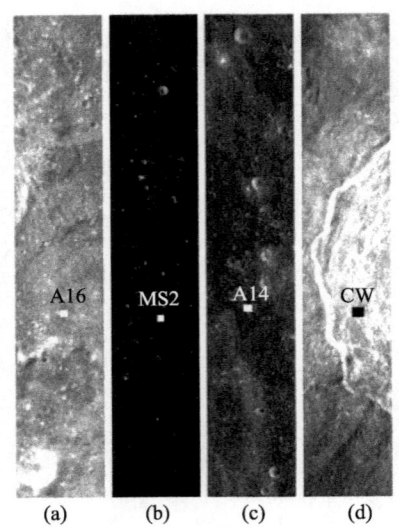

图8 四个交叉校正检验区域的位置示意图

底图为各自的757 nm灰度影像，(a)为Apollo 16(A16)在2225轨影像的位置；(b)为Mare Serenitatis (MS2)在2220轨影像的位置；(c)为Apollo 14(A14)在2871轨影像的位置；(d)为Copernicus撞击坑壁(CW)在2875轨影像的位置

表3 四个交叉校正检验区域的影像位置和相对应的望远镜光谱

位置	望远镜光谱	IIM 轨道号	列号	行号	像元
Apollo 16 (A16)	HC0028	2225	66~75	11157-11166	10×10
Mare Serenitatis (MS2)	HB0916	2220	68~76	6920-6929	9×10
Apollo 14 (A14)	HC0750	2871	55~65	11492-11501	11×10
Copernicus crater wall (CW)	HA0338	2875	72~88	9654-9668	17×15

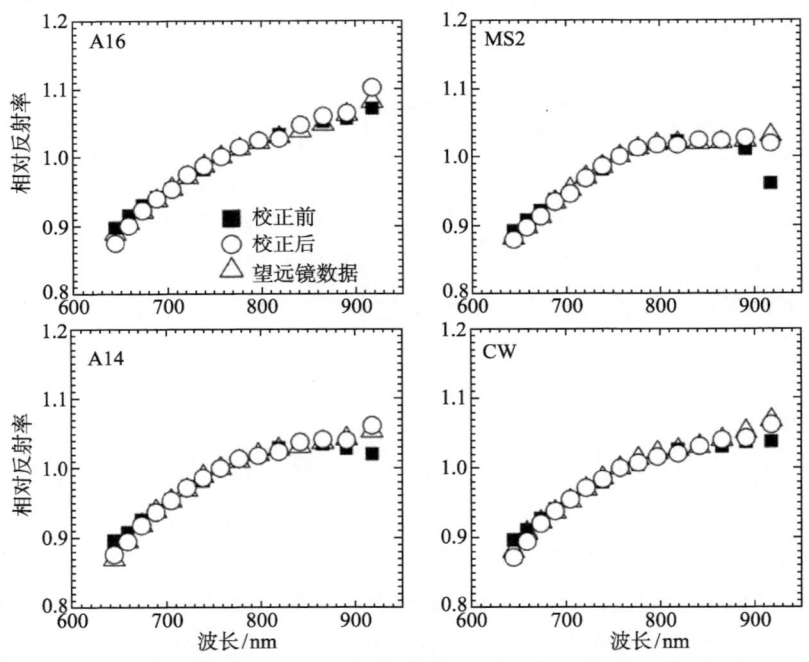

图9 四个交叉校正检验区域的相对IIM反射光谱和望远镜光谱

A16为Apollo 16绝对定标场的光谱，MS2为Mare Serenitatis的光谱，A14为Apollo 14的光谱，CW为Copernicus撞击坑壁的光谱

图10为4个交叉校正检验区域的绝对反射率曲线。Mare Serenitatis (MS2)的反射率非常低，在反射率交叉校正前，其光谱特征表现出强烈的1000 nm吸收带，指示该地区富含铁镁质硅酸盐，与该地区的成熟度不符。从绝对定标前后的光谱曲线看，所有区域的反射率在短波段有所削弱，在891和918 nm处得到加强。

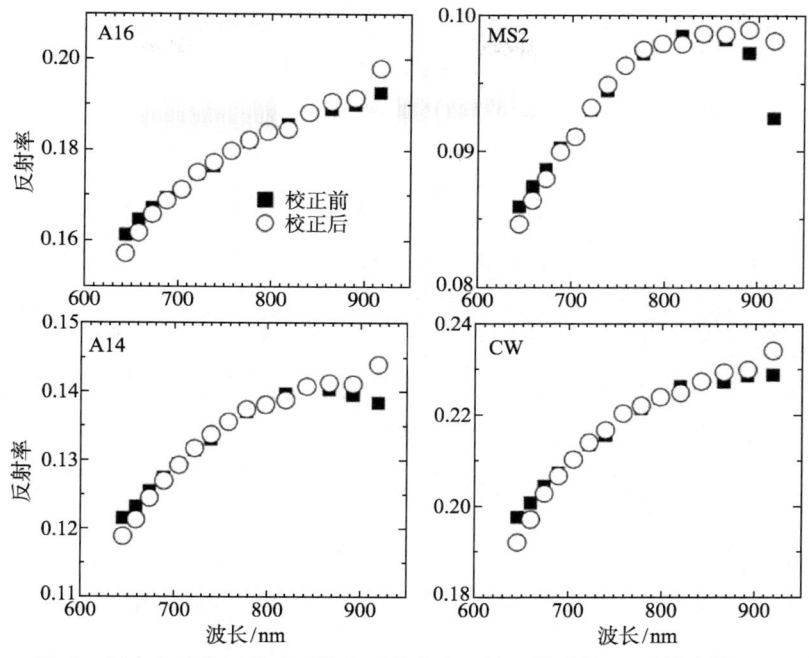

图10　四个交叉校正检验区域在反射率交叉校正前后的IIM反射光谱

A16为Apollo 16绝对定标场的光谱，MS2为Mare Serenitatis的光谱，A14为Apollo 14的光谱，CW为Copernicus撞击坑壁的光谱

从图9和10可以看出，利用本文得到的增益和漂移因子对其他区域的反射率校正结果与望远镜光谱非常接近，说明该方法适合全月球的光谱校正。

5.4　月球的反射率影像

通过上述方法可以对整个月球的光谱进行标定，将辐亮度转换成反射率数据。本节仅以4个典型的月表地质单元为例说明问题。主要选择Tycho撞击坑、Copernicus撞击坑、Mare Serenitatis、和Apollo 14作为样本进行讨论。图11(图版Ⅶ)为上述4个区域的反射率RGB合成影像(R为658 nm，G为757 nm，B为918 nm)。Tycho撞击坑的反射率较高，与其形成的年龄较小有关，在撞击坑的中央峰、底部和侧壁有明显的1000nm吸收峰(图11(a))。Copernicus撞击坑的反射率稍低于Tycho撞击坑，在其撞击坑内部可见有玄武岩的吸收

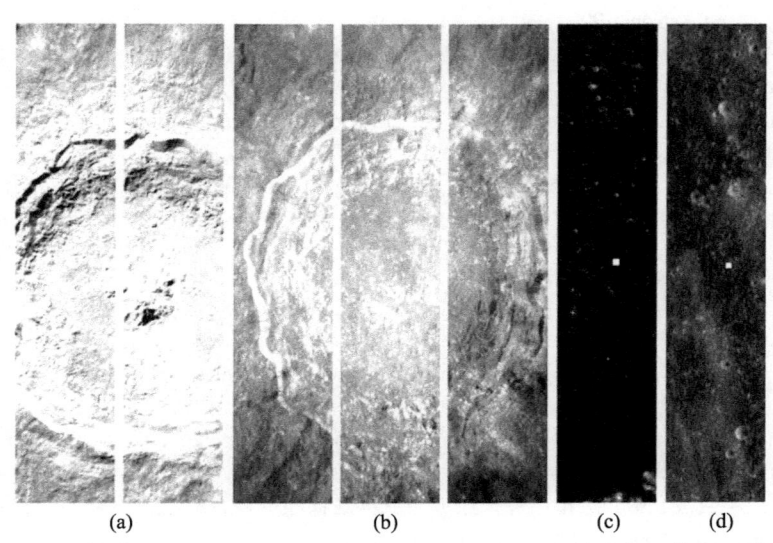

图11　Tycho撞击坑(a)、Copernicus 撞击坑(b)、Mare Serenitatis 月海(c)和Apollo 14着陆点(d)的RGB合成影像图(R为658 nm，G为757nm，B为918nm)

特征(图11(b))。月海和高地的反射率形成鲜明的对比，如月海Serenitatis反射率低，高地Apollo 14着陆区反射率高(图11(c)和(d))。从细节看，月海Serenitatis附近可见大量的小撞击坑，拥有1000nm吸收峰，相比高地Apollo 14着陆区在1000nm处几乎没有吸收峰(图11(c)和(d))。

6 结论

(i) 从IIM数据的改进定标结果看(图1)，条纹去除过程消除了平场效应。在条纹去除后进行反射率转换避免了坏线、坏点和平场效应对反射率定标的影响。

(ii) Aristarchus区域进行反射率交叉校正可以获得IIM反射率在近红外波段的增益和漂移因子。根据计算得到的增益和漂移因子对月球其他地质单元的反射率进行交叉校正，校正后的光谱与望远镜光谱匹配很好(图9和10)，表明该校正因子可以有效地校正月球的反射率。

(iii) 交叉校正后的光谱有效地纠正了891 nm和918 nm波段的反射率异常(图9和10)，对于定量反演月表的FeO和TiO_2有重要影响。

致谢 本次工作使用的2C级IIM数据由中国科学院国家天文台月球与深空探测科学应用中心提供；IIM数据的绝对定标方法得到南京大学吴昀昭老师的大力帮助，在此一并表示感谢。

参 考 文 献

[1] Ouyang Z Y, Li C L, Zou Y L, et al. Chang'E-1 Lunar mission: An overview and primary science results. Chin J Space Sci, 2010, 30(5): 392—403

[2] Ouyang Z Y, Li C L, Zou Y L, et al. Primary scientific results of Chang'E-1 lunar mission. Sci China Earth Sci, 2010, 53(11): 1565—1581

[3] Li C L, Liu J J, Ren X, et al. The global image of the Moon obtained by the Chang'E-1: Data processing and lunar cartography. Sci China Earth Sci, 2010, 53(8): 1091—1102

[4] Li C L, Ren X, Liu J J, et al. Laser altimetry data of Chang'E-1 and the global lunar DEM model. Sci China Earth Sci, 2010, 53(11): 1582—1593

[5] Wang Z Z, Li Y, Zhang X H, et al. Calibration and brightness temperature algorithm of CE-1 Lunar Microwave Sounder (CELMS). Sci China Earth Sci, 2010, 53(9): 1392—1406

[6] Ling Z C, Zhang J, Liu J Z, et al. Preliminary results of FeO mapping using imaging interferometer data from Chang'E-1. Chin Sci Bull, 2011, 56(4): 376—379

[7] Wu Y Z, Xu X S, Xie Z D, et al. Absolute calibration of the Chang'E-1 IIM camera and its preliminary application. Sci China Ser G-Phys Mech Astron, 2009, 52(12): 1842—1848

[8] Chen S B, Meng Z G, Cui T F, et al. Geologic investigation and mapping of the Sinus Iridum quadrangle from Clementine, SELENE, and Chang'e-1 data. Sci China Phys Mech Astron, 2010, 53(12): 2179—2187

[9] Liu F J, Qiao L, Liu Z, et al. Estimation of lunar titanium content: Based on absorption features of Chang'E -1 interference imaging spectrometer (IIM). Sci China Phys Mech Astron, 2010, 53(12): 2136—2144

[10] Zhang L Y, Li C L, Liu J Z, et al. Data processing of imaging interferometer of the Chang'E project. In: Proceedings of the International Lunar Conference. Toronto, Canada, American Astronautical Society, 2005

[11] Wu Y Z, Zhen Y C, Zou Y L, et al. Global absorption center map of the mafic minerals on the Moon as viewed by CE -1 IIM data. Sci China Phys Mech Astron, 2010, 53(12): 2160—2171

[12] Wu Y Z, Zou Y L, Zheng Y C, et al. A preliminary experience in the use of Chang'E-1 IIM data. Planet Space Sci, 2010, 58: 1922—1931

[13] Tan B X, Li Z Y, Chen E X, et al. Preprocessing of EO-1 Hyperion hyperspectral data (in Chinese). Remote Sensing Info, 2005, 6: 36—41[谭炳香, 李增元, 陈尔学, 等. EO-1 Hyperion 高光谱数据的预处理. 遥感信息, 2005, 6: 36—41]

[14] Ohtake M, Matsunaga T, Yokota Y, et al. Deriving the absolute reflectance of lunar surface using SELENE (Kaguya) multiband imager data. Space Sci Rev, 2010, 154: 57—77

[15] 刘建忠, 张文喜, 凌宗成, 等. 嫦娥一号干涉成像光谱数据光度校正与反射率换算. 绕月探测工程探测数据应用研究进展论文集. 北京：中国科学院探月工程总体部. 2009, 200—204

NASVD 方法在 CE1-GRS 谱线分析中的应用研究

杨 佳[1]　葛良全[1]　熊盛青[2]

（1. 成都理工大学地学核技术省重点实验室，成都 610059；
2. 中国国土资源航空物探遥感中心，北京 100083）

摘 要　直接从谱形特征上对嫦娥一号伽玛射线谱仪(CE1-GRS)的 3 级谱线数据进行分析，难以确定月表元素的种类。提出采用噪声调整的奇异值分解(NASVD)方法对 CE1-GRS 谱线进行定性分析。分析结果表明，该方法能够识别出的月表可能元素包括 U、Th、K、Fe、Ti、Si、O、Al、Mg 和 Ca 等 10 种元素，也能够通过各观测谱线对应于第一主成分的幅度绝对值大小反映对应月表区域的总放射性活度强弱。

关键词　嫦娥一号伽玛射线谱线　定性分析　NASVD 方法

伽玛射线谱仪是搭载在嫦娥一号探月卫星上的有效载荷之一，主要用于获取月表元素的分布特征，为月球地质解译提供依据[1-2]。

嫦娥一号伽玛射线谱仪(CE1-GRS)是由中国科学院紫金山天文台研制的。采用的主探测器和反符合探测器均为大尺寸的 CsI(Tl)晶体[3]。CE1-GRS 在轨运行的主要性能指标为：可探测的射线能量范围为 290 keV~11.2 MeV；测量道数为 512 道，每一道能量为 22.14 keV；能量分辨率是 13.58%@609keV；反符合效率低于 70%。探测数据对月表的覆盖范围达到 100%。

1　3 级谱线数据特点

CE1-GRS 的 3 级数据产品中包含月表不同经纬度范围区域对应的 512 道伽玛能谱数据。这些数据是经过能量标定、增益校正、死时间校正、反符合处理以及轨道高度归一化处理后得到的。

CE1-GRS 每个单点谱的采样时间为 3 秒。为了统一对月表各区域的采样时间并降低探测数据中的统计涨落影响，本文采用各区域探测数据的 3 秒平均伽玛谱线进行分析。图 1 显示的是月球赤道某 5°×5° 区域对应的 0.324~9.002 MeV 能量范围的 3 秒平均谱线。月表其他区域对应的平均谱线也具有相似的谱形特征。

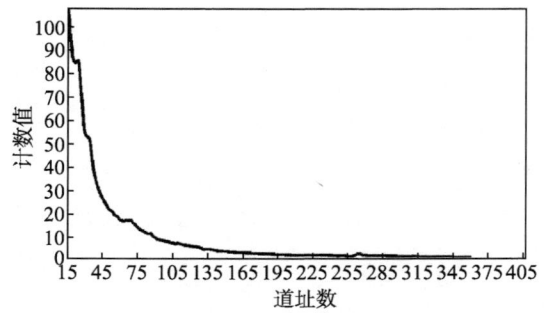

图 1　CE1-GRS 所测月球赤道某 5°×5°区域对应的 3 秒平均谱线

本文原载于《核电子学与探测技术》，2010，Vol.30，No.1，145~150。

从图 1 中只能看出三个明显峰形，如图 2 所示，它们分别位于第 24 道、第 33 道以及第 66 道附近。从 CE1-GRS 的能量分辨率及峰形推测这些峰都包含多种元素的计数贡献。根据提供的能量与道址对应关系知：^{40}K 的 1.461 MeV 特征伽玛射线位于第 66 道附近的重峰里；第 24 道对应的能量值范围为 500.813~522.952 keV，由此推断第 24 道的峰为 511keV 的湮没辐射峰。而第 33 道附近对应的能量范围约为 611~810 keV，难以确定其可能元素的贡献。因此考虑采用解谱方法对 CE1-GRS 谱线进行定性分析，以获取月表元素的种类。

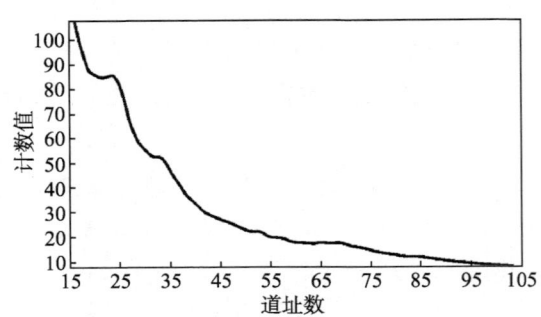

图 2　全谱中的第 15~105 道范围 3 秒平均谱线

2　NASVD 方法在解谱中的应用

2.1　NASVD 方法原理

噪声调整的奇异值分解(Noise Adjusted Singular Value Decomposition，NASVD)方法是由加拿大的 Hovgaard 提出，并被用于核应急领域中寻找未知的放射源、核动力碎片[4]以及航空伽玛能谱数据的降噪处理[5-6]。

NASVD 方法的两个重要特点是：第一，奇异值分解(SVD)方法被用于提取谱线主成分；第二，为了使观测谱线中的绝大多数信号能够集中体现在低序主成分中，先采用噪声的先验模型将各输入谱线的每道计数率的方差调整至单位方差，然后再进行奇异值分解。

2.2　核心算法

由于 CE1-GRS 谱线数据集的所有谱线都具有相似的谱形特征，并考虑到对 0.324~9 MeV 能量范围(第 15~407 道范围)的谱线感兴趣，因此对月表所有经纬度范围区域对应的 1790 条 3 秒平均伽玛谱线构成的谱线数据集矩阵 $A_{1790 \times 393}$ 进行分析。

设计算法步骤如下：

(1) 读入观测谱线矩阵 $A_{m \times n}(m \geq n)$；

(2) 对 $A_{m \times n}$ 进行方差调整，得到单位方差矩阵 A_{NA}；

(3) 对矩阵 A_{NA} 进行奇异值分解：$A_{NA} = U'T'V'^{T}$；

(4) 对 V' 和 $U'T'$ 进行逆变换：

主成分矩阵 $S = \sqrt{\text{diag}(S_{1unit})}V'$，$S$ 的列向量依次为各主成分(特征向量)，是对应特征值的大小按降序排列；

幅度矩阵 $B = \sqrt{\text{diag}(C_1)}U'T'$；

(5) 将谱线数据集 $A_{1790 \times 393}$ 表示为各条观测谱线对应的 393 层谱线叠加：$A = BS^{T} = \sum_{i=1}^{399} b_i s_i^{T}$，其中 b_i 表示

B 的第 i 个列向量，s_i^T 表示 s^T 的第 i 个行向量。将低序层谱线 $b_i s_i^T$ 中的峰信号对相应各条谱线进行定性分析。

设原始观测谱线数据矩阵为 $A=(\alpha_{ji})_{m\times n}$，实现方差调整的算法如下：

(1) A 各列元素求和：$s_1^T=(s_{11},s_{12},\cdots,s_{1n})$；

(2) 归一化处理：$s_{1\text{unit}}^T = s_1^T \Big/ \sum_{j=1}^{n} s_{1j}$；

(3) A 各行元素求和：$C_1=(c_{11},c_{12},\cdots,c_{m1})^T$；

(4) 求出 A 对应的平均谱矩阵 $C_1 S_{1\text{unit}}^T$；

(5) 求出单位方差矩阵 $A_{NA}=\left[\dfrac{a_{ij}}{\sqrt{(C_1 S_{1\text{unit}}^T)_{ij}}}\right]_{m\times n}$。

3 实验及结果

采用以上算法得到的 CE1-GRS 谱线数据集对应的前 8 个主成分如图 3 所示。由于代表信号的主成分(通常为前 3~8 个主成分)应显示出具有相关性的谱形状信息[6]。从图中可以看出：前 5 个主成分都显示出具有相关性的谱形状。而第六及以上主成分中，并没有显示出具有相关性的谱形状信息而主要表现出噪声。

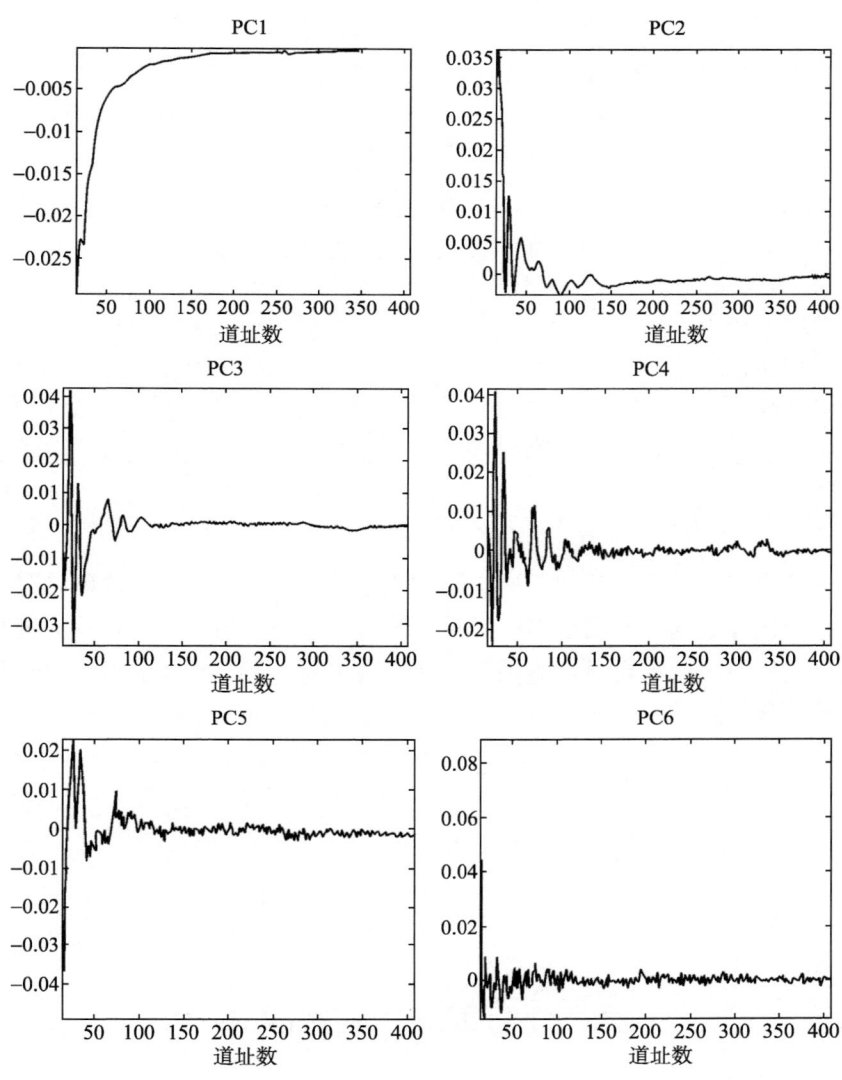

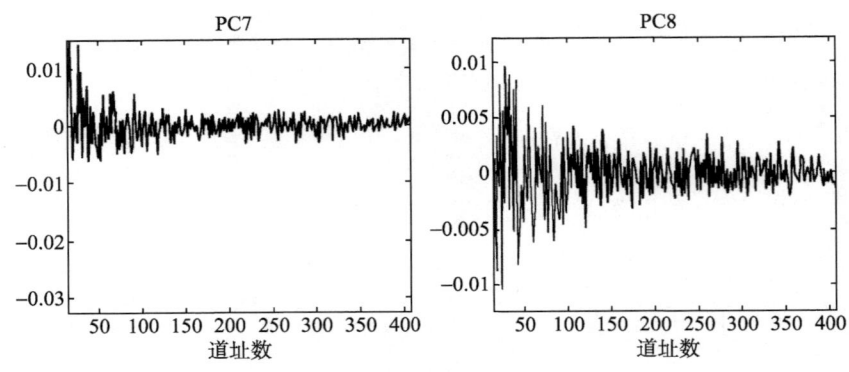

图 3 CE1-GRS 谱线数据集的前 8 个主成分

图 4 显示了该谱线数据集对应的前 200 个特征值，图中的纵坐标是各特征值的对数值。一般来说，数值大的特征值对应的主成分代表信号，数值小的特征值对应的主成分代表噪声[6]。从图 4 中可以看出，曲线迅速下降后几乎变成一条直线，这说明代表信号的低序主成分对应的方差大，而表示噪声的主成分的方差几乎为一常数，即每一个噪声成分对输入谱线的方差贡献大致相等。

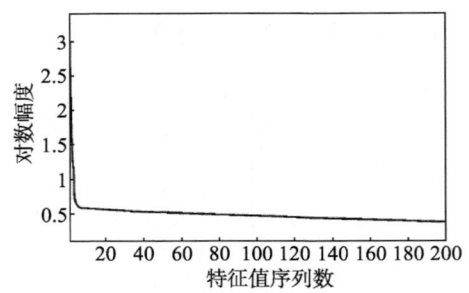

图 4 CE1-GRS 谱线数据集的前 200 个特征值

由以上分析可确定采用前 5 个主成分对 CE1-GRS 谱线数据集进行定性分析。

第一主成分(PC1)体现 CE1-GRS 数据中的绝大多数信号，是谱线数据集的平均谱形状，从中可以看出观测谱线的三个明显峰形。第二主成分(PC2)代表除去第一主成分的贡献后，谱线中绝大多数信号的谱形状，以此类推。$b_i s_i^T$ 的各行是各条观测谱线的平均谱，$b_i s_i^T$ ($i = 2,3,4,5$) 的各行是各条观测谱线的第 i 层谱线。$\sum_{i=1}^{5} b_i s_i^T$ 的各行则是降噪后的各条谱线。

图 5 显示了图 1 的前 5 层谱线。易知，前 5 层谱线中的峰是峰信号，这些峰信号不一定都来自于元素信息，也可能包含干扰峰等其他影响因素。可见，峰信号与观测谱线中存在的元素特征伽玛射线信号之间满足必要非充分关系。因此需要通过鉴别各条观测谱线前 5 层谱线中的各个峰信号对应的能量值是否等于特定元素的特征伽玛射线能量值来对 CE1-GRS 谱线进行定性分析。

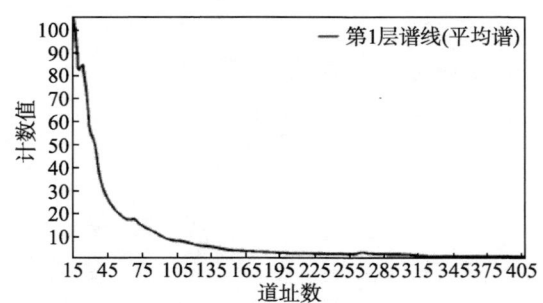

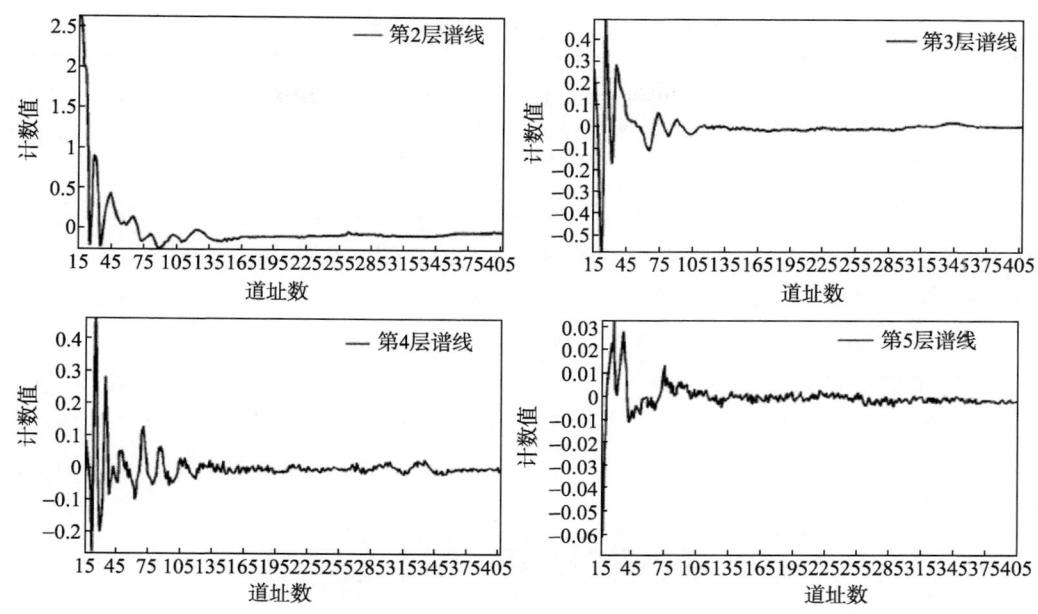

图 5 图 1 的前 5 层谱线

Reedy(1978)给出了最完整的月表发出伽玛射线的能量值及其粒子注量率列表[2]。文中描述了月表主要伽玛射线源(O、Mg、Al、Si、K、Ca、Ti、Fe、Th 和 U)、其他 22 种元素以及在行星、彗星或宇宙飞船中可能产生伽玛射线的任何一种元素所发出的能量大于 0.2MeV 的伽玛射线能量值，并分别计算了这些伽玛射线未经任何相互作用而从月表逃逸出来的粒子注量率[1]。文献[2]中的表 2.9 列出了采用轨道伽玛射线能谱测量技术探测到十种元素发出的最强烈的伽玛射线能量值。

将各条观测谱线前 5 层谱线中的各个峰信号对应的能量值与上述文献列出的所有伽玛射线能量值进行比较，以获取 CE1-GRS 探测到的月表各区域元素种类。表 1~4 列出了图 5 中满足"峰信号对应的能量值等于文献[1-2]列出的伽玛射线能量值"的道址序列数及对应的可能元素。各表中列出的可能元素考虑了月表发出伽玛射线的粒子注量率因素。如果某伽玛射线的粒子注量率过小，则不考虑其对应元素。各元素产生伽玛射线的反应类型、粒子注量率及假定丰度等参数详见文献[1-2]。由这四个表可见，图 1 对应的月表区域被识别出的可能元素有：U、Th、K、Fe、Ti、Si、O、Al、Mg 和 Ca。

表 1 图 5 中第 2 层谱线的峰信号与对应的可能元素

道址数	17	45	57	66	79~84	99~109	115~134
可能元素 (E/MeV)	U(0.352)	Th(0.965); Th(0.969); Ti(0.983)	U(1.238); Fe(1.238)	K(1.461)	Si(1.779); U(1.765); U(1.730); Al(1.809); Si(1.809); Mg(1.809); Fe(1.811); Al(1.779)	Al(2.210); U(2.204); Si(2.235)	Th(2.615); O(2.741); Al(2.754); Mg(2.754); Si(2.839); Si(2.754); Ca(2.793)

表 2 图 5 中第 3 层谱线的峰信号与对应的可能元素

道址数	16	25~29	36~37	50~51	72~74	89~92	291~320	327~364
可能元素 (E/MeV)	Th(0.338)	U(0.609)	Th(0.795); U(0.768)keV; Th(0.583); Ca(0.771)	U(1.120)	Th(1.588);Ca(1.611); Al(1.634);Si(1.634); Mg(1.634)	Ca(1.970); Ca(1.943)	Ca(6.420);O(6.917); Ti(6.419);Ti(6.762)	Al(7.724); Fe(7.631); Fe(7.646)

表 3 图 5 中第 4 层谱线的峰信号与对应的可能元素

道址数	34	41~42	47~57	68~70	83~89	102~115	278~350
可能元素 (E/MeV)	Th(0.727)	Th(0.911)	U(1.238); U(1.120); Al(1.014); Fe(1.238); Si(1.014)	Th(1.496); U(1.509)	Al(1.809); Si(1.809); Mg(1.809); Fe(1.811); Ca(1.943)	Si(2.235); U(2.448); Al(2.230); Fe(2.523); Ca(2.230)	O(6.129); O(6.917); O(7.117); Ti(6.419); Ti(6.762); Ca(6.420); Fe(7.631); Fe(7.646); Al(7.724)

表4 图5中第5层谱线峰信号与对应的可能元素

道址数	33~38	69~82
可能元素(E/MeV)	Th(0.795); Th(0.727)U(0.768); Ca(0.771)	Si(1.779); U(1.765); U(1.730); Th(1.588) Al(1.720); Ca(1.611); Al(1.634); Si(1.634)

对月表其他所有区域对应的各条观测谱线的前5层谱线进行分析,结果表明,该方法能够识别月表各区域的可能元素包括 U、Th、K、Fe、Ti、Si、O、Al、Mg 和 Ca 等10 种元素。

鉴于各条观测谱线的第一层谱线分别为月表各区域的平均谱,谱形完全相同,区别仅在于各平均谱的计数值不同。并且各谱线对应于 PC1 的幅度值比对应于 PCi(i>1)的幅度值大 1~2 个数量级。因此,各观测谱对应于 PC1 的幅度绝对值能够反映对应月表区域的总放射性活度。图6 是各谱线对应于 PC1 的幅度值构成的火柴杆图,即幅度矩阵 B 的第 1 个列向量 b_1。从图中能够看出:谱线对应于 PC1 的幅度绝对值越大,则对应月表区域的总放射性活度越强。

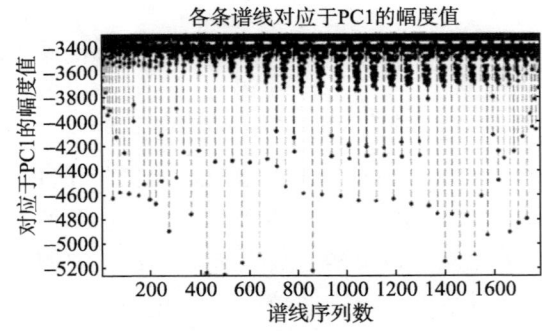

图6 各谱线对应于 PC1 的幅度值构成的火柴杆图

4 结论

本文针对从 CE1-GRS 的 3 级谱线数据的谱形特征上进行月表元素识别存在难度的问题,提出采用噪声调整的奇异值分解(NASVD)方法对 CE1-GRS 谱线进行定性分析。分析结果表明,该方法能够识别的月表可能元素包括 U、Th、K、Fe、Ti、Si、O、Al、Mg 和 Ca 等 10 种元素。并且各观测谱对应于 PC1 的幅度绝对值大小能够反映对应月表区域的总放射性活度强弱。

参 考 文 献

[1] Reedy R C. Planetary gamma-ray spectroscopy [C]//Merrill R B. Proc. Lunar Planet. Sci. Conf. 9th. New York: Pergamon Press,1978: 2961—2984

[2] Jolliff B L, Wieczorek M A, Shearer C K and Neal C R. New Views of the Moon [M]. NY: Mineralogical Society of America. 2006:172—208

[3] 舒双宝, 常进, 蔡明生等. 月球伽玛射线谱仪的研制及其性能[J]. 天文学报, 2006, 47(2):218—223

[4] Hovgaard J. A new processing technique for airborne gamma-ray spectrometer data (Noise adjusted singular value decomposition) [C]// [s. n.]. Proceedings of Sixth Topical Meeting on Emergency Preparedness and Response. San Francisco: ANS,1997: 123—127

[5] Hovgaard J, Grasty R L. Reducing statistical noise in airborne gamma-ray data through spectral component analysis [C]// Gubins A G. Proc. of Exploration 97: 4th Decennial Conf. on Mineral Exploration. Toronto: Prospectors and Developers Association of Canada, 1997: 753—764

[6] Guidelines for radioelement mapping using gamma ray spectrometry data [R]. Vienna, Austria: Intern at ion al Atomic Energy Agency Technical Documents, 2003

中红外光谱在月球探测中的应用

刘 剑[1,2] 欧阳自远[1] 李春来[3] 邹永廖[3] 胥 涛[4]

(1. 中国科学院 地球化学研究所，贵阳 550002；2. 中国科学院 研究生院，北京 100039；
3. 中国科学院 国家天文台，北京 100012；4. 海南大学 理工学院，海口 570228)

摘 要 硅酸盐矿物的中红外光谱一般具有明显的 CF 特征和 Reststrahlen 吸收带，这些特征都与硅酸盐的组成密切相关，可以作为判别其组成的指示特征。月球表面的主要矿物有辉石、斜长石、钛铁矿和橄榄石，其中除了钛铁矿以外，都属于硅酸盐的范畴。如果充分利用月球表面硅酸盐的红外光谱特征，即 CF 特征和 Reststrahlen 吸收带，将可以用于探测 VIS-NIR 光谱无法完成的探测目标：钙长石、橄榄石、石英和碱性长石等。如果在嫦娥工程的后续探测器(如轨道器、月球车或者软着陆平台)上搭载中红外光谱仪对月表物质进行探测，将有利于成功实现对月球表面的矿物和岩石的精确探测，这也将是我国首次采用中红外光谱探测仪对月球进行系统的矿物与岩石探测。

关键词 中红外光谱 月球 矿物 岩石 光谱分析

所有物质，只要其温度超过绝对零度，就会不断向外发射能量，这种由于物体中的分子、原子受到热激发而发射电磁波的现象称为热辐射，热辐射具有连续的辐射谱。对于行星，其热辐射主要位于红外区，即热红外光谱。热红外光谱位于可见光与微波之间，波长范围为 0.78~500 μm。热红外波段按波长一般分为三个谱区：近红外(0.78~3 μm)、中红外(3~25 μm)以及远红外(25~500 μm)。对于月球表面物质而言，其红外光谱主要位于中红外谱区。热红外辐射不仅与物质的表面物理状态有关，而且是物质内部组成结构和温度的函数。

对于行星，短波长区热辐射的影响一般可以忽略；在长波长区，反射光谱的贡献可以忽略；但是在中间过渡区反射光谱和热辐射对探测器都有贡献。而且物体离太阳越远，热辐射越弱，且辐射峰值向长波方向移动；反射光谱的强度也相应变弱。

红外光谱已经成功地用于地球遥感探测，如 AIRS 探测仪、ASTER 探测器和 ATSR 探测仪等。红外光谱也成功地应用于火星探测，如 Mariner 6/7 的 IRS 探测仪、Mariner 9 的 IRIS 探测仪、Viking 1/2 的 IRTM 探测仪、Mars Global Surveyor 的 TES 探测仪和 Mars Odyssey 的 THEMIS 探测仪等。但至今都没有利用中红外探测仪对月球表面物质组成进行系统探测。然而中红外光谱对于月表硅酸盐矿物的探测有较大的应用潜力，因此研究利用中红外光谱识别月表物质组成的理论和方法就显得比较重要。

1 矿物与岩石的中红外光谱特征

月表主要矿物是辉石、斜长石、钛铁矿和橄榄石，其中除了钛铁矿以外，都属于硅酸盐的范畴，其中，辉石的成分变化较大，在月海玄武岩中主要是单斜辉石，非月海玄武岩及高地岩石中主要是斜方辉石；斜长石为月海玄武岩及高地岩石的主要矿物，包括从钠长石向钙长石过渡所有的过渡型斜长石系列，斜长石一般呈白色粒状集合体或板晶产出；橄榄石在不同岩石中的含量不同，如在月海玄武岩及橄长岩

本文原载于《矿物学报》，2006, Vol. 26, No. 3, 435~440。

内含量较高，是作为主要的矿物相产出，而在一些岩石中含量较低，是以次要矿物产出。如果利用中红外光谱对月表物质进行探测，就有必要先了解与月表有关的矿物的中红外光谱特征。因此在此介绍一些典型的月表硅酸盐矿物的中红外光谱特征。

1.1 硅酸盐矿物的中红外光谱特征

在硅酸盐的晶体结构中，每一个硅原子都被四个氧原子所包围，四个氧原子分布在硅原子的四个角顶上，构成硅氧四面体，它是硅酸盐的基本单位。硅氧四面体在结构中可以单独存在，也可以以不同的方式相互联结，从而形成多种复杂的络阴离子。硅酸盐的基频振动模式由 Si-O 伸展以及弯曲模式支配，伸缩振动对应 8~12 μm 谱区，弯曲振动对应 15~25 μm。

硅酸盐矿物的中红外光谱有一个最为明显的特征就是具有 CF(Christiansen Frequency)特征。CF 特征表征反射率最小值(发射率最大值)的光谱特征。CF 特征，一般认为是在单个基频分子振动波段之前，折射指数发生快速变化(即反常色散)引起的，其折射指数接近于其周围介质的折射指数，使后向散射最小化。CF 特征发生的位置相对基频振动略微靠近短波处，因此吸收相对较弱。由于后向散射和吸收都比较弱，红外辐射较易通过样品，导致反射率最小或发射率达到最大值。

Conel[1]最先指出硅酸盐光谱中与 CF 特征相关的特征波长可以用做研究矿物组成的指示特征。Salisbury 与 Walter[2]的研究结果也表明物质的 CF 波长可以作为确定其组成的判别依据,同时强调了该技术用于中红外遥感的潜力。

硅酸盐矿物在 8.5~12.0 μm 之间有一 Reststrahlen 吸收带，来自 Si-O 键的伸缩振动，其具体位置与物质组成相关。在 16.5~25 μm 间还有一稍弱的强吸收带，该吸收带与 Si-O-Si 弯曲振动相联系。在这两个波段之间，粒子处于体积散射区域，在反射光谱上产生一个最大值，在发射光谱上产生一个最小值，Salisbury 与 Walter[2]指出该峰对应的波长与矿物组成密切相关，并称之为透射特征 TF(Transparency Feature)。而且，CF 特征的波长与透射特征 TF 的波长彼此相关，都可以作为判断行星表面硅酸盐组成的依据。

Salisbury[3]的研究则表明透射特征并不适用于月球表面的中红外光谱探测，因为透射特征在真空环境下几乎不存在。因此对于月球的中红外光谱只有 CF 特征和 Reststrahlen 吸收带适用于判别矿物的组成。但是，对于某特定物质，CF 特征的波长并不固定，例如，在真空中与空气中，CF 特征的波长就不同，其位置会产生位移。

1.2 岩石的中红外光谱特征

岩石的红外光谱为矿物的混合物光谱。与混合物的反射光谱需要复杂的理论模型来分解不同，在中红外光谱区，岩石的红外光谱是其矿物组分光谱的简单线性组合，与丰度成正比。但是 CF 特征的位置不再与某个光学常数相关联。当物质组成从长英质向超基性过渡时，CF 特征向长波方向移动，研究表明 CF 特征与总体化学组成密切相关联[2,4]。

在多数情况下，造岩矿物的光谱特征都可以在岩石光谱上表现出来。尤其是清晰的强谱带，例如石英的谱带。在有些情况下，一些尖锐的谱带密集在一起，拓展成一个单一的谱带。研究表明，组成岩石的矿物越简单，越好鉴定，例如，长英质岩石光谱里的造岩矿物谱带比镁铁质岩类光谱容易识别。

基性岩(辉长岩、辉绿岩和玄武岩)的合成光谱：20 μm 附近有一个宽的极低值谱带，它与 10~11 μm 谱区的主要极低值谱带同时并存，而且光谱曲线也显示出斜长岩不很强的宽谱带。辉长岩，其主要吸收带不对称，它的极低值在 11 μm 附近，而辉绿岩和玄武岩的极低值较接近 10 μm。基性岩的 CF 特征，一般位于 8 μm 或更长的波段区。

超基性岩(富橄榄石橄榄岩和富辉石橄榄岩)的合成光谱：其两个端元组分的光谱曲线彼此十分不同。

它们的 CF 特征波长最大，而且主要吸收带低值所处位置的波长也比其他岩石类型长。

2 中红外光谱在月球探测中的应用研究

为了充分利用中红外光谱的优点，实现中红外光谱对月球表面物质的探测，国外许多研究人员利用月球样品进行了实验室中红外光谱研究。这些基础研究为我国嫦娥工程利用中红外光谱进行月表物质探测奠定了良好的基础。

2.1 月球样品的中红外光谱特征

Nash[5]在实验室里测量和研究了 7 份具有代表性的月球样品(分别代表角砾岩、月壤和月球火成岩)位于 4~12 μm 之间的红外光谱。通过对这些典型月球样品的研究可以了解月球物质的中红外光谱性质。实验室研究结果如图 1 所示，所有样品的红外光谱中最为明显的一个特征就是 CF 特征(反射率最小，发射率最大)都出现在 8 μm 附近，而且随着铁镁质的成分的增加，CF 的位置渐渐地往长波方向移动；在 4.5~6.0 μm 存在一个比较弱的吸收特征，这是由于分子振动引起的谐波；在 8.5~12.0 μm 之间有一个 Reststrahlen 吸收特征，Reststrahlen 吸收带是物质组成的特征。根据以上特征，可以知道利用中红外光谱对月球表面物质进行探测是可行的。

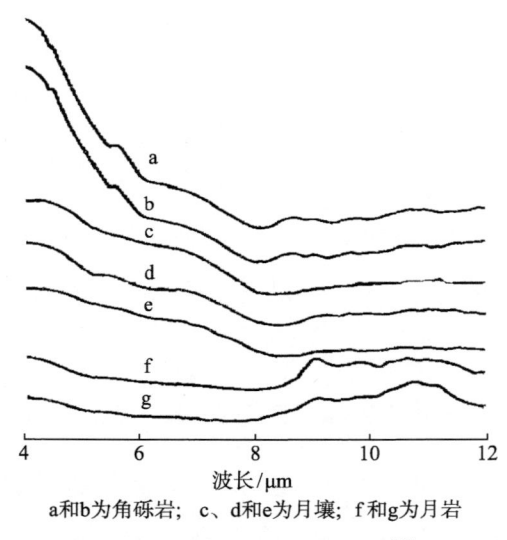

a和b为角砾岩； c、d和e为月壤； f和g为月岩

图 1 月球样品的反射光谱[5]

Betts 等[6]研究了另外 6 个代表性 Apollo 月壤样品和从 Apollo 11 月海玄武岩样品(样品号：10058)中分离出的单矿物的中红外光谱。研究的光谱区间为 2.2~25 μm。研究结果表明这些月壤样品的中红外光谱极为相似，但是也显示了一些不相似的特点，尤其是在 4.5~6.5 μm 之间的光谱过渡区和 CF 特征区(反射率最小的 8 μm 附近)。Apollo 16 月壤样品的光谱与 Nash[5]所研究的来源于同一采样点的角砾岩的光谱非常地相似。同时，他们还研究了从编号为 10058 的 Apollo 11 月海玄武岩样品中分离出的单矿物的中红外光谱。研究结果为：该样品的主要矿物组分(斜长石、辉石、橄榄石和钛铁矿)的光谱特征能够清楚地被识别出来。可见，Nash[5]和 Betts 等[6]的实验室月球样品研究表明，利用中红外光谱对月球表面进行探测，从理论上完全可以实现。

具体应用中红外探测月球表面物质时，还必须了解影响中红外光谱的各种物理因素。影响中红外光谱的各种因素有：取向效应，如结晶取向会影响观测到的光谱；吸附的月尘，较大粒子上的月尘会引起异常的散射，使光谱形状发生改变；月表的覆盖物；光谱对比度与粒度；充填作用和热梯度等等。这些物理特征和环境因素都会对中红外光谱产生影响。

通常实验室样品的中红外光谱被用来研究如何确定岩石的组成。早期研究表明岩石的物理状态及其所处的环境会影响它们的光谱特征。特别是对于细小的粉尘状物质，Reststrahlen 吸收带的光谱对比度将会显著地消失[7]。中红外区光谱比其他谱区对矿物组成(如斜长岩长石)更加敏感。Betts 等[8]在前人工作的基础上，分析了更多的月壤样品，这些样品的组成具有更广泛的代表性，而且覆盖的取样点和地质单元更多。为了验证 CF 特征与地球上硅酸盐矿物的相关性是否同样适用于月球组分的识别，他们将月壤样品的 CF 参数与代表岩石物质组成的一些参数，如 SCFM 指数联系起来。SCFM 是反映月球总体组成的指标，定义为 $SiO_2/(SiO_2+CaO+FeO+MgO)$[4]。

图 2 表明光谱参数(CF)以及矿物组成参数(SCFM)之间有松散但显著的相关性。由于月球探测的主要对象为月壤，所以上述实验表明 CF 参数可以大致区分月球表面的矿物组成，但是只有显著的矿物组成变化才能有效地区分开来。在图 2 中存在数据点的分群现象，所有 Apollo 14 和 Apollo 16 的测量值位于左上方，这是因为它们的镁铁质含量要低一些，长石质含量要高一些。Apollo 17 样品又分为三小块。处于最低处的 Apollo 17 样品相对图上的其他点比较特殊，事实上正是如此，该样品(74220)取自月壤的橙色玻璃，有些研究者认为它们是易碎的碎屑状岩石而不是月壤。Apollo 17 的另外两组样品与地质单元相对应，右下方的两个点对应的样品来自地质单元 Valley Floor，左上方的三个点来自 Light Mantle 单元。所以，在热红外区光谱中使用 CF 值，可以可靠地识别一般的矿物和地质单元，但是要做清晰明确的判断，需要 CF 有显著的差异。在遥感探测和现场分析获得的月球(包括水星)的中红外光谱中，CF 特征可能是最显著的光谱特征。

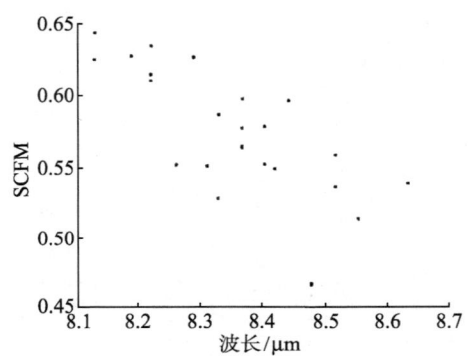

图 2　一系列月壤样品 CF 特征的波长与 SCFM 指标的关系图[8]

为了更好理解中红外光谱特征与粒度以及组成之间的相关性，Cooper 等[9]分析了 35 个岩石样品及其对等的粉末状样品。他们发现固体结晶质岩石的 CF 特征与相对应的粉末样品不同，而且差异的大小与岩石类型相关。粒状火成岩的 CF 特征对于岩石类型的确定都比较有用，而且他们也证明了 SCFM 指数[4]更适合作为组成指数。对于某特定物质，CF 特征的波长并不固定。例如，在真空中以及空气中，CF 特征的波长就不同。然而，Salisbury 和 Walter[2]已经指出在空气中测得的特定物质的 CF 波长与真空中的测量值有系统相关性，而且与火成岩的 SCFM 指数有很好的相关性。

2.2　中红外光谱在月表物质探测中的应用

上个世纪 60 年代以来，许多研究者开始利用地基望远镜和气球搭载的光谱仪进行中红外试验和研究，取得了很多成果。其光谱范围为 7~13.5 μm，空间分辨率约为 5~300 km。这些基础研究工作为如何利用矿物中红外光谱的 CF 特征和 Reststrahlen 吸收带进行月球表面物质组成的探测奠定了坚实的基础。

Kozlowski 等人[10]通过望远镜对月表正面五个区域进行了中红外光谱(7.5~11.4 μm)的研究。这五个区域中有 3 个位于 Copernius 撞击坑中，另外 2 个位于月海中。Copernius 撞击坑中的 3 个区域的中红外光谱表明在8 μm附近存在有与硅酸盐相关的CF特征，在10.5 μm附近存在一明显的、比较深的低反射率特征，这是橄榄石存在的明显的光谱特征，这结果与以前对Copernius撞击坑的研究结果是相符合的；而位于月海中的2个区域的光谱在8.1~8.9 μm之间存在有一个斜长石光谱特征。

Bell等[11]利用地基望远镜对Tycho(第谷)进行成像光谱研究，其中红外光谱区间为8.3~13.3 μm。此外Bell等[12]利用地基望远镜还对Schickard、Baade和Inghirami撞击坑附近的区域进行成像光谱研究，主要研究4~8 μm这一过渡区(光线由体散射转变到面散射)的光谱特征。这两次观测研究都采用成像光谱技术，这为将来进行遥感器成像光谱探测奠定了良好的基础。

Apollo 17的在轨指挥舱曾搭载了热辐射红外辐射计。该仪器进行了从1.2 μm到约70 μm的月球表面辐射探测。其空间分辨率约为2.2 km。Apollo 17热辐射计的探测结果可以用于确定最小块状物质的大小及其丰度，如Arisarchus撞击坑，它的中心在日出前温度高达120~130 K，然而它的围壁以及周围平地温度为100 K左右，根据推测，该撞击坑的内部有11%~16%为基岩露头。

Clementine搭载了一个长波红外探测器(LWIR)，波段中心为8.75 μm，带宽1.5 μm。空间分辨率：极地为200 m，赤道为55 m。相机主要用于极地夜晚成像，其他区域可以选择性地高分辨率成像。Lawson等[13]将LWIR测得的DN值按Plank定律转换为亮度温度，从而获取了月球表面的亮度温度。根据亮度温度，可以研究月球表面的粒度、密度等物理信息。Lawson等[14]利用LWIR数据证实，如果采取垂直观测，采用朗伯近似月球表面的热辐射是可行的。Lawson等[15]还发现在白天影响月球表面温度变化最主要的因素是反照率(受单粒子散射反照率支配)以及太阳入射角。

2.3 中红外光谱在月表物质探测中的优点

对月球而言，中红外光谱区域包含有硅酸盐矿物至关重要的光谱特征，如果利用中红外光谱特征进行遥感探测或者现场分析，其效果可能是利用短波(可见光—近红外)进行探测无法比拟的。相对于短波光谱而言，利用中红外光谱进行探测具有如下的优点：

对于斜长石，尤其是含铁量不高的钙长石，VIS-NIR光谱不适合于探测月表的钙长石的分布和丰度，但是利用中红外光谱进行探测，就完全可以实现对钙长石的丰度和分布的填图(因为在5~15 μm的光谱区间存在对钙长石十分敏感的光谱特征)；中红外光谱对月表橄榄石中镁铁的含量极为敏感，可以用于探测月表橄榄石的化学组成。中红外光谱同样可用于探测无铁的矿物，但是如果利用UV-VIS-NIR光谱探测石英和碱性长石很难实现。

3 结论

月球表面物质的中红外光谱包含有硅酸盐矿物重要的光谱特征：CF特征和Reststrahlen吸收带。CF特征和Reststrahlen吸收带可以用于判别硅酸盐矿物的组成。如果利用中红外光谱特征进行遥感探测或者现场分析，其效果是利用短波(可见光—近红外)进行探测无法取代的。利用中红外光谱特征进行探测，可以成功实现对斜长石(尤其是钙长石)、橄榄石、石英和碱性长石探测。

因为中红外光谱对硅酸盐矿物的探测具有独特优势，许多研究者利用地基望远镜和气球搭载的中红外光谱仪对月球表面进行探测和研究，已奠定了坚实的基础。Apollo 17搭载的热辐射红外辐射计和Clementine搭载的长波红外探测器(LWIR)，它们利用红外遥感器进行了很成功的尝试，但是都局限于对月表物质的热惯量等物理性质进行相关的探测，利用热惯量研究暴露年龄、粒度和密度等。而正式利用中红外光谱进行月球表面物质识别的遥感探测器至今还没有。

我国已经启动了首个月球探测工程——嫦娥一期工程，嫦娥一期工程将发射一颗绕月探测卫星对月表和地月空间环境进行探测。嫦娥工程是一个系统工程，如果在后续探测器中，如轨道器、月球车或者软着陆器，搭载相应的中红外光谱探测仪对月表物质进行遥感探测或现场分析，那么将更加有利于我国成功实现对月球表面矿物和岩石的准确探测，也将是月球上首次采用中红外光谱探测仪进行矿物与岩石的探测。

参 考 文 献

[1] Conel J. Infrared emissivities of silicates: experimental results and a cloudy atmosphere model of spectral emission from condensed particulate mediums[J]. *Journal of Geophysical Research*, 1969, 74: 1614—1634

[2] Salisbury J W, et al. Thermal infrared(2.5–13.5μm) spectroscopic remote sensing of igneous rock types on particulate planetary surfaces[J]. *Journal of Geophysical Research*, 1989, 94: 9192—9202

[3] Salisbury J W. Thermal infrared(8 to 14μm) spectroscopic remote sensing of rock type on particulate planetary surfaces[J]. *Lunar and Planetary Science Conference* ⅩⅨ, *Houston: Lunar and Planetary Institute*, 1995: 1021—1022

[4] Walter L S, et al. Spectral characterization of igneous rocks in the 8-12 μm region[J]. *Journal of Geophysical Research*, 1989, 94: 9203—9213

[5] Nash D B. Infrared reflectance spectra(4-12 μm) of lunar samples[J]. *Lunar and Planetary Science Conference XXII, Houston: Lunar and Planetary Institute*, 1991: 957—958

[6] Betts B H. Infrared reflectance spectra of selected lunar soils and mineral concentrates[J]. *Lunar and Planetary Science Conference* ⅩⅩⅥ, *Houston: Lunar and Planetary Institute*, 1995: 113—114

[7] Aronson J R, et al. (1979) Infrared spectra of lunar soils and related optical constants [J]. *Lunar and Planetary Science Conference 10*[J]. *Houston: Lunar and Planetary Institute*, 1979, 1787—1795

[8] Betts B H. Mid-infrared reflectance spectra of lunar samples: Christiansen frequency correlations with composition[J]. *Lunar and Planetary Science Conference* ⅩⅩⅧ, *Houston: Lunar and Planetary Institute*, 1997: 1844—1845

[9] Cooper B L, et al. Midinfrared spectral features of rocks and their powders[J]. *Journal of Geophysical Research*, 2002, 107(4): 5017

[10] Sprague A L, et al. The Moon mid-infrared (7.5–11.4 μm) spectroscopy of five selected regions[J]. *Lunar and Planetary Science Conference* ⅩⅩⅢ, *Houston: Lunar and Planetary Institute*, 1992: 1343—1344

[11] Bell J F, et al. Imaging spectroscopy of the Moon in the mid-infrared: 8.3 to 13.3 microns image cubes of TYCHO[J]. *Lunar and Planetary Science Conference* ⅩⅩⅦ, *Houston: Lunar and Planetary Institute*, 1995: 97—98

[12] Bell J F, et al. Mid-infrared(5.0-7.0 microns) imaging spectroscopy of the moon from the KAO[J]. *ASP Conference Series*, 1995, 73: 341—344

[13] Lawson S L, et al. Jakosky Brightness temperatures of the lunar surface: the Clementine long-wave infrared global data set[J]. *Lunar and Planetary Science Conference* ⅩⅩⅦ, *Houston: Lunar and Planetary Institute*, 1999: 15—29

[14] Lawson S L, et al. Brightness temperatures of the lunar surface: Calibration and global analysis of the Clementine long-wave infrared camera data[J]. *Journal of Geophysical Research*, 2000, 105(2): 4273—4290

[15] Lawson S L, et al. Lunar surface thermophysical properties derived from Clementine LWIR and UVVIS images[J]. *Journal of Geophysical Research*, 2001, 106(11), 27911—27932

月球紫外-可见-近红外反射光谱的基本特征及解析方法

胥 涛[1,2]　David T Blewett[3]　李春来[4]　王世杰[1]

（1. 中国科学院地球化学研究所，贵阳 550002；2. 中国科学院研究生院，北京 100039；3. NOVASOL. Honolulu. Hawaii 96813，USA；4. 中国科学院国家天文台，北京 100012）

摘　要　月球紫外-可见-近红外反射光谱(强度，反照率以及吸收特征)是研究月球表面物质组成的重要手段。文章简要介绍了月球表面反射光谱的基本特征、形成机制以及现有月球反射光谱数据及其解析方法。这些方法主要有光谱参数、连续背景扣除、高斯反卷积分析以及主成分分析。还讨论了高光谱探测器在月球探测中的应用前景、已有方案以及获取新的月球样品对月球反射光谱研究的重要性。

关键词　基本特征　解析方法　反射光谱　月球

1　月球表面反射光谱的基本特征

夜晚我们看到的来自月球的光线并不是发出月球本身，而是月球反射来自太阳的光线。太阳光线与月球表面物质接触时，会发生散射、反射、透射、吸收以及折射等物理光学作用。在作用过程中，本来表现为各向同性(辐照强度与角度无关)的入射太阳辐射会在某些性质上发生改变，例如，月球表面不同区域的反射特性不一致，月海区相对富镁铁等暗色矿物，对光线的吸收比较强，于是成为夜晚见到的月球上的暗色斑块；与之相反，月球高地富长石等浅色物质，反照率比较高，这些区域在夜晚就显得比较亮。

除了在反照率等方面发生改变，在某些特定的波长，会出现强度不一的吸收特征，特别是在俗称为"指纹区"的红外谱区。这些吸收特征与特定的矿物岩石有关，所以可以根据月球的反射光谱(强度，反照率以及吸收特征)，反演月球表面的物质组成。月壤样品 62231 是现在月球光学遥感采用的光谱标准。图 1 中

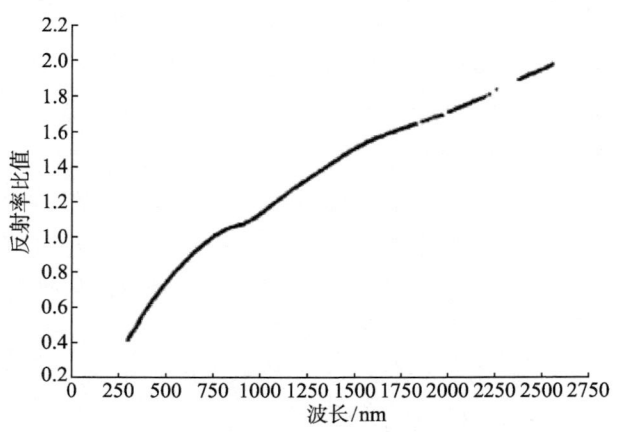

图 1　阿波罗 16 月壤样品 62231 的二向反射光谱
入射角 i=30°，发射角 e=0°，相对 1020 nm 光谱比值的平均值

本文原载于《地球与环境》，2004，Vol. 32，No. 3-4，27~33。

62231的二向反射光谱随着波长的增长，反射率增加，即变"红"。

月球除了反射入射太阳辐射，本身还向外发出热辐射。但是，在紫外-可见-近红外谱区，主要是反射的太阳辐射，而且反射的形式是漫反射。太阳辐射的主要能量集中在 0.2~3 μm，其中可见光区(0.38~0.76 μm) 约占总能量的一半。研究月球表面的反射光谱一般测量 0.4~2.6 μm 间的光谱曲线。

2 现有的月球反射光谱数据

美国的 McCord 等利用夏威夷岛 Mauna Kea 山上的夏威夷大学望远镜(口径 2.2 m) 获取月球正面一些小的地质单元(直径 3~10 km) 的可见近红外反射光谱[1] (0.65~2.5 μm)。该望远镜所处的海拔高度有 4200 m，空气吸收低，特别是在水的红外吸收区，所以可以获得高质量的连续反射光谱数据。

他们获得的反射光谱只有约5%是高质量的，分辨率约 3 km，其余约 1/3 的数据是用 60 cm 的望远镜获得的，所获得的数据主要是针对组成相对均一的月海区月壤。在一定程度上可以弥补空间分辨率的不足。所用望远镜的光谱分辨率在 0.6~1.4 μm 间为 10nm，在 1.35~2.5 μm 间为 20 nm。

处理获得的原始光谱数据需要采用光谱标准，他们采用的校正标准是 Apollo16 附近的一块区域。具体而言采用月壤样品 62231 对应的区域。在 1980 年以前一般采用 MS2(Mare Serenitatis)作为标准。使用月壤作为校正标准，比采用像太阳一样的天体要准确得多[2]。处理时首先获得被观测区域相对 Apollo 16 区域的通量比值(公式(1))，然后将该比值乘以阿波罗 16 样品 62231 的实验室光谱。美国一般使用 Adams 获得的 62231 的方向半球反射光谱。

$$R = \frac{flux_A}{flux_{Apollo16}} \quad (1)$$

公式(1)中的比值可以视为测量相对反射率，采用这种比值形式可同时提供大气及仪器校正。但仍然不能完全消除观测误差以及重复测量导致的统计误差。阿波罗 16 样品 62231 的实验室光谱测定是很准确的。美国国家宇航局资助的布朗大学 RELAB 光谱实验室采用海伦板(美国国家标准)作为实验室标准，它被校正成绝对反射体标准，在近红外反射比为 97%~99%。按上述处理流程得到的不是反照率数据，所得到的月球反射光谱数据相对 1.02 μm 处的反射率作了归一化处理，因为在该波长没有明显的大气吸收。

美国布朗大学的 RELAB 实验室按 PDS[3]规范整理了月球反射光谱数据，并发布到网上[4]。这些数据的公开发表将方便月球光谱的分析工作。

3 月球光谱的吸收机制

物质分子的运动具有多种形式，需要从外界获取的能量也各不相同，所以会在不同的波长位置出现吸收峰，形状特征也会彼此不同。按成因这些吸收谱线可分为以下四类[5]。

第一类是分子平动引起的吸收，它的特征是非量子化，所导致的吸收波段是连续的；第二类是分子转动引起的吸收，它对应 RF 区域或者叠加到振动跃迁(热红外区域的精细结构)，例如，液态水在射频区有吸收；第三类是分子振动引起的吸收，它位于热红外辐射区。例：H_2O，基频(2.9、3.1、6.1 μm)。倍频以及组合频峰(0.94，1.1，1.4，1.9 μm)；OH 基团，基频(2.8 μm)，基波 1.4 μm；SO_2 0.28,4.1 μm；硅酸盐 9.0~12.5 μm；碳酸盐 7μm；硝酸盐 7~8 μm；硫酸盐 9μm；磷酸盐 9.3 μm。

最后一类是电子跃迁在光学区域引起的吸收。这类谱线是矿物中的过渡金属离子吸收产生的，根据这些谱线识别月球表面的矿物类型，而且这些吸收特征正好位于月球的反射光谱能量区间。下面将详细讨论。

3.1 晶体场跃迁

过渡金属离子(Fe，Ti，Mn，Cr 等)具有未填满的 d 电子轨道。在没有外加电场的状态下，这些离子的结构是稳定的。但在晶格中，周围的离子会使电子外壳变形，导致能级分裂。例:八面 O^{2-} 配位体中的亚铁，它的能级将被分裂成两个子能级。这两个子能级间的能量差大约相当于波长为 1 μm 的光子，所以包含亚铁离子的硅酸盐矿物在 1 μm 附近有一吸收带。晶格的大小会改变能量间隔，在斜方辉石中，吸收带中心在 1 μm 附近。钙是一种大离子，它的大量存在会使晶格变大，所以高钙单斜辉石亚铁离子的波长大于 0.95 μm。橄榄石在 1 μm 附近有三个重叠得非常紧的波段；斜方辉石约 1.8 μm，单斜辉石约 2.3 μm。尖晶石只有一个 2 μm 的波段，在斜长石中，Fe^{2+} 作为不纯物会在 1.25 μm 附近引入一个弱的吸收带。

3.2 电荷转移跃迁产生的吸收

电子从某个离子的一个能级跃迁到另一个离子的能级。如硅酸盐中的: $Fe^{2+}\text{-}O^{2-} \rightarrow Fe^{3+}\text{-}O^{3-}$，波长约为 260 nm。在普通辉石以及玻璃中，会发生 Fe-Ti 电荷转移跃迁，波长为 340~400 nm。

3.3 共价键-导带跃迁

当入射光子能量等于或大于矿物禁带宽度时，绝大多数入射光子被吸收，使价带顶部的电子跃迁到导带，能级差异对应紫外波长。该光谱区域对分析不是特别有用。因为太阳在 200nm 以下，光强比较弱，而且大多数物质在 200 nm 以下有吸收。

3.4 色心产生的吸收

色心是产生颜色的缺陷中心，它通常与晶格的点缺陷相关。在矿物生成时，经常形成各种点缺陷。诸如空位，置换或填隙离子，以后晶体受力，受放射线辐照以及其他作用也能产生新的点缺陷。这些点缺陷导致矿物内局部结构的电中性的确定，是形成色心的潜在原因。色心的光吸收是色心中的电子(或空穴)以电偶极矩跃迁的形式跃迁到一个被束缚的激发态(陷在导带的陷阱中)所引起的[6]。

4 月球表面的岩石类型及其光谱特征

月球表面主要由富钙的斜长岩长石、富钙以及贫钙辉石组成。但在有些地区，橄榄石和钛铁矿也是重要组分。在扩展可见光谱区域(从近紫外到近红外)，这些矿物的反射性质主要受 Fe 和 Ti 的支配。Fe 和 Ti 是过渡金属元素，在橄榄石与辉石的晶体结构中，它们的 d 轨道没有填满，会在这些矿物的波谱上产生吸收特征。因为 Fe 的宇宙丰度大概是 Ti 的 300 倍。Ti 只在钛铁矿等富钛的氧化物以及某些含铁的玻璃的光谱中比较显著，亚铁在紫外波段产生一个强的紫外吸收，并在近红外 1 μm 附近有一个弱的吸收。随着矿物中亚铁含量的增加，或者月壤中含亚铁的矿物相增加，这两个重叠的吸收带复合起来，使总反射率降低。与此同时，1 μm 附近的吸收带变深，而且在紫外边缘尖锐的吸收边缘会导致 UV 波长的吸收率相对连续背景吸收率的比值降低[5]。

5 月球表面主要过程及其对月球反射光谱的影响

月球的空气非常稀薄，表面的结晶质岩石以及角砾岩(在大的撞击中产生)直接暴露于宇宙户。月球表面沐浴在太阳风粒子中。太阳风质子注入矿物以及玻璃颗粒中，深度约为数十 nm。陨石则是影响月球表面的第二种重要粒子。它们以全宇宙速度撞击月球表面，导致表面物质的粉碎、熔融、蒸发，还会制造出通过玻璃质连接在一起的聚集体颗粒，称为聚集体。风化过程(即月壤成熟)也会通过矿物和玻璃中亚铁的还原(可能有氢的辅助)产生微小(纳米态)的单质铁(Fe^0)的圆形包裹体。亚微观的金属铁对月球表面的反射性质有强烈的影响——削弱总体反射率，减小 1 μm 波段的对比度，使光谱曲线的坡度变化随着波长的增加而变大,即变"红"。这些光谱效应已经被用来将月壤中亚铁与单质铁的光谱效应(依赖于月壤的成熟度)区分开。钛是月球表面另一种对反射光谱有重要影响的元素。富钛的不透明矿物钛铁矿($FeTiO_3$)的光谱没有明显的吸收特征,从 UV 到 NIR 光谱曲线比较平坦,所以富钛铁矿的月壤的反射率相对较低,而且光谱曲线的的坡度相对一般表现为光谱"红"的月球物质要小(即变"蓝")。钛铁矿和铁的竞争效应是 Lucey 等开发的钛与铁定量技术的基础[7-10]。

6 月球反射光谱数据的解析

6.1 连续光谱的解析

月球连续光谱一般在地面通过望远镜观测或在实验室测定。它的空间分辨率低,光谱分辨率高,局限于月球正面。月球连续光谱与遥感数据获得的反射光谱不同,前者与使用的测量仪器有关,一般为单向—半球反射光谱,后者一般为二向反射光谱,这两类数据如果需要对照,需要校正因子。

连续光谱的解析是通过一些光谱参数来实现的,研究人员已经从对月海玄武岩样品的化学分析以及光谱测定中,提取出了一些与化学矿物组成相关的光谱参数,包括 UV/VIS、反照率、1 μm 波段特征以及 2 μm 波段特征,并详细研究了它们之间的相互关系,见表 1。

表 1 月球反射光谱参数与月海玄武岩各种组分之间的关系[11]

UV/VIS	反照率 (0.62μm)	1μm 波段特征		2μm 波段特征	
		组分	吸收峰深度	组分	吸收峰深度
负相关:玻璃紫外吸收(负相关:FeO—TiO_2)	正相关:斜长石含量	玻璃	正相关:玻璃中的 FeO 正相关:玻璃含量	辉石组分	正相关:辉石含量
负相关:单质铁(Fe^0)	负相关:玻璃吸收	辉石组分	正相关:辉石含量		负相关:不透明矿物 (负相关:TiO_2—FeO)
负相关:反照率	负相关:不透明矿物 (负相关:TiO_2—FeO)	橄榄石	正相关:橄榄石含量		
	负相关:Fe^0		负相关:不透明矿物 (负相关:TiO_2—FeO)		

在月壤中发现的很多组分在 1 μm 附近有吸收特征,这是由于它们所含的六配位中的亚铁,它们对 1 μm 波段的影响按逐渐递减的顺序排列如下:均一玻璃、辉石、橄榄石、斜长石。均一的玻璃有一宽的、以 1.05 μm 为中心的电子跃迁吸收,在纯的玻璃中,它的强度直接与 FeO 含量相关。对于月球上的普通辉石,1 μm 波段的最低点在波长上作为化学组成的函数,在 0.91~0.98 μm 之间变化,辉石的亚铁波段比玻璃的要窄。橄榄石则有一单独的非对称宽峰,中心位置在 1.03~1.06 μm 间,由于橄榄石的吸收相对玻璃与辉石要弱,所以月壤中少量的橄榄石对该波段的贡献很容易被其他物质给掩盖。对于斜长岩长石,它的 FeO 含量低于 1%,有一宽吸收峰,中心在 1.25~1.30 μm 间(与钙长石含量有关),该斜长石峰会在 1 μm 吸收峰的长波一侧引起弯曲,但在月壤中很难看到,这是因为它很弱,很容易被其他更强的吸收物质掩盖。各种矿物的吸收特征如图 2。

对于 1 μm 吸收峰,它的深度与所有在 1 μm 波段有吸收的物质的相对丰度有关。然而,对于准确理解观察到的月壤 1 μm 吸收的深度,需要搞清楚成熟月壤中的聚集体以及不透明物质对该波段的影响。

为了更好地利用连续光谱数据,有人提出了连续背景扣除以及高斯解卷积。月球表面的月壤在太阳风以及微陨石撞击下,不断"成熟",在反射光谱上表现为使光谱不断变"红"且特征吸收峰变窄,这会使光谱分析变得困难。于是有人提出通过归纳总结月球反射光谱的曲线特征,用某种数学方法拟合连续背景,然后按该方法扣除该背景。

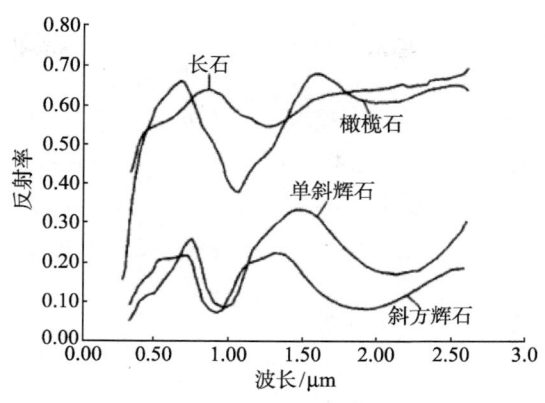

图 2 月球主要矿物的反射光谱

(注：数据来自 Brown University RELAB 实验室)

根据连续光谱的一般特征，起初将从 1000 nm 到 2000 nm 的与反射光谱相切的直线作为连续背景，即线性扣除法；此后，Hiroi 等提出了双线性扣除[12]，即作两条与反射光谱相切的直线，一条从 700 nm 到 1500 nm，另一条从 1500 nm 到 2500 nm。如果这两条切线在 1500 nm 附近彼此重合，它们的接触点移动到它们的中点；如果它们不重合，就用相应空缺波长区域的反射光谱连接起来。最近为了更好地模拟连续背景，Hiroi 又提出了下面的数学公式[13]

$$C(\lambda) = c_{-1}/\lambda + c_0 + c_1\lambda \tag{2}$$

λ 是波长，$C(\lambda)$ 是连续背景函数，c_{-1}、c_0 以及 c_1 是常数；可以通过代入 750 nm、1500 nm 以及 2600 nm 三点的 λ 以及对应反射率求解方程获得。

从反射光谱中扣除背景后，可以通过高斯解卷积[14]得到各吸收特征的光谱参数。于是可以利用现有的矿物波谱知识作光谱分析。具体做法是将扣除连续背景后的反射光谱转换为它的自然对数，然后分解成改进过的高斯函数以及一扁平的背景。所得到的高斯曲线代表某矿物的单独的吸收特征。扣除连续背景时常常会附带扣除一些信号，所以需要以某种方式加以弥补。上述方法称为 MGM 方法。得到的吸收带参数有吸收中心，宽度以及强度。美国布朗大学 RELAB 光谱实验室利用 Matlab 语言、Idl 语言以及 VMS Fortran 语言实现了 MGM 方法的基本功能，可供参考。

以上方法是从光谱曲线的形状出发，而不是从连续背景的真正成因出发，去导出一个扣除方法。所以不可避免有误差。但是不可否认 MGM 方法是目前分析月球光谱数据的有力工具。

6.2 光学遥感数据的解析

美国 1994 年发射的克莱门汀 UV-VIS 相机只工作在 0.415、0.75、0.90、0.95 与 1.00 μm 这五个波段，所以相对连续光谱，在非常重要的 1 μm 波段数据要少得多。

为了从有限的数据获得同样有价值的信息，Tompkins 与 Pieters[15]在利用克莱门汀数据研究月球撞击坑中央峰的组成时，提出用关键比值(Key Ratio)表示 1 μm 吸收特征的强度，它的做法是将 0.75 μm 波段的反射率与 1 μm 附近三个波段(0.90，0.95，1.00)的反射率相比，由于矿物组成决定了这三个长波波段(0.90、0.95、1.00)中的哪一个最接近吸收的最低点，于是将其中最强的一个比值作为关键比值。从关键比值的定义可以看出，它可以有效地反映 1 μm 波段的相对强弱，而且可以根据关键比值采用哪个长波波段来近似估计吸收最低点的位置，也就可以区分不同的矿物组成。如辉石的关键比值所采用的长波波段定在 1 μm 内，即 0.90 或 0.95 μm，而对于玻璃与斜长石，由于它们 1 μm 吸收的中心在 1 μm 外，所以关键比值定义为 0.75 μm/1.00 μm。

此外，他们还提出了光谱曲率(Spectral Curvature)的概念，定义为克莱门汀"五色"光谱在 0.75、0.90、

0.95 μm 这三个点形成的角度大小。它可以有效地表示 1 μm 吸收特征在那些对区分镁铁质矿物非常关键的波段处的弯曲情况。随着 1 μm 吸收最低点向长波方向移动，光谱曲率增大。研究表明光谱曲率与镁铁质矿物的类型密切相关，特别是对辉石的组成敏感，它可以用于区分低钙与高钙辉石。

总之，利用关键比值与光谱曲率这两个参数，可以从有限的遥感数据中得到同样有价值的信息。但这对原始探测数据的校正提出了很高的要求，因为小的校正误差就可能导致完全错误的解析。

美国克莱门汀探测器获得的多光谱数据可以从网上获取(PDS NODE)，格式为 pds 格式。相应的分析工具有美国地质调查局开发的免费的 ISIS 系统。

6.3 混合物的光谱分析

我们获得的月球反射光谱通常代表一些矿物的混合物。一般使用多变量分析技术来研究混合物的光谱解析[16]，如主成分分析[17](PCA)就被用于研究月球表面反射光谱的多组分特性。研究时可以将某些组分视为端员组分(Endmembers)，并将月球的反射光谱理解为这些组分光谱特征的非线性组合。通过对整个数据集作 PCA 分析，可以确定在光谱图上存在多少线性独立的变量源，被混合的端员组分可能是什么，是什么物理过程影响组分的混合以及对从遥感数据区分端员组分，并可以帮助月球探测方案设计者确定采用什么光谱通带是最合适的？确定端员组分的标准是：端员组分必须位于 PCA 图的最高点且端员组分必须是利用现有信息可以识别的。

7 总结

现有月球紫外-可见-近红外的分析方法已比较成熟，但是对月球连续光谱的背景的成因还不是很清楚，需要进一步的实验以及理论探讨，这对月球光谱的精确解析很重要。

月球样品的研究是遥感探测的重要基础，月岩以及月壤样品对于反射光谱的校正与解析至关重要，目前，拥有样品的国家仅美国和俄罗斯，依赖他们大量提供样品是不现实的。中国目前已经启动探月计划，我们认为中国应该综合月球现有研究的热点以及中国的国情，适时提出有较高科学价值的取样方案。另一方面，从科学的角度来说，美国以及前苏联登月所取样品的地域代表性以及化学、矿物学以及岩石学组成的代表性都远远不够[18]，迫切需要有新的样品来增加研究活力。这也有助于丰富我们对月球物质反射光谱特性的认识。

高光谱技术在光谱分辨率上接近地面现代望远镜，而空间分辨率又大大优于望远镜，所以在未来的行星探测中，可以预见高光谱成像技术将得到积极的应用。美国已在陆地卫星上使用了高光谱相机(如美国 EO-1 卫星上的高光谱成像仪 HY-PERION，光谱范围 400~2500 nm，波段数 220 个，波段连续，地面分辨率 30m)，中国目前也已经具备了研制高光谱相机的实力，"神舟"三号飞船搭载的中分辨率成像光谱仪，有 34 个波段(严格地说应该属于"准高光谱"，因为光谱并不连续)，完全可以将该技术移植到月球卫星遥感上，需要解决的是数据的传输与储存。

到目前为止，美国尚未在对月遥感中使用高光谱技术，但已有许多科学家提出了在近期应用该技术的方案。Lelong[19]就提出了使用更多波段的成像光谱技术(表 2)。该表选取的光谱通道，是在分析许多月球样品光谱的基础上提出来的。采用该方案，可以有效识别并区分月球表面的主要成岩矿物。

表 2 在紫外-可见光谱波段可作的矿物识别以及使用的主要波段

光谱区间(μm)	矿物识别目标
0.3；0.4；0.55；0.75	光学成熟度(optical maturation)；二氧化钛含量；钛铁矿的存在
0.75；0.90；0.95；0.98；1.00；1.05；1.10	1μm 附近 Fe^{2+} 吸收带的定位；斜方辉石、单斜辉石以及橄榄石的区分
1.05；1.10；1.20；1.30；1.60；1.80	1.1—1.3μm 范围内 Fe^{2+} 吸收带的定位；斜长岩端员组分之间的区分
1.80；1.95；2.00；2.10；2.20；2.30；2.50	2μm 附近 Fe^{2+} 吸收波段的定位；斜方辉石以及单斜辉石的区分

欧空局已于 2003 年九月成功发射了 SMART-1 探测器，它搭载了一台 SIR 红外光谱仪，工作区间为 0.9-2.4 μm，光谱分辨率达到 0.06 μm，空间分辨率为 300 m。该仪器将用于区分月球表面不同的组分。该仪器也是欧空局未来星际探测的重要工具，如对火星的探测。这台仪器不仅空间分辨率高，而且具有很高的光谱分辨率。如果它能顺利到达月球，将给我们带来一个全新的视野，极大促进对月球起源与演化的认识。

致谢 本文得到了李德先同学、傅平青同学的帮助，特此感谢！

参 考 文 献

[1] McCord T B, Clark R N, Hawke B R, et al. Moon: Near-infrared Spectral Reflectance, a First Good Look[J]. J. Geophys. Res, 1981.86(B11):10883—10892
[2] Pieters C M. The moon as a spectral calibration standard enabled by lunar samples[EB/OL]. http://wufs.wustl.edu/geodata/mk88-1-120cvf-3 -rdr-120color-vl/mkls_0001/document/pieter99.pdf .1999
[3] National Aeronautics and Space Administration (NASA). Planetary Data System. http://pds.jpl.nasa.gov/.2004-01/2004-03.
[4] Washington University in St. Louis. PDS Geosciences Node. http://wufs.wustl.edu/missions/lunarspecj/2000-12/2004-04
[5] Blewett D T. Lunar remote sensing and spectroscopy grab-bag, prepared for B. R. Hawke's Lunar Geology class[R]. Dept. of Geology and Geophysics, University of Hawaii. October, 2000
[6] 陈丰，林传易，张慧芬，等. 矿物物理学概论[M]. 北京：科学出版社. 1995
[7] Blewett D T. Lucey P G. Hawke B R.et a l. Clementine images of the lunar sample-return stations: Refinement of FeO and TiO, mapping techniques[J]. J. Geophys. Res. 1997. 102(E7): 16319—16 325
[8] Lucey P G. Blewett D T. Hawke B R. Mapping the FeO and TiO_2, content of the lunar surface with multispectral imagery [J]. J. Geophys. Res. 1998. 103(E2):3679—3699
[9] Lucey P G. Blewett D T, Jolliff B L. Lunar iron and titanium abundance algorithms based on final processing of Clementine UVVIS data[J]. J. Geophys. Res. 2000. 105(E8): 20297—20306
[10] Lucey P G, Blewett D T, Taylor G J, et al. Imaging of lunar surface maturity[J]. J. Geophys. Res. 2000. 105(E8): 20377—20386
[11] Pieters C M. Mare basalt types on the front side of the Moon: A summary of spectral reflectance data[J]. Proceedings of the Ninth Lunar and Planetary Science Conference. 1978: 2825—2849
[12] Horoi T, Pieters C M. Modified Gaussian Deconvolution of reflectance sepectra of lunar soils[J]. Lunar and Planetary Science XXIX. 1998: 1253
[13] Sunshine J M, Pielers C M, Pratt S F. et al. Absorption Band Modeling in Reflectance Spectra: Availability of the Modifed Gaussian Model. Lunar and Planetary Science XXX. 1999:1306
[14] Hiroi T, Pieters C M, Noble S K. Improved scheme of modified gaussian deconvolution for reflectance spectra of lunar soils[J/CD]. Lunar Planet. Sci., XXXI. Houston. TX. 2000: 1548
[15] Tompkins S, Pieters C M. Mineralogy of the lunar crust: Results from Clementine[J]. Meteoritics&.Planetary Science. 1999, 34: 25—41
[16] Johnson P E, Smith M O, Adams J B. An Application of a Strategy for Interpretation of Multispectral Reflectance Data from Planetary Surfaces[J]. Lunar and Planetary Science XV. 1981: 407—408
[17] Blewett D T, Hawke B R, Lucey P G, et al. Remote sensing and geologic studies of the Schiller-Schickard region of the Moon[J]. J. Geophys. Res., 1995. 100(E8): 16959—16 977
[18] Dunkin S. Heather D. New views of the Moon[EB/OL]. http';/physicswcb. org/article/world /l 2/7 /8.1999
[19] Camille C D. Lelong. Institute of Planetology. Geology and Mineralogy through the LunarSat project[EB/OL]. http://ifp.uni-muenster.de/sohl/euromoon/spectrum/mineralogy.html.2003-09/2004-03

A Preliminary Experience in the Use of Chang'E-1 IIM Data

Wu Yunzhao[1] Zheng Yongchun[2] Zou Yongliao[2] Chen Jun[3] Xu Xisheng[3]
Tang Zesheng[4] Xu Aoao[4] Yan Bokun[5] Gan Fuping[5] Zhang Xia[6]

(1. School of Geographic and Oceanographic Sciences, Nanjing University, Nanjing 210093, China;

2. National Astronomical Observatories, Chinese Academy of Sciences, Beijing 100012, China;

3. State Key Laboratory for Mineral Deposits Research, Department of Earth Sciences, Nanjing University, Nanjing, 210093, China;

4. Collaborative Research Laboratory on Lunar and Planetary Exploration, Macao University of Science and Technology, Taipa, Macao;

5. Institute for Remote Sensing Method, China Aero Geophysical Survey and Remote Sensing Center for Land and Resources, Beijing 100083, China;

6. Institute of Remote Sensing Applications, Chinese Academy of Sciences, China)

Abstract The Interference Imaging Spectrometer(IIM) onboard the first lunar satellite of China, Chang'E-1, has acquired 84% of the area between south and north latitude 70°. To contribute to its usability, this paper presents our preliminary experience in the use of IIM data. Firstly, we provide one practicable method for the on-orbit correction of the inhomogeneity of sensor response. Secondly, aiming at the problem that the spectral range of IIM does not cover the absorption peak of the mafic mineral completely, we explore a method to approximate the absorption band center for IIM data. A strong correlation between the absorption band center and the wavelength at which the first derivative equals to 0 (i.e., stagnation point) was revealed. Based on the corrected data and the correlation, the absorption band center of several large craters was mapped. The distribution of rocks and minerals shown in the map of absorption band center for Aristarchus and Copernicus is in agreement with previous studies but with much finer structure. Horizontal and vertical lithologic diversity was detected in Zucchius crater. This paper demonstrates the potential of IIM data for the identification of lunar rocks due to its high spatial and spectral resolution. In a future study we will produce a global map of the absorption band center with greater accuracy and it is expected that this global map will provide complementary information for other hyperspectral data such as SP on KAGUYA or M^3 on Chandrayaan-1.

Key words Chang'E-1 IIM Calibration Absorption band center Moon

1 Introduction

The first lunar satellite of China, Chang'E-1 (CE-1), was launched on 24th October 2007 and ended its 16-month mission by impacting the Moon on 1st March 2009. The Interference Imaging spectrometer (IIM) which covers wavelengths from 480 to 960nm with 32 channels onboard Chang'E-1 was aimed to acquire the mineral composition of the lunar surface. IIM has acquired 84% of the area between south and north latitude 70° during its one-year mission. The final data provided by the Ground Segment for Data, Science and Application (GSDSA) of China's Lunar Exploration Program, was Level 2C. The IIM 2C data have been systematically processed in several steps as described in Zhang et al. (2005). However, it should be processed further before it is used. Two factors hamper the direct application of IIM data: (1) the IIM 2C data have not been converted into reflectance; (2) the sensor response was inhomogeneous from left to right so that the same lunar surface exhibits different spectra along the line direction(see details below), which causes apparent, but unreal compositional variations. Moreover, the spectral range of IIM does not cover the absorption peak of the mafic minerals completely, which limits its ability and feasibility for identifying minerals. It is necessary to investigate the method for improving its usability. To contribute to its usability, in this paper we present our preliminary study of Chang'E-1 IIM data, including the calibration, flat-field correction and preliminary application. In this study all the processes were performed using the Lunar Geochemical Mapping System(LGMS v1.0), which was designed specifically for CE-1 optical images by our group.

2 Data introduction and calibration

All scientific data probed by CE-1 were received, processed, archived and distributed by GSDSA, which was located in the National Astronomical Observatories, Chinese Academy of Sciences. The processing flowchart of CE-1's IIM data was described in detail in Zhang et al. (2005). The IIM 2C data are in radiance format with dark-current subtraction, flat-field correction, convention to radiance. The photometric normalization corrections of IIM 2C data were performed with the empirical photometric function proposed by (Hillier et al., 1999) and the parameters generated by the GSDSA to convert the data into the standard geometry ($i=30°$, $e=0°$). The instrument is a Sagnac-based pushbroom Fourier transform imaging spectrometer and maps the lunar surface with a swath of 25.6 km from the polar orbit of 200 km altitude. The performance index of IIM data is shown in Table 1.

Table 1 Key IIM measurement characteristics

Image width	25.6 km
Spatial resolution	200 m(nadir)
Spectral range	480–960 nm
Spectral resolution	325.5 cm^{-1}
Radiometric sampling	12 bit
MTF	⩾0.2
SNR	⩾100
Number of pixels	128

The reflectance spectra of Apollo 16 soil 62231 measured in the laboratory and the corresponding area of undisturbed (mature) soil of Apollo 16 landing site has been often used for the calibration of lunar remote sensing data (Pieters, 1999; Pieters and Pratt, 2000; Matsunaga, et al., 2008). The IIM data used in this study have been photometrically normalized to the same photometric geometry as the bidirectional reflectance measurement of 62231. This study used the bidirectional reflectance measurement of 62231 to convert IIM data into reflectance.

The location used for the calibration of telescopic and Clementine spectra in the previous study is commonly located at the point to the west of the landing site that is the third vertex of an equilateral triangle which has North Ray and South Ray craters as the other two vertices (Blewett et al., 1997; Pieters et al., 2008). Unfortunately, this site is located just between CE orbit 2224 and 2225, i.e., IIM does not cover this site. By comparing the IIM data of the two orbits with high spatial resolution and multispectral Clementine data, we selected one site from Orbit 2225, which is close to the previous calibration standard site, whose spectral feature is similar with the calibration standard site, whose surface is homogeneous with free of crater rays, and whose viewing and illumination geometry is similar with that of laboratory reflectance spectra of Apollo 16 soil 62231 (Table 2), as the calibration site to convert IIM data into reflectance (Fig.1). Considering the pushbroom design of IIM and the artifacts caused by the variations in the pixel-to-pixel sensitivity of the detector columns, a 5×17 pixel box in the middle of IIM columns (Column: 69–73; Line: 11151–11167) was averaged to supply the spectrum.

Table 2 Imaging information of the calibration site area used in this study

Coordinate	8.79°S, 14.9°E
Imaging date	2008-05-21
Imaging time(UTC)	01:32
Instrument incidence angle	0°
Sun incidence angle	28°
Sun azimuth angle	290°
Phase angle	28°

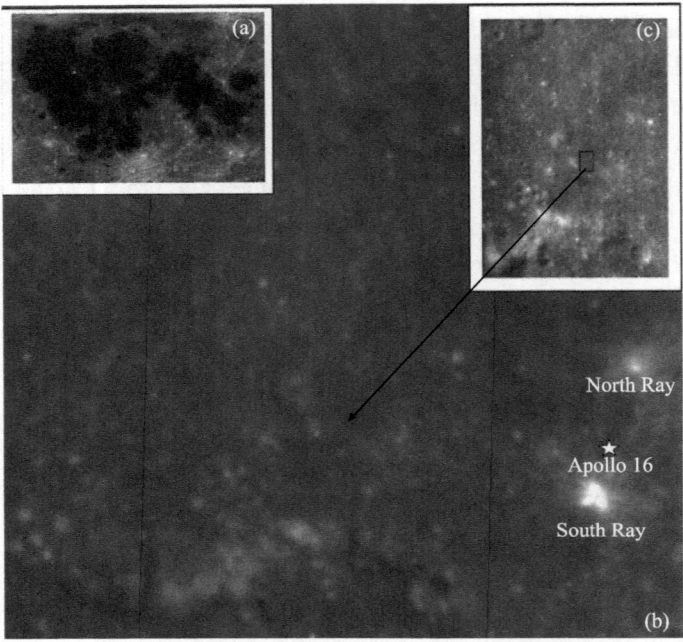

Fig.1 Calibration site of IIM data. (a) shows the covering region of Orbit 2225. (b) is Clementine 750 nm albedo data with 100 m spatial resolution for the Apollo 16 region. The calibration site area used in this study is indicated with an arrow. (c) is IIM data with the red box as the calibration site area

3 Correction for the inhomogeneity of sensor response

Fig.2(a)–(d) (Plate VII) show the composite images of 918 nm (R), 757 nm (G) and 618 nm (B) of the converted IIM reflectance data for several different areas. Fig.2(a) represents a low latitude area (calibration site) of the southern hemisphere and Fig.2(b) represents a high latitude area (C. Mayer crater) of the northern hemisphere. The composition of both areas is highland anorthosite. Fig.2(c) and (d) represent compositionally diverse area (Aristarchus crater). Although the composition and the geographic location vary for these areas, they share in common a hue that changes from left to right. Fig.3 shows the reflectance of the dashed line shown in Fig.2 for 918, 757 and 618 nm. For each plot the dashed line covers an area of approximately uniform composition along the line direction. Hence, the spectral profile should be flat from left to right. However, as shown in Fig.3 (Plate VIII) for each plot the reflectance is variable, especially for the longer bands (e.g. 918 nm), with reflectance increasing from left to right. This effect is due to the inhomogeneity of the sensor response and is independent of the character of the lunar surface and geographic location.

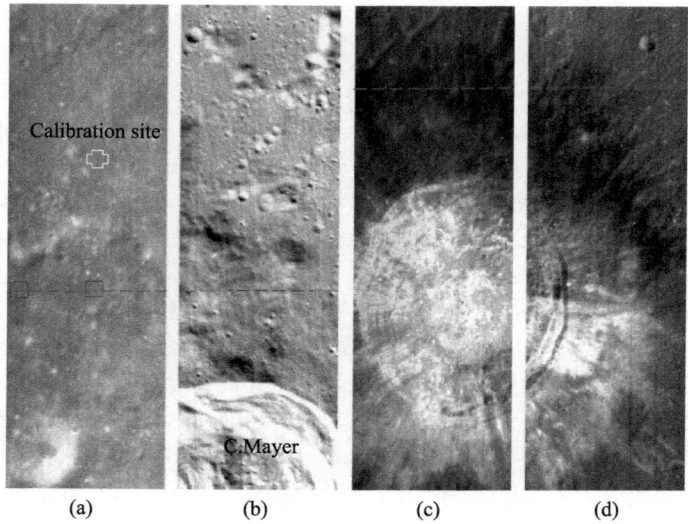

Fig.2 Composite image of 918 nm (red), 757 nm (green) and 618 nm (blue) of IIM data for calibration site (a), the crater C. Mayer (b) and Aristarchus area (c and d). For each plot the composition of the surface from left to right crossed by the dashed line is approximately the same

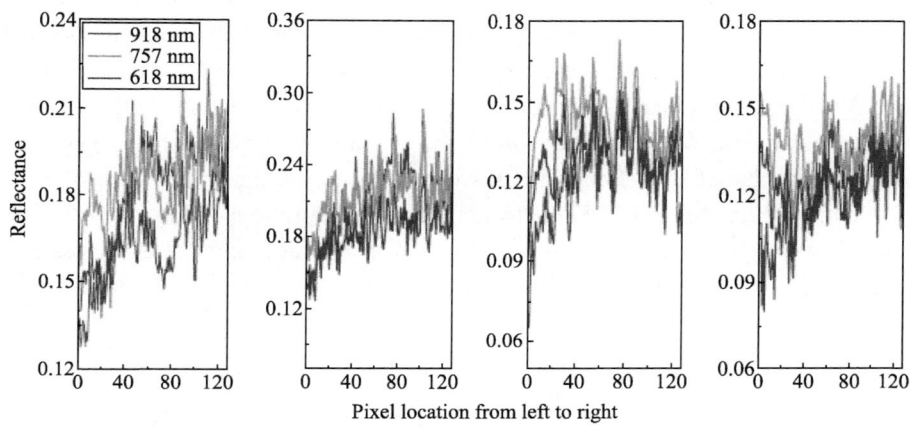

Fig.3 Reflectance profile of the dashed lines shown in Fig.2 parts A–D, for 918, 757 and 618 nm

The inhomogeneity of the sensor response must be corrected before the data are used for geological research. From the on-orbit flat-field correction point of view, in this study we propose a correction method named reflectance normalization, based on the spectral features of the lunar surface. The area covering the calibration site used in this study was selected for the flat-field correction. The correction procedures include the following steps:

(1) The absolute reflectance of soil 62231 was scaled to unity near 750 nm, named scaled reflectance (Fig.4a).

(2) Selection of correction standard line: because we want to correct the inhomogeneity along the line direction, our assumption is that the surface of the Moon for the whole line is the same. In this case the spectral variation along the line direction is caused only by the sensor response without the compositional effects. The width of the image is about 25.6 km and it is infeasible to find one line with the same surface. In order to avoid the lateral variation of lunar soil composition, three lines near the calibration site were carefully selected. The subset of each line with the same surface was combined into one correction standard line (Fig.4b).

(3) Normalize the reflectance of each pixel of the standard line for each band to the scaled reflectance (Fig.4c).

(4) Obtain the correction factor by quadratic polynomial fitting for the values derived above. Then the artifacts of the IIM data can be corrected by multiplying the correction factor (Fig.4d). In order to acquire the

correction factor with higher accuracy, some outliers with very high or low reflectance, caused perhaps by abnormal composition or instrumental noise, were removed before the polynomial fitting.

The correction does not just mean homogenizing of the hue along the line direction with the random adjustment of the reflectance. Two factors ensured that the corrected reflectance of each pixel has physical meaning, i.e., the reflectance of the surface. Firstly, the area for the selection of the correction standard line was the same as the land surface used to convert IIM 2C data into reflectance, i.e., the Apollo 16 landing site. Secondly, the correction factor was derived from the fitting of the reflectance of the correction standard line which has the same scaled reflectance of soil 62231.

Fig.4(d) shows the correction factor derived from the above method for three longer bands. In the space domain the left side of the correction factor is higher than the right side, which suggests that the sensor response in the left side is lower than that in the right. In the spectral domain, the correction factor for the long bands is larger than that for the short bands, which suggests that the inhomogeneity of the sensor response for the long bands is more serious. To validate whether the correction factor is suitable for other areas, all the plots of Fig.2 were corrected. As shown in Fig.5(a)–(d) (Plate VIII), the artifacts of the hue have been removed after the flat-field correction, and the spectra of the three NIR bands were homogenous along the line direction (Fig.6, Plate VIII).

In order to validate the final IIM data processed above, reflectance spectra of IIM were compared with the Earth-based telescopic spectra (Pieters and Pratt, 2000) for several relatively uniform areas, in which the topographic variation is relatively small. Fig.7 shows the final processed IIM spectra with 25×25 blocks of 200 m pixels (5×5 km) averaged for Apollo 16 region, MS-2, Aristarchus Plateau2 and Aristarchus Plateau3. It should be noted that the signal-to-noise ratio (SNR) of the shortest and longest bands of IIM (Bands 1 and 32) is very low and they were not shown in Fig. 7, i.e., 30 IIM points in each plot. As shown in Fig. 7, the remaining first six bands (Bands 2–7) are problematic for unknown reasons. Moreover, to ensure the quality of the data are as good as possible for the application the remaining first three bands (Bands 8–10) were also removed due to their low SNR. Therefore, only 21 bands (Bands 11–31) with a spectral range of 571–918 nm were used in the following study. In our experience using 21 bands for IIM application can ensure as good quality as possible and as many bands as possible. Fig.7 shows that reflectance spectra of the IIM and Earth-based telescopes match well in their mutual spectral range.

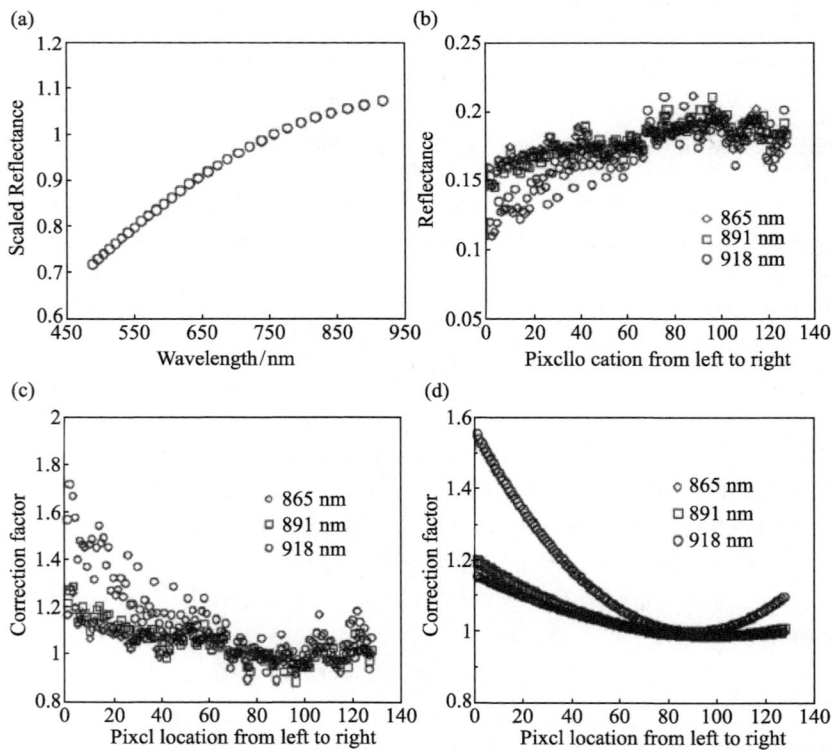

Fig.4 Derivation of the correction factor for the inhomogeneity of sensor response of the IIM data for three longer bands

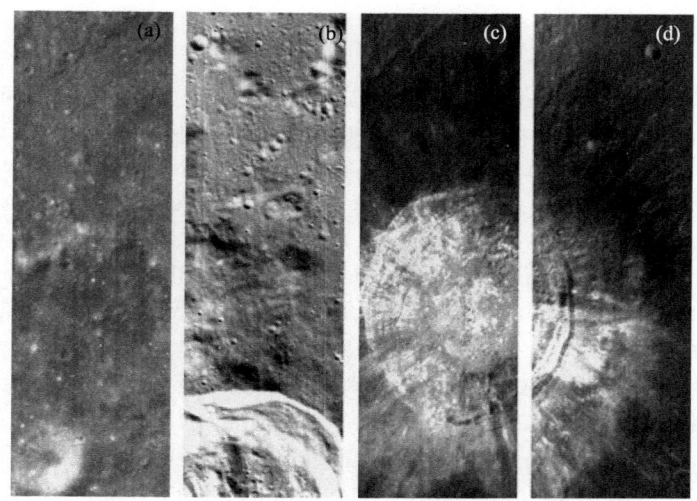

Fig.5 Composite image of the same area as Fig.2 but with the corrected data

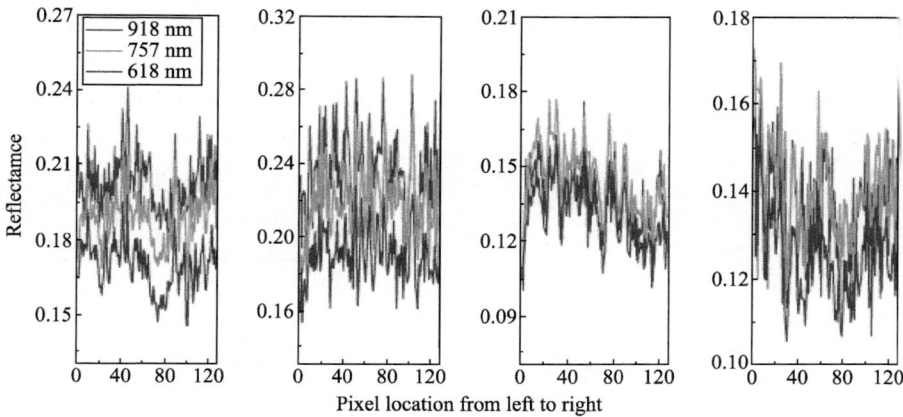

Fig.6 Reflectance profile same as Fig.3 but with the corrected data

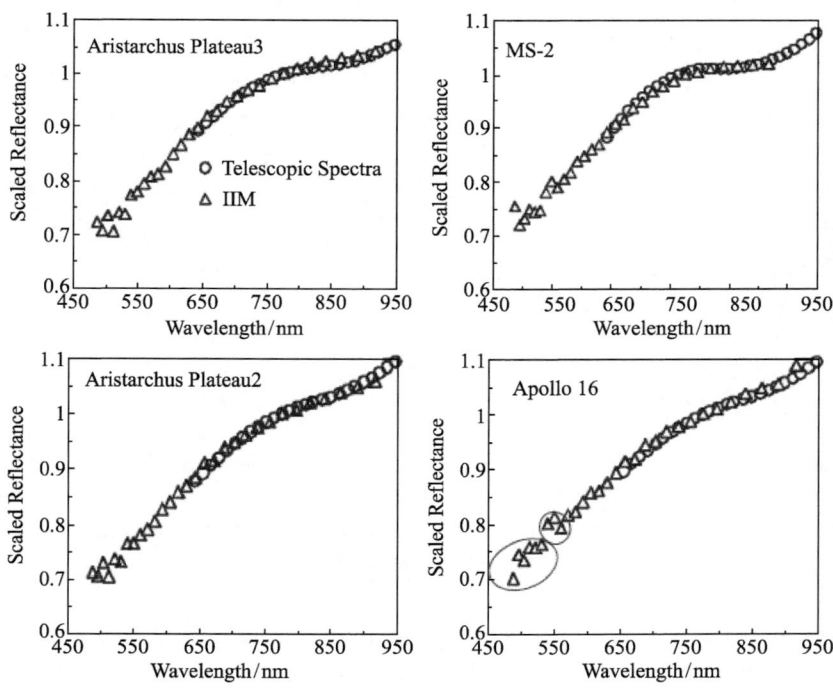

Fig.7 Comparison of the IIM reflectance spectra and Earth-based telescopic spectra for several relatively uniform areas. All spectra are scaled to unity at 776 nm. The ellipses denote the bands that will be removed during the application

4 Application of IIM data

The reflectance spectra of the lunar surface were affected by both composition and maturity. To investigate the method of extending the spectral range of IIM data to acquire the absorption band center of the mafic minerals, fresh exposures were considered because they are less affected by space weathering and revealed the minerals of the stratigraphy more clearly than older craters. The Aristarchus crater is the first choice because it is very new, has geological and spectral diversity, and has many Earth-based telescopic spectra of high quality. All the telescopic spectra of the Aristarchus crater except for that of south floor were used in this study. The floor of the crater may contain more melt than other sites in the crater. Hence it was excluded because the spectral feature of the original minerals was significantly altered.

The Earth-based telescopic spectra for 5 fresh exposures in the Aristarchus crater are shown in Fig.8(a). They all exhibit characteristic absorption band centered between 900 and 1000 nm. This diagnostic Fe^{2+} electronic transition absorption varies with the composition (both elements such as Fe, Mg, Ca and minerals such as pyroxene and olivine) of the lunar regolith (Adams, 1974; Cloutis and Gaffey, 1991; Sunshine et al., 1990). The wavelength of the absorption band center is the key parameter to identify minerals with reflectance spectra. For example, even the Mg-number, which serves as an important petrologic discriminator, was detected based on the variation of absorption band center (Okuno et al., 2008). Unfortunately, IIM cannot cover the absorption peak of the mafic minerals completely due to its narrow spectral range. However, next we try to extend the spectral range of IIM to acquire the absorption band center of the mafic minerals based on the first derivative of the spectra curve.

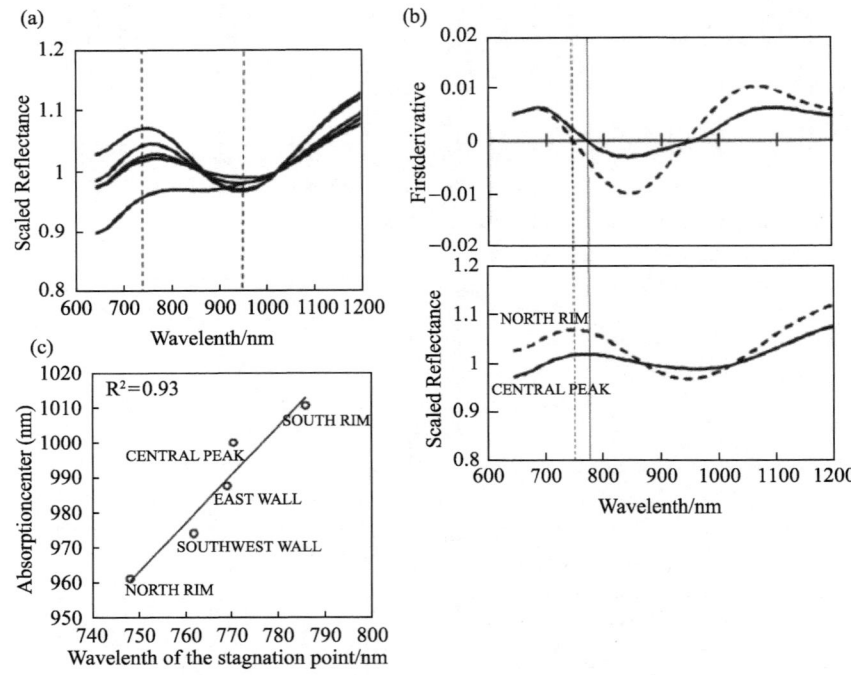

Fig.8 (a) Reflectance spectra for Aristarchus crater obtained with Earth-based telescopes (scaled to unity at 1020 nm; Pieters and Pratt, 2000). (b) Two representative spectra in Aristarchus crater and their first derivative. The two dashed lines indicate the wavelength of the stagnation point (i.e., at which the first derivative equals to 0). (c) The relationship between the absorption band center and the stagnation point

The absorption band center for the telescopic spectra was derived after a continuum, which is a straight line tangent to the spectrum at the peaks near 750 and 1500 nm, estimated and removed. The first derivative, which corresponds to the slope of the spectral curve, was computed using the Savitzky–Golay method (Savitzky and

Golay, 1964). Because the Earth-based telescopic spectra are noisy, they are smoothned using the cubic spline fitting procedure. As shown in Fig.8(b), the reflectance increases from shorter bands (e.g. 650 nm) towards the longer bands. At the wavelength of the stagnation point, i.e., the first derivative equals to 0, neither the reflectance increases nor decreases, i.e., it seems stagnant. As the wavelength continuing increase, the first derivative becomes negative from positive. Therefore, the wavelength of the stagnation point may be related to the wavelength of the absorption band center. Fig.8(c) shows the relationship between the absorption band center and the stagnation point. As stated above, it can be seen that the stagnation point was highly correlated with the absorption band center ($R^2=0.93$). Similarly positive correlation can also be found for the mixtures with pure minerals such as orthopyroxene and clinopyroxene simulated in the laboratory (Sunshine et al., 1990). Hence, the key parameter to identify minerals, absorption band center, could be mapped with this correlation and the stagnation point derived from the IIM data.

Figs.9 (Plate IX) and 10 show the map of the absorption band center produced using the IIM data and the formula in Fig.8 overlay the 776 nm IIM basemap, and the corresponding 776 nm IIM brightness image. The pixels exposing the basemap in Fig.9 are those without stagnation point within the band range of IIM, which means that their absorption band center is very long and are probably anorthositic rocks, pyroclastic deposits or impact melts (e.g. Lucey et al., 1986; Pieters, 1993). No obvious instrumental effects are observed in these derivative images, which is perhaps a test for our on-orbit flat-field correction method. Our result for Aristarchus crater is approximately consistent with those mentioned in Lucey et al. (1986) and also related to the mineral parameter map derived from M^3 data (Mustard et al., 2000). The absorption band center of the south rim and Point A is the longest. It corresponds to olivine depicted by previous research (e.g., Lucey et al., 1986, Le Mouelic et al., 1999, Mustard et al., 2000) the absorption band center of which is longer than pyroxene. Generally, the floor has longer absorption band center than the walls, which is consistent with the melt sheet of the floor. Minor differences between ours and Lucey's results exist. For example, the west wall has the shortest absorption

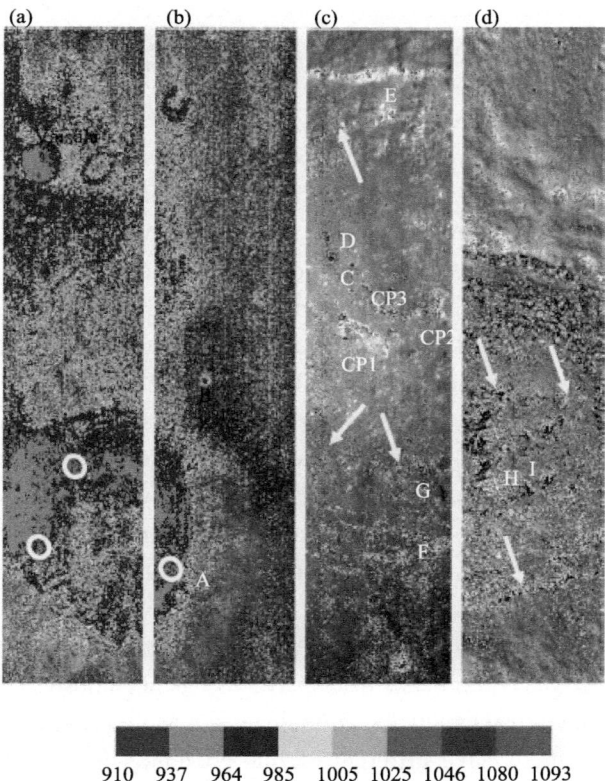

Fig.9 Map of the absorption band center for Aristarchus (a and b), Copernicus (c) and Zucchius (d). The ellipses denote the location of the telescopic spectra

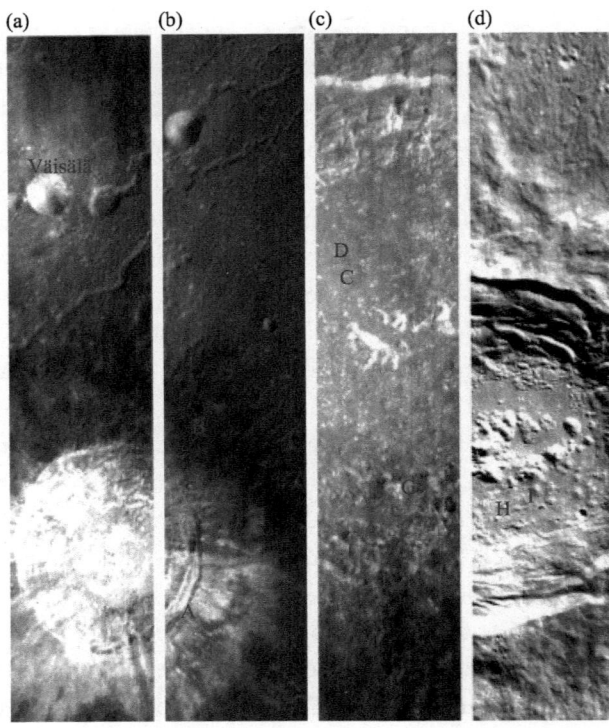

Fig.10 IIM 776 nm image showing the same location as Fig.9

band center for IIM while for Lucey's research the northwest wall has the shortest absorption band center. Although our results are based on the Earth-based telescopic spectra, we still believe that our results are more accurate because the spatial resolution of the earth-based spectra was low and the telescopic spectra of the Northwest wall, Southwest wall and East wall were all located at the boundary between two distinct absorption band centers (ellipse in Fig.9a).

The north wall also possesses a dichotomy with a longer absorption band center on the top and shorter absorption band center on the bottom. The dichotomy is consistent with the color ratio composite map derived from M^3 data (Mustard et al., 2000). Unlike the west wall, which is uniformly bright, the bottom of the north wall is brighter than the top (Fig.10a). As observed with the Lunar Orbiter images, the brighter bottom of the north wall seems unlike outcrops that were exposed by landslides. The dichotomy of the north wall needs to be further researched with other hyperspectral data. Le Mouelic et al. (1999) identified olivine-rich area on the southeastern rim and recently Mustard et al. (2000) confirmed this. As shown in Fig. 9(b) the area (marked with Point A) with an occurrence of olivine stated in Le Mouelic et al. (1999) and Mustard et al. (2000) does have a longer absorption band center. However, it is not clear that Aristarchus crater has as much olivine as described in Le Mouelic et al. (1999) and Mustard et al. (2000). Due to the mixture between fresh rays and surrounding stratigraphy, the absorption band center has a concentric distribution with longer wavelength centers outside the crater, e.g. Points B, C and D in Fig.9. Important to the spectral interpretation is the spectral mixture between the fresh material and the surrounding substrate. On the top of Fig. 9(a), the 8km Väisälä crater also exhibits lithologic heterogeneity. The absorption band center of Väisälä crater is longer for the northeast wall and has a shorter center for southwest wall.

The olivine composition of the central peaks of Copernicus has been observed by the Earth-based telescopic spectra (Pieters, 1982; Lucey et al., 1991) and Clementine data (Le Mouelic and Langevin, 2001), who found that the three central peaks of Copernicus have different olivine content, with the easternmost central peak (CP2) being the richest in olivine. Fig.9(c) shows that from right to left (CP2 to CP3 to CP1), the absorption band center moves to shorter wavelengths, which means that olivine is the richest in the CP2 and poorest in the CP1. Our results are in good agreement with previous observations but provide much finer structure, which testifies to the accuracy of detection and our process method of this instrument. Although telescopic spectra have a high spectral resolution, the spatial resolution is limited. Clementine multispectral images are limited to its spectral resolution and cannot determine the specific absorption band center. Our results show the advantage of IIM with its high spatial and spectral resolution, e.g. the different pyroxene types of C, D and CP1 were clearly identified which had not been discriminated before.

The two small craters C and D reveal fresh materials bearing Ca-rich pyroxene of the northern floor. Generally the floor of Copernicus crater does not have a stagnation point while the floor of Aristarchus has a stagnation point within the spectral band of IIM, which suggests that the pristine floor of Copernicus was more altered (more impact melts or space weathered) than that of Aristarchus. This is consistent with the fact that Copernicus is larger and older than Aristarchus. Our results show not only a difference of the minerals between the three central peaks, but also the compositional diversity between the south and north wall, which confirms the results by Pieters et al. (1994). The localized abundance of iron-bearing minerals along the southern walls was revealed in the absorption band center map. The absorption band center of Points F and G in the south crater is shorter than CP1 and a little longer than Points C and D. In general, the south wall is more mafic than the north wall. The north wall is more feldspathic than the south wall due to its fresh exposure but no stagnation point and its high brightness.

Zucchius crater (61.4°S; 51.3°W) provides another example showing the spatial heterogeneity of the lunar crust derived from the IIM (Fig.9d). Contrary to the former two craters, which have central peaks of with longer wavelength absorption band centers, the absorption band center of the Zucchius central peaks is shorter, with Points H and I being the shortest, indicating low-Ca orthopyroxene. Although much shading exists in the north wall due to the rugged terrain and low solar elevation angle, the absorption band center of the north and south wall is similar. On the one hand, it suggests that the difference is insensitive to the topography effect; on the other hand, it also suggests that the compositional heterogeneity of the Zucchius walls is smaller than the former two craters. It further indicates that the compositional stratigraphy of the lunar highland crust is relatively simple compared with the nearside stratigraphy around mare, at least in the horizontal direction.

Both the south and north walls of Zucchius crater show longer absorption band centers than the central peak. It suggests that the upper stratigraphy is gabbroic and the lower stratigraphy is noritic. Pieters (1993) summarized the stratigraphic information of the lunar nearside curst derived from the telescopic spectra. The upper crust of the lower left nearside (third quadrant) exhibits high-Ca pyroxene gabbroic unit such as Bullialdus (20.7°S; 22.2°W) and Tycho (43.4°S; 11.1°W) crater. Our results also show that Zucchius, which is close to the southwest of the lunar nearside, also exhibits gabbroic upper crust. The lower crust exposed in the central peaks shows the compositional diversity including gabbroic and noritic units (Fig.9d). The lower crusts of Bullialdus and Tycho also have different compositions though they have the same upper crust. These results suggest that the upper crust is relatively simple compared with the lower crust in the highland areas. Tompkins and Pieters (1999) also found that many central peaks contain multiple lithologies with Clementine data. This study suggests that the lower crust is more heterogeneous than the upper crust, which may be related to the global fractionation of the magma ocean.

Even though the formula shown in Fig.8 is not always accurate at the global scale, for local area such as one crater, e.g., within the swath of IIM image (25.6 km), the relative wavelength of the absorption band center is suitable for the relationship between the absorption band center and stagnation point. As shown in Fig. 9(a)–(c), the distribution of the absorption band center is consistent with the mineral distribution detected by previous research for both Aristarchus and Copernicus, which indicates that our method is feasible to clarify the relationship between different lithologies and to detect heterogeneities at small scales. This method is insensitive to the topography and illumination geometry which can be confirmed by the concentric distribution of absorption band center for some small craters, e.g. Points B, C and D. It also indicates that the similar absorption band center for the south and north wall of Zucchius crater was not caused by the topography but they have consistent composition. Moreover, it can also contribute to the finding of small craters which have low contrast with surroundings. For example, Point J is difficult to find in the brightness image (Fig.10c) but can be easily shown in the absorption band center map.

The method shown in this study is empirical. Indeed the absorption band center of the lunar spectra is the complex convolution of individual absorption band centers of the mafic minerals and the continua. It should be noted that the identification of rock types is somewhat simplistic using only the absorption band center. In this preliminary study we focused on the extrapolation of IIM spectral range to improve its usability. Despite this, the consistent results for Aristarchus and Copernicus crater with previous researchers indicate the effectiveness of our method. Moreover, the concentric distribution of longer absorption band centers outside the small craters may have important implications. For example, the relationship may be used to assess the mixture between rays, ejecta and the surrounding substrate. At any rate, this method has a potential for the spatial mapping of minerals on the Moon when integrated with M^3 or SP which have broad spectral range.

5 Conclusions

To contribute to its usability of IIM data, in this paper we discussed our preliminary experience in the use of Chang'E-1 IIM data. We firstly provided one feasible method for the on-orbit correction of the inhomogeneity of sensor response, and then revealed the relationship between the absorption band center and the wavelength at which the first derivative equals to 0. The consistency between the distribution of the absorption band center and the mineral distribution detected by previous research for both Aristarchus and Copernicus crater indicates the usability of IIM data and that our method is feasible. This method extends the spectral range of IIM and helps to clarify the relationship between different lithologies. Various mafic minerals were mapped with finer structure than previous research for Aristarchus and Copernicus crater. The lithologic diversity of the Zucchius crater central peak was identified. High resolution IIM spectra confirm the overall heterogeneity of the lunar crust both horizontally and vertically. This paper demonstrates the potential of IIM data for the identification of lunar rocks with its high spatial resolution and large number of spectra. In future study we will build a more accurate model for predicting the absorption band center with IIM data and produce a global map of absorption band centers. We hope this global map can be helpful to research the composition of the Moon together with other lunar explorers.

Acknowledgements The authors thank the reviewers for their many helpful comments. The level 2C data of CE-1's IIM were provided by the Ground Segment for Data, Science and Application (GSDSA) of China's Lunar Exploration Program. We are most grateful for the data description from researchers of GSDSA including Prof. Ouyang Ziyuan, Prof. Li Chunlai, Dr. Zhang Wenxi and others. We would also like to thank Prof. Bruce Hapke from University of Pittsburgh, Prof. Paul Lucey from University of Hawaii at Manoa, Dr. Andrew Ball from the Open University, Dr. Ian Garrick-Bethell from University of California, Santa Cruz and Mr. Thomas Chiasson from Nanjing University, for helpful suggestions and English language support on the manuscript. This work were jointly supported by grants from the open fund of the State Key Laboratory for Mineral Deposits Research, Naning University (2008-II-03), the National Natural Science Foundation of China (no. 40904051 and 40701125), the Macau Science and Technology Development Fund (Grants: 003/2008/A1 and 018/2010/A), the Project of China Geological Survey (1212010811050) and the National High Technology Research and Development Program of China (863 Program: 2008AA12A213 and 2010AA122203).

References

[1] Adams, J.B., 1974. Visible and near-infrared diffuse reflectance spectra of pyroxenes as applied to remote sensing of solid objects in the solar system. J. Geophys. Res. 79, 4829—4836

[2] Blewett, D.T., Lucey, P.G., Hawke, B.R., Jolliff, B.L., 1997. Clementine images of the lunar sample-return stations: refinement of FeO and TiO_2 mapping techniques. J. Geophys. Res. 102 (E7), 16319—16325. doi:10.1029/97JE01505

[3] Cloutis, E.A., Gaffey, M.J., 1991. Pyroxene spectroscopy revisited spectral-compositional correlations and relationships to

[4] Hillier, J.K., Buratti, B.J., Hill, K., 1999. Multispectral photometry of the Moon and absolute calibration of the Clementine UV/Vis camera. Icarus 141, 205—225

[5] Le Mouelic, S., Langevin, Y., 2001. The olivine at the lunar crater Copernicus as seen by Clementine NIR data. Planet. Space Sci. 49, 65–70

[6] Le Mouelic, S., Langevin, Y., Erard, S., 1999. The distribution of olivine in the crater Aristarchus inferred from Clementinc NIR data. Geophys. Res. Lett. 26, 1195—1198

[7] Lucey, P.G., Hawke, B.R., Horton, K., 1991. The distribution of olivine in the crater Copernicus. Geophys. Res. Lett. 18, 2133—2136

[8] Lucey, P.G., Hawke, B.R., Pieters, C.M., Head, J.W., McCord, T.B., 1986. A compositional study of the Aristarchus region of the Moon using near-infrared reflectance spectroscopy. J. Geophys. Res. 91, D344—D354

[9] Matsunaga, T., Ohtake, M., Haruyama, J., et al., 2008. Discoveries on the lithology of lunar crater central peaks by SELENE spectral profiler. Geophys. Res. Lett. 35, L23201. doi:10.1029/2008GL035868

[10] Mustard, J.F., Pieters, C.M., Isaacson, P.J., Head, J.W., Klima, R.L., Petro, N.E., Staid, M., Sunshine, J.M., Runyon, C. Taylor, L.A., 2010. Compositional characteristics of the Aristarchus crater from (M^3) data. 41st Lunar Planet. Sci. Conf. 2000

[11] Okuno, H., Yamanoi, Y., Saiki, K., 2008. The absorption-peak map of Mare Serenitatis obtained by a hyper-spectral telescope. Earth, Planets and Space 604, 425—431

[12] Pieters, C.M., 1982. Copernicus crater central peak: lunar mountain of unique composition. Science 215, 59—61

[13] Pieters, C.M., 1993. Compositional diversity and stratigraphy of the Lunar crust derived from reflectance spectroscopy. In: Pieters, C., Englert, P. (Eds.), Remote Geochemical Analysis: Elemental and Mineralogical Composition. Cambridge University Press, Cambridge, pp. 309—339

[14] Pieters, C.M., 1999. The Moon as a spectral calibration standard enabled by lunar samples, LPI contribution 980. In: Proceedings of the new views of the Moon II: understanding the Moon through the integration of diverse datasets. Lunar Planet. Inst., Flagstaff, Ariz., 22—24 September

[15] Pieters, C.M., Head, J.W., Isaacson, P., Petro, N., Runyon, C., Ohtake, M., Foing, B., Grande, M., 2008. Lunar international science coordination/calibration targets (L-ISCT). Adv. Space Res. 42, 248—258

[16] Pieters, C.M., Pratt, S., 2000. Earth-based near-infrared collection of spectra for the Moon: a new PDS data set. 31st Lunar Planet. Sci. Conf., 2059

[17] Pieters, C.M., Staid, M.I., Fischer, E.M., Tompkins, S., He, G., 1994. A sharper view of impact craters from clementine data. Science 266, 1844—1848

[18] Savitzky, A., Golay, M.J.E., 1964. Smoothing and differentiation of data by simplified least squares procedures. Anal. Chem. 36, 1627—1639

[19] Sunshine, J.M., Pieters, C.M., Pratt, S.F., 1990. Deconvolution of mineral absorption bands: an improved approach. J. Geophys. Res. 95, 6955—6966

[20] Tompkins, S., Pieters, C.M., 1999. Mineralogy of the lunar crust: results from Clementine. Meteor. Planet. Sci. 34, 25—41

[21] Zhang, L.Y., Li, C.L., Liu, J.Z., Liu, J.J. Data processing of imaging interferometer of the Chang'E project In: Proceedings of the International Lunar Conference 2005, Toronto, Canada

Absolute Calibration of the Chang'E-1 IIM Camera and its Preliminary Application

Wu Yunzhao[1,2]   Xu Xisheng[3]   Xie Zhidong[3]   Tang Zesheng[1]

(1. Collaborative Research Laboratory on Lunar and Planetary Exploration, Macao University of Science and Technology, Taipa, Macao, China;
2. School of Geographic and Oceanographic Sciences, Nanjing University, Nanjing 210093, China;
3. Department of Earth Sciences, Nanjing University, Nanjing 210093, China)

Abstract The interference imaging spectroradiometer (IIM) onboard the first lunar satellite of China "Chang'E-1" can now provide approximately global high spectral and spatial resolution reflectance spectra of the Moon. It is the essential instrument with which to accomplish one of the four missions of the first lunar satellite of China. As the current data provided by the Lunar Exploration Program Center and National Astronomical Observatories (NAOC) are not reflectance and the sensor response is inhomogeneous in the line direction, users can not use the current data directly. Moreover, due to the narrow band range, IIM data cannot cover the absorption peak of the mafic minerals of the Moon completely, which limits its ability for identifying minerals. The main objective of this study is to describe the methods for absolute calibration, correction and acquiring the absorption center of minerals for IIM data. The results from our study show that in the space domain the sensor response decreases toward the left, and in the spectral domain the response of the longer bands is more inhomogeneous than that of the shorter bands. After the calibration and correction, the reflectance of IIM matches the earth-based telescopic spectra well, which suggests the possible use of the processed data in the geological research. A high correlation was found between the absorption center and the wavelength at which the first derivative equals 0, i.e., the so-called Stagnation Point in the mathematical sense. In the end, we show a preliminary applied study of the two craters with diameter larger than 35 km using the calibrated data. The spectra of IIM data show that the lunar crust has compositional diversity within the km scale. Pure anorthosite may be found on the wall and floor of the Aristarchus crater with the map of absorption center, which indicates that anorthosite is ubiquitously present within the lunar crust. IIM, with its capacity to acquire lunar composition at the regional and global scale, will contribute to the research of lunar origin and evolution.

Key words   Chang'E-1   IIM   Calibration   Radiometric Correction   The Moon

Knowledge of the abundance and distribution of the rocks and minerals across the lunar surface has thrown light on the understanding of the nature and origin of the Moon[1,2]. The samples returned by six Apollo and three Luna programs have provided invaluable information for understanding the compositions of the Moon. However, the lunar sample collection is a relatively small and non-random representation of the Moon's crust. Remote sensing, which provides a global perspective that the lunar sample collection alone cannot, is still the major means by which to acquire the compositions of the Moon.

Remote sensing within the visible and near infraredregion has been widely used to acquire the compositions and elements of the Moon[3-8]. All the three lunar satellites launched recently: SELENE, CE-1, and Chandra yaan-1 are equipped with imaging spectroradiometers, which will contribute to the lunar research. Imaging spectroradiometer is the simultaneous acquisition of spatially coregistered images in many spectrally contiguous bands. It is the essential instrument with which to finish one of the four missions of the first lunar satellite of

China[9]. Now CE-1 has successfully fulfilled the tasks of exploring the moon. With the IIM data processed completely and distributed to the users, it is necessary to analyze and interpret the data as soon as possible, and to understand the Moon using our own data.

We acquired the IIM data (Level 2C) from the Lunar Exploration Program Center and National Astronomical Observatories (NAOC) in April. The 2C data have been systematically processed in several steps. However, data should be further processed before it can be used. Two factors hamper the application of IIM data. 1) The units of the 2C data of IIM are not reflectance but radiance; 2) the sensor response is inhomogeneous from the left to the right, so that the same lunar surface exhibits different spectra along the line direction (see details below), which hence has caused multi-resolution for the same composition. Moreover, the spectral range of IIM is very narrow and hence cannot cover the absorption peak of the mafic minerals of the Moon completely, which limits its ability for identifying minerals. This study aims to describe the methods for the absolute calibration, correction and acquiring the absorption center of minerals for IIM data.

1 Absolute calibration of IIM data of CE-1

For the details of 2C data users can read the data instruction from NAOC. Here is a brief summary of the 2C data. The 2C data used in the study are in radiance format with dark-current subtraction, flat-field correction, convention to radiance, and photometric normalization corrections to standard geometry ($i = 30°$, $e = 0°$). Table 1 shows the performance index of IIM data.

Table 1 Key IIM measurement characteristics

Image width	Spatial resolution	Spectral range	Spectral resolution	Radiometric sampling	Number of pixels
25.6 km	200 m (nadir)	480 – 960 nm	32 (9.6 nm@543.5 nm, 13.1 nm@632.8 nm, 20 nm@783.8 nm, 22.5 nm@831.2 nm)	12 bit	128

Previous research showed that the use of lunar soil as a calibration standard was more accurate than using stellar model[10]. Fortunately, now the return of lunar samples allows their spectra and properties to be measured accurately in earth-based laboratories, which greatly improves the calibration accuracy. The lunar samples are used as a spectral calibration standard, because the sample is representative of well-developed soils of the area, and the calibration target is a relatively homogeneous area with no nearby units of a significantly different material. Pieters et al. reasoned that the very mature soil 62231 was representative of mature soils throughout the Apollo 16 region, and they calibrated the telescopic and Clementine spectra using soil 62231[11]. The IIM data used in this study have been photometrically normalized to the same photometric geometry as the bidirectional reflectance measurement of 62231, i.e. $i = 30°$, $e = 0°$. This study uses the bidirectional reflectance measurement of 62231 to convert IIM data into reflectance.

The location where the previous studies calibrated telescopic and Clementine spectra lies at the point to the west of the landing site, the third vertex of an equilateral triangle, which has North Ray and South Ray craters as the other two vertices[12,13]. Unfortunately, this site is just located between Orbit 2224 and 2225, i.e., IIM does not cover this site. By comparing the IIM data of the two orbits with high spatial resolution and multispectral Clementine data, we chose one site as the calibration site to convert IIM data into reflectance (Fig. 1) from Orbit 2225 which is close to the previous calibration standard site. The spectral feature at this site is similar to that of the calibration standard site, and the surface is homogeneous, free from crater rays. The absolute spectral calibration of IIM data follows these steps.

a) Selection of calibration site: a 5×17 pixel area(Column: 69-73; Line: 11151-11167).

b) Translate 62231 spectrum into IIM 32-filter spectrum using the Gaussian curve as spectral response function. This spectrum is denoted as the standard $R_{standard}$.

c) Derive IIM reflectance for any lunar area: $R = DN_{IIM}/DN_{Ap16} \times R_{standard}$.

Fig. 2 shows the spectral curve before and after the conversion into reflectance. In this study the two endpoints, i.e., 480 and 946 nm, have been eliminated due to their low Signal Noise Ratio (SNR).

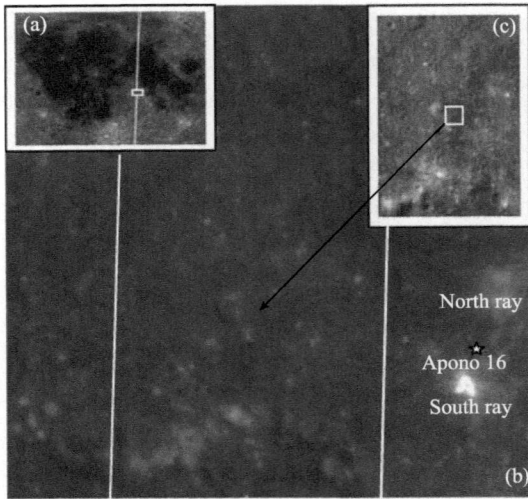

Fig. 1　Calibration site of IIM data. (a) The covering region of Orbit 2225; (b) Clementine 750 nm albedo data with 100 m spatial resolution for the Apollo 16 region. The calibration site area used in this study is indicated with an arrow; (c) IIM data with the box as the calibration site area

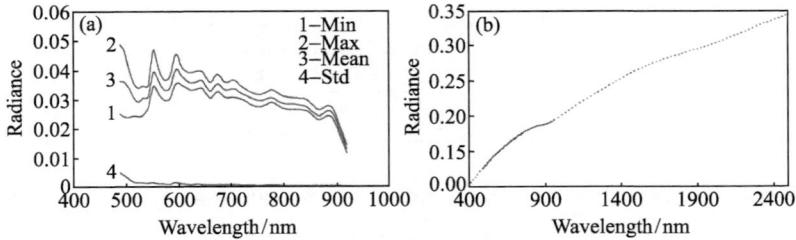

Fig. 2　L2C data of IIM (a) and the converted reflectance (b) of the calibration site.

2　Correction for the inhomogeneity of sensor response

Figs. 3(a) and (c) (Plate IX) show the true color composite images of basaltic Mare (MS2) and crater Petavius areas. Although the composition and the degree of impacting by meteorolites are different for the two areas, they have one thing in common: The hue changes from left to right. The shading is more obvious in the ratio image. Figs.3(b) and (d) are the ratio composite images of 891/918 nm (R), 918/891 nm (G), and 757/918

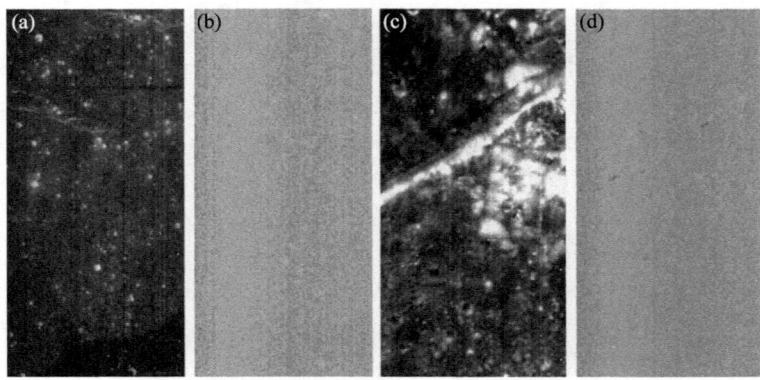

Fig. 3　True color image of IIM data for MS2 (a) and Petavius (c) area, and the ratio composite image of 891/918 nm (R), 918/891 nm (G), 757/918 nm(B) of the two areas ((b) and (d))

nm (B) of the two areas, respectively. It can be seen that left 1/5 portion is pink while the other 4/5 portion is green. Yet the true surface character such as craters and rima cannot be exhibited. Fig. 3 indicates that the shading is not related to the character of the land surface, nor to the geographic location, which provides the basis for the correction.

To further understand the mechanism of the shading along the line direction, we have extracted two spectra of the MS2 area (Orbit 2220) and the line profile of 918, 891, and 865 nm (red line in Fig. 3(a)). Fig. 4(a) shows the two spectra of left and middle portion (red box in Fig. 3(a)) which share the same composition (basalt) and structure. As shown in Fig. 4(a), the two spectra of the left and middle portion are different especially for longer bands, though they have the same composition. Fig. 4(b) shows that from left to right all the reflectance of the three bands increases, of which the longest band (918 nm) increases the most, which suggests that the long band is the most inhomogeneous.

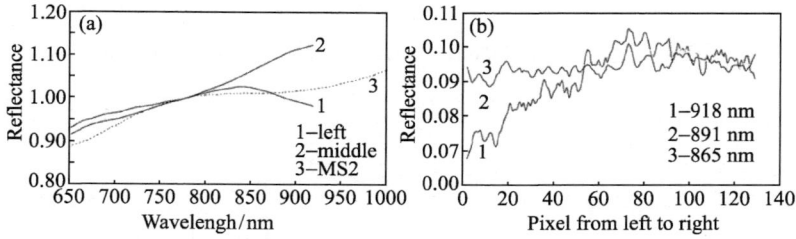

Fig. 4 (a) Uncorrected IIM and telescopic reflectance for the left and middle portion of MS2 area (red box in Fig. 3(a)); (b) reflectance of 918, 891, and 865 nm of the dashed line shown in Fig. 3(a)

The area covering the calibration site used in this study is chosen for the correction of the shading of IIM data. The correction procedures include the following steps.

(1) The absolute reflectance of 62231 was scaled to unity near 750 nm, named scaled reflectance.

(2) Selection of correction standard line: Three lines each containing 128 pixels near the Apollo 16 calibration site were carefully chosen. Because we wanted to correct the inhomogeneity along the line direction, we assumed the composition of the surface for the whole line, i.e., 128 pixels, was the same. Then only the sensor response can cause the spectral variation along the line direction without the compositional effects. So the standard line must be carefully chosen. In this study the correction standard line was chosen with the help of highly spatial resolution and multispectral Clementine data.

(3) Normalize the reflectance of each pixel of the standard line for each band to the scaled reflectance.

(4) Obtain the correction factor by the quadratic polynomial fitting for the values derived above. Then the shading of the IIM data can be corrected using the correction factor.

The correction does not just mean homogenizing the hue along the line direction with the random adjustment of the reflectance. Two factors can ensure that the corrected reflectance of each pixel has the physical meaning, i.e., reflectance of the surface. Firstly, the area for selecting the correction standard line is the same place as the land surface used to convert IIM 2C data into reflectance, i.e., Apollo 16 landing site area. Secondly, the correction factor is derived from fitting the reflectance of the correction standard line which is corrected to the reflectance of 62231.

3 Correction result

Fig. 5 shows the correction factor derived using the above method. In the space domain the left side of the correction factor is higher than the right side, which suggests that the sensor response on the left side is lower than that in the right. In the spectral domain, the correction factor for the long bands is larger than that for the short bands, which suggests that the inhomogeneity of the sensor response for the long bands is more serious. A comparison of the reflectance before (Fig. 4(a)) and after the correction (Fig. 6) suggests that the sensor response along the line direction tends to homogeneity. To validate whether the correction factor is suitable for other areas, we have corrected the highland area with plagioclase composition. As shown in Fig. 7(a) and (b), the shading has been removed after the correction, and the spectra of the three NIR bands are homogenous along the line direction (Fig. 7(c)).

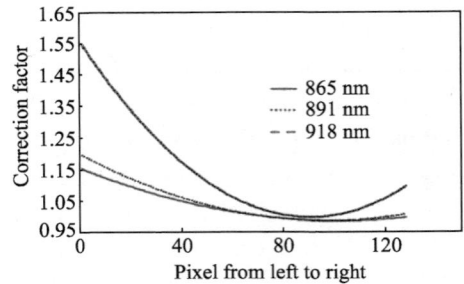

Fig. 5   Correction factor for the inhomogeneity of sensor response of the IIM data

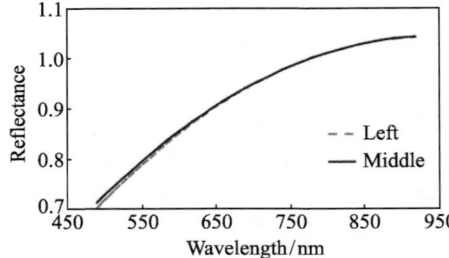

Fig. 6   The corrected reflectance of the left and middle portion of MS2 area (red box in Fig. 3(a))

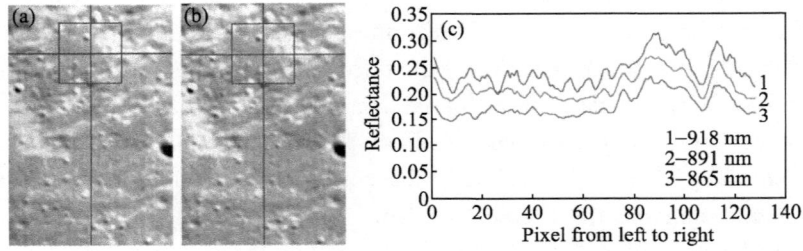

Fig. 7   The uncorrected (a) and corrected reflectance (b) for the high latitude area with anorthosite composition and the corrected reflectance for 918, 891, and 865 nm of the line profile (the line in (a) and (b))

4   Preliminary application of IIM data

The currently favored hypothesis for the formation of the ancient lunar crust is flotation of plagioclase feldspar in a global magma ocean. Testing the hypotheses of lunar origin requires the knowledge of the global composition of the lunar surface, in particular that of the material from the low crust. Craters larger than 35 km in diameter contain central peaks. The materials that make up the peaks are regarded as from the most deep-seated unit of the crater, and maybe coming from the low crust or up mantle. Pieters et al by researching the composition of many central peaks using the earth-based telescopicspectra found that the crustal stratigraphy was diverse[14]. However, the research of lunar stratigraphy is limited by the low spatial resolution data of the earth-based telescopic spectra. Further research is needed using the highspatial resolution spectrometer.

We have examined the composition of the crater Petavius using the corrected spectra of IIM data (Orbit 2495). The previous study showed that Petavius had excavated material with no mafic minerals and was thus an anorthosite. The corrected spectra of IIM show that besides the anorthosite which has no absorption, the mafic composition also exists because there is absorption near the longer bands (Fig. 8). The two central peaks are close to each other with a distance of 6 km, which suggests that the lunar crust has compositional diversity not only in the vertical direction but also in the horizontal level.

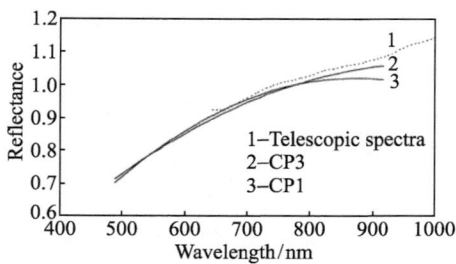

Fig. 8 The corrected reflectance for two individual central peaks of the crater Petavius and the telescopic spectra

The spectral range of IIM is only 480–960 nm and isless able to identify rock and minerals than telescopic spectra whose spectral range is wider. However, the spatial resolution of the telescopic spectra is low. Some minor compositions may not exhibit their own spectral feature due to the low contents within the instantaneous field of view. This study suggests that IIM data have the merits of high spatial resolution, in spite of its narrow spectral range.

IIM data have 32 bands, and strong correlations have been observed between the spectral data of each band. Therefore, large amounts of redundancy exist between these bands. Principal Component Analysis (PCA) can transform many correlated bands into a smaller number of uncorrelated variables called principal components. The PCA method is used to classify the rock types of the Aristarchus area (Orbit 2897 and 2898). Fig. 9 shows the image of the first three principal components with PC1 (R), PC2 (G), and PC3 (B) before and after the correction, respectively. The area with same components from left to right can be used to identify the difference between the classification results of the uncorrected andcorrected IIM data. The area marked by the dashed line in Fig. 9 has the same composition from left to right by checking the Clementine data. The telescopic spectra show that the component of this area is dark mantle blanketing (DMD). Fig. 9 shows that this area is classified into two types on the uncorrected data (left figure) and yet only one type on the corrected data (right figure). Hence, the correction improved the accuracy of the classification.

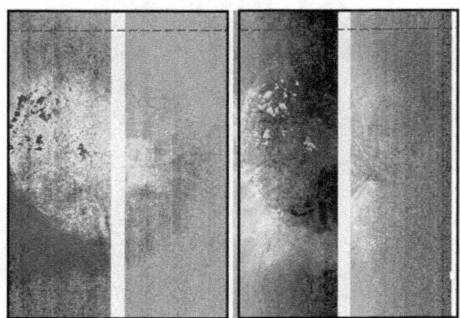

Fig. 9 Left: The composite images for PC1(R), PC2(G), and PC3(B)of the Aristarchus area before (left) and after (right) the correction

The characteristic absorption bands of the principal mafic minerals constituting the lunar crust are at around 1000 and 2000 nm. However, the valid spectral range of IIM is only between 488 and 918 nm, which limits its ability for identifying the minerals. Based on the earth-based spectra[15], we found a high correlation between the absorption center and the wavelength at which the first derivative equals to 0, i.e., so-called Stagnation Point in mathematical sense (Fig. 10). Hence, the key parameter to identify minerals, absorption center, could be known with this correlation and the stagnation point derived from the IIM data. Fig. 11 (Plate X) shows the map of the absorption center produced using the corrected IIM data and the formula shown in Fig. 10. The absorption center of the northwest wall is the shortest, while the south wall is the longest among the walls. Their compositions are probable orthopyroxene and clinopyroxene, respectively. Ohtake et al. found that anorthosite exists on the walls of the Aristarchus crater using the Multi-band Imager (MI)[16]. In comparing with MI, IIM has higher spectral resolution and hence can get the stagnation point of the reflectance spectra. As shown in Fig. 11, besides the south wall of Aristarchus the floor probably contains anorthosite. Anorthosite was previously suggested to be less abundant or absent in the Procellarum area[17]. This study provides a new samplefor the global distribution of anorthosite, and hence support the models of lunar magma ocean evolution.

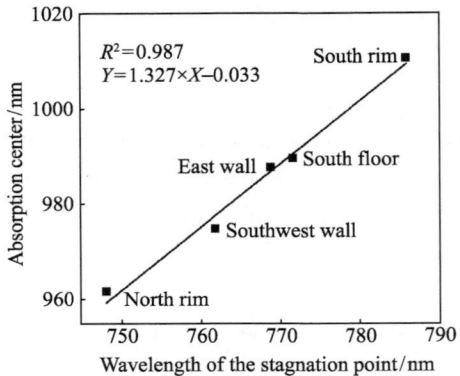

Fig. 10 The relationship between the absorption center and the wave length at which the first derivative equals 0 derived from the telescopic spectra for the Aristarchus crater

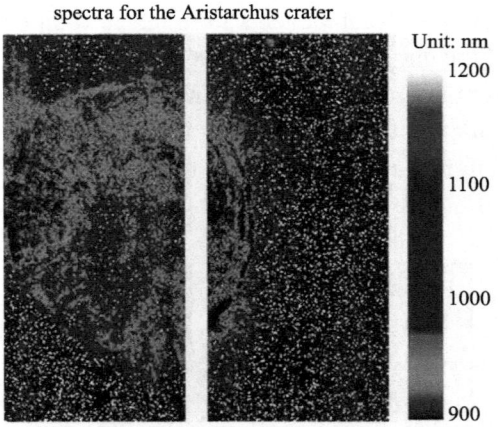

Fig. 11 Map of the absorption center of the Aristarchus crater

5 Conclusions

IIM can acquire the composition of the lunar surface with high spatial and spectral resolution. Just because the current data provided by the NAOC have not been converted into reflectance and the sensor response is inhomogeneous, these data cannot be used directly. Moreover, due to the narrow band range of IIM data, the absorption peak of the mafic minerals of the Moon can not be covered completely by the IIM data, which limits its ability for identifying minerals. This paper has explored the methods of absolute calibration and correction of the inhomogeneity of the sensor response in the line direction with the results that in the space domain the sensor response decreases toward the left, and in the spectral domain the longer bands respond more inhomogeneous than the shorter bands. The corrected spectra match the telescopic spectra well, which suggests that after the correction the IIM data can be used in geological investigation. Moreover, a high correlation was found between the absorption center and the wavelength at which the first derivative equals to 0. Two craters with a diameter larger than 35 km, Petavius and Aristarchus, have been investigated using the corrected IIM data. The derived rock compositions for the individual sets of central peaks show that the lunar crust is compositionally diverse even at the very small scale, both at the vertical and horizontal direction. IIM will contribute to a better understanding of the lunar science by its ability to acquire data with high spatial and spectral resolution.

Acknowledgements The 2C IIM data are provided by the Lunar Exploration Program Center and NAOC. We are most grateful to researchers of NAOC including Prof. Ouyang Ziyuan, Prof. Li Chunlai, Dr. Zhang Wenxi, and others for their data introductions.

References

[1] Ouyang Z Y. Introduction to Lunar Science (in Chinese). Beijing: China Astronautic Publishing House, 2005
[2] Xiao L, Ronald G, Zeng Z X, et al. Methodology, achievements and prospects of comparative planetary geology (in Chinese). GeolSciTech Inf, 2008, 27(3): 1—13
[3] Lucey P G, Taylor G J, Malaret E. Abundance and distribution of iron on the Moon. Science, 1995, 268: 1150—1153[doi]
[4] Pieters C M, Mustard J F, Sunshine J M. Quantitative mineral analyses of planetary surfaces using reflectance spectroscopy. In: Dyar M D, McCammon C, Schaefer M W, eds. Mineral Spectroscopy: A Tribute to Roger G. Burns. Houston: Geochemical Society, 1996. 307—325
[5] Xu T, Li C L. Abundances and distribution of elements on the Lunar surface (in Chinese). Chin J Space Sci, 2001, 21(4): 332—340
[6] Xu T, David T B, Li C L, et al. Fundamental characteristics of UV-VIS-NIR reflectance spectra and methods of their interpretation(in Chinese). Earth Environ, 2004, 32(3-4): 27—33
[7] Xue B, Yang J F, Zhao B C. The study of spectral feature of major minerals on the Lunar surface (in Chinese). ProgGeophys, 2004,19(3): 717—720
[8] Chevrel S D, Pinet P C, Daydou Y, et al. The Aristarchus Plateau on the Moon: Mineralogical and structural study from integrated Clementine UV-VIS-NIR spectral data. Icarus, 2009, 199: 9—24[doi]
[9] Xue B. Study on information processing and application of CE-1 in- terference imaging spectrometer (in Chinese). Doctor Dissertation. Xi'an: Xi'an Institute of Optics and Precision Mechanics, ChineseAcademy of Sciences, 2006
[10] Pieters C M, Pratt S. Earth-based near-infrared collection of spectra for the Moon: A new PDS data set. Lunar Planet Sci XXXI, 2000: 2059
[11] Pieters C M. The moon as a calibration standard enabled by lunar samples. New Views of the Moon II: Understanding the Moon Through the Integration of Diverse Datasets, Flagstaff. 1999
[12] Pieters C M, Head J W, Isaacson P, et al. Lunar international science coordination/calibration targets (L-ISCT). Adv Space Res, 2008, 42: 248—258[doi]
[13] Blewett D T, Lucey P G, Hawke B R, et al. Clementine images of the lunar sample-return stations: Refinement of FeO and TiO_2 mapping techniques. J Geophys Res, 1997, 102: 16319—16326[doi]
[14] Pieters C M. Compositional diversity and stratigraphy of the Lunar crust derived from reflectance spectroscopy. In: Pieters C, Englert P, eds. Remote Geochemical Analysis: Elemental and Mineralogical Composition. Cambridge: Cambridge University Press, 1993.309—339
[15] Lucey P G, Hawke B R, Peters M C, et al. A compositional study of the Aristarchus region of the Moon using near-infrared reflectance spectroscopy. J Geophys Res, 1986, 91: 344—354
[16] Ohtake M, Matsunaga T, Haruyama J, et al. The global distribution of pure anorthosite on the Moon. Nature, 2009, 461: 236—241
[17] McEwen A S, Robinson M S, Eliason E M, et al. Clementine observations of the Aristarchus region of the Moon. Science, 1994, 266: 1858—1862

第五部分 伽马与X射线谱仪

嫦娥一号卫星 X 射线谱仪的性能模拟

曹学蕾 王焕玉 张承模 张家宇 彭文溪 杨家伟 梁晓华
汪锦州 高旻 陈勇

（中国科学院高能物理研究所粒子天体物理中心，北京 100049）

摘 要 本文主要讲述了嫦娥一号卫星(CE-1)X 射线谱仪(LOXIA)的性能模拟。包括，探测效率的模拟；探测器能谱响应的模拟；角度响应的模拟三个方面，并对模拟结果进行分析，为谱仪的研制提供了参考。

关键词 Geant4 探测效率 角度响应 能谱响应

X 射线谱仪是中国探月计划嫦娥一号卫星的主要载荷之一[1]，仪器的研制目的是测量月表物质的γ射线和荧光 X 射线谱，探测有用元素的含量和分布。科学目标是在一定时间内(≥12 个月)通过对月球表面 X 射线荧光的(轨道高度≤200 km)探测，获得 Mg、Al 和 Si 等元素的分布情况，同时对可能存在的月球天然放射性元素所产生的 X 射线进行观测[2]。

本文详细讨论了 X 射线谱仪的探测效率，角度响应，能谱响应的模拟，给出了蒙特卡罗模拟计算结果，对该谱仪的设计与定标以及试验数据处理都有非常重要的意义。

1 探测器结构

LOXIA 的探测器部分由 16 路高能 X 射线探测器单元与 4 路低能 X 射线探测单元构成，探测器类型为 Si-PIN 型。单路高能 X 射线探测器灵敏探测面积为 100 mm^2，探测能区为 10~60keV；低能 X 射线探测器灵敏探测面积为 25mm^2，测量能区为 1~10keV。总探测面积为 1700mm^2，测量能区 1~60keV[2]。

2 模拟程序的建立

2.1 Geant4

Geant4 是由欧洲核子研究中心开发的一个模拟软件。该软件可精确地模拟粒子穿过介质时发生的各种作用以及产生的一些结果。Geant4 提供了大量的物理模型，来描述粒子与物质的相互作用。Geant4 采用 C++语言编写的。在编程中由于其模块化的特点，使得程序的移植性非常好[3-4]。

模拟过程包括下面几个方面，探测器建模，物理过程的确定，初始事件的产生，数据的读出等。

2.2 探测器建模

探测器模型，按照 X 射线谱仪实际尺寸构造，如图 1 所示。探测器上面板尺寸：202 mm×202 mm×5 mm，中间安装两准直器。

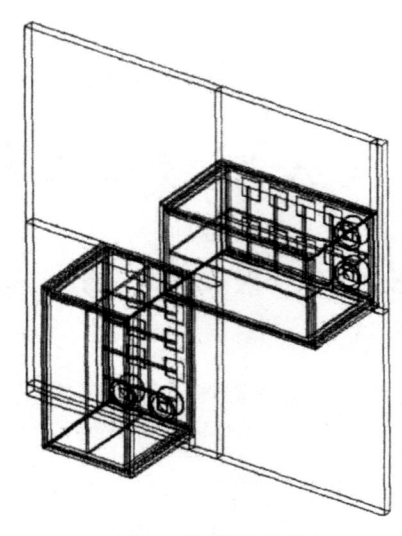

图 1 探测器建模

准直器部分由内到外依次为碳纤维板、磁铁、软铁、铝壁。准直器中间，探测器前面加一层 Al 膜反射可见光。支撑膜的支架材料用碳纤维。四路低能探测器单元前面有屏蔽铍窗。灵敏探测器的确定：将 16 路高能 X 射线探测器，以及 4 路低能 X 射线探测器定义为灵敏探测器。

2.3 低能模型

Geant4 提供的电磁作用的物理模型有两个，标准模型、低能模型。标准模型适合于模拟能量高的粒子相互作用。且标准模型对能量比较高的粒子作用模拟做了最优化设计；而低能模型，更多的考虑了原子结构的影响。且对模拟的范围做了拓展，低能模型适用的有效能量区间为 250 eV 到 100 GeV，原子序数 Z 从 1 到 99[5]。

X 射线荧光过程跟原子结构有直接的关系，所以实际模拟中采用了低能模型。

3 探测效率模拟

3.1 X 射线的吸收

X 射线作用于探测器时，可能发生光电效应、康普顿效应，以及热效应等。X 射线会消失或者改变动量以及能量。这一过程称为吸收。

X 射线可能通过光电效应损失其全部能量，也可能通过康普顿效应损失其部分能量。也可能通过多次散射，在探测器中损失其全部能量。实际模拟计算时，以全能损失的 X 射线光子数与总的入射光子数的比值定义为探测器对 X 射线全能峰的探测效率。文中以下所指探测效率皆为全能峰探测效率。

X 射线作用于探测器，其被吸收的程度与 X 射线穿过的厚度成正比。穿过厚度越大，X 射线被吸收的越多。所以，在一定范围之内，对于入射能量一定的一束 X 射线来说，探测器的厚度与探测效率成正比。

3.2 谱仪的探测效率

对于经过特殊封装的 X 射线谱仪系统来说，系统的探测效率，除了与探测器本身的厚度有关，还与探测器的封装有关。X 射线谱仪，探测器有前置 Be 窗，准直器上安装了一层 Al 膜。系统的探测效率会受到 Al 膜以及 Be 窗的影响。

光电效应的反应截面大致与入射光子能量的 7/2 次方成反比[6]。

$$\sigma \propto Z^5 \cdot (h\nu)^{-\frac{7}{2}}$$

当入射 X 射线能量比较低的时候，X 射线与谱仪的探测效率主要受 Be 窗以及 Al 膜的光电吸收作用影响。随着入射 X 射线能量的增加，X 射线对 Be 窗以及 Al 膜的透过率增加，谱仪系统的探测效率增加，如图 2 所示。当入射 X 射线能量比较高的时候(>10keV)时，由于 Al 膜与 Be 窗很薄，X 射线对 Al 膜以及 Be 窗的透过率已经接近 100%，此时谱仪系统的探测效率主要受探测器厚度的限制，探测效率随 X 射线能量增加而降低，如图 3 所示。

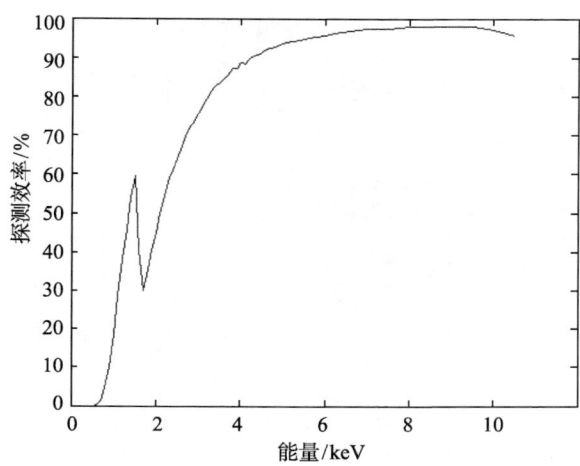

图 2　X 射线谱仪 0.5-10keV 能段探测效率的模拟

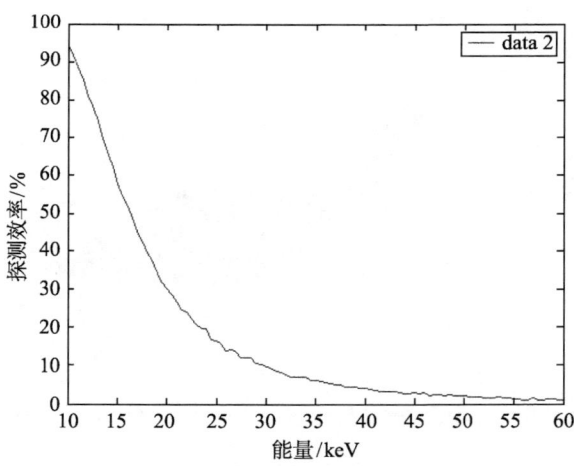

图 3　X 射线谱仪 10-60keV 能段探测效率的模拟

当 X 射线的入射能量达到激发某元素某一能级的临界值时，X 射线的吸收急剧增大，这一临界值称为吸收限。从图 2 中可以非常清楚的看到 Al 的 K 吸收限在 1.559 keV。所以受此影响，对于 Si 的 K 线，探测效率约为 30%。

4　角度响应模拟

4.1　入射角

入射角定义为，X 射线入射方向与法线方向的夹角。入射角度变大时，X 射线与探测器物质的作用距离 d 增大，Al 膜 Be 窗也会随着入射角度变化而变化，各个方面都会引起探测效率的变化。

X 射线谱仪由于其准直器的限制，探测器的视场角<55°，所以实际模拟时，入射角模拟范围 0°-56°。

4.2 角度响应模拟结果

能量比较低时，探测效率主要受封装 Be 窗以及 Al 膜的影响，能量比较高时，探测效率主要受探测器厚度的影响。从图 4 及图 5(图版 X)可以看出，入射能量大于 10 keV 时，随着入射角度的增加，探测效率明显的增大。而在 2keV 时探测效率随着入射角度的增加而明显减小。

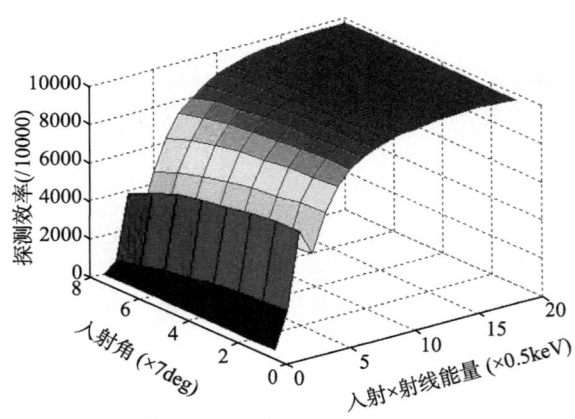

图 4　探测效率随能量以及入射角的变化关系(1)

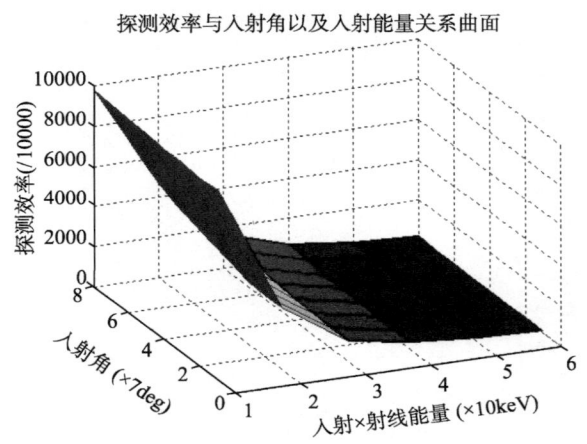

图 5　探测效率随能量以及入射角的变化关系(2)

5 能谱响应模拟

5.1 能量分辨率的拟合

对实验测得系统对几种 X 射线源的能量分辨率进行拟合，得到能量与能量分辨率的关系。然后再用模拟的方法可以得到系统对于各能量的 X 射线的能谱响应。

系统的能量分辨率受两种因素的影响，一是探测器的固有能量分辨；二是谱仪电子学系统。由探测器固有能量分辨引起的全能峰的展宽用 $\Delta E1$ 来表示，由电子学造成的全能峰的展宽用 $\Delta E2$ 来表示。

$$\Delta E1 \propto \delta N \propto \sqrt{N} \propto \sqrt{E} \tag{1}$$

其中 E 为入射 X 射线的能量，$\overline{N}$ 电荷数的平均值，δN 为 $\overline{N}$ 的标准差。

对相同的电子学系统来说 $\Delta E2$ 可以看作是不变的。结合误差传递公式，系统对能量为 E 的全能峰半高宽 *FWHM* 满足下面的关系：

$$FWHM = \sqrt{KE + \Delta E2^2} \tag{2}$$

其中 *K* 为常数。

将实验测得的 Fe55 源 5.9keV 全能峰的半高宽以及能量，Cu 的 Kα 线 8.04keV 的全能峰半高宽以及能量分别代入公式(2)，可以求出常数 K 跟 $\Delta E2$ 的值。

对其中一单路低能探测单元测得结果进行拟合，得到全能峰半高宽与能量的关系如下：

$$FWHM = 1.7479 * \sqrt{E} + 117.24 \tag{3}$$

5.2 能谱模拟的结果

将关系式(3)写入模拟程序，然后对 Fe55 源的 X 射线进行能谱模拟得到下面的结果(图 6、图 7)。

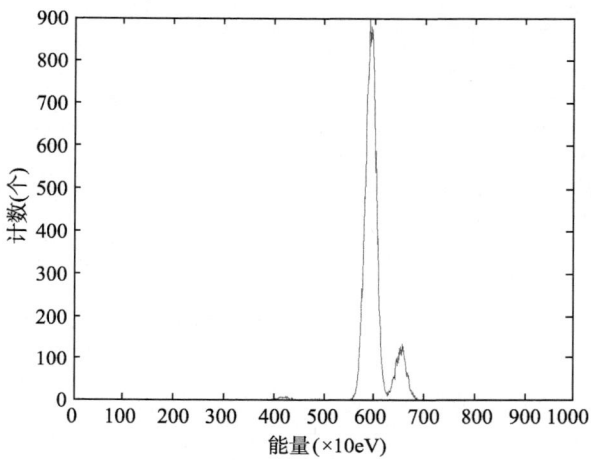

图 6 模拟得到的 Fe55 放射源能谱

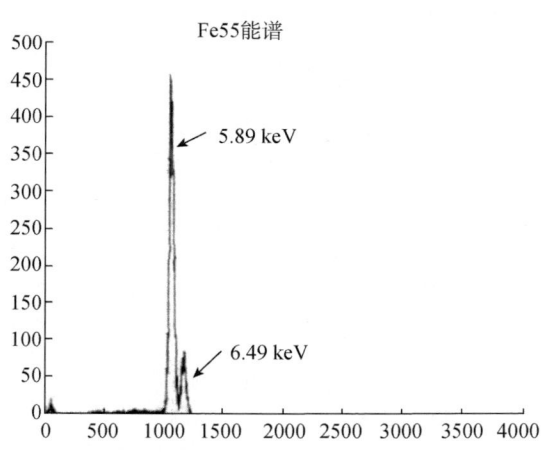

图 7 实验测得的 Fe55 放射源的能谱

将图 6 与图 7 比较：Fe55 源中 5.89keV 与 6.49keV 两种 X 射线的相对分支比进行比照，实验测得的 5.89keV 与 6.49keV 相对分支比为 6.77，模拟得到的结果为 6.59。模拟结果与实测结果偏差为 2.7%。

当入射光子因发生光电效应而被探测器吸收而损失掉全部能量时，Si-PIN 探测器可能被激发产生 Si 的荧光光子。因为探测器体积有限，所以荧光光子可能从探测器边缘逃逸，而造成边缘效应，这样在测得的能谱中会有一个逃逸峰，其中心能量：hv-Ek，其中，hv 是入射光子的能量，Ek 是逃逸 Si 的荧光光子的能量。所以对 Fe55 源的响应能谱来说应该在 4.41keV 处有一个逃逸峰。不过由于 Si 的荧光产额只有 0.049，所以逃逸峰占的比例非常小。从图 6 对 Fe55 的模拟中可以看到逃逸峰的存在。与图 7 实验测得的结果完全相符。

6　总结与讨论

通过模拟计算得到了嫦娥一号卫星 X 射线谱仪的探测效率曲线，角度响应曲面。

在对探测效率进行模拟计算时，发现 Al 膜对于低能 X 射线，包括 Mg，Al，Si 三种主要元素的 K 系荧光线吸收严重，因此在探测器的设计改进中做了修改，将低能 X 射线前端的 Al 膜去掉。

在对能谱响应的模拟计算中，验证了逃逸峰的存在。

参 考 文 献

[1] 探月工程总体部, 2005, 嫦娥一号卫星有效载荷正样研制任务总要求
[2] 王焕玉, 2005, CE-1 有效载荷 X 射线谱仪正样设计报告
[3] Geant4-a simulation toolkit, Nuclear Instruments and Methods in Physics Research A 506(2003)250-303S. Agostinelli etc
[4] Geant4 toolkit for simulation of HEP experiments Nuclear Instrument s and Methods in Physics Research A 502(2003)666-668 V. N. Ivanchenko
[5] Physics Reference Manual, 2001, CERN, 15-75, http:// geant4.web.cern.ch/geant4/
[6] 王绶琯，周又元. X 射线天体物理学[M], 北京：科学出版社, 1999: 76—110

嫦娥一号卫星 X 射线谱仪磁屏蔽设计与模拟计算

张家宇 王焕玉 张承模 杨家卫 梁晓华 汪锦州 曹学蕾
高旻 崔兴柱 彭文溪

(中国科学院高能物理研究所粒子天体物理重点实验室，北京 100049)

摘要 介绍了嫦娥一号卫星 X 射线谱仪准直器磁屏蔽设计的基本思路，通过磁场点测实验得到了谱仪准直器的真实磁场分布，并且根据实验结果对谱仪准值器的磁屏蔽效果进行了模拟计算，由模拟结果可知该准直器对电子能起到较好的屏蔽作用。

关键词 X 射线谱仪，准直器，磁屏蔽

X 射线谱仪是我国"嫦娥一号"探月卫星的一个有效载荷，它可探测月表元素受太阳 X 射线或宇宙射线激发产生的 X 射线荧光，并能对太阳 X 射线辐射进行监测，通过数据反演法可获得月表主要元素的含量和分布，以确定月表岩石类型和资源分布，并为月球探测和检验月球形成与演化模型提供重要信息。该 X 射线谱仪有两个探测器阵列，均由 8 个高能探测器单元(面积 1cm^2、探测能区 10~60 keV)和 2 个低能探测器单元(面积 25 mm^2、能区 1~10 keV)组成。在 X 射线谱仪的整个任务阶段，宇宙中弥漫的各种带电粒子引起的信号会使谱仪难以分辨某些元素特征谱线。为尽可能减小此类干扰，每个探测器阵列配备有相应的磁屏蔽准直器，用以实现谱仪的成像功能,并降低带电粒子给谱仪造成的本底噪声。谱仪准直器的尺寸为 80 mm×44 mm×50 mm，由外至内为铝、软铁、铷铁硼和碳纤维板等材料。其中，铷铁硼永磁铁产生的磁场和准直器的铝反射膜，使带电粒子在到达 X 射线探测器前就被偏转或屏蔽，这极大地增强了系统的抗辐照能力，减少了系统的本底噪声和死时间。铷铁硼永磁铁外配有软铁，以减少磁场外漏。带准直器探测器阵列的模型结构见图 1。

1 磁场测量

根据 X 射线谱仪准直器的实际尺寸，由理论推算可知，对 10 keV 电子，若施加 200 G 的横向磁场，则电子最多飞行 24.22 mm 就被偏转到准直器壁上，从而达到对空间电子辐射进行有效屏蔽的目的[1]。同时，采用 E^{-2} 幂率谱，利用 Geant4 模拟软件对 1~80 keV 电子的磁屏蔽效果的模拟计算可知，对于 X 射线谱仪准直器，350 G 的磁场就可完全偏转 80 keV 以下的电子[2]。综合上述磁场对带电粒子（主要是电子）磁屏蔽方面的考虑，以及谱仪系统对重量和对周围其他设备电磁干扰方面的考虑，最后确定 X 射线谱仪准直器磁屏蔽磁体的净质量不大于 450 g，每组磁体的厚度为 3~4 mm，磁体间的磁场强度要尽量均匀，平均强度在 300 G 左右。据此，中国科学院电工所设计制造了 X 射线谱仪的准直器铷铁硼永磁铁。2006年 12 月，我们利用我所加速器中心的磁场点测机和一维霍尔探头等磁场测量设备，测量了 X 射线谱仪准

本文原载于《核技术》，2008，Vol.31，No.6，423~426。

直器的磁场分布，以得到准直器内部及其附近区域的真实磁场分布。磁场点测实验的磁场点测机包括一维霍尔探头及安装模具，特斯拉计(型号：DTM-151)；其他设备有光学定位仪、不锈钢底座、专用照明灯光、测试记录系统等。

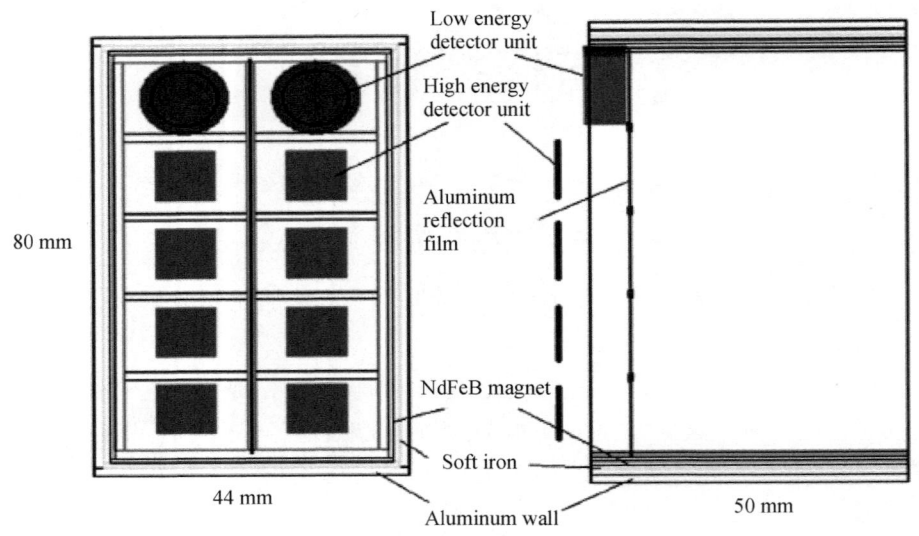

图 1　带准直器探测器阵列的模型结构图

测量实验包括下列步骤：①把准直器及临近区域划分为等步长(2 mm)空间网格(图 2)；②用光学定位法确定准直器的中心位置，即磁场点测时的相对坐标原点；③将一维霍尔探头安装在不同形状的模具上，用磁场电测机使探头在不同方向上按划分的步长移动(其中两维程序控制，一维手动)，测量各格点的磁场分量。

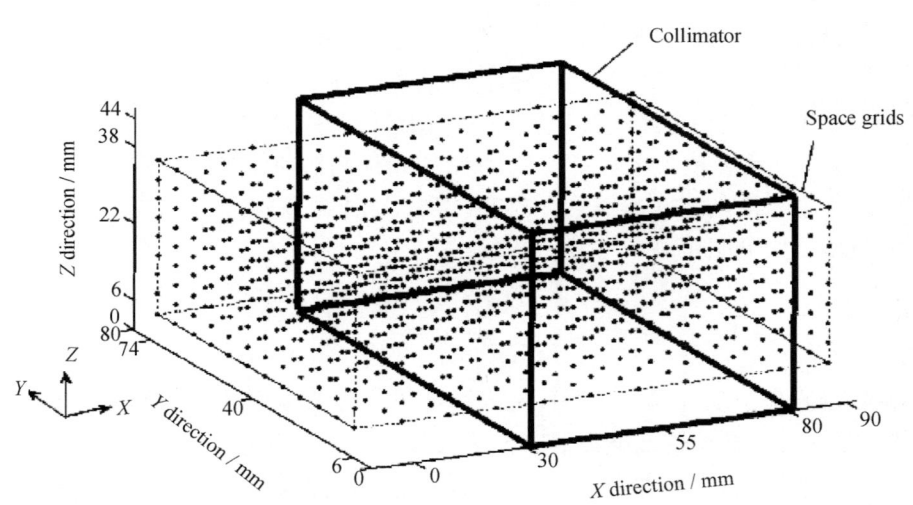

图 2　准直器磁场点测实验空间网格划分示意图

对准直器及其附近的 68 mm×32 mm×90 mm 的区域(格点数 27370)进行了磁场测量。图 3 为测量区域总磁场强度在各中截面上的分布。

测量结果表明，磁场主要被限制在准直器区域内。准直器外部区域的磁场强度<100 G；准直器中心区域的磁场分布均匀性较好，平均磁场强度 > 300 G，均达设计要求。

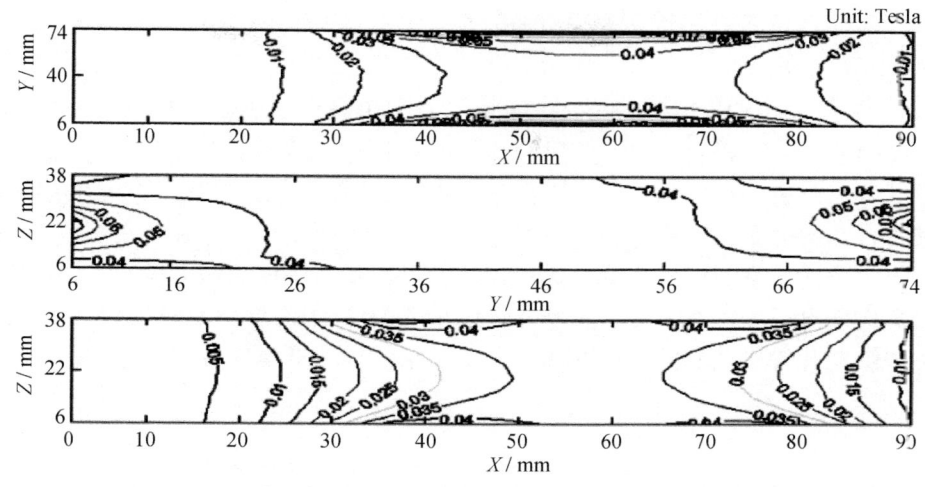

图 3 总磁场强度在测量区域各中截面上的分布图(位置坐标参照图2)

2 模拟计算

测得谱仪准直器的磁场分布后，就可由模拟计算对准直器的磁屏蔽能力进行评估。因为质子和重粒子的质量比电子大得多，数百高斯的磁场对于数 keV 质子的偏转效果已不明显，对于它们的屏蔽是依靠准直器的铝反射膜，故准直器磁屏蔽效果的模拟计算针对电子进行。

模拟采用的是 CERN 开发的 Geant4 模拟计算程序系统[3]，模拟过程包括下列步骤：① 根据准直器及其组成情况建模，主要涉及下述组成部分的材料与尺寸：准直器、探测器、入射窗与反射层，以及各组成部分间的相对位置关系等；② 测量的磁场区域小于准直器，须对测得磁场值进行外推，但考虑到最外层区域不大，且磁场较强，故外推时认为准直器的最外层磁场与最外一组测量值相等；③ 把磁场测量值按照 Geant4 的坐标系重新排列，生成数据文件，作为磁场的输入文件；④ 读入磁场输入文件，并利用插值法近似得到各测量点附近的磁场值，从而在准直器区域内建立起近似的真实磁场；⑤ 选取不同能量的入射电子，模拟无磁场和有磁场情况下准直器的磁屏蔽效果，并定义磁屏蔽效率 η：屏蔽效率(η)=[1−有磁场时到达探测器的电子述(N_b)/无磁场时到达探测器的电子述(N_{b0})]×100%。

在 40~400keV 电子能区内对准直器的磁屏蔽效果进行模拟计算，每间隔 20 keV 入射电子 10^6 个，2π 立体角各向同性入射，入射粒子的位置为在准直器上方的均匀入射。模拟计算的结果见图 4。

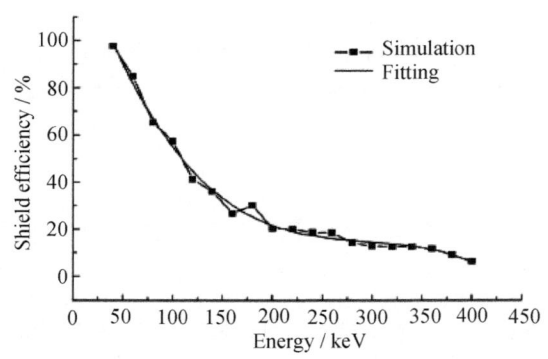

图 4 对不同能量电子的屏蔽效率

由模拟结果可知，准直器对不同能量电子的屏蔽效率差别较大，对于<40 keV 的电子，其屏蔽效率接近 100%；对于能量>400 keV 的电子，其屏蔽效率则小于 6%。对模拟结果进行拟合可发现，在 40 keV ⩽

$E \leqslant 400$ keV 能区，准直器对电子的屏蔽效率(η)与入射电子的能量(E)之间的关系可用式(1)近似表示(χ^2=5.89831, R^2=0.99302)：

$$\eta(E) = 142.0886 - 1.23474E + 0.00409E^2 - 0.00000464446E^3 \tag{1}$$

月球辐射环境中的电子主要由来自太阳风的低能电子和与太阳活动有关的高能电子组成。其中太阳风低能电子的能量为十余 eV 到上百 eV；与太阳活动有关的高能电子，其能量范围可从数 keV 到 MeV 量级，但是其能谱符合幂率分布，高能电子随着能量的增加而迅速减少，尤其是太阳耀斑期的高能电子，其能谱符合双幂率谱形式，高能成分下降更为迅速。通过参考部分探测器的探测结果，可构造近似的太阳宁静期和太阳耀斑期电子谱，对谱仪准直器在太阳宁静期和耀斑期的磁屏蔽效率进行模拟，从而对谱仪准直器在太阳宁静期和耀斑期时的磁屏蔽能力进行评估。模拟计算所选用的入射电子谱和模拟结果如表 1 所示。

表 1 准直器对不同分布入射电子的屏蔽模拟*

太阳活动	电子能量范围与谱指数	屏蔽效率 η
宁静期 (Quiet periods)	Energy range, 2~100 keV Spectrum index, $\delta = 2.5$ 谱指数参考 WIND[4]探测器 1995-2-22(1130-1330UT)探测结果 Referring to spectrum index results of WIND detector[4]	83.05
耀斑期 (Flare periods)	Energy range, 8~1000keV； Spectrum index, $\delta=1.1$(8~50 keV), $\delta=3.1$(50~1000 keV) 谱指数参考 WIND[4]探测器 1995-4-22 (1230-1630UT)探测结果 Referring to spectrum index results of WIND detector[4]	74.22
	Energy range, 2~1000 keV； Spectrum index, $\delta=2.5$(2~10keV), $\delta=1.1$(10~100keV), $\delta=2.8$(100~1000keV) 谱指数参考 ISEE-3[5]探测器 1978-9-23(1130-1336UT)探测结果 Referring to spectrum index results of ISEE-3 detector[5]	58.53

* 每种入射谱的入射电子数都取 10^6. The number of incident electrons for each kind spectrum is 10^6。

由上述模拟结果可知：准直器磁场对于典型的太阳宁静期和太阳耀斑期入射电子的屏蔽效率分别达到了 83%和 58%以上。因此，由上述模拟结果可知，通过谱仪准直器磁场是能够对月球辐射环境中由太阳耀斑引起的高能电子进行有效屏蔽的。

3 结论

磁场测量试验的结果表明，X 射线谱仪准直器磁场达到了预期的设计要求。由以此测量结果为基础进行的模拟计算可知，X 射线谱仪准直器对于月球辐射环境中的电子能够起到有效的磁屏蔽作用。

参 考 文 献

[1] 曹学蕾. 高性能空间 X 射线谱仪研究. 博士论文. 中国科学院高能物理研究所, 2006
  CAO Xuelei. The study of high performance X-ray detector, Ph.D theses, Institute of High Energy Physics, Chinese Academy of Sciences
[2] 梁晓华. 半导体 X 射线探测器性能研究. 硕士论文. 中科院高能物理研究所, 2003
  LIANG Xiaohua. The study of the performance of the semiconductor X-ray detector. Master theses, Institute of High Energy Physics, Chinese Academy of Sciences
[3] http://geant4.web.cern.ch/geant4/G4UsersDocuments/UsersGuides/ForApplicationDeveloper/html/index.html
[4] Lin R P. Wind observations of suprathermal electrons in the interplanetary medium, Space Science Reviews, v. 86, Issue 1/4, p. 61—78 (1998)
[5] Lin R P. Energetic Electrons Accelerated in Solar Particle Events, AIP Conference Proceedings, 2000, 528(1): 32—38

空间 X 射线成像谱仪系统及其软件研制

王焕玉　张承模　陈　勇　梁小华　曹学蕾　汪锦州　杨家卫
高　旻　张家宇　彭文溪　崔兴柱

（中科院高能物理所，北京　100049）

摘　要　嫦娥工程是我国重大空间项目之一，是我国开展深空科学探测的标志性工程。空间 X 射线成像谱仪系统是嫦娥一号卫星的重要科学载荷之一。它包括太阳 X 射线监测器、X 射线成像谱仪和电子学电路系统三部分。本文介绍空间 X 射线成像谱仪的工作原理、软件结构、主要性能和测量结果。

关键词　嫦娥一号卫星　太阳 X 射线监测器　X 射线成像谱仪

月球作为地球的唯一卫星，表面保存着31亿年以前的地质活动记录，探测月球对人们认识地球、太阳系的起源和演化历史有重要意义。地球矿物资源的日益匮乏也使人类把目光投向了月球。月球有丰富的矿藏，月球上稀有金属的储藏量比地球还多。在月球广泛分布的岩石中，蕴藏有丰富的钛、铁、铀、钍、稀土、镁、磷、硅、钠、钾、镍、铬、锰等矿产。岩石中含有地球中全部元素和60种左右的矿物，其中6种矿物是地球没有的。据估计，仅月海玄武岩中含有可开采利用的钛金属至少就有100万亿吨，而克里普岩中稀土元素、钍、铀的资源量分别约为 6.7 亿吨、8.4 亿吨和 3.6 亿吨[6]。

本文介绍的嫦娥一号卫星空间 X 射线成像谱仪系统[1](以下简称空间 X 射线成像谱仪)是月球探测卫星的重要载荷之一。它的主要目标是探测月表元素受太阳 X 射线或宇宙射线激发产生的 X 射线，通过数据处理获得月表主要元素的含量和分布，为月球的开发利用提供有关资源分布的数据，为深入研究月球的组成、月球地质演化和热历史提供相关资料。

1　空间 X 射线成像谱仪系统组成及其工作原理

1.1　空间 X 射线成像谱仪系统结构

X 射线成像谱仪由三部分组成，包括面对月球的 X 射线探测阵列、太阳监测器和相关的电子学系统(简称电控箱)。如图 1 所示。

X 射线成像谱仪用于观测月球方向发生的 X 射线事例，软 X 射线能段事例(1 keV-10 keV)由阵列的低能 X 射线探测器观测，硬 X 射线能段(10~60 keV)的事例则由高能 X 射线探测器观测。

月球表面的 X 射线主要来自两方面。一方面来自受太阳 X 射线或者高能宇宙线激发月球表面元素产生的 X 射线，另一方面来自蕴藏在月球上的天然放射性元素(同位素)核反应过程所产生的 X 射线。当距月球 200 km 环月轨道上的 X 射线谱仪视场内，月表有 X 射线产生时，被由 x 和 y 两方向探测器阵列组成的 X 射线谱仪进行成像观测。对有效能段内的不同月表元素或天然放射物质产生的特征 X 射线会得到不同的能谱[2]。

本文原载于《核电子学与探测技术》，2008，Vol. 28，No. 2，215~222。

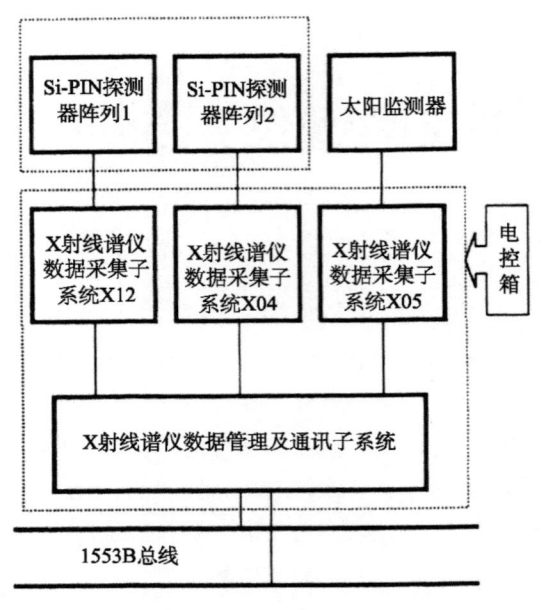

图 1 系统构成图

宇宙中的带电粒子(如 e、p、α)给 X 射线观测带来相当大的影响,有时带电粒子引起的信号使得某些元素产生的 X 射线特征谱线分辨十分困难。为了尽可能地避免因此带来的干扰,X 射线谱仪系统中设计了专门的磁铁和屏蔽措施,使得带电粒子在达到 X 射线谱仪(阵列)之前就被偏转,减少了系统噪声信号和本底、以及系统的死时间,同时,增强了系统抗辐照的功能。

太阳监测器指向太阳,监测太阳 X 射线辐射状况,用以配合月表 X 射线观测,进而获得元素的绝对丰度分布[3]。

电子学系统在 X 射线谱仪系统软件的配合下,对上述探测器获得的信号进行采集、处理和记录,以服务请求方式将记录的数据按规定的格式传送到有效载荷数管分系统,并保证在进站后下传到地面数据处理中心。

1.2 X 射线谱仪电子学系统

CE-1 月球探测卫星 X 射线成像谱仪的电子学系统如图 2 所示。它由以下几部分组成,即月球 X 射线阵列探测器电子学,太阳监测器电子学,数据管理与通讯电路,DC-DC 电源电路、高压产生器和系统状态监测电路等。月球 X 射线阵列探测器电子学和太阳监测器电子学包括电荷灵敏放大器、脉冲成形器、脉冲幅度甄别器、主放大器、峰值保持器和带有内置模数变换器的微控制器,但在月球 X 射线阵列探测器电子学中还包括通道选择电路和通道编码电路。来自太阳的 X 射线照射月球的表面时,月壤中的不同元素受激发产生不同的特征 X 射线。当它们到达月球 X 射线阵列探测器(包括 X、Y 两个探测器)时,在其后面的电荷灵敏前置放大器输出与入射 X 射线能量成正比的电压信号,由于 Si-PIN 探测器的信号比较弱,经过前置放大器的信号必须要成形之后才能进入主放大器和峰值保持电路,同时经过甄别器处理给出线性门开启信号、译码信号和中断申请信号。中断信号通知微处理器 80C196 启动 ADC 转换和率计电路对 X 射线事例记录和存储。同时,太阳监测器电路也开始工作,其工作过程与 X 射线阵列探测器相似,将太阳 X 射线的流量和能量信息记录到它自己的双端口 RAM 存储器中。X、Y 月球 X 射线阵列探测器和太阳监测器分别产生一个中断信号,通知数据管理与通讯电路取走它们当前记录的信息。数据管理与通讯电路有 80C186CPU 和 1553B 总线接口等组成[4]。它的功能除了以中断方式响应 X、Y 月球 X 射线阵列探测器和太阳监测器的请求外，还承担着将获取的数据打包送往1553B总线控制器，以及下传和上下行指令的传送等任务。

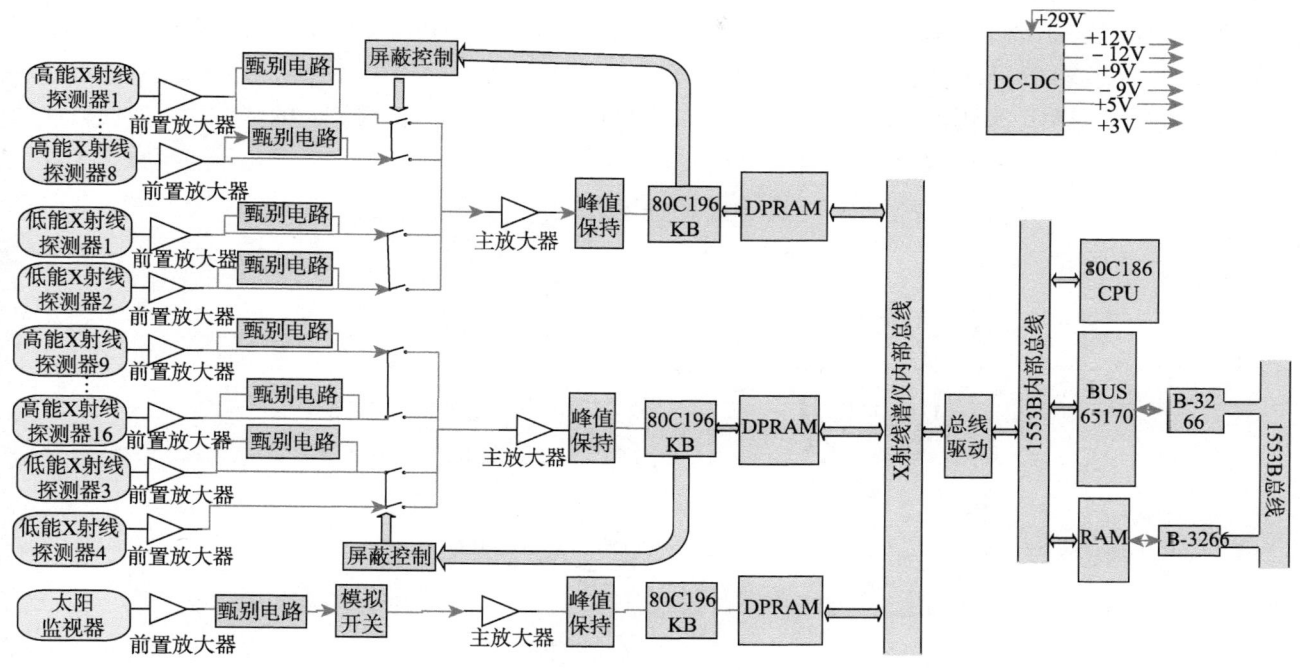

图 2　空间 X 射线谱仪电子学系统框图

图中 X、Y 分别表示 X 向和 Y 向探测阵列，PA 表示前置放大器，SHAPING 表示成形电路，DL 表示延迟，K、A 表示主放大器，PH 表示峰保电路，ENCODE 表示译码电路，DPRAM 表示双端口 RAM，SD 表示太阳监测探测器。

2　数据采集与管理软件

X 射线成像谱仪系统由三个数据采集子系统和一个数据管理子系统构成，是包含 4 个 CPU 的分布式多处理器系统。与此相应，它的软件包含三个数据采集软件(X 射线谱仪数据采集软件 1A、X 射线谱仪数据采集软件 1B 和 X 射线谱仪数据采集软件 2)和一个数据管理软件(X 射线谱仪数据管理及通讯软件)。各软件运行时分别被写在三个数据采集电路和数据管理及通讯电路的程序存储器(E^2PROM)中，软件与硬件之间的对应关系如表 1 所示。由于 X 射线谱仪数据采集子系统 X 和 X 射线谱仪数据采集子系统 Y 是完全相同的两个子系统，所以 X 射线谱仪数据采集软件 1A 和 X 射线谱仪数据采集软件 1B 也是完全相同的两个软件，只是名称上分为两个软件，实际为同一个软件的两个复制件。

表 1　软件与硬件对应关系表

序号	软件名称	硬件名称	备注
1	X 射线谱仪数据采集软件 1A	X 射线谱仪数据采集子系统 X	Si-PIN 探测器阵列 1(即相对于插槽面板 X 向，数据标识为 1111H)
2	X 射线谱仪数据采集软件 1B	X 射线谱仪数据采集子系统 Y	Si-PIN 探测器阵列 2(即相对于插槽面板 Y 向，数据标识为 2222H)
3	X 射线谱仪数据采集软件 2	X 射线谱仪数据采集子系统 X05	与太阳监测器相对应
4	X 射线谱仪数据管理及通讯软件	数据管理及通讯子系统	1553B BUS 接口电路

数据采集软件1运行于Si-PIN探测器数据采集系统，主要采集月球表面的X射线荧光事例(包括低能和高能两种事例)；数据采集软件2运行于太阳监测器数据采集系统，主要用来采集太阳X射线空间辐射能谱和时间信息；数据管理及通讯软件主要完成科学数据管理和与总线控制器的数据通信。三个软件之间的关系是X射线谱仪系统的内部关系。主要体现在各软件接口和通讯协议上，根据具体任务需要，协议由X射线谱仪内部自己确定。

每个软件都采用结构化程序的设计方法，增加各模块的聚合度，减少模块间的耦合，以方便软件维护；同时又力求软件模块的划分简洁清晰。程序结构如下面图3。

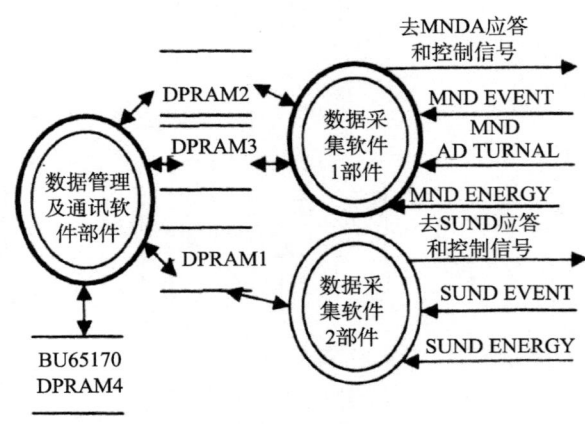

图3　软件系统数据流图

3　主要性能指标

月球探测X射线成像谱仪系统的主要性能指标如下，

X射线谱仪的主要技术指标

A. 探测器类型：　　　Si-PIN 半导体
B. 探测器有效面积：　17 cm^2
C. 探测能区：　　　　1~60 keV
D. 分辨率：　　　　　≤9%@59.5 keV
月面本征分辨率：　　170 km×170 km

太阳X射线监测器的主要技术指标

E. 探测器类型：　　　Si-PIN 半导体
F. 探测器有效面积：　0.25 cm^2
G. 探测能区：　　　　1~10 keV
分辨率：　　　　　　≤300 eV@5.95 keV
H. 总功耗：　　　　　11.41 W
I. 总重量：　　　　　5.47 kg
J. 工作模式：　　　　逐事例+率计+遥控操作。

4　小结

月球X射线成像谱仪系统已经在轨运行，研制期间通过了模样、初样和正样样机三个研制阶段，通过了嫦娥一号月球探测卫星有效载荷系统的桌面联调以及整星测试，完成了软件的代码走查和测试，并且完成了空间环境模拟试验和地面性能检测和性能标定。图 4 给出了对玄武岩、斜长角闪岩、能谱线性和放射性同位素 Am-241 的荧光 X 射线谱线的测试结果[5]。结果表明月球 X 射线成像谱仪系统的性能已经满足科学目标要求，为深空 X 射线探测奠定了技术基础。

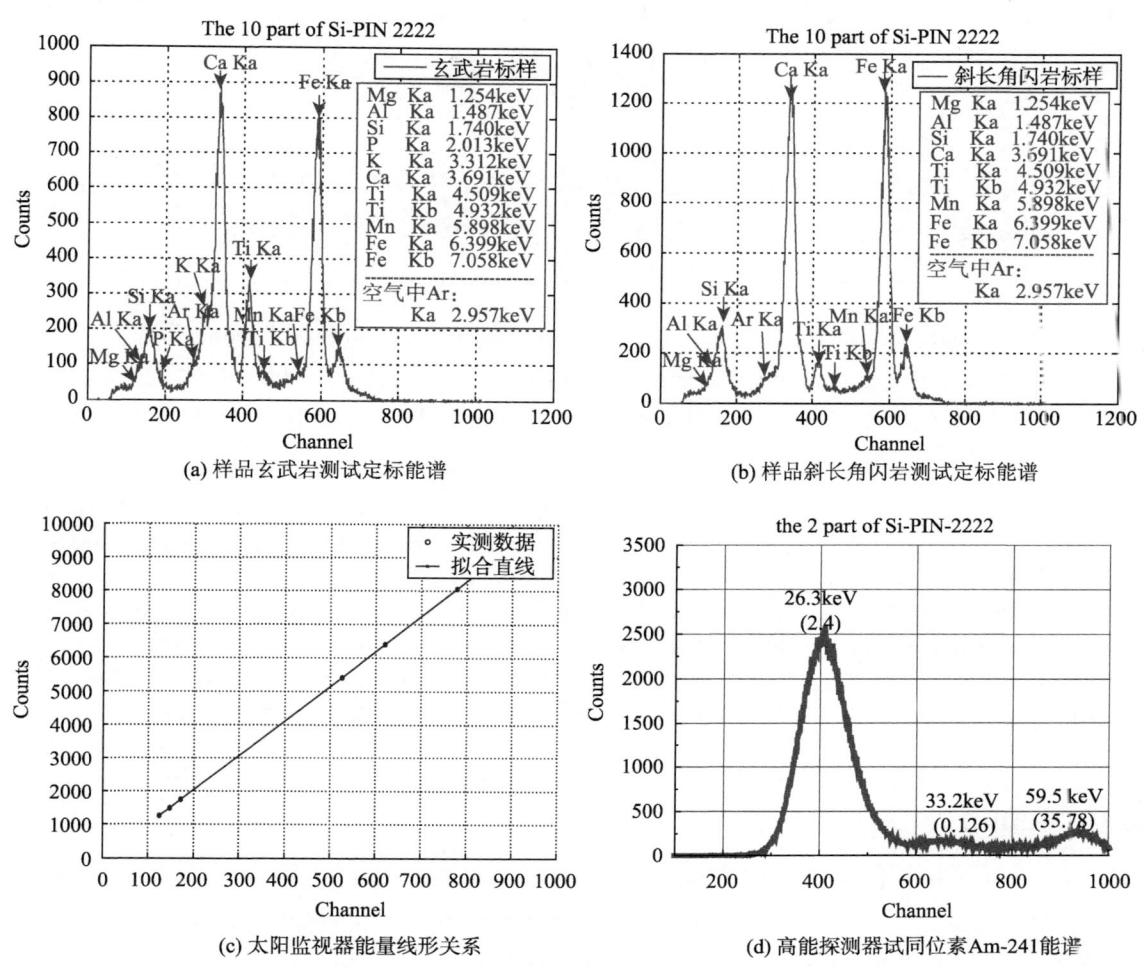

图 4 玄武岩和斜长角闪岩荧光谱线测试图

应该指出的是，在本仪器系统研制中我们采用了粒子物理研究中专用模拟软件对探测器的磁屏蔽性能、探测效率、各种太阳活动情况下的有关月表元素的探测效果进行了蒙特卡罗计算，为探测器的系统设计和性能测试提供了帮助。

参 考 文 献

[1] 王焕玉等. 嫦娥一号卫星 X 射线谱仪初样设计报告. 2003
[2] 吉昂，陶光仪等. X 射线荧光光谱分析. 北京：科学出版社，2003
[3] Clark P E, Trombka J I. J. Geophys. Res., 1977, 102, 16361
[4] Y Q Ma, Y H Wang, C M Zhang, et al. 27th ICRC, Hamburg, Aug, 2001
[5] 核素图表组. 核素常用数据表. 北京：原子能出版社，1977
[6] 欧阳自远，李春来，邹永廖等. 月球手册. 2004

Gamma-ray Detector on Board Lunar Mission Chang'E-1

J.Chang[1]   T. Ma[1]   N. Zhang[1]   M.S. Cai[1]   Y.Z. Gong[1]   H.S. Tang[1]
R.J. Zhang[1]   N.S.Wang[1]   M. Yu[1]   J.P. Mao[1]   Y.L. Zhou[2]   J.Z. Liu[2]
A.A. Xu[3]   L.G. Liu[3]

(1. Purple Mountain Observatory, 2 West Beijing Road, Nanjing 210008, China;

2. National Astronomical Observatory, 20A Datun Road, Beijng 100012, China;

3. MACAU University of Science and Technology, Avenida Wai Long, Taipa, Macau, China)

Abstract   Chang'E-1 is the first Chinese lunar mission which was launched in 24th October 2007. Gamma-ray spectrometer (GRS) is included in the payload of Chang'E-1. Specific objectives of the GRS are to map abundances of O, Si, Fe, Ti, U, Th, K, and perhaps, Mg, Al, and Ca, to depths of about 20 cm. There are remarkable advantages for GRS application to remote sensing elemental materials over the entire lunar surface: large effective area, high energy resolution and good ability for background rejection. The design of GRS is described and its performance in the laboratory is presented. The GRS performance both in calibration and in lunar orbit are introduced in this paper.

Key words   Lunar mission   Chang'E-1   Gamma-ray spectrometer   Elemental abundance

1 Introduction

The Chang'E-1 (CE-1) is the first Chinese lunar mission. The science objectives of CE-1 are as following:
- To make a detailed map of the moon, or to obtain 3-dimensional image of the lunar surface
- To observe chemical element distribution on the surface of the moon
- To probe the depth of lunar soil, or regolith
- To explore the lunar space environment

To accomplish these four tasks, Chang'E-1 carries six kinds of payloads, including a total of 25 devices installed in five instrument systems
- Optical Imaging System
- Laser Altimeter
- Gamma/X-Ray Spectrometers
- Microwave Detector
- Space Environment Monitor System

The GRS onboard CE-1 satellite plans to map the Moon surface by polar orbiter. The flight model of the instrument is shown in Fig. 1. The orbiter altitude is 200km. Gamma-ray spectrometer (GRS) has been proven to be powerful instrument for remote measuring the abundance of chemical elements, like C, O, Mg, Al, Si, K, Ca, Fe, Th and U on the planetary surface.[1] Several missions planned by combining high energy gamma ray detectors to explore the planetary surface composition, such as APOLLO,[2] LUNAR PROSPECTOR before,[3] SELENE (KAGUYA)[4] and CE-1 later for the Moon, MARS Odyssey for the Mars and NEAR Shoemaker for the asteroid Eros.[5,6] This paper describes a gamma-ray spectrometer (GRS) onboard CE-1. Its outside and crosssectional view are shown in Fig. 2. The GRS is designed to make global mapping by remote sensing elemental materials over the entire lunar surface. This mission goal is to give the major elements' distribution, such as U, Th, K, Fe, Ti, Si, O, Al, Mg, Ca and so on. The gamma-ray fluxes from the moon are expected to be low and many days of accumulation will be necessary to provide statistically significant measurements. We hope to answer fundamental questions about the nature, origin, evolution of the moon by the measurements.

Fig.1　The flight model of GRS onboard CE-1

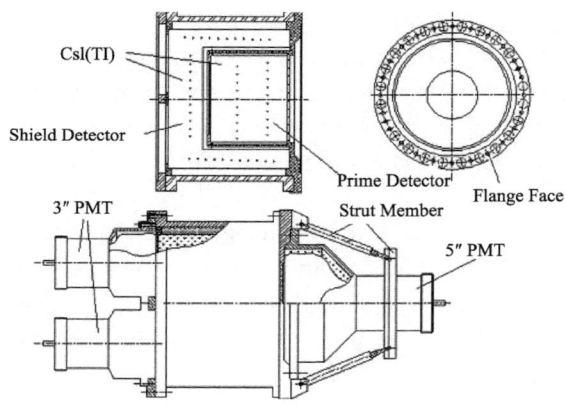

Fig.2　The structure of the gamma-ray spectrometer (GRS)

2　Instrument overview

The GRS has a large geometrical factor of about 134.5 cm^2 sr and is attached to spacecraft body so as to fulfill mechanical and thermal interface condition. Active thermal control system for the GRS is not employed. Considering the constraint of the weight and cost, the GRS main detector is surrounded by thick anticoincidence scintillator to reduce spacecraft background without boom. GRS consists of two parts; gamma-ray detector (GRD), and electronics control box (ECB). The GRS and the GRD are connected by two wire harnesses. The GRD includes central scintillation detectors and surrounding anticoincidence detector, and photomultiplier tubes (PMT). The high voltage electronics box and preamplifiers are also installed on the GRD. The data processing unit is included in the ECB which conducts gating by anticoincidence signal, pulse height analysis, telemeter coding, command signal decoding and supplying power. The instruments characteristics are showed in Table 1.

Table 1 Characters of GRS instrument

Component	Specifications
Prime detector	CsI(Tl) scintillator Φ11.8cm×7.8cm
Shield detector	CsI(Tl) scintillator cup Φ17.8cm×10.8cm
Energy range	0.3-10MeV 512 channel spectra for prime detector 256 channel spectra for shield detector
Energy resolution of prime detector	8.2% fwhm @ 662keV 12% fwhm @ 511keV
Energy resolution of shield detector	12% fwhm @ 662keV
Anticoincidence Background rejection	>97% above 5 MeV
Telemetry data rate	12kbps
Power consumption	8.7W
Mass	32kg

3 Instrument design

The prime Gamma-ray detector (GRD) is a 12cm diameter × 7.6cm height CsI(Tl) crystal. It is surrounded the side and back by a well type CsI(Tl) crystal shield with thickness of about 3cm (20cm diameter × 12cm height). The shielding CsI(Tl) counter generate signals by absorption of Compton scattering of gamma- rays from spacecraft and by penetrating charged particles, which inhibit coincident signals from central CsI(Tl) counter. Two gamma-ray spectra are collected simultaneously: the prime detector spectrum and shield detector spectrum. Although high purity Ge detector has the excellent energy resolution, it needs cooling system and is degraded by possible serious radiation damage in a year. The GRS can be operated at room temperature with simple thermal regulation system, maintaining good energy resolution, which was tested in a temperature chamber.

In order to achieve the best energy resolution, many tests were made to select the best combination of CsI(Tl) crystal and PMT. It is important to use stable HV (high voltage) modules. The signal gains of the PMT are not particularly sensitive to temperature, but are very sensitive to voltage variations. Each PMT was supplied HV from each HV module for redundancy. Each HV module has 7 voltage levels, which can be selected easily by the command. Fig. 3 is block diagram of GRS detector electronics.

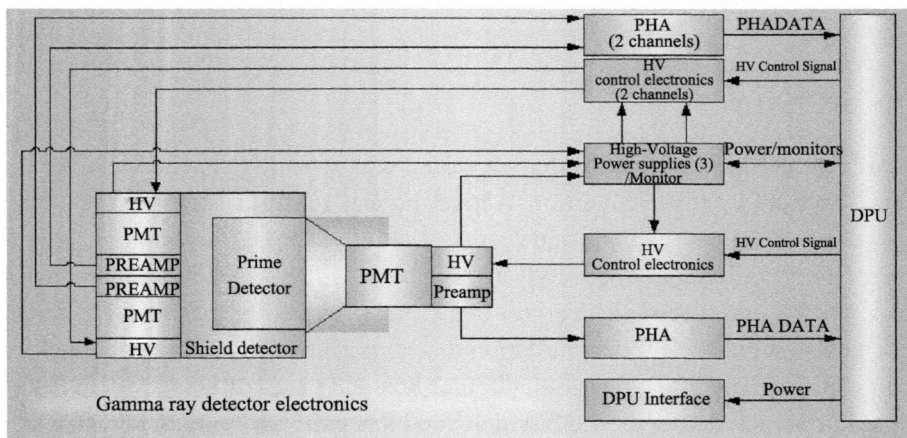

Fig.3 Function block diagram of the GRS including the gamma-ray scintillation detectors (main and shield) with photomultiplier tubes (PMT), preamp modules, the pulse shapers and pulse-height analyzers (PHA) etc. Spectra formed in the data processing unit (DPU), which also provides the spacecraft interface

The DPU (data process unit) records PHS (pulse height spectrum) every 3s. The integration time is adjustable to 1s, 2s and 3s. The GRS produces two spectra: shield detector spectrum with 256 channels, prime detector spectrum with 512 channels. The energy range is from 0.3 to 10 MeV.

4 Calibration results

The calibration for the GRS was conducted by evaluating the detector performance in laboratory. The detector response was measured using calibrated gamma- ray sources such as ^{57}Co, ^{137}Cs, and ^{60}Co at 121.54keV, 661.81keV, 1.17MeV, and 1.33MeV respectively. The experiments were numerically simulated to evaluate detection efficiency as a function of the energy using GEANT4 software. Here the efficiency was defined as the ratio of the counting rate to the incident flux from each source calculated from source intensity and subtended solid angle which was narrow enough to approximate parallel beam.

Fig. 4 shows the simulated (line) and measured (pluses) efficiency of CE-1 GRS as a function of energy. The PMT of the central counter was attached to the entrance surface of the crystal, which reduced detection efficiency.

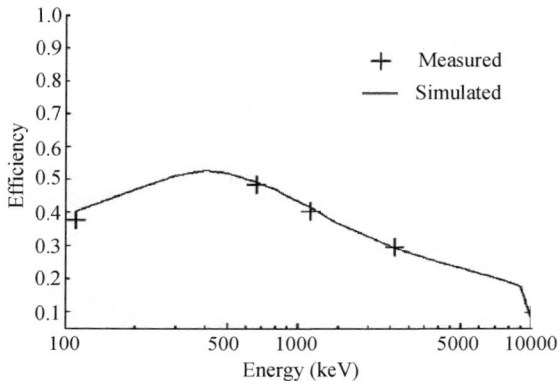

Fig. 4 Simulated (line) and measured (pluses) detection efficiency of CE-1 GRS for normal incidence as a function of energy

Figure 5 shows measured (dot) and fitted (line) energy resolution of CE-1 GRS as a function of energy.

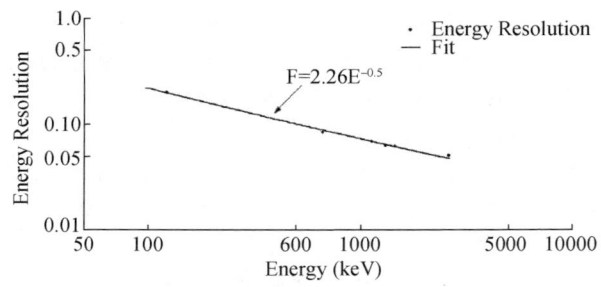

Fig. 5 Measured (dot) and fitted (line) energy resolution of CE-1 GRS as a function of energy

Figure 6 shows the lab gamma-ray background spectrum measured by CE-1 GRS.

5 Operation of the GRS in lunar orbit

GRS started the normal observation since Nov. 28 2007. Sample spectrum from the GRS that illustrate the excellent quality of the data are shown in Fig. 7. It can be seen that typical gamma-ray lines from lunar surface are clearly identified in the spectrum. The data analysis shows that the main performance parameters of GRS met all the requirements of design. We therefore expect to attain all the scientific objectives.

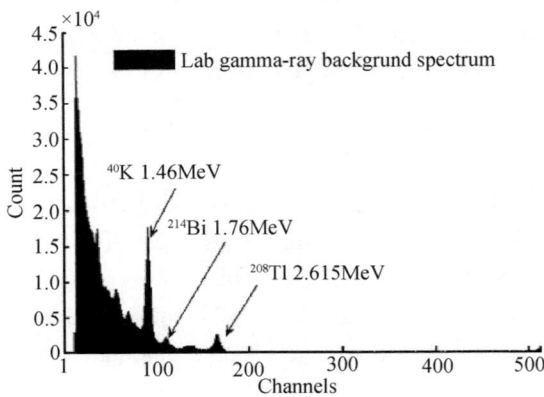

Fig. 6 Measured gamma-ray background spectrum in lab by CE-1 GRS

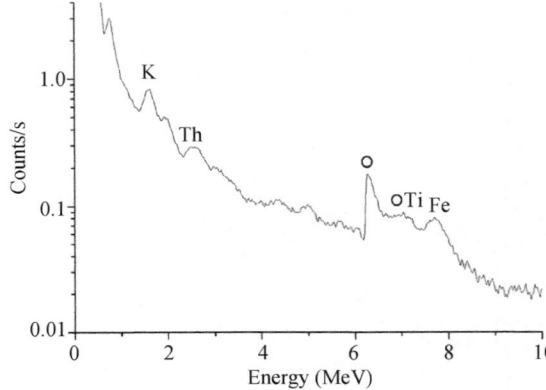

Fig. 7 Energy spectrum measured during the first 8 h of 28 Nov. 2007 by the GRS in lunar mapping orbit that illustrates the quality of data

6 Conclusion

A good energy resolution gamma ray detector (GRD) with a large geometrical factor is presented. The calibration results and observation in the orbit show that the performance of GRS meets all the requirements of science. The measurement provided by the CE-1 GRS will make contributions to our understanding of the elemental composition and the evolution of moon.

Acknowledgment The author thanks all GRS team members for their outstanding performance in developing this instrument. This work was supported by NSFC (No. 10573039, 10333040).

References

[1] L. G. Evans, R. C. Reedy and J. I. Trombka: in Remote Geochemical Analyses: Elemental and Mineralogical Composition, edited by C.M. Pieters and P. A. J. Englert, CambridgeUniv. Press, New York (1993) 167
[2] J. I. Trombka, C. S. Dyer, L. G. Evans, M. J. Bielefeld, S. M. Seltzer, A. E. Metzger: Astrophys. J 212 (1977) 925
[3] W.C. Feldman, B. K. Barraclough, K. R. Fuller, D. J. Lawrence, S. Maurice, M. C. Miller, T. H. Prettyman, A. B. Binder: Nucl. Instrum & Meth. A422 (1999) 562
[4] N. Hasebe et al: To be published in J. Phys. Soc. Jpn. Suppl. (in this volume)
[5] W. V. Boynton et al.: Space Sci. Rev. 110 (2004) 37
[6] J.O. Goldsten et al.: Space Sci. Rev. 82 (1997) 169

Time Series Data Correction for the Chang'E-1 Gamma-ray Spectrometer

Zhang Liyan[1,2] Zou Yongliao[1] Liu Jianzhong[1] Liu Jianjun[1] Shen Ji[1]
Mu Lingli[1] Ren Xin[1] Wen Weibin[1] Li Chunlai[1]

(1. National Astronomical Observatories, Chinese Academy of Sciences, Beijing 100012, China;
2. Graduate University of Chinese Academy of Sciences, Beijing 100049, China)

Abstract The main goal of the gamma-ray spectrometer (GRS) onboard Chang'E-1 (CE-1) is to acquire global maps of elemental abundances and their distributions on the moon, since such maps will significantly improve our understanding of lunar formation and evolution. To derive the elemental maps and enable research on lunar formation and evolution, raw data that are received directly from the spacecraft must be converted into time series corrected gamma-ray spectra. The data correction procedures for the CE-1 GRS time series data are thoroughly described. The processing procedures to create the time series gamma-ray spectra described here include channel processing, optimal data selection, energy calibration, gain correction, dead time correction, geometric correction, orbit altitude normalization, eliminating unusable data and galactic cosmic ray correction. Finally, descriptions are also given on data measurement uncertainties, which will help the interested scientists to understand and estimate various uncertainties associated with the above data processing.

Key words instrumentation spectrographs (gamma-ray spectrometer) — gamma-rays observations—methods data analysis

1 Introduction

The gamma-ray spectrometer (GRS) is one of the main payloads onboard Chang'E-1 (CE-1). Its main scientific objectives are to analyze the abundances and distributions of useful elements on the lunar surface, to study the distribution of lunar resources like rocks, minerals, etc., and to find enriched areas of utilizable resources together with the imaging interferometer and X-ray spectrometer onboard CE-1, in order to improve our understanding of lunar formation and evolution. To achieve the above-mentioned scientific objectives, it is important that raw data that are received directly from CE-1 GRS be converted into fully corrected time-ordered gamma-ray spectra that are used to directly derive elemental maps. In this paper, we provide a thorough description of the data correction processes for the CE-1 GRS time series data, which will provide a foundation for future Chinese lunar gamma-ray experiments that need to carry out similar processing steps and help interested scientists to understand and estimate the data correction procedures and methods, and various uncertainties associated with the processing for better using the corrected data to carry out research on lunar surface composition, lunar origin and evolution.

The data correction processing methods described in the paper to create the CE-1 GRS time series gamma-ray spectra include channel processing, the optimal data selection from two data acquisition stations, energy calibration, gain correction, dead time correction, geometric correction, orbit altitude normalization, eliminating unusable data and galactic cosmic ray correction. Before time series data correction is described, a brief review of the CE-1 GRS sensor operation and data collection modes is provided. We also review and summarize the additional auxiliary data sets that are used for the CE-1 GRS data correction. In the last section of the paper, descriptions are given of data measurement uncertainties inherent in the time series data before and after the data processing.

2 CE-1 GRS sensor and support information

2.1 Review of the CE-1 GRS Sensor

CE-1 GRS was designed and developed by the Purple Mountain Observatory, Chinese Academy of Science (CAS). It consists of a gamma-ray detector (GRD) and electronics control box (ECB) (Chang et al. 2009) as shown in Fig. 1 (a) and (b). As shown in Fig. 1 (b), GRD includes a central prime crystal, surrounding anticoincidence crystal, one 5″ photomultiplier tube (PMT) and two 3″ PMTs (Chang et al. 2009). The prime crystal is an 11.8cm diameter by a 7.8cm long cylinder of CsI (Tl) crystal. It measures gamma-rays with energies from 0.3 to 11.2MeV. The measured energy resolution is 8.27% at 0.662MeV and operates in $E^{-1/2}$ (where E = gamma-ray energy). The side and back of the prime crystal is surrounded by a well type CsI (Tl) crystal shield with thickness of about 3cm. The 5″ PMT collects gamma-rays from the prime crystal, and the two 3″ PMTs collect gamma-rays from the anticoincidence crystal. In lunar orbit, the efficiency of the anticoincidence system's background rejection is almost 70%.

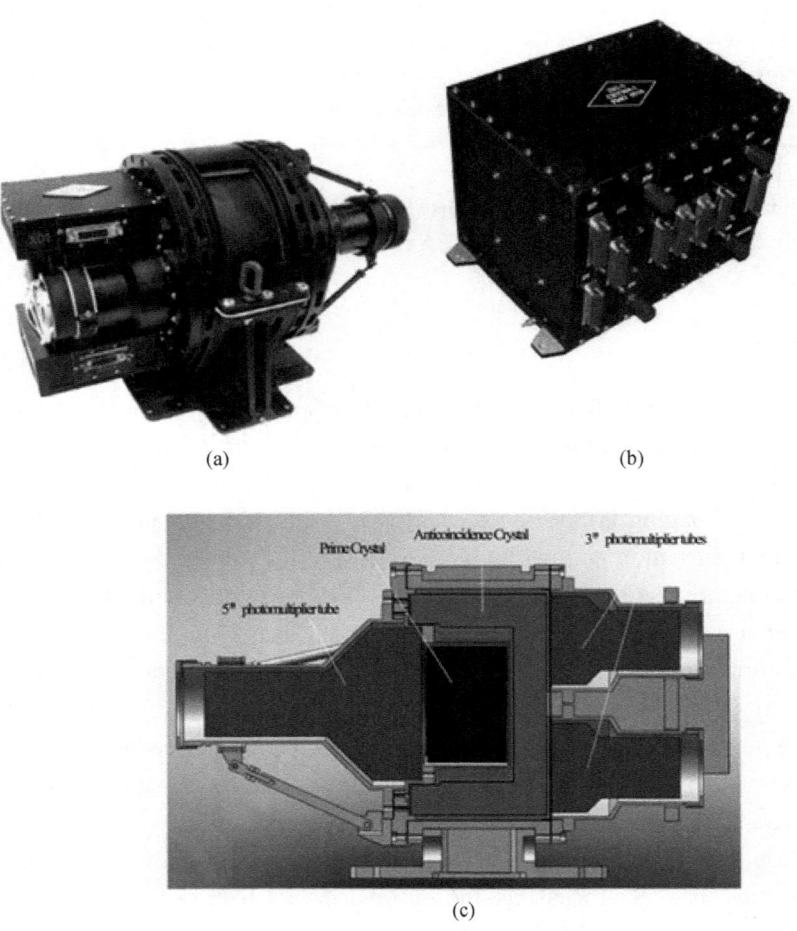

Fig.1 (a) GRD, (b) the electric control box and (c) the structure of GRD

CE-1 operates in a circular lunar polar orbit with inclination of approximately 90 degrees with respect to the lunar equatorial plane at a height of about 200 km above the lunar surface, and it adopts a three-axis stabilized attitude control. GRS is installed in the corner of the spacecraft, and cannot be shielded by any large-sized or high-quality component of the spacecraft within the range of −60°~+60° in its field of view. GRS is an uncollimated, omnidirectional gamma-ray detector. It includes one prime detector and one anticoincidence detector. The prime detector collects a spectrum with 512 channels every three seconds originating from the moon and other non-lunar sources. The anticoincidence detector collects a spectrum with 256 channels every second originating from the absorption of Compton-scattered gamma-rays from the spacecraft, and whose primary objective is to reduce the spacecraft's background gamma-rays striking the prime detector. The data collection circuit of GRS uses the anticoincidence procedure to eliminate the anticoincidence signals and measure the lunar gamma-ray spectra. The collected data are then transmitted to the ground from CE-1 and are used to obtain the elemental abundances on the lunar surface through data analysis.

GRS started its first observation of the lunar gamma-ray spectra at 19:58 on 2007 Nov. 27 (UTC). During the duration of the CE-1 mission, the working times of GRS are shown in Table 1.

Table 1 Working Times of CE-1 GRS

No.	Beginning time (UTC)	End time (UTC)
1.	2007–11–27 T19:58	2007–12–01 T02:22
2.	2007–12–04 T01:27	2008–01–26 T22:56
3.	2008–01–30 T00:59	2008–02–06 T00:38
4.	2008–05–15 T11:28	2008–07–03 T02:30
5.	2008–07–03 T02:30	2008–07–25 T11:20
6.	2008–11–21 T06:16	2008–11–21 T22:29

2.2 Auxiliary data

In addition to the two fundamental spectra from the prime detector and the anticoincidence detector, a number of other auxiliary CE-1 data products and data sets are used in the GRS analysis. They include the engineering and telemetry parameters of GRS, spacecraft ephemeris and attitude data, GRS installation parameters on the spacecraft and ground calibration data of GRS.

The engineering and telemetry parameters for the GRS analysis include the sensor temperature and the high voltage. The sensor temperature is a measurement of GRS housing temperature and is measured once per second. The high voltage (HV) measurements monitor the combined HV on each of the three PMTs once per second, that include one PMT of the prime detector and two PMTs of the anticoincidence detector. During the mission, the HV could be changed to keep the gain for each PMT within an acceptable range.

The spacecraft's ephemeris data include spacecraft positions and speeds of X, Y and Z axes in the J2000 geocentric coordinate system. The spacecraft's attitude data include three Euler attitude angles that are yaw, roll and pitch angle. The spacecraft ephemeris and attitude data are used to calculate lunar positions of measured spectra and GRS sensor observation angles.

The GRS installation parameters include the instrument coordinates and angles of the X, Y and Z axes in the spacecraft body's coordinate system.

The ground calibration data include the laboratory calibrations and numerical simulations of GRS. The ground calibrations are used to acquire 1) detector energy linearity, resolution and range, 2) effective area and detection efficiency of the GRS detector as a function of energy, 3) gain calibration, 4) angle calibration, 5) temperature influence on the energy linearity and resolution, 6) dead time calibration, 7) anticoincidence counting rate influence on the dead time of the prime detector, and 8) efficiency of anticoincidence background rejection.

3 Primary processing

In this section, we describe the primary processing that is carried out on the CE-1 GRS data. This processing includes steps (Fig. 2) such as channel processing, optimal data selection, system correction, geometric corrections, orbit altitude normalization, unusable data elimination and galactic cosmic ray correction. The final result is a fully corrected time series data set that is ready for spectral analysis, mapping, and abundance calculations (Metzger et al. 1974, 1977, 1979; Metzger 1993; Lawrence et al. 1998, 2004; Prettyman et al. 2006).

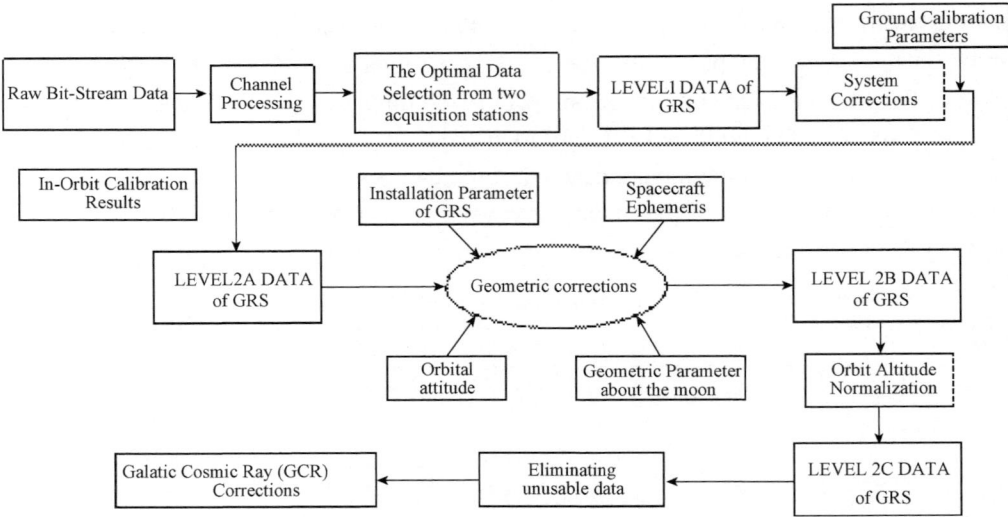

Fig.2　Procedure of time series data correction for CE-1 GRS

3.1 Channel processing

Raw data transmitted from CE-1 are comprised of CCSDS (Consultative Committee for Space Data Systems) data frames, so the first step is channel processing (Fig. 3), which includes frame synchronization, descrambling, Reed-Solomon decoding, virtual channel data extraction and GRS data packet acquisition. It aims to acquire GRS raw packet data from CCSDS data frames.

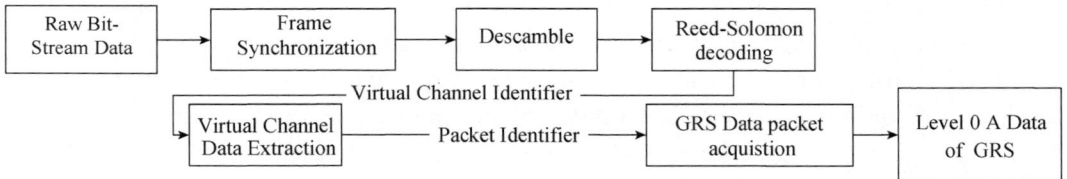

Fig.3　Channel processing procedure

1) Frame Synchronization

Raw data that are received directly from CE-1 are comprised of CCSDS data frames. Frame synchronization is achieved with the aid of a sync pattern, which is either injected periodically into the data stream (continuous transmission) or appended at the beginning of each packet (packet transmission). Frame synchronization of the CE-1 raw data is achieved with the aid of a sync pattern, which is appended at the beginning of each packet, and a sync pattern code is used to extract CCSDS data frames from raw bit-steam data.

2) Descrambling

To assure the adequate change of channel density and avoid continuous 0 or 1 values for a long duration, data bits of each frame are scrambled at the transmitting end of a communication system with a pseudo-random sequence in the high-speed multi-channel multiplexer of the payload data management system (PDMS) onboard

CE-1. At the receiving end of the Ground Research and Application System (GRAS), data bits of each frame are descrambled using the same pseudo-random bit pattern to recover the original data.

3) Reed-Solomon Decoding

The moon is approximately 384 400 km from the earth, so channel errors are inevitably produced due to various kinds of interference factors. Thus, in the high-speed multi-channel multiplexer of the PDMS onboard CE-1, Reed-Solomon codes for correcting errors are used to reduce the channel errors. At the receiving end of the GRAS, a Reed-Solomon decoder extracted from the data stream is used to correct errors in the data frames.

4) Virtual Channel Data Extraction

The Virtual Channel Data Units (VCDUs) are created to transmit eight sets of raw data acquired by CE-1. At the receiving end of the GRAS, virtual channel data are extracted on the basis of a virtual channel identifier in the CCSDS data frame.

5) GRS data packet acquisition

After virtual channel data extraction, GRS data packets are extracted based on a GRS packet identifier in the CCSDS data frame from the virtual channel.

3.2 Optimal data selection

Optimal data selection (Fig. 4) includes the following steps:

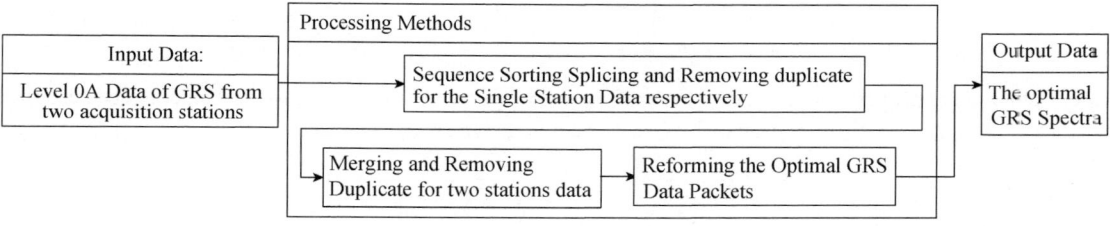

Fig.4　Processing procedure for optimal data selection

(1) The first step is where the processes of sequence sorting, splicing and removing duplicates are carried out on the GRS data packets from two data acquisition stations (the Miyun Station and the Kunming Station), respectively.

(2) The second step is where merging, removing duplicates and forming the optimal data packets are carried out on the GRS data packets from two data acquisition stations.

(3) The last step is that the optimal data packets are disassembled and reformed according to the structures of GRS data packets and scientific data blocks to acquire the optimal GRS spectra.

3.3 System corrections

Using the results of the lab calibration and in-orbit calibration, a series of corrections are made to the optimal GRS spectra and are therefore referred to as system corrections. Below, we describe the three major system corrections: energy calibration, gain and dead time corrections.

3.3.1 Energy calibration

The goal of energy calibration is to convert channel spectra acquired by GRS to energy spectra. By measurements in the laboratory, we found that the relationship between channel and energy of the GRS detector is linear, so the linear relationship can be determined by fitting known characteristic energy values and their corresponding channel values acquired by measuring and analyzing gamma-rays emitted by the different characteristic energies. By using results from the experiment where the GRS measures gamma-ray spectra emitted by the calibrated gamma-ray sources (i.e., ^{57}Co, ^{137}Cs, and ^{60}Co) (Chang et al. 2009) and the radioactive sources in the laboratory (i.e., ^{40}K and ^{208}Tl), we can acquire gamma-ray spectra of the six characteristic energies that are 122, 662 keV, 1.17, 1.33, 1.46 and 2.615 MeV, respectively. The corresponding channel values of these six full energy peaks are obtained by the Gauss fitting of their characteristic energy spectra. The parameters a and b in the following equation are solved by fitting the above six pairs (channel versus energy) of data. Thus, the linear fitting result and the deviations between the measured and fitted energy are shown in Fig. 5(left) and (right), respectively.

$$E1 = a \times \text{channel} + b, \qquad (1)$$

where $E1$ is the energy value, channel is the corresponding channel value of $E1$, and a and b are two fitting parameters that will be solved. By applying Equation (1), channel spectra observed by GRS are converted to energy spectra.

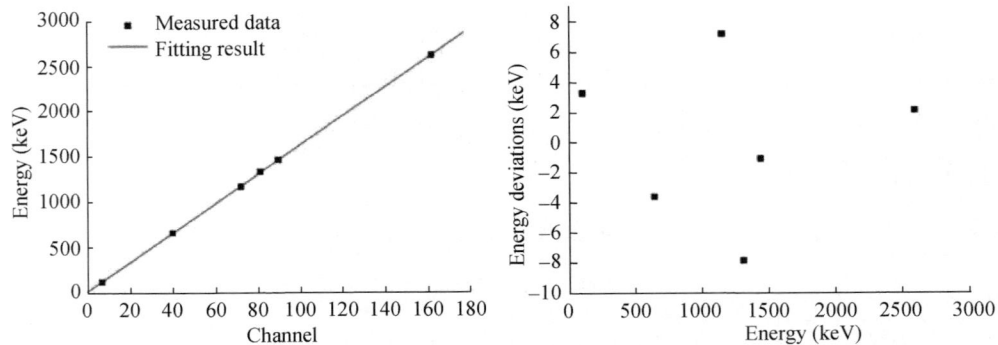

Fig.5 *Left*: plot of channel versus energy, *right*: energy deviations versus energy. By applying Equation (1), channel spectra observed by GRS are converted to energy spectra

3.3.2 Gain correction

Gain variations are caused by PMT changes and temperature changes. They can also result in channel shifts, so gain variations have to be monitored and corrected. The CE-1 GRS gain changes are monitored by analyzing the variation of the three characteristic peak positions including the 0.511 MeV annihilation peak, 6.13 MeV oxygen peak and 10.830 MeV nitrogen peak throughout the mission. Three peaks are clearly defined during the spacecraft's orbital period (127 minutes), and all the spectra detected in the lunar orbit are accumulated according to the spacecraft's orbital period. The three peak positions are obtained by performing Gauss fitting on the accumulated spectra for every spacecraft's orbital period. If the time variations of the peak positions are clear, we have to carry out gain corrections.

According to the above monitoring methods, we monitor the gain variations throughout the mission. We found that channel shift did not occur in the spectra detected by CE-1 GRS before 09:26:02 UTC on 2008 June 12. Since the static random access memory (SRAM) of the GRS failed at 02:30:27 UTC on 2008 July 3, gain corrections had to be carried out for spectra detected by CE-1 GRS during the period between 09:26:02 UTC on June 12 and 02:30:27 UTC on 2008 July 3.

The methods of gain corrections include 1) three peak positions are obtained for every 127-minute accumulated spectra throughout the mission; 2) the three peak positions for the first orbit of accumulated spectra are defined as the corresponding reference peak positions; 3) a linear relation is derived between the three reference peak positions and the corresponding peak positions for the other orbits of accumulated spectra; 4) the channels of spectra observed in other orbits are corrected by being recorded in the reference channels by using the linear relation built in the previous step; 5) the reference gain and offset values are obtained by linear fitting of the three pairs (reference channel vs. energy) for the first orbits of accumulated spectra; 6) the gain corrections for the other orbits of spectra are carried out by using the reference gain and offset values.

3.3.3 Dead time correction

Dead time of the prime detector can be caused by two reasons, which are cases where the prime detector is busy processing an event or the anticoincidence detector is busy processing an event. Based on the Lunar Prospector data analysis, we find that the count rate collected by the prime detector surrounding the moon is less than 1000 counts per second. By the GRS calibration tests in the laboratory, we find that the characteristic peak area at ^{208}Tl (abbreviation of thallium) has little change when the count rate collected by the prime detector is less than 1000 gamma-ray counts per second. Hence, we can conclude that dead time caused by the prime detector being busy processing an event does not need to be corrected.

Dead time caused by the anticoincidence detector means that the spectra collected by the prime detector are affected by an anticoincidence square wave. In the ground calibration tests, the relative change of the peak area at ^{208}Tl acquired by the prime detector is investigated when the different count rates are measured by the anticoincidence detector. Then the relation between the count rate of the anticoincidence detector and the relative change of the peak area at ^{208}Tl is derived by the use of the experimental data. In the GRS data processing, the relative change is solved by using this relation derived from the ground tests and the count rate measured by the anticoincidence detector. Then, the counts observed by the prime detector divided by the relative change are the dead time corrected counts.

3.4 Geometric corrections

The goals of geometric corrections are to obtain lunar positions of measured spectra and observation angles of the GRS detector by building the observation equation using the following information such as the spacecraft ephemeris and attitude data, GRS installation parameters on the spacecraft, geometric parameters about the moon, etc. By using geometric corrections, we can obtain the geometric information including lunar surface longitude and latitude, orbit height, instrument incidence angle and azimuth angle at the moment of data acquisition.

The core of geometric positioning arithmetic is 1) to build the spatial triangle (Fig. 6) using three points including the CE-1 spacecraft's centroid, the moon's centroid and the lunar surface's observation point of GRS, which is formed by the GRS observation vector, the CE-1 position vector and the position vector of the lunar surface observation point, 2) to build the observation equation, and 3) to solve it.

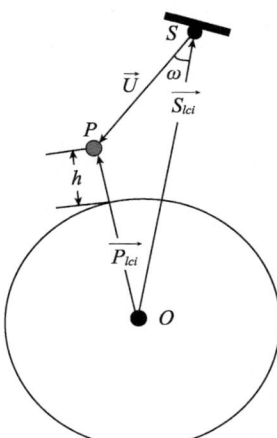

Fig.6 Spatial triangle formed by the observation vector (U), the CE-1 position vectors (S) and the position vectors of the lunar surface's observation point (P)

The processing procedure for geometric corrections is shown in Fig. 7. It mainly includes the three steps as follows:

1) Data Preparation

Besides the system-corrected GRS spectra, data needed in geometric corrections include GRS installation parameters, spacecraft ephemeris and attitude data, geometric parameters about the moon, etc.

2) The conversion of coordinate systems

Because the above input data needed for geometric corrections are defined under different coordinate systems, these input data must be converted into data under the identical coordinate system (CS). GRS installation parameters are defined under spacecraft body CS, spacecraft ephemeris and attitude data are defined under flat equatorial geocentric CS, and the lunar surface positions of the measured spectra obtained by geometric corrections are defined under lunar geodetic CS. To build the spatial triangle as shown in Fig. 6, we must first carry out a series of coordinate conversions on the input data under different CSs so that the GRS observation vector and CE-1 position vector are converted into the vector under the lunar solid CS, respectively. The series of coordinate conversions are shown in Fig. 7.

3) Building the observation equation and deriving geometric information

After coordinate conversions, the observation equation is built based on the spatial triangle as shown in Fig. 6 to solve the lunar positions and the relevant geometric parameters. In this paper, the process of establishing and solving the observation equation is abridged.

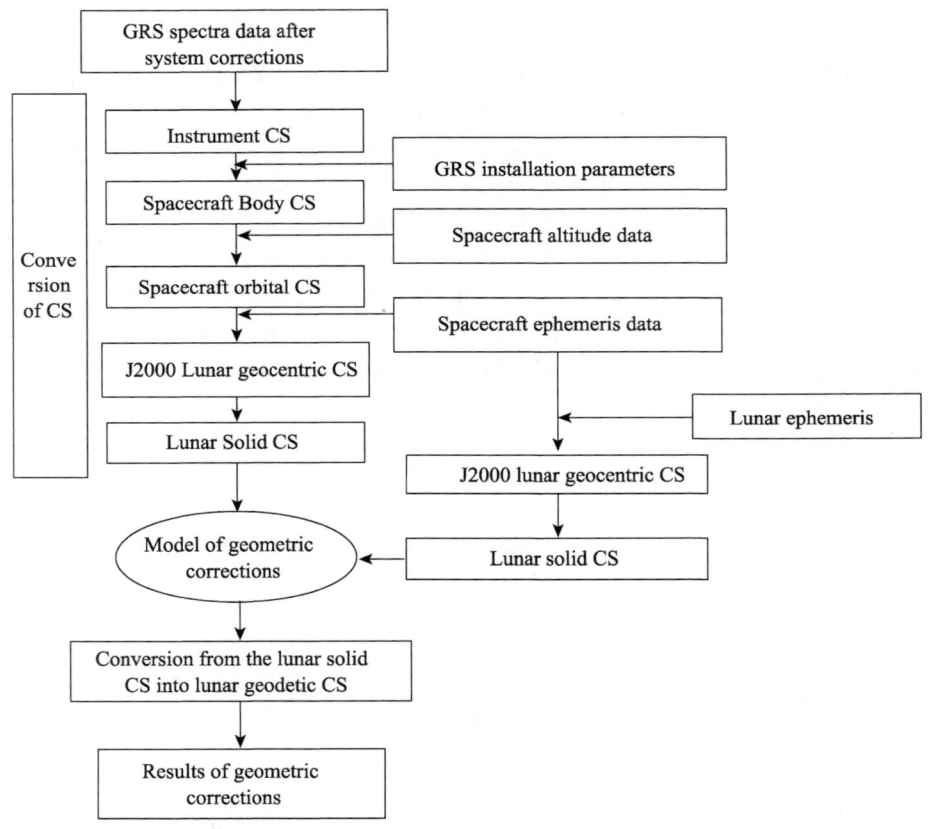

Fig.7 Process of geometric corrections

3.5 Orbit Altitude Normalization

Since the flux of the lunar gamma-rays is different at various spacecraft heights, we need to make corrections for variations of the lunar gamma-ray flux resulting from variations of the spacecraft height. In addition to the gamma-rays coming from the moon, there is a nonlunar component of gamma-rays coming from the galaxy (Lawrence et al. 2004). This nonlunar component does not vary with spacecraft height. It is measured using the background data that CE-1 GRS measured surrounding the moon. The variation of the lunar gamma-rays' flux acquired by the GRS at various spacecraft heights results from the variation of the Moon's solid angle as shown in Fig. 8. The orbit altitude normalization performs the Moon's solid angle correction. Below, we describe how the moon's solid angle correction is carried out.

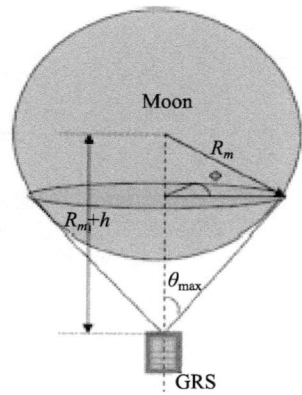

Fig.8 The moon's solid angle of GRS

If we assume that the moon has a spherical surface, then the fractional solid angle relative to the surface varies with the spacecraft's altitude, h, and can be expressed using the following relation

$$\Omega(h) = \int_0^{2\pi} d\phi \int_0^{\theta_{max}} \sin\theta d\theta = \int_0^{2\pi} d\phi \int_{\cos\theta_{max}}^1 d(\cos\theta) = 2\pi(1 - \cos\theta_{max}) \qquad (2)$$

where θ is the angle from the sub-spacecraft point to a visible location on the moon, and $\theta_{max} = \sin^{-1}\left(\dfrac{R_m}{R_m + h}\right)$ is the maximum value that θ can have for a given altitude. For simplification across data sets, we have normalized the solid angle of Equation (2) to the solid angle at the lunar surface, i.e., $\Omega'(h) = \Omega(h)/\Omega(0)$. The full solid angle correction, including the non-lunar background, is then

$$\frac{C(h) - C_{background}}{C(0) - C_{background}} = \frac{\Omega(h)}{\Omega(0)} = \Omega'(h) \qquad (3)$$

where $C(h)$, $C_{background}$ and $C(0)$ are the uncorrected, background and solid angle corrected spectral counts, respectively.

Equation (3) now becomes

$$C(0) = \frac{C(h) - [1 - \Omega'(h)]C_{background}}{\Omega'(h)} = \frac{C(h) - \sqrt{1 - \left(\dfrac{R_m}{R_m+h}\right)^2} C_{background}}{1 - \sqrt{1 - \left(\dfrac{R_m}{R_m+h}\right)^2}} \qquad (4)$$

3.6 Eliminating unusable data

As for the orbit altitude normalized spectral data, the spectral data observed by GRS under the safe mode (that is 0 as the HV level) need to be eliminated to carry out the elemental abundance inversion. In the defining convention we adopt for the spectral quality status, if the first character is identified as "W" in quality status composed of the eight characters in one spectrum, this spectrum also needs to be eliminated. In addition, the following abnormal spectral data should be eliminated from the above nominal data sets: spectral measurements that show data are clearly off-limits (i.e., the spectrum measured at 06:49:09.060 UTC on 2008 December 28 in Fig. 9) and spectral measurements with geometric positioning errors (i.e., the geometric positioning results of the spectrum in Table 2).

Table 2 One spectrum with abnormal geometric information

Field	Time	GRS Instrument status	Longitude	Latitude	Instrument incidence angle	Instrument azimuth angle	Spectrum with 512 channels	Quality status
Value	2007-12-07 T06:03:24.534Z	431	*	*	*	*	...	0X00EEFF

Notes: * (=9999.9999) represents abnormal data.

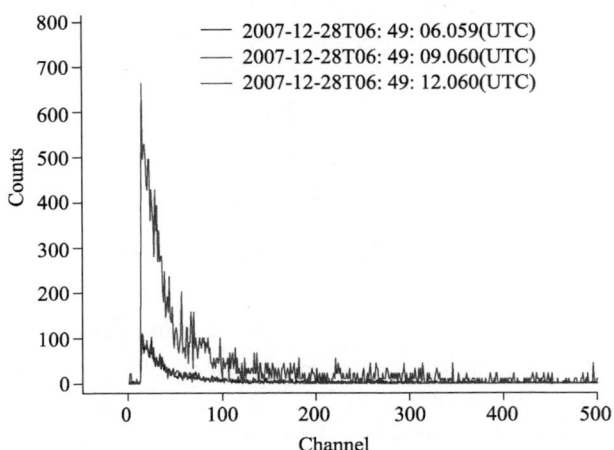

Fig.9 One abnormal spectrum and two normal spectra

3.7 Galactic cosmic ray corrections

Since most of the primary and background gamma-rays being observed with the CE-1 GRS originate from galactic cosmic rays (GCRs) (the one exception is gamma-rays produced by radioactive decay), we need to monitor and correct for the time variability of the GCR flux. After much trial and error, we decided to monitor the time-varying GCR flux by measuring the counting rate in the 6.13 MeV oxygen gamma-ray line. This oxygen line is a good indicator of the GCR flux (Lawrence et al. 2004) because it is produced by inelastic scattering of fast neutrons, which are the direct product of the interaction of the GCR protons with the moon. In addition, since oxygen abundances should be mostly constant over the moon (Haskin & Warren 1991), the time variation of the 6.13 MeV oxygen line should be dominated by GCR flux variations.

Since there are oxidants in the remaining propulsion fuel in the spacecraft after maneuvering into orbit around the moon, the 6.13 MeV oxygen gamma-rays come from the remaining propulsion fuel in addition to the lunar surface. During the period of the GRS data acquisition, less than 20 kg of fuel was consumed and the vast majority of consumed fuel was used in spacecraft maneuvers. Below, we carry out the comparative analysis on variation of the 6.13 MeV oxygen line before and after spacecraft maneuvers. The variation of the 6.13 MeV oxygen line before and after the orbit maintenance is shown in Fig. 10. The variation of the 6.13 MeV oxygen line is shown in Fig. 11 before and after the spacecraft's attitude trim. From Figs. 10 and 11, we can find that there are little variations of the 6.13 MeV oxygen line in two spacecraft maneuvers. Therefore, we decided to monitor the time-varying GCR flux by measuring the counting rate in the 6.13 MeV oxygen line.

In order to apply the GCR corrections, we have smoothed the 6.13 MeV line's counting rate over 127 minutes (one orbit of GRS spectral data) and normalized them to the counting rate seen at the beginning of the orbital mission on 2007 December 29. The processes of the GCR corrections include, 1) the 6.13 MeV line's counting rate during the first orbit of the valid spectra of the GRS orbital mission is smoothed and selected as the reference data; 2) the 6.13 MeV line's counting rates during the other orbits of the valid spectra of the GRS orbital mission are smoothed and defined as the uncorrected data; 3) the relation between uncorrected data and the reference data is defined as a linear relation; 4) the uncorrected data are corrected using the above linear function built in step 3).

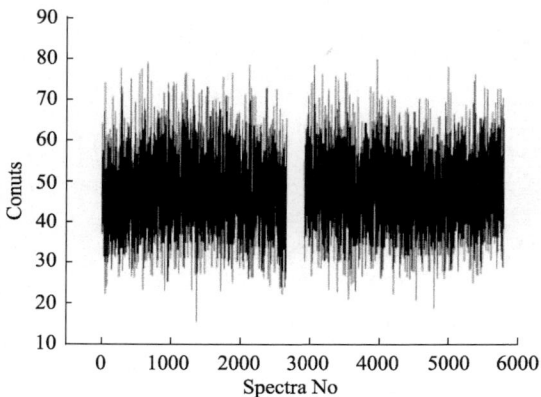

Fig.10   Variation of the 6.13 MeV oxygen line before and after the orbit maintenance

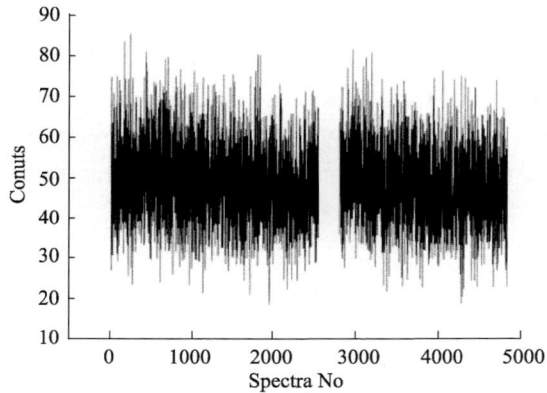

Fig.11   Variation of the 6.13 MeV oxygen line before and after spacecraft attitude trim

Fig. 12 (Plate XI) shows the uncorrected and corrected counts of the 556th~705th orbit of GRS data. From Fig. 12, the corrected data show better correction results.

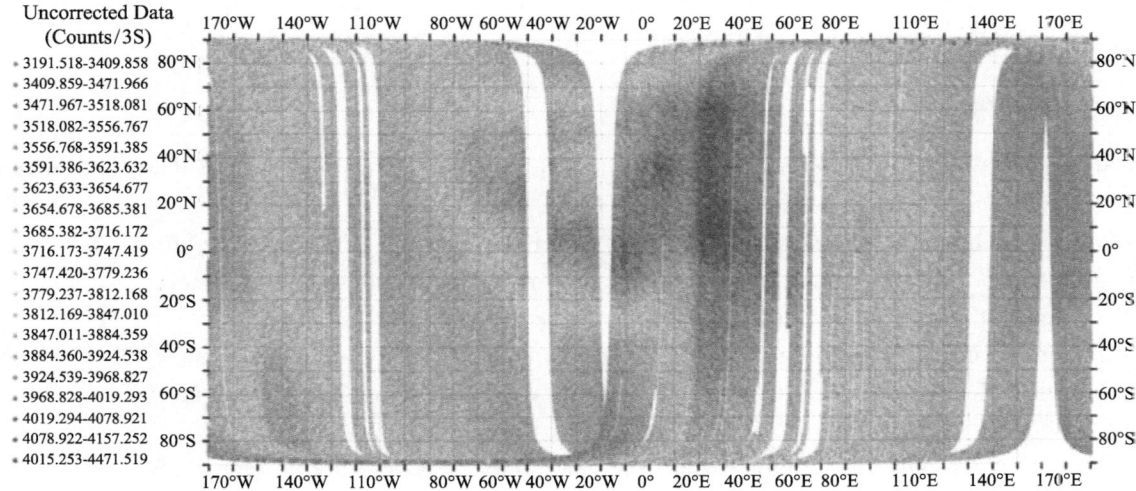

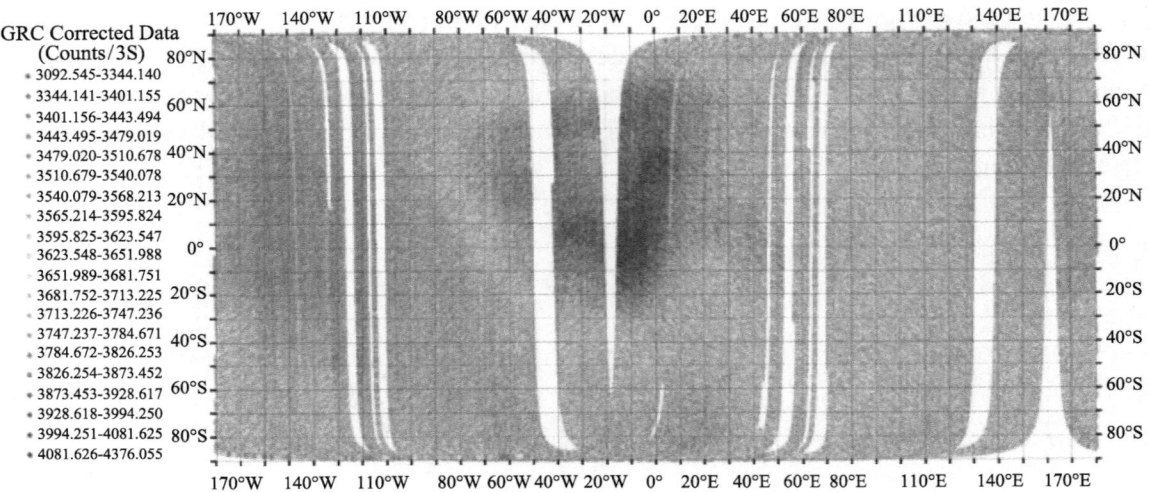

Fig.12 *Top panel*: the uncorrected count distribution; *bottom panel*: the GCR corrected count distribution

4 Data uncertainties

Besides the uncertainties due to Poisson statistics, non-Poisson uncertainties are responsible for the measured spectral uncertainties. All of the time series corrections described in this paper are carried out in order to reduce non-Poisson uncertainties in the CE-1 GRS data. The ultimate goal is to reduce the uncertainties to the level of Poisson uncertainties. One way to quantify uncertainties (Lawrence et al. 2004) is to map the data onto equal-area pixels and measure the standard deviation for each pixel using the following equation

$$\sigma_{\text{measured}} = \sqrt{\frac{\sum_{i=0}^{N_{\text{spectra}}} (S_i - S)^2}{N_{\text{spectra}} - 1}} \quad (5)$$

where N_{spectra} is the number of accumulated spectra in each pixel, S is the mean counts measured per three seconds in each pixel, and S_i is the measured counts in a given 3s spectrum i. The fractional standard deviation is then $\sigma_{\text{measured}}/S$.

By examining a plot of measured uncertainty of uncorrected data versus the Poisson uncertainty, we can see what impact the non-Poisson uncertainties have on the raw data observed by CE-1 GRS. From a plot of measured uncertainty of the time series corrected data versus the Poisson uncertainty, we can also see to what extent the non-Poisson uncertainties in raw data observed by CE-1 GRS are reduced by the time series data correction (Lawrence et al. 2004).

5 Conclusions

In this paper, detailed discussions are given on the time series data correction procedures and methods for the CE-1 GRS.

(1) We discuss how to acquire GRS raw packets from raw bit-stream data transmitted from CE-1 by channel processing. In channel processing, the methods of frame synchronization, descrambling, Reed-Solomon decoding and GRS raw packet acquisition are described.

(2) The methods of the optimal GRS raw packet selection are discussed because GRS raw packets transmitted from CE-1 are simultaneously acquired by two data acquisition stations (Miyun station and Kunming Station).

(3) The methods of energy calibration, gain and dead time corrections are presented based on the results of

the lab calibration and in-orbit calibration.

(4) We discuss how to acquire lunar positions of the measured spectra and observation angles of the GRD. They are acquired by building the observation equation using information such as spacecraft ephemeris and attitude data, GRS installation parameters on the spacecraft, geometric parameters about the moon, etc.

(5) The method for correcting variations in the lunar gamma-ray flux resulting from variations of the spacecraft's orbit altitude is described.

(6) We discuss the method for correcting the time variability of the GCR flux. After much trial and error, we selected the 6.13 MeV oxygen line to monitor the time-varying GCR flux, and better results of GCR corrections were obtained.

Acknowledgements The authors would like to thank all the CE-1 GRS team members for their outstanding work. The authors are very thankful for the constructive comments and suggestions from the two anonymous reviewers. This work is supported by the National High Technology Research and Development Program of China (Grant Nos. 2008AA12A212 and 2010AA122202) and the National Natural Science Foundation of China (Grant Nos. 41040031 and 40904024).

References

[1] Chang, J., Ma, T., Zhang, N., et al. 2009, in Proc. Int. Workshop Advances in Cosmic Ray Science, J. Phys. Soc. Jpn., 78 (Suppl. A), 26
[2] Haskin, L. A., & Warren, P. 1991, Lunar chemistry, in Lunar Sourcebook, eds. Heiken, G. H., et al., (New York: Cambridge Univ. Press), 357
[3] Lawrence, D. J., Feldman, W. C., Barraclough, B. L., et al. 1998, Science, 281, 1484
[4] Lawrence, D. J., Maurice, S., & Feldman, W. C. 2004, Journal of Geophysical Research (Planets), 109, 7
[5] Metzger, A. E. 1993, Composition of the Moon as determined from orbit by gamma-ray spectroscopy (New York: Cambridge Univ. Press), 341
[6] Metzger, A. E., Haines, E. L., Etchegaray-Ramirez, M. I.,& Hawke, B. R. 1979, in Lunar and Planetary Science Conference Proceedings, ed., N. W. Hinners, 10, 1701
[7] Metzger, A. E., Haines, E. L., Parker, R. E., & Radocinski, R. G. 1977, in Lunar and Planetary Science Conference Proceedings, ed. R. B. Merril, 8, 949
[8] Metzger, A. E., Trombka, J. I., Reedy, R. C., & Arnold, J. R. 1974, in Lunar and Planetary Science Conference Proceedings, 5, 1067
[9] Prettyman, T. H., Hagerty, J. J., Elphic, R. C., et al. 2006, Journal of Geophysical Research (Planets), 111, 12007

Automatically Smoothing the Spectroscopic Data by Cubic B-Spline Basis Functions

Zhu Menghua[1] Liu Lianggang[1] Zheng Mei[2] Qi Dongxu[1] Zheng Caimu[1]

(1. Space Exploration Laboratory, Macao University of Science and Technology, Macao, China;
2. Unit 61541 of the PLA, Beijing, China)

Abstract In the present paper, a new criterion is derived to obtain the optimum fitting curve while using Cubic B-spline basis functions to remove the statistical noise in the spectroscopic data. In this criterion, firstly, smoothed fitting curves using Cubic B-spline basis function are selected with the increasing knot number. Then, the best fitting curves are selected according to the value of the minimum residual sum of squares (RSS) of two adjacent fitting curves. In the case of more than one best fitting curves, the authors use Reinsch's first condition to find a better one. The minimum residual sum of squares(RSS) of fitting curve with noisy data is not recommended as the criterion to determine the best fitting curve, because this value decreases to zero as the number of selected channels increases and the minimum value gives no smoothing effect. Compared with Reinsch's method, the derived criterion is simple and enables the smoothing conditions to be determined automatically without any initial input parameter. With the derived criterion, the satisfactory result was obtained for the experimental spectroscopic data to remove the statistical noise using Cubic B-spline basis functions.

Key words Spectroscopic data Automatic smoothing Spline function

Least-square fitting method with B-spline functions is a helpful tool in reducing the statistical noise in the spectroscopic data, such as gamma-ray spectrum. It can be used to obtain a smoother fitting curve compared with the methods such as convolution method[1-8] and the Fourier transformation[9-13]. In addition, this method can eliminate the noise efficiently, especially in the case that the noise and noise-free curve have approximate frequency. Another praiseworthy characteristic of this method is that, it can give an explicit expression of the result curve which is convenient for the calculation of the derivative and area[14].

However, the optimum fitting curve can not be obtained using this method on the noise spectroscopic data with ignorance of the number of initial knots and their positions. The different determination can influence the result of the fitting curve apparently and until now, this problem has not been well addressed in the literatures. The general method described by Reinsch[15,16] can be used to obtain the optimum fitting curve only when the number of the initial knots is known. In this method, the smoothing curve can be determined automatically in the interval $[a, b]$ with m channels by minimizing the value of among all fitting curves of the noisy data such that.

$$\sum_{x=1}^{m} \{[S^*(x) - F(x)]W_x\}^2 \leqslant C \tag{1}$$

Herein, W_x is the weight assigned to channel x, and C is a smoothing parameter which controls the extent of smoothing. When $C=0$, no smoothing is carried out and the noisy data are interpolated, and as C increases, the degree of smoothing increases. It should be noted that this condition has little meaning in the noise elimination especially in the reduction of statistical noise, since the degree of smoothing required in the statistical noise elimination of spectra is far greater than the normal requirement, which improve the difficulty of determining the value of C and W_x before calculating[17]. Sophisticated determination of the number of the initial knots and their positions in the processing always gives uncompleted or distorted elimination and cannot be fit for different spectra.

To avoid this problem, we present a simple method to get the optimum smoothing curve automatically while using least square method with Cubic B-spline basis functions. The details are described as below.

Firstly, in the range of interest channels, $a \leqslant x \leqslant b$ where x is the channel number, five channels x_0, x_1, x_2, x_3 and x_4 with equal interval that satisfy $a < x_0 < x_1 < x_2 < x_3 < x_4 \leqslant b$ are selected as knots to fit the noisy data. With this set of channels as center knots, the general cubic B-splines fitting curve has the unique representation of the form

$$S(x) = \sum_{j=0}^{4} c_j \phi_j \tag{2}$$

where ϕ_j is the fundamental spline functions.

The method of least squares assumes that the best-fit $S^*(x)$ curve with selected channels has the minimal sum of the least square error from the measured spectra $F(x)$.

$$\|\delta_i\|^2 = \sum_{x=1}^{m}[S_i^*(x) - F(x)]^2 = \min[S_i(x) - F(x)]^2 \tag{3}$$

$$S_i(x) = C_{i,0}\phi_0(x) + C_{i,1}\phi_1(x) + \cdots + C_{i,n}\phi_n(x), \quad n \leqslant m \tag{4}$$

where m is the number of the channels of the spectra in the interval $[a, b]$, and n is the number of the selected channels, In addition, the initial values of C_i are set to be the average count value of the adjacent channel of each selected knot.

Then, new channels are selected in the middle of each interval and the fitting as above is repeated until the interval is equal to 1. Least square method gives the best-fit curve minimal sum of the least square error from the noisy data at each time and with the increase of channels selected, all the fitting curves form a finite set. Anyone from this set is the best fitting curve corresponding to selected channels.

However, as described by Naoki Saitou[18], we don't recommend the minimum residual sum of squares (RSS) of fitting curve with noisy data as the criterion to determine the best fitting curve, because the value of RSS decreases to zero as the number of selected channels increases and minimum value gives interpolated result without smoothing effect. What we do is calculating the RSS of two adjacent fitting curves $S_i^*(x)$ and $S_{i+1}^*(x)$

$$\|\epsilon_i\|^2 = \sum_{x=1}^{m}[S_{i+1}^*(x) - S_i^*(x)]^2 \tag{5}$$

where i is the fitting number with selected channels (2^i+1) as knots. Since the noise in the Fourier domain is in frequency order and the fitting curves in the fitting set are also in order corresponding to the number of selected channels, the value of $\|\epsilon_i\|^2$ can be considered as the bias error of $S_i^*(x)$ relative to $S_{i+1}^*(x)$. The lower this value, the closer these two adjacent curves will be. This value can also indicate that the curve $S_{i+1}^*(x)$ contains more complex components than $S_i^*(x)$, which can be considered as more noise in $S_{i+1}^*(x)$. Because all the curves $S_i^*(x)$ are the best-fit curves, minimum RSS means that these two adjacent curves have little difference and little noise $S_i^*(x)$ contained that can estimate noisy data approximately. However, sometimes, with the fitting curve set, minimum RSS is not unique, corresponding to more than one fitting curves. In this case, according to Reinsch's first condition that the smoothest spline curve has minimum value of $\int_b^a [S_i^{*\prime\prime}(x)]^2 \mathrm{d}x$ in the interval $[a, b]$, the result curve can be obtained.

The experimental data as shown in Fig.1 (a) is measured from ^{60}Co source with 8193 channels by Na I detector for more than 2 hours. According to the method described above, the values of $\|\epsilon_i\|^2$ are calculated as shown in Fig.1 (b). Because the selected channel number is 5 in the first fitting compared with 8193 in the last fitting, the range of $\|\epsilon_i\|^2$ is from $\|\epsilon_2\|^2$ to $\|\epsilon_{12}\|^2$. What can be seen from this sub-figure is that two minimum values about $\|\epsilon\|^2$ satisfy the condition where $i=6$ and $i=9$ in the fitting set. The curves $S_6^*(x)$ and $S_9^*(x)$ corresponding to two minimum values are shown in Fig.1 (d) and Fig.1 (e), respectively. The values of $\int_1^{8193}[(S_6^*)''(x)]^2 dx$ and $\int_1^{8193}[(S_9^*)''(x)]^2 dx$ are shown in Fig.1 (c) which can indicate the smoother fitting curve clearly.

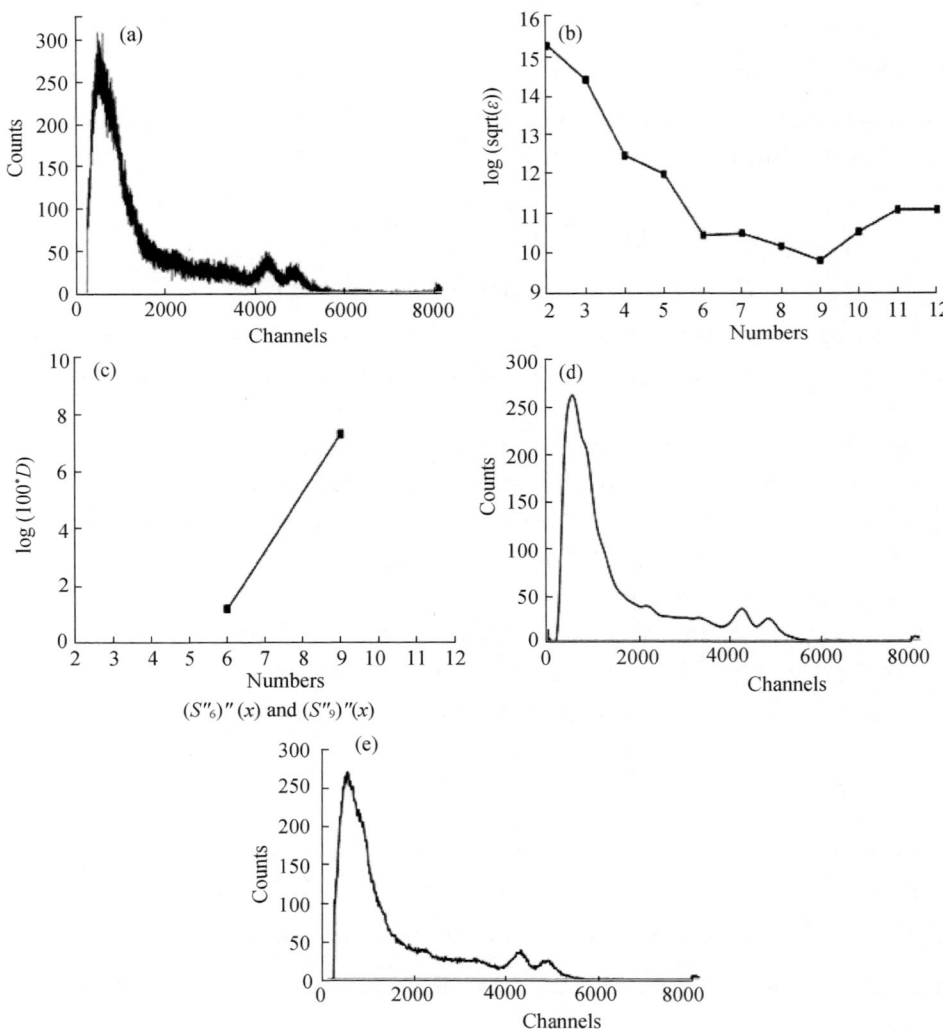

Fig.1 Gamma-ray spectrum of ^{60}Co (Na I) and its fitting result using B-spline basis functions

(a): Experimental spectra of ^{60}Co source; (b): Logarithm of $\|\epsilon_i\|^2$ with i from 2 to 12;

(c): Square integral values of $(S_6^*)''(x)$ and $(S_9^*)''(x)$; (d): Fitting curve where $i=6$; (e): Fitting curve where $i=9$

As one can see from the illustration as above, the optimum smoothing fitting curve can be determined automatically without any input value while using least square method with Cubic B-spline basis functions to remove the statistical noise in gamma-ray spectra.

We gratefully acknowledge the assistance of Herman Zhu for technical support. This work also would not have been possible without financial support of the Science and Technology of Development Fund of Macao (China) and NSF-China project.

References

[1] Edwards T H, Willson P D. Applied Spectroscopy, 1974, 28: 541
[2] Gorry P A. Analytical Chemistry, 1990, 62: 570
[3] Evans S, Hiorns A G. Surface and Interface Analysis, 1986, 8: 71
[4] Black W W. Nuclear Instruments and Methods, 1969, 71: 317
[5] Mariscotti M A. Nuclear Instruments and Methods, 1967, 50: 309
[6] Blaauw M. Nuclear Instruments and Methods, 1993, A336: 273
[7] Routti J T, Prussin S G. Nuclear Instruments and Methods, 1969, 76:109
[8] Koskelo M J, Aarnio P A, Routti J T. Nuclear Instruments and Methods, 1981, 190: 89
[9] Blinowska K J, Wessner E F. Nuclear Instruments and Methods, 1974, 118: 597
[10] Inouye T, Harper T, Rasmussen N C. Nuclear Instruments and Methods, 1969, 67: 125
[11] Kekre H B, Madan V K. Nuclear Instruments and Methods, 1986, A245: 542
[12] Kekre H B, Madan V K, Bairi B R. Nuclear Instruments and Methods, 1989, A279: 596
[13] Hampton C V, Lian B, McHarris W C. Nuclear Instruments and Methods, 1994, A353:280
[14] ZHU Menghua, LIU Lianggang, QI Dongxu, et al. Chinese Ph C, 2009, 33: 24
[15] Reinsch C H. Numerische Mathematik, 1967, 10: 177
[16] Reinsch C H. Numerische Mathematik, 1971, 16: 451
[17] Proctor A, Fay M J, Hoffmann D P, et al. Applied Spectroscopy, 1990, 44: 1052
[18] Saitou N, Iida A, Gohshi Y. Spectrochimica Acta, 1983, 38: 1277

第六部分　微波探测仪

嫦娥 1 号卫星微波探月技术机理和应用研究

姜景山[1]　王振占[1]　李 芸[1,2]

(1. 中国科学院空间科学与应用研究中心,北京 100190；2. 中国科学院研究生院,北京 100190)

摘　要　微波探测仪是嫦娥1号卫星有效载荷之一,主要用于测量不同深度的月壤微波辐射亮温,进而反演月壤厚度的信息并对月球的 ^{3}He 资源量和分布进行评估。这是国际上第一次利用被动微波遥感探测器在月球轨道直接测量月表亮温信息。因此,对月壤辐射传输模型的研究是极其必要的。文章分析了月球微波探测的机理和存在的问题,并给出了初步解决途径。

关键词　嫦娥1号微波探测仪　遥感　月壤　微波辐射传输

1　前言

微波探测仪是嫦娥1号卫星有效载荷之一,主要用于探测全月表及不同深度的月壤微波辐射亮温,反演月壤厚度的信息并进而对月球的 ^{3}He 资源量和分布进行评估。微波探测仪的科学目标：利用微波信号对月球表面物质的穿透传播特性,从表征月球物质微波辐射的亮温数据中,获取月球月壤的厚度信息；获得月球表面的微波辐射特性参数,为研究月球的微波背景构建"微波月亮",获得月球黑夜及月球两极的微波遥感信息[1]。

微波探测仪是国际上第一次利用被动微波遥感探测器在月球轨道直接测量亮温来探测月表信息。因而,对月壤辐射传输模型的研究是实现这一目标的重要基础。从20世纪70年代开始,就开始了对地球表面辐射传输的研究[2-4],但由于水的存在,微波穿透深度很小,所以在地球上,微波传输通常不包含土壤厚度和下层岩石信息。月球上基本没有水的存在,3 GHz 频率可以穿透深度达 5 m 以上[5],所以从嫦娥1号微波探测器接收的亮温数据能够反映月壤热传导和结构特性。

研究月壤厚度对未来月球探测、登月与月球资源开发均具有十分重要的意义。首先,月壤是对月遥感探测的直接目标,它包含了大量月球化学信息,包括其化学和矿物组成、月壤形成和演化等；其次,由于在月壤形成过程中,伴随太阳风的作用,^{3}He 不断注入月壤层中,从而积聚了大量的核资源,确定月壤的厚度以及分布是对 ^{3}He 储量进行估算的必要前提[6-8]。

2　国内外研究现状及分析

研究月壤厚度的方法可归纳为直接和间接两种[9]。直接测量方法包括 Apollo 和 Luna 探测计划中实施的月表钻探实验。这种直接钻探的方法仅能测量出降落区较薄的月壤层厚度,具有很大的局限性。因而,在月壤厚度的直接测量中,大部分结果来源于月震实验和多频电磁探测。直接的方法还可以利用月球卫星图像以及遥感数据进行推算月壤厚度。随着遥感技术的不断发展,Shkuratov 和 Bondarenko 提出了利用地基雷达遥感数据对月壤厚度进行反演的方法[10]。间接方法主要是基于撞击实验,通过对撞击坑形态和直径分布频率的分析,推出月壤的厚度。

从月球轨道用被动微波遥感的方法获得全月球月壤厚度数据，至今还没有先例。迄今为止只有欧洲的月球计划中有微波遥感器的规划。将微波遥感器用于月球卫星的对月观测是中国的科学家首次提出并在嫦娥1号上实施。欧洲计划的"欧洲大学生探月计划"中将采用这种方法[11]。

目前，国内许多单位对利用微波探测仪亮度温度(以下简称亮温)科学探测数据进行月壤厚度反演表现出浓厚的兴趣，包括：中国科学院空间中心、复旦大学、国家天文台、地球化学研究所、遥感应用研究所、吉林大学等。复旦大学金亚秋教授和美国的A. K. Fung教授应中科院空间中心的委托分别建立了月壤微波传输模型：金亚秋应用WKB方法开发了1个3层模型结合光学观测数据来反演月壤厚度[12,13]；Fung用一种多层模型产生8个点位5个频率(2 GHz，3 GHz，7.8 GHz，19.5 GHz，37 GHz)的深度从1~9 m的近似辐射传输，并包含了月壤散射贡献。

中科院空间科学与应用研究中心从月球微波探测仪立项之初就开始了对月壤微波辐射特性和月壤介电特性的研究[14]，并进行了大量月壤特性模拟和测试工作[15,16]。另外这个研究小组还开展了详细的月壤微波辐射特性验证试验，分别于2007年2月在北京唐家岭的小牛房进行了对于不同湿度的沙子和泡沫的微波辐射特性测量试验，试验结果验证了理论模型的正确性和仪器性能的可靠性，同时结合SMOS计划，在塔克拉玛干沙漠进行了地面试验。此外还从微波辐射原理出发，利用非相干法推导了针对月壤的介电特性、温度特性等进行了月壤温度剖面、月表多波段辐射亮温的理论模拟[17,18]，其结果已经用于微波探测仪数据的预处理和初步应用上。

3 月表微波探测面临的问题分析

3.1 月表微波探测目前要解决的问题

在月壤厚度的研究方法中，利用月震数据、多频电磁探测数据、撞击坑的形态和分布规律等推算月壤厚度，存在很多局限性，数据的获取都比较困难，难以实现，误差也比较大，并且只能应用于特定的局部区域。月壤厚度的数据目前还仅限于20世纪Apollo和Luna计划获得的9个采样点，虽然取自不同的地质类型和地貌类型，但仅限于月球正面，而且采样点过稀，不具有代表性。

利用微波探测仪从月球轨道遥感月壤厚度的核心问题是，建立月壤厚度与探测仪接收的亮温之间的函数关系，而这个关系建立的根据就是微波在介质中的辐射传输特性的模拟。为了模拟这个过程，需要建立的有关参数模型包括：

(1)月表层温度剖面的模型；
(2)月壤介电特性随着深度和地理位置变化的变化规律；
(3)月壤的比热、密度等物理参数的变化规律；
(4)月表地形等粗糙度参数化以及对微波辐射的影响；
(5)月表层土壤和岩石的结构变化规律。

目前，所有关于这些参数的资料和数据有限，尤其没有针对微波辐射测量的资料，所以需要利用在地球上的微波辐射传输的模拟和验证结果，结合这些参数的已知数据分析，把嫦娥1号微波探测仪数据和其他有效载荷数据进行有机结合，来分析表面的微波辐射规律，进而探索月壤特性的变化规律。

3.2 解决途径分析

3.2.1 全月月壤温度剖面和亮温的模拟

月表亮温是月表温度剖面、介电特性剖面、表面粗糙度等参数综合作用的结果，对这些参数模拟结果的准确程度决定了最终结果的准确度。在以往的各种理论研究中，不管是中国的还是其他国家的研究，都只是使用了这些参数的简单模型，例如，把温度只是简单模拟为两个温度，一个代表表层温度，一个代表所有下层的温度。显然这种简化对于计算结果影响很大，因为表层的温度波动对于3 GHz这样穿透深度很大的通道来讲，亮温的不确定性很大，也就是对于月壤反演的影响很大。在充分考虑了对于月表微波亮温影响的各种因素：通过对微波辐射传输理论的研究，建立月壤微波辐射传输模型，并利用热传导方程解得的月壤温度剖面，在考虑表面地形和微波探测仪天线方向图特性的情况下，对全月亮温进行模拟。最关键的一点是可以利用微波探测仪的数据进行比对和验证。

利用月球表面的亮温数据，可以反演月球参数的反演，如月壤厚度等。反演算法建立的基础是微波辐射传输方程。微波在介质中传播除了受到介质的衰减以外，还与介质发生作用，介质本身还辐射能量。对于层状介质来讲，如果存在介质突变面，则在界面除了反射入射到的能量外，还有部分能量透射到另外的介质中去。如果介质中存在与入射电磁波波长相当的粒子，那么电磁波在其中传播的时候还要产生散射。总的来讲，有两种方法可以用来模拟介质的微波辐射[18]：一种是相干的方法(coherent approach)，这种方法基于Maxwell方程和波动逸散理论，它同时考虑反射振幅和相位的影响，相干的方法必须用Maxwell方程的一个解来计算电磁场矢量，进而获得辐射强度。这种方法认为介质是水平均匀的，介电常数只是厚度的函数，在该层内忽略散射的影响，其介电常数在该层内认为不变。但是这种方法由于考虑相位的影响，所以分层厚度就会对计算结果产生依赖性。另外一种是非相干的方法(incoherent approach)，这种方法基于辐射传输理论，只考虑振幅，忽略相位的影响。这种方法可以直接计算辐射强度，但是其成立的前提是假设介质中存在大量的与波长尺度量级相当的散射体。这些散射体的随机分布使得介质中两点之间的波传播产生随机分布的相位因子。这样传播过程就变成一个不均匀的过程，可以用波的功率密度来描述。为了求解月壤的微波辐射传输方程，需要分析月壤参数：月壤密度、月壤的介电常数、基岩的介电常数、月壤以及基岩的温度分布、比热和热导率等对于微波辐射特性的影响。需要建立如下变化规律。

(1)月壤体积密度(g/cm^3)与月壤深度的相关关系。

(2)确定月球表面任何一层的介电常数，进而确定该层的吸收系数，那么整个月壤的微波辐射亮温就可以通过积分来确定。

(3)确定基岩的介电常数。虽然下伏基岩的介电常数变化一般不大，但是由于基岩温度和反射率的变化，会对最终亮温的计算产生很大的影响，尤其在月壤厚度较小的时候。

(4)确定月壤的比热随温度变化而变化。

(5)确定月壤的热导率：热导率不但是温度的函数，也随着深度的变化而变化。

如果上述参数的变化规律确定，那么我们就可以首先利用光学和其他月球探测器数据，结合热传导方程，分析月面表层温度的变化。得到月壤和基岩的温度分布。然后，利用月壤剖面温度，以及月壤的介电常数和基岩的介电常数，根据均匀和非均匀的辐射传输模型对月表亮温进行模拟，并对模拟结果进行分析和比较。最后，分析表面地形对于亮温的影响，结合天线方向图进行分析，并利用模拟结果建立亮温与月壤厚度之间的关系。

3.2.2 微波探测仪数据预处理

嫦娥1号卫星所携带的微波探测器是国际上首次利用微波探测器来测量全月球表面(包括月球背面)的微波亮温。对其数据的有效处理，可以获得全月球白天和黑夜的微波亮温数据。星载微波辐射计预处理是指从原始数据到亮温数据产生的全过程。主要内容包括原始数据的地理定位，辐射定标，天线亮温的冷空间溢出和交叉极化订正，像元地面分辨率均匀化处理等内容。由于月球轨道的特殊性，我们重点分析冷空定标的数据有效性以及旁瓣等的影响。

在发射前，低端定标选择没有任何干扰的冷空背景(2.7 K)，但是卫星发射后，由于轨道和太阳等星体的关系，导致冷空信号的干扰。所以需要根据数据特点和天线方向图分析，确定太阳相对冷空的位置，判断太阳在冷空天线的后面、在主波束范围内、在旁瓣范围内等情况。利用天线方向图计算旁瓣的冷空贡献。

其次，由于嫦娥1号微波探测器在月球轨道运行过程中，其对月观测的天线旁瓣会照射到冷空间和主波束以外，因此为了准确反演出主波束的月表亮温，需要对这些贡献进行评估和计算。另外，由于安装精度、天线指向以及卫星姿态等的影响，按照设计极化方式接收的目标辐射可能包含其他极化方式的辐射，这样就会产生交叉极化的影响，因此需要交叉极化修正。亮温算法就是建立天线温度与月表亮温的关系。最后，还要利用表面的特征点对测量数据进行外定标。目前，包含月壤厚度月表的特征点只有几个登月点的数据，对这些数据的分析，结合微波探测仪的实测数据，会有助于提高定标精度。

3.2.3 月壤信息反演技术研究

根据理论模拟的结果，利用不同频率亮温和月壤厚度的关系建立月壤厚度反演的算法，并利用实测亮温对全月月壤厚度进行反演计算。

此外，通过厚度信息与亮温之间的关系，可以分析极地和赤道地区厚度的差异，以及两极地区介电特性与水存在的可能性。还可以进行月表未知信息的探索。

4 月表微波探测的初步研究结果

目前，已经从微波辐射传输的原理出发，结合国际上最新的月表土壤和岩石的科学信息，充分利用嫦娥1号卫星微波探测器和其他有效载荷的数据，进行月表亮温的理论模拟和应用研究，推导出月壤厚度反演的多层介质辐射传输模型和表面温度剖面模型。通过与微波探测仪数据的比对，验证模型，改进算法，从而分析月球表面的亮温、厚度与地理位置等的关系，取得了初步成果。

首先，利用热传导方程，研究了全月表层温度剖面的变化规律。图1为月球赤道一昼夜表面温度的变化，可见表面温度呈现周期性变化，图1的时间起点是月球该地区的日落时刻。夜晚和白天的时间各是地球上的近似15天。图1的前面一半是黑夜，后面一半是白天。日落后温度迅速下降，在日出前温度最低，正午温度最高。白天的温度起伏达到250 K左右，而夜晚温度也有30~40 K的变化。

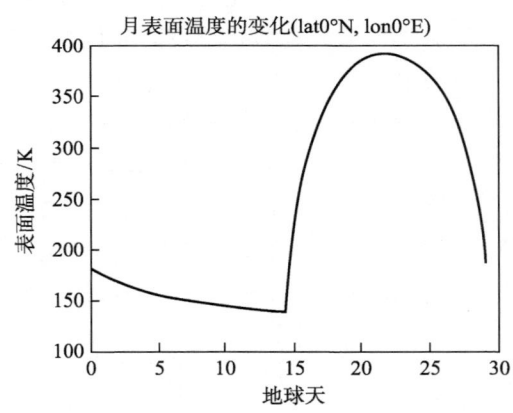

图1 月球赤道一昼夜表面温度变化

图2为月球赤道一昼夜两个时刻的温度剖面比较。可以看出赤道地区表面20 cm以内的月壤温度受太阳的影响而剧烈波动，最大差异达到200 K以上，因此表层的温度梯度在亮温的昼夜波动中起着决定作用。

极地与赤道最大的不同就是表面温度的周期长，大约为赤道地区的12倍。而最高和最低温度仅为157.93 K和64.97 K。在其他地区，月球表面温度和剖面温度的变化趋势同低纬度地区相似。

为了分析表层温度的昼夜变化对于CELMS接收亮温的影响，根据模拟的温度剖面，利用多层辐射传输模型[5]进行亮温模拟，获得全月亮温的昼夜变化规律，不同频率、厚度与亮温的关系。图3给出了月球赤道一昼夜的亮温变化，月壤厚度是1 m，图4给出的是月球赤道在月壤厚度为3 m时一昼夜的亮温变化。

整个低纬度地区结果非常相似。4个通道的亮温在一昼夜的不同时刻大小出现交替，3 GHz 的昼夜起伏很小，因为其主要反映深层的温度和介电特性，而 37 GHz 起伏最大，说明其受到表层温度的影响剧烈。纬度不同，太阳照射引起的表层温度分布不同，因此亮温也随着纬度的升高而减小。

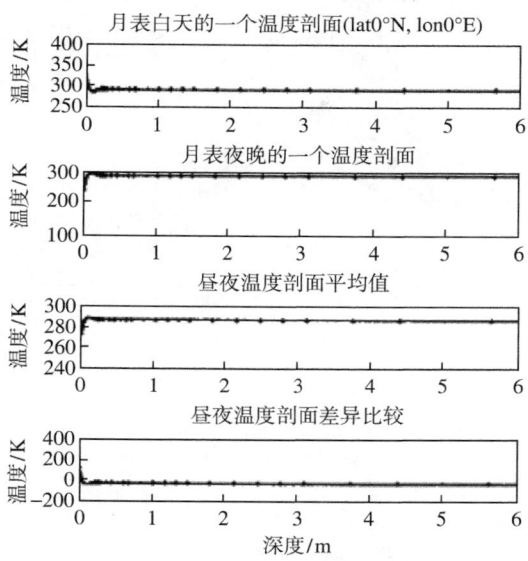

图 2　月球赤道一昼夜的剖面温度及平均值比较

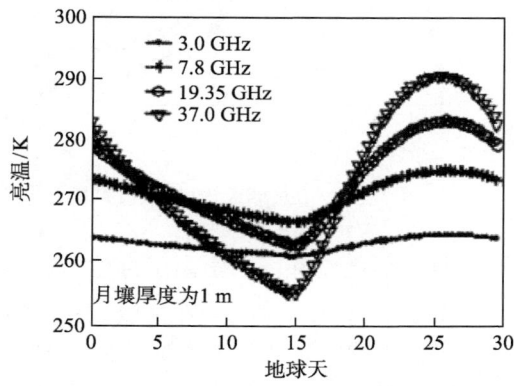

图 3　月球赤道一昼夜亮温变化(月壤厚度 1m)

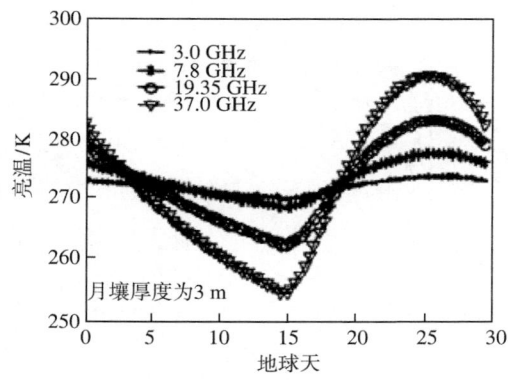

图 4　月球赤道一昼夜亮温变化(月壤厚度 3m)

图 5 和图 6 是北极地区不同月壤厚度下的亮温变化情况，亮温变化与低纬度地区的变化规律差异明显，

它们分别代表了月球一昼夜北极月壤厚度为 0.5 m 和 5 m 的亮温变化,这是因为每个频率的穿透深度不同,而北极表面温度变化相对较小,底面温度又非常低的缘故。

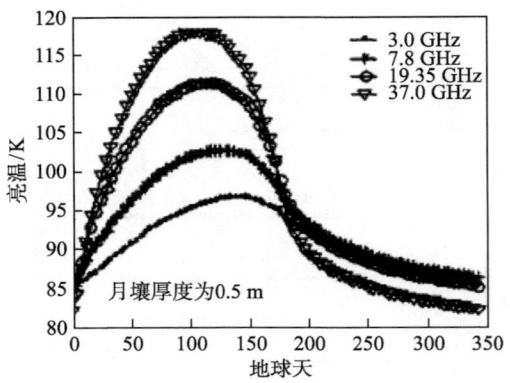

图 5　月球北极一昼夜亮温变化(月壤厚度 0.5 m)

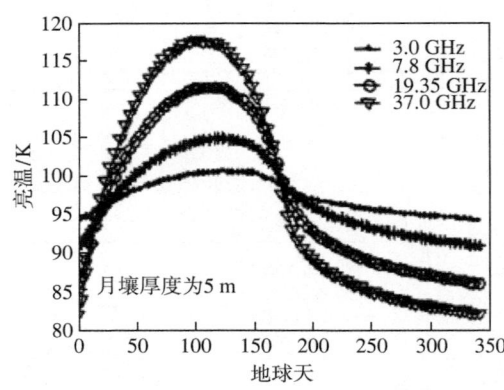

图 6　月球北极亮温一个月球天内的变化(月壤厚度 5 m)

为了分析月壤厚度与亮温变化的规律,图 7 和图 8 给出了月球赤道不同频率亮温随月壤厚度的变化。图 7 是月球赤道正午时刻的亮温变化趋势,月壤厚度从 0.2~6 m,4 个频率亮温各不相同,并且随频率的增加而增加,随深度的增加而增加。而图 8 是月球赤道黎明时刻的亮温变化趋势,4 个频率相互交叉,且在 50 cm 以下非常接近。在黎明时刻,3.0 GHz 的亮温在月壤厚度超过 2.5 m 后达到 4 个频率的最大值。

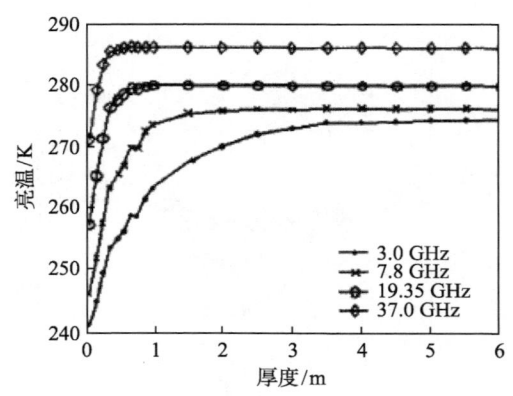

图 7　月球赤道在月球正午时刻亮温随厚度的变化

通过理论模拟结果发现,当考虑表面温度梯度和月岩的物理特性的变化,利用多层辐射传输模型[5]

进行亮温模拟的亮温与厚度的关系时,3 GHz 穿透深度达 5~6 m,7.8 GHz 达到 2 m 左右,19.35 GHz 约为 1 m,而 37 GHz 的穿透深度不超过 50 cm。

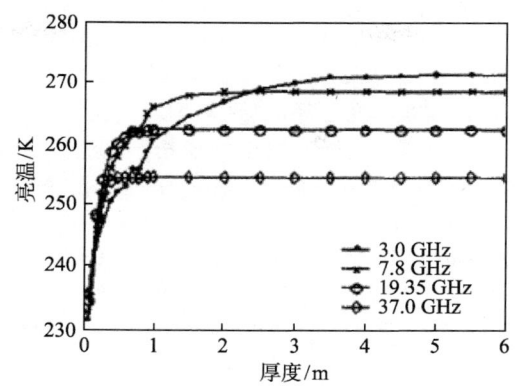

图 8　月球赤道在月球黎明时刻亮温随厚度的变化

5　结语

随着遥感技术的不断发展,高精度的定量遥感已经在很多领域内得到广泛应用,其中,微波遥感由于其能够穿透次表层,不受太阳光照条件的影响等特点,在新一轮探月热潮中备受国际关注。利用微波辐射测量技术,获取月球表面月壤的厚度信息,从而得到月球表面年龄及其分布特征,结合返回的月壤分析数据,可以计算月球表面 ^{3}He 和其他气体的含量、资源分布及储量。根据月球表面的亮温分布,用微波辐射测量的方法进行月球含水的可能性探测具有吸引力,这也是当前月壤厚度研究的一种新方法。

文章分析了月球微波探测的机理和所面临的问题,并给出了初步解决途径。为了对于辐射传输模型进行验证,在微波探测仪发射之前,进行了详细的地面试验,目的是在地面上利用微波探测仪对已知介电特性的介质进行测量,进而验证测量亮温与目标厚度的关系。通过对泡沫和干沙的 3 次测量,获得了一个泡沫厚度增加的亮温变化结果,一个沙子厚度增加亮温变化结果以及沙子的厚度逐渐减小的亮温曲线。由于亮温变化是介质介电特性、温度特性等多个参数综合作用的结果,因此对于获得的亮温随着厚度变化的关系需要用微波辐射传输理论进行验证[11]。结果表明,理论和实际测量的一致性很好,天线温度相关性可达到 90% 以上。

参 考 文 献

[1] Sun Huixian, Dai Shuwu, Yang Jianfeng, et al. Scientific objectives and payloads of Chang-E-1 lunar satellite [J]. Journal of Earth System Science, 2005, 114(6): 789—794

[2] Tsang L, Njoku E, Kong J A. Microwave thermal emission from a stratified medium with nonuniform temperature distribution [J]. Journal of Applied Physics, 1975, 46(12): 5127—5133

[3] Njoku E, Kong J A. Theory for passive microwave remote sensing of near-surface soil moisture [J]. Journal of Geophysical Research, 1977, 82(20): 3108—3118

[4] BurkeW J, Schmugge T, Paris J F. Omparison of 2.8—21 cm microwave radiometer observations over soils with emission model calculations [J]. J. Geophysical research, 1979, 84(C1): 287—294

[5] Wang Zhenzhan, Li Yun, Jiang Jingshan, et al. Wave radiative transfer model applied to lunar soil remote sensing [J]. ICEF2008, Chongqing, China

[6] Freanch B M. Lunar Sourcebok: A User's Guide to the Moon [M]. Cambridge university Press, 1991

[7] 胥传东, 郭伟, 张晓辉, 等. "嫦娥 1 号"(CE-1) 微波探测仪地面定标及验证实验[J]. 第八届国际月球探测和利月大会, 2006

[8] 郑永春. 模拟月壤研制与月壤的微波辐射特性研究[D]. 北京: 中国科学院地球化学研究所, 2005

[9] 李雄耀, 王世杰, 陈丰, 等. 月壤厚度的研究方法与进展 [J]. 矿物学报, 2007, 27(1): 64—68
[10] Yurij G. Shkuratov, Nataliya V. Bondarenko. Regolith layer thickness mapping of the moon by radar and optical data [J]. Icarus, 2001, 149(2): 29—338
[11] Jiang Jingshan. Development of microwave remote sensing technology in China and microwave sounding the lunar soil from China lunar satellite CE-1 [J]. ICMMT 2008, An invited key note speech. 2008, 4: 21—24
[12] 金亚秋. 空间微波遥感数据验证理论与方法[M]. 北京: 科学出版社, 2005
[13] 法文哲, 金亚秋. 月球表面多通道辐射亮度温度的模拟与月壤厚度的反演[J]. 自然科学进展, 2006, 16(1), 86—94
[14] 蓝爱兰, 张升伟. 利用微波辐射计对月壤厚度进行研究[J]. 遥感技术与应用, 2004, 19(3): 154—158
[15] 李涤徽, 吴季, 姜景山, 等. 模拟月壤微波介电特性的初步研究[J]. 遥感技术与应用, 2005, 20(1): 141—147
[16] 李涤徽, 姜景山, 吴季, 等. 模拟月壤微波介电特性的实验研究与统计分析[J]. 科学通报, 2005, 50(10): 1040—1049
[17] Wang Zhenzhan, Li Yun, Jiang Jingshan, et al. Methods of remote sensing Lunar surface by Microwave sounder on CE-1[J]. 2008 International Conference on Microwave and Millimeter Wave Technology, Nanjing, China. 2008, 4: 21—24
[18] Ulaby, F T, Moore R K, Fung A K. Microwave Remote Sensing, vol. I: Active and Passive[M]. Addison-Wesley Publishing Company, 1981

利用微波辐射计对月壤厚度进行研究

蓝爱兰　张升伟

(中国科学院空间科学与应用研究中心，北京　100080)

摘　要　介绍了运用并矢 Green 函数和起伏逸散定理来计算平面分层媒质的辐射亮温，同时利用最小二乘法对多通道辐射计的模拟测量结果进行处理得到分层媒质厚度的方法。将该方法应用于月壤厚度的反演研究，在假定月壤如平面分层结构模型的情况下，得到了其厚度的反演结果，并对反演误差原因进行分析。

关键词　月壤厚度反演　分层媒质　并矢 Green 函数　微波辐射计

1　引言

从 17 世纪初伽利略发明望远镜算起，人类进行科学探测月球的历史已有 300~400 年。在这几百年的探月史中，出现了两次高峰，第一次高峰期是 20 世纪 50 年代末至 70 年代初的美苏登月竞赛；进入 90 年代后，人类迎来了第二个探月高潮，但此时的探月目的已由"征服月球"变为"利用月球"。

月球上不具备化学风化作用、流水和冰蚀等产生粒径均匀沉淀物的条件，但由于陨石的不断撞击,在月球表面上覆盖着一层由岩石碎屑、粉末、角砾、撞击熔融玻璃等物质组成的、结构松散的混合物，我们称之为月壤。月壤中富含 ^{3}He 等稀有气体，是月球的重要资源。而 ^{3}He 是一种可长期使用的、清洁、安全和高效的核聚变发电的燃料，与氘-氚(D-T)聚变反应释放中子不同，氘-氦(D-^{3}He)聚变反应释放出质子，因此其反应所需防护的设施、材料和环保条件较 D-T 反应简便且廉价。如果能弄清月壤在月球表面的精确厚度、^{3}He 在月表不同月壤区以及月壤不同层位的确切情况，对未来开发利用月壤中的 ^{3}He，为人类提供一种可长期使用的、清洁、安全和高效的核聚变发电的燃料，为解决人类目前所面临的能源危机提供了一种可能。但目前对于月壤中 ^{3}He 的含量及其分布还知之甚少，因此，对月壤厚度分布进行研究意义深远。

根据前人的研究结果[1]，我们可以近似认为月球表层是由月壤和月岩组成的分层结构。虽然有些资料认为月球表层是由月尘、月壤和月岩三部分组成，但由于月尘的化学成分与月壤大致相同，且厚度较小，我们将其视为月壤的一部分。下文中介绍了一种求解分层媒质亮温的方法，并采用最小二乘法对多通道辐射计的测量数据进行比较分析，反演出月壤的厚度。

2　测量原理

由于微波波长较长，具有一定穿透能力，可以获得月球次表层的信息；且微波辐射计能全天时工作，因此我们采用微波辐射计对月球进行探测。微波辐射计是一种被动微波遥感器，主要用于测量物质自身的热辐射。当上半空间为自由空间时，我们可以近似认为辐射计所测得的亮温就是被测物的亮温。

在微波波段，黑体(发射率 $e=1$)的辐射能量与物理温度之间的关系可由 Plank 定律的 Rayleigh-Jeans 近似式表示，而对于灰体($0 < e < 1$)，则一般用亮温表示：

$$T_B = eT_0 \quad (1)$$

其中：T_0是物质的物理温度。

由于物质的热电磁辐射实际上是物质吸收的电磁波能量的再发射，即物质辐射微波能量的能力也就是物质吸收微波能量的能力，所以发射率e可近似用吸收系数k_a表示，即$e=k_a$。在均匀媒质中，若体散射可以忽略不计，则有：

$$e=k_a=2k'' \tag{2}$$

其中：波数$k=\omega\sqrt{\mu\epsilon}$。

设一半空间媒质的电导率为σ、复介电常数为ϵ，则其功率损耗率为：

$$\alpha=\omega\sqrt{\mu\epsilon}\left[\frac{1}{2}\left(\sqrt{1+\frac{\sigma^2}{\omega^2\epsilon^2}}-1\right)\right]^{\frac{1}{2}} \tag{3}$$

因此，在深度z ($z<0$)处的媒质的热辐射对表面亮温的贡献为：

$$T_B'(z)=T_B(z)e^{2\alpha z} \tag{4}$$

由公式(2)~(4)可知，对同一种介质，频率越高，发射率越大，在介质中的传输损耗也越大，穿透能力较弱；而低频的发射率虽小，但在介质中的传输损耗小，穿透深度大。

物质的复介电常数ϵ可以表示为：

$$\varepsilon=\varepsilon'-j\varepsilon''=\varepsilon_r'\varepsilon_0-j\frac{\sigma}{\omega} \tag{5}$$

对于弱损耗介质，因为$\sigma\ll\omega\varepsilon'$，所以，

$$\alpha\approx\frac{\pi f}{c}\frac{\varepsilon_r''}{\sqrt{\varepsilon_r'}}\approx k'' \tag{6}$$

而对于强损耗介质，则有$\sigma\ll\omega\varepsilon'$，因此

$$\alpha\approx\sqrt{\frac{\omega u\sigma}{2}} \tag{7}$$

3 分层媒质亮温的计算模型

非均匀物质在自然界中广泛存在，且大部分非均匀物质的介电系数等物理性质是随机变化的，很难用确定的函数表达式描述，这加大了对其各方面进行研究的难度。为解决这个问题，我们将非均匀介质层沿介电系数变化方向分成许多小薄层，只要小薄层的厚度足够小，即可将其视为均匀层。也就是说，非均匀介质层可等效为由许多均匀小薄层组成的分层媒质，由此，可将对非均匀媒质的研究转变成对均匀分层媒质的研究。因此，对分层媒质的研究在遥感学上占有很重要的地位。而随着遥感学的发展，求解分层媒质亮温的方法也越来越多，以下主要介绍利用并矢 Green 函数和起伏逸散定理求解分层媒质亮温的方法。

无界均匀空间的并矢 Green 函数为[2]：

$$\bar{\bar{G}}(\vec{r},\vec{r}')=-\hat{z}\hat{z}\frac{\delta(\vec{r}-\vec{r}')}{k_0^2}\begin{cases}\frac{i}{8\pi^2}\int d\vec{k}_\rho\frac{1}{k_z}\left[\hat{v}(k_z)\hat{v}(k_z)+\hat{h}(k_z)\hat{h}(k_z)\right]e^{i\vec{k}\cdot(\vec{r}-\vec{r}')}, & z>z'\\ \frac{i}{8\pi^2}\int d\vec{k}_\rho\frac{1}{k_z}\left[\hat{v}(-k_z)\hat{v}(-k_z)+\hat{h}(-k_z)\hat{h}(-k_z)\right]e^{i\vec{k}\cdot(\vec{r}-\vec{r}')}, & z<z'\end{cases} \tag{8}$$

其中：水平极化矢量$\hat{h}=\frac{\hat{k}\times\hat{z}}{|\hat{k}\times\hat{z}|}$，垂直极化矢量$\hat{v}=\frac{1}{k_0}\hat{h}\times\vec{k}$，$\hat{k}$表示波的传播方向矢量。

$$\vec{\kappa}=\hat{x}k_x+\hat{y}k_y-\hat{z}k_z$$

由无界均匀空间的并矢 Green 函数，我们可以推导出图1所示分层媒质的并矢 Green 函数。

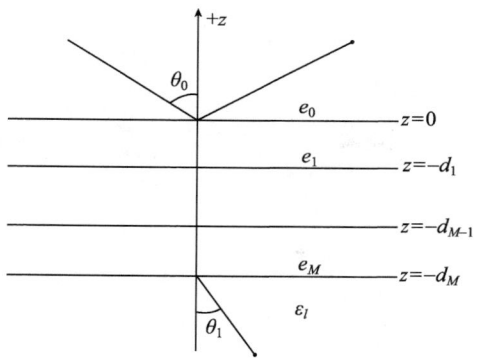

图 1 平行分层媒质模型

设点源(z')位于 0 区域,(即有 $z < z'$)则 0 区域中的场由上行波和下行波组成,由式(8)可得该区域中并矢 Green 函数为[2]:

$$\bar{\bar{G}}_{00}(\vec{r},\vec{r}') = \frac{i}{8\pi^2}\int d\vec{k}_\rho \frac{1}{k_z}\{[R_h\hat{h}(k_z)e^{i\vec{k}\cdot\vec{r}} + \hat{h}(-k_z)e^{i\vec{\kappa}\cdot\vec{r}}]\hat{h}(-k_z)e^{-i\vec{k}\cdot\vec{r}'}$$
$$+ [R_v\hat{v}(k_z)e^{i\vec{k}\cdot\vec{r}} + \hat{h}(-k_z)e^{i\vec{\kappa}\cdot\vec{r}}\hat{h}(-k_z)e^{i\vec{\kappa}\cdot\vec{r}}]\}, \quad z < z' \tag{9}$$

根据边界 $z = -d_l$, $l = 0, 1, \cdots, M$ 处边界条件的匹配,得

$$\bar{\bar{G}}_{l0}(\vec{r},\vec{r}') = \frac{i}{8\pi^2}\int d\vec{k}_\rho \frac{1}{k_z}\{\hat{h}(k_z)e^{-i\vec{k}\cdot\vec{r}'}[A_l\hat{h}(k_{lz})e^{-i\vec{k}_l\cdot\vec{r}} + B_l\hat{h}(-k_z)e^{i\vec{\kappa}_l\cdot\vec{r}}]$$
$$+ \hat{v}(-k_z)e^{-i\vec{k}_l\cdot\vec{r}'}[C_l\hat{v}(k_{lz})e^{i\vec{\kappa}_l\cdot\vec{r}} + D_l\hat{v}(-k_{lz})e^{-i\vec{k}_l\cdot\vec{r}}]\} \tag{10}$$

其中:下标 $l0$ 表示观测点在区域 l,源点在区域 0。

在下垫区域 t,有

$$\bar{\bar{G}}_{t}(\vec{r},\vec{r}') = \frac{i}{8\pi^2}\int d\vec{k}_\rho \frac{1}{k_z}\{[T_h\hat{h}(-k_{lz})e^{i\vec{k}_l\cdot\vec{r}}\hat{h}(-k_z)e^{-i\vec{\kappa}_l\cdot\vec{r}}$$
$$+ T_v\hat{v}(-k_{lz})e^{i\vec{\kappa}_l\cdot\vec{r}}\hat{v}(-k_z)e^{-i\vec{k}\cdot\vec{r}'}\} \tag{11}$$

其中:系数 A_l, B_l ($l = 0, 1, \cdots, M$)、反射系数 R_h 和 R_v 及透射系数 T_h 和 T_v 均可由分层媒质界面上切向场连续的边界条件得到[3]。

根据 Green 函数的对称关系 $\bar{\bar{G}}_{l0}(\vec{r},\vec{r}') = \bar{\bar{G}}_{0l}^T(\vec{r}',\vec{r})$,可得:

$$\bar{\bar{G}}_{0l}(\vec{r},\vec{r}') = \frac{i}{8\pi^2}\int d\vec{k}_\rho \frac{1}{k_z}e^{i\vec{k}\cdot\vec{r}}\{\hat{h}(k_z)[A_l\hat{h}(-k_{lz})e^{-i\vec{\kappa}_l\cdot\vec{r}'} + B_l\hat{h}(k_{lz})e^{-i\vec{k}_l\cdot\vec{r}'}]$$
$$+ \hat{v}(k_z)[C_l\hat{v}(-k_{lz})e^{-i\vec{\kappa}_l\cdot\vec{r}'} + D_l\hat{v}(k_{lz})e^{-i\vec{k}_l\cdot\vec{r}'}]\} \tag{12}$$

对于图 1 所示的 M 层媒质,设第 l 层中热源为 $\vec{J}_l(\vec{r},\omega)$,$\vec{r}$ 在第 l 层($l = 0, 1, \cdots, M+1$)。在微波波段 $h\omega \ll BT$ 的条件下,由起伏逸散定理得到热源相关函数的期望为:

$$\langle \vec{J}_l(\vec{r},\omega)\vec{J}_l^*(\vec{r}',\omega') \rangle = \frac{4}{\pi}\omega\varepsilon_l''(z)BT_l(z)\bar{\bar{I}}\delta(\vec{r}-\vec{r}') \tag{13}$$

其中:B 为 Boltzmann 常数,$T_l(z)$ 为第 l 层的物理温度分布,$\varepsilon_l'(z)$ 为第 l 层介电常数的虚部,$\bar{\bar{I}} = \hat{x}\hat{x} + \hat{y}\hat{y} + \hat{z}\hat{z}$。

由辐射强度和辐射亮温的定义,极化的辐射亮温可写成:

$$T_{Bp}(\hat{k},\omega) = \frac{(2\pi\cdot c)^3}{2B\omega^2}\varepsilon_0\int_0^\infty d\omega'\int_0^\infty k^2 dk\int d\vec{k}'\{\hat{p}\cdot\langle\vec{E}(\vec{k},\omega)\vec{E}^*(\vec{k}',\omega')\rangle\cdot\hat{p}\exp[(\vec{k}-\vec{k}')\cdot\vec{r} - i(\omega-\omega')t]\} \tag{14}$$

其中：$\hat{p} = \hat{h}, \hat{v}; c = (\mu_0 \epsilon_0)^{\frac{1}{2}}$。

由分层媒质的并矢 Green 函数式（12），有：

$$\vec{E}(\vec{r}, \omega) = \sum_{l=1}^{M+1} \int_{-\infty}^{+\infty} d\vec{r}_p' \int_{d_l}^{d_{l-1}} dz' \overline{\overline{G}}_{0l}(\vec{r}, \vec{r}') \cdot \vec{J}(\vec{r}') \tag{15}$$

其中 $\vec{r} \in 0$ 区域，$\vec{r}' \in$ 第 l 层，$d_{M+1} \to \infty$。

将(15)式代入(14)，并由：

$$\vec{E}(\vec{k}, \omega) = \int d\vec{r} \vec{E}(\vec{r} \omega) e^{-j\vec{k} \cdot \vec{r}}$$

可得到：

$$T_{Bh}(\hat{k}, \omega) = k^3 \cos\theta \sum_{l=1}^{M+1} \int_{d_l}^{d_{l+1}} dz' \frac{\epsilon_l''(z')}{\epsilon_0} T_l(z') \left| \frac{1}{k_z} [A_l \hat{h}(-k_{lz}) e^{ik_{lz}z'} + B_l \hat{h}(k_{lz}) e^{-ik_{lz}z'}] \right| \tag{16a}$$

$$T_{Bv}(\hat{k}, \omega) = k^3 \cos\theta \sum_{l=1}^{M+1} \int_{d_l}^{d_{l+1}} dz' \frac{\epsilon_l''(z')}{\epsilon_0} T_l(z') \left| \frac{1}{k_z} [C_l \hat{v}(-k_{lz}) e^{ik_{lz}z'} + D_l \hat{v}(k_{lz}) e^{-ik_{lz}z'}] \right| \tag{16b}$$

其中：θ 为区域 0 的观测角。

当分层媒质中 $\epsilon(z)$ 和 $T(z)$ 为连续廓线函数时，辐射亮温表达式(16)可以写成权重积分形式[2]。

在引言中，我们已经提到月球表层的结构可以近似认为由月壤和月岩组成的分层结构，因此其亮温与厚度的关系可以用式(16a，b)来描述(图2)。

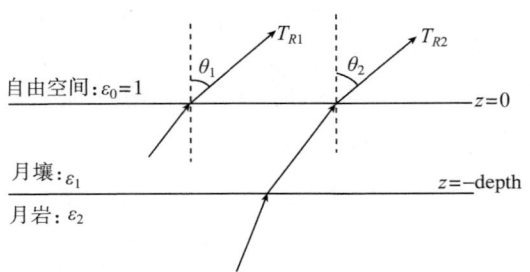

图 2 月球表层的理想结构模型

根据前人对各月壤月岩样本介电常数的研究结果[1]可以发现：各样本的介电常数不尽相同。文中月壤的介电常数选用给出的样本 15601 的介电常数 ϵ_1= 3.22(1+ 0.00251j)(Gold et al.,1973)[1]，月岩则选用样本 12002 的介电常数 ϵ_2=7.8(1+ 0.056j)(Katsube and Collett，1971)[1]。

迄今为止，我们对月球表层温度随深度的变化情况了解甚少。地球上，表面温度对土壤温度的影响不超过 1m；而月壤的物性与干土接近，也应具有这个特点。以下我们分别对温度廓线为指数分布和线性分布两种情况进行讨论。

月壤温度随深度指数变化时，设温度廓线 $T(z)$(z 的单位是 cm)：

$$T(z) = \begin{cases} (T_0 - T_2) e^{z/20} + T_2, z > -100 \\ T_2, z \leqslant -100 \end{cases} \tag{17}$$

而月壤温度随深度线性变化时则设温度 $T(z)$(z 的单位是 cm)：

$$T(z) = \begin{cases} \dfrac{(T_0 - T_2)z}{100} + T_0, z > -100 \\ T_2, z \leqslant -100 \end{cases} \tag{18}$$

其中：T_0 为表面温度，T_2 为 1m 以下部分的温度。

设 T_0= 300 K 和 T_2= 200 K，观测角 θ= 0°，根据式(16)和(17)、(18)可得温度为指数分布和线性分布时

月面亮温随月壤厚度的变化情况。

由图3可见：

(1) 月面亮温与月壤厚度之间不是一一对应关系，即一个亮温值可能对应多个厚度值，这是由式(16)中 $e^{ik_{1z}z}$ 和 $e^{-ik_{1z}z}$ 的存在所引起的；

(2) 温度随深度变化不同时，月面亮温随月壤厚度变化呈相同的变化趋势；

(3) 频率为 3 GHz 的热辐射的穿透能力比 7.8 GHz 强，即辐射计的探测深度会随频率的增加而减小。

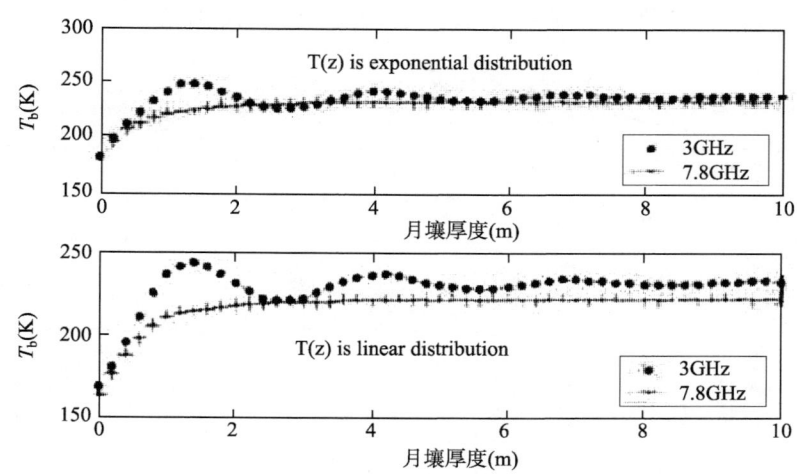

图 3 月面的亮温随月壤厚度的变化情况

4 厚度反演

从上文我们已知不同频率的辐射计的探测深度不同，且亮温与厚度之间并非一一对应关系，因此我们采用多通道辐射计对月球进行探测，并通过对多通道微波辐射计的测量数据进行比较分析而获得月球表层不同深度的信息。

为了使月壤厚度反演问题得到简化，我们选取辐射计的观测角 θ=0°，则水平极化亮温与垂直极化亮温一致，即 $T_{Bh} = T_{Bv}$。我们选择多通道辐射计的频率分别为 3 GHz 和 7.8 GHz。

假设月壤温度为指数变化，由式(16、17)可计算出第 i 通道(i=1，2，3)在月壤厚度为 d_j 时的辐射亮温 $T_{Bi}(d_j)$，而对应第 i 通道的观测值为 T_{BQi}(由于没有实际测量数据，我们用相应的理论值加噪来代替)。然后，利用最小二乘法对这些数据进行比较，从而得到月壤的厚度 $d\text{-}inverse$。即寻找 $d\text{-}inverse$，使得在 $d_j = d\text{-}inverse$ 时，

$$\sum_{i=1}^{N}(T_{Bi}(d_j) - Tboi)^2$$

达到最小。根据最小二乘法得到不同误差条件下月壤厚度反演结果与真实值之间的关系如图4所示。

由图4可见：

(1) 反演精度受测量误差影响很大，测量误差越小，反演值与模拟真实值之间的偏差也越小；

(2) 由于深处的亮温经衰减后对辐射计的贡献比较小，受测量误差的影响较大；

(3) 由图3可知 7.8 GHz 辐射计的探测深度比较小，因此在月壤厚度比较大的情况下，7.8 GHz 辐射计几乎不起作用，而亮温与厚度之间不是一一对应关系，这就导致反演值与真实值之间的误差并非单纯随厚度的增加而增大，而是无规律变化。

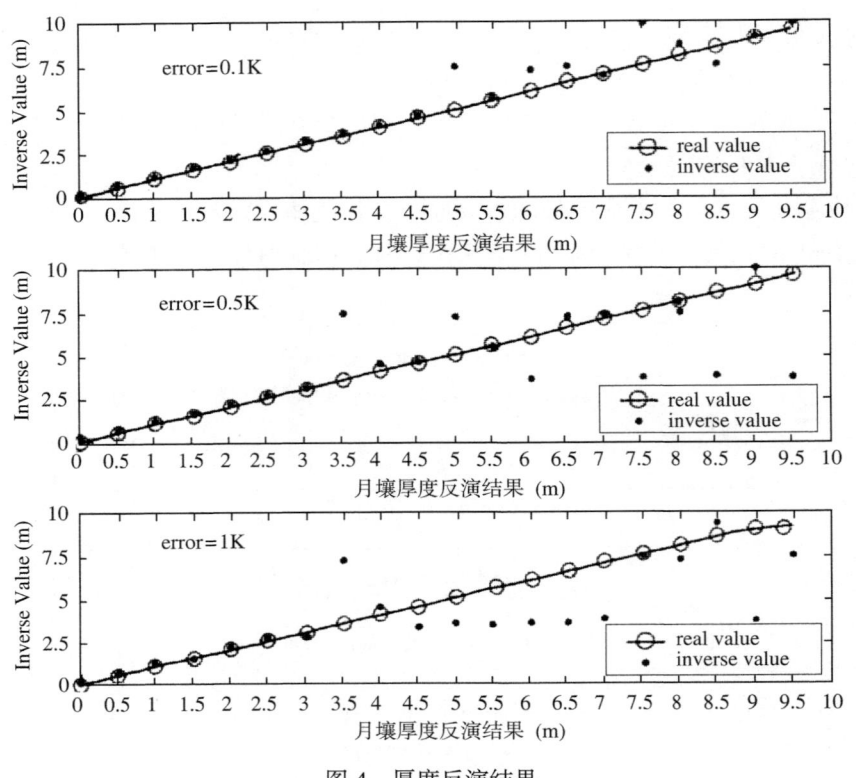

图 4 厚度反演结果

5 结语

从上文可以看出,如果我们所假定的月球表层模型与真实月球表层相符,那么,通过双通道辐射计对其进行测量,然后再利用最小二乘法对测量数据进行处理,从而反演得到月壤厚度。但由于 7.8 GHz 辐射计的探测深度小于 3 GHz 辐射计的探测深度,而月面亮温与月壤厚度之间并非一一对应,这导致了在月壤厚度大于 7.8 GHz 辐射计的探测深度后的反演结果不理想;另外,反演结果还受测量误差的影响,我们可以通过提高辐射计分辨率来减小测量误差,从而提高反演精度。

致谢 本论文得到复旦大学波散射和遥感中心的金亚秋教授的指导,在此表示感谢!

参 考 文 献

[1] Heiken. Lunar Sourcebook: A User's Guide to the Moon [M]. Cambridge University Press,1991.
[2] 金亚秋. 电磁散射和热辐射的遥感理论 [M]. 北京:科学出版社,1993.
[3] F T 乌拉比. 微波遥感 [M]. 北京: 科学出版社,1987.
[4] 张祖荫,林士杰. 微波辐射测量技术及应用 [M]. 北京:电子工业出版社,1995.
[5] 戴振铎,鲁述. 电磁理论中的并矢格林函数 [M]. 武汉:武汉大学出版社,1996.
[6] 刘钦圣. 最小二乘问题计算方法 [M]. 北京:北京工业大学出版社,1989.
[7] Bardaiti F,Solimini D,Tognolatti P. Effect of Electrical and Properties of Materials on Microwave Brightness Temperature of the Trrestrial Crust J. IGARSS'86,1989,7: 605~610.

月表温度剖面对于"嫦娥一号"卫星微波探测仪探测亮温影响的模拟研究

李芸[1,2]　王振占[1]　姜景山[1]

(1. 国家863计划微波遥感技术实验室，中国科学院空间科学与应用研究中心，北京 100190；
2. 中国科学院研究生院，北京 100049)

摘要 月球表层的温度剖面是月球表面微波遥感的一个重要参数。利用从Apollo, Luna等登月试验带回的样品得到的有关月表物质物理特性的知识，通过求解热传导方程，分析了月面表层温度的时间和空间的变化规律，模拟产生了全月的600 cm温度分布结果。结果表明：月球表层温度剖面的变化绝大部分集中在表层20 cm的范围，除了在两极地区，温度的昼夜变化波及到约1 m深度，大于这个深度温度基本上稳定不变。温度的波动很大程度上导致了"嫦娥一号"微波探测仪(CELMS)的不同通道亮温的变化。本文的研究结果证明了温度剖面对于CELMS亮温的影响，从另外一个角度证明了CELMS测量数据的正确性，为遥感数据的解译和科学目标的反演提供了依据。

关键词 "嫦娥一号"微波探测仪(CELMS)　月表温度　月球表层温度剖面　热传导方程　亮温模拟

"嫦娥一号"卫星微波探测仪(CE-1 Lunar Microwave Sounder，CELMS)是世界上首次在月球轨道进行探测的微波辐射计，用于获取月表的亮温，进而反演出月壤的厚度信息。月壤的微波辐射亮温包含了月表以下各层亮温贡献的总和。频率不同探测的深度不同，从而反映的月壤厚度不同。厚度信息的体现主要在于：不同厚度内月壤的物理特性，如温度、介电常数、密度、热导率、比热等参数不同，反应到微波辐射传输模型中，就是不同层的发射率、透过率和物理温度不同，导致这些层内的辐射亮温和到达表面亮温的比例不同。因此通过分析微波探测仪接收到的亮温，就可以反演辐射来源于哪些深度，进而反演这些深度内月壤层的物理特性。

由于这些物理参数对于微波辐射计接收的亮温影响是不同的，同时这些参数也是CELMS的探测目标，因此对这些参数的影响机制进行分析是必要的，也是必需的。

本文主要分析不同深度月壤层的温度变化对于微波亮温的影响。由于月表没有大气，太阳构成了月表层温度变化的能量来源。随着太阳的升降，月表温度发生剧烈的变化。由于月壤内几乎没有空气，月壤内部的温度传输主要取决于热传导。这样表面的剧烈温度变化必然通过热传导作用与下面一定深度内的月壤发生热交换，因此造成表面温度分布的不均匀，同时造成表层温度分布的波动。本文就是利用热传导方程来模拟这些表面和剖面温度变化情况，进而分析这个温度变化对于微波辐射的影响，为月壤厚度信息反演提供依据和先验信息。

1 月表温度及其剖面的获取方法概述

月球表面温度是研究月壤的重要参数之一。月表温度的已知程度及其准确度直接关系到对于遥感目标

本文原载于《中国科学 D辑：地球科学》，2009, Vol.39, No.8, 1045~1058。

的实现,尤其是月壤厚度的反演。因为月表的微波辐射不但是表面发射率的函数,也是表面温度的函数。月表温度也是研究月球热进化过程的必要边界条件。月壤颗粒的尺寸、体密度(或者孔隙度)、热导率以及岩石的丰度(含量)在月表温度特性中起到至关重要的作用。到目前为止,除了利用 Apollo 登月点带回的样品进行分析以外,没有直接获取的参数。

目前获取月球表面温度的方法大致有两种[1]:(1)直接的月球表面温度测量;(2)间接的方法,包括利用遥感测量的手段,包括绕月飞行器的遥感测量以及地基的遥感测量;利用登月器带回来的月壤样品的热物理性质推算的月表温度;利用通过样品得到的热物理参数的模型,通过求解热辐射或热传导方程得到最终的月球表面温度物理模型,等等。

Apollo 计划的 15 和 17 号是到目前为止仅有的两次在登月点成功地对月球表面温度进行测量的直接试验,测量结果表明,Apollo 15 登月点的最高温度为 374 K,最低温度为 92 K,Apollo 17 登月点的最高和最低温度均比 Apollo 15 登月点高出大约 10 K[2]。但是 Wieczorek 和 Huang[3]通过对 Apollo 15 和 Apollo 17 登月点的表面和次表面的温度数据进行重新分析,认为月球表面温度与 18.6 年的月球轨道周期密切相关,而这个信号在以前的研究中都被忽略,因此所获得的有关热流数据的可靠性就不得而知了。

通过绕月飞行器的遥感测量数据可以建立表面物质的密度、比热、热导率、热发射率等参数之间的关系,以及这些参数对热辐射亮温的贡献,Horai 和 Fujii[4]通过分析温度和孔隙间的气压对于热导率和弥散系数的影响,建立了热导率、体密度和温度之间的经验关系,发现这个关系式与月壤样品的数据一致。Racca[5]通过分析现存的有关月表探测器的温度信息,建立了两个表面温度的数学模式,这个模型与试验结果吻合的很好。

到目前为止,地基遥感观测,主要是基于红外和微波波段的观测,通过解译获得的亮温数据反演得到月球表面实际温度。Price 等[6]利用 MSX 卫星获得了 4.3 μm 谱段月表月食过程的热图像,空间分辨率 45 km,发现很多热温度点,这些温度点与火山坑对应,另外在阴影和半阴影区的温度比高原地区温度高。MSX 观测到的月食过程月表温度的不均匀冷却的现象揭示了各种月表特征的热物理特性的差异。

地基遥感观测主要集中在月球正面的赤道地区,空间分辨率较差[7]。Mitchell 等[8]利用 Hat Creek 射电天文台的 BIMA 的 0.3 cm 毫米波干涉仪以及波长范围 1.3~20.5 cm 的 VLA 超大阵列观测的数据,来分析水星表面的热辐射,发现水星表面的土壤比月表阴暗区的月壤透过率高至少 2~3 倍,而比月表高地的透过率至少高出 40%,它认为原因可能是水星表面 Fe,Ti 的丰度与月表有较大差异所致。

为了更好地了解月壤参数之间的相互作用,Urquhart 和 Jacksky[9]建立了一个白天的表层温度的热模型,其中考虑了温度与比热和热导率的相关性。他们认为这些参数之间的相关性对于反演月壤信息非常关键,原因是月表温度昼夜之间的差异非常巨大。研究发现,虽然不能完全独立地研究粒子尺寸、密度和热导率等参数,但是这些参数之间的关系,以及这些参数对于月表温度的影响还是能够非常确切地描述:随着月壤的体密度增大,月球夜晚月壤的表面温度越高;而粒子的尺寸增大导致月球夜晚月壤表面温度降低。单独通过红外遥感不能得出唯一的表面参数特性,尤其是不能遥感粒子的尺寸。另外岩石丰度对于确定表面的热惯导来讲不是一个大的影响因素。

利用通过样品得到的热物理参数的模型,通过求解热辐射或热传导方程得到最终的月球表面温度物理模型是目前广泛使用的方法[10~14]。这种方法主要根据求解热辐射和热传导方程,利用目前已知的关于月壤物理化学特性的研究结果,来模拟月表层温度的分布和变化规律。本文就是利用这种方法,结合从样品分析数据得到的月壤介电特性、热导率和比热等的最新研究结果,来模拟月表层的温度随着经纬度、时间以及日月距离的变化。

2 热传导方程求解月表温度剖面的方法

2.1 热传导方程

月球表面温度物理模型的理论基础是半有限固体的热传导理论,月球表面物质同时受到来自外部的太阳辐射和月球内部热流两个热源的作用，据 Stefan-Boltzmann 定律，月球表面物质在受到热源作用的同时按温度的四次方向外界辐射热量。根据能量守恒，月球表面物质所吸收的热量等于它向外界辐射的能量，得到温度和太阳辐照度以及内部热流之间的关系，从而可以对月球表面温度进行求解。热传导方程[15]可写为

$$\rho(x,T)C(x,T)\frac{\partial T}{\partial t} = \frac{\partial}{\partial x}\left[K(x,T)\frac{\partial T}{\partial x}\right] + Q(x,T) \tag{1}$$

或者写成差分形式：

$$\rho(x,T)C(x,T)\frac{\partial T}{\partial t} = \frac{\left[K(x,T)\frac{\partial T}{\partial x}\right]_{x+\frac{\Delta x}{2}} - \left[K(x,T)\frac{\partial T}{\partial x}\right]_{x-\frac{\Delta x}{2}}}{\Delta x} + Q(x,T) \tag{2}$$

其中，$\rho(x,T)$ 表示密度；$C(x,T)$ 表示比热；$K(x,T)$ 表示热导率；$Q(x,T)$ 表示部分透明介质由于吸收太阳的辐射而产生的源项，这里不考虑这个辐射，因此 $Q(x,T)=0$。$\rho(x,T)$，$C(x,T)$ 和 $K(x,T)$ 都是表层深度 x 和温度 T 的函数。边界条件如下：

(i) 在表面：

$$K_s\frac{\partial T}{\partial x}|_s = \frac{I_\theta}{4\pi r^2}(1-A_b)\sin^+\phi - \varepsilon\sigma_B T_s^4 + J_0 \tag{3}$$

其中，$K_s(\partial T/\partial x)$ 表示传入次表面的能量；$\partial T/\partial x$ 表示表面的温度梯度，K_s 为表面的热导率；$\varepsilon\sigma_B T_s^4$ 表示表面辐射的(红外)能量，ε 为红外表面发射率(一般设定为 0.90~1.0 之间)，σ_B 为 Stefan-Boltzman 常数 5.6703×10^{-8} Wm^{-2}·K^{-4}，T_s 为表面温度；$I_\theta/4\pi r^2$ 表示到月表上的总辐照度，I_θ 为太阳辐照度 3.827×10^{26}W，r 为月亮-太阳之间的距离 1.496×10^{11} m；A_b 为热辐射反照度 0.12；ϕ 为太阳高度角，$\sin^+\phi$ 表示当太阳在水平线以下，$\sin^+\phi=0$，否则 $\sin^+\phi=\sin\phi$；J_0 表示月球内部发射的热通量，这里可以忽略。

(ii) 在热平衡深度 Z_0：

$$\frac{\partial T}{\partial x}|_d = \frac{-J_0}{K_d} \ll 1 \tag{4}$$

其中，K_d 为在 Z_0 时的温度梯度。根据 Mitchell 等[8]，由于 J_0 只有大约 20 erg·cm^{-2}·s^{-1} (1 erg=10^{-7}J)，所以 J_0 可以忽略，它对于亮度变化没有影响。另外模拟 Mitchell 等[8]结果表明当 $x>1.35$ m 时，$\Delta T<0.1$ K，所以可以选择大于这个深度的任意深度作为下边界，此时 $(\partial T/\partial x)|_{Z_0}=0$。

如果假设太阳光是一个点源，其光线在月表面某点的散射是均匀的；表面的热辐射反照度是一个常数；月表不存在大气，光不发生漫射，那么，就可以根据偏微分方程的数值求解方法获得热传导方程的数值解。利用前向差分法可以计算任意时刻、任意位置的月表温度剖面，包括表面的温度。在计算中，把表面到 6 m 深度分为 45 层，图1给出分层与深度的对应关系。其中第 44 层和 45 层在计算时令它们的温度相等。

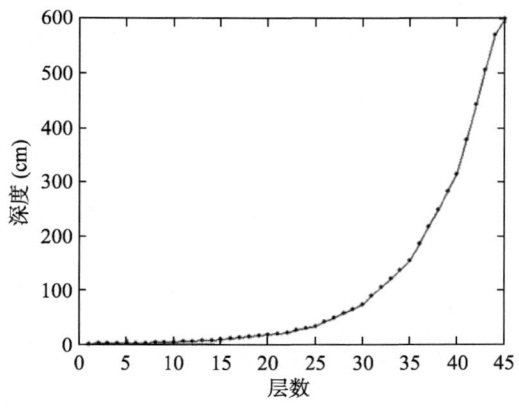

图 1 分层与深度的对应关系

这种模型(以下简称多层模型)主要考虑表层温度的剧烈波动，以及月壤密度、热导率、比热等参数随着深度的变化，与Vasavada等[10]的两层模型相比，更加真实地模拟月表参数随着深度的分布特征。

2.2 月表物理参数的模型

研究表明，月壤密度与温度的关系很弱，因此$\rho(x,T)=\rho(x)$，月壤体积密度(g/cm^3)与月壤深度(cm)的关系[16]表示为

$$\rho = 1.92\frac{z+12.2}{z+18} \quad (5)$$

过去一些人认为月表的比热是一个常数，但是Horai等[4]研究发现C随温度变化而变化。Jones等[17]给出实验关系：

$$C(T)=c_1T^3+c_2T^2+c_3T+c_4 \quad (6)$$

其中，$c_1=1.13112\times10^{-8}$，$c_2=1.21176\times10^{-5}$，$c_3=5.72364\times10^{-3}$，$c_4=0.189972$[1]。Urquhart等[9]通过对于前人的数据在70~400 K范围拟合，得到关系式的系数，其中$c_1=1.24\times10^{-9}$，$c_2=-1.96\times10^{-6}$，$c_3=1.19\times10^{-3}$。单位cal·g^{-1}·K^{-1}。经过比较可以看出二者的差异不明显。

在处理月表热导率的时候为了简单一般认为它是常数，范围为1×10^{-3}~30×10^{-3} W/(m·K)。而实际上热导率不但是温度的函数，也随着深度的变化而变化。Linsky[18]给出一个关系式：

$$K(x,T) = K(x) + \bar{\varepsilon}_M\sigma T^3(x)S(x) \quad (7)$$

其中，$\bar{\varepsilon}_M$为红外发射率，$S(x)$为辐射表面的有效平均间距，表示介质间不连续的间隔。在Mitchell等[8]中，

$$K = K_C\left[1+\chi\left(\frac{T}{T_{350}}\right)^3\right] \quad (8)$$

K_C表示光导率，χ是比值，在Vasavada等[10]中，使用两层模式：在顶层2 cm，$K_C=9.22\times10^{-4}$ Wm^{-1}·K^{-1}，$\chi=1.48$；底层：$K_C=9.3\times10^{-4}$ Wm^{-1}·K^{-1}，$\chi=0.073$。Urquhart等[9]引用Watson的测试数据，认为热导率受温度的调制，尤其对于表面温度剧烈变化的月球表面。但是Urquhart等[9]认为由于月岩的热导率随温度变化的比例小得多，所以可以认为是常数。给出月壤的热导率的形式为

$$K=a+bT^3 \quad (9)$$

其中系数a，b从Urquhart等[9]中得到。为了进行模拟计算，根据Horai等[4]的数据和Urquhart等[9]的Apollo11, 12, 15登月点的数据，通过拟合得到Watson方程(9)的系数变化规律方程(10)和(11)，拟合结果如图2。

$$a = 1.0957\rho^2 - 1.1117\rho + 0.6797 \quad (10)$$
$$b = -0.0067\rho^2 + 0.1465\rho + 0.0433 \quad (11)$$

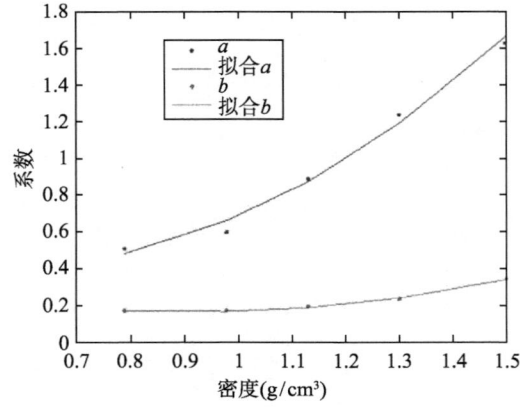

图2 Watson方程系数随着月壤密度的变化

3 模型模拟结果及分析

3.1 表面温度全月分布

图 3(图版 XI)给出模拟的在位置(0°, 0°)正午时刻全月表面温度的分布图,可见在正对地球一侧是白天,表面温度的分布从中心向两极逐渐减小。同时,随着经度的增大温度也在减小。背对太阳一侧是夜晚,温度也呈逐渐过渡的趋势。从太阳落下地平线开始,整个黑夜温度都在缓慢下降,直到黎明后太阳升起,温度迅速上升。因此从温度分布上看左右不是对称的。另外南北极温度明显不同,因为这里模拟的结果表明:北极是白天,南极是黑夜。黑夜的最低温度可达到 50 K。可见昼夜温差非常大。

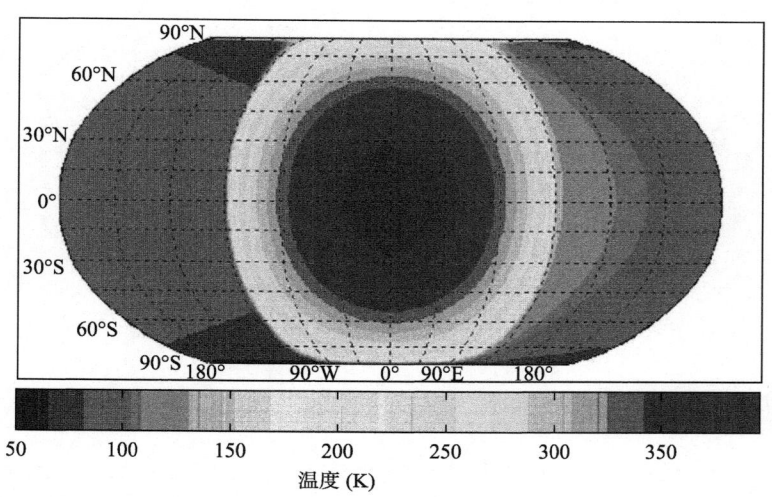

图 3 月球赤道正对地球中心点正午时刻全月的温度分布模拟结果

图 4 给出南北纬 40°相同时刻表面温度的差异比较。可见由于存在太阳高度角的差异,表面温度也存在一定的差别,尤其在白天太阳升起和落下的时刻,由于时间上的差异,导致差异达到 15 K 以上。

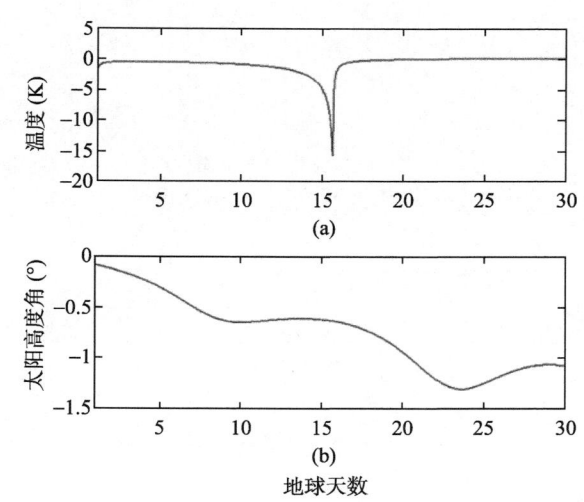

图 4 南北纬 40°相同时刻表面温度的差异比较
(a) 南北纬 40°相同时刻表面温度差;(b) 南北纬 40°相同时刻太阳高度角差

图 5 给出南北极相同时刻表面温度的比较。可见南北极的昼夜是相反的。白天最高温度差异 2.4 K,

北极略高。晚上基本没有差异。

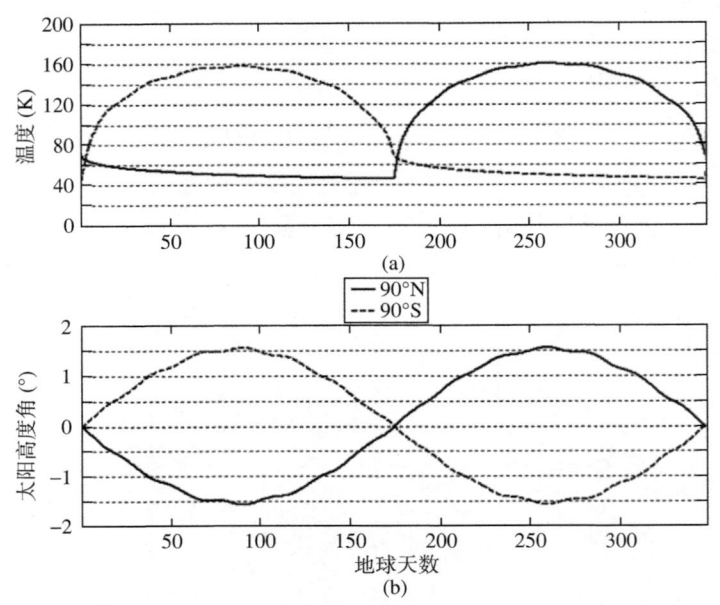

图 5　南北极相同时刻表面温度的比较
(a) 南北极表面温度比较；(b) 南北极太阳高度角比较

3.2　底面温度全月分布规律

为了分析月球月壤层温度的分布，根据月表热流的分布特点，本文模拟了全月 6 m 处的温度分布，如图 6 所示(图版 XI)，可见在这个深度，温度基本不受昼夜变化的影响，只是随着纬度的增大而减小。在赤道，底面温度在 239 K 左右，而极地约为 95 K 左右。

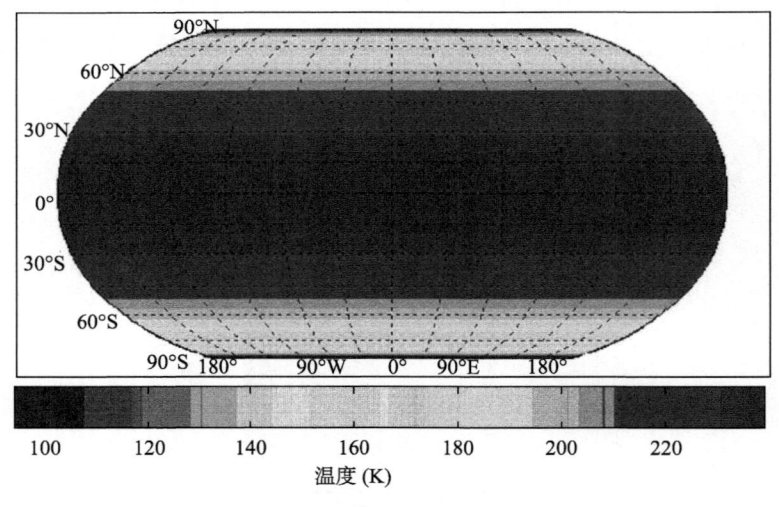

图 6　全月 6 m 深度的温度分布

3.3　特定点表面温度昼夜变化

为了分析特定点表面温度的昼夜变化，分别模拟了赤道、北纬 40°和 80°，北极的昼夜温度以及 Apollo 15 和 17 登月点的温度。图 7 给出赤道地区、北纬 40°，80°地区的表面温度的变化。可见二者的变化趋势是一致的，只是最高的温度和最低的温度不同，在赤道、北纬 40°和 80°最高温度分别可达 396.3，372.7 和 264.5K，最低温度分别为 87.8，84.8 和 70.3 K。图 7(b)给出对应的太阳高度角。太阳高度角的负角是根据太阳和月亮的轨道计算的，其实这时的高度角相对月表来说为 0。太阳高度角的最大值和最小值与月表温度的最大值和最小值不是对应的，一般来说，温度的最大值出现在太阳高度角最大值之后，而最低温度出现在太阳升起前。

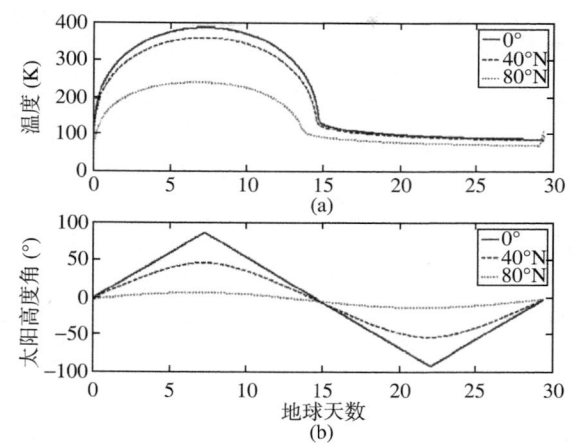

图 7　表面温度随着太阳高度角的变化
(a) 一个月球昼夜内表面温度变化；(b) 一个月球昼夜内太阳高度角变化

太阳在低纬度地区的变化周期相当于地球的 29.5 天。而在高纬度，尤其在 85°以外，周期逐渐加长，在北极，一个月球天近似等于地球的一年。如图 8 所示，横坐标是地球月。由于极地绕太阳周期的增大，导致表面温度出现随太阳距离变化的波动。在下一节将讨论这个问题。北极的最高温度约 160 K 左右，而最低温度根据目前模拟结果可能小于 50 K。模拟的误差取决于对于极地月壤结构、物理性质等的已知程度。

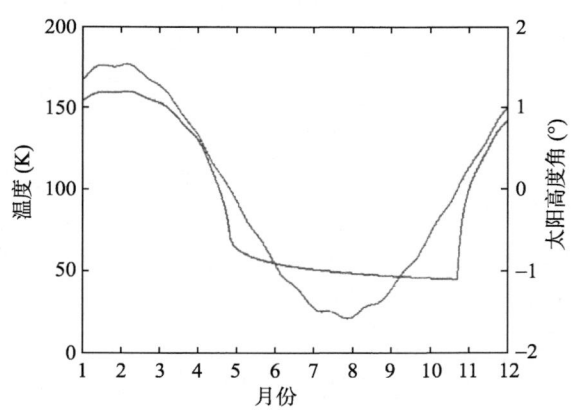

图 8　北极表面温度随着太阳高度角的变化

Apollo 15 和 17 号登月点是到目前为止仅有的两次在登陆点成功地对月球表面温度进行测量的直接试验，测量结果如图 9，Apollo 15 登月点的最高温度为 374 K，最低温度为 92 K，Apollo 17 登月点的最高和最低温度均比 Apollo 15 登月点高出大约 10 K[2]。为了说明模拟的准确性，本文分别模拟了 Apollo 登月点的昼夜温度变化。因为在这两个登月点当年的美国宇航员分别安置了测温仪器，观测表层热流的变化，获得了一些结果。图 10 显示在 Apollo 15 登月点最高温度 370 K 左右，最低温度在 80 K 以上，Apollo 17 登月点最高温度在 380 K 以上，最低在 90 K 以上，可见与实测结果非常一致。

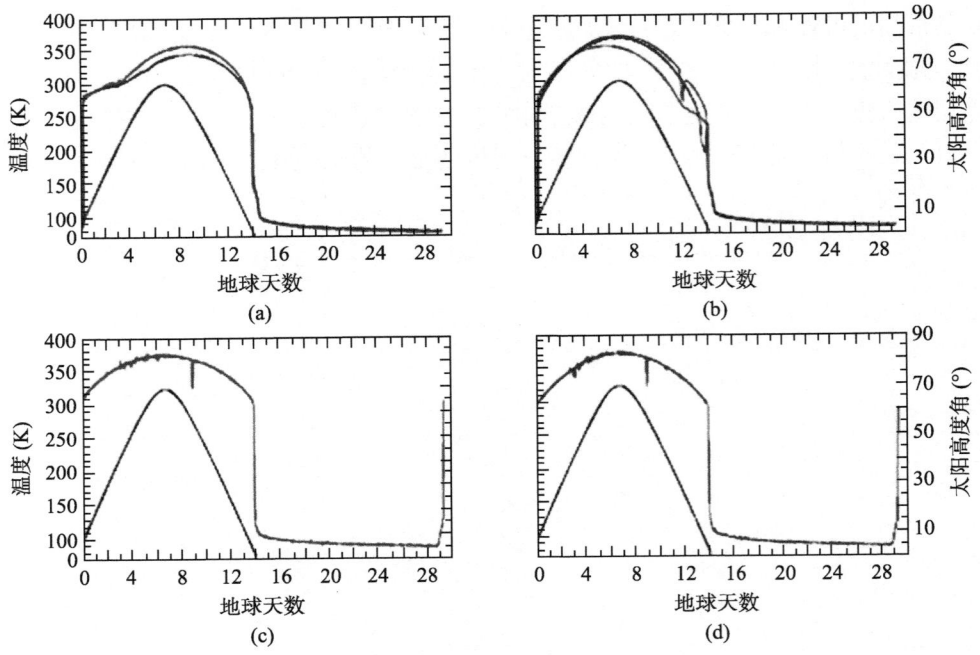

图 9 给出的 Apollo 15 和 Apollo 17 的表面温度变化[3]
(a), (b)为 Apollo 15; (c), (d)为 Apollo 17

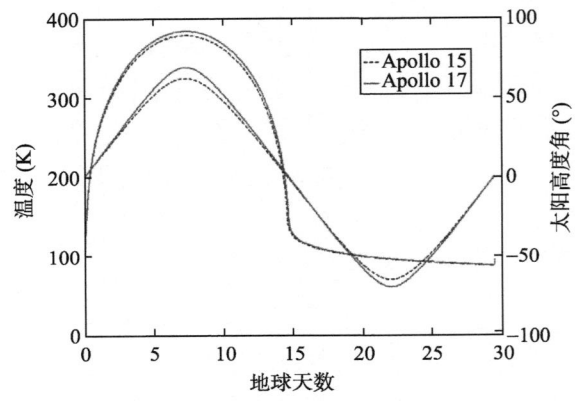

图 10 Apollo 15, 17 登月点的表面温度随着太阳高度角的变化

Wieczorek 和 Huang[3]通过对 Apollo 15 和 Apollo 17 的表面和次表面的温度数据进行重新分析,认为月球表面温度与 18.6 年的月球轨道周期密切相关,而这个信号在以前的研究中都被忽略,因此所获得的有关热流数据的可靠性就不得而知了。图 11 给出模拟的 Apollo 15 登月点在 20 年内表面温度在太阳高度角最大、最小情况下的波动,可见这个波动可能由于月球与太阳的距离不同引起表面温度的不同,这一点在高纬度地区比在低纬度地区更加明显。

3.4 剖面温度的昼夜变化

根据目前的研究成果,月壤表面是一层由非常细的尘埃颗粒组成,就是月尘层。这层月尘起到的作用相当于一个保温毯,使里面的温度和外面的温度交换缓慢,由于没有空气,热交换只能通过这一层的传导到里面月壤,而月壤的温度也通过这一层向外辐射。月尘层的厚度约为 2 cm。因此月表的温度变化的剧烈区域就表现在浅层的剧烈变化,温度梯度非常大。

图 12 给出赤道、北纬 80°和北极的剖面温度的分布,分别对应与太阳高度角最大和最小的情况,但不

是温度最高和最低值。可以看出,除了极区,剖面温度的波动范围基本上在 20 cm 以内的表层,而极区由于日照周期的延长(约为低纬度地区的 12 倍),导致温度波动的深度可以达到 1 m,如图 12 所示。在任何一个地区,底层的温度基本上是不变的,也就是说,当深度大于 1 m,温度就是一个恒定值。

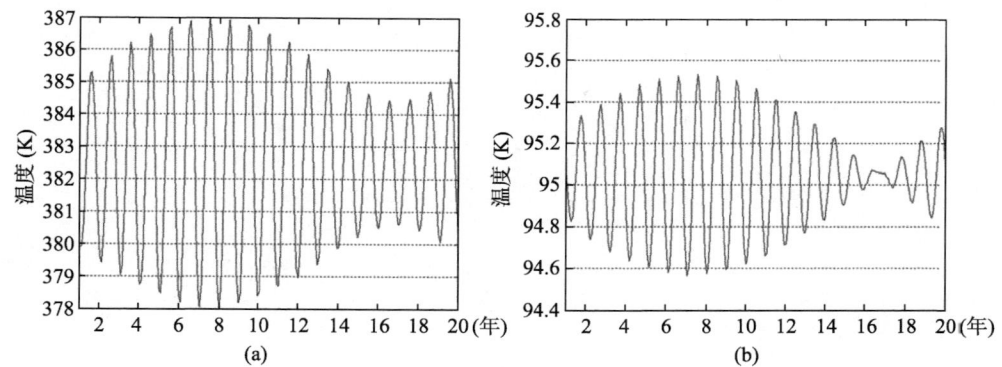

图 11　20 年内 Apollo 15 点表面最高、最低温度的变化
(a) Apollo 15 登月点最高表面温度变化;(b) Apollo 15 登月点最低表面温度变化

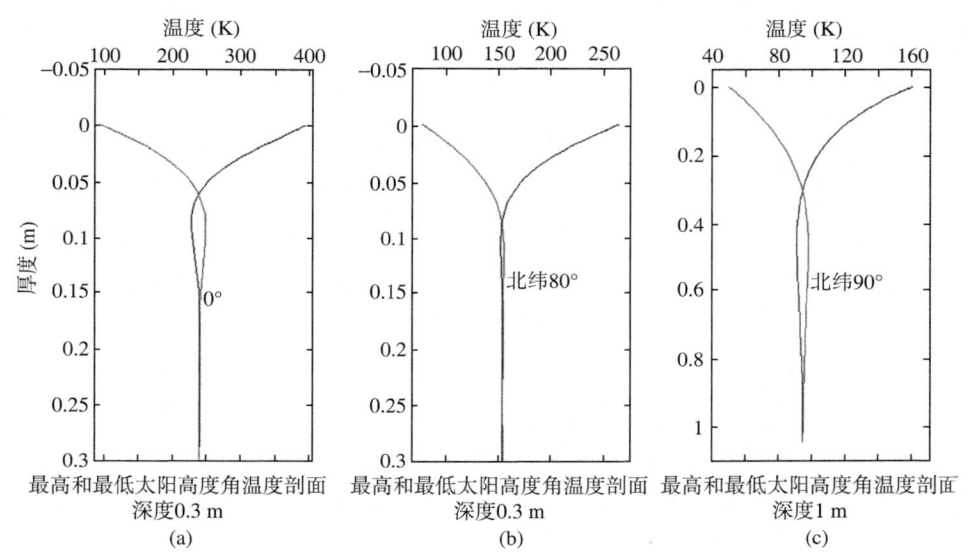

图 12　赤道地区(a)、北纬 80°(b)和北极地区(c)剖面温度的变化

南北纬相同时刻的剖面温度由于太阳高度角的差异也出现不同,如图 13 所示。图 13 中比较了南北纬 40°在最高温度点和最低温度点时刻剖面温度的差异,可见南北纬的温度差异主要在表层 70~80 cm 以内,从此深度以下基本没有明显区别。

3.5　温度与日月距离

根据方程(3),如果太阳的辐照度 I_θ 近似认为常数,那么日月距离的作用体现在月表上的总辐照度 $I_\theta/4\pi r^2$ 随着距离的变化而变化,另外随着距离的变化,太阳高度角也在发生变化。这样热传导方程的月表边界条件发生周期性改变。因此,体现在温度剖面,尤其是表面温度随着这个波动的变化在长期时间范围内,呈现周期性。图 14 和 15 分别给出赤道、北纬 80°一年内太阳高度角变化引起的表面温度的变化。赤道地区,由于太阳高度角一年之内变化很小,所以表面温度变化不大,而北纬 40°高度角引起的表面温度的变化就已经明显,最高温度的不同月份之间的差异可达到 7~8 K。而北纬 80°的差异达到 25 K 以上。

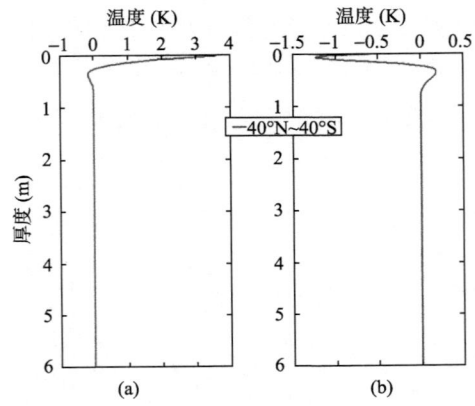

图 13 相同时刻南北纬相同纬度的温度剖面差异比较
(a) 最高表面温度时南北纬 40°的温度剖面差异；(b) 最低表面温度时南北纬 40°的温度剖面差异

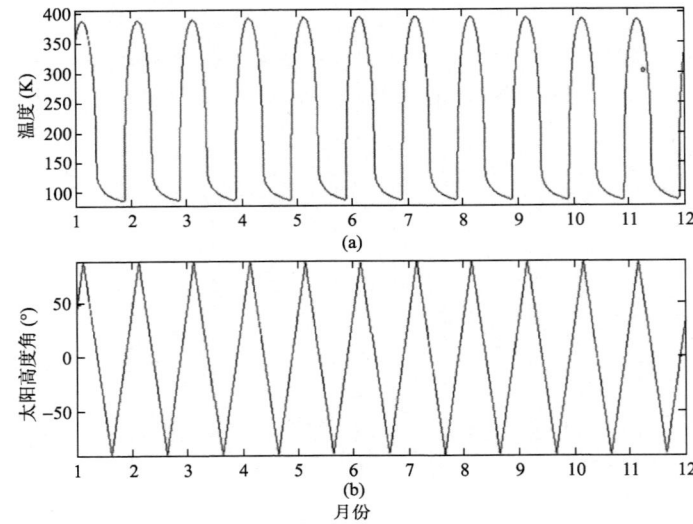

图 14 赤道地区一年内太阳高度角变化引起的表面温度的变化
(a) 一年内表面温度变化；(b) 一年内太阳高度角变化

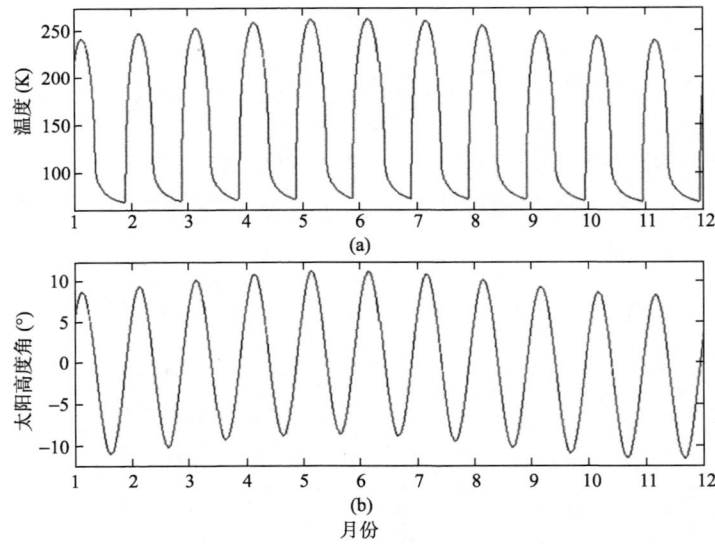

图 15 北纬 80°一年内太阳高度角变化引起的表面温度的变化
(a) 一年内表面温度变化；(b) 一年内太阳高度角变化

日月距离导致了表面温度的变化,同样也导致了剖面温度的变化。这里就不再赘述。

4 温度变化导致的亮温变化

月表温度的周期变化以及表层温度的波动,在 CELMS 的亮温上表现了亮温随着时间的特殊变化规律以及在特定时刻亮温随着厚度的变化而变化。图 16~19 给出月表不同地区在一个月球昼夜内的亮温变化情况的比较。这些亮温的比较是基于给定月壤厚度的假设前提下的。计算过程中,对于给定厚度以下认为是月岩的半无限空间,考虑月岩的辐射和对于上面月壤下行辐射的反射[19]。为了比较,这里只给出了赤道和北纬 80°在月壤厚度为 1 和 6 m 两种情况下的亮温变化。

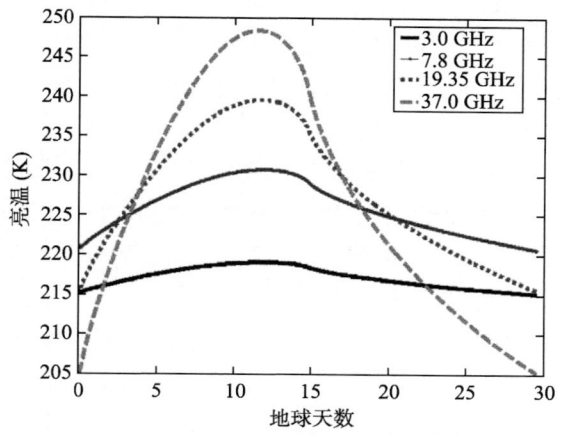

图 16 月球赤道月壤厚度为 1 m 地区的不同频率亮温随着时间的变化

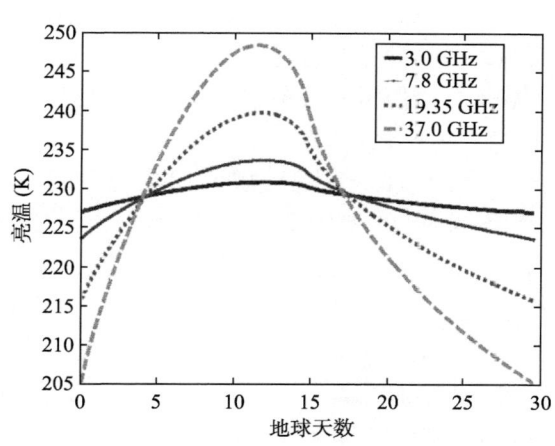

图 17 月球赤道月壤厚度为 6 m 地区的不同频率亮温随着时间的变化

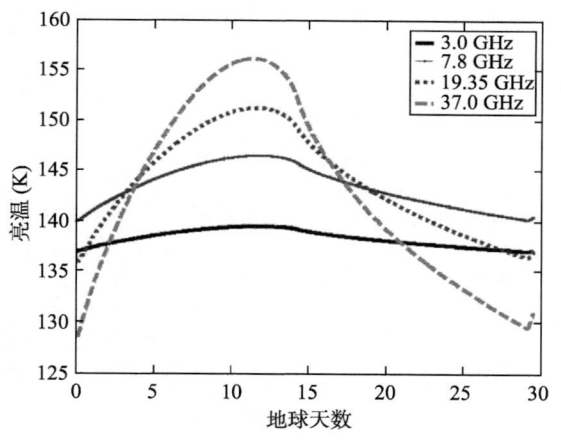

图 18 月球北纬 80°月壤厚度为 1 m 地区的不同频率亮温随着时间的变化

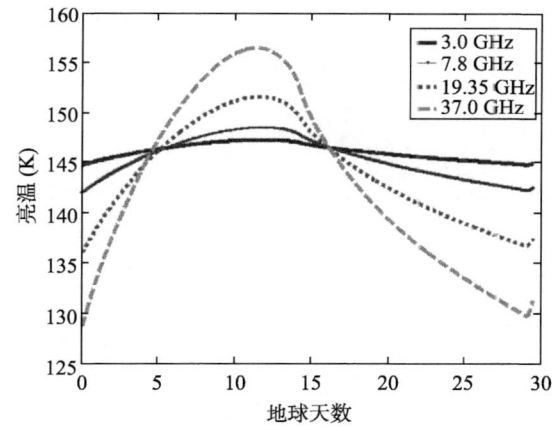

图 19 月球北纬 80°月壤厚度为 6 m 地区的不同频率亮温随着时间的变化

从图 16~19 可以看出:

(1)不同频率的亮温变化规律不同。从 4 个图形明显看出频率的相应差异,一天之内频率不同,亮温变化的起伏特点不同,37.0 GHz 起伏最大,3.0 GHz 最小,7.8 和 19.35 GHz 居中。

(2) 不同厚度的亮温变化规律不同。在 1 m 厚度下,赤道地区(如图 16),当太阳上升到地平线以上,4 个频率的亮温都快速上升,由于夜晚的降温,使得温度上升的起点不同,37.0 GHz 亮温最低,因为表面温度的剧烈下降对于 37.0 GHz 的影响最大,37.0 GHz 反映表层 0.5 m 以内的温度和介电特性的变化。

而 3.0 和 19.35 GHz 亮温比较接近，这是体现了温度剖面、穿透深度、底层反射和辐射的一种折衷特性。3.0 GHz 穿透月壤深度可达 5 m 以上，这样，当达到 1 m 的月岩界面，亮温是月岩自身的辐射、月岩的反射以及上伏月壤的辐射三者贡献之和，1 m 内的月壤对 3.0 GHz 的贡献比 19.35 GHz 小很多，因此反射的月壤贡献也很小，主要的亮温来自 1 m 以下月岩的贡献。19.35 GHz 的亮温基本上是表层 1 m 月壤的贡献。这个亮温比 37.0 GHz 的大。上述现象也从图 20 得到证明。在北纬 80，表面温度和剖面温度更低，这样 3.0 GHz 的亮温就变得比只有表层贡献的 19.35 GHz 的略高。7.8 GHz 的亮温之所以比 3.0 GHz 的高，也是基于上述原因，就是说，7.8 GHz 的探测深度约为 2.0 m，因此它接收的亮温除了 1.0 m 月壤的贡献，还包括月岩的贡献。月岩的贡献在温度相同的情况下随着频率升高而增大。月壤的贡献也是如此，因此 7.8 GHz 的亮温比 3.0 GHz 在 1 m 月壤的情况下要高。

在 6 m 月壤厚度下的亮温与 1.0 m 的变化规律是不同的(图 19 和 21)，原因是每个频率探测的只有月壤的贡献，而没有月岩的贡献，因此这时的亮温的差异反应了温度剖面和穿透深度的关系。频率越低，穿透深度越深，越能反应底层的温度；频率越高，越反应的表面的温度。因此在日出前，37.0 GHz 最低，而中午 37.0 GHz 最高。

(3) 不同纬度的亮温变化规律不同。不同纬度的表层温度分布是不同的，因此亮温的变化也是不同的。

(4) 不同时间的亮温变化规律不同。一日之内的不同时间，由于厚度和温度剖面的差异，4 个通道的亮温大小关系呈现交替变化的趋势。所以在分析月表亮温测量结果的时候，不能简单地凭借频率之间的特定情况下的大小关系来判断测量的有效性，而是应该结合测量的具体时间和厚度来分析。

为了更好地分析厚度和测量亮温的关系，图 20~23 分别给出赤道和北纬 80º 在最高表面温度和最低表面温度下亮温随着厚度的变化趋势图，可以看出：

在最高温度下(如图 20 和 21)，4 个频率亮温都随着厚度的增加而增大，37.0 GHz 的亮温基本上在 0.5 m 以后不再变化，19.35 GHz 的亮温在 1.0 m 以后不再有变化，而 7.8 GHz 的响应厚度约达 2~2.5 m，3.0 GHz 可达到 5.5~6.0 m。亮温的大小关系与频率的大小近似正比关系。

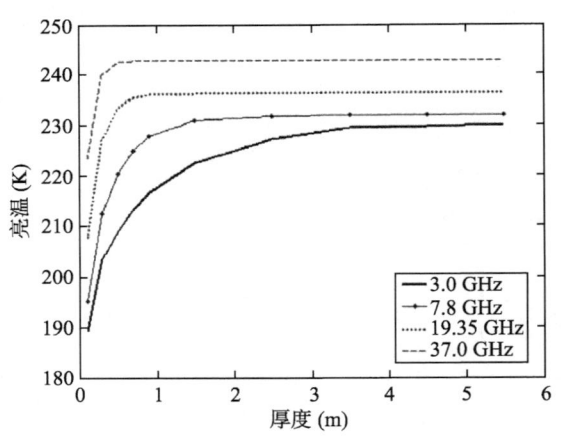 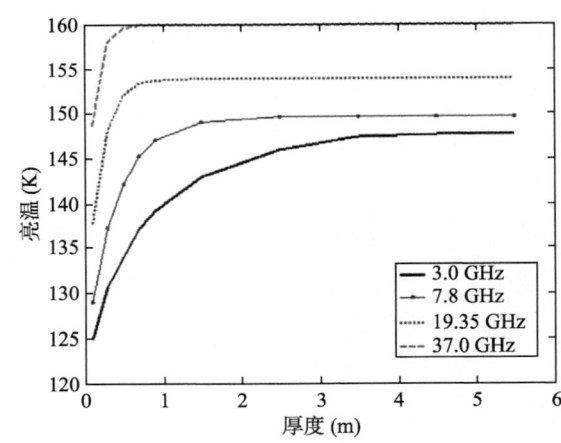

图 20 月球赤道在表面温度最高时刻亮温随厚度的变化　　图 21 月球 80°N 在表面温度最高时刻亮温随厚度的变化

在最低温度下(如图 22 和 23)，亮温随着厚度增大的趋势是一致的，但是，由于温度剖面的变化，在厚度较低的情况下，亮温之间的变化比较剧烈，而且相对大小出现交替。对于基本裸露的月岩地区(月壤厚度不大于 10 cm)，37.0 GHz 的夜晚温度最低，3.0 GHz 最高，但是差异不是很大。但是随着厚度的逐渐增加，约在 50 cm 左右，4 个通道亮温出现交替，频率从高到底，亮温升高梯度逐渐减慢，3.0 GHz 上升最慢，因此它的亮温从 10 cm 的最高到比 7.8 GHz 的低，然后比 19.35 GHz 的低，最后比 37.0 GHz 的低，最终达到最低，但这时的厚度也只有 50 cm 左右，然后又逐渐超过 37.0，19.35，7.8 GHz，最后当厚度达到 2 m 以上，3.0 GHz 的亮温达到 4 个频率中的最高。

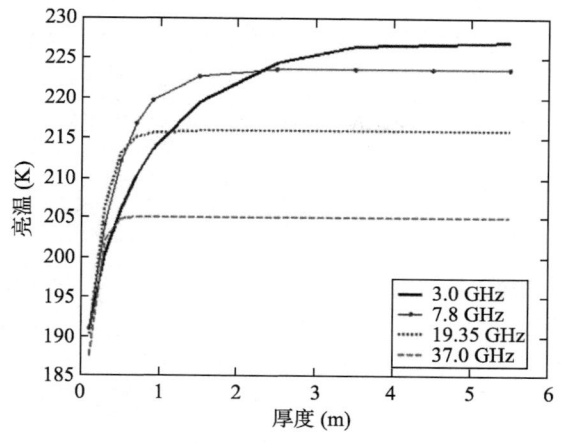 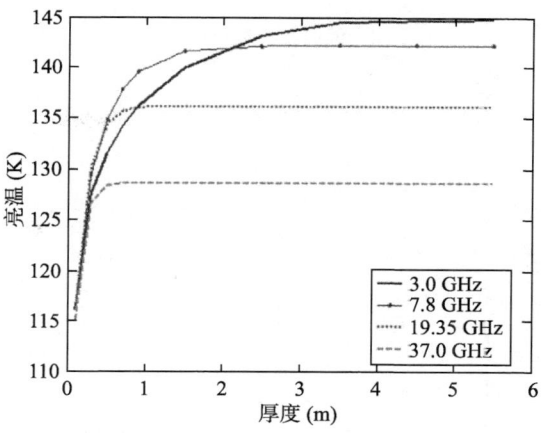

图 22 月球赤道在表面温度最低时刻亮温随厚度的变化　　图 23 月球北纬 80°在表面温度最低时刻亮温随厚度的变化

5 结论

本文利用热传导多层模型，根据月球表层已知的物理参数模式，模拟了月表层温度的分布，结果表明：

(1) 全月任一时刻表面温度分布是不均匀的，不但呈现白天南北纬的相同纬度差异，其中南北极的差异最大，而且夜晚随着太阳落下以后时间的差，温度分布也是不同的。

(2) 全月的底面温度(一般来讲大于 1 m 以上，本文模拟的最大深度 6 m)分布只随着纬度的分布而不同，南北纬的差异很小，与经度关系不明显。

(3) 表层温度的剖面分布变化剧烈。但是这个变化受月球表面昼夜差异的影响，一般来讲，在±85°(最大约到±88°)以内的纬度表层温度的波动集中在 20 cm 以内，只有两极地区由于日照周期长而出现波动深度增加到近 1 m 的现象。

为了分析温度剖面对微波辐射计亮温的影响，文章还进行了 CELMS 4 个通道亮温的模拟，结果表明温度的波动是不同时刻亮温变化的主要原因，随着月壤厚度的不同，不同时刻 4 个频率亮温之间的大小关系呈现不同的交替变化，这有助于分析 CELMS 的数据内在规律和特点。

因此，本文的研究结果证明了温度剖面对于 CELMS 亮温的影响，从另外一个角度证明了 CELMS 测量数据的正确性，为遥感数据的解译和科学目标的反演提供了依据。

但是，模拟结果还存在一些不确定性，主要原因是月表参数的不确定性引起的，如密度、比热和热导率等，这些参数的模型是根据有限的测量结果得出的。在全月的代表性有待进一步评估。为此，本文比较了用多层模型和两层模型(使用的参数和简化方法不同)得出的表面温度和剖面温度模拟结果之间的差异。如图 24 和 25 所示。可见虽然两种模型都可以模拟表面温度的变化趋势，但是两层模型要比多层的结果偏高。原因是多层模型尽可能详细地考虑温度在不同层之间的传输差异，更接近真实情况。

文中还对日月之间的距离对于表面和剖面温度的影响进行了分析。利用平均日月距离、两层模型的结果与多层模型并考虑日照变化的比较结果如图 26 所示。图 26 给出的是北极地区模拟的全日温度变化情况的比较，实线表示两层模型没有考虑日月距离的变化，虚线表示用多层模型考虑日月距离的变化。图 26 的左半侧表示北极夜晚，可见差异比较明显，原因是表层参数的不同和模型的精度不同。除此以外，在右半侧的白天，尤其是正午时刻，由于太阳和月球之间的距离的变化，导致表面温度的波动更大(图中虚线的最大值处)。由于目前关于月表层温度的知识稀少，上述结论有待进一步验证。

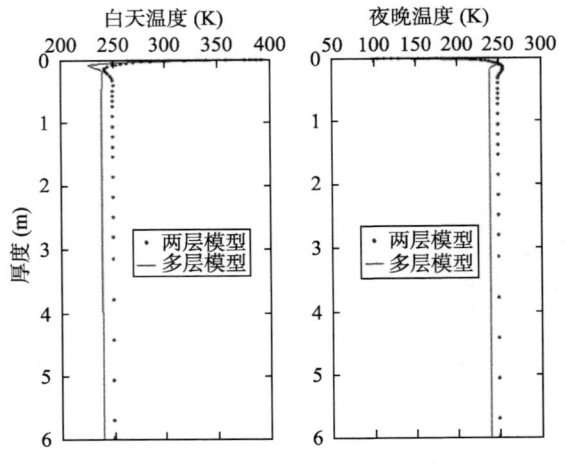

图 24　不同模型模拟的赤道表层温度剖面的差异比较

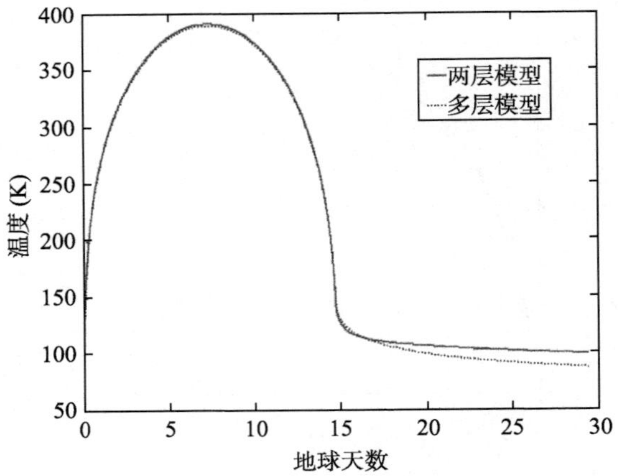

图 25　不同模型模拟的赤道表面温度的差异比较

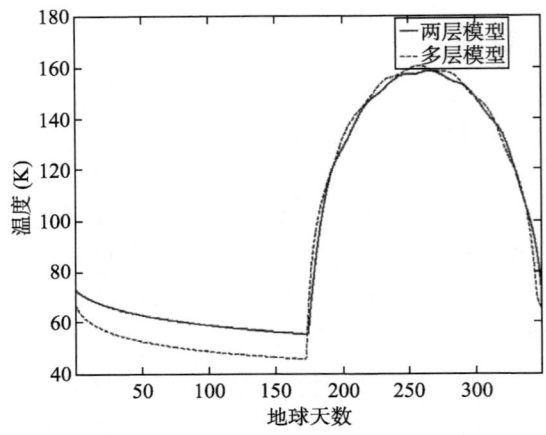

图 26　不同模型是否考虑日月距离变化而导致的北极表面温度的比较

致谢　感谢审稿专家提出的宝贵意见。

参 考 文 献

[1] 李雄耀, 王世杰, 程安云. 月球表面温度物理模型研究现状. 地球科学进展, 2007, 22(5): 480—485
[2] Lucas J W, Conel J E, Hagemeyer W A. Lunar surface thermal characteristics from surveyor 1. J Geophys Res, 1967, 72: 779—789
[3] Wieczorek M A, Huang S. A reanalysis of Appllo 15 and 17 surface and subsurface temperature series. Lunar Planet Sci, 2006, XXXVII

地基雷达技术及其在太阳系天体探测中的应用

郑 磊 苏 彦 郑永春 李春来 赵 攀

（中国科学院 国家天文台，北京 100012）

摘 要 人类利用雷达波进行天文研究距今已有40多年历史了。它是一种发射雷达波到目标天体，通过分析其回波特性来进行天文探测的技术。本文从地基雷达在太阳系天体探测中的应用出发，分析了地基雷达相比其他探测手段的优点；介绍了地基雷达的基本工作原理；给出了近年来地基雷达的发展情况和探测成果；最后从现有条件出发，探讨了我国开展地基雷达探测的设想。

关键词 地基雷达 深空探测 反射

1 引言

地基雷达是指依托地面大口径天线作为发射和接收天线的雷达探测系统。它利用雷达与目标的相对运动，通过检测回波信号的时间延迟和多普勒频移参数，来实现对目标的成像。

人类利用地基雷达开展对太阳系内天体的探测始于二战后。1946年1月，美军的一个地面站发送雷达波到月球，在不到3s之后收到了来自月球的回波。这是人类利用雷达探测太阳系天体首次收到的回波。此后地基雷达由于其灵活、主动、经济的特点，成为人类进行天体探测的重要手段之一。

地基雷达对太阳系内天体的观测可以满足多种需求。它可以完成天体表面的地形测绘、探测天体表层特性(厚度、介电常数、地质结构等)以及实现对永久阴影区的探测等，在天体探测中具有其独特和重要的作用。

地基雷达有着其他探测手段难以比拟的优点：(1)高分辨率：目前对月观测的分辨率最高可达20m[1]。这远高于地基光学望远镜的分辨率，甚至比嫦娥一号CCD相机的分辨率(120m)要高。(2)对探测目标是空间解析的，即可以获得目标的距离、速度、大小形状等方面的信息。(3)主动的多波段极化探测，与被动的探测手段不同，它通过发射特定的极化调制信号，接收时利用信号的相关，可以去除接收信号中的噪声成分。从而可以精确地将目标天体的几何特性、物理特性和观测量联系起来，使得探测结果更加可信。(4)具有一定穿透性，可以对天体表层下的结构进行探测。(5)灵活性与经济性。一套地基雷达观测系统建成后，就可以对多个天体进行探测。整套系统的建设、运行费用相对于昂贵的航天项目来说是很经济的。此外，地基雷达可以进行全天候和全天时观测。历史上有证据证明小行星撞击地球曾经给地球上的生命带来灭顶之灾。如果将来一旦需要监视对地球可能带来危害的近地小行星，地基雷达完全可以胜任。但地基雷达也存在一定的缺点：(1) 信号衰减快，信号强度与距离的4次方成反比，这就需要具有高信号发射功率和灵敏的接收系统。(2) 数据处理算法复杂，还需要对被观测天体的相对运动规律有精确的先验知识。

由此可见，地基雷达系统是进行天体探测的一种灵活有效的手段，可应用于天文研究和军事领域等方面。与发射探测卫星的耗资巨大相比，利用地基雷达不仅能在节约经费的情况下得到与前者同量级分辨率的观测结果，而且通过优化数据处理方法，能使分辨率得到提高。该技术适用范围广，不局限于探测单一

本文原载于《天文学进展》，2009，Vol.27，No.4，373~382。

类型的天体，适宜运用于各种天体的研究工作。虽然在我国利用地基雷达进行天体探测工作尚属空白，但综合这些优点可以预见，地基雷达必将广泛地应用在我国的天体探测中，并且将会大大推动我国未来深空探测工作的发展。

2 地基雷达成像原理

地基雷达向空中预定方向发射脉冲电磁波，电磁波以近似直线的路径传播，当遇到目标物时就会发生散射，返回的散射电磁波被雷达接收，被称为回波信号，该过程也被称为后向散射。

地基雷达成像原理的核心就在于"定位"。即在回波信号中将来自不同"区域"的目标散射信号区分开来，这个"区域"的大小即为地基雷达的分辨率。

地基雷达工作时，可以采用同一天线发射和接收雷达波，称为收发同置；也可采用不同的天线分别作为发射和接收天线，称为收发分置。为了便于理解其基本原理，下面以收发同置的情况进行说明。

由于地基雷达在成像时利用回波信号的时间延迟和多普勒频移参数来进行"定位"，通常在延迟多普勒坐标系(x,y,z)下进行讨论。它可以由天体的直角坐标系(x'',y'',z'')经过两次旋转得到，如图1所示。首先以O点为圆心旋转坐标轴，使得x''轴指向观测者的方向，得到坐标系(x',y',z')。然后再以x'为轴，旋转$y'z'$平面，使得z'轴与天体的视旋转轴重合，即得到此时的延迟多普勒坐标系(x,y,z)。整个过程可表示为：

$$\begin{bmatrix} x \\ y \\ z \end{bmatrix} = \begin{bmatrix} 1 & 0 & 0 \\ 0 & \cos\gamma & -\sin\gamma \\ 0 & \sin\gamma & \cos\gamma \end{bmatrix} \begin{bmatrix} \cos\lambda_{srp}\cos\theta_{srp} & \sin\lambda_{srp}\cos\theta_{srp} & \sin\theta_{srp} \\ -\sin\lambda_{srp} & \cos\lambda_{srp} & 0 \\ -\cos\lambda_{srp}\sin\theta_{srp} & -\sin\lambda_{srp}\sin\theta_{srp} & \cos\theta_{srp} \end{bmatrix} \begin{bmatrix} x'' \\ y'' \\ z'' \end{bmatrix},$$

其中λ_{srp}，θ_{srp}分别表示x'轴在$x''y''$平面上的投影与x''轴和x'轴的夹角，γ则表示z'轴与z轴的夹角[2]。

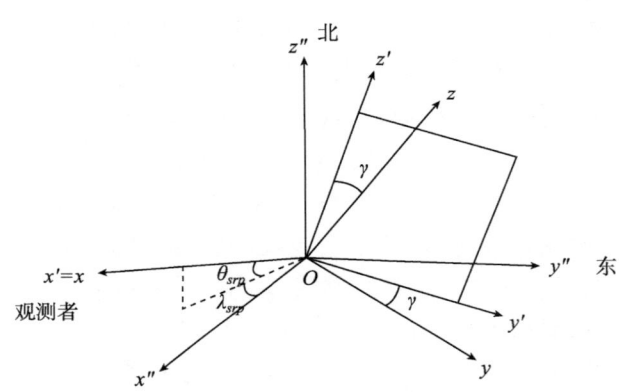

图 1　延迟多普勒坐标系与天体直角坐标系的转换关系

图2为天体延迟多普勒坐标系的示意图。在图中取x轴与天体表面的交点为SRP点，这是天体(设为月球)表面距雷达距离最近的点，取其到雷达的距离为R。则对于月面上任意一点$P(x_p,y_p,z_p)$，则P点回波信号时间延迟Δt可由$\Delta t \approx \frac{2}{c}\left[R+r-x_p\right]$得到。$P$点回波信号多普勒频移$\Delta f$则为：$\Delta f = \frac{2\pi w}{\lambda}y_p$。其中$r$为天体半径，$\omega$为天体自转频率。由上述两式可知，在天体延迟多普勒坐标系中，在与y-z平面平行的平面上以其与x轴的交点为圆心在天体表面上画圆，可得在天体表面上的一组圆，每个圆上的各点到雷达的距离相等，即对应的回波信号的时间延迟相同；类似地，在与x-z平面平行的平面上以其与y轴的交点为圆心，也能得到一组圆，每个圆上各点的多普勒频移量相同。这两组圆就像经纬线，区分出天体表面上各区域的反射信号，从而实现对天体表面的"定位"。

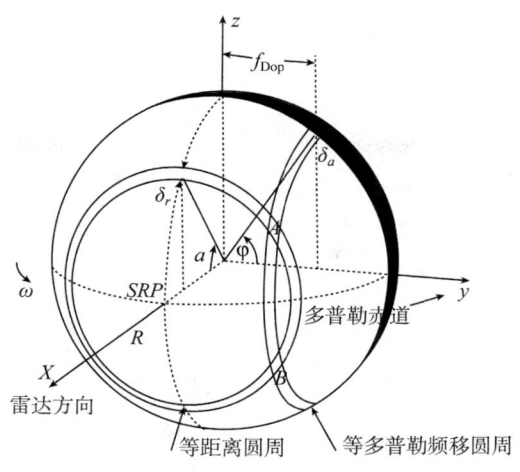

图 2 延迟多普勒坐标系示意图

值得注意的是,天体上相对于 x-y 平面南北对称的两点(A,B),其回波信号具有相同的时间延迟和多普勒频移,则在接收端无法区分,这种现象称为"南北模糊"。对波束宽度较窄的发射天线,可以对南北半球分别进行探测,从而避免南北模糊。否则只能采用双天线接收,利用干涉处理的方法来消除南北模糊[3,4]。

从上面的讨论可知,准确地提取回波信号中的时间延迟和多普勒频移参数是地基雷达成图的关键。

设载频信号为 $\exp(j2\pi f_c t)$,脉冲信号以重复周期 T 依次发射,即发射时刻为 $t_m = mT$ (m 为脉冲个数),称为慢时间。在每个脉冲内,以发射时刻为起点的时间用 t_k 表示,称为快时间。快时间用来计量单个脉冲传播的时间,而慢时间则反映发射了多少个脉冲,这两个时间与时间 t 的关系为 $t_k = t - t_m$,因而发射信号可以表示为:

$$s_i(t_k, t_m) = a_r(t_k)\exp\left[j2\pi\left(f_c t + \frac{1}{2}\gamma t_k^2\right)\right],$$

其中 $a_r(t_k) = \begin{cases} 1, & \left|\dfrac{t_k}{T_p}\right| \leq \dfrac{1}{2} \\ 0, & \left|\dfrac{t_k}{T_p}\right| > \dfrac{1}{2} \end{cases}$,$f_c$ 为载频,T_p 为脉宽,γ 为调频率。这是雷达波常用的线性调频信号,可以起到脉冲压缩的作用。观测区域中任意一点 P 的回波信号的基频信号(去载频)可表示为:

$$s(t_k, t_m) = a_r(t_k - \alpha)\exp\left[j\pi\gamma(t_k - \alpha)^2\right]\exp[-j2\pi f_c \alpha]。$$

其中 $\alpha = \dfrac{2R(t_m)}{c}$ 表示回波信号的时间延迟,c 表示光速,$R(t_m)$ 是 P 点距雷达的瞬时距离(一般来说,脉冲周期非常短,因此瞬时距离可近似认为与 t_k 无关)。

首先进行距离匹配滤波,提取出 P 点距离信息,也就是时间延迟参数。其匹配函数为:

$$s_r(t_k) = a_r(t_k)\exp\left(-j\pi\gamma t_k^2\right)。$$

得到信号:$s(t_k, t_m) = A\mathrm{sinc}[\Delta f_r(t_k - \alpha)]\exp[-j2\pi f_c \alpha]$。其中,$A$ 为距离压缩后点目标信号的幅度,Δf_r 为线性调频信号的频带。sinc 函数的表达式为 $\mathrm{sinc}(a) = \dfrac{\sin(\pi a)}{\pi a}$,它在 $\pi a = 0$ 时取最大值,因此通过检测距离匹配滤波输出信号的峰值时刻,则可提取出时间延迟量[5]。

根据雷达与观测目标的相对运动规律,进行距离补偿,就可以将不同脉冲中,来自同一区域的信号序列组合在一起,进行傅里叶变换。通过对其频域的分析,则可提取出多普勒频移参数[4]。即实现了对目标天体表面的定位。最后根据延迟多普勒坐标系和直角坐标系的转换关系,完成对天体表面的成图。

3 研究进展

二次大战以后,地基雷达逐渐发展成为天文观测的一种重要手段。到目前为止,地基雷达已经被应用在月球、火星、金星、水星以及小行星探测中,得到了许多探测成果。特别是美国的金石太阳系雷达(GSSR,Goldstone Solar System Radar)和设在波多黎各的Arecibo天文台地基雷达,取得了令人瞩目的成果。表1中列出其部分参数。

表1 国外主要地基雷达探测系统参数

名称	发射天线口径/m	频段/MHz	发射功率	接收天线
Arecibo	305	430 2380	1MW	Arecibo 305m Green Bank 100m
Golden stone	70	8560	500kW	DSS-13 34m Green Bank 100m VLA 天线阵

3.1 月球探测

月球是地基雷达探测应用最早也是最多的天体。早在20世纪六七十年代就进行了地基雷达月球地形测绘。Schubert等人开展了月面撞击坑形貌学研究,如坑环高度与坑深的比值、坑深度与直径的比值[6];Schaber等人根据雨海盆地异常低的雷达反射率,研究了盆地内的岩浆流[7]。后来,随着月球探测中激光高度计和立体相机的成功应用,地基雷达在对月球正面中低纬度地区的地形测绘中的应用价值逐渐丧失,目前已经很少开展。但是由于月球两极存在永久阴影区,这些阴影区是可见光波段探测手段无法观测的,而微波探测手段不受光照影响。因此,这些区域是当前地基雷达月面探测的重点。

地基雷达可以用来绘制月球两极的地形图,特别是在未来最有希望建立月球基地的南极,这项工作尤为重要。2008年2月27日,NASA公布了由金石太阳系雷达获得的迄今为止最高分辨率(20m/像素)的月球南极地形图(图3)[8]。该图像比Clementine月球探测器获得的分辨率高50倍。它揭示了月球南极的地形比以前认知的更加高低不平,为学者选取着陆点提供了非常有价值的参考信息。

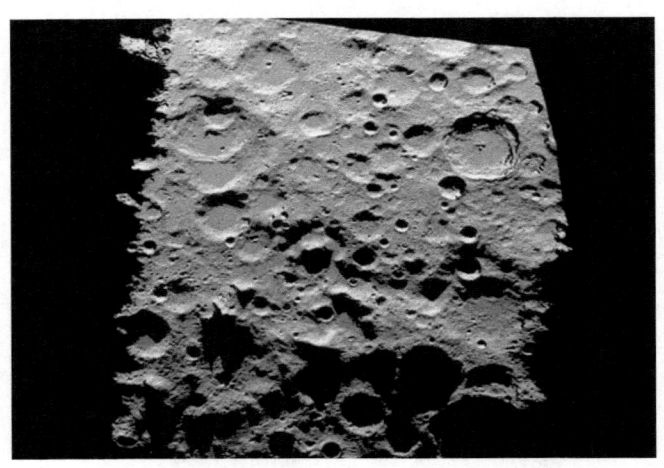

图3 金石太阳系雷达获取的月球南极地形图[8]

地基雷达也可用于两极的月壤特性研究以及寻找水冰的工作。图4是利用Arecibo地基雷达发射、Green Bank 100m射电望远镜(GBT,美国西弗吉尼亚州)接收的月球南极雷达圆极化率图。探测波长70 cm,图中将雷达阴影区的圆极化率设为0,整个区域圆极化率从0.5(黑色)到1.1(白色)不等。由于圆极化率与天体表面的岩石丰度相关,根据图4就能推演出极区的月壤岩石丰度特性[9]。如图中Zucchius, Hausen, Moretus,

Schomberger 等 4 个环形山附近出现低圆极化率的圆晕，说明这些地区的喷发物中石块含量较低。

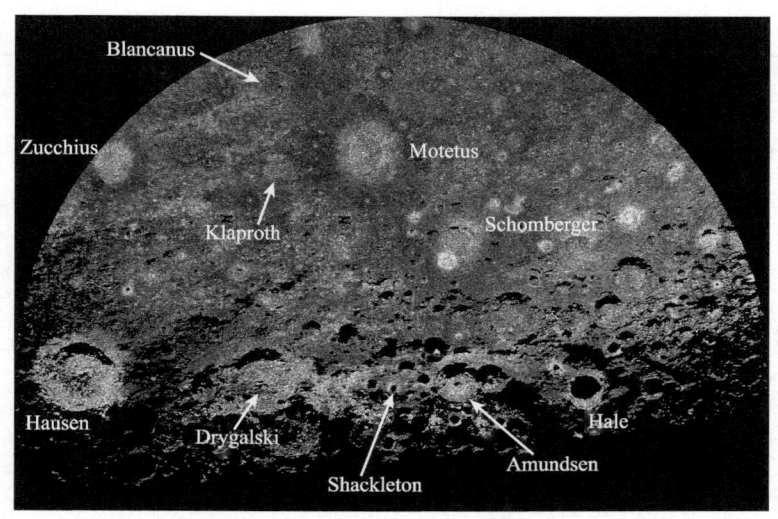

图 4　月球南极地区的雷达极化率图[9]

研究人员也能通过比较目标区域回波特性与水冰回波特性的异同，来寻找大面积的水冰。1992 年，美国康纳尔大学的 Stacy 等人利用 Arecibo 天文台 S 频段雷达波，以 125m 空间分辨率、双极化方式对月球极地进行观测，搜寻月球极地永久阴影区的水冰。探测中没有发现任何一块面积大于 $1km^2$ 的区域存在高雷达后向散射截面和高圆极化率，即没有发现月球极地存在大面积分布的水冰[10]。

地基雷达还能用于月壤厚度的探测。Shkuratov 等人根据 1981–1984 年 Arecibo 天文台获得的 70cm 波长雷达后向散射全极化回波马赛克影像，结合 Clementine 月球探测器对月面铁和钛分布的光谱数据，获得了月球正面月壤厚度分布[11]。

3.2　探测火星

随机长编码延迟-多普勒技术以及干涉延迟-多普勒技术为地基雷达的火星探测提供了可能[12]。美国深空网下属的金石站 70 m 天线 1986 年完成升级改造以后，除 1997 年因火星大冲而中断观测外，从未间断过火星观测。利用火星雷达观测数据可以提取火星表面高程、反射率、粗糙度等信息[12]，其经度方向的空间分辨率达 20 km，纬度方向达 150 km。这些观测数据提供了火星地形和地质物理等有关的直接信息。通过地基雷达对火星表面的探测，可以了解在火星上建立无人着陆场的危险性和火星车的通行条件，可以估算火星表面物质的介电常数等[13]。目前这些数据已经成功应用于 Viking, Pathfinder, 2003 Mars Exploration Rovers 等火星探测活动的着陆点选择、评估和验证[14,15]。

3.3　探测金星

由于金星表面被浓密的大气层遮蔽，光学观测对其无能为力，因此一直不能准确测量金星的自转周期和半径。1961 年，地基雷达首次获得了金星的雷达回波，并揭示了金星的自转周期为 243d。1981 年，Arecibo 天文台的地基雷达获得的图像显示，金星上存在明亮的环状结构，其分布密度很小，这些环状结构被认为是火山喷发或小天体撞击形成的，但其分布密度明显小于月球和木卫一 Io[16]。1999 年，Arecibo 天文台地基雷达再次对金星进行探测，其分辨率为 2 km。发现了在 1997 年麦哲伦探测器观测到且已被命名的金星上口径超过 5 km 的撞击坑外的两个新撞击坑，口径分别为 41 km 和 19 km[17]。这两个撞击坑刚好处在麦哲伦探测器覆盖区域的缝隙中，可见地基雷达可以为其他的探测手段进行验证，两者互为补充。

表 2 中列出了不同探测手段获得的金星表面雷达图像的空间分辨率。随着雷达器件以及数据处理算法

的改进，金星地基雷达探测的分辨率上还存在很大的提高空间。

表 2　不同探测手段获得的金星表面雷达图像的空间分辨率

探测平台	探测手段	空间分辨率
Pioneer Venus Orbiter	雷达高度计	>10km
Arecibo 天文台	地基雷达	2km
麦哲伦号探测器	雷达	110m

3.4　探测水星

1965 年 4 月，当水星接近地球在内合(Inferior conjunction)的位置时，Arecibo 天文台利用地基雷达的观测结果，发现水星的自转周期是 59d。这项惊人的发现，使得大家对于水星的自转运动有了崭新的认识。

3.5　地基雷达探测新技术

近年来，新发展的地基雷达天文探测技术——雷达散斑变换干涉测量技术(RSDI, Radar Speckle Displacement Interferometry)，可以测量月球的物理天平动，估算月球的瞬时自旋轴的倾角及其随时间的变化[18]。特别是，RSDI 是探索水星天平动和倾斜度问题的最有希望的地基天文技术之一[19]。

2001—2002 年，雷达散斑变换干涉测量技术在美国和欧洲已经成功得到应用。通过地基雷达的散斑变换干涉测量，确定了水星的极轴[20]。

另外，地基雷达也是水星永久阴影区的重要探测手段。1999 年 Harmon 等利用 Arecibo S 波段雷达波获得了水星北极地区的雷达图像(图 5)[21]。观测波长：12.6 cm；空间分辨率：1.5 km；图幅大小为 450 km。图中明亮的地区被认为是永久阴影区的水冰沉积。

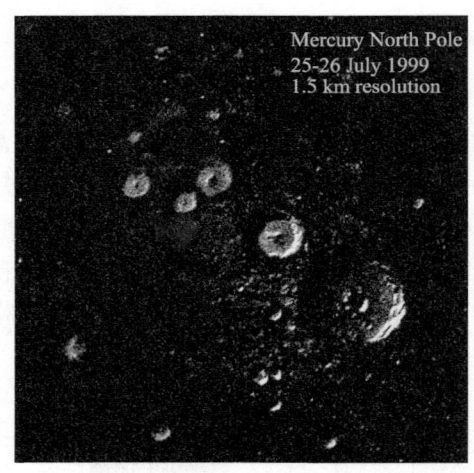

图 5　Arecibo S 频段雷达波获得的水星北极地区的雷达图像[21]

3.6　小行星探测

小行星表面的雷达反射率与近表层物质的密度相关，而极化率和与波长同尺度的近表层结构相关。因此，利用地基雷达可以观测小行星的表面性质[22]；同时还能精确测定近地小行星的形状，进而提高近地小行星轨道的预测精度。目前观测的小行星主要有 340 个，其中 222 是近地小行星，118 个位于小行星带。地基雷达探测可以获得近地小行星的高分辨率成像和三维地形特征。Arecibo 和 Goldstone 雷达系统获得的小行星的影像足以与航天器搭载的最高空间分辨率相媲美[23]。在过去几年里，美国金石太阳系雷达每年都能新发现近地小行星。

从上述国外的探测成果来看,地基雷达在天体探测中应用的面非常广,而且适用于各种天体,是非常实用的一种探测手段。

4 国内地基雷达天体探测

地基雷达的天体探测在我国尚属空白,而现在正是开展地基雷达天体探测的一个良好契机。我国成功实现了月球探测,又正在与俄罗斯合作进行火星探测,深空探测活动处于蓬勃开展的时期。但仅依靠周期相对较长的通过运载火箭发射卫星进行探测,是无法满足深空探测的迫切需要的。地基雷达技术作为深空探测的一个灵活有效的手段,必将广泛应用在深空探测中,并大大推动我国深空探测的发展。

现在我国已基本具备了进行地基雷达天体探测的条件。首先我国的机载与星载雷达探测技术均已比较成熟,拥有利用雷达数据成像的经验,为地基雷达的数据处理打下了一定的基础。其次在发射和接收设备方面,我国现有部分地基雷达设备,以前主要运用在导航、预警或电离层探测等方面,稍加改造后就可运用在天体探测中。云南曲靖 29 m 天线和发射机正在进行改造,预计建成后工作频率为 500 MHz,发射功率约 2 MW,脉冲宽度 20 μs,可以获得较强的雷达回波,能基本满足地基雷达观测的需要;另外,随着我国近年来在大口径射电望远镜研制方面的突破,现在已建成北京密云的 50 m 射电望远镜以及云南昆明的 40m 射电望远镜,均可作为接收天线,或者将其改造成为发射天线。特别是目前正在贵州兴建的 500 m 射电望远镜,更是以后地基雷达探测的理想设备。因此,在当前环境因素下,我国地基雷达天体探测的开展在技术上可以逐步开展,可以先寻求国际合作,申请其他国家现有的地基雷达系统进行月球探测,利用我国自主研发数据处理系统,对其探测数据进行处理;作为验证手段,将得到的探测结果与我国嫦娥工程得到的探测结果进行比对。在数据处理技术日益成熟的基础上,再着手改造发射和接收设备,独立进行地基雷达的天体探测;在探测目标上,也采取由易趋难的原则,先从月球探测入手,重点在月球极地(特别是永久阴影区)的高分辨率地形测绘和月壤特性(包括月壤厚度、介电常数、次表层结构等)的反演。既可为我国现有的月球探测数据提供验证,也可以为建立月球基地的选址提供重要参考。通过对月球的探测达到技术成熟后,就可以逐步对其他天体进行探测。例如对火星表面地形测绘,可以获得较高的精度,为火星车的探测路线规划提供借鉴;对近地小行星(包括空间碎片)的观测和监视是近年来地基雷达观测的热点,可以获得小行星形状、表面性质、运行轨道等参数;近地小行星阿芬塞斯(Apophis)将在 2029 年和 2036 年与地球相遇,并有可能撞击地球,引起世界各国的广泛关注。通过地基雷达探测阿芬塞斯的形状、大小、旋转状态、估算其运行轨道,评估撞击地球的可能性,将为地基雷达探测其他小天体提供经验借鉴;将来水星、金星、土星环、流星、彗星和小行星等太阳系各类天体,以及地球电离层等都可以成为地基雷达的探测目标。

当然,在发展地基雷达天体探测技术的道路上还将会面临很多困难。目前存在的主要难点是还没有地基雷达探测太阳系天体的数据处理经验,技术储备也十分有限;尤其是在数据处理设备的研制上,要求精度高,需要精确解析探测目标的相对规律;另外还有,人员储备有限,需要逐步培养和建立地基雷达探测的科研团队;地基雷达需要进口大功率发射机等;但是大势所趋,只要坚持由易趋难,渐进发展的原则,通过实践掌握关键技术,相信地基雷达技术终将会在我国天文探测工作中大放异彩,为深空探测服务。

参 考 文 献

[1] Hensley S. Radar Conference, RADAR'08, IEEE, 2008: 1.
[2] Bruce A, Campbell, Donald B, Campbell, et al. Transactions on Geosciences and Remote Sensing. IEEE, 2007, 45(12): 4032
[3] Campbell B A. Radar Remote Sensing of Planetary Surfaces. UK: Cambridge University Press, 2002: 97
[4] Stacy N J S. PhD Dissertation. Ithaca: NY Cornell Univ, 1993

[5] 保铮，邢孟道，王彤. 雷达成像技术，北京：电子工业出版社. 2004：124
[6] Schubert G, Lingenfelter R E, Terrile R J. Icarus, 1977, 32(2): 131
[7] Schaber G C, Thompson T W, Zisk S H. Earth, Moon and Planets. 1975, 13: 395
[8] http://www.nasa.gov/exploration/home/022708.html, 2008
[9] Campbell B A, Campbell D B. Icarus, 2006, 180(1): 1
[10] Stacy N J S, Campbell D B, Ford P G. Science. 1997, 276(6): 148
[11] Shkuratov Y G, Bondarenko N V. Icarus, 2001, 149(2): 329
[12] Haldemann A F C, et al. Workshop on Concepts and Approaches for Mars Exploration, TX: Houston, 2000
[13] Moore H J, Jakosky B M. Icarus, 1989, 81(1): 164
[14] Haldemann A F, Larsen K W, Jurgens R F, et al. AGU Fall Meeting Abstracts. 2001, 42, 0568
[15] Jurgens R F, Haldemann A F C, Larsen K W, et al. Bulletin of the American Astronomical Society, 2001, 33: 1103
[16] Cutt J A, Thompson T W, Lewis B H. Icarus, 1981, 48(3): 428
[17] Burba G A, Schaber G G, et al. Lunar and Planetary Science, 1999, 283: 2062
[18] Holin I V. The Moon Beyond 2002: Next Steps in Lunar Science and Exploration, 2002: 25
[19] Holin I V. Solar System Remote Sensing, 2002: 33
[20] Holin I V. Meteoritics & Planetary Science, 2002, 37: A65
[21] Harmon J K. Perillat P J, Slade M A. Icarus, 2001, 149(1): 374
[22] Binzel R P, A'Hearn M, Asphaug E, et al. Planetary and Space Science, 2003, 51(7-8): 443
[23] Margot J L, Nolan M C. AAS/Division for Planetary Sciences Meeting, 1999: 31

天然掺杂铁氧体的电磁参数调控机制分析及其在吸波材料中的应用

郑永春[1,2]　王世杰[2]　冯俊明[2]　李春来[1]　欧阳自远[1,2]　刘建忠[1]　李晓彪[2]

(1.中国科学院国家天文台，北京　100012；
2.中国科学院地球化学研究所环境地球化学国家重点实验室，贵阳　550002)

摘　要　对不同地质产状类型天然铁氧体的电磁参数测量结果发现，岩浆岩型(A 型)天然铁氧体的电磁参数明显区别于其他类型铁氧体，是一种可用于微波吸收材料的天然矿物。该矿物经选矿工艺制备后，在 8~12GHz 频率范围内，ε'=58.60，ε''=10.0，μ'=1.2~1.5，μ''=1.0~1.2，属于具有一定磁导率、低介电常数、高电磁损耗的磁性物质，其粉体适合作为吸波材料的磁性吸收剂。对 A 型天然铁氧体的矿物组成和化学成分的分析结果表明，与单一磁铁矿或二元型铁矿物组成的天然铁氧体相比，A 型天然铁氧体中的杂质元素和另相矿物含量明显高于其他天然铁氧体，其中的杂质成分对电磁参数起着重要的调控作用，这种调控作用有利于降低吸波材料的反射系数，并提高电磁波吸收效率。而且，A 型铁氧体中的这种天然掺杂导致的结构缺陷及其对电磁参数的影响效应是人工制备的铁氧体难以实现的。材料应用试验与吸波性能测量结果表明，以 A 型天然铁氧体作为粉体填料与天然橡胶制成的吸波片材在 8~12 GHz 频率范围内的电磁波吸收频宽和反射系数均优于传统的羰基铁粉与橡胶制成的吸波片材，完全可以替代羰基铁粉应用于微波吸收材料的开发，而且具有原料丰富、制备工艺简单、成本低、吸波性能稳定的比较优势。

关键词　天然掺杂　铁氧体　复介电常数　复磁导率　微波吸收材料

微波吸收材料(microwave absorbing material，MAM)又称雷达波吸收材料(radar absorbing material，RAM)，是一种能够吸收和衰减入射的电磁波，并将电磁能转化成热能或其他形式的能量而耗散掉，从而使电磁波的反射、散射和透射都很小的功能材料，包括吸波胶片、吸波涂层、泡沫吸波材料等几种类型。

雷达在工作时，由于周围物体的多重反射、杂乱回波及彼此干扰而影响系统的正常工作，应用吸波材料可以消除雷达魅影、降低雷达截面积、改善天线方向图，从而抑制这些干扰[1]。同时，微波吸收材料在消除微波设备使用中的环境干扰或内部吸收屏蔽以防止微波泄漏、微波暗室吸波材料、抗电磁波干扰(electro-magnetic interference，EMI)遮蔽材料、隐身技术、电磁辐射防护材料等方面都有广泛的应用。随着现代雷达与微波电子技术的飞速发展，微波吸收材料在国民经济和国防建设中的应用领域也越来越广。

石墨、碳精粉、SiC、SiN、磁性高分子、金属超细粉、功能陶瓷、铁氧体(ferrites)等是主要的吸波材料填料，尤以铁氧体占最重要地位[2]。铁氧体是氧化物软磁性材料的总称，主要分为尖晶石系(spinel)、六方晶系(hexagonal)和石榴石系(garnet)三类，属于磁介质型吸波材料。目前，我国在微波吸收材料的研制和生产中仍广泛地使用着从 20 世纪 60 年代开始的羰基铁粉作为吸波材料中的吸收剂。由于该材料吸收性能不够理想，产品性能受合成工艺影响，从而使产品的合格率不够稳定。

我国幅员辽阔，地质地理条件复杂多样，天然矿物资源种类繁多，储量丰富。通过对天然矿物微波电

本文原载于《中国科学　E 辑：技术科学》，2006, Vol.36, No.5, 550~559。

磁参数的研究，从中筛选出具有微波吸收性能的优势矿物，利用矿物学和地球化学的研究手段，分析其矿物组成、矿物结构、杂质成分、粒度、粒形等因素对其微波电磁参数的影响效应，阐明天然矿物的电磁参数调控机制。在此基础上，采用材料超细粉碎和精选纯化技术，去除不利于微波吸收的因素，保留微波吸收性能突出的优势矿物和组分。然后，以非金属有机高分子树脂(如环氧树脂、热塑料、天然橡胶等)为基体，均匀分散地填充吸波矿物粉体；由介电常数和损耗较低的特殊纤维(如石英纤维、玻璃纤维等)进行增强，制成复合材料。这种复合材料既能吸收电磁辐射、减弱电磁波散射又能承受一定的荷载，而且比普通金属材料重量轻、机械强度高。

在自然界的天然矿物中，存在着多种铁氧体矿物。通过对这些矿物的电磁参数测量，发现某些矿物具有开发成为微波吸收材料的潜力。但大多数学者认为，由于矿物在自然界中形成时的化学热力学和化学动力学条件存在差异，导致天然矿物大多含有不同程度的杂质成分，其结构和组成也存在着较大可变性，从而造成吸波材料的性能不稳定。因此迄今为止，国内外仅有少数几种基于天然矿物开发的微波吸收材料[3,4]。本文提出利用我国丰富的天然铁氧体矿物，通过超细粉碎、精细选矿、人工制备以强化性能等加工，制成吸波性能良好的粉体填料。该粉体填料电磁性能优良，且原料丰富、成本低廉、制备工艺简单，在微波吸收材料中具有良好的开发应用前景，可广泛应用于国民经济和国防装备各领域。

1 材料吸收电磁波的两个条件

1.1 吸波材料与自由空间达到阻抗匹配

在吸波材料研制时，需考虑两个基本条件。首先，吸波材料的表面与自由空间要求达到阻抗匹配，使入射的电磁波在自由空间/吸波材料界面不产生反射而进入介质内部。当电磁波入射至厚度为 d 的铁氧体材料表面时，一部分电磁波在材料表面发生反射，另一部分经材料衰减吸收后透射出去。只有当电磁波透入材料内部时，材料内部的各种吸收机制才能发挥作用。当入射波为平面电磁波，则吸收体归一化的输入阻抗 Z_{in} 为

$$Z_{in} = \left(\frac{\mu_r}{\varepsilon_r}\right)^{1/2} \tanh\left[j\left(\frac{2\pi f d}{C}\right)\left(\frac{\mu_r}{\varepsilon_r}\right)^{1/2}\right] \tag{1}$$

其中，$\mu_r=\mu'-j\mu''$ 为吸波材料的复磁导率，μ' 和 μ'' 分别为复磁导率的实部和虚部；$\varepsilon_r=\varepsilon'-j\varepsilon''$ 为吸波材料的复介电常数，ε' 和 ε'' 分别为复介电常数的实部和虚部，$j=\sqrt{-1}$。d 为吸波材料厚度，f 和 C 分别为入射电磁波的频率和光在真空中的传播速度。当电磁波由阻抗为 Z_0 的自由空间入射至阻抗为 Z_{in} 的吸波材料表面时，将发生部分反射，其反射系数如下式：

$$R.L.(dB) = 20\lg\left|\frac{(Z_{in}-1)}{(Z_{in}+1)}\right| \tag{2}$$

为使反射系数为零，就需要使吸波材料的输入阻抗 $Z_{in}=1$，即吸波材料与自由空间之间达到阻抗匹配。根据(1)和(2)式，阻抗匹配与否取决于6个参数，即 ε', ε'', μ', μ'', f 和 d。

理论上讲，微波吸收材料应尽可能满足 $\mu_r=\varepsilon_r$，以利用最薄的吸波材料达到最佳的吸收效果。但在宽频带范围内 $\mu_r=\varepsilon_r$ 的材料是不存在的。因此，为了达到阻抗匹配和反射系数最小化，应尽量满足：

$$\frac{\mu_0}{\varepsilon_0} = \frac{\mu_r}{\varepsilon_r} \tag{3}$$

其中，$\mu_0=4\pi\times 10^{-7}\text{N}\cdot\text{A}^{-2}$，为自由空间的磁导率，$\varepsilon_0=8.85419\times 10^{-12}\text{F/m}$，为自由空间的介电常数，即 $\frac{\mu_0}{\varepsilon_0}=376.7\Omega$。因此，具有吸波材料开发潜能的天然铁氧体矿物，应满足微波频率下 ε_r 越小，而 μ_r 越大，

这样才能使材料的电磁波反射系数尽可能小。

1.2 吸波材料的电磁波吸收机制

在吸波材料的反射系数最小化的前提下，电磁波入射进入吸收体，这时需要考虑微波吸收材料的第二个基本条件，即物质的电磁波吸收性能最好，电磁能衰减量最大。已有研究表明，对具有双复特性的铁氧体材料而言，损耗主要来源于磁矩进动过程中阻尼作用所产生的各种磁损耗、电极化和传导所产生的介电损耗。因此，用损耗角正切表示的微波吸收材料的损耗为

$$\tan\delta = \tan\delta_m + \tan\delta_e \tag{4}$$

其中，$\tan\delta_m$ 为磁损耗角正切，$\tan\delta_e$ 为介电损耗角正切。

$$\tan\delta_m = \frac{\mu''}{\mu'}, \quad \tan\delta_e = \frac{\varepsilon''}{\varepsilon'} \tag{5}$$

由(4)和(5)式可知，微波吸收材料的 ε''(表征材料的介电损耗)和 μ''(表征材料的磁损耗)越大，吸波材料的损耗就越大，电磁波吸收性能就越好。

公式(4)中，介电损耗角正切 $\tan\delta_e$ 为

$$\tan\delta_e = \frac{\varepsilon''}{\varepsilon'} = \frac{\dfrac{\sigma}{f\varepsilon_0} + \varepsilon_r\sin\alpha}{\varepsilon'} \tag{6}$$

其中，σ 为吸波材料的电导率，f 为工作频率，α 为电感滞后角。

由(6)式可知，材料的电导率 σ 也会影响其介电损耗，电导率越高，介电损耗角正切越大，吸波效率越高。通过提高铁氧体的电导率 σ，降低介电常数 ε'，就可提高吸波材料的介电损耗[5]。

因此，反射系数最小化要求吸波材料与自由空间达到阻抗匹配，微波吸收效率最大化要求材料的电磁损耗最大化。综合考虑两个方面的要求，通过提高材料的 ε'' 和 μ''，同时适当降低 ε' 和 μ'，将使吸波材料的微波吸收性能得到提升。

2 天然掺杂铁氧体的电磁参数调控机制分析

2.1 地质产状对天然铁氧体电磁参数的影响

在天然铁氧体矿物中，常见的两大类型为 $M^{2+}Fe_2O_4$ 型和 Fe_2O_3 型[6]。我们测量了采自国内不同地区、不同地质产状类型的天然铁氧体矿物的电磁参数(表 1，2)。复介电常数测量采用谐振腔微扰法[7,8]完成，测量中心频率为 9.37 GHz，复磁导率测量由南京大学信息物理系完成[9]。

表 1 不同地质产状类型 $M^{2+}Fe_2O_4$ 型天然铁氧体矿物的微波电磁参数

类型	地质产状	复介电常数		复磁导率		矿物组成与结构
		ε'	ε''	μ'	μ''	
A	岩浆岩型	58.60	10.0	1.2~1.5	1.2	呈多元型铁氧体矿物混晶结构，并含少量硫化物矿物
B	沉积变质型	68.93	2.70	1.0	<1.0	单一磁铁矿
C	热液型	122.70	3.01	1.0	<1.0	单一磁铁矿
D	接触带型	83.28	1.90	1.0	<1.0	二元型铁矿物(磁铁矿+钛铁矿)结构，杂质少
E	岩浆岩型	111.70	2.80	1.0	<1.0	单一磁铁矿

从不同地质产状类型 $M^{2+}Fe_2O_4$ 型和 Fe_2O_3 型天然铁氧体的微波电磁参数比较可以看出，岩浆岩型

$M^{2+}Fe_2O_4$ 型天然铁氧体矿物(以下简称 A 型)的 ε'' 和 μ'' 明显高于其他产状类型的铁氧体矿物,可作为吸波材料粉体填料的首选物质。我们进一步研究了天然铁氧体中的杂质元素和杂质矿物对其电磁参数的影响,以分析其电磁参数调控机制。

表 2　不同地质产状类型 Fe_2O_3 型天然铁氧体矿物的微波电磁参数

类型	地质产状	复介电常数		磁导率
		ε'	ε''	μ'
a	沉积型	13.60	1.16	< 1.0
b	沉积型	8.32	1.49	< 1.0
c	岩浆岩型	9.40	0.54	< 1.0
d	不详	10.89	0.21	< 1.0
e	热液型	13.40	0.54	< 1.0
f	变质型	12.04	0.36	< 1.0

2.2　天然铁氧体中杂质元素对电磁参数的影响

化学成分对天然铁氧体的电磁参数起着重要的调控作用,尤其是矿物中所含的微量杂质元素对铁氧体的电磁参数影响很大。表 3 为我国不同地质产状类型 $M^{2+}Fe_2O_4$ 型铁氧体的化学成分。

表 3　不同地质产状类型 $M^{2+}Fe_2O_4$ 型天然铁氧体矿物的主量元素组成(质量分数)

类型	SiO_2	TiO_2	Al_2O_3	Fe_2O_3	FeO	MgO	MnO	CaO	V_2O_5	P_2O_5	S	合计
A	1.52	14.39	2.63	78.84	−	2.43	0.32	0.47	0.72	−	0.34	101.57
B	0.44	0.12	0.15	66.90	30.84	0.01	−	0.07	−	−	0.13	98.69
C	0.15	−	−	69.16	30.90	0.08	0.18	−	−	−	−	100.47
D	−	7.51	−	91.17	−	−	−	−	0.70	0.10	0.001	99.48

由表 3 可知,虽然不同地质产状类型 $M^{2+}Fe_2O_4$ 型天然铁氧体的主要化学成分都为 Fe_2O_3 和 FeO,但其中所含 Mg,Ca,Al,Mn,V 和 S 等杂质元素的含量存在较大差异。结合表 1 中铁氧体的电磁参数数据,可以发现杂质元素的参与明显影响着铁氧体矿物的复介电常数和复磁导率。随着杂质元素含量的增加,铁氧体的 ε' 减小,ε'' 增大,μ' 与 μ'' 也所有变化,从而更有利于阻抗匹配和电磁波吸收。A 型天然铁氧体矿物的杂质元素含量大大高于其他类型铁氧体,其电磁参数也最优。

由电子探针分析结果可知(5 个点的平均值,表 4),A 型天然铁氧体中的基质部分含有 3%(质量分数)左右的 Al_2O_3 和 MgO 等,其他杂质元素主要寄宿在含镁钛铁矿和尖晶石中。多种价态杂质元素的金属离子进入 $M^{2+}Fe_2O_4$ 型天然铁氧体矿物取代 M^{2+}(图 1~4),这些低价或高价态的低极化率离子进入晶体结构,形成晶体缺陷结构并引起缺陷电导效应,从而导致天然铁氧体矿物的 ε' 减小,而 ε'' 增大。

表 4　A 型天然铁氧体的电子探针分析结果

	化学成分/%(质量分数)							矿物含量/%(体积分数)
	FeO	Al_2O_3	TiO_2	V_2O_5	MgO	MnO	合计	
铁氧体基质部分	81.76	2.98	12.03	0.72	3.02	−	100.51	95
含镁钛铁矿	41.67	−	50.57	−	5.82	1.15	99.21	1~2
尖晶石	18.35	55.42	5.43	0.53	20.50	0.29	100.52	2~3

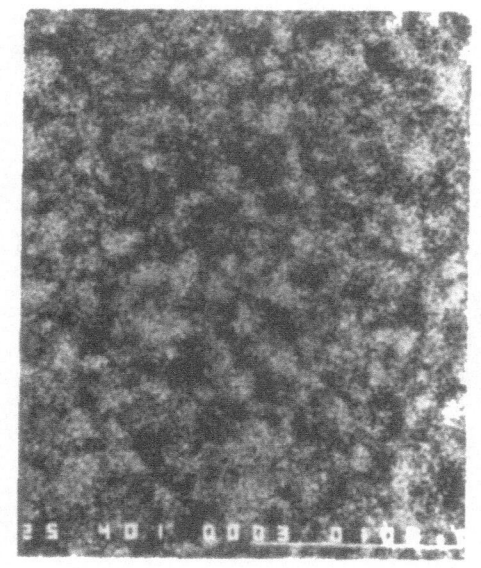 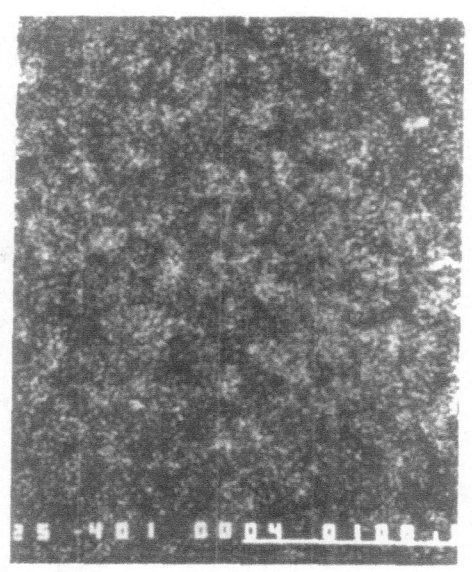

图 1 A 型天然铁氧体中 Fe^{3+} 离子分布(放大倍数: ×540 倍)　图 2 A 型天然铁氧体中 Ti^{4+} 离子分布(放大倍数: ×540 倍)

 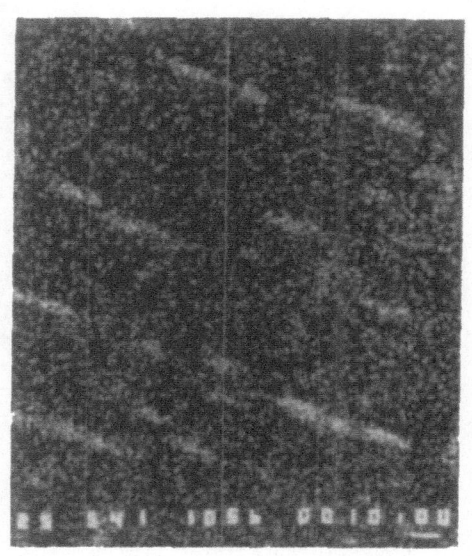

图 3 A 型天然铁氧体中 Al^{3+} 离子分布(放大倍数: ×540 倍)　图 4 A 型天然铁氧体中 Mg^{2+} 离子分布(放大倍数: ×540 倍)

2.3　天然铁氧体中杂质矿物对电磁参数的影响

天然铁氧体型吸波粉体的选矿流程是：先从初始铁氧体矿石中脱去脉石和硫化矿物，分离出以粒状钛铁矿和钛磁铁矿复合矿物相为主的铁(钒)精矿产品，再经过深加工和表面处理，最终得到吸波材料的粉体填料(图 5)。岩矿鉴定结果表明，铁氧体吸波粉体中最主要的含铁矿物为钛磁铁矿，以粒状集合体形态产出，颗粒直径多为 0.2~0.6 mm，实际上是具固溶体分解结构的复合矿物相，包括磁铁矿、钛铁晶石(粒径约 0.5 μm)、镁铝尖晶石(粒径约 1~4 μm)、钛铁矿片晶(粒径约 15 μm)、微细粒磁黄铁矿以及类质同象的含钒、铬、镓等矿物组成。磁铁矿约占吸波粉体的 75%(体积分数)，杂质矿物以微细颗粒均匀镶嵌在磁铁矿中，含量约占 25%(体积分数)，其中尖晶石、含镁钛铁矿等约占 2%~3%(体积分数)(图 6)。尖晶石属低介电常数矿物，ε' 约为 5~6(实测值)，而钛铁矿的 ε' 实测值约为 31，均低于磁铁矿的 ε'[7,8]。因此，尖晶石、含镁钛铁矿等另相杂质矿物的存在使 A 型天然铁氧体矿物的 ε' 显著降低。同时，这些杂质矿物也是钛、钒、钴、镍、镓、铬等杂质金属元素的载体矿物。

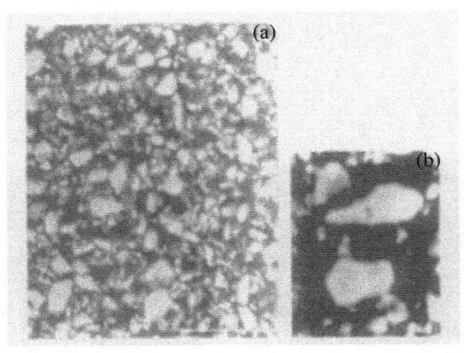

图 5 A 型天然铁氧体吸波粉体的颗粒形态
(a)放大 400 倍，(b)放大 1600 倍

图 6 A 型天然铁氧体中的复合矿物相结构
黑色条带为尖晶石，灰色基质为磁铁矿

进一步研究表明，天然铁氧体中杂质矿物的电导率明显不同于其中磁铁矿的电导率(表5)，尤其是钛铁矿、黄铁矿和磁黄铁矿的电导率明显高于磁铁矿的电导率，甚至高2个数量级。根据Lichtenecker混合公式，多相混合体系的电磁参数与其中每一相的电磁参数及其在混合体系中所占体积百分比成正相关关系[7,8,10~12]。根据公式(6)，这些高电导率杂质矿物的存在提高了天然铁氧体的电导率，铁氧体的介电损耗相应增大，从而使A型铁氧体的电磁参数优于二元型铁矿物和单一型磁铁矿，更适合作为吸波材料的粉体填料。

表 5 A 型天然铁氧体复合矿物相中部分单矿物的电导率　　　　　　　　　(单位：$\Omega^{-1} \cdot cm^{-1}$)

矿物	磁铁矿	钛铁矿	黄铁矿	钛铁晶石	磁黄铁矿	含钒尖晶石
分子式	$Fe^{2+}Fe_2^{3+}O_4$	$FeTiO_3$	FeS_2	$TiFe_2^{2+}O_4$	$Fe_{1-x}S(x=0~0.17)$	$M^{2+}Fe_2O_4$(M为金属元素)
电导率	$10~10^2$	$10^2~10^4$	$0.67×10^5~3.33×10^5$	未测量	$10^2~10^4$, 10^6	10^2

3　A 型天然铁氧体复合材料与羰基铁粉复合材料的吸波性能比较

根据公式(3)，从复介电常数分析，A 型天然铁氧体的 ε'=58.60，低于羰基铁粉的 ε'=80，从而更有利于阻抗匹配，降低入射电磁波在自由空间与吸波材料界面产生的反射；A 型天然铁氧体的 ε''=10.0，大于羰基铁粉的ε''=8.0，从而使前者的介电损耗大于后者，更有利于电磁波吸收。从复磁导率比较，A型天然铁氧体的$\tan\delta_m \approx 1.0$，羰基铁粉的$\tan\delta_m \approx 1.25$，两者基本接近。

为比较A型天然铁氧体吸波粉体和羰基铁粉的微波吸收性能，我们开展了材料应用试验。采用相同工艺、相同粉体含量与天然橡胶混合制成复合吸波片材，分别进行了吸收频宽和功率反射系数的扫频(8~12 GHz)测量，测量结果如图7所示。

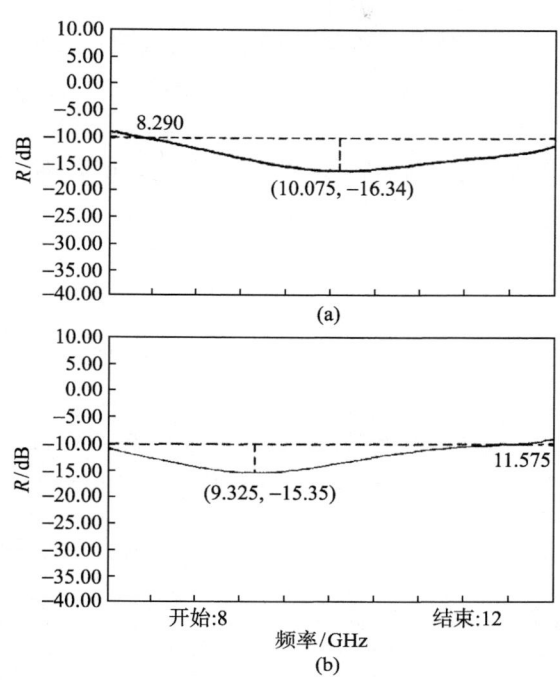

图7 A型天然铁氧体粉体(a)和羰基铁粉(b)与天然橡胶制成的复合吸波片材的吸收频宽和功率反射系数
测量频率：8~12 GHz，极化方式：VV

材料应用试验及8~12 GHz的宽带扫频测量结果表明，采用A型天然铁氧体作为粉体填料制成的复合吸波片材在8.9~9.6 GHz频率范围内的功率反射系数小于0.5%~1.0%，比采用羰基铁粉制成的复合吸波片材下降了1~2倍，说明A型铁氧体具有较好的吸波性能。测量结果还表明，电磁波反射衰减在-10 dB以下(即相当于电磁波吸收率大于90%)的吸收频宽，采用A型铁氧体作为粉体填料制成的复合吸波片材比采用羰基铁粉制成的复合吸波片材拓宽了0.5 GHz。两种片材的峰值吸收效率接近，最高吸收峰对应的频率也比较接近。

更为重要的是，A型天然铁氧体吸波粉体原料丰富、制备工艺简单、成本低、吸波性能稳定，为微波吸收材料的研制与生产拓宽了原料来源，完全可以取代羰基铁粉作为吸波粉体，在国民经济和国防装备所需微波吸收材料开发中具有明显的成本优势和广泛的应用前景。

4 结果与讨论

化学成分与矿物组成简单的天然铁氧体矿物，如单一型磁铁矿和二元型铁矿物，其中的杂质元素和杂质矿物含量很少，一般ε'较大，ε''较小，原因在于其更接近于理想的$M^{2+}Fe_2O_4$型铁氧体，缺少杂质组分对其电磁参数进行调控，因此难以开发成为有应用前景的微波吸收材料。而A型(岩浆岩型)天然铁氧体矿物由于含有极化率较低的杂质元素和介电常数较低的另相杂质矿物，使其ε'明显低于其他地质产状类型的天然铁氧体矿物；已有研究表明，铁氧体介电损耗主要来源于电偶极子取向极化弛豫和界面极化弛豫效应，而这两种效应均与进入铁氧体的杂质有关,即矿物晶格中混入高价或低价离子、以及铁氧体中含有高电导率另相杂质矿物,其结果都将导致天然铁氧体的电导率 σ 增加,从而使其介电损耗增大,有利于电磁波吸收。因此,A 型天然铁氧体中杂质元素与另相杂质矿物的存在对其微波电磁参数起着重要调控作用,而且,这种天然掺杂引起的电磁参数调控效应是人工制备的铁氧体很难实现的。

采用相同工艺分别将 A 型天然掺杂铁氧体粉体和羰基铁粉与天然橡胶制成复合吸波片材,电磁波吸收频宽和反射系数的测量结果表明,在 8~12 GHz 频率范围内,A 型铁氧体粉体复合吸波片材的电磁波吸收频宽和吸收率(功率反射系数)均优于传统的羰基铁粉复合吸波片材,而且吸波性能稳定,完全可以替代羰基铁粉作为吸波材料的粉体填料。更重要的是,A 型天然掺杂铁氧体具有原料丰富、成本低、制备工艺简单的比较优势,在微波吸收材料开发中具有广泛的应用前景。

致谢 感谢中国科学院国家天文台邹永廖研究员、中国科学院地质科学院地质研究所李兆丽博士在实验测试和论文成文过程中的帮助和支持。

参 考 文 献

[1] Amin M B, James J R. Techniques for utilization of hexagonal ferrite in radar absorbers. Radio Electron Eng, 1981, 51 (5): 209—225
[2] Petrov V M, Gagulin V V. Microwave absobing materials. Inorg Mater, 2001, 37 (2): 93—98
[3] 娄明连, 阚涛. 用铁砂研制铁氧体电波吸收材料. 安徽大学学报(自然科学版), 1998, 22 (1): 36—41
[4] 云月厚, 邰显康, 李国栋, 等. 富含稀土白云选铁尾矿制备的微波吸收材料特性研究. 稀土, 2003, 24 (2): 68—70
[5] 科夫涅里斯特 Ю К, 拉扎列娃 И Ю, 拉瓦耶夫 А А. 微波吸收材料. 蔡德录, 刘承钧, 译. 北京: 科学出版社, 1985.1—80
[6] Barsoum M W. Fundamentals of Ceramics. Material Science Series, New York: The McGraw-Hill Companies Inc, 1997. 668
[7] Zheng Y C, Wang S J, Feng J M, et al. Measurement of the complex permittivity of dry rocks and minerals: Application of polythene dilution method and lichtenecker's mixture formulae. Geophys J Int, 2005, 163 (3): 1195—1202
[8] Zheng Y C, Wang S J, Ouyang Z Y, et al. Measurement of the dielectric properties of volcanic scoria and basalt at 9370 MHz. Acta Geol Sin-Engl, 2005, 79 (2): 801—805
[9] 王相元, 盛玉宝. 羰基铁复合材料的复介电常数和复磁导率与体积浓度关系. 宇航材料工艺, 1989, (4): 47—50
[10] Lichtenecker K. Die dielektrizitats konstante naturlicher und kunstlicher mischkorper. Phys Z, 1926, 27: 115—158
[11] Lichtenecker K, Rother K. Die herleitung des logarithmischen mischungsgesetzes aus allgemeinen prinzipien der stationaeren stroemung. Phys Z, 1931, 32: 255—260
[12] Zakri T, Laurent J P, Vauclin M. Theoretical evidence for `lichtenecker's mixture formulae' based on the effective medium theory. J of Phys D Appl Phys, 1998, 31: 1589—1594

空间微波环境对"嫦娥一号"微波探测仪在轨定标影响分析

崔海英　王振占　张晓辉　董晓龙

(国家863计划微波遥感技术实验室，中国科学院空间科学与应用研究中心，北京　100190)

摘　要　"嫦娥一号"(CE-1)卫星利用微波探测仪(Chang'e Lunar Microwave Sounder-CELMS)对月球探测在国际上尚属首次。CELMS星上定标时采用冷空作为其低温定标源。当定标天线指向冷空获取宇宙背景辐射亮温时，宇宙中一些离月球距离较近的星体，如太阳、地球以及一些对月张角较大且辐射强度较高的射电源，如银心、天鹅座、金牛座等在某些时段均可进入其波束范围，是影响CELMS准确定标的主要因素。从分析CE-1的轨道入手，结合冷空星体的分布，研究了各星体相对于CELMS定标天线波束的入射角，进而得到了空间微波环境对CELMS探测结果的影响：太阳辐射对冷空天线的影响最大，其次是地球和银心，所有这些影响综合可达10 K量级。

关键词　嫦娥一号　月球微波探测仪　在轨定标　轨道分析

1　引言

CELMS是CE-1卫星的有效载荷之一，共有4个工作频率：3 GHz、7.8 GHz、19.35 GHz和37 GHz。其科学目标是测量不同深度的月壤微波辐射亮温，进而反演月壤厚度的信息并对月球的氦-3资源量和分布进行评估。从月球轨道用被动微波遥感的方法获得全月球月壤厚度信息，至今还没有先例[1]。

CELMS采用实时的两点定标技术，定标热源采用接收机内部的匹配负载，冷源采用冷空宇宙背景。通常情况下，冷空背景辐射亮温在微波波段约为2.7 K。但CELMS冷空定标天线安装在卫星的前进方向，天线波束较宽，宇宙中一些离月球(包括月球本身)距离较近的星体，如太阳、地球以及一些对月张角较大且辐射强度较高的射电源，如银心、天鹅座、金牛座等，会对冷空天线的接收产生影响，造成利用冷空背景作为定标标准时产生误差[2]，因此需要仔细分析冷空背景变化对冷空背景辐射亮温的影响。

由于CELMS处在月球这一全新的空间微波环境中，与以往地球卫星的微波探测仪冷空定标误差结果如何，是非常值得研究的问题。本文从分析CE-1的轨道入手，通过冷空星体的分布，对空间微波环境进行综合考虑，得到所有星体对于冷空天线总的亮温贡献，为CELMS天线在轨定标提供研究依据，为CE-1卫星探月取得新成果、新发现打好基础。

2　CE-1轨道特点

CE-1卫星于2007年10月24日成功发射并于2007年11月7日准确进入环月圆轨道，经过一段时间的卫星调姿、通信链路测试等工作后，于2007年11月27日CELMS加电工作。CE-1卫星设计寿命为1年，使命轨道的标称参数为[3]：

半长轴：1932.850 km(平均高度 194.65 km)
偏心率：0
倾角：90°
周期：127.088 min。

CE-1 卫星相对于标称轨道的保持和控制的精度要求是：高度 ± 25 km，倾角 ± 5°。

月球平均半径为 1 738 km，自转角速度为 2.661699×10⁻⁶ rad/s，一个轨道周期月球赤道转过 35.27 km，在 27.3 天内微波探测仪可以将全月球覆盖两遍。月球的赤道与黄道的夹角约为 1.54°，两者接近共面，因此卫星极轨道面与黄道面基本垂直。由于环月轨道在惯性空间中的方位基本不变，因此一年内太阳相对卫星轨道面正好旋转一圈，如图 1 所示，其中 β 表示太阳矢量与卫星轨道面的夹角。

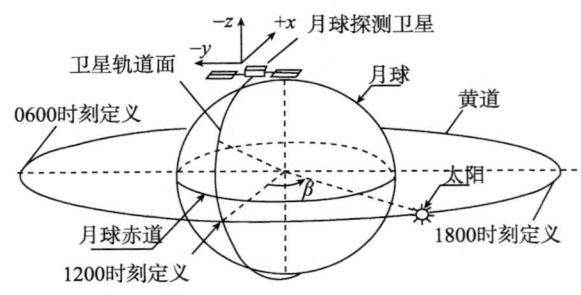

图 1 太阳、月球、卫星三者之间关系

太阳矢量与轨道面夹角是指太阳光在轨道面上的投影与太阳光的夹角。考虑轨道面的法线方向，则有：

$$\beta = \arcsin(\cos i \sin \delta_{\dot{Y}} - \sin i \cos \delta_{\dot{Y}} \sin \alpha_{\dot{Y}} \cos \Omega + \sin i \cos \delta_{\dot{Y}} \cos \alpha_{\dot{Y}} \sin \Omega) \tag{1}$$

其中：i 和 Ω 是 CE-1 卫星的轨道倾角和升交点赤经，$\alpha_{\dot{Y}}$ 和 $\delta_{\dot{Y}}$ 分别是太阳的赤经和赤纬。在 CE-1 卫星寿命期内，β 角的变化曲线如图 2 所示。

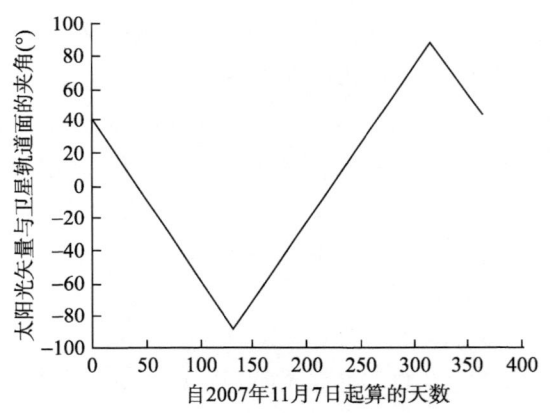

图 2 CE-1 寿命期内太阳与卫星轨道面夹角的变化曲线

基于卫星能源考虑，CE-1 环月阶段在轨运行具有正飞和侧飞两种状态：当太阳矢量与卫星轨道面夹角 β 在 −45°≤β≤45° 范围时采用正飞姿态，正飞姿态定义与卫星轨道坐标系 $Sxyz$ 一致，即 z 轴由卫星质心指向月心，y 轴指向轨道面的负法向，x 轴在卫星轨道面内与 z 轴垂直并指向卫星运动方向。CE-1 卫星外形及载荷所在坐标系平面如图 3 所示。正飞时姿态条件满足有效载荷对月面立体成像要求，同时通过帆板跟踪太阳满足电源供电要求。当太阳矢量与卫星轨道面夹角 β 在 −45°≤β≤45° 范围以外时，采用侧飞状态，此时姿态条件不满足有效载荷对月面立体成像要求，太阳帆板固定朝向+x 方向，满足电源供电要求。不论正飞还是侧飞时，对月观测天线波束指向始终与+z 一致；而对冷空天线，正飞状态时其波束指向与+x 轴方向一致，侧飞状态时与+y 轴方向一致。由于侧飞状态微波探测仪不工作，因此本文只考虑正飞状态下的微波探测仪的空间微波环境。时间系统采用 UTC 时间。

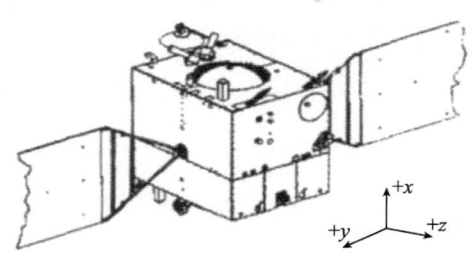

图 3 CE-1 外形图

3 月球轨道冷空星体分布及其对于 CELMS 冷空定标天线的影响

微波探测仪星上定标时采用冷空作为其低温定标源，定标天线指向冷空获取宇宙背景辐射亮温。图 4 是 2007 年 4 月 15 日在月球上看到的天空星体分布图，随着冷空天线的在轨运行，这些星体的辐射都有可能进入到冷空天线的主瓣范围，成为影响准确定标的主要因素。而这些星体由于自身微波辐射强度、与月球之间的距离以及对月张角的差异，对冷空天线的贡献是不同的。表 1 给出了主要星体在太空中的射电源位置、角径和亮温。从中可以判断，一些星体，如太阳、地球以及几个辐射较强的星座，如天鹅座、仙后座、猎户座、金牛座和银心等都可能对冷空天线定标有影响。

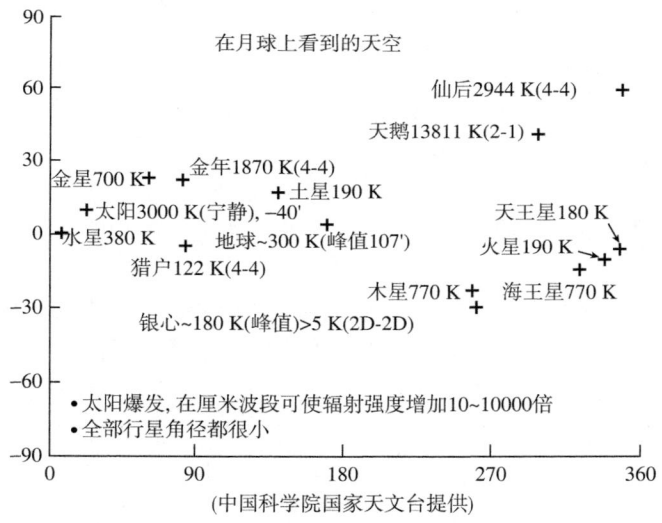

(中国科学院国家天文台提供)

图 4 2007 年 4 月 15 日月球上看到的天空

表 1 月球上天空星体射电源位置、角径和亮温表

	赤经(°)	赤纬(°)	角径	亮温(K)
太阳	22.80	9.60	大约 40'	4 000
地球	170.25	3.80	大约 107'	180
银心	260.75	−29.00	2°×2°	300
金牛	82.50	22.00	4'×4'	1 870
仙后	350.50	58.55	4'×4'	2 944
天鹅	299.25	40.58	2'×1'	13 811
猎户	83.25	−5.50	4'×4'	1 122

注：中国科学院国家天文台提供。

另外由于 CELMS 冷空天线指向卫星前进方向,通过计算在 194.65 km 的平均轨道高度上看到月表水平线的角度,来判断月球本身是否会进入冷空天线的波束范围内,影响定标的结果。CELMS 的冷空定标天线的波束宽度可达 ±30°。

把星体看作是一个均匀的辐射体,根据探测仪与星体的相对位置计算星体相对于天线的入射角,即星体与冷空天线的波束指向之间的夹角 ϕ。再通过对应天线的辐射方位图就可以得到星体对冷空定标源产生的亮温影响。

3.1 太阳

太阳是太阳系的能源中心,太阳射电辐射来自太阳的外层大气等离子体。CE-1 寿命期处于太阳活动极小年,因此,主要考虑宁静太阳射电和太阳缓变射电产生的总辐射。从图 1 中我们可以看出,当卫星运行到月球赤道附近时,太阳与冷空天线的波束指向几乎垂直,而运行至其他位置,入射角 ϕ 随着 β 角的变化而变化:当 β 角为 0°附近,在两极附近 ϕ 角最小,很有可能进入天线波束范围内,当 β 角为 ±90°附近,ϕ 角都接近 90°,太阳不会进入冷空天线的波束范围内。图 5 是当 $\beta=0°$(2007 年 12 月 18 日)附近入射角 ϕ 的变化曲线,图 6 是该时段卫星星下点纬度变化曲线。可以看出,基本上当卫星运行在南纬 70°以上,太阳就有可能进入天线波束范围。同时要考虑月球本身遮挡问题,如果卫星运行在南纬 70°以上且当地为白天的星下点时,太阳才会进入冷空天线的波束范围内。

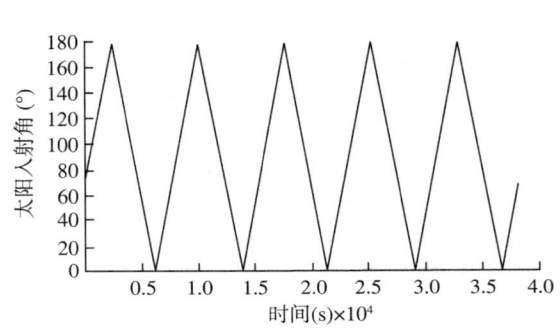

图 5　太阳入射角 ϕ 的变化曲线

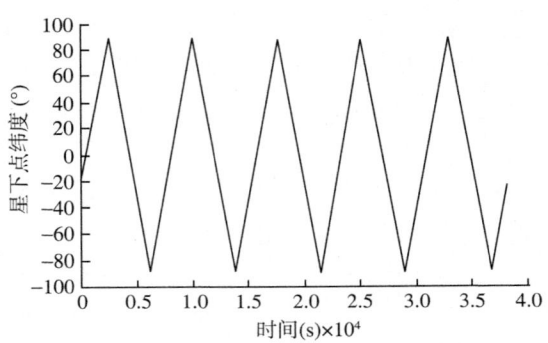

图 6　卫星星下点纬度变化曲线

通过星体在冷空天线的入射角及出现在天线方向图上的位置,就可以通过方向图的加权积分计算这个星体在天线上产生的亮温贡献。图 7 是太阳引起的冷空背景温度的变化曲线。可见,在这段时间的月球轨道上,由于太阳引起的冷空亮温变化最大可达到 5 K 以上。

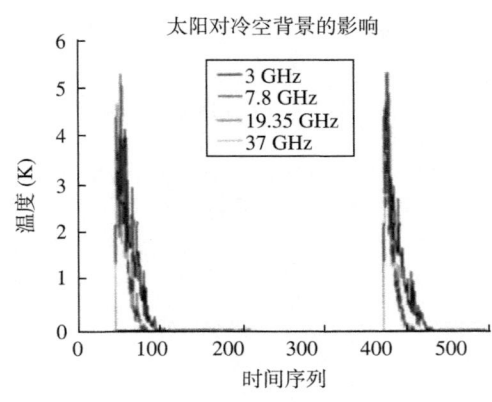

图 7　太阳引起的冷空背景温度变化曲线

3.2 地球

卫星绕月飞行,月球绕地运动,地球也有可能进入冷空天线的波束范围内。考虑到月球本身遮挡问题,当卫星运行在月球的背面时,地球是不可能进入冷空天线波束范围内的,这可以通过计算卫星的星下点轨迹月面经度来判定。图 8 是 2007 年 12 月 10 日 8 时始 5 个轨道周期的入射角 ϕ 变化,图 9 和图 10 是该时段卫星星下点纬度和经度变化曲线。从图中可以看出,此时卫星在月面零经度子午面附近飞行,当运行在北纬 70°以上、零经度线附近时,地球就可以进入冷空天线波束范围内。

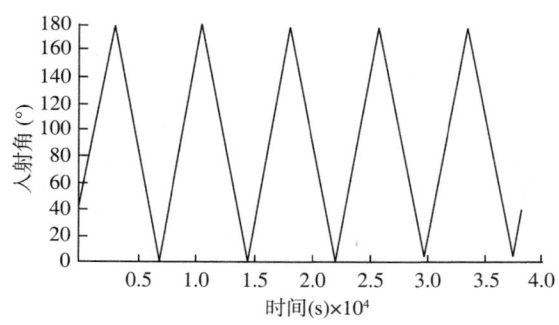

图 8 地球入射角 ϕ 的变化曲线

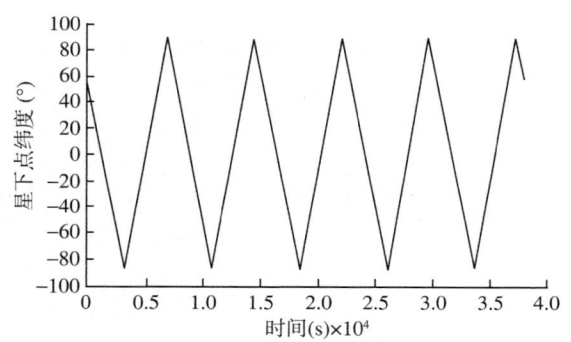

图 9 卫星星下点纬度变化曲线

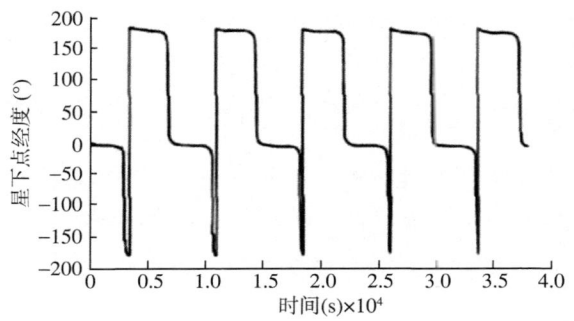

图 10 卫星星下点经度变化曲线

此时,地球引起的冷空背景温度的变化如图 11 所示,最大变化可达 3 K 以上。

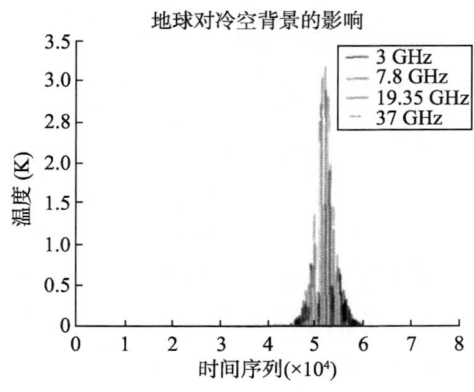

图 11 地球引起的冷空背景温度的变化曲线

3.3 月球

月球也有可能进入冷空天线的波束范围内,如图 12 所示。月球进入冷空天线波束范围的临界值 α 为:

$$\alpha = 90° - \arcsin \frac{R_M}{R_M + H} \tag{2}$$

其中：R_M 为月球平均半径，H 为轨道平均高度。经过计算得临界值 α 为 25.94°，这个夹角小于天线主波束范围。即使考虑卫星姿态控制误差，基本上在微波探测仪工作期间月球始终都处在天线波束范围内。月面温度对冷空背景亮温始终存在影响，其大小随月面温度的变化而变化。根据模拟的天线方向图来估算的某一特定时刻月球引起的冷空背景温度变化如图 13 所示。可以看出，月球对 3 GHz 冷空天线温度的影响最大，其余影响较小，这是由于 3 GHz 天线波束范围最宽的缘故。

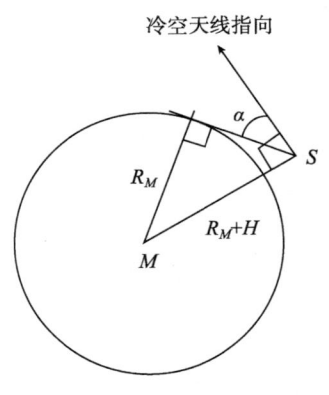
图 12 月球进入冷空天线示意图

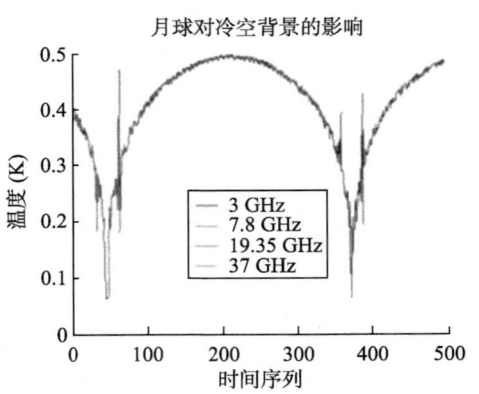
图 13 月球引起的冷空背景温度的变化

3.4 银心、金牛、仙后、天鹅、猎户

除了太阳、地球和月球外，还有一些星体有可能进入冷空天线的波束范围内。图 14 中全部行星的角径都很小，可忽略，但一些对月张角大的强射电源星座，比如银心、金牛、仙后、天鹅和猎户星座，需要逐一分析这些射电源对冷空天线定标产生的影响。

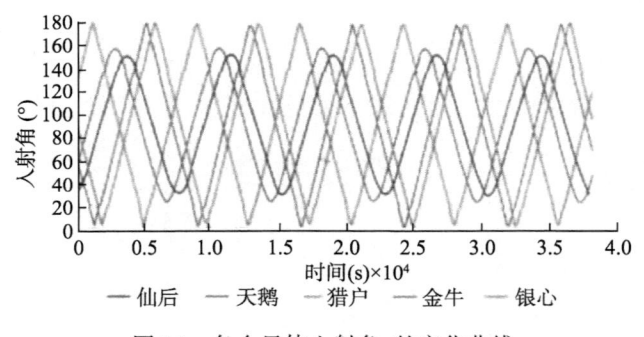
图 14 各个星体入射角 ϕ 的变化曲线

环月轨道在惯性空间中的方位基本不变，而星座在惯性空间中的位置变化十分缓慢，在 CE-1 号的寿命期内，基本保持不动。因此只要计算一个轨道周期的角度，就基本可以知道整个寿命期内这些星体进入冷空天线波束范围的情况。图 14 分别是 UTC 时间 2007 年 12 月 27 日 0 时始 5 个轨道周期内银心、金牛、仙后、天鹅、猎户进入天线波束范围的情况，此时卫星星下点轨迹如图 15 所示。各个星体引起的冷空背景温度的变化如图 16 所示。

从图 16 中可以看出，仙后、天鹅和猎户星座对定标天线温度影响已经很小，因此对角径很小的行星，均可不予考虑。

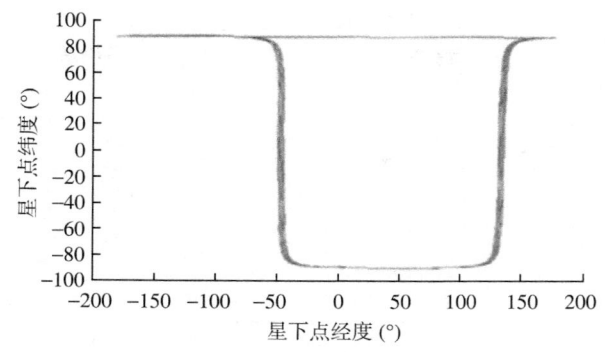

图 15　卫星星下点轨迹

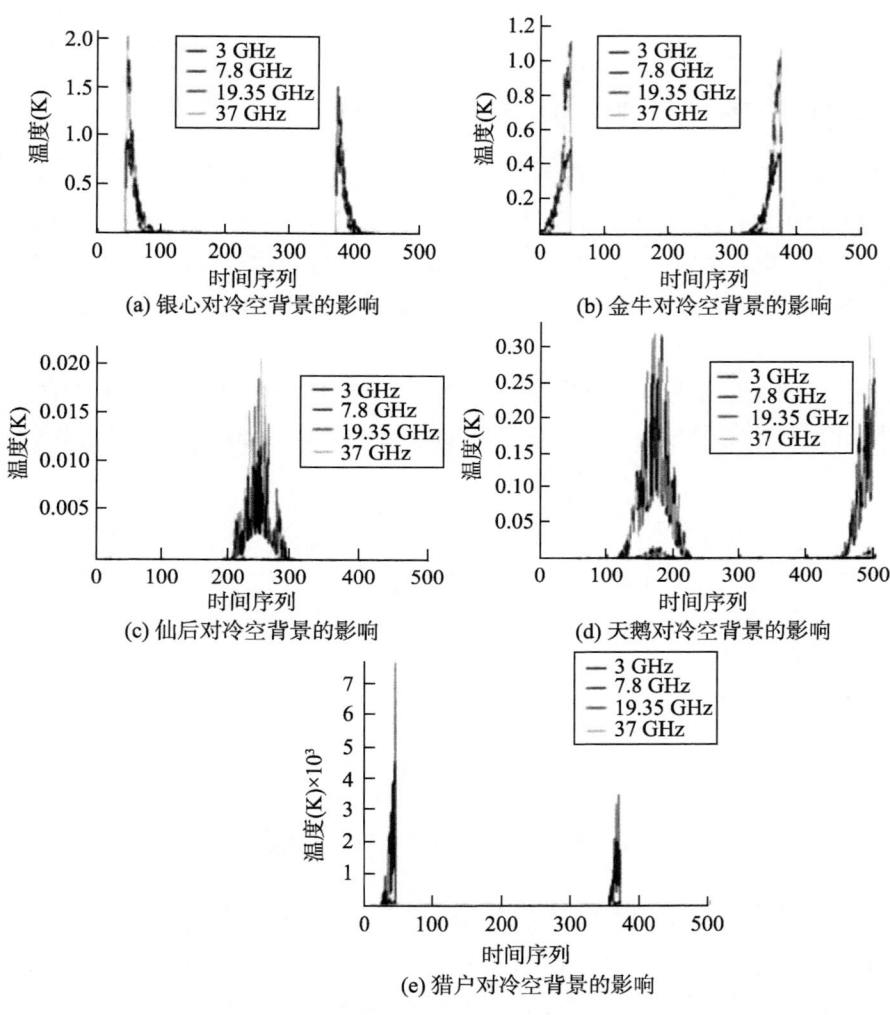

(a) 银心对冷空背景的影响

(b) 金牛对冷空背景的影响

(c) 仙后对冷空背景的影响

(d) 天鹅对冷空背景的影响

(e) 猎户对冷空背景的影响

图 16　各个星体引起的冷空背景温度的变化曲线

3.5　冷空背景温度的模拟结果

综合考虑上述太阳、地球、月球及银心、金牛、仙后、天鹅和猎户星座的宇宙背景影响因素，模拟出的冷空背景温度如图 17 所示，可以看出，冷空背景的变化导致定标天线接收的亮温变化可达到 9 K 多。总体来讲，4 个频率的天线接收到的亮温变化大致相当，只是 3 GHz 天线波束较宽，其冷空背景较不纯净，比其他 3 个频率的冷空温度高，这其中大部分是由于月球的影响造成的。

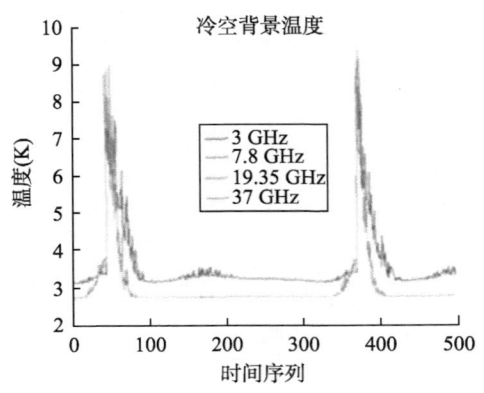

图 17 冷空背景温度模拟曲线

4 日月距离变化及其对月球表面温度的影响

月球周围没有大气,月面物质的热容量小且导热率很低,因此月球表面温度变化很大。太阳光直接照射月球赤道上,白昼时温度可达130℃,黑夜温度可降至-180℃,月球上不会出现诸如地球上的明显四季变化。但日月间距离的变化,会对月表温度有影响。根据日月位置的近似计算公式[4],通过坐标转换,可以得到太阳和月亮在历元J2000.0平赤道地心系下的位置坐标,则月亮相对太阳的位置矢量为:

$$r_{日月} = r_{月} - r_{日} \tag{3}$$

自2007年11月7日CE-1进入工作状态开始的日月距离如图18所示,在一年的寿命期内,大约在2008年1月8日附近月亮距太阳最近,2008年7月18日月亮距太阳最远,远近距离之差约为572万km。

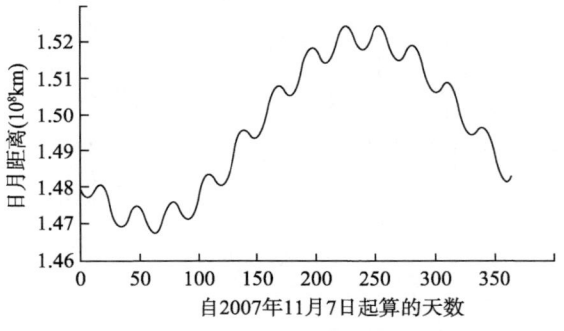

图 18 CE-1寿命期内日月距离变化曲线

在李芸[5]撰写的论文《月表温度剖面对于CELMS探测亮温影响的模拟研究》中给出了模拟的Apollo15登月点在20a内表面温度在太阳高度角最大、最小情况下的波动。在太阳高度角最大的情况下,Apollo15登月点最高表面温度变化可达9 K,这是由于太阳的距离不同引起表面温度的不同,这一点在高纬度地区比在低纬度地区更加明显。

5 结语

CELMS是世界上第一台月球轨道微波辐射计,用于探测月表的亮温分布。CELMS星上定标时采用冷空宇宙背景作为其冷源。宇宙背景中离月球较近的星体,如太阳和地球,以及对月张角大且辐射强度较高的射电源,如银心、天鹅座、金牛等星座都会进入冷空天线的波束范围内,影响定标的结果。本文研究结果表明:太阳辐射对冷空天线的影响最大,这其中包括日月距离的影响,其次是地球和银心。另外,随着卫星的前进，冷空天线由于波束宽度较大会接收到月球自身的辐射。所有这些影响综合可达 10 K 量级。天线的指向对于测量的冷空背景亮温影响很大。

致谢 中国科学院国家天文台相关人员为本文的研究提供了数据支持，特此致谢！

参 考 文 献

[1] Jiang Jingshan, Wang Zhenzhan, Li Yun. Study on Theory and Application of CE-1 Microwave Sounding Lunar Surface [J]. Engineering Sciences, 2008, 10 (6): 16—22

[姜景山, 王振占, 李芸. 嫦娥1号卫星微波探月技术机理和应用研究[J]. 中国工程科学, 2008, 10(6): 16—22]

[2] Plonski M, Smith C. Algorithm Theoretical Basis Document (ATBD), for the Conical-Scanning Microwave Imager/Sounder(CMIS) Environmental Data Records (EDRs) [R]. Volume 17: Temperature Data Record and Sensor Data Record Algorithms, Version 2.0, 2001, Atmospheric and Environmental Research, Inc. 131 Hartwell Avenue Lexington, MA 02421—3126

[3] Yang Weilian, Zhou Wenyan. Orbit Design for Lunar Exploration Satellite CE-1[J]. Spacecraft Engineering, 2007, 16(6): 16—24

[杨维廉, 周文艳. 嫦娥一号月球探测卫星轨道设计[J]. 航天器工程, 2007, 16(6): 16—24]

[4] Liu Lin. Orbit Theory of Spacecraft [M]. Beijing: National Defense Industry Press, 2000: 255—258

[刘林. 航天器轨道理论[M]. 北京：国防工业出版社, 2000, 255—258]

[5] Li Yun, Wang Zhenzhan, Jiang Jingshan. Simulation on Influence of Lunar Surface Temperature Profiles on CE-1 Lunar Microwave Sounder Brightness Temperature [J]. Science in China Series D, 2009, 39(8): 1—14

[李芸, 王振占, 姜景山. 月表温度剖面对于CELMS探测亮温影响的模拟研究[J]. 中国科学 D 辑, 2009, 39(8): 1—14]

基于 SVM 和"嫦娥一号"数据的月球表面亮温分布

周明星 周建江 汪 飞

(南京航空航天大学电子信息工程,南京 210016)

摘 要 月球表面的微波辐射亮度温度分布与月表物质的物理化学和地理分布特性密切相关。为了分析月球表面微波辐射亮温的分布特点,利用支持向量机(SVM)方法对嫦娥一号(CE-1)绕月卫星搭载的微波辐射计获得的 2C 级亮温数据建立回归分析模型,并利用粒子群算法优化 SVM 回归模型,建立了月球表面不同地理位置的 4 个频率通道(3GHz,7.8GHz,19.35GHz,37GHz)的微波辐射亮温与时间的关系,获得了这 4 个频率的微波辐射亮温在月表很窄时间段的全球分布,因而显示出了更多的细节特征。最后对这些特征进行了描述并对影响月球表面亮温的因素进行了讨论。

关键词 亮度温度 支持向量机 粒子群 微波辐射计 CE-1

0 引言

月球表面的微波辐射亮度温度是月表物理化学特性的直接反映,包含了表面温度、介电常数、月壤厚度以及表面地形等的综合信息[1]。对月球表面微波辐射亮度温度的研究开始于上世纪 40 年代[2],到了 60 年代末和 70 年代初进入了一个高潮时期,在这期间,许多科学家从各个方面对月球表面亮度温度进行了研究,所用的微波波长几乎覆盖了整个微波频段。但自从人类第一次登上月球后,科学家们的视线逐渐转移到对月球样品的研究上,慢慢地不再重视对亮度温度的研究。随着人类第一轮月球探测的结束,对月球表面亮度温度的研究被完全遗忘了。伴随着新一轮探月高潮的来临,对月球表面亮度温度的研究重新引起科学家的重视。但比起以前来说,现在科学家对月球表面微波辐射亮度温度(亮温)的研究采用了更加先进的手段,并且将研究的重点放在探测月球水冰、月壤厚度和月球资源上[3]。我国的嫦娥一号(CE-1)绕月卫星发射后,引起了国内外学者对月球表面微波辐射亮温研究的兴趣[1,4-5]。

微波辐射计(MRM)是 CE-1 号绕月卫星的有效载荷之一,它有 4 个工作频率(3.0 GHz,7.8 GHz,19.35 GHz 和 37 GHz),用于探测月球表面微波辐射亮温,研究月壤厚度分布并在此基础上评估月球上 ^{3}He 资源的含量。这是人类第一次在月球轨道上用被动微波遥感技术探测月壤厚度。经过一年多的在轨飞行,MRM 获得了大量的亮温数据。MRM 亮温数据以 PDS(Planetary Data System)标准存储,2C 级轨道数据包括了数据采样时间、4 个频率通道的亮温、星下点太阳入射角和方位角、月球表面经纬度,以及轨道高度等信息。根据这些信息可以得到各频率通道亮温在月球表面的分布。

从目前对 CE-1 亮温数据处理的初步结果来看,虽然已有的亮温数据可以覆盖全月表面,但获得的全月表面亮温分布都是对月球上一天当中一个比较宽的时间范围内的亮温数据在月球表面进行投影得到的[1,4-5]。然而,由于月球表面亮温受月球表面物理温度的影响,同一地点不同时刻的亮温差别明显。例如,图 1 为 CE-1

卫星上的微波辐射计对 Apollo11 登月点地区在月球上一天(地球上的 29.53 天)当中的实测亮温。

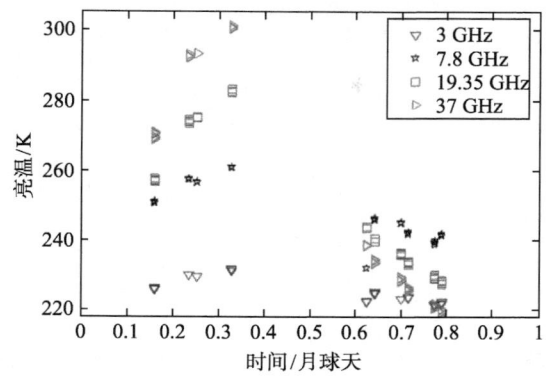

图 1　Apollo11 登月点亮温随时间的分布

　　从图 1 中可以发现，频率越高，亮温在月球上一天中的变化就越大。经过对 2C 级亮温数据整理发现，用越窄的时间范围(比如接近正午的很短时间段)的数据得到的亮温分布图越能反映出月球表面亮温随纬度的分布规律以及地形、月表物质介电性能等对亮温的影响。这是因为，月球表面的微波辐射亮度温度与月表物质的物理化学性质以及地理位置密切相关。其中，月球表面微波辐射亮温受月球表面温度的影响最大，而月球表面温度不仅随纬度变化，同时在月球一天当中，同一地点的温度随时间的变化很剧烈，如果用不同时刻的亮温数据绘制全球亮温分布图，那么温度的剧烈变化引起的亮温的剧烈变化就会掩盖其他诸如地理位置、地形、月表物质介电性能等对亮温的影响，使得亮温分布图没有规律，在越窄的时间范围内，月球表面的温度变化就越小，时角范围足够小时，月球表面温度可以看作只随月球纬度缓慢变化，而不随时间变化，这样就使得亮温随月球纬度的变化的规律表现的更加明显，同时也使其他因素对亮温的影响更好地体现出来。另外，在反演月壤厚度的过程中，相比直接利用不同时刻的亮温数据反演月壤厚度对需要计算不同时刻的月球表面温度，如果得到特定时刻亮温在全月表面的分布，在反演月壤厚度时就只需要计算这一时刻的月球表面的温度，这将使月壤厚度的反演变得更加方便。然而，以图 1 为例，对于月球表面局部范围，从 CE-1 获取的实测亮温数据不能够覆盖一天当中整个时段，所以从已经得到的 2C 级数据当中，并不能直接获取到很窄的某一时间范围内的全月表面亮温分布。针对月球表面局部范围内，微波辐射计实测亮温数据有限的特点，本文尝试利用 SVM[6-7]回归分析方法从现有的 2C 级亮温数据中获取某一很窄时间范围的全月表面亮温分布，并讨论亮温分布的一些规律。

　　SVM(Support Vector Machine)是 Vapnik[6]等人提出的一种以统计学习理论为基础的新型机器学习方法。这种方法起先是针对分类问题提出的，随着 Vapnik 引入不敏感损失函数，SVM 已推广到非线性系统的回归估计中，并表现出了极好的学习性能[8]，能够较好地解决小样本、非线性等问题。本文所用的就是这种用于曲线拟合的回归型支持向量机(SVR)。根据 CE-1 上微波辐射计的分辨率，本文将月球表面划分为 360×360 的投影网格。不同于以往对亮温数据的处理，本文对每个投影网格所对应的范围内的 2C 级亮温数据建立 SVM 回归模型，并利用粒子群优化算法[9]优化 SVM 回归模型当中的参数，得到每个网格区域 4 个频率通道(3 GHz，7.8 GHz，19.35 GHz，37 GHz)的微波辐射亮温与时间的关系，以获得到全月表面在很窄时段内的亮温分布，进而分析亮温分布特点并对影响月球表面亮温的因素进行讨论。

1　回归型的 SVM

　　回归问题的数学描述为：根据给定的训练集 $T = \{(x_1, y_1), \cdots, (x_l, y_l)\}$，其中 $x_i \in \mathbf{R}^n$ 表示输入，$y_i \in \mathbf{R}$ 表示输出，$i = 1, \cdots, l$ 表示样本个数，寻找 $\mathbf{R}^n$ 上的一个实值函数 $f(x)$，以便使用 $y = f(x)$ 来推断任一

输入 x 所对应的 y 值[7]。对于线性情况，用 $f(x) = w\cdot x + b$ 拟合样本，w 是权向量。对于非线性情况，通过非线性变换 $\phi(\cdot)$ 将输入空间映射到一个高维特征空间，并在这个特征空间用线性函数 $f(x) = w\cdot\phi(x) + b$ 拟合样本，从而将非线性问题转换为线性问题。SVR 问题等价为线性约束二次规划的优化问题[10]：

$$\min\left(\frac{1}{2}\|w\|^2 + C\sum_{i=1}^{l}(\xi_i + \xi_i^*)\right)$$
$$\text{s.t}\begin{cases} y_i - ((w\cdot x_i) + b) \leq \varepsilon + \xi_i^* \\ ((w\cdot x_i) + b) - y_i \leq \varepsilon + \xi_i \\ \xi_i \xi_i^* \geq 0 \end{cases} \tag{1}$$

其中，最小化 $1/2\|w\|^2$ 意味着使 VC 维最小，即学习机的复杂性越小。ε 是拟合误差，用它做约束条件来控制支持向量个数和泛化能力。这样，上述最优化问题体现了结构风险最小化准则[10]，从而得到的回归结果具有良好的泛化能力。ξ_i，ξ_i^* 是当约束条件不能实现时引入的松弛变量，$C>0$ 为惩罚系数，它是模型复杂度和估计误差之间的一个平衡量。利用拉格朗日乘子法，得到对偶最优化问题：

$$\min_{\alpha,\alpha^*}\frac{1}{2}\sum_{i=1}^{l}\sum_{j=1}^{l}(\alpha_i - \alpha_i^*)(\alpha_j - \alpha_j^*)K(x_i, x_j)$$
$$\varepsilon\sum_{i=1}^{l}(\alpha_i + \alpha_i^*) - \sum_{i=1}^{l}y_i(\alpha_i - \alpha_i^*) \tag{2}$$
$$\text{s.t} \sum_{i=1}^{l}(\alpha_i - \alpha_i^*) = 0,\ 0 \leq \alpha_i,$$
$$\alpha_i^* \leq C,\ i = 1,\cdots,l$$

其中 α_i，α_i^* 为拉格朗日乘子，$K(x_i, x_j) = \phi(x_i)^T\cdot\phi(x_j)$ 为核函数。由这个约束问题的最优解得到回归函数：

$$f(x) = \sum_{i=1}^{l}(\alpha_i - \alpha_i^*)K(x_i, x) + b \tag{3}$$

其中，b 由下式计算[8]：

$$b = \frac{1}{N_{\text{NSV}}}\left\{\sum_{0<\alpha_i<C}\left[y_i - \sum_{x_i\in SV}(\alpha_j - \alpha_j^*)K(x_j, x_i) - \varepsilon\right]\right.$$
$$\left. + \sum_{0<\alpha_i^*<C}\left[y_i - \sum_{x_j\in SV}(\alpha_j - \alpha_j^*)K(x_j, x_i) + \varepsilon\right]\right\} \tag{4}$$

式(4)中下标 NSV 表示标准支持向量[8]，N_{NSV} 表示标准支持向量数量，SV 表示所有支持向量。

2 回归模型的建立与参数优化

2.1 CE-1 号 MRM 亮温数据分析

CE-1 绕月卫星上 MRM 获取的 2C 级亮温数据以 PDS 格式存储。根据数据中包含的数据采样时间、星下点太阳入射角和方位角、月球表面经纬度等信息可以算得数据采样点在月球一天中所对应的时间。为了更加准确地描述这个时间信息，本文引入"时角"概念[11]，由于月球的周期运动，可以选择观测位置的任意时刻作为月球上一天的起点，对应时角 0°。这样，月球在一昼夜中旋转一周，观测位置的时角就由 0° 增加到 360°。根据模拟的月球表面亮温变化的特点[12]，即日出时刻为亮温最低点，所以本文选择月球上观测点日出时刻(温度最低时刻)作为起始时刻。

尽管 CE-1 号卫星的 MRM 在绕月飞行期间测得的亮温数据可以多次覆盖月球表面，但对亮温数据统计结果表明：对于特定区域，实测亮温数据不能够覆盖一天当中整个时段。图 1 中 MRM 测得的 Apollo11 登月点的亮温数据就是一个示例。这样，如果限定某个较小的时角范围，用已有的亮温数据就不能得到全月表面在这一时段的亮温分布。因此，如果缺少特定区域特定时刻的亮温数据，就要用该区域其他时刻的数据来拟合该时刻的数据。经过计算所有数据采样点对应的时角信息，发现数据采样时间多数集中在接近白天正午时刻(太阳直射当地，对应时角 90°)和午夜时刻(对应时角 270°)，如图 2 所示。为了获得最小的亮温拟合估值误差，应当选择数据采样点最多的时刻来获得最好的数据支持，因此，选择数据量最多的正午时刻和午夜时刻来获取全月表面亮温分布相比选用其他时段更为合理。

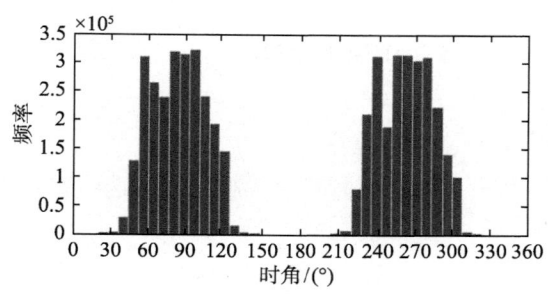

图 2　数据采样点的时角信息统计

2.2　建立预测模型

为了获取全球范围内特定时刻的亮温分布，首先要选择合适的投影网格作为像素点，然后计算该时刻每个网格所对应范围内的亮温。CE-1 上 MRM 的分辨率是 30–50km，CE-1 卫星星下点轨迹相对月面向西运动，每轨在赤道位置向西运动大约 1°左右，约 30.3 km，而在纬度变化方向，平均在 1°范围内有 11 个采样点。本文选择 360×360 的网格进行全球亮温投影，来使亮温在纬度方向显示出更多的细节。

因为 SVM 具有适用于解决小样本、非线性等问题的特点[10]，所以本文尝试用 SVM 的方法来预测每个网格素点特定时刻的亮温。实验中采用的是台湾大学林志仁教授提供的 SVM 工具箱[13]。核函数选择径向基函数 $K(x_i, x) = \exp(-\|x - x_i\|^2/\sigma^2)$。

2.3　SVM 参数的优化

不敏感系数 ε、惩罚因子 C 和核函数中的 σ 对回归模型的训练结果起着重要作用。本文中 ε 取常数 10^{-3}，用粒子群优化算法(PSO)来选取最优的 C 和 σ。

PSO 源于对鸟群群体运动行为的研究[9]，PSO 求解优化问题时，将每个问题的解看作是 d 维搜索空间中的一个微粒(点)，称作粒子。每个粒子都以一定的速度在搜索空间中飞行，并且有一个适应度来评价该粒子的优劣。为了寻找最优解，每个粒子根据当前自己找到的最优解(p_{best})和全局最优解(g_{best})来改变自己的移动方向。对于第 j 维的第 i 个粒子，它的速度和位置按如下公式来更新[14-15]：

$$v_{ij}^{t+1} = w \cdot v_{ij}^t + c_1 r_1 \cdot \left(p_{\text{best},ij}^t - p_{ij}^t\right) + c_2 r_2 \left(g_{\text{best},ij}^t - p_{ij}^t\right) \tag{5}$$

$$p_{ij}^{t+1} = p_{ij}^t + \beta \cdot v_{ij}^t \tag{6}$$

其中，t 是进化的次数，v_{ij} 是第 j 维第 i 个粒子的速度，p_{ij} 是第 j 维第 i 个粒子的位置，w 是个惯性权重系数，用以控制局部最优和全局最优对速度的更新，r_1 和 r_2 是[0, 1]之间的随机数，β 是位置更新口用来控制速度权重的常数，通常取 1，c_1 和 c_2 是加速因子，是非负常数。

用 PSO 对 SVM 参数优化时，粒子包括两部分：C 和 σ。优化过程如下：

(1) 随机产生"N"个粒子，形成粒子群，并初始化每个粒子的初速度。

(2) 用训练集训练 SVM，并评估每个粒子的适应度。本文选用 k 次交叉验证结果作为适应度。在 k 次交叉验证中，将训练集分割成 k 个子样本集，一个单独的子集被保留作为验证模型的数据，其他 $k-1$ 个子样本集用来训练。交叉验证重复 k 次，每个子样本验证一次，平均 k 次的结果得到一个单一估测。k 次交叉验证适应度函数定义如下[14]：

$$f_{\text{cross validation}} = \frac{1}{k}\sum_{i=1}^{k}|\overline{e}_i|\times 100\% \tag{7}$$

$$\overline{e}_i = \frac{1}{m}\sum_{j=1}^{m}\left|\frac{y_j - \hat{y}_j}{y_j}\right| \tag{8}$$

其中 y_j 是真实值，$\hat{y}_j$ 是估计值，m 是子集中样本个数，$\hat{e}_i$ 是子集平均相对误差。

(3) 根据适应度更新 p_{best} 和 g_{best}。

(4) 根据公式(5)和(6)更新粒子的速度和位置。

(5) 重复步骤(2) – (4)，当达到最大迭代次数或最优解在一定迭代次数内变化小于设定值时终止更新，并输出最优的 C 和 σ。

在完成用 PSO 对 SVM 参数优化后，用得到的最优的 C 和 σ 重新训练 SVM 回归模型，求解回归方程，得到每个像素点所需要的预测结果。

3 月表特定时刻亮温分布

如上所述，本文选择 360×360 的网格进行全球亮温投影。利用 SVM 回归模型对每个网格像素点所覆盖范围内的 4 个频率通道的亮温数据进行训练，得到该像素点 4 个通道亮温与所定义的时角的关系，从而得出每个像素点特定时刻的亮温值，经过投影后，便得到了特定时刻的全球亮温分布。

图 3 为 4 个频率通道的亮温在月球上正午时刻的分布，按照本文对时角的定义，此时对应的时角是 90±1°。图中(a)，(b)，(c)，(d)分别是 3 GHz，7.8 GHz，19.35 GHz，37 GHz 的四个频率的亮温分布。

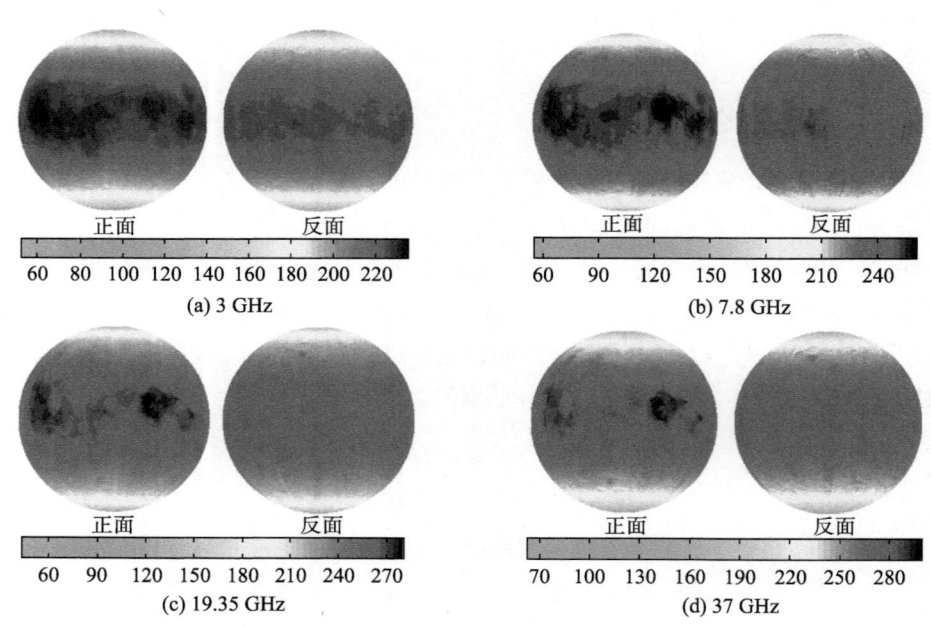

图 3　月球正午时分四个频率通道的全球亮温分布

亮温分布图显示，月球表面各频率亮温随纬度升高而降低，这是因为影响亮温的月球表面物理温度从赤道向两极逐渐降低的缘故。频率越高，全月表面亮温分布差异越大，3 GHz 亮温两极与赤道最大差异约为 170 K，而 37 GHz 亮温的最大差值达到了 230 K。这是因为频率越高，电磁波对月壤的穿透深度就越小，这 4 个频率中，37 GHz 的电磁波对月壤穿透深度较小，这一频率通道的亮温主要表现了月球表面最表层物质的属性，因为月球表面各地同一时刻表面物理温度差异大，所以亮温受温度的影响差异也大。而 3 GHz 的电磁波对月壤穿透深度较大，而较深层的月壤温度变化较小，所以 3 GHz 的亮温分布差异也较小。

另外，随着频率的升高，地形对亮温的影响表现的越来越明显，3 GHz 通道亮温的分布主要表现出亮温随纬度的变化而产生的差异，因为 3 GHz 电磁波穿透深度大，所以地形对亮温的影响表现不明显，只有模糊的轮廓。亮温分布图显示频率越高，体现出来的细节变化越多，在 37 GHz 亮温分布图上某些撞击坑已经清晰可见，如：朗哥芒坦坑(49.6°S, 21.8°W)、麦金拉斯坑(50.5°S, 6.3°W)、达朗贝尔坑(50.8°N, 163.9°E)、亚佛加德罗坑(63.1°N, 164.9°E)、季可夫坑(62.3°N, 171.7°E)等以及南北纬 45°附近的一些撞击坑。图 4 和图 5 更清楚地表现出了这个特点，并且投影图像所呈现出来的特征也越接近 Clementine UV/VIS 图像的特征[16]。

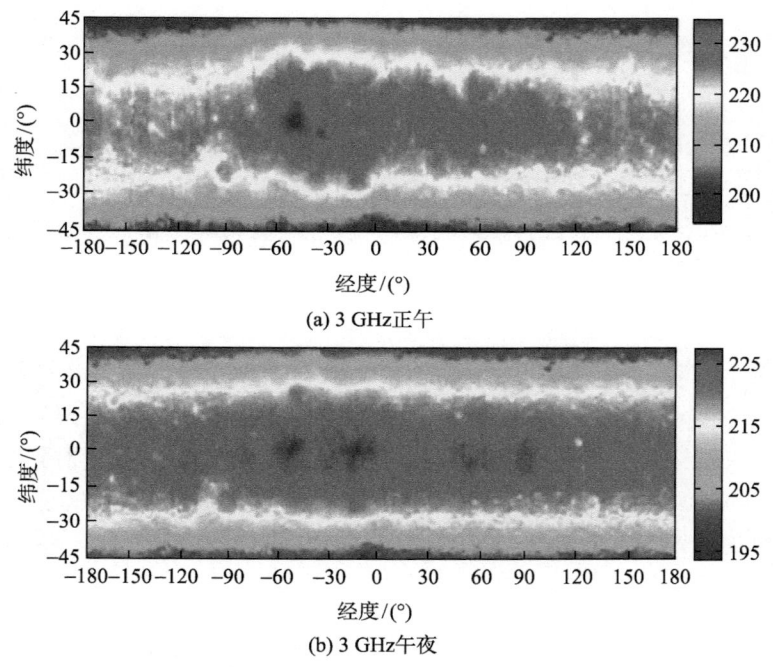

图 4　南北纬 45°范围内 3GHz 的亮温分布

图 4 和图 5 分别是月球表面南北纬 45°之间 3 GHz 和 37 GHz 的亮温投影。其中，图 4(a)是月球正午(时角 90±1°)3 GHz 亮温分布，图 4(b)是月球午夜(时角 270±1°)3 GHz 亮温分布。图 5(a)是月球正午 37 GHz 亮温分布，图 5(b)是月球午夜 37 GHz 亮温分布。

从这两个图上可以看出，对于 3 GHz，白天和夜晚亮温变化较小，而对于 37 GHz，白天和夜晚亮温差别较大，这个特点和模拟结果一致[12]，即频率越高，亮温的昼夜差别就越大。

另外，午夜的 37 GHz 亮温投影比正午的 37 GHz 亮温投影显示出更多的细节。这是因为，正午时刻 37 GHz 亮温在投影范围内差别较大，约从 240 K 变化到 290 K，而午夜时刻 37 GHz 亮温在这个范围内的投影变化区间约为 200 K 至 235 K，在同一频率和相同的投影方式下，亮温变化范围比较小的则显示出更多的细节。另一方面，由于地形的起伏，白天太阳照射时受地形影响形成的阴影区也会影响亮温对地形特征的体现。对于这些细节与月表地形、温度、月壤的铁钛含量，密度等的关系还需做进一步研究。

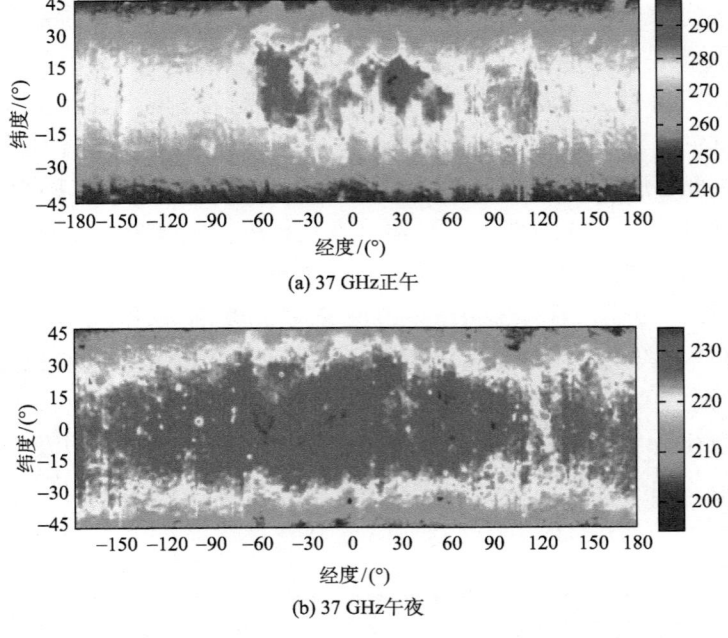

(a) 37 GHz正午

(b) 37 GHz午夜

图 5 南北纬 45°范围内 37 GHz 的亮温分布

此外，在图 5 中可以看出，有些地区的亮温在正午时刻相对于周围地区要高，而在夜晚这些地区的亮温反而比周围地区低，比如北纬 40°至南纬 10°，西经 20°至 60°，这一范围内这种现象就很明显。这应该与当地月壤的性质有关，使得这些地区白天温度很高，夜晚温度又降得很低。需要进一步研究月表物质的性质才能得出更加准确的结论。

4 讨论

文中对 ε，C 和 σ 进行优化处理以保证回归精度，因为本文是将月球表面网格化以后对每一网格所包含的区域的 4 个频率的亮温数据进行回归估计，最后得到这 4 个频率的微波辐射亮温在月表很窄时间段的全球分布，因此，每个网格像素点的回归精度与该像素点的实测亮温数据有关。图 6 为 Apollo11 登陆点 4 个频率通道亮温的实测值与回归值的对比，本文以此为例来说明用 PSO 优化的 SVM 模型得到的结果与实测数据的误差。

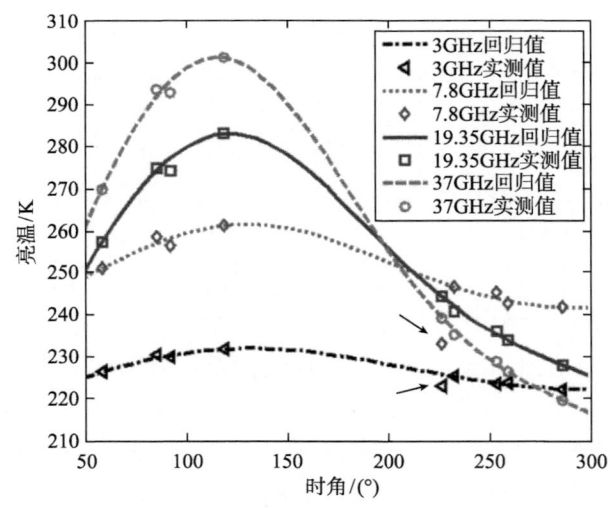

图 6 4 个频率通道亮温的实测值与回归值的对比

图6中以三角形、菱形、方块和圆圈分别表示3 GHz、7.8 GHz、19.35 GHz和37 GHz这4个频率通道的实测亮温,分别用点虚线、点线、实线和虚线来表示这4个频率亮温的回归曲线。图中箭头所指的两个亮温值是3 GHz和7.8 GHz通道在时角为226°时的两个实测值,这两个值偏离邻近时刻测量的亮温值和回归曲线都比较大,这个时间是在月球当地的夜晚且在午夜之前的时刻,按照模拟的月球表面亮温变化规律,特定地点在相邻时刻的亮温不应该发生突变,并且亮温在整个夜晚是逐渐降低的[12],而箭头所指的两点亮温实测值却明显小于所测的相邻时刻的亮温,并且时间晚于它的各点亮温测量值都比该点亮温测量值大并逐渐降低。因此可以判断箭头所指的这两点测量值本身存在着比较大的误差,是一个异常点。除箭头所指的这两点外,其余各点实测值和回归值的误差较小。3 GHz、7.8 GHz、19.35 GHz和37 GHz这4个频率通道实测值和用PSO优化的SVM模型得到的回归值之间的平均误差分别是0.23 K、0.66 K、0.66 K和0.72 K。其中3 GHz通道的误差最小,小于微波辐射计的温度分辨率0.5K,而反演月壤厚度主要用3GHz的亮温值,因此,在今后的工作中可以尝试把用这种方法得到的特定时刻的亮温分布用于估计月球表面月壤厚度的分布。

与图2中的统计结果一致,因为在0°~50°以及300°~360°的这些时段里,装载在CE-1号上的微波辐射计没有采样到当地的亮温数据,造成这些时段里面没有数据支持,而这是一个很宽的时段,因而没办法得到这个时段里较准确的数据拟合,所以图6中没有给出0°~50°以及300°~360°的结果。这是一个遗憾,如果有这些时段的采样数据,就能得到当地亮温在一天当中较完整的变化规律。

另外,由于2C级亮温数据本身的一些误差和异常值的存在,使得部分数据点亮温误差比较明显,且这些误差点的分布比较有规律,往往是同一经度上不同纬度的亮温投影都与周围亮温分布有明显差异,这个在7.8GHz通道比较明显。这就说明,这些误差是由2C级亮温数据点本身的误差引起的。根据局部范围内亮温在同一时刻不会发生突变的特点,可以找出这些误差点,并对它们进行修正,进而修正对应的2C级数据的误差。修正数据采样点误差对以后利用亮温数据来反演月壤厚度有重要意义,因此在今后的工作当中,需要研究对数据误差进一步修正的方法。

5 结论

本文将月球表面划分为360×360的投影网格,对每一网格对应区域亮温建立SVM回归模型,并用PSO优化算法优化SVM回归模型的参数,获得了这4个频率的微波辐射亮温在月表特定时刻的全球分布,亮温分布图显示,月球表面各频率亮温随纬度升高而降低。频率越高,全月表面亮温分布差异越大,月表地形对亮温的影响表现的越来越明显;频率越高,同一地点亮温的昼夜差别就越大;对于同一频率,午夜的亮温投影比正午的亮温投影显示出更多的细节,对于这些细节与月表地形、温度、月壤的铁钛含量,密度等关系将在今后的工作中做进一步研究;有些地区呈现出白天亮温越高,夜晚亮温就越低的特征,解释这一现象需要进一步研究月表物质的性质。另外,还可根据分布图上显示出来的亮温异常点对2C级亮温数据进行误差修正。

误差分析表明,用PSO优化算法优化SVM回归模型得到的亮温与实测值误差较小,其中3GHz的亮温回归值的平均误差最小,因此在今后的工作当中将尝试将本文结果用于反演月壤厚度。

致谢 本文感谢月球探测工程中心提供2C级亮温数据,感谢中国科学院北京天文台郑永春老师提供的帮助以及审稿专家认真细致的修改。

参 考 文 献

[1] 王振占, 李芸, 姜景山. 用"嫦娥一号"卫星微波探测仪亮温反演月壤厚度和3He资源量评估的方法及初步结具分析[J]. 中国科学D辑: 地球科学, 2009, 39(8): 1069—1084

Wang Zhen-zhan, Li yun, Jiang Jing-shan. Lunar surface dielectric constant, regolith thickness and helium-3 abundance distributions retrieved from microwave brightness temperatures of CE-1 Lunar microwave sounder[J]. Sci China Ser D-Earth Sci, 2009, 39(8): 1069—1084

[2] Dicke R H, Beringer R. Microwave radiation from the sun and the moon, Astrophys[J]. Acta Automatica Sinica, 1946, 103(3): 375—376

[3] Wu Ji, Li Di Hui. Microwave brightness temperature imaging and dielectric propertiesof Lunar soil[J]. Earth Syst. Sci, 2005, 114(6): 627—632

[4] Zheng Y C, Bian W, Su Y, et al. Brightness temperature distribution of the moon: result from Chinese Chang'E-1 Lunar Orbiter[C]. Goldschmidt Conference 2009, Davos, Switzerland, June 21—26, 2009

[5] Fa W Z, Jin Y Q. Analysis of microwave brightness temperature of lunar surface and inversion of regolith layer thickness: Primary results of Chang'E-1 multi-channel radiometer observation[J]. Science in China Series F: Information Sciences, 2010, 53(1): 168—181

[6] Vapnik V, The nature of statistical learning theory[M]. New York: Springer, 1995

[7] 邓乃扬, 田英杰. 数据挖掘中的新方法：支持向量机[M]. 北京：科学出版社, 2006: 77—78

[8] 杜树新, 吴铁军. 用于回归估计的支持向量机方法[J]. 自动化学报, 2003, 15(11): 1580—1663
Du Shu-xin, Wu Tie-jun. Support vector machines for regression[J]. Journal of System Simulation, 2003, 15(11): 1580—1663

[9] Kennedy J, Eberhart R. Particle swarm optimization[C]. IEEE International Conference on Neural Networks, vol. 4. 1942—1948, Perth, Australia, Nov 27—Dec 1, 1995

[10] Huang X X, Shi F H, Gu W, et al. SVM-based fuzzy rules acquisition system for pulsed GTAW process[J]. Engineering Applications of Artificial Intelligence, 2009, 22(8): 1245—1255

[11] 郗晓宁, 王威, 高玉东. 近地航天器轨道基础[M]. 长沙：国防科技大学出版社, 2003: 20—36

[12] Zhou M X, Zhou J J, Wang F. Analysis and simulation of microwave brightness temperature on Lunar surface[C]. 60th International Astronautical Congress, Daejeon, South Korea, Oct 12—16, 2009

[13] Chang C C, Lin C J. LIBSVM-a library for support vector machines[EB/OL]. [2010]. http://www.csie.ntu.edu.tw/~cjlin/libsvm/

[14] Fei S W, Wang M J, Miao Y B, et al. Particle swarm optimization-based support vector machine for forecasting dissolved gases content in power transformer oil[J]. Conversion and Management, 2009, 50(6): 1604—1609

[15] Pedrycz W, Park B J, Pizzi N J. Identifying core sets of discriminatory features using particle swarm optimization[J]. Expert Systems with Applications, 2009, 36(3): 4610—4616

[16] U.S. Geological Survey. Clementine 750nm mosaics warped to ULCN2005[DB/OL]. [2009—03—21]. http://webgis.wr.usgs.gov/pigwad/down/moon_warp_clementine_750nm_b2.htm

Microwave Brightness Temperature Imaging and Dielectric Properties of Lunar Soil

Wu Ji[1]   Li Dihui[1]   Zhang Xiaohui[1]   Jiang Jingshan[1]
A T ALTYNTSEV[2]   B I LUBYSHEV[2]

(1. Center for Space Science and Applied Research, Chinese Academy of Sciences, P.O. Box 8701, Beijing 100080, China;
2. Institute of Solar Terrestrial Physics, Siberia Branch of Russian Academy of Sciences, P.O. Box 4026, Irkutsk 664033, Russia;
1、2. China–Russia Joint Research Center on Space Weather)

Abstract   Among many scientific objectives of lunar exploration, investigations on lunar soil become attractive due to the existence of He^3 and ilmenite in the lunar soil and their possible utilization as nuclear fuel for power generation. Although the composition of the lunar surface soil can be determined by optical and γ/X-ray spectrometers, etc., the evaluation of the total reserves of He^3 and ilmenite within the regolith and in the lunar interior are still not available. In this paper, we give a rough analysis of the microwave brightness temperature images of the lunar disc observed using the NRAO 12 meter Telescope and Siberian Solar Radio Telescope. We also present the results of the microwave dielectric properties of terrestrial analogues of lunar soil and, discuss some basic relations between the microwave brightness temperature and lunar soil properties.

Key words   Microwave brightness temperature   Dielectric constants of lunar soil   Microwave radiometer

1 Introduction

China plans to launch a lunar exploration orbiter, Chang'E-1, in the coming few years. A microwave radiometer will be on this mission to collect lunar brightness temperature data for determining the thickness of the lunar soil and other soil characteristics.

The Moon is the only and the nearest natural celestial body revolving around the Earth. In view of its uniqueness, it was the first object of study and has been very attractive since ancient times. Study of the lunar microwave brightness temperature began in the 1940s, represented by the work of Dicke and Beringer (1946) and Piddington and Minnett (1949). They discovered the approximate sinusoidal change of lunar brightness temperature with the lunar phase. In the 1950s, Troitskii made further study of lunar brightness temperature and quantified some of the lunar radio emission properties and regularities discovered by earlier researchers (Hagfors et al 1969). With the establishment of the large-scale radio astronomical telescope in the 1960s and early 1970s, many scientists joined the study of lunar brightness temperature. The wavelengths used in observation ranged from 168 cm to 1 mm, covering nearly all the microwave frequencies. But these studies stopped when Neil Armstrong set his foot on the lunar surface at 02:56:15 UT on 21st July 1969. From then on, a great number of scientists from different disciplines were involved in the study of lunar material collected by former Soviet Union robots and Apollo astronauts. With the completion of the first round of lunar exploration by human beings, the study of lunar microwave brightness temperature was completely forgotten. Accompanied by a new upcoming era of lunar exploration and the development of science and technology, observation of the lunar microwave brightness temperature has again become important for exploration of water-ice, characterizing thickness of lunar soil and lunar resources. In this paper, we first explain two significant lunar microwave brightness temperature images observed with the ground-based radio astronomical telescopes in recent years and then describe our study of the microwave dielectric properties of lunar soil, and finally, discuss the possible relations between microwave brightness temperature and depth of the lunar soil.

2   Lunar microwave brightness temperature

For a long time, visible light and infrared have been the preferred wavelengths for lunar observations though microwave band was also used occasionally. Because of the synchronization of the Moon's rotation and revolution, visible and infrared light cannot 'see' the dark side of the Moon, but microwave can. As we know, it has the 'ability' to 'see' things in the dark and there is no need for solar illumination. Moreover, it can penetrate a certain depth under the surface of dielectric material. Therefore, microwaves can be used for the Moon's all-time observation, and study of the Moon's microwave brightness temperature can probably be useful in obtaining information of the depth and structures of the lunar soil. Fig. 1 shows the microwave brightness temperature image of the Moon's surface observed on April 1st,1994 using the NRAO 12 meter Telescope at Kitt Peak, Arizona at a frequency of 90 GHz (3.3 mm). Figure 2 is its nearly corresponding optical black and white image. The radio surface of Fig. 1 appears different from the familiar optical surface shown in Fig. 2. In comparison to the optical image (Fig. 2), the microwave image of the Moon (Fig. 1) roughly reflects the topographic characteristics of mare and highlands, the protrusion extending S–N in the middle of the figure nearly corresponds to the lunar highland terrain. Figure 3 is the microwave brightness temperature image of the lunar disc with 20 arc-sec angular resolution observed on October 27th, 2001 using the Siberian Solar Radio Telescope (SSRT). The lunar phase was nearly 70% during the observation period. Different from figure 1, the brighter region in Fig. 3 (Plate XII) is approximately in accordance with the mare area in the upper right of the optical moon shown in Fig. 2. We expected that Fig. 1 would show the temperature features of the lunar surface and that Fig. 3 would reflect the properties up to a certain depth under the surface.

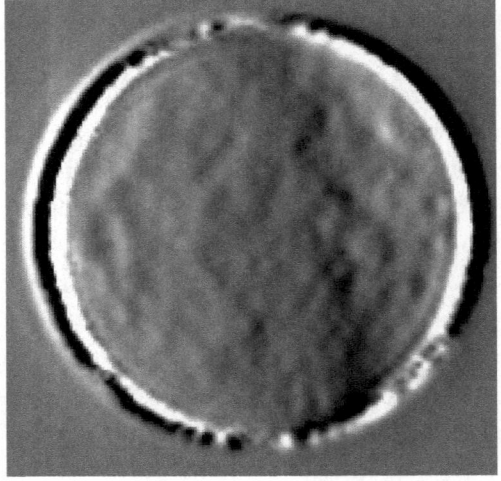

Fig.1   Microwave brightness temperature image of the Moon at 90 GHz

Fig.2   Optical image of the Moon

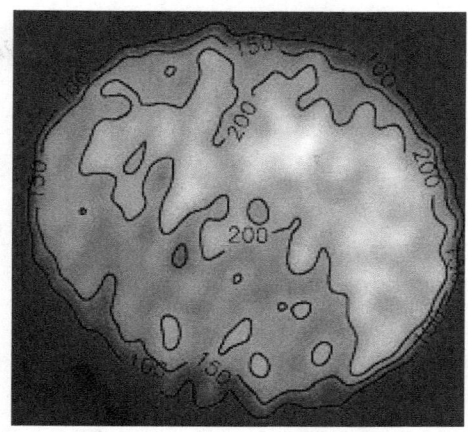

Fig.3 Microwave brightness temperature image of the Moon at 5.7 GHz

3 Dielectric properties of the lunar soil simulators

The microwave dielectric constant is an important parameter for interpretation of the lunar brightness temperature images. In order to understand the microwave brightness temperature data expected from Chang'E's microwave radiometer, nine terrestrial basalts and anorthosites were selected based on the similarity of their chemical compositions to various lunar soils, as determined by X-ray fluorescence. We measured the relative dielectric constants of these lunar analogues at five densities (0.8, 1.0, 1.2, 1.4, 1.6 g/cm^3) over the range of 0.5–20 GHz with open-ended coaxial line model on the HP8722C Network Analyzer. The measurement accuracy of the model is ±0.05 for both real part ε' and loss tangent $\varepsilon'/\varepsilon''$, and the error caused by measurement operation is generally ±0.02. All the measurements are given in an experiment report (Li Dihui et al 2004) and each of them is the average of 3 to 5 sets of measurement values in different positions. Changes of the real parts ε' and imaginary parts ε'' of the relative complex dielectric constants with density and frequency are shown in figures 4 and 5, where different curves represent the samples with different contents of Ti and Fe oxides.

3.1 Quantitative characteristics of the relative complex dielectric constants

Variance tests and polynomial regression indicate that density, frequency and composition contribute to the relative complex dielectric constants. Density is the most dominating factor followed by frequency whereas composition shows a complicated situation. Averages of the real parts of the relative complex dielectric constants at five densities 0.8, 1.0, 1.2, 1.4 and 1.6 g/cm^3 are 2.1560, 2.4425, 2.7098, 3.0729 and 3.4288 respectively and those of the imaginary parts are 0.0844, 0.1024, 0.1208, 0.1627 and 0.1922 respectively. The maximum of the real parts is approximately 1.6 times the minimum and the maximum of the imaginary parts is about 2.3 times the minimum. In the range of frequency from 500 MHz to 20 GHz, the maximum difference between the real parts is 0.8816 and that between the imaginary parts is 0.738.

3.2 Relations between relative complex dielectric constant and density

Real parts and imaginary parts of the relative complex dielectric constants are strongly affected by the sample density. Figure 4 shows the projections of the relative dielectric constants of different analogues vs. density at different microwave frequencies on a planar rectangular coordinate system. Different curves represent terrestrial analogues of lunar samples having different Fe and Ti concentrations (for both, real and imaginary parts). It shows that the real parts have a quicker increase with increased density at the same frequency and all the samples fall in a narrow belt. Regression analyses indicate that a linear relation exists between the real parts and the density over the range of 0.5–20 GHz and the correlation coefficients average 0.989. The imaginary parts show a similarity to the real parts over the range of > 0.5–10 GHz, but begin to change irregularly at about 10

GHz. Accordingly, the imaginary parts under 10 GHz are linearly correlated with the density with the average correlation coefficient of ~0.94, which drops to 0.69 in the range of all measuring frequencies.

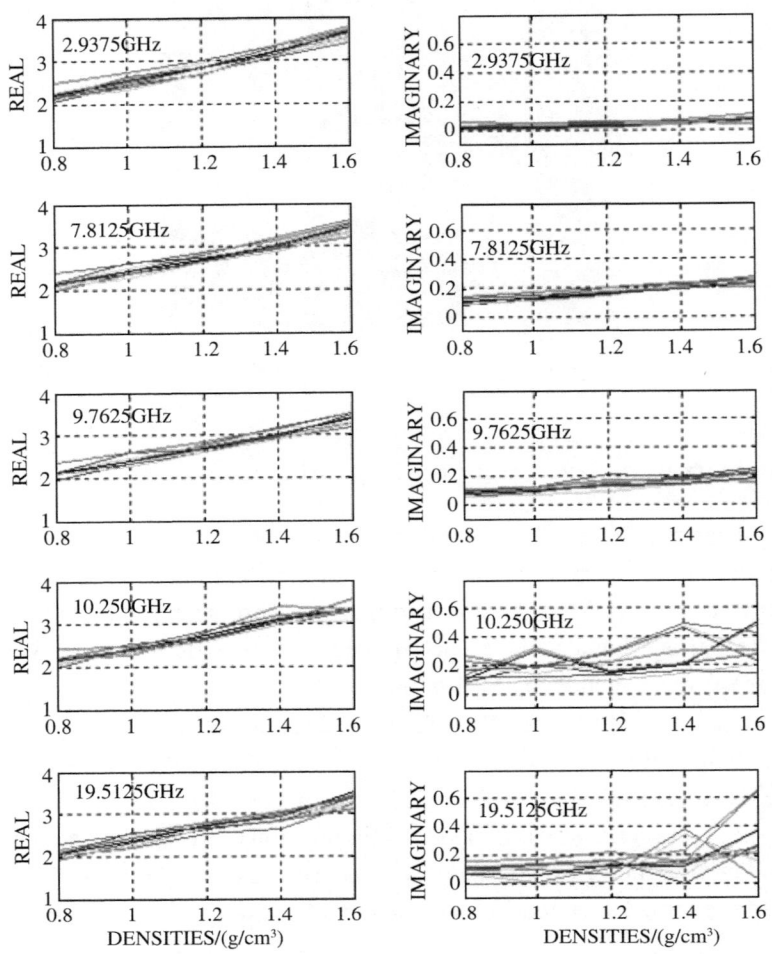

Fig.4 Changes of relative complex dielectric constant (real and imaginary part) with density (the colored lines in the diagrams represent samples with different weight percentages of TiO_2 content: blue −1.0545%, dark green −1.3589%, red −2.4877%, cyan −2.8983%, pink −3.6799%, light brown −3.9355%, black −7.2076%, light blue −8.2427%, light green −11.1473%)

3.3 Relations between relative complex dielectric constant and frequency

Figure 5 shows the projections of the relative dielectric constants of different simulators vs. microwave frequency at different densities on a planar rectangular coordinate system. Real parts and imaginary parts of the relative complex dielectric constants demonstrate totally opposite changes with frequency. With the increase of microwave frequency, the real parts tend to decline whereas the imaginary parts rise. Regression analyses indicate that real parts are also linearly but negatively correlated with microwave frequency in the range of 500 MHz to 20 GHz, and the correlation coefficients average 0.73, obviously lower than those between the real parts and density. Similar to the changes observed with density, the projection curves of the imaginary parts vs. frequency extend smoothly over the 0.5–10 GHz range, but fluctuate in 10–20 GHz, with two to three jumps up and down. The calculations indicate that most of the imaginary parts still vary linearly with increased microwave frequency, but the correlation becomes weaker as indicated by the average correlation coefficient of about 0.34.

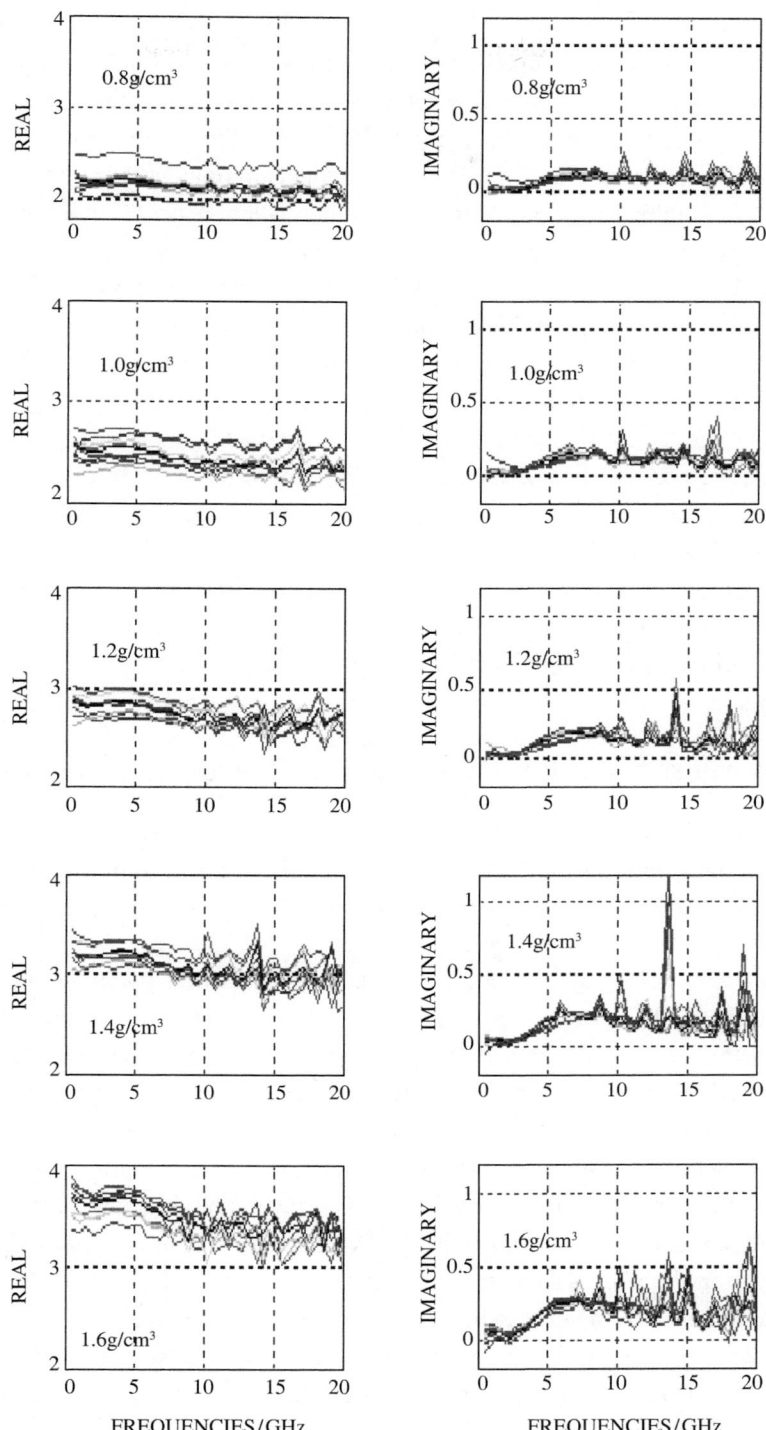

Fig.5 Changes of relative complex dielectric constant with frequency (see figure 4 for the meaning of the colored lines in the diagrams)

3.4 Relations between relative complex dielectric constant and composition

Analysis of measurements reveals that the relative complex dielectric constants and TiO_2 or Fe-oxides or TiO_2 + Fe-oxides or several main element oxides content do not follow a simple relation and TiO_2 or Fe-oxides or TiO_2 + Fe-oxides do not contribute to the dielectric constants. The correlation coefficients between TiO_2 contents and the relative complex dielectric constants average 0.5 for the real part and 0.24 for the imaginary part. Multiple regressions analysis was made by taking the % wt. of major element oxides of SiO_2, Al_2O_3, CaO, MgO, Na_2O, TiO_2 and ΔFe as the arguments of chance variables and each real or imaginary part of the relative dielectric constants as their dependent variable. The results are in accordance with a multivariate linear function. In view of the extremely diversified combinations of material chemical compositions and the entirely different microstructures among the combinations, the effect of the composition on the relative dielectric constants is probably much more complicated than those of density and frequency, and the precise mathematical and physical links between relative dielectric constant and composition need to be studied more.

It is interesting that the dielectric properties of lunar soil analogues are close to those of true lunar samples (Heiken *et al* 1991), but vary differently from those of terrestrial rocks (Ulaby *et al* 1990). Thus we believe that the real lunar soil is realistically simulated in dielectric properties to the lunar soil analogues used in this study.

Understanding of the microwave dielectric properties of materials is important for the application of remote sensing techniques. Some workers have confirmed the correlations between the microwave dielectric constants of a material and its microwave brightness temperature (Lai *et al* 1982; Teng *et al* 1984; Xiao 1988). For example, they have successfully distinguished between water and land, some distinct types of rocks, different structural formations and iron deposit bodies by means of such correlations. Therefore, it is possible that further careful study of microwave dielectric constants of materials and their corresponding microwave brightness temperature data such as shown in figures 1 and 3 will be helpful in determining the regolith thickness, composition, structure and other physical parameters.

4 Discussion on microwave brightness temperature and lunar soil

As we have discussed above, different media have different microwave brightness temperatures, and the microwave brightness temperatures contain information of media density, composition and structure, etc. The physical relation of microwave brightness temperature (B_t) and medium is given by

$$B_t = e(d, \varepsilon', \varepsilon'')T$$

where T is the physical temperature of a medium and can be measured by other means, $e(d, \varepsilon', \varepsilon'')$ is the microwave emissivity of the medium varying between one and zero, a function of d (density), ε' is the real part of the dielectric constant of the medium, ε'' the imaginary part of the dielectric constant, where the polarization and viewing angles are assumed to be constant and therefore dependence on these parameters is not discussed here.

A metallic medium is generally considered to be 'cold'. Therefore, it has a low e value, close to 0, whereas a dense non-conductive dielectric medium is considered 'warm', with a high e value close to 1.0. So, a non-conductive dielectric medium has a high value in ε' and correspondingly shows a high microwave brightness temperature, and a highly conductive medium has high ε'' value, with low microwave brightness temperature. In view of its stronger effect on the ε' of a dielectric medium, the density, in fact, plays the role of an amplifier of microwave brightness temperature. The denser the medium, the larger the e value and the higher the microwave brightness temperature. The microwave frequency, however, just shows the opposite effect.

Lunar soil is a non-conductive medium. As inferred from Lai *et al* (1982) and our dielectric measurements, in an area with similar composition, the lunar soil with a higher brightness temperature should indicate its higher density. In particular, the lunar rocks have higher ε' or high density. Therefore, areas where the regolith layer is not thick or has rocks directly below or is covered with lunar rocks, will appear brighter in a microwave brightness temperature image than the area covered withthicker regolith.

The lower frequency would result in a higher microwave brightness temperature (as shown in figure 5) after absolute calibration. The slope of the measurements from low to high frequency band can also reflect the density of the area. Larger slope reflects higher density as shown in figure 5 and can be used in determining the density.

Limited by the measurement conditions, the mechanism of dependence of composition on dielectric properties of a medium is still not understood. We, however, believe that lunar soil composition (especially with high content of TiO_2 or Fe-oxides) plays an important role in the microwave brightness temperature in view of the rough correspondence between the brighter regions seen in figure 3 and its correspondence with the mare area, in view of the higher dielectric constant values obtained by Chung *et al* (1971).

5 Conclusions

The microwave brightness temperature images of the Moon possibly contain new information about properties of lunar rocks and regolith which are not available in optical images.

Our studies on the microwave dielectric properties of lunar soil analogues and the analysis of the microwave brightness temperature images suggest that denser lunar soil will appear brighter on a microwave brightness temperature image and lower frequency band brightness temperature will correspond to a higher response from lunar soil. These two characteristics will enable us to extract properties of lunar soil with different densities from images in different frequency bands.

References

[1] Chung D H, Westphal W B and Simmons G. 1971. Dielectric behavior of lunar samples: Electromagnetic probing of the lunar interior; *Geochim. Cosmochim.Acta Suppl.*2.2381—2390

[2] Dicke R H and Beringer R. 1946. Microwave radiation from the sun and the moon; *Astrophys. J.* 103(3): 375—376

[3] Hagfors T, Green J L and Guillen A. 1969. Determination of the albedo of the Moon at a wavelength of 6 meter; *Astron. J.*74: 1214—1219

[4] Heiken G H, Vaniman D T and French B M. 1991. Lunar Sourcebook: a user's guide to the moon; (New York, Port Chester, Melbourne, Sydney: Cambridge University Press) 536—552

[5] Lai Zhaosheng, Xiao Jinkai and Feng Junming. 1982. Dielectric constant determination of some minerals and rocks under microwave frequencies; In: Researches on Mineral Physics and Mineral Materials (Beijing: Science Press) 108—114

[6] Li Dihui *et al*. 2004. Dielectric properties of several powdered metal-oxide mixtures and lunar soil simulants at varied microwave frequencies.Center for Space Science and Applied Research; *Chinese Academy of Sciences, Experimental Report*, 1—48 (in Chinese)

[7] Piddington J H and Minnett H C. 1949. Microwave thermal radiation from the moon; *Aust. J. Sci. Res.*A2: 63

[8] Teng Xuyan, Xiao Jinkai, Shi Changging, Lai Zhaosheng, Peng Hongxian and Yang Bolin. 1984. Passive microwave radiometry in the Gobi-Desert region; *Remote Sensing of Environment*, 15: 37—46

[9] Ulaby F T *et al*. 1990. Microwave Dielectric Properties of Dry Rocks; *IEEE Transactions on Geoscience and Remote Sensing* 28(3): 325—336

[10] Xiao Jinkai. 1988. Dielectric property research of minerals and rocks and its significance on remote sensing; *Environmental Remote Sensing*, 3: 135—146 (in Chinese)

The Analysis of Affections to the Cold Space Calibration Source of Chang'E-1 Payload Microwave Detector

Zhang Huiya[1,2]　Zhang Xiaohui[1]　Yang Junli[1,2]

(1.Center for Space Science and Applied Research, Chinese Academy of Sciences, No.1 Nanertiao, Zhongguancun, Haidian District, Beijing 100080, China;
2.Graduate School of Chinese Academy of Sciences, Beijing 100036, China)

Abstract　Chang'E-1 (CE-1) will be the first satellite of China in lunar orbit. Its Microwave Detector makes real-time and periodical calibration at high and low temperature points. The low brightness temperature calibration source will be provided by calibration antenna which is pointing towards the cold space.

This paper focuses on analyzing solar radiation affections to the cold space calibration source while the satellite is in lunar orbit. We get the following results:

(1) Solar radiation affections are maximal when the satellite is right facing the sun and the maximal brightness temperatures which close to the original antenna temperature T are 86.3200, 85.6700, 31.0020, 29.8905 K, respectively.

(2) The incidence of solar radiation affections in forward-flight is smaller than in sideward-flight, and the affections fluctuate rapidly in forward-flight when reaching to and leaving from the peak value. But when in sideward-flight, they are smoothly descending all the time.

(3) The solar radiation affections to channel 1 (3 GHz) and channel 2 (7.8 GHz) are almost as three times high as to the other two channels.

(4) The affections to E plane, H plane and 45° plane of the same channel are quite similar.

These results not only provide important theoretical reference to Microwave Detector's calibration, but also would be useful when observing the earth in future.

Key words　CE-1　Microwave Detector　Cold space　Calibration　Solar radiation

1 Introduction

Microwave Detector is one of the main payloads of ChangE-1 lunar satellite. It is a multi-band microwave radiometer with operation frequencies of 3.0, 7.8, 19.35 and 37.0 GHz, respectively. Generally, the basic calibration way of satellite-borne microwave radiometer includes three steps (Xiao Zhihui et al., 2000): ① pre-launch ground calibration, ② on-orbit calibration and ③ absolute and analogy calibration. The on-orbit calibration ensures the stable work of microwave radiometer in outer space, and it always executes periodically through two points.

Different satellites have different structures, and carry out periodical calibration in different ways. There are two tested structures (Xiao Zhihui et al., 2000). First, the instrument uses switches for connecting to observation signals and calibration signals by turning in a cycle, such as SMMR (the Seasat Scanning Multichannel Microwave Radiometer) (Njoku et al., 1980; Swanson and Riley, 1980), MSR (Microwave Scanning Radiometer) (Morita et al., 1989), ATSR/MWR (Along-Track Scanning Radiometer/Microwave Radiometer) (Bernard et al., 1993;Eymard et al., 1994) and TMR (TOPEX/Poseidon Microwave Radiometer) (Janssen et al., 1995; Ruf et al., 1995). Second,

---

本文原载于 ADVANCES IN SPACE RESEARCH, 2008, Vol.42, 350~357。

during the scan, the entire antenna assembly rotates as a unit, and the stationary hot-load and cold-sky reflectors are viewed as the feeds pass beneath them, so the radiometer view hot/cold calibration target once per scan, such as SSM/I (the Defense Meteorological Satellite Program operational Special Sensor Microwave Imager) (Colton, 1999; Hollinger et al., 1990), MIMR (the Multifrequency Imaging Microwave Radiometer) (Menard and Reynolds, 1991; Thornbury, 1990) and AMSR (Advanced Microwave Scanning Radiometer).[1,2]

Microwave Detector is designed as the second structure. The hot reference source is provided by a match-load detected in real-time and the cold source is provided by a special horn antenna viewing the cold space. Base upon observations from the NASA's COBE, space provides a uniform brightness temperature of 2.73 K with a variation of not more than 100μK.[3] Actually, because of the absolute outside affections such as solar radiations, the cold source is not as exact as 2.73 K. In order to avoid the solar affections effectively, most satellite-borne microwave radiometers run in following two ways. In the first way, the data which were affected by the sun were not used in calibration process (Njoku et al., 1980; Swanson and Riley, 1980). In the other, avoid the sun running into the antenna's main-beam and side-lobe (Janssen et al., 1995; Ruf et al., 1995). Neither of the two ways deals with the affected data directly.

According to the instruction of CE-1 satellite's orbit and its poses in flight, the calibration antenna is pointing to the satellite's orbit tangent while in forward-flight, and pointing to the sun while in sideward-flight. Thus, most cold observation data are affected by the sun. If we put away all these affected data, it will be a great loss in the field of data processing and research work. So this paper focuses on analyzing solar radiation affections to the cold space calibration source while the satellite is in lunar orbit.

2 Analysis

According to the principle that two points determine a straight line, we can calibrate through two points while the linearity of the microwave radiometer receiver is ensured. If one of the two points is unknown, we could not draw the line. So if one of the calibration sources is uncertain, we could not calibrate easily.

Microwave Detector has two calibration branches and one observation branch (Fig.1), and uses switches to choose the calibration mode or observation mode periodically. The calibration branches connect to the high and low reference source which are key points of calibration equation, respectively. They affect the veracity and accuracy of the whole calibration system directly.

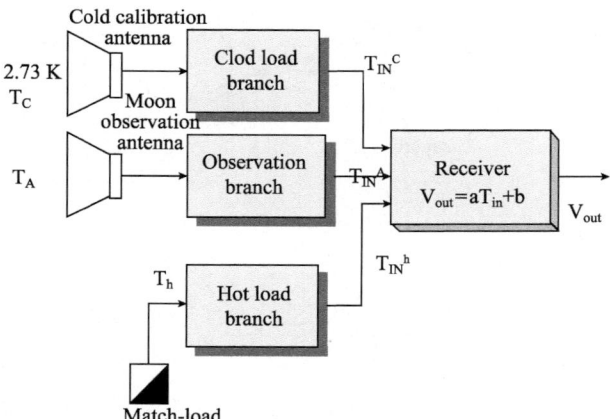

Fig.1 The simple diagram about one channel of Microwave Detector

2.1 Optimum observation condition—the satellite is in the shadow of the moon

Learnt from the ubiety of the sun, moon and satellite, moon has a shadow zone (Ouyang Ziyuan, 2005). While the satellite moves into the centre of the shadow, the solar radiations are ten thousandth of the sun constant.

According to the character of Microwave Detector's antenna (Figs.6–8), the patterns are with very low sidelobes and the antenna is low loss. Under such circumstances, we can use 2.73 K as the brightness temperature of the cold source.

2.2 Worst observation condition—the sun is in the view of the calibration antenna

2.2.1 The radiate state of the sun

If the satellite is out of the moon's shadow, the sun will be in the view of the cold calibration antenna in some periods. The solar radiations come from atmosphere outside the sun. The range of the wavelength extends all over the radio band. So far, base upon the observations, solar radiations have three totally different components (Zhao Renyang et al., 1997): quiet solar radio, stable-change solar radio and solar radio outbreak. CE-1 will be launched next year when solar activity is minimal, so in this paper we only analyze the total radiate flux produced by quiet and stable-change solar radio. According to the radio observations of the annular solar eclipse on September 23, 1987 (Zhou Shurong et al., 1990), we can get antenna temperature which generate by solar flux. The angle between the cold calibration antenna boresight and a vector from the spacecraft to the sun is called sun-cold horn angle. Solar radiation affections to the cold source can be concluded by sun-cold horn angle and normalized antenna radiation pattern.

2.2.2 The flight-pose adjusting of the satellite

The following three equipments of CE-1 satellite have fixed directions: sun sail board points to the sun, observation equipment points to the moon and control-communication equipment points to the earth. In order to get enough solar energy in the flight, satellite flies in forward and sideward ways. If the sun incidence angle is larger than 45°, the satellite moves on steadily with fixed pose. Otherwise, sun-limb cannot get enough solar energy and the satellite has to change its flying pose. Fig.2 shows two sets of antennas of Microwave Detector. Generally, in the satellite's coordinate (Fig.2), Z axis in the direction of observation antenna, X axis is the direction of cold calibration antenna and Y axis is the direction of sun limbs. The drawings of the satellite flying pose are shown in Figs.3 and 4. When in forward-flight, the X axis (in Fig.3) in the satellite's coordinate is pointing to the orbit's tangent (also the speed direction of the satellite). When in sideward-flight, the satellite turn 90° and the Y axis (in Fig.4) in the satellite's coordinate is point to the orbit's tangent. While the sun-angle β (the angle between orbit plane and the sun) is in the range of [−45°, 45°] and [135°, 225°], the satellite moves on with forward-flight pose (Fig.3); while β is in the range of [45°, 135°] and [225°, 315°], the satellite moves on with sideward-flight pose (Fig.4).

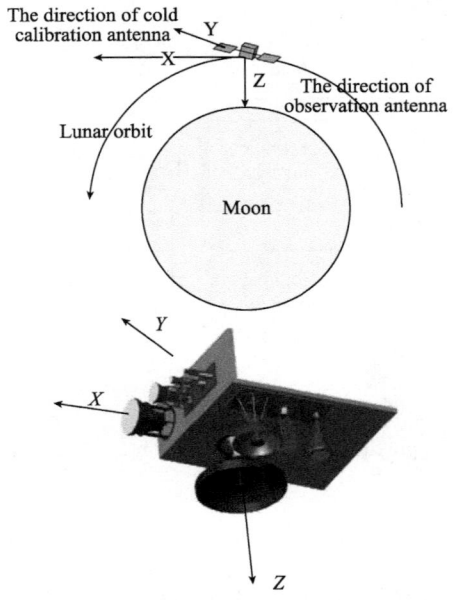

Fig.2 Two sets of antennas of Microwave Detector

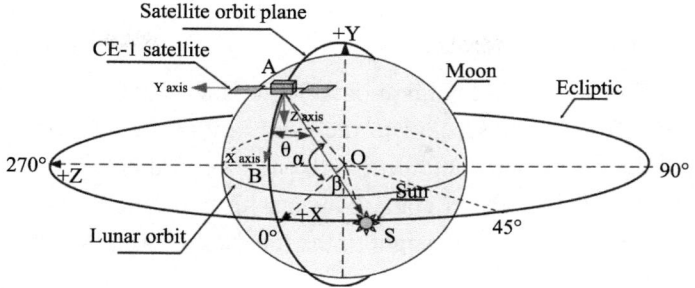

Fig.3 The ubiety among the sun, the moon and the satellite while the satellite is in forward-flight

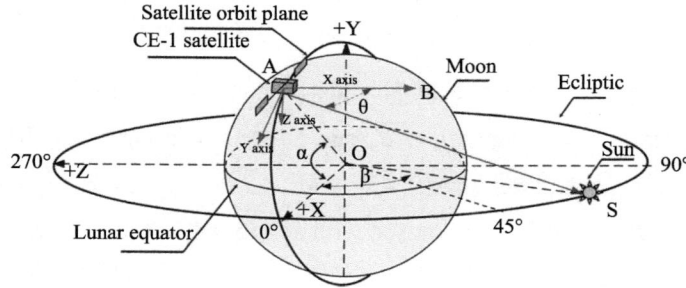

Fig.4 The ubiety among the sun, the moon and the satellite while the satellite is in sideward-flight

2.2.3 Antenna temperature

Antenna received power in unit frequency is given by

$$P = SA \tag{1}$$

where S is the flux of solar radiation, A is the effective area of antenna. The relation between A and G (antenna gain) is written as (Zhao Renyang et al., 1997)

$$A = \frac{\lambda^2}{4\pi} G \tag{2}$$

Substituting Eq. (2) for A in Eq. (1) and rearranging

$$P = S \frac{\lambda^2}{4\pi} G \tag{3}$$

Solving Eq. (3) for T (antenna temperature)

$$T = P/2k = S \frac{\lambda^2}{8\pi k} G \tag{4}$$

where k is Boltzmann constant, and $k=1.38 \times 10^{-23}$ J/K.

Both 2007 and 1987 are the beginnings of solar activity cycles; it means that the sun is in minimal activity. We use the observation data in 1987 as a reference. According to the information from Zhou Shurong et al. (1990), the reference solar flux is listed in Table 1 (The wavelengths in parenthesis of the second column which are close to the wavelengths of Microwave Detector were measured in 1987.). The frequencies, wavelengths, gains, effective areas of antenna and antenna temperatures calculated by Eq.(4) for all the four channels are also listed in this table.

Table 1 Parameters for calculating antenna temperature

Frequency f/GHz	Solar flux S (s.f.u)	Wavelength λ (cm)	Gain G (dB)	Effective area of antenna A (m^2)	Antenna temperature T (K)
3	72(10.6 cm)	10	16.25	0.3312	86.4000
7.8	264(3.2 cm)	3.846	18.87	0.0896	85.6700
19.35	790(1.46 cm)	1.550	17.70	0.0111	31.8190
37	2347(0.86 cm)	0.811	18.33	0.0035	29.8905

2.2.4 Sun-cold horn angle

When the sun is in the view of the cold calibration antenna, the angle θ between the cold calibration horn antenna boresight and a vector from the spacecraft to the sun is called sun-cold horn angle (Figs.3 and 4). We suppose the sun turns around the moon, and compute θ in the moon coordinate.

Firstly, we suppose there is no deflection angle between lunar orbit plane and lunar polar plane. As we known, the radius of the moon, $R=1738$ km; the height of the lunar orbit, $h = 200$ km, then satellite location $A(x_A, y_A, z_A)$ in the moon coordinate can be written as

$$\begin{cases} x_A = r\cos\alpha \\ y_A = r\sin\alpha \\ z_A = 0 \end{cases} \tag{5}$$

where α is the angle between the satellite and lunar equator, and $r=R+h=1938$ km. So we have $\overrightarrow{OA}=(r\cos\alpha, r\sin\alpha, 0)$. Because $\overrightarrow{OA}$ is always plumbing with $\overrightarrow{AB}$ and $\overrightarrow{AB}$ is pointing to the direction of orbit tangent, we have $\overrightarrow{AB}=(r\sin\alpha, -r\cos\alpha, 0)$ in Fig.3 (where $\beta \in [0, 45]$) and $\overrightarrow{AB}=(0, 0, -1)$, in Fig.4 (where $\beta \in [45, 90]$).

Secondly, we suppose lunar equator is superposed with ecliptic, then the sun's location S (x_S, y_S, z_S) in moon coordinate can be written as

$$\begin{cases} x_S = R_S \cos\beta \\ y_S = 0 \\ z_S = -R_S \sin\beta \end{cases} \tag{6}$$

so $\overrightarrow{AS}=(R_S\cos\beta-r\cos\alpha, -r\sin\alpha, -R_S\sin\beta)$.

At last, when the satellite is in forward-flight pose (Fig.3)

$$\cos\theta = \frac{R_S \sin\alpha \cos\beta}{\sqrt{R_S^2 - 2R_S r \cos\alpha \cos\beta + r^2}} \tag{7}$$

where $\alpha \in [0°, 360°]$, $\beta \in [0°, 45°]$.

When the satellite is in sideward-flight pose (Fig.4)

$$\cos\theta = \frac{R_S \sin\beta}{\sqrt{R_S^2 - 2R_S r \cos\alpha \cos\beta + r^2}} \tag{8}$$

where $\alpha \in [0°, 360°]$, $\beta \in [45°, 90°]$.

2.2.5 Results

With antenna temperature T, sun-cold horn angle θ and normalized antenna radiation pattern, the relationship between α, β and ΔT of channel 1 can be illustrated in Fig.5 (other channels are similar). For the short angle step we chose, we get a number of ΔT. All minimum ΔT is less than 0.001K, and according to the calibration precision ΔT_A and the sensitivity of all the antennas which are showed in Table 2, it is unnecessary to analyze the minimum ΔT. So we only list the maximum ΔT of forward-flight ($\beta = 0°$), forward-flight ($\beta = 45°$), sideward-flight ($\beta = 45°$), sideward-flight ($\beta = 90°$), the whole forward-flight and sideward-flight periods in Table 3.

After analyzing the tables and figures, we finally come to the following main conclusions,

(1) In Fig.5 (Plate XII), ΔT has a small incidence and fluctuates rapidly when reaching to and leaving from the peak value in forward-flight. The figure is like a hill, the flat parts of which are the affections that can be neglected. But when in sideward-flight, ΔT is descending smoothly and the incidence is large. The figure is like a pensile carpet which means the affections to cold source are always great during the whole sideward-flight period. There are several reasons for these. First, the direction of calibration antenna is pointing towards lunar orbit tangent which is the same as $-Z$ axis, so θ does not change with α in sideward-flight. Second, the average

distance between the sun and moon is much larger than the average distance between the moon and the satellite, so θ changes with β a little which makes ΔT descending smoothly. But in forward-flight, θ changes both with α and β. Most of the time the satellite is in the shadow of the moon and the sun runs into side-lobe of calibration antenna, so ΔT has a peak value region.

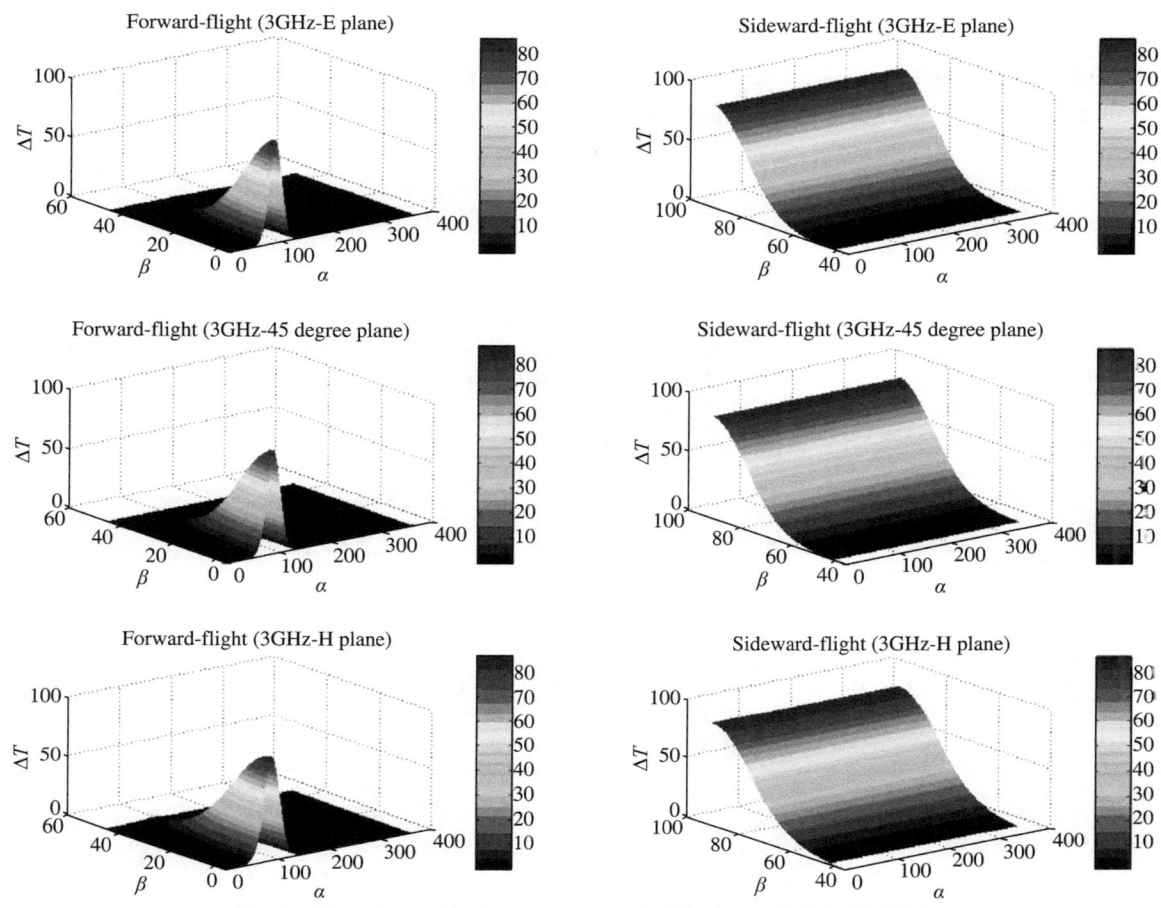

Fig.5 The relationship between α, β and ΔT in forward-flight (3GHz)

Table 2 The performance of Microwave Detector

Frequency (GHz)	Sensitivity (K)	Linearity	ΔT_A (K)
3.0	0.30	0.99957	0.4840
7.8	0.14	0.99991	0.4257
19.35	0.11	0.99984	0.1613
37	0.17	0.99990	0.3722

Table 3 Maximum ΔT(K) of all channels in forward-flight and sideward-flight

Flying pose	Forward 0°	Forward 45°	Forward	Sideward 45°	Sideward 90°	Sideward
Frequency 3 GHz E plane	86.0250	0.3688	86.0250	0.0785	86.0250	86.0250
H plane	86.3200	1.2567	86.3200	1.2567	86.3000	86.3200
Frequency 7.8 GHz						
E plane	85.6300	0.1249	85.6300	0.0178	85.6300	85.6300
H plane	85.6700	0.0703	85.6700	0.0704	85.6700	85.6700
Frequency 19.35 GHz						
E plane	30.5130	0.1518	30.5130	0.0301	29.1665	30.5130
H plane	31.0020	0.0804	31.0020	0.0069	29.5385	31.0020
Frequency 37 GHz						
E plane	29.8905	0.0691	29.8905	0.0075	29.5010	29.8905
H plane	29.8905	0.0478	29.8905	0.0430	29.0825	29.8905

(2) After learning from Table 3, we know that the affections to channel 1 (3 GHz) and channel 2 (7.8 GHz) are greater than channel 3 (19.35 GHz) and channel 4 (37 GHz). Although the solar flux in channel 1 and 2 are smaller, the wavelengths and effective areas of antenna are larger than the others. It causes that the antenna temperatures in channel 1 and 2 are almost as three times high as the others.

(3) The affections to E plane, H plane and 45° plane in the same channel have similar characteristics. It is determined by the antenna, radiation pattern (Figs.6–8). Fig.6 (Plate XIII) is a 3D emulational model of channel 1 (3 GHz)'s antenna radiation pattern. Figs.7 and 8 show the pattern of E plane and H plane in channel 1, respectively. Learnt from these pictures, we have known that the antenna radiation pattern is really similar to each other.

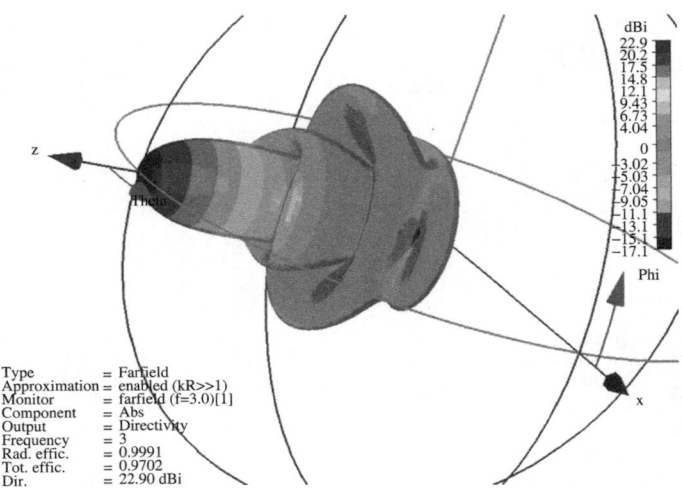

Fig.6   The 3D emulational model of channel 1 (3 GHz)'s antenna radiation pattern

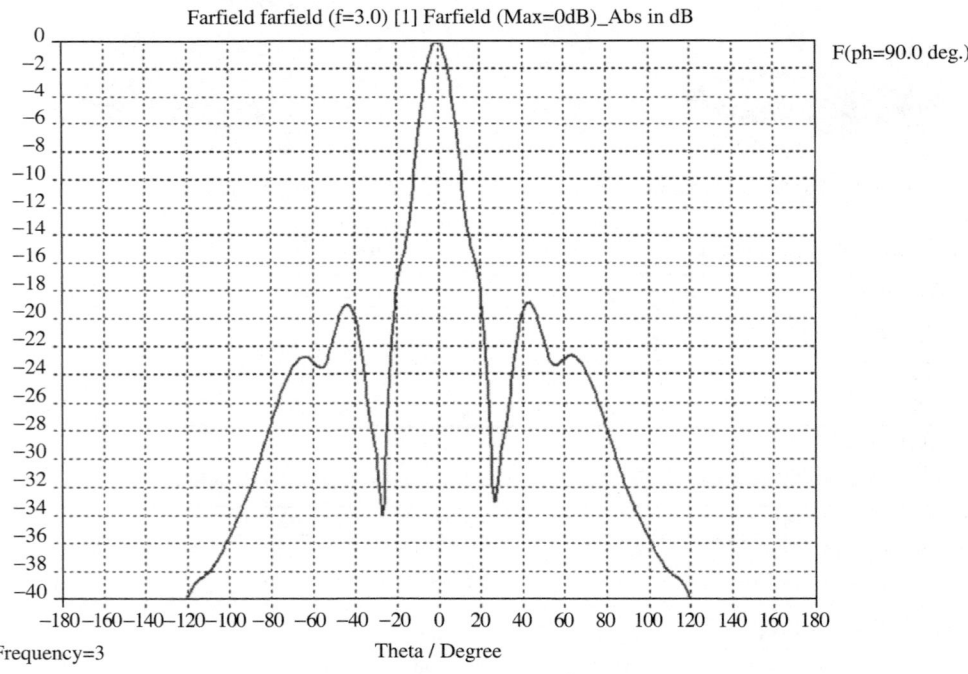

Fig.7   The antenna radiation pattern of E plane in channel 1 (3 GHz)

(4) In the same channel, the affections in forward-flight are increasing at first and then decreasing when reaching to the peak value, but the affections in sideward-flight are decreasing smoothly all the time. When β is expanding in forward-flight, the satellite is getting away from the location right facing the sun and ΔT is

descending; when β is expanding in sideward-flight, the satellite is near to the location right facing the sun and ΔT is ascending. Thus, the solar radiation affections to the cold source are maximal when the satellite is right facing the sun and the maximum ΔT is close to the above antenna temperature in Table 1.

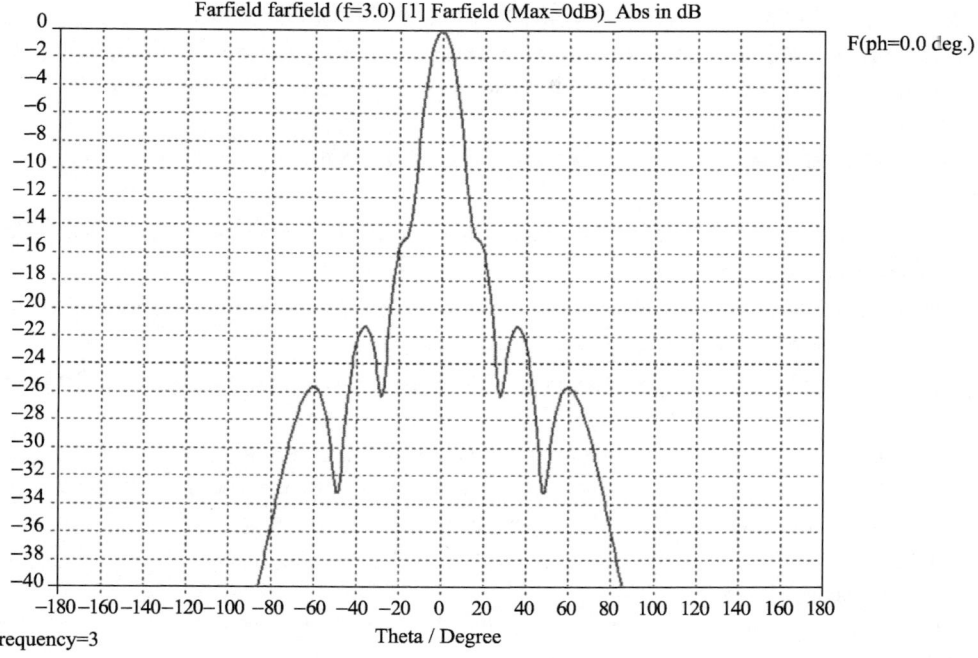

Fig.8 The antenna radiation pattern of H plane in channel 1 (3 GHz)

3 Conclusion

Generally, the satellite-borne low temperature calibration source is provided by the calibration antenna pointing towards the cold space. Because the sun sometimes runs into the view of the antenna, it becomes the primary affection to the cold source while the microwave radiometer is calibrating. So far, people usually do not use the affected data or avoid the sun from running into the view of the antenna effectively. In this paper, we provide the direct analysis of solar radiation affections.

According to the radio observations of the annular solar eclipse on September 23, 1987 (Zhou Shurong et al., 1990), we can then get antenna temperature which is generated by solar flux. We suppose the moon and the satellite are static, and get the angle between the cold calibration antenna boresight and a vector from the spacecraft to the sun (sun-cold horn angle) in lunar coordinate. Solar radiation affections related with α and β to the cold source can be concluded through sun-cold horn angle and normalized antenna radiation pattern. Solar radiation affections to the cold source are maximal while the satellite is right facing the sun. The maximal temperatures which close to antenna temperatures are 86.32, 85.67, 31.2, 29.89 5 K, respectively.

This paper introduces the qualitative analysis of solar radiation affections to the cold source. The results not only provide important theoretical reference to the satellite-borne real-time calibration when observing the moon, but also would be useful when observing the earth.

References

[1] Bernard, R., Le, C.A., Laurence, E., et al. The microwave radiometer aboard ERS-1: Part I. Characteristics and performances [J]. IEEE Trans. Geosci. Remote Sens. 31 (6), 1186—1198, 1993

[2] Colton, M.C. Intersensor calibration of DMSP SSM/Is:F-8 to F-14,1987–1997 [J]. IEEE Trans. Geosci. Remote Sens. 37 (1), 418—439, 1999

[3] Eymard, L., Le, C.A., Tabary, L. The ERS-1 microwave radiometer [J].Int. J. Remote Sens. 15 (4), 845—857, 1994
[4] Hollinger, J.P., Peirce, J.L., Poe, G.A. SSM/I instrument evaluation [J]. IEEE Trans. Geosci. Remote Sens. 28 (5), 781—790, 1990
[5] Janssen, M.A., Ruf, C.S., Keihm, S.J. TOPEX/Poseidon microwave radiometer (TMR):II antenna pattern correction and brightness temperature algorithm. IEEE Trans. Geosci. Remote Sens. 33 (1), 138—146, 1995
[6] Morita, I.A.K., Motomura, K., Suzuki, T. Improvement in the spatial accuracy of the MOS-1 microwave scanning radiometer by signal processing [R]. Proceedings of 1989 International Symposium on Noise and Clutter Rejection in Radars and Imaging Sensors, 341—346, 1989
[7] Menard, Y., Reynolds, M. The design of the ESA multiband imaging microwave radiometer MIMR [R]. Proceedings of IGASS'91, 2359—2363, 1991
[8] Njoku, E.G., Stacey, J.M., Barath, F.T., et al. The seasat scanning multichannel microwave radiometer (SMMR): instrument description and performance. IEEE Oceanic Eng. OE-2(2), 100—115, 1980
[9] Ruf, G.S., Keihm, S.J., Janssen, M.A. TOPEX/Poseidon microwave radiometer (TMR): I instrument description and antenna temperature calibration. IEEE Trans. Geosci. Remote Sens. 33 (1), 125—137, 1995
[10] Thornbury, A. The multifrequency imaging microwave radiometer (MIMR) [R]. Proceedings of IGASS'90, 1597—1600, 1990
[11] Zhao, Renyang, Jin, Shengzhen, Fu, Qijun Solar Radio Microwave Outbreak. Scientific Press (in Chinese), 1997
[12] Swanson, P.N., Riley, A.L. The seasat scanning multichannel microwave radiometer (SMMR): radiometric calibration algorithm development and performance. IEEE J. Oceanic Eng. OE-5(2), 116—124, 1980
[13] Zhou, Shurong, Zuo, Xiao, Yongsheng, Li Observations and Studies of China on annular solar eclipse of September 23, 1987. Scientific Press (in Chinese), 1990
[14] Xiao, Zhihui, Zhang, Zuyin, Guo, Wei A review: the calibration of ground-based, airborne and satellite-borne microwave radiometers. Remote Sens. Technol. Appl. 15 (2), 113—120 (in Chinese), 2000
[15] Ziyuan, Ouyang Introduction to Lunar Science. China Aerospace Press (in Chinese), 2005

第七部分 其他

嫦娥一号卫星的制导、导航与控制

黄江川　张洪华　李铁寿　宗　红

(北京控制工程研究所，北京 100080)

摘　要　嫦娥一号卫星是中国首颗月球卫星。卫星制导、导航与控制(GNC)任务复杂多变，对系统实时性、可靠性和精度要求较高。文章介绍嫦娥一号卫星 GNC 系统组成、控制方法、系统特点和典型飞行结果。

关键词　嫦娥一号卫星　制导、导航与控制　月球　系统

1 引言

中国第一颗大型月球探测航天器嫦娥一号卫星于 2007 年 10 月 24 日成功发射。2007 年 10 月 31 日，嫦娥一号卫星在预定时间和预定地点进入预定的地月转移轨道，2007 年 11 月 5 日，嫦娥一号卫星在近月点进入预定的绕月轨道，2007 年 11 月 7~18 日，嫦娥一号成功完成对月定向和三体指向控制在轨测试。

嫦娥一号卫星 GNC 系统完成了许多复杂任务。在调相轨道，GNC 系统执行一系列姿态机动和轨道控制，使卫星在适当时间转入地月转移轨道。在地月转移轨道，GNC 系统保证卫星对太阳定向，并执行几次轨道中途修正，使卫星捕获预定环月轨道起始点。在月球轨道捕获阶段，GNC 系统执行几次轨控发动机点火，使卫星捕获月球轨道并进入标称环月轨道。在环月轨道，GNC 系统使卫星本体对月球定向、太阳帆板对太阳定向、定向天线对地球定向。

本文概要介绍嫦娥一号卫星 GNC 系统组成、控制方法、系统特点和典型飞行结果。

2 卫星运动模型与控制目标

嫦娥一号卫星是带有挠性太阳帆板、大型充液贮箱和中心刚体的复杂运动体，卫星运动包括刚体平动与转动、挠性振动、液体晃动等。

引入坐标系："O_i"代表惯性坐标系，"O_b"代表卫星本体坐标系，"O_d"代表卫星目标坐标系。设从"O_b"系旋转到"O_d"系的欧拉轴单位矢量为 $\boldsymbol{k}=[k_1 k_2 k_3]^T$，欧拉角为 ϕ，则"O_d"系相对于"O_b"系的姿态可用单位四元数表示：

$$\Delta q = \begin{bmatrix} \Delta \boldsymbol{q}^{-T} & \Delta q_4 \end{bmatrix}^T$$

其中：$\Delta \bar{\boldsymbol{q}} = [\Delta q_1\ \Delta q_2\ \Delta q_3]^T$，$\Delta q_1 = k_1 \sin\frac{\varphi}{2}$，$\Delta q_2 = k_2 \sin\frac{\phi}{2}$，$\Delta q_3 = k_3 \sin\frac{\phi}{2}$ 是四元数的矢量部分，$\Delta q_4 = \cos\frac{\phi}{2}$ 是四元数标量部分，满足 $|\Delta q_4|^2 + \|\Delta \bar{\boldsymbol{q}}\|^2 = 1$。

记"O_b"系相对于"O_i"系的旋转角速度为 ω_s，"O_d"系相对于"O_i"系的旋转角速度为 ω_d，则"O_d"系相对于"O_b"系的角速度为 $\Delta\omega = \omega_d - \omega_s$。卫星目标系相对本体系运动学可以表示为：

本文原载于《空间控制技术与应用》，2008, Vol.34, No.1, 29~32。

$$\Delta \dot{\bar{q}} = \frac{1}{2}\Delta\omega\Delta q_4 + \frac{1}{2}\Delta\tilde{\omega}\Delta\bar{q} \tag{1a}$$

$$\Delta \dot{q}_4 = -\frac{1}{2}\Delta\omega^{\mathrm{T}}\Delta\bar{q} \tag{1b}$$

其中：$(\cdot)$表示相对的时间导数，$\Delta\tilde{\omega}$ 是角速度 $\Delta\omega$ 的反对称阵。

卫星动力学模型可以简写如下[1,2]：

$$M\ddot{X} + F_{\mathrm{tr}}\ddot{\eta}_{\mathrm{r}} + F_{\mathrm{tl}}\ddot{\eta}_{\mathrm{l}} = P_{\mathrm{s}} \tag{2}$$

$$I_{\mathrm{s}}\dot{\omega}_{\mathrm{s}} + \tilde{\omega}_{\mathrm{s}}I_{\mathrm{s}}\omega_{\mathrm{s}} + F_{\mathrm{sr}}\ddot{\eta}_{\mathrm{r}} + F_{\mathrm{sl}}\ddot{\eta}_{\mathrm{l}} + R_{\mathrm{has}}\dot{\omega}_{\mathrm{la}} + R_{\mathrm{ras}}\dot{\omega}_{\mathrm{ra}} = T_{\mathrm{s}} \tag{3}$$

$$I_{\mathrm{la}}\dot{\omega}_{\mathrm{la}} + F_{\mathrm{la}}\ddot{\eta}_{\mathrm{l}} + R_{\mathrm{has}}^{\mathrm{T}}\dot{\omega}_{\mathrm{s}} = T_{\mathrm{la}} \tag{4a}$$

$$I_{\mathrm{ra}}\dot{\omega}_{\mathrm{ra}} + F_{\mathrm{ra}}\ddot{\eta}_{\mathrm{r}} + R_{\mathrm{ras}}^{\mathrm{T}}\dot{\omega}_{\mathrm{s}} = T_{\mathrm{ra}} \tag{4b}$$

$$\ddot{\eta}_{\mathrm{r}} + 2\zeta_{\mathrm{r}}\omega_{\mathrm{r}}\dot{\eta}_{\mathrm{r}} + \omega_{\mathrm{r}}^2\eta_{\mathrm{r}} + F_{\mathrm{tr}}^{\mathrm{T}}\ddot{X} + F_{\mathrm{sr}}^{\mathrm{T}}\dot{\omega}_{\mathrm{s}} = 0 \tag{5a}$$

$$\ddot{\eta}_{\mathrm{l}} + 2\zeta_{\mathrm{l}}\omega_{\mathrm{l}}\dot{\eta}_{\mathrm{l}} + \omega_{\mathrm{l}}^2\eta_{\mathrm{l}} + F_{\mathrm{tl}}^{\mathrm{T}}\ddot{X} + F_{\mathrm{sl}}^{\mathrm{T}}\dot{\omega}_{\mathrm{s}} = 0 \tag{5b}$$

式中 X 是卫星平动位置列阵，ω_{s} 是卫星本体角速度列阵，$\tilde{\omega}_{\mathrm{s}}$ 是角速度列阵的反对称阵，M 是卫星质量阵，I_{s} 是卫星惯量阵，P_{s} 是作用在卫星上外力列阵，T_{s} 是作用在卫星上力矩列阵，ω_{la}、ω_{ra} 分别是左、右太阳帆板角速度列阵，ω_{l}、ω_{r} 分别是左、右太阳帆板模态频率对角阵，η_{l}、η_{r} 分别是左、右太阳帆板模态坐标阵，ζ_{l}、ζ_{r} 分别是左、右太阳帆板模态阻尼系数阵，I_{la}、I_{ra} 分别是左、右太阳帆板惯量阵，F_{tl}、F_{tr} 分别是左、右太阳帆板振动对本体平动的柔性耦合系数阵，F_{sl}、F_{sr} 分别是左、右太阳帆板振动对本体转动的柔性耦合系数阵，F_{la}、F_{ra} 分别是左、右太阳帆板振动对自身转动的柔性耦合系数阵，R_{las}、R_{ras} 分别是左、右太阳帆板转动与卫星转动的刚性耦合系数阵，T_{la}、T_{ra} 分别是左、右太阳帆板上的控制力矩列阵。式(2)描述了卫星平动动力学，式(3)描述了卫星转动动力学，式(4)描述了卫星帆板转动动力学，式(5)描述了卫星挠性振动动力学。

卫星姿态控制目标就是使得本体系相对惯性系姿态与目标系相对惯性系姿态一致。姿态控制的目标姿态在不同阶段有不同取法。在巡航姿态，卫星 X_{s} 轴对太阳定向并可以设定偏置俯仰角和角速度绕太阳矢量方向旋转，此时目标坐标系 X_{d} 轴取为指向太阳方向；在轨控阶段，卫星目标坐标系取为轨控点火目标姿态；在环月阶段，目标坐标系取为环月轨道坐标系。姿态控制目标可以简述为，根据卫星运动学和动力学构造控制力矩使得"O_{b}"系跟踪"O_{d}"系并抑制挠性振动。

卫星轨道控制目标描述如下：根据卫星运动学和动力学构造控制力使得卫星在预定时间和预定地点获得预定速度增量，实现轨道控制。

3 系统组成与控制方法

3.1 系统组成

嫦娥一号卫星 GNC 系统的敏感器包括太阳敏感器、星敏感器、紫外月球敏感器、速率积分陀螺和加速度计；执行机构包括飞轮装置、推力器、帆板驱动装置、天线驱动装置和轨控发动机；控制器包括控制计算机、应急计算机、配电器和二次电源。GNC 系统的软件包括控制计算机系统软件、应用软件，应急软件和部件 LTU 软件。LTU 通过内部总线与控制计算机相连，构成计算机控制网络。控制系统的这种分布式体系结构保证 GNC 分系统高效、可靠、实时实现嫦娥一号卫星的控制功能和性能。

3.2 巡航期间的姿态控制

在卫星环月运行之前，除了轨控阶段，卫星运行于巡航姿态。姿态确定是利用太阳敏感器的输出给出太阳矢量方向在卫星本体系的表示，然后根据太阳敏感器的安装矩阵计算卫星偏航角和俯仰角。巡航姿态角速度的确定是利用速率积分陀螺的输出,然后根据陀螺的安装矩阵计算卫星三轴姿态角速度。巡航姿态的控制分为太阳捕获和太阳定向两个阶段:在太阳捕获阶段,根据0-1式太阳敏感器输出,利用相平面控制算法,通过推力器点火驱使卫星旋转使太阳矢量进入数字太阳敏感器视场;在太阳定向阶段,通过数字太阳输出和陀螺输出外推,根据系统动力学,利用相平面控制算法和PID控制算法,通过推力器点火和飞轮转动保证卫星X_s轴指向太阳。

巡航姿态控制的特点是卫星既可以绕X_s轴慢旋,也可以使X_s轴绕俯仰轴偏置并绕太阳矢量慢旋。这种运动状态一方面可满足卫星总体测控需求,另一方面可有效避免推力器喷气对卫星轨道的影响。

巡航姿态控制在轨飞行结果见图1、图2。图中描述了从太阳捕获到太阳定向过程中星体对太阳指向的变化和三轴角速度的变化。由图可见:卫星准确捕获太阳并以高精度和高稳定度跟踪太阳。

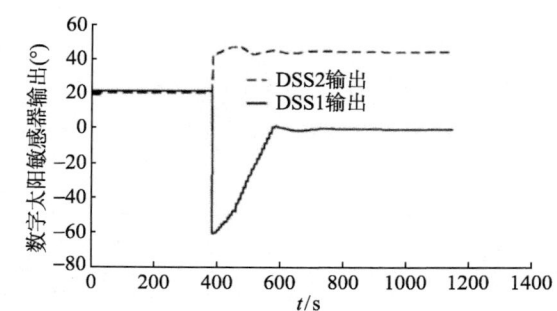

图1 巡航姿态控制期间数字太阳敏感器(DSS)的输出曲线

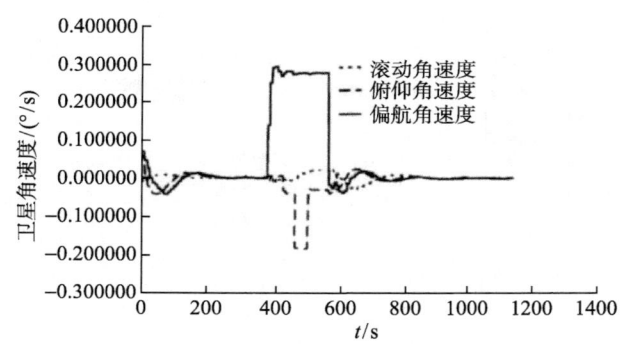

图2 巡航姿态控制期间三轴角速度曲线

3.3 轨道控制

奔月轨道的特点是预先设定地月转移标称轨道。实际轨道与标称轨道的初始微小偏差经5天的飞行放大,可能导致卫星撞月或离月。因此,变轨的高精度控制成为一大技术难点。而要实现变轨的高精度控制就要面对复杂的卫星对象。正如上节动力学描述,严格意义上讲,嫦娥一号卫星刚体平动与转动、挠性振动和液体晃动互相耦合,在快速机动过程又有三轴非线性耦合影响,控制系统稍有疏忽就可能引发多种运动与控制系统相互作用从而导致系统不稳定。

要保证轨控精度,卫星从太阳定向姿态就必须实施姿态快速机动转到轨控定向姿态,其中凸现三轴耦合的非线性问题;在490N发动机点火期间,轨控定向要高精度维持预定惯性指向,其中凸现推力偏斜干扰、挠性振动和液体晃动的抑制问题;490N发动机必须在预定时间点开机并且在预定速度增量点关机,其中凸现点火时间精准问题。这些问题涉及一系列复杂的姿态机动控制、姿态维持控制和变轨制导控制,与此同时强调及时(实时性)、准确(高精度)和可靠(可靠性)。

为此，GNC系统创造性地设计了星上网络控制系统，提出了在线规划调度和新型控制方法，高标准实现了变轨控制过程中的姿态控制和轨道控制。其中：卫星姿态确定利用了星敏感器与陀螺联合定姿算法；卫星姿态机动利用了基于四元数的高品质相平面控制算法；卫星姿态维持利用了基于四元数的"PID+滤波器"算法以及数字化脉宽调制算法；卫星导航利用了高精度加速度计；卫星制导利用了高精度、高可靠关机策略。为保证系统可靠性，还创造性设计了自主故障诊断和系统重构以及自主变轨恢复方案。

变轨控制期间第三次近地点加速的在轨飞行结果见图3、图4。由图表明，卫星在预定时间完成姿态机动和姿态保持，进而在预定时间进入轨控点火阶段并保持轨控定向姿态。在轨数据显示轨控精准，因此，原先拟定的三次中途修正减少到一次，大大节省了宝贵的推进剂，为后续新的任务实施提供了良好条件。

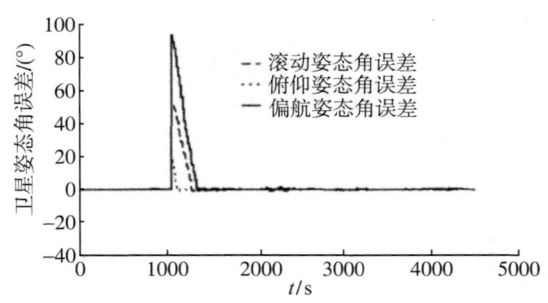

图3 变轨控制期间的三轴姿态角误差

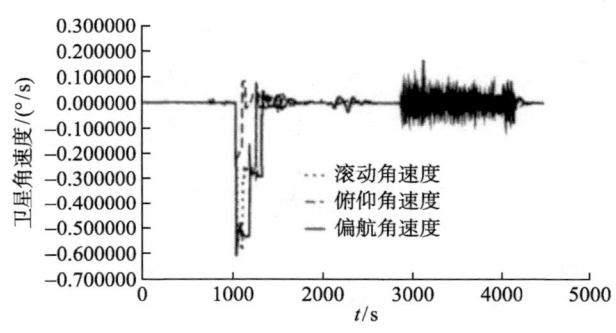

图4 变轨控制期间的三轴姿态角速度

3.4 环月期间的卫星控制

在环月期间，日、地、月相对于卫星的运动关系变化复杂，卫星控制面临诸多技术挑战。卫星本体对月定向，其技术难点是卫星轨道的实时计算和怎样利用对月定姿敏感器；太阳帆板跟踪太阳，其技术难点是太阳相对卫星轨道面以年为周期变化，不能照搬地球卫星太阳同步轨道帆板跟踪太阳方法；定向天线跟踪地球，其难点是地球轨道的实时计算和双轴驱动的控制方法。

为此，在环月期间，姿态确定利用了星敏感器结合星上轨道外推以及紫外月球敏感器结合太阳敏感器综合定姿两套方案；本体对月定向姿态控制利用了基于相平面的喷气控制结合基于PID算法的飞轮控制方法；太阳帆板对日定向和定向天线对地定向，则利用"两次垂直转动可以保证第三轴指向任意方向的基本原理"；在此基础上，姿态控制系统根据创造的实用算法，并基于帆板驱动装置和双轴天线驱动装置，实现了定向控制目的。这些方法使得卫星三体指向同时定向成为可能。

环月期间卫星控制的在轨飞行结果见图5~图8。由图可以看出，卫星本体高精度、高稳定度对月定向，帆板对日定向，定向天线对地定向。

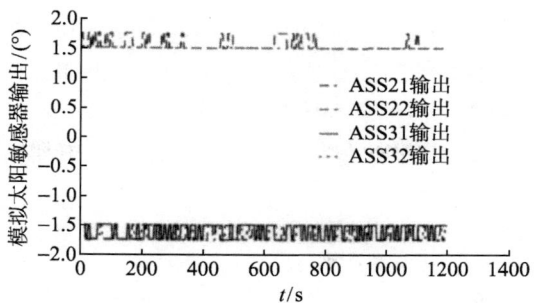

图 5　环月控制期间卫星帆板上的模拟太阳敏感器(ASS)输出曲线

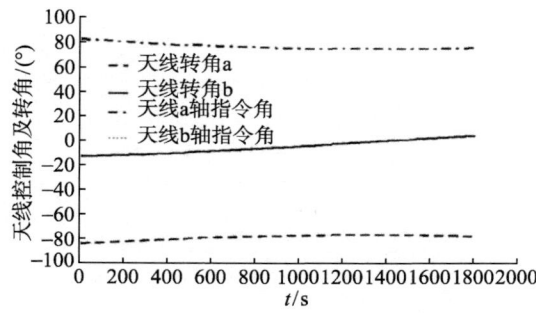

图 6　环月期间卫星控制的天线指令角和实际转角曲线

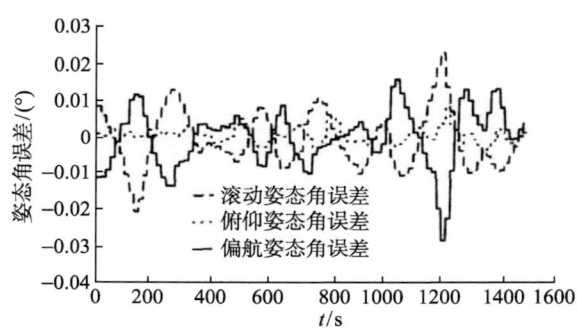

图 7　环月期间卫星控制的三轴姿态角偏差曲线

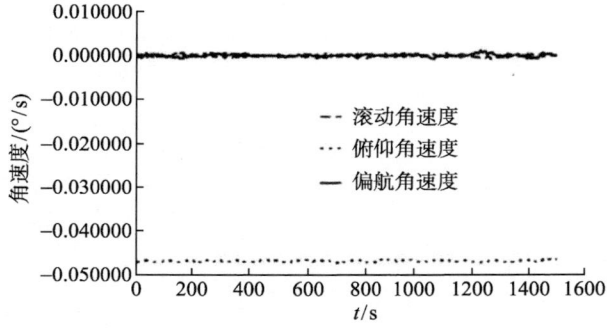

图 8　环月期间卫星控制的三轴姿态角速度曲线

4　结论

嫦娥一号卫星 GNC 系统任务复杂多变，它对系统实时性、可靠性和精度要求较高。本文概要介绍了"嫦娥一号"卫星 GNC 系统组成、控制方法、系统特点和典型飞行结果。"嫦娥一号"卫星 GNC 系统创造了几个第一：第一个高精度变轨控制系统，能够在国内第一次按地面指令时序自主完成复杂的 490N 发动机变轨控制；第一个三体定向控制系统，能够实时、高精度实现帆板跟踪太阳、定向天线跟踪地球、卫星本体有载荷一面跟踪月球；第一个飞控仿真与支持系统；第一次实现奔月轨道及其控制的高精度仿真；第一个月球紫外敏感器；第一个双轴天线驱动装置；第一个成功使用高精度加速度计于变轨控制中。

参 考 文 献

[1] John W, Kenneth K D. The attitude control problem [J]. IEEE Transactions on Automatic Control, 1991,36(10): 1148—1162

[2] 屠善澄. 卫星姿态动力学与控制[M]．北京：宇航出版社, 1998

嫦娥一号卫星热设计及计算分析

侯欣宾　邵兴国　徐　丽　贾　宏

(中国空间技术研究院，北京　100094)

摘　要　嫦娥一号卫星整星主结构以东方红三号卫星平台为基础。由于其复杂的飞行阶段和飞行姿态，以及月球表面特殊的温度分布，卫星表面的外热流非常复杂、变化剧烈，给整星的热控设计带来很大困难，使得整星的热控方案与东方红三号卫星有很大不同。文章着重分析了卫星的特点，并且给出了主要的热控设计方案，最后给出了整星的热分析模型，并对计算结果进行了分析。分析结果表明目前的设计方案满足了卫星在各种工况下的温度指标，实现了总体提出的热控要求。

关键词　嫦娥一号卫星　热控设计　热分析

1　概述

嫦娥一号(CE-1)卫星是我国第一个月球探测器，设计寿命为1年。它将实现我国第一次绕月飞行探测，预计发射时间为2007年。CE-1卫星采用东方红三号(DFH-3)卫星平台(见图1)，外形为2 200mm×1 720mm×2 000mm 的箱形结构，分为上舱和下舱两大部分。其中上舱为在东方红三号平台通信舱的结构基础上，在+Z和-Z方向增加两块剪切板，主要用于对月探测有效载荷设备的安装和部分卫星平台设备的安装。下舱主要安装蓄电池、电源控制器及控制分系统设备。推进系统基本维持东方红三号平台设计方式。

CE-1卫星在热控方面飞行工况复杂、外热流变化剧烈(卫星所有6个面在不同时期都会受到太阳直接照射，日照期和阴影期月球表面红外热流变化非常大)、设备功耗变化大，有效载荷的温度要求苛刻等特点。热控技术被列为关键技术之一。本文通过对CE-1卫星热控特点的分析，综合考虑设备的温度要求和复杂多变的外热流分布，在计算分析的基础上进行了卫星的整体热控设计。通过对计算温度结果进行分析，目前的设计方案可以满足总体提出的温度指标要求。

本文在对CE-1卫星进行概要介绍的基础上，给出了CE-1卫星的整体热控设计方案，并且进行了整星的外热流和传热网络瞬态计算分析，给出了整星温度计算分析结果。

2　CE-1卫星简介

2.1　CE-1卫星坐标系定义

CE-1卫星采用三轴稳定姿态，整星坐标系(见图1)定义为：
坐标原点O：对接锥下法兰框上的三个定位销钉所确定的理论圆心；
X轴：过坐标原点，垂直于星箭分离面，沿卫星的纵轴方向，指向上舱方向为正；
Z轴：过坐标原点，位于星箭分离面内，指向对月方向为正；

本文原载于《航天器工程》，2006, Vol.15, No.4, 21~26。

Y 轴：位于星箭分离面内，与 X 轴、Z 轴构成右手系。

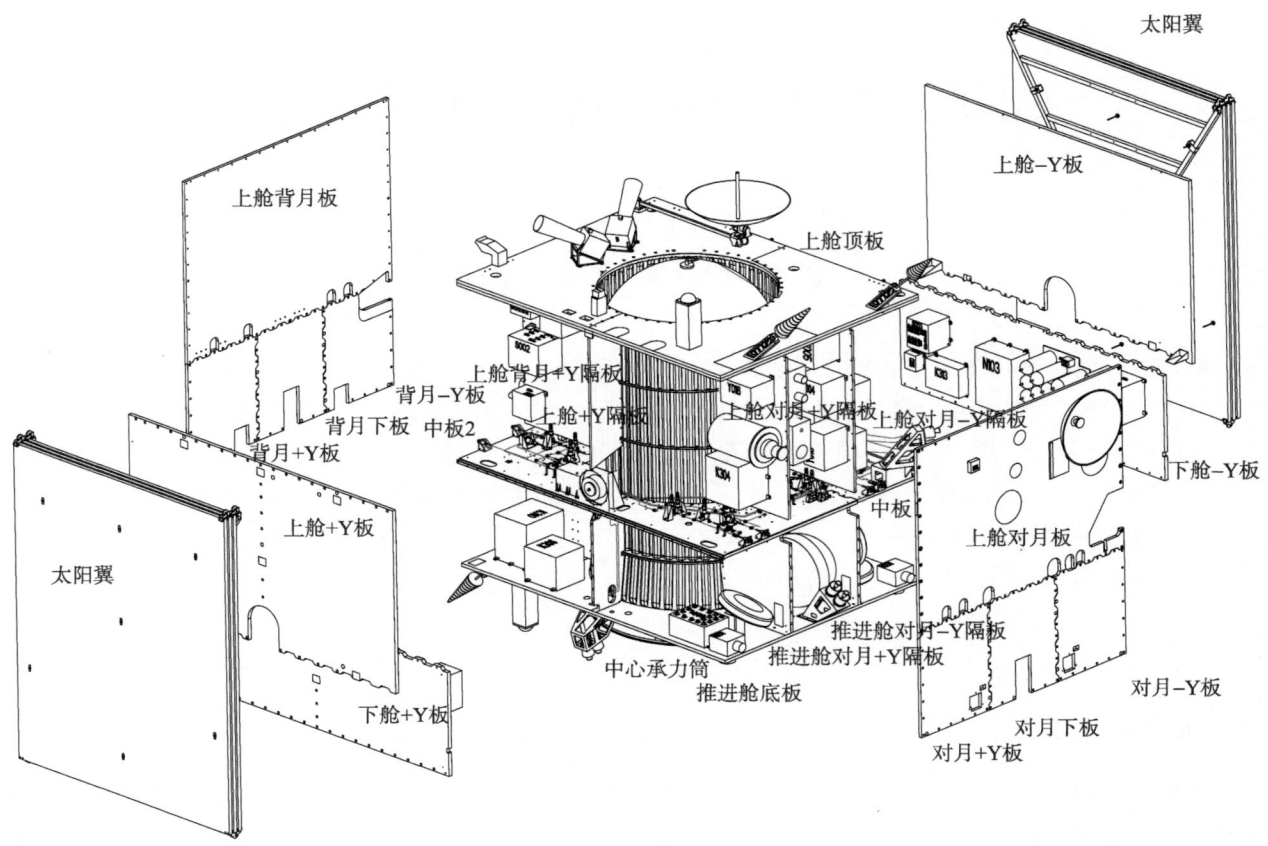

图 1　CE-1 卫星主结构示意图

2.2　CE-1 卫星轨道参数

CE-1 卫星轨道参数为：
月球的赤道与黄道的夹角　1.5°
月球半径　1 730km
轨道高度　200km±25km
轨道倾角　90°
轨道周期　127min

2.3　CE-1 卫星飞行阶段及姿态控制方式

为满足 CE-1 卫星不同阶段任务的要求，卫星的姿态指向主要分为：
(1) 　调相轨道阶段：巡航姿态，基本为 $+X$ 对日定向；
(2) 　地月转移轨道阶段：巡航姿态，$+X$ 向对日定向；
(3) 　环月轨道阶段：
为保证能源供应，根据太阳角的变化采用正飞姿态和侧飞姿态，分别定义为：正飞姿态和侧飞姿态

1) 正飞姿态(见图 2)：

$+Z$ 轴沿卫星与月球中心的连线，指向月球中心；
$+X$ 轴沿当地水平面指向前进方向；

Y轴垂直卫星轨道面，XYZ成右手坐标系。

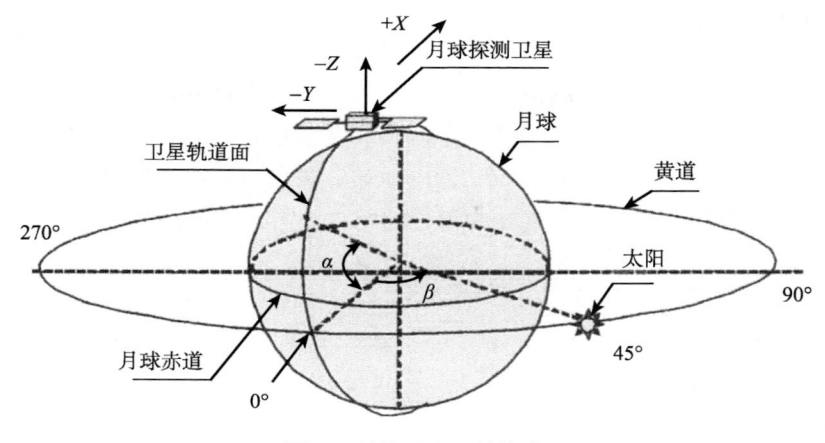

图2　环月正飞卫星姿态

2) 侧飞姿态（当太阳入射光与电池帆板的夹角小于45°时，卫星绕Z轴旋转90°）：

$+Z$轴沿卫星与月球中心的连线，指向月球中心；

$+X$轴沿卫星轨道面法线指向太阳一侧；

XYZ成右手坐标系($+Y$轴沿卫星轨道切线方向，与速度方向相同或相反)

2.4　典型设备的温度要求

卫星内部主要电子设备的温度范围一般为–10℃~+45℃，主要有效载荷设备——激光高度计和光学成像系统(安装于对月$+Y$隔板上)的主要部件温度要求在20℃附近(需要进行特殊热控设计)，蓄电池的正常工作温度范围为–5℃~+20℃。

3　CE-1卫星热控设计方案

3.1　飞行阶段分析

通过对CE-1卫星各个飞行姿态外热流的分析，可对CE-1卫星飞行分为两大阶段：一是奔月阶段，此阶段由于没有月球红外和反照热流，且太阳直接入射卫星的+X面，其他表面都处于直接向空间散热状态，所以对应卫星的极端低温工况；二是环月阶段，卫星在不同的飞行姿态下所接受的太阳入射热流、反照热流、月球红外热流都不相同，具体属于低温或高温工况与飞行姿态和散热面的选择有很大关系。

3.2　热控设计方案

3.2.1　热控散热面的选择

CE-1卫星的飞行工况复杂、外热流变化剧烈，受到重量和能源的限制，热控必须采用合理的被动热控设计维持整星的温度均衡，散热面的选择显得尤为重要。

1) 上舱

通过对卫星的外热流分析可知，在环月阶段，月球表面红外热流的周期性变化十分剧烈(月表最高温度为400K，最低温度为100K左右)，为了减小红外热流变化对卫星温度，特别是有效载荷温度的影响，将受红外热流影响最大的卫星+Z面全部包覆多层。卫星的+X、+Y、–Y表面在不同的时刻都会受到太阳入射热流的影响，为了保证星内设备的散热，特别是有效载荷设备的低温需求，均开设一定面积的光学二次表面镜(OSR)散热面，作为+Z/–Y、+Z/+Y 隔板设备的主要散热面。–Z 表面作为–Z/–Y、–Z/+Y 隔板设备的主要散热面，选用 OSR 作为散热面涂层。

2) 下舱

制导、导航和控制(GNC)设备功耗相对稳定，主要安装在下舱底板上。根据设备功耗分布，开设一定的 OSR 散热面，作为下舱 GNC 设备的主要散热面。蓄电池和电源控制器的功耗随工况变化较大，蓄电池布置于卫星下舱±Y面板上，为保持蓄电池对温度较为苛刻的要求，需在下舱±Y板开设大面积的 OSR 散热面。

3.2.2 其他部分的热控措施

其他部分的热控措施有：

(1) 卫星内部采用喷漆处理，强化辐射换热；
(2) 一般仪器设备均进行高发射率处理；
(3) 对特殊温度要求有效载荷设备进行隔热设计，进行主动控温；
(4) 对于大功率的设备采用安装结构板预埋热管的方式强化换热；
(5) 蓄电池组外表面采用隔热设计，蓄电池组和电源控制器安装底板预埋热管强化换热，低温工况采用电加热器补偿；
(6) 在结构板及部分设备表面粘贴主动加热器，补偿低温工况。

4 CE-1 卫星热分析模型

4.1 计算工况选择

根据 CE-1 卫星的轨道定义，在正常飞行状态下包括 12 个主要工作状态。根据对称性分析，共有 6 个典型工况，即对日定向、环月轨道正飞 0°、环月轨道正飞 45°、环月轨道侧飞 45°，环月轨道侧飞 90°、环月轨道正飞 225°(其中角度代表图 2 中的太阳角 β)。

对于进入环月阶段的 5 个工况，根据外热流的不同(红外热流、太阳常数和热控涂层退化)，分别计算了两个极端工况，即低温工况和高温工况，具体参数选择见 4.2 节。

4.2 参数选择

4.2.1 热计算参数选择

表 1 给出对应每个工况的热流计算参数选择[1]，红外热流参考 4.2.2 节计算的结果。

表 1 各种工况热流计算参数选择

	对日定向	环月正飞 0°		环月正飞 45°		环月侧飞 45°		环月侧飞 90°		环月正飞 225°	
		低温	高温	低温	高温	低温	高温	低温	高温	低温	高温
太阳辐照度/(W/m^2)	1 309	1 309	1 399	1 309	1 399	1 309	1 399	1 309	1 399	1 309	1 399
反照热流/(W/m^2)	0	95	102	95	102	95	102	95	102	95	102
OSR 太阳吸收率	0.135	0.135	0.16	0.135	0.16	0.135	0.16	0.135	0.16	0.135	0.16

4.2.2 月球表面红外热流分析

假设卫星受到的月表红外热流近似为星下点红外热流，不考虑月表导热的因素影响，月球日照面每一微面积的红外热流与吸收的太阳热流能量平衡，则该星下点的月表红外热流为：

$$q = S\cos\beta(1-\rho)\cos\alpha \tag{1}$$

式中，S 为太阳辐照度，α 为卫星星下点的纬度，β 为太阳角(参见图2)，ρ 为月球反照率取 0.073[3]。根据该假设，日照区月表红外热流分布呈纬度角 α 的余弦分布，阴影区则为0。

图3为文献[3]给出的根据 Apollo 11 的探测数据得到的月球表面温度曲线。可以看出月表日照最高温度为400K左右，最低为100K左右。考虑到红外热流密度与表面温度的四次方成正比，则日照期红外热流密度与表面温度的四次方成正比，则日照期红外热流近似呈余弦分布，阴影期可以近似简化为定温直线。这与公式(1)反映的理想月表红外热流基本相符。

本文计算采用的月表红外热流，见图4。其中实线为根据公式(1)得到的理论值，虚线为计算分析采用的近似阶梯热流曲线。根据周期的能量平衡计算得到各工况对应 q_1 和 q_2，分别见表2，其中阴影热流根据阴影区温度取 $q_3=50W/m^2$，日照面 $q_2=2q_1$。考虑设计余量，计算中各高温工况红外热流分别增大10%、低温工况分别减小10%，环月轨道侧飞90°由于处于阴阳交界面处，红外热流误差较大，分别取高温200 W/m² 和低温50 W/m²。

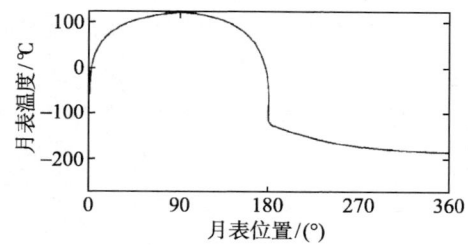

图3 月球表面的温度分布[3]

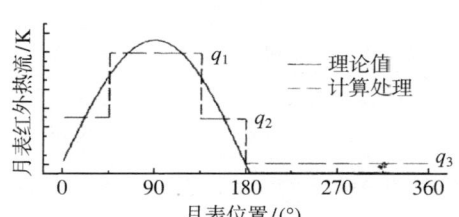

图4 月球表面瞬态红外辐射热流分布示意图

表2 红外热流瞬态计算参数选择

	0°轨道		45°轨道		90°轨道	
	高温	低温	高温	低温	高温	低温
q_1/ (W/m²)	572	429	302.5	394.5	200	50
q_2/ (W/m²)	1 144	858	605	789	200	50
q_3/ (W/m²)	50	50	50	50	200	50

5 CE-1 卫星热分析结果

5.1 各工况部分设备计算结果分析

典型工况温度计算结果见表3。表中选取了主要安装板的典型设备（非瞬态热功耗设备）在各工况下的低温值（低温小工况的低温值）和高温值（高温小工况的高温值）。通过计算分析得出，对日定向、侧飞90°（低温）工况对应卫星的低温工况。正飞45°（高温）、侧飞45°（高温）和正飞225°（高温）对应卫星的高温工况。在目前的计算中，环月轨道阶段大部分设备温度都维持在 0~30℃。蓄电池和数传设备等瞬态功耗设备温度变化见5.2和5.3小节。本文对有效载荷不做详细分析，主要保证周围的环境温度。

表3 典型工况典型设备温度/℃

设备名称	安装板	奔月轨道	环月正飞0°		环月正飞45°		环月侧飞45°		环月侧飞90°	
			低温	高温	低温	高温	低温	高温	低温	高温
天线驱动线路	上舱顶板	−3.6	13.9	19.8	15.9	20.8	15.3	21.3	11.5	18.9
帆板驱动机构	上舱中板	2.8	11.3	17.1	14.6	20.1	12.2	17.4	8.6	14.0

续表

设备名称	安装板	奔月轨道	环月正飞0°		环月正飞45°		环月侧飞45°		环月侧飞90°	
			低温	高温	低温	高温	低温	高温	低温	高温
遥控单元	上舱−Z/−Y隔板	−0.7	11.1	18.0	13.1	18.9	10.6	17.2	4.4	11.2
加热控制器B	上舱−Z/+Y隔板	5.0	16.1	21.4	19.5	22.1	16.0	20.9	9.7	15.3
载荷配电器	上舱+Z/−Y隔板	−2.4	9.0	15.7	12.1	16.8	11.2	17.0	10.3	18.8
γ/χ电控箱	上舱+Z/+Y隔板	−0.5	14.4	21.0	20.7	24.9	13.5	19.1	11.9	20.7
陀螺线路	下舱+Y板	2.0	5.9	11.8	14.4	22.8	6.8	12.0	2.7	5.4
动量轮线路	下舱−Y板	−0.9	8.5	15.5	10.6	17.7	9.0	14.8	3.9	9.6
陀螺组合件	下舱底板	17.5	23.6	30.4	25.7	33.1	19.7	25.3	17.9	21.4

5.2 蓄电池及电源控制器温度结果分析

蓄电池和电源控制器均为电源系统重要设备。蓄电池为氢镍蓄电池，充放电热耗变化大，其正常工作温度范围为−5℃~+20℃。通过计算分析，正飞45°轨道和侧飞90°轨道对应环月阶段蓄电池的极端高温和极端低温工况，温度曲线见图5、图6，其中N102为蓄电池，N103为电源控制器。正飞45°工况由于太阳入射+Y面夹角达到45°，使得+Y侧蓄电池温度很到，接近18℃，而−Y侧蓄电池需要补偿100W，以维持两组蓄电池5℃的温差要求。在此工况下，电源控制器最高温度为30℃。侧飞90℃工况两侧蓄电池分别需要补偿100W和150W以控温到0℃以上。

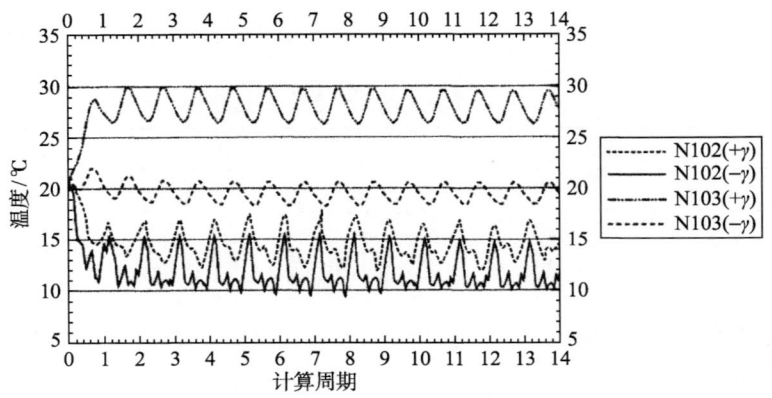

图5 蓄电池及电源控制器温度变化(正飞45°高温工况)

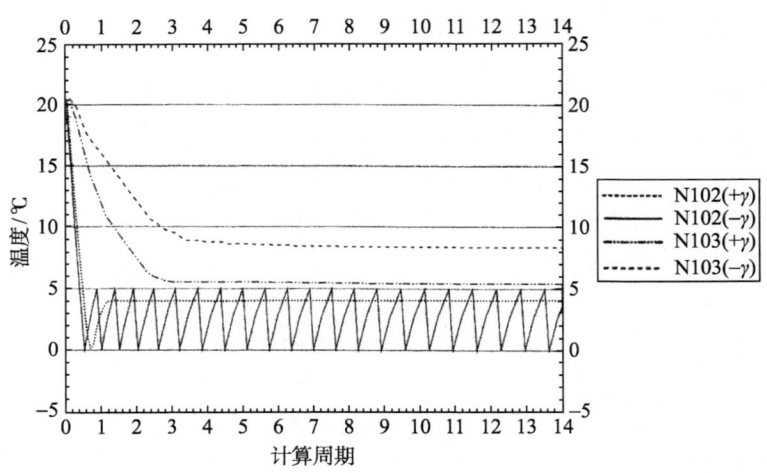

图6 蓄电池及电源控制器温度变化(侧飞90°低温工况)

5.3 瞬态大功率设备温度结果分析

CE-1 卫星的数传设备（如 X 频段固放和 S 频段固态功率放大器等）具有大功率瞬态工作的特点，热控采取了安装板预埋热管和增加补偿加热器的措施。计算采用的工作时间为每圈连续工作 36min。从图 7 中可以看出由于设备的大功率瞬态工作特点，36min 工作期引起的设备温度波动值达到 18℃。所以在确定设备的工作时间时，应考虑设备的初始温度和可能产生的温度波动值，防止设备过热。

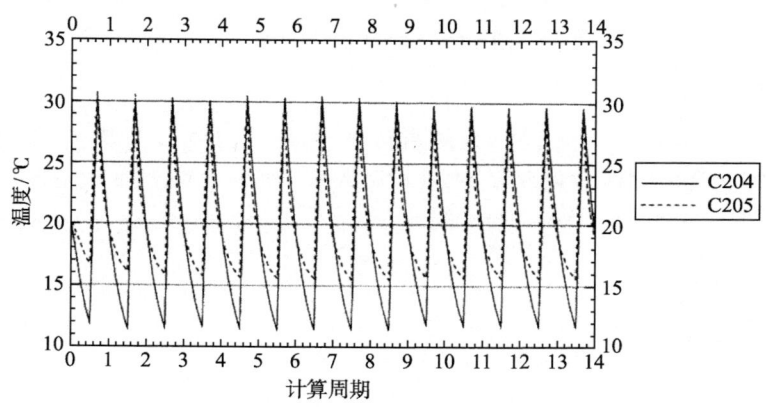

图 7　X 频段及 S 频段固态功率放大器温度变化(正飞 45°高温工况)

6　结论

本文通过介绍 CE-1 卫星复杂的飞行阶段、飞行姿态以及月球表面特殊的温度分布，给出了主要的热控设计方案。重点介绍了月表红外热流的处理方法和卫星的散热面的设计思路，并对整星的热分析模型计算结果进行了分析。设计基本满足了卫星在各种工况下的温度要求。但对于我国第一个月球探测器，在设计经验方面尚有很多不足，为了保证设计的准确性，需要进一步完善一下工作：

(1) 月表红外热流的合理性分析；
(2) 蓄电池热设计详细分析；
(3) 加强有效载荷设备的热设计；
(4) 通过合理的整星试验检验热控设计的合理性和准确性。

参 考 文 献

[1] 闵桂荣，郭舜. 航天器热控制[M]. 北京：科学出版社, 1998 [Min Guirong, Guo Shun. Spacecraft thermal control [M]. Science Press, 1998
[2] NASA Space vehicle design criteria-Lunar surface models.[R]. NASA SP—8023, 1969, 5
[3] Gilmore D G. Spacecraft thermal control handbooks [M]. The Aerospace Press, 2002
[4] Adorjan A S. Directional behavior of thermal emission from a rough lunar surface[R]. AIAA-71—0480

"嫦娥一号"月球探测卫星真空热试验的初步思路

马有礼　景甫林

(北京卫星环境工程研究所，北京　100029)

摘　要　文章介绍了月球环境条件、设计月球探测器需要考虑的环境因素，以及绕月飞行的"阿波罗"服务舱(service module)真空热试验的外热流模拟情况，对中国"嫦娥一号"月球探测卫星的真空热试验提出了初步设想。

关键词　月球环境　月球探测　卫星　外热流模拟

1　引言

月球距地球约 3.8×10^5km，是距离地球最近的自然卫星。由于人类渴望深入了解月球，并利用月球为人类服务，因而已经开展了一系列的月球探测活动。其中最典型的是在 1969 年至 1972 年间，美国宇航局先后 6 次成功地完成了"阿波罗(Apollo)"飞船登月探测任务，并取得了一定的收获。但要实现利用月球资源(如可作为人类的代替能源的储量丰富的 ^{3}He)和从月球上向深空其他星球发射航天器，还需要做大量的工作。因此，在积极实施火星探测计划的同时，美国提出了重返月球的计划。据此计划，定于 2015~2020 年之间，美国航天员重返月球，除了进行各种探测研究外，还要在月球上建立太空基地，以便在月球上组装航天器，从月球向深空其他星球发射航天器(比如向火星发射载人飞船)，达到节约能源和资金的目的。欧空局、日本和印度在此之前也相继制定了月球探测计划。

中国从 2004 年起，也正式启动不载人的"探月工程"。工程分为"绕、落、回"三个实施阶段：第一阶段是力争在 2006 年 12 月发射第一颗绕月飞行的"嫦娥一号"月球探测卫星，主要任务是获取月球表面三维立体影像，分析月球表面有用元素的含量和物质类型的分布特点，探测月球的土壤特性和地球至月球的空间环境等；第二阶段是计划在 2010 年前向月球发射无人探测装置，进一步开展月球探测工作；第三阶段，在 2020 年前完成月球土壤样品的采集工作。

要完成"探月工程"计划，对这些航天器进行真空热试验是必不可少的。本文先就绕月飞行探测卫星真空热试验的相关问题，特别是外热流模拟问题谈谈设想。由于迄今为止，对绕月飞行的"嫦娥一号"月球探测卫星情况知之甚少，因此还不能做到有针对性。随着卫星研制工作的进展，对真空热试验的要求和实现方法才能逐步细化。正因为如此，本文若有不妥之处，欢迎给予指正。

2　月球的环境条件

有关月球的环境条件已在文献[1]中做了详细叙述，这里就其中与完成绕月探测卫星真空热试验关系密切的两种环境条件略加说明。

本文原载于《航天器环境工程》，2004, Vol. 21, No.2, 1~7。

2.1 月球表面高、低温及其剧烈变化

由于月球上空没有大气层对太阳辐射的吸收与屏蔽,所以月球表面直接暴露于太阳的辐射之下,致使月球表面在白天的最高温度达到 390K,而月球表面的热量在黑夜又毫无阻挡地迅速向深空间辐射出去,使月球表面温度快速下降,最低温度可达到 105K,白天和黑夜温差悬殊。不仅如此,从白天向黑夜过渡的过程中,依据月球纬度的不同,温度下降的变化率可达到 190~240K/h。当然,在月球从黑夜转为白天时,月球表面温度也有类似的反变化。

正因为如此,不仅在设计月球登陆器(含月球车)时必须考虑采取适当的热控措施使各种仪器设备正常工作,而且在设计绕月飞行的月球探测器时同样要考虑这些问题。因为月球表面的高温、低温及其迅速变化同样要影响到绕月飞行探测器的温度,特别是探测器朝向月球的一面受到的影响会更大。

除此以外,另一个热源就是太阳直接照射月球探测器和月球表面的反照,太阳对月球探测器的辐照度与太阳对地球航天器的辐照度差不多,最大也是 1 个太阳常数。其值随月球探测器运行的轨道位置不同而变化。当然,当绕月的月球探测器进入月球阴影或背向太阳的一面,太阳辐照度为零。所以,太阳辐射与月球表面温度对月球探测器的影响必须同时考虑。

2.2 真空

月球表面及其上空几乎没有大气存在,其压力低到 10^{-11}Pa。在这么高的真空环境中,如果飞船载人登上月球,飞船内部必须充有 $8×10^4$Pa 能维持生命的大气,还需配备对突然失压紧急处理的设施。

月球探测器长期工作于极高真空,其材料和元器件的电、热、机械性能会受到明显的影响。这些影响主要有:

a. 某些表面材料的加速蒸发能导致表面结构、表面导电率、高压击穿、表面吸收比和发射率等特性的改变;

b. 材料在高真空下放出的气体中,可能有一部分再凝结在探测器的低温光学表面上产生污染,影响光学系统正常工作;

c. 超高真空可能使相对运动部件的接触部位形成干摩擦和冷焊现象;

d. 高真空不利于探测器的散热。

虽然真空带来以上不利影响,但也存在一些有利因素:

a. 真空虽然不利于月球探测器的散热,但却有利于探测器的保温,节约能源这个优点在探测器的某些部位是有用的;

b. 在高真空环境下由于没有氧化作用,因此能增加金属材料的疲劳寿命。有研究表明:在 $4×10^{-3}$Pa 的真空环境中,316 个型号不锈钢的疲劳寿命是空气中的 25 倍到 77 倍,镍的疲劳寿命是空气中的 9 倍;

c. 月球表面没有风、雨、雪的影响,对设计也是有益的。

3 设计月球探测器需要考虑的环境因素

早在 20 世纪 60 年代初,美国宇航局在一篇工程报告中列举了设计月球探测器结构需要考虑的环境因素。现在看来,仍有很好的参考价值。表1列出了月球探测器发射前、发射、飞行、在月球上着陆和在月球上空运行全过程中需要考虑的各种环境因素。

表1 月球探测器的环境因素

序号	环境	发射前	发射	飞行	月球上着陆	月球上空运行(夜间)	月球上空运行(白天)
1	加速度		√		√		
2	噪声		√		√		
3	大气	√					
4	霉菌	√					
5	湿度	√					
6	真空			√	√	√	√
7	雨淋	√					
8	盐雾	√					
9	飞沙、灰尘	√			√	√	√
10	冲击、随机		√		√		
11	冲击、地震					√	√
12	空间电磁波辐射			√	√	√	√
13	空间流星			√	√	√	√
14	空间粒子辐射			√	√	√	√
15	空间辐射			√	√	√	√
16	温度(高)	√	√	√	√	√	√
17	温度(低)			√	√	√	√
18	振动	√	√	√	√	√	√

4 Apollo服务舱真空热试验的外热流模拟

我们介绍以下内容，主要是为了有所借鉴。

Apollo服务舱运行在距月球96.54km高的轨道上，它的直径为3.98m，高为5.49m。它的真空热试验在美国宇航局载人航天器中心的空间环境模拟室A中完成。该模拟室的直径为19.8m，高为36.6m，热沉温度约为90K，模拟室压力1.33×10^{-4}Pa，在顶部外边装有太阳模拟器，在直径为4m的辐照面范围内，辐照度可在646~1507W/m²范围内调节，光谱波长范围为0.25~3.0μm。为了完成Apollo服务舱的真空热试验，还专门研制了一个大型红外模拟器。

4.1 外热流模拟装置

Apollo服务舱的真空热试验中，同时使用了两种外热流模拟装置，即：红外模拟器和太阳模拟器。红外模拟器安装在围绕试件的180°弧形面上，而太阳模拟器光束则照射在另一个180°的弧形面上。由于太阳模拟器的使用比较成熟，而红外模拟器是为完成试验专门研制的，所以仅对后者加以介绍。

4.2 红外模拟器的设计要求

a. 红外辐射热流必须能覆盖围绕服务舱180°的弧形面，且各向同性，以模拟月球热辐射的方向性；

b. 各个加热区的热流应能独立控制，以便模拟各种月球轨道(赤道、近地极和45°月球轨道)下的热流剖面；

c. 色温必须不超过315.6℃，并且要使月球表面的位置与每一个加热区对应一致；

d. 每个加热区的热响应时间应尽量短，以模拟瞬态热流；

e. 每个加热区必须保持服务舱表面的热流控制在441.7~1388.1W/m²范围内(作者认为，此处低热流的要求是不合适的)；

f. 红外模拟器应有提升机构,当模拟月球黑夜面的热流时,将红外模拟器提升到试件的斜上方,以使试件能看到用液氮冷却的热沉,当需要加热时再将红外模拟器降到原位。

4.3 红外模拟器加热源的选择

为了试验,曾研究了3种辐射加热源,即:带有反射器的红外灯(钨丝石英灯);粘贴在背面涂有黑漆的1.6mm厚铝板上的薄膜型电加热器;很薄的镍铬合金片。

经过分析比较和必要的试验验证后认为:

a. 红外灯阵不能满足辐射热流各向同性要求,色温538℃太高,并随输入功率的变化而变化,所以放弃了选用红外灯作为加热源;

b. 对于粘贴在1.6mm厚铝板上的薄膜型加热器,经试验后发现它的热响应太慢,由-17.8℃升到148.9℃需要10min,由148.9℃降到-17.8℃需要15min。这种热响应速度不能满足模拟瞬态热流要求,还担心胶黏剂放气形成污染,再加上成本高,所以也放弃了这种加热源;

c. 镍铬合金片的色温能满足要求,便于加工,成本低,采取措施后有快速热响应,所以最后决定选用含80%镍、20%铬的镍铬合金片作为加热源。

4.4 红外加热笼的设计

按照要求,红外加热笼设计成180°半圆形的弧面,直径为4.88m,高为5.49m,距离试件表面为0.457m。分成18个加热区,每个区由4个长5280mm、宽95mm、厚0.15mm的镍铬合金片串联而成,它们之间的距离为6.4mm。在每个加热区的加热片上布有6~12对热电偶,且与加热片之间有良好的电绝缘,加热片朝向试件的一面喷涂黑漆。这种设计思路与国内常用的加热笼设计思路有着较大的差距。这种设计方法更接近加热板,所以必然导致对热沉遮挡太大,从而不能实现低温热流的模拟。为了能模拟相应于月球表面黑夜期间接近于零的外热流,又专门设计了加热笼提升机构。该机构由电动机、一个变速比为80:1的密封齿轮箱(内充大气并进行温控)、绞盘、缆绳和必要的支撑杆等组成。当需要模拟低热流时用提升机构在1min内能将红外笼提升到试件的斜上方,以便试件看到热沉,热流迅速减小。当需要加热时,再用1min的时间,将红外笼降到原位置。无论是提升还是下降到规定的位置,其重复性都要很好。试件、红外加热笼和提升机构的具体位置见图1和图2所示。

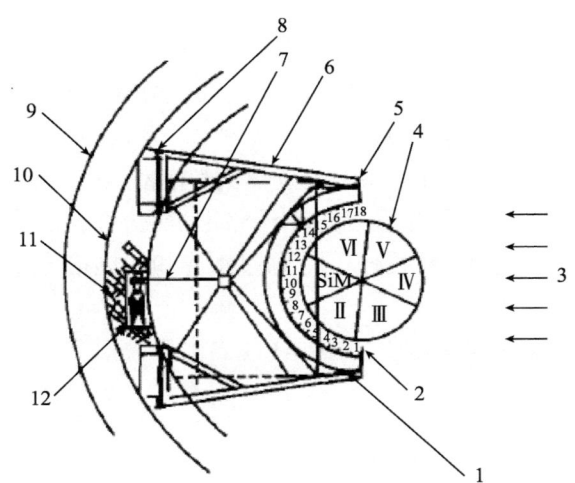

图1 红外加热笼提升机构

1. 加热笼结构(外径为51mm的管子); 2. 红外模拟器加热区; 3. 太阳光束; 4. Apollo服务舱(直径为3960mm); 5. 铰链点; 6. 提升杆; 7. 提升缆绳; 8. 铰链点; 9. 空间模拟室壁; 10. 热沉; 11. 梯子; 12. 提升电机

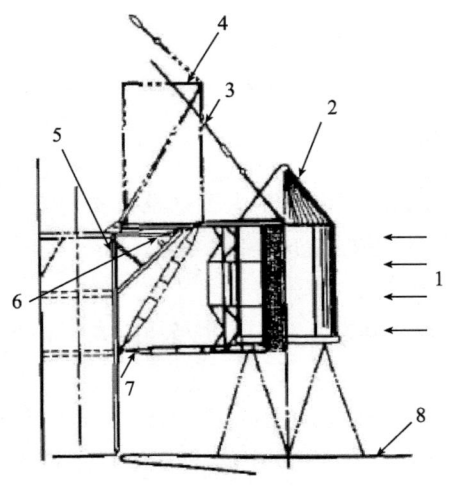

图 2 红外加热笼在空间模拟室内的安装情况
1. 太阳光束；2. Apollo 服务舱；3. 提升缆绳；4. 红外模拟器提升后的位置；
5. 支撑结构；6. 支撑臂；7. 稳定杆；8. 月球平台

4.5 红外加热笼的性能

4.5.1 控制系统

红外加热笼的温度及其提升机构都采用计算机控制，并研制了专用的软件包，能进行开环和闭环控制，软件包具有非常灵活的输入选择，这些选择主要有：

a. 选择 1~99 个不同要求的温度剖面；
b. 选择功率控制信号修改的时间间隔；
c. 选择超前准备的调节参数，以保证红外加热笼的热响应满足要求；
d. 使用多个系数来修改每一个控制区要求的温度剖面；
e. 定义计算热电偶平均温度的热电偶数量，用于闭环控制；
f. 确定提升或下降机构的控制参数；
g. 选择控制、记时和显示方式。

4.5.2 红外加热笼的标定

由于加热片对热沉的遮挡严重，再加上热流计的热响应又比较慢，所以试验中不采用热流计的测量结果作为闭环控制的反馈量，而采用加热笼的加热片温度为控制量。因此，必须在试验前进行加热笼的标定，找出试件表面吸收的热流与加热片温度之间的关系，建立热数学模型。在标定中还要测出模拟室的背景热流。

在标定中使用了两种热流计，一种是涂覆高发射率黑漆的热流计，直接粘贴在试件表面上。另一种热流计也涂覆高发射率黑漆，并安装在水冷的铝板上，通过水冷和加热相结合，使热流计粘贴在铝板部位的温度维持在常温。

通过标定得出了用于控制的热数学模型，还测出了来自太阳模拟器、不冷却的壁面以及其他辐射源的背景热流大约为 47W/m^2。

4.5.3 红外加热笼的性能

红外加热笼对加热和冷却的响应表示在图 3 上。由图可见，加热区的平均温度可达到 150℃，但降温只能降到 -100℃，这不能满足相应于月球黑夜期间所要求的热流水平。当将红外加热笼提升到试件斜上方后，大约在 90s 内，热流从最大的 1387W/m^2 降到 15.8W/m^2，基本上满足了模拟低热流的要求。

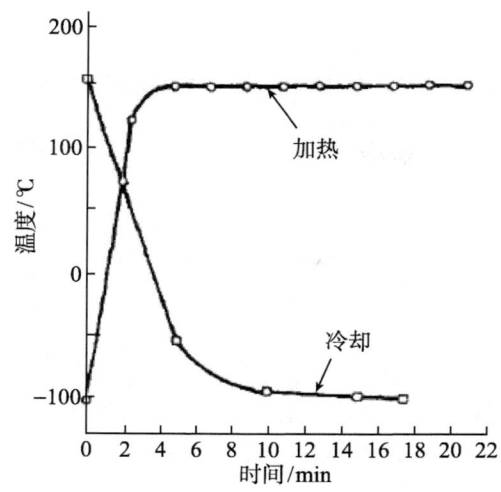

图3 全尺寸红外模拟器加热区的热响应

5 "嫦娥一号"月球探测卫星真空热试验的初步思路

5.1 "嫦娥一号"卫星的整体概况

"嫦娥一号"月球探测卫星预计发射后先围绕地球以大椭圆轨道运行几圈,再从远地点转移出地球轨道进入向月球飞行的巡航轨道,经过9天左右的飞行后,进入绕月球两极飞行的椭圆形轨道,最后变成距月球约200km高的圆形月球轨道。在此轨道上完成各种预定的探测任务。

5.2 对"嫦娥一号"卫星真空热试验预计的技术要求

a. 试验在载人航天器空间环境试验设备中进行,模拟室压力和热沉温度与以往卫星要求相同;

b. 按照"东方红三号"平台卫星真空热试验的思路,仍然使用L型支架作为"嫦娥一号"月球探测卫星的试验支架(使用太阳模拟器除外),支架与卫星的对接环处要有热控措施,支架应能调节水平,可能还要在支架水平部位的上方增设液氮冷却的冷板;

c. 温度测量通道和电源数量与使用"东方红三号"平台卫星的真空热试验相比,可能有所增加;

d. 外热流模拟装置可能是红外加热笼或红外灯阵与薄膜型电加热器的组合使用,设计时要考虑满足卫星经过地球轨道、巡航轨道、绕月运行轨道及其相关过渡轨道所遇到的太阳辐射、月球反照和红外辐射的热流模拟要求,其中要特别考虑低热流和瞬态热流的模拟要求;

e. 试验过程中要注意防止有机物对卫星敏感部位(如光学系统)的污染。

5.3 外热流模拟

要完成"嫦娥一号"月球探测卫星真空热试验,真空度和热沉温度已不是问题。最难的还是外热流模拟问题。

5.3.1 外热流模拟的特点

a. 试验工况多,要模拟各个飞行阶段的外热流;

b. 由于月球表面高温、低温及其剧烈变化对绕月探测卫星的影响,所以不仅要模拟比较大的热流(包括太阳辐射和反照热流),而且还要模拟很低的热流,以及模拟变化大的瞬态热流;

c. 对热流控制系统的要求高。

5.3.2 外热流模拟装置的选择

国内常用的外热流模拟装置有薄膜型电加热器、红外加热笼和红外灯阵，预计在进行绕月探测卫星真空热试验时，新建的太阳模拟器也将投入使用。

1) 薄膜型电加热器

对于探测卫星外部敷设多层隔热组件的部位，可能选用薄膜型电加热器来模拟这些表面吸收的外热流，它具有热流模拟精度高和易于模拟瞬态热流的优点。

2) 红外加热笼

经过精心设计的红外加热笼可以用来模拟"嫦娥一号"月球探测卫星的外热流，但要注意以下几个问题：

a. 红外加热笼加热带的工作温度不宜太高，经过热设计和必要的试验验证，如果加热带的温度比较高(允许的温度待定)，在热平衡试验中使用敏感面涂黑漆的绝热型热流计来测量热流可能带来较大的测量误差。为此，有可能改用吸收式热流计来测量吸收热流，但最好的方法还是尽量避免加热带温度过高；

b. 红外加热笼工作系统(包括：加热笼、热流计、电源和控制)的热响应比较慢，不能满足瞬态热流的模拟要求，试验前可能要进行标定，从而确定合理的热流控制方法。红外加热笼的标定不一定像 Apollo 服务舱那样标定出加热带温度与热流之间的关系，而可以标定出每个加热区的电流与热流之间的关系，试验过程中可以分台阶进行开环控制；

c. 为了模拟很小的热流，需要将加热笼移出被加热表面。因此，可能要设计一套提升(下降)加热笼的机构或移动加热笼的机构。其过去的经验和技术可以用于本移动机构的设计。当然，如果像百叶窗那样，能将每个加热区转动成与被加热面成垂直状态，也可以达到迅速降温的目的，但转动机构相对复杂一些，国内在红外加热笼系统中还尚未使用过。

3) 红外灯阵

在采用"东方红三号"平台卫星的真空热试验中使用了红外灯阵模拟卫星南面板和北面板的吸收热流。在"嫦娥一号"月球探测卫星真空热试验中有可能继续采用红外灯阵模拟某些面的吸收热流。为了模拟很小的热流，可能还需要提升(下降)或移动红外灯阵的机构。采用热屏等温型热流计测量吸收热流。

4) 太阳模拟器

能否将红外模拟器与太阳模拟器配合起来使用，要取决于月球探测卫星使用热管的情况。即使允许月球探测卫星在运动模拟器上作必要的运动，也会给红外模拟器模拟外热流带来麻烦。如果单独使用太阳模拟器而不配合使用红外模拟器，看来要完成外热流模拟任务比较困难。因此，现在虽不能排除配合使用太阳模拟器的可能，但这种可能性很小。

6 结束语

在使用以上谈及的几种外热流模拟装置时会各有优缺点，由于种种条件限制，本文目前还不能给出倾向性很明确的结论性意见，文中的内容只是进行"嫦娥一号"月球探测卫星真空热试验的初步思路，仅供参考。

参 考 文 献

[1] 肖福根，庞贺伟. 月球地质形貌及其环境概述[J]. 航天器环境工程, 2003—2
[2] Skinner J, Wallin P, Wolff M. The design and application of an infrared simulator for thermal vacuum testing [R]. NASA SP—298
[3] Alvin Smith. Early concepts for lunar structures [R]. AIAA 92—1030

嫦娥一号卫星定向天线动力学仿真分析

荣吉利[1]　李　健[1,2]　徐天富[1]

(1.北京理工大学宇航学院，北京　100081；2.广西工学院汽车工程系，柳州　545006)

摘　要　采用大型有限元软件 MSC.NASTRAN，对嫦娥一号卫星定向天线压紧与展开两种工作状态下的模态与压紧状态下的加速度动力学响应，进行数值计算分析。针对双轴定向天线结构特点，结合双轴电机试验参数，引入扭簧单元对双轴电机结构进行模拟，并将天线压紧状态下的固有频率和关键点的加速度响应计算结果与试验结果进行了对比分析，两者具有较好的一致性，证明了有限元模型建立及扭簧处理方法的正确性。研究了天线展开状态下双轴刚度与系统一阶模态之间的关系，提出了符合天线技术要求的轴向刚度的临界值，即应大于 0.035N/m。计算模型、方法对相关的工程研究和计算有一定的参考价值。

关键词　嫦娥一号卫星　定向天线　动力学分析　有限元　数值仿真　扭转弹簧

1 引言

嫦娥一号(CE-1)月球探测卫星采用了"三体定向"技术，即卫星在环月飞行时，要保持卫星对月球定向，太阳电池翼对太阳定向，以及天线对地球定向。卫星相对于地球没有固定的对地面，环月时每一时刻天线对地球面的指向在卫星坐标系中均不相同，为了实现对地通信，需要采用高增益可转动定向天线进行对地跟踪，即天线的反射面要始终对准地球，以确保卫星与地球时刻保持通信联系[1-2]。随着通信容量的提升，以及传输速率要求的提高，国内外在星载天线设计方面已经有了长足的发展[3-7]。CE-1 卫星机械转动定向天线是国内卫星上首次采用的双轴大角度机械扫描天线，是保证卫星能成功进入绕月轨道的关键部件之一。卫星在发射阶段、飞行阶段、到达月球轨道等任务过程中都需要天线的支持。更重要的是，CE-1 卫星绕月工作时间较长，卫星飞行轨道的变化、飞行姿态的调整，感知与地面的距离，这些来自地面指挥系统的信号都必须通过天线接收，计算机才能做出相应的指令；同时，卫星搜集到月球的所有信息都将通过天线发送回地球。可以说，定向天线在 CE-1 任务中扮演了极其重要的角色，其工作状态的正常与否，将直接影响到此次探月任务的成败。因此，定向天线也是专家极为关心的问题之一[8-10]。

2 系统组成

定向天线(见图 1)由以下几部分组成：射频部分，包括反射面、馈源、旋转关节、高频电缆组件；机构部分，包括双轴驱动机构、压紧释放机构、展开机构；结构部分，包括反射面支架、展开臂、系统安装板；此外，还包括展开机构的控制器和天线的低频控制线束。压紧释放机构用来将天线系统可靠地压紧在卫星侧板上，使系统在卫星的发射段可以承受较大的载荷而不被损坏，并且满足天线系统压紧状态的模态要求；在卫星入轨后，顺利地解除对天线系统的约束，使天线展开机构可以正常运动。展开机构采用步进电机加谐波减速器作为动力源，由电机驱动铰链运动，当锁定机构完成锁定时触发微动开关,电机停止转动。天线展开后,可以通过控制双轴驱动机构的转动角度实现反射面的定向,从而完成天线对地球的定向。

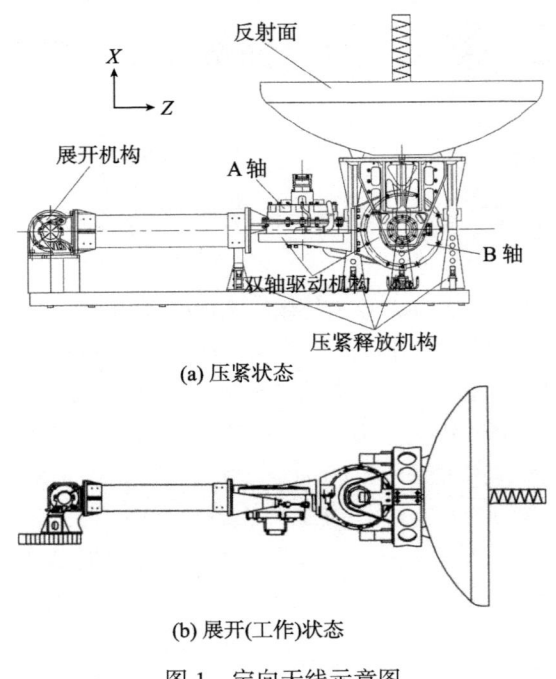

图 1 定向天线示意图

3 有限元模型

3.1 双轴驱动机构

双轴驱动机构是天线的关键部分。在卫星工作状态下,双轴驱动部件的扭转刚度相对较小,因此,在进行卫星天线动力学仿真分析时,双轴驱动机构力学参数(如约束方式、刚度等)的设置,对数值计算的准确性具有至关重要的作用,需要在仿真分析前先行测量转轴的刚度。

由于各转轴之间只有相对转动的运动趋势,并且两轴的扭转刚度已经通过试验获得,为真实描述电机及其转轴的质量特性,将电机等效为一梁单元,并赋予梁单元与电机及其转轴相同的质量特性;同时,也将两轴的固定部分与转盘用此梁单元来连接,并仅释放沿转轴方向的转动自由度。根据试验测量结果,采用 MSC.NASTRAN 软件中的弹簧单元来模拟转轴的扭转特性,即将转轴的扭转刚度直接在弹簧刚度系数中进行定义。双轴的有限元模型如图2所示。从图中可以看到,锁定组件与 A 轴固定部分的连接处有一些共用单元,而 A 轴的中部是一个梁单元,用于模拟电机及相对运动的约束。B 轴的建模方式及参数定义与 A 轴完全一样。

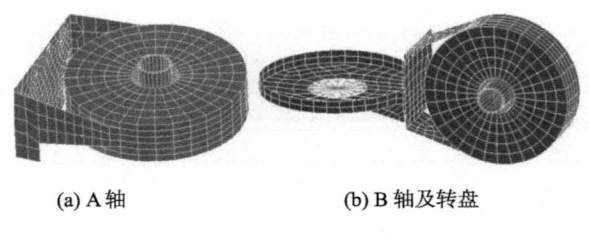

图 2 有限元模型

3.2 边界条件

由于本文最主要是对定向天线部分进行整体模态分析及响应分析,因此必须要确定边界条件。对于压紧状态,要考虑系统安装板及压紧座的影响,根据现实情况,在计算整体模态时,将约束点的 6 个方向的自由度全部约束住,在进行压紧状态的响应分析时,将加速度激励直接加在约束点上。为了便于计算,将系统安装板上的指定位置的约束点,用一个多点约束(MPC)单元全部连接到某一节点上,这样,只要对这一指定节点的各个方向自由度进行约束或加载,即可达到对系统安装板上各点的约束及加载。虽然该方法会适当提高安装板的刚度,但数值计算结果表明,天线前五阶模态振型均集中在反射面、馈源等部件上,与安装板相距较远;另外,将计算结果与每个约束点单独约束、加载对比,发现反射面与馈源顶端响应值基本保持不变,因此,可以采用 MPC 方法将安装板上的约束点连接后统一加载。天线压紧状态下的有限元模型,如图 3 所示。整个天线系统共有 41 171 个单元,对于展开状态,由于此时系统安装板已经不是主要考虑对象,因此可以将其忽略,而直接考虑底座与展开部分。另外,展开状态的约束,就是直接将底座上螺钉孔内各节点的 6 个自由度全部约束。

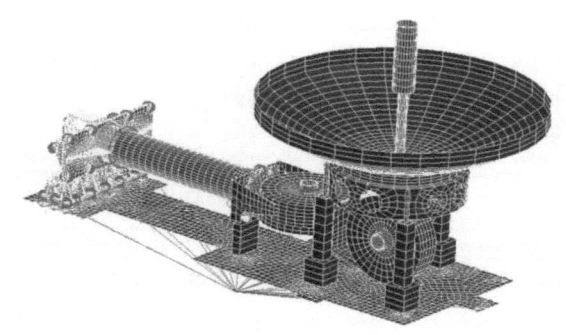

图 3　天线压紧状态下的有限元模型

4　计算结果分析

4.1　压紧状态固有频率分析

当天线处于压紧状态时,其第一阶频率约为 92Hz,振型为沿展开臂方向,即沿 Z 轴方向的摆动,最大振幅位于反射面柱面边缘外侧(90°)及馈源顶端。第二阶频率为 103Hz 左右,振型表现为垂直于展开臂及底板方向的横向摆动,最大振幅同样位于反射面柱面边缘外侧(0°)及馈源顶端。第三阶频率约为 115Hz,振型为反射面的扭转振动。第四阶频率为 124Hz,振型为展开臂与馈源沿 Y 方向的横向摆动,最大振幅仍位于馈源顶部。第五阶频率为 130Hz,振型为反射面与馈源沿 Z 方向的振动。与第一阶振型不同的是,第五阶振型的反射面与馈源并不是同步振动,两个部件的振动应该有一个相位差;而第一阶振动是反射面的振动还带动馈源运动,这就是两者比较显著的差别。由于试验测量时仅扫到 100Hz 的频率,因此,试验结果仅包含前两阶模态,分别为 89Hz 和 100Hz。将数值计算结果与试验结果进行对比,计算结果略微偏高,但误差均控制在 3%~4%。这说明,计算结果与试验结果无论是在趋势还是数值上,都吻合得较好。

4.2　压紧状态动力学响应分析

本文主要是对压紧状态的正弦振动试验进行模拟,表 1 分别给出了 X、Y、Z 方向上所加的正弦加速度的大小。根据压紧状态馈源顶端振幅大于其他测点的情况,研究 X、Y、Z 方向正弦加速度激励作用下,馈源顶端的动力响应,计算结果如表 2 所示。

表1 正弦振动试验量级

X方向		Y方向		Z方向	
频率/Hz	量级	频率/Hz	量级	频率/Hz	量级
10~20	6.2mm	10~20	5.0mm	10~20	5.0mm
20~35	$10g_n$	20~35	$8g_n$	20~35	$8g_n$
35~65	$15g_n$	35~65	$12g_n$	35~65	$12g_n$
65~100	$8g_n$	65~90	$6g_n$	65~80	$6g_n$
		90~100	$5g_n$	80~100	$5g_n$

表2 馈源顶端方向加速度响应计算结果

激励方向	模态阻尼	X方向数值		Y方向数值		Z方向数值	
		共振点/Hz	响应值/g_n	共振点/Hz	响应值/g_n	共振点/Hz	响应值/g_n
X	0.02	>105	16.32	>105	23.20	>105	89.08
	0.03	>105	16.32	>105	18.51	>105	78.09
	0.05	>105	16.32	>105	14.67	>105	69.20
Y	0.02	102	4.74	102	342.24	102	228.54
	0.03	101	3.10	101	207.57	101	169.51
	0.05	99	1.78	99	104.60	99	119.80
Z	0.02	95	0.42	95	75.69	95	223.81
	0.03	90	0.10	90	23.72	90	181.30
	0.05	89	0.24	98	42.97	98	145.03

由于在 X 方向的共振频率并不在 105Hz 以内，因此只提取了馈源顶端的峰值。从计算结果看，其加速度响应峰值在 X 方向受模态阻尼的影响较小，而 Y 方向与 Z 方向受模态阻尼的影响较为明显，表现为随着模态阻尼的增加，其加速度响应峰值明显减小。与此同时，各点响应值的数值计算结果较试验测试结果普遍偏小。因此，选择正确的模态阻尼对计算结果的正确性具有重要的现实意义。

由于 Y 方向的固有频率在 102Hz 左右，因此在 Y 方向激励下的响应峰值基本就出现在这一区域内。随着模态阻尼的增加，馈源顶端加速度响应值迅速降低，降幅分别为 39.3%和 69.4%。可见，在 Y 方向激励下，模态阻尼对一些关键点的影响还是非常大的。根据测试结果，在 Y 方向激励作用下，反射面边缘外侧0°与90°方向的值分别为 $84.6g_n$ 与 $119.7g_n$。馈源顶端 Y 方向加速度值为 $350.4g_n$，在 Z 方向激励作用下，馈源顶端 Y 方向加速度值为 $78.4g_n$，而对应的数值计算结果分别为 $342.2g_n$ 与 $75.6g_n$。结果表明，仅模态阻尼为 0.02 时与试验值相对较为接近。随着模态阻尼的增加，响应值必然减小，此时与试验值相差会较大，因此，从这点上讨论，模态阻尼取为 0.02 较为合适。

与 Y 方向激励情况类似，Z 方向的共振频率在 89~92Hz 范围内。馈源顶端加速度响应数值测试结果为 $179.56g_n$。将计算结果与试验结果对比发现，在模态阻尼为 0.02 时，加速度响应值偏大；而取 0.03 时，两者吻合较好；取 0.05 时，计算结果与试验结果相差较大，数值明显偏小。

从响应分析的结果看，计算结果与试验结果基本能吻合。对 Y 方向的响应分析可以看到，当模态阻尼取为 0.02 时，测点与试验数据吻合得较好。但是，随着阻尼的增加，其响应值会有较为明显的下降。而对 Z 方向各测点的响应分析来看，模态阻尼取为 0.02 时，关键点的响应值偏大；而取 0.03 时，吻合较好。考虑到天线在压紧状态时主要是受到垂直安装板以及沿 Y 方向的横向"摆动"，即主要沿 X 方向与 Y 方向振动，因此，综合考虑，在动力学响应分析时，可以将模态阻尼取为 0.02。

4.3 展开状态固有频率分析

展开状态共包括 15 种工况，即 A 轴(靠近展开臂)分别旋转 0°、45°、90°、135°、180°(共 5 种工况)与

B轴(连接反射面)分别旋转0°、90°、180°(共3种工况)之间的两两组合。天线一阶模态计算结果如表3所示。由于双轴驱动机构主要是由电机提供其刚度，而电机转轴相对于卫星其他部件而言，其刚度较低，因此，天线展开状态15种工况中第一阶模态振型均为从AB两轴连接部分、B轴、反射面直到馈源顶端各部件整体围绕A轴转动并带动展开臂轻微摆动。结果表明，A轴扭转刚度相对较低。而从计算结果看，各工况的一阶模态频率均大于3.0Hz，处于3.3~4.5Hz之间。

表3　天线展开状态15种工况下系统一阶模态频率计算结果

B轴转角/(°)	0					90					180				
A轴转角/(°)	0	45	90	135	180	0	45	90	135	180	0	45	90	135	180
一阶模态频率/Hz	4.43	4.22	4.12	4.22	4.45	3.50	3.38	3.31	3.37	3.51	4.45	4.24	4.12	4.21	4.44

4.4　双轴电机刚度对展开状态一阶模态的影响

在天线展开工作状态下，双轴驱动主要是由相应的电机提供的，而根据数值计算结果，定向天线的第一阶模态又表现为沿A轴方向的扭转振型，因此，有必要研究双轴电机刚度与系统一阶模态之间的关系。对比上述15种工况中系统一阶模态，取一阶模态最小值所对应的工况(即A、B轴转角均为90°)进行研究，计算结果如图4所示。从图中可以看出，系统一阶模态频率随着扭转刚度的增加而增加，当双轴电机刚度大于0.2N/m时，两者基本呈线性关系。根据定向天线一阶模态指标要求[11]，为了保证系统一阶模态频率大于1.0Hz，通过对数值计算结果离散点的拟合，认为双轴电机刚度不能小于0.035N/m，因此可将此值作为保证天线一阶模态频率大于1.0Hz要求所对应的电机扭转刚度的临界值。

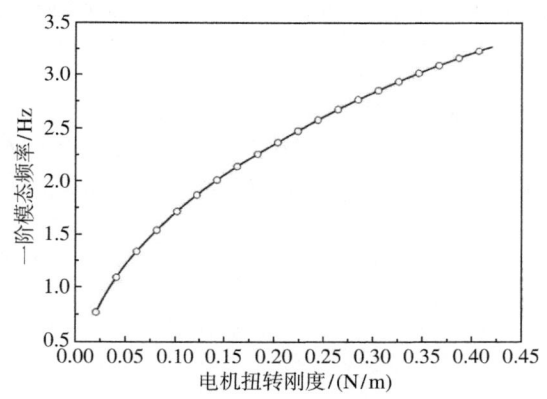

图4　电机扭转刚度与天线一阶模态频率之间的关系

5　结论

本文以CE-1卫星定向天线为研究对象，采用MSC.NASTRAN有限元软件，对天线压紧与展开两种工作状态下的固有频率及压紧状态下的动力学响应进行分析。根据计算结果与试验结果的对比分析可以看出，定向天线压紧状态的频率值与试验值误差控制在5%以内，计算结果较试验测试结果略高。振型方面，一阶振型出现于沿展开臂方向；二阶振型与展开臂在同一平面内，并垂直于展开臂方向，出现摆动的情况；三阶振型出现的是反射面的扭转模态。从振型上看，馈源顶部振幅较大。

从定向天线展开工作状态的15种工况模态分析结果看，由于双轴驱动部件仅靠电机控制，其扭转刚度相对其他部件的刚度要低，因此第一阶模态出现在沿A轴轴向的扭转上，且频率值也较低，在3.3~4.5Hz

范围。

针对双轴电机刚度与卫星低阶模态的影响，还研究了双轴电机刚度值变化与天线一阶模态频率的关系。从计算结果可以看出，天线的一阶模态频率会随着双轴电机刚度的增加而增加，若要保证天线一阶模态频率满足大于 1.0Hz 的技术要求，则双轴电机刚度应大于 0.035N/m。

为了避免压紧状态下某些部件由于较大的动力响应值损坏，应该适当地加大阻尼，以减小部分点的响应值。从计算结果看，纵观整个模型各节点，压紧状态下响应值不论是 Y 方向还是 Z 方向，最大值均出现在馈源顶端，而且最大值可达到约 $350g_n$，为有效避免过高的加速度响应，还是应当适当加入阻尼而抑制过高的响应。另外，反射面边缘外侧两个方向的值相对也较大，这些都是值得注意的。

参 考 文 献

[1] Lecha J, Woods C. Design development and testing of the X-ray timing explorer high gain antenna system [C] // The 29th Aerospace Mechanisms Symposium. Washington: NASA, 1995: 193—207

[2] Stephen R. Topex high-gain antenna system deployment actuator mechanism [C] // The 25th Aerospace Mechanisms Symposium. Washington: NASA, 1991: 205—212

[3] Jan H, Enrico S, Michael S, et al. A compact, light-weight high data-rate antenna system for remote-sensing orbiter s and space exploration [J]. Acta Astronautica, 2009, 65: 1738—1744

[4] Hiroaki T. Design optimization studies for large-scale contoured beam deploy able satellite antennas[J]. Acta Astronautica, 2006, 58: 443—451

[5] Shogen K, Nishida H, Toyama N. Single shaped reflector antennas for broadcasting satellites [J]. IEEE Transaction on Antennas and Propagation, 1992, 40 (2): 178—187

[6] Cherrette A R, Lee S W, Acosta R J. A method for producing a shaped contour radiation pattern using a sing le shaped reflector and a single feed [J]. IEEE Transaction on Antennas and Propagation, 1989, 37(6): 698—706

[7] Akira M, Akio T, Naokazu H, et al. Technology status of the 13m aperture deployment antenna reflectors for Engineering Test Satellite VIII[J]. Acta Astronautica, 2000, 47: 2—9

[8] 叶培建, 饶炜, 孙泽洲, 等. 嫦娥一号月球探测卫星技术特点分析[J]. 航天器工程, 2008, 17(1): 7—11

[9] 欧阳自远. 月球科学概论[M]. 北京: 中国宇航出版社, 2005

[10] 叶培建, 孙泽洲, 饶炜. 嫦娥一号月球探测卫星研究综述[J]. 航天器工程, 2007, 16(6): 9—15

[11] 孙大媛. 一种卫星机械转动定向天线的研制[J]. 航天器工程, 2007, 16(6): 46—50

绕月探测工程地面接收站通用解调处理机性能测试

郑 磊　苏 彦　李春来　张洪波

(中国科学院国家天文台，北京 100012)

摘 要 介绍了绕月探测工程数据接收地面站使用的通用解调处理机的基本组成和工作原理；讨论了针对误码率、极化合成、帧同步和滑步容错等指标要求的测试原理和方法，并对其测试结果进行了分析。为卫星通信的基带设备的测试提供了参考。

关键词 绕月探测　卫星通信　解调

发展空间科学，开展深空探测，是我国的航天科技发展目标。绕月探测工程作为我国开展深空探测的第一步，在政治、军事、科学、技术等诸多相关领域都将产生深远的影响，是我国航天事业的又一个重要里程碑。

在此背景下，结合其他国家深空探测的经验，我们在工程一开始就把质量和可靠性放在首位，以确保任务的顺利完成。要求整个系统的设计都应在满足工程应用的基础上，以方便运行、维护及操作为着眼点，重点考虑深空探测的特点，采用成熟技术，突出可靠性，并兼顾系统的扩展能力，适应未来发展的需要。

通用解调处理机作为数据接收环节中必不可少的基带处理设备，对其各项指标都要进行严格和详细的设计、调试和检测，保证工程任务的顺利完成。

文章针对通用解调处理机在绕月探测工程中承担的任务、设备组成和工作原理、设计指标以及我们怎样其是否符合设计指标的要求等各方面进行介绍。

1 工作原理和设计指标

1.1 工作原理

数据接收地面站的主要任务是完成卫星下传数据的接收和记录存储，并将数据实时或事后传送到地面应用系统总部。卫星下传 S 频段信号经天线接收后，送入接在天线馈源右旋和左旋输出口的低噪声放大器；放大后的射频信号由光端机经光路送至下变频输入口，再通过中频开关矩阵将变频后的中频信号分别送至两套通用解调处理机。图 1 中模拟仿真器可以模拟卫星调制信号输出，在系统中作测试用。

通用解调处理机在数据接收中的主要任务是在卫星下行信道中接收前端送来的两路中频模拟信号，经解调、CCSDS 处理后得到卫星下传的位流数据(RS 译码前数据)和成帧数据(RS 译码后数据)。通用解调处理机除了可以满足绕月探测工程任务要求外，还充分考虑了系统的可扩展性，本文主要围绕绕月探测工程的任务要求进行讨论。

由图 1 知，通用解调处理机分为通用解调器和 CCSDS 处理器两大部分。

本文原载于《天文研究与技术》，2007，Vol. 4，No. 3，258~265。

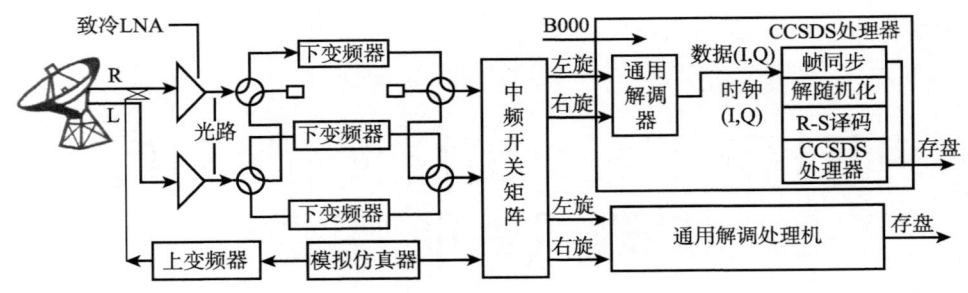

图1 数据接收系统框图(部分)

1.1.1 通用解调器

通用解调器同时接收前端下行信道送来的两路中频模拟信号,分别经输入匹配隔离、AGC放大和带通滤波得到恒定电平的带限信号,然后进行高速A/D采样,得到两路中频数字信号。此中频数字信号与本地数字载波进行正交混频和低通滤波整形,得到I、Q两路正交基带信号。对这两路基带信号进行码同步、信道译码和码型转换,最终得到I、Q两路串行数据。(当工作在BPSK调制体制时,只有I路输出)。

当工作在分集合成模式时,通用解调器将两路输入信号先进行最大比合成,再完成合成信号的解调处理。

从工作流程看(图2),通用解调器可以划分为5大功能模块:(中频)信道模块、分集合成模块、载波解调模块、码同步模块和译码模块。

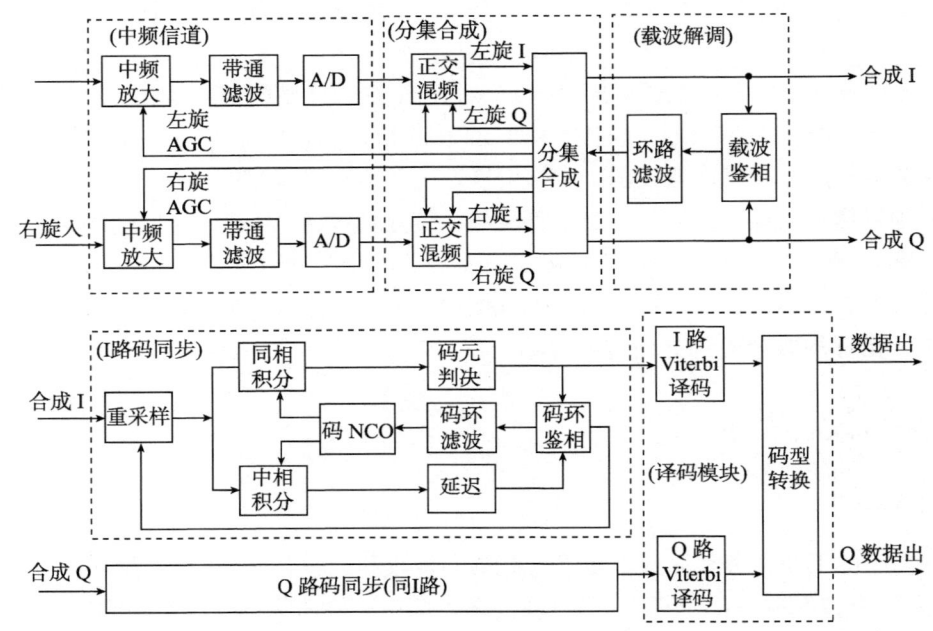

图2 通用解调器框图

中频信道接收前端送来的70MHz中频信号,首先进行AGC放大,然后经过模拟带通滤波,滤除信息带宽之外的频率分量,避免采样时发生频谱混叠[1]。依据带通采样定理进行A/D转换,将模拟信号转换为数字信号,送后端分集合成模块。根据后端送来的反馈控制信号控制放大增益,实现自动增益控制。

分集合成模块采用对称分集锁相环和AGC/AM加权最大比合成器技术。利用共模环跟踪输入信号的公共频率和相位变化,差模环跟踪两路输入信号之间的频率和相位变化,并对称控制两路输入信号的对应的本振NCO频率和相位,保证跟踪的连续性。将两个环的输出作为控制信号,输出无相位阶跃,能够实现快衰落信号的最佳分集合成,避免信号深衰落造成的数据丢失。

载波解调模块将分集合成模块送来的两路合成后的I和Q信号进行载波鉴相,鉴相结果经过环路滤波后,得到共模控制量,送至分集合成单元,完成共模环闭环过程。

码同步模块采用数据转换跟踪环(DTTL),从前端送来的基带数字信号中提取码同步时钟,进行最佳码元判决,得到串行数据流。

译码模块除了完成Viterbi译码(CCSDS标准),还实现I、Q两路数字信号的并串转换(BPSK除外)。

1.1.2 CCSDS处理器

CCSDS(Consultative Committee for Space Data System)标准是在美国1983年以前所提出的国内测控网标准基础上,以空间站为对象模式提出的国际性宇航标准。旨在利用不同国家的空间和地面设备,使用一个共同的数据系统标准进行空间数据的传输与获取,以促进空间科学的相互合作与发展。它建立了数据包和虚拟信道的概念,对不同信息源、不同速率、不同性质的多种信息数据统一管理[3]。

CCSDS处理器接收通用解调器送来的数据和时钟,完成帧同步、解随机化、RS译码和VCDU提取,并给出译码质量信息,然后存盘。

帧同步器以三态方式工作:搜索态、校核态、锁定态。三态逻辑关系如图3所示。

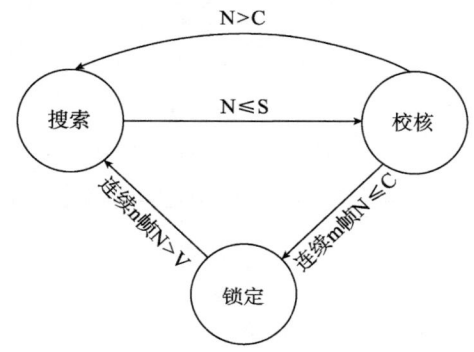

图3 帧同步算法示意图

N:误码数;S:搜索容错;C:校核容错;V:锁定容错;m:校核帧数;n:锁定帧数

帧同步器接收来自码同步器的串行数据流和码同步时钟,检出串行信号中的帧同步码组。完成数据帧格式的提取,产生帧同步脉冲、字同步脉冲及帧锁定状态指示。

1.2 设计指标

设计指标可分为功能指标和性能指标两大部分。

通用解调处理机的基本功能指标包括:编码方式,调制体制,调制码率,工作频率,帧格式、状态指示,存盘、记录、回放、虚拟信道、业务类型和业务等级各方面。设计时应跟卫星发送调制信号和任务要求相匹配,比较直观,本文重点介绍性能指标。

绕月探测工程数据接收地面站承担的任务是接收绕月探测一号卫星S频段下行数传信道的数据,在信道参数满足的条件下,满足3Mbps码速率、误码率优于$1×10^{-6}$的接收要求。

考虑到深空探测数据传输路径上的各种干扰、空间信道的衰减和卫星飞行速度产生的影响等,导致信道裕量有限,所以采用了误码率性能好的BPSK调制体制[1],并对解调处理的各个环节性能要求都较高。主要性能指标有:

- 通用解调器在3Mbps码速率下,误码率为$1×10^{-3}$~$1×10^{-7}$时信噪比偏离理论值不超过0.7dB。
- 分集合成采用最大比合成算法,要求:
  两路等信噪比时,合成增益优于2.5dB;
  两路不等信噪比时,合成信噪比优于$10\log(E_{b1}/n_1+E_{b2}/n_2)-0.5$dB;

其中：E_{b1}、E_{b2} 是：单路中单位比特的平均信号能量；
n_1、n_2 是：单路中噪声的单边功率谱密度。

- 帧同步容错能力：0~3 bit(可设)。
- 抗码位滑动能力：0~±3 bit (可设)。

2 测试项目、原理、方法与结果分析

为了提高设备的可靠性和有效性，在研制、生产、验收三个阶段中都应对其进行相应的测试。某些内部单元的测试，例如中频信道的 AGC 范围和时间常数，载波同步模块的载波捕获带宽、载波同步带宽、载波捕获时间和捕获/同步多普勒频率变化率，以及码同步模块的码环带宽和码同步捕获/同步带宽等，在研制和生产阶段完成测试。设备安装后，进行系统性的测试，验证设备是否满足设计要求和工程需要，本文介绍了设备安装后的系统性指标和功能测试。在测试中使用的测试设备见表1。

表1 调试和测试中使用的测试设备

设备名称	型号	个/套数
频谱仪	E4440A	1
中频噪声源	YM2-50	2
模拟仿真器	自研	1
实时误码率测试软件	自研	1
可调衰减器	0-11dB(1dB 步进)	2
低损耗连接电缆和转接头		若干

2.1 误码率测试

利用系统的有线长环测试链路进行测试，连接如图4。

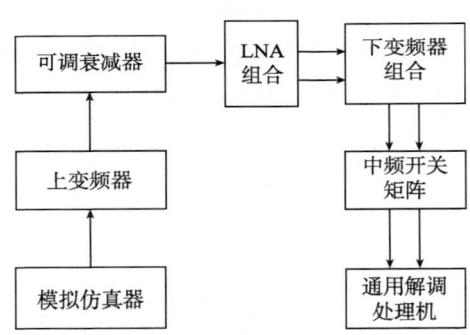

图4 长环测试系统框图

在误码率测试中，我们需要检测通用解调处理机入口处的信噪比，由于很难直接测量信噪比，一般通过测量载噪比，来换算为信噪比。载噪比和信噪比的转换公式如下[2]：

$$(C/n_0)_{dB} = (E_b/n_0)_{dB} + (R_b)_{dB} \tag{1}$$

其中：C：载波平均功率，E_b：单位比特的平均信号能量，n_0：噪声的单边功率谱密度，R_b：信号传输码率，C/n_0：表示载噪比，E_b/n_0：表示信噪比。

首先，对系统进行中频定标。发送单载波，利用频谱仪测出载噪比的值，在信号传输码率已知的情况下计算出信噪比值。其次，统计系统误码率。发送调制波，统计在此信噪比值下的传输误码率。

误码率统计可利用模拟仿真器发送已知数据，则接收端比对接收的每一帧数据即可统计出误码率，实

际测试时编写了实时误码率测试软件来统计误码率，其界面见图5。

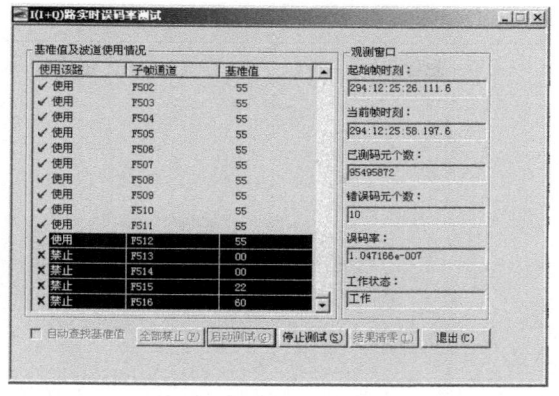

图 5 实时误码率测试界面

2.1.1 接收门限值

接收门限值是通用解调处理机在执行任务时，满足一定误码率条件下相对应的载噪比最低门限值。考虑 BPSK 调制体制，由公式 2[2]：

$$P_e = Q\left(\sqrt{\frac{2E_b}{n_0}}\right), \tag{2}$$

其中 P_e：误码率；E_b：单位比特的平均信号能量；n_0：噪声的单边功率谱密度，

$$Q(x) = \int_x^\infty \frac{1}{\sqrt{2\pi}} e^{-y^2/2} \mathrm{d}y,$$

可以求得在不同误码率的情况下，对应的信噪比理论值。然后根据公式(1)，计算出满足 $P_e=10^{-6}$ 时不同码速率的载噪比理论值。

在中频定标，将其载噪比值标到稍高于理论值。然后发送 3Mbps 和其他码速率调制波，精调可变衰减器，测得统计误码率满足 $P_e=1\times10^{-6}$ 时的中频载噪比值即为接收门限值。测试结果见表 2。

表 2 不同码速率下接收门限值的测量值

码速率	3Mbps	1.5Mbps	750kbps	375kbps	187.5kbps
1×10^{-6} 时 C/n_0 理论值	75.2	72.2	69.2	66.2	63.2
统计码元数	1×10^8	1×10^8	1×10^8	1×10^8	1×10^8
对应的误码情况	1.8×10^{-7}	2.5×10^{-7}	4.2×10^{-7}	2.6×10^{-7}	1.6×10^{-6}
C/n_0 实测值	76.2	73.2	70.2	67.8	64.2

注：由于可调衰减器的步进最小值为 1dB，所以在实测时没有严格调准误码率为 10^{-6}，可以多测几个点拟合出接收门限值(我们只关心 3Mbps 的情况，可以在表 3 中求出准确值，这里不再赘述)。

2.1.2 信噪比和误码率关系测试

在 3Mbps 码速率下，测试不同信噪比值对应的误码率值。

由公式(1)可得 BPSK 调制体制下，给定误码率下对应的信噪比理论值。

实测值和理论值分别见表 3。将实测值与理论值比较，该设备满足误码率为 $1\times10^{-3}\sim1\times10^{-7}$ 时信噪比偏离理论值不超过 0.7dB 的设计要求。

表3 信噪比和误码率关系

E_b/n_0(db)			P_e
实测值	理论值	差值	
12.1	11.7	0.4	$2.7×10^{-8}$
11.1	10.6	0.5	$7.4×10^{-7}$
10.1	9.6	0.5	$8.3×10^{-6}$
9.1	8.6	0.5	$6.3×10^{-5}$
8.1	7.7	0.4	$3.0×10^{-4}$
7.1	6.7	0.4	$1.1×10^{-3}$

2.2 最大比合成测试

在系统测试中利用信噪比与误码率的对应关系，来验证极化合成性能是否满足指标要求(见图6)。

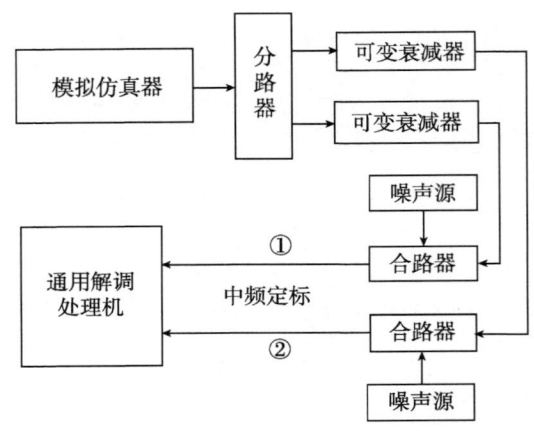

图6 分集合成测试系统框图

在合路器输出端定标，调整可变衰减器和噪声源的值，使两路信号的信噪比相等。

模拟仿真器发送3Mbps码速率调制波。首先只选用单路信号输入模式，统计此时误码率P_{e1}。然后选用最大比合成模式，增加两路可调衰减器值，调整此时误码率值$P_{e2}=P_{e1}$。这时衰减器增加的值即为所测最大比合成增益(表4)。

表4 最大比合成测试

		载噪比(dB)	误码率
等信噪比	左旋	76.25	$1.8×10^{-7}$
	右旋	76.14	$1.7×10^{-7}$
	合成		$3.7×10^{-7}$
不等信噪比	左旋	74.25	$2.4×10^{-5}$
	右旋	71.14	$2.0×10^{-2}$
	合成		$7.4×10^{-7}$

注：等信噪比测试时，左右旋两路各加3dB衰减后进行最大比合成。合成后误码率略高于单路时，可判断等信噪比时最大比合成增益接近3dB。不等信噪比测试时，按设计要求，此时合成信噪比应优于$10\log(E_{b1}/n_1+E_{b2}/n_2) - 0.5=10.6$dB。10.6dB对应的误码率值应略差于$1×10^{-6}$(门限信噪比值11dB)，而测试值优于$1×10^{-6}$，可以判断其在这点上满足指标要求。

接着可以调整可变衰减器值，在两路不等信噪比时进行测试，测试方法与上面相同。

改变衰减器值再进行测试，测试结果均满足指标要求，这里不再列出。

2.3 起始有效帧数

通用解调处理机对接收到的数据进行载波捕获、码同步和帧同步后存盘,而捕获和同步的过程将导致数据的丢失。为了考虑在执行任务期间,地面站接收下行数据的完整性,需要知道记录数据的起始有效帧数,以制定卫星和地面站的发送和接收策略。这也是工程上我们比较关注的一项指标。

可利用系统内的中频自检链路进行测试,连接如图7。

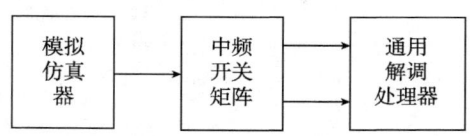

图 7 中频自检系统框图

在强信号时进行测试(定标在远高于门限值,可近似看作没有误码)发送端传输数据时,会加入帧计数码按帧数依次进行编号。接收端查看存盘文件中第一帧的帧计数码即可知起始有效帧计数值为多少。

3Mbps 码速率下,帧长 512 字节,字长 8bit,起始有效帧数为 9 帧,即在载波捕获的情况下,经过 12ms 记录第一帧数据。

由于在参数设定时,设备内部就完成了载波捕获,等待接收数据,所以上面测试值并没有包含载波锁定时间,但也反映了实际任务时设备的反应时间。

2.4 容错测试

容错功能用来避免传输时不可预知的帧同步码误码对数据接收带来的影响。利用模拟仿真器的位流数据模飞功能(可以将满足 CCSDS 格式的位流数据直接调制后发送),我们可以在位流数据的帧同步码组中人为加入误码,在强信号下进行容错模拟测试。系统连接如图7。

模拟仿真器以固定码(55)调制,发送 3Mbps 码速率调制波,通用解调处理器存盘得到标准文件。在此文件中将连续 600 帧(为了能在界面显示上有足够的时间)的同步码组中加入 n 比特误码。生成容错测试文件。

在不同的容错参数下,将容错测试文件放到模拟仿真器中位流数据模飞,观察帧同步状态指示是否正确。

实测时:采用的容错测试文件详见图 8(2bit 误码,1ACFFC1D→7ACFFC1D)。

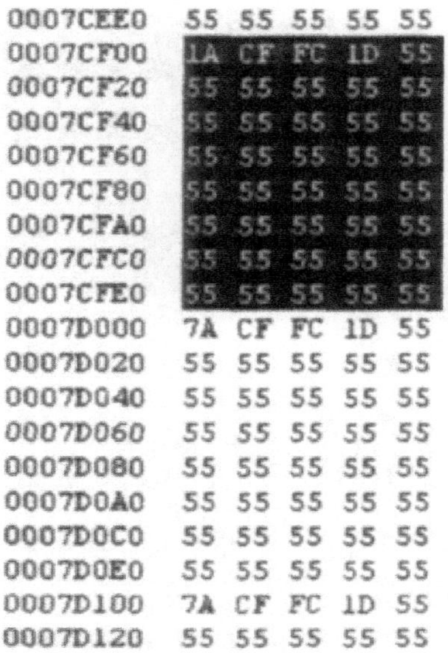

```
0007CEE0    55 55 55 55 55
0007CF00    1A CF FC 1D 55
0007CF20    55 55 55 55 55
0007CF40    55 55 55 55 55
0007CF60    55 55 55 55 55
0007CF80    55 55 55 55 55
0007CFA0    55 55 55 55 55
0007CFC0    55 55 55 55 55
0007CFE0    55 55 55 55 55
0007D000    7A CF FC 1D 55
0007D020    55 55 55 55 55
0007D040    55 55 55 55 55
0007D060    55 55 55 55 55
0007D080    55 55 55 55 55
0007D0A0    55 55 55 55 55
0007D0C0    55 55 55 55 55
0007D0E0    55 55 55 55 55
0007D100    7A CF FC 1D 55
0007D120    55 55 55 55 55
```

图 8 同步码头 1~7h

容错参数：搜索容错：1bit 校核容错：0bit 锁定容错：0bit
校核帧数：3 帧 锁定帧数：3 帧

测试中出现帧同步失锁(见图 9)，符合设定参数。其他测试结果这里不再一一赘述，在下面将要进行的滑步测试我们可以进一步的验证帧同步算法。

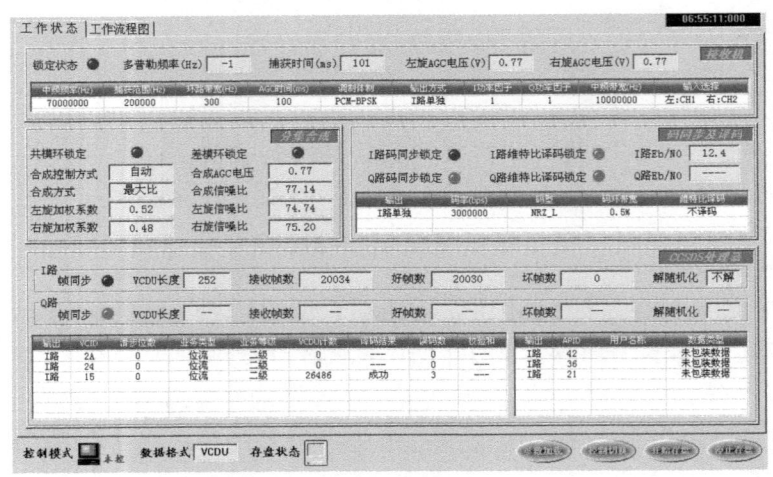

图 9 容错测试帧同步失锁

2.5 滑步测试

滑步功能是用来避免数据传输时不可预知的帧长度变化对数据接收带来的影响。跟容错测试的原理一样，我们人为的生成滑步测试文件，利用位流数据模飞，在强信号下，通过中频自检链路进行滑步模拟测试。系统连接如图 7。

在标准文件中选取任意一帧使其数据少或多 1~3 bit，生成滑步测试文件。分别设置滑步容错位数为 1~3 bit，模拟仿真器位流数据模飞滑步测试文件，查看通用解调处理机存盘数据，验证其是否符合滑步容错算法，同时还可验证帧同步算法。

实测时：采用的滑步测试文件详见图 10(多 1bit，则 0101_0101 0101→0101 0010 1010)。

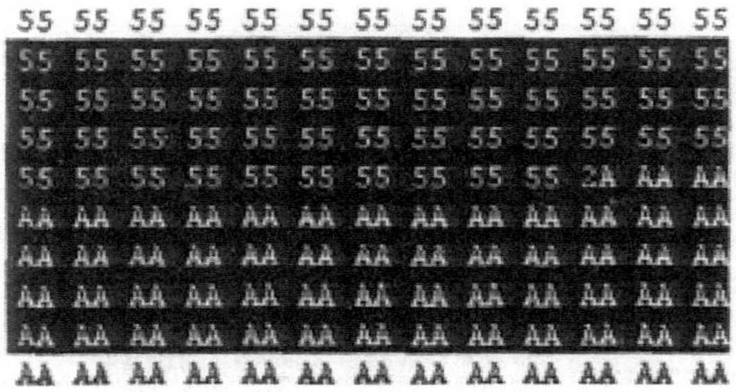

图 10 滑步测试文件(多 1bit)

当容错设为 0bit 时，存盘数据从多 1bit 的那帧开始错了三帧半。这是因为接收端检测不出多了 1bit，则开始就错了半帧；接下来的每一帧都错开了 1bit，导致帧同步码出错，连续 3 帧(锁定帧数 3)后从锁定态转到搜索态，直到再次帧同步锁定后才有新的数据存盘。所以总共错了 3 帧半。

当容错设为 1bit 时，存盘数据在只多 1bit 的位置错 1 帧。这时接收端检测出多了 1bit，则在多 1bit 的那帧的帧同步码后去掉 1bit，因此前半帧 0101 0101→1010 1010，后半帧 0101 0010 1010→1010 0101 0101，见图 11。

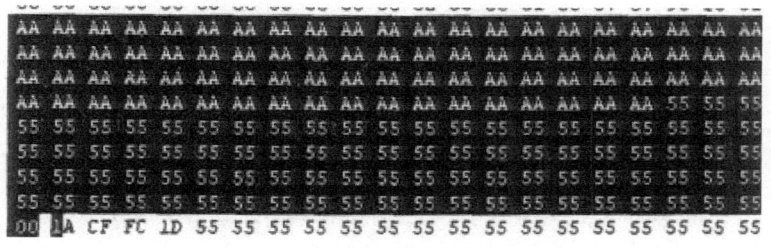

图 11　滑步测试存盘数据(容错 1，多 1bit)

3　测试注意事项

- 测试时应注意信号电平不能过高，以免烧坏设备。开启设备前可把衰减器调至最大。
- 所有设备严禁带电插拔，更改连接时应先断电。特别是低噪声放大器。
- 统计误码率时错误码元数应大于 100 个。
- 噪声源设定值应大于系统底噪。
- 极化合成测试时应选择合适的误码率定标点，以免误码率过低，检测时间长，或检测不到。

4　结束语

随着卫星通信的蓬勃发展，调制解调设备的应用将会日趋广泛。为了适应各个领域的需求，解调设备的设计将会多元化，相应的性能测试则会越来越重要。本文在结合自身实践经验的基础上，较为系统的介绍了绕月探测工程通用解调处理机的各项性能指标测试，希望能为今后解调设备的设计测试提供一些参考。

参　考　文　献

[1] Bernard Sklar. Digital Communications Fundamentals and Applications Second Edition [M]. 北京：电子工业出版社，2002.
[2] 曹志刚，钱亚生，现代通信原理[M]. 北京：清华大学出版社，1992.271—280
[3] Recommendation for Space Data System Standards Telemetry Channel Coding [M].CCSDS 101. 0-B-5BlueBook,2001.3.

基于 SAN 的绕月探测工程数据存储系统架构的设计与实现

张舟斌[1,2]　左　维[1]　李春来[1]

(1. 中国科学院国家天文台，北京　100012；2. 中国科学院研究生院，北京　100049)

摘　要　在分析了两种网络存储技术优缺点的基础上，针对绕月探测工程地面应用系统数据管理的要求，提出了数据存储系统架构的基本技术方案，并讨论了如何实现系统的高可靠性、高可用性和虚拟化的存储管理。

关键词　绕月探测　SAN　存储系统

月球探测是目前国际空间科学中的一个热点，在未来 5～15 年内，月球探测是国际深空探测活动的重要目标，月球已成为未来航天大国争夺战略资源的焦点。开展月球探测工作是我国迈出航天深空探测第一步的重大举措，将成为我国空间科学和空间技术发展的第三个里程碑。

作为绕月探测工程五大系统之一的地面应用系统，将主要负责绕月卫星在轨运行的日常管理、探测数据的接收、存储备份、处理和科学研究与应用，是绕月探测工程科学探测目标能否实现的关键，也是绕月探测工程科学价值的最终体现。

数据是地面应用系统最宝贵的资源，因此如何构建一个安全可靠的数据存储平台是地面应用系统中的一个非常重要的工作内容。在存储系统的设计过程中，我们采用了当前具有广泛应用前景的网络存储技术来构建地面应用系统的海量数据存储系统，把设计重点放在如何提高信息存储的可靠性和可用性这一关键性的问题，同时兼顾系统的可扩展性，以适应未来发展的需要。

文章针对系统技术方案的选择、数据存储系统架构的实现，以及如何实现存储系统的可靠性及高可用性等方面进行论述。

1　网络存储技术

网络存储系统基本上可以分为 SAN(Storage Area Network，存储区域网络)和 NAS(Network Attached Storage，网络附加存储)两种类型，目前，这两种网络存储技术被广泛应用于企业级数据存储解决方案中，以满足集中式超大容量、数据共享、远程数据访问、虚拟化存储服务的需求。

1.1　SAN 介绍

根据 SNIA(Storage Networking Industry Association，存储网络工业协会)给出的 SAN 定义："SAN 是一个网络，其主要目的是在计算机和存储元素之间以及存储元素之间传输数据"。SAN 通过专用的交换设备和环路设备连接若干存储设备和备份设备组成一个单独的数据网络，它独立于服务器网络系统之外，但又允许任何服务器连接至任何存储设备(见图 1)。SAN 采用一种串行的、无阻塞的、专门为数据传输而优化设计的 FCP(Fiber Channel Protocol，光纤通道协议)作为存储访问协议，实现以前无法或很难实现的应用模

本文原载于《天文研究与技术》，2007，Vol.4，No.4，343~348。

式。光纤通道很好地融合了通道和网络技术的优势,可在所连接的设备之间,提供点到点的直接连接或交换的点到点连接,避免了大流量数据传输时容易发生的阻塞和冲突,具有更高的带宽、更长的连接距离、更好的安全性和扩展性。

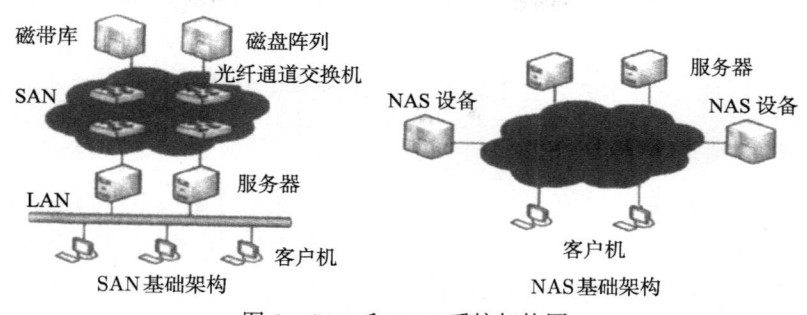

图 1 SAN 和 NAS 系统架构图

1.2 NAS 介绍

NAS 的定义是:"NAS 是一个存储单位,连接在网络上提供文件访问",在 NAS 存储结构中,存储系统不再通过 I/O 总线附属于某个特定的服务器,它完全独立于网络中的主服务器,被定义为一种特殊的专用数据存储服务器(见图 1)。NAS 系统拥有一个专用的服务器,该服务器上安装着一个瘦操作系统和优化的文件系统,并专门服务于文件请求,以提高系统性能和不间断的用户访问。NAS 解决方案通常配置为作为文件服务的设备,由工作站或服务器通过网络协议(如 TCP/IP)和应用程序(如网络文件系统 NFS 或者通用 Internet 文件系统 CIFS)来进行文件访问,因为 NAS 设备支持 NFS 和 CIFS 协议,故异构平台之间的文件可以通过这两种协议,而达到真正的文件共享[1]。

2 存储系统的设计与实现

2.1 方案的选择

地面应用系统的核心任务是要提供数据的存储、管理、备份与服务,所需存储的数据主要包括:原始数据、遥测数据、轨道数据、快视数据、科学研究数据产品等,预计一年在线存储数据总量高达 25.1T,因此存储系统必须要有极大的存储能力、良好的可扩展性、快速的数据访问性能和高效的系统整体性能。我们在大量调研了网络存储技术之后,并结合当前比较成熟的海量数据存储解决方案,认为相比于 NAS 系统,SAN 解决方案更适用于我们的数据存储系统。主要表现在以下三个方面:

2.1.1 系统存储能力及可扩展性

NAS 系统只能通过提高单个控制器的处理能力和增大单个控制器控制的存储设备来扩展,它的可扩展性受到设备容量的限制,并且新增的 NAS 设备与原有的设备不能集成为一体,形成一个连续的文件系统。随着系统中 NAS 设备的逐渐增加,也会给系统的管理带来一定的难度。

SAN 存储系统则具有更好的可靠性和扩展性。SAN 所采用的光纤通道标准定义了三种传输拓扑结构以实现系统的灵活扩展。通过光纤交换机将服务器和存储设备互联,使 SAN 具有几乎可无限扩大的存储能力,并且所有存储资源都处在同一个专用数据网中,可方便地对存储资源进行划分和管理。

2.1.2 数据访问的性能

NAS 采用 NFS、CIFS 网络文件系统协议在网络上提供文件级的数据访问功能。当计算机通过网络从 NAS 访问数据时,需采用软件的方式对网络协议栈进行处理,这将会耗费一定的 CPU 资源。同时,传统网络并不是为在可靠的链路上传输大批量数据而设计的,这些都将对系统的整体效率产生一定的影响。从根本上来讲,NAS 系统不能满足大容量连续数据传输的要求,不适用于数据库或大型计算系统,而更适用于文件长度比较短、对传输速度要求不高的应用。

SAN 则是针对海量和面向数据块的数据传输,服务器和存储设备之间的协议是专为数据密集型存储所设计,访问共享存储效率十分高。并且它把数据处理和通信管理的任务放在 HBA 上进行处理,具有非常低的 CPU 时间占用。因此,SAN 更适合于集中的存储备份和关键任务数据库等应用。

2.1.3 备份任务对系统性能的影响

NAS 是基于网络的存取设备,NAS 环境中的数据备份不是集中化的,因此仅限于使用直接连接设备(如专用磁带机或磁带库)或者基于网络的策略。在该策略中,设备上的数据将通过用户的网络进行备份,给网络带来了沉重的负担,严重时甚至会引起网络瘫痪。

SAN 是构建于光纤的专用数据网络,可以极高的提高带宽(新的 FC 标准可使带宽达到 4GB),并且 SAN 中的数据备份可以将备份流量局限在数据网络内部,能极大地降低网络的负载,做到真正的 LANFree 备份。

2.2 存储系统实现的拓扑结构

地面应用系统的数据存储系统是由 3 台 IBM P570 服务器,2 台 EMC DS4700M 光纤交换机,1 台 EMC CX700 磁盘阵列和 1 台 STK L700E 磁带库组成,其拓扑结构如图 2 所示。其中,各主业务服务器均分别连接至两台光纤交换机上,将 EMC 磁盘阵列连接至两台光纤交换机实现各类数据的高速在线存取,将 STK 磁带库连接至其中一台光纤交换机实现各类数据的高速近线归档备份,连同那些用以存储陈旧、短期内不会用到的数据的脱机磁带一起,构建成在线存储、近线存储、离线存储相结合的三级层次存储体系。

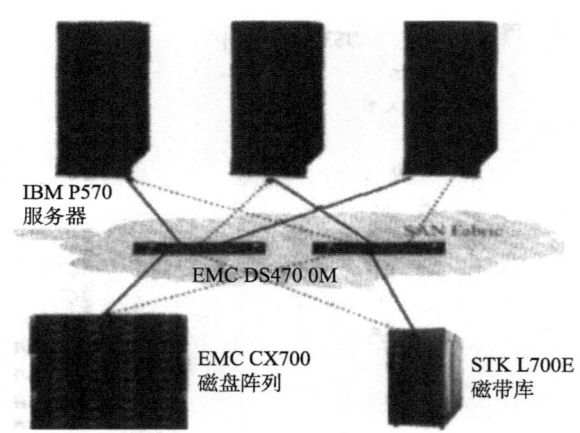

图 2 存储系统拓扑图

在此数据存储网络中,所有链路均采用光纤连接的方式,每条链路速率可达 2G/4Gbps。磁盘阵列和各业务主机均采用两条链路的方式连接至光纤交换机,以实现链路的冗余备份及负载均衡功能。这样,链路均正常时系统会将数据流量均匀分布在两条链路上,提高了系统的整体性能,而当其中的一条链路发生故障时,系统就会自动将全部的数据流量加载到另一条链路上,有效地避免了单点故障,提高了系统的可用性。

2.3 存储系统的数据可靠性与安全性

数据是最宝贵的资源,数据的可靠性与安全性问题是计算机系统亟需解决的首要任务。在地面应用系统存储系统的设计与实现过程中，我们从系统容错以及对数据的安全访问角度，采取了多种技术手段以保证数据的高可靠性和安全性。

在存储系统中，磁盘失效的潜在危险随着所使用的磁盘数的增长而线性增长，而磁盘矢效将直接导致数据的不可用。因此，我们采用 RAID5 技术来实现提供容错功能的冗余磁盘系统，以提高系统的可靠性。RAID5 通过条带化存储和奇偶校验两个措施来实现其冗余和容错的目标，在数据写入时，通过在一组排列的数据上执行异或操作来计算这组数据的校验值，并同时将数据和校验值分散存储在所有成员磁盘上。当系统中的某一磁盘驱动器发生故障时，配置的热备盘就会替换掉失效磁盘，并且通过对剩余成员磁盘中的数据执行异或操作以重建失效磁盘上的数据，同时将数据恢复至热备盘中。图 3 显示了已实现 RAID5 的磁盘资源的划分。

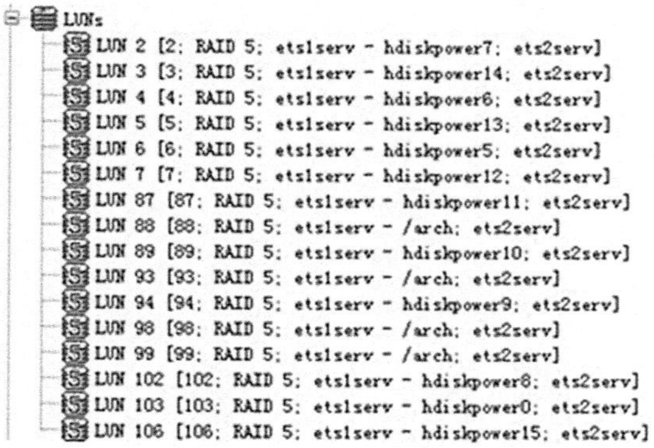

图 3 LUN 配置图

SAN 的结构允许任何服务器可以连接到任何存储阵列，存储设备被当作服务器的本地设备进行访问，文件系统和数据的维护在服务器端完成。因此，必须采用有效的数据隔离手段，以防止各服务器互相破坏彼此的文件系统和数据。为此，我们对磁盘资源做了 LUN(Logic Unit Number，逻辑单元号)屏蔽和交换分区的划分，以实现数据的安全访问机制。

光纤通道设备以逻辑单元号(LUN)的方式提供数据资源，LUN 屏蔽技术通过将存储资源上的某些 LUN 指定给特定的服务器，使这些 LUN 只能被有访问它们权限的系统看到，以此保证数据的一致性。在地面应用系统的存储系统中，我们通过创建存储组(Storage Group)的方式来实现数据隔离(见图 4)，存储组中的

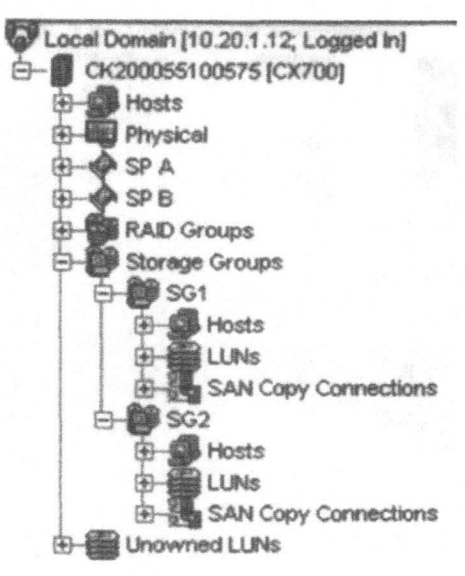

图 4 存储组

主机只能访问属于同一个存储组的 LUN 资源,从而控制了 LUN 的访问权限。除此之外,我们还在光纤交换机层级设置了交换分区。交换分区的工作原理是通过向交换机的 SNS(Simple Name Server,简单名字服务器)注册主机 HBA 卡的 WWN(World Wide Name,全局名),从而将不同主机配置成几个不同的逻辑组,以实现交换机连接的所有主机端口级的屏蔽,确保只有同一逻辑组内的成员才能互相通信及访问特定的存储资源,从而避免可能发生的数据丢失或破坏[2]。同时,交换分区的划分也有利于减少数据包在光纤交换机内的广播,减少光纤交换机端口之间的干扰,从而提高设备的访问速度。图 5 显示了系统所建立的交换分区。

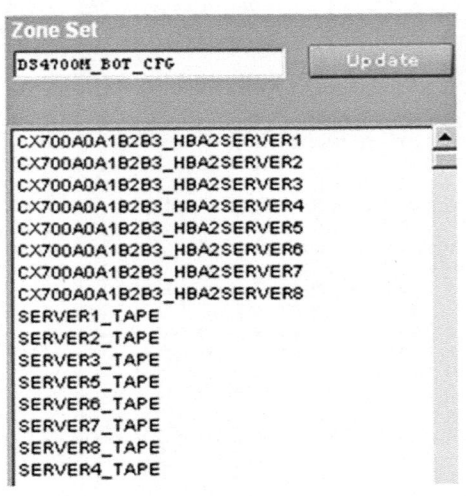

图5 Zone 的配置

2.4 存储系统高可用服务器集群

地面应用系统的服务器由 2 台 IBM P570 整合式服务器、1 台 IBM P570 仿真测试开发服务器组成。每台整合式服务器通过分区的手段在逻辑上成为运行管理、数据处理、数据管理服务器双机,提供主要的业务处理能力,各业务服务器双机通过安装 IBM HACMP(High Availability Cluster Multi-Processing,高可用集群多处理软件)以主/备方式构成高可用集群系统,以保证主机系统安全可靠、连续不间断的工作[3](见图 6)。

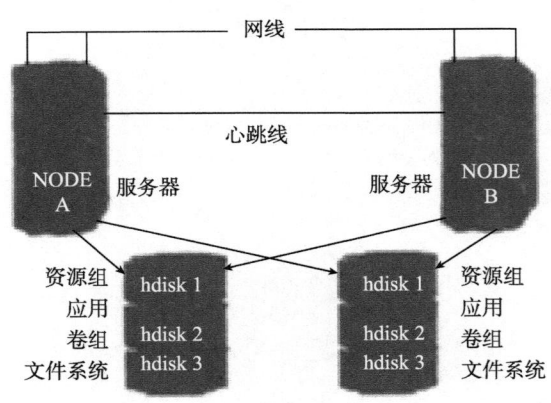

图 6 HACMP 集群

每组业务服务器通过设定浮动的服务 IP 地址对外提供服务,双机之间通过 RS232 标准连线相互连接串口实现心跳检测,当主服务器出现系统级、数据库级或应用级故障时,在监视器连续没有收到的心跳信号达到一定数目后，则认为主服务器失效，集群系统即开始启用系统恢复功能，将服务 IP 地址浮动到备份服务器上，并接管失效主服务器的资源，如服务 IP 地址、卷组和文件系统，同时也可以通过客户化的 SHELL 程序来启动相应的应用程序，实现透明的服务切换。资源的接管根据集群中节点的优先级进行分配，主服务器的优先级高于备份服务器，因此在主服务器修复重新启动后，将重新接管资源。此外，我们还为每台服务器配置了主备两块网卡以应对网卡失效时造成的服务中断，当节点的服务网卡发生故障时，HACMP 侦测到后会将备份网卡的备份 IP 切换成浮动的服务 IP，而将原服务网卡的 IP 切换成启动 IP，以恢复被短暂中断的服务。

2.5 虚拟化的存储管理

根据数据不同的重要性、访问频次等指标，地面应用系统的数据存储系统设计成在线存储、近线存储和离线存储相结合的三级层级存储结构，数据存储和管理需要涉及到磁盘阵列和磁带库两种不同类型的设备。由于应用程序不能直接访问磁带库设备，因此我们利用提供的 Veritas 和 HSM(Hierarchical Storage Management，分级存储管理)软件接口将磁盘阵列与磁带库虚拟成一个逻辑存储体，把磁带库作为磁盘阵列的存储空间扩展，并对所存储的数据进行了三级编目，分别对应磁盘阵列文件系统目录、磁带库分区目录和已出库磁带数据目录，为应用程序提供透明的数据访问。数据编目信息通过建立数据文件目录索引的方式存放于数据库中，用以实时记录数据文件的目录信息，包括文件名称、大小、路径、存放物理设备等等，并且使目录信息随着文件的产生、删除、迁移等操作随时保持更新。虚拟文件系统结构如图 7 所示。通过这种层次化的虚拟存储管理手段，根据事先制定的策略，得以将那些非重要性且访问频次低的数据迁移到相对廉价的磁带库中，一方面节约了数据存储的成本，另一方面也大大减少了本地磁盘的空间占用率，提高了整个系统的存储功能。

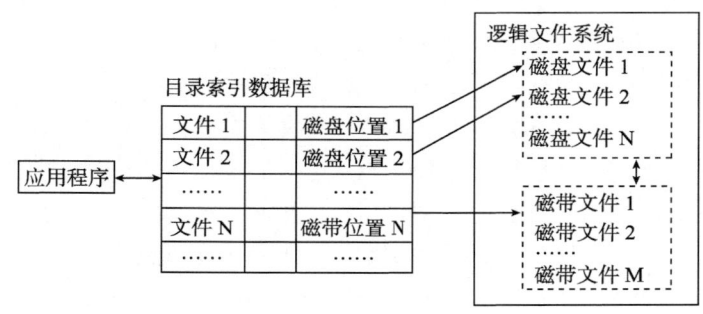

图 7　虚拟文件系统

3　结束语

随着信息的爆炸式增长以及数据重要性的提高，如何构建一个安全可靠的信息存储中心一直是业界研究发展的重点。本文结合绕月探测工程中数据存储系统的解决方案，较为系统地介绍了如何构建一个安全可靠、性能良好、稳定运行的信息存储中心，为企业级信息存储提供了一个参考。

参 考 文 献

[1] Farley M. Building Storage Networks [M]. 北京：机械工业出版社，2001.
[2] 闫光星，罗宁，白英彩. 存储区域网络的数据安全访问[J]. 微型电脑应用，2004, 20(6): 11—12.
[3] 李杰飞，侯燕燕，李卫东. 高可用群集多处理系统及在防地减灾计算机网络的应用[J].地震, 2001, 21 (4): 129—135.

基于复用的软件构架评估方法及其在嫦娥工程中的应用

胡智新[1,2]　李春来[1]　欧阳自远[1]

(1. 中国科学院国家天文台，北京　100012；2. 中国科学院研究生院，北京　100039)

摘　要　针对目前软件构架评估方法未考虑复用构架评估知识的局限性，提出一种新的、基于构架评估知识复用的软件构架评估模式(包括软件构架评估方法元模型和应用框架)。运用一致的评估元模型，建立高层模型，辅助构架评估决策；并在应用框架支持下，系统地复用评估历史数据，将 SAEM 的活动整合为一个系统的、可复用的、可管理的过程。该方法已成功应用于中国探月工程地面应用系统的软件构架评估，降低了深空探测航天复杂系统的研制风险。

关键词　软件构架　软件构架评估　软件构架评估模式　场景

0　引言

软件构架(software architecture，SA)(Kruchten，1995；IEEE，1998)是软件系统最初设计决策的体现和风险承担者交流的手段，决定着软件系统所有的质量属性和项目结构。软件构架评估(evaluating software architecture，ESA)(Li and Henry，1993)是在项目的竞争需求和构架决策映射到代码之前对决策进行公布和评估，判断 SA 是否满足项目所期望的质量属性要求，将决策文档化，是保证软件质量的必要条件。成功的软件产品开发和演化依赖于 SA 的设计或选择是否恰当。因此，尽早评估 SA 至关重要。

ESA 分为提问技术(如调查问卷、检查表、场景等)、度量技术(如仿真、试验等)和混合技术(提问技术和度量技术相结合)(Abowd et al.，1996；Bass et al.，1998)。Parnas and Weiss(1985)提出了 ESA 的基础，Carnegie Mellon 的软件工程研究所(software engineering institute，SEI)则将软件构架评估引入软件开发周期，针对不同的焦点，提出一系列基于场景的软件构架评估方法(software architecture evaluation methods，SAEM)。目前，广泛应用于软件工程实践中的主流 SAEM 有软件构架分析方法(software architecture analysis method，SAAM)、构架权衡分析方法(architecture tradeoff analysis method，ATAM)、成本-收益分析方法(cost benefit analysis method，CBAM)、中间设计的主动评审(active reviews for intermediate designs，ARID)、质量属性专题讨论法(quality attribute workshop，QAW)以及软件性能工程(software performance engineering，SPE)(Kazman et al.，1994，1996，2001；Abowd et al.，1996；IEEE，1998；Kazman，1998；Smith and Woodside，1999；Clement，2000；Williams and Smith，2003；Clement et al.，2002；Barbacci，2003a，b；Moore et al.，2003)。

本文分析基于场景的主流 SAEM，存在未从复用的角度考虑软件构架评估知识的管理和应用的局限性，缺乏 SAEM 元模型、应用框架支持 ESA 的高层模型辅助决策和评估知识的复用。结合中国探月工程地面系统工程实践，本文提出一种新的、基于构架知识复用的软件构架评估模式——软件构架评估方法元模型和软件构架评估应用框架(以下简称"评估元模型"和"应用框架")。目的：(1)提出一致的评估元模

本文原载于《地球科学——中国地质大学学报》，2006，Vol.31，No.3，384~388。

型，建立高层模型，辅助构架评估决策，并指导复用构架评估数据；(2)在应用框架支持下，系统地复用评估历史数据，将 SAEM 的活动整合为一个系统的、可复用的、可管理的过程。提高 ESA 效率和质量，降低风险和成本。

1 基于场景的软件构架评估方法对比

本节分析广泛应用于软件工程实践的、基于场景的主流 SAEM(表 1)。由于未考虑软件构架知识的系统性复用，基于场景的 SAEM 存在以下局限性：①ESA 缺乏高层模型(high-level model)辅助决策。ESA 是一个风险度很高的软件过程，不同的 SAEM 关注焦点、方法、适用状况、应用时机和结果精确度等存在较大差异，需要软件组织构建高层模型进行辅助决策：采用何种 SAEM，何时实施，及如何复用构架评估知识和历史数据；②缺乏基于评估知识复用的 SAEM 元模型和应用框架。目前的 SAEM 只是关注于单一 SA 的评估，只是方法的重用，缺乏从软件过程性管理的整体性考虑。此外，构架评估是一个人员密集型过程，评估效果过多依赖于评估人员素质。因此需建立基于评估知识复用的、一致的 SAEM 元模型和应用框架支持 ESA。

表 1 基于场景的软件构架评估方法对比分析

	SAAM	ATAM	ARID	CBAM	SPE	QAW
涉及的质量属性	可修改性、变化性和功能	所有质量属性	设计方法的适宜性	成本和收益	性能	所有质量属性
分析对象	构架文档	构架方法和样式，构架文档	组件的接口规范	关键属性、ATAM评估结果	构架文档、性能需求文档	多个尚未完成的构架文档
评估时机	在组件已经分配到各个模块中以后	在构架设计方法选定后的软件开发各阶段	系统开发早期或概念阶段	构架设计方法已定，且 ATAM 评估已完成	已确定组件的性能约束条件时	多个软件构架完成之前
采用方法	头脑风暴，场景分析、演练	头脑风暴，构架分析，确定敏感点、权衡点	头脑风暴	量化收益和成本值	构建、调整和验证性能模型，量化静态分析场景	求精质量属性,比较构架优劣
复杂度	简单	复杂	简单	较复杂	复杂	较复杂
支持复用	否	否	否	否	否	否

2 软件构架评估元模型和应用框架

由于软件构架评估过程是基于组件开发的基本活动，多数软件开发组织开发、维护的产品，在特定领域的构架评估存在共性，因此，针对目前 SAEM 未从复用的角度考虑软件构架知识的系统性复用，软件组织不能以构架评估的知识和历史数据指导以后的构架评估过程的局限性，本文提出一种新的、基于构架评估知识复用的软件构架评估模式——评估元模型和应用框架，将 SAEM 的活动整合为一个系统的、可复用的、可管理的过程。利用评估元模型和应用框架支持 ESA 高层决策，复用构架评估历史数据，提高 ESA 的效率与质量。

2.1 软件构架评估元模型

在分析基于场景的主流 SAEM 的基础之上，本文抽象 SAEM 的基本要素,构建新的评估元模型(图 1)。目的：组织 ESA 各要素，指导软件组织根据待评估 SA 的特点、资源约束等因素，在评估应用框架的基础上，在评估过程中参考可复用历史数据库(以下简称"历史数据库")的应用实例(场景、模型、构架决策或风险问题等)，决策采用何种 SAEM、何时实施(图 2)，并在评估结束后增加新的数据至历史数据库，支持其演化。通过复用 ESA 知识和历史数据，提高构架评估质量和效率，降低决策的风险和代价。评估元模型的应用方法和步骤见 3.3 节。

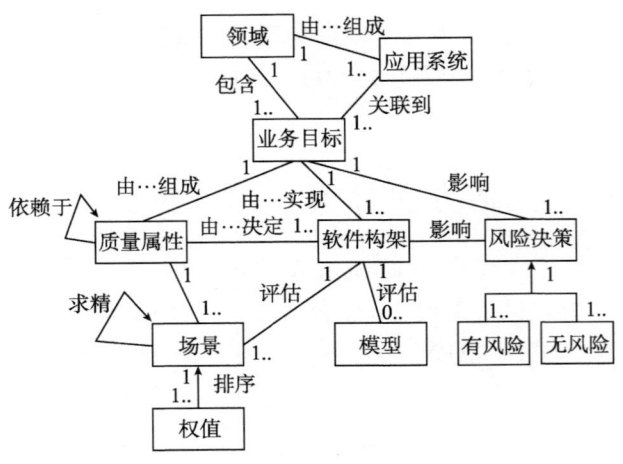

图 1　软件构架评估元模型

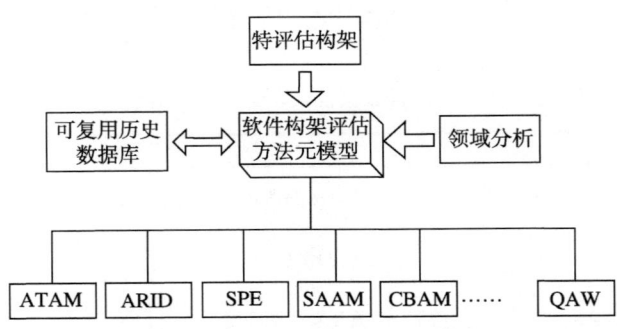

图 2　软件构架评估元模型应用概念图

2.2　软件构架评估应用框架

从基于组件化的软件过程管理考虑，本文提出基于历史数据复用的评估应用框架，将过程管理、数据复用和技术支持整合(图3，斜线阴影部分是目前 SAEM 的范围)。目的：ESA 在统一的应用框架内，基于评估元模型，与管理机制融合，运用 SAEM 库、可复用历史数据库(以下简称"历史数据库")和工具集，提供统一平台，支持面向特定领域的、大粒度的构架评估历史数据复用，实现构架评估的工程一体化。

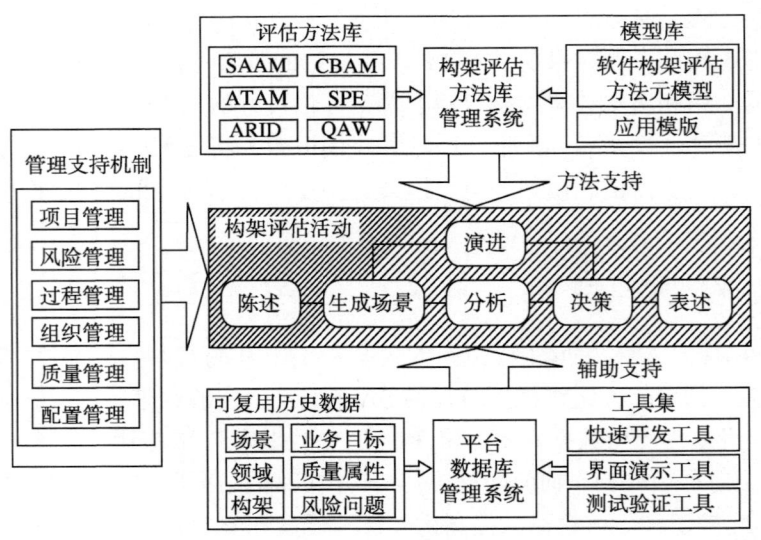

图 3　软件构架评估应用框架的过程模型图

通过评估应用框架和评估元模型的支持，复用了构架评估历史数据，特别是复用已有的、经实践验证的构架决策，可提高构架评估的质量和效率，降低风险和代价，提高评估结果的可信度。

评估应用框架的主要应用步骤：

(1) 分析。根据待评估 SA 的业务目标和领域特性，通过领域分析和评估元模型，结合历史数据库，分析其质量属性和拟采用的构架方法，在项目管理的指导下选择 SAEM。

(2) 实施。基于评估元模型，通过工具集支持，复用历史评估数据，实施选定的构架评估方法：①在生成场景时，基于评估元模型，从历史数据库中，提取与待评估构架质量属性相关的场景为种子场景，在此基础上扩展或裁剪，生成待评估构架的场景；②在工具集的支持下应用步骤①中生成的场景，结合历史数据库的场景验证案例，验证待评估构架方法是否满足质量属性要求，运用风险管理识别、分析风险问题，并与历史数据库对照；③决策。根据验证结果和风险问题(构架对关键质量属性的满足程度、重要场景的排序等)，参考历史数据库的构架决策历史数据，进行构架评估决策。如果发现构架方法不能满足业务目标，或风险度过高，需调整构架方案并返回步骤(1)重新评估。

(3) 演进。形成构架评估最终报告。基于评估元模型，提取评估过程中产生的数据(包括业务目标、质量属性、构架方法、场景、测试案例、模型、风险问题、构架决策和最终报告)至历史数据库，对历史数据库进行配置管理，维护其完整性和一致性，促进其演进。

在实施过程中，同时实施项目管理、质量管理和过程管理，控制评估过程、质量、进度、成本和决策。

3 案例应用

本节介绍在中国探月工程的地面应用系统(ground segment for data and application system，GSDAS)研制中，利用本文提出的软件构架评估方法模式成功进行了 GSDAS 的 ESA，降低了深空探测航天复杂系统的研制风险(评估过程的细节情况由于保密缘由进行了更改和省略)。

3.1 项目背景

嫦娥工程(2003~2007)是中国研制和发射月球探测器开展月球科学探测，是对月球表面的地貌、地形、地质构造、元素分布、物质类型与月表环境进行全球性、整体性与综合性的探测，加深对月球、地-月系和太阳系起源与演化的认识，对有开发利用前景的月球能源与资源的分布和规律进行探测与研究。GSDAS 是嫦娥工程五大系统之一。GSDAS 负责嫦娥工程科学目标；嫦娥 1 号卫星(CE-1)的业务运行及有效载荷的科学探测计划制定；协作处理海量的月球地质、地理、遥感、空间等科学数据，组织科学研究等任务。

3.2 分析阶段

首先，建立评估小组，提出日程安排，做好准备工作；其次，评估负责人介绍评估元模型和应用框架的方法、过程和技巧；项目决策者介绍 GSDAS 的上下文和系统开发的主要业务目标、关键质量属性(集中于性能、可靠性、可操作性和可维护性)；构架师对 SA 做出描述(包括三级分层结构、客户机/服务器方法、管道/过滤器方法)。应用领域分析和评估元模型，结合历史数据库，分析 GSDAS 的质量属性和拟采用的构架方法，此次 SAEM 选择 ATAM。

3.3 实施阶段

(1) 基于评估元模型，从历史数据库中提取与待评估 SA 质量属性相关的场景为种子场景，根据关键质量属性在此基础上扩展或裁剪，生成 GSDAS 的质量效用树(utility tree)(表2)，设定场景优先级。本文对场景采用 9 级权值，把审查的关注重点放在了(H，H)上，其次是(H，M)、(M，H)、(M，M)其余不考虑。

表 2　标注了优先级的 GSDAS 效用树子集

第 2 层质量属性	第 3 层质量属性求精	第 4 层质量属性场景	重要性	难度	累加和
性能	P1：0 级数据处理时间为毫秒级	P1.1：0 级数据的处理时间<0.1s	H	H	(H,H)
	P3：准实时处理和显示快视图像	P3.1：处理和显示快视图像时间<5s	L	H	(L,H)
可靠性/可用性	Ra1：系统过载或崩溃数据不丢失	Ra1.1：双机实时切换，恢复故障机<5min	H	H	(H,H)
	Ra4：数据处理、存储的故障时间	Ra4.1：数据处理、存储或备份引起的故障时间每周< 1h。处理机、网络及存贮资源的使用、I/O 通道的吞吐量的余量>30%	H	M	(H,M)
可维护性	M1：对一个子系统的更改不要求改变其他子系统	M1.1：对数据管理子系统的修改不会影响其他子系统及数据的查找、存储	M	H	(M,H)
	M5：对操作系统、数据库、COTS 产品的升级时间缩短 50%	M5.1：1 天完成 AIX 操作系统的升级	M	M	(M,M)
		M5.2：1 天完成应用软件包的升级	M	L	(M,L)
可演化性	E1：系统支持 50 个大规模数据请求的输出	E1.1：系统能够每天处理 100 个不同的数据输出请求（>1G 数据/每个请求）	M	M	(M,M)

(2) 在工具集的支持下应用步骤(1)中生成的场景，结合历史数据库的场景验证案例，描述并分析高优先级场景的 SA 设计方法，确定出 SA 的有风险决策、无风险决策、敏感点和权衡点，运用风险管理识别、分析风险问题，并与历史数据库对照。验证 GSDAS 的构架方法是否满足质量属性要求。其后，评估人员集体讨论形成更大的场景集合，通过投票表决确定这些合成场景的优先级，并使用合成场景再次分析 GSDAS 的构架方法，未产生以前步骤中没有发现的高优先级场景。

(3) 决策阶段。根据步骤(2)的验证结果，结合历史数据库，判定 GSDAS 采用的 SA 可满足其业务目标。确定了 GSDAS 的 5 个权衡点并录入文档，形成评估最终报告，评估小组向 GSDAS 风险承担者表述评估结果。

3.4　演进阶段

基于评估元模型，提取 GSDAS 评估过程中产生的数据至历史数据库，对历史数据库进行配置管理，维护其完整性和一致性，支持其演进。

4　总结

与传统的 ESA 相比，本文提出新的、基于构架评估知识复用的软件构架评估方法模式——软件构架评估元模型和应用框架，在统一平台支持下，有效复用了面向特定领域的、大粒度的构架评估要素的历史数据，特别是复用已有的、经实践验证的构架决策数据，提高了构架评估的质量、效率和可信度，降低了风险和代价。

从细化软件构架评估模式的应用前景来看，未来的研究方向和突破点为：①增强评估应用框架的灵活性、稳定性和可扩展性，更好地支持复用；②集成软件构架评估模式于软件产品线开发中，优化软件组织的资源。

参 考 文 献

[1] Abowd, G., Bass, L., Clements, P., et al., 1996. Recommended best industrial practices for system architecture evaluation. Technique Report, CMU/SEI-96-TR-025.
[2] Barbacci, M.,2003a. Using the architecture tradeoff analysis method (ATAM) to evaluate the software architecture for a product line of avionics systems: A case study. Technique Report, CMU/SEI-2003-TN-012, SEI, Carnegie Mellon University. http://www.sei.cmu.edu/publications/documents/03.reports/03tn012.html.

[3] Barbacci, M., 2003b. Quality attribute workshops(QAWs). Third Edition. Technique Report, CMU/SEI-2003-TR-016, SEI, Carnegie Mellon University. http://www.sei.cmu.edu/publications/documents/03. reports/03tr016. html.
[4] Bass, L., Clement, P., Kazman, R., 1998. Software architecture in practice. Addison-Wesley, Reading, MA.
[5] Clement, P., 2000. Active review for intermediate designs. Technique Report, CMU/SEI-2000-TN-009, SEI, Carnegie Mellon University. http://www.sei.cmu.edu/publications/documents/0.reports/00tn009.html.
[6] Clement, P., Kazman, R., Kelein, M., 2002. Evaluating software architectures: Methods and case studies. Addison-Wesley, MA. IEEE, 1998. IEEE glossary of software engineering terminology, 610.12—1990
[7] Kazman, R., Abowd, G., Bass, L., et al., 1994. SAAM:A method for quality through formal technical review. In: Proceedings of the 16th International Conference on Software Engineering, Sorrento, Italy, May, 113—122.
[8] Kazman, R., Abowd, G., Bass, L., et al., 1996. Scenario-based analysis of software architecture. *IEEE Soft-ware*, 13(6): 47—55.
[9] Kazman, R., 1998. The architecture tradeoff analysis method. In: Proceedings of the Fourth International Conference on Engineering of Complex Computer Systems(ICECCS98).
[10] Kazman, R., Asundi, J., Klein, M., 2001. Quantifying the costs and benefits of architectural decisions. In: Proceedings of the 23rd International Conference on Software Engineering (ICSE23), Toronto, Canada, May,297—306
[11] Kruchten, P.B., 1995. The 4+1view model of architecture. *IEEE Software*, 12(6): 42—50
[12] Li, W., Henry, S., 1993. Object-oriented metrics that predict maintainability. Systems and Software, 23(2): 111—122
[13] Moore, M., Kazman, R., Klein, M., et al., 2003. Quantifying the value of architecture design decisions: Lessons from the field. In: Proceedings of the 25th International Conference on Software Engineering (ICSE25), Portland, Oregon, May.
[14] Parnas, D.L., Weiss, D., 1985. Active design review: Principles and practices. In: Proceedings of the 18th International Conference on Software Engineering.
[15] Smith, C.U.,Woodside, M., 1999. Performance validation at early stages of software development. *The Journal of Systems and Software*. http://www.perfeng.com/papers/smitwood.pdf.
[16] Williams, L.G., Smith, C. U., 2003. PASASM: A method for the performance assessment of software architecture. In: Proceedings of the Workshop on Software and Performance (WOSP2002), Rome, Italy, July.

月球探测计划中影像数据的格式

李瑞玲　刘建忠　李春来

(中国科学院国家天文台，北京　100012)

摘　要　月球影像数据是研究月球的地貌、地表形态，绘制全月地形图以及进行各种相关科学研究的基础。早期的 Lunar orbiter, Apollo, Surveyor 基本是以模拟影像格式传回地球，经过扫描后以照片形式发布。Clementine 的数据是通过 PDS 中心，以 PDS 的形式发布的。影像数据的保存与发布格式应遵循以下原则：数据格式要便于经常更新和补充；能够在各种平台和各不同目的使用人群中方便地使用；要符合国际标准，方便处理，能够与以前多次探测计划得到的数据结合使用。为中国嫦娥工程的影像数据格式提出建议。

关键词　影像数据　数据格式　月球探测

0　引言

月球是距离地球最近的天体，长期以来一直是人类空间探测的首选目标。美国和前苏联在 1958~1976 年间开展了多次以月球探测为中心的空间探测计划，1994 年和 1998 年美国又分别发射了"克莱门汀号"和"月球勘探者号"探测器，每次探测活动都返回了大量照片和月球资料。丰富的月球影像数据，为研究月球的地貌、地表形态，月面地形图的绘制，月球表面物质特征及其区域分布等科学研究提供了珍贵的资料。1989 年美国提出"重返月球"的计划以来，欧洲、日本、印度等国都提出了各自的月球探测计划，我国也启动了"嫦娥"工程的月球探测计划[1]。月球影像数据的保存与发布格式，不仅关系到月球影像数据资源能否得到充分利用，也是数据能够进行不断更新和补充的前提。本文综述了多次探测计划中影像数据的保存及发布格式，也对我国"嫦娥工程"中影像数据的保存和发布格式提出了建议。

1　各次月球探测计划中影像数据的存储及发布格式

1.1　Lunar Orbiter 系列任务及影像数据格式

Lunar Orbiter 系列是一个月球环绕轨道探测计划，1966~1967 年在轨，共发射 5 个探测器，全部成功。该项目设计完成三个方面的基本任务，即轨道摄影、绘制月球重力场和月球空间环境评价[2]。其中轨道摄影是探测计划重点要完成的任务。轨道摄影的主要目的是获取详细的高分辨率照片，为阿波罗登月着陆点的选择提供依据，也为地球上的诸多观测点进行科学研究提供翔实的资料图片。Lunar Orbiter 在轨期间共返回了 1654 幅高质量照片，其中 840 张为阿波罗计划选择着陆点服务，其余的照片拍摄了月球正面和 95%的月球背面的照片，并收集了预先选定的 36 个地区的中、高分辨率图像。

在 Lunar Orbiter 探测器中，同时装备了两台照相机，同步操作但视域和分辨率不同。摄像系统包括相机、胶片、胶片处理器和经过通信系统传输到地面的读出系统。摄像系统在胶片上同时记录两帧数据：广角影像(80 mm 镜头)和窄角影像(160 mm 镜头)，结果影像存储在读出系统中。影像数据通过原始胶片在

本文原载于《地球物理学进展》，2006，Vol.21，No.4，1155~1160。

读出系统中进行机载光学扫描，形成一系列带状的模拟数据，以模拟影像的格式发送到地球。在地球上接收到视频信号后，每帧照片上的数据要重新写到新的胶片上，通过相机进行曝光，重建影像[3]。为了数据能够长期保存和便于利用，原始胶片上的数据都经过数字化处理，以 JPEG 的黑白照片格式保存。从胶片上打印出来的照片手工镶嵌成高分辨率和中等分辨率图像广泛发布。

Lunar Orbiter 的影像数据都是硬拷贝的黑白带状照片，或将视频信号直接记录在磁带上制作影像。但以这种格式保存的影像数据在重新扫描后分辨率会比原始数据降低，而且照片产品常会出现数据缺失或重复，影响使用。另外，这个过程要耗费大量人力和时间，重建数据过程是一个极为复杂的过程，保存和使用都不方便。

1.2 Surveyor 系列任务及影像数据格式

Surveyor 系列是美国在 1966～1968 年间实施的不载人登月计划，包括了 7 次登月任务，是美国探测器首次安全登上月球。Surveyor 的主要任务是获取月面的近景图像，确定地形条件是否有利于安全着陆。每个 Surveyor 探测器都配备了电视摄像机，共获得 86000 多张高分辨率照片。得到的影像提供了有关探测器附近的地形情况，以及这些地区撞击坑的数量、分布特征以及土壤力学、月表组成物质等方面的详细资料[4]。

Surveyor 探测器携带了电视摄像机用来获取影像数据。安装在一个可以调节方位和高度的镜子下，摄像机的操作完全依靠地面指令。将摄像机获取的影像转换为标准的电视信号，在地球上将电视影像通过扫描显示器显示。每帧输入的电视画面可以作为一帧识别图像被地面接收，并以与输入图像相同的速度适时显示出来。图像记录在一个视频磁带记录仪上或胶片上。直接从原始的负片上制作负片或在探测器上运用机载的程序将模拟影像转为数字化影像，结果图像可以直接打印[5]。

因此，Surveyor 探测中得到的影像数据以两种形式保存：记录在磁带上的模拟影像和在探测器上将模拟影像转为数字化影像之后的照片。将模拟影像转为数字影像后，影像的分辨率比原始的模拟影像有所降低，但在图像发布时一般经过去噪声、去条带等处理，比原始的模拟影像对比度更强。

1.3 Apollo 系列任务及影像数据格式

Apollo 探测从 1963～1972 年，历时 10 年。Apollo 7 和 9 的任务是环月飞行以检测登月舱的性能，没有返回月球数据；Apollo 8 和 10 检测环月时各部件的性能，返回了月面照片；Apollo 13 因故没有登上月球，但也返回了照片。Apollo 11，12，14，15，16，17 都实现了载人登月并安全返回，这六次任务中返回了大量有科研价值的影像数据和约 400kg 的月球样品，并做了大量包括月壤结构、月震、月球空间环境等在内的实验[6]。

Apollo 的各次任务中都装备了不同镜头的相机，得到的数据直接通过相机胶片曝光或在地面应用一些技术对胶片重新曝光得到，都是彩色或黑白的照片[7]。因为得到的数据都以模拟影像格式或微缩胶片形式存储，使用起来极为不便。这些照片都精心保存于 NASA 的数据库中，要得到这些照片必须通过查询缩微胶片等方式查看哪些照片可以再制。因此，大多 Apollo 的照片几乎从来没有机会向公众展示。LPI(Lunar and Planetary Institute)利用 NASA 提供的影像对 Apollo 的胶片进行扫描，创建数字化的文件制成影像图的目录数据库，成为现在方便使用的，在线的数字化资源[8]。每幅照片数字化为 24 位的彩色照片，Targa 格式。Targa 格式图像再进一步处理，生成 JPEG 格式的图像文件，经过进一步加工后以 JPEG 格式保存。但影像目录库中的影像分辨率较低，无法直接用于科学研究。为了方便 Apollo 数据在未来月球探测中使用，现在正在进行的月球数据工程将原来很少用的照片、硬拷贝文件、缩微胶片等数据重新数字化，保存为 CDF 格式的文件[9]，可以在线浏览和使用。

Apollo 探测中得到的影像都是硬拷贝的彩色或黑白照片，由于资料极其珍贵，这种照片格式的影像使用时有许多限制，大大增加了影像利用的难度。另外，这种照片格式的影像若保存时间过长，会发生变形，色彩和黑白对比度也会发生变化，彩色照片易褪色，黑白照片的色阶也会发生变化，使照片质量下降，不利于影像的长期保存，使用也极为不便。

1.4 Clementine 探测任务及影像数据格式

Clementine 探测器于 1994 年 1 月发射，主要任务是检测传感器和飞船的性能，并进行月球科学观测，在轨期间进行了两个月的月球测绘。Clementine 探测器上携带了六个有效载荷获取影像数据。两台星象跟踪仪相机[10]，另外还有紫外、长波红外和近红外照相机、激光高度计，主要用于成像，成像传感器采用 CCD 技术。

Clementine 探测器在轨的两个月中获取的影像范围覆盖了月面 99%以上的地区，是月球探测中首次获得全月的数字化数据。探测器机载携带了三台计算机，一个 32 字节的精简指令集计算机处理器用来做图像处理。数据处理单元有独立的微型程序化相机，操纵图像压缩系统，控制数据流。在图像处理过程中，数据存在数据记录器中，以后通过 128km/s 的下行线传到地面站[11]。

影像在探测器上用芯片进行过压缩，是经过离散余弦变换的压缩格式，在存储时也采用这种格式。Clementine 的影像数据存储在工程数据集中，都是未经处理和校正的原始数据，根据 PDS 标准重新定义和组织数据格式。完整的 Clementine 影像数据集包括探测器获取的原始影像数据、相关的辅助数据及解压处理软件[11]。数据解压软件可在 SUN/UNIX，IBM/PC 及 Macintosh 计算机平台中运行，最后以 PDS，GIF，TIFF 格式输出解压后的影像[12]。工程数据集中的产品由 Clementine 科学组对数据格式进行标准评估和认证后，保存在光盘上，由 PDS 中心发布到各科研团体。因为这些数据都是按 ISO 9660 的标准进行存储的，和许多计算机平台，如 UNIX，IBM/PC 等兼容。这些数据可用于工程方面，或直接进行科学分析，也可以用来构筑其他科学研究的产品。

模拟格式的影像数据灰度和亮度是固定值，很难改变。当相邻地区的影像需要拼接时色彩很难匹配，尤其是在一些特征明显的地区，影响对数据特征的分析。而数字化存储的影像经过处理相邻地区可以进行无缝拼接。

用 PDS 格式来存储影像，在月球影像存储方面是一个极大的进步。一方面，月球影像数据可以和其他行星数据采用相同的数据存储格式，方便数据间的转换和交流；另一方面，PDS 格式在数据存储和数据描述方面有许多优点，而且这种格式的影像都是数字化的，有利于数据的长期利用和存储。

1.5 Galileo 及 Smart-1 探测的主要任务及影像数据格式

Galileo 是一个木星探测器，1989 年发射，先后两次飞过月球。Galileo 携带的固体相机(SSI)采用 CCD 技术，飞临月球时对月表进行了多波段成像[13]，获取的数据对研究月表成分和月表形态研究有重要意义。影像采用 VICAR 的格式，每幅影像都附有单独的 PDS 头文件。

SMART-1 是欧洲空间局于 2003 年发射的第一个月球探测器，主要目的是检测在未来深空探测中利用的太阳电推进技术以及探测器和载荷的其他新技术，同时进行全面的月球科学观测[14]。Smart-1 携带了 AMIE 小型高分辨率相机进行月球地形成像，将再次获得月球南极的高分辨率图像，现在已经传回了月球影像。

Smart-1 计划采用 PDS 格式来组织和发布所获取的数据，以期能和以前月球探测中获取的数据资源整合利用。

2 常用影像数据格式及其特点

2.1 HDF(Hierarchical data format)格式

HDF 是美国国家高级计算应用中心(National Center for Supercomputing Application)为了满足各种领域

研究需求而研制的一种能高效存储和分发科学数据的新型数据格式[15]。HDF 能够存储不同种类的科学数据，包括栅格图像数据、多维数组、指针及信息说明等文本数据。这种数据格式的主要优点有：

(1) 是一种自描述性文件格式，能够提供数据说明的全面信息。还提供命令方式，分析现存 HDF 文件的结构，并即时显示图像内容。科学家可以用这种标准数据格式快速熟悉文件结构[16]，并能立即着手对数据文件进行管理和分析。

(2) 文件格式可移植性强(独立于操作平台)，可以存储并处理大数据量，文件可以在不同平台间传递而不用转换格式。

(3) 文件的可扩展性较强，可加入新数据模式，与其他标准格式兼容性较强。

(4) 从目前可用的软件来说，HDF 格式的文件可以用 PCI、ENVI、ERDAS 等多种遥感图像处理软件和地理信息系统软件读取和处理。

HDF 格式的缺点是天文数据分析包不支持这种格式[15]，因此要进行快速处理比较困难；而且仅从存储的字节很难反映数据的内容。

这种格式被广泛用于目前国外各种卫星传感器的标准数据格式，包括 NASA 已发射的 Landsat-7 号卫星，MODIS 的一级数据等。

2.2 FITS(Flexible Image Transport System)格式

FITS 格式起源于七十年代后期，由 NASA 和 IAU 联合开发，是目前国际上广泛用于天文学数据分析、存储和转换的一种格式[17]，能通过在线形式发布。

这种数据格式的主要优点有：

(1) 存储量大，能存储许多新型的、结构复杂的数据，适合于许多新的，专业性强的天文仪器数据的存储。

(2) 数据格式标准化程度高，可移植性强，可以包括许多关于图像的描述性数据。

(3) 存储的文件除图像数据外，还支持多维数组、光谱数据及文本文件等。

但 FITS 格式的数据在使用中也有明显的不足，主要表现在：

(1) 数据格式没有压缩形式，数据量极大。在天文学光谱数据方面应用更为广泛。

(2) 数据的头文件编辑时遵循一定规则，用普通的文本编辑语言无法进行编辑，因此，无法进行数据内容的增补和修改。要求用专业的天文学应用程序进行处理，对适合使用的人群有限制。

2.3 PDS(The Planetary Data System)格式

是 NASA 的空间科学组发布的一种数据存储格式，现在已经成为全球科学研究中常用的基本数据资源[18]。PDS 用来存储和发布来自于 NASA 的行星探测、天文观测及实验室的数据。所有的 PDS 产品都可通过在线目录方便获取，并经常进行更新和升级，保证使用者能方便使用。

PDS 数据的主要优点有：

(1) 这种数据格式是为了方便行星数据交流而设计的，可以通过网上电子交流；数据的存储使用 CD，降低了大容量文件存储中的成本和存在的风险。

(2) PDS 使用一定标准存储和描述数据，设计时考虑了以后使用者在不熟悉原始数据格式，用不同的计算机平台的情况下都可以方便地分析和使用数据。这些标准中包括了数据结构、数据描述的内容、发布介质的设计等方面。

(3) PDS 中的数据都是按照 ISO 9660 的标准组织和存储的，这种存储格式利于数据的稳定性并能保证数据长期可用。

(4) 数据的存储使用了压缩格式，减少了数据量。同时数据集中还提供了关于卫星姿态、星历数据、解压软件等各种文件说明信息，方便进行数据处理。

(5) 从现有的软件来看，PDS影像数据可以用ISIS软件读取和处理，NASA的网站上提供了NASA View程序显示影像，另一程序MapMaker可将输出影像转为GIF，TIFF，JPEG等格式。PDS中心提供了数据转换的代码，方便各种软件识别数据。

火星、水星的一级数据，月球探测器Clementine的紫外/可见光相机数据、激光高度计获取的地形数据等都以这种格式存储，MODIS的0级数据也以这种格式存储。

2.4 CDF(Common Data Format)及VICAR(Video Image Communication and Retrieval)格式

CDF是一种存储和管理多维数据的自描述性文件格式，是一个科学数据管理包[19]，可存储和管理矢量、标量以及多维数据。数据的共享需要ISTP/IACG程序(网上可下载程序)，数据也可以通过IDL语言组织和读取。

VICAR用于存储从地球及探测器上获取的行星数据，在设计和使用上与FITS和PDS格式相似[20]。文件包括了以VICAR格式存储的影像及这种影像的处理程序。目前的许多探测项目中用PDS格式文件作为VICAR影像的头文件，PDS负责VICAR影像文件的发布。缺点是以这种格式存储的影像文件只能用VICAR文件中提供的软件来处理。

3 中国的探月工程中影像数据格式的选择

我国启动月球探测计划"嫦娥"工程以来，与探月计划一期工程科学目标相关的科学研究得到了我国科学家的极大关注[21~23]，嫦娥一期工程的首要科学目标是获取月球表面的三维影像，精细划分月球表面的基本构造和地貌单元；进行月球表面撞击坑形态、大小、分布、密度等的研究，为类地行星表面年龄的划分和早期演化历史研究提供基本数据；划分月球断裂和环形影像纲要图，勾画月球地质构造演化史；为月面软着陆区选址和月球基地的位置优选提供基础资料[1]。卫星上搭载有效载荷可见光立体相机和激光高度计来实现这一科学目标。可见光立体相机用来获取不同轨道、不同时段及不同角度的影像数据，用得到的影像进行月球三维影像的制作和月球地形、地貌图的绘制。如何系统组织和管理这些影像数据不仅关系到影像能否充分利用，也是数据质量的一个体现。因此，影像数据格式的选择是月球探测数据存储和发布过程中极为重要的工作。

我国是首次进行月球探测，相关的各方专家和公众都极为关注，获取的数据既要能够满足科学家进行科学研究的需求，也要能够为普通公众提供一个了解月球和月球探测意义及其成果的平台。因此，选择的影像数据的格式不仅要能够为科研提供基础资料，还要方便数据的适时浏览和显示。另外，选取的数据格式应能够满足我国月球探测对数据存储和管理的要求。具体来说，在影像格式选择过程中可以遵循以下原则：

(1) 影像数据格式要便于经常更新和补充。月球探测中获取的数据不仅要用于现在的科学研究，也为未来月球探测和行星探测提供数据库资源，所以数据格式选择要能保证数据长期使用，并能够根据不同的需求进行修改和补充。

(2) 数据的可移植性强，能够支持不同的操作平台及操作系统；尽可能支持当前的多种遥感图像处理软件和GIS软件，不受或减少计算机软硬件系统对数据处理和应用方面的限制，利于数据资源得到最大程度的利用。

(3) 数据格式要遵循一定惯例，符合国际数据组织标准，方便处理，能够与以前多次月球探测计划得到的科学数据以及行星数据结合使用。

(4) 所使用的数据格式应具有压缩格式，尽量减少下传数据的数据量，保证数据质量。

通过对不同探测计划中影像数据格式发展过程的总结，结合一些常用的影像数据格式的特点，"嫦娥"工程中影像数据可能采用的格式有目前国际上较常用的 PDS 或 HDF 格式，考虑到数据格式的统一性，且天文学的影像更多采用 PDS 组织和发布数据，"嫦娥"工程影像数据的标准格式可以采用 PDS 格式。

从 PDS 数据格式的特点可以看出，这种格式基本能够满足数据格式对数据保存、处理、发布等多方面的要求。在我国的月球探测中使用这种影像格式，优势还表现在：

(1) 以前的各次月球探测已经发布和即将发布的数据都采用这种格式，在数据格式的组织和管理方面已有比较成功的经验。使用 PDS 格式可以借鉴其他国家在数据管理和发布方面的成功经验，又保证了数据格式与国际月球探测接轨，利于数据的管理及共享和交流。

(2) PDS 中心针对这种格式开发了多种适时显示的软件，保证了数据在不经过任何处理、不用转换格式的条件下能够适时显示和浏览影像，方便普通公众及时了解关于我国月球探测的进展和成果。

(3) 由于对数据理解的不统一和数据分析软件的多样性，目前还没有统一、成熟和共同遵循的数据格式[24]。PDS 中心有专用的转换代码，针对不同软件提供影像的头文件，目前的多种商业软件都能方便处理，为多种来源的月球数据的集成提供了可能。

4 结语

从 20 世纪 50 年代末美苏掀起第一次月球探测的高潮到美国提出"重返月球"的构想，由于空间探测技术和理论的不断深入，月球探测取得了许多重大成果，月球探测中获取的影像数据格式也经历了漫长的发展过程。早期的 Lunar Orbiter，Apollo，Surveyor 基本是以模拟影像格式传回地球，经过扫描后以照片形式发布，使用者获得的都是硬拷贝的照片文件。由于各次探测中任务的变化以及影像获取系统的差异，不同探测器获得的照片数据很难与其他任务中得到的数据进行坐标匹配，也很难进行数字化处理；而且这种数据最大的缺点是随着时间的推移，照片会出现变色、褪色以及变形的情况，不利于数据的长期保存和使用。Clementine 的影像数据是以 PDS 的形式发布的，影像都以数字化方式保存。以这种格式存储方便数据的交流与应用，也利于数据的保存和更新。通过对常用影像数据格式特点的分析，我国嫦娥工程中影像数据可以采用 PDS 作为标准格式发布。以这种格式作为影像数据的存储和发布格式，能够满足数据方便使用、不断更新等多方面的特点。

参 考 文 献

[1] 欧阳自远，我国月球探测的总体科学目标与发展战略[J]. 地球科学进展，2004，19(3)：351—358.
[2] Thomas P. Hansen. Guide to lunar orbiter photography, NASA, SP - 242, 1970.
[3] Gaddis L R, Sucharski T, Becker T, et al. Cartographic processing of digital lunar orbiter data, Lunar and Planetary Science XXXII, 2001, 1892.
[4] The surveyor program, http://www.lpi.usra.edu/expmoon/surveyor/surveyor.html.
[5] surveyor, http://nssdc.gsfc.nasa.gov/planetary/lunar/surveyor.html.
[6] National aeronautics and space administration. Apollo program summary report. 1975, 3. 1—3. 63.
[7] Grant Heiken (Editor), David Vaniman, Bevan M. French. Lunar sourcebook: a user's guide to the moon [M]. Cambridge university press, New York, 1991, 595—599.
[8] Apollo image atlas, http://www.lpi.usra.edu/.
[9] David R. Williams, Edwin J. Grayzeck. The lunar data project-restoration of Apollo data for future lunar exploration [J]. Lunar and Planetary Science XXXVII, 2006, 1187.
[10] Edwards K E, Colvin T R, Becker T L, et al. Global digital mapping of the moon[J]. LPS XXVII, 1996, 335—336.
[11] Stewart Nozette, Rustan P, Pleasance L P, et al. The Clementine Mission to the moon: scientific overview [J]. Science, 1994, 266: 1835—1839.
[12] Eliason E M. Clementine mission: The archive of image products and data processing capabilities [J], LPS XXVI, 1995, 369—370.

[13] Oberst J, Wahlisch M, Zhang W, *et al*. New data on lunar topography derived from Galileo and Clementine stereo images [J]. LPS XXVII, 1996, 973—974.
[14] Foing B H, Racca G D, Grande M, *et al*. ESA'S smart'1 mission at the moon: fir t results, status and next steps [J]. LPS XXXVI, 2005, 2404.
[15] Jennings D G, McGlynn T A, Jordan J M. Investigating HDF as an astronomical transport and archival format [A]. Astronomical data analysis software and systems III ASP conference series[C]. 1994, (61): 526—529.
[16] Members of the MODIS Characterization Support Team. MODIS Level 1B Product User's Guide [M], NASA/Goddard Space Flight Center, 2003, 6—43.
[17] Hanisch R J, Farris A, Greise E W, et al. Definition of the flexible image transport system [J]. A& A, 2001, 376: 359—380.
[18] Jet Propulsion Laboratory. Planetary data system standards reference [J]. California Institute of Technology Pasadena, 2002.
[19] The common data format, http://cdf.gsfc.nasa.gov/.
[20] Anderson D, Mann M. VICAR image processing using unix, xwindows, and cdroms [J]. LPSC, XX, 1989, 17—18.
[21] 薛彬，杨建锋，赵葆常. 月球表面主要矿物反射光谱特性研究[J]. 地球物理学进展，2004，19(3)：717—720.
[22] 陈俊勇，宁津生，章传银，等. 在嫦娥一号探月工程中求定月球重力场[J]. 地球物理学报，2005，48(2)：275—281.
[23] 李泳泉，刘建忠，欧阳自远，等. 月球磁场与月球演化[J]. 地球物理学进展，2005，20(4)：1003—1008.

绕月探测工程卫星数据的存储与管理

朱 兰[1,2]　左 维[1]　李春来[1]

(1. 中国科学院国家天文台，北京 100012；2. 中国科学院研究生院，北京 100049)

摘　要　从数据的类型和产品格式说明开始，对绕月探测工程中的科学探测数据的管理方法进行了研究。通过数据存储架构和软件结构的分析，指出数据存储的软硬件环境；通过数据存储和备份策略的分析，指出数据存储的可靠性。这些存储与管理方法将为更好地管理数据并向外界提供检索和获取服务提供保障。

关键词　绕月探测　存储架构　存储策略　备份策略

2007年10月24日，嫦娥一号探测卫星的发射圆了中国人几千年的奔月梦；
2007年11月26日，第一幅图的公布，标志着绕月探测工程取得了圆满成功。

这一成功的开始，为大量接收来自月球和地月空间环境的数据打下了坚实的基础，同时也创造了良好的条件。在未来的一年里，将有大量的数据从嫦娥一号卫星上源源不断地传到地球。作为嫦娥工程五大系统之一的地面应用系统(GSDSA)将把这些宝贵的第一手资料妥善存储、管理，并且经过深加工后，向全世界的科学家和天文爱好者公布。

加拿大档案管理员 Arthur George Doherty 指出："档案是一个国家所有财产中最为珍贵的财产，是传给下一代的礼物。档案保管的好坏真实地反映出我们的文明程度"。嫦娥一号卫星传回的科学数据便是一种不可或缺的、重要的国家科研档案，相对其他科学数据而言，其获得的代价与成本更为巨大，其潜在价值和作用将远远超出最初的设想[1]。

如何存储并管理这每天29 G，一年将近25 T 的数据，是首先要解决的问题。为此搭建一个安全且高效的硬件环境作支撑，在其上铺设管理这些数据的应用软件，使硬件与应用软件相辅相成，将确保绕月探测工程接收到的海量数据安全可靠的存储和使用。

1　GSDSA 的基本情况

地面应用系统(GSDSA)是一个实时、多功能的科学探测卫星业务运行与科研应用支持与服务系统。与卫星、火箭、测控、发射场系统一起组成了嫦娥工程的五大系统。它既是工程的起点，又是工程的归宿，是科学目标和科学价值的最终体现，对于月球探测工程至关重要。地面应用系统是目前唯一具有深空(月球)探测器数据接收、空间有效载荷管理、地面系统运行控制、空间探测数据处理、数据管理和科学应用与研究能力的航天工程和科学应用研究的业务系统，也是我国月球与行星科学的研究中心。其核心任务是接收全部的月球探测科学数据，进行探月数据的预处理、科学应用和研究，并最终完成绕月探测工程既定科学目标所规定的任务。

地面应用系统由运行管理分系统(OMS)、数据接收分系统(DAS)、数据预处理分系统(DPS)、数据管理分系统(DMS)、科学应用与研究分系统(SAS)等五个分系统组成。其中数据接收分系统是拥有我国目前口径最大、接收能力最强，并集深空探测卫星数据接收和射电天文观测与研究为一体的密云地面站和昆明地

本文原载于《天文研究与技术》，2008，Vol.5，No.4，365~372。

面站，分别建有 50 m 口径(密云)和 40 m 口径(昆明)的天线系统。

地面应用系统的两架天线就像是嫦娥工程的两只照向太空的眼睛，确保来自月球的信息安全完整地回到地球。

2 数据的范围

2.1 数据类型

嫦娥一号卫星上搭载了 8 种有效载荷，包括 CCD 立体相机、干涉成像光谱仪、激光高度计、γ 射线谱仪、X 射线谱仪、微波探测仪、太阳高能粒子探测器和太阳风离子探测器。这些载荷采集到的数据连同载荷本身的工程数据共同组成了卫星下传到地面的原始数据。这些数据可分为：

(1) 科学数据产品：有效载荷工作产生的探测数据。

(2) 工程数据产品：有效载荷工作时的状态数据。

经数据预处理程序后，地面应用系统将生成不同级别的数据产品，这些级别定义见表 1。

表 1 科学数据产品分级表

数据级别	英文名称	数据描述	备注
原始数据	Raw data	卫星下行数据，经位同步的数据流	位流数据
	Frame data	卫星下行数据经帧同步，解扰，RS 译码的数据帧序列	
0 级数据	Level 0A	经帧同步，解扰，RS 译码的有效载荷数据源包或 IIM 数据帧	
	Level 0B	在 Level 0A 的基础上，经有效载荷数据源包排序，优化拼接，去重复，去源包包头，生成有效载荷科学数据块	
1 级数据	Level 1	在 Level 0B 级的基础上，经物理量转换，并以探测时间 1 轨为单位进行数据分幅的数据，生成的相关辅助数据文件、质量文件、描述文件	
2 级数据	Level 2	在 Level 1 的基础上经过辐射，几何，光度等校正的数据	根据有效载荷不同分为 2A、2B、2C 等级别。
3 级数据	Level 3	在 Level 2 的基础上，经过深加工获得的数据产品	

2.2 数据产品格式说明

每个级别的产品都有各自的数据格式，其中有三类基本的数据产品格式：

(1) PDS 格式数据产品：1、2 级科学数据产品、卫星参数数据(轨道和姿态)和定标数据等产品采用 PDS(Planetary Data System 行星数据系统)格式。

(2) Microsoft EXCEL 格式：工程数据产品采用 Microsoft EXCEL 格式。

(3) 其他格式：原始数据、0 级数据采用二进制文件格式；质量文件、报告文件和说明文件等采用文本文件格式。

3 数据管理系统的体系结构

3.1 数据存储架构

地面应用系统的核心任务是提供数据的存储、管理、备份与服务，所需存储的数据主要包括：原始数据、遥测数据、轨道数据、快视数据、科学研究数据产品等，一年在线存储数据总量有 25.1 T。因此存储系统必须有极大的存储能力与良好的可扩展性、快速的数据访问性能和高效的系统整体性能。

我们采用了当前具有广泛应用的 SAN(Storage Area Network，存储区域网络)网络存储技术来构建地面应用系统的数据存储系统。根据 SNIA(Storage Networking Industry Association，存储网络工业协会)给出的

SAN 定义："SAN 是一个网络，其主要目的是在计算机和存储元素之间以及存储元素之间传输数据。地面应用系统的数据存储系统是由 3 台 IBM P570 服务器，2 台 EMC DS4700M 光纤交换机，1 台 EMC CX700 磁盘阵列和 1 台 STK L700E 磁带库组成。其中，各主业务服务器均分别连接至两台光纤交换机上，将 EMC 磁盘阵列连接至两台光纤交换机实现各类数据的高速在线存取，将 STK 磁带库连接至其中一台光纤交换机实现各类数据的高速近线归档备份，连同那些用以存储陈旧、短期内不会用到的数据的脱机磁带一起，构建成在线存储、近线存储、离线存储相结合的三级层次存储体系(如图 1 所示)。

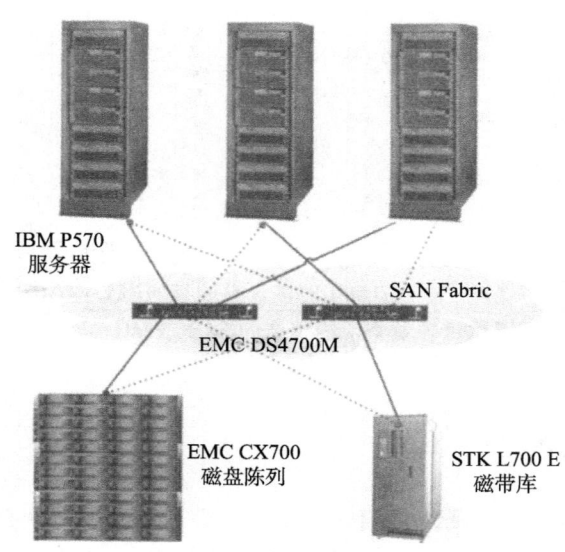

图 1　存储系统拓扑图

在此数据存储网络中，所有链路均采用光纤连接的方式，每条链路速率可达 2G/4Gbps。磁盘阵列和各业务主机均采用两条链路的方式连接至光纤交换机，以实现链路的冗余备份及负载均衡功能。这样，链路均正常时系统会将数据流量均匀分布在两条链路上，提高了系统的整体性能，而当其中的一条链路发生故障时，系统就会自动将全部的数据流量加载到另一条链路上，有效地避免了单点故障，提高了系统的可用性。

3.2　软件功能结构

数据管理系统中的应用软件(DMSA)采用 B/S(Browser/Server 浏览器/服务器)结构，在这种结构下，客户机上只要安装一个浏览器，用户界面完全通过浏览器实现一部分事务逻辑，主要事务逻辑在服务器端实现。浏览器通过服务器同数据库进行数据交互。

DMSA 负责制定存档业务运行计划，调度数据管理业务的运行，对 GSDSA 产生的各类数据进行统一管理，提供统一的用户管理和监控检索界面，提供信息和产品发布的渠道。主要任务包括：①数据存储、管理与服务；②系统业务运行管理；③信息与产品发布。其功能结构见图 2。

服务器端用于完成对数据的存档、下载、管理及通信等功能的实现。底层使用 Oracle 数据库作为数据存储的平台。Oracle 数据库因其较好的开放性，较强的可伸缩性和并行性成为目前使用较广的关系型数据库。

客户端则实现与用户的交互。

主控台用于监视 DMS 服务器的运行状态及运行日志等关键信息，还可以控制 DMS 服务器，比如停止或者启动指定的永久进程、恢复或中止指定的作业等。

系统运行管理网站完成对 DMSA 系统运行参数的管理，包括用户管理、各类数据的配置、监视以及检索显示等功能。

数据订购网站，完成给需要数据服务的用户提供相关服务的功能。包括：用户登录、新用户注册、数据产品信息检索、数据产品订购、数据服务在线帮助、订单查询和跟踪、新闻发布等子功能。

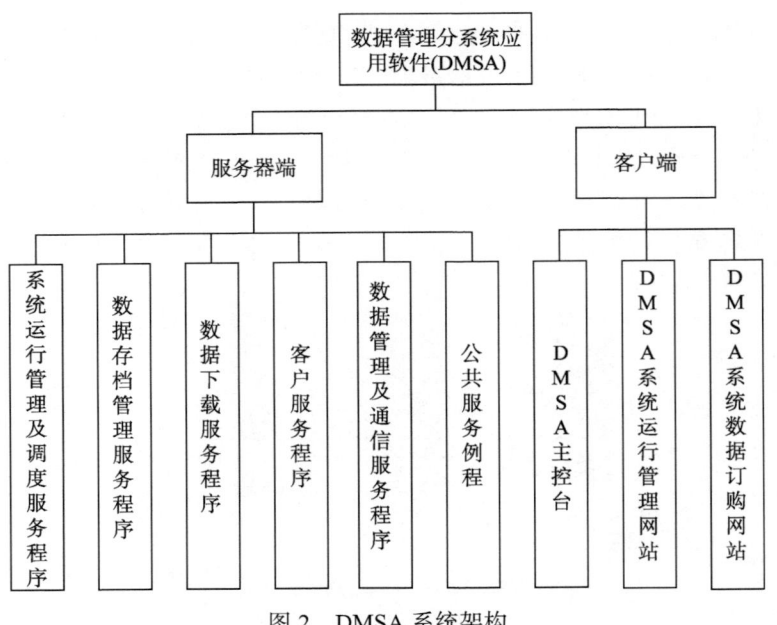

图 2 DMSA 系统架构

4 数据的存储备份与获取

嫦娥工程中所获得的科学数据将利用地面应用系统搭建的硬件与软件环境进行科学的管理并最终面向全世界的科学家和普通用户。

4.1 数据存储策略

4.1.1 数据生命周期管理

数据管理从数据产生到数据最终被废弃贯穿数据的整个生命周期,因此数据管理也就是对数据生命周期的管理,根据数据所处的生命周期阶段的不同,数据存储的位置也不同。在生命周期初期,数据经常被存取,也是数据最有价值的时期,因此将数据储存在成本最高的在线存储设备上;当数据进入生命中期时,其价值也相对随之下降,此时将数据移至近线磁带存储;当数据进入生命周期末期时,则移往离线设备存放,一旦需要时可以随时将之回存到在线设备。如图 3 数据管理层次结构图所示,数据管理应用软件是建

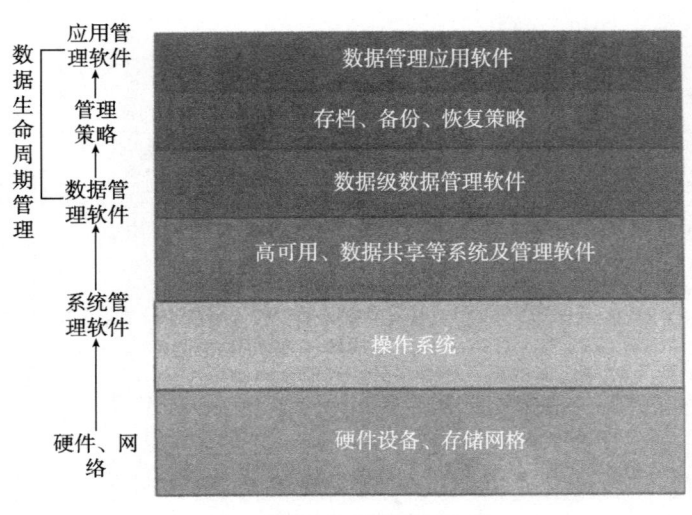

图 3 数据管理层次结构图

立在设备管理商用软件和存档备份策略之上的更高一级的应用，数据管理软件、数据管理策略和数据管理应用软件相互配合，才能最终共同完成对数据生命周期的管理。

4.1.2 数据编目

从数据编目角度看，探月卫星资料可划分为元数据和数据。元数据指经提取的、能够对某具体的数据对象或过程控制进行抽象描述的关系型数据，它包括数据属性信息、业务监控信息、配置管理参数等。数据泛指探月卫星探测资料和产品数据集，由文件系统、二级存储数据、离线数据构成。数据编目采用多级编目结构，元数据作为主编目，下设三级子编目，分别对应数据的文件系统目录、二级存储数据目录和离线数据管理目录。使用数据库构建元数据主编目，文件系统目录构建数据一级子编目，自动磁带库的磁带分区目录构建数据二级子编目，已出库磁带数据目录构建数据三级子编目。客户通过一级数据映射，存储访问文件系统中的在线数据，通过二级数据映射，存储访问磁带库中的近线数据，通过三级数据映射，访问已出库的离线数据。主编目和一、二、三级子编目的编目信息均存放于数据库中。

4.1.3 磁盘阵列资源分配

磁盘阵列(Disk Array)中将存放使用最为频繁的在线存储数据，这些数据在盘阵上可以被方便地调出使用。地面应用系统的磁盘阵列可划分为四类逻辑区：

第一类逻辑区为 OMS、DPS、DMS、SAS 的专用区，存放各种应用程序模块、控制文件、日志文件、配置参数等等。

第二类逻辑区为数据库存储区，存放数据库库体。

第三类逻辑区为业务系统数据区，存放各分系统生产的数据以及数据处理所需的各种中间文件。

第四类逻辑区为业务数据在线存储区，又分为四个区域：

® 临时存储区：存放各分系统送达的需存档的数据及其他交互数据；

® 永久存储区：存放配置参数文件等需永久保存的小文件；

® 滚动存储区：根据需要滚动存放一个时间序列的原始数据和各类产品文件，以及由于探月卫星资料应用领域的不断拓展，各种新的处理方法与算法的研究工作的进行，可能经常调用的大量历史资料；

® 数据下载区：存放根据用户请求从磁带上调回来的数据文件，这些文件保存一段时间，以方便被重复使用。在磁盘阵列上各存储区存储的数据见表2。

表2 各存储区数据存储列表

存储区域	存储数据类型	备注
临时存储区	抽样数据、遥测数据、轨道数据、兴隆图像数据、科学计划、探测计划、业务运行计划、原始数据、0级数据、1级数据、2级数据、3级数据、4级数据、科学研究数据和科普数据等所有要存档的数据	所有需存档的数据都要由生产者将数据放在临时存储区，数据管理分系统得到数据到达通知后从临时存储区复制数据
永久存储区	一些数据量小且经常会使用到的文件存放在永久存储区，方便使用	在存档处理时由临时存储区复制到永久存储区
滚动存储区	遥测数据、轨道数据、兴隆图像数据、科学计划、探测计划、业务运行计划、原始数据、0级数据、1级数据、2级数据、3级数据、4级数据、科学研究数据和科普数据等要在磁带库中近线存储的数据	存档处理时由临时存储区复制到滚动存储区；定时将滚动存储区数据批量存入磁带库；当数据在滚动存储区存储时间超过其在线存储时限时删除该数据
数据下载区	根据用户需要从近线磁带库回调回来的所有数据文件	根据数据清理规则进行定期清理

4.1.4 磁带库资源分配

磁带库(Tape Library)是近线存储介质，存放使用率偏低的数据。根据数据的级别，将磁带库划分为若干存储池，每个存储池可动态定义若干磁带。按照数据级别、生产系统将数据分类存入事先定义的存储池中。磁带库划分的存储池见表3。

表3 各存储区数据存储列表

存储池	存储数据
原始数据存储池	原始数据(卫星下行数据,未经任何处理)
0级数据存储池	0级数据(包括0A和0B的数据)
1级数据存储池	1级数据(格式重整后的数据)
2级数据存储池	2级数据(辐射、几何、光度等校正数据)
科研数据存储池	3级数据、4级数据、科学研究数据和科普数据
测控运行数据存储池	遥测数据、轨道数据等辅助数据集,科学计划、探测计划、业务运行计划等
卫星图像数据存储池	兴隆图像数据等

4.1.5 数据存储策略

每天根据运行管理软件(OMSA)发送的业务运行计划生成存档计划和产品归档计划,然后根据数据预处理软件(DPSA)每轨产品生产计划更新相应的产品归档计划。再根据存档计划每天按时对计划内的各种数据进行分类存档。要存档的数据主要包括:原数据、0级数据、1级数据、2级数据、抽样数据、辅助数据等。这些数据由各分系统产生之后直接送达临时存储区,并向OMSA发送数据到达消息,OMSA确认后向数据管理软件(DMSA)发送数据到达通知,DMSA收到通知后向OMSA发送应答信息,同时将数据从临时存储区转入永久存储区和滚动存储区,并进行质量检验和元数据提取,完成数据的在线归档;每天根据存档计划进行数据的集中近线存档(如图4)。其他存档数据,如3级数据、科学研究数据和科普数据等,由SAS产生之后向DMS发送存档请求,DMS进行手动批量存档。

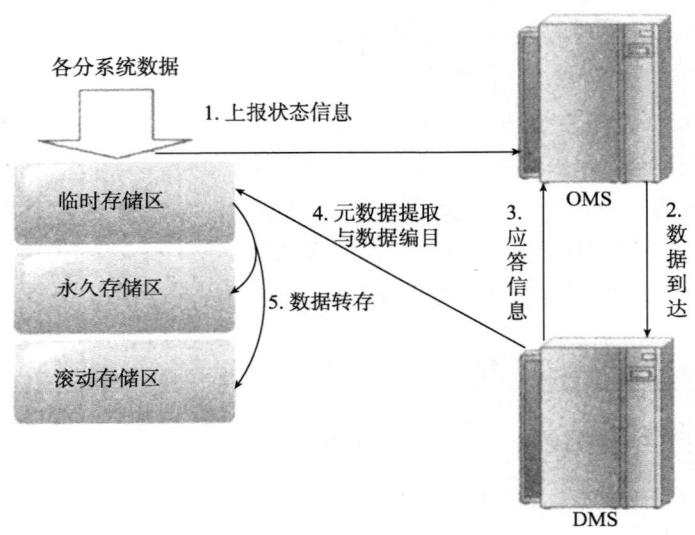

图4 数据存储策略

4.2 数据备份策略

数据的备份工作作为一项长期需要进行的重要工作,备份策略的制定需要仔细考虑。对Oracle数据库中的数据,采用NBU(Netbackup网络存储)系列产品进行专门的备份工作。

NBU是企业级备份管理软件,包括服务器/客户端(Server/Client)软件:

(1) 在需要连接存储设备(如磁带库或光盘库)的服务器上安装服务器(Server)软件。

(2) 在需要提供数据进行备份的机器上安装客户端(Client)软件。

Server和Client之间通过TCP/IP Sockets通讯。

磁盘阵列中的专用区、数据库存储区、永久存储区和滚动存储区中的数据以及 GSDSA 总部的 UNIX 服务器和 PC 服务器的系统文件和应用文件，采用全自动磁带库作为数据的备份介质，根据备份数据分类存储需求，将磁带库中所有磁带驱动器定义成一组或几组资源。按照预案进行指定文件系统、数据库与数据的全备份、增量备份或差分备份。制定备份窗口、时间和备份方式，全备份每周一次，增量备份或差分备份每天一次。由于备份主机多，数据量大，采用分组分期备份方式完成所有数据的全备份。并根据备份的数据量、数据增量、备份窗口等因素，制定可行的备份日程表。

4.3 数据获取方式

当数据全部传回地球以后，嫦娥工程的科学数据将陆续向全世界的用户开放。用户根据权限不同可分为直接用户、间接用户和大众用户。

(1) 直接用户指的是 GSDSA 的操作和管理人员，这些用户通过数据管理软件(DMSA)提供的主控台控制系统运行，并通过 B/S 界面监视和检索系统运行情况；

(2) 间接用户指的是卫星研制单位、科学家群体及 GSDSA 的其他分系统，这些用户自动或直接通过 B/S 界面检索他们所需要的数据，并下载所需数据；

(3) 大众用户是指普通的非专业用户，这类用户可通过登录 GSDSA 的对外网站学习相关的科普知识，了解卫星的运转情况及相关新闻，下载一些对外开放的小量数据。

数据管理系统应用软件的客户端提供了网站形式的数据产品检索服务，该系统基于 J2EE 平台构建，采用三层体系结构，用户仅需使用 Web 浏览器就可以完成数据的检索、浏览、定购和下载。同时还提供给用户不同的数据获取方式，包括在线 FTP 下载和离线介质转储。

5 结束语

对于嫦娥工程科学数据的管理，我们利用存储介质实现数据的在线、近线和离线存储，使用 Oracle 数据库和 NBU 备份管理软件实现数据的高效备份和可靠管理，这些工作为今后的科学实验以及更多月球产品的开发提供了有力保障，也为广大数据用户提供了使用数据的可靠来源。数据的科学组织与存储，也为今后的数据产品格式的多样化和复杂化，提供了可扩展的基础。

参 考 文 献

[1] 凌晓良，LEE Belbin，张洁. 澳大利亚南极科学数据管理综述[J]. 地球科学进展，2007,22(5):532—539.

单频干扰下 BPSK 接收性能恶化分析及应用

郑 磊[1]　苏 彦[1]　朴廷彝[1]　李 斌[2]　李春来[1]

(1. 中国科学院国家天文台，北京 100012；2. 中国科学院上海天文台，上海 200030)

摘 要 为了提高现有单频干扰下 BPSK 接收性能恶化分析方法的准确性和实用性，本文引入干扰信号相位作为随机变量，得出了新的分析方法。该方法首次准确量化了单频干扰下 BPSK 接收性能的恶化程度，普遍适用于 BPSK 调制体制下的卫星通信系统。给出了 MATLAB 数值仿真和现场实测结果。将其应用到探月工程嫦娥一号(CE-1)任务中，对 CE-1 通信系统受日本 SELENE 月球探测器信号干扰进行了准确的定量分析，为地面接收站的抗干扰设计以及中日双方探测器频率协调工作提供了依据。

关键词 通信　干扰　BPSK　嫦娥

0 引言

BPSK 调制体制由于其较高的可靠性，在深空探测卫星通信中被广泛采用[1]。随着各国深空探测活动的日益频繁，干扰下的 BPSK 接收性能恶化分析为工程抗干扰设计提供了理论依据，发挥着越来越重要的作用。针对单频干扰的情况，文献[2]将干扰信号折算成接收频带内的等效噪声分量进行分析，这也是目前工程分析中普遍采用的方法，但是单频干扰信号和信道高斯白噪声的合成分量并不满足高斯分布，因此其分析方法并不严谨，其结果也不准确；文献[3-4]将干扰信号相位作为确定量展开讨论，但是由于干扰源和接收站的距离变化以及空间传输的不确定性等原因，其干扰相位多为随机量，因此这种方法仅适用极特殊的情况，不具一般性；文献[5]虽然将干扰信号相位作为随机变量引入了误码率分析中，却仅对其进行了定性分析，未能将其量化。

因此，为了得到准确量化并更具工程实用性的分析方法，本文将干扰信号相位作为随机变量进行定量分析，研究出改进的单频干扰下 BPSK 接收性能恶化分析方法，并给出 MATLAB 仿真和在探月工程嫦娥一号卫星的数据接收地面站的实测结果。

1 理论分析

图 1 是 BPSK 接收机受单频干扰的一般模型[1]。

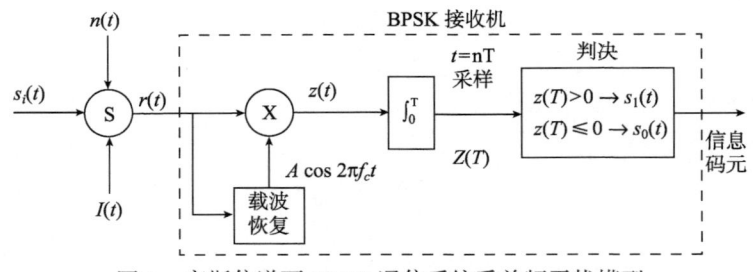

图 1　高斯信道下 BPSK 通信系统受单频干扰模型

图中 $s_i(t)$ 表示幅度为 A，载频为 f_c，码速率为 R_b 的 BPSK 调制信号(码元宽度 $T=1/R_b$)

$$s_i(t) = \begin{cases} s_1(t) = A\cos 2\pi f_c t & "1" \\ s_2(t) = -A\cos 2\pi f_c t & "0" \end{cases} \tag{1}$$

其中 $A = \sqrt{2E_b/T}$，E_b 表示每个码元周期内 $s_i(t)$ 的能量。

$I(t)$ 表示幅度为 A_i，频率为 f_i 的单载波信号(f_i 在接收系统频带内)

$$I(t) = A_i \cos(2\pi f_i t + \varphi) \tag{2}$$

其中 φ 是在 $(0, 2\pi)$ 均匀分布的随机变量[6]。

$n(t)$ 表示信道的加性高斯白噪声，双边功率谱密度为 $N_0/2$ [7]。

上述三个信号同时被送入 BPSK 接收机进行相干解调和检测。当发送 $s_1(t)$ 时，相干解调后输出信号为

$$\begin{aligned} z(t) &= \frac{A^2}{2}(1 + \cos 4\pi f_c t) \\ &+ \frac{AA_i}{2}\begin{pmatrix} \cos[2\pi(f_c+f_i)t+\varphi] + \\ \cos[2\pi(f_c-f_i)t-\varphi] \end{pmatrix} + n(t)A\cos 2\pi f_c t \end{aligned} \tag{3}$$

令 $\Delta f = f_c - f_i$，展开讨论。

1) $\Delta f = 0$ 时

注意到 $f_c \ll 1$ 和 $f_c + f_i \gg 1$，因此在 $t = nT$ 时(n 是自然数)时刻的采样值为

$$Z(T) = \frac{A^2}{2}T + \frac{AA_i}{2}T\cos\varphi + A\int_0^T n(t)\cos 2\pi f_c t\, dt \tag{4}$$

其中第一项为所需的信号分量，第二项为单频信号干扰项，第三项为信道高斯白噪声分量。

令随机变量 $X = \cos\varphi$，

$$Y = \frac{A^2}{2}T + A\int_0^T n(t)\cos 2\pi f_c t\, dt$$

由概率论计算可得，X 的概率分布函数为

$$F(x) = \begin{cases} 0 & x < -1 \\ \frac{1}{\pi}(\pi - \arccos x) & -1 \leqslant x \leqslant 1 \\ 1 & x > 1 \end{cases} \tag{5}$$

Y 满足均值 $u = \frac{A^2}{2}T$，方差 $\sigma_0^2 = E_b N_0/2$ 的正态分布。

因此发送 $s_1(t)$ 时的误码概率 $P(e/s_1)$ 为

$$P(e/s_1) = P(Z < 0) = P\left(\frac{AA_iT}{2}X + Y < 0\right)$$

令 $b = \frac{AA_i}{2}T$，X、Y 的概率密度函数为 $\varphi_X(x)$，$\varphi_Y(y)$。由于 X、Y 独立，则

$$P(e/s_1) = \iint\limits_{bx+y<0} \varphi_X(x)\varphi_Y(y)\,dxdy$$

最后可得

$$P(e/s_1) = Q\left(\frac{b+u}{\sigma_0}\right) + \int_{-b}^{b}\frac{1}{\pi}\left(\pi - \arccos\left(-\frac{y}{b}\right)\right)\cdot\frac{1}{\sigma_0\sqrt{2\pi}}\exp\left(-\frac{1}{2}\left(\frac{y-u}{\sigma_0}\right)^2\right)dy \tag{6}$$

其中

$$Q(x) = \frac{1}{\sqrt{2\pi}} \int_x^\infty \exp\left(-\frac{1}{2}(u)^2\right) du$$

同理可得，当发送 $s_0(t)$ 时的误码概率为

$$P(e/s_0) = Q\left(\frac{b+u}{\sigma_0}\right) + \int_{-b}^{b} \frac{1}{\pi} \arccos\left(-\frac{y}{b}\right) \cdot \frac{1}{\sigma_0 \sqrt{2\pi}} \exp\left(-\frac{1}{2}\left(\frac{y+u}{\sigma_0}\right)^2\right) dy \tag{7}$$

2) $\Delta f \neq 0$ 时

令 $c = \frac{AA_i}{2} \frac{\sin(\pi \Delta f T)}{\pi \Delta f}$，与(1)中推理近似，可得，发送 $s_1(t)$ 时的误码概率为

$$P(e/s_1) = Q\left(\frac{c+u}{\sigma_0}\right) + \int_{-c}^{c} \frac{1}{\pi}\left(\pi - \arccos\left(-\frac{y}{c}\right)\right) \cdot \frac{1}{\sigma_0 \sqrt{2\pi}} \exp\left(-\frac{1}{2}\left(\frac{y-u}{\sigma_0}\right)^2\right) dy \tag{8}$$

发送 $s_0(t)$ 时的误码概率为

$$P(e/s_0) = Q\left(\frac{c+u}{\sigma_0}\right) + \int_{-c}^{c} \frac{1}{\pi} \arccos\left(-\frac{y}{c}\right) \frac{1}{\sigma_0 \sqrt{2\pi}} \cdot \exp\left(-\frac{1}{2}\left(\frac{y+u}{\sigma_0}\right)^2\right) dy \tag{9}$$

由于发送 $s_1(t)$ 和 $s_0(t)$ 的统计概率相等，综合式(6-9)，接收误码率为

$$P_e = Q\left(\frac{d+u}{\sigma_0}\right) + \frac{1}{2}\left(\int_{-d}^{d} \frac{1}{\pi}\left(\pi - \arccos\left(-\frac{y}{d}\right)\right) \cdot \frac{1}{\sigma_0 \sqrt{2\pi}} \exp\left(-\frac{1}{2}\left(\frac{y-u}{\sigma_0}\right)^2\right) dy \right.$$
$$\left. + \int_{-d}^{d} \frac{1}{\pi} \arccos\left(-\frac{y}{d}\right) \cdot \frac{1}{\sigma_0 \sqrt{\pi}} \exp\left(-\frac{1}{2}\left(\frac{y+u}{\sigma_0}\right)^2\right) dy \right) \tag{10}$$

其中 $\Delta f = 0$ 时，$d = \frac{AA_i}{2} T$；$\Delta f \neq 0$ 时，$d = \frac{AA_i}{2} \frac{\sin(\pi \Delta f T)}{\pi \Delta f}$。

性能恶化量 D 表示为：

$$D = \lg(P_e/P_0) \tag{11}$$

其中 $P_0 = \sqrt{2E_b/N_0}$ 表示无干扰时的误码率。结合(10)~(11)式，我们便可以计算出系统受此单频干扰时的性能恶化量。

在 MATLAB 中采用数值运算的方法对上述结果进行分析。其结果见图 2~图 4。图 2 绘出了当信噪比一定时，不同干扰频率和强度对应的性能恶化量。可以看出，单频干扰对 BPSK 接收系统造成的性能恶化量动态范围很大。严重的时候可以使误码率恶化 5 个量级，轻微时几乎没有影响，其大小由干扰信号的频率和强度决定。

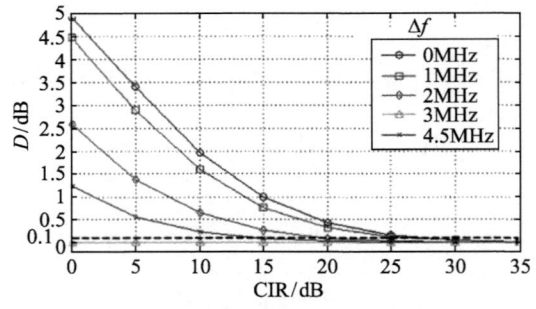

图 2 接收性能恶化量理论计算曲线

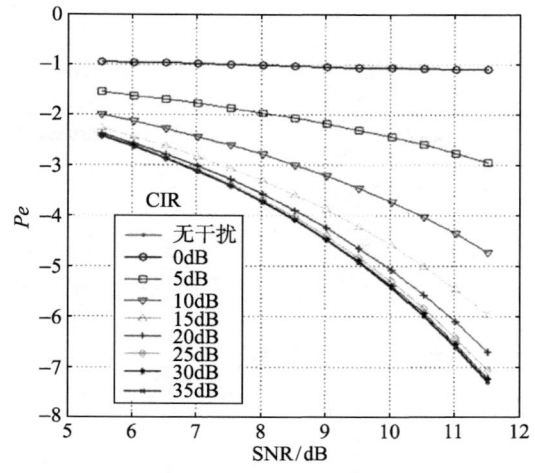

图 3 误码率随载干比变化理论计算曲线

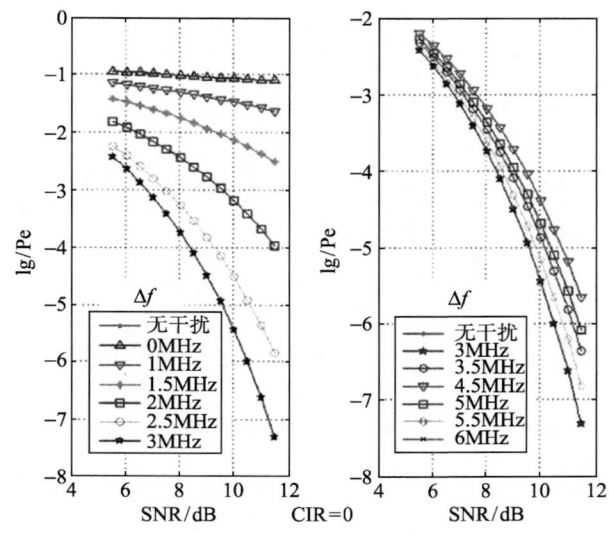

图 4 误码率随干扰频率变化理论计算曲线

为了进一步研究单频干扰对系统接收性能的影响规律，图 3 绘出了当干扰信号频率一定时，不同载干比对应的误码率曲线。可以看出，当载波与干扰信号频率差一定时，随着载干比的升高，即干扰信号强度越小，干扰信号对系统接收性能的恶化影响越小。

图 4 绘出了当干扰信号强度一定时，不同干扰频率对应的误码率变化曲线。可以看出，当载干比一定时：

(1) 干扰信号频率对接收性能的影响，以信号载频为中心按对称分布。
(2) $\Delta f = n/T$ (n 为自然数)时，即干扰频率落在信号两能量瓣之间时，其对信号接收无影响；
(3) 在信号能量瓣内，干扰信号偏离能量瓣中心频率越远，对接收误码率的影响越小。

2 MATLAB 仿真

在 SIMULINK 中进行仿真分析，采用伪随机序列发生器产生周期为 2^{23} 的伪随机序列，利用乘法器进行 BPSK 调制；之后加上单频干扰和高斯白噪声，进行相干解调和积分判决；最后统计误码率[8]。所得结果见图 5、图 6。

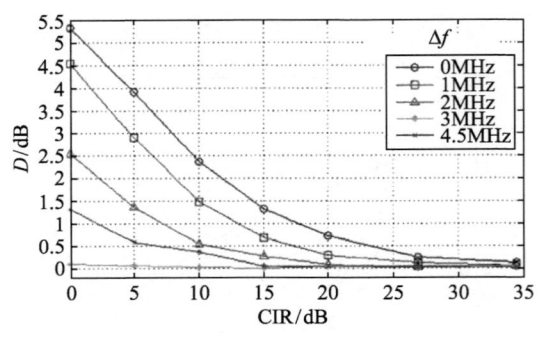

图 5 接收性能恶化量仿真曲线

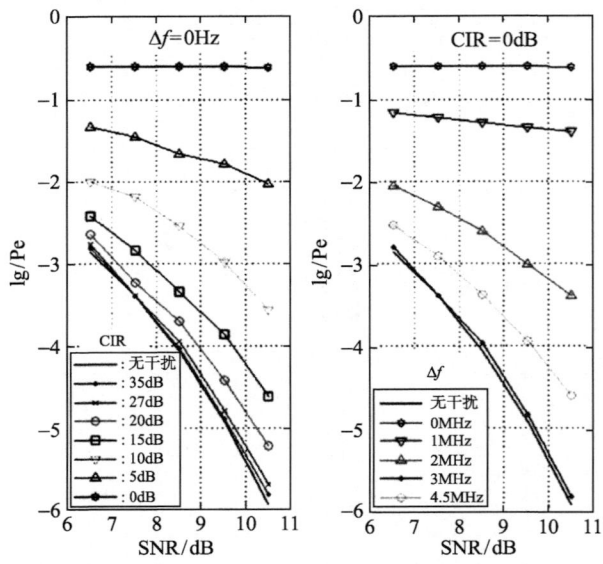

图 6 误码率随载干比和干扰频率变化仿真曲线

与理论分析结果相对应，图 5 绘出了当信噪比一定时，不同干扰频率和强度对应的系统接收性能恶化量。图 6 左图绘出了当干扰频率一定时，不同载干比对应的误码率变化曲线；右图绘出了当干扰信号强度一定时，不同干扰频率对应的误码率变化曲线。

将仿真和理论结果进行比对，当系统信噪比在 6~11dB，载干比在 0~35dB 以及载波与干扰频率差在 0~4.5MHz 范围变化时，两者的各种曲线在变化趋势上吻合。在相同情况下的误码率数值上，仿真值略大于理论值，但是两者偏差较小，误码率始终处在同一量级，且随着载干比和载波与干扰频率差的增大，两者逐渐收敛。考虑仿真计算中各项近似处理造成的误差，说明理论分析与仿真结果吻合。

3 现场实测

探月工程"嫦娥一号(CE-1)"探测器的下行数据采用 BPSK 调制体制，其数据接收工作由国家天文台密云和昆明两个地面接收站完成。本文利用配置在密云站的 BPSK 接收机进行了现场实测，测试系统连接见图 7。

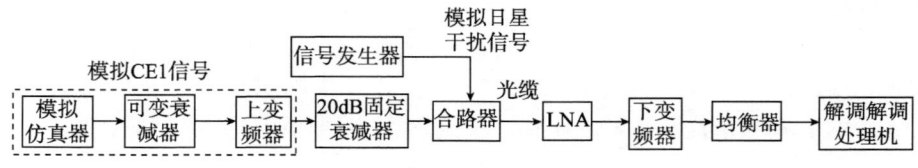

图 7 现场实测系统连接框图

图中模拟仿真器产生 BPSK 调制信号，信号发生器选用 Agilent 4438C，产生单频干扰信号。链路中在合路器前加入 20dB 固定衰减器来降低测试标较设备对接收系统引入的噪声。分别测试在不同信噪比时，接收系统在不同干扰信号强度和频率下的误码率。

为了便于比较，图 8 中将实测结果与理论和仿真结果罗列在一起，左图绘出了当干扰频率一定时，不同载干比对应的误码率变化曲线；右图绘出了当载干比一定时，不同干扰频率对应的误码率变化曲线。

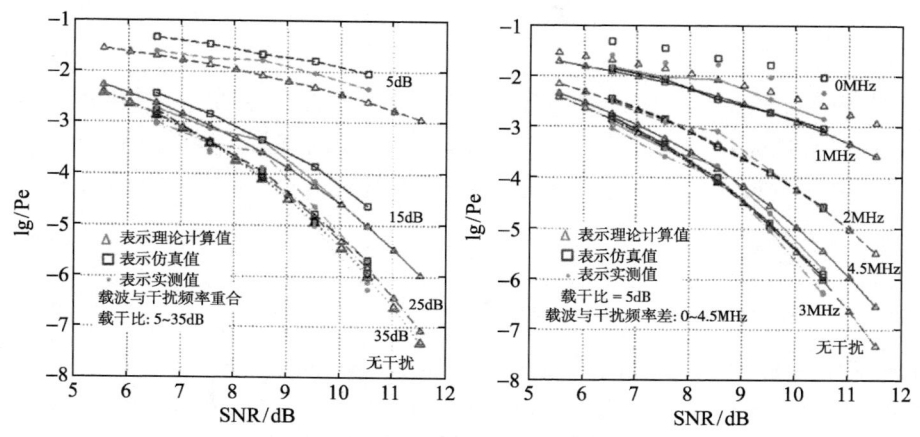

图 8 误码率随载干比和干扰频率变化实测曲线

从图中可以看出，理论、仿真和模拟实测的结果在变化趋势上吻合，且随着载干比和载波与干扰频率差的增大，三者逐渐收敛。而在相同干扰情况下，理论值与实测值偏差不超过一个量级，特别是当载干比大于 20dB 或频率差大于 2MHz 时，三者的值几乎重合在一起。说明该新方法准确的反应了实际工程中，单频干扰对 BPSK 接收机造成的影响。

4 工程应用

2007 年 9 月，日本研制的 SELENE 月球探测器发射成功。它由一颗母星和两颗子星——"中继星"(RSAT)和"月球重力场分布测量卫星"(VARD)组成[9]。同年 10 月底，我国研制的探月工程"嫦娥一号(CE-1)"月球探测器顺利升空。

CE-1 的 TD3 波束采用 BPSK 调制体制，中心频率 2290MHz，带宽 6MHz，其作为探测数据下行的唯一途径，对整个探月工程至关重要。而日星的 VS3 单载波波束频率落在了 TD3 传输频带内，可能会对 CE-1 数据接收系统造成不可接受的性能恶化。表 1 给出了双方探测器信号的主要参数。

表 1 CE-1 和 SELENE 波束主要参数

	日星 VS3 波束	CE-1 TD3 波束
发射天线增益 G	5 dBi	18 dBi
频率范围	2287.238 MHz~2287.388 MHz	中心频率 2290 MHz
发射类别	20K0NON(必要带宽=20kHz，无调制载频(N)，无调制(O)，无信息(N))	6M00GXX 必要带宽 6MHz BPSK 调制
EIRP	21.5 dBm	59 dBm

为了评估日星信号对 CE-1 下行数据接收造成的影响，我们应用该误码性能恶化分析方法。考虑最坏的情况，当系统信噪比处于 10^{-6} 接收门限时，取载干比 37.5dB，频率差 2.7MHz 代入(10-11)式中，求得误码率的恶化量为 0.0001dB，非常微小，不必对系统另进行抗干扰设计。

探测器发射后，根据 CE-1 和日星的轨道预报进行计算，可知在北京时间 2008 年 1 月 7 日 11:50~12:00，日本 VRAD 子卫星与 CE-1 的角距离小于 0.1°，而密云站接收波束宽度为 0.18°，因此在这段时间内，日星 VS3 波束落在了密云站接收波束内。同时，当天 11:28~12:28 密云站正在执行当日的第二轨 CE-1 下行数据接收任务，任务中利用解调设备上的实时误码率测试工具，对接收数据的帧同步码进行了误码率统计，其结果可以反映整个数据接收的误码率情况。实测的误码率为 3×10^{-7} 与往日相近，未见恶化，表明日星 VS3 信号对 CE-1 数据接收造成的影响非常微小。该结果与分析结果吻合，再次验证了新方法的正确性。

5 结论

综上所述，本文引入干扰信号相位作为随机变量进行研究，首次得到准确量化并更具工程应用性的单频干扰下 BPSK 接收性能的恶化分析方法。通过比较理论、仿真和现场实测结果，验证该方法准确可信。将其成功应用在了探月工程日星干扰的分析和评估中，为地面接收站的抗干扰设计以及中日卫星频率协调工作提供了依据。

参 考 文 献

[1] Sklar B. Digital communications fundamentals and applications, second edition[M]. 北京：电子工业出版社, 2002.

[2] 刘剑锋. DS/BPSK 扩频系统单频干扰方法研究及其仿真实现 [J]. 现代电子技术, 2008(01): 110—112 [Liu Jian-feng. Research on the method of single-tone jamming of DS／BPSK spread spectrum system and implementation of simulation [J]. Modern Electronics Technique, 2008(01):110—112.

[3] 眭惠巧. 对 BPSK 相干接收最佳干扰的研究 [J]. 无线电通信技术, 2001, 27(6): 33—34. [Gui Hui-qiao. Research on the best interference to the BPSK coherent reception system [J]. Radio Communications Technology, 2001, 27(6): 33—34.

[4] 方立, 匡镜明, 吕昕. 干扰环境下 BPSK 接收机跟踪—检测联合性能分析 [J]. 电视技术, 2001(03): 33—36. [Fang Li, Kuang Jing-ming, Lü Xin. Union tracking and detection performance of BPSK receiver in the presence of tone interference [J]. Television Technology, 2001(03): 33—36.

[5] 刘志华, 高慧敏, 宋玉凤. 单音信号对 BPSK 的干扰效果分析 [J]. 无线电工程, 2008, 38(8): 18—21. [Liu Zhi-hua, Gao Hui-ming, Song Yu-feng. Analysis of single-tone interfering effects on BPSK system [J]. Radio Engineering, 2008, 38(8): 18—21.

[6] 张邦宁, 魏安全, 郭道省. 通信抗干扰技术[M]. 北京：机械工业出版社, 2006(03).

[7] 曹志刚, 钱亚生. 现代通信原理[M]. 北京：清华大学出版社, 1992.

[8] 李贺冰. Simulink 通信仿真教程[M]. 北京：国防工业出版社, 2006.

[9] 郑永春. "月女神"探测器揭开日本月球探测的新篇章 [J]. 国际太空, 2007(11): 14—16. [Zheng Yong-chun. SELENE detector the epoch of Japanese lunar exploration [J]. International Outer Space, 2007(11): 14—16.

嫦娥一号卫星星地时差校正量计算方法研究

任金彬　刘俊泽　朱光明

(北京航天指挥控制中心，北京　100094)

摘　要　影响月球探测卫星星地时差的因素较多，文章介绍了其计算模型，并对其中的模糊距离进行了详细分析。通过对试验数据的分析，运用最小二乘法，分别建立了集中校时量和均匀校时量计算模型。实际结果表明：运用此两种校时模型和方法可以方便有效地控制嫦娥一号卫星的星地时差，并使其在一段时期内保持在 5ms 范围以内。

关键词　嫦娥一号卫星　长期管理　星地时差　校时

1　引　言

嫦娥一号卫星的 25kHz 时钟信号由测控数传分系统中的统一频率源提供，由数据管理分系统产生卫星基准时间并分配给用户，保证星地时间同步运行[1]。由于空间环境和设备等多种因素的影响，卫星时间相对于地面时间会发生漂移，时间偏差过大时将影响到星上自主执行指令的时间准确性[2]。出于保证部分对月观测设备数据定位精度的需要，在卫星环月飞行期间，星地时钟同步误差要求更高，要始终小于 5ms，这就需要地面对星地时差进行实时监视并分析统计其变化规律，以实施可靠有效的校正。本文在分析确定星地时差直线拟合方程的基础上，主要研究了集中校时量和均匀校时量计算模型，依据这两个校时模型对嫦娥一号卫星实施的两次校时结果表明，校时模型方便可靠，可满足嫦娥一号卫星的星地时差精度要求。

2　星地时差计算模型与分析

2.1　星地时差计算模型

嫦娥一号任务中的时差计算分为两种：实时计算和事后计算。实时计算指每收到一帧星上时间计算一次星地时差；事后计算指对一段时间内的星地时差数据进行统计分析，估计或外推一个特定时刻的星地时差值及时差漂移率。实时计算结果主要用于监视，事后计算结果主要作为校时判断和校时量计算的依据。

星地时差 ΔT 的计算公式如下：

$$\Delta T = T_f - T_s - (\Delta t_1 + \Delta t_2) - \Delta t_3 - t_{EB90} \tag{1}$$

各参数的物理意义及说明如下：

(1) T_f 为地面测站时间，是每帧遥测数据的帧头信息，可实时处理。

(2) T_s 为星上时间，工程遥测参数，是相对于 2004 年 1 月 1 日的累积秒，可实时处理。

(3) $\Delta t_1 + \Delta t_2$ 为星地设备固定时延，由测控系统提供[3]。其中，Δt_1 为星上产生帧同步前沿至发射出去固定时延；Δt_2 为地面接收至解调出帧同步后沿固定时延。不同的遥测码速率和遥测编码方式下星地设备固定时延不同，以正样对接测量数据为准。

本文原载于《航天器工程》，2008，Vol.17，No.6，84~87。

(4) Δt_3 为空间传输时延，等于 R/C。其中，R 表示测控站(船)接收到某帧遥测数据时卫星到测站的斜距；C 为电磁波传输速率，取 299 792 458.0m/s。

(5) t_{EB90} 为帧同步 EB90 传输时延。遥测码速率是 512bit/s 时，t_{EB90}=31.25ms；遥测码速率是 256bit/s 时，t_{EB90}=62.5ms。

2.2 模糊距离计算模型

嫦娥一号卫星作为执行月球探测任务的探测器，进入环月段以前卫星到测站的斜距 R 已超过 300000km，即大约电磁波在 1s 所传播的距离。目前，嫦娥一号卫星所使用的两个主要测站喀什站和青岛站均使用 S 频段统一系统(USB 系统)测距，其测距终端通过比对收、发测距信号提取距离信息[4]。因此，在 USB 系统跟踪目标过程中，当目标斜距刚好超过电磁波 1s 所传播的距离 C 时，USB 系统的测距将出现跳点，需要求解模糊距离[5]。卫星到测站的斜距 R 的完整计算模型与如下：

$$R = \begin{cases} R_W & (R \leqslant 299\ 792\ 458\text{m}) \\ R_W/2 + 299\ 792\ 458 & (R > 299\ 792\ 458\text{m}) \end{cases} \tag{2}$$

式(2)中，R_W 表示测站发送的外测距离。另外，若测站双捕或失锁时的外测数据与其他数据相差较大，导致计算的星地时差严重偏离正常值，应予以剔除。

3 校正量计算模型

3.1 星地时差直线拟合方程

将测量数据以地面测站时间为横坐标，以计算的星地时差为纵坐标描绘在坐标纸上，并把数据点描绘成测量曲线[6]。从星地时差点随时间的分布可以看出，星地时差随时间基本上是线性变化的，可用线性方程表示，即

$$y = a_0 + a_1 x \tag{3}$$

根据最小二乘法[7]，应使各数据点与拟合直线之间的残差平方和最小，根据所有测量数据可得

$$u = \sum [y_i - (a_0 + a_1 x_i)]^2 \tag{4}$$

式(4)中，y_i 表示星地时差；x_i 表示当前时间距初始时间的时间长度，即

$$x_i = t_i - t_0 \tag{5}$$

式(5)中，t_i 为地面测站时间。式(4)分别对 a_0 和 a_1 求偏导数得

$$\frac{\partial u}{\partial a_0} = -2(y_1 - a_0 - a_1 x_1) - 2(y_2 - a_0 - a_1 x_2) - \cdots - 2(y_n - a_0 - a_1 x_n)$$

$$\frac{\partial u}{\partial a_1} = -2x_1(y_1 - a_0 - a_1 x_1) - 2x_2(y_2 - a_0 - a_1 x_2) - \cdots - 2x_n(y_n - a_0 - a_1 x_n)$$

为满足公式(4)，其必要条件是

$$\frac{\partial u}{\partial a_0} = 0, \quad \frac{\partial u}{\partial a_1} = 0$$

整理后得

$$n a_0 + \left(\sum x_i\right) a_1 = \sum y_i \tag{6}$$

$$\left(\sum x_i\right)a_0 + \left(\sum x_i^2\right)a_1 = \sum x_i y_i \tag{7}$$

联立(6)式和(7)式求解得

$$a_0 = \frac{\sum y_i \sum x_i^2 - \sum x_i \sum x_i y_i}{n\sum x_i^2 - \left(\sum x_i\right)^2} \tag{8}$$

$$a_1 = \frac{n\sum x_i y_i - \sum x_i \sum y_i}{n\sum x_i^2 - \left(\sum x_i\right)^2} \tag{9}$$

将式(8)和式(9)代入式(3)，即得到用最小二乘法拟合的线性方程。

3.2 集中校时量计算模型

将式(5)代入式(3)得

$$y_i = a_0 + a_1(t_i - t_0) \tag{10}$$

为确保星上时间精度优于 5ms，在时差大于或等于 5ms 时，需对星上时间实施集中校时。将某一合理范围内的任意时刻 t_i 代入式(10)，即可估计或外推出该时刻的星地时差 ΔT。则该时刻的集中校时量 X 需满足下面的条件

$$|\Delta T - X| < \varepsilon \tag{11}$$

对于嫦娥一号卫星，为留有余量建议 ε 取 4，即将时差校正在 4ms 以内。若统计信息表明时差漂移率在增大，则 $X=\Delta T+4$ 时均匀校时次数最少；若统计信息表明时差漂移率在减小，则 $X=\Delta T-4$ 时均匀校时次数最少。

当 $\Delta T>5$ 时，拨快星上时钟；当 $\Delta T<-5$ 时，拨慢星上时钟。

3.3 均匀校时量计算模型

由直线中斜率的物理意义可知，a_1 即为星地时差漂移率，为方便起见通常将其量纲转化为 ms/d。若 a_1 是依据未进行均匀校时的测量数据拟合的结果，则均匀校时量 Y 为

$$Y = \frac{24\times 3\,600}{a_1} \tag{12}$$

若 a_1 是依据已进行均匀校时的测量数据拟合的结果，则均匀校时量 Y 为

$$Y = \frac{24\times 3\,600}{a_1 + a_p} \tag{13}$$

式(13)中，a_p 为上次校时前的时差漂移率。

当 $Y>0$ 时，每 $|Y|$s 星上时钟拨快 1ms；当 $Y<0$ 时，每 $|Y|$s 星上时钟拨慢 1ms。

4 校时案例

图1是 2007 年 11 月 12 日 13:40:00~15:00:00 期间的时差数据分布图，以及按照上述计算模型设计的时差统计分析软件绘制的拟合曲线图。图中 a_0 表示开始时刻的时差拟合值，a_1 表示这段时间内通过最小二乘法估计得时差漂移率。由于这之前卫星未进行任何校时控制，星上时间每天约慢 66ms。

计划在 2007 年 11 月 12 日 15:54:00 先通过集中校时将星地时差校正到 0，再实施均匀校时。由式(10)可计算出校时时刻的星地时差 ΔT 约为 1 199ms，集中校时量取 1 199ms。将 a_1=66ms/d 代入式(12)，计算得均匀校时量 Y=1 309s。

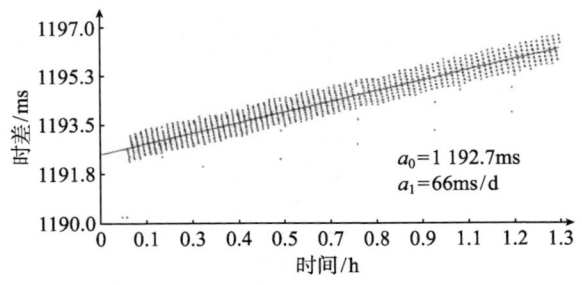

图 1 均匀校时前星地时差分布及拟合直线图

图 2 是 2007 年 11 月 12 日 15:54:00 至 14 日 22:16:20 按此方案校时后的星地时差分布及拟合直线图。从图 2 可以看出，校时时刻的实际星地时差为 0.2ms，校正后的时差漂移率约为 0.5ms/d。若时差漂移率比较稳定，每 10 天作一次集中校时控制即可满足星地时差的精度要求，校正结果比较理想。

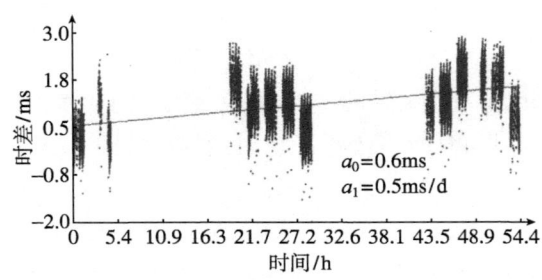

图 2 均匀校时后星地时差分布及拟合直线图

5 关于时差漂移率的变化问题

在嫦娥一号卫星运行的 4 个多月中，通过长时间对卫星时差漂移率的观测，我们发现卫星时差漂移率并不是始终稳定在一个固定范围内，有时会出现较大的变化。表 1 为不同时间段内抽取的各天时差漂移率的统计情况。

表 1 2007 年 11 月-2008 年 3 月时差漂移率统计表

阶 段	日 期	时差漂移率/(ms/d)	阶 段	日 期	时差漂移率/(ms/d)
2007 年 11 月	18 日	64.0	2008 年 2 月	12 日	65.5
	19 日	63.5		13 日	64.1
	20 日	64.2		15 日	64.9
2007 年 12 月	18 日	63.3		23 日	72.9
	19 日	65.3		24 日	73.9
	21 日	65.6		26 日	73.0
2008 年 1 月	17 日	63.9	2008 年 3 月	18 日	74.7
	18 日	63.8		21 日	72.7
	20 日	60.4		23 日	72.5

从表 1 中可以发现卫星时差漂移率在 2008 年 2 月中旬出现了明显的较大变化，2 月中旬之前的漂移率稳定在 65ms/d 左右，但在 2 月 23 日以后漂移率则保持在 73ms/d 附近。针对这一明显变化，我们重点分析这阶段卫星状态发生的变化。由于在 2 月 18 日后随着卫星轨道面与太阳光线夹角继续增大，卫星进入了持续近两个月的连续全阳照飞行阶段，即卫星未进入月影范围，处于持续可见阳光状态。可见，由于 2 月 18 日前后卫星所处光照条件发生了显著变化，卫星内部热环境也随之发生明显变化，环境温度的变

化是影响时钟稳定度的主要因素，因此，可以初步推断这是导致时差漂移率变化的主要原因。

6 结束语

在运用最小二乘法对试验数据分析和处理的基础上，本文分别建立了集中校时量和均匀校时量计算模型。实际结果表明：运用本文的校时模型和方法可以方便有效地控制嫦娥一号卫星的星地时差，并使其在一段时期内保持在 5ms 范围以内，为嫦娥一号卫星顺利开展各种试验项目提供了保障。

参 考 文 献

[1] 刘蕴才，导弹卫星测控系统工程：下册[M]. 北京：国防工业出版社, 1996, 6.
[2] 刘蕴才. 导弹卫星测控系统工程：上册[M]. 北京：国防工业出版社, 1996, 6.
[3] 李邦复. 遥测系统：上册[M]. 北京：宇航出版社, 1987, 10.
[4] 范剑峰. 载人飞船工程概论[M]. 北京：国防工业出版社, 2000, 8.
[5] 刘蕴才. 导弹航天测控总体[M]. 北京：国防工业出版社, 2000, 1.
[6] 李邦复. 遥测系统：下册[M]. 北京：宇航出版社, 1987, 10.
[7] 周生国. 机械工程测试技术[M]. 北京：北京理工大学出版社, 1998, 4.

嫦娥一号卫星热控系统及其特点

向艳超　邵兴国　刘自军　谭沧海

(北京空间飞行器总体设计部，北京　100094)

摘　要　对嫦娥一号卫星热控系统的设计进行了论述，并在此基础上提出了嫦娥一号卫星热控系统的设计特点，可为我国后续深空探测器热控系统的设计提供借鉴。

关键词　嫦娥一号卫星　热控系统　系统设计

1　前言

嫦娥一号卫星是我国第一颗绕地球飞行后经过地-月转移轨道进入绕月使命轨道的航天器，其经历的热环境较近地轨道的航天器所经历的热环境更复杂、恶劣；同时星上数传设备的热耗波动大，载荷设备的温度指标要求高。为了确保嫦娥一号卫星首发成功，热控分系统在设计中采用了以下原则，即：深入分析和统筹考虑星上资源、技术指标要求和外界热环境，重视从系统的角度进行综合设计；充分利用主动控温措施，提高热控系统在轨调节能力和手段；坚持以被动热控措施为主，以主动热控措施为辅的设计原则，充分发挥被动热控措施、尤其是热管的作用；坚持以成熟技术为主，适当采用新技术，提高热控系统的环境适应性。嫦娥一号卫星在轨运行结果表明：热控分系统的设计理念是正确的。

本文对嫦娥一号卫星热控系统设计及其特点进行了描述。

2　卫星总体特点

嫦娥一号卫星的主体结构继承了东方红三号卫星的结构，即中心承力筒加蜂窝板的板式结构，太阳翼采用单自由度对称双翼布局。卫星由长征三号甲运载火箭送入近地点200km，远地点51 000km、倾角31°、周期为16h的超地球同步轨道，之后卫星经历调相轨道、地-月转移轨道，最后进入轨道高度为200km的圆形极月使命轨道。途中卫星经过1次远地点加速、3次近地点加速、1次中途修正、3次近月点制动共计8次轨控。

卫星在一年的寿命期间内，β角(太阳矢量与轨道面的夹角)在 0°～360°范围内变化，为保证太阳翼发电，卫星采用了正飞和侧飞两种飞行姿态。当β角在0°～45°、135°～225°及315°～360°范围内时卫星采用正飞姿态运行；当β角在其他范围内时，卫星采用侧飞姿态。图1a，1b给出了卫星在正飞姿态和侧飞姿态时卫星、太阳和月球之间的相对关系。

另外，由于太阳、地球及月球的相对运动，在2008年2月21日及2008年8月21日，将出现月食现象。由于在月食期间，卫星没有了外热流，同时星上设备依靠蓄电池组供电，电源使用受到限制，因此嫦娥一号卫星热控系统需要配合卫星总体需求，确保星上设备满足月食低温要求。

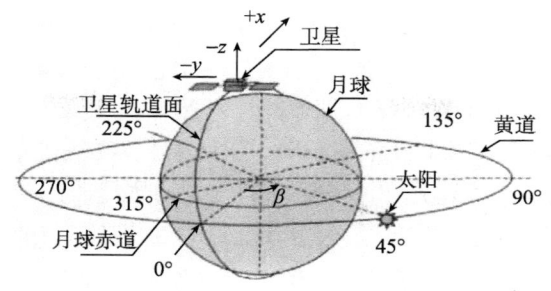

(a) 环月正飞太阳、月球、卫星关系

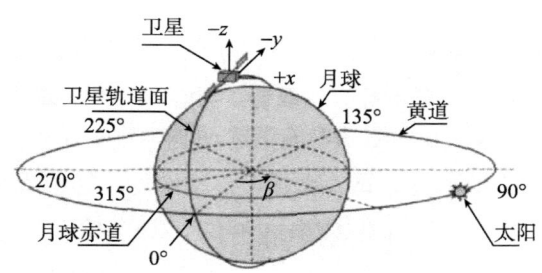

(b) 环月侧飞太阳、月球、卫星关系

图 1 太阳、月球、卫星关系

3 热控分系统的功能需求分析

为了保证嫦娥一号卫星在轨运行期间星上设备的温度环境满足要求，热控系统必须具备以下功能：

(1) 能够适应地、月的热环境，尤其是月球的恶劣热环境。月面光照区最高温度约123℃，阴影区最低温度低于–180℃；光照区的最大月球红外辐射强度是地球红外辐射强度的6倍多；

(2) 能够适应由于星上设备大的工作模式变化所引起的星上热环境的变化，尤其是短期工作的数传设备，解决设备工作时的散热和不工作时的保温之间的突出矛盾；

(3) 保证激光高度计、CCD相机和光谱仪等载荷设备对热环境的严格要求；

(4) 保证单组蓄电池中电池单体间的温差小于3℃，两蓄电池组间的温差小于5℃；

(5) 在月食期间，保证星上设备的温度满足低温工作、存贮要求，尤其保证蓄电池组的工作温度满足–10℃以上。

4 热控系统设计

根据热控系统的功能需求和卫星经历的空间热环境，热控系统在进行设计时基本思路是统筹考虑星上资源(主要是电源)、设备温度要求和热环境等因素，采取适当的措施和手段能够减小月球红外辐射对卫星的影响，充分可控地利用星上资源，发挥被动热控措施尤其是热管的作用，在保证系统可靠性的基础上使整个热控系统的性能达到相对最优。

4.1 OSR散热面及多层布局

在卫星在轨运行的一年里，太阳相对于月球将围绕月球转一周，相应地卫星各个舱板都能受到太阳的辐照，且外热流变化大。因此嫦娥一号卫星没有类似近地轨道卫星上的外热流稳定的散热面。经过详细的分析可知，在嫦娥一号卫星的6个舱面中，+Z面由于受到强的月球红外的辐射，无法设置任何散热面，故全部用多层隔热组件覆盖，以隔离月球红外辐射对卫星热控系统的影响；–Z面仅受太阳辐照，适合于设置OSR涂层散热面，是星上吸收外热流最稳定的散热面；+Y面、-Y面既受到太阳辐照，又受到较强的月球红外辐照，但在同一时刻只有一面能够受到太阳辐照。因此，其外热流在整个寿命周期内变化大；+X面、-X面也是既受到太阳辐照，又受到较强的月球红外辐照，且在一个轨道周期里外热流变化大。

综合考虑6个面的外热流特点及星上设备的布局，嫦娥一号卫星热控系统最终在除+Z面的其他5面上都布置了OSR散热面。基本的散热通道是：服务舱内的设备的热耗通过其+Y、-Y面的OSR散热面排散；载荷舱+Z隔板上设备的热耗通过+X面及载荷舱+Y、-Y面的OSR散热面排散；载荷舱-Z隔板上设备的热耗通过-Z面的OSR散热面排散；-X面仅布置了个别单机的局部OSR散热面。

4.2 热管布局

热管是热控设计中的重要部件，嫦娥一号卫星热控系统中共使用了32根热管，其中包括9根外贴热管、23根预埋热管。在嫦娥一号卫星热管布局时，不但充分继承了近地轨道卫星上热管应用的经验，即通过预埋或外贴等方式，利用热管实现舱板的等温化设计；而且根据卫星外热流的特点及星上设备温度控制需求，利用槽道热管实现了下舱+Y、-Y舱板间的热耦合，扩展了热管网络的应用范围[1]。图2给出嫦娥一号卫星上利用周向热管进行大范围内热耦合的实例，即用3根轴向槽道热管外贴在蓄电池安装板外侧，外贴热管与舱板内的预埋热管形成热管网络，从而形成+Y板、-Y板间强的热耦合。+Y板、-Y板间热耦合保证蓄电池组间的温差要求，同时也降低了光照侧蓄电池组的温度，减少蓄电池组散热面面积，为蓄电池度过月食提供了基本保证。

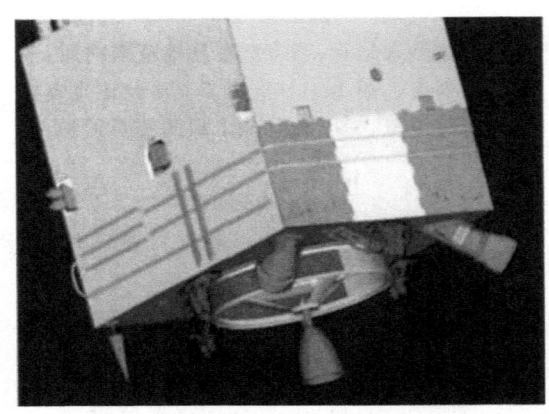

图2 周向热耦合热管布局图

在充分发挥常规槽道热管作用的同时，根据星上外热流变化不甚剧烈的特点及CCD立体相机热控需要，大胆创新，研制并成功在CCD立体相机的热控中应用相变材料热管。相变材料热管构形详见图3所示，在中间圆形腔体内充装液氨，作为常规热管使用；两边两个腔体内充装相变材料，腔体中的肋片起到增强热管与相变材料热耦合的作用。相变材料热管的使用平抑了剧烈变化的外热流给CCD立体相机焦面波动造成影响，改善了CCD立体相机的热环境，为CCD立体相机的正常、可靠工作提供了基础保障。

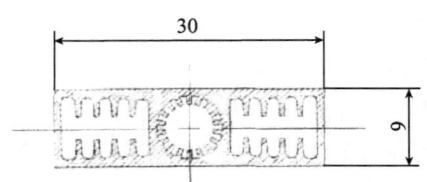 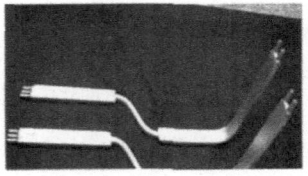

图3 相变材料复合热管

4.3 主动控温设计

嫦娥一号卫星热控系统的主动控温设计继承了东方红四号平台卫星的主动控温方法,并在此基础上进行了改进,以适应测控、有效载荷等短缺工作设备对热控制的要求及能源限制,最终形成了高适应能力的智能主动控温系统。嫦娥一号卫星主动控温系统在系统实现上充分利用星上的数据管理设备、遥测遥控设备,加上热控系统研制的执行部件———加热控制器,形成智能主动控温系统的物理结构,详见图4所示。

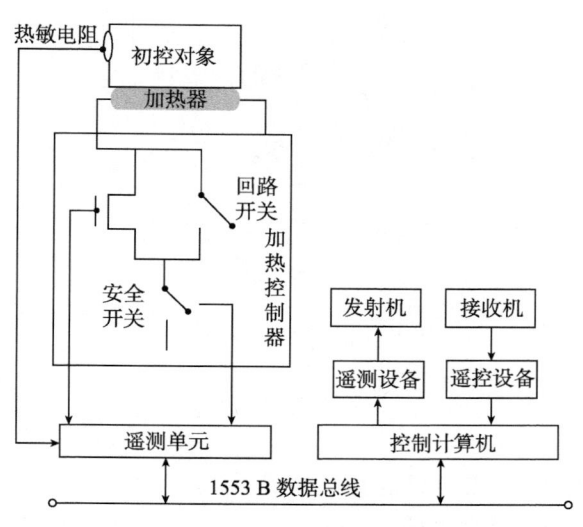

图 4　智能主动控温系统示意图

为了提高热控系统环境适应性和在轨管理的能力,嫦娥一号卫星主动控温系统充分发挥数管计算机大的数据处理能力,利用软件驱动加热控制器实现对加热回路的控制。嫦娥一号卫星主动控温系统具有以下特点:

(1) 利用计算机实现了对加热回路状态的批处理集中设置,以适应卫星在正常飞行、变轨阶段、以及月食阶段星上能源供给的限制,实现可控地利用星上的能源;

(2) 实现了多个热敏电阻的联合控温,提供了被控对象的温度均匀性和控温系统的可靠性;

(3) 在蓄电池组温度控制上实现了跟踪控温功能,为保证蓄电池组间的温差要求提供了保证;

(4) 能够对加热回路的状态设置,如:加热回路开关状态、控温热敏电阻使用、控温阈值、热敏电阻数据有效范围等参数通过遥控进行修改,在轨管理能力及故障应急能力显著增强。

嫦娥一号卫星在轨运行结果表明:主动控温系统的设计运行稳定,性能满足设计要求。

5　热平衡试验验证[2]

为了验证嫦娥一号卫星热控系统设计的正确性和功能满足设计要求,在完成热控系统的设计及实施后,需要进行卫星的整星热平衡试验。近地轨道航天器外热流变化相对缓慢,在地面进行热平衡试验验证时一般采用轨道平均外热流,即用稳定的轨道平均外热流模拟轨道的外热流。而嫦娥一号卫星的外热流变化剧烈,热流变化范围可达 $3W/m^2 \sim 1300W/m^2$ 因此采用轨道平均外热流方法,将不能反映在轨的真实情况。为了充分验证嫦娥一号卫星的热控系统的功能和性能,嫦娥一号卫星的热平衡试验主要进行准瞬态试验、辅以个别稳态试验。

5.1　外热流模拟方法

嫦娥一号卫星热平衡试验在 KM6 真空环境模拟器中进行,卫星试验时的姿态为 $-Y$ 朝上。卫星外热流采用红外灯阵、红外笼和薄膜加热器组合模拟。其中卫星+Y面、-Y面、+X面、-Z面的散热面采用红外灯阵模拟外热流，多层隔热组件表面采用薄膜加热器模拟外热流。-X面的三块散热面受安装空间限制采用了红外笼模拟外热流，多层隔热组件外热流用电加热器模拟。+Z面为对月面，全部包覆了多层隔热组件，多层隔热组件表面采用红外灯模拟外热流。图5给出了卫星热试验时红外灯阵的布局示意图。

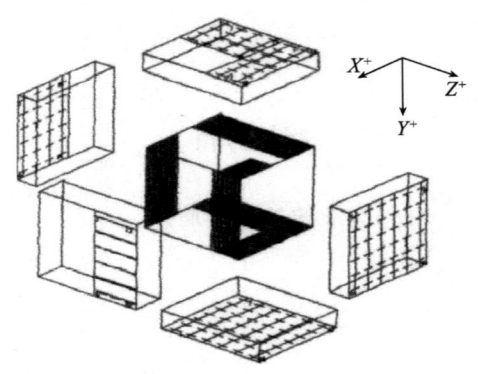

图 5　红外灯阵布局示意图

5.2　热流测量方法

在嫦娥一号卫星热平衡试验中散热面的外热流主要采用红外灯阵进行模拟。由于红外灯阵发出的热辐射具有可见光成分，不同的热控涂层对此具有不同响应特性。为了减小吸收热流的测量误差，提高试验的有效性，在试验中共计采用了5种不同敏感面的热流计，其中以OSR敏感面热流计作为OSR散热面的控制热流计(-X面使用黑漆敏感面热流计作为OSR散热面控制热流计)；采用聚酰亚胺敏感面热流计作为+Z面多层隔热组件吸收热流的控制热流计。同时为了提高绝热型热流计的测量范围和精度，在试验实施时改进了热流计的结构和安装方式，通过在绝热型热流计的背面粘贴一层单面镀铝膜和采用非接触安装方式，减少了热流计背面漏热，提高了热流测量精度。

5.3　瞬态试验验证

嫦娥一号卫星在轨运行时外热流变化剧烈，因此在热平衡试验中主要进行瞬态试验来验证热控系统的功能和性能。在瞬态试验过程中，为了消除由于热流计对外热流变化响应慢而带来的热流测量不准的问题，在进行热平衡试验前，利用模拟星对热流计进行标定，从而得到热流与红外灯阵灯丝电流间的关系，如图6所示。图6中RLJ 28～RLJ 33为热流计的编号。

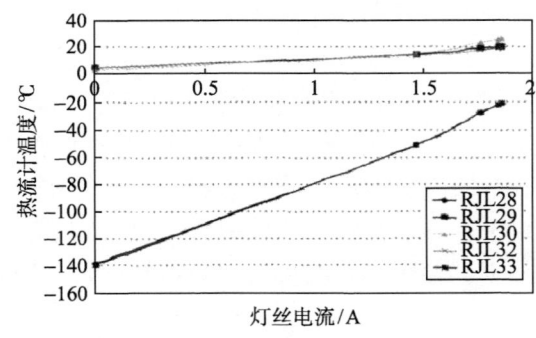

图 6　热流计温度与灯阵电流曲线

在瞬态试验过程中，利用能量积分平衡的方法，分台阶模拟瞬态外热流。外热流按照由热流-灯丝电流曲线获得的电流值直接施加，热流计仅用于监视外热流。嫦娥一号卫星所进行的瞬态试验充分验证了热控系统功能和性能满足设计要求。

6 热控系统设计的特点

嫦娥一号卫星热控系统是在借鉴国外深空探测器热控系统设计方法和充分考虑国内技术现状基础上形成的[3]。较近地轨道航天器热控系统，嫦娥一号卫星的热控系统设计具有显著的特点，初步形成了我国深空探测器热控系统设计的基本思想。具体表现如下：

(1) 嫦娥一号卫星热控系统设计时综合考虑并有效、可控地利用星上能源。热控系统充分利用星上智能主动控温系统，在星上电源充足时充分利用电源，在电源紧张时控制对电源的使用，把有限的电源用来保证卫星关键设备的温度。同时，智能主动控温系统的使用提高了热控系统适应复杂、多变的外界热环境的能力。

(2) 嫦娥一号卫星热控系统设计时充分利用成熟的被动热控措施，尤其是充分利用热管的作用，并适当采用新技术。在热控系统设计时，不但实现了利用槽道热管进行箱式卫星相对两舱间的热耦合，提高了蓄电池组控温品质；而且对常规槽道热管进行创新，研制并成功应用了相变材料热管，为CCD立体相机的热控制提供了坚实的基础[4]。

(3) 嫦娥一号卫星热控系统设计时留有一定的设计预留量。由于嫦娥一号卫星需要经历地-月转移轨道环境、月球环境等未知的热状态的考验；另外，在新的未知环境条件下，热控材料的热控特性和材料的退化特性也是未知的，所有的未知因素增加了探测器热控系统的设计风险。因此需要增加热控系统的设计余量。嫦娥一号卫星热控系统的设计余量主要通过使用智能主动控温措施系统实现，提高热控系统的调节能力。

7 结束语

嫦娥一号卫星经历了调相轨道、地-月转移轨道、环月轨道等复杂、多变的外界热环境，卫星受到的外热流变化非常大。作为卫星的关键基础支持分系统，热控分系统在继承传统设计思想的基础上，更加重视系统设计的理念；在充分利用成熟技术的基础上，勇于创新，大胆使用新技术。在轨飞行结果表明：热控系统指标满足总体要求，与设计预期值吻合得较好。嫦娥一号卫星热控系统设计的理念和技术为我国后续深空探测器的热控设计奠定了坚实的基础。

参 考 文 献

[1] 邵兴国，向艳超，谭沧海. 热管在嫦娥卫星的应用及试验验证[J]. 航天器工程，2008，17 (1)：63—67
[2] 邵兴国，刘自军，谭沧海. 航天器热试验外热流实施和测试探索[J]. 航天器工程，2007，16 (4)：94—98
[3] 向艳超，吴燕，邵兴国. 深空探测器热控系统设计及其特点[J]. 航天器工程，2007，16 (6)：82—86
[4] 耿利寅，邵兴国. 嫦娥卫星光学成像探测系统焦平面组件热设计[C]. 成都：第七届空间热物理会议，2005

嫦娥一号绕月探测器轨道投入过程实时监测判定的原理与技术实现

平劲松[1]　王明远[1,2]　史　弦[1,2]　简念川[1,2]

（1. 中国科学院上海天文台，上海　200030；2. 中国科学院研究生院，北京　100049）

摘　要　深空探测任务中的轨道机动是保证探测器进入预定轨道的关键环节，也是实际测控任务中的重点和难点。在轨道机动过程中，探测器通过点火产生自身加速度，此过程会造成飞行状态不稳定，使得对卫星机动过程的预测和判定变得更加复杂。针对这些问题，结合中国第一个深空探测任务嫦娥一号(CE-1)卫星，对其轨道机动段，特别是近月点入轨制动这一关键弧段，提出了基于视向速度对探测器飞行状态进行实时监测估计的原理和方法，进一步建立了相应的实时监测系统，并应用于实际工程任务，同时对该系统的表现进行评估，为未来深空探测中的类似问题提供了一种有效的解决方法。

关键词　嫦娥一号，轨道机动，近月点制动，实时监测系统

1　引言

在深空探测器的跟踪与控制中，对轨道机动段的实时监测是至关重要的环节，对实时监测系统的稳定性、可靠性和实时性都提出了较高的要求。以往深空任务均采用尽可能少的测量数据对探测器轨道机动过程中的状态进行评估。例如，在NASA的土星探测器Cassini-Huygens进入绕土星轨道的制动过程中，地面站利用接收到的探测器信号频率对其状态进行估计[1]；在NASA的月球探测器Lunar Prospector的绕月轨道投入制动段，跟踪控制部门利用深空探测网接收到的探测器双程视向速度(2-Way Doppler)观测量与事先生成的模型进行比较，来确定探测器的状态[2]；日本的月球探测器KAGUYA于2007年发射，在其绕月轨道投入过程中，日本航天局(JAXA)使用了Usuda站的双程视向速度观测量进行监测，同时美国喷气推进实验室(JPL)使用三程视向速度(3-Way Doppler)观测量与预测值的残差对其进行了监视[3]。

中国第一个月球探测器嫦娥一号(CE-1)于2007年10月24日在西昌卫星发射中心发射。在进入工作轨道之前，对探测器成功实施了8次轨道机动，首先通过一次远地点加速和三次近地点加速，将探测器送入地月转移轨道，再经过一次地月转移轨道中途修正机动使探测器到达月球时的轨道参数符合设计要求，最后经过三次近月点制动，探测器被月球引力场捕获，使探测器成为200km高的绕月极轨卫星。上述轨道机动过程中，第一次近月点制动，即绕月轨道投入(Lunar Orbit Insertion, LOI) 制动尤为关键，其直接决定了整个任务的成败，因此，对该弧段的监测必须做到实时、准确。本文针对任务中的这一关键弧段进行理论和仿真分析，提出了一种方法，利用尽可能简单的模型和尽可能少的外部测量数据对变轨过程中卫星状态进行确认，同时预测变轨后卫星的轨道，并介绍了该方法在实际任务中的应用情况及结果。

本文原载于《空间科学学报》，2011，Vol. 31，No. 3，330~337。

2 理论分析

2.1 CE-1 绕月轨道投入制动过程力学模型

CE-1 在最后一次近地点点火以后进入地月转移轨道,经过微小的中途修正机动后,到达近月点时处于双曲轨道状态,此时探测器运动的基本公式为

$$\ddot{r} = F_0 + F_\varepsilon$$
$$F_0 = -\frac{GM}{r^2}\left(\frac{r}{r}\right), \quad (1)$$
$$F_\varepsilon = \sum_{k=1}^n F_k(r,\dot{r},t,\varepsilon^k)。$$

其中,r 为探测器的位置矢量,r 为 r 的模,$\dot{r}$ 为 r 的一次导数,$\ddot{r}$ 为 r 的二次导数;F_0 为中心天体引力,其中 G 为万有引力常数,M 为中心天体的质量;F_ε 包含了影响探测器运动的其他力学因素,包括中心天体的非球形引力摄动、第三体引力摄动、大气阻力摄动、太阳辐射压摄动等. 由于月球的影响球半径为 6.61×10^4 km,CE-1 的月心距离超过该半径时,取地球为中心引力体,考虑月球和太阳的第三体摄动;CE-1 进入月球影响球半径后,取月球为中心引力体,考虑地球和太阳的第三体摄动。

通常情况下,探测器推进系统通过两种方式对探测器轨道实施机动,即脉冲式和持续式。脉冲式机动是指探测器单次、大推力的机动,探测器的瞬时速度变化为 Δv;持续式机动中,通常轨道机动持续一段时间,在这段时间内探测器的质量、运动方向都会发生改变。实际上任何探测器都有质量,不可能实现速度的瞬时变化。脉冲式机动模式简单明了,需要的参数远少于持续式机动,适合用于实时监测。本文所介绍的实时监测系统在实际任务执行时是利用脉冲机动模式对卫星状态进行分析的。

2.2 测量数据的选择及观测方程

在 CE-1 任务中,测控系统由统一的 S 波段(Unified S-Band, USB)系统和甚长基线干涉测量(Very Long Baseline Interferometry, VLBI) 系统组成。前者提供探测器的距离观测量和双程视向速度观测量,并接收探测器的遥测信号,发送遥控指令;后者提供探测器信号到达两个 VLBI 测站的时间延迟和延迟率,从而得到探测器在天球坐标系中的角位置信息。在这些数据中,由 USB 系统接收的探测器遥测数据,包括加速度计等数据,是用于确定轨道机动过程中卫星状态的最直接依据但由于在轨道机动过程中,卫星将经历复杂的状态调整,这些数据的稳定性和准确性极有可能受到影响。因此,需要利用外部测量数据,即测距、测速或测角数据,对卫星变轨过程中的状态进行实时估计。

根据设计轨道,LOI 制动过程中探测器位于距离地球约 38×10^4 km 的近月点处,其轨道面相对地球测站处于正视(face-on) 状态,即探测器与测站的连线几乎垂直于轨道平面,此时测距数据对卫星状态的变化不敏感。而 VLBI 系统涉及 4 个天线数据的采集、传输和处理,因此从接收信号到产生最终数据有几分钟的时间延迟,不适合用于监测对实时性要求较高的 LOI 制动过程。因此,选取探测器的双程视向速度观测量作为 LOI 制动段的基础监测数据,该数据由 USB 测站提供,数据采样频率为 1Hz,通过网络 TCP/IP 协议实时传输至前述实时监测系统所在的计算机平台。

双程视向速度观测量与探测器的位置状态间存在如下关系[4]:

$$\dot{\rho}(t) = \dot{r}(t) + \omega r_s \cos\delta_0 \sin(\omega t + \phi + \lambda + \alpha_0)。 \quad (2)$$

其中,$\dot{\rho}(t)$ 为探测器与测站间的距离差分即双程视向速度观测量,$\dot{r}(t)$ 为探测器与地心间的距离差分,ω 为地球的自转角速度,r_s 为测站到地球自转轴的距离,δ_0 为卫星赤纬,t 为从初始历元开始计算的时间,ϕ 为初始历元相位,λ 为测站经度,α_0 为探测器赤经。该公式从理论上确定了双程视向速度数据与探

测器的位置状态存在一一对应的关系,因此可使用该数据对探测器制动过程中及制动后的状态进行定性分析。

3 监测系统分析与设计

3.1 测站双程视向速度观测量与探测器总速度增量 Δv 的关系

模拟仿真部分利用了美国 AGI 公司的 Satellite Tool Kit (STK)软件包。在实际 LOI 制动过程中,跟踪测量数据是由位于青岛的 USB 测站提供的。图 1 给出了在脉冲制动和持续制动两种情况下探测器相对于测站的双程视向速度模拟观测量。图 1 中,实线表示脉冲制动对应的结果,制动时刻为卫星到达近月点时刻;虚线表示持续制动对应的结果,开始时刻为探测器到达近月点前 10 min,并保持持续制动中的总速度增量 Δv 与对应脉冲制动相同。除初始轨道和机动时间外,模拟持续制动还需输入星上发动机参数、卫星质量和卫星燃料质量。

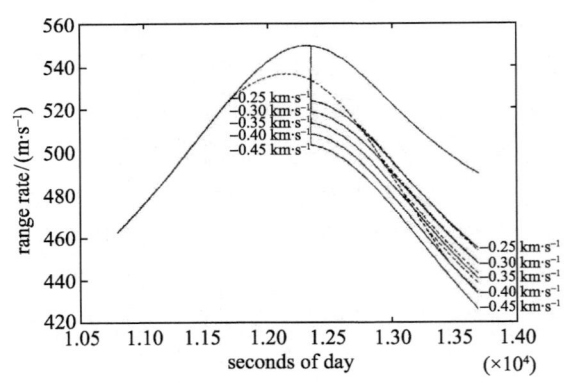

图 1 LOI 制动过程中探测器相对于测站的视向速度变化
虚线对应持续制动,实线对应脉冲制动

由图 1 可以看出,脉冲制动导致双程视向速度观测量在制动时刻发生跃变,而持续制动使观测量平缓变化,Δv 在−0.35km/s 附近时,脉冲模式与持续模式制动方法得到的双程视向速度观测量相对接近。另外,无论是脉冲制动还是持续制动,不同 Δv 导致的双程视向速度观测量在制动后(脉冲制动)或制动过程中(持续制动)都处于近似平行状态,这使得可以利用线性插值的方法对实际 Δv 进行估计,模型 Δv 的步长越小,线性插值得到的结果就越精确。

3.2 探测器总速度增量与制动后轨道的关系

LOI 制动过程主要由星上主发动机完成。在进行姿态调整后,主发动机点火,产生与速度方向相反的加速度,从而达到减速制动的目的。以脉冲制动为例,图 2 显示了不同 Δv 对应的制动后轨道。细实线表示被月球捕获的制动后轨道,虚线表示由于|Δv|过小导致飞掠月球,以及由于|Δv|过大导致探测器落月的制动后轨道。

若将制动后轨道倾角约束在 90°±1°范围内,将轨道近月点高度约束在 200km±15km 范围内,可得 Δv 上限为 $\Delta v_{max} = -0.2525$km/s;将轨道远月点高度约束在 200km±15km 范围内(即探测器直接进入绕月圆轨道),可得 Δv 下限为 $\Delta v_{min} = -0.8203$km/s。根据设计轨道,CE-1 在 LOI 制动后应进入周期约为 12h 的绕月轨道,对应的 Δv 为 $\Delta v_{des} = -0.3456$km/s。

进一步考虑 Δv 与制动后轨道的定量关系,对脉冲制动和持续制动两种模式进行分析的结果如图 3 所示,其中轨道半长轴及轨道周期与 Δv 间用 6 次多项式拟合,轨道倾角与 Δv 间用 2 次多项式拟合。由图 3

可以发现,当|Δv|大于约 0.35km/s 后,利用两种模式得到的 Δv 与制动后轨道关系近乎重合。由于 CE-1 的设计 Δv 约为–0.346km/s,因此可利用较为简单的脉冲制动作为实时监测系统的估计模式。

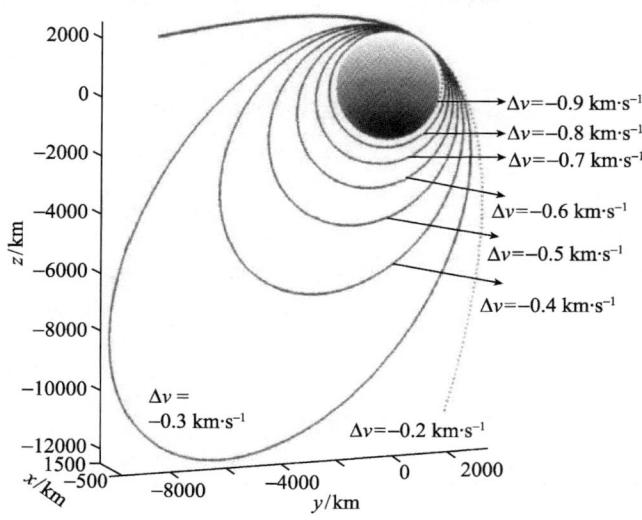

图 2　月固坐标系下 LOI 制动中 Δv 与制动后轨道对应关系(粗实线对应制动前嫦娥一号的地月转移轨道,虚线描述制动导致速度变化太大或太小时引起探测器坠毁或飞掠月球的状态,细实线表示介于两者之间的状态)

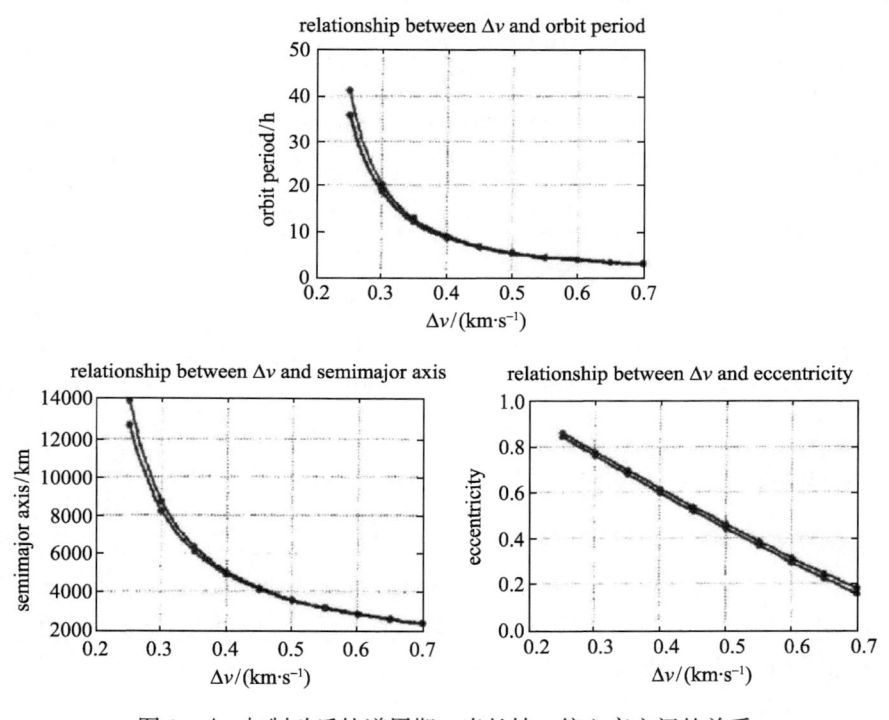

图 3　Δv 与制动后轨道周期、半长轴、偏心率之间的关系

3.3　LOI 制动过程实时监测系统设计

基于以上分析,利用双程视向速度观测数据,设计建立针对探测器 LOI 制动过程的实时监测系统。该系统的设计思想如下。

(1) 制动开始前,根据制动前的定轨结果和设计制动参数,针对不同 Δv 的情况,对制动过程进行模拟,产生一组相应双程视向速度观测量的模型值,同时产生不同 Δv 与制动后轨道的轨道参数间的多项式拟合结果。

(2) 制动过程中，将双程视向速度实际测量值与上一步中产生的一组模型值进行比较，通过插值等方法得到当前探测器 Δv 的估计，并根据得到的 Δv 估计值对制动后探测器投入轨道的过程进行预测。具体输入/输出参数列于表 1。

表 1 实时监测系统输入/输出参数

输入参数	制动前最后一次定轨结果 $(t, \boldsymbol{r}, \dot{\boldsymbol{r}})$ 设计制动起始时刻 t_{db} 设计制动结束时刻 t_{de} 设计 Δv, Δv_d 测量站在地固坐标系中的位置 (x_s, y_s, z_s) 实时 2-Way Doppler 测量数据的 $v_r(t)$
输出参数	探测器 Δv 估计值 Δv_p 投入轨道半长袖预测值 a_p 投入轨道偏心率预测值 e_p 投入轨道倾角预测值(月固坐标系) i_p 投入轨道周期预测值 T_p

综合 3.1 和 3.2 的结果，可以得到测站双程视向速度观测量与探测器总速度增量 Δv 之间的关系，即假设某时刻的测站双程视向速度实测值为 $v_r(t)$，根据一组不同的 Δv 得到的测站双程视向速度模型值为 $v'_{ri}(t)$，则根据该时刻的测量结果利用线性插值，计算预测 Δv 为

$$\Delta v_p(t) = \frac{v_r(t) - v'_{rb}(t)}{v'_{ra}(t) - v'_{rb}(t)}(\Delta v_a - \Delta v_b) + \Delta v_b 。 \tag{3}$$

其中，$v'_{ra}(t)$ 和 $v'_{rb}(t)$ 为与 $v_r(t)$ 最接近的两个模型值，Δv_a 和 Δv_b 为其对应的探测器速度增量 Δv。根据模拟结果，对 Δv、制动后轨道的半长轴、偏心率、倾角、轨道周期进行多项式拟合得到的系数分别为 P_a, P_e, P_i 和 P_T，则制动后的轨道参数可由下式预测得到：

$$a_p = \sum_{n=0}^{6} P_a^n \Delta v^n, \quad e_p = \sum_{n=0}^{2} P_e^n \Delta v^n, \\ i_p = \sum_{n=0}^{2} P_i^n \Delta v^n, \quad T_p = \sum_{n=0}^{6} P_T^n \Delta v^n 。 \tag{4}$$

4 实时任务应用结果及分析

CE-1 卫星于 2007 年 11 月 5 日到达近月点，制动前最后一次定轨历元为 11 月 3 日 16:00UTC，此时卫星距离月心 1.24×10^5 km，超过月球影响球半径，因此在探测器的月心距缩小到 6.61×10^4 km 之前以地球为引力中心，地球重力场取 WGS84 模型[5]的前 8 阶次，太阳、月球作为点质量产生第三体摄动；探测器的月心距缩小到 6.61×10^4 km 之后以月球为引力中心，月球重力场取 LP165P 模型[5-6]的前 50 阶次，太阳、地球作为点质量产生第三体摄动。

实时监测系统的工作流程如图 4 所示，该系统由两部分组成，即实时数据传输部分和实时数据处理部分。

在实时数据传输部分中，喀什、青岛两 USB 测站的双程视向速度测量数据通过北京航空航天指挥控制中心实时传输至上海 VLBI 数据处理中心，传输通信协议为高级数据链路控制协议(High Level Data Link Control protocol, HDLC)。其是面向比特的同步通信协议，为全双工点对点操作提供完整的数据透明度，且支持对等链路。这种特性保证了测站与数据处理中心的数据传输能够流畅进行。VLBI 数据处理中心的数据传输 I/O 配置项接收到数据后，在解码保存的同时，通过 TCP/IP 协议将含有 USB 数据的数据帧转发

至实时数据处理部分，并解码保存为秒文件进行后续使用。上述数据传输过程均为实时传输，时间消耗在 ms 量级，符合实时要求。

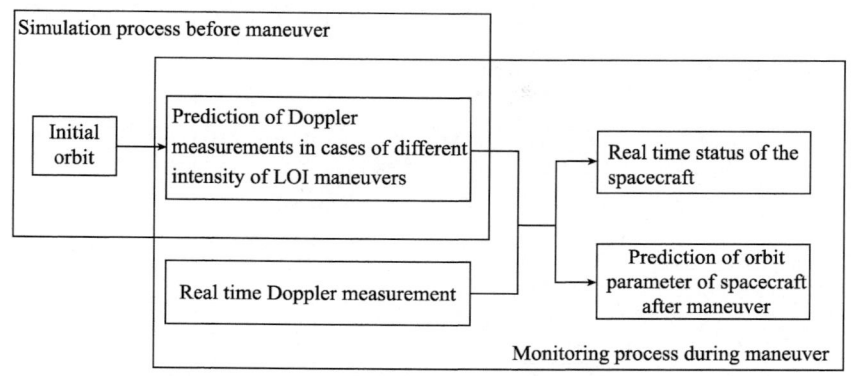

图 4 实时监测系统工作流程

4.1 制动开始和结束时刻的判定

因为双程视向速度观测量对卫星速度非常敏感，所以该观测量可作为实时判断制动点火是否正常开始和结束的依据。对于每一时刻 t_i，利用当前秒和前一秒的双程视向速度观测数据进行一次多项式拟合，得到由两点所构成直线的斜率 $k(t_i)$，同时对 t_{i-50} 到 t_{i-1} 的斜率序列求平均值 $M(t_i)$，并得到该序列的均方差 $R(t_i)$。若

$$|k(t_i) - M(t_i)| > R(t_i) \tag{5}$$

则认为机动开始；在机动开始后某一时刻的斜率若再出现满足式(5)的变化，则认为机动结束。图 5 显示了对 CE-1 实际 LOI 制动前后 1h 内对双程视向速度测量数据进行斜率分析的情况，从图 5 可以看出，用斜率判断制动开始和结束时刻是较为有效的方法。利用该方法判断的制动开始时刻为当天的 03:15:12.5 UTC，制动结束时刻为 03:36:27.5 UTC，制动过程共 1275s。

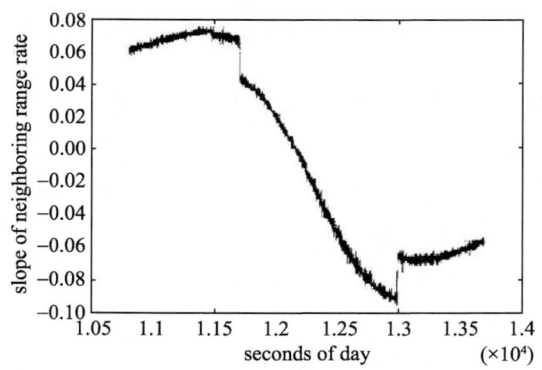

图 5 LOI 制动前后 1h 内双程视向速度数据的斜率变化

4.2 探测器总速度增量 Δv 及制动后轨道参数的估计

Δv 的预测包括定性和定量两部分工作，前者是指对速度增量进行临界判定，即判断点火量是否能使探测器进入安全的轨道范围，既不会飞掠月球，也不会落月(如 3.2 节所述)；后者是指在点火过程中对 Δv 的大小进行判断。

图 6(a)显示了采用脉冲制动模型的情况下，双程视向速度模型观测量与实测数据的比较情况。其中，

虚线表示无制动情况下的双程视向速度模型值序列；点线表示根据3.2中约束条件得到的对应于Δv临界值的双程视向速度模型值序列；细实线表示在临界范围内对Δv进行细分后得到的一组双程视向速度模型值序列，范围为–0.30~–0.50km/s，步长为25m/s；粗实线表示实测双程视向速度序列。

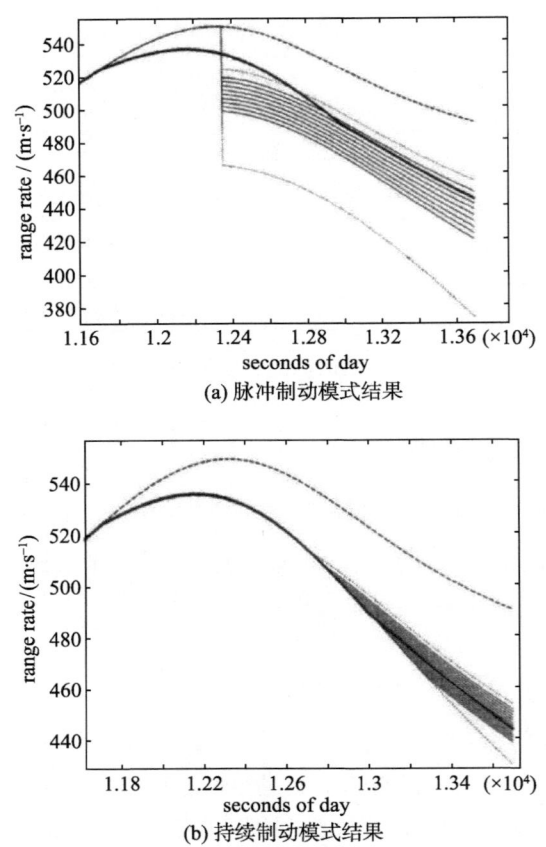

图 6　LOI 制动过程模型观测量与实际观测量的比较

虚线表示无制动情况下的模型值序列，点线表示临界情况下的模型值序列；细实线表示在临界范围的一组模型值，粗实线表示实测双程视向速度序列

如图6(b)中粗实线所示的实测双程视向速度值以每秒1个数据点的速度增加。在判定制动开始后，若在制动时间到达临界值上限所对应时刻之前判断制动结束，则认为点火意外中止，将及时上报飞行指挥中心；当实测数据显示制动正常进入临界值上下限，则开始利用式(3)所示的插值方法对制动量进行估计；得到Δv后，再根据式(4)估计制动后的轨道参数。

4.3　结果及分析

为了比对两种制动模式，事后利用持续制动模式对LOI制动过程重新进行了分析，分析方法与脉冲制动模式相同，结果如图6(b)所示，利用脉冲和持续两种模式在制动结束时刻估计得到的Δv，以及对应的制动后轨道参数列于表2。

表 2　Δv 及制动后轨道参数估计结果

制动过程模式	Δv/(m·s⁻¹)	制动后轨道参数		
		轨道半长径/km	轨道偏心率	轨道倾角/(°)
脉冲制动	320.65	7125.4	0.72651	90.223
持续制动	352.93	6194.4	0.69063	90.351

制动结束后，CE-1 测轨系统利用制动后半小时的观测数据进行短弧定轨，得到制动结束时刻的轨道参数；另外，在制动过程中，VLBI 测轨分系统测角计算配置项还利用 VLBI 时延、时延率，加上距离观测量和双程视向速度观测量，得到了探测器的准实时单点定位结果。将事后短弧定轨所得轨道参数作为标准，与脉冲制动、连续制动、单点定位这三种估计模式得到的制动后轨道参数进行比较，各轨道根数之差的百分比结果列于表 3。

表 3　各种估计方法结果与短弧定轨结果之差　　(%)

	脉冲制动	持续制动	单点定位结果
轨道半长径	15.72	0.5978	−0.5064
轨道偏心率	6.335	1.083	−0.2379
轨道倾角	−0.0445	0.0976	0.0245
轨道升交点经度	0.0050	−0.027	−0.0025
近点幅角	−0.0420	−1.94	−0.220

从表 1 可以看出，单点定位结果与短弧定轨结果最为接近，但单点定位结果所需观测量较多，且有一定时间延迟，其稳定性与实时性不能构成理想的监测系统。对于确定轨道大小和形状的半长轴和偏心率两个根数，持续制动模式和单点定位的结果比较好，对制动后轨道的半长径和偏心率的估计的误差(相对于短弧定轨结果)都在 1%左右；但是持续制动模式对轨道的倾角、升交点经度和近点幅角的估计偏差较大，这是由于持续制动时间较长，无法准确还原探测器轨迹在制动时间段内的变化，导致轨道的空间定向根数估计值产生偏差。

脉冲制动模式得到的结果虽然与其他结果都存在较大的差异，但其结果已经能够满足实时监测的要求。

5　结论与讨论

利用双程视向速度测量数据对中国第一颗月球卫星 CE-1 的 LOI 制动过程进行实时监测并对制动后轨道参数进行估计；在脉冲制动和持续制动两种模式下对该过程进行模拟，并阐述了在实际监测过程中使用脉冲制动模式进行模拟的理由。将实际应用结果与单点定位及短弧定轨方法得到的结果进行比较发现，虽然使用持续制动对观测量进行估计的结果与短弧定轨结果及单点定位结果更为接近，但是利用脉冲制动对观测量进行估计的误差也都在允许范围之内。另外由于利用脉冲制动模式进行估计所需参数少，在卫星状态较不稳定时也可使用，因此更适用于实时监测。

在未来的深空探测任务中，对类似关键弧段的监测也可使用这种基于双程视向速度测量数据的监测与判定技术，特别是针对小天体探测器以及平动点探测器。

参 考 文 献

[1] Asmar S W, Johnston D V, Maize E, and Mitchell R T. Critical monitoring of the Cassini Saturn orbit insertion maneuver [R]//8th International Conference on Space Operations, Montreal, Canada, 2004
[2] Lozier D, Galal K, Folta D, Beckman M. Lunar Prospector mission design and trajectory support [R]. *Spacef. Dyn.*, 1998, **100**(1):297—312
[3] Haw R J, Mottinger N A, Graat E J, Jefferson D C, Park R, Menom P, Higa E. Kaguya orbit determination from JPL [R]// AIAA/AAS Astrodynamics Specialist Conference and Exhibit, Honolulu Hawaii, 2008
[4] Breidenthal J C, Komarek T A. Deep Space Telecommunication Systems Engineering [M], New York: Plenum Press, 1983. 123—178
[5] National Imagery and Mapping Agency. Department of Defense World Geodetic System 1984, Its Definition and Relationships with Local Geodetic Systems [R]. Reston, 2000
[6] Konopliv A S, Asmar S, Carranza E, Sjogren W L, Yuan D N. Recent gravity models as a result of the lunar prospector mission [J]. *Icarus* 2001, **150**(1):1—18

月球重力场对"嫦娥一号"近月轨道的影响

曹建峰[1,2,3]　黄　勇[1]　胡小工[1]　陈　明[3]

(1. 中科院上海天文台，上海　200030；2. 中国科学院研究生院，北京　100049；
3. 北京航天飞行控制中心，北京　100094)

摘　要　2008 年 12 月 6 日"嫦娥一号"卫星开始了为期半个月的变轨试验，卫星距离月球表面最近处约为 15 km，这在国内尚属首次。试验期间，国内 USB 和 VLBI 测控网进行了跟踪测量，获取了卫星不同飞行高度的测轨资料。通过对变轨试验期间的 USB 和 VLBI 测量数据的定轨计算，分析了月球重力场误差对于绕月低轨卫星的影响，计算表明，尽管目前的月球重力场模型高阶项由于没有月球背面的测量数据而不准确，但对绕月低轨卫星的定轨精度提高仍然有重要帮助。分析了 VLBI 数据对绕月低轨卫星定轨的贡献，比较了 USB 数据单独定轨以及 USB 和 VLBI 联合定轨两种情况，结果表明 VLBI 数据的加入可有效提高定轨精度。该工作对于我国后续月球探测工程具有一定的借鉴意义。

关键词　月球重力场　"嫦娥一号"　精密定轨　USB　VLBI

0　引言

2007 年 10 月 24 日 18 时 30 分，我国首颗绕月探测卫星"嫦娥一号"(CE-1)在西昌卫星发射中心顺利升空，按照设计轨道历经调相 24 小时轨道段，调相 48 小时轨道段，地月转移轨道段，月球捕获轨道段，最终进入月面高度约 200 km 的环月飞行轨道段[1]。"嫦娥一号"卫星的主要科学目标是：获取月球表面三维影像；分析月球表面有用元素及物质类型的含量和分布；探测月壤厚度；探测地月空间环境。CE-1 顺利进入月球轨道并传回月球三维影像，标志着我国首次探月工程获得圆满成功，是我国由近地空间探测到深空探测的重要转折。

在我国后续月球探测任务中，将发射嫦娥二号卫星并展开月球探测二期工程。后续月球探测卫星的轨道高度将比目前"嫦娥一号"的 200 km 低，可能会在 100 km 左右。对于测控系统，由此将带来月球重力场误差对轨道计算的影响问题。

月球重力场一直都是月球探测的焦点之一，其研究始于 20 世纪 60 年代，目前所公布的月球重力场模型几乎都综合了历史上不同飞行器如 Apollo, Clementine, Luna, Lunar Prospector(LP)的跟踪数据。美国喷气试验中心(Jet Propulsion Laboratory，JPL)的 JGL165p 月球重力场模型阶次为 165，空间分辨率约为 30 km[2]。但是所有这些月球重力场模型都存在着相同的问题，由于月球的公转和自转周期相同，月球始终只有半面朝向地球，导致地面无法观测到月球背面，月球背面的重力异常只能通过拟合推估间接求出，极大影响了月球背面的月球重力场精度[2-5]。

日本于 2007 年 9 月实施的 SELENE 探月计划，主要目标是提高月球重力场的精度，特别是背面重力场的精度。为此，他们特别设计了一个环月中继卫星，当环月主飞行器飞到月球背面时，将与环月中继卫星保持多普勒测量链路的通畅，其对应的观测量也从双程多普勒(地面天线→环月主飞行器→地面天线)变成了四程多普勒(地面天线→环月中继卫星→环月主飞行器→环月中继卫星→地面天线)[6]。目前，日本的研究人员利用 SELENE 测量数据以及历史数据解算的最新月球重力场模型 SGM90d[3]，但是此模型尚未公开发布，目前我们轨道计算采用的模型仍为 JGL165p。

CE-1 的任务轨道为 200 km 高度的极轨近圆轨道。在完成各项科研试验任务后，2008 年 12 月 6 日至 20 日期间 CE-1 进行了为期半个月的变轨试验。试验期间，卫星经历了由 200 km×200 km 轨道先后变轨成为 200 km×100 km，100 km×100 km，100 km×15 km 轨道，最后再抬升为 100 km×100 km 的圆轨道，试验任务顺利圆满完成，获取了大量有效测量数据。此次试验任务第一次将卫星控制到距离月面仅 15 km，并成功将轨道抬升为 100 km×100 km 轨道，为探月工程后续落、回等阶段科研任务工作的顺利开展提供了有利条件。

甚长基线干涉测量(Very Long Baseline Interferometry，VLBI) 技术诞生于 20 世纪 60 年代，具有高精度、高分辨率以及多用途等特点，已经在天文地球动力学和大地测量中得到了富有成果的应用[7]。VLBI 观测技术是对无线电测距测速技术的有益补充，在国外深空探测中得到了广泛的应用[7]。绕月探测工程是 VLBI 技术第一次成功应用于我国的航天任务。刘迎春等[8]，黄勇等[9]和王威等[10]仿真计算表明，USB 与 VLBI 联合定轨可有效提高定轨精度。王宏等处理了嫦娥一号的奔月段和环月段数据，验证了仿真计算的结论[11]。

CE-1 变轨试验中，综合使用了 USB 与 VLBI 两种测量手段。本文利用此次变轨试验得到的 CE-1 在不同轨道高度的测量数据，进行轨道计算，着重分析月球重力场误差对于 CE-1 近月轨道的影响，另外还比较了 USB 数据单独定轨以及 USB 和 VLBI 联合定轨两种情况，分析了 VLBI 数据对绕月低轨卫星定轨的贡献。

1 试验概况及轨道计算参数

CE-1 变轨试验准备工作从 2008 年 12 月 6 日开始,并于 9 日进行了轨道控制,卫星轨道高度从 200 km×200 km 降至 100 km×100 km，此轨道状态一直持续到 18 日再次进行轨控，轨道高度降至 100 km×15 km，由于卫星近月点高度较低，维持了大约 30 小时，20 日再次调整至 100 km×100 km 轨道。

参与试验的 USB 测站有青岛和喀什，VLBI 测站为上海、北京、昆明和乌鲁木齐。试验期间，青岛、喀什以及 VLBI 测站每天都进行了观测。试验期间测量数据的时间分布见图 1，本文的分析计算都基于图 1 的测轨数据进行。

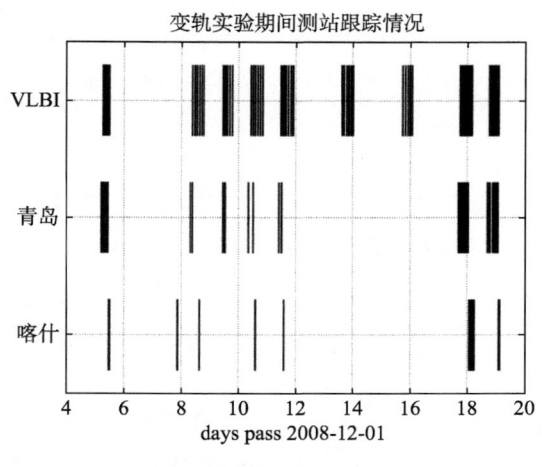

图 1 试验期间测站跟踪情况

本文轨道计算分析处理过程共涉及四类观测数据。USB 系统 12 m 天线上行、18 m 天线下行的测量体制总称为三程测距测速系统。对应的测量数学模型为：

(1) 三程测距：

$$\rho = \frac{|\vec{r} - \vec{R}_1(t_t)| + |\vec{r} - \vec{R}_2(t_r)|}{2}$$

其中 $\vec{r}$ 为信号转发时刻飞行器位置矢量，$\vec{R}_1(t_t)$ 为信号发射时刻上行站位置矢量，$\vec{R}_2(t_r)$ 为信号接收时刻下行站位置矢量。

(2) 三程测速：

$$\dot{\rho} = \frac{\rho(t_2) - \rho(t_1)}{T_c}$$

其中 T_c 为计数间隔，$t_2 = t + \frac{T_c}{2}$，$t_1 = t - \frac{T_c}{2}$。$\rho(t_1), \rho(t_2)$ 分别为 t_1，t_2 时刻的三程测距[12]。

由上海 25 m 天线、北京 50 m 天线、昆明 40 m 天线和乌鲁木齐 25 m 天线组成了 VLBI 测轨分系统，获得了时延及时延率数据。VLBI 测定的是同一信号到达两个天线的时间延迟以及时间延迟的时间变化率。设 $\vec{r}(t), \vec{R}_1(t_1), \vec{R}_2(t_2)$ 分别是信号发射时飞行器的位置矢量，该信号到达第 1 站时的台站位置矢量，该信号到达第 2 站时的台站位置矢量。由于到达各站的时间是不相等的，则 VLBI 时延和时延率的观测模型可简单地表示为（c 为光速）：

$$\tau = \frac{1}{c}(\rho_1 - \rho_2) = \frac{1}{c}(|\vec{r}(t) - \vec{R}_1(t_2)| - |\vec{r}(t) - \vec{R}_2(t_2)|)$$

$$\dot{\tau} = \frac{1}{c}(\dot{\rho}_1 - \dot{\rho}_2) = \frac{1}{c}\left(\frac{(\vec{r}(t) - \vec{R}_1(t_1)) \cdot (\dot{\vec{r}}(t) - \dot{\vec{R}}_1(t_1))}{\rho_1} - \frac{(\vec{r}(t) - \vec{R}_2(t_2)) \cdot (\dot{\vec{r}}(t) - \dot{\vec{R}}_2(t_2))}{\rho_2}\right)[13]$$

轨道计算所采用的软件为上海天文台基于美国航天局哥达德飞行中心综合定轨软件 GEODYN II 研制的探月和深空探测数据处理软件系统。该系统特别针对 VLBI 数据类型增加了数据预处理。关于该数据处理系统的详细算法和流程请参考文献[9]。计算过程中应用到的动力学模型和测量见表 1。

表 1 软件轨道确定策略

项目	模型
坐标系	月心 J2000 天球坐标系
月球重力场模型	JGL165P1
N 体摄动	太阳及大行星
行星历表	JPL DE403
太阳辐射压	固定面质比
相对论影响	Schwarzschild
解算参数	6 个轨道参数+ 光压系数+ 测距系统差
数据使用及权重设置	测距: 3m; 时延率: 0.01 cm/s
大气延迟修正	Marini-Murray 模型

定轨计算中各类观测量的相对权重根据以往定轨经验和实际情况，本文分析中使用 USB 测距与 VLBI 时延率观测量进行轨道计算，而将 USB 测速与 VLBI 时延观测数据作为定轨精度的检验。另外，测轨数据残差 RMS，系统差解算值和光压系数解算值，也是判断定轨内符精度的重要指标。

在定轨计算中，对 100 km× 100 km 轨道和 100 km× 15 km 轨道分别使用 64 阶次和 165 阶次的重力场模型进行轨道计算，并比较定轨结果。分析结果表明，对 100 km× 100 km 轨道，二者相差不大，测轨数据的拟合情况均良好。而对 100 km× 15 km 轨道，用 64 阶次重力场，定轨数据残差水平明显变差，但是如果采用完整的 165 阶次月球重力场模型，定轨数据拟合情况显著改善。

2 月球重力场误差

目前的月球重力场模型都存在着相同的问题,由于没有月球背面测轨数据的支持,其重力异常只能通过分析背面重力对正面轨道的影响间接求出。相比于正面的重力场,月球背面重力场的精度要下降约 3 倍[14]。本文进行轨道计算时使用的是 JGL165P1 月球重力场模型,同样有类似问题。

JGL165p 月球重力场基于分析 LP 的正面双程 Doppler 和测距数据得到。LP 主要科学目标之一是对月球重力场的测定,Konoplv 等通过对 LP 的高精度测轨,解算出的月球重力场揭示了 3 个新的质量瘤,而且通过提高月球极惯量矩(与月球重力场的二阶系数等价)的精度,对月球内核的半径和化学成分进行了约束。同时指出,由于 LP 没有月球背面的观测数据,得到的 20 阶以上的重力场球谐展开系数可能是不可靠的[15]。

从重力场本身的形式误差来看,月球重力场形式误差与地球重力场形式误差有显著的差别。地球重力场的形式误差随阶数增加而变大,而月球重力场的形式误差随阶数增加而变小。地球重力场的误差是较合理的,其原因在于,通过飞行器轨道的变化来探测重力场时,总是对应于较大空间尺度质量分布的低阶球谐系数比对应于较小空间尺度质量分布的高阶球谐系数更容易反映在轨道变化中,因而低阶系数的精度应该高于高阶系数的精度。月球重力场高阶项误差主要反映的是对其误差的先验 Kaula 约束,而不完全是测轨数据的贡献[14]。

从信噪比的角度来看,地球重力场直到 150 阶,其形式误差仍小于重力场本身的值,而对于月球重力场,约从 80 阶开始,形式误差就与重力场本身的量级相当了。地球重力场的高精度是在发展了众多的测量新技术,如低轨卫星对高轨卫星的跟踪、低轨卫星之间的跟踪、非保守力加速仪测量等后获得的,而月球重力场的建立主要依赖的是对环月飞行器的地面测轨。地球和月球重力场精度的差异也表明提高月球重力场的精度仍然是值得深入研究的重要科学问题[15]。

3 轨道计算及分析

本节主要分析:
(1) "嫦娥一号"卫星变轨试验不同轨道段的定轨结果,并进行精度评估;
(2) 月球重力场误差对不同高度轨道卫星的影响,包括星历预报和定轨。

3.1 100 km×100 km 轨道

CE-1 在 100 km×100 km 圆轨道飞行了大约 9 天,USB 与 VLBI 设备持续跟踪了 4 天(12 月 9 日至 12 日)。24 小时左右 CE-1 就要进行一次动量轮卸载,对于卸载过程无法进行精确的动力学建模,因此本文未进行 4 天长弧精密定轨,而是以卸载时刻作为分割点,分段进行轨道计算。

分别使用 64×64,165×165 阶次重力场进行一天星历积分,分析月球重力场模型对于 100 km×100 km 星历预报的影响,星历偏差见图 2,在横向(T)上星历偏差约 50 m、径向(R)、法向(N)上偏差约 20 m。64×64 阶次的重力场已经完全满足工程百米的精度要求,但是计算量只有 165×165 阶次的 15%,大大节省了计算时间,有利于快速轨道预报。

进一步进行定轨分析,选取 12 月 9 日 08:00-18:00 数据进行定轨分析,定轨策略见表 1。使用 64×64 阶次重力场进行轨道计算,时延率数据残差的 RMS= 0.46 ps/s,测距数据残差的 RMS= 2.1 m。测距和时延率数据 O-C 残差见图 3。

如果使用 165 阶次月球重力场进行轨道计算,时延率数据残差的 RMS= 0.43 ps/s,测距数据残差的 RMS= 2.1 m,残差图与 64 阶次情况基本一致。

计算表明,使用 64 与 165 两种阶次的重力场进行计算,数据拟合情况及数据噪声水平大致相当,得到的星历偏差见图 4,最大偏差约 30 m,使用 64 阶次的重力场进行轨道计算基本可以满足工程精度要求。

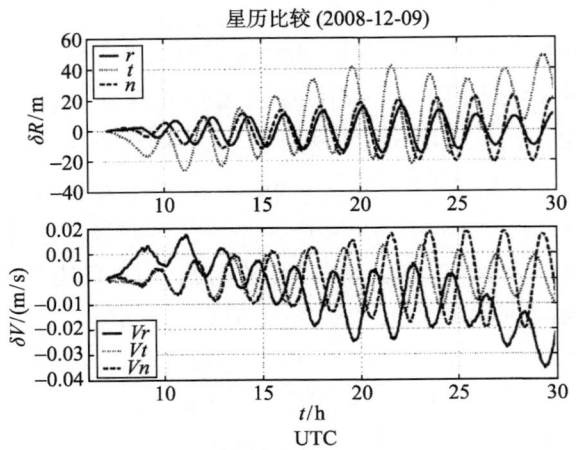

图 2 不同阶次重力场模型的星历预报比较

(分别使用 64 与 165 阶次重力场进行一天星历积分，比较重力场对 100 km× 100 km 轨道预报的影响，上图为位置差，下图为速度差)

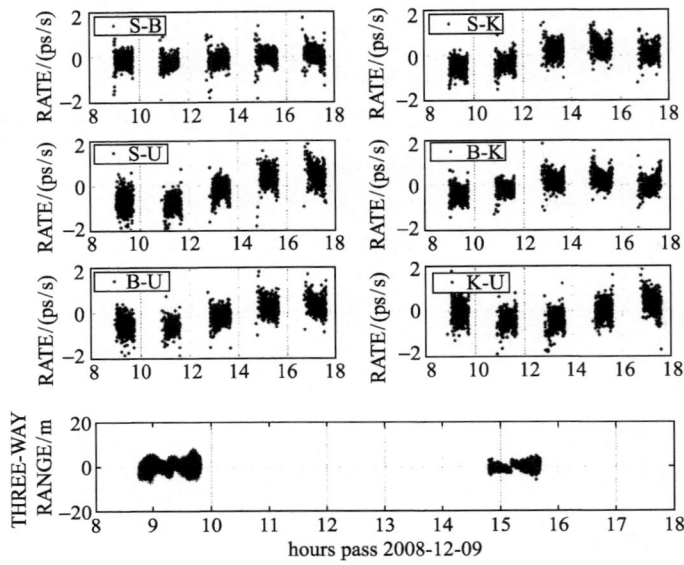

图 3 64 阶次重力场模型定轨后的 USB 与 VLBI 时延率数据残差

(S、B、K、U 分别代表上海、北京、昆明和乌鲁木齐)

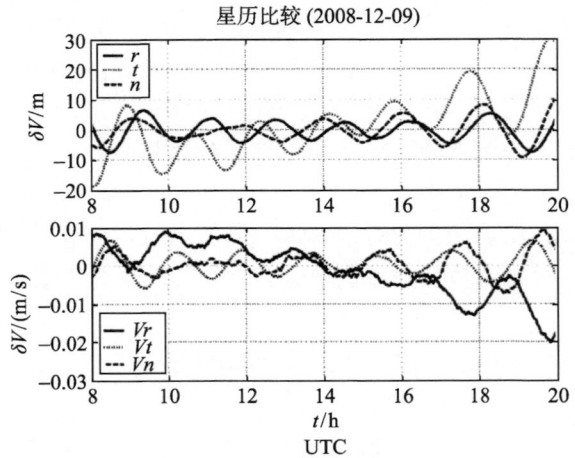

图 4 64 与 165 阶次重力场轨道计算星历比较

极轨卫星近月距的变化关键在于轨道偏心率 e 的变化[16]。计算表明，在 100 km× 15 km 的轨道状态下，CE-1 卫星的半长轴，偏心率与近月距的变化很快，卫星的半长轴无长期变化，其变化周期主要为轨道周期，但偏心率有长周期变化，导致一天内近月距下降了近 7 km。当近月距小于月球半径时可能就会撞上月球。

3.2 100 km× 15 km 轨道

对 100 km× 15 km 轨道，同样分别使用 64 与 165 阶次的月球重力场模型进行一天的星历积分，比较月球重力场模型对卫星星历预报的影响。结果表明影响非常明显，在横向方向上偏差达到千米量级，此误差已经远远超出工程精度指标。

进一步进行定轨计算，与 100 km× 100 km 轨道类似，以卸载时刻为分割点，利用 12 月 18 日控后 10 小时测轨数据进行轨道计算。轨道计算策略见表 1。使用 64 阶次重力场进行轨道解算，定轨后 O-C 残差图见图 5。时延率数据残差的 RMS= 1.03 ps/s, 测距残差的 RMS= 9.3 m。图 5 可以看出定轨数据残差时间序列明显呈非随机状态，而且震荡幅度比较大，初步怀疑是月球重力场模型误差所致。

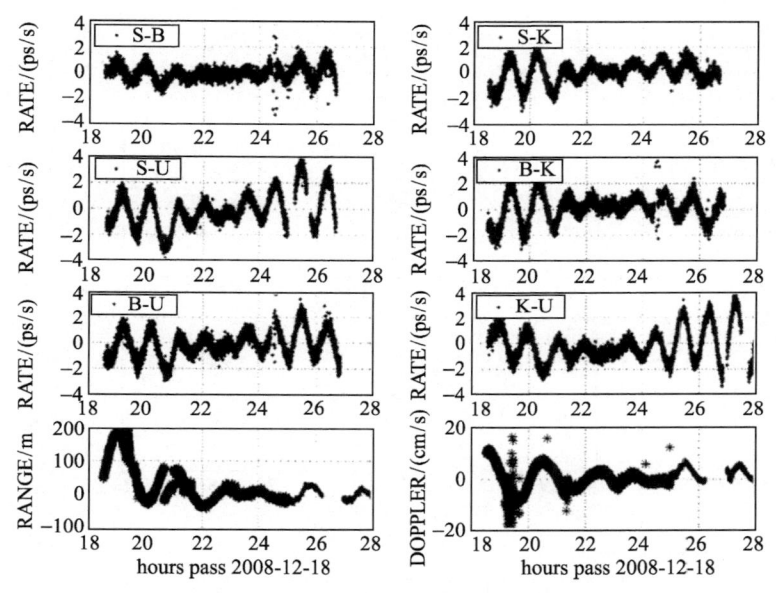

图 5 64 阶次重力场定轨残差图

USB 测速数据未参加定轨，可以作为定轨精度的检验。对 64 阶次重力场解算的轨道进行校验，从图 5 中右下最后一幅子图可以看出，测速资料拟合明显不好，也反映出轨道解算不够准确。

既然使用 64 阶次重力场定轨结果不好，尝试使用 165 阶次重力场进行轨道计算，其他动力学模型不变，结果表明定轨后数据拟合情况明显改善，参见图 6,时延率残差的 RMS = 0.58 ps/s,测距残差的 RMS= 3.3 m。

比较图 5、图 6 可以发现使用 165 阶次重力场后测轨数据的拟合情况明显比 64 阶次好。同样利用 USB 测速数据进行定轨精度的校验，测速资料拟合情况明显变好，见图 6 中右下最后一幅子图，27~28 小时为喀什站，测速约有 2 cm/s 的系统差。

比较 64 与 165 阶次重力场分别解算的卫星轨道，星历偏差见图 7，在定轨弧段内，轨道最大偏差约为 400 m，其中横向方向最大，约为 300 m，径向和法向方向约 100 m。高阶月球重力场对轨道影响非常明显，这既反映在观测数据的拟合上，也反映在解算的星历上。

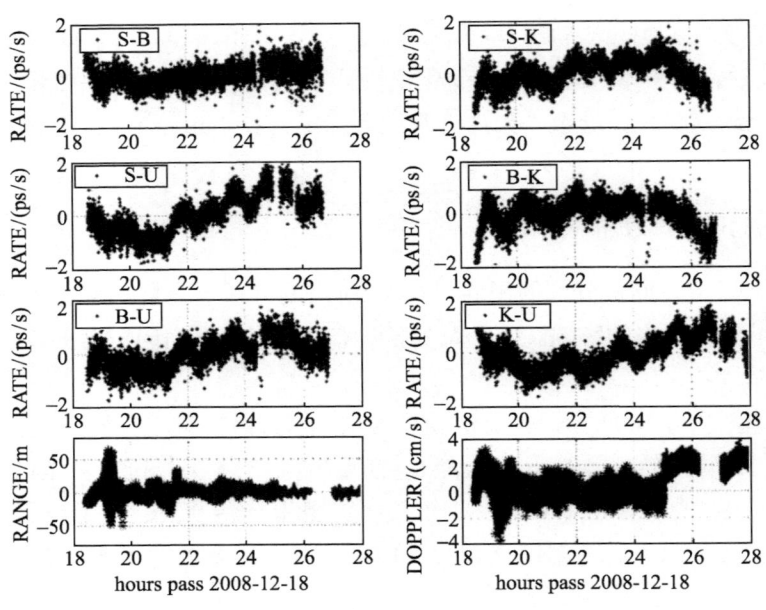

图 6 165 阶重力场解算轨道残差图

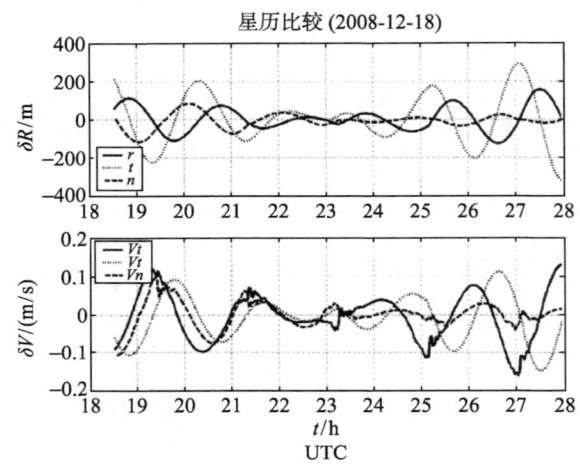

图 7 不同重力场轨道解算星历比较

3.3 VLBI 对定轨的贡献

仿真计算表明，USB 距离观测对径向(R)约束较好，VLBI 对沿迹(T)和法向(N)敏感，USB 与 VLBI 联合定轨可提高定轨精度。对我国具体情况，依靠国内 USB 站进行测距和 VLBI 资料也可以达到较高定轨预报精度[6]。

由于缺乏独立于 USB 和 VLBI 的测轨数据，难以评估定轨结果的真实外符合精度。为了验证 VLBI 测轨数据对 100 km×15 km 环月轨道确定的贡献，我们使用重叠弧段法来验证定轨精度。

100 km×15 km 周期约 2 小时，此轨道阶段持续跟踪的时间为 12 小时。我们分别使用 18 日 18：00-26：00 数据，与 18 日 22：00 至 19 日 6：00 的观测资料进行轨道计算，参与轨道计算观测资料为 USB 测距和 VLBI 时延率，定轨策略同表 1，使用 165 阶次的重力场模型。定轨弧段均为 4 圈，搭接弧段为 2 圈，比较解算的轨道，在搭接弧段内横向偏差约 20 m，径向和法向偏差优于 5 m，这与 200 km 圆轨道阶段计算结果相一致；如果单独使用 USB 测距和测速资料进行搭接弧段精度比较，在搭接弧段内横向偏差与 80 m，径向与法向约 20 m，比 200 km 圆轨道偏差稍大，可参见表 2。通过搭接弧段比较，可以认为 USB

与VLBI数据的联合定轨较单用USB数据定轨可信，VLBI观测资料对定轨精度的提高有帮助。

表2 搭接弧段定轨星历比较

定轨数据	定轨/搭接弧长(周期)	星历偏差		
		R (m)	T (m)	N (m)
USB	4/2 圈	20	80	20
USB+VLBI	4/2 圈	5	20	5

使用USB与VLBI资料联合定轨与单独使用USB资料轨道计算，并比较定轨星历，参见图8，在定轨弧段内，星历偏差在横向方向上约120 m，径向上约20 m，这比200 km圆轨道阶段星历比较偏差稍大，所以VLBI数据的参与对轨道计算精度仍约有百米的提高，VLBI的贡献更加明显。进行搭接弧段比较采用的定轨数据仅有4圈弧长，如果增加跟踪弧段，使用USB与VLBI联合定轨精度上仍可进一步提高。

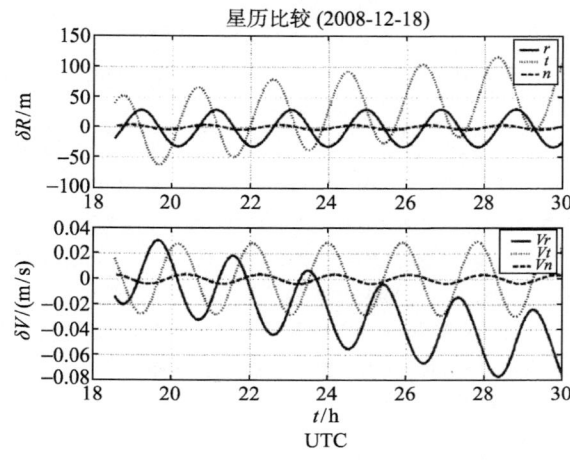

图8 USB数据定轨与USB、VLBI联合定轨星历比较

4 结论

月球重力场摄动影响会导致偏心率变大，有可能使卫星轨道迅速变扁，缩短卫星寿命，在100 km×15 km轨道阶段反映尤为明显。本文对CE-1变轨试验期间的测轨数据进行处理分析，比较分析不同阶段的定轨精度、预报精度，重力场对定轨预报精度的影响。根据对变轨期间测轨数据处理计算的结果，我们可以对月球低轨卫星定轨和轨道预报精度得出以下一些结论：

(1) 尽管月球重力场的高阶谐系数并不可靠，但考虑高阶数的重力场对低轨卫星定轨精度的提高仍有帮助。65阶以上高阶月球重力场对月球低轨卫星影响明显，对100 km×100 km轨道进行一天轨道积分变化仅几十米，但对100 km×15 km可以达到数千米。100 km×15 km轨道阶段进行轨道解算时，必须考虑足够的重力场阶数(条件允许可以考虑165阶次)，否则无法对观测资料很好地进行拟合，轨道计算偏差较大，定轨弧段内为百米，预报精度则更差。

(2) VLBI参与轨道计算可提高定轨精度，在100 km×15 km轨道阶段在5~6圈的定轨弧段内仍可有效提高定轨精度。

参 考 文 献

[1] 杨维廉，周文艳. 嫦娥一号月球探测卫星轨道设计[J]. 航天器工程，2007，16(6)：16—24 [YANG Wei-lian, ZHOU Wen-yan.Orbit design for lunar exploration satellite CE -1[J]. Spacecraft Engineering, 2007, 16(6) : 16—24.]

[2] Konopliv A S, Binder A B, Hood L L, et al. Improved gravity field of the moon from lunar prospector [J]. Science, 1998, 281: 1476—1480.

[3] Noriyuki N, Takahiro I, et al. Farside gravity field of the moon from four-way doppler measurements of SELENE (Kaguya) [J]. Science, 2009, 323: 900—905.

[4] Frank W, Gll E, Montenbruck O, et al. Lunar prospector orbit determination and gravity field modeling based on weilheim 3-way doppler measurements[J]. J. of the Braz. Soc. Mechanical Sciences, 1999: 280—286.

[5] 鄢建国, 平劲松, 李斐, 等. 应用 lp165p 模型分析月球重力场特征及其对绕月卫星轨道的影响[J]. 地球物理学报, 2006, 49(2): 408—414. [YAN Jian-guo, PING Jin-song, LI Fei, et al. Character analysis of the lunar gravity field by the LP165P model and its effect on lunar satellite orbit[J]. Chinese Journal Of Geophysics, 2006, 49(2): 408—414.]

[6] 平劲松, 河野裕介, 河野宣之, 等. 日本 SELENE 月球探测计划和卫星间多普勒跟踪的数学模型[J]. 天文学进展, 2001, 19(3): 354—364. [PING Jin-song, Kono Y, Kawano N, et al. SELENE Mission: Mathematical model for SST dopler measurements [J]. Progress In Astronomy, 2001, 19(3): 354—364.]

[7] 叶叔华, 黄珹. 天文地球动力学[M]. 济南: 山东科学技术出版社, 2000. [YE Shu-hua, HUANG C. Astrogeodynamics[M]. Ji'nan: Shandong Science and Technology Press, 2000.]

[8] 刘迎春, 张飞鹏, 董晓军. 月球探测卫星的轨道支持[J]. 飞行器测控学报, 2003, 22(1): 15—19. [LIU Ying-chun, ZHANG Fei-peng, DONG Xiao-jun. Orbital support to lunar exploration spacecraft[J]. Journal of Spacecraft TT&C Technology, 2003, 22(1): 15—19.]

[9] 黄勇. "嫦娥一号"探月飞行器的轨道计算研究[D]. 上海: 上海天文台, 2006. [HUANG Yong. Orbit determination of the first Chinese lunar exploration spacecraft CE-1[D]. Shanghai: Shanghai Astronomical Observatory, 2006.]

[10] 王威, 胡小工, 黄勇, 等. 影响奔月飞行器定轨精度的误差源分析[J]. 飞行器测控学报, 2005, 24(1): 44—50. [WANG Wei, HU Xiao-gong, HUANG Yong, et al. Analysis of error sources for trans-lunar spacecraft orbit determination [J]. Journal of Spacecraft TT&C Technology, 2005, 24(1): 44—50.]

[11] 王宏. 我国嫦娥一号卫星初步定轨结果分析[J]. 深空探测研究, 2008, 6(2): 37—40. [WANG Hong. Preliminary analysis of CE-1 orbit determination[J]. Deep Space Exploration, 2008, 6(2): 37—40.]

[12] 汤锡生, 陈贻迎, 朱民才. 载人飞船轨道确定和返回控制[M]. 北京: 国防工业出版社, 2002. [TANG Xi-sheng, CHEN Yi-ying, ZHU Min-cai. Manned Spacecraft Orbit Determination and the Return of Control[M]. Beijing: National Defense Industry Press, 2002.]

[13] Theodore D, Moyer. 深空网导航数据的测量和计算公式[M]. 刘迎春译. 北京: 清华大学出版社, 2006. [Theodore D, Moyer. Formulation for Observed and Computed Values of Deep Space Network Data Types for Navigation [M]. Beijing: Tsinghua University Press, 2006.]

[14] Konopliv A S, Binder A B, Hood L L, et al. Improved gravity field of the moon from lunar prospector [J]. Science, 1998, 281: 1476—1480.

[15] 胡小工, 黄珹, 黄勇. 环月飞行器精密定轨的模拟仿真[J]. 天文学报, 2005, 46(2): 186—195. [HU Xiao-gong, HUANG Cheng, HUANG Yong. Simulation of precise orbit determination for a lunar orbiter [J]. Acta Astronomical Sinice, 2005, 46(2): 186—195.]

[16] 王歆, 刘林. 目标天体极轨道卫星的轨道寿命[J]. 宇航学报, 2001, 22(5): 62—65. [WANG Xin, LIU Lin. Lifetime of polar-orbit satellite[J]. Journal of Astronautics, 2001, 22(5): 62—65.]

嫦娥一号绕月卫星对月球重力场模型的优化

鄢建国[1,2,3]　平劲松[2,4]　KOJI Matsumoto[3]　SANDER Goossens[3]
唐歌实[4]　李斐[1]　刘俊泽[4]　李金岭[2]

（1. 武汉大学测绘遥感信息工程国家重点实验室，武汉　430070；
2. 中国科学院上海天文台，上海　200030；
3. 日本国立天文台 RISE 项目组，水泽　0230801，日本；
4. 北京航天指挥控制中心飞行动力学重点实验室，北京　100094）

摘　要　月球重力场是揭示月球内部结构和物质组成的重要信息,探测月球重力场仍然是绕月探测任务中的重要科学目标之一。在已有月球重力场模型基础上，利用嫦娥一号探测数据，并结合"月女神"一号探测器、月球勘察者(LP)及早期月球探测器轨道跟踪数据，本文解算得到了高精度月球重力场模型 CEGM02(100 阶次)，在 CEGM01 月球重力场模型基础上对模型进行了优化。对新模型的分析结果表明，嫦娥一号卫星轨道跟踪数据的融入，使得对月球重力场长波长部分的解算精度有显著提高，相比于 SGM100h 模型在 5 阶以内精度提高约 2 倍，在 10 阶以内有明显贡献，在 20 阶内都有贡献。初步判断这是由于嫦娥一号卫星轨道动量轮卸载的频度不足"月女神"的 1/4，而同时轨道相对较高所导致。文中结合 CEGM02 和激光测月观测结果解算了月球平均转动惯量 0.393446(±0.000006)，对月球内部构造研究提供了更强的约束。

关键词　CEGM02　嫦娥一号　月球重力场模型　功率谱　转动惯量

　　月球重力场是月球科学的一个重要部分，是研究月球物理性质及内部结构、月球的起源和演化等科学问题的主要的手段[1]。高精度月球重力场模型给出的月球主转动惯量可以约束月球内核的大小及状态，弥补月震观测的不足对月核研究带来的限制或缺陷[2]。利用卫星重力方法解析月球卫星轨道信息和重力场模型的同时，还可以解算月球固体潮 Love 数 k_2 项，进而结合激光测月(LLR)数据估算月球对地球潮汐力的弹性响应，作为确定月球内核状态的另一个重要的约束条件[3]。

　　对月球重力场的高精度探测研究具有重要的理论意义和应用价值，是当今国际月球探测的重要科学目标之一。由于绕月卫星受到了从内到外、从正面到背面的月球全部物质质量的引力作用，卫星重力方法可以方便地实现对月球重力场的测量。地球卫星重力测量的历史和经验表明，利用这种方法进行重力场探测时，使用连续自由飞行在位于不同轨道上的多颗探测器，能够实现重力场的高精度、高分辨率、全球覆盖的测量。在对月球重力场进行探测时，历史上比较完备的探测是日本于 2007 年 8 月发射的探月计划"月女神"一号和早期美国发射的 LP 探测器，它们是专门以月球重力场和测月学探测为主要目标的月球探测器。尽管全部的月球物质质量都对 LP 产生引力作用，考虑到测控观测量多在月球正面获得，这时卫星对月球背面重力场短波长成分不敏感，LP 只实现了全月球重力场和正面高分辨率重力场的观测。而"月女神"一号在上述基础上，利用四程卫星中继无线电测量技术，首次实现了对月球背面月球重力场短波长成分的直接测量。基于"月女神"一号得到的月球重力场模型 SGM90d[4,5]的精度和分辨率，相比于 LP 系列模型[6]在月球背面有显著的改进，并发现了月球背面的环状包围的质量异常区。利用 LP 和"月女神"探测器联合获得的月球重力场异常特征为研究月球壳幔结构、均衡补偿状态以及月球二分性等科学问题提供了新的依据[4]。最近，松本晃志利用"月女神"一号整个任务期间的四程多普勒测量数据及历史上月球探测器跟踪数据得到了更高精度的月球重力场模型SGM100h，并对月球背面的特征盆地进行了更细致的划分[7]。

尽管LP和"月女神"一号在月球重力场领域取得了前所未有的结果，由于两个计划的探测主要是通过对100km高度的极轨道卫星的测量得到的，而且"月女神"卫星平均6~9h的一次动量轮卸载，使得她的自由飞行轨道不完全满足实现高精度、高分辨率、全球覆盖的月球卫星重力场测量的条件。在目前得到的重力场模型中，只用到了极少量的800km以上飞行的极轨道探测器的观测数据。为此，最新的SGM100h模型在多方面仍然需要继续优化，包括（i）实现高空间分辨率或高于100阶次的模型；（ii）实现更高精度的长波长或低阶的模型；（iii）实现对月球边缘重力场的高精度探测；（iv）消除65阶次以上的球谐函数重力场模型系数的强相关性以进一步弱化算法带来的数学效应。为此获取更高精度和更高分辨率的月球重力场模型仍然是当前和近未来月球探测计划的重要或首要目标[8,9]。比如，于2009年6月发射的美国新千年重返月球的第一个探测计划LRO(Lunar Reconnaissance Orbiter)[8]，将在50km高的圆极轨道在轨运行一年，其主要科学目标就是获取月面高精度高分辨率地形及月球重力场，根据已有的测高数据解算得到的地形模型空间分辨率为100m，为目前分辨率最高的地形模型[10]。另外，2011年美国将发射的月球探测计划GRAIL(Gravity Recovery and Interior Laboratory)，将主要用于月球重力场探测。这一计划将采用地球重力场探测计划GRACE的卫星-卫星跟踪测量模式[11]，用于高精度月球重力场模型探测，预期精度将比日本"月女神"一号计划得到的重力场模型SGM90d提高3个数量级[9]。

嫦娥一号绕月探测卫星于2007年10月24日发射，经过调相段、地-月转移段、月球捕获段和环月飞行段等多次轨道调整后，成为绕月飞行的极轨圆轨道卫星，轨道平均高度为200km[12]。嫦娥一号卫星任务期间积累了大量的测距测速跟踪数据，这类测控数据可以用来进行或参与月球重力场的解算。在任务前的仿真分析表明，使用嫦娥一号测控数据可以有效反演50阶次以内的月球全球重力场，而对更高阶次的重力成分不太敏感。与飞行在100km高度的"月女神"一号主卫星比较，除了采用了不同几何构型的轨道之外，嫦娥卫星飞行过程中动量轮卸载的频度是每24~36h一次，一个完整的自由飞行弧段可以包括18圈之多，是"月女神"的4~6倍。作者利用这些数据，采用了与"月女神"一号SGM90d和SGM100h重力场模型解算时同样的定轨与重力分析软件以及相同的时间、地球与天球参考架系统，独立地解算了月球重力场模型[13]，验证了数据的有效性。在此基础上本文进一步结合"月女神"一号、LP及历史已有的跟踪数据，进行了高精度月球重力场模型解算，对现有月球重力场模型进行了优化。首先简单介绍了绕月探测卫星测量数据的情况，继而对月球重力场模型解算和结果进行了分析，并进一步探讨了该模型在月球内部构造研究中的潜在应用。

本文计算采用的解析软件为美国航天局戈达德飞行中心(GSFC/NASA)授权中国科学院上海天文台和日本国立天文台使用的GEODYNII/SOLVE[14,15]探测器轨道和中心引力体重力场分析软件。与"月女神"团队解析SGM90d和SGM100h重力场模型使用的软件一致，并且采用了一致的天体力学与天体测量模型以及参数作为输入条件。

1 数据和模型

嫦娥一号月球探测器是中国独立发射的第一颗月球探测卫星，其主要目标是获取月球表面的三维影像和月面物质分布。该卫星在入轨成为绕月卫星之后，嫦娥一号的跟踪测量主要由中国卫星测控网的双程测距测速和中国科学院的天文甚长基线干涉测量(VLBI)网联合实现。本文使用的数据主要为青岛和喀什两测控站的双程测距测速数据，数据采样率为1s。测控任务设计其中，测距标称精度为2m，测速标称精度为

10mm·s^{-1}。然而事实上，经过精密定轨标定了测控系统偏差后的残差表明，嫦娥一号月球探测器的测距精度为1m，测速精度为3mm·s^{-1}。参考LP100h和SGM100h解析中的经验，在后续月球重力场解算中，对嫦娥一号卫星跟踪数据的权进行了一定程度的弱化，并将数据进行了10s平均，定轨解算弧段长度为一天。使用3mm·s^{-1}的权重发现并不对结果有改进或破坏。正常任务期间，2007年12月每天观测弧段约为6h；2008年1和2月每天观测弧段约为3h；3和4月为侧飞期间，没有安排测控观测；2008年5~11月每天观测弧段约为4h[13]。图1为测量数据的月面覆盖情况。

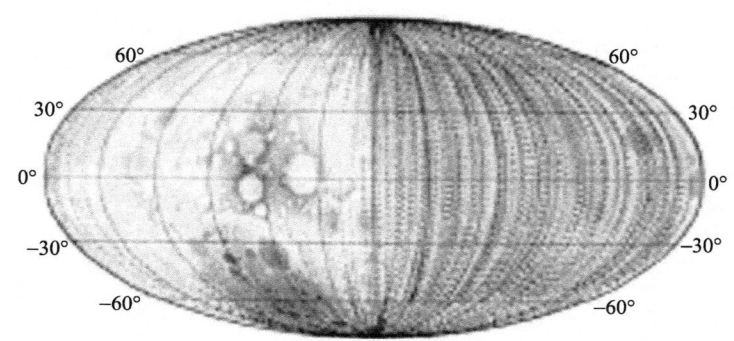

图1 嫦娥一号跟踪数据月面覆盖图
左侧为月球背面，右侧为月球正面

为了检验和确认上述嫦娥一号跟踪数据用于重力场解析的有效性和结果模型的特征，本文中方作者基于动力法精密定轨解算月球重力场模型的原理及策略。在"嫦娥一号"测控数据精度和覆盖均有限的条件下，独立使用嫦娥一号月球探测器6个月的在轨运行双程测距测速跟踪数据，成功得到了50阶次月球重力场模型CEGM01[16]。并通过多种方式，如重力场模型频谱特性、实测数据定轨残差、月球重力场异常特征、与地形的相关性及导纳值，对解算得到的CEGM01月球重力场模型进行了分析评价，分析了相应的物理特性和效果。结果证明了模型解算过程的有效性。文献[16]还确认，单独使用嫦娥一号的测距测速跟踪数据解析月球低阶重力场时，18阶次以上的模型误差与模型本身的差异不显著，而18阶次以下的部分，模型误差比CEGM01模型本身小很多，表明上述测控数据对低阶重力场部分更敏感，有希望在融合使用更多的历史测控数据解析月球重力场中发挥作用，该项工作为我国开展后续的月球重力场探测试验打下了良好的基础。

本文使用的"月女神"一号数据包括中继星、VRAD星及主卫星的双程测距测速数据，中继星与主卫星的四程测速数据，其中"月女神"一号主卫星双程测距测速数据由日本宇航局的全球地面网(GN)11m口径天线获得。位于日本中部臼田深空站的64m口径天线也参与了主卫星的部分时段的跟踪。中继星和VRAD星的双程测距测速数据以及中继星-主卫星-地面测站间的四程测速数据仅由臼田深空站64m口径天线获得。本文直接采用文献[9]的定权值。

解算中用到的历史数据还包括美国1960年代发射的Apollo15与Apollo16子卫星、Lunar Orbiter I-V及Celementine等探测器的轨道跟踪数据，均由美国JPL/NASA深空网跟踪站获得。此外还包括LP正常任务段跟踪数据。上述各探测器跟踪数据的类型、权重、数据量及解算中平均弧段长度等如表1所示[6,7]。数据的权重主要根据测量数据的精度而定，同时采用了LP月球系列重力场模型求解中数据的定权方式[6]。

绕月卫星精密定轨和月球重力场模型解算中用到重力场球谐函数表达形式[1,17,18]：

$$V(r,\varphi,\lambda) = \frac{GM}{r} = \left[\sum_{n=0}^{N_{\max}} \sum_{m=0}^{n} \left(\frac{R}{r}\right)^n \bar{P}_{nm}(\sin\varphi)(\bar{C}_{nm}\cos m\lambda + \bar{S}_{nm}\sin m\lambda)\right] \quad (1)$$

表1 月球重力场解算中各探测器跟踪数据取权基本情况

探测器名称	跟踪数据类型	数据量	测量数据权重	平均弧段长度
嫦娥一号	双程测距	224240	1m	24h
	双程测速	224190	10mm·s^{-1}	
"月女神"一号 主卫星	四程测速	78004	1mm·s^{-1}	12h
	双程测速(臼田)	1786747	1mm·s^{-1}	
	双程测速(GN)		2mm·s^{-1}	
"月女神"一号 中继星	双程测距	150470	5m	2.6d
	双程测速	159225	1mm·s^{-1}	
"月女神"一号 VLBI星	双程测距	150470	5m	2.4d
	双程测速	42502	1mm·s^{-1}	
	双程测距	35567	5m	
LO-I	双程测速	36735	4.5mm·s^{-1}	12h
	三程测速	3867	4.5mm·s^{-1}	
LO-II	双程测速	42376	4.5mm·s^{-1}	12h
	三程测速	5370	4.5mm·s^{-1}	
LO-III	双程测速	15175	4.5mm·s^{-1}	12h
	三程测速	1957	4.5mm·s^{-1}	
LO-IV	双程测速	19390	4.5mm·s^{-1}	12h
	三程测速	1780	4.5mm·s^{-1}	
LO-V	双程测速	13397	4.5mm·s^{-1}	12h
	三程测速	2260	4.5mm·s^{-1}	
Apollo 15 子卫星	双程测速	28986	4.5mm·s^{-1}	8h
	三程测速	16522	4.5mm·s^{-1}	
Apollo 16 子卫星	双程测速	15459	4.5mm·s^{-1}	8h
	三程测速	15584	4.5mm·s^{-1}	
Clementine	双程测速	354020	3mm·s^{-1}	2d
	双程测速	5091	6m	
LP	双程测速	3155182	2mm·s^{-1}	2d

式中r, λ, φ分别是月心球坐标系下月理坐标系中的向径和经纬度，GM为月球引力常数，R为参考球半径(文中取为1738.0km)，$\bar{C}_{nm}$和$\bar{S}_{nm}$为待估计的完全正则化斯托克斯参数，这里也称为位系数，$\bar{P}_{nm}$为n阶m次正则化连带勒让得函数。

这里重力场的解算通过动力法定轨解析得到，即探测器精密定轨和重力场模型解算同时完成，其中全局参数包括2~100阶次的重力场位系数和月球二阶引力位Love数k2，局部参数包括各弧段探测器初始轨道根数、每个站每个跟踪弧段测距测速系统偏差、用于模制卫星动量轮卸载的经验加速度在轨道面径向/沿迹/法向3个方向上的常数分量以及太阳光压系数。在正常任务期间嫦娥一号卫星的姿态调整频率相对较低，约一天1次，并且在此处的解算中尽量剔除了姿态调整时间段的数据。而在"月女神"一号的任务期间，主卫星姿态调整频率为一天3~4次，导致卫星轨道的连续弧段变短，每个弧段在经度方向上的覆盖范围比嫦娥一号卫星小3~4倍，对重力场中的长波长解析带来限制。文中解算策略为，对每个探测器每个弧段均生成观测量对全局参数和局部参数的法方程矩阵，对各探测器的所有弧段的法方程矩阵单独融合为一个法方程矩阵，在此基础上将各探测器的法方程矩阵融合得到总的法方程矩阵，最后通过分块求逆的方法得到重力场位系数的估值[17]。

解算过程中利用到的惯性系为月心J2000，历表为JPL DE421[17]。考虑到的动力学模型有先验重力场模型、太阳光压、N体摄动及地球扁率的间接摄动等，计算中选取的先验重力场模型为SGM100g[7]。

2 解算结果及分析

为了判定解析的有效性,本文作者先利用分析软件系统独立地重现了 SGM90d 和 SGM100h 的重力场位系数的估值和模型误差的估值,再分 3 次先后引入 2,6,10 个月的嫦娥一号测量数据并入法方程矩阵,利用上述解析过程对 100 阶次的重力场进行解析,发现重力场位系数在得到改进的同时,模型误差随着引入嫦娥一号测量数据的增加,逐步减小。图 2 给出了包括最后一步引入全部 10 个月嫦娥一号测距测速数据的解算模型与先验重力场模型的功率谱密度曲线,即阶方差 σ_n (sig)与误差阶方差 δ_n(sigvar)曲线图,二者的计算公式分别为[18]:

$$\sigma_n = \sqrt{\frac{\sum_{m=0}^{n}(\bar{C}_{nm}^2 + \bar{S}_{nm}^2)}{2n+1}}; \quad \delta_n = \sqrt{\frac{\sum_{m=0}^{n}(\sigma_{\bar{C}_{nm}}^2 + \sigma_{\bar{S}_{nm}}^2)}{2n+1}}, \tag{2}$$

式中,$\bar{C}_{nm}$,$\bar{S}_{nm}$ 是 n 阶 m 次完全正则化的位系数,$\sigma_{\bar{C}_{nm}}$,$\sigma_{\bar{S}_{nm}}$ 分别是相应位系数的方差,用于表示重力场在频率域中的强度,误差阶方差则可以反映重力场的误差。

图 2 中(a)图给出了 CEGM02(100 阶次)与 SGM100h(100 阶次)的阶方差和误差阶方差,同时给出了 LP100K(100 阶次)模型作为比较。选取 LP100K 进行比较的原因在于,LP100K 模型解算中利用到了 LP 所有正常任务段的数据和历史跟踪数据,与 SGM100h 和 CEGM02 模型解算中的历史数据源和解析软件、流程、天体力学和天体测量模型与参数都一致。通过与 LP100K 的比较,可以更清楚地显示出月球背面跟踪数据和嫦娥一号轨道跟踪数据在重力场模型解算中的贡献。(b)图给出了分别利用嫦娥一号跟踪数据解算得到的模型位系数阶方差曲线,可以明显看出随着数据量的增加,低阶项位系数精度有明显提高。

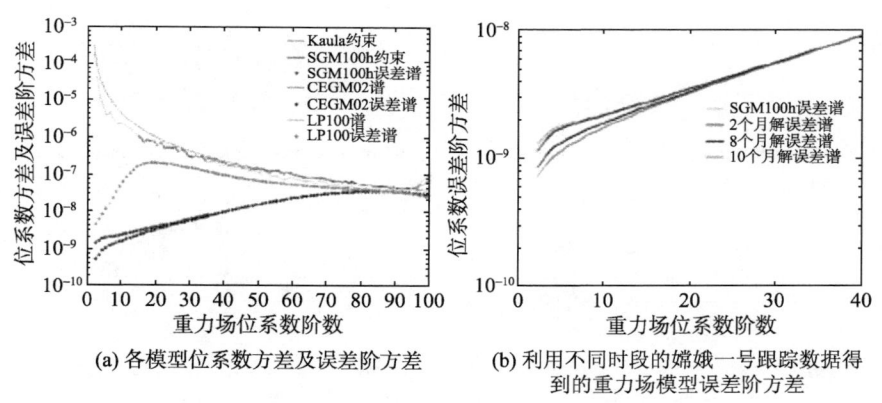

(a) 各模型位系数方差及误差阶方差

(b) 利用不同时段的嫦娥一号跟踪数据得到的重力场模型误差阶方差

图 2 重力场位系数阶方差及误差阶方差曲线图
图中截断至 40 阶

SGM100h 的解算用到 SELENE 所有的四程测速和双程测距测速数据,以及历史跟踪数据,包含有丰富的重力场信息。嫦娥一号卫星轨道高度为 200km,相比于 SELENE 主卫星和 LP 正常任务段 100km 高的轨道,对从观测量中有效分离重力场长波长信息有利。CEGM02 模型对 20 阶次以下的部分有改善,20 阶次以上部分主要来自 LP 和"月女神"一号卫星的贡献。

嫦娥一号卫星在 200km 高的轨道上飞行十个月积累的跟踪数据主要包含了来自月球重力场长波波段即中低阶次的引力效应。由图 2 中的模型误差阶方差曲线图可以看出,相比于 SGM100h,CEGM02 模型,在 5 阶以内精度提高约 2 倍,在 10 阶以内有明显贡献。"月女神"一号四程测速数据提供的月球背面重力场信息可以直接解算到 70 阶次,进行更高阶次的解算则需要引入 Kaula 约束[7]。本文解算中采用了与 SGM100h 模型解算一致的 Kaula 约束常数值 3.6×10^{-4}。

图 3 和图 4(图版 XIII)分别给出了利用三个重力场模型位系数协方差矩阵计算得到的重力场异常和月球大地水准面误差分布,分别截止到 10 阶次和 25 阶次。可以看出,对重力场长波段误差而言,SGM100h

和 CEGM02 比 LP100K 有显著改进，融入嫦娥一号跟踪数据对重力场长波段的精度改进也较为显著。LP100K、SGM100h 和 CEGM02 截断至 10 阶次的重力场异常误差的 rms 值分别为 0.38，0.021 和 0.005mgal，大地水准面误差的 rms 值分别为 0.49，0.03 和 0.007m。这一数据与图 2 中位系数误差阶方差曲线反映的低阶项位系数精度的改进一致。

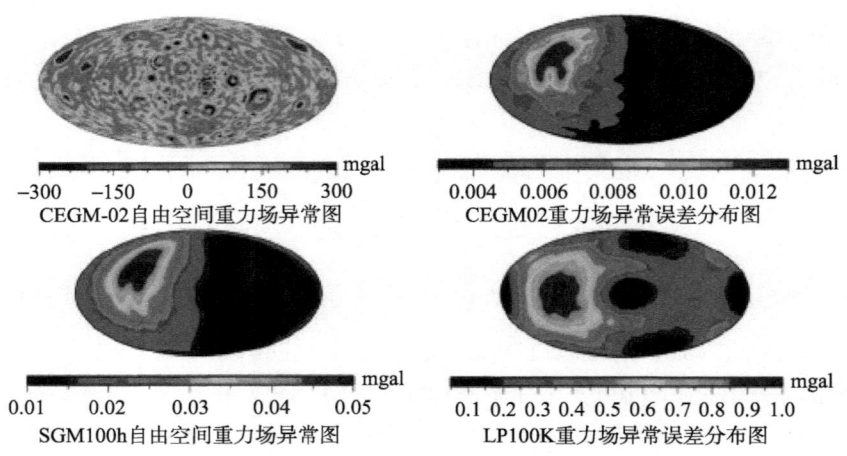

图 3　CEGM02 模型重力场异常以及各模型截断至 10 阶次的重力场异常误差分布图
重力场异常图对称中心为(180°E, 0°N)；误差图的对称中心为(270°E, 0°N)，左边为月球背面、右边为月球正面

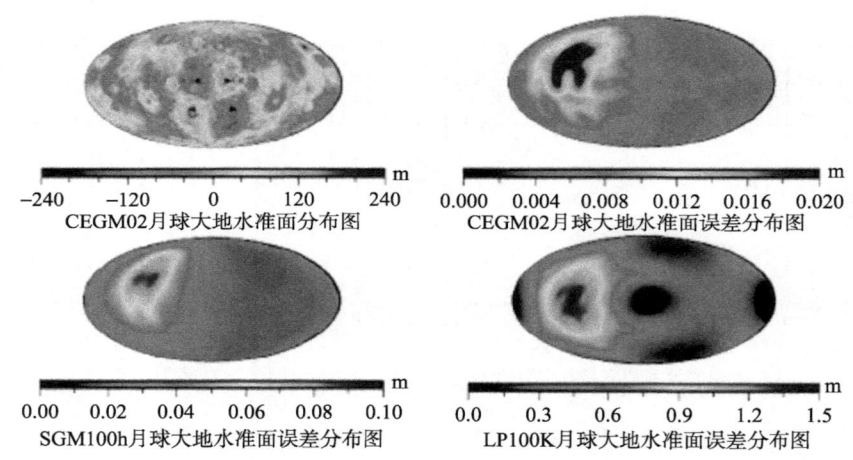

图 4　CEGM02 模型月球大地水准面以及各模型截断至 10 阶次的月球大地水准面误差分布图
月球大地水准面分布图对称中心为(180°E, 0°N)；误差图的对称中心为(270°E, 0°N)，左边为月球背面、右边为月球正面

为了检验模型在应用中的可靠性和有效性，这里采用 LP 卫星的重建轨道与基于不同重力场模型的外推轨道比较的方法实施。比较弧段为 LP 正常任务段 1998 年 5 月 21~22 日两天弧段 1 与同年 6 月 23~24 日两天弧段 2。比较时，首先得到两个弧段的重建轨道(不同重力场模型得到的重建轨道的差异极小，可忽略)，然后利用不同的重力场模型把弧段 1 得到的精密星历外推至 6 月 23 日与弧段 2 重建轨道的星历进行比较。表 2 给出了基于不同重力场模型外推结果与重建轨道在轨道面径向、沿迹及法向 3 个正交方向上差值的 RMS 值。相比于 LP100K 和 SGM100h，CEGM02 在沿迹及法向方向上的 RMS 存在明显的改进。对 LP 探测器，由于没有大气阻力作用，轨道沿迹方向上的误差主要由重力场误差引起，而长时段弧段外推轨道误差则由重力场低阶项误差引起，轨道差异在沿迹方向上的减小说明了嫦娥一号轨道跟踪数据对月球重力场低阶次系数解算精度的改进效果是可靠的。

表 2 不同模型预报轨道与重建轨道差异的比较

轨道矢量方向	径向(m)	沿迹(m)	法向(m)
LP100K	3.16	1563.25	50.80
SGM100h	9.55	1556.46	15.44
CEGM02	8.50	1366.64	15.33

通过卫星跟踪数据解算得到的高精度月球重力场模型可以提供月球转动惯量信息。Apollo 计划期间在月面放置了 4 个月震仪，运行期间接收到了大量月球深震信息，通过对 P 波和 S 波走时信息的分析，可以加强对内部构造特别是月壳的了解。由于 S 波强烈衰减，无法提供月球更深部的信息[2]。目前对月核成分和半径的了解主要通过月球转动惯量和物理天平动信息[19,20]。LLR 可以得到月球天平动参数 β 和 γ，结合月球重力场 2 阶项位系数信息，可以解算得到月球转动惯量及平均转动惯量 A，B，C 和 I，计算公式如下[2]：

$$\beta = \frac{C-A}{B}, \quad \gamma = \frac{B-A}{C}$$
$$J_2 = \frac{2C-A-B}{2MR^2}, \quad C_{22} = \frac{B-A}{4MR^2} \quad (3)$$
$$I = \frac{A+B+C}{3}$$

式中 M 为月球质量，R 为月球平均半径，J_2 和 C_{22} 分别为月球重力场二阶项位系数，其中 J_2 取为二阶项带谐位系数的负值。

根据 LLR 观测得到的天平动信息，结合不同重力场模型 2 阶项位系数，表 3 给出了平均转动惯量的计算值。

表 3 不同模型的平均惯性矩

	平均惯性矩	方差(5 倍形式误差)
LP100K[6]	0.393120	0.00016
SGM100h[5]	0.363493	0.00008
CEGM02	0.393446	0.00006

由表 3 可以看出，CEGM02 解算得到的平均转动惯量相比于 SGM100h，精度略有改进，计算结果与 Williams 给出的值比较接近。平均转动惯量给出月球密度径向分布积分约束，是用于推断月核成分和大小的一个重要地球物理量[21]。根据 CEGM02 平均惯性矩的误差信息，假设月核半径为 300km[22]，月核密度的解算误差可以约束至 $0.08 \text{gm} \cdot \text{cm}^{-3}$。

3 讨论与结论

通过对嫦娥一号、"月女神"一号、LP 及历史积累月球探测器跟踪数据的融合处理和分析，解算得到了 100 阶次的新月球重力场模型 CEGM02。解算结果表明，增加嫦娥一号轨道跟踪数据后，对现有月球重力场模型中低阶次位系数解算精度有显著改进。通过阶方差及误差阶方差计算，CEGM02 相比于 SGM100h 在 5 阶项以内精度上改进达 2 倍，在 10 阶以内有明显贡献，在 20 阶内都有贡献。通过重力场异常及 LP 长期轨道预报误差检验表明本文模型是合理可靠的，可以作为我国未来月球探测中的参考模型使用。

通常情况下进行卫星重力场探测时，卫星飞行高度越低对重力场响应越敏感。以上解析过程还表明，同时采用飞行在不同轨道的探测器进行重力场反演时，分离出引力效应中的长波部分的能力还与卫星自由飞行弧段的长短有关。比如在"月女神"任务中，由于主卫星的自由飞行弧段长度只有 6~9h，如果仅仅使用主卫星的双程和四程多普勒数据，在重力场低阶上有较大的损失。这时候，如果引入连续不间断自由飞行在100×2400km高的中继卫星多普勒观测量,尽管观测机会不多,使得低阶重力场的精度可以提高2倍多。后者的测量数据总量不多,每周两次,每次数小时。简单地延续上述分析过程,加入200km高度的嫦娥一号卫星的测量数据后,由于卫星连续自由飞行的弧长可以达到24~36h,对重力场低阶项有进一步的改进。

在整理本文的过程中,本文得到的具有自主知识产权的CEGM02模型,从2010年10月开始已经成功用于我国嫦娥-2号探测卫星的测控任务事前分析和任务实施阶段的卫星轨道测控中。

在后续工作中将考虑使用包括"月女神"一号和嫦娥一号的VLBI观测在内的全部数据,预计可以继续提高月球重力场模型精度,目前正在开展这方面的工作。另外,利用模型的2阶项的位系数,并结合LLR给出的天平动信息,计算了月球平均惯性矩,分析得到了月核密度误差约束,这些可作为月球内部构造研究方面的参考。

致谢 嫦娥一号轨道跟踪数据是由中国卫星测控网和VLBI网获得并提供,"月女神"一号跟踪数据由日本国立天文台RISE/SELENE项目组提供,历史探测器轨道跟踪数据由NASA的PDS网站提供。文中的计算分析工作分别在中国科学院上海天文台和日本国立天文台的服务器上完成。GEODYNII/SOLVE软件的使用得到了GSFC/NASA的支持。

参 考 文 献

[1] Kaula W M. The gravity field of the Moon. Science, 1969, 166: 1591—1598

[2] Williams J G, Boggs D H. Lunar core and mantle. What does LLR see? In: Proceedings of the 16th International Workshop on Laser Ranging, Poznan: Poland, 2009. 101—120

[3] Williams J G. A scheme for lunar inner core detection. Geophys Res Lett, 2007, 34: L03202

[4] Namiki N, Takahiro I, Matsumoto K, et al. Farside gravity field of the Moon from four-way Doppler measurement of SELENE(Kaguya). Science, 2009, 323: 900—905

[5] Goossens S, Matsumoto K, Ishihara Y, et al. Analysis of tracking data and results from Kaguya (SELENE) satellites for lunar gravity field estimation. Eos Trans, AGU, Fall Meet Suppl, 2008, 89(53): Abstract P31B-1400

[6] Konopliv A S, Asmar W, Carranza E, et al. Recent gravity models as a result of the lunar prospect mission. Icarus, 2001, 150: 1–18

[7] Matsumoto K S, Goossens Y, Ishiharay, et al. An improved lunar gravity field model from SELENE and historical tracking data: Revealing the farside gravity features. J Geophys Res, 2010, 115, E06007, doi: 10.1029/2009JE003499

[8] Mazarico E, Lemoine F G, Neumann G A, et al. Preparations for Lunar Reconnaissance Orbiter gravity and altimetry missions. Eos Trans, AGU, Fall Meet Suppl, 2008, 89(53): Abstract P31B-1401

[9] Zuber M T, Smith D E, Alkalai L, et al. Outstanding questions on the internal structure and thermal evolution of the moon and future prospects from the grail mission. In: 39th Lunar and Planetary Science Conference (Lunar and Planetary Science XXXIX), League City, Texas, 2008. 1074

[10] Smith D E, Zuber M T, Neumann G A, et al. Initial observations from the Lunar Orbiter Laser Altimeter. Geophys Res Lett, doi:10.1029/2010GL043751, in press

[11] Tapley B D, Bettadpur S, Watkins M et al. The gravity recovery and climate experiment: Mission overview and early results. Geophys Res Lett, 2004, 31: L09607, doi:10.1029/2004GL019920

[12] Ouyang Z Y. Scientific objectives of Chinese lunar exploration project and development strategy (in Chinese). Adv Earth Sci, 2004, 19(3): 351–358 [欧阳自远. 我国月球探测的总体科学目标与发展战略. 地球科学进展, 2004, 19(3): 351—358]

[13] Yan J G, Ping J S, Li F, et al. Chang'E-1 precision orbit determination and lunar gravity field solution. Adv Space Res, 2010, 46: 50—57

[14] Rowlands D D, Marshall J A, Mccarthy J, et al. GEODYN II System Description. Vols. 1-5. Contractor Report, Hughes STX Corp. Greenbelt, Maryland, 1997

[15] Ullman R E. SOLVE Program: Mathematical Formulation and Guide to User Input. Hughes/STX Contractor Report, Contract NAS5-31760. NASA Goddard Space Flight Center, Greenbelt, Maryland, 1994

[16] Yan J G, Li F, Ping J S, et al. Lunar gravity field model CEGM-01 based on tracking data of Chang'E-1 (in Chinese). Chin J Geophys, 2010, 53(12): 2843—2851[鄢建国, 李斐, 平劲松, 等. 基于"嫦娥一号"跟踪数据的月球重力场模型CEGM-01. 地球物理学报, 2010, V53(12): 2843—2851]

[17] Heiskanen W A, Moritz H. Physical geodesy. Bull Géodésique, 1967, 86(1): 491—492

[18] Kaula W M. Theory of Satellite Geodesy. Applications of Satellite to Geodesy. Waltham, Mass: Blaisdell Pub. Co., 1966

[19] Dickey J O, Bender J E, Faller J E, et al. Lunar laser ranging: A continuing legacy of the Apollo program. Science, 1994, 265: 482—490

[20] Williams J G. DE421 lunar orbit, physical librations, and surface coordinates. JPL IOM 335-JW, DB, WF-20080314-001, March 14, 2008

[21] Wieczorek M A, Jolliff B L, Khan A, et al. The constitution and structure of the lunar interior. Rev Mineral Geochem, 2006, 60: 221—364

[22] Ananda M P, Ferrari A J, Sjogren W L. An improved lunar moment of inertia determination: A proposed strategy. Earth Moon Planets, 1977, 17: 101—120

嫦娥一号绕月探测卫星精密定轨实现

陈 明[1,2]　唐歌实[1]　曹建峰[1,2]　张 宇[1]

(1. 北京航天飞行控制中心，北京　100094；2. 中国科学院上海天文台，上海　200030)

摘　要　对探月任务精密定轨技术进行了论述，分析了轨道确定过程中的关键技术问题。基于 SMART-1 探月卫星测轨数据，对精密定轨软件系统进行了测试验证，3d 数据弧段定轨结果精度优于百米。在嫦娥一号任务实施过程中，各轨道段轨道的计算结果准确，卫星成功进入环月使命轨道，特别是原计划三次中途修正仅执行了一次，为卫星节约了宝贵的燃料。与外部星历互差的结果表明，整个任务阶段定轨精度在百米量级，环月段定轨精度约数十米。实施结果表明，该文给出的定轨技术理论正确，关键技术解决有效，完全满足探月任务工程测控和科学研究的需要。

关键词　嫦娥一号探月卫星　绕月探测　精密定轨　测试验证　精度评估

人类对月球的探测活动从 20 世纪 50 年代末开始至今已有近半个世纪的历程，并取得了丰硕的成果[1,2]。自 20 世纪 90 年代以来，逐渐形成了第二次探月高潮[3]。随着在月球及深空探测领域的不断深入发展，国外在深空航天器测量技术、航天器定轨定位等方面进行了深入的研究，并已形成了完整的方法及体系[4]。美国在该领域走在世界的前列，随着 SMART-1 等探月及深空任务的成功执行，欧空局也建立起了相应的测量和信息处理体系。我国的科研人员也对探月相关问题进行了研究[5,6]，还有部分科研人员在国外开放的软件平台上对利用国外的深空探测任务获取的测量数据进行了分析。但由于缺乏探月任务支持，国内尚无独立自主的可用于支持探月任务的测量及定轨系统。21 世纪初，我国提出了适合自己国情的探月计划，整个计划可分为绕月飞行、月面软着陆、采样返回三个阶段。一期工程即绕月探测工程于 2004 年立项启动[7]。为增加测控覆盖率，探月任务中还与欧空局进行了国际合作。这种不同测轨模式、不同测量网间的信息融合技术是轨道计算中要解决的重要问题之一。轨道计算精度对于探月任务的成功实施至关重要，嫦娥奔月遭遇的四大难题中就有两个与轨道相关[7]。

1　探月卫星精密定轨技术

探月卫星精密定轨技术包括估值方法、积分技术、参数求解、动力学模型、观测模型等部分。本文采用的估值方法为加权最小二乘贝叶斯估值方法[8]，采用的积分技术为定步长 Adams-Cowell 多步法和变步长 RKF7(8)相结合的积分技术[8]。可求解的参数包括历元时刻卫星的位置速度、大气阻尼系数(仅在包含近地调相段时考虑)、太阳光压系数、动量轮卸载量、测量数据系统偏差等参数。

1.1　坐标系与星历表

环月卫星相关的坐标系包括月心天球坐标系、月心固连坐标系、月心瞬时真天球坐标系，其具体定义与坐标系间的转换参见文献[9]。本文定轨系统选用的历表为 JPL 发布的 DE 系列历表[9]。考虑到月球非球形引力场采用 JGL165P1 模型,而此模型与 DE403 历表相一致,故在使用中一般选用 DE403 历表。

1.2 运动方程与变分方程

卫星总的加速度可描述为:

$$\ddot{\vec{r}} = \vec{a}_{PM} + \vec{a}_{NS} + \vec{a}_{IO} + \vec{a}_{TID} + \vec{a}_{D} + \vec{a}_{SR} + \vec{a}_{TAC} + \vec{a}_{T} + \vec{a}_{REL} + \vec{a}_{WOL} \tag{1}$$

于是,卫星的运动方程就可以简记为:

$$\ddot{\vec{r}} = f(\vec{r}, \dot{\vec{r}}, t, p), p = (\vec{r}^0, \dot{\vec{r}}^0, p^*)^T \tag{2}$$

其中,$\vec{r}$ 为卫星的位置向量;$\vec{p}$ 为模型参数向量;$\vec{r}_0$、$\dot{\vec{r}}_0$ 为卫星在 J2000.0 惯性坐标系中某历元时刻的位置和速度矢量;p^* 为与摄动力有关的模型参数;$\vec{a}_{PM}$ 为由 n 个天体产生的质点引力加速度;$\vec{a}_{NS}$ 为由天体引力位的非球形部分产生的非球形引力加速度;$\vec{a}_{IO}$ 为由地球和月球相互间的非球形部分吸引力产生的加速度;$\vec{a}_{TID}$ 为由天体潮汐引起的加速度;$\vec{a}_{D}$ 为由气动力产生的加速度(该项加速度仅在调相段存在);$\vec{a}_{SR}$ 为由太阳辐射压产生的加速度;$\vec{a}_{TAC}$ 为由卫星姿态控制系统调整过程引起的加速度;$\vec{a}_{T}$ 为由卫星发动机推力产生的加速度;$\vec{a}_{REL}$ 为由相对论效应引起的加速度;$\vec{a}_{WOL}$ 为由卫星动量轮卸载引起的加速度。

对于嫦娥一号卫星测控过程,调姿与轨控过程在定轨过程中一般不予考虑。卫星动量轮卸载是通过喷气调姿实现的,这里对其进行单独建模解算。考虑到动量轮喷气卸载持续时间较短,因而可假定其等效为惯性空间 x、y、z 方向上的常值作用力模型,因此有:

$$\vec{a}_{WOL} = (a_x \ a_y \ a_z)^T = (a_{x_0} \ a_{y_0} \ a_{z_0})^T \tag{3}$$

其中,a_{x_0}、a_{y_0}、a_{z_0} 为常值,可作为已知量使用或者作为动力学方程中的待估参数求解。卸载作用时间为卸载持续时间。模型偏导数为:

$$\frac{\partial \vec{a}_{WOL}}{\partial (a_{x_0}, a_{y_0}, a_{z_0})} = \frac{\partial (a_x, a_y, a_z)}{\partial (a_{x_0}, a_{y_0}, a_{z_0})} = \boldsymbol{I}_{3\times 3} \tag{4}$$

由于动量轮卸载的相关信息可能包含在星上遥测数据中,因此卸载作用力可由遥测数据进行估算,该估算值可作为卸载作用力参数估值的初值使用;若无该信息(如测控区外卸载情形),则可以设置其初值为零,再进行解算。

在轨道确定过程中,还需要计算状态变量 $\vec{r}(t)$、$\dot{\vec{r}}(t)$ 的变分[10]。

1.3 观测量与积分中心

对于 CE-1 月球探测卫星,测量系统包括 USB 和 VLBI 测量系统。USB 测量系统得到的观测量为三程测距和(ρ_{RT})、双程测距(ρ)、测距变率($\rho_>$)、方位角(A)和仰角(E)。由于距离遥远,角度测量量的精度很低,对轨道约束很弱,在定轨中不用。VLBI 跟踪系统得到的观测量为时延、时延率。

对 USB 测量系统,其所测得的斜距是从发射机经卫星到地面接收机的往返信号传输时间及对应的时标来计算的。双程测距和三程测距的主要区别是双程测距信号的接收和发射在同一个测站,而三程测距信号的接收和发射在不同的测站。具体处理时可统一成三程测距进行处理。

由于在从地球到月球的飞行过程中,积分中心需在地心与月心间转换,特别是当积分中心为月心时,状态量是相对于月心,而观测量是基于地面测站,观测偏导数可以认为是相对于地心惯性系的状态量。假定卫星的地心位矢为 $\vec{r}$,月心的地心位矢为 $\vec{r}_{EL}$,卫星的月心位矢为 $\vec{r}_L$,忽略相对论效应的影响,则其关系可表示为 $\vec{r} = \vec{r}_{EL} + \vec{r}_L$。设 Y 为观测量,则观测偏导数可表示为 $\partial Y/\partial \vec{r}_L = (\partial Y/\partial \vec{r}) * (\partial \vec{r}/\partial \vec{r}_L) =$

$\partial Y/\partial \vec{r}$，因而不必区分观测偏导数是相对于地心还是相对于月心。但在计算观测量时，还需要考虑地心坐标系与太阳系质心坐标系间的转换关系，这是由于月球星历是在质心坐标系下描述的[11]。

1.4 估计模型与轨道确定

轨道确定的基本过程是对来自观测模型的一组参数的估值进行微分改正，以便测量的观测数据和该模型计算的对应量之差值的加权平方和为最小。这其中涉及的关键问题包括动量轮卸载、自旋对测量的影响、测量数据加权、测量量系统差的求解、测量数据的选用与初轨计算等。

由于嫦娥一号卫星自主飞行过程中的姿态保持采用动量轮控制的方式，而动量轮卸载采用喷气卸载的方式，会带来质心方向的加速度，这个问题在定轨过程中需要考虑，具体模型如前文所述。在实际应用中，CE-1卫星在调相轨道段、地月转移轨道段动量累积较少，动量轮卸载主要安排在轨控前与调姿过程相结合。在环月飞行阶段，动量累积加剧，一般1~2 d会卸载一次。解算弧段为2~3 d，解算卸载次数为1~2次，更长的弧长对提高定轨精度无明显的意义。

卫星自旋对测量的影响也是定轨要考虑的问题之一。一般来说，自旋不会引起定轨结果的系统性偏差，但会影响到残差的水平，进而影响到对定轨结果精度的评估。CE-1卫星在调相和转移段为慢自旋状态，其对测量量的影响很小，可不必考虑。初轨计算问题是轨道确定的重要组成部分，在没有轨道先验信息或轨道先验信息较弱时，仅利用测轨数据来获取具有一定精度的轨道初值，供轨道改进使用[10]。

1.5 定轨软件系统

本文描述的定轨软件系统为北京航天飞行控制中心针对CE-1卫星任务而研发的地面支持系统的一个组成部分。定轨软件系统包括测量数据实时接收与监视、测量数据预处理、初轨计算、轨道改进、轨道评估与选优、轨道预报等几个部分。以数据、曲线图、预报残差等形式实时监视测量数据的接收情况。测量数据预处理则主要完成对测量数据的预处理功能。初轨计算则主要完成初始轨道的计算工作，为后续的轨道改进提供所需的初始状态。初轨计算一般仅需在入轨及轨控后进行。轨道改进完成卫星精密轨道确定的工作。轨道评估与选优完成对轨道精度的评估与选优工作。轨道预报完成对后续一个时段内的轨道外推计算的功能，这是编制后续飞控计划及组织飞控实施的依据。由此可见，定轨软件系统在整个飞控任务实施过程中起着基础而重要的作用。

2 定轨验证

探月任务精密定轨软件系统是以上文所述的技术为理论基础，针对探月任务的特点并兼顾其他卫星定轨需求完全独立自主开发的一套系统。探月卫星精密定轨计算结果的测试与验证是定轨系统要解决的重要问题之一。对定轨系统的测试验证包括内部测试验证和外部测试验证。内部测试所需的数据通过仿真产生，内部测试验证主要解决软件与方案的一致性及系统的自洽性问题。外部测试验证主要是将定轨计算结果与外部相对成熟软件的计算结果进行比较及对实测数据进行处理分析。

测轨系统于2006年5月组织对绕月飞行的SMART-1卫星(http://sci.esa.int/science-e/www/object/)进行了跟踪测轨试验，本次测轨试验获取了包括USB系统与VLBI系统在内的全部观测类型。本次测轨试验定轨结果的评估采取了内符合与外符合相结合的评估方式。内符合主要采取本定轨软件重叠弧段定轨结果星历偏差比较的方式，外符合采取本定轨软件计算结果与ESA事后精密星历的比较的方式。本次测轨数据重合弧段星历的比较结果表明，在搭接弧段内，位置偏差最大约100 m，主要集中在沿迹方向上。与ESA提供的事后精密星历比较可知，4 d弧段数据定轨，星历最大偏差约120 m，主要集中在沿迹线方向上。

3 嫦娥一号卫星任务中的应用

3.1 测控任务中的定轨及精度评估

嫦娥一号卫星于 2007 年 10 月 24 日 18:05 发射升空，历经绕地调相段、地月转移段、月球捕获段等轨道阶段，于 2007 年 11 月 07 日 08:35 进入环月使命运行轨道，在完成了一系列的在轨测试工作后，进入为期一年的科学探测任务工作阶段。2008 年 11 月，嫦娥一号卫星完成全部预定任务，随即转入扩展任务阶段。2009 年 3 月 1 日，嫦娥一号卫星在地面控制中心的操作下完成了在月面的硬着陆。

参加本次跟踪测轨任务的系统包括 USB 测量系统与 VLBI 测量系统。USB 测量系统包括国内测量站和国外测量站两个部分，国内站主要是青岛站和喀什站，部分其他测量站参与了调相轨道段及地月转移轨道段早期的测量工作；国外站主要是 ESA 的库鲁站等。VLBI 测量系统主要由国内 4 个 VLBI 测量站组成，包括上海佘山站、北京密云站、昆明站及乌鲁木齐南山站等。VLBI 测量系统的数据处理中心在上海天文台。任务实施中，USB 测量数据实时获取并传送至北京飞控中心；VLBI 测量数据则首先传送至 VLBI 数据处理中心，经相关处理后得到时延、时延率数据。

图 1、图 2 为 CE-1 任务期间地月转移段部分弧段定轨数据残差图。统计分析表明，USB 测距残差 RMS 约为 1.5 m，时延 RMS 约为 3.5 ns，时延率 RMS 约为 0.5 ps/s。

CE-1 卫星任务定轨精度事后评估采用外部星历比较的方法。本文用于比较的外部结果由软件 GEODYNII 生成。将任务期间计算的轨道星历与外部事后星历直接进行比较，并以二者的最大偏差作为衡量定轨精度的依据。任务中，各关键弧段定轨及预报星历精度评估如表 1 所示。

分析表 1 结果可知，嫦娥一号卫星自入轨开始至顺利进入工作轨道期间，全程定轨位置精度在百米量级，速度精度在 1 cm/s 量级。轨控特征点预报位置精度在百米量级，速度精度在 10 cm/s 量级，误差最大值为第一次近月点预报。近月捕获点附近的轨道精度较低是由轨道特性所决定的。第一次近月制动点为以月球为中心引力体的双曲轨道的近月点，第二次近月制动点为以月球为中心引力体的大椭圆轨道的近月点，这些位置处的卫星状态变化迅速，轨道确定和预报的精度相对较低。由于高精度的轨道计算与轨道预报结果使得原计划三次地月转移中途轨道修正仅执行了一次，为卫星节约了大量的燃料。

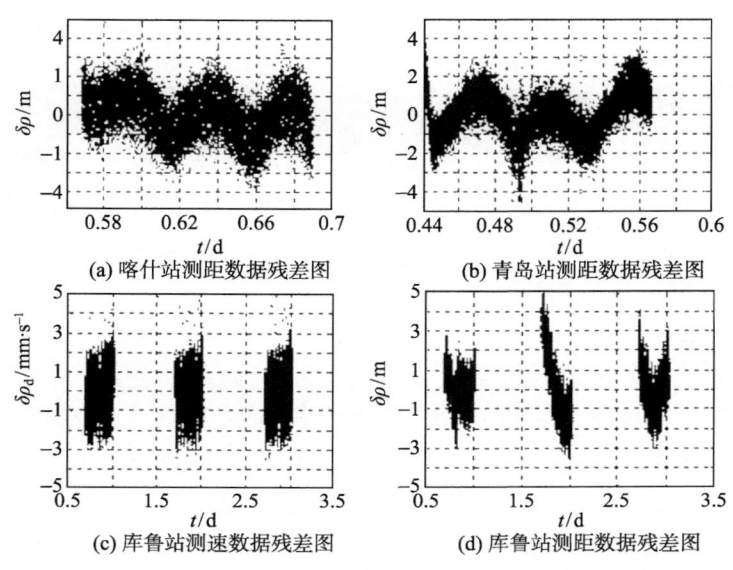

图 1　地月转移部分测量弧段 USB 数据定轨残差图

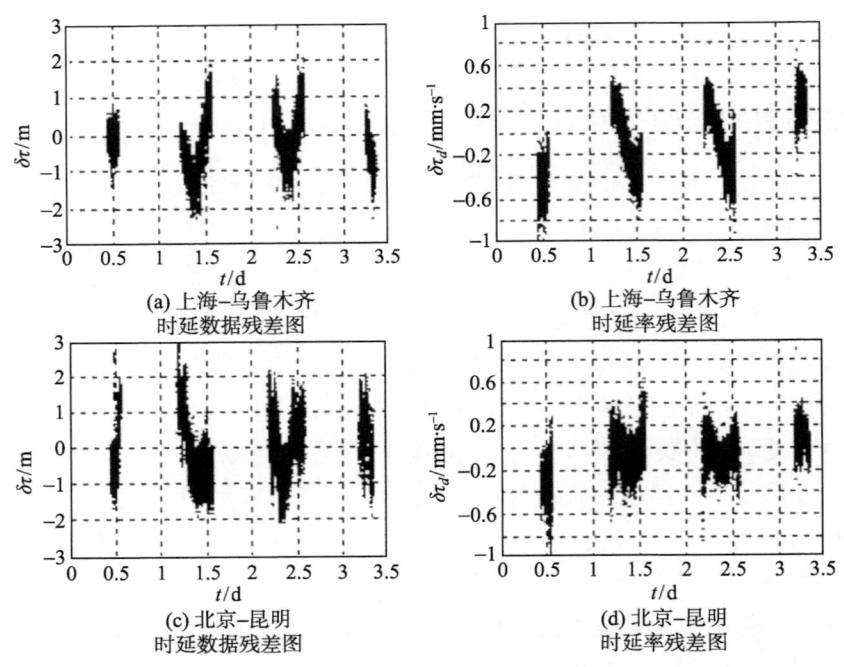

图 2 地月转移部分测量弧段 VLBI 数据定轨残差图

表 1 定轨及预报精度比较

定轨弧段及预报目标点	定轨弧段星历		预报目标点星历	
	位置偏差最大值 /m	速度偏差最大值 /m·s⁻¹	位置偏差最大值 /m	速度偏差最大值 /m·s⁻¹
第二次近地点控前 3h 第二次近地点	100	0.05	130	0.1
第三次近地点控前 3h 第三次近地点	100	0.03	180	0.13
中途修正点前 6h 中途修正点	100	0.025	140	0.004
第一次近月制动前 8h 第一次近月制动点	500	0.01	750	0.6
第二次近月制动前 6h 第二次近月制动点	300	0.1	100	0.1

3.2 环月段定轨精度分析

环月轨道为本次探月任务的使命轨道，这里给出正常环月一个时段的轨道计算结果比对情况，用以说明环月段的定轨精度情况。环月段轨道运行高度为 200 km 的圆极轨道，取北京时间 2007 年 11 月 10 日 7:00 至 18:00 时段作为分析对象，计算结果与 GEODYNII 计算结果的星历偏差如图 3 所示。由图 3 可见，定轨星历偏差小于 40 m，其中径向偏差约 10 余 m。

3.3 动量轮卸载影响分析

嫦娥一号卫星的动量轮卸载会产生喷气推力，进而对轨道产生影响。以 2007 年 11 月 10 日 7:00 至 18:00 间的观测弧段为例，动量轮卸载时刻为 9:01。在不考虑动量轮卸载影响的条件下，定轨数据残差如图 4(a) 所示。由图 4(a) 可知，9:01 前后，数据拟合有明显的差异，其后的数据拟合残差明显优于之前的数据拟合结果。这种差异主要是由于动量轮卸载造成的，后段的数据拟合优于前段是由于后段数据在弧段长度和数据量上占有明显的优势所致。

同样的测轨数据，但在定轨过程中考虑卸载的影响，卸载加速度如式(3)所示，定轨过程中解算该加速度。数据残差如图 4(b)所示，可以看出，数据拟合程度显著好转，特别是卸载前的数据，这表明式(3)能够较好地描述卸载过程。

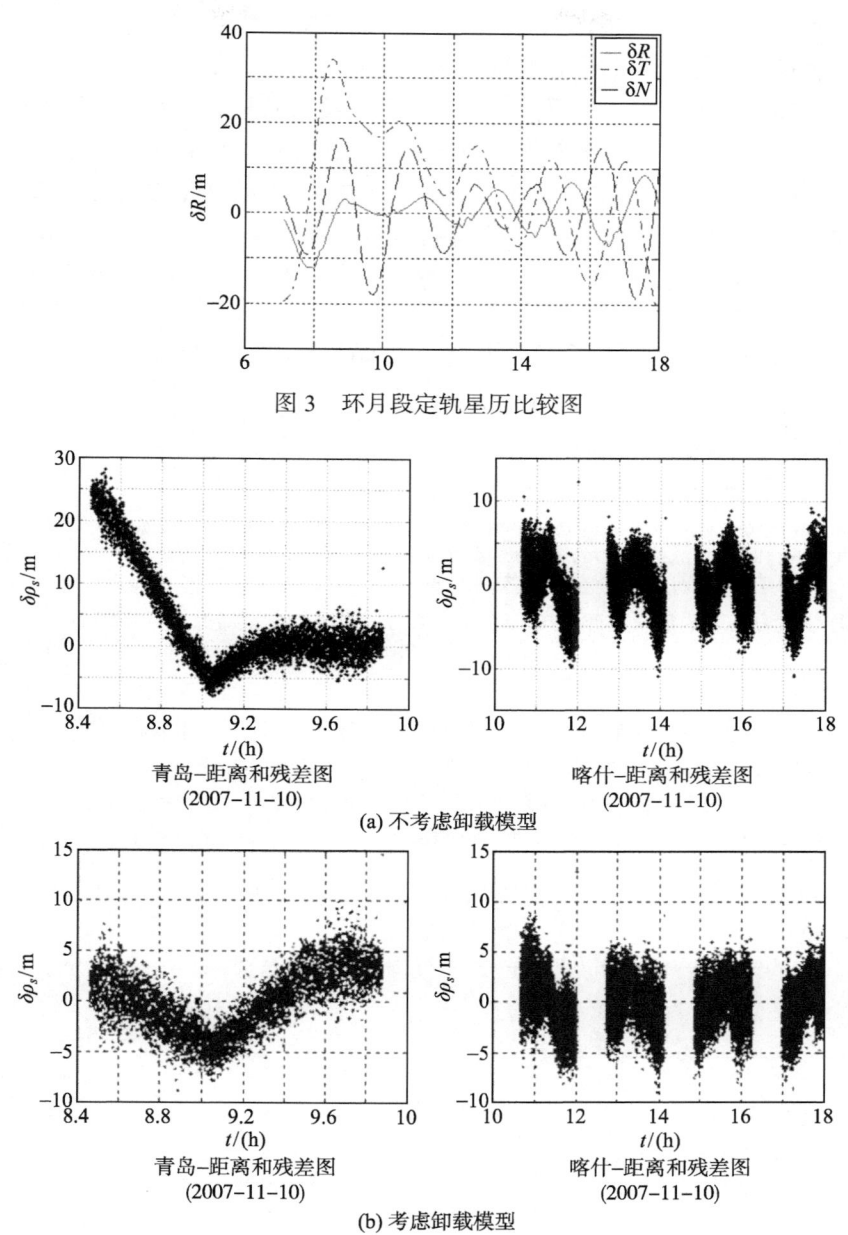

图 3　环月段定轨星历比较图

图 4　环月段青岛站与喀什站距离和数据残差图

轨道确定系统除应用于支持嫦娥一号卫星任务工程测控支持外，还可实现对测量数据的事后精密定轨。精密定轨精度受限于测量设备能够达到的精度。环月卫星定轨精度与引力场精度密切相关，在使用新的更精确的月球引力场的条件下，定轨精度将会进一步提高。精密定轨为嫦娥一号卫星的科学应用提供了星历支持。如月球地形模型 CLTM-S01[12]的计算即基于本轨道确定系统提供的精密星历。

4　结语

嫦娥一号绕月探测卫星精密定轨对嫦娥一号绕月探测卫星的测控支持是我国月球探测任务的首次实

现，也是我国迈向深空探测的第一步。探月卫星的定轨技术将为我国后续的月球探测任务及其他深空探测任务提供重要的技术基础。

提高定轨精度将依赖于三个方面的工作，一是提高测量精度；二是提高测量数据修正精度；三是提高定轨模型精度。与国际上的先进水平相比，我们现在的测量精度还有很大的提升空间，应该加强在这方面的工作。同时探索新的高精度的探月卫星的测量手段和方法也是十分有意义的[13]。采用新的手段和方法以进一步提高测量数据的修正精度，包括中性大气折射修正精度及电离层时延修正精度，也是提高定轨精度需要完成的一个重要工作。提高定轨模型精度，特别是相对论框架下的动力学与观测模型的精度，细化各种力学因素的建模分析；进一步提高定轨精度，以适应深空探测的需要，进行诸如月球/行星引力场的研究等深空探测领域的工程和科学研究，将为卫星精密定轨技术的深入发展带来广阔的前景。

致谢 感谢上海天文台的平劲松博士、胡小工博士和黄勇博士对本文工作的帮助。

参 考 文 献

[1] 郗晓宁，曾国强，任萱，等. 月球探测器轨道设计[M]. 北京：国防工业出版社，2001
[2] 吴伟仁，刘晓川. 国外深空探测的发展研究[J]. 中国航天，2004(1)：26—30
[3] 陈钦丽. 世界各国探月计划一览[J]. 航天员，2006, (5)：24—27
[4] Moyer T D. Formulation for Observed and Computed Values of Deep Space Network Data Types for Navigation, Monograph 2[C]. Deep Space Communications and Navigations Series, NASA/JPL, 2000
[5] 刘经南，魏二虎，黄劲松，等. 月球测绘在月球探测中的应用[J]. 武汉大学学报·信息科学版，2005, 30(2)：95-100
[6] 李斐，鄂建国. 月球重力场的确定及构建我国自主月球重力场模型的方案研究[J]. 武汉大学学报·信息科学版，2007, 32(1)：6—10
[7] 文广. "嫦娥奔月"遭遇四大难题[J]. 科学之友，2006 (5)：43
[8] 汤锡生，陈贻迎，朱民才. 载人飞船轨道确定和返回控制[M]. 北京：国防工业出版社，2002
[9] 陈明，唐歌实. 月球星历及坐标转换的实现[C]. 飞行力学与飞行试验学术交流年会，西宁，2004
[10] 曹建峰，陈明，唐歌实. 探月工程中初轨计算方法的研究[C]. 航天测控技术研讨会，西宁，2006
[11] Moyer T D. Mathematical Formulation of the Double-Precision Orbit Determination Program [R]. Technical Report 32-1521971, NASA/JPL, 1971
[12] 平劲松，黄倩，鄂建国，等. 基于嫦娥一号卫星激光测高观测的月球地形模型 CLTM-S01[J]. 中国科学(G 辑)，2008, 38(11)：1601—1612
[13] 刘基余. "嫦娥"卫星绕月飞行轨道的激光测定法[J]. 武汉大学学报·信息科学版，2005, 30(10)：870—880

深空探测用数字开环多普勒技术初步研制及其在嫦娥一号探测任务中的应用

简念川[1]　尚　堃[1,2]　张素君[1,2]　王明远[1,2]　史　弦[1,2]
平劲松[1]　鄢建国[5]　唐歌实[3]　刘俊泽[3]
邱　实[1]　冯礼和[1]　张　华[4]　王　震[4]

(1. 中国科学院上海天文台，上海 200030；2. 中国科学院研究生院，北京 100049；
3. 北京航天飞行控制中心，北京 100094；4. 中国科学院国家天文台乌鲁木齐天文站，
乌鲁木齐 830011；5. 武汉大学测绘遥感信息工程国家重点实验室，武汉 430070)

摘　要　基于数字无线电技术，开发了用于卫星视线方向速度测量的无线电开环多普勒测量方法和技术原理样机。并在嫦娥一号卫星测轨任务中进行了观测试验，利用卫星转发的 S 波段载波信号进行开环多普勒测量。结果显示，在 1 s 积分情况下，嫦娥一号卫星的开环多普勒测量精度 RMS 达到 3 mm/s(1σ)，这已经与目前使用的 USB(Unified S-Band, 统一 S 波段)测速数据的精度水平相当。这个测量精度主要受制于上行站原子钟的短期稳定度。进一步通过两站开环差分测量的办法，可以有效地消除信号发射时的原子钟频率漂移和不稳定性，使得测量精度 RMS 提高到 1 mm/s(1σ)。开环多普勒数据和差分数据已经开始尝试用于嫦娥一号卫星的定轨，数据的精度评估和科学应用也将逐步展开。这项技术的研制成功将对我国未来深空探测有重要的意义。

关键词　开环多普勒　差分开环多普勒　嫦娥一号　深空探测

深空探测器的无线电通讯系统主要有两个作用：(i) 对航天器的指令和控制，通过从地球向航天器传输信号，即上行链路；(ii) 将遥测信号返回地面，通过从航天器向地球传输信号，即下行链路。这种无线电通信系统主要应用于两个方面：无线电科学和深空探测器导航。无线电科学是指利用在深空探测任务中测量到的无线电波的振幅，相位，极化等特性进行的科学研究，其应用领域包括：行星质量估计和质量分布，行星电离层、大气层、和环的测量，行星形状和表面的研究，太阳风的观测，以及检验广义相对论等[1,2]。而深空航天器导航，主要目的是通过无线电测距和地球航天器之间信号的多普勒频移测速获得探测器的位置和速度信息，对于月球或行星距离的深空航天器的导航定位是通过不同的无线电和光学手段来完成的。从 1970 年开始，根据行星际航天器飞行轨道的不同阶段，一般是利用某一种数据类型(测距或测速)的组合来进行跟踪观测，其中也包括了 VLBI(very long baseline interferometry)观测数据。到 20 世纪 80 年代，深空无线电跟踪基本上是依赖于测距(Range)和多普勒(Doppler)测速系统。2000 年以后，在不断追求高精度和低成本的基础上更强调了系统的稳定性和实时性，快速修正、即时响应的跟踪模式对光学和无线电跟踪手段都提出了需求[3]。

根据通讯系统的上行链路和下行链路的构型可以将台站对航天器的跟踪划分为单程(One-way)、双程(Two-way)和三程(Three-way)模式[3]，而根据通讯链路是否闭合分为闭环(Closed-loop)和开环(Open-loop)模式。单程跟踪模式是指由航天器上携带的晶体振荡器产生下行信号，地面台站接收并跟踪信号，这种模式

没有上行链路。双程跟踪模式是由地面台站向航天器发射上行链路，航天器接收上行信号并通过星上转发设备产生一个相干的下行信号，同一个地面台站将接收下行链路，并将这个信号与发射时的参考频率相比较，这种模式有上行和下行链路，并且信号发射源和接收源为相同的台站。三程跟踪模式与双程类似，区别只在于信号发射和接收由两个不同的台站来完成。双程和三程的区别如图1所示。另外，通信系统所采用的频段是在国际上统一划分的。表1是ITU(international telecommunication union)划分的用于深空通信的频率范围[4]。对于双程或三程多普勒跟踪测量，航天器必须发射一个同上行信号相干的下行信号，即对上行信号乘上一个转发比。表2给出了美国深空网DSN(Deep Space Network)使用的，也是CCSDS测控通信体制建议的不同波段常用的星地链路锁相转发比[4]。

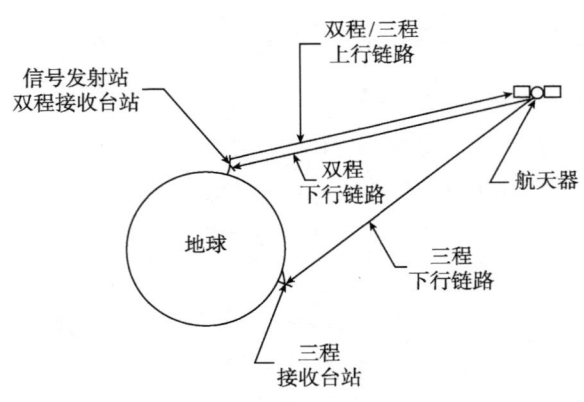

图1 双程或三程数据发射、接受模式示意图[5]

表 1 深空通信的上行和下行频率

频段	上行频率/MHz	下行频率/MHz
S	2110~2120	2290~2300
X	7145~7190	8400~8450
Ka	34200~34700	31800~32300

表 2 航天器收发机的转发比

上行频段	下行频段	转发比(下行/上行)
S	S	240/221
S	X	880/221
S	Ka	3344/221
X	S	240/749
X	X	880/749
X	Ka	3344/749

在这3种基本的跟踪模式中，双程测量是一种闭环跟踪模式，依靠一个台站同时完成信号发射和接收，这样的通讯链路是闭合的。而单程和三程为开环跟踪模式，其通讯链路两端开放。开环模式与闭环模式在深空探测器测控控制方面的最根本区别在于，开环测量模式要求信号发射源(台站或卫星)和接收源(另一个台站)采用相互独立的频率标准源，在获取Doppler时，增加了测量中的不稳定性和频率源带来的误差。开环测量方法对测量技术和数据处理方法都提出了更高的要求。早期的深空探测导航系统多以闭环测量为主，如美国的阿波罗号、海盗号、旅行者号等，到后来发展为完善的USB(Unified S-Band，统一S波段)测量系统。20纪80年代后期，星上晶体振荡器稳定度有较大提高，使得单程开环测量模式变为可能[6]。开环测量包括单程/三程多普勒(One-way/Three-way Doppler)、差分单程/三程多普勒(Differenced One-way/Three-way Doppler, DOD/DTD)和差分单程/三程测距(Differenced One-way/Three-way Range, DOR/DTR)等测量模式。相对于传统闭环测量，开环测量模式优势主要在于以下几点：对于单程开环测量，

其优势主要在于不需要上行链路,只需要卫星发射下行信号,这样可以节省资源消耗,更容易在遥远距离的深空测量中实现。而对于三程开环测量,虽然与双程相比仍然需要上行信号,但在某些深空任务中,由于飞行器距离非常遥远,当信号从卫星转发回地球时发射站已经不在视线范围内了,这时无法做到双程闭环跟踪,只能由另外一个测站来接收信号,即必须采用开环三程的跟踪模式;同时这种类似 VLBI 的开环三程测量方式在对卫星横向速度的分辨率上有明显的改进;另外如果上行台站和下行台站沿东西方向分布,还可以有效的延长可视时间。近年来美国 DSN 深空网对伽利略号、火星勘测者号等深空探测器都应用了开环测量模式,开环测量在美国深空网的应用中得到了长足的发展,特别是三程测速模式和 VLBI 测量模式,在其主导的深空测控任务中发挥了巨大的作用。

由两个接收站独立的开环测量所组成的开环差分观测模式能提供精度更高的观测值[7,8],包括差分单程/三程多普勒(DOD/DTD)和差分单程测距(DOR)。差分测量的优点在于可以很好地消除开环观测中信号发射源(台站或卫星)的系统频率漂移等误差源,缺点在于差分多普勒速度测量消除了大部分的飞行器径向速度信息,单独使用此观测量不能得到好的定轨结果,在实际的应用中需结合非差分观测量来定轨[9]。

我国首颗月球探测卫星——嫦娥一号,于 2007 年 10 月 24 日在西昌卫星发射基地发射,2007 年 11 月 7 日进入环月轨道,工作一年来发回大量科学探测数据。嫦娥一号的导航主要依赖于中国 USB(Unified S-Band,统一 S 波段)测控网的测距和测速数据,以及中国 VLBI 网的 VLBI 数据。USB 的测量方式是由上行台站发射 S 波段的上行信号到嫦娥卫星,嫦娥卫星接收并通过星上转发设备转发下行信号到原台站,继而获得卫星的距离和速度信息。这是一种典型的闭环测量模式。

在嫦娥一号任务之后是即将到来的中俄联合火星探测任务。该任务计划于 2009 年 9 月或 10 月发射一颗中俄联合火星探测器。届时将由俄罗斯的运载火箭搭载俄罗斯的"Phobos-Grunt"火卫一探测器和中国的 YH-1 火星探测器共同发射。预计于 2010 年 8 月或 9 月到达火星,随后展开相应的科学探测。YH-1 的导航仍然依靠高精度的测距和测速资料。但与嫦娥一号的情况不同,由于中国的深空网尚未建成,国内的上行台站没有足够的发射功率可以把上行信号发射到火星距离(0.5 AU~2 AU),因此 YH-1 的通信和测量必须采用基于星上晶振的单程开环方式。相应的在 YH-1 探测器上配备了星载接收机和发射机的独立通信体制,而没有通常的基于锁相环技术的转发器。这样在 YH-1 任务中,单程开环模式将代替传统的双程闭环模式首次应用于中国的深空探测,即星上发射下行信号,中国的 VLBI 网的天线将直接接收单程信号,继而提取卫星距离和速度信息。嫦娥一号任务为研发开环测量技术提供了一个很好的平台。因此在嫦娥一号任务期间,我们利用三程开环模式来测试单程开环模式,即国内上行站发射信号到卫星,卫星接收并转发成下行信号,然后利用中国的 VLBI 网的天线接收这些信号。为此,分别在中国 VLBI 网的上海佘山站、乌鲁木齐南山站和云南昆明站安装了新的数据采集记录设备,实现远程控制并对基带转换器输出的 4 路基带信号直接进行 A/D 采样,把 VLBI 原始电压信号记录到硬盘中,并通过 TCP-IP 协议传回到用户端,进行事后数据处理,最后提取出开环多普勒(速度)数据和开环差分多普勒(速度)数据,为将来的 YH-1 任务中的开环测量做准备。

1 开环多普勒测量原理

1.1 开环多普勒公式

电磁波的多普勒效应是由于发射器和接收器的相对速度以及所处的不同的引力场造成的两者钟的原时不同而引起的。真空中电磁波的速度始终是光速。如果忽略引力场的变化,只考虑狭义相对论效应,在一阶近似情况下,单程多普勒的表达式可以近似表示为[5]

$$f_R^{1-way} \approx \left(1 - \frac{v}{c}\right) f_T \tag{1}$$

其中 f_T 表示电磁波的发射频率，f_R 表示接收频率。v 表示发射器和接收器的相对速度，定义两者相对远离时速度为正，此时接收到的频率相比发射的频率变低(即红移)。在这种近似下，电磁波的多普勒计算公式和介质波多普勒效应的严格公式是一致的。严格的电磁波多普勒计算公式如下[5]：

$$f_R = \frac{1-\frac{v}{c}\cos\theta}{\sqrt{1-\left(\frac{v}{c}\right)^2}} f_T \tag{2}$$

其中 θ 为电磁波波矢和台站和卫星相对速度的夹角。

对于三程多普勒的情况，设发射站上行频率为 f_T，卫星接收频率为 f_S，此时卫星和发射站的相对速度为 v_1，则两者的单程多普勒频率表达式为

$$f_S \approx \left(1-\frac{v_1}{c}\right)f_T \tag{3}$$

下面考虑三程的情况。假设星上的转发比设为 M，则经卫星转发后的发射频率为 $f_S \cdot M$，接收站接收到的频率为 f_R，此时接收站和卫星的相对速度为 v_2，则两者的多普勒表达式为

$$f_R \approx \left(1-\frac{v_2}{c}\right)f_S \cdot M \tag{4}$$

因此总的三程多普勒频率表示为

$$f_R \approx \left(1-\frac{v_1}{c}\right)\left(1-\frac{v_2}{c}\right)f_T \cdot M \approx \left(1-\frac{v_1+v_2}{c}\right)f_T \cdot M \tag{5}$$

这里忽略了光速的二阶小量。

把两个相对速度的平均值定义为三程多普勒的速度，即有

$$f_R^{3\text{-way}} \approx \left(1-\frac{2v_{3w}}{c}\right)f_T \cdot M \tag{6}$$

$$v_{3w} \approx -\frac{1}{2}\cdot\frac{f_R^{3\text{-way}} - f_T \cdot M}{f_T \cdot M}\cdot c \tag{7}$$

(7)式就是三程多普勒在一阶近似下的转换公式。即只要记录下了信号到达接收站的频率 f_R，并已知上行频率 f_T 和转发比 M，就可以换算到三程速度 v_{3w}。这就是开环多普勒速度测量的基本原理。

1.2 差分多普勒公式

考虑站 1 和 2 同一时刻观测 t_R 接收到两个单程多普勒频移 $f_R^{1\text{-way},S1}$ 和 $f_R^{1\text{-way},S2}$，相同的接收时刻对应着不同的发射时刻 t_{T1} 和 t_{T2} 通过(1)式构造差分单程多普勒测量表达式为

$$\begin{aligned} D\_f_R^{1\text{-way}} &= f_R^{1\text{-way},S2}(t_R) - f_R^{1\text{-way},S1}(t_R) \\ &= \left(1-\frac{v^{S2}}{c}\right)f_T(t_{T2}) - \left(1-\frac{v^{S1}}{c}\right)f_T(t_{T1}) \end{aligned} \tag{8}$$

考虑站 1 和 2 观测不同的三程多普勒频移 $f_R^{3\text{-way},S1}$ 和 $f_R^{3\text{-way},S1}$，通过(6)式构造差分三程多普勒测量表达式为

$$\begin{aligned} D\_f_R^{3\text{-way}} &= f_R^{3\text{-way},S2}(t_R) - f_R^{3\text{-way},S1}(t_R) \\ &= \left(1-\frac{2v_{3W}^{S2}}{c}\right)f_T(t_{T2})M - \left(1-\frac{2v_{3W}^{S1}}{c}\right)f_T(t_{T1})M \end{aligned} \tag{9}$$

目前多普勒测量中主要的误差源来自于信号发射器的频率标准源(单程对应的是星载晶振而三程对应的是地面原子钟)带来的误差。表征频率标准源性能的主要指标有频率准确度、频率稳定度和频率漂移率。频率准确度是频率标准源输出频率的实际值与标称值的相对偏差。频率稳定度表示频率标准源输出频率因受随机噪声影响产生的随机起伏特性。频率漂移率是指受元、器件老化影响后它的输出频率随时间的线性变化关系。发射器频率标准源的准确度对多普勒观测值的影响表现为系统性，可通过差分方法消除。下面的分析给出了频率标准源对多普勒测量造成的影响。先给出频率标准源的简单演化模型：

$$f_T = f_{T_0} + Bt + A \cdot rand(t) \tag{10}$$

f_{T_0} 为标称值，B 为线性演化因子，A 为相位噪声对应的频率随机分布，可以用阿伦方差来衡量。

对于单程多普勒测量，频率标准源一般为星上的晶振，其频率准确度对多普勒观测值的影响为

$$\Delta f_R^{1\text{-way}} = \Delta f_T \left(1 - \frac{v}{c}\right) \approx \Delta f_T \tag{11}$$

对于三程多普勒测量，频率标准源是地面的原子钟，其对多普勒观测值的影响为

$$\Delta f_R^{3\text{-way}} = \Delta f_T M \left(1 - \frac{2v_{3W}}{c}\right) \approx \Delta f_T M \tag{12}$$

下面考虑差分方法对频率准确度的影响。对于单程差分多普勒测量，晶振准确度对多普勒观测值的影响为

$$\Delta D\_f_{R_{1,2}}^{1\text{-way}} \approx \Delta f_T(t_{T2}) - \Delta f_T(t_{T1}) \tag{13}$$

t_{T1} 和 t_{T2} 对应相同接收时刻对应的信号发射时的不同时刻。

对于差分三程多普勒测量，原子钟准确度对多普勒观测值的影响为

$$\Delta D\_f_{R_{1,2}}^{3\text{-way}} \approx \Delta f_T(t_{T2})M - \Delta f_T(t_{T1})M \tag{14}$$

从上面的分析结果可以看出：如果不考虑其他因素的影响，频率标准源自身的误差和单程多普勒观测误差相当，三程多普勒的观测误差是频率标准源自身误差的 M 倍。而差分多普勒取决于在两次发射时刻间隔 t_{T1} 和 t_{T2} 内频率标准源的准确度，当 t_{T1} 和 t_{T2} 相差不大时，可以认为消除了频率标准源长期漂移对观测量带来的影响，模拟数据表明地月距离的测量中此时间间隔量级约为 0.01 s。因此，非差分的多普勒精度取决于频率标准源的准确度，而差分多普勒测量的精确取决于时刻间隔 t_{T1} 和 t_{T2} 内频率标准源的准确度。这样看来，差分多普勒精度要高于非差分多普勒精度，主要在于消除了频率标准源准确度的误差。

2 嫦娥一号开环多普勒测量实验

2.1 嫦娥一号卫星信标结构

嫦娥卫星在 S，X 波段都有下行信号。X 波段信标设计为 VLBI 准白噪声信号，没有稳定的频点，无法进行多普勒测量。S 波段在 2210 MHz 附近安排有单程下行和上行转发信号，理论上可以进行单程和三程开环测量。然而实验结果表明，单程信号的瞬时稳定度较差：频率稳定度约为 1 Hz，对应卫星自身晶振瞬时稳定度为 4.5×10^{-10}，推算到速度精度约为 0.1 m/s。考虑到这个精度对工程定轨没有贡献，因而没有对嫦娥卫星进行单程多普勒实验。而嫦娥卫星地面上行站为喀什和青岛站，配备有较为稳定的信源，三程多普勒数据分析表明上行站频率标准源短稳约为 1.31×10^{-11}，对应于接收端多普勒精度为 20 mHz，推算的

速度精度为 3 mm/s，可以支持工程定轨。因此本文中的实验针对的是嫦娥卫星 S 波段的三程多普勒信号进行的观测。

2.2 数据采集观测设备与采集流程

利用中国 VLBI 网的射电天线资源，分别在上海佘山站、乌鲁木齐南山站和云南昆明站安装了数据采集记录设备。

数据 A/D 采样设备采用了 ADLINK 公司生产的 NUDAQ PCI-9812/9810 高速数据采集板卡和日本 NICT 开发的(K5/VSSP32)VLBI 数据采集卡。ADLINK 公司的 NUDAQ PCI-9812/9810 是基于 32 位总线的高性能数据采集卡，可在短时间内连续不间断地高速数据采集，并将数据流送到计算机硬盘中，最大的采样速率可达到 20 MbPs。

对这块板卡进行了 2 次开发，增加了外部时钟预触发和触发模块，并引入了相干的外部频标信号[10]，还在短期连续采集间断记录的基础上，增加了数据长期连续采集和记录的功能。ADLINK PCI-9812/9810 的布局如图 2 左所示。

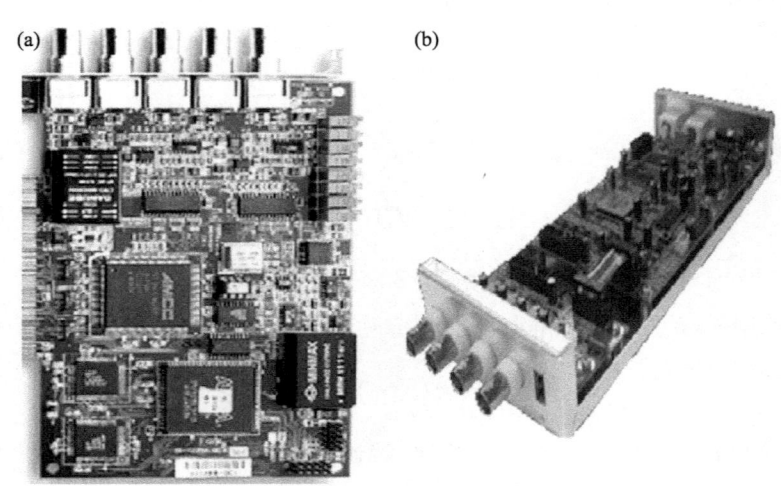

图 2 数据 A/D 采集卡
(a) ADLINK PCI-9812/10 板卡布局；(b) K5/VSSP32 数据采集板卡

日本 NICT 设计开发的 K5/VSSP32 数据采集卡设计为 1，4 通道两种工作模式 VLBI 数据采集设备，单通道采样频率最高可达 64 MbPs，四通道可以达到 256 Mbit/s，量化率最高可达 8 bit。这块板卡是成熟的用于 VLBI 以及其他广泛领域的商业板卡。板卡如图 2 右所示。

航天器发射的信号首先经天线终端接收得到射频(Radio frequency，RF)信号，然后降频到中频(Intermediate frequency，IF)信号，最后输入基带转换器(BBC)得到基带信号，飞行器信号一般放置在基带 500 kHz 左右。远程控制上述 A/D 采集卡对 1~4 路 BBC 输出的基带信号进行 A/D 采样，把数据记录到硬盘中。采取事后处理的方式完成开环多普勒信号的提取，数据记录流程如图 3 所示。

2.3 开环多普勒速度提取方法

与射电源的天文观测不同的是，射电望远镜观测到的卫星信号信噪比较大，信号的能量也较高。利用前面介绍的设备记录到的嫦娥一号卫星的下行基带信号的结构如图 4 所示，左边是直接记录的原始数据，右边是 1 s 原始数据的频谱图，其中采样频率为 4 MHz，记录带宽为 2 MHz。

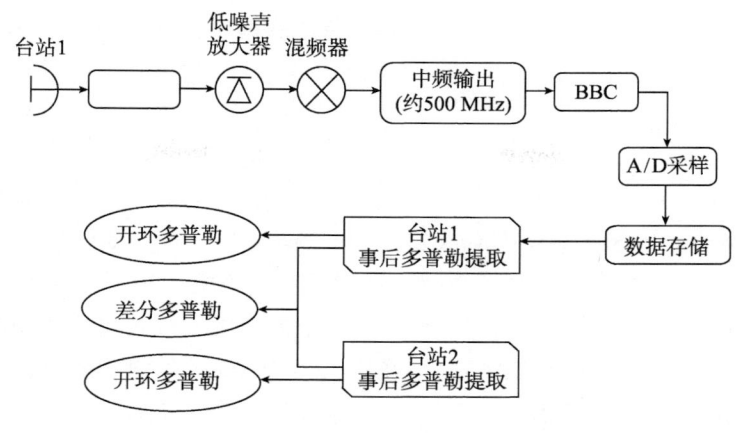

图 3 开环多普勒数据采集流程示意图

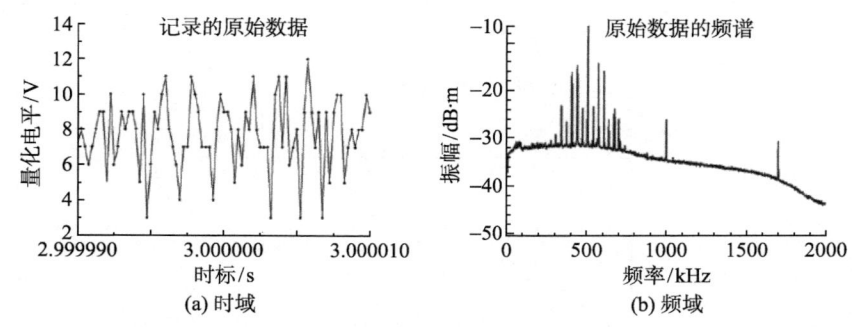

图 4 A/D 采样的嫦娥一号卫星的下行基带信号结构

开环多普勒频率数据提取采用了两种方法，在相同的积分时间内两种方法得到的多普勒值完全吻合：

(1) 对原始数据滤波，下变频后对一段时间内的相位变化计数，相位计数对时间的平均即得到基带原始多普勒值；

(2) 对原始信号滤波下变频后，利用多项式拟合得到的相位对积分时间求平均，从而求得基带原始多普勒值。经实际数据检验。

开环差分多普勒频率数据的提取也采用了两种方式：

(1) 时域方法，对两台站记录的原始基带时域信号进行混频或相关处理，然后提取差分三程多普勒；

(2) 频域方法，对两台站记录的基带信号独立提取三程多普勒值，然后直接做差得到差分三程多普勒。

经实际数据检验，如果不考虑舍入误差两种算法结果并无差别，但在具体实现中由于原始数据为海量数据，显然第二种方法实现起来比较简便，我们的数据处理基于第二种方法。

将得到的开环多普勒频率和开环差分多普勒频率，按照(1)~(9)式给出的开环速度和开环差分速度转换关系，可以将积分多普勒频率转换为积分多普勒速度。在对嫦娥卫星的具体实验中，给出的是 1 s 积分的三程速度 V_{3W} 和 $V_{3W}^{S2} - V_{3W}^{S1}$。下面的数据处理结果都是给出 1 s 积分的开环三程多普勒速度和差分三程多普勒速度，不再给出基带频率。可以按照同样的方法给出不同时间间隔积分的开环三程多普勒速度和差分三程多普勒速度。

3 嫦娥一号开环多普勒测量结果

3.1 开环三程多普勒实验结果

针对嫦娥卫星的开环三程多普勒的观测从 2008 年 5 月展开，首先在上海佘山站安装 A/D 采样并同时开发事后算法,在一年的时间内先后将该系统安装到乌鲁木齐南山站和云南昆明站,整个系统在嫦娥一号卫星任务期间对其进行了多次观测。这里以2008年12月18日的嫦娥卫星观测为例,说明开环三程多普勒的数据情况。

2008年12月18日的佘山站三程多普勒观测是目前为止持续时间最长的一次开环多普勒实验。此时的嫦娥卫星高度近月点约为17 km,远月点约为100 km。观测接收台站为上海佘山站,上行站为喀什站和青岛站。开环三程多普勒的有效观测大约持续9 h,中间更换过上行站。由于任务期间USB测控网同时提供闭环双程多普勒数据,因此这里给出开环和闭环数据的对比情况。

图5(a)给出了大约从UTC 18日18时到19日24时,上海佘山站观测到的开环三程多普勒速度观测结果,单位为m/s,这里已剔除一些明显的粗数据。图中大约19.05日处出现的数据错位现象是由于上行站更换的原因。图5(b)给出了相同时间范围内的USB双程多普勒观测结果作为对比。三程多普勒观测的有效数据大约有9 h,双程多普勒的有效数据大约有13 h。

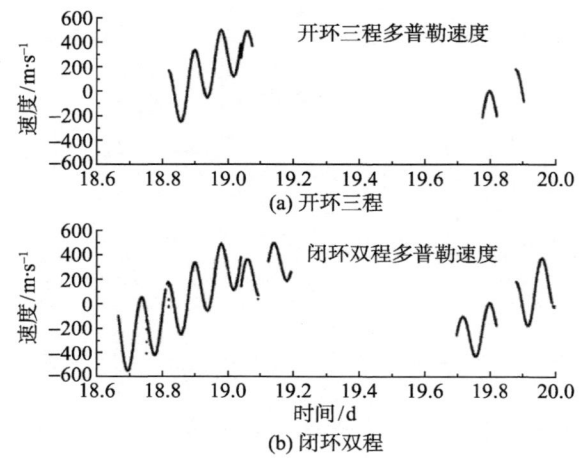

图5　嫦娥卫星多普勒速度观测结果

对上述数据直接做分段多项式拟合,用拟合后残差的RMS(Root Mean Square)表示数据的记录精度。图6给出了分段多项式拟合的残差值(上图为三程多普勒,下图为双程多普勒),单位为cm/s。从图中可以看出,大部分数据的残差都比较小,在±1 cm/s以内,少部分数据的残差比较大,这是不同观测弧段之间台站本身的记录连接问题引起的,在开环和闭环多普勒数据中有类似的现象。剔除大于1σ的野值数据统计RMS[11,12]。

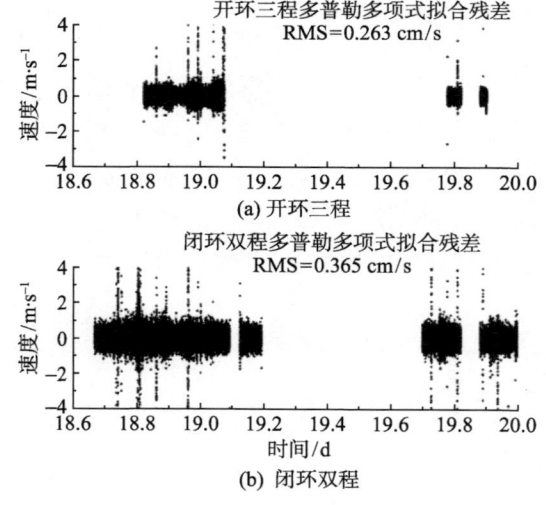

图6　多普勒速度分段多项式拟合残差

结果显示，开环三程多普勒速度残差的 RMS(1σ)约为 0.263 cm/s，闭环双程多普勒速度残差的 RMS(1σ)约为 0.365 cm/s，三程多普勒速度的精度略高于双程多普勒。

将上述开环多普勒速度用于定轨迭代，同样可以评估数据的精度。这里使用的定轨软件为 NASA GFSC 提供的 GEODYN II 软件。定轨过程如下：先将卫星发布的精密轨道作为初轨，然后仅用三程多普勒观测数据进行初轨改进，再用改进后的轨道推算三程多普勒的理论值，将这个理论值减去三程多普勒的观测值就得到定轨残差值。这种做法最大程度的扣除了系统差的影响，可以直接反映出数据的随机差。图 7 中的上图给出了仅用开环三程多普勒数据的定轨残差值，所取的月球重力场为 100×100 阶，图 7 中的下图同时给出了仅用闭环双程多普勒数据的定轨残差值。从图 7 中可以看出，残差并不呈现正态分布，而是有明显的结构。这种结构的低频部分与图 5 中多普勒信号的周期性相关，这显然是由于嫦娥的轨道周期(约 2 h)造成的；另外，残差中的高频变化则很有可能来自于月球高阶重力场的贡献，这为通过视线速度反演月球重力场提供了可能[13,14]。并且开环三程多普勒定轨残差和闭环双程多普勒定轨残差有明显的相关性，这说明两者在定轨残差的水平上是相符的。需要说明的是，这种处理办法仅用于评价数据精度，并不能评估三程数据对定轨的贡献程度，这主要是由于目前数据量较少，无法有效的进行定轨迭代。对开环数据全面的定轨精度评估将在未来的工作重点之一。

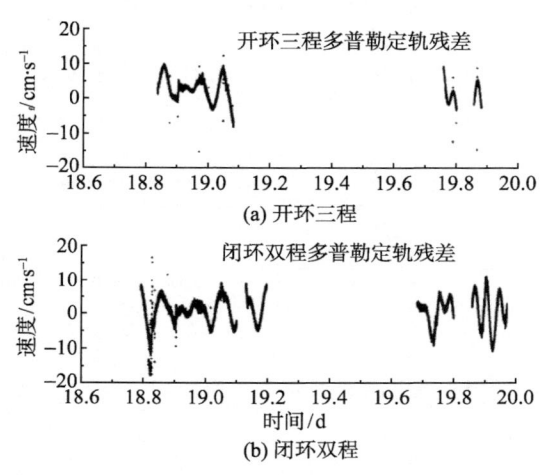

图 7　多普勒速度定轨残差

3.2　开环差分三程多普勒实验结果

2008 年 8 月 29 日，上海佘山站和乌鲁木齐南山站同时进行了嫦娥卫星的开环三程多普勒观测实验，每个站分别记录了两段开环数据。将两站的三程多普勒进行差分处理，首次得到了嫦娥卫星的开环差分三程多普勒结果。由于目前的定轨软件尚不支持开环差分数据，因此利用自编软件由嫦娥轨道计算出了差分的理论值，用观测值减去理论值(O-C)得到残差，统计这个结果以反映差分数据对精度的提高。表 3 给出了差分前和差分后系统差和随机差的统计结果。图 8 给出了 29 日第一段数据的差分结果。

表 3　嫦娥卫星开环差分多普勒残差统计

差分三程多普勒统计结果 2008 年		系统误差 (mean)/cm·s⁻¹	随机差 (std)mm·s⁻¹
8月29日 第一弧段	佘山	4.3	3.3
	南山	4.5	3.5
	差分结果	0.4	1.0
8月29日 第二弧段	佘山	5.2	3.2
	南山	5.4	4.0
	差分结果	0.39	1.2

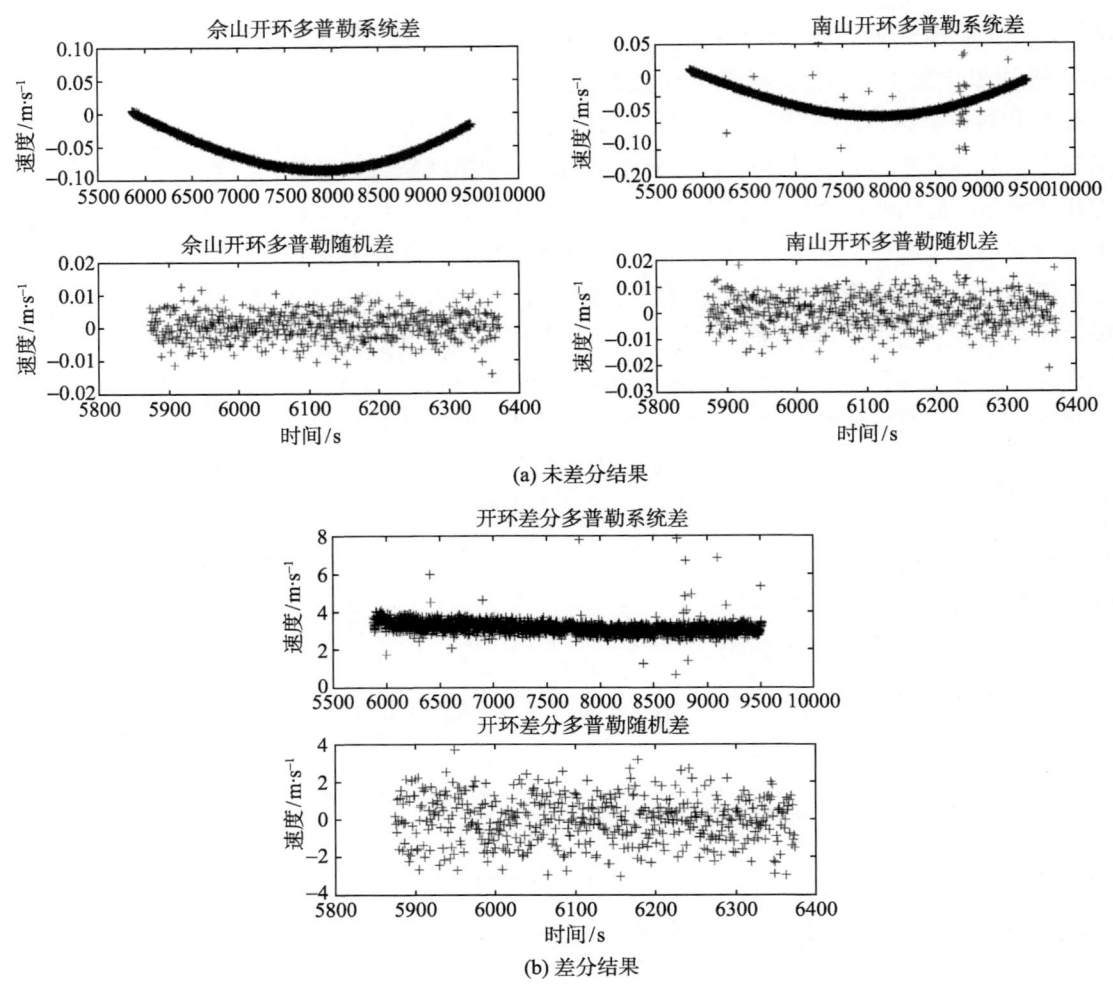

图 8　差分多普勒速度残差(2008 年 8 月 29 日第一弧段)

3.3　开环多普勒结果精度总结

自 2008 年 5~12 月，对嫦娥一号卫星先后进行了 20 多次开环多普勒观测。统计结果表明：在 1 s 积分的情况下，开环三程多普勒测量随机差约为 3 mm/s，精度约为 10^{-11}；差分三程多普勒随机差约为 1 mm/s，精度约为 $3×10^{-12}$。三程多普勒的测量结果随机误差和上行站频率标准源瞬时稳定度(10^{-11})是一致的，而差分三程多普勒的精度由于消除了频率标准源准确度的误差，因此高于非差分多普勒的精度。

从上述结果可以看出差分多普勒可以很显著地提高测量精度，降低随机误差。但是，受频率标准源稳定度的影响，差分三程多普勒精度局限于 1 mm/s，可以满足一般的工程定轨需求，但用来做精密定轨或无线电科学则远远不够。国际上地面原子钟短稳可以达到 10^{-16}，星载晶振可达 10^{-14}，反映到多普勒测量可以达到每秒微米的精度。因此，还需大力提高国内频率标准源指标，以满足目前深空探测的需求。

4　开环多普勒测量的应用和展望

开环多普勒测量将会是未来 YH-1 任务的主要测量模式。对 YH-1 卫星的轨道测量将采用 VLBI 和单程多普勒联合测量模式，萤火卫星将采用 X 波段三点频无调制信号，标称频率分别为：8402，8424 和 8446 MHz。星上频率稳定度可以达到 10^{-12}，与嫦娥地面原子钟的稳定度(10^{-11})相比，YH-1 的开环多普勒测量精度有希望提高一个量级，达到 0.1 mm/s，这将有效地提高工程定轨精度。

在无线电科学研究方面，开环多普勒数据可以用于行星重力场的反演，可以为火星大气掩星观测提供高精度的多普勒或相位输入观测量。通过双频的多普勒观测，可以对行星电离层、大气层、行星环以及磁场等进行探测和科学研究[2]。

根据嫦娥一号卫星的观测结果，1 s 积分的三程多普勒 RMS 为 3 mm/s，300s 积分的 RMS 可达到 0.0577 mm/s。2002 年 Cassini 飞行器的单程多普勒 300s 积分的 RMS 为 0.0022 mm/s[15]，比嫦娥一号的结果好 20 多倍，除了数据处理方法的不同，也和 Cassini 飞行器上晶振的稳定度和巡航段所处的引力场环境有关。如果多普勒的测量能够达到这个精度，就可以在太阳系内开展引力理论验证，如 PPN 参数检验等[15]，这也是高精度多普勒数据在未来深空探测中的应用之一。

5 结论

本文利用嫦娥一号卫星任务期间中国 VLBI 网的卫星跟踪数据，开发了针对深空飞行器开环多普勒速度提取的算法。根据对嫦娥一号卫星开环多普勒的处理结果，1 s 积分的非差分多普勒残差精度达到 3 mm/s，差分多普勒精度达到 1 mm/s。

进一步研究针对月球以及火星距离的更低信噪比、更低幅度的微弱信号的探测器，改进软件算法和效率，改进接收机的性能以及地面信号发射器的稳定度，提高多普勒计算的精度是下一步要继续开展的工作。

致谢 感谢总装 USB 测控网和中国科学院 VLBI 系统对观测实验的支持。感谢 NASA GFSC 提供的 GEODYN 软件平台。

参 考 文 献

[1] Armstrong J W. Low-Frequency Gravitational Wave Searches Using Spacecraft Doppler Tracking. Pasadena: JPL Publication, 2002
[2] Asmar S W, Renzetti N A. The deep space network as an instrument for radio science research. Pasadena: JPL Publication, 1993. 80—93
[3] Thornton C L, Border J S. Radiometric Tracking Techniques for Deep-Space Navigation. Pasadena: JPL Publication, 2002. 10
[4] Yuen J H. Deep Space Telecommunications Systems Engineering. New York: Plenum Press, 1983. 123—178
[5] Ruggier C J. Frequency and Channel Assignments. DSMS Telecommunications Link Design Handbook. Pasadena: JPL Publication, 2000. 5—7
[6] Ellis J. Deep Space Navigation With Noncoherent Tracking Data. TDA Progress Report, 1983. 12: 42—74
[7] O'Reilly B D, Chao C C. An Evaluation of QVLBI OD Analysis of Pioneer 10 Encounter Data in the Presence of Unmodeled Satellite Accelerations. DSN Progress Report, 1974. 12: 22—42
[8] Rourke K H, Ondrasik V J. Application of Differenced Tracking Data Types to the Zero Declination and Process Noise Problems. DSN Progress Report, 1971. 12: 32—1526
[9] Bhaskaran S. The Application of Noncoherent Doppler Data Types for Deep Space Navigation. TDA Progress Report N95—32221: 54—65, 1995
[10] Ping J S, Frank W, Yusuke K, et al. High frequency components in LP Doppler data. J Planet Geod, 2001, 36: 15—22
[11] Ping J S, Yusuke K, Frank W, et al. A method of getting velocity information from S/C VLBI observation. J Beijing Norm Univ (Nat Sci), 2000, 36: 769—774
[12] 史林，赵树杰. 数字信号处理. 北京：科学出版社，2007
[13] 丁玉美，高西全. 数字信号处理. 西安：西安电子科技大学出版社，2001
[14] 杨小牛，楼才义，徐建良. 软件无线电原理与应用. 北京：电子工业出版社，2001. 21—30
[15] Bertotti B, Iess L, Tortora P. A test of general relativity using radio links with the cassini spacecraft. Nature, 2003, 425: 374—376

嫦娥一号卫星双轴天线轨迹规划

王淑一[1,2]　宗　红[1,2]　李铁寿[1,2]

(1. 北京控制工程研究所，北京　100190；
2. 空间智能控制技术国家级重点实验室，北京　100190)

摘　要　针对嫦娥一号卫星双轴天线指向地球的控制要求，给出双轴天线框架角的计算方法。根据地球可见条件及双轴天线转动角度和角速度的限制，确定星上双轴天线自主跟踪地球的范围。最后，在轨飞行试验结果验证了该方法的有效性。

关键词　嫦娥一号卫星　双轴天线　框架角

嫦娥一号卫星带有一副两自由度可转动的高增益抛物面天线(简称双轴天线)[1]，用于环月期间科学探测有效载荷数据传输和星上遥测数据传输。在环月轨道的正常运行模式中，该天线将指向并跟踪地球[2]。2007年11月9日11时21分，嫦娥一号卫星进行了双轴天线自主跟踪地球试验：控制双轴天线自主跟踪地球。

嫦娥一号卫星的双轴天线构形[3]设计为X-Y方式，见图1。从星体到反射器经过的第一框架转轴A轴与星体+X轴平行，第二框架转轴B轴与星体X轴垂直，天线反射器理论电轴与B轴垂直。

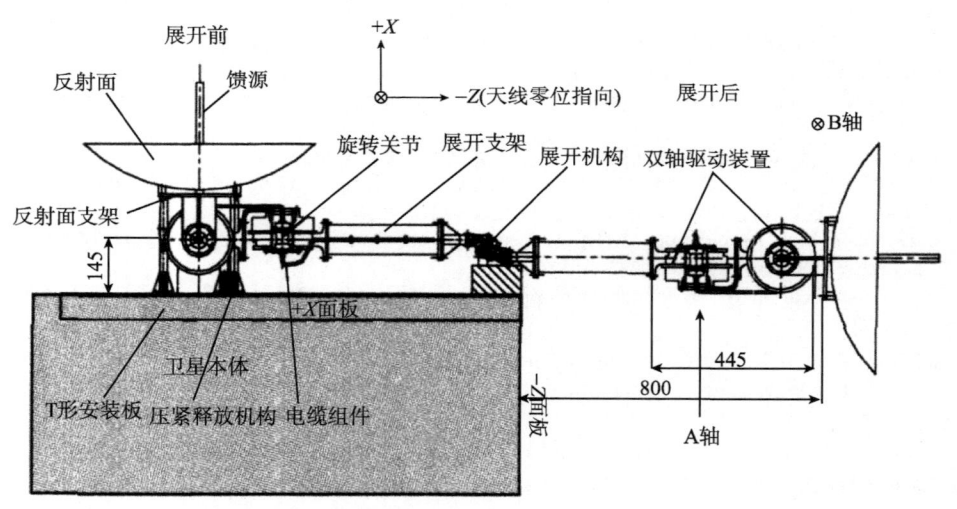

图1　双轴天线展开前状态和展开后零位状态

X-Y安装方式的天线零位状态为：A轴转动到使得B轴与星体+Y轴平行、B轴转动到使得天线反射器理论电轴与星体-Z轴平行。

图1为展开前/后零位状态示意图，A轴转角α：零位状态下，按右手螺旋定则，绕星体+X轴旋转为正；B轴转角β：零位状态下，按右手螺旋定则，绕星体+Y轴旋转为正。两个转轴的机械转角范围为$-90°\leqslant\alpha$，

本文原载于《空间控制技术与应用》，2009，Vol.35，No.3，18~22。

β≤90°。α 和 β 称为双轴天线的框架角。

1 框架角的计算

记本体坐标系单位矢量为 x_b，y_b，z_b，原点在卫星质心处；天线反射器坐标系单位矢量为 x_a，y_a，z_a，原点在反射器质心处（图2）。z_a 为理论电轴负向单位矢量。当 A 轴处于零位时，y_a 与 y_b 平行。x_a 与 y_a、z_a 构成右手坐标系，当 A、B 轴均处于零位时，x_a 与 x_b 平行。由此有

$$z_a = -x_b \sin B + y_b \sin\alpha \cos\beta - z_b \cos\alpha \cos\beta \tag{1}$$

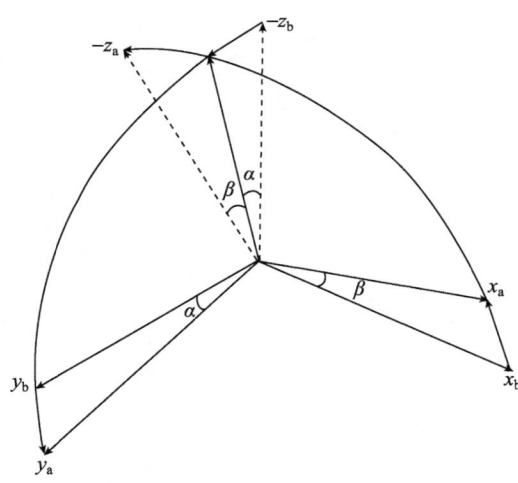

图 2 本体坐标系、天线坐标系和框架角定义

记卫星质心到地心的单位矢量为 E，E 在本体坐标系的分量定义为

$$E_{bx} = E \cdot x_b, \quad E_{by} = E \cdot y_b, \quad E_{bz} = E \cdot z_b。$$

为使天线理论电轴指向地心，要求 $-z_a = E$，则框架角 α，β 应满足

$$\begin{cases} -\sin\beta = E_{bx} \\ \sin\alpha \cos\beta = E_{by} \\ -\cos\alpha \cos\beta = E_{bz} \end{cases} \tag{2}$$

即

$$\begin{cases} \beta = -\arcsin E_{bx} \\ \alpha = \arctan\left[-\dfrac{E_{by}}{E_{bz}}\right] \end{cases} \tag{3}$$

2 双轴天线跟踪范围的确定

2.1 地心矢量计算

为确定框架角，首先需要确定地心矢量。这里约定惯性系为月心地球赤道惯性坐标系，坐标系原点在月心，坐标轴单位矢量为 x_i，y_i，z_i，平面 $x_i y_i$ 与地球赤道平行，z_i 与地轴平行并指北，该惯性系下地心单位矢量 E 可以根据地球星历计算得到。

定义轨道坐标系单位矢量为 x_o，y_o，z_o，$x_o z_o$ 平面与轨道平面重合，z_o 指向月心。地心矢量 E 在轨道坐标系三分量定义为

$$E_{ox} = \boldsymbol{E} \cdot \boldsymbol{x}_o, \quad E_{oy} = \boldsymbol{E} \cdot \boldsymbol{y}_o, \quad E_{oz} = \boldsymbol{E} \cdot \boldsymbol{z}_o。$$

根据轨道星历计算可计算出任意时刻 $\boldsymbol{x}_i, \boldsymbol{y}_i, \boldsymbol{z}_i$ 到 $\boldsymbol{x}_o, \boldsymbol{y}_o, \boldsymbol{z}_o$ 的坐标变换阵 $\boldsymbol{A}_{OI}$，从而有

$$\begin{bmatrix} E_{ox} \\ E_{oy} \\ E_{oz} \end{bmatrix} = \boldsymbol{A}_{OI} \begin{bmatrix} E_{x_i} \\ E_{y_i} \\ E_{z_i} \end{bmatrix} \tag{4}$$

用 3-1-2(z-x-y 顺序)转序的欧拉姿态角 φ, θ, ψ 描述本体系相对于轨道系的姿态，则从 $\boldsymbol{x}_o, \boldsymbol{y}_o, \boldsymbol{z}_o$ 到 $\boldsymbol{x}_b, \boldsymbol{y}_b, \boldsymbol{z}_b$ 的坐标变换阵[4] $\boldsymbol{A}_{BO}$ 为

$$\boldsymbol{A}_{BO} = \boldsymbol{R}_y(\theta)\boldsymbol{R}_x(\varphi)\boldsymbol{R}_z(\psi) \tag{5}$$

式中，

$$\boldsymbol{R}_y(\theta) = \begin{bmatrix} \cos\theta & 0 & -\sin\theta \\ 0 & 1 & 0 \\ \sin\theta & 0 & \cos\theta \end{bmatrix},$$

$$\boldsymbol{R}_x(\varphi) = \begin{bmatrix} 1 & 0 & 0 \\ 0 & \cos\varphi & \sin\varphi \\ 0 & -\sin\varphi & \cos\varphi \end{bmatrix},$$

$$\boldsymbol{R}_z(\psi) = \begin{bmatrix} \cos\psi & \sin\psi & 0 \\ -\sin\psi & \cos\psi & 0 \\ 0 & 0 & 1 \end{bmatrix}。$$

由此算出地心矢量 $\boldsymbol{E}$ 在本体系的分量为

$$\begin{bmatrix} E_{bx} \\ E_{by} \\ E_{bz} \end{bmatrix} = \boldsymbol{A}_{BO} \begin{bmatrix} E_{ox} \\ E_{oy} \\ E_{oz} \end{bmatrix}$$

2.2 地球可见的充分条件

地球可见条件指卫星至地球的视线不受月球遮挡的条件。若不考虑卫星姿态、天线框架角的变化及其限制，则当地心矢量 $\boldsymbol{E}$ 与月心方向 $\boldsymbol{z}_o$ 的夹角 θ_{EM} 大于月球半张角 ρ_M 时地球可见。故地球可见的充分条件为

$$\theta_{EM} \geqslant (\rho_M)_{\max} = \arcsin\left(\frac{R_M}{R_M + 175}\right) = 65.3° \tag{6}$$

式中，$R_M = 1738\text{km}$ 为月球半径。嫦娥一号卫星的轨道高度 $h = (200 \pm 25)\text{km}$。

地球可见充分条件可以表述为

$$E_{oz} = \cos\theta_{EM} \leqslant \cos65.3° = 0.4178 \tag{7}$$

为分析方便，本文引入另外两个参数来描述 $\boldsymbol{E}$ 与轨道系的关系。记 $\boldsymbol{E}$ 与轨道平面的夹角为 σ($-90°\leqslant\sigma\leqslant 90°$)，$\boldsymbol{E}$ 在轨道法向 $-\boldsymbol{y}_o$ 一侧时 σ 为正。$\boldsymbol{E}$ 在轨道平面内投影绕轨道负法向 $\boldsymbol{y}_o$ 相对于 $-\boldsymbol{z}_o$ 的转角为 τ，见图3。

参数 τ 可以看成无量纲的时间(卫星在轨道上绕 $-\boldsymbol{y}_o$ 轴运动)，则

$$\begin{cases} E_{ox} = -\cos\sigma\sin\tau \\ E_{oy} = -\sin\sigma \\ E_{oz} = -\cos\sigma\cos\tau \end{cases} \tag{8}$$

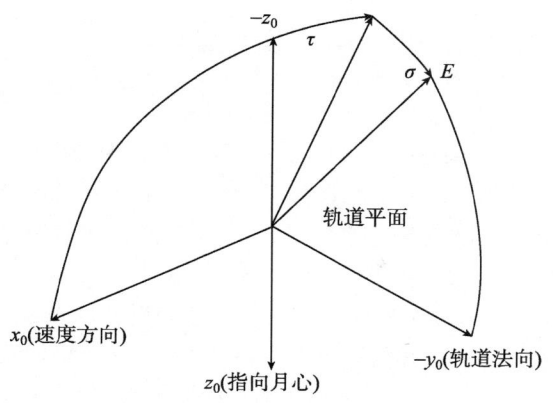

图 3 轨道坐标系和 σ, τ 角定义

由于地球方向变化角速度比轨道角速度慢得多,在一个轨道周期中 σ 近似为常数。因此地球矢量 $\boldsymbol{E}$ 在以卫星质心为球心的天球上的轨迹是以 $\boldsymbol{y}_o$ 为中心轴,$90°-|\sigma|$ 为半径的小圆。利用式(8),地球可见的充分条件(7)成为

$$E_{oz} = -\cos\sigma\cos\tau \leqslant 0.4178$$

根据几何关系可得到地球全轨道可见的充分条件

$$|\sigma| \geqslant \arcsin\left(\frac{R_M}{R_M+175}\right) = 65.3° \tag{9}$$

或者

$$|E_{oy}| \geqslant \frac{R_M}{R_M+175} = 0.9085 \tag{10}$$

显然,当地球在轨道面内时($\sigma=0°$),不可能全轨道可见;当地球在轨道面法向时($\sigma=\pm90°$),则全轨道地球可见。

2.3 框架角的限制条件

卫星正飞姿态下三轴欧拉角为小角度,则本体坐标系 x_b, y_b, z_b 与轨道坐标系 x_o, y_o, z_o 基本重合,故 $E_{bx}=E_{ox}$,$E_{by}=E_{oy}$,$E_{bz}=E_{oz}$。结合式(2)和(8)得框架角 α、β 与 σ、τ 的关系为

$$\begin{cases} \beta = \arcsin(\cos\sigma\sin\tau) \\ \alpha = -\arcsin\dfrac{\sin\sigma}{\cos\beta} \end{cases} \tag{11}$$

根据圆轨道性质,τ 近似匀速变化,即 $\dot\tau=$ 常数,$\ddot\tau=0$,故

$$\begin{cases} \dot\beta = \dot\tau\dfrac{\cos\sigma\cos\tau}{\cos\beta} \\ \dot\alpha = \dot\beta\dfrac{\sin\alpha\sin\beta}{\cos\alpha\cos\beta} \end{cases} \tag{12}$$

$$\begin{cases} \ddot\beta = \dfrac{\dot\beta^2\sin\beta - \dot\tau^2\cos\theta\sin\tau}{\cos\beta} \\ \ddot\alpha = \dfrac{(\dot\alpha^2+\dot\beta^2)\sin\cos\beta + 2\dot\alpha\ddot\beta\cos\alpha\sin\beta + \ddot\beta\sin\alpha\sin\beta}{\cos\alpha\cos\beta} \end{cases} \tag{13}$$

经三角运算和代数运算可得

$$|\dot{\beta}|=|\dot{\tau}|\frac{|\cos\sigma\cos\tau|}{\sqrt{(1-\cos^2\sigma\sin^2\tau)}},$$

$$|\dot{\alpha}|=|\dot{\tau}|\frac{|\sin\sigma\cos\sigma\sin\tau|}{1-\cos^2\sigma\sin^2\tau}。$$

取极限后可知，当 $\sigma=0$，$\tau=90°$时，$|\dot{\alpha}|$ 很大，而 $|\dot{\beta}|=|\dot{\tau}|$。此时若要跟踪目标，需要框架角 α 有很大的角速度，此种现象类似于过顶盲区[5]问题称为框架锁定。为了避免框架锁定，需要适当缩小天线跟踪工作范围。

忽略姿态误差，以地心方向 E 与轨道平面的夹角 σ 为参变量，计算 α、β、$\dot{\alpha}$、$\ddot{\alpha}$ 随 τ 变化的函数曲线。环月圆轨道的周期取为127min，轨道角速度为 $0.0472(°)/s$。计算结果见图4~图8。

当地球在轨道平面内($\sigma=0°$)时，α 恒为 0 （见图4），β 匀速改变(见图5)；

当地球在轨道平面法向($\sigma=90°$)时，α 恒为 $-90°$(见图4)，β 恒为 $0°$(见图5)；

同理，当地球在轨道平面负法向时($\sigma=90°$)时，α 恒为 $90°$，β 恒为 $0°$；

当地球接近轨道平面(σ 接近 $0°$)时，α 在 $\tau=\pm90°$附近有剧烈变化(见图6)；

当 $|\sigma|>15°$时，在 $-85°\leqslant\tau\leqslant85$ 范围内实施天线跟踪地球控制，框架角 α 的角速度小于 $0.2(°)/s$(见图7)，角加速度小于 $4\times10^{-4}(°)/s^2$(见图8)，而框架角 β 的角速度小于 $0.05(°)/s$，角加速度小于 $2\times10^{-4}(°)/s^2$，图略。

同样可以计算出：当 $|\sigma|<15°$时，在 $-80°\leqslant\tau\leqslant80°$范围内实施天线跟踪地球控制，框架角 α 的角速度小于 $0.2(°)/s$，框架角加速度小于 $8\times10^{-4}(°)/s^2$。

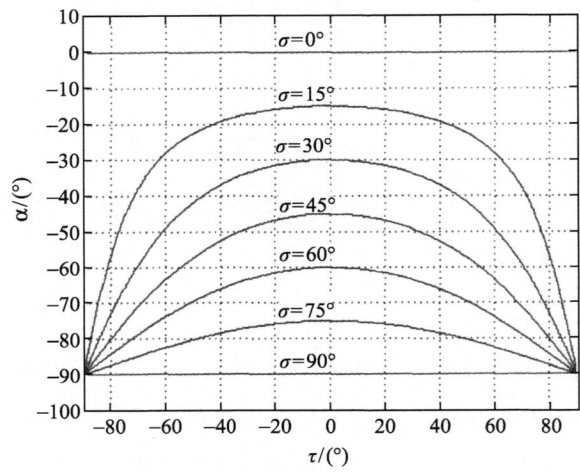

图4　α 随 τ 变化曲线($\sigma=0°$, 15°, 30°, …, 90°)

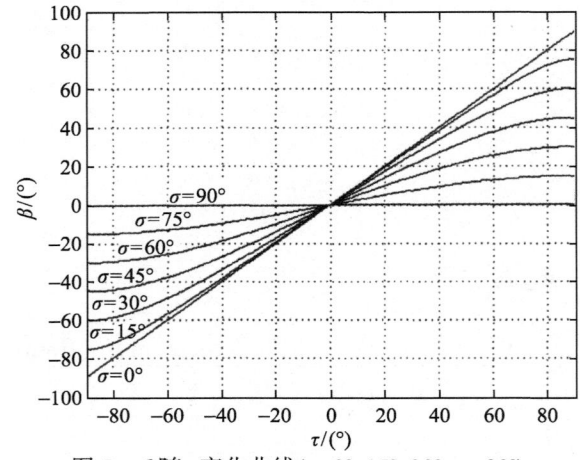

图5　β 随 τ 变化曲线($\sigma=0°$, 15°, 30°, …, 90°)

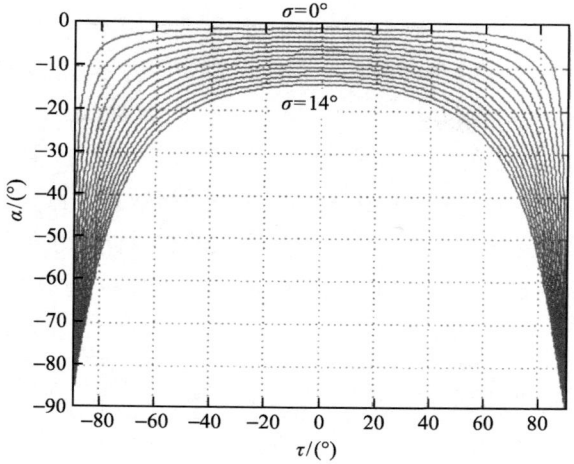

图 6　α 随 τ 变化曲线($\sigma=0°, 1°, 2°, \cdots, 14°$)

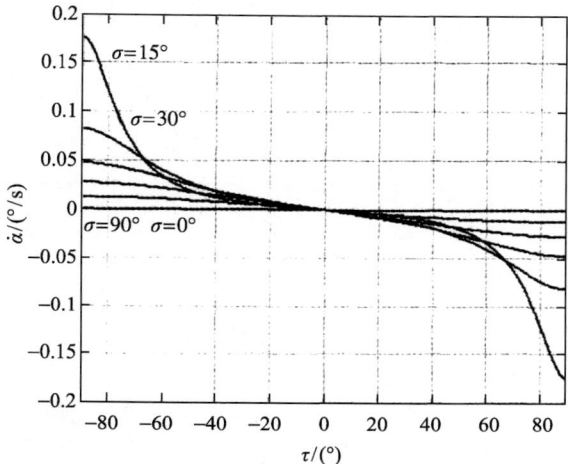

图 7　$\dot{\alpha}$ 随 τ 变化曲线($\sigma=15°, 30°, \cdots, 90°$ 和 $0°$)
(左边由上至下，右边由下至上)

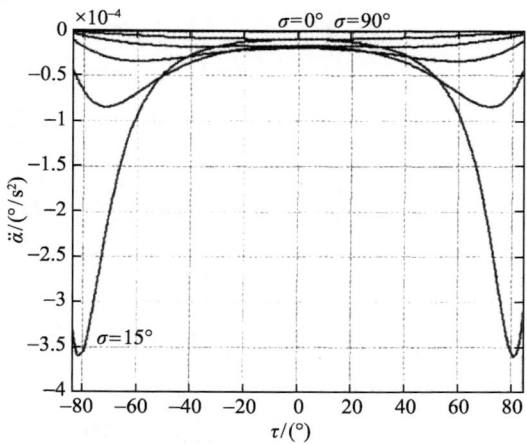

图 8　$\ddot{\alpha}$ 随 τ 变换曲线($\sigma=15°, 30°, \cdots, 90°$ 和 $0°$)
(由下至上)

3 飞行试验结果

根据嫦娥一号卫星地球可见的充分条件并避免框架角锁定现象的出现，选取 $E_{oz}<0.4178$ 或 $|E_{oy}|>0.9085$ 的条件下才进行天线跟踪地球的控制。

2007 年 11 月 9 日，卫星转入星光环月模式[2]，并进行了定向天线自主跟踪地球试验。北京时间 11 时 21 分定向天线自主跟踪地球，向地面传回了卫星和载荷的遥测数据，跟踪精度满足使用要求。

星上根据地面注入的卫星轨道平根数拟合值，按照简化的二体问题进行轨道外推，实时计算地心方位。在双轴天线跟踪上地球前后，遥测下传的地心方位 E_o 的曲线如图 9 所示。

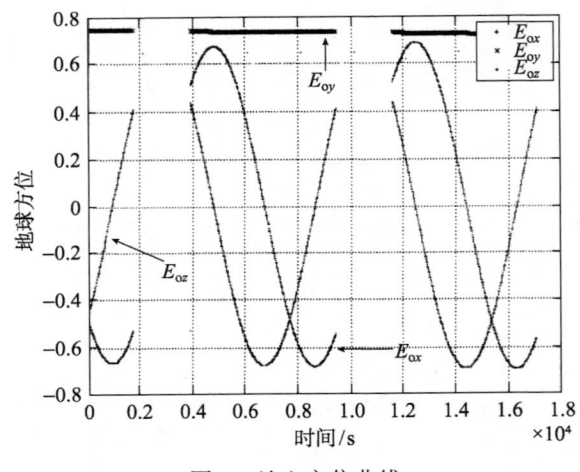

图 9　地心方位曲线

根据卫星实时遥测数据，双轴天线转角输出曲线见图 10。

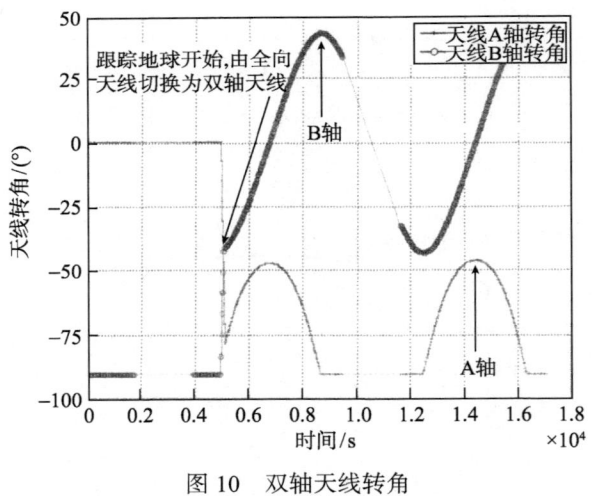

图 10　双轴天线转角

从双轴天线跟踪曲线可知：该弧段中 $E_{oy}=0.77$ 左右，当 $E_{oz}<0.4178$ 时地球可见，在可见条件下双轴天线能够跟踪上地球并下传遥测数据。

图 11 为某跟踪弧段内，双轴天线跟踪地球时 B 轴的转角误差曲线，考虑机构误差[6]后，其指向精度能够满足使用要求。

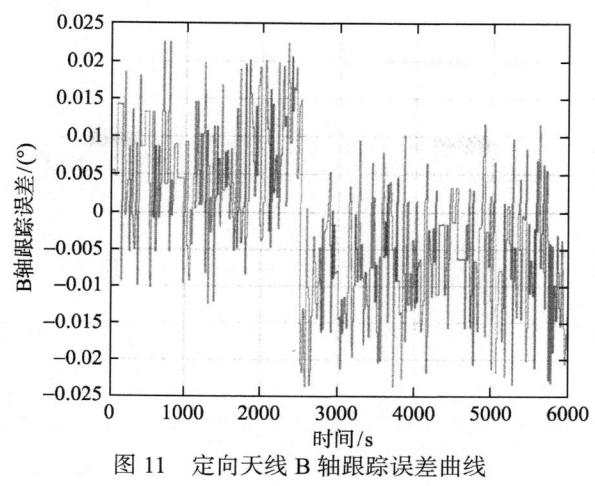

图 11 定向天线 B 轴跟踪误差曲线

4 结论

当地球方向与轨道平面夹角大于 15°时，在 $-85° \leqslant \tau \leqslant 85°$ 范围内实施天线跟踪地球控制，则最大跟踪角速度不超过 0.2(°)/s，角加速度不超过 $4 \times 10^{-4}(°)/s^2$；当地球方向与轨道平面夹角小于 15°时，在 $-80° \leqslant \tau \leqslant 80°$ 范围内实施天线跟踪地球控制，则最大角速度不超过 0.2(°)/s，角加速度不超过 $8 \times 10^{-4}(°)/s^2$，双轴天线跟踪的物理条件能够满足这两个约束；结合地球可见的充分条件，给出了嫦娥一号卫星双轴天线对地跟踪范围，并在嫦娥一号卫星上得到应用。从在轨遥测数据分析：星上双轴天线控制律能很好地控制双轴天线跟踪地球，跟踪误差满足使用要求。

参 考 文 献

[1] 叶培建, 饶炜. 嫦娥一号月球探测卫星技术特点分析[J]. 航天器工程, 2008, 17(1): 7—11
[2] 宗红, 李铁寿, 王大轶. 月球卫星 GNC 系统方案设想[J]. 航天控制, 2005, 23(1): 2—6
[3] 孙大媛. 一种卫星机械转动定向天线的研制[J]. 航天器工程, 2007, 16(6): 46—50
[4] 章仁为. 卫星轨道姿态动力学与控制[M]. 北京: 北京航空航天大学出版社, 1998
[5] 董小萌, 张平. 两轴稳定平台的过顶盲区问题[J]. 北京航空航天大学学报, 2007, 33(7): 811—815
[6] 孙京, 马兴瑞, 于登云. 星载天线双轴定位机构指向精度分析[J]. 宇航学报, 2007, 28(3): 545—550

"嫦娥一号"卫星轨控标定方法研究与实现

唐歌实[1,2]　陈莉丹[2]　刘勇[2]

(1. 北京航空航天大学宇航学院，北京　100191；　2. 北京航天飞行控制中心，北京　100094)

摘　要　在航天测控任务中，对轨控效果进行标定并合理利用可以实现更为精准的轨道控制。提出了一种综合利用控前控后精密轨道、轨控过程遥测姿态数据、遥测加速度计测量数据对沉底发动机、轨控发动机、加速度计刻度系数进行标定的方法；介绍了该方法在中国首次月球探测任务中的应用情况；最后分析了标定结果对定轨及定姿精度的敏感程度，从而在理论上进一步说明在后续深空探测中利用精密轨道进行轨控标定的可行性和重要性。

关键词　轨道控制　标定　月球探测　卫星

"嫦娥一号"(CE-1)卫星于 2007 年 10 月 24 日顺利升空，经过调相段、转移段、捕获段共 8 次轨道控制，于 11 月 7 日成功进入环月使命轨道。由于在任务中采用了将前次轨控标定结果用于下次控制注入的闭环控制策略，卫星实现了更为精准的轨道控制。

1　轨控标定方法

在 CE-1 任务中，轨控过程由沉底小发动机和轨控大发动机先后开机完成，沉底过程的加入使整个轨控过程被分成相互独立的两段，并且沉底速度变化没有计入轨控速度关机的判断过程，这给精确标定带来了一定困难；另外，深空探测距离远、遥测码速率低，难以准确获得两类发动机开关机时刻也成为制约标定精度的瓶颈。针对深空探测任务的特殊性，本文提出了一种综合利用控前控后精密轨道、轨控过程遥测姿态数据、遥测加速度计测量数据对沉底发动机、轨控发动机、加速度计刻度系数进行标定的方法，取得了很好的标定效果。

1.1　轨控过程动力学模型

对轨控过程的标定首先要建立 CE-1 卫星轨控过程的动力学模型

$$\ddot{r} = -\frac{\mu}{r^3}r + P + F(F_b, k_f, \gamma, \Phi, \phi) \tag{1}$$

式中 r 为卫星相对中心天体的位置矢量；μ 为中心天体引力常数；P 为各种摄动力；F 为轨控过程发动机产生的推力，其为发动机标称推力 F_b、发动机推力系数 k_f、姿态角滚动 γ、俯仰 Φ、偏航 ϕ 的函数。根据任务经验及仿真分析[1]，在 CE-1 任务中，当卫星与月心距离大于 $6×10^4$ km 时以地球为中心天体，摄动主要考虑地球非球形引力、引力场阶数 32×32，太阳、月球质点引力，大气摄动、光压摄动，大行星摄动、在与月心距离小于 $6×10^4$ km 时以月球为中心天体，摄动主要考虑月球非球形引力、引力场阶数 64×64，太阳、地球质点引力，光压摄动，大行星摄动。在轨控过程数值积分时采用 RKF78 单步法。

1.2 轨控过程标定的寻优方法

1.2.1 轨控过程标定的目标函数

CE-1 任务的沉底过程由 4 台标称推力为 F_1 的小发动机执行，速度增量在 4 m/s 左右。定义沉底过程目标函数为

$$J_s(F) = \Delta V_s - \Delta V_c \tag{2}$$

式中 ΔV_s 为根据动力学模型外推到沉底结束时的卫星速度增量；ΔV_c 为卫星沉底过程实际速度增量，由于沉底速度增量较小，加速度计刻度系数对其影响忽略不计，所以可以通过求解遥测加速度计数据得到沉底实际速度增量，迭代沉底过程使目标函数最小。

定义由标称推力为 F_2 大发动机执行的轨控过程目标函数为

$$J_{m1}(F) = e^T \Lambda e \tag{3}$$

$$或 J_{m2}(F) = a_m - a_c \tag{4}$$

式中 $e = [x, y, z, \dot{x}, \dot{y}, \dot{z}]^T - [x_c, y_c, z_c, \dot{x}_c, \dot{y}_c, \dot{z}_c]^T$，$[x, y, z, \dot{x}, \dot{y}, \dot{z}]^T$ 是由数值外推得到的关机时刻位置和速度、$[x_c, y_c, z_c, \dot{x}_c, \dot{y}_c, \dot{z}_c]^T$ 是由控后精密轨道得到的关机时刻位置和速度；Λ 为 6×6 的对角形的权矩阵，对速度的权系数取 1，对位置的权系数要根据轨道不同合理选取[2]；a_m 为数值外推得到的关机时刻轨道半长轴；a_c 为控后精密轨道得到的关机时刻轨道半长轴。

可以看出：J_{m1} 以星历最接近为目标，适应普遍情况，而 J_{m2} 以轨道半长轴为目标，物理意义直观且不用选取位置的权系数，对于拱点控制的标定具有明显优势。

1.2.2 轨控过程标定的寻优方法

在对沉底或是轨控过程标定的过程中，都要通过迭代控制过程对目标函数求解最小值，这里采用二次插值法来对目标函数寻优[3]。令发动机推力系数 $k = \dfrac{F}{F_b}$，其中 F_b 为标称推力，则目标函数式(2)~(4)可转化为关于 k 的函数 $J(k)$。

因为发动机的推力系数在 1 附近，所以令 $k_0 = 1.0$，$k_1 = 1.5$，$k_2 = 0.5$，则可使二次插值法的初始条件成立

$$\left. \begin{array}{l} J(k_0) < J(k_1) \\ J(k_0) < J(k_2) \end{array} \right\} \tag{5}$$

用二次函数来拟合 $J(k)$ 曲线：

$$h(k) = a_0 + a_1 k + a_2 (k)^2 = J(k) \tag{6}$$

可得

$$\left. \begin{array}{l} h(k_0) = a_0 + a_1 k_0 + a_2 (k_0)^2 = J(k_0) \\ h(k_1) = a_0 + a_1 k_1 + a_2 (k_1)^2 = J(k_1) \\ h(k_2) = a_0 + a_1 k_2 + a_2 (k_2)^2 = J(k_2) \end{array} \right\} \tag{7}$$

经计算，当 $k = \bar{k}$ 时，二次函数取得最小值

$$\bar{k} = \frac{\{[(k_0)^2 - (k_2)^2]J(k_1) + [(k_2)^2 - (k_1)^2]J(k_0) + [(k_1)^2 - (k_0)^2]J(k_2)\}}{2[(k_0 - k_2)J(k_1) + (k_2 - k_1)J(k_0) + (k_1 - k_0)J(k_2)]} \tag{8}$$

判断式(9)是否成立，

$$|k_0 - \bar{k}| < \varepsilon \tag{9}$$

式中 ε 为迭代收敛精度，为一小量。如果不成立，则比较 $J(k_0)$ 与 $J(\bar{k})$ 的大小，重新选择 k_0、k_1、k_2 迭代

轨控过程，直到式(9)成立。则最终$\overline{k^*}$为使$J(k)$最小的发动机推力系数，从而求出轨控过程大发动机开机产生的速度增量ΔV_m，求得加速度计刻度系数$k_a = \dfrac{\Delta V_z}{\Delta V_m}$，$\Delta V_z$为注入速度增量与遥测控后剩余速度增量的差。

1.3 轨控标定流程

CE-1卫星轨控标定主要分为遥测数据的处理、小发动机沉底过程标定和大发动机轨控过程标定3个部分，具体流程如图1所示。

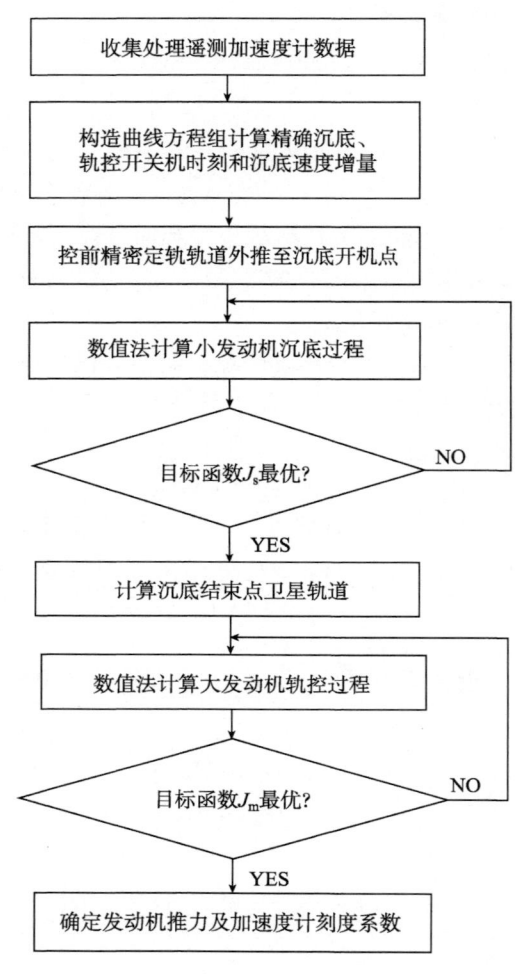

图1 轨控标定流程

2 CE-1卫星轨控标定的实现

2.1 轨控标定算例

下面以调相段第一次近地点控制为例具体说明轨控标定的过程。

2.1.1 加速度计遥测数据的提取和拟合

图2显示的为第一次近地点加速轨控过程加速度计遥测下传数据经过事后初步处理的结果。在低遥测码速率情况下直接从遥测数据判断轨控特征点会带来较大误差，任务中采用了分段数据拟合的方法解决了

这一问题。将轨控过程曲线分为控前恒星定向、沉底过程、轨控过程、控后太阳捕获四段，选择适当跨度的数据分别进行最小二乘曲线拟合，因为轨控过程中发动机推力近似恒定，所以得到直线方程组

$$\left.\begin{array}{l} f_1(t) = m_1 t + n_1 \\ f_2(t) = m_2 t + n_2 \\ f_3(t) = m_3 t + n_3 \\ f_4(t) = m_4 t + n_4 \end{array}\right\} \tag{10}$$

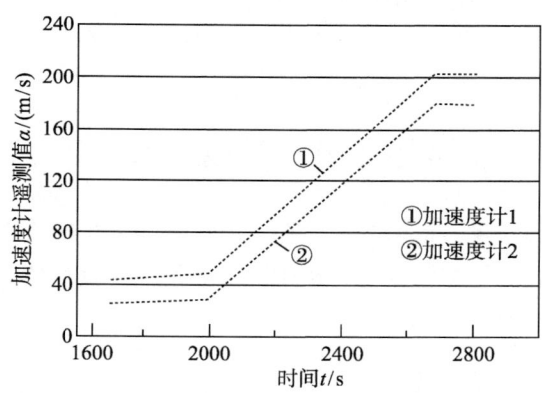

图 2　第一次近地点轨控过程加速度计遥测数据

由 f_1、f_2 联立可求得沉底开始特征点 t_1；f_2、f_3 联立可求得轨控开机特征点 t_2；f_3、f_4 联立可求得轨控关机特征点 t_3；然后将 t_1、t_2、t_3 分别代入方程可求得特征点所对应的加速度计测量值 ΔV_1，ΔV_2，ΔV_3 从而可以精确得出沉底速度增量 ΔV_c。

沉底开始特征点：$t_1 = 1762.8591$ s　　$\Delta V_1 = 25.8131$ m/s
轨控开机特征点：$t_2 = 2003.0624$ s　　$\Delta V_2 = 29.5575$ m/s
轨控关机特征点：$t_3 = 2683.0156$ s　　$\Delta V_3 = 179.6793$ m/s

2.1.2　沉底过程标定

求得精确特征点后，可得沉底实际速度增量 $\Delta V_c = \Delta V_2 - \Delta V_1 + d_a \cdot (t_2 - t_1) = 3.8205$ m/s，d_a 为标定加速度计 2 零漂值，经拟合 $d_a = \frac{m_1 + m_4}{2}$。然后按第一节所述方法以速度增量最接近为迭代目标对沉底过程进行标定，得到沉底发动机实际推力 $F_s = 9.55$ N，这样就实现了沉底过程的剥离。

2.1.3　轨控过程标定

用标定的 F_s 外推沉底过程，得到轨控发动机开机点的精确轨道，按第一节所述方法以轨道半长轴最接近为迭代目标对轨控过程进行标定，得到轨控发动机实际推力 $F_m = 501.5583$ N、轨控过程实际速度增量 $\Delta V_m = 150.0211$ m/s、加速度计 2 刻度系数 $k_a = 1.0022$。

2.2　轨控标定结果的应用

在调相段第一次远地点和第一次近地点控制时，控制偏差分别为 1.7 % 和 0.3 %，在利用精密定轨轨道对前两次控制进行了标定和分析后，认为卫星加速度计 2 刻度存在约 0.2 % 的系统偏差，所以在第二次近地点加速时将系数 1.002 运用到控制注入中，即：实际注入控制量 = 理论计算轨控过程控制量 × 1.002，表 1 是第二次近地点控制的情况。

表 1 第二次近地点控制情况

参数	理论	实际
控后轨道半长轴/m	66 964 826.076	66 958 733.065
沉底速度增量/ (m/s)	4.041 2	4.096 0
轨控速度增量/ (m/s)	166.327 5	166.254 1
变轨总速度增量/ (m/s)	170.368 2	170.350 1

经计算卫星第二次近地点控制的速度增量偏差仅为 0.01 %，在后续任务中利用精密轨道和实际姿态对卫星控制过程进行标定，不断修正沉底发动机推力和加速度计刻度系数，使后续卫星控制偏差均保持在万分之几的量级，实现了精准控制。

3 CE-1 卫星加速度计刻度系数的敏感度分析

本文提出的标定方法以外测精密轨道为基础，所以定轨精度直接影响标定结果的可信度；同时标定过程要使用卫星定姿结果数值积分，定姿精度也会影响标定精度。下面以第二次近月制动为例说明 CE-1 任务中定轨与定姿精度对加速度计刻度系数标定结果的影响。

3.1 对定轨偏差的敏感度分析

第二次近月制动在近月点实施，控制以轨道半长轴为目标，所以这里在控前控后精密轨道分别加入半长轴偏差进行标定计算，结果如图 3 所示。

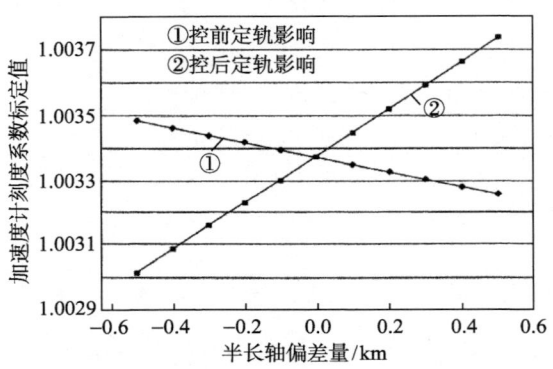

图 3 第二次近月点制动定轨偏差对标定结果影响

可以看出，标定的加速度计刻度系数值与定轨偏差大致成线性关系。第二次近月点制动的实际加速度计刻度系数标定结果为 1.0033，控前控后精密轨道偏差 ± 500 m 时影响标定结果偏差最大 0.05 %，根据事后对 CE-1 任务定轨精度的分析，近月制动阶段定轨半长轴偏差优于 100 m，所以在近月制动阶段定轨偏差对加速度计刻度系数标定结果仅影响到 0.01 %。

对于拱点控制的一般情况，不考虑姿态偏差，可以推出近似关系[4]：

$$\delta_k = \frac{\mu/2a_0^2 v_0 + \mu/2a_1^2 v_1}{\Delta V} \delta_a \tag{11}$$

式中 δ_k 为加速度计刻度系数标定误差；δ_a 为定轨偏差；μ 为中心天体引力常数；a_0、v_0 为控前轨道半长轴和速度；a_1、v_1 为控后轨道半长轴和速度；ΔV 为本次控制的理论速度增量。可以看出，对于某次特定控制，刻度系数标定误差与定轨偏差成正比，而对于不同的控制，刻度系数偏差还与中心天体、轨道半长轴以及本次控制量大小有关。CE-1 任务调相段定轨精度大大优于环月段，再结合控制量等因素，经分析，除调相段远地点控制、中途修正外，其他历次控制定轨精度均可以保证刻度系数标定偏差在 0.01 % 以下。

3.2 对姿态偏差的敏感度分析

在标定中,要利用卫星的实际姿态来确定轨控过程中发动机推力在惯性空间的精确方向进行轨道数值积分,所以卫星定姿精度也会影响标定结果。CE-1 卫星星敏在相对于惯性空间的角速度较稳定时定姿精度在角分以下量级,并且在轨控过程中扰动很小。经计算,如果轨控过程星敏可用,利用星敏定姿结果进行标定,则定姿偏差影响加速度计刻度系数的标定结果偏差最大仅为 0.000 01 %,完全可以忽略。即使在轨控过程中出现星敏不可用或定姿数据无法下传,利用理论姿态进行标定,由于轨控过程大部分时间卫星姿态均能稳定在理论姿态 0.2° 以内(见图4),标定结果也仅偏差 0.02 %。

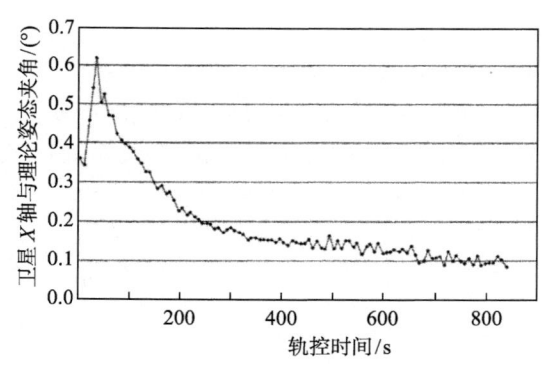

图4 第二次近月点制动过程卫星 X 轴与理论姿态偏差

4 结束语

本文首先介绍了利用控前控后精密轨道、轨控过程遥测星敏姿态数据、遥测加速度计测量数据对沉底发动机、轨控发动机、加速度计刻度系数进行标定的方法;以 CE-1 任务第一次近地点控制为例说明了如何采用对遥测数据进行曲线拟合的方法来克服遥测数据更新速率慢带来的误差,实现沉底过程速度增量的准确剥离和各特征点的确定,最终实现对控制的精确标定,将标定的加速度计刻度系数引入后续控制使控制精度提高了一个量级;最后分析了标定的加速度计刻度系数对定轨精度和定姿精度的敏感度,说明在 CE-1 任务中定轨与定姿精度影响加速度计刻度系数在万分之一以下,完全可以对卫星千分之一以上的刻度系数系统偏差进行捕获和精确标定,从而大大提高控制精度。

在以后的深空探测任务中,轨道控制可谓是"失之毫厘,谬以千里",提高控制精度有着重要意义。从 CE-1 任务控制的成功实施可以看出,虽然深空探测任务中测控系统面临巨大压力,距离远、测定轨难,但仍可以利用以精密轨道为基础对轨控过程进行标定的方法来提高控制精度,为任务的成功实施提供更坚实的保障。

参 考 文 献

[1] ERIC CARRANZA, ALEX KONOPLIV, MARK RYNE. Lunar Prospector Orbit Determination Uncertainties Using the High Resolution Lunar Gravity Models [C]. AAS/AIAA Astrodynamics Specialists Conference, 1999.
[2] 韩冬. GNC 对地面测控的计算要求[R]. 航天科技集团公司中国空间技术研究院报告, 2006.
[3] 魏权龄, 王日爽, 徐兵. 数学规划引论[M]. 北京: 北京航空航天大学出版社, 1991.
[4] 肖业伦. 航天器飞行动力学原理[M]. 北京: 宇航出版社, 1995: 64—65.

嫦娥-1卫星绕月捕获分析与快速判断

史弦　简念川　鄢建国　平劲松

(中国科学院上海天文台，上海 200030)

摘　要　月球和行星环绕探测器的轨道投入和确认是整个任务中的关键步骤之一。以 CE-1 任务为参考，本文分析了卫星绕月捕获的正常制动范围，针对入轨的快速判断提出了多种实时任务中的解决方法。并对这些方法进行了仿真分析。比较实测制动量和根据设计轨道或入轨前测量轨道得到的理论制动量与入轨参数数据库，可以方便快速地确认卫星入轨过程和状况，避免了事后定轨所产生的滞后效应。

关键词　嫦娥-1　绕月探测器　捕获

1　问题提出

嫦娥-1 在发射升空约 9 天后将被投入绕月调相轨道。月球轨道器在从地月转移轨道进入绕月轨道的过程中，近月点附近的制动减速对轨道器能否正确入轨起到了至关重要的作用。迅速准确地判断 CE-1 探测器近月点附近制动和制动结束后的绕月入轨状态是测控系统的关键任务，同时还对进一步的测控(轨道控制和姿态控制)提供参考依据。具体问题和任务包括：

(1) 正常制动情况下卫星减速总量的上下限范围是多少？

(2) 上行制动 CMD 信号发出后不论是否正常收到来自卫星的制动点火 TLM 反馈信号，如何迅速地从轨道遥测数据中判定是否开始制动？

(3) 正常开始制动到结束制动时，速度增量是多少？

(4) 正常开始制动到结束制动时，卫星进入的轨道参量是多少？

尽管在事后可以通过测量数据分析得到上述信息，任务期间要求迅速给出各问题的准确答案。

2　近月点捕获制动总量分析

按照制动设计，对应不同制动量，轨道器将被月球捕获进入不同的绕月轨道，如果刹车量不足，探测器将飞掠月球。对刹车量下限估计如下。

2.1　飞行器近月点捕获判据

- 逃逸速度判据：设计轨道高度为 200km，对应的逃逸速度为 2.25km/s，飞行器的速度必须小于此。
- 月球引力范围判据：取月球引力范围为 43000km，则对应于轨道高度 200km 的情况，达到月球引力范围边缘的轨道半长径为：

$$a = \frac{43000\text{km} + 1738\text{km} + 200\text{km}}{2} = 22469\text{km}$$

因此，制动后的轨道半长径不能大于此值。

本文原载于《深空探测研究》，2006，Vol.4，No.3，9~15。

- 第二次到达近月点的轨道倾角判据：设计轨道倾角为90°，考虑卫星第二次回到近月点时的倾角，不能太大地偏离90°。
- 第二次到达近月点的轨道高度判据：设计轨道高度为200km，考虑卫星第二次回到近月点时的轨道高度，不能太大地偏离200km。
- 轨道周期判据：考虑到中国上空观测条件的限制，第一次制动后进入的大椭圆轨道周期以12小时或24小时为佳。

2.2 模拟计算

利用STK分析软件可以方便得到不同制动量对应得到的轨道参数，见表1。

表1 不同近月点制动量对应的轨道根数

DelV/(km/s)	E	T/s	A/km	I/Degree	Hp/km
0.21	0.91484969	467317	30044.808	106.49856969	4713.121
0.22	0.91482718	346018	24590.131	100.86775640	1127.465
0.23	0.90341541	267941	20735.834	95.21186219	445.094
0.24	0.88914181	214468	17875.963	92.09572883	290.123
0.25	0.87399960	176281	15685.516	90.57935508	250.171
0.26	0.85855976	148063	13963.420	89.83586249	239.066
0.27	0.84301992	126596	12578.865	89.45848713	235.971
0.28	0.82746153	109857	11444.035	89.26072217	235.225
0.29	0.81192367	96524	10498.343	89.15515817	235.158
0.30	0.79642765	85712	9698.926	89.09906578	235.251
0.31	0.78098629	76805	9014.740	89.07060911	235.345
0.32	0.76560796	69366	8422.828	89.05815537	235.396

根据以上数据作图1，蓝色的点为对应某一制动量的仿真数据，红色曲线为6阶多项式拟合结果。

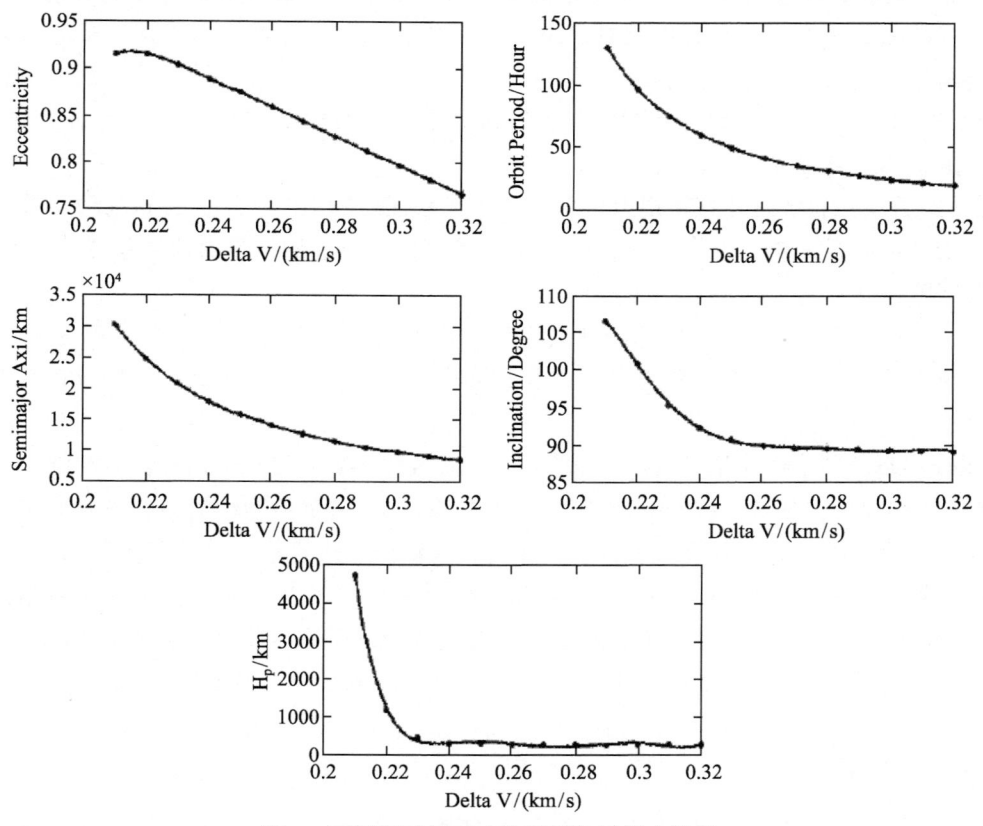

图1 不同制动量对应轨道根数及拟合结果

针对捕获判据的模拟分析结果
- 逃逸速度判据：根据模拟结果，飞行器到达近月点时的卫星速度为1.755042km/s，小于200km高度的逃逸速度2.25km/s。
- 月球引力范围判据：根据上文求得的轨道半长径最大值22469km和表1中的仿真结果，制动量必须大于0.23km/s。
- 第二次到达近月点的轨道倾角判据：若认为±5°是可以接受的倾角偏差范围，则根据表1中所列结果，制动量必须大于0.23km/s。
- 第二次到达近月点的轨道高度判据：若认为±50km为可以接受的轨道高度偏差范围，则根据表1中所列结果，制动量应大于0.25km/s。

轨道周期判据：若制动后进入的轨道为24小时轨道，则由表1可见制动量需大于0.29km/s，对应于24小时轨道的制动量为0.29930km/s，接近0.3km/s。对应于12小时轨道的制动量为0.356km/s。

综上所述，满足嫦娥-1卫星工程测控需求的正常的制动范围在0.25km/s-0.35km/s。制动时间为近月点前后各15分钟的范围内。

3　对策方法

在飞行测控中，对上述制动过程的定性和定量判断也变得至关重要。由于制动过程中轨道剧烈变化，解决上述问题的方法分为两类，即利用实时分析比对方法和制动前后利用短弧定轨来快速确定轨道及其变化。前者效率高、实时性强、给测控预留了进一步的判断和应对时间；后者精度略高，但是最快也要在制动结束后30分钟左右给出结果，很可能耽误了实时判断和应对的时机，导致任务出现高风险。本文重点讨论实时分析比对方法。

在实时分析比对和判断模式下，针对上述情况，对制动过程进行预先分析，分两级建立数据库，在实时测控中利用两种以上的方式进行分析比对，确定制动过程、状态和结果，将是可行且高效的方法。

第一级数据库包括两个部分：

(1) 利用CE-1地月转移设计轨道和制动绕月限制条件，预报产生总制动量与最终投入轨道的参数库，定性和定量分析制动量与投入轨道的对应关系；

(2) 利用发射后CE-1地月转移轨道上最后一次中途修正后确定的卫星轨道和制动绕月限制条件，预报产生总制动量与最终投入轨道的参数库，产生制动量与投入轨道的定量对应关系供进一步判断使用。

第二级数据库包括三部分，或三类方法：

(1) 利用卫星遥测数据TLM(制动点火时间、制动结束时间、单位之间制动量)得到近月点前后制动时间与制动总量的关系，利用预报得到的总制动量与最终投入轨道参数的关系，进一步得到制动时间与投入轨道参数的关系；

(2) 针对陆地USB测站喀什站，预报一系列不同的制动量情况下测站得到的轨道器制动前后和制动期间的双程视向Doppler测速数据和Doppler增量；建立Doppler增量与制动量的关系数据库；进一步通过制动量与最终投入轨道的参数库，建立测站视向Doppler数据与最终投入轨道参数的对应关系；

(3) 针对陆地VLBI测控网，预报一系列不同的制动量情况下VLBI网得到的轨道器制动前后和制动期间的单向差分速度数据和增量；建立差分速度增量与制动量的关系数据库；进一步通过制动量与最终投入轨道的参数库，建立VLBI网单向差分速度数据与最终投入轨道参数的对应关系。

4　CE-1制动分析

轨道器的制动方式可分为两种——脉冲制动和连续制动。脉冲制动是指轨道器只在某一时间点上进行制动，表现为速度的瞬时减少；连续制动是指轨道器在一段时间内进行持续的制动以达到一定的减速量，下面分别针对两种制动模式进行分析。真实情况通常是后者，然而实践表明，对前者的分析更便捷简单，可以为制动前后轨道变化提供可靠的参考。

4.1 脉冲型制动

选取减速方向为轨道器速度反方向(Antivelociry Vector)；脉冲制动时间点选为轨道器到达近月点时间；减速量表现在飞行器整体速度标量(Magnitude)的减少上。按照制动设计，对应不同制动量，轨道器将被月球捕获进入不同的绕月轨道，如果刹车量不足，探测器将飞掠月球。

减速量大于270m/s时，探测器将被投入到近月点高度约200公里，周期小于24小时的绕月轨道。图2中给出了近月点时刻5个制动0.7km/s，0.6km/s，0.5km/s，0.4km/s和0.3km/s量分别对应的绕月轨道示意图。图中红色表示的为地月转移轨道，白色表示的为绕月轨道，各条轨道由内至外与制动量一一对应。

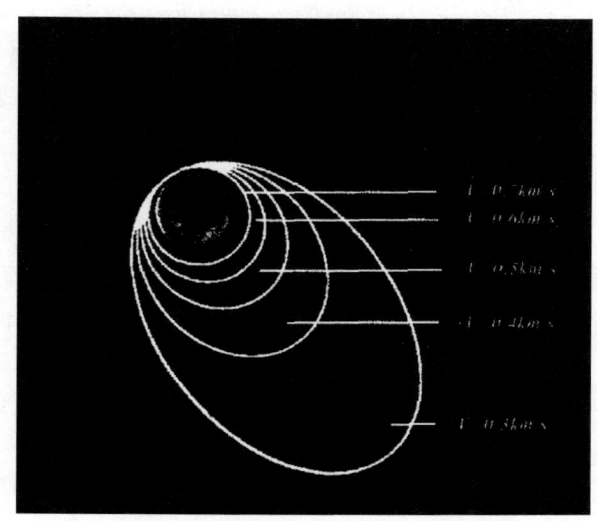

图2 不同制动量对应的CE-1探测器绕月轨道

进一步，以CE-1设计轨道近月点时刻的吻切轨道为例，预报分析不同制动量对应的绕月轨道得到如下结果：轨道周期与制动量的关系，轨道半长轴与制动量的关系，以及轨道偏心率与制动量的关系。这些关系可以方便地用一元多项式拟合。六次拟合曲线和仿真预报值的这些关系图示如下，下面三图中蓝色点为对应不同制动量的仿真轨道根数值，红色曲线为多项式拟合得到的轨道根数随制动量变化曲线。拟合得到的三个多项式为：

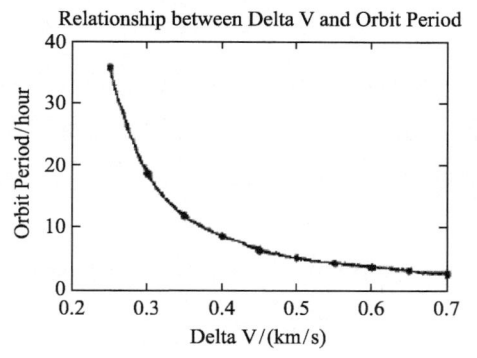
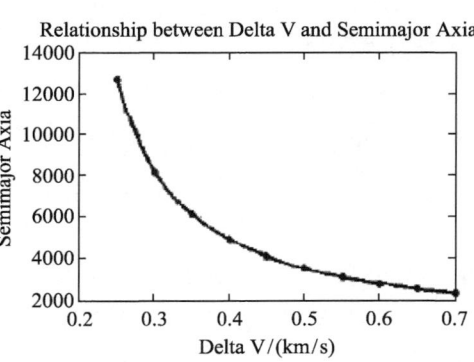

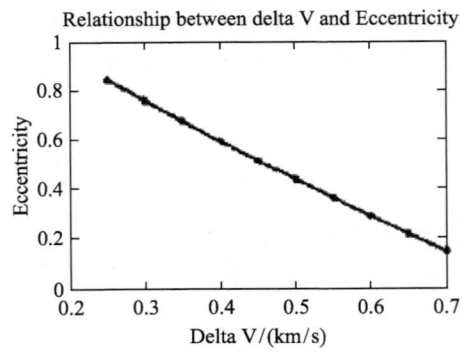

图 3 脉冲制动模式下轨道根数与制动量关系

4.2 连续制动

在实际情况中,一般使用连续制动方式把探测器投入绕月轨道。还是选取减速方向为轨道器速度反方向;制动时间为近月点前后约半小时。同样地对于连续制动模式下不同的制动量进行轨道仿真,得到三个吻切轨道根数与制动量的关系:即轨道周期与制动量的关系,轨道半长轴与制动量的关系,以及轨道偏心率与制动量的关系。

作为比较,对应与相同的总制动量,图中还给出了相应的脉冲制动情况下的轨道根数与制动量的关系,并利用一元多项式进行拟合。六次拟合曲线和仿真预报值的这些关系图示如下,上面三图中蓝色点为对应不同制动量的仿真轨道周期值,红色曲线为多项式拟和得到的轨道周期随制动量变化曲线。从图中可以看出,在制动量大于 400m/s 时,脉冲制动和连续制动的结果相差不大,用脉冲制动的分析可以方便地取代连续制动的结果。

利用设计轨道,我们已经建立了一套分析方法和第一级数据库。对于地-月转移轨道中途修正的情况,在得到工程或精密轨道后,可以简单套用上述方法生成更精密的数据库。

5 CE-1 绕月制动入轨判断方法流程

基于第三节产生的一级数据库,可以有三种方法快速判断卫星入轨状态:

(1) 在连续点火制动情况下利用卫星制动遥测数据得到近月点前后制动时间与制动总量的关系,结合预报得到的总制动量与最终投入轨道参数的关系,可以得到制动时间与投入轨道参数的关系,进而可以实时快速地预报制动过程、状态和制动结果。图 4 给出了从点火制动开始时刻、固定单位时间内的制动量前提条件下制动时间周期与制动总量的关系,以及制动时间与最终绕月入轨轨道参数的关系。这里给出的是程序软件依据经验给出的一个示意关系。真正的关系要根据 CE-1 卫星的发动机参数和制动控制参数,参考上述分析流程来精确确定。

(2) 针对陆地 USB 测站喀什站,预报一系列不同的制动量情况下测站得到的轨道器制动前后和制动期间的双程视向 Doppler 测速数据和 Doppler 增量;建立 Doppler 增量与制动量的关系数据库;进一步通过制动量与最终投入轨道的参数库,建立测站视向 Doppler 数据与最终投入轨道参数的对应关系。

以 USB 喀什站为例,图 5 显示了针对不同的制动量,在制动前后半小时内,轨道器相对喀什站视向速度的变化曲线。制动量取值从 0.3km/s 到 0.7km/s,步长为 0.1km/s。

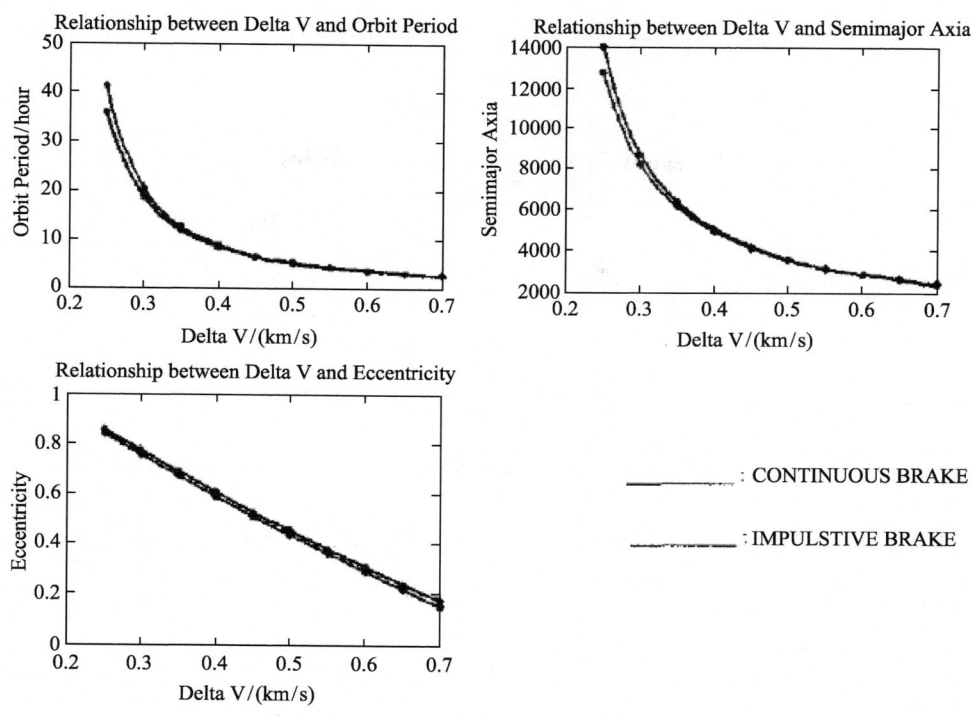

图 4 两种制动方式下轨道根数与制动量关系比较

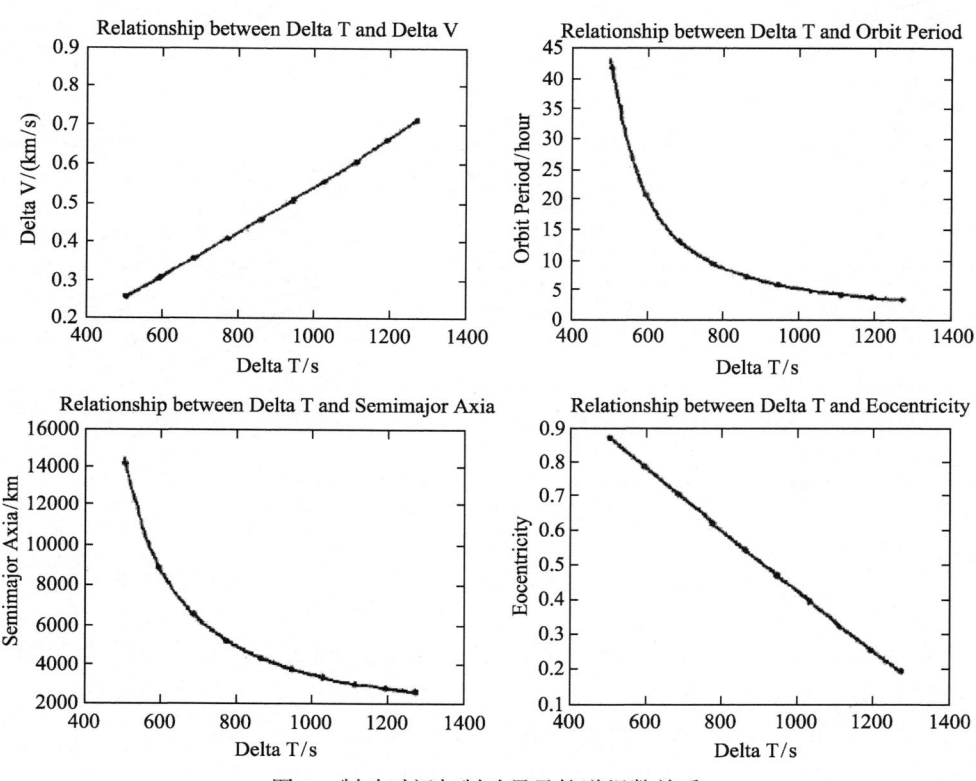

图 5 制动时间与制动量及轨道根数关系

设计制动总量约 0.27km/s。低于 270m/s 卫星入轨风险很高。0.3km/s 到 0.5km/s 为比较合理的制动量区间，下图对这一区间进行了细化分析。

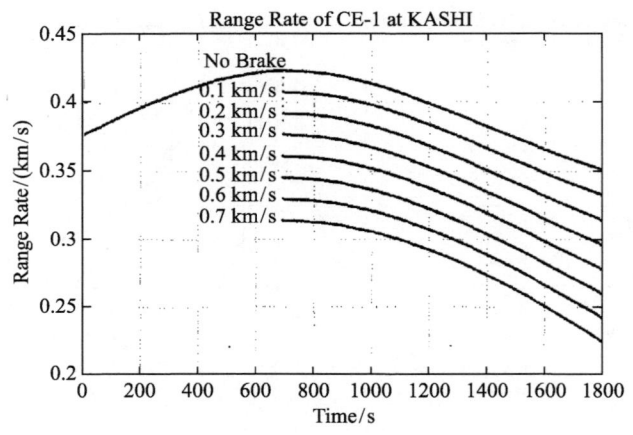

图 6-1 脉冲制动中制动前后轨道器相对于喀什站视向速度变化曲线

图 6-2 中制动量分别取 0.3km/s，0.33km/s，0.35km/s，0.37km/s，0.4km/s，0.43km/s，0.45km/s，0.47km/s，0.5km/s；时间为 28 Apr16:35-28 Aprl6:50。

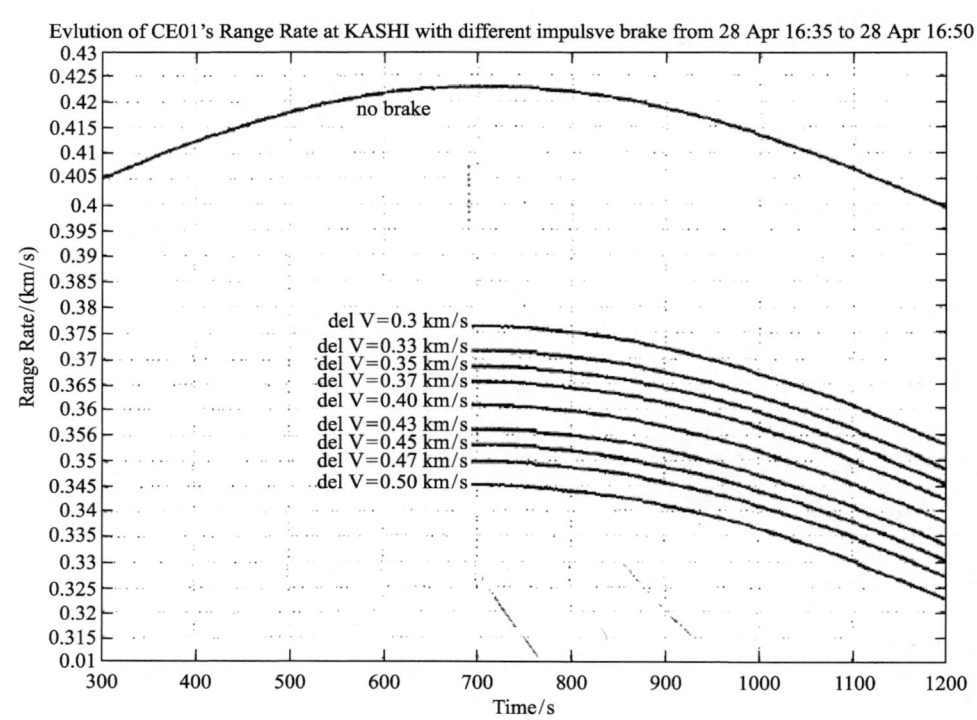

图 6-2 脉冲型制动模式下，制动量为 0.3km/s、0.5km/s 的视向速度变化曲线

更真实地预报计算了对应于制动量连续变化的视向速度随时间变化的情况，并与脉冲型制动进行了比较。地面测站为喀什站。如图 6-3 所示。

对制动量为 0.3km/s 到 0.7km/s 的情况进行细化分析，得到：

这样，根据上述探测器视向速度观测量变化曲线得到相应的制动量，再根据制动量与入轨参数的关系曲线，便可确认轨道器入轨状况。图 7 给出了根据卫星入轨前后视向速度变化量与入轨参数的关系数据库产生的曲线。

(3)针对陆地 VLBI 测控网，预报一系列不同的制动量情况下 VLBI 网得到的轨道器制动前后和制动期间的单向差分速度数据和增量；建立差分速度增量与制动量的关系数据库；进一步通过制动量与最终投入轨道的参数库，建立 VLBI 网单向差分速度数据与最终投入轨道参数的对应关系。中国 VLBI 网目前拥有上海、乌鲁木齐、北京和昆明四个台站，其分布如下图。

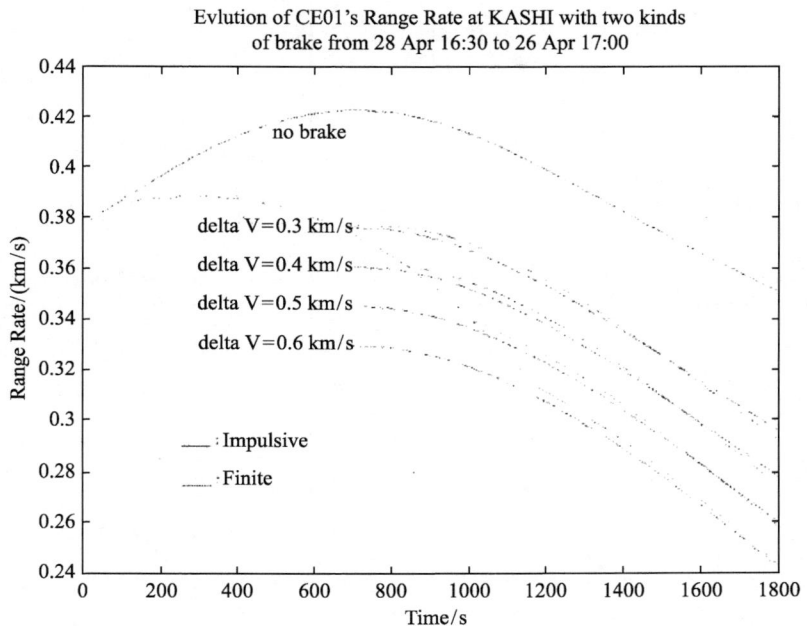

图 6-3　两种制动模式下轨道器相对于喀什站视向速度变化曲线比较

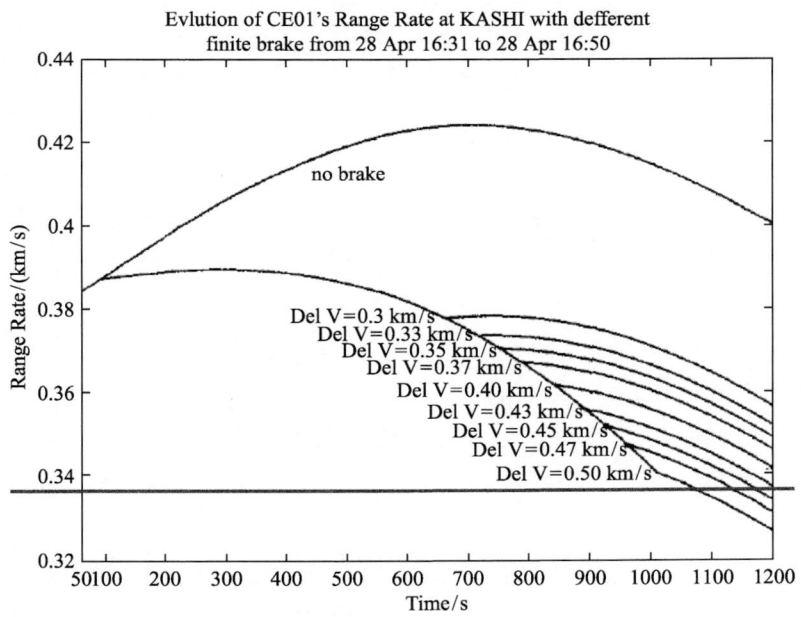

图 6-4　连续制动中制动量为 0.3km/s、0.5km/s 的视向速度变化曲线

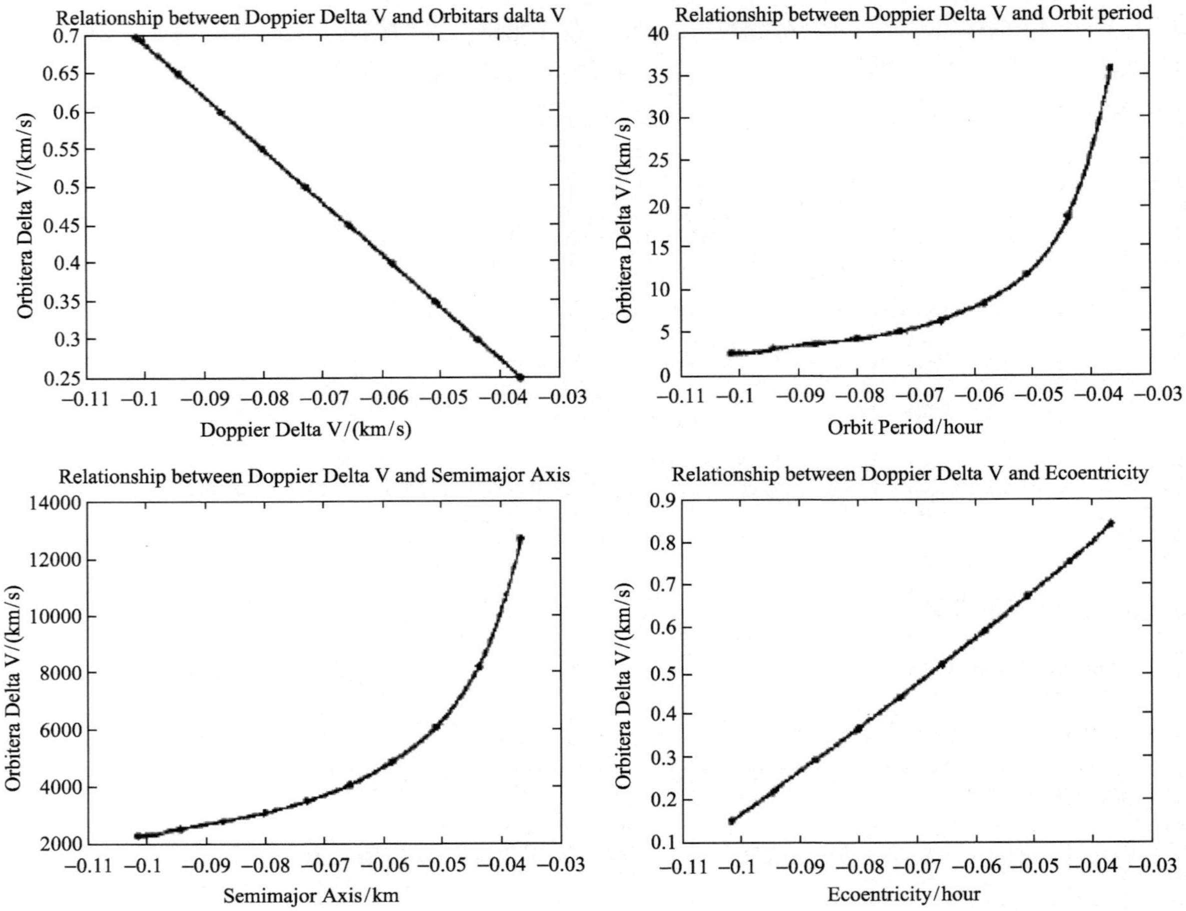

图 7 卫星相对喀什站视向速度变化量与制动量及轨道根数关系

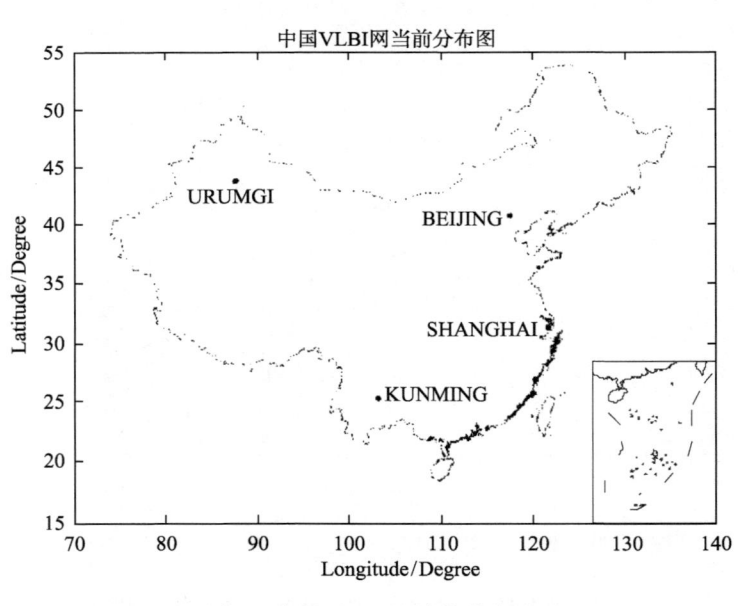

图 8 中国 VLBI 网当前分布图

参考视向 Doppler 的情况，以上海-乌鲁木齐基线为例，可以预报得到在 CE-1 绕月入轨前后的差分速度与制动量的关系如图。

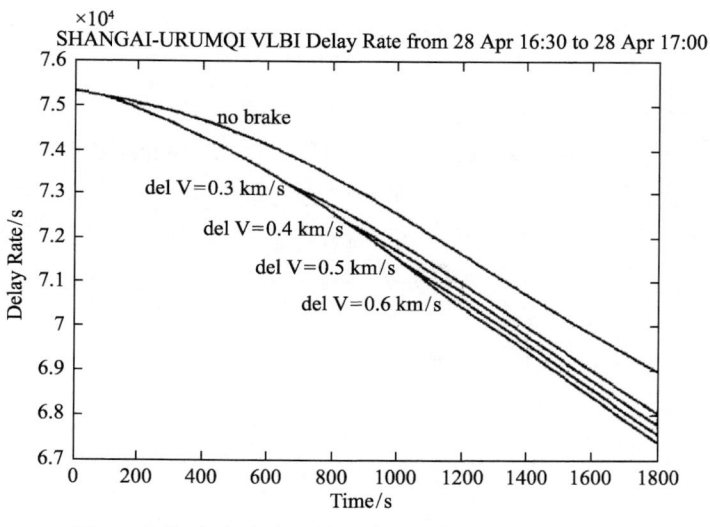

图 9 上海-乌鲁木齐基线速度差分与制动量关系

同样，给出差分速度与制动量及轨道参数的关系：

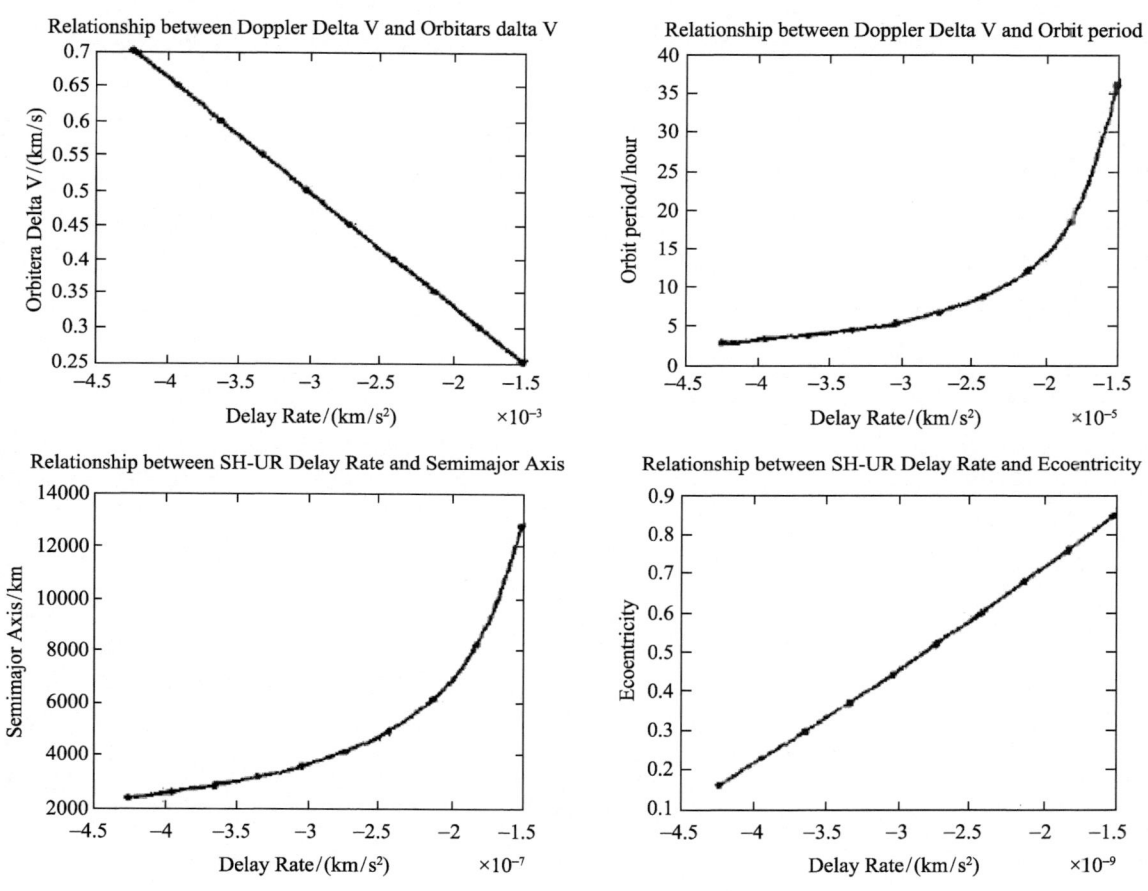

图 10 上海-乌鲁木齐基线 VLBI 差分速度与减速量及轨道根数关系

在实际应用中，根据实测数据(制动遥测参数、视向 Doppler 速度、差分 VLBI 延迟率等)，比对这些数据与入轨参数的关系库或曲线，可以方便地知道卫星的入轨状况。除了轨道半长轴、轨道周期、轨道偏心率以外，其他参数可以根据制动平面方便得到。这里不再赘述。

6 小结

月球和行星轨道器的实时入轨确认是整个探测任务中的关键步骤之一。本文针对该课题提出了多种解决方法并对这些方法进行了仿真分析。首次以 CE-1 任务为参考建立了根据设计轨道得到的理论制动量与入轨参数数据库。在实际任务中通过实时比较实测制动量和该数据库的对应关系，可以方便快速地确认卫星入轨过程和状况，避免了事后定轨所产生的滞后效应。

致谢 本文得到了中国科学院知识创新工程、百人计划和 CE-1 工程的支持。

嫦娥一号卫星的地月转移变轨控制

宗 红 王淑一 韩 冬 王大轶 李铁寿 张洪华 黄江川

(北京控制工程研究所，北京 100080)

摘要 文章阐述了嫦娥一号卫星地月转移阶段(从星箭分离到进入使命轨道)的高可靠、高精度自主变轨控制方案，介绍了飞行轨道、轨控策略及控制参数优化、星上自主变轨控制的系统设计和相关参数的地面标定等，给出了在轨飞行试验的验证结果。

关键词 嫦娥一号卫星 地月转移 轨道控制 自主变轨控制

1 引言

嫦娥一号卫星于北京时间 2007 年 10 月 24 日 18 时 05 分 04 秒由长征三号甲运载火箭从西昌卫星发射中心发射升空。经过一次远地点变轨和三次近地点变轨，嫦娥一号于 10 月 31 日进入地月转移轨道，并于 11 月 5 日准确按计划完成第一次近月制动，成为中国第一颗月球卫星。又经过两次变轨后，她终于到达离月面 200km 的通过月球两极上空的圆形工作轨道。

嫦娥一号卫星与月球轨道交会过程中地月转移变轨控制至关重要，特别是第三次近地点加速和第一次近月点制动两次关键变轨，其控制窗口具有唯一性和短暂性，必须保证按飞行计划及时、准确地完成各次变轨控制。为此，嫦娥一号卫星采用星上自主定姿、自主姿态控制、自主开/关变轨发动机、自主故障检测以及快速恢复轨控的自主变轨控制方案，由地面配合进行轨控参数优化及推力标定和加速度计标定，并采取保证变轨精度的系统设计，出色地完成了地月转移过程中的各项轨控任务。

文中所指的地月转移阶段是指从星箭分离开始到进入使命轨道的整个过程。本文介绍了嫦娥一号卫星地月转移阶段的飞行轨道和变轨策略、轨道控制大系统、星上自动变轨控制的设计、变轨控制参数的计算和标定、保证变轨精度的其他措施以及飞行验证结果。

2 地月转移飞行轨道及控制要求

2.1 在地月系统中的标称飞行轨道

嫦娥一号卫星的地月转移阶段，包括调相轨道、地月转移轨道和绕月轨道。飞行轨道如图 1 所示[1]。

嫦娥一号卫星由长征三号甲运载火箭送入近地点 200km、远地点 51000km 的大椭圆轨道(超 GTO)。调相轨道任务是将超 GTO 轨道变为远地点约 400000km 的地月转移轨道。星箭分离后，卫星在周期 16 小时的超 GTO 轨道上运行一圈半后，在远地点做一次小的轨道机动，将轨道近地点高度变为 600km，再运行一圈半，在近地点进行第一次大的轨道机动，将轨道周期变为约 24 小时，接着运行 1~3 圈后，进行第二次近地点变轨，将轨道周期变为约 48 小时。运行 1 圈后，在调相轨道运行结束到达最后一个近地点时，进行第三次近地点变轨，使卫星进入地月转移轨道。

本文原载于《空间控制技术与应用》，2008，Vol.34，No.1，44~50。

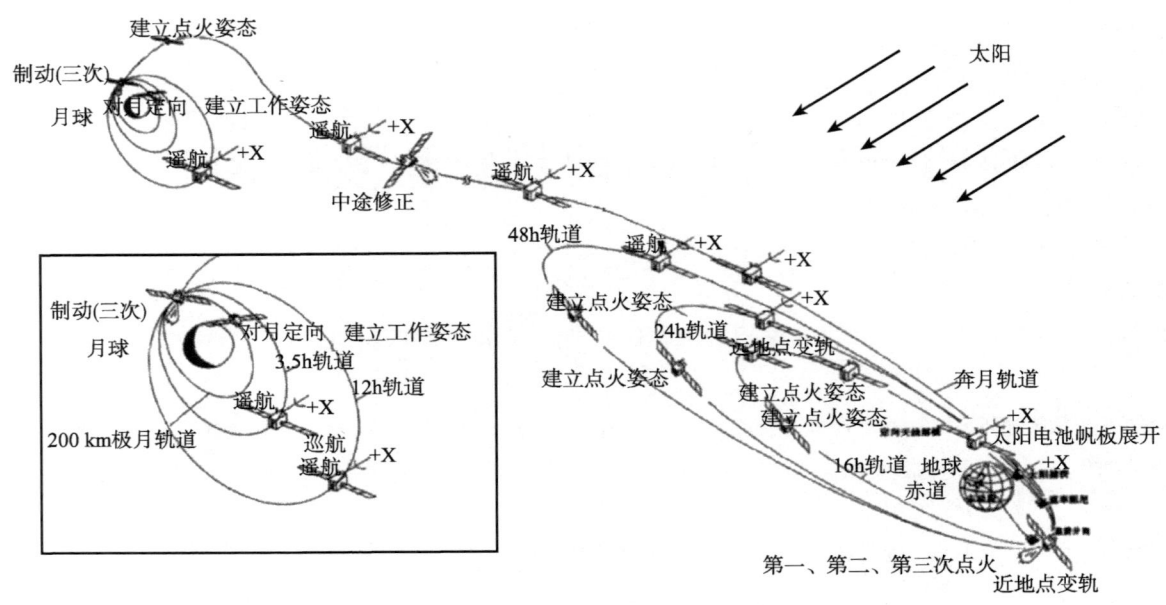

图 1 嫦娥一号卫星飞行轨道示意图

地月转移轨道共飞行 114 小时,是接近燃料消耗最少的转移轨道。在转移轨道飞行途中一般都需进行若干次轨道修正,正常情况下是 2~3 次,一次在离开近地点后的 24 小时以内完成,最后一次是在到达近月点前的 24 小时以内完成。

卫星到达近月点后,为了使其变为绕月飞行的月球卫星,需要在近月点进行 3 次减速机动。依次将轨道周期变为 12 小时、3.5 小时和 127 分钟,最终进入使命轨道。

2.2 对轨道控制的要求

科学探测要求嫦娥一号卫星的工作轨道为高度 200km±5km,相对于月球赤道的倾角为 90°±5°。为了到达这一工作轨道,嫦娥一号飞行过程中要经历 8~10 次轨道控制,包括 1 次远地点变轨、3 次近地点变轨、1~3 次中途修正和 3 次近月点制动。

在进入地月转移轨道时,很小的初始速度误差就会导致到达近月点时出现几千千米的位置误差,初始速度误差越大,轨道修正所需的燃料就越多。此外,由于卫星在近地点的高度低、速度快,若轨道控制的误差较大,就会导致近地点位置发生变化,这时地面就不能保证连续的测控条件,因此嫦娥一号的轨道控制必须足够精确。

地月转移轨道的入口和第一次近月点制动都具有唯一性。地月转移轨道入口要求必须在特定的时间从特定的位置上进入转移轨道,否则不能按照预定计划与月球交会;第一次近月点制动则要求必须在近月点附近进行减速,否则卫星将飞离月球。如果这两次变轨中任何一次失利,要想重新到达月球附近就需要花费大量的燃料和时间,甚至根本无法实现。

为确保变轨按计划及时执行,考虑到恶劣情况,在没有地面测控支持时,卫星也要具有一定自主变轨的能力。

基于上述考虑,对嫦娥一号卫星的轨道控制提出了精确性和及时性的要求,同时也要具备一定的自主性。

3 星地大回路轨道控制

嫦娥一号卫星轨道控制由星上和地面共同完成。星上部分主要是制导、导航和控制(GNC)分系统及推进分系统的相关设备，包括姿态敏感器（星敏感器和陀螺）、计算机和执行机构（变轨发动机和姿态控制推力器）等，并依靠测控数传分系统同地面保持上、下行通信联系。地面部分主要包括各测控站（船）的跟踪、遥测和遥控设备以及位于北京的航天飞行控制中心。

卫星飞行各阶段，由地面进行精确的轨道测量。获取测距、测速数据和甚长基线干涉（VLBI）系统的测角数据。飞控中心处理观测数据，确定轨道参数。由此在地面制定轨道控制策略，计算优化的变轨控制参数，并适时将有关数据注入星上计算机。卫星根据注入的变轨控制参数自动地进行姿态机动、变轨姿态保持、发动机开关机和巡航姿态恢复等控制过程。轨道机动完成后，地面再进行轨道测量和确定、对轨控效果进行标定和评估以及制定后续的轨控策略，从而构成星地大回路轨道控制。

4 变控制参数计算和推力标定

如前所述，在变轨控制实施过程中，在地面进行的轨控参数计算和轨控后相关参数标定是必不可少的步骤。

4.1 变轨控制的约束条件

变轨控制主要考虑以下约束条件：
(1) 燃料消耗：嫦娥一号卫星携带的燃料量有限，所能提供的总速度增量受到限制；
(2) 测控范围：虽然卫星具有一定的自主变轨能力，但是出于安全性考虑，仍希望整个变轨过程都在地面实时监视下进行，而轨道控制时这样的条件不一定能够满足，因此制定轨控策略时要考虑这一次轨控及后续轨控的地面测控条件；
(3) 卫星能源：卫星轨控过程中，可能处于地球或月球阴影中，或者帆板不一定能够对准太阳，此时卫星依靠电池供电，这要求轨道控制整个过程不能超过电池供电的最长时间。

在不同的阶段和不同的情况下，这些约束条件的重要性不同。在调相轨道阶段主要考虑测控范围的限制；在近月制动时主要考虑卫星能源；在发生故障而需要重新设计轨控策略时主要考虑燃料消耗约束。

4.2 点火姿态和角速度的选择

轨控参数计算时，根据轨控的目标，在满足测控条件的约束下对轨控开机时刻、轨控姿态和轨控时长等参数进行优化，使得在达到目标轨道的同时消耗燃料最少。同时星上还具备匀速转动变轨的能力，轨控过程中推力方向在空间按一定的角速度旋转，可以进一步减少轨控的燃料消耗。

为了保证轨控过程中星敏感器不受日光、月光和地气光干扰，轨控姿态在保证+X（变轨推力）方向的情况下，可以绕+X轴旋转一定的角度以寻找合适的姿态。

4.3 有限推力轨道控制参数计算

在制定轨道控制策略时，按照脉冲变轨方式进行计算，计算过程中只有一个未定变量，即速度增量的大小。轨道控制策略确定后再按照有限推力方式计算当前这一次轨控的控制参数。有限推力轨控时需要确定两个变量：轨控开机时刻、轨控关机时刻/开机时长。

轨控开机时刻的选择：一般情况下，有限推力变轨以近地（月）点为中点，前后各取一半的点火时间。也可以调整开机时刻使轨道的近地点幅角达到期望的目标，这时就要采用牛顿迭代法来调整开机时刻。

轨控关机时刻的选择：首先由脉冲变轨给出关机时刻的初始值，然后再用数值迭代的方法进行精确计算。迭代时要考虑测控条件的约束，若不满足测控条件的约束，则对关机时刻进行修正，以满足轨控时的测控约束。对于近月点制动，关机时刻比较容易确定，只需从开机时刻开始，数值积分到轨道半长轴满足预定目标即可。

4.4 加速度计和发动机推力的在轨标定

加速度计测量量本身包含零位偏差和脉冲当量误差，如不考虑这些偏差将影响轨道控制精度。为保证轨控的准确性，需要对加速度计进行标定并给予补偿。

加速度计的在轨标定分为两个方面，一是对加速度计零位偏差的标定。每次轨控前，统计卫星没有喷气的时间段内加速度计的数据，给出平均值，作为加速度计的零位偏差，以便在轨控中对使用加速度计数据计算的卫星速度增量进行补偿。二是利用定轨数据对加速度计的刻度系数进行标定。每次变轨结束后，依据地面测、定轨后给出的变轨过程中的速度增量 δV 和变轨过程中利用加速度计累计的速度增量，计算加速度计脉冲当量标定系数，在下一次变轨策略计算中对卫星变轨速度增量进行补偿，以提高轨控精度。

卫星入轨后，发动机推力会随着推进剂贮箱温度和压力等参数的变化呈现出不同的特性。若采用同样的推力进行轨控策略的计算，势必会带来较大的计算误差，影响变轨精度。所以在每次变轨后对推力器的推力标定是高精度变轨必不可少的步骤。目前国内外均有许多种推力标定的方法。嫦娥一号卫星主要根据加速度计数据对发动机推力进行标定，同时根据卫星贮箱压力温度等参数进行适当修正。

5 星上自主变轨控制的设计

为了解决深空探测中卫星的变轨问题，保证准确、及时、可靠地完成卫星的变轨控制，在嫦娥一号卫星的变轨控制中，采用了一种星上自主地进行的姿态确定、姿态机动以及自主变轨的变轨程序。

5.1 变轨控制飞行程序

月球探测卫星大部分轨道控制利用 490N 大推力发动机完成，少量中途轨道修正以及环月运行后轨道维持控制采用 10N 小推力发动机进行。根据每种发动机使用特点，制定了 490N 变轨准备子程序，变轨控制程序，10N 变轨准备、变轨控制以及 10N 轨道维持子程序。保证各种轨道控制准时、可靠。

5.2 与变轨控制有关的工作模式设计

GNC 分系统设计了四种工作模式，用于卫星变轨准备和变轨控制：

(1) 恒星捕获：在卫星建立轨控点火姿态之前，利用敏感器信息预估卫星的惯性姿态并进行卫星姿态控制；

(2) 惯性调姿：用于建立卫星轨控点火姿态，实现卫星三轴大角度姿态机动；

(3) 恒星定向：在卫星建立轨控点火姿态之后保证卫星的稳态控制；

(4) 轨控定向：进行轨控发动机开、关机控制，确定卫星的点火姿态，并进行点火期间的姿态稳定控制。

5.3 预置控制参数的自主变轨控制程序

在卫星建立轨控点火姿态之前，卫星进入恒星捕获模式，对陀螺漂移和加速度计零位偏差进行标定；根据程控指令，卫星自主转入惯性调姿阶段，建立卫星轨控点火姿态；调姿到位后，卫星自主转入轨控前的姿态稳定阶段，利用星敏感器对卫星姿态进行滤波修正；程控时间到，卫星自主进入轨控定向模式，星上自主控制轨控发动机开关机。轨控发动机关机后，卫星稳定一段时间，自主转入太阳定向模式，恢复卫星巡航姿态。

5.4 建立点火姿态的再定向机动

近地轨道卫星的姿态机动多为单轴姿态机动，或是三轴小角度的姿态控制。嫦娥一号卫星在轨道控制前，需要将卫星从对日定向的巡航姿态调整到轨道控制所需的点火姿态。这种姿态调整可能是任意姿态的调整,为此采用四元数方式同时进行三轴姿态机动。控制律设计中在考虑调姿时间有限制这一条件的同时还考虑了根据卫星调姿姿态设置不同调姿角速度的方法,这样既满足时间要求又可适当地减小轴间耦合。

5.5 点火姿态的测定

点火姿态主要通过陀螺数据估计确定。当卫星姿态角速度较大时,在适合星敏感器测量的条件下,自动引入星敏感器信息修正卫星点火姿态。

5.6 轨控发动机的开机和关机控制

轨控开机采取预先注入开机时间的自主程控点火,轨控关机采用速度增量及时间双保险关机的控制方法。当加速度计信号积分值达到预定速度增量数值时,计算机发出关机指令。时间关机控制方法是指当点火时间累计值达到预定点火时长数值时,计算机发出关机指令。这一控制逻辑能够保证,在正常情况下使用速度关机方法,实现高精度变轨;加速度计异常情况下使用时间关机方法,防止错过控制窗口或引发灾难性故障。

5.7 点火姿态的稳定控制

考虑到轨控发动机点火的干扰力矩大,可能激发液体推进剂晃动和太阳帆板挠性振动,设计了基于脉宽调制(PWM)的 PID 和滤波校正的纯数字化的喷气姿态控制律[2],有效地保证了变轨期间卫星姿态控制精度。同时尽可能地减少了推力器脉冲工作次数。图 2 给出了稳定控制原理框图。

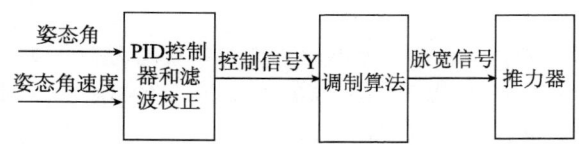

图 2 轨控期间姿态稳定控制原理框图

5.8 关机后巡航姿态的自主恢复

以往静止轨道卫星变轨结束后,均由地面控制卫星进行太阳捕获,建立巡航姿态。嫦娥一号卫星在变轨发动机关机并稳定一段时间后,自主地进入太阳定向模式,利用卫星上安装的太阳敏感器进行太阳捕获和太阳定向控制,自主地恢复巡航姿态。

5.9 自主故障检测、处理和恢复

目前静止轨道卫星的变轨控制多采用地面控制的方式,在变轨过程中若出现姿态控制或推进系统异常情况,由地面进行故障诊断,然后采取措施关闭轨控发动机,中止轨道控制。中低轨道卫星具有一定的自主轨控能力,但在轨控过程中出现姿态异常等现象,仍然能自主退出轨控,待地面排除故障后,再择机进行轨控。这种控制策略,对于轨控窗口有唯一性要求的月球探测卫星却不再适用。

嫦娥一号卫星变轨控制过程中,是由星上自主检测卫星姿态信息,当检测出故障后,紧急关闭变轨发动机,自主地进行故障处置。同时,星上自主控制进入一个过渡模式,待卫星的姿态角速度被适当阻尼后,自主地重新转入轨控准备阶段,重新设置轨控流程。根据重新设定的时间,卫星自主地转入相应的工作阶段,恢复轨控,并根据所需要的控制量完成变轨开、关机。

6 保证变轨精度的其他措施

对于嫦娥一号卫星或其他对轨道要求严格的航天器,必须考虑到姿态控制喷气(含动量轮卸载)构成其

轨道运动的摄动力，可能会影响轨道确定精度和变轨控制参数计算精度，最终影响变轨精度。为此，在嫦娥一号卫星 GNC 系统设计中采用了减少喷气和计量喷气的措施，还研究了多种工作模式下的喷气影响分析和补偿方法。

6.1 巡航姿态的动量轮控制

以往卫星太阳定向模式下多采用喷气姿态控制。非力偶式安装的推力器工作时，不但产生姿态控制力矩，而且产生使卫星质心速度改变的推力，使卫星轨道发生复杂变化。为了减小卫星姿控的喷气量，嫦娥一号卫星的太阳定向巡航姿态采用动量轮控制方式，有效地减少了喷气对轨道的扰动。但是，由于受到环境干扰力矩的作用，会造成动量轮角动量饱和。星上采用三轮零动量工作方式，增加系统存储角动量的能力，减少喷气卸载次数。

6.2 巡航姿态的慢旋方式和停旋控制

卫星在巡航姿态下处于不同的轨道段，所受的干扰力矩不同。调相段主要受重力梯度力矩的影响，地月转移段主要受太阳光压力矩的影响，月球捕获段主要受月球引力的影响。在上述阶段卫星处于惯性定向姿态，干扰力矩引起不同程度的动量积累。在巡航姿态下采用地面控制卫星慢旋与星上自主启旋的方式，使卫星绕对日定向轴(+X轴)慢旋，抵消大部分干扰力矩的影响，减少巡航姿态长期运行中的喷气卸载。在卫星每次变轨前，地面控制卫星停旋在适当相位，以减少卫星惯性调姿控制的喷气量。

6.3 动量轮喷气卸载的计量纳入轨道预报模型

嫦娥一号卫星 GNC 分系统设计，采用了提供遥测通道将姿态控制推力器的工作时间和工作次数下传到地面，结合姿态敏感器数据可以在地面较为准确地计量每一个遥测周期内喷气推力的大小和方向的方法。根据遥测数据获得喷气加速度在航天器惯性坐标系中的分量，定轨时计算并计入了喷气摄动，以补偿姿控喷气对轨道确定精度的影响。该方法显著地提高了卫星轨道确定的精度，补偿流程如图 3 所示。

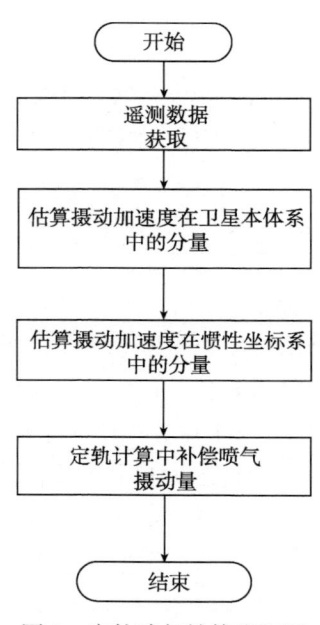

图 3 定轨喷气补偿流程图

为了防止动量轮卸载发生在地面不可见(无遥测)弧段，GNC 系统设计了强制卸载手段。飞控过程中，安排在地面可见弧段(在飞出测控区之前的一段时间)进行强制卸载，通过地面注入卸载指令，强制动量轮卸载有效地避免了航天器在不可测控弧段喷气卸载。

6.4 变轨前后姿控喷气纳入变轨参数计算模型

由于在巡航姿态下进行变轨前的测、定轨，计算出轨控参数后才进行姿态机动建立点火姿态，而姿态机动采用喷气控制完成，轨控过程中姿态控制和轨控后恢复巡航姿态的姿态机动也会喷气。在变轨参数计算时如果不考虑这些喷气，会对变轨精度产生影响。

根据数学仿真和飞行遥测数据，可以估算不同条件下姿态机动时的喷气摄动，并生成数据表。在轨控参数计算中，根据航天器本体相对于目标姿态的误差四元数查表求得对应的航天器姿态机动产生的速度增量，据此在轨控计算中修正轨控量，从而提高轨控精度，补偿计算流程如图4所示。

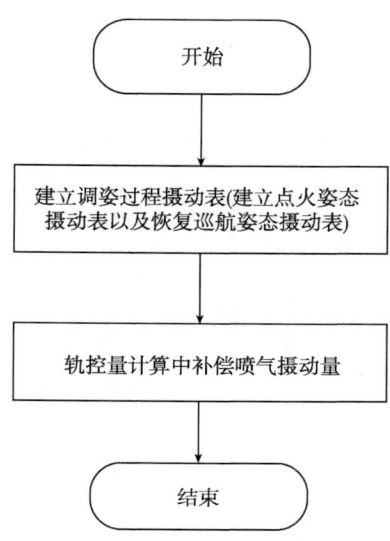

图 4 轨控参数计算中补偿喷气流程图

7 飞行试验结果

嫦娥一号卫星于2007年10月24日18时29分38秒入轨，入轨后的近地点高度为200km，远地点高度为51000km。入轨后当天就对加速度计进行了标定，标定出了两个加速度计的零位偏差。

7.1 调相轨道阶段

星箭分离后，嫦娥一号卫星进入周期为15.8小时的超地球同步转移轨道。经过大约1天(绕地球一圈半)，于10月25日17时55分进行了一次远地点变轨。卫星从对太阳定向的巡航姿态开始，自动进行大角度姿态机动，建立点火姿态；发动机点火关机后，卫星自动恢复巡航姿态；整个过程中地面没有向卫星发出任何遥控指令。以后每次变轨都要重复这一过程。因此，这是对于新设计的变轨控制流程的一次在轨试验，同时也将轨道的近地点高度提高到600km，改善了后续近地点变轨操作时的观测条件。

在远地点变轨后大约1天，即26日17时33分和29日17时49分，成功地进行了第一次近地点变轨控制。由于近地点附近地面观测时间短，且易受轨道误差影响，故使用了前述的自动轨道控制流程。整个过程中地面未向卫星发出任何遥控指令。变轨完成后，卫星轨道周期从15.8小时变为24小时，远地点高度从51000km变为72000km。

按照卫星飞行程序，在第一次近地点变轨后大约3天，进行了第二次近地点变轨，轨道周期变为48小时，远地点高度升至120000km，超过了以往中国人造地球卫星所到达过的高度。

卫星继续飞行约2天后，在10月31日17时15分开始进行了第三次近地点加速。完成变轨后进入了与月球交会的地月转移轨道，这一次轨控特意提前了3分钟，这样就将轨道的近地点幅角减小了0.23°，使得中途修正的速度增量减小了一半。地月转移轨道入口点时刻为10月31日17时25分4.7秒，比设计的地月转移轨道入口时刻仅提前了23.3秒。

7.2 地月转移轨道阶段

由于实现了在发射窗口前沿发射，并且轨道控制精度较高，所以取消了原计划在进入地月转移轨道第17小时所进行的第一次中途轨道修正，在进入转移轨道后第41小时进行了一次小的修正，修正量为4.84m/s。这次中途修正之后，根据定轨的结果，卫星到达近月点的高度为211km，轨道倾角为90.17°，近月点时刻为11月5日11时25分48秒，仅仅比标称轨道晚了20秒到达近月点。因此再一次取消了原计划在到达第一个近月点前24小时进行的第三次中途修正。

卫星在转移轨道上共飞行约114小时，于11月5日11时15分开始进行第一次近月点制动变轨，11时36分33秒变轨发动机关机，卫星进入周期为12小时2分的绕月轨道，成为一颗月球卫星。

7.3 近月点制动轨道阶段

第一次近月点制动后的环月轨道周期为12小时145秒。在11月6日11时21分开始进行第二次近月点制动，制动后的环月轨道周期为3.5小时104秒。在11月7日8时24分开始进行第三次近月点制动，制动后将卫星的远月点高度降低到187.66km，成为了近月点，原来的近月点成为了远月点，高度为213.2km，满足高度200km±25km、倾角90°±5°的要求。至此嫦娥一号卫星进入了预定工作轨道。

8 结论

嫦娥一号卫星GNC分系统的高可靠、高精度自主变轨控制，保证了卫星从绕地球运行轨道顺利转移到预定的绕月球运行轨道，为中国首次月球探测工程的圆满成功作出了重要贡献。本文从变轨参数优化和参数标定、星上自动变轨控制和系统设计保障措施等方面详细描述了嫦娥一号卫星地月转移阶段的变轨控制过程以及高精度的变轨控制方法，同时给出了在轨飞行试验的验证结果，可供后续深空探测系列卫星的轨道控制系统设计参考。

参 考 文 献

[1] 宗红，李铁寿，王大轶. 月球卫星GNC系统方案设想[J]. 航天控制，2005，23(1)：2—6
[2] 王寨，李铁寿，王大轶. 探月卫星变轨时的姿态控制研究[J]. 航天控制，2005，23(1)：11—14

瞬时状态归算用于嫦娥一号卫星关键轨道段监测

李金岭　郭　丽　钱志瀚　平劲松

(中国科学院上海天文台，上海 200030)

摘　要　我国绕月探测工程嫦娥一号卫星(嫦娥一号)采用了联合 S 波段测距测速与 VLBI 跟踪的测轨模式。我们实现了对跟踪资料的一种实时处理方法，获得了嫦娥一号卫星的瞬时状态(位置和速度)时间序列，用于关键轨道段监测。本文介绍该实时处理方法以及在嫦娥一号卫星关键轨道段监测中的具体应用，以资批评借鉴。

关键词　月球探测　轨道机动　轨迹监测　瞬时状态　嫦娥一号卫星

我国月球探测计划嫦娥一号工程于 2004 年 1 月正式立项[1,2]，于 2007 年 10 月 24 日发射首颗探月卫星(嫦娥一号)。该卫星累计完成了 494 d 的飞行，并于 2009 年 3 月 1 日成功实现了受控撞月，为我国后续探月计划积累了宝贵的技术和工程经验。我国联合 S 波段测距测速(USB)及 VLBI 系统对该卫星进行了长期跟踪观测，本文介绍在部分关键轨道段，如月球捕获段和落月段等轨道机动段，利用瞬时状态(ISV)(位置和速度)归算方法对卫星轨迹的实际监测情况。

深空目标的轨道分析和 ISV 归算既有相通之处，也存在显著区别[3~7]。主要的区别在于，定轨分析一般要求有足够长的跟踪弧段，通过状态转移矩阵将整个弧段中不同时刻的观测资料联系起来统一求解，强调精确模制跟踪弧段中飞行器的受力状态。而在 ISV 归算中，待求参数是各个给定时刻目标的位置与速度时间序列，对跟踪弧段的长度没有苛刻要求，不需要模制飞行器的受力状态，从几何角度实时监测跟踪数据的质量和实现快速轨迹测定。尤其是在轨道机动段、月面软着陆以及月面行进等过程中，由于目标的受力状态不易精确模制，一般可采取 ISV 归算方法[8~11]实现轨迹快速测定。

1　ISV 方法简介

现代观测资料的参数解析一般遵循这样的基本操作过程，即对待求参数赋予尽量准确的初值，依照有关模型计算观测量的理论值、以及理论值对待求参数的偏导数，将观测值与理论值之差对待求参数初值的改正值进行泰勒展开，通过误差方程解算获得参数初值的改正值，进而获得经观测资料更新之后的参数值。形式地表示如下[12,13]：

$o - c = \Sigma \frac{\partial c}{\partial X} \mathrm{d}X$ 为误差方程，经解算获得 dx，o 为观测值，c 为待求参数取初值时由观测模型计算得到的理论值，X 为待求参数集合，$\partial c/\partial X$ 为由观测模型推演得到的理论值对待求参数的偏导，$o-c$ 为观测值与理论值之差，dX 为解算参数，为待求参数初值的改正值。X+dX 为经观测更新的参数取值。

本节从观测模型、理论值与偏导数的计算、观测量的预处理和误差方程的约束等方面，对 ISV 方法予以简要介绍。

1.1 观测模型

嫦娥一号卫星跟踪资料包括 USB 的双程、三程测距和对应的积分 Doppler 测速[14]，以及 VLBI 的时延、时延率等[4]。

1.1.1 USB 测距

$$\rho = (|\boldsymbol{r}_s(t) - \boldsymbol{r}_1(t_1)| + |\boldsymbol{r}_s(t) - \boldsymbol{r}_2(t_2)|)/2 \tag{1}$$

t_1 和 t_2 分别为上行信号发出、下行信号接收的时刻，$r_1(t_1)$ 和 $r_2(t_2)$ 为测站位矢，$r_s(t)$ 为信号转发时刻 t 时飞行器的位矢，ρ 为观测时刻 t_2 时的测距值。若信号的发射和接收由同一测站完成时为双程测距，由不同测站完成时为三程测距。

1.1.2 USB 积分 Doppler 测速为

$$\dot{\rho} = (\rho_2(t) - \rho_2(t - \Delta t))/\Delta t \tag{2}$$

Δt 为积分时间，$\rho(t_2)$，$\rho(t_2 - \Delta t)$ 表示两个时刻的测距，$\dot{\rho}$ 即为 t_2 时刻的积分 Doppler 测速。

1.1.3 VLBI 时延 τ 和时延率 $\dot{\tau}$

$$\tau = \frac{1}{c}(|\boldsymbol{r}_2(t+tr+\tau) - \boldsymbol{r}_s(t)| - |\boldsymbol{r}_1(t+tr) - \boldsymbol{r}_s(t)|) \tag{3}$$

c 为真空中的光速，tr 为信号波前自飞行器至参考测站的传输时间，r_1，r_2 和 r_s 分别表示参考测站和远端测站以及卫星的位矢，τ 即为 $t+tr$ 时刻的观测时延。时延率为

$$\dot{\tau} = \Delta\tau/\Delta t \tag{4}$$

Δt 为积分时间，依照相关约定，$\dot{\tau}$ 的观测时刻取为积分的中间时刻。

(1)~(4)式还应附加传输介质、仪器差、钟差等修正，本文所用的嫦娥一号卫星跟踪资料均已经采取了各项误差的相应处理措施，精度满足工程需求[5,6]。

1.2 理论值计算

如(1)~(4)式所示，计算跟踪观测量的理论值，需要基于对应时刻飞行器和测站的状态矢量，其中飞行器状态为待求参数，测站状态为已知参数。某时刻飞行器的状态矢量初值可由初始轨道根数预测，一般在惯性参考架中进行。给定时刻的测站状态矢量可由参考时刻国际地固参考架(ITRF)中的数值推算得到，经由潮汐、板块运动等地球物理效应改正之后，转换至惯性参考架。

飞行器和测站的状态矢量由某一信号波前相互关联，对于 t 时刻的飞行器状态矢量，与之对应的测站状态矢量的参考时刻是唯一确定的，反之亦然。两个参考时刻之差即为信号的传输时间(光行时)。无论给定星端还是给定站端在某时刻的状态矢量，计算其所对应的另一端状态矢量的参考时刻，此过程可迭代进行。以下讨论星端和站端状态矢量以及信号传输时间的迭代计算，进而获得观测量的理论值和参数偏导。

1.2.1 星端状态

轨道预报是空间环境监测和实时跟踪测量中的一个重要环节，通常采用分析方法进行[15]。尽量提高预报精度，这是提高参数解算精度和参数解的收敛速度的重要保障。以地球卫星为例，在地心天球参考架中，运动方程表示为

$$\ddot{\boldsymbol{r}} = \boldsymbol{f} \tag{5}$$

式中 $\ddot{\boldsymbol{r}}$ 为卫星的加速度，$\boldsymbol{f}$ 为作用力，包括中心天体引力、地球形状摄动、日月及行星摄动、大气阻力、太阳光压等。如果各种作用力精确已知，则由参考时刻 t_0 的状态 $\boldsymbol{r}_0$ 和 $\dot{\boldsymbol{r}}_0$ 可以推得任意时刻 t 的状态 $\boldsymbol{r}_0$ 和 $\dot{\boldsymbol{r}}_0$，即(5)式需要如下初始条件，

$$\begin{cases} r_0(t_0) = r_0; \\ \dot{r}_0(t_0) = \dot{r}_0. \end{cases} \quad (6)$$

一般情况下，卫星的初始状态 r_0 和 $\dot{r}_0$ 无法预先精确可知，只能得到参考值 r_0^* 和 $\dot{r}_0^*$。在定轨分析中，需要通过迭代计算对该参考值进行逐步精化，以获得参考时刻高精度的状态。在 ISV 归算中，由初轨的参考值 r_0^* 和 $\dot{r}_0^*$，预报卫星在某时刻 t 的状态 r_0^* 和 $\dot{r}^*$，由误差方程的解算得到对预报值的改正，进而迭代获得经观测量修正的精确值。

1.2.2 站端状态

飞行器瞬时状态解算中，测站状态被视为已知，因而必须尽量确保精度，以降低其误差对飞行器状态解算精度的影响。测站在观测历元的状态由 ITRF 中参考历元的量导出，经多种地球物理效应的改正，由地固系转换到惯性系，进而与飞行器的状态初值一道，迭代计算信号传输的光行时。

1.2.3 光行时与理论值的迭代计算

如上所述，空间飞行器和测站的状态在参考时刻上是由信号同一波前而相互关联的，两个参考时刻之差即为光行时。以 VLBI 观测时延为例，信号波前在星端发出、站端接收。基线两端测站接收到同一波前的时间差即为 VLBI 时延观测量。在软件编制中我们设计了光行时如下步骤的迭代计算方式：

(1) tr 为光行时，初值取为 0，即 $tr0=0$。
(2) 预报给定时刻 t 时飞行器在惯性参考架中的位矢 $r_s(t)$。
(3) 计算 $t+tr0$ 时刻参考测站在惯性参考架中的位矢 $r_1(t+tr0)$。
(4) 更新光行时估值 $tr = (|r(t+tr0)| - |r_s(t)|)/c$。并令 $dtr = |tr - tr0|$，以及 $tr0 = tr$。当 dtr 小于某门限设置时，将 tr 作为光行时最终估值，并进行后续步骤的计算。否则，返回步骤(3)，重新计算 $r_1(t+tr0)$ 及更新 tr 的估值。
(5) 令 $tr1=tr$，表示某波前自飞行器至参考测站的光行时。
(6) 令 τ 为基线测量时延，初值取为 0，即 $\tau 0=0$。
(7) 计算 $t+tr1+\tau$ 时刻远端测站在惯性参考架中的位矢 $r_2(t+tr1+\tau)$。
(8) 更新估值 $\tau = (|r_2(t+tr0)| - |r_s(t)|)/c - tr1$。并令 $d\tau = |\tau - \tau_0|$，以及 $\tau 0 = \tau$。当 $d\tau$ 小于某门限设置时，将 τ 作为时延的最终估值，并进行后续步骤的计算。否则，返回步骤(vii)，重新计算 $r_2(t+tr1-\tau)$ 及更新 τ 的估值。

由上述迭代得到的 τ 即为 VLBI 观测时延的理论计算值，进而依据积分时间等约定得到时延率理论计算值 $\dot{\tau}$，以及不难组合得到双程与三程测距和积分 Doppler 等观测量的理论计算值 ρ 和 $\dot{\rho}$。

如此设计的光行时和观测量理论值的迭代计算过程，收敛速度快、精度有保障，其有效性已经得到仿真计算和实测资料解析的验证。

1.3 参数偏导

仅以 VLBI 时延对卫星坐标 x_s 的计算为例，由(3)式不难得到，

$$\frac{\partial (c\tau)}{\partial X_s} = \left(\frac{X_1 - X_s}{R_1} - \frac{X_2 - X_s}{R_2} \right), \quad (7)$$

其中 R_1 为卫星至参考测站的距离，x_1 为参考测站的坐标，其他类推。

1.4 观测量的内插

仍以 VLBI 时延为例，一条基线某时刻的观测量是两端台站对同一波前的测量结果，不同基线即便在

同一时刻的观测量却不一定对应于同一信号波前。归算飞行器的瞬时状态, 可以存在两种方式, (a)归算波前发出时刻 t 的飞行器状态; (b)归算 t 时刻某基线观测量所对应的 $t-tr$ 时刻(tr 表示信号光行时)的飞行器状态。在(a)方式时, 与 t 时刻(波前)对应的各基线观测量不一定存在, 但可经由各基线观测数据序列的内插而得到。(b)方式时, 以某基线 B 观测量的参考历元 t 为节点, 即使观测量按时间均匀采样, 所得飞行器状态的参考历元 $t-tr$, 其间隔将因星站几何构型的相对变化而不同。况且, B 以外的其他基线与 $t-tr$ 历元波前所对应的观测量不一定存在, 因而也需要由观测序列进行内插。可见, 无论采用(a)和(b)中的任一方式, 都需对基线观测序列进行内插。

定轨计算中, 解算参数为参考历元的飞行器状态, 不同基线不同时刻的观测量将分别对应于不同时刻的飞行器状态, 但都可以表示为参考历元状态的函数(状态转移矩阵), 因而无需对观测资料序列做内插处理。在 ISV 归算中, 各时刻的卫星状态都被作为解算参数而分别求解, 强调各时刻瞬时状态与观测量的一一对应关系, 而不是通过状态转移矩阵一次性统一解算整个弧段的所有观测资料。定轨分析中强调飞行器轨道运动过程中的受力约束, ISV 归算不考虑这样的约束, 仅强调与观测网的几何关系, 在处理方式上不同于定轨分析, 因而有其独到之处。尤其对于轨道机动段跟踪资料的处理, 在难以精确模拟受力情况时, 仅强调几何关系将更适用于实时、快速的工程性需求。

常规的观测资料内插方法有多种类型, 如切比雪夫多项式插值、拉格朗日多项式插值、牛顿插值法、3 次样条插值等。3 次样条插值方法基于插值函数一阶导数平滑、二阶导数连续的假设, 首先一次性对整个数据序列在各采样节点构建二阶导数序列, 之后可方便地得到数据序列之内任意引数处的函数内插值。模拟和实验观测数据处理的检验表明, 采用该插值方法具有快速和高精度的特点。

1.5 地心距和速率约束

对月球卫星, 我国的 VLBI 网的相应立体角仅 0.5°左右, 意味着即使有了 4 条基线的观测量, 即存在 3 个独立的时延观测量, 误差方程仍会存在较强的相关性, 不能很好求解卫星的三维位置, 径向分量的确定精度要远低于横向分量。为此, 可与测距资料联合处理, 或采用地心距约束,

$$|r_s|=|r_{s0}|$$

即假定在 t 时刻飞行器的地心距 $|r_s|$ 与初轨预报值 $|r_{s0}|$ 相等。对上式线性化, 组成误差方程中的一式,

$$\frac{X_s}{|r_{s0}|}\mathrm{d}X_s + \frac{Y_s}{|r_{s0}|}\mathrm{d}y_s + \frac{Z_s}{|r_{s0}|}\mathrm{d}Z_s = 0, \tag{8}$$

地心距约束的先验误差由初轨误差确定。

类似地, 对时延率观测量, 为改善误差方程的条件和提高速度矢量的解算精度, 可综合处理 Doppler 测速资料, 或引入地心距速率约束,

$$|\dot{r}_s|=|\dot{r}_{s0}|$$

即假设在 t 时刻飞行器相对于地心的速率 $|\dot{r}_s|$ 与初轨预报值 $|\dot{r}_{s0}|$ 相等。线性化得

$$\frac{\dot{X}_s}{|\dot{r}_{s0}|}\mathrm{d}\dot{X}_s + \frac{\dot{Y}_s}{|\dot{r}_{s0}|}\mathrm{d}\dot{Y}_s + \frac{\dot{Z}_S}{|\dot{r}_{s0}|}\mathrm{d}Z_s = 0 \tag{9}$$

类似地, 先验误差可由初轨误差确定。

2 ISV 归算的应用举例

嫦娥一号卫星测控系统分别于 2005 年 3 月 17 日至 20 日, 2006 年 5 月 29 日至 6 月 2 日以及 2007 年 1 月 22 日至 24 日组织了我国 USB 与 VLBI 网对探测一号卫星和欧空局 Smart-1 环月卫星的跟踪联测。通

过实验观测的实时与事后资料处理演练,对飞行器ISV归算在方法和软件上均得到了不同程度的改进,尤其是提高了软件的实时跟踪资料处理能力和参数的解算精度。嫦娥一号卫星于2007年10月24日发射升空之后,我国USB、VLBI网进行了长期的跟踪监测,ISV归算方法在跟踪资料质量的实时监测和关键弧段的快速轨迹测定方面,获得了成功应用。

2.1 2006年Smart-1实验观测

2006年5月29日至6月2日,我国USB,VLBI网对欧空局的Smart-1环月卫星进行了联合测轨试验。当时,Smart-1卫星对我国地区可视期约12 h,中间穿插河外射电源观测以进行钟差校正。VLBI观测每5 s给出一组各基线的时延和时延率观测量,USB每秒提供一组测距和测速观测量。但是,由于当时的测距测速存在较大的系统差,对于实时的ISV归算无法判断并予以处理,因此采用了地心距和速率的约束方式。

图1给出了6月2日实时解算的VLBI时延拟合残差,拟前残差是观测时延与初轨预报值之差,如图1中黑点所示,拟后残差是参数解算后的观测时延残差,如图1中灰点所示。从中可见,拟前残差在上下5 m之内波动,相当于十几个纳米水平,表明跟踪观测以及软件对观测量的预报精度都达到了较好的水平。拟后残差在1 m水平,明显好于拟前残差,表明ISV归算是成功的。

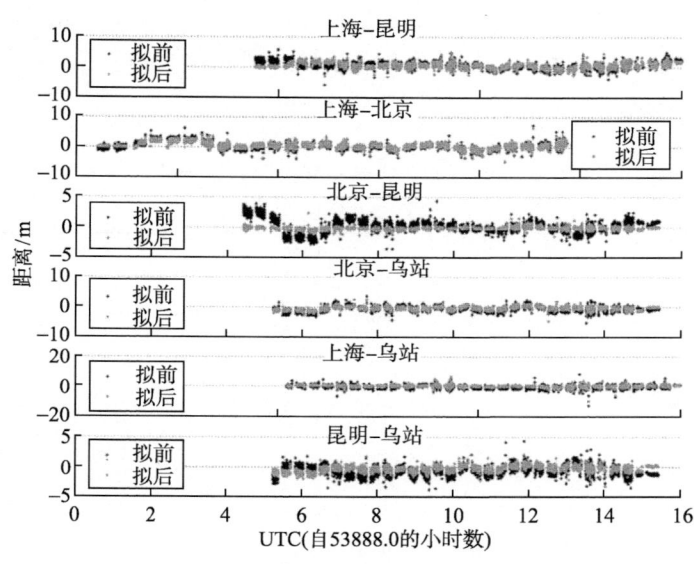

图1 Smart-1卫星观测时延的拟前、拟后残差分布

图2是实时ISV归算结果与ESA事后重建轨道在地心赤经赤纬方向的比较,可见二者符合较好,差值序列的标准偏差约为0.1 as,在月球距离相当于186 m。这肯定了VLBI观测量的精度以及ISV归算方法和软件的可靠性。

2.2 嫦娥一号卫星月球捕获段轨迹监测

嫦娥一号卫星的设计轨道依次是地球停泊轨道、地月转移轨道和绕月轨道。从地月转移轨道进入绕月轨道的过程中,近月点附近的减速制动是关键环节。对近月点附近飞行器轨道的快速与准确识别,是测轨系统的主要任务,可用于判断飞行器是否成功入轨,以及用于轨道进一步调控的基础依据。在嫦娥一号卫星近月点附近的减速制动阶段,卫星的位置并不发生突变,但速度却会因制动而在瞬间发生突变。通过解算位置矢量时间序列可以用于轨道识别,但需一定时段的数据积累。若能同时获得一定精度下的ISV,将之转换为轨道根数,则可用于轨道的快速识别。

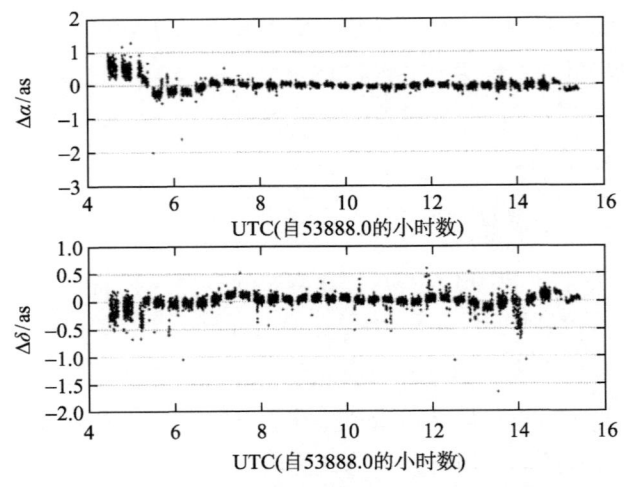

图 2 Smart-1 卫星实时 ISV 归算与 ESA 重建轨道的比较

2007 年 11 月 5 日,嫦娥一号卫星在近月点附近实施了减速制动,我们利用 VLBI 跟踪时延、时延率和 USB 的测距测速,实时归算得到了嫦娥一号卫星的 ISV 时间序列,并转换为月心天球参考系中的瞬时轨道根数,如表 1 所示。从中可见,捕获段的轨道机动前后历时约 30 min,轨道偏心率的演变表明,卫星轨迹从双曲线依次演变为抛物线到椭圆,表明卫星已经被月球成功捕获。任务执行期间,我们以轨道偏心率的演变情况快速做出了正确的判断,滞后时间仅约 5 min。

表 1 2007 年 11 月 5 日的嫦娥一号卫星月球捕获段轨迹监测

时刻(UTC)	半长径/km	偏心率
双曲		
03:13:01	6618.708	1.29441
03:13:31	6641.511	1.29340
03:14:01	6664.319	1.29238
近抛物线		
03:17:07	3406.991	1.08194
成功捕获		
03:20:53	2990.028	0.84911
03:20:58	3016.474	0.84918
03:42:01	6205.055	0.68567
03:42:30	6154.900	0.68305
03:43:00	6155.691	0.68311

2.3 嫦娥一号卫星环月段轨道机动监测

嫦娥一号卫星被月球成功捕获之后,又连续进行了多次轨道机动以进入工程设计的绕月轨道,我们利用 ISV 归算方法成功进行了轨迹演变监测。表 2 为 2007 年 11 月 6 日嫦娥一号卫星轨道机动的一次监测举例。从中可见,虽然轨道机动过程仅历时约 15min,但轨道半长径、偏心率的演变非常显著,分别从约 6000 km 至 2700 km,从 0.68 至 0.28。轨道倾角和升交点角距分别保持在约 87°和 265°,说明此次轨道机动主要是轨道形状的变化,由扁到近圆。

表 2 2007 年 11 月 6 日嫦娥一号卫星轨道机动监测

时刻(UTC)	半长径/km	偏心率	倾角/(°)	升交点角距/(°)
03:21:14	6015.141	0.67682	87.524	265.446

续表

时刻(UTC)	半长径/km	偏心率	倾角/(°)	升交点角距/(°)
03:21:39	5543.681	0.64883	87.522	265.431
03:22:08	4785.971	0.59599	87.496	265.293
03:22:23	4618.108	0.58076	87.500	265.293
03:22:43	4418.871	0.56143	87.503	265.290
03:26:29	3360.283	0.42134	87.555	265.223
03:26:49	3237.410	0.39931	87.559	265.221
03:27:33	3214.633	0.39420	87.561	265.277
03:28:37	3071.672	0.36547	87.573	265.279
03:28:46	3056.361	0.36189	87.581	265.279
03:29:11	2993.481	0.34900	87.578	265.278
03:29:26	2971.901	0.34367	87.583	265.281
03:31:43	2600.635	0.24986	87.607	265.299
03:31:58	2597.407	0.24883	87.605	255.310
03:32:18	2592.536	0.24780	87.603	255.326
03:32:33	2594.346	0.24800	87.602	255.339
03:32:57	2591.983	0.24783	87.594	255.357
03:33:02	2593.461	0.24816	87.593	255.361
03:33:17	2595.940	0.24898	87.589	255.373
03:33:46	2601.796	0.25105	87.578	265.396
03:35:34	2681.237	0.27258	87.535	265.481
03:36:43	2704.682	0.27867	87.531	265.484

2.4 嫦娥一号卫星受控撞月段轨迹监测

嫦娥一号卫星经过连续 494 d 的飞行,于 2009 年 3 月 1 日成功实现了"受控撞月",为我国后续月球探测计划积累了技术与工程经验。图 3 是利用 USB,VLBI 跟踪测量,经 ISV 归算得到的嫦娥一号卫星落月段轨迹测定,分别为月球主轴系中的月面经度、纬度和高度(月球参考半径 1738 km)。在地心天球坐标系至月球主轴系的转换中采用了 DE405 历表及其天平动参数。从图 3 可见,在 UTC07h45m 前后嫦娥一号卫星轨迹的月面经纬度表现出明显的趋势性改变,与轨道机动时段(UTC7h36m 至 7h56m)很好对应。由于各轨迹点均为独立归算得出,彼此的连续性和整体的平滑度表明,对轨迹的跟踪观测与 ISV 归算是相当成功的。这对于实现探月二期着陆器落月轨迹测定,从方法、软件到信心等各方面都是有力的支持。

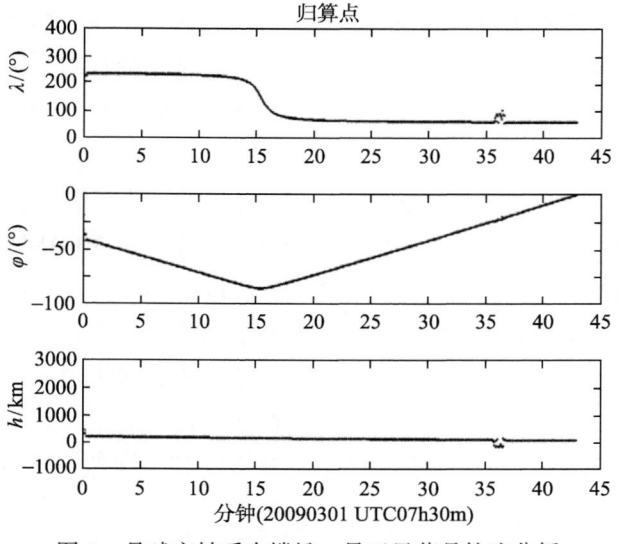

图 3　月球主轴系中嫦娥一号卫星落月轨迹分析

进一步分析得到了嫦娥一号卫星落月点坐标如表3所示,其中解算值是由跟踪资料经ISV分析所得最后信号发出时刻的嫦娥一号卫星月面坐标,其误差为一倍形式误差。平滑值是对点位坐标序列经平滑拟合后的结果,其误差为点位坐标序列的标准偏差。

表3 嫦娥一号卫星落月点坐标分析

	解算值	平滑值
时刻	20090301	UTC8h13m06.514s
东经/(°)	52.2760 ± 0.0018	52.2732 ± 0.0040
南纬/(°)	1.6407 ± 0.0031	1.6440 ± 0.0091
月面高/km	−3.30 ± 0.06	−3.56 ± 0.18

关于撞月时刻,VLBI跟踪资料互相关处理过程中显示,测站跟踪信号消失的时刻约为UTC8h13m7.8s,此时对应的地月距离约为378336 km,信号传输时间约为1.26 s,由此推断的嫦娥一号卫星撞月时刻应在UTC8h13m6.5s附近,这与表3中通过ISV分析给出的时刻是相符的。另外,表3中平滑值的误差所对应的月面切向误差为0.30 km,三维定位误差为0.35 km。考虑到此处定位分析的步长设置为0.01 s,若节点时刻的误差为3倍步长,即0.03s,由点位坐标平滑序列所得落月点坐标在月面经、纬和高度方向上的误差约为(±0.0001)°,(±0.0017)°和(±0.00) km。再有,ISV分析中所用VLBI数据的互相关积分时间为1 s,各基线最后数据点的时标可能存在约(±0.25) s的偏差,由点位坐标平滑序列估算得到的3个方向上的对应变化分别为(±0.0006)°,(±0.014)°和(±0.01) km。综合考虑点位坐标序列的波动、ISV分析中的步长设置和互相关积分时间等因素,落月点坐标的误差在月面经、纬和高度方向分别为(±0.0040)°,(±0.0168)°,(±0.18) km。对应的月面切向误差为0.52 km,三维定位误差为0.55 km。若VLBI互相关时延的时标误差为±0.5 s,对应的月面切向、三维定位误差将达0.90和0.92 km。可见,对于快速运动的目标,缩短VLBI互相关积分时间有利于提高定位精度。综合以上分析可以推断,若探月二期中对着陆器的跟踪方式仍然采用射电测距测速和VLBI技术,即使维持目前的技术指标,仍有保障获得着陆器月面1 km的定位精度。

3 讨论

ISV归算对卫星跟踪弧段的长度无苛刻要求,主要从几何角度实现对跟踪数据质量的监测和实现快速轨迹测定,较为适用于轨道机动段的跟踪测量,这已经被我国探月工程一期的相关实验观测所证实。规划中的我国探月工程二期将实现月面软着陆、释放月球车,并对选定的月面区域进行巡视勘察。对着陆器落月段轨迹的监测、月面着陆点坐标的确定以及对巡视器月面勘察轨迹的测定等,ISV归算方法有望继续发挥作用。

需要明确,ISV归算只能在同一信号波前具有足够独立观测量的基础上进行。嫦娥一号卫星采用了USB与VLBI联合跟踪的测量模式,这是实现ISV归算的根本保障。VLBI技术对于嫦娥一号卫星,尤其是在短弧情况下的精密定轨,其作用也是明显的。2007年11月8日至30日的嫦娥一号卫星在轨测试期间,基本维持了每天一次的动量轮卸载。以一天的跟踪数据进行轨道分析,预测12 h,并与第二天的定轨结果进行比对。作为示例,表4列出了5段重叠轨道起始点(零时)坐标差在RTN(轨道面径向、沿迹和法向)方向上的投影。从中可见,运用USB与VLBI跟踪数据的联合定轨结果,在3个方向上坐标差的均方差,要明显小于仅用USB跟踪数据的定轨情况。VLBI技术在我国探月一期项目中的成功应用,为后续探月工程及其他深空探测奠定了良好的基础。

表 4 不同资料解算所得部分重复弧段的差值情况统计 (单位/m)

日期	USB+VLBI			USB		
	径向	沿迹	法向	径向	沿迹	法向
24/25	12.09	36.70	14.38	14.76	149.16	233.49
25/26	3.45	−43.10	−116.50	50.04	−43.90	−63.48
26/27	1.39	−123.35	−19.20	3.9	−196.44	−121.72
27/28	10.83	−10.35	4.02	10.07	250.37	42.97
28/29	1.64	14.87	−0.87	1.28	−446.26	449.87
均值	5.88	−25.05	−23.63	16.01	−57.41	106.23
均方差	5.17	62.47	53.32	16.74	277.54	232.70

致谢 感谢嫦娥一号卫星 VLBI 测轨分系统 VLBI 中心提供本文分析数据。

参 考 文 献

[1] 欧阳自远. 我国月球探测的总体科学目标与发展战略. 地球科学进展, 2004, 19(3): 355—357
[2] 欧阳自远. 月球探测进展与我国的探月行动. 自然杂志, 2005, 27(4): 187—188
[3] 王家松, 陈建荣, 马鹏斌, 等. USB 与 VLBI 联合确定"探测一号"卫星轨道. 飞行器测控学报, 2006, 25(1): 31—33
[4] Sovers O J, Fanselow J L. Astrometry and geodesy with radio interferometry: Experiments, models, results. Rev Mod Phys, 1998, 70(4): 1393—1454
[5] 黄勇. "嫦娥一号"探月飞行器轨道计算研究. 博士学位论文. 上海: 中国科学院上海天文台, 2006
[6] 郭丽. 基于 VLBI 跟踪观测的空间飞行器瞬时状态矢量归算. 博士学位论文. 上海: 中国科学院上海天文台, 2007
[7] 乔书波, 李金岭, 孙付平. VLBI 在探月卫星定位中的应用分析. 测绘学报, 2007, 36(3): 262—268
[8] Li J L, Guo L, Zhang B. The Chinese VLBI network and its astrometric role. In: Jin W J, Platais I, Perryman M A C, eds. Proceedings IAU Symposium No. 248, 2007 Oct 15-19, Shanghai. Cambridge: Cambridge University Press, 2008. 182—185
[9] Guo L, Li J L, Qiao S B, et al. Monitoring the lunar capture of Chang E-1 satellite by real-time reduction of the instantaneous state vectors. In: Jin W J, Platais I, Perryman M A C, eds. Proceedings IAU Symposium No. 248, 2007 Oct 15-19, Shanghai. Cambridge: Cambridge University Press, 2008. 194—195
[10] Li J L. The contribution of CVN to the CE-1 mission. In: Finkelstein A, Behrend D, eds. Measuring the Future. Saint Petersburg: Nauka, 2008. 193—198
[11] 郭丽, 李金岭. s7123a 观测实验测角归算结果讨论. 飞行器测控学报, 2007, 26(6): 30—35
[12] Press W H, Teukolsky S A, Vetterling W T, et al. Numerical Recipes in Fortran 77: The Art of Scientific Computing. London: Cambridge University Press, 1992
[13] 徐士良. FORTRAN 常用算法程序集. 北京: 清华大学社出版社, 1995
[14] 李济生. 人造卫星精密轨道确定. 北京: 解放军出版社, 1995
[15] 刘林. 人造地球卫星轨道力学. 北京: 高等教育出版社, 1999

环月卫星可见时段的计算和分析

马茂莉 郑为民 李金岭 王广利

(中国科学院上海天文台,上海 200030)

摘 要 针对CE-1卫星精确的撞月时刻与撞月点坐标,首先通过探测器载波信号的本地相关处理技术,精确分析了载波信号在VLBI各测站的消失时刻,进而推算了卫星的撞月时刻;通过实时单向多普勒频移测量的事后分析,核实了卫星撞月过程中的飞行姿态演化;最后结合VLBI互相关时延与测距资料,经定位归算确定撞月点坐标。分析表明,CE-1 卫星撞月时刻的误差为±5μs,撞月点坐标月面切向和三维定位误差分别约为 0.274km 和 0.319km(1σ)。

关键词 嫦娥一号卫星;撞月时刻与坐标;本地相关测量;VLBI;多普勒频移测量;月球探测

0 引言

我国月球探测计划嫦娥工程分为"绕、落、回"3 个发展阶段[1],首颗月球探测卫星"嫦娥一号"(CE-1)于 2007 年 10 月 24 日在西昌卫星发射中心发射,累计完成 494 天的飞行后,于 2009 年 3 月 1 日在地面测控网控制下以月面撞击方式完成硬着陆。我国统一 S 波段测控系统(USB)和甚长基线干涉测量(VLBI)测轨分系统对撞月过程进行了全程跟踪。根据中国探月网资料[2],CE-1 卫星撞月时刻为北京时间 2009 年 3 月 1 日 16:13:10,世界协调时 08:13:10(UTC08:13:10),撞月点坐标为月面东经 52.36°,南纬 1.50°。VLBI 分系统在上海佘山站配置的实时高精度多普勒测量系统现场获得了 CE-1 载波信号消失时刻为 UTC08:13:08;VLBI 相关处理中心对 4 个 VLBI 测站跟踪资料的事后互相关处理表明,测站跟踪信号的消失时刻约为 UTC08:13:07.08[3]。文献[3]综合 VLBI 及 USB 跟踪数据,分析确定的卫星撞月时刻为 UTC08:13:06.514,撞月点坐标为东经 52.2732°,南纬 1.6440°,月面高−3.65km(月球参考半径 1738km),误差在月面经、纬、高度方向分别为 ± 0.0040°, ± 0.0091°, ± 0.18km,对应的月面切向、三维定位误差为 0.52km 和 0.55km。文献[2]与文献[3]给出的撞月时刻相差 3.486s,撞月点坐标东经与南纬分别相差 2.6km 和 4.4km。文献[3]分析认为,撞月点坐标之差主要源于撞月时刻的估算误差。由此可见,精确分析确定卫星的月面撞击时刻,对于确定撞月点坐标、检验轨控精度等有重要作用。

在落月过程中,CE-1 卫星发射 F_1、F_2 测控载波信号以及数传信号。在此后的一年时间内,我们分别用多普勒测量方法、VLBI 相关处理[4]和载波本地相关测量方法对测控信号或载波信号进行了分析。由于卫星硬着陆时,星载信标机将停止发射信号。本文以此为前提,根据载波信号消失时刻确定卫星撞月瞬时时刻。2009 年 3 月 1 日,设置于佘山站的实时高精度多普勒测量系统监测落月过程中 F_1 信标频率变化的全过程。VLBI 中心依据该系统实时提供的卫星多普勒频率消失时刻,现场获得了信号消失的时刻(精度为 0.5s)。撞月过程结束后,我们以 VLBI 事后处理方式获得了 F_1 及 F_2 载波的时延、时延率,并根据有效数据点消失时刻,进一步获得了精度更高的信号消失时刻。由于多普勒测量和 VLBI 互相关处理中的数据点步长为 1s,因而所估算的信号消失时刻可能存在 0.5s 的误差。

本文原载于《宇航学报》,2011,Vol.32,No.3,477~481。

VLBI 定位分析时利用 VLBI 时延及 USB 测距数据，归算出卫星在撞月过程中的轨迹，并将最后轨迹点的时刻作为卫星撞月时刻。VLBI 互相关处理时设定的 1s 步长导致时延的时标存在 0.5s 的误差。定位归算时以离散方式处理，也存在 0.01s 的步长，故所确定的撞月时刻存在 5ms 的误差。另外，定位归算要求同一波前至少存在 3 个时延测量值，即信号的波前至少被 3 个测站成功接收，否则将导致法方程系数矩阵秩亏，无法获得参数解。在不考虑步长设置的前提下，定位归算的最后时刻点有可能超前于撞月时刻。综上所述，文献[3]通过定位归算导出的撞月时刻存在至少 0.1s 的不确定性，更精确的撞月点坐标因而也有待于重新计算[3]。

　　本文通过载波本地相关测量技术，将测站接收数据与依据理论构建的"参考信号"进行互相关处理，根据信号消失前后幅度的变化以估算测站卫星信号消失时刻。本地构建的参考信号时刻与氢原子钟同步，具有极高精度。将各个测站的信号消失时刻扣除钟差、电离层和大气时延改正及光行时后，得到精确的卫星撞月时刻，并结合卫星多普勒速度变化，分析了卫星撞月过程中的飞行姿态演化。最后综合 VLBI 时延及 USB 测距数据，经定位归算确定撞月点坐标。

1 采用载波本地相关测量技术估算撞月时刻

　　CE-1 卫星受控撞月过程中发射 F_1、F_2 测控信号和数传信号，VLBI 测轨分系统佘山、密云、昆明、南山站(分别简写为 SH、BJ、KM、UR)全程跟踪了撞月过程，并记录了 F_1、F_2 测控信号(表 1)。撞月当天 UTC08:13:06 的 F_1 及 F_2 的频谱见图 1，频率分辨率为 15Hz，峰值最强的信号分别为对应于 F_1、F_2 载波信号。

表 1　F_1、F_2 频率及 VLBI 台站可视时间
(UTC2009 年 3 月 1 日)

信号	中心频率/MHz	SH、BJ、KM、UR	
F_1	2210.XX	06:21-06:56，07:29-08:13	共 81min
F_2	2234.XX	06:32-06:50，07:29-08:13	共 74min

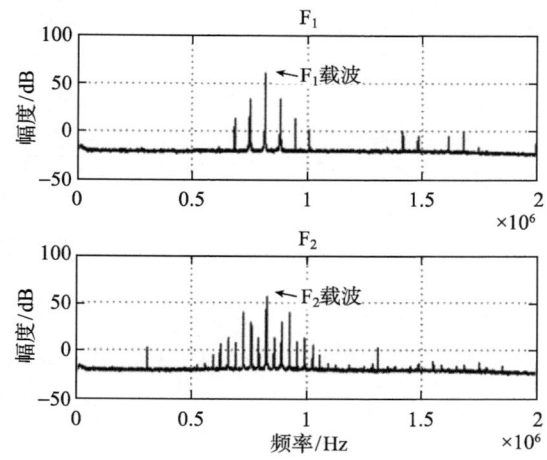

图 1　UTC08:13:06 时的 F_1 及 F_2 测控载波信号频谱图

　　设测站接收的载波信号频率为 ω，采样时间间隔为 T_s，测站终端采集的信号幅度为 $A(nT_s)$，附加相位延迟为 $\phi(nT_s)$，噪声为 $N(nT_s)$，满足均值为 0 的高斯分布，则载波信号可表示为

$$C_{PR} = A(nT_s)\cos(\omega nT_s + \phi(nT_s)) + N(nT_s) \tag{1}$$

虚构参考信号，与载波信号相关[5-6]。设 C_{PR} 为两者的互相关函数，则

$$C_{\mathrm{PR}} = \sum_{n=0}^{N}[A(nT_s)\cos(\omega nT_s + \phi(nT_s)) + N(nT_s)]\exp(\mathrm{j}\omega_0 nT_s) \qquad (2)$$

简写为

$$C_{\mathrm{PR}} = A\exp(-\mathrm{j}\phi)/2 + N \qquad (3)$$

其中 $A=2|C_{\mathrm{PR}}|$，$\phi=-\mathrm{phase}(C_{\mathrm{PR}})$。载波本地相关测量技术原理见图 2。

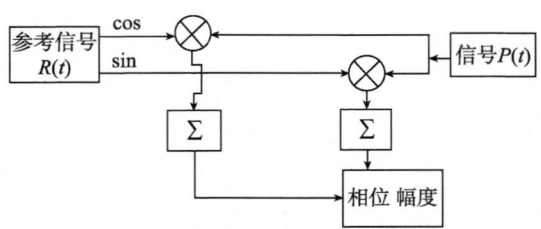

图 2 载波本地相关测量技术原理图

当载波信号消失时，相关结果的幅度和相位均为噪声。在实时单向多普勒测量及文献[3]估算的撞月时刻基础之上，本文用滑动积分的处理方法，重点分析 UTC08:13:07s 至 8s 的数据，得到 F_1、F_2 信号相关幅度见图 3。取积分时间为 50μs，滑动步长为 10μs。

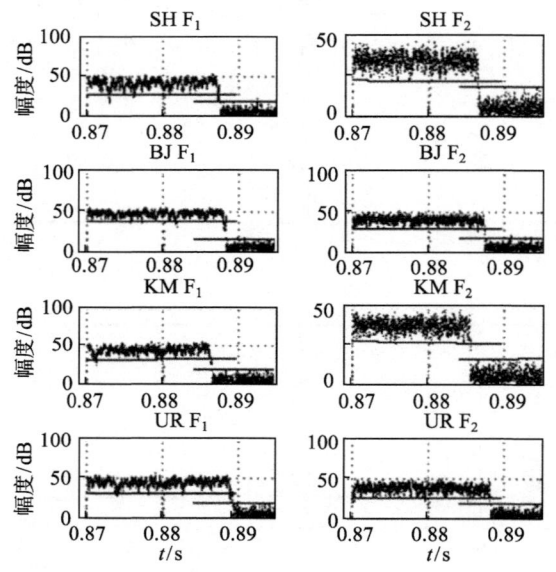

图 3 撞月过程中 VLBI 四台站 F_1、F_2 信号幅度变化

图 3 中，横坐标为时间，单位秒，数据点间隔为 10μs。可见，在载波信号消失过程中，本地相关幅度下降明显，达 10～20dB。分别对 F_1、F_2 信号本地相关幅度和噪音的幅度进行直线拟合，并标出其点位波动 3σ 分界线(图 3 横线)。两条分界线之间的数据对应于信号在各测站消失的过程。对于 F_1 及 F_2 信号，佘山、密云、昆明、南山站两条 3σ 分界线内分别有 14、27、16、29 及 1、5、2、1 个数据点。将这些数据点对应的时刻分别作平均，即对应信号消失时刻。获得 F_1、F_2 信号测站消失时刻为 UTC08:13，秒及秒以下时刻见表 2。

表 2 VLBI 测站 F_1、F_2 信号消失时刻   (单位:秒)

波段	SH	KM	UR	BJ
F_1	7.887350	7.886330	7.889065	7.888165
F_2	7.886480	7.885450	7.888200	7.887300

测站钟差[7]、电离层[8]和大气时延改正[8]及光行时[9]数据见表 3。将各个测站的信号消失时刻扣除钟差、电离层和大气时延改正及光行时后，推算得到对应的卫星撞月时刻如表 4 所示。钟差、电离层与大气时延改正均为实测值，其误差在纳秒量级。光行时为依据历表、地球定向、月球天平动、测站坐标和撞月点坐标等的理论估算值，误差不超过纳秒量级，因此在微秒量级上由各站本地相关估算的撞月时刻，其误差主要由相关测量的误差引起。

表 3 测站钟差、电离层和大气时延改正及光行时

	SH	KM	UR	BJ
钟差/μs	27.39	46.5	2.507	71.36
电离层/ns	5.6284	9.2259	4.4941	4.3579
大气时延/ns	8.6970	6.3358	7.1337	8.6764
光行时/s	1.24068788	1.23964132	1.24243010	1.24148650

表 4 由测站信号消失时刻推算的卫星撞月时刻　　　　　　　　　　（单位:秒）

波段	SH	KM	UR	BJ
F_1	6.6466347	6.6466422	6.6466324	6.6466071
F_2	6.6457647	6.6457622	6.6457674	6.6457421

由表 4 可知，

(1) 由佘山、昆明、南山站本地相关估算的撞月时刻吻合较好，差别在微秒量级，误差 ±5μs；

(2) 密云站撞月时刻与其他三站差别较大，F_1、F_2 信标消失估算的差别分别在 30μs、23μs 左右，测量误差约 ±20μs。故本文取佘山、昆明、南山站由 F_1、F_2 信号消失估算的撞月时刻均值为卫星撞月时刻，分别为 6.646636s 及 6.645765s，误差 ±5μs；

(3) F_2 信标机撞月时刻比 F_1 信标机超前 0.871ms，在百微秒水平上差异显著；

(4) 载波本地相关测量技术估算的卫星撞月时刻相比于定位归算[3]滞后约 0.13s。

2　卫星撞月姿态分析

佘山站利用实时高精度多普勒测量系统现场测量了 F_1 信标的多普勒频率。经事后分析，将 F_1 信标多普勒频率转化为多普勒速度，即测站视线方向上卫星的视向退行速度，见图 4。

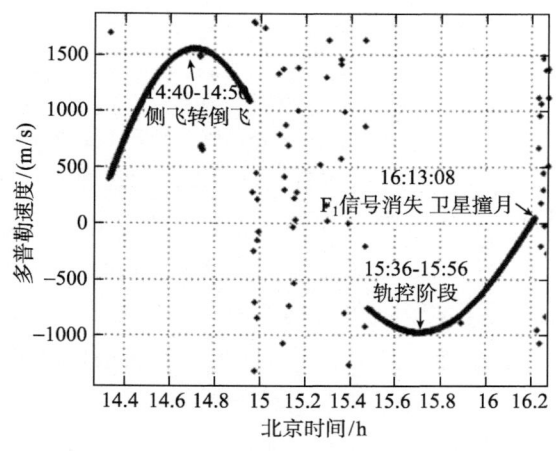

图 4 CE-1 卫星落月过程中视向速度变化

图 4 中数据点步长为 1s。设定卫星远离测站的速度为正，由图 4 可见，北京时间 14:40 至 14:50 卫星

视向速度最大,开始由侧飞状态逐步转变为倒飞,15:00 至 15:20 期间 F_1 信标机对月,测站信号消失(表现为噪声点)。此后,卫星再次逐步转变为侧飞阶段,至 15:40 左右视向速度达最小。15:35 至 15:55 为轨控减速段,F_1 信号于北京时间 16:13:08 消失,表明卫星撞月。自 15:40 左右视向速度最小至 16:13 左右的撞月阶段中,F_1 信标机逐步转为对地球方向,同时,位于卫星另一端的 F_2 逐步对月球方向。因此由多普勒频移测量结果可判断,F_2 信标机先撞月,F_1 信标机后撞月。这与上文本地相关分析结论一致。

3 撞月点坐标分析

综合 VLBI 时延及 USB 测距数据,对 CE-1 卫星落月段 UTC08:07:00 之后的跟踪数据,以 0.01s 的步长进行定位归算,以确定卫星的落月点坐标。定位归算轨迹点拟合的数学关系在月经、纬、高三个方向分别为:

$$\lambda = 0.001483t^2 - 0.161317t + 53.345015(°)$$
$$\varphi = 0.006023t^2 + 3.276960t - 25.242792(°) \quad (4)$$
$$h = 0.0437t^2 - 3.4002t + 18.4074(km)$$

其中 t 为自 UTC08:06:00 起算的分钟数。三个方向上点位坐标序列的标准偏差分别为 0.003546°、0.008317°、0.163535km,对应的月面切向定位误差 0.274km,三维定位误差 0.319km。

定位归算的最后资料点时刻为 UTC08:13:6.514[3],对应于(4)式的 t 为,

$$t = UTC08:13:06.514 - UTC08:06:00$$
$$= 7.10856(min) \quad (5)$$

由表 4,取 F_1、F_2 的撞月时刻分别为 6.646636s、6.645765s,对应的 t 为 7.110777min、7.110763min。由(4)式估算的撞月点坐标见表 5。可见,F_1、F_2 撞月时刻对应的撞月点坐标之差在米级水平;本地相关所得结果相比于定位归算滞后约 0.13s,对应的撞月点坐标在月面切向和三维方向之差分别约为 0.15km、0.224km。

表 5 撞月点坐标估算结果

撞月时刻(UTC08:13)	月经/(°)	月纬/(°)	月面高度/km
定位分析 6.514s	52.273222	−1.643973	−3.554894
本地相关 F_1 6.646636s	52.272911	−1.636518	−3.561054
本地相关 F_2 6.645765s	52.272913	−1.636565	−3.561015

4 结论

载波信号的本地相关分析表明,F_2 信标比 F_1 信标提前约 0.871ms 消失,即 F_2 先于 F_1 撞月。在多普勒频移测量的卫星姿态分析中也得到相同的结论。取 F_2 信标消失时刻为卫星撞月时刻,即北京时间 2009 年 3 月 1 日 16:13:06.645765 (UTC08:13:06.645765),精度 ±5μs;撞月点坐标为东经 52.2729°,南纬 1.6366°,月面高−3.56km,对应的月面切向、三维定位误差分别为 0.274km 和 0.319km。

本地相关测量能够精确地估算卫星撞月时刻,但不能独立提供撞月点坐标。利用 VLBI 互相关时延和 USB 测距资料的定位分析可以较准确地监测卫星撞月轨迹,但将最后轨迹点对应的时刻作为卫星撞月时刻,超前于实际撞月时刻。本文综合了本地相关与定位分析的特点,由前者精确估算撞月时刻,由后者获得撞月轨迹的经验拟合,进而组合导出撞月点坐标。撞月点坐标百米级的误差主要源于定位分析中数据的点位波动。估算表明,CE-1 卫星撞击月球时在月固系中的速度为 1697.2m/s,投影到月纬、月经和高程方向的速度依次为 1694.1m/s、−53.5m/s 及−87.4m/s。当撞月时刻准确到 10μs 时,理论上由撞月时刻导致的

定位误差在厘米量级。

参 考 文 献

[1] 欧阳自远. 我国月球探测的总体科学目标与发展战略［J］. 地球科学进展, 2004, 19(3): 355—358. [Ou Yang Zi-yuan. Scientific objectives of Chinese lunar exploration project and development strategy[J]. Advances in Earth Science, 2004, 19(3) : 355—358]

[2] 嫦娥一号卫星受控撞月成功[OL]. 中国探月: 嫦娥一号撞月专题, 2009[2010-7-12]. http://www.clep.org.cn/index.asp?modelname=2009/zt_luoyue/zt_cely_ttxw_content&recno=1

[3] 李金岭, 郭丽, 钱志翰, 等. 瞬时状态归算用于嫦娥一号卫星关键轨道段监测[J]. 中国科学, 2009, 39(10): 1393—1399. [Li Jin-ling, Guo Li, Qian Zhi-han, et al. The application of the instantaneous states reduction to the orbital monitoring of pivotal arcs of the Chang'e-1 satellite [J]. Science China Ser G, 2009, 39(10) : 1393—1399]

[4] 郑为民, 舒逢春, 张冬. 应用于深空跟踪测量的 VLBI 软件相关处理技术[J]. 宇航学报, 2008, 29(1), 18—23. [Zheng Wei-min, Shu Feng-chun, Zhang Dong. Application of software correlator to deep sapce VLBI tracking [J]. Jounal of Astronautics, 2008, 29(1) : 18—23]

[5] 杨艳, 郑为民. VLBI 相位校正信号提取的软件实现方法[J]. 中国科学院上海天文台年刊, 2006, 27: 107—117. [Yang Yan, Zheng Wei-min. Software realization method of extraction VLBI phase calibration signal [J]. Annals of Shanghai Astronomical Observatory Chinese Academy of Sciences, 2006, 27: 107—117]

[6] 郑君里, 应启珩, 杨为理. 信号与系统[M]. 北京: 高等教育出版社, 1999: 341—362

[7] 朱陵凤, 李超, 刘利, 等. 基于国产氢原子钟的钟差预报方法研究[J]. 大地测量与地球动力学, 2009, 29: 148—151. [Zhu Ling-feng, Li Chao, Liu Li, et al. Research on methods for predicting clock error based on domestic hydrogen atomic clock [J]. Journal of Geodesy and Geodynamics, 2009, 29: 148—151]

[8] 曲伟菁, 朱文耀, 宋淑丽, 等. 三种对流层延迟改正模型精度评估 [J]. 天文学报, 2008, 49: 113—122. [Qu Wei-jing, Zhu Wen- yao, Song Shu-li, et al. The evaluation of precision about Hopfield, Saastamonien and EGNOS tropospheric delay correction model [J]. Acta Astronomica Sinica, 2008, 49: 113—122]

[9] 赵铭. 天体测量学导论[M]. 北京: 中国科学技术出版社, 2006: 313—330

"嫦娥"卫星绕月飞行的星载激光定轨法

刘基余

(武汉大学测绘学院,武汉 430079)

摘 要 依据地月激光测距的成功实践和我们对卫星激光定轨的基础研究,笔者提出用地面对"嫦娥"卫星作激光测距的方法,高精度地测定"嫦娥"卫星绕月飞行时的实时轨道参数。为此,需要:a. 给"嫦娥"卫星装备无电功耗需求的激光后向反射镜阵列,以便对它进行星载激光定轨测量;b. 给"嫦娥"卫星装备 GPS 信号接收机,实现离地环行轨道米级精度的自主定轨,确保"嫦娥"卫星准确进入地月转移轨道,并为星载激光定轨提供初始值。

关键词 "嫦娥"卫星 星载 GPS 测量 自主定轨

0 引言

欧阳自远院士在《我国月球探测一期工程的科学目标》一文中指出:"我国月球探测活动的第一步将首先从环月卫星的遥感探测开始,通过利用光学、微波和能谱等探测器对月球表面三维影像、有用元素、月壤厚度和地月空间环境进行全球性、综合性和整体性探测。"据悉,"嫦娥"卫星将携带立体相机、成像光谱仪、激光高度计、微波辐射计、太阳宇宙射线检测器和低能离子探测器等多种科学仪器绕月飞行探测。前三种仪器主要用于为月球"画像"。依笔者之见,即使三种仪器能够采集到精细的画像数据,如果没有精确的卫星绕月飞行实时在轨点位作数据处理基准,就很难精确画出月球的像;因此,"嫦娥"卫星绕月飞行探测定轨将成为能否为月球精确画像和确定有用元素分布位置的技术关键。

据笔者所知,许多国家都非常重视获取卫星遥感图片的位置基准数据。例如,20 世纪 80 年代中期,法国为了给 SPOT 遥感卫星定轨,建立了一个专用的"星载多普勒无线电定轨定位系统"(DORIS 系统)。该系统是一个与 GPS 相反的"信标上行"系统,它不像 GPS 系统那样,由 GPS 卫星发送导航定位信号,而是由地面播发站向卫星播发无线电信标,星载 DORIS 接收机接收该无线电信标,进而测得 2036.25MHz 和 401.25MHz 的双频多普勒频移,依此而解算出该颗卫星的在轨实时位置。又如,欧洲空间局(ESA)为了测定 ERS 欧洲遥感卫星轨道而建立了一个"精确距离及其变率测量系统"(PRARE 系统);它包括星载微波收发机、地面微波转发站、地面主控站和地面标校站。20 世纪 90 年代以来,许多国家的对地观测卫星,多采用星载 GPS 测量定轨,其定轨精度已达到了 cm 级。对于"嫦娥"卫星而言,其轨道测定就较地球卫星复杂一些。我们知道,月球是在一个椭圆轨道上环绕地球飞行的自然天体,其轨道倾角为 5.145°,偏心率是 0.0549,近地点为 356400km,远地点为 406700km。由此可见,当"嫦娥"卫星在距离月球 200km 的极月圆轨道上飞行时,采用基于地面设备一般的卫星测轨方法,是难以精确测得距离地球 38 万 km 之遥的卫星轨道参数,而很难达到理想的预期目标。为此,本文对如何实现"嫦娥"卫星的高精度轨道测定,提出下述建议,供决策参考。

本文原载于《航天器工程》,2006,Vol. 15,No. 1,9~13。

1 激光测月的成功实践为"嫦娥"卫星激光定轨奠定了基础

1969年7月21日,美国Apollo(阿波罗)11宇航员Neil Armstrong和Buzz Aldrin实现了人类有史以来的第一次成功登月,并在月球表面上安置了一个用于测量距离的激光后向反射镜阵列。同年8月1日,美国加州大学的Lick天文台利用美国第二架口径最大的120inch望远镜和10mμs、1J激光脉冲,成功地测量了地球和月亮之间的距离为 383911.218km ± 0.045km。其后,美国和原苏联分别在月球表面上还布设了Apollo-14,Apollo-15和Luna-17,Luna-21等四个激光后向反射镜阵列。且知,Apollo-11至Apollo-14的距离为1250km,Apollo-11至Apollo-15的距离为970km,Apollo-14至Apollo-15的距离为1100km,Apollo-11至Luna-21的距离为760km,Apollo-15至Luna-17的距离为820km。上述月球激光后向反射镜阵列的主要参数如表1所示。目前,国际上有下列台站进行着经常性的地月激光测距(统称为激光测月):美国Texas州的McDonald天文台,美国Hawaii州的Haleakala天文台,法国的Grasse观测站;澳大利亚的Orrorral观测站和德国的Wettzell观测站。

表1 月球激光后向反射镜阵列的主要技术参数

名称	Apollo-11	Luna-17	Apollo-14	Apollo-15	Luna-21
设置者	美国	原苏联和法国	美国	美国	原苏联和法国
设置日期	1969.7.21	1970.11.01	1971.2.05	1971.7.31	1973.6.15
阵列尺寸(cm^2)	46×46	—	46×46	104×61	—
棱镜数量	100块	14块	100块	300块	14块

地月激光测距成果不仅用于探求天体物理学中的重大课题之一——引力常数是否随时间而变化,而且具有下述作用:

(1) 研究天体演化和地球动力学。不断提高激光测月精度的目的,不是为了测得某时刻地月之间的绝对距离,主要是为了能够精确地测定地月距离在几个月,甚至几年内的变化率。引起距离变化的主要原因是地球和月球自转的不均匀性,月球轨道参数的不精确性,大陆的漂移。从激光测月的长期资料中,便可探讨这些重大问题。

(2) 编算精密月亮历表。在地月激光测距问世以前,天文学家们一直利用方位测量来决定月亮的轨道参数。因其观测精度低,致使所求得的月亮轨道参数比较粗糙。如月球轨道平均半径(地月质量中心的平均距离)的误差高达±3~4km,偏心率误差为±10^{-7}。用这些不精确的常数计算出来的月球相对于地球的位置误差可达几km。如果用精确的地月激光测距值和理论值进行比较,其差值可以认为是一些常数不精确引起的,因此,可以用地月激光测距成果来改正这些常数,而算得精密的月亮历表。

(3) 研究月亮物理天平动。月球存在一种绕其重心的运动,它是一些小振幅的周期摆动,后者被称为月亮物理天平动。其最大摆动仅为±0.27″,它的周期是一年。周期为一个月的摆动,仅达±0.06″。如此小的摆动,就要求高精度的测量,地月激光测距则能满足这种高精度要求。通过月亮物理天平动的研究,可以探求月球动力学特性和月球内部质量分布,改进地月系统的某些参数,而算得精密的月亮历表。

(4) 研究地球自转和地极移动。地月激光测距所用的时刻,受到地球自转和地极移动的影响,而引起地月激光测距误差,因此,从地月激光测距成果中可以推求出地球自转和地极移动的改正量。

2 多台地面激光测距仪同时测量可为"嫦娥"卫星精确定轨

正如上述,地月激光测距,从20世纪60年代末期开始,一直延伸到今天。其测距精度已从刚开始的几十米提高到了厘米级,而能够精确测定月球远离地球的速率为3.8cm/yr。因此,笔者建议,采用多台地面激光测距仪同时测量至绕月飞行的"嫦娥"卫星距离,进而精确解算出"嫦娥"卫星在轨实时位置,为测绘月球表面起伏图的图像处理提供基准数据。

图 1 表示多站激光测距定轨的基本构成及其工作原理。在地面三个 SLR 测站(A, B, C)上，各设置一台测程可达 410000km 的卫星激光测距仪；它们的站坐标是精确已知的，且分别为 X_a, Y_a, Z_a, X_b, Y_b, Z_b 和 X_c, Y_c, Z_c，时元 t "嫦娥"卫星绕月飞行的在轨点位坐标记作 $X_{CH}(t)$, $Y_{CH}(t)$, $Z_{CH}(t)$。下述数学论证表明，只要各台激光测距仪同时测得至"嫦娥"卫星的距离，就能够精确解算出时元 t "嫦娥"卫星绕月飞行的在轨三维坐标 $X_{CH}(t)$, $Y_{CH}(t)$, $Z_{CH}(t)$；其测量精度不低于米级。"嫦娥"卫星绕月飞行的激光定轨的难点是，测程可达 410000km 的卫星激光测距仪。我国现有的武汉、上海、长春、北京和昆明等五个激光测卫站上的卫星激光测距仪，其最大测程仅达 20000km 左右，都不具备测量"嫦娥"卫星绕月飞行的能力，必须经过重大改造。对于绕月飞行轨道上的激光测距照准问题，是依靠卫星激光测距仪的"全程"（离地环行轨道→地月转移轨道→绕月飞行轨道）激光测距来实现的(详见下文所述)，不需要第三者为绕月飞行轨道上的激光测距提供轨道预报。

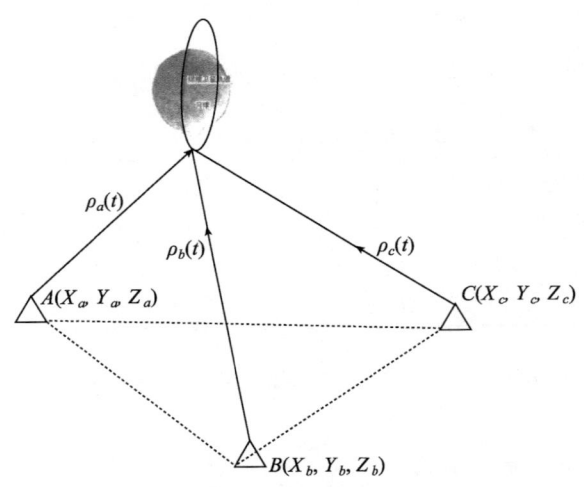

图 1 多站激光测距定轨原理图示

当地面三个 SLR 测站(A, B, C)上的卫星激光测距仪在时元 t 测得"嫦娥"卫星绕月飞行时的距离 $\rho_a(t)$, $\rho_b(t)$, $\rho_c(t)$，则有

$$\begin{aligned}\rho_a(t) &= \sqrt{[X_a - X_{CH}(t)]^2 + [Y_a - Y_{CH}(t)]^2 + [Z_a - Z_{CH}(t)]^2} \\ \rho_b(t) &= \sqrt{[X_b - X_{CH}(t)]^2 + [Y_b - Y_{CH}(t)]^2 + [Z_b - Z_{CH}(t)]^2} \\ \rho_c(t) &= \sqrt{[X_c - X_{CH}(t)]^2 + [Y_c - Y_{CH}(t)]^2 + [Z_c - Z_{CH}(t)]^2}\end{aligned} \tag{1}$$

上式经过线性化，而求得"嫦娥"卫星绕月飞行时的三维坐标修正矩阵：

$$X = A^{-1} B \tag{2}$$

式中，

$$X = [\Delta X_{CH}(t) \ \ \Delta Y_{CH}(t) \ \ \Delta Z_{CH}(t)]^T$$

$$B = \begin{bmatrix} \rho_{a0} - \rho_a \\ \rho_{b0} - \rho_b \\ \rho_{c0} - \rho_c \end{bmatrix}$$

$$A = \begin{bmatrix} \dfrac{X_a - X_{CH0}}{\rho_{a0}} & \dfrac{Y_a - Y_{CH0}}{\rho_{a0}} & \dfrac{Z_a - Z_{CH0}}{\rho_{a0}} \\ \dfrac{X_b - X_{CH0}}{\rho_{b0}} & \dfrac{Y_b - Y_{CH0}}{\rho_{b0}} & \dfrac{Z_b - Z_{CH0}}{\rho_{b0}} \\ \dfrac{X_c - X_{CH0}}{\rho_{c0}} & \dfrac{Y_c - Y_{CH0}}{\rho_{c0}} & \dfrac{Z_c - Z_{CH0}}{\rho_{c0}} \end{bmatrix}$$

$$\rho_{a0} = \sqrt{(X_a - X_{CH0})^2 + (Y_a - Y_{CH0})^2 + (Z_a - Z_{CH0})^2}$$
$$\rho_{b0} = \sqrt{(X_b - X_{CH0})^2 + (Y_b - Y_{CH0})^2 + (Z_b - Z_{CH0})^2}$$
$$\rho_{c0} = \sqrt{(X_c - X_{CH0})^2 + (Y_c - Y_{CH0})^2 + (Z_c - Z_{CH0})^2}$$

若给"嫦娥"卫星装备 GPS 信号接收机，不仅能够实现离地环行轨道米级精度的自主定轨，而且能够为激光定轨提供解算所需的初始值 X_{CH0}, Y_{CH0}, Z_{CH0}，进而求得。

$$\left.\begin{array}{l}X_{CH}(t) = X_{CH0} + \Delta X_{CH}(t) \\ Y_{CH}(t) = Y_{CH0} + \Delta Y_{CH}(t) \\ Z_{CH}(t) = Z_{CH0} + \Delta Z_{CH}(t)\end{array}\right\} \quad (3)$$

卫星激光定轨，已在高达20000km的导航卫星（GPS/GLONASS 卫星）和高达1000km左右的对地观测卫星上，成功地进行了工程实用，例如1992年8月10日，美国航空航天局（NASA）和法国国家空间研究中心（CNES）联合发射的 Topex/Poseidon 海洋测高卫星。它要求其径向误差在13cm 以内，而采用了星载 GPS 定轨和激光定轨并行的实施方案；其结果表明，径向误差为3~4cm，法向误差为5~10cm，切向误差为9~16cm。2001 年10 月7 日，美国航空航天局和法国国家空间研究中心又联合发射的 Topex/Poseidon 海洋测高卫星的后续卫星——JASON-1。该颗卫星的有效载荷与 Topex/Poseidon 卫星相似，仅将星载 Monarch GPS 信号接收机改换为性能更优的 BlackJack GPS 信号接收机。2002年3月17日，美国航空航天局和德国地学研究中心（GFZ）合作发射了两颗重力测量与气候科研设备卫星，分别命名为 GRACEA 卫星和 GRACEB 卫星，而构成 GRACE 卫星重力测量系统。GRACEA 卫星和 GRACEB 卫星相距220km，它们的轨道高度为500km，于近极共面轨道上飞行。GRACE 卫星和德国的 CHAMP 卫星一样，均装备了 BlackJack GPS 双频接收机和星载激光后向反射镜阵列，实施星载 GPS 测量和激光并行定轨。上述卫星激光定轨的工程实践表明，卫星在轨实时点位坐标的测量精度达到了 cm 级。因此，笔者提出的用多台地面卫星激光测距仪同时测量至绕月飞行的"嫦娥"卫星距离，进而精确解算出"嫦娥"卫星在轨实时位置，是能够达到 m 级甚至更高的测量精度。

3 星载 GPS 接收机用于离地环行轨道精确自主定轨可为星载激光定轨提供初始值

1978 年2 月22 日第一颗 GPS 试验卫星的入轨运行，开创了以导航卫星为动态已知点的无线电导航定位的新时代。GPS 卫星所发送的导航定位信号，是一种可供无数用户共享的空间信息资源。陆地、海洋和空间的广大用户，只要持有一种能够接收、跟踪、变换和测量 GPS 信号的接收机，就可以全天时、全天候和全球性地测量运动载体的七维状态参数和三维姿态参数，其用途之广，影响之大，是任何其他无线电接收设备望尘莫及的。21 世纪初叶，GPS 卫星全球定位系统的空间部分和地面监控系统，都将实施现代化，并计划于2005 年开始增设第三导航定位信号（L5），而形成用3 个 GPS 信号（L1，L2，L5）同时进行导航定位的新格局。GPS 现代化，不仅使全球广大用户能够用 GPS 动态载波相位测量获得厘米级精度的3 维实时点位坐标，而且能够用 C/A 码伪距测量解获得米级的单点定位精度。

在 DDKIN GPS 动态数据处理软件的基础上[4-6]，研究成功了 SARTOD 卫星自主定轨软件；并用 CHAMP 卫星 GPS 测量数据进行了试算。其结果表明，达到了米级的卫星自主定轨精度，亦即，算得 CHAMP 卫星实时自主定轨的三维位置误差分别为：RMSX = 2.86m，RMSY = 2.87m，RMSZ = 2.21m，如图2 所示。因此，笔者建议，给"嫦娥"卫星装备 GPS 信号接收机，当其离开地面时，便实施星载 GPS 测量自主定轨（称之为离地环行轨道定轨），以确保"嫦娥"卫星准确进入地月转移轨道，并为卫星激光测距仪从离地环行轨道开始的星载激光定轨提供初始值；此后，卫星激光测距仪一直保持在激光测距工作状态，直至绕月飞行轨道。这种"嫦娥"卫星的"全程"激光测距，既能够精确定"嫦娥"卫星的离地环行轨道→地月转移轨道→绕月飞行轨道的轨道参数，又不必担忧"嫦娥"卫星在绕月飞行轨道上的入射激光能否准确返回地面问题。

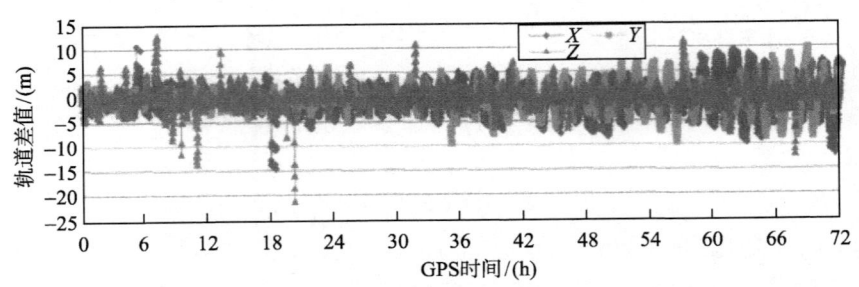

图 2　星载 GPS 测量自主定轨的实测成果(由王甫红博士解得)

参 考 文 献

[1] 欧阳自远，李春来，邹永廖，刘建忠，徐琳. 我国月球探测一期工程的科学目标. 航天器工程，Vol14，No1，2005，P1—5

[2] 刘基余. "嫦娥"卫星绕月飞行的激光定轨建议. 卫星应用简报，2004 年第 23 期，1—4

[3] 刘基余. "嫦娥"卫星绕月飞行轨道的激光测定法. 武汉大学学报（信息科学版），第 30 卷，第 8 期，2005 年

[4] 刘基余. GPS 卫星导航定位原理与方法. （北京）科学出版社，2003 年 8 月，PP433

[5] 刘基余. Evaluation Practice on Airborne GPS Data Quality (SCI: IDS No.: 535MX). Survey Review, V0136, No284, April 2002, P463—469

[6] 刘基余，陈小明，李德仁等. GPS Kinematic Carrier Phase Measurements for Aerial Photogrammetry (SCI: IDS No.: VL895). ISPRS Journal of Photogrammetry and Remote Sensing. Vol51, No5, 1996, P230—242

"嫦娥一号"卫星的调相轨道设计

杨维廉

(中国空间技术研究院,北京 100094)

摘　要　中国第一颗月球探测卫星"嫦娥一号"的飞行轨道的设计中采用了调相轨道,在"长征三号甲"运载火箭提供的超地球同步转移轨道与地月转移轨道之间增加了一段由周期为24h和48h轨道构成的环绕地球飞行的调相轨道。为了将几条不同的轨道精确地拼接起来,必须考虑地球引力场对轨道的摄动影响。克服这个难点的做法是基于经典的轨道摄动理论,先将整段调相轨道设计为考虑地球引力场 J_2 项影响的平轨道,在与运载的发射轨道拼接时,先将运载的包括短周期摄动的瞬时轨道转换为平轨道,在与地月转移轨道拼接时将调相轨道转换成拼接点的瞬时轨道。由于采用了平轨道的处理方法使得轨道控制策略的表述十分简明并易于操作。

关键词　月球探测　地月转移轨道　调相轨道　轨道摄动　卫星

1　引言

"嫦娥一号"(CE-1)卫星飞行轨道中包含了调相轨道段,它将运载火箭"长征三号甲"提供的超 GTO(地球同步转移轨道)与地月转移轨道连接起来。超 GTO 的近地点 200km,远地点 51000km,周期约 16h。卫星在调相轨道段利用自身的推进系统作四次轨道机动,将卫星送入地月转移轨道。卫星与运载分离后在超 GTO 上运行三圈,在第二个远地点作第一次小的轨道机动,将近地点提高到 600km。后三次都是大的近地点机动,先将轨道周期变成 24h,运行三圈后变成 48h,最后一次机动将卫星送入飞行时间 114h 的地月转移轨道[1]。

在探月飞行中采用调相轨道的做法,在 20 世纪 90 年代已经出现,它不仅可以在运载的载荷能力不够时起助推作用,还可以减小转移轨道中途修正所需的速度增量以及扩大发射窗口[2-4]。

在具体设计调相轨道时要解决把几条不同的轨道精确地拼接起来的问题,其核心是如何处理轨道摄动问题。在"嫦娥一号"卫星飞行轨道设计中我们找到一个简便而有效的解决办法。这个方法分两步走,第一步先不考虑轨道摄动的影响,认为所有的轨道都是不变的椭圆轨道,它们都处在同一个轨道平面内;第二步利用经典的轨道摄动理论分析解,通过轨道的平根数进行更精确的拼接。

2　忽略摄动影响的调相轨道设计

因为飞行过程的所有轨道机动都是在近地点或远地点进行,如果不考虑摄动的影响,飞行过程中所涉及的所有椭圆轨道的倾角、升交点赤经 Ω 和近地点幅角 ω 都可以取相同的值。这就使得问题大大地简化。

由于"长征三号甲"运载火箭最初主要是针对发射地球同步轨道来设计的,星箭分离后的轨道近地点是在赤道附近,也就是近地点幅角是 180°左右,因此地月转移轨道的近地点幅角变化范围也受到很大的限制,发射月球卫星的日期也要根据这个限制范围来选择。

在发射日期选择后再任选该日的某个时刻 T_L(例如中午 12 点整)作为近地点时刻来解算出一条地月转

移轨道。这条轨道的倾角31°、近地点高度600km是预先选定的,升交点赤经 Ω 和近地点幅角 ω 是解算出来的。

超GTO的6个轨道参数中,近地点高度、远地点高度(对应半长轴和偏心率)和轨道倾角是由运载能力及发射方式确定的,是轨道设计的输入条件;近地点幅角也可以根据卫星轨道的要求来调节;升交点的位置是相对于地球固连坐标系表述的,它是由发射轨道本身确定,不受卫星轨道的约束,卫星轨道升交点赤经的要求可以通过发射时间的变化来满足。因此在地月转移轨道确定后,运载方的工作是对超GTO进行调节使近地点幅角满足卫星轨道的要求。

在超GTO确定后,就可以把整条飞行轨道拼接起来,先计算出星箭分离的时刻 T_S:用 P_1 来表示星箭分离时的卫星轨道周期,P_2 表示远地点机动后的周期,P_{24} 表示24h轨道的周期,P_{48} 表示48h轨道的周期,于是

$$T_S = T_L - 1.5(P_1 + P_2) - 3P_{24} - P_{48} \tag{1}$$

需要注意的是,运载方所给的轨道参数是相对于地球固连坐标系,升交点位置是用地理经度表示的,它与地球的旋转无关。如果将它作为 T_S 时刻的瞬时轨道则需将升交点位置转换成对应时刻的赤经 α。在一般情况下它是与所要求的升交点赤经 Ω 是不同的,于是就需要对分离时刻 T_S 进行修正,也就是修正转移轨道近地点时刻 T_L。如果经度和时刻的单位分别是度和小时,则时刻的修正量应为

$$\Delta T = (\Omega - \alpha)/15 \tag{2}$$

得到这个修正量以后就要重新解算 $T'_L = T_L + \Delta T$ 时刻的一条新的地月转移轨道,然后基于这条新轨道再重复上述的过程进行迭代直到满足要求。

3 考虑 J_2 摄动的轨道模型

根据经典的轨道摄动理论,如果只考虑地球引力场中的运动,卫星在任意时刻的瞬时(密切)轨道根数可以表示为

$$\left.\begin{aligned}
a &= a^* + \delta a(a^*, e^*, i^*, \Omega^*, \omega^*, M^*) \\
e &= e^* + \delta e(a^*, e^*, i^*, \Omega^*, \omega^*, M^*) \\
i &= i^* + \delta i(a^*, e^*, i^*, \Omega^*, \omega^*, M^*) \\
\Omega &= \Omega^* + \delta \Omega(a^*, e^*, i^*, \Omega^*, \omega^*, M^*) \\
\omega &= \omega^* + \delta \omega(a^*, e^*, i^*, \Omega^*, \omega^*, M^*) \\
M &= M^* + \delta M(a^*, e^*, i^*, \Omega^*, \omega^*, M^*)
\end{aligned}\right\} \tag{3}$$

式中带星号的根数为包括长周期变化的轨道根数,也称为拟平根数,右边部分第二项是短周期摄动项,其量级为 $o(J_2)$,$J_2 = 0.00108263$,是地球引力场二阶调和项的系数。拟平根数又可以表示为

$$\left.\begin{aligned}
a^* &= \bar{a} \\
e^* &= \bar{e} + \delta e^*(\bar{a}, \bar{e}, \bar{i}, \bar{\Omega}, \bar{\omega}) \\
i^* &= \bar{i} + \delta i^*(\bar{a}, \bar{e}, \bar{i}, \bar{\Omega}, \bar{\omega}) \\
\Omega^* &= \bar{\Omega} + \delta \Omega^*(\bar{a}, \bar{e}, \bar{i}, \bar{\Omega}, \bar{\omega}) \\
\omega^* &= \bar{\omega} + \delta \omega^*(\bar{a}, \bar{e}, \bar{i}, \bar{\Omega}, \bar{\omega}) \\
M^* &= \bar{M} + \delta M^*(\bar{a}, \bar{e}, \bar{i}, \bar{\Omega}, \bar{\omega})
\end{aligned}\right\} \tag{4}$$

式(4)中带杠的根数为平根数,右边的第二部分是长周期摄动项,是慢变化部分。需要注意,半长轴 a 在这样的引力场中不存在长周期摄动。平均轨道根数可以表示为

$$\left.\begin{aligned}&\bar{a}=\bar{a}_0\\&\bar{e}=\bar{e}_0\\&\bar{i}=\bar{i}_0\\&\bar{\Omega}(t)=\bar{\Omega}(t_0)+\dot{\bar{\Omega}}(\bar{a}_0,\bar{e}_0,\bar{i}_0)(t-t_0)\\&\bar{\omega}(t)=\bar{\omega}(t_0)+\dot{\bar{\omega}}(\bar{a}_0,\bar{e}_0,\bar{i}_0)(t-t_0)\\&\bar{M}(t)=\bar{M}(t_0)+\dot{\bar{M}}(\bar{a}_0,\bar{e}_0,\bar{i}_0)(t-t_0)\end{aligned}\right\} \quad (5)$$

在本设计中只考虑 J_2 的一阶长期摄动和一阶短周期摄动,于是轨道模型可以简化为

$$\left.\begin{aligned}&a=\bar{a}+\delta a(\bar{a},\bar{e},\bar{i},\bar{\Omega},\bar{\omega},\bar{M})\\&e=\bar{e}+\delta e(\bar{a},\bar{e},\bar{i},\bar{\Omega},\bar{\omega},\bar{M})\\&i=\bar{i}+\delta i(\bar{a},\bar{e},\bar{i},\bar{\Omega},\bar{\omega},\bar{M})\\&\Omega=\bar{\Omega}+\delta\Omega(\bar{a},\bar{e},\bar{i},\bar{\Omega},\bar{\omega},\bar{M})\\&\omega=\bar{\omega}+\delta\omega(\bar{a},\bar{e},\bar{i},\bar{\Omega},\bar{\omega},\bar{M})\\&M=\bar{M}+\delta M(\bar{a},\bar{e},\bar{i},\bar{\Omega},\bar{\omega},\bar{M})\end{aligned}\right\} \quad (6)$$

一阶长期摄动的变率为

$$\left.\begin{aligned}&\dot{\bar{\Omega}}=-\frac{3}{2}J_2\left(\frac{R_\mathrm{e}}{P}\right)^2\bar{n}\cos i\\&\dot{\bar{\omega}}=-\frac{3}{4}J_2\left(\frac{R_\mathrm{e}}{P}\right)^2\bar{n}(1-5\cos^2 i)\\&\dot{\bar{M}}=\bar{n}\left[1-\frac{3J_2}{4}\left(\frac{R_\mathrm{e}}{p}\right)^2\eta(1-3\cos^2 i)\right]\\&p=\bar{a}(1-\bar{e}^2),\quad \eta(\sqrt{1-\bar{e}^2})\end{aligned}\right\} \quad (7)$$

式(7)中的 R_e 是地球赤道半径,n 是平均运动。短周期摄动采用 Hill 变量的表达式[5]

$$\left.\begin{aligned}&\delta\dot{r}=-\frac{\gamma Gp}{4r^2}\left\{(1-3c^2)\left[(1+\eta)^{-1}+\eta\frac{r^2}{p^2}\right]e\sin f+2s^2\sin 2u\right\}\\&\delta r=\frac{\gamma p}{4}\left\{(1-3c^2)\left[1+(1+\eta)^{-1}e\cos f+2\eta\frac{r}{p}\right]+s^2\cos 2u\right\}\\&\delta G=\frac{\gamma Gs^2}{4}\left[3\cos 2u+3e\cos(2u-f)+e\cos(2u+f)\right]\\&\delta u=-\frac{\gamma}{8}\left\{6(1-5c^2)(f-l)+4\left[1-6c^2+(1-3c^2)(1+\eta)^{-1}\right]e\sin f\right.\\&\quad\left.+(1-3c^2)(1+\eta)^{-1}e^2\sin 2f-(1-7c^2)\sin 2u-2(2-5c^2)e\sin(2u-f)+2c^2e\sin(2u+f)\right\}\\&\delta H=0\\&\delta h=\frac{1}{4}\gamma c\left[6(M-f-e\sin f)+3\sin 2u+3e\sin(2u-f)+e\sin(2u+f)\right]\end{aligned}\right\} \quad (8)$$

式(8)中的小参数 $\gamma=J_2\left(\dfrac{R_\mathrm{e}}{P}\right)^2$,$c=\cos i$,$s=\sin i$,Kepler 根数与 Hill 变量之间的变换是

$$r = \frac{p}{1+e\cos f}, \quad u = \omega + f, \quad h = \Omega, \quad \dot{r} = \sqrt{\frac{\mu}{p}} e\sin f, \quad G = \sqrt{\mu p}, \quad H = G\cos i \tag{9}$$

4 考虑摄动的调相轨道设计

星箭分离时的 GTO 轨道和地月转移轨道都是瞬时轨道，因为瞬时轨道的周期是随时间变化的，在它们之间直接用瞬时的调相轨道来拼接是困难的，克服这个困难的有效方法是采用平根数。具体做法是先将 GTO 轨道瞬时根数变成平根数，随后的 24h 和 48h 的轨道均采用平根数来描述，在与地月转移轨道拼接前将 48h 轨道的平根数再转变成瞬时根数。

"嫦娥一号"卫星在飞离地球前的三次大的轨道机动都将在近地点进行，对于轨道机动期间卫星工作状态的监测主要是由"远洋"测量船来执行的，将轨道的近地点尽可能安排在同一地区是十分有利的，为此对 24h 和 48h 的轨道进行了特殊的设计，对于 24h 轨道要求它的近点周期等于一个交点日；对于 48h 轨道要求它的近点周期等于两个交点日，满足这个要求后就能使卫星到达近地点时基本上处在同一位置。近点周期是卫星在平均轨道上从近地点到近地点的时间间隔，交点日是平均轨道升交点相对地球旋转一圈的时间。如果用 T_A 表示近点周期，D_n 表示交点日，ω_e 表示地球自转的速率，则有

$$T_A = \frac{2\pi}{\dot{M}}, \quad D = \frac{2\pi}{\omega_e - \dot{\Omega}} \tag{10}$$

基于所采用的轨道模型，24h 轨道的要求是：$T_A = D_n$，即要求

$$\bar{n}\left\{1 - \frac{3}{4}J_2\left(\frac{R_e}{p}\right)^2\left[\sqrt{1-e^2}(1-3\cos^2 i) + 2\cos i\right]\right\} = \omega_e \tag{11}$$

式中 ω_e 是地球自转速度。对于 48h 轨道的要求是：$T_A = 2D_n$，即要求

$$\bar{n}\left\{2 - \frac{3}{2}J^2\left(\frac{R_e}{p}\right)^2\left[\sqrt{1-e^2}(1-3\cos^2 i) + \cos i\right]\right\} = \omega_e \tag{12}$$

方程(11)、(12)中当倾角 i 确定后，还有两个未知量 a 和 e，有无穷多组解，再对近心距提出具体要求后可以获得唯一解。要求平均近心距 $\bar{r}_p = 6\,980.155$ km，于是 24h 轨道的有关平根数是

$$\bar{a} = 42158.240, \quad \bar{e} = 0.8344296, \quad \bar{r}_p = 6980.155\text{km}, \quad \bar{i} = 30.989° \tag{13}$$

式中下标 p 表示近地点。交点日是 86 134.332s，即 23h55min34s，而不是 24h。48h 轨道的相应平根数是

$$\bar{a} = 66928.771, \quad \bar{e} = 0.8957077, \quad \bar{r}_p = 6980.155\text{km}, \quad \bar{i} = 30.989° \tag{14}$$

交点日是 86 150.157s，即 23h55min50s；近点周期是 47h51min40s，而不是 48h。

可以把这两条平轨道连成一条作为不变的"标准构件"，去设计不同日期发射的调相轨道。

GTO 是地球同步转移轨道，它的远地点高度是地球同步卫星的高度约 36 000km，超 GTO 是指远地点高度超过这个高度的椭圆轨道。"长征三号甲"运载火箭现有的载荷能力可以将"CE-1"卫星发射到倾角 31°，近地点高度 200km，远地点高度 51 000km 的超 GTO 上。在调相轨道设计时，可以保持这三个参数基本不变。对 GTO 需要调整的是近地点幅角 ω，不同日期发射由于月球位置的变化要求近地点幅角 ω 作相应的变化。

这里以 2007 年 4 月发射的调相轨道为例来介绍具体的设计过程。

运载提供的超 GTO 的参数如表 1 所示。

表 1 运载提供的超 GTO 的参数

a/km	e	i/(°)	ω/(°)	Ω/(°)	f/(°)
31 978.596	0.7942931	31	178.779	−20.2547	16.1683

表 1 中 Ω 的数值是地理经度，它与发射时刻没有关系。首先将这组瞬时根数转换成对应的平根数，并计算出近点周期 T_A，这组星箭分离点的轨道表明此时卫星已不在轨道的近地点位置，故需要求出卫星过近地点后已飞过的时间 ΔT。如果把分离时刻记为 $T_0 = 0$，则卫星到达第二个远地点的时刻 $T_1 = 1.5 \times T_A - \Delta T$。由于具体的发射时刻还未选定，把升交点赤经的平根数暂记为 Ω_0，表 2 给出了超 GTO 的平根数。

表 2 超 GTO 的平根数

$\bar{a}$ /km	$\bar{e}$	$\bar{i}$ /(°)	$\bar{\omega}$ /(°)	$\bar{\Omega}$ /(°)	$\bar{f}$ /(°)
31840.442	0.7933379	30.989	178.750	Ω_0	16.1864

$T_A = 56\,533.224\text{s} = 15\text{h}42\text{min}13\text{s}$，$\Delta T = 180.449\text{s}$，$T_1 = 1.5 \times T_A - \Delta T = 84\,619.051\text{s}$，轨道的近点周期是 15h42min13s。由这组平根数进一步可以算出：

$\bar{r}_p = 6\,580.213\text{km}$，$\bar{r}_a = 57\,100.672\text{km}$，$\bar{v}_p = 10.4227\text{km/s}$，$\bar{v}_a = 1.2011\text{km/s}$（其中下标 a 表示远地点）。

在时刻 T_1 到达远地点时需进行一次小的轨道机动将近地点提高到 600km，这里要求机动后 $\bar{r}_p = 6\,980.155\text{km}$，因此远地点速度应增加到 $\bar{v}_a = 1.2332\text{km/s}$，所需的脉冲速度增量是 $\Delta v = 0.0321\text{km/s}$。表 3 是机动后的轨道平根数。

表 3 远地点机动后的轨道平根数

历元 T_1/s	$\bar{a}$ /km	$\bar{e}$	$\bar{i}$ /(°)	$\bar{\omega}$ /(°)	$\bar{\Omega}$ /(°)	$\bar{f}$ /(°)
84619.051	32040.414	0.78214529	30.989	179.122	$\Omega_0 - 0.219$	180

这条轨道的近点周期是 $T_{A1} = 57\,067.423\text{s} = 15\text{h}51\text{min}07\text{s}$。近地点幅角和升交点赤经因地球引力场 J_2 摄动已发生了变化。

第一次大的轨道机动是在近地点，机动的结果变成 24h 的轨道，它的周期是 23h55min34s。机动的时刻应该是 $T_2 = T_1 + 1.5 \times T_{A1} = 170\,220.186\text{s}$，机动前的近地点速度 $\bar{v}_p = 10.0881\text{km/s}$，机动后应增加到 10.2350km/s，所需的脉冲速度增量是 $\Delta v = 0.1469\text{km/s}$，第一次近地点机动后的平根数在表 4 中列出。

表 4 第一次近地点机动后的平根数

历元 T_2/s	$\bar{a}$ /km	$\bar{e}$	$\bar{i}$ /(°)	$\bar{\omega}$ /(°)	$\bar{\Omega}$ /(°)	f /(°)
170220.186	42158.240	0.8344296	30.989	179.401	$\Omega_0 - 0.417$	0

本文按卫星在 24h 轨道运行两圈来设计，两个交点日共计 172 268.664s。下一个近地点机动的时刻是 $T_3 = 342\,488.850\text{s}$，机动后变成 48h 轨道，交点日是 86 150.157s，近地点的速度应该增加到 10.4045km/s，所需的脉冲速度增量是 $\Delta v = 0.1695\text{km/s}$，表 5 所列是机动后的平根数。

表 5 第二次机动后的平根数

历元 T_3/s	$\bar{a}$ /km	$\bar{e}$	$\bar{i}$ /(°)	$\bar{\omega}$ /(°)	$\bar{\Omega}$ /(°)	f /(°)
342488.850	66928.771	0.8957077	30.989	179.789	$\Omega_0 - 0.665$	0

在 48h 轨道上运行一圈需 172 300.314s，因此到达地月转移轨道近地点时刻是 $T_4 = 514\,789.164\text{s}$，表 6 给出对应的平根数。

表 6 地月转移轨道近地点时刻的轨道平根数

历元 T_4/s	$\bar{a}$ /km	$\bar{e}$	$\bar{i}$ /(°)	$\bar{\omega}$ /(°)	$\bar{\Omega}$ /(°)	f /(°)
514789.164	66928.771	0.8957077	30.989	179.971	$\Omega_0 - 0.781$	0

据此可知，从星箭分离到转移轨道近地点的总时间是 142h59min49.16s。

到此为止，既没有确定星箭分离的具体时刻，也没有确定卫星进入地月转移轨道的具体时刻，但已经知道了从星箭分离到进入转移轨道的总时间间隔。为了找出确切的分离时刻，可以先选择一个转移轨道时刻以及相应的轨道升交点赤经，然后计算分离时刻及分离轨道的升交点赤经，这两个升交点赤经的差应该是 0.781°，为此需要按第一节介绍的做法作若干次迭代。最后算出星箭分离时刻应该是 17 日 23:43:16，超 GTO 轨道的升交点赤经是 180.283°，转移轨道时刻是 23 日 22:43:05，对应升交点赤经 180.504°。表 7 给出所有关键点的平根数，其中的升交点赤经是根据摄动理论推算的，在 4 月 23 日 22:43:05 进行最后一次近地点机动前的轨道升交点赤经是 180.502°，它与所要求的值 180.504° 已经非常接近。几个关键点的平根数及根据摄动理论计算的瞬时根数如表 7、表 8 所示。

表 7 几个关键点的平根数

历元	$\bar{a}$ / km	$\bar{e}$	$\bar{i}$ /(°)	$\bar{\omega}$ /(°)	$\bar{\Omega}$ /(°)	$\bar{f}$ /(°)
4/17 23:43:16	31840.442	0.7933379	30.989	178.750	181.283	16.186 4
4/18 23:13:35	32040.414	0.7821453	30.989	179.093	181.064	180
4/19 23:00:16	42158.240	0.8344296	30.989	179.401	180.866	0
4/21 22:51:25	66928.771	0.8957077	30.989	179.789	180.618	0
4/23 22:43:05	66928.771	0.8957077	30.989	179.971	180.502	0

表 8 几个关键点的瞬时根数

历元	a / km	e	i /(°)	ω /(°)	Ω /(°)	f /(°)
4/17 23:43:16	31978.596	0.7942931	31.000	178.779	181.282	16.168 3
4/18 23:13:35	32037.234	0.7821219	30.989	179.093	181.064	180
4/19 23:00:16	42385.989	0.8353661	31.000	179.401	180.866	0
4/21 22:51:25	67509.683	0.8966329	30.999	179.789	180.618	0
4/23 22:43:05	67509.689	0.8966329	30.999	179.971	180.502	0

前面所做的一切都是基于 J_2 摄动的模型，而忽略了日、月的摄动，这种忽略对 48h 的轨道影响较大，应修正这个影响。为此用卫星 21 日 22:51:25 进入 48h 的瞬时轨道根数作为初值进行精确的推算，作用力的模型包括地球引力场 70 阶的模型及日、月的引力，推算到 23 日 22:43:05 的轨道根数如表 9 所示。

表 9 精确推算的轨道根数

历元	a / km	e	i /(°)	ω /(°)	Ω /(°)	f /(°)	根数类型
4/23 22:43:05	67490.010	0.896549	30.983	179.982	180.496	7.155	瞬
4/23 22:43:05	66923.512	0.895646	30.973	179.972	180.496	7.161	平

这组精确的数据说明卫星提前到达近地点。根据这组数据还可以算出卫星飞离近地点已有 84s。据此推算出更为精确的过近地点时刻是 22:41:41，对应时刻的轨道的瞬根数和平根数如表 10 所示。

表 10 卫星到达近地点时刻的轨道根数

历元	a / km	e	i /(°)	ω /(°)	Ω /(°)	f /(°)	根数类型
4/23 22:41:41	67503.430	0.896570	30.983	179.982	180.496	0	瞬时
4/23 22:41:41	66923.512	0.895646	30.973	179.972	180.496	0	平均

经过这种修正后的这个时刻及相应的瞬时轨道根数就是作最后一次近地点机动将卫星送入地月转移轨道的根据。这时的地心距是 6 981.862km，速度是 10.405 6km/s，基于这条轨道的瞬时根数计算出的地

月转移轨道的瞬时根数在表 11 中列出。

表 11 地月转移轨道的瞬时根数

历元	a/km	e	i/(°)	ω/(°)	Ω/(°)	f/(°)
4/23 22:41:41	212857.337	0.967199336	30.983	179.983	180.485	0

这条转移轨道近地点的地心距是 603.725km，近地点速度是 10.597 6km/s，因此最后一次近地点机动所需的速度增量是 $\Delta v_p = 0.192$km/s。

5 结束语

本文给出了一种基于经典轨道摄动理论的分析解来设计调相轨道的方法，这种方法可以十分方便地满足对近地点位置的要求，而且为轨道控制策略的制定提供了基础的理论模型，并使得轨道控制策略的表述十分简明且易于操作。由于将主要的轨道摄动影响都包括进去，使得整条调相轨道具有足够的拼接精度。

参 考 文 献

[1] 杨维廉，周文艳. 嫦娥一号月球探测卫星轨道设计[J]. 航天器工程，2007，16(6)：16—34
[2] DUNHAM D, JEN S, UESUGI K, et al. A Launch Window Study for Geotail's Double Lunar Swing by Trajectory [R]. IAF Paper 90-309, 1990
[3] UESUGI K, MATSUO H, KAWAGUCHI J, et al. Japanese First Double Lunar Swing by Mission-Hiten [R]. IAF Paper 90—343, 1990
[4] CARRING D, CARRICO J, JEN J, et al. Trajectory Design for The Deep Space Program Science Experiment (DSPSE) Mission [R]. AAS 93—260, 1993
[5] AKSNES K. On the Use of the Hill Variable in Artificial Satellite Theory: Brouwer's Theory [J]. Astron& Astrophys, 1972, 17: 70—75

基于大倾角卫星轨道跟踪数据的月球重力场模型仿真解算

李斐[1]　鄢建国[1,2]　平劲松[2]　叶叔华[2]　唐歌实[3]

(1. 武汉大学测绘遥感信息工程国家重点实验室，武汉　430079；
2. 中国科学院上海天文台，上海　200030；
3. 总装备部北京指挥控制中心，北京　100000)

摘　要　本文针对已有月球探测任务主要为极轨的特点，仿真分析了大倾角轨道卫星跟踪数据在月球重力场解算中的贡献。文中针对极轨道、77°倾角和极轨道结合77°倾角轨道三种情况各三个月的轨道跟踪数据进行了月球重力场模型仿真解算，通过重力场功率谱、基于解算模型位系数协方差矩阵的重力异常及月球大地水准面误差以及精密定轨等手段对解算模型进行了精度评价。结果表明结合大倾角的轨道可以较为明显地改进月球重力场模型的计算精度。

关键词　月球重力场　大倾角　月球重力场异常误差　月球大地水准面误差

1　引言

月球重力场是月球内部质量非均匀分布的一种反映，是研究月球内部构造以及月球演化的重要数据源[1]。通过在频率域对月球地形和重力场进行相关与导纳分析，可以估计月球壳幔弹性厚度[2]，有助于了解月球浅部结构；月球重力场低阶项位系数则对了解月球深部构造特征提供了很强的约束，低阶项位系数和月球天平动信息是确定月核密度、大小及状态的主要约束条件[3]。

利用卫星轨道跟踪数据进行月球重力场模型解算的研究始于对1966年前苏联月球探测计划Luna 10轨道跟踪数据的分析，后续的月球探测计划Lunar Orbiter系列任务和Apollo 15/16，以及20世纪90年代发射的Clementine和Lunar Prospector(LP)探测器的轨道跟踪数据进一步提高了月球重力场模型解算精度。Clementine探测器为一极轨大偏心率轨道，有利于改进月球重力场低阶位系数的精度[4]，LP为一极轨圆轨道卫星，正常任务段平均轨道高度为100 km，扩展任务段平均轨道高度为30 km，适合于解算月球重力场中高阶位系数，并有利于提高月球正面重力场的精度和空间分辨率[5]。

由于月球自转和公转周期近似相等，导致月球只有一面对着地球，使得对月球卫星的直接观测仅限于月球正面区域。为了克服这一局限，日本月球探测计划SELENE通过高轨-低轨卫星跟踪卫星模式首次直接探测了月球背面重力场[6]。这一计划的成功实施极大地改进了当前月球重力场模型的精度，特别是精化了月球背面重力场，给出了月球背面典型盆地区域重力异常的环状特性，即中心区域为大的负重力异常，外层由正的重力异常包围。这些特征对于进一步了解月球正面和背面壳幔结构的二分性以及月球热演化历史提供了新的信息[6]。

用于月球重力场模型解算的主要轨道跟踪数据均来自对极轨道卫星的跟踪，这一倾角的轨道可以最大

本文原载于《地球物理学报》，2011，Vol. 54，No. 3，666~672。

程度地覆盖月球表面，但不利于降低月球重力场特定阶次位系数之间的相关性。地球重力场的解算面临类似问题。为了降低地球重力场位系数之间的相关性，解算 EGM96 重力场模型时综合了历史上不同倾角的卫星轨道跟踪数据[7]。图 1 给出了不同倾角情况下月球重力场模型误差引起的轨道误差在轨道面径向、沿迹及法向的分布情况。可以看出，仅用极轨卫星轨道跟踪数据解算得到的月球重力场模型，对极轨这一轨道的卫星具有较好的轨道预报精度，随着倾角偏离极轨的程度越大，轨道预报精度越差[8,9]。

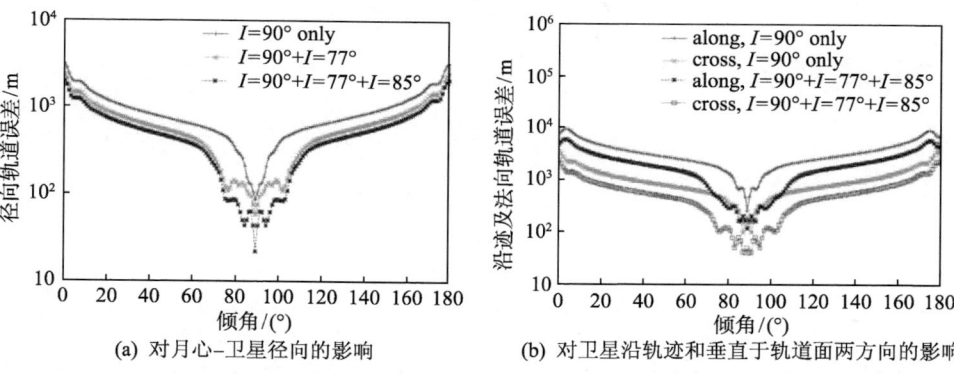

图 1 2 个月极轨、2 个月极轨+2 个月 77°倾角、2 个月极轨+2 个月 77°倾角+2 个月 85°倾角轨道得到的月球重力场对卫星轨道精度的影响

我国"嫦娥一号"正常任务期间积累了大量轨道跟踪数据，并成功用于月球重力场模型解算[9,10]，原拟定在任务后期进行变轨，以一定倾角轨道进行月球地形及重力场探测，由于星上机械故障这一计划未能实施。"嫦娥二号"已于 2010 年 10 月发射，在正常任务期间有望调整轨道倾角进行跟踪测量。本文针对现有月球重力场模型解算中主要用到极轨卫星这一情况，仿真分析了利用 90°倾角卫星 90 天时段长度、77°倾角卫星 90 天长度以及综合 45 天 90°倾角卫星与 45 天 77°倾角卫星三种情况下月球重力场仿真解算的情况（90°倾角解算得到的模型记为 B，77°倾角解算得到的模型记为 C，90°与 77°综合解算得到的模型记为 D，见后文），对各解算模型进行了位系数功率谱分析、基于解算模型位系数协方差矩阵进行了月球重力异常和月球大地水准面误差分析以及精密定轨分析。这一研究主要针对我国后续月球探测计划的科学目标，以提出更有特色的月球重力场探测模式。

2 仿真计算设置

仿真过程中选择 LP100J[5]模型为真实模型。利用 LP100J 模型生成三种观测模式下的真实观测量，观测类型包括青岛和喀什两个站的双程测速观测量，以及中国 VLBI 网(上海，北京，昆明，乌鲁木齐)四个台站的时延和时延率观测量。VLBI 时延的精度设定为 3 ns，时延率的精度设定为 1 ps/s，USB 双程测速的精度为 1 mm/s。仿真解算中在真实观测量的基础上增加高斯白噪声，其标准偏差值为测量数据精度。仿真时段为 2010 年 5 月 1 日至 2010 年 8 月 1 日。数据仿真中考虑到月球的遮掩，测站地面高度角设定为 10°。

考虑到我国后续月球探测的轨道设计，三种仿真模式中轨道高度均选取为 100 km，极轨和 77°倾角轨道下的偏心率均为 0.0005，接近于圆轨道。模拟计算中用到的动力学模型包括月球非球形引力摄动、太阳和地球的引力摄动、木星、水星、金星等大行星的摄动、固体潮摄动、地球扁率的间接效应、太阳光压摄动以及相对论效应。计算过程中采用的坐标系是 J2000.0 月心天球坐标系，时间系统为 UTC 协调世界时，日月等天体的位置采用 JPL DE405/LE405[11]历表计算。计算中采用了 GSFC/NASA/USA 授权使用的轨道分析软件 GEODYN II/SOLVE[12,13]。

月球重力场模型解算采用动力法[4,5]，即在精密定轨的同时解算重力场位系数。各模型求解的阶次设定为 50×50，该阶次对应的月球表面水平分辨率为 100 km[14]。由于仿真观测数据只有月球正面的观测数据，在求解中需要加入 Kaula 先验约束以平滑重力场参数的求解[4,5]。基于动力法求解月球重力场位系数涉及非线性问题的线性化，为了提高解算精度，解算需要迭代处理。文中迭代次数为 2 次，迭代收敛准则为观测量残差达到先验误差水平。进行精密定轨及重力场参数解算时采用的先验重力场模型为在 LP100J 的基础上加上模型各位系数对应的 3 倍方差。考虑到轨道运行过程中的姿态调整，解算弧段长度选取为 1 天，以避免动量轮卸载因素对精密定轨及重力场解算的影响。

3 月球重力场模型功率谱分析

基于球谐函数展开的月球重力场模型位系数的阶方差信息可以反映重力场在频域中的信号强度(图 2 中 sig)。阶方差的计算公式为[15]

$$\sigma_n = \sqrt{\frac{\sum_{m=0}^{n}(\bar{C}_{nm}^2 + \bar{S}_{nm}^2)}{2n+1}} \tag{1}$$

式中 $\bar{C}_{nm}, \bar{S}_{nm}$ 是正则化的位系数，n 为阶，m 为次。对阶方差信息进行分析判断的常用工具是 Kaula 准则。Kaula 准则是对重力场模型位系数统计规律的一个近似描述，即正则化的位系数具有零均值，标准偏差与阶数 n 的平方成反比[16]，其数学表达形式为

$$\bar{C}_{nm}, \bar{S}_{nm} \sim 0 \pm \frac{\text{const}}{n^2} \tag{2}$$

计算中常数 const 取值为 3.0。在月球重力场模型确定中引入 Kaula 准则的目的主要是作为一个正则化因子，用来克服由于缺乏月球背面轨道跟踪数据数据导致重力场求解时的不稳定性，同时对高阶位系数的计算起到一平滑作用[17]。Kaula 准则用曲线 $3.0 \times 10^{-4} / n^2$ 进行描述，其中 n 对应重力场模型位系数的阶数。

月球重力场模型的误差阶方差用来反映重力场位系数误差在频域中的强度(图 2 中 sigvar)，可以说明重力场解算模型的精度水平，其计算公式为[16]

$$\delta_n = \sqrt{\frac{\sum_{m=0}^{n}(\sigma_{\bar{C}_{nm}}^2 + \sigma_{\bar{S}_{nm}}^2)}{2n+1}} \tag{3}$$

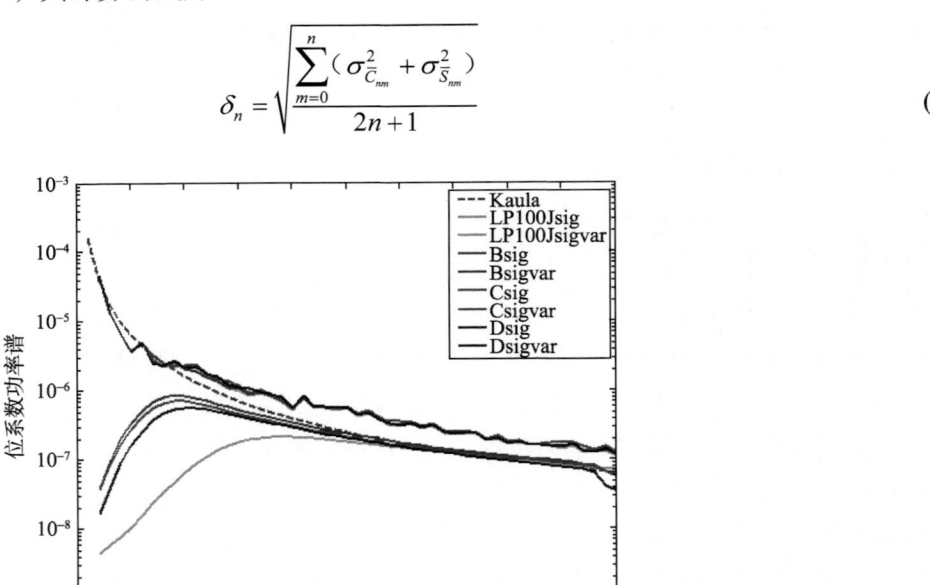

图 2　Kaula 曲线，极轨、77°倾角以及极轨和 77°倾角综合解算得到的重力场、真实重力场与先验重力场模型位系数阶方差及误差阶方差曲线

图2给出了解算模型与先验模型的功率谱曲线图，图中B,C和D分别表示仅用90°倾角卫星、仅用77°倾角卫星、以及综合两颗卫星解算得到的重力场模型。由图2可以看出，三种计算模式下得到的重力场模型与真实模型均比较接近，且模型D得到的阶方差与先验真实模型更为接近。相比于先验真实模型LP100J，三个解算模型对中长波部分位系数没有显著改进。这是因为LP100J解算中综合了历史高轨道跟踪数据，包括Clementine和Apollo及Lunar Orbiter系列探测器，这些探测器的轨道跟踪数据对改进中长波重力场信息起到了重要作用。从图2中可以看到不同倾角轨道跟踪数据对重力场位系数相对改进的程度，77°倾角卫星跟踪数据解算的模型相比于极轨卫星而言，对5阶次到15阶次位系数精度有比较明显的改进，这可能是因为77°倾角的月球卫星轨道相比于极轨有利于降低特定阶次扇谐系数和田谐系数的相关性。融合极轨与77°倾角卫星的跟踪数据联合解算的重力场模型，则对2至15阶次的位系数精度有显著改进，这主要是由于综合不同倾角卫星的轨道跟踪数据，降低了特定阶次位系数之间的相关性，从而提高了位系数的解算精度。

4 月球重力异常及月球大地水准面误差分析

基于误差传播理论，利用解算模型得到的位系数协方差矩阵进行了月球重力异常及月球大地水准面误差分析。月球重力异常和月球大地水准面的计算公式分别为[15]

$$\Delta g(\phi,\lambda) = 10^5 \times \sum_{n=2}^{n_{\max}}\sum_{m=0}^{n} \frac{\mu}{a^2}(n-1) \times [\bar{C}_{nm}\cos(m\lambda)+\bar{S}_{nm}\sin(m\lambda)]\bar{P}_{nm}(\sin\phi) \quad (4)$$

$$\Delta N(\phi,\lambda) = a\sum_{n=2}^{n_{\max}}\sum_{m=0}^{n}[\bar{C}_{nm}\cos(m\lambda)+\bar{S}_{nm}\sin(m\lambda)]\bar{P}_{nm}(\sin\phi) \quad (5)$$

其中a是月球平均球半径，(ϕ,λ)为月面地理经纬度，μ为月球引力质量常数，n和m分别为重力场位系数的阶与次，$n_{\max}$为模型的最大截断阶数，$\bar{P}_{nm}$为连带勒让德函数。(4)(5)两式可以简化为线性关系式：

$$y = Hx \quad (6)$$

式中x表示位系数参数向量，y表示月球重力异常或者月球大地水准面，两者之间通过矩阵算子H联系。令通过模型解算得到的重力场位系数协方差矩阵为$Q(x)$，则通过误差传播定律有

$$\sigma^2(y) = H^{\mathrm{T}}Q(x)H \quad (7)$$

其中$\sigma(y)$为y的方差。基于三种解算模式得到的月球重力异常及月球大地水准面误差计算情况如表1所示。由表1可以明显看出，综合极轨和77°倾角卫星进行解算的情况具有最小的月球重力异常和月球大地水准面误差，由77°倾角卫星解算模型得到的月球重力异常和大地水准面精度相比于90°倾角卫星解算模型均有提高，这与图2中关于各模型位系数误差阶方差的区别一致。

表1 月球重力异常和月球大地水准面RMS值

	重力异常误差/mGal	大地水准面误差/m
极轨卫星解算模型	34.08	19.02
77°倾角卫星解算模型	31.87	16.77
极轨+77°倾角卫星	27.88	14.54

图3至图5（图版XIV）给出了基于三个解算模型的位系数协方差矩阵计算得到的月球正面重力异常误差分布图。图3中的误差分布呈比较明显的倾斜条纹形式，这主要是由于卫星的轨道特征，导致卫星轨道跟踪数据星下点覆盖呈倾斜条纹形式。图中红色部分较大的重力异常误差区域是由于没有跟踪数据覆盖所引起，77°倾角的轨道无法覆盖两极，导致在两极区域有较大的重力异常误差。图4中重力异常分布呈现明显的过两极的条带特性，在高纬度及两极具有较小的重力异常误差，这是由于极轨卫星跟踪数据星下

点分布在两极与中高纬度地区更为密集。图 5 给出的综合极轨和 77°倾角卫星得到的重力异常误差分布相比于图 3 和图 4 均有所改进,图 5 中重力异常误差分布的倾斜条带效应相比于图 4 有较为显著的弱化,重力异常精度有明显提高。这从月球重力场特征量误差分布的角度证实了倾角卫星对改进月球重力场模型精度的贡献。

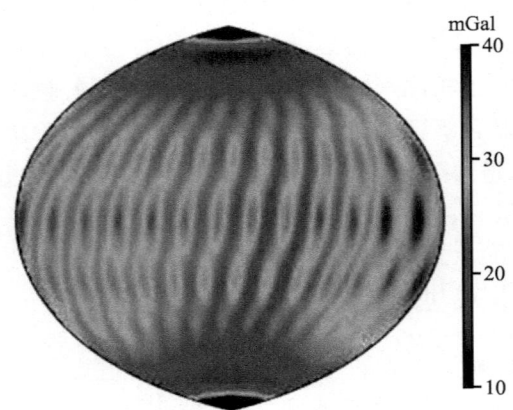

图 3　77°倾角卫星解算重力场模型对应的月球正面重力异常误差分布

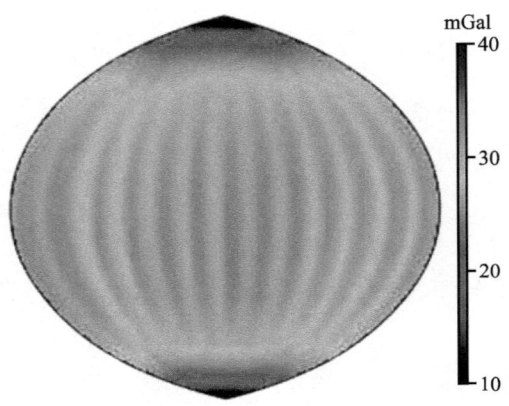

图 4　极轨卫星解算重力场模型对应的月球正面重力异常误差分布

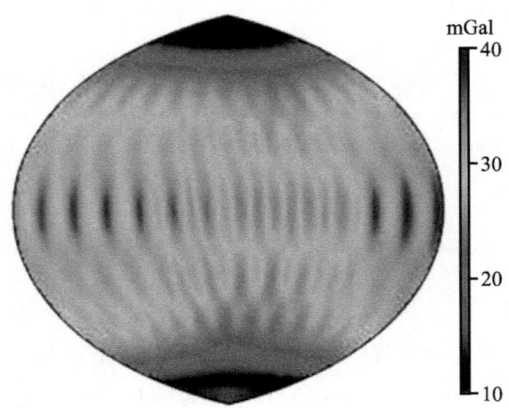

图 5　极轨卫星与 77°倾角卫星联合解算重力场模型对应的月球正面重力异常误差分布

5 定轨误差分析

为了进一步验证倾角卫星在提高月球重力场模型解算精度中的贡献, 利用不同重力场模型对77°倾角卫星进行了精密定轨。精密定轨中涉及的其他动力学模型及参考框架与第2节中一致。考虑的重力场模型包括带有误差的先验重力场模型(A)、极轨卫星解算得到的模型(B)、77°倾角解算得到的模型(C)以及综合极轨卫星和77°倾角卫星解算得到的模型(D)。随机选取了12个弧段, 弧段按照天数进行编号, 精密定轨后给出的参考量为双程测速残差。结果如表2所示。

表2 不同重力场模型精密定轨后双程测速残差 RMS 值 (单位: cm/s)

弧段编号	模型			
	A	B	C	D
6	3.658	2.264	0.106	0.102
9	8.33	5.985	0.102	0.100
12	50.560	3.328	0.105	0.103
33	3.492	2.107	0.104	0.101
36	8.208	5.325	0.102	0.100
39	21.8	2.614	0.101	0.099
45	4.367	5.949	0.114	0.102
49	16.052	3.654	0.100	0.099
55	6.489	0.926	0.104	0.099
67	5.411	4.115	0.105	0.102
78	14.053	3.820	0.101	0.100
84	3.299	1.243	0.108	0.103

由表2可以看出, 基于带有误差的先验重力场模型进行精密定轨时, 具有较大的残差, 以极轨卫星跟踪数据解算得到的重力场模型相比于先验重力场模型有所改进, 但误差仍然较为显著, 这说明基于极轨卫星得到的重力场模型不能较好地应用到其他倾角卫星轨道上, 这与图1中给出轨道预报误差分布一致。利用77°倾角卫星跟踪数据解算得到的重力场模型具有较好的精密定轨结果, 这是因为该重力场模型完全是由这一卫星轨道的跟踪数据解算所得, 故具有较好的拟合程度。综合极轨和77°倾角卫星跟踪数据解算得到的重力场模型可以得到最好的轨道残差值, 并能恢复到先验测量精度水平。这主要是因为综合极轨和77°倾角卫星的轨道跟踪数据, 通过降低特定阶次位系数之间的相关性, 提高了重力场模型解算精度。表2的数据一方面说明了整个仿真计算过程的准确性, 另外也说明为了满足未来某一特定倾角月球卫星精密定轨的需求, 需要综合不同倾角卫星的轨道跟踪数据进行综合解算以得到高精度月球重力场模型。

考虑到仿真解算中已知真实轨道信息, 为了更客观地评价重力场模型精度, 利用三个不同模型对77°倾角卫星进行了精密定轨。表3给出了三种情况下解算轨道与真实轨道的差值。由表3可以看出, 综合极轨和大倾角卫星跟踪数据解算得到的重力场模型(D)具有最好的定轨精度。基于三个不同重力场模型进行精密定轨后, 位置误差的均值分别为221.75米, 5.82米, 1.63米, 对应的标准偏差为286.68米, 4.49米, 0.94米, 可以看出模型D的定轨精度相比于B, 改进超过两个量级, 相对于C改进达5倍。仅根据极轨卫星轨道跟踪数据解算得到的重力场模型(B)对倾角卫星定轨具有较大的误差, 这与表2得出的结论一致。部分弧段中仅用倾角卫星得到的轨道(C)和(D)具有较为接近的结果, 这些弧段主要是轨道面接近通视的情况, 即轨道面垂直于地面测站到探测器连线方向, 这一几何构型下具有更长的观测时段, 可以对轨道起到较好的约束效果。

表3 不同重力场模型精密定轨结果　　　　　　　　　　　　　　　　　(单位：m)

弧段编号	X			Y			Z			Position		
	B	C	D	B	C	D	B	C	D	B	C	D
6	62.636	−1.24	0.80	70.169	−1.108	0.432	141.97	−5.502	1.814	170.30	5.748	2.03
9	−8.639	−0.157	−0.009	−41.670	0.121	−0.187	47.168	0.542	0.331	63.529	0.577	0.393
12	22.666	1.185	−1.074	−31.070	−2.732	2.217	14.911	0.474	0.005	41.249	3.015	2.464
33	−17.985	−7.033	−0.165	39.588	−5.301	−0.948	0.56	−5.361	−0.449	43.486	10.311	1.062
36	1.052	−0.028	0.034	7.015	0.106	0.113	66.662	1.225	0.795	67.038	1.230	0.804
39	10.427	0.205	−0.146	−31.783	−1.031	0.557	−59.380	1.581	−0.528	68.153	1.899	0.782
45	135.62	11.458	−3.187	−59.824	0.639	−0.267	−68.886	4.406	−0.909	163.45	12.293	3.3253
49	−185.33	−1.996	−0.516	−354.08	−4.239	−1.209	−155.21	−1.182	−0.237	428.73	4.833	1.336
55	633.11	5.828	1.938	−432.37	−4.237	−1.578	693.62	4.571	1.737	1033.8	8.533	3.044
67	48.670	0.111	−0.153	−107.31	−1.139	0.459	361.36	7.360	−1.960	380.09	7.448	2.019
78	−4.526	0.120	0.098	−37.022	0.697	0.617	29.20	−0.575	−0.639	47.368	0.912	0.893
84	−145.66	12.850	−1.316	47.196	−1.867	−0.079	−15.553	1.934	−0.531	153.90	13.128	1.421

6 结论

本文针对我国后续月球探测计划可能改变倾角，在近极轨道附近以大倾角轨道运行，综合这一倾角轨道的跟踪数据进行重力场模型解算有利于降低月球重力场模型特定阶次位系数之间的相关性，以起到提高月球重力场模型解算精度的作用。另外已有重力场模型的解算数据主要来自于极轨卫星，这一模型对其他倾角的卫星具有较大的轨道预报误差。基于上述考虑，本文针对仅用极轨卫星，仅用77°倾角卫星以及综合极轨卫星和77°倾角卫星三种情况，进行了月球重力场模型仿真计算。从重力场位系数功率谱、月球重力异常及大地水准面误差、精密定轨等方面对三个解算模型进行了精度评价。结果表明综合极轨卫星和77°倾角卫星得到的月球重力场模型，相比于只使用一种倾角的卫星解算得到的模型，在中低阶次位系数精度有显著改进，由此可以得到更为精确的月球重力异常及月球大地水准面分布，并在定轨精度上有大幅提高。本文计算结果可以对我国后续月球探测计划用于月球重力场探测提供一定程度的参考。

致谢　本文所提及的GEODYNII轨道分析软件是经GSFC/NASA/USA授权中国科学院上海天文台使用，并且在上海天文台的计算机工作站上完成。

参 考 文 献

[1] Khan A, Mosegaard K, Williams J G, et al. Does the Moon possess a molten core? Probing the deep lunar interior using results from LLR and lunar prospector. *J.Geophys. Res.*, 2004, 109, E09007, doi: 10.1029/2004JE002294

[2] Crosby A, McKenzie D. Measurements of the elastic thickness under ancient lunar terrain. *Icarus*, 2005, **173**(1):100—107

[3] Hanada H, Iwata T, Namiki N, et al. VLBI for better gravimetry in SELENE. *Advances in Space Research*. 2008, **42**(2): 341—346

[4] Lemoine F G, Smith D E, Zuber M T, et al. A 70th degree lunar gravity model (GLGM-2) from Clementine and other tracking data. *J.Geophys. Res.*, 1997, **102** (E7):16339—16359

[5] Konopliv A S, Asmar S W, Carranza E, et al. Recent gravity models as a result of the Lunar Prospector mission. *Icarus*, 2001, **150**(1):1—18

[6] Namiki N, Iwata T, Matsumoto K, et al. Farside Gravity Field of the Moon from Four-Way Doppler Measurements of SELENE (Kaguya). *Science,* 2009, **323**(5916): 900—905

[7] Lemoine F G, Kenyon S C, Factor J K, et al. The development of the joint NASA GSFC and the National Imagery and Mapping Agency (NIMA) geopotential model EGM96. NASA Technical Paper NASA/TP-1998-206861. Goddard Space Flight Center, Greenbelt., 1998

[8] Rosborough G W, Tapley B D. Radial, transverse and normal satellite position perturbations due to the geopotential. *Celestial Mechanics*, 1987, **40**(3—4): 409—421

[9] 鄢建国, 李斐, 平劲松等. 基于嫦娥一号跟踪数据的月球重力场模型 CEGM01. 地球物理学报, 2010, **53**(12):2843—2851
Yan J G, Li F, Ping J S, et al. Lunar gravity field CEGM01 based on Chang'E-1 orbital tracking data. *Chinese J. Geophys.* (in Chinese)**,** 2010, **53**(12):2843—2851

[10] Yan J G, Ping J S, Li F, et al. Chang'E-1 precision orbit determination and lunar gravity field solution. *Advances in Space Research*, 2010, **46**(1):50—57

[11] Standish E M. JPL Planetary and Lunar Ephemerides, DE405/LE405. Interoffice Memorandum, 1998, 312.F-98-048.1—18

[12] Rowlands D, Marshall J A, Mccarthy J, et al. GEODYN II system description. Vols.1—5, contractor report, Hughes STX Corp., Greenbelt, MD,1997

[13] Ullman R E. SOLVE program: mathematical formulation and guide to user input, Hughes/STX Contractor Report, Contract NAS5-31760. NASA Goddard Space Flight Center, Greenbelt, Maryland, 1994

[14] 陈俊勇, 宁津生, 章传银等. 在"嫦娥一号"中求定月球重力场. 地球物理学报, 2005, **48** (2):275—281
Chen J Y, Ning J S, Zhang C Y, et al. On the determination of lunar gravity field in the Chinese first lunar prospector mission. *Chinese J. Geophys.* (in Chinese), 2005, **48** (2):275—281

[15] Heiskanen W A, Moritz H. Physical Geodesy. San Francisco: Freeman,1967

[16] Kaula W M. Theory of Satellite Geodesy. Waltham, mass: Blaissell Publishing Company,1966

[17] 鄢建国, 平劲松, 李斐等. 应用 LP165P 模型分析月球重力场特征及其对绕月卫星轨道的影响, 地球物理学报, 2006, **49**(2):408—414
Yan J G, Ping J S, Li F, et al. Character analysis of the lunar gravity field by the LP165P model and its effect on lunar satellite orbit. *Chinese J. Geophys.* (in Chinese), 2006, **49**(2):408—414

射电望远镜指向误差的广义延拓插值修正方法

孔德庆[1,2]　施浒立[1,2]　张喜镇[1]　张洪波[1]

(1. 中国科学院国家天文台，北京 100012；2. 中国科学院研究生院，北京 100039)

摘　要　提出了一种新的基于广义延拓插值的射电望远镜指向误差修正模型，在线性修正模型的基础上对观测域进行球面的 Delaunay 三角剖分，利用高精度的广义延拓插值逼近方法生成不分项的整体分布曲面。此方法可在分片边界点上满足插值条件，并充分利用分片插值区域的周围结点(包括内点)信息，实现分片最佳拟合。应用 50 m 天线指向实验数据进行分析，结果表明该模型与线性修正模型相比，指向预测精度提高了 30.3%，从而能更有效地提高望远镜的指向精度。

关键词　射电望远镜　指向误差　广义延拓插值

为了提高信号探测的灵敏度，射电望远镜正朝着大口径、高频段方向发展。指向精度是射电望远镜最重要的性能指标之一，一般要求指向偏差小于天线半功率宽度(HPBW)的 10%[1]。在大型射电望远镜建造的过程中，结构的硬件校准并不能完全解决望远镜的指向问题，还需进一步通过软件校准才能满足精度要求。

射电望远镜指向误差修正主要采用的方法[2]：通过对已知精确方向目标源的指向测量实验，获得指向偏差实验样本。建立指向偏差数学模型，用实验样本数据离线求得回归函数。利用回归函数，在线预测天线原始指令角的指向偏差，原始指令角减去该偏差得到校准指令角，使天线更准确地指向原始指令角方向，从而提高天线指向精度。

目前，国内外天线指向校正大多采用线性指向模型(linear pointing model, Linear-PM)[2,3]。这种模型存在两点不足：一是部分忽略了非线性偏差；二是外部因素对天线指向的影响难以考虑周全。另一种方法是将测量得到的误差数据直接存入天线控制计算机中，经插值进行实时改正[4,5]；这一方法虽然考虑到了一些误差源，但几何逼近性较弱，在各区域边界上可出现函数的台阶性突变。针对上述方法的不足，笔者提出了一种新的基于广义延拓逼近法的天线指向修正模型。

1　指向误差的线性修正模型

大型射电望远镜一般采用地平座架，具有方位 A 和俯仰 E 两个自由度，指向偏差按可分项偏差分解为：①方位轴不垂直；②俯仰轴与方位轴不正交；③电轴与俯仰轴不正交；④天线结构受载变形引起的偏差；⑤轴角编码器偏差等。在线性修正中，由于各偏差分量都比较小，总偏差为各分项偏差的代数和。经过对指向实验数据样本的最小二乘求解，可得出作为待定系数的各分项度量值。式(1)为一线性化方法[5]：

$$\begin{cases} \delta A_l \cos \Delta E = C_1 \cos \Delta E + C_3 + C_4 \cos \Delta A \sin \Delta E + C_5 \sin \Delta A \sin \Delta E + C_6 \sin \Delta E \\ \delta E_l = C_2 - C_4 \sin \Delta A + C_5 \cos \Delta A + C_7 \cos \Delta E + C_8 \cos^3 \Delta E \end{cases} \quad (1)$$

式中 δA_l 和 δE_l 分别是方位和俯仰的线性模型修正函数，ΔA 为方位角误差，ΔE 为俯仰角误差，C_1 为方位角编码器零点误差，C_2为俯仰角编码器零点误差，C_3为光轴或电轴与俯仰轴不正交误差，C_4为方位轴东西向偏斜引起的误差，C_5为方位轴南北向偏斜引起的误差，C_6为俯仰轴与方位轴心的不正交误差，C_7为电轴的重力变形误差一次项，C_8为电轴重力变形误差的高次项。

线性修正模型不足之处有：部分忽略了非线性误差，大口径射电望远镜的误差源非常复杂，误差描述函数很难描述所有的指向误差，所以这种方法对指向精度的提高有一定的限制；分解误差的分项描述函数不考虑相位关系，也会使误差辨识与实际情况不相符；忽略了各分量之间的耦合作用；在模型建立过程中，采用的线性化等近似处理损失了部分精度；各分项度量值反映了实验时外部环境因素对指向的偏差，但当天线在不同于实验环境因素下工作时，由此模型预测的指向偏差往往会有较大的误差[2]。

2 基于广义延拓插值的指向误差修正

利用望远镜指向误差在线性修正基础上的整体校正修正方法，可以消除线性误差识别和修正法残差较大的缺点，获得非常高的指向精度。方法的不同之处是：将误差数据不采用分项误差辨识与改正，而是利用高精度的广义延拓插值逼近方法[6]生成不分项的整体分布曲面，产生整体误差修正数据。广义延拓逼近法是在插值法和拟合法的基础上，集两者长处于一体的一种广义插值逼近方法，它使根据测量数据生成的误差修正数据精度极高。

2.1 射电望远镜指向测量域的 Delaunay 三角剖分

为了利用广义延拓插值法，需对测量域进行区域剖分。在所有可能的三角网中，Delaunay 三角网在二维拟合方面表现最为出色，因此常常被用于不规则三角网(TIN)的生成。Delaunay 三角网具有"空外接圆"性质，任何一个三角形的外接圆均不包含其他数据点；在所有可能形成的三角网中，Delaunay 三角网中的三角形的最小内角是最大的[7]。这样就有效地保证了 TIN 三角网是最接近等角或等边的最优三角网，可以减少由狭长的、内角尖锐的三角形产生的潜在数字精度问题。

射电望远镜指向测量域在进行 Delaunay 三角剖分时，可以将每个区域看作半径长度为固定值的球面三角形，其球面角由下式计算：

$$\begin{cases} \cos A = \cos a - \cos b \cos c \\ \cos B = \cos b - \cos a \cos c \\ \cos C = \cos c - \cos a \cos b \end{cases} \tag{2}$$

式中 A, B, C 为球面三角形球面角，a, b, c 为与球面角对应的球面三角形的边。跟据平面域的 Delaunay 三角剖分方法[7]可建立球面的三角剖分。

2.2 指向误差的广义延拓逼近模型

望远镜指向方位偏差和俯仰偏差模型可表示为

$$\delta A = \delta A_l + \delta A_g, \quad \delta E = \delta E_l + \delta E_g \tag{3}$$

式中 δA_g, δE_g 为方位、俯仰偏差线性修正后的残差的广义延拓插值函数。

将观测域 D 作三角剖分，得到 m 个互不重叠的子区域 $D_1, D_2, \cdots, D_m$，每个三角形子域 D_e 上的结点编号为 1,2,3，其延拓域 D'_e 上有 $s(s>3)$ 个结点，如图 1。

设单元 D_e 上的方位残差的广义逼近函数为[6]

$$\begin{aligned} \delta A_{ge}(A,E) &= a_1 + a_2 A + a_3 E + a_4 A^2 + a_5 AE + a_6 E^2 \\ (A,E) &\in D_e, e = 1, 2, \cdots, m \end{aligned} \tag{4}$$

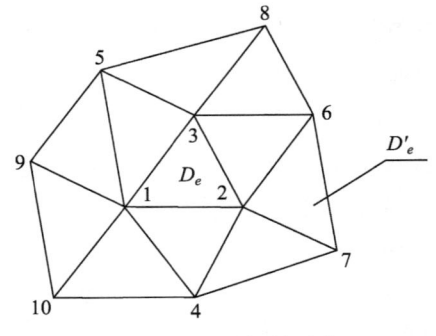

图 1 二维延拓域示意

则待定系数 $a_1, a_2, \cdots, a_6$ 由下式确定：

$$\begin{cases} \min I(a_1,a_2,\cdots,a_6) = \sum_{k=4}^{s}(a_1+a_2A_k+a_3E_k+a_4A_k^2+a_5A_kE_k+a_6E_k^2-U_k)^2 \\ \text{s.t.} U_i = a_1+a_2A_i+a_3E_i+a_4A_i^2+a_5A_iE_i+a_6E_i^2, \quad i=1,2,3 \end{cases} \quad (5)$$

式中 A_k 和 H_k 分别是节点 k 处的方位和俯仰值，$U_i(i=1,2,\cdots,m)$ 为节点处方位偏差在线性修正后的残差，即

$$U_i = \mathrm{d}A_i - \delta A_i$$

其中 $\mathrm{d}A_i$ 为节点 i 处实测的方位和俯仰偏差。

引入拉格朗日乘子 $\lambda_1, \lambda_2, \lambda_3$，应用拉格朗日乘子法求解问题(5)，可得到如下的关于 $a = \{a_1,a_2,a_3,\cdots,a_6,\lambda_1,\lambda_2,\lambda_3\}^\mathrm{T}$ 的线性代数方程组：

$$\begin{bmatrix} C & 0 \\ A & C^\mathrm{T} \end{bmatrix} \begin{bmatrix} a \\ \lambda \end{bmatrix} = \begin{bmatrix} M_0 \\ M_1 \end{bmatrix} \quad (6)$$

其中

$$a = [a_1,a_2,\cdots,a_6]^\mathrm{T}, \quad \lambda = [\lambda_1,\lambda_2,\lambda_3]^\mathrm{T}, \quad M_0 = [U_1,U_2,U_3]^\mathrm{T}$$

$$M_1 = \left[\sum_{i=4}^{s}U_i, \sum_{i=4}^{s}A_iU_i, \sum_{i=4}^{s}E_iU_i, \sum_{i=4}^{s}A_i^2U_i, \sum_{i=4}^{s}A_iE_iU_i, \sum_{i=4}^{s}E_i^2U_i\right]^\mathrm{T}$$

而分块矩阵 **0** 为零矩阵。

$$C = \begin{bmatrix} 1 & A_1 & E_1 & A_1^2 & A_1E_1 & E_1^2 \\ 1 & A_2 & E_2 & A_2^2 & A_2E_2 & E_2^2 \\ 1 & A_3 & E_3 & A_3^2 & A_3E_3 & E_3^2 \end{bmatrix}$$

$$A = \begin{bmatrix} \sum_{i=4}^{s}1 & \sum_{i=4}^{s}A_i & \sum_{i=4}^{s}E_i & \sum_{i=4}^{s}A_i^2 & \sum_{i=4}^{s}A_iE_i & \sum_{i=4}^{s}E_i^2 \\ \sum_{i=4}^{s}A_i & \sum_{i=4}^{s}A_i^2 & \sum_{i=4}^{s}A_iE_i & \sum_{i=4}^{s}A_i^3 & \sum_{i=4}^{s}A_i^2E_i & \sum_{i=4}^{s}A_iE_i^2 \\ \sum_{i=4}^{s}E_i & \sum_{i=4}^{s}A_iE_i & \sum_{i=4}^{s}E_i^2 & \sum_{i=4}^{s}A_i^2E_i & \sum_{i=4}^{s}A_iE_i^2 & \sum_{i=4}^{s}E_i^3 \\ \sum_{i=4}^{s}A_i^2 & \sum_{i=4}^{s}A_i^3 & \sum_{i=4}^{s}A_i^2E_i & \sum_{i=4}^{s}A_i^4 & \sum_{i=4}^{s}A_i^3E_i & \sum_{i=4}^{s}A_i^2E_i^2 \\ \sum_{i=4}^{s}A_iE_i & \sum_{i=4}^{s}A_i^2E_i & \sum_{i=4}^{s}A_iE_i^2 & \sum_{i=4}^{s}A_i^3E_i & \sum_{i=4}^{s}A_i^2E_i^2 & \sum_{i=4}^{s}A_iE_i^3 \\ \sum_{i=4}^{s}E_i^2 & \sum_{i=4}^{s}A_iE_i^2 & \sum_{i=4}^{s}E_i^3 & \sum_{i=4}^{s}A_i^2E_i^2 & \sum_{i=4}^{s}A_iE_i^3 & \sum_{i=4}^{s}E_i^4 \end{bmatrix}$$

求解式(6)得 a，代入式(4)得单元 D_e 上的分片光滑逼近函数 δA_{ge}。把单元子域的逼近函数拼合起来，可得全域上的高精度逼近函数

$$\delta A_g = \bigcup_{e=1}^{m} \delta A_{ge} \quad (7)$$

同理，可得俯仰残差的广义延拓逼近 δE_g。

图 2 为指向误差线性修正和广义延拓插值修正示意图。

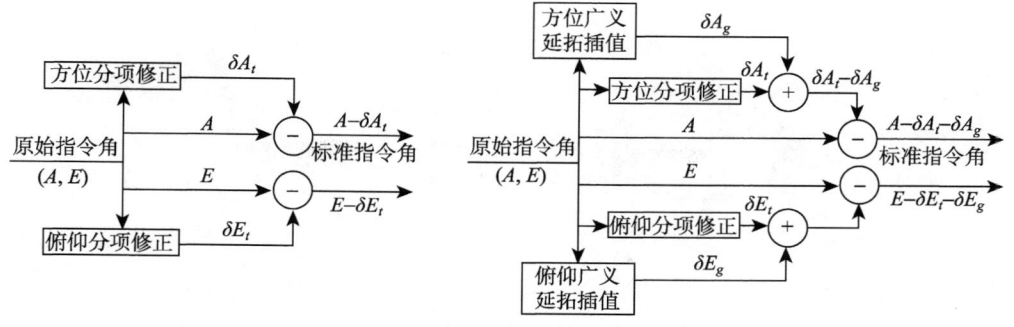

图 2 线性修正和广义延拓插值软件校准模块

3 计算结果与比较

3.1 实验样本说明

笔者对中国科学院国家天文台新建设的 50 m 射电望远镜,在 2006 年 7 月份进行的全天域指向校准实验数据进行了基于广义延拓插值的指向校准模型的计算与分析。该实验选取已知方向的有较强辐射的射电点源,借助 VLBI 数据接收系统接收辐射信号,运用天线控制软件引导天线跟上源、偏开方位、扫描、偏开俯仰、扫描,进行扫描法测量。实验具有有效数据样本 157 个,测量点天区分布如图 3 所示。

3.2 试验数据处理结果

观测点在球面内的 Delaunay 三角剖分结果如图 4 所示。图 5 为方位、俯仰指向误差修正曲面,从图中可以看出,基于广义延拓插值的修正技术可以实现观测域中各片的光滑连接,并能体现区域的细节,降低线性校正时的非线性误差。

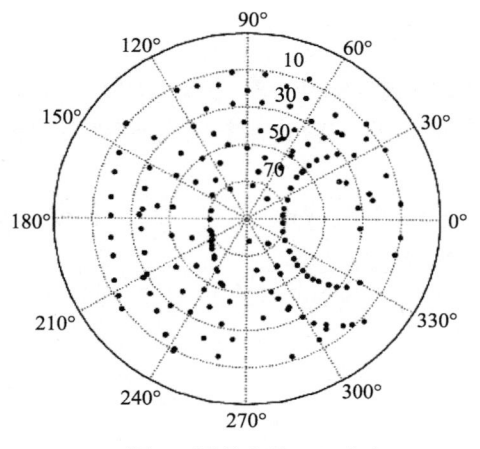

图 3 测量点的天区分布

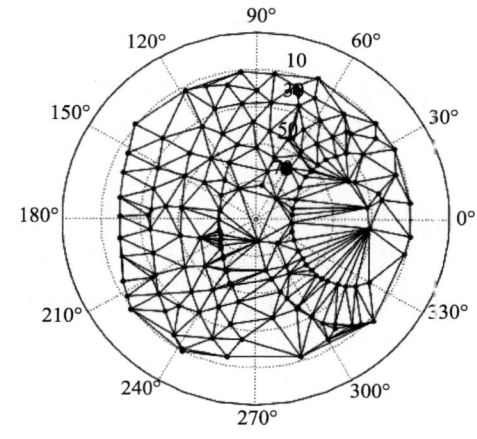

图 4 测量域的 Delaunay 三角剖分

将观测数据随机分为辨识样本和预测样本,其中辨识样本用于修正模型参数的辨识,预测样本用于对修正模型性能进行验证,而不参与修正模型参数的辨识。射电望远镜指向误差的广义延拓插值修正和线性修正效果如表 1 所示。从表中可以看出,基于广义延拓插值的修正后误差(rms)明显优于线性修正,在不同的辨识样本点数下,前者修正后误差比后者修正后误差分别减少 30.3%,20.5% 和 15.6%;同时,由上述推导可知在辨识样本点附近,基于广义延拓插值修正后的误差为观测误差。由此可知,广义延拓插值修正法在观测域上可以获得非常高的指向精度。另外,广义延拓插值修正误差随着辨识样本点的减少而增大。所以,测量数据应尽可能得多,并应分布均匀。

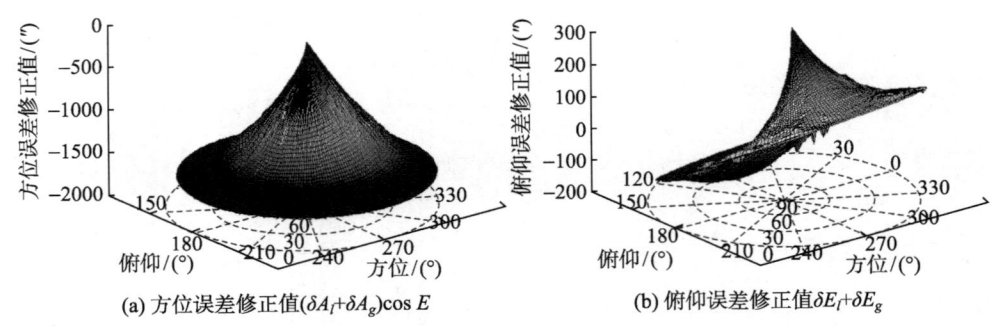

图5 方位、俯仰指向误差修正曲面

表 望远镜指向误差修正效果比较 (单位″)

项目	广义延拓插值修正误差	线性修正误差	未校准前误差	辨识样本点数	预测样本点数
方位	12.0	17.8	1586.9		
俯仰	13.9	19.5	173.5	147	20
合成	18.4	26.4	1596.3		
方位	14.6	16.2	1601.6		
俯仰	12.1	17.6	154.1	127	40
合成	19.0	23.9	1609.0		
方位	13.6	16.0	1565.6		
俯仰	15.3	18.4	157.3	107	60
合成	20.5	24.3	1573.5		

注：1) 表中修正误差为通过辨识样本辨识出的修正模型在预测样本点处相对实测值的偏差。
2) 每组中两种校正方法所用的样本点相同。

4 结束语

笔者提出的基于广义延拓插值的射电望远镜指向校准模型将测量域进行最优三角剖分，在每个区域内分别对指向残差进行修正，理论上能够全面考虑外部因素对天线指向的影响，并能部分减少指向的非线性偏差。

从对实验数据的计算与分析结果看，广义延拓插值修正模型能够较好地预测天线指向偏差，精度比传统的线性修正模型更高。为提高误差修正精度，应进行全方位的射电观测。

考虑多维(三维及以上)模型，广义延拓插值修正模型应可用于综合环境影响因素方面的指向修正，理论推导及实验验证有待进一步研究。

参 考 文 献

[1] Levy R. Structural Engineering of Microwave Antennas [M]. New York：IEEE Press, 1996
[2] Meeks M L. The Pointing Calibration of the Haystack Antenna [J]. IEEE Antennas and Propagation, 1968, 1(16):746—751
[3] 王绶琯，吴盛殷. 射电天文方法[M]. 北京：科学出版社, 1988
[4] Ott M. Witzel A. Quirenbach A. An Updated List of Radio Flux Density Calibrators [J]. Astronomy and Astrophysics.1994, 284(1):331—339
[5] 袁慧仁，彭云楼. 天线参数的射电天文测量[M]. 北京：电子工业出版社, 1987
[6] 施浒立，颜毅华. 工程科学中的广义延拓逼近法[M]. 北京：科学出版社, 2005
[7] Dawy'er R A. A Fast Divide-and-conquer Algorithm for Constructing Delaunay Triangulation [J]. Algorithmica,1987,(2):137—151

"嫦娥"卫星绕月飞行轨道的激光测定法

刘基余

(武汉大学测绘学院，武汉 430079)

摘 要 依据地月激光测距的成功实践和对卫星激光定轨的基础研究，提出了用地面对嫦娥卫星作激光测距的方法，高精度地测定嫦娥卫星绕月飞行时的实时在轨位置，论述了多站激光定轨和单站激光定轨的解算数模。

关键词 嫦娥卫星 绕月飞行 激光定轨 卫星激光测距

我国的探月计划"嫦娥工程"将分为"绕、落、回"三步走：第一步"绕"是在距离月球200 km的极月圆轨道上，"嫦娥一号"卫星对月球进行绕月探测，获取月球三维影像图；第二步"落"是向月球发送月球探测系统，当它安全降落在月球表面后，实施行走区域探测；第三步"回"是先后向月球发送月球返回探测系统和载人登月探测系统。

2004年4月12日，欧阳自远院士在中国地质大学空间技术科学研讨会上说，我国"嫦娥一号"卫星的首要目的是为月球"画像"，也就是要通过各种手段获取月球表面影像和立体图像。据悉，"嫦娥一号"卫星将携带立体相机、成像光谱仪、激光高度计、微波辐射计、太阳宇宙射线检测器和低能离子探测器等多种科学仪器，绕月飞行探测。前三种仪器主要用于为月球"画像"。依笔者之见，即使三种仪器能够采集到精细的画像数据，如果没有精确的卫星绕月飞行实时在轨点位作数据处理基准，也很难精确地画出月球的像。因此，"嫦娥一号"卫星绕月飞行探测定轨将成为能否为月球精确画像的关键技术。

1 "嫦娥"卫星激光定轨的技术基础

1969年7月21日，美国Apollo（阿波罗）-11宇航员Neil Armstrong和Buzz Aldrin实现了人类有史以来的第一次成功登月，并在月球表面上安置了一个用于测量距离的激光后向反射镜阵列。同年8月1日，美国加州大学的Lick天文台利用美国第二架口径最大的120 in望远镜和10 ns、1 J激光脉冲，成功地测量了地球和月亮之间的距离为（383911.218 ± 0.045）km。其后，美国和前苏联分别在月球表面上布设了Apollo-14、Apollo-15和Luna-17、Luna-21等4个激光后向反射镜阵列。且知，Apollo-11至Apollo-14的距离为1250 km，Apollo-11至Apollo-15的距离为970 km，Apollo-14至Apollo-15的距离为1100 km，Apollo-11至Luna-21的距离为760 km，Apollo-15至Luna-17的距离为820 km。目前，国际上有下列台站经常进行着地月激光测距(统称激光测月)：美国Texas州的McDonald天文台、美国Hawaii州的Haleakala天文台、法国的Grasse观测站、澳大利亚的Orrorral观测站和德国的Wettzell观测站。地月激光测距成果不仅用于探求天体物理学中的引力常数是否随时间而变化，而且还用于研究天体演化与地球动力学，编算精密月亮历表，研究月亮物理天平动和探究地球自转与地极移动规律。这也为笔者的下述建议奠定了坚实的技术基础（详见文献[1]）。

本文原载于《武汉大学学报：信息科学版》，2005，Vol. 30，No. 10，870~872。

2 "嫦娥"卫星的 SLR 多站观测激光定轨法

如上所述,地月激光测距从 20 世纪 60 年代末期开始一直延伸到今天,其测距精度已从刚开始的几十 m 提高到了 cm 级,而能够精确测定月球远离地球的速率为 3.8 cm/a,因此笔者建议,采用多台地面激光测距仪同时测量至"嫦娥"卫星的距离,从驻留轨道开始,对它作全程飞行的激光测距,依此精确解算出"嫦娥"卫星在轨飞行的全程实时位置,为测绘月球表面起伏图的图像处理提供基准数据。

图 1 表示多站激光测距定轨的基本构成及其工作原理。在地面 3 个 SLR 测站(A, B, C)上,各设置一台测程可达 410000 km 的卫星激光测距仪,它们的站坐标是精确已知的,且分别为(X_a, Y_a, Z_a)、(X_b, Y_b, Z_b)和(X_c, Y_c, Z_c),时元 t,"嫦娥"卫星绕月飞行的在轨点位坐标记作($X_{CH}(t)$, $Y_{CH}(t)$, $Z_{CH}(t)$)。下述数学论证表明,只要各台卫星激光测距仪同时测得至"嫦娥"卫星的距离,就能够精确解算出时元 t 时"嫦娥"卫星绕月飞行的在轨三维坐标($X_{CH}(t)$, $Y_{CH}(t)$, $Z_{CH}(t)$),其测量精度不低于 m 级。"嫦娥"卫星绕月飞行的激光定轨的难点是测程可达 410000 km 的卫星激光测距仪。我国现有的武汉、上海、长春、北京和昆明等 5 个激光测卫站上的卫星激光测距仪,其最大测程仅达 20000 km 左右,都不具备测量"嫦娥"卫星绕月飞行的能力。它们必须经过重大改造,才可具有激光测月能力。

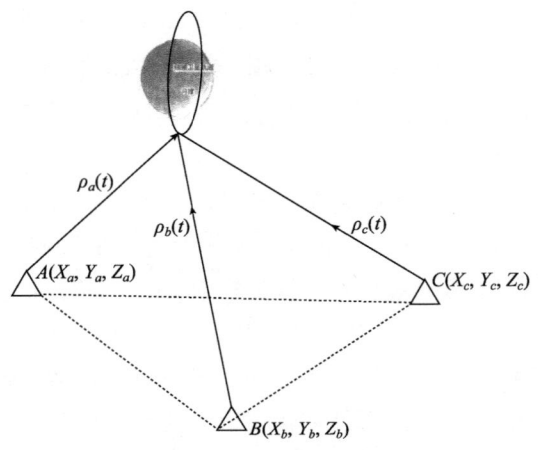

图 1 多站激光测距定轨原理图示

地面 3 个 SLR 测站(A, B, C)上的卫星激光测距仪在时元 t 测得的"嫦娥"卫星绕月飞行时的距离 $\rho_a(t)$、$\rho_b(t)$、$\rho_c(t)$ 分别为:

$$\begin{aligned}\rho_a(t) &= \sqrt{[X_a - X_{CH}(t)]^2 + [Y_a - Y_{CH}(t)]^2 + [Z_a - Z_{CH}]^2} \\ \rho_b(t) &= \sqrt{[X_b - X_{CH}(t)]^2 + [Y_b - Y_{CH}(t)]^2 + [Z_b - Z_{CH}]^2} \\ \rho_c(t) &= \sqrt{[X_c - X_{CH}(t)]^2 + [Y_c - Y_{CH}(t)]^2 + [Z_c - Z_{CH}]^2}\end{aligned} \quad (1)$$

对式(1)进行线性化,求得"嫦娥"卫星绕月飞行时的三维坐标修正矩阵为:

$$X = A^{-1}B \quad (2)$$

式中,

$$X = [\Delta X_{CH}(t) \ \Delta Y_{CH}(t) \ \Delta Z_{CH}(t)]^{\mathrm{T}}$$

$$A = \begin{bmatrix} \dfrac{X_a - X_{CH0}}{\rho_{a0}} & \dfrac{Y_a - Y_{CH0}}{\rho_{a0}} & \dfrac{Z_a - Z_{CH0}}{\rho_{a0}} \\ \dfrac{X_b - X_{CH0}}{\rho_{b0}} & \dfrac{Y_b - Y_{CH0}}{\rho_{b0}} & \dfrac{Z_b - Z_{CH0}}{\rho_{b0}} \\ \dfrac{X_c - X_{CH0}}{\rho_{c0}} & \dfrac{Y_c - Y_{CH0}}{\rho_{c0}} & \dfrac{Z_c - Z_{CH0}}{\rho_{c0}} \end{bmatrix}$$

$$B = [\rho_{a0} - \rho_a \quad \rho_{b0} - \rho_b \quad \rho_{c0} - \rho_c]^T$$

$$\rho_{a0} = \sqrt{(X_a - X_{CH0})^2 + (Y_a - Y_{CH0})^2 + (Z_a - Z_{CH0})^2}$$

$$\rho_{b0} = \sqrt{(X_b - X_{CH0})^2 + (Y_b - Y_{CH0})^2 + (Z_b - Z_{CH0})^2}$$

$$\rho_{c0} = \sqrt{(X_c - X_{CH0})^2 + (Y_c - Y_{CH0})^2 + (Z_c - Z_{CH0})^2}$$

若给嫦娥卫星装备 GPS 信号接收机，不仅能够实现驻留轨道 m 级精度的自主定轨，而且能够为激光定轨提供解算所需的初始值(X_{CH0}, Y_{CH0}, Z_{CH0})，进而求得：

$$\begin{cases} X_{CH}(t) = X_{CH0} + \Delta X_{CH}(t) \\ Y_{CH}(t) = Y_{CH0} + \Delta Y_{CH}(t) \\ Z_{CH}(t) = Z_{CH0} + \Delta Z_{CH}(t) \end{cases} \quad (3)$$

卫星激光定轨已在高达 20 000 km 的导航卫星（GPS/GLONASS 卫星）和高达 1 000 km 左右的对地观测卫星上成功地进行了工程实用。如 1992 年 8 月 10 日，美国航空航天局（NASA）和法国国家空间研究中心（CNES）联合发射的 Topex/Poseidon 海洋测高卫星，它要求其径向误差在 ± 13 cm 以内，而采用了星载 GPS 定轨和激光定轨并行的实施方案。其结果表明，径向误差为 3~4 cm，法向误差为 5~10 cm，切向误差为 9~16 cm。2001 年 10 月 7 日，NASA 和 CNES 又联合发射了 Topex/Poseidon 海洋测高卫星的后续卫星——JASON-1，该颗卫星的有效载荷与 Topex/Poseidon 卫星相似，仅将星载 Monarch GPS 信号接收机改换为性能更优的 Black Jack GPS 信号接收机。2002 年 3 月 17 日，NASA 和德国地学研究中心（GFZ）合作发射了两颗重力测量与气候科研设备卫星，分别命名为 GRACE A 卫星和 GRACE B 卫星，从而构成 GRACE 卫星重力测量系统。GRACE A 卫星和 GRACE B 卫星相距 220 km，它们的轨道高度为 500 km，而于近极共面轨道上飞行。GRACE 卫星和德国的 CHAMP 卫星一样，均装备了 Black Jack GPS 双频接收机和星载激光后向反射镜阵列，实施星载 GPS 测量和激光并行定轨。上述卫星激光定轨的工程实践表明，卫星在轨的实时点位坐标的测量精度达到了 cm 级。因此，笔者提出的用多台地面激光测距仪同时测量至绕月飞行的"嫦娥"卫星距离，进而精确解算出"嫦娥"卫星在轨实时位置，是能够达到 m 级甚至更高的测量精度的。

3 "嫦娥"卫星的 SLR 单站观测激光定轨法

若一个 SLR 测站不断地测得至"嫦娥"卫星的站星距离，则可以求得下列误差方程式：

$$V = \frac{\partial \rho}{\partial a}\Delta a + \frac{\partial \rho}{\partial e}\Delta e + \frac{\partial \rho}{\partial i}\Delta i + \frac{\partial \rho}{\partial \Omega}\Delta\Omega + \frac{\partial \rho}{\partial \omega}\Delta\omega + \frac{\partial \rho}{\partial M_0}\Delta M_0 + (\rho^c - \rho^m) \quad (4)$$

式中，a 为卫星轨道的长半轴；e 为卫星轨道的偏心率；i 为卫星轨道的倾角；Ω 为卫星轨道的升交点赤径；ω 为卫星的近地点角距；M_0 为卫星在时元 t_0 的平近点角。

依式(4)可以构建下列 6 个法方程式：

$$[k(a)k(a)]\Delta a + [k(a)k(e)]\Delta e + [k(a)k(i)]\Delta i + [k(a)k(\Omega)]\Delta\Omega + [k(a)k(\omega)]\Delta\omega$$
$$+ [k(a)k(M_0)]\Delta M_0 + [k(a)l]\Delta l = 0$$
$$[k(e)k(a)]\Delta a + [k(e)k(e)]\Delta e + [k(e)k(i)]\Delta i + [k(e)k(\Omega)]\Delta\Omega$$
$$+ [k(e)k(\omega)]\Delta\omega + [k(e)k(M_0)]\Delta M_0 + [k(e)l]\Delta l = 0$$
$$[k(i)k(a)]\Delta a + [k(i)k(e)]\Delta e + [k(i)k(i)]\Delta i + [k(i)k(\Omega)]\Delta\Omega +$$
$$[k(i)k(\omega)]\Delta\omega + [k(i)k(M_0)]\Delta M_0 + [k(i)l]\Delta l = 0$$
$$[k(\Omega)k(a)]\Delta a + [k(\Omega)k(e)]\Delta e + [k(\Omega)k(i)]\Delta i + [k(\Omega)k(\Omega)]\Delta\Omega +$$
$$[k(\Omega)k(W)]\Delta W + [k(\Omega)k(M_0)]\Delta M_0 + [k(\Omega)l]\Delta l = 0$$

$$\begin{aligned}&[k(W)k(a)]\Delta a+[k(\omega)k(e)]\Delta e+[k(\omega)k(i)]\Delta i+[k(\omega)k(\Omega)]\Delta\Omega+\\&[k(\omega)k(\omega)]\Delta\omega+[k(\omega)k(M_0)]\Delta M_0+[k(\omega)l]\Delta l=0\\&[k(M_0)k(a)]\Delta a+[k(M_0)k(e)]\Delta e+[k(M_0)k(i)]\Delta i+[k(M_0)k(\Omega)]\Delta\Omega+\\&[k(M_0)k(\omega)]\Delta W+[k(M_0)k(M_0)]\Delta M_0+[k(M_0)l]\Delta l=0\end{aligned} \quad (5)$$

式中，

$$\partial\rho/\partial M_0=k(M_0)=(a/r)^2\sqrt{1-e^2}k(\omega)+\frac{1}{r}ae^2k(r)\sin E;$$

$$\partial\rho/\partial a=k(a)=\frac{r}{a}k(r)-\frac{3n}{2a}(t-t_0)k(M_0);$$

$$\partial\rho/\partial e=k(e)=\frac{1}{1-e^2}\left(2\sin f+\frac{e}{2}\sin 2f\right)k(\omega)-ak(r)\cos f;$$

$$\partial\rho/\partial\omega=k(\omega)=\frac{1}{\rho}[(XY_S+X_SY)\cos i+(XZ_S-X_SZ)\cos\Omega\sin i+(YZ_S-Y_SZ)\sin\Omega\sin i];$$

$$\partial\rho/\partial\Omega=k(\Omega)=\frac{1}{\rho}(Y_SX+X_SY);$$

$$\partial\rho/\partial i=k(i)=\frac{Z_S}{\rho}(-X\sin\Omega+Y\cos\Omega-Z\cot i);$$

$$l=\rho^c-\rho^m$$

依式(5)可解得"嫦娥"卫星 6 个轨道参数（根数）的修正量，进而求得"嫦娥"卫星的 6 个轨道参数：$a=a_0+\Delta a$，$\Omega=\Omega_0+\Delta\Omega$，$e=e_0+\Delta e$，$\omega=\omega_0+\Delta\omega$，$i=i_0+\Delta i$，$M_0=M_{00}+\Delta M_0$。

4 结语

笔者依据地月激光测距的成功实践和对卫星激光定轨的基础研究，提出了采用多台地面激光测距仪同时测量至"嫦娥"卫星的距离，从驻留轨道开始，对它作全程飞行的激光测距，依此精确解算出"嫦娥"卫星在轨飞行的全程实时位置，特别是绕月飞行的"嫦娥"卫星的三维坐标，为测绘月球表面起伏图的图像处理提供基准数据，并论证了多站和单站激光定轨的解算数模。依据现有的国外激光测月成果推论，并考虑换算到月球坐标系的精度损失，"嫦娥"卫星在轨实时位置的激光定轨精度不会低于 m 级。

参 考 文 献

[1] 刘基余. "嫦娥"卫星绕月飞行的激光定轨建议. 卫星应用简报，2004, (23):1—4
[2] Bender P L, Currie D G, Dicke R H, et al. The Lunar Laser Ranging Experiment. Science, 1973, 182(4109): 229—238
[3] Lunar Laser Ranging. http://www.lpi.usra.edu/expmoon/Apollo 14/A14 Experiments LRRR. htm, 2004
[4] 刘基余. GPS 卫星导航定位原理与方法. 北京：科学出版社，2003—433
[5] Liu J Y. Satellite Laser Ranging Errors. The 6th International Workshop on Laser Ranging Instrumentation, Antibes, France, 1986

嫦娥一号卫星热控设计中热管的应用及验证

邵兴国　向艳超　谭沧海

(北京空间飞行器总体设计部，北京　100094)

摘　要　为克服由于月球热环境的特殊性给热控设计带来的困难，尤其是卫星度过月食的极端状态条件，首次采用了舱外两舱热耦合热管、相变材料热管技术，为最终嫦娥一号卫星热控状态满足总体的技术要求发挥了关键作用。由于两舱热耦合技术的采用，两舱的热能量得到了相互补偿，因此减少了整星散热面，减少了热补偿功率需求，提高了月食结束时蓄电池的温度，使热控技术方案成为相对优化的方案。文章对热管技术在嫦娥一号卫星热控设计中的应用进行了总结，并给出了热分析及热平衡。

关键词　嫦娥一号卫星　热管　舱段间热耦合　应用

1 前言

嫦娥一号卫星的热环境较地球卫星热环境要恶劣得多，其数传设备热耗波动大、有效载荷工作温度要求苛刻、能源系统资源紧张等，均给热控系统设计带来很大困难，同时也带来了新的挑战。

卫星在特定轨道、姿态条件下运行时，在太阳、月球红外热流的共同作用下，卫星不存在有外热流相对稳定舱面可用作散热面。另外，卫星在运行过程要经历月食，月食期间，由于能源系统的限制，热控系统可用的资源很少，因此，卫星还要经受月食低温环境的考验[1]。为了适应月球热环境和星上资源的限制，满足星上设备对热控的需求，热控系统采用了特殊的热管系统，形成±Y舱板热管耦合网络系统，成功解决了卫星的热控难题。

本文着重介绍±Y舱热管耦合网络系统和成像光谱仪使用热管的设计和试验。

2 ±Y舱热管耦合网络设计

嫦娥一号卫星绕月球两极飞行，在一年的寿命期间内，β角(太阳矢量与轨道面的夹角)在0°~360°变化；另外，卫星在轨运行期间需要根据能源的需要进行姿态调整，即当β角在−45°~＋45°、−135°~＋225°范围内时卫星采用正飞姿态运行(见图1a)；当β角为其他角度时，卫星采用侧飞姿态运行(见图1b)。

根据卫星的飞行姿态，卫星＋Y舱面(或−Y舱面)与阳光矢量的最大夹角为45°，在一侧最大光照条件下，两侧舱板吸收热流之差达213 W/m²(见图2)；而在月食期间，蓄电池组的电补偿功率有限，电池需要保温。所有这些都对安装在±Y舱板内侧的工作温度要求严格的蓄电池组的热控设计提出了严峻挑战。

如此大的外热流差异，造成两侧蓄电池组的热环境有很大的不平衡，当一侧的电池舱受到太阳照射和月球红外热流的共同影响时，温度会很高，而另一侧的电池处于太阳的阴影区，月球红外影响也相对较小，温度会很低，需要进行电功率补偿，图3给出了未安装热管时的蓄电池组的温度；而当卫星进入月食过程时，能源系统不能提供足够的电功率维持电池自身要求的温度指标，因此需要尽可能减少散热面面积。

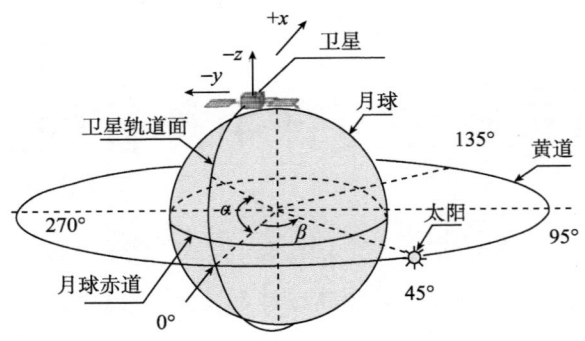

(a) 环月正飞太阳、月球、卫星关系

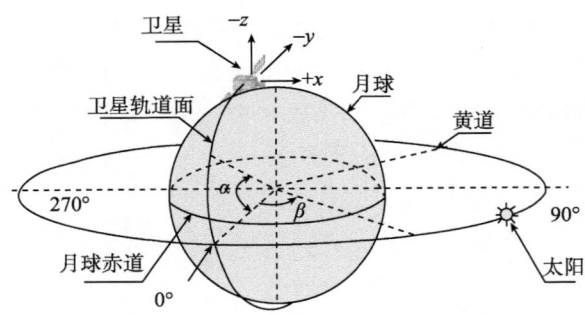

(b) 环月侧飞太阳、月球、卫星关系

图 1　太阳、月球、卫星关系

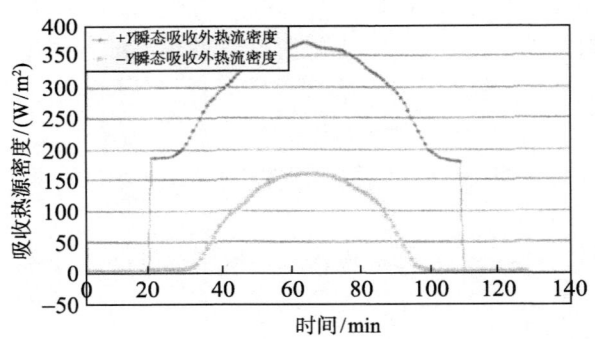

图 2　±Y 舱外热流

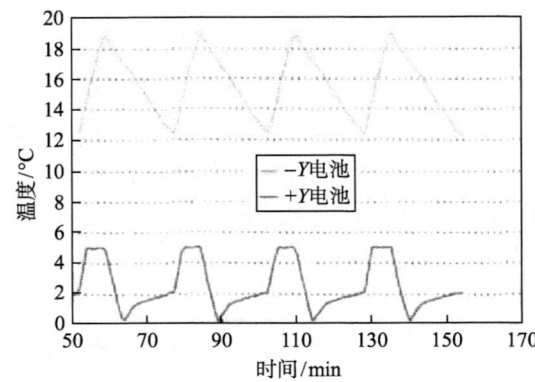

图 3　未装热管时两舱电池温度

综合上述考虑，利用热管进行±Y 舱板的热耦合设计，形成两舱板热管耦合体系，改善蓄电池组在高温工况下温度水平，同时减少散热面，解决月食过程的降温问题。

2.1 轴向槽道热管方案

热管耦合网络技术可以采用环路热管技术和传统的轴向槽道热管技术两种方案。热耦合设计的目的是解决电池舱的高温问题和月食期间的低温问题，不同的技术途径达到的目标一致，但各自的技术特色不同，按照技术的成熟度和可靠性因素考虑，最终选用轴向热管方案，即用3根外贴轴向槽道热管将+Y、-Y舱板热耦合在一起，详见图4所示。采用轴向槽道热管进行热耦合的不利因素是在地面试验验证时，由于重力的影响，会导致一个方向的性能不能得到试验验证。

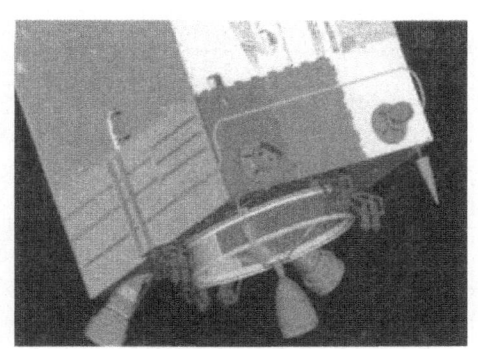

图4 热管耦合方案

2.2 热耦合设计热分析

针对上述热管耦合系统进行了热分析。图5给出了高温工况下蓄电池组的温度变化曲线，表1给出了不同工况下蓄电池组的温度结果。

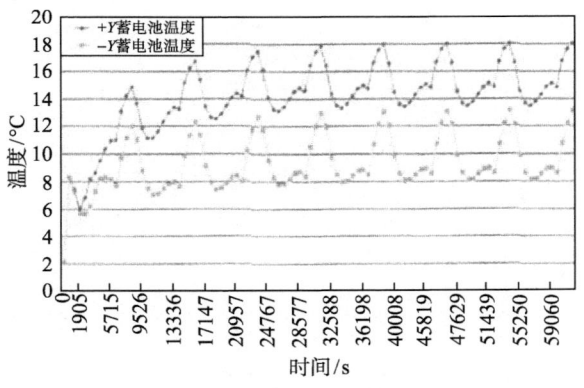

图5 增加热管后电池温度变化

表1 计算结果　　　　　　　　　　(单位：℃)

温度点	高温工况			月食工况
	MAX	MIN	温差	结束温度
1	18.4	13.9	4.5	-10.4
2	19.3	14.8	4.7	-8.3
3	17.9	13.4	4.5	-10.7
4	16.9	12.4	4.5	-10.2
5	16.4	11.9	4.5	-11.6

从分析结果可以看出，在高温工况下，+Y、−Y 蓄电池组之间的温差明显减小，蓄电池组自身的温度波动也有较为明显的改善；在月食工况，由于在热管耦合后减小了蓄电池组的散热面，因此蓄电池出月食的温度也由原来的−18℃提高到−10℃左右。分析结果表明蓄电池组的热控设计方案是有效的。

2.3 整星热平衡试验结果

正样真空热试验两侧蓄电池均为正样产品，蓄电池在主动控温作用下，正常工况时，蓄电池组低温被限制在 5℃以上，蓄电池组工作在较好的温度环境中；月食结束时蓄电池温度高于−10℃，详见图 6。表 2 给出了热平衡试验中蓄电池组的温度。

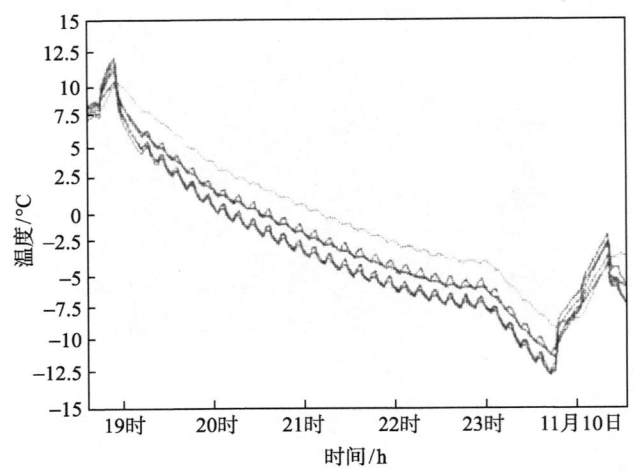

图 6　电池温度波动状态(月食)

表 2　±Y 电池整星热试验的温度值　　　　　　　　　　(单位：℃)

−Y 测点	低温工况 1	低温工况 2	高温工况 1		高温工况 2		+Y 测点	低温工况 1	低温工况 2	高温工况 1		高温工况 2	
			高温	低温	高温	低温				高温	低温	高温	低温
1	5.5	6.7	7.6	4.9	10.1	6.3	1	5.1	6.7	10.3	6.0	11.3	7.8
2	7.0	8.2	8.8	5.8	10.0	6.1	2	6.0	7.8	10.9	6.6	10.7	7.5
3	7.4	8.5	9.5	6.4	9.8	6.1	3	6.0	7.9	10.6	6.3	10.5	7.1
4	6.7	7.9	8.5	5.8	9.5	5.8	4	5.1	7.0	10.0	5.4	10.2	6.5
5	6.7	7.9	8.8	6.1	8.5	4.9	5	5.4	7.0	9.4	6.3	9.7	6.3
单体温差	1.2	1.5	1.9	1.5	1.6	1.4	单体温差	0.9	1.2	1.6	1.2	1.5	1.5

2.4 热管工作情况

热平衡试验过程中，三根 U 型热管工作正常。图 7 给出了±Y 舱板中对应的预埋热管的温度变化曲线，从图中可以看出两舱板中的预埋热管的温差小于 3℃，这说明耦合热管网络系统作用显著。图 8、图 9 给出了热平衡试验中两组蓄电池组的温度变化曲线，从图中可以看出，由于耦合热管网络系统的作用，两组蓄电池组的温差明显减小，温度波动变缓，蓄电池组的热环境得到明显改善。

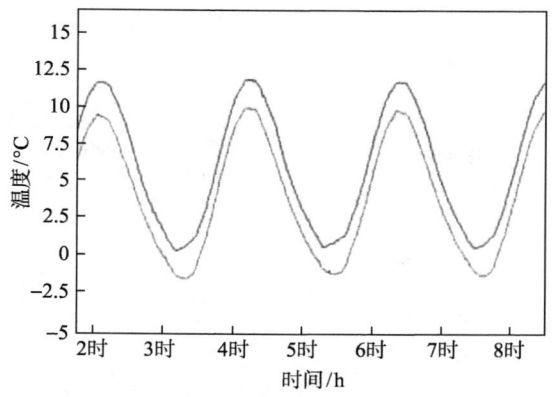

图 7 ±Y舱板内预埋热管温度波动状态

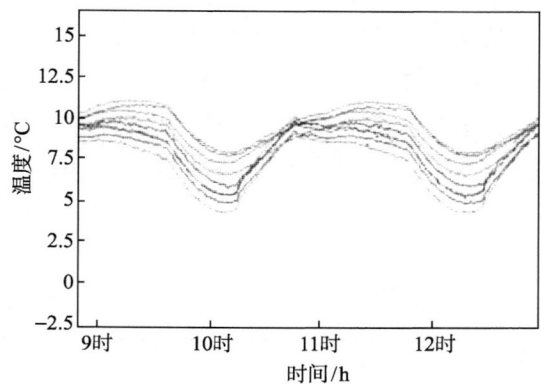

图 8 电池温度波动状态($\beta=-22.5°$)

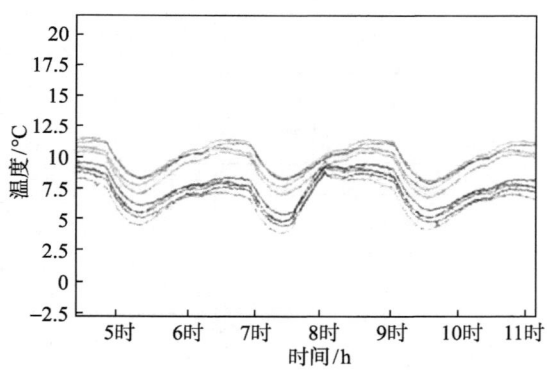

图 9 电池温度波动状态($\beta=-45°$)

3 X频段固态放大器热控修正设计及试验

3.1 修正原因及改进方案

X频段固态放大器是短期工作设备，由于设备短期热耗较大，造成设备短期工作时温度波动剧烈，峰值接近设备允许的工作温度上限，图10给出了设计修正前热平衡试验过程中该设备温度变化曲线，其峰值接近50℃。

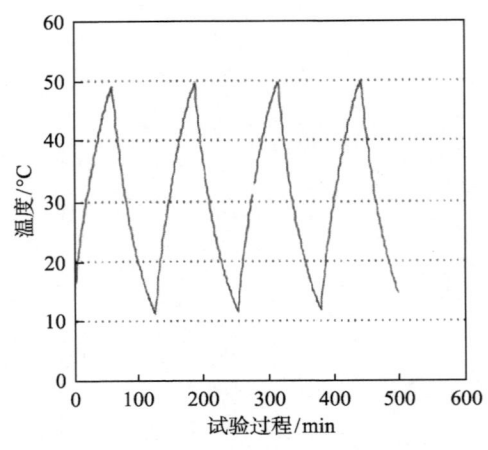

图 10　更改前 X 频段固态放大器温度波动

为了能够有效地平抑设备的温度波动幅度，同时又不增加设备不工作时的热补偿功率，在后续的修正设计中使用了热管，即在设备下方增加了两根 9×18 的预埋热管，热管安装在舱板内侧，舱板外侧包覆多层隔热组件，如图 11 所示。通过热管良好的导热性能，增强设备与卫星机构及其内部的热耦合关系，增大设备的热惯性，降低设备温度波动幅度；同时由于没增加散热面面积，因此没有增加设备不工作时的热补偿功率。

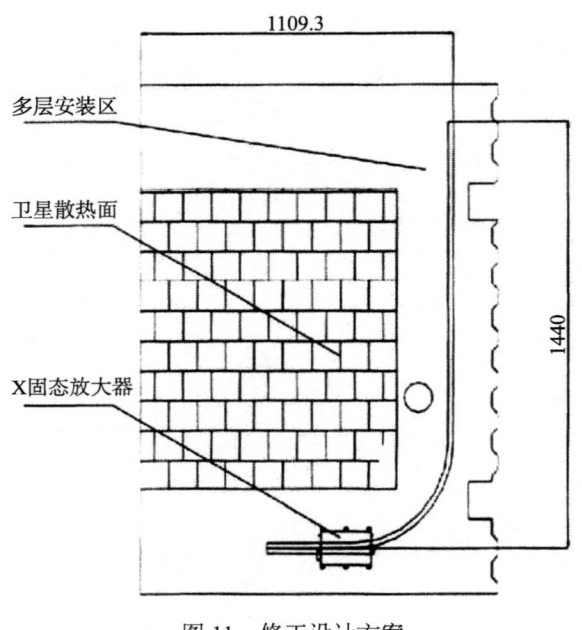

图 11　修正设计方案

3.2　试验验证

修正后的设计参加了热平衡试验，图 12 给出了 X 频段固态放大器的温度变化曲线。从图中可以看出，增加热管后设备温度波动幅度由更改前的超过 30℃降低到更改后的不足 5℃，设备的温度波动得到了有效抑制，这表明了增加预埋热管其改进设计的有效性。

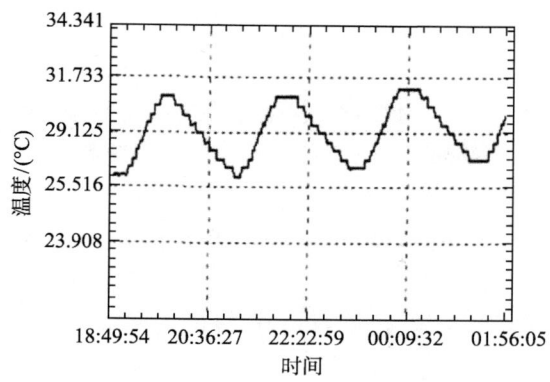

图 12 更改后试验中 X 频段固态放大器温度波动

4 相变材料热管复合技术应用

在嫦娥一号卫星上，为了规避月球红外热流的影响，安装在对月板处的载荷设备的散热面设在+X 板上，利用热管将 X 板的散热面和散热设备热耦合进行设备的温度控制。+X 板散热面在外热流的作用下，温度有很大的波动(孤立散热面的温度波动 20℃至–20℃)，造成被控区域温度波动幅度较大，高温时温度过高，低温时需要电功率补偿。因此需要采用增大热容设计方法，使被控对象温度波动过大的现象得到纠正。

载荷热控措施最终选定了热管-相变材料方案，既有热管功能，也能够充装相变材料。具体实施如下：相变工质选用正十二烷，相变温度为–10℃；相变材料的充装量不低于 60g；相变材料与热管管体之间应有尽量大的热导[2]。最终完成的相变材料复合热管界面及热管外形见图 13，在中间圆形腔体内充装液氨，作为常规热管使用；两边两个腔体内充装相变材料，腔体中的肋片起到增强热管与相变材料热耦合的作用。

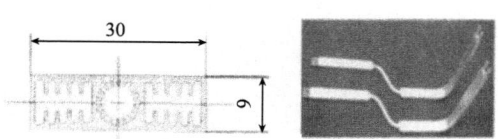

图 13 相变材料复合热管设计和研制

相变材料复合热管的性能在产品热真空试验和载荷舱热平衡试验中得到了验证[3]，图 14 给出了相变材料热管产品在热真空试验时的温度曲线，从图中可以看出热管温度在设定的温度点有明显的温度阻尼作用。采用热管相变材料方案后，载荷的温度波动水平控制在可以接受的温度范围内，满足设计要求并有一定余量[4]。

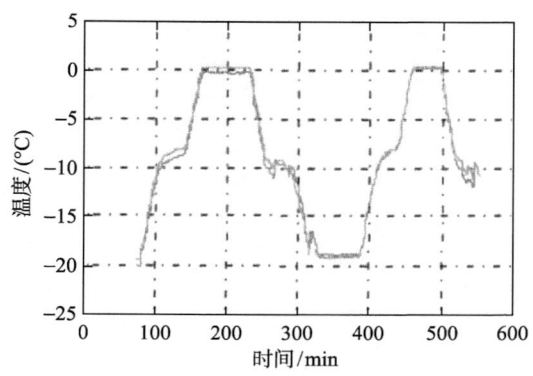

图 14 相变材料热管热试验温度曲线

5 结论

轴向槽道热管工作可靠、性能稳定,为嫦娥一号卫星在复杂外热流环境条件下确保仪器设备正常温度环境起到了较为突出作用。

(1)采取轴向槽道热管两相对舱板间的热耦合技术,为首次在此类卫星上使用,解决了±Y舱内蓄电池组在轨出现的高温问题,同时也为卫星成功度过月食的低温环境提供了根本保证;

(2)相变材料复合热管是热管应用领域中的一次新的尝试,它具有良好热传导性能和控温性能的显著特点,将会使其在未来的空间飞行器热控设计中得到广泛的应用。

参 考 文 献

[1] 曹剑峰. 月食过程月表太阳辐射及温度变化模拟[J]. 航天器工程,2006,15(4)
[2] Gilmore D G. Spacecraft thermal control handbook[M]. EI Segundo CA: the Aerospance Corporation Press,2002
[3] 郭霖,等. 相变材料热管性能的试验研究[C]. 第10届全国热管会议,贵阳,2006
[4] 耿利寅,邵兴国. 嫦娥一号卫星光学成像探测系统焦平面组件热设计[C]. 第七届空间热物理会议,成都,2005

嫦娥一号月球卫星缩短阴影时间的分析与实现

李革非[1,2]　韩潮[1]

（1. 北京航空航天大学宇航学院，北京 100083；2. 北京航天飞行控制中心，北京 100094）

摘要　通过对"嫦娥一号"月球卫星在 2008 年 2 月 21 日月食期间的卫星阴影时间进行分析，明确了卫星轨道参数升交点经度和卫星相位与卫星阴影的密切关系，得出了通过月食前调整卫星相位以缩短卫星阴影时间的方法。"嫦娥一号"卫星利用该方法实施的调相轨道控制，使卫星阴影时间缩短了约 1.5 h，有效地保障了月食期间的卫星安全。

关键词　月食　月球卫星　卫星阴影时间　轨道调相

"嫦娥一号"月球卫星在轨一年的环月飞行期间，月球会发生 2 次月食，分别在 2008 年 2 月 21 日和 2008 年 8 月 17 日。

月食期间"嫦娥一号"卫星将随月球一起进入地影区，出现长时间无光照情况，对整个卫星平台的正常运行造成影响。由于月食时间卫星处于阴影时间较长，为保证卫星安全，希望尽量缩短卫星处于阴影的时间。下面，通过对 2008 年 2 月 21 日月食期间的卫星阴影进行分析，得到缩短卫星阴影时间的解决方法。

1 阴影计算方法

1.1 锥形阴影模型

阴影是指星体或卫星因其他星体遮挡而无法受太阳照射的情形。遮挡物为地球时，阴影称为地影；遮挡物为月球时，阴影称为月影。特别的，月球受地球遮挡出现阴影称为月食，太阳受月球遮挡出现阴影称为日食。

较精确的阴影模型是锥形阴影模型。

不考虑地球大气衰减以及地球扁率效应，从卫星上看太阳、地球、月球的视半径分别为

$$\alpha_s = \arcsin\left(\frac{a_s}{r_{sa}}\right), \quad \alpha_e = \arcsin\left(\frac{a_e}{r_{ea}}\right), \quad \alpha_m = \arcsin\left(\frac{a_m}{r_{ma}}\right)$$

其中，a_s, a_e, a_m 分别为太阳、地球、月球的半径；r_{sa}, r_{ea}, r_{ma} 分别为卫星到太阳、地球、月球的距离。视日心与地心的夹角 θ_{se} 和视日心与月心的夹角 θ_{sm} 分别为

$$\theta_{se} = \arccos\left(-\frac{\vec{r}_{ea} \cdot \vec{r}_{sa}}{|\vec{r}_{ea}||\vec{r}_{sa}|}\right), \quad \theta_{sm} = \arccos\left(\frac{\vec{r}_{ma} \cdot \vec{r}_{sa}}{|\vec{r}_{ma}||\vec{r}_{sa}|}\right)$$

对于地影有：当 $-\vec{r}_{ea} \cdot \vec{r}_{sa} \leq 0$ 时，卫星不在地影中；否则

(1) $\theta_{se} \geq \alpha_e + \alpha_s$ 时，卫星不在地影中；

(2) $\theta_{se} \leq |\alpha_e + \alpha_s|$ 时，卫星在本影或伪本影中；

本文原载于《中国科学 E 辑：技术科学》，2009，Vol.39，No.3，562~567。

(3) $|\alpha_e-\alpha_s|<\theta_{se}<(\alpha_e+\alpha_s)$时，卫星在半影中。

同样，对于月影有：当$\bar{r}_{ma}\cdot\bar{r}_{sa}\leq0$时，卫星不在月影中；否则

(1) $\theta_{sm}\geq\alpha_m+\alpha_s$时，卫星不在月影中；
(2) $\theta_{sm}\leq|\alpha_m+\alpha_s|$时，卫星在月本影或伪月本影中；
(3) $|\alpha_m-\alpha_s|<\theta_{sm}<(\alpha_m+\alpha_s)$时，卫星在月半影中。

1.2 半影和伪本影的阴影因子

将卫星是否在地球或月球的阴影之内用阴影因子F表示，其定义为

(1) $F=1$，卫星在地影和月影之外；
(2) $F=0$，卫星在地球或月球的本影之内；
(3) $0<F<1$，卫星在地球或月球的半影或伪本影之内，F根据太阳被蚀面积的大小计算。

当卫星在地球半影中时，

$$A_e = \alpha_s^2 \cos^{-1}\left(\frac{\beta}{\alpha_s}\right) + \alpha_e^2 \cos^{-1}\left(\frac{\theta_{se}-\beta}{\alpha_e}\right) - \theta_{se}\sqrt{\alpha_s^2-\beta^2}$$

其中，$\beta = \dfrac{\theta_e^2 + \alpha_s^3 - \alpha_e^2}{2\theta_{se}}$

当卫星在地球伪本影中时，$A_e = \pi\alpha_e^2$

当卫星在月球半影中时，

$$A_m = \alpha_s^2 \cos^{-1}\left(\frac{\beta}{\alpha_s}\right) + \alpha_m^2 \cos^{-1}\left(\frac{\theta_{sm}-\beta}{\alpha_m}\right) - \theta_{sm}\sqrt{\alpha_s^2-\beta^2}$$

其中，$\beta = \dfrac{\theta_m^2 + \alpha_s^2 - \alpha_m^2}{2\theta_{sm}}$

当卫星在月球伪本影中时，$A_m = \pi\alpha_m^2$。则阴影因子F为

$$F = 1 - \frac{\max(A_e, A_m)}{\pi\alpha_s^2}$$

阴影因子也称为太阳强度系数。阴影因子越小，太阳照射强度越弱；阴影因子越大，太阳照射强度越强。

2 月食期间的月球卫星阴影分析

2.1 阴影与卫星轨道升交点和相位的关系

月食期间，太阳、地球和月球三者基本处于一条直线上，太阳-地球-月球连线与卫星环月轨道平面的夹角以及卫星在轨道上的位置对卫星处于地影的时间长度均有影响。

太阳-地球-月球连线与轨道平面的夹角影响地影时间长短的变化幅度。夹角为0°或180°时，即连线与轨道平面平行时，无论卫星处于轨道任何位置，卫星基本上与月球同步进入地影，地影时间长短的变化幅度小；夹角为90°或270°时，即连线与轨道平面垂直时，卫星所处轨道的不同位置将使卫星进出地影的时间不同，地影时间长短的变化幅度大。

因此，环月卫星处于地球阴影的时间取决于月食期间环月轨道的升交点赤经和卫星的轨道相位。

另外，卫星绕月飞行期间，在大部分时间内每圈都有月球遮挡造成的月影。月影的时间长度与日月连线与轨道平面的夹角有着十分密切的关系。对于月球卫星200 km的极轨圆轨道，当日月连线与轨道平面平行时，每圈的月影时间最长，在127 min的轨道周期内约有45 min处于月影中；当日月连线与轨道平面

夹角大于63.74°或日月连线与轨道平面法线夹角小于26.26°时，卫星飞行一圈均处于阳照中；其他情况下月影时间随夹角不同而变化。

卫星在月食期间的最长阴影是地影和月影的综合。

2.2 分析算例

下面计算分析"嫦娥一号"月球卫星在2008年2月21日月食期间的阴影时间长度，阴影时间长度是地影和月影的叠加结果。按太阳强度系数0.5计算等效地影区间，即当太阳强度系数小于0.5时，将半影区间归算到阴影区中；当太阳强度系数大于0.5时，将半影区间归算到阳照区中。月影则包括了月球遮挡的本影和半影。地影和月影叠加时，若地影和月影之间间隔时间小于30 min，则将该间隔归算至阴影区间。这样得到卫星在月食期间的最大阴影时间长度。

以2008年2月21日零点(北京时间)为起算时刻，卫星在J2000惯性系的升交点经度从0°以10°为间隔变化至360°，卫星平近点角从0°以10°为间隔变化至360°，计算卫星在2008年2月21日月食期间的阴影时间长度，从而考察月球卫星在月食期间的阴影时间长度与轨道升交点和相位的变化关系。

下面给出了4组图，分别为升交点经度Ω=70°，150°，240°，340°的阴影时间长度随初始平近点角的变化。图1～图4中，DSE为卫星的地影时间，DSEM为地影叠加月影的时间。

图1～图4中表明(在2008年2月21日的月食期间):

(1) 当卫星升交点经度Ω=70°时，日地连线与卫星轨道平面法线的夹角为168.2°，即日地连线近似垂直于轨道平面。当初始轨道平近点角从0°到360°变化时，长地影与短地影之间有跳变，地影最长时间约14460 s(约4 h 1 min)，地影最短时间约9810 s(约2 h 43 min)，两者相差约1 h 18 min，长地影的相位范围约225°。由于日地月连线近似垂直于轨道平面，因此无月影。

(2) 当卫星升交点经度Ω=150°时，日地连线与卫星轨道平面法线的夹角为85.5°，即日地连线近似平行于轨道平面。当初始轨道平近点角从0°到360°变化时，长地影与短地影之间变化幅度较小，地影最长时间约13320 s(约3 h 42 min)，地影最短时间约9890 s(约2 h 45 min)，两者相差约57 min。由于日地月连线近似平行于轨道平面，月影时间最长，约45 min。图中，地影和月影叠加后的最长阴影时间约17960 s(约5 h)，最短阴影时间约12470 s(约3 h 28 min)，两者相差约1 h 32 min，叠加后长阴影的相位范围约100°。

(3) 当卫星升交点经度Ω=240°时，日地连线与卫星轨道平面法线的夹角为10.7°，即日地连线近似垂直于轨道平面，当初始轨道平近点角从0°到360°变化时，长地影与短地影之间有跳变，地影最长时间约13330 s(约3 h 42 min)，地影最短时间约8360 s(约2 h 19 min)，两者相差约1 h 23 min，长地影的相位范围约300°。由于日地月连线近似垂直于轨道平面，因此无月影。

(4) 当卫星升交点经度Ω=340°时，日地连线与卫星轨道平面法线的夹角为86.9°，即日地连线近似平行于轨道平面，当初始轨道平近点角从0°到360°变化时，长地影与短地影之间变化幅度较小，地影最长时间约13520 s(约3 h 45 min)，地影最短时间约9830 s(约2 h 44 min)，两者相差约1 h 1 min。由于日地月连线近似平行于轨道平面，月影时间最长，约有45 min。图中，地影和月影叠加后的最长阴影时间约18010 s(约5 h)，最短阴影时间约10370 s(约2 h 53 min)，两者相差约2 h 7 min，叠加后长阴影的相位范围约195°。

上述图1～图4给出了不同轨道升交点造成的日地连线与卫星轨道平面法线夹角的2种特殊情况，即日地连线近似垂直于轨道平面和日地连线近似平行于轨道平面。日地连线近似垂直于轨道平面时，长地影时长与短地影时长相差约1 h 20 min，由于无月球阴影，卫星阴影随轨道相位变化简单，呈现长-短阴影区间单层变化。日地连线近似平行于轨道平面时，长地影时长与短地影时长相差约1 h，由于存在月球阴影，地影和月影叠加后使得卫星阴影随轨道相位变化复杂，呈现长-短阴影区间多层变化。

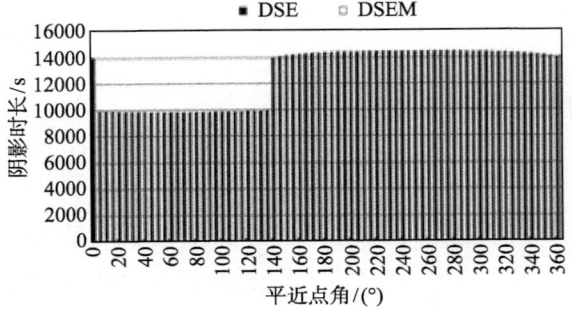

图 1 2008-2-21 月食日月球卫星阴影时间长度($D=70°$)

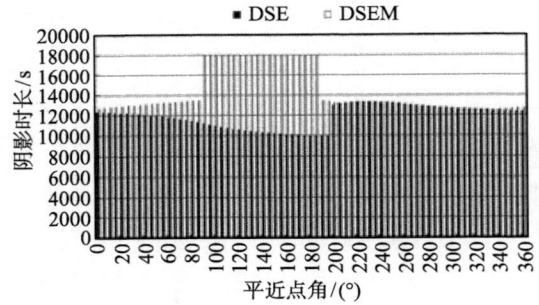

图 2 2008-2-21 月食日月球卫星阴影时间长度($D=150°$)

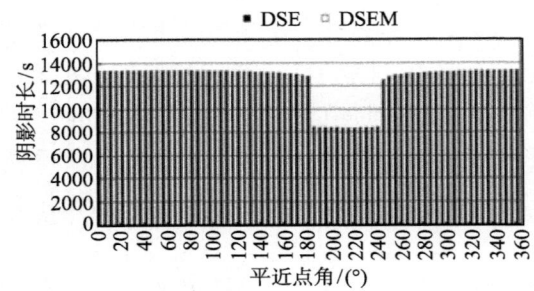

图 3 2008-2-21 月食日月球卫星阴影时间长度($D=240°$)

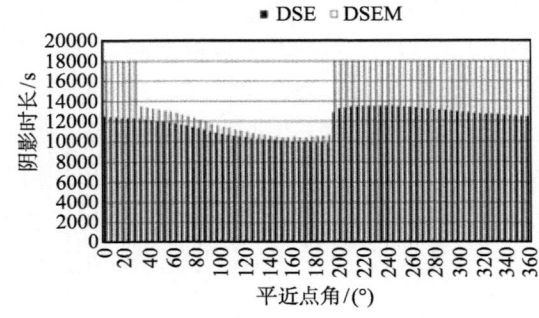

图 4 2008-2-21 月食日月球卫星阴影时间长度($D=340°$)

3 缩短阴影的调相轨道控制

3.1 调相轨道控制分析

月食期间，卫星阴影时间与卫星升交点和卫星相位具有密切的关系。2008年2月21日月食期间，初始轨道不同的相位会使卫星长地影与短地影的时间长度有较大的变化，长地影与短地影之间有跳变，地影最长时间与地影最短时间相差约1h以上。因此，当卫星初始相位使得卫星出现长阴影时间时，通过在月食前调整卫星相位，使卫星长阴影时间缩短为短阴影时间。

轨道相位调整需通过调整轨道半长轴实现。

对于圆轨道，轨道角速度与轨道半长轴的变化关系为

$$\Delta n = -\frac{3}{2} \cdot \frac{n \Delta a}{a}$$

轨道速度与轨道半长轴的变化关系为

$$\Delta v = -\frac{1}{2} \cdot n \Delta a$$

轨道角速度引起轨道相位的变化为

$$\Delta u = \Delta n \cdot \Delta t$$

对于高度200 km环月圆轨道，角速度为

$$n = \sqrt{\frac{\mu_m}{a^3}} = 8.20713e-4 (\text{rad}/s)$$

假设轨道半长轴调整5 km，则轨道角速度改变为

$$\Delta n = 3.17613e-6 (\text{rad}/s)$$

速度增量为

$$\Delta v = 2.05178 (\text{m/s})$$

假设轨道半长轴调整10 km，则轨道角速度改变为

$$\Delta n = 6.35226e-6 (\text{rad}/s)$$

速度增量为

$$\Delta v = 4.10356 (\text{m/s})$$

由于阴影与相位之间关系的复杂性，相位最大调整量为180°。

(1) 轨道半长轴调整5 km时，相位调整所需的时间为

$$\Delta T = \frac{\Delta \phi}{\Delta n} = 989125(\text{s}) \approx 129 (\text{圈})$$

即，为缩短长阴影期为短阴影期，最大相位调整为180°时，若半长轴改变约5 km(升高或降低)，需提前129圈(约11天)进行相位调整，需速度增量2.052 m/s。

(2) 轨道半长轴调整10 km时，相位调整所需的时间为

$$\Delta T = \frac{\Delta \phi}{\Delta n} = 494562(\text{s}) \approx 65 (\text{圈})$$

即，为缩短长阴影期为短阴影期，最大相位调整为180°时，若半长轴改变约10 km(升高或降低)，需提前65圈(约5天)进行相位调整，需速度增量4.104 m/s。

3.2 调相轨道控制实施

月食调相轨道控制分2个步骤：①月食阴影相位区间计算；②月食调相控制计算。

3.2.1 月食阴影相位区间计算

月食阴影相位区间计算是根据初始轨道计算月食日阴影时间,确定月食日短阴影的相位区间$[u_s, u_e]$。

(1) 初始轨道积分至月食日零点,计算月食日零点的轨道相位u_0。使u的变化范围为$u_0 \to u_0+360°$,变化间隔为5°;

(2) 对不同u计算地影、月影,并进行地影、月影叠加,记录月食阴影文件;

(3) 由于阴影关于相位呈多层分布,自动确定短阴影的相位区间算法复杂,因此根据月食阴影文件人工确定短阴影对应的相位区间$[u_s, u_e]$。月食日零点的目标相位取为:

$$u_m = u_s + \frac{u_e - u_s}{2}$$

3.2.2 月食调相控制计算

月食调相控制计算是根据初始轨道和月食日零点目标相位u_m,月食调相轨控时刻T_C,计算月食调相轨道控制脉冲轨控量。其中月食调相轨控时刻T_C可人工指定,也可根据降低轨道或抬高轨道确定在近月点或远月点作为调相轨控时刻。处理过程如下。

(1) 初始轨道积分至月食日零点T_E,计算月食日零点相位u_0。

计算相位调整量: $\Delta u = u_m - u_0$。

(2) 初始轨道积分至轨控时刻: T_C, ORB_{C1}。

轨控点到月食日零点的时间: $\Delta t = T_E - T_C$。

调整相位的半长轴变化量: $\Delta a = -\frac{2a}{3n\Delta t}\Delta u$。

速度增量: $\Delta v = \frac{n}{2}\Delta a$。

脉冲轨控: $ORB_{C1} \to ORB_{C2}$。

ORB_{C2}积分至月食日零点,计算u_{m1}。

相位偏差: $\delta u_m = u_{m1} - u_m$。

根据δu_m计算δa, $\delta a = -\frac{2a}{3n\Delta t}\delta u$。

$\Delta a = \Delta a + \delta a$,迭代计算直到相位偏差$\delta u_m$满足要求。

4 "嫦娥一号"卫星缩短阴影时间

4.1 2008年2月21日月食期间的卫星阴影计算

根据"嫦娥一号"卫星轨道,以2008年2月21日0点(北京时间)为起算时刻。卫星平近点角从0°开始、以5°为间隔变化至360°,计算卫星不同相位在2008年2月21日月食期间的阴影时间(如图5)。

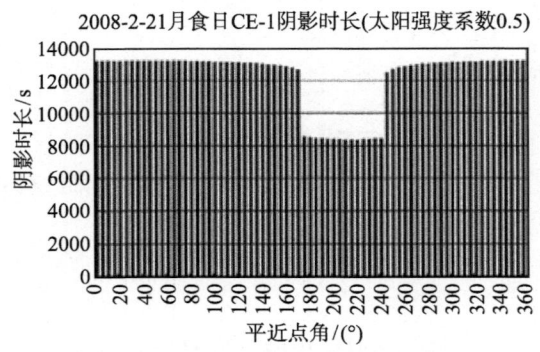

图5 2008-2-21月食期间 CE-1 阴影时长

2008年2月21日月食期间,卫星升交点经度$\Omega=265°$,日地连线与卫星轨道平面法线的夹角约为23.8°,卫星只有地影没有月影。

采用太阳强度系数$K=0.5$进行月食期间的地影计算。当初始轨道平近点角从0°到360°变化时,长地影与短地影之间有跳变,地影最长时间约13253 s(约3 h 41 min),地影最短时间约8392 s(约2 h 20 min),两者相差约1 h 21 min,长地影的相位范围约290°,短地影的相位范围约70°(从175°到245°).

4.2 调相轨道控制计算

根据2008年2月21日月食期间的卫星阴影计算结果,按太阳强度系数$K=0.5$计算,月食日初始相位调整至[175°,245°]范围内,可以使卫星处于短阴影时间。选择初始相位调整210°进行调相轨道控制(实际相位调整−150°)。按照调相轨控后的轨道进行复核计算,卫星在月食日处于短阴影时段。根据2008年2月21日对"嫦娥一号"卫星的测控表明,调相轨道控制后,卫星实际的阴影时间与计算的基本一致。

下面图6和图7给出的是初始相位调整DM=0°时的最长阴影图示和初始相位调整DM=210°时的最短阴影图示。

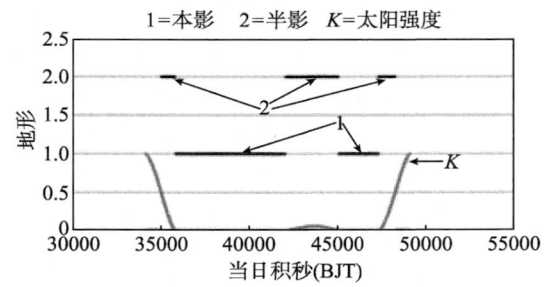

图6 2008-2-21,CE-1 初始相位调整 DM=0 度的地影

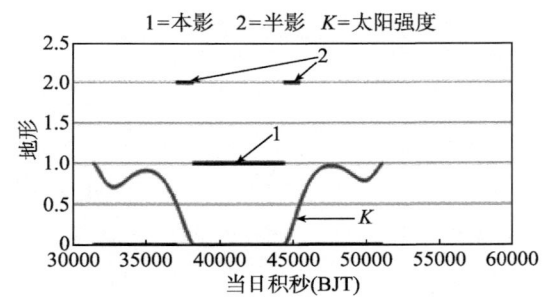

图7 2008-2-21,CE-1 初始相位调整 DM=210 度的地影

相位调整前,卫星地影本影区间有2段,半影区间有3段。中间半影区间的太阳强度系数近似为0,半影区间等同于本影区间;两端半影区间之外,卫星完全进入阳照区。

相位调整后,卫星地影本影区间只有1段,半影区间有2段。两端半影区间之外,卫星仍处于地影半影区间,但太阳强度系数均大于0.7,满足太阳强度系数大于0.5的要求,等同于卫星处于阳照区。

通过相位调整,卫星阴影时间缩短了约1.5 h。

5 结论

(1) 月食期间,月球卫星处于阴影的时间长度与卫星升交点经度和卫星相位具有密切的关系。日地连线近似垂直于轨道平面时,卫星不同相位形成的长地影时长与短地影时长相差约1 h20 min,由于无月球阴影，卫星阴影随相位变化简单，呈现长-短阴影区间单层变化。日地连线近似平行于轨道平面时，卫星不同相位形成的长地影时长与短地影时长相差约 1 h，由于存在月球阴影，地影和月影叠加后使得卫星阴影随相位变化复杂，呈现长-短阴影区间多层变化。

(2) 在月食前，通过轨道控制改变卫星轨道半长轴调整卫星相位，是缩短卫星长阴影时间的有效方法。2008 年 2 月 21 日月食期间，"嫦娥一号"月球卫星进行的调相轨道控制使得卫星阴影时间缩短了约 1.5 h，为卫星安全渡过月食期提供了必要的保障。

参 考 文 献

[1] 郗晓宁, 曾国强, 任萱, 等. 月球探测器轨道设计. 北京: 国防工业出版社, 2001
[2] 李济生. 人造卫星精密轨道确定. 北京: 解放军出版社, 1995
[3] 汤锡生, 陈贻迎, 朱民才. 载人飞船轨道确定和返回控制. 北京: 国防工业出版社, 2002
[4] Bernard K, Jay M, Karen R. Mission design of the clementine space experiment. AAS Paper 95—124
[5] Soyka M T. Clementine: contingency options for the phasing loops used in the lunar transfer. AAS Paper 95—126
[6] David L, Ken G, David F, et al. Lunar prospector mission design and trajectory. AAS Paper 98—323

嫦娥一号卫星数据高可靠性保护设计

叶志玲　张　猛　郭　坚　赵　蕾

（北京空间飞行器总体设计部，北京　100094）

摘　要　针对我国第一个月球探测器的特点，对星载网络数据保护做了专门的设计。文章着重介绍了嫦娥一号卫星星载数据高可靠性保护的设计方案与具体实现情况。

关键词　嫦娥一号卫星　数据　可靠性　保护

1　数据保护概念及方法

数据保护包含两个方面[1]：一是保护数据不泄漏，防止未经授权的访问；二是保护数据不丢失。通过对底层数据进行加密来实现星载数据的不泄漏，本文不做重点介绍。

保护数据不丢失，可分为多个保护级别。这些级别是根据数据的可用性包括恢复时间目标(RTO)，即使系统恢复所需要的时间和恢复点目标(RPO)，即可接受的数据丢失量来划分的。保护级别越高，RTO和RPO也就越少；不过实施的相对成本也就越高。这些级别分别是备份、本地复制、远程复制和实时连续复制。

备份是为了在系统出现故障时进行数据恢复，包括磁带备份和磁盘备份等。

本地复制包括快照和克隆。快照是数据在某个时间点(拷贝开始的时间点)的映像。它是基于指针、节省空间的逻辑拷贝，通常要求少于30%的源卷容量，速度较快。克隆是数据的完整复制，是真实的考贝，它的过程较慢，而且每个副本需要与源卷容量相同的存储空间。

远程复制为业务连续性和灾难备份提供了强有力的保证，通过远程站点故障切换确保数据和系统的可用性，在生产中心停机后数分钟内，数据能够恢复，业务继续运营。

持续数据保护(Continuous Data Protection, CDP)，它的关键词是持续。就给定的数据集而言，CDP提供恢复点的连续体，能够存取任何时间点上的数据，而不仅仅针对那些由快照流程预先确定的特殊时刻。CDP允许应用恢复到特定的时间点之前，而不是恢复到预先确定的时间点上。恢复点在时间发生后选定并动态重建。它提供了粒度无限的恢复点(RPO)。

随着数据不断地激增以及集中化存储的趋势，数据访问开始面临着一个新的困境：当大量用户访问集中数据时，有限的带宽可能会造成严重的影响。于是，具有分布式的网格存储日益受到关注，它将网络视为一台超级计算机，充分利用分布在不同地理位置上的存储资源，构建一个动态响应用户需求的虚拟数据中心资源池，实现存储资源的全面共享。网格存储具有更高的容错与冗余度、在负载波动的情况下有更好的性能、以及更低的成本等优点。

而卫星内部的多台星载计算机即构成了分布式的网络拓扑结构，在数据保护设计上应该充分利用分布在不同地理位置上的存储资源，实现存储资源的全面共享。接下来以嫦娥卫星为例具体介绍星载计算机是如何利用网络资源实现数据高可靠性的保护设计的。

本文原载于《航天器工程》，2008，Vol.17，No.1，53~56。

2 星载网络数据保护环境

嫦娥一号星载网络由数管分系统和其他分系统组成,其中数管分系统由中心计算机 CTU (Central Terminal Unit)、远置单元 RTU A-D (Remote Terminal Unit A-D)和遥控单元组成。其他分系统由制导、导航与控制计算机和有效载荷总线控制器等计算机组成,其中4台远置单元与其他分系统的计算机通过两条冗余的 1553B 总线[2]与中心计算机连接;遥控单元通过串行数据接口与中心计算机连接,此外,遥控单元与其他计算机之间还有直接指令接口,遥控单元和远置单元 A 之间还有串行数字量接口。二级分布的星载网络拓扑结构如图1所示。

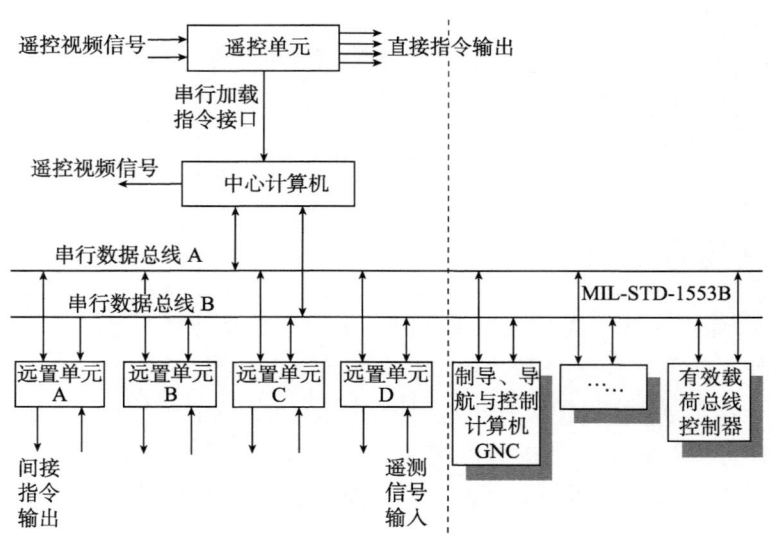

图 1 星载网络拓扑结构图

数管分系统是嫦娥一号卫星的一个重要分系统。数管分系统的主要任务包括[3]卫星的遥测、遥控、热控自主管理、蓄电池组放电电流平衡控制自主管理、定向天线展开自主管理、为其他分系统提供数据注入和重要数据保护、为电源控制器提供数据交换以及整星应急控制等。它又是一个软件密集型系统,软件在分系统中占的分量非常重[4],其中中心计算机软件最复杂、功能最强大,控制着整星的数据流,因此对数据进行高可靠性的保护就显得尤其重要。

3 高可靠性数据保护设计分析

空间环境复杂多样,以往卫星型号在对重要数据进行保存时并没有充分利用星载网络资源,仅仅选择一台计算机对本机的数据进行保存,可靠性不高;也仅仅采用周期性刷新重要数据的方法,方式单一,当重要数据变更,不能实时地刷新保存过的重要数据,直到刷新周期到,才能有效刷新数据,影响数据的真实性。在星载计算机网络中任两台计算机同时出现瞬时故障,都有可能丢失重要数据,从而影响到整星持续稳定的工作。

月球环境更加陌生,为了防止未知的异常情况威胁卫星正常工作,作为整个卫星数据管理的核心系统——星载计算机,采用以往重要数据保存的方法已不能满足月球探测的需要,对于整星的重要数据要利用更加有效手段进行强有力的保护,从而保证整星持续、正常、稳定的工作。

将星载网络视为一台超级计算机,提出一种星载数据保存与恢复的新方法:在数据保护设计上应该充分利用分布在不同地理位置上的存储资源,实现存储资源的全面共享。网格存储具有更高的容错与冗余度,在负载波动的情况下也将具有更好的性能。根据 RTO 和 RPO 的不同来划分,将数据进行分级保护。充分利用现有资源进行分级备份存储,在尽量减少总线冗余数据量的基础上,提高数据保护的可靠性,同时实时刷新变化的数据。该方法对数据保护过程中出现的异常情况进行考虑,提高了数据保护的真实性。更适应于航天领域中的月球及深空探测类型航天器,从而保证探测航天器在整个飞行工作阶段的持续稳定工作。

4 高可靠性数据保护具体实现

为了实现新设计,在嫦娥一号中对于整星的重要数据利用分类保存及备份存储的方法,充分利用了星载网络的优势,方法流程如图 2 所示,本机为计算机网络数据管理的核心单元中心计算机,它机为受本机控制和管理的从属计算机,由远置单元 A、B、C、D,制导、导航与控制计算机 GNC(Guide Navigation Control),有效载荷总线控制器组成,本机与它机之间通过数据总线通讯;本机上电或复位时,本机首先恢复星上时及本机数据,然后本机取它机数据;本机把恢复后的数据及星上时进行保存;当本机数据发生变化时,它机将变化的数据进行保存;当它机请求恢复它机数据时,本机及时恢复它机数据,同时将本机数据恢复到它机。该方法对数据保护过程中出现的异常情况进行了考虑,任一台计算机出现了瞬时故障,都不会造成数据的丢失,增强了数据保护的可靠性。实时的数据刷新机制提高了数据保护的真实性。

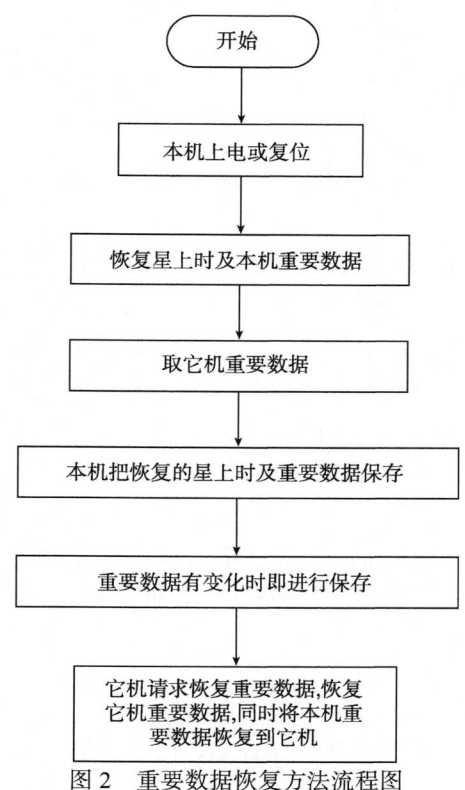

图 2 重要数据恢复方法流程图

又由于本机中重要数据的数据量大,利用所有它机进行全部数据的备份存储,总线数据冗余通信量大且会占用过多的内存数据存储空间,针对这一问题将本机数据进行分类备份实时存储。如图 3 所示,将本机中心计算机 CTU 重要数据按照 RTO、RPO 的不同分为三类,分别为运行状态重要数据、延时指令重要数据、自主热控重要数据;其中,运行状态重要数据为中心计算机 CTU 除自主热控功能外的运行状态数据,此数据是保障中心计算机正常运行最基本的数据,将它保存在制导、导航与控制计算机、有效载荷总线控制器、远置单元 RTU A、远置单元 RTU B 中;自主热控重要数据为自主热控功能相对应的重要数据,包括热控工作模式、高低温阈值、回路工作状态、热敏电阻状态等,保存在制导、导航与控制计算机、有效载荷总线控制器、远置单元 RTU A、远置单元 RTU B 中;延时指令重要数据为尚未执行的延时指令,延时指令重要数据保存在远置单元 RTU C、远置单元 RTU D 中。它机重要数据包括制导、导航与控制计算机重要数据、有效载荷总线控制器重要数据，它机重要数据保存在本机中心计算机 CTU 中。从图 3 可见每一类重要数据分别管理，分别保存和恢复，减少了总线通信的数据量，提高了效率的同时备份存储，提高了数据的可靠性。

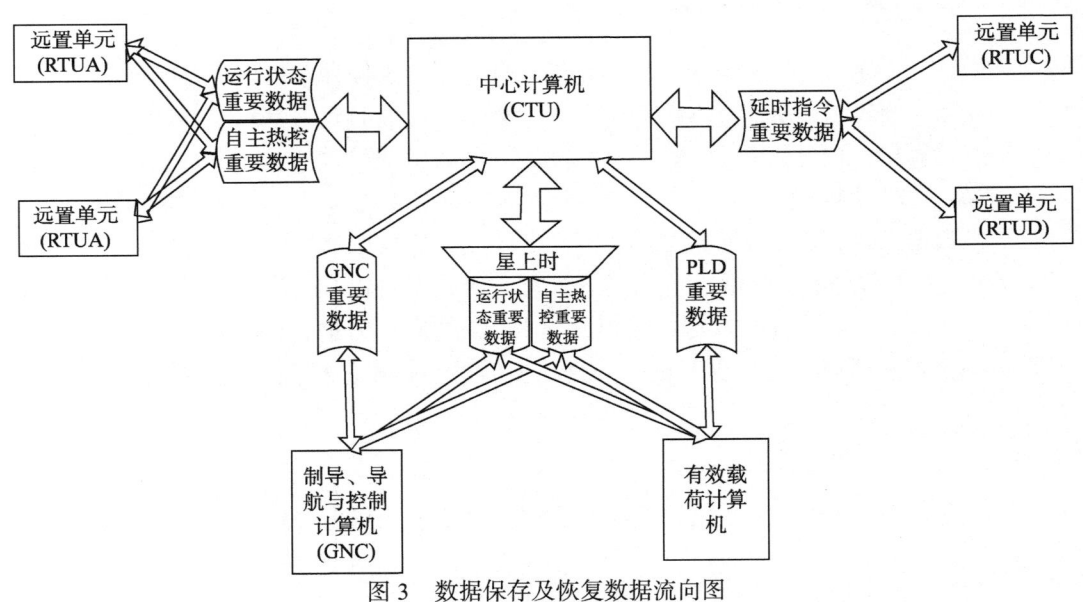

图 3　数据保存及恢复数据流向图

星上时间对于数管分系统、制导、导航与控制分系统、有效载荷分系统等都是非常关键的参数，其中，数管分系统需要根据它来发送延时指令和删除延时指令，制导、导航与控制分系统也要根据它来确定控制的时刻。但是星上时间又是实时变化的，在对其进行保存、恢复时，利用以往添加校验和的方法已不能对时间的有效性进行验证了，因此把对运行状态重要数据的恢复与对时间码的恢复结合起来实现，通过运行状态重要数据恢复的情况间接判断恢复回来的星时的有效性。为了保证恢复星上时间的精度，取时间的操作均为两次，第一次取星上时间的目的是通知 GNCC 或有效载荷计算机下次还要取时间，在收到该消息后，GNC 或有效载荷计算机立即将保存的星上时间更新，CTU 软件在 125ms 后再次取星上时间，这样再加上一个 125ms 的差值，即是准确的星上时间。而在平时则可通过集中校时和均匀校时来确保星上时间码的准确性。此种方法在整星测试中已得到充分的验证，达到了理想的效果。

5　结论

在当今社会，信息是如此重要，丢失了物质没什么，但是如果失去了信息，那么一切都将消失。而在卫星中数据即是一种描述信息的信息，更需要我们想尽一切办法来保护。本设计方案通过实际应用验证，说明是一种行之有效的利用网络资源对卫星数据进行保护的方法。在接下来的设计工作中我们还将继续对本设计方法做尽一步的完善与提升，来满足日益复杂的星载数据管理工作。

参 考 文 献

[1]　于希国，叶毓睿. 谈数据保护技术保护数据不泄漏不丢失[J]. 硅谷动力，2007—04. http//enet.com.cn
[2]　GJB289A-97. 数字式时分制指令/响应型多路传输数据总线[S]. 国防科学技术工业委员会，1997
[3]　谭维炽，顾莹琦. 空间数据系统[M]. 北京：中国科学技术出版社，2004
[4]　Mazze C，Scheffe A，et al. ESA's software engineering standard- the foundation for reliable software [J]. ESA Bulletin，1992 (69)

"嫦娥一号"任务全球地形实时仿真技术研究

陈宏敏[1,2]　战守义[1]　刘　涛[2]　唐歌实[2]　张　伟[2]

（1. 北京理工大学计算机学院，北京　100081；2. 北京航天飞行控制中心，北京　100094）

摘　要　针对"嫦娥一号"任务中航天器绕地球快速漫游的需求，设计了一种全球地形实时仿真技术。该技术采用建立静态层次细节模型和连续层次细节模型相结合的方法。运用逻辑四叉树结构组织全球海量数据的静态层次细节模型，满足海量地形数据的快速访问，并设计了数据更新算法，用以更新不同时相的地形数据；设计可见区域多细节层次数据块的快速确定算法实现全球地形的实时漫游，运用动态缓存调度策略提高了地形的绘制速度。实验表明，运用该技术实现了微机上全球地形的实时漫游。

关键词　航天飞行实验；全球地形仿真；四叉树；实时漫游

1　引言

航天飞行器视景仿真是将飞行器及其周围的场景映射到一个虚拟仿真环境，通过控制计算机的实时数据驱动飞行器的飞行状态，为故障检测、实时监控等领域提供强有力的分析依据。航天发射是一个充满各种难以预知风险的过程，复杂的实时数据使人们进行飞行状态判断时往往难以及时做出最佳选择，从而引发严重后果。采用可视化仿真技术，可以利用实时数据为飞行机制提供一个直观清晰的表现方式，缩短人们对结果进行分析和响应的时间，以更直观地、逼真地反映飞行器的飞行状况，为人们的正确判断和决策提供有利的时机和依据[1]。

"嫦娥一号"任务的整个过程包括发射段、以地球为中心的调相轨道飞行、地月系之间的奔月飞行以及进入环月工作轨道的飞行。发射段是运载火箭发射升空到星箭分离飞行，调相段是从星箭分离到卫星第三次近地点变轨的飞行。发射段和调相段航天器绕地球飞行，发射段航天器距地球较近，调相段航天器距地球较远。当航天器距地球较近时，地球场景需要高分辨仿真，以免失真；当航天器距地球较远时，地球场景只需用较低分辨率仿真，用以提高效率。为仿真这样一个飞行场景，需建立全球范围内的多分辨率地形场景，以满足不同飞行过程中不同分辨率全球场景的仿真。

全球地形几何和纹理的数据量巨大，可达到几百 GB 乃至几个 TB，这些数据不可能一次性读入闪存，需要利用访问速度较慢的硬盘存储数据。若海量地形数据的存储和地形的显示在单机上，要考虑数据从硬盘到内存的交换；若海量地形数据存储在服务器端，要考虑网络的带宽，进行服务器端数据到显示端数据的调度。内存中的地形数据还需从内存到图形卡显存的数据传输。内存到图形卡是即时模式，传输速度比较快，但是硬盘和内存之间进行数据交换或通过网络交换数据，速度很慢，已成为大规模场景基于外存(Out-of-Core)的实时绘制的一个瓶颈，设计有效的海量数据组织和调度算法，对于提高全球范围内场景的实时绘制速度，具有非常重要的意义。细节层次(Level Of Details，LOD)技术是提高海量三维场景数据绘制的有效途径，它通过预先建立具有多个 LOD 的离散模型，或者依据视点参数实时动态构建具有不同 LOD 的多分辨率模型，达到提高绘制效率的目的。前一种称为静态层次细节模型，后一种称为连续层次细节模型。

地形多细节层次模型分为基于不规则地形的多分辨率模型[3,4]和基于规则格网的多分辨率模型[5,6]。

在不规则地形的简化方面 Hoppe 做出了杰出的工作:他首先提出了通用的用于模型简化的累进网格法,通过每次选择性地将一条边缩为一点,实现了从原始模型到最简模型的过渡,反方向则可进行最简模型的复原和模型的渐进传输。然后,他又在此基础上提出了视点相关累进网格法,使这种方法适合于地形的显示,并可以根据需要控制生成三角形的数目。但由于不规则网格多分辨率模型的几何关系比较复杂,不便于数据的实时存取,这显然不适合大区域地形数据的可视化。

规则地形简化的代表作是 Duchaineau 等的 ROAM(Real-time Optimal Adaptive Meshes)算法,其核心是引入了两个优先级队列:分裂队列用来控制简化的顺序,合并队列用来获得帧之间的相关性。由于按优先级顺序进行分裂,因此可以精确地控制三角形的数目。但为了得到优先级队列,它必须在每一帧为每个三角形计算与视点相关的最大误差。虽然这个计算并不复杂,但由于是自底向上地全部计算,因此它是整个算法最耗时的部分。在这方面,Lindstrom,Rottger 等也做出了杰出的工作。

针对以上方法的分析,基于规则格网的多分辨率模型适合海量地形数据的实时绘制,因此以规则网格数据为基础,提出一种静态层次细节模型和连续层次细节模型相结合的全球海量地形数据组织与绘制方法,以实现航天飞行在全球范围内实时漫游。

2 框架

海量数据可视化的核心是基于视相关的地形场景多分辨率模型[2],该模型是将在该视点确定的视域下的几何和纹理数据按离视点的远近组成多细节层次的数据集合。对于全球地形可视化系统,仅仅考虑基于视相关的地形场景多分辨率模型是不够的,还包括海量地形几何和纹理数据的管理,以方便所要求数据的快速存取,因此先建立全球海量数据静态层次细节模型,在绘制时根据视点建立连续层次细节模型,用以提高海量地形数据的绘制。

框架图如图 1 所示,包括海量原始数字高程模型(Digital Elevation Model,DEM)数据和数字正射影像(Digital Orthography Model,DOM)数据的静态层次细节模型和基于视相关的地形场景连续层次细节模型两部分。前者主要是预处理部分,后者属于实时绘制部分。

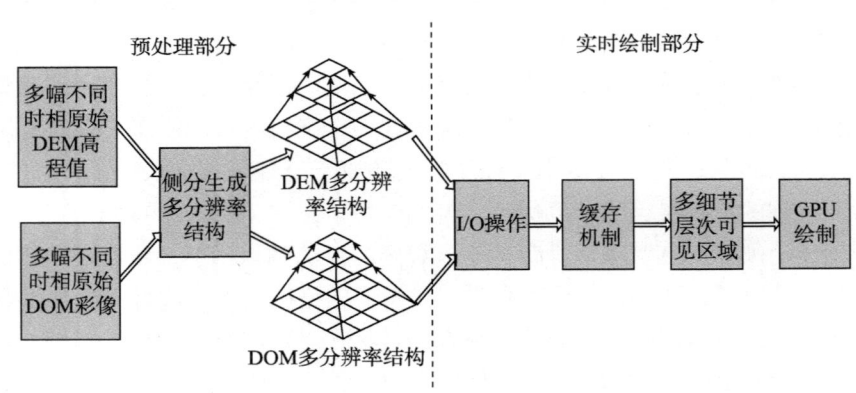

图 1 全球地形漫游框架图

(1) 预处理部分。该部分能够把不同时相的原始地形数据进行剖分生成四叉树多分辨率管理结构,该结构中上一层节点数据是下一层节点数据的父节点,下一层节点数据是上一层节点数据的子节点,父节点数据覆盖场景范围是子节点覆盖范围的四倍,父节点数据分辨率比子节点数据分辨率低一倍。为提高 I/O 读取效率,四叉树节点文件不是按四叉树结构目录层次存放,而是每一层目录并行存放,防止节点数据文件的存放路径过深。

(2) 实时绘制部分。根据当前视点对不可见区域数据进行剔除，组成一个多边形可见区域。根据误差判断确定可见区域的多细节层次(LOD，Level Of Detail)数据的节点序列，根据该序列通过 I/O 操作从四叉树结构中读取相应节点数据文件。为避免频繁的 I/O 操作，建立节点数据缓存机制，该缓存建立在内存区，把最近用到的节点数据存放其中，这样绘制数据时首先从数据缓存去寻找节点数据，未找到的数据从硬盘的四叉树地形数据结构中寻找。由于可见区域数据是多细节层次，不同细节层次节点数据拼接时有缝隙，还需对相邻不同细节层次节点数据进行拼接。当前得到的多细节层次的地形数据就是基于视相关的地形场景多分辨率模型数据。最后 GPU(Graphics Processing Unit)根据可见区域内的多细节层次数据进行绘制。

3 全球海量数据的静态层次细节模型

3.1 地形数据的四叉树存储结构

地形数据主要包括 DEM 数据和 DOM 数据，采用效率最高的基于规则网格的地形数据。因为基于不规则网格模型通常要把整个模型一次性读入内存进行处理，显然无法满足海量地形数据的可视化。

基于规则网格的 DEM 数据和 DOM 数据都可以存储为位图形式的数据，可以使用一种结构进行存储和读取。采用四叉树结构存储地形数据有效地和全球经纬坐标统一，在四叉树结构中，最上一层为第 0 层，数据覆盖全球，只有一个节点数据，第一层数据为第 0 层数据横向二叉剖分为 2 个节点，分别表示东西半球，第二层数据为第一层节点数据四叉剖分为 8 个节点数据，以下各层节点数据是上一层节点数据的四叉剖分，其结构图如图 2。

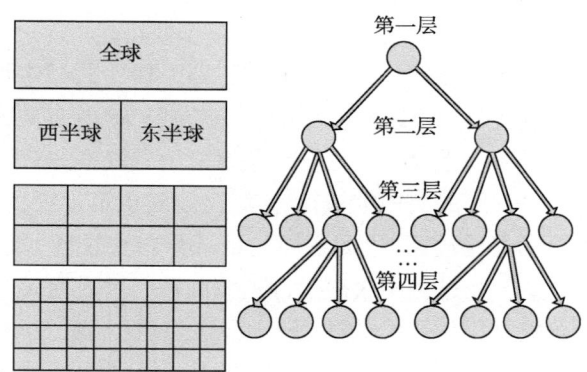

图 2 全球四叉树结构

四叉树数据存储结构中，数据节点覆盖一定范围内的数据，DEM 数据节点采样点为 $(2n+1)\times(2n+1)$ 个，通常 n 取 5；DOM 数据采样点为 $2m\times2m$ 个，通常 m 取 8。这样每一层节点数据的经纬坐标可以通过以上数据计算得到。DEM 和 DOM 数据存储的区别仅在于数据节点，DEM 为 33×33 的双精度高程数据，DOM 为 256×256 的影像值，为提高存储空间的利用率，可对 DOM 数据节点进行压缩。

假设层(layer)、块号(xblock, yblock)、单元格(xcell, ycell)。层表示数据块节点的层次；各层节点处数据块内部由单元格构成。分别建立块号和单元格坐标系，对某层块和单元格都可在这层相应坐标系确定一个坐标。每块由 $N*N$ 单元格组成，N 在 DEM 数据记为 demN、影像数据记为 imgN、矢量数据记为 vecN。分割后相应层的块数和单元格数与层次的关系如公式(1)-(4)：

纵向块数： $BNx = 2^{layer}$ (1)

横向块数： $BNy = 2^{layer-1}$ (2)

纵向单元格数： $GNx = 2^{layer+N}$ (3)

横向单元格数： $GNy = 2^{layer+N-1}$ (4)

块号和单元格坐标系与大地坐标系之间对应关系，单元格和块与大地坐标系的关系如公式(5)-(6)：

单元格与大地坐标系对应关系：

$Cs = 360./1 \ll (layer+N)$(度) (5)

块与大地坐标系对应关系：

$Bs = 360./(1 \ll layer)$(度) (6)

这样可对这三个坐标系进行转换。

每个节点数据只有几十 KB 到几百 KB，绘制时只需把视域覆盖的节点数据读入内存，避免海量数据不能一次性读入内存的问题。四叉树结构每一层数据代表一个分辨率，整个四叉树存储结构是一个多分辨率的存储结构，根据距视点的远近调入相应分辨率的数据，提高绘制效率。每个节点数据是以位图形式存储，还可进行数据压缩，减少硬盘存储空间。

3.2 地形数据的逻辑四叉树组织结构

为提高四叉树地形数据的存取效率，四叉树组织结构采用逻辑四叉树结构，而非物理四叉树结构。

定义 1：物理四叉树，数据是按严格意义上的四叉树进行存储，父结点数据覆盖范围与四个子结点数据覆盖范围相同，子结点数据的分辨率是父结点数据一倍，每个节点有相应的数据文件。读取某个节点数据直接访问相应四叉树目录下的节点数据即可，操作比较方便。

定义 2：逻辑四叉树，各结点数据仍然按照四叉树结构生成，父结点数据覆盖范围与四个子结点数据覆盖范围相同，子结点数据的分辨率是父结点数据一倍。但数据不是按严格意义上的四叉树进行存储，整个四叉树全部结点数据可以存储在一个文件中，也可以存放在多个文件中，在文件头创建每个节点数据的索引以方便快速查询。

构建物理四叉树读取结点数据比较方便，只需生成其在四叉树中的相应目录即可找到，而逻辑四叉树非严格意义上的四叉树，须从索引表中查找结点数据的存储位置。物理四叉树的层次比较深，结点文件的目录层次比较多，读取文件的时间耗费比较多，对于一个第二十层数据的存取显示，就必须要获得前面二十层目录数据的相关信息。因为每一层的文件数量比较多(比如对于第十层，有 $2\times4^{10-1}$ 个文件)，所以各层之间在磁盘上的分配一般相距得都比较远，也即各层的目录分布在磁盘上间隔比较大的不同地方。所以，获取二十层的目录信息，正常最少要经过二十次磁盘寻道、旋转、读数据块的时间，设一次的大致时间是 10ms，二十次则大约为 200ms，若数据文件包含的数据大小为 64×64，显示一整屏地形需要 1280×800/(64×64) = 250 个数据文件，考虑一次只更新屏幕数据的 10%，按照上述极端的算法则共需要耗时 250×200×10% = 5s。显然，这么大的延迟对于与目前要求的实时显示相距甚远，影响地形漫游的实时性，因此采用逻辑四叉树。逻辑四叉树中结点数据存放的路径较浅，若一层目录可存放 8 层节点数据文件，3 层目录可存到非常高分辨率的数据节点层次，可存放物理四叉树第 24 层，减少了访问目录的时间。

在逻辑四叉树中共有三类文件，层索引文件、块索引文件和数据文件，层索引文件包含每层数据块索引的位置，块索引文件包含每块数据存放的位置，通过索引表可以迅速的查找到所需数据的信息，以便迅速访问到数据如图 3 所示。

3.3 逻辑四叉树结构的构建

由于渠道、资金等问题，一次不可能获得全球高分辨率的地形数据，每次得到的是某一分辨率、某一区域、某一时相的数据，因此，需通过算法管理数据的构建和更新。通常最新时相的数据更换老时相的数据，高分辨率的数据更换较低分辨率的数据，根据这一原则设定本次数据入库是否更换当前数据库中的节点文件。通常获得 DEM 数据和 DOM 数据是一块一块的，并非正好和逻辑四叉树中 DOM 和 DEM 所要求的数据块节点大小一致，而且原始数据块通常是单一分辨数据，无法满足四叉树的多分辨率结构，因此需对原始数据进行重采样、合并、抽稀处理以达到四叉树结构的要求。

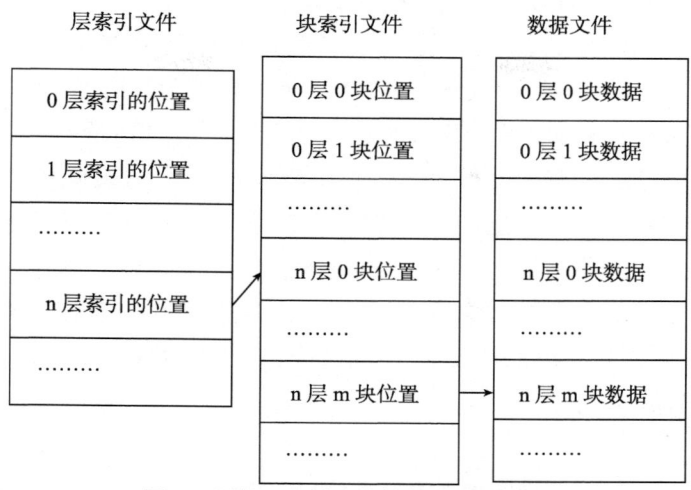

图 3　逻辑四叉树存储结构的数据访问

因此逻辑四叉树结构的构建算法如下：

step1：确定当前原始数据的分辨率和坐标信息；

step2：根据分辨率确定该数据对应四叉树的层次 layer，并从该层开始入库；

step3：根据坐标信息、layer 确定原始数据覆盖的节点；

step4：处理每个节点数据，从原始数据重采样该节点的数据，查找库中该层数据的索引表，判断该节点数据是否存在，若存在，根据更换标志进行更新，不存在，直接入库，并在索引表中修改该节点的存在标志；

step5：通过对上一层高分辨地形数据进行抽稀，依次更新库中较低分辨率层次数据。

4　逻辑四叉树组织数据的实时绘制

4.1　可见区连续多分辨率模型的快速确定

在任务中，每次成像的可见面片只占全球几何面片很微小一部分，只绘制可见区域的数据节点使全球地形的实时漫游成为可能。因此海量数据的实时绘制算法的核心是建立基于视相关的地形场景连续多分辨率模型，即可见区域多细节层次数据块的确定，算法如下：

step1：根据视点和视线方向建立视域四棱锥，确定该四棱锥与地球球面相交组成的多边形区域 G。该多边形区域边界由经纬坐标表示。并且确定从大地坐标系和屏幕坐标系的转换矩阵 T。

step2：建立区域判断数据链表 LG，误差判断数据链表 LE，可见数据链表 LV。各链表的基本单元是层(layer)、块号(xblock, yblock)信息。把逻辑四叉树结构的顶层数据节点放入判断数据链表 LG 中。

step3：从 LG 中取出一块数据节点，判断该节点是否在可见区域 G 范围内，若不在 G 范围内，将该数据块从 LG 中剔除并返回到该步的开始。若在 G 范围内，把该数据节点放入 LE 中。当 LG 为空时，算法结束。

step4：从 LE 中取出一块数据节点，对该数据块通过转换矩阵 T 确定该数据块在屏幕上的投影面积 S，若 S 大于某一给定面积阈值 $S0$，取其对应的四个子节点数据放入 LG 中作为判断数据块，若 S 小于某一给定面积阈值 $S0$，将该数据块放入可见数据链表 LV 中。当 LE 为空时执行 step3。

根据该算法得到一个能够反映层次细节模型的多分辨率可见数据结点集合 LV。根据 LV 获取相应的 DEM 和 DOM 数据块，用以绘制。LV 的数据节点离视点远则分辨率低，离视点近则分辨率高。屏幕投影面积阈值 S_0 是经验值，取决于任务要求显示的精度要求。

可见节点集合是一个多分辨率数据节点集合，不同分辨率数据节点之间会产生裂缝，消除裂缝的方法通常有两种[3]：将高分辨率数据节点的顶点移到相邻低分辨率数据节点边上；对低分辨率数据节点进行分裂，使其与高分辨率数据节点有相同边界。第一种方法的优点是其三角形的数目在消除裂缝过程中不会增多。但它至少会带来下列问题：①需要进行插值产生新的数据点来弥合裂缝；②由于产生了新的数据点，原来的误差计算准则可能被破坏；③产生 T 型连接。第二种方法可以避免 T 型连接，但若相邻数据节点分辨率相差过大会出现畸形三角形，通常的做法是限制相邻地块的分辨率差不超过 1。采用第二种方法，分裂低分辨率的数据节点，使其与高分辨率节点共边，用以消除不同分辨率数据节点拼接时的裂缝问题，在建立可见节点集合块时限制相邻数据节点分辨率不超过 1，效果良好。

4.2 动态缓存调度策略

在航天飞行漫游过程中，前后两帧图像具有帧相关性，在内存区域开辟一片缓存区保存最近访问的数据，不需要每一帧的所有节点数据从硬盘中读取，防止频繁访问硬盘，影响地形的绘制速度。

这种方法是以牺牲内存的方法提高绘制效率，地形数据包括 DEM、DOM 数据，因此分配两块缓存区域保存 DEM、DOM 数据。缓冲区的大小不能过大，过大影响内存的使用效率，并且大到一定程度并不能有效提高绘制的速度。在实验中，DEM 缓存区为 150 块节点数据的大小，DOM 缓存区为 70 块节点数据的大小。缓存区的更新机制采用最近使用原则，把最近使用的数据节点的优先级置于最高，当缓存区数据节点数目超过额定值，把优先级最低的数据删除，实现缓存区的更新。因此数据存储区分为显存区、数据缓存区、逻辑四叉树存储区如图 4 所示，显存区的节点数据直接用 GPU 绘制。当视点变动引起视域内的数据变动，显存区的数据不能覆盖整个视域，须从数据缓存区中查询不在显存区的数据节点；若数据缓存区也不存在视域内需求的节点数据，通过访问硬盘中的逻辑四叉树数据存储区中获得数据。实验证明，运用该方法可提高 50%左右的绘制效率。

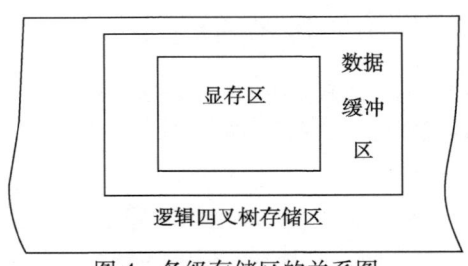

图 4 各级存储区的关系图

5 实验数据

我们将上述算法应用到嫦娥一号任务测控可视化过程中，利用 PC 平台实现了火箭和卫星的绕地球实时仿真飞行。任务中硬件配置为 P4 2.8GHz CPU、1GB 内存、NVIDIA Geforce 7300 512MB 显卡、120G 硬盘，软件配置中操作系统为 Windows XP，编程工具使用 VC++ 6.0 和 OpenGL，屏幕分辨率为 1280×1024。实验数据是 20GB DOM 数据、5GB DEM 数据。在该地形中进行 3 次不同地域漫游，得到的实验结果如表 1。平均每秒达到 30 帧/s 左右，达到实时效果。实验效果图如图 5 所示。

表1 实验结果

帧数	花费时间/ms	帧速率/(1/s)
410	12424	33 帧
1128	38367	29.4 帧
573	16371	35 帧

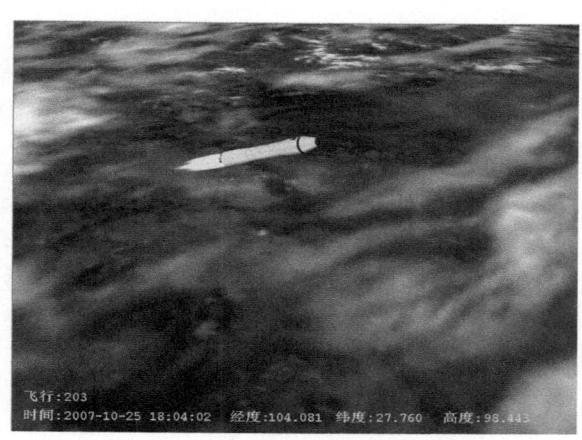

图5 嫦娥一号卫星任务火箭飞行效果图

6 结论

提出了一种面向嫦娥一号卫星任务的全球地形实时仿真技术，该技术的特点是以全球地形实时仿真为研究对象，而当前通用的实时地形仿真技术大多数是以大区域的地形仿真为目的，该技术着重从海量数据组织、可见区域多细节层次模型的确定两方面对全球地形实时仿真寻求突破，提出了一种静态层次细节模型和连续层次细节模型相结合的全球海量地形数据组织与绘制方法，利用逻辑四叉树组织和管理全球海量地形数据的静态层次细节模型，大大提高了海量数据的存取效率；设计了可见区域连续多细节层次数据块的快速确定算法和动态缓存调度策略，大大提高了地形绘制速度，在单机实验中每秒绘制30帧/s左右，达到实时效果。

在今后的研究中，将进一步提高地形的数据量，海量地形数据将存放服务器端，通过网络存取数据，并要达到实时漫游效果。

参 考 文 献

[1] 柴毅, 史晶晶, 冯大龙. 基于Vega的航天发射场视景仿真系统实现[J]. 计算机仿真, 2007, 24(6): 62—65, 211
[2] 谭兵, 徐青, 周杨. 大区域地形可视化技术的研究[J]. 中国图像图形学报, 2003, 8(5): 578—584
[3] De Floriani, Paola Meggillo. Multiresolution models for topo-graphic surface description [EB/OL]. (1996-07) [2008—04] http://www.disi.unige.it/person/DeflorianiL/publications.html.
[4] Hoppe H. Smooth view-dependent level-of-detail control and its application to terrain rendering [EB/OL]. (1998-10) [2008—04]. http://research.microsoft.com/en-us/um/people/hoppe/
[5] Mark Duchaineau, Murray Wolinsky. ROAM Terrain: Real-time optimally adapting meshes [EB/OL]. (1997-10) [2008—04]. http://www.vterrain.org/LOD/Papers/
[6] Renato Pajarola. Large scale terrain visualization using the restricted quadtree triangulation [EB/OL]. (1998-10) [2008—04]. http://www.vterrain.org/LOD/Papers/

星载毫米波辐射计地面定标实验

张 东 郭 伟

(中国科学院空间科学与应用研究中心，北京 100080)

摘 要 对 CE-1(嫦娥一号)卫星有效载荷微波探测仪初样样机 8mm 微波辐射计进行了地面定标实验。测量了接收机的线性度和灵敏度；建立了射频前端微波辐射传输模型；应用该模型并结合测试结果，建立了失配条件下辐射计天线温度的定标方程。实验结果表明，接收机线性度为 0.9999，灵敏度为 0.17K，定标误差小于 0.85 K。为 CE-1 正样样机的研制以及科学目标的实现提供了定标参考。

关键词 嫦娥一号；微波辐射计；定标

1 引言

CE-1 卫星为我国第一颗环月飞行探测卫星。有效载荷微波探测仪由 4 个频段的全功率型微波辐射计组成来实现对月球表面的微波辐射特征测量。这 4 个频段分别为：3.0 GHz、7.8 GHz、19.35 GHz 和 37 GHz[1]。定标模式设计为星上高、低温两点定标。低温定标参考为冷空的黑体辐射，通过指向冷空的喇叭天线接收后经由传输线(波导或同轴电缆)、开关及隔离器传输给接收机；高温定标参考为测温匹配负载，通过开关及隔离器与接收机连接。观测天线指向星下点进行非扫描式观测。这 3 个支路通过开关的切换周期性地执行观测模式和定标模式。

为了保证在轨观测中实现高精度定标，首先需要在研制阶段执行严格的地面定标与测试。目前，微波探测仪已经完成了初样样机的研制。本文针对其中的 37 GHz 辐射计开展了地面定标实验，测量了接收机的线性度和灵敏度，建立了射频前端微波辐射传输模型和失配条件下辐射计的定标方程。地面实验结果证明了初样样机达到了预期的设计性能，为 CE-1 正样样机的研制以及将来的在轨定标提供了参考。

2 定标方法

37 GHz 辐射计工作框图如图 1 所示。

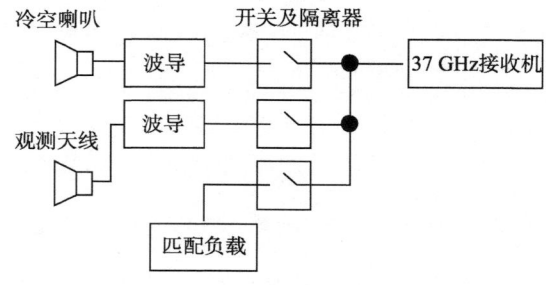

图 1　37 GHz 辐射计工作框图

本文原载于《遥感技术与应用》，2005，Vol. 20，No. 4，439~442。

在轨工作期间，通过开关在冷空喇叭、观测天线、匹配负载通道之间的切换周期性地执行定标模式和观测模式。其中，低温定标参考为冷空喇叭接收到的冷空黑体辐射，高温定标参考由热敏电阻测温的匹配负载提供。为了实现高精度准确定标，首先要保证接收机的线性度和高灵敏度，这需要在发射前的地面定标实验中进行严格的测试；其次，由于天线与波导、波导与开关、匹配负载与开关之间不可能完全匹配，波导、开关及隔离器还存在一定的插入损耗，因此，天线口面接收到的亮温与传输到接收机输入端口的等效亮温之间需要建立辐射传输模型，来表征真实的辐射传递过程并建立定标方程；最后，将接收机的输出电压代入到定标方程中，确定观测天线接收到的来自月球表面的微波辐射亮温。

为了保证在轨定标的准确性，要求地面定标实验按照设计的在轨定标模式执行，考核发射前地面定标的准确性，为在轨定标提供参考。由于地面定标中很难获取准确的冷空黑体辐射参考，因此，低温定标参考采用了液氮制冷的黑体定标源。这是与在轨定标的唯一的差别。二者仅存在参考亮温值的不同，不会影响地面定标实验对在轨定标的验证。

3 接收机线性度及灵敏度测试

对 CE-1 微波探测仪初样接收机线性度和灵敏度的测试，通过在图 1 中匹配负载和开关隔离器之间设置一个可调衰减器，改变接收机的输入亮温，测定接收机相应的输出电压。此时，接收机的等效输入噪声温度[1]T_{IN} 为：

$$T^{IN} = \frac{T_N}{L} + \left(1 - \frac{1}{L}\right) \times T_0 \tag{1}$$

其中：T_N 为匹配负载的温度，L 为可调衰减器的衰减系数，T_0 为开关隔离器的温度。利用数据分析软件拟合出测得的输入、输出对应的各个点，对测定数据进行分析即可得到接收机的线性度和灵敏度。测试结果如图 2 所示。

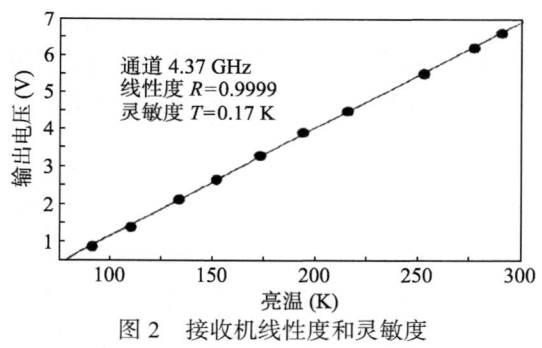

图 2 接收机线性度和灵敏度

由图 2 可见，37 GHz 辐射计接收机的线性度为 0.9999，灵敏度为 0.17 K，达到了初样设计指标，能为地面或在轨定标提供高精度的保证。

4 射频前端辐射传输模型

为了获得高精度定标，建立失配条件下射频前端辐射传输模型[3]，如图 3 所示。以对月观测支路为例，图中 α_w 和 α_s 分别是波导和开关隔离器的功率传输系数，T_w 和 T_s 分别是波导和开关隔离器的物理温度，Γ_{wl} 和 Γ_{wr}、Γ_{sl} 和 Γ_{sr} 分别是 2 个参考面处从左和右看时的功率反射系数；T_A 是对月观测天线的输出亮温。冷空观测和匹配负载支路的相关量亦标于图 3 中。

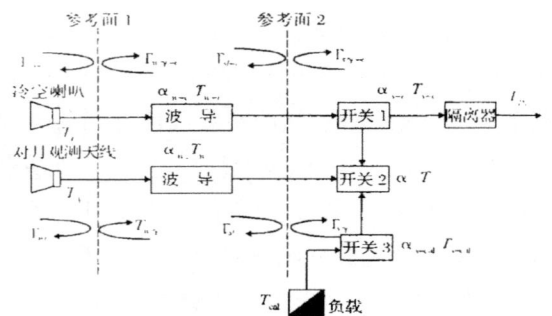

图 3 微波辐射计射频前端模型

以对月观测支路为例,分析接收机的输入亮温 T_{IN} (T_{IN} 在冷空、场景观测天线和匹配负载 3 个支路分别为 T'_A、T'_C 和 T'_{CAL})。T_A 经过射频前端的的输出 T'_A 由 2 部分组成:一部分是输入亮温 T_A 经过射频前端衰减后的直接输出;另一部分是射频前端各部件的辐射亮温对 T'_A 的贡献,其中包括在两个参考面上反射形成的贡献(忽略经过两次反射后对 T'_A 的贡献)。分析得到式(2):

$$T'_A = T_A(1-\Gamma_{wl})\alpha_\omega(1-\Gamma_{sl})\alpha + T_w(1-\alpha_w)(1-\Gamma_{sl})\alpha_s(1+\alpha_v\Gamma_{wr}) + T_s[(1-\alpha)+\alpha_s\Gamma_{sr}^{1/2} \\ + (1-\Gamma_{sl})^{1/2}(1-\Gamma_{sr})^{1/2}wr^{1/2}\alpha_w e^{j2\theta_A^2}] \tag{2}$$

其中:$\theta_1 = \frac{2\pi}{\lambda}l_A$ 为对月观测支路两个参考面之间的电相位,l_A 为对月观测支路两个参考面之间的电长度。对冷空支路,分析得到式(3):

$$T'_c = T_c(1-\Gamma_{wl-c})\alpha_{w-c}(1-\Gamma_{sl-c})\alpha_{s-c} + T_{w-c}(1-\alpha_{w-c})(1-\Gamma_{sl-c})\alpha_{s-c}(1+\alpha_{w-c}\Gamma_{wr-c}) + T_{s-c}[(1-\alpha_{s-c}) \\ + \alpha_{s-c}\Gamma_{sr-c}^{1/2} + (1-\Gamma_{sl-c})^{1/2}(1-\Gamma_{sr-c})^{1/2}\Gamma_{wr-c}^{1/2}\alpha_{w-c}e^{j2\theta_c^2}] \tag{3}$$

其中:$\theta_0 = \frac{2\pi}{\lambda}l_c$ 为冷空观测支路两个参考面之间的电相位,l_c 为冷空观测支路两个参考面之间的电长度。对匹配负载支路,分析得到式(4):

$$T'_{CAL} = T_{CAL}\alpha_{-CAL} + T_{s-CAL}(1-\alpha_{-CAL}) \tag{4}$$

由此,建立了 CE-1 微波探测仪初样各支路的射频前端传输模型,也即得出了各支路输入亮温和接收机接收亮温的关系,通过此关系即可消除在定标中由于射频前端各元件之间不匹配及其插损所带来的误差。

5 实验结果

本定标实验建立在接收机定标和连接网络定标基础上,通过冷空定标天线照射低温定标参考和来自匹配负载之路的高温定标参考建立天线输出亮温和接收机输出电压的定标方程。匹配负载定标如图 3 中匹配负载支路所示,为 300 K 记做 T^h_{CAL};低温由液氮制冷实现,为 77.4 K 记为 T^c_{CAL};在高、低温定标源情况下接收机的输出设为 V^h_{out} 和 V^c_{out},以上的上标 c 和 h 分别代表冷和热。

根据式(3)和(4)得:

$$T^h_{IN} = T^h_{CAL}\alpha_{-CAL} + T^h_{s-CAL}(1-\alpha_{s-CAL}) \tag{5}$$

$$T^C_{IN} = T^C_{CAL}(1-\Gamma_{wl-c})\alpha_{w-c}(1-\Gamma_{sl-c})\alpha_{s-c} + T_{w-c}(1-\alpha_{w-c})(1-\Gamma_{sl-c})\alpha_{s-c}(1+\alpha_{w-c}\Gamma_{wr-c}) + T_{s-c}[(1-\alpha_{s-c}) \\ + \alpha_{s-c}\Gamma_{sr-c}^{1/2} + (1-\Gamma_{sl-c})^{1/2}(1-\Gamma_{sr-c})^{1/2}\Gamma_{wr-c}^{1/2}\alpha_{w-c}e^{j2\theta_c^2}\Gamma^2] \tag{6}$$

又因为接收机的输入和输出有:

$$V_{out} = a(T_{IN}+b) \tag{7}$$

则有:

$$V_{\text{out}}^h = a(T_{\text{IN}}^h + b) \tag{8}$$

$$V_{\text{out}}^c = a(T_{\text{IN}}^c + b) \tag{9}$$

把式(5)和(6)分别代入式(8)和(9)，结合冷空观测支路射频前端辐射传输中功率传输和反射系数(见表1)即得接收机定标方程系数 a 和 b。由此确定接收机定标方程为：

$$V_{\text{out}} = 0.02907 \times T_{\text{IN}} - 1.77711 \tag{10}$$

式(2)中的 T'_A 作为式(10)中的 T_{IN} 代入可得：

$$T_A = a_1 V_{\text{out}} - a_2 T_w - a_3 T_s - b' \tag{11}$$

其中：$a_1 = 1/(aD)$；

$a_2 = (1-\alpha_w)(1-\Gamma_{sl})\alpha_s(1+\alpha_w\Gamma_{wr})/D$；

$a_3 = [(1-\alpha_s) + \alpha_s\Gamma_{sr}^{1/2} + (1-\Gamma_{sl})^{1/2}(1-\Gamma_{sr})^{1/2}\Gamma_{wr}^{1/2}\alpha_w e^{j2\theta}\Gamma^2]/D$；

$b' = b/D$；

$D = (1-\Gamma_{wl})\alpha_w(1-\Gamma_{sl})\alpha_s$。

结合对月观测支路射频前端传输模型中功率传输系数和反射系数(见表1)，最后可得定标方程系数见表2。

表 1　射频前端功率传输和反射系数

观测支路	Γ_{sr}	Γ_{sl}	Γ_{wr}	Γ_{wl}	α_w	α_s
测量值	0.0192	0.00039	0.0049	0.0128	0.9231	0.8479
冷空支路	Γ_{sr-c}	Γ_{sl-c}	Γ_{wl-c}	Γ_{wl-c}	α_{w-c}	α_{s-c}
测量值	0.0203	0.00059	0.0038	0.0095	0.9318	0.8370
负载支路	α_{s-cal}					
测量值	0.8512					

表 2　定标方程系数

a	b	a_1	a_2	a_3	b'
0.02907	−61.1321	44.553	0.0742	0.2546	−79.177

CE-1 微波探测仪地面定标的验证在微波暗室中进行。由于实验设施的限制，目前只能采取高、低温两点的验证，即对月观测天线对准液氮制冷定标源；对月观测天线观测暗室中的黑体，用温度计对黑体同步测温。将微波探测仪的测量温度(即通过定标方程的反演温度)与液氮制冷定标源标准温度以及暗室黑体温度计的测量温度进行比较(见图4)，验证定标的准确性和可靠性，经过分析，得出本实验的定标误差小于0.85 K。

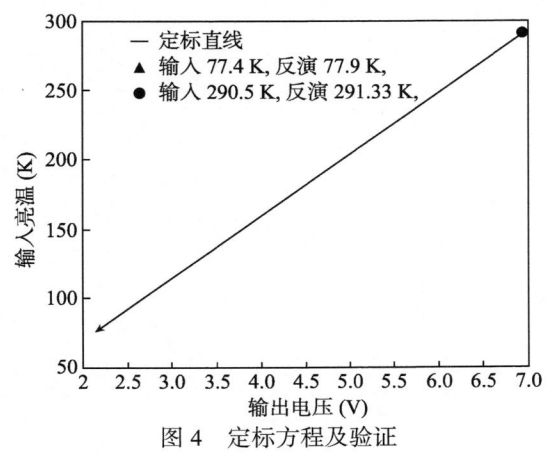

图 4　定标方程及验证

6 结语

本文对 CE-1 微波探测仪初样 37 GHz 辐射计系统进行了详细的地面定标实验，得到了定标方程。在轨时亦采用高、低温两点定标法：低温定标参考为冷空喇叭接收到的冷空黑体辐射，高温定标参考由热敏电阻测温的匹配负载；和地面定标唯一的不同是低温定标参考源不同，故可以通过地面定标对在轨定标进行验证，同时可以为 CE-1 微波探测仪正样的研制及科学目标的实现提供参考，并为正样定标中误差的修正打下基础。

参 考 文 献

[1] 月球探测卫星有效载荷微波探测仪初样方案设计报考报告[R]. 空间中心，2003
[2] 乌拉比 F T，穆尔 R K，冯健超. 微波遥感[M]. 北京:科学出版社，1982
[3] Christopher S R, Stephen J K, Michael A J. Instrument Description and Antenna Temperature Calibration [J]. IEEE Transactions on Geosciences and Remote sensing，1995, 22(1): 125—137
[4] 姜景山，王文魁，都亨.空间科学与应用[M]. 北京:科学出版社，2001

月球软着陆点的选择与几个预选点的初步对比分析

熊盛青[1]　闫柏琨[1]　甘甫平[1]　王振超[2]

（1. 中国国土资源航空物探遥感中心对地观测技术工程实验室，北京　100083；
2. 中国地质大学(北京)，北京　100083）

摘　要　月球软着陆探测是中国二期探月工程的主要目标，软着陆点的选择是工程实施与科学目标能否顺利完成的关键之一。克里普岩对于研究月球的起源和演化有重要意义，但由于该岩石被玄武岩所覆盖，难以利用轨道探测器开展全面深入的研究。撞击坑是研究月表以下物质成分的窗口，在克里普岩区选择条件合适的撞击坑开展软着陆探测有助于对克里普岩的深入研究。在克里普岩区选取 Copernicus、Kepler 及 Aristarchus 3 个撞击坑作为预选着陆点，并利用嫦娥一号 CCD 数据、LIDAR 数据以及 Clementine UV/VIS/NIR 数据从月形月貌特征和物质组成两个方面对预选点进行了初步对比分析，以期为我国二期探月工程提供参考与依据。

关键词　月球　软着陆探测　遥感

1　引言

月球软着陆探测可为月球研究提供最直接的第一手资料。迄今为止，美国与前苏联已向月球成功发射了 19 颗载人或不载人软着陆探测器，收集了大量的实测数据与月岩(壤)样品，为月球遥感探测提供了坚实的基础，深化了对月球的研究。

中国探月工程整体分为绕、落、回 3 个阶段[1,2]，其中第一个阶段，即绕月探测工程，已圆满完成，嫦娥一号月球卫星搭载了多个不同用途的传感器，获取了月球与月球空间的环境数据；第二个阶段，即月球软着陆探测，开始提上了议事日程，着陆点的选择是完成任务面临的首要问题。本文在总结月球软着陆探测已取得的成果和存在的问题基础上，结合月球科学研究的需要，提出了以 3 个克里普岩区的撞击坑作为预选着陆点，综合利用嫦娥一号 CCD 数据、LIDAR 数据及 Clementine UV/VIS/NIR 数据从月形月貌特征和物质组成两个方面对预选点进行初步对比分析，以期为我国二期探月工程提供参考与依据。

2　月球软着陆探测回顾

前苏联始于 20 世纪 50 年代的 Luna 登月计划开启了人类对月球软着陆探测的时代。截至 1976 年，Luna 计划共有 8 个月球车成功登陆月球。前苏联成功登月之后，美国于 1966 年开展了 Surveyor 登月计划，1966~1968 年期间，该计划共有 5 个月球车成功登陆月球，除最后一个月球车外，其余登月的主要目的是为后续的载人登月奠定科学基础。在无人登月成功之后，美国开展了 Apollo 载人登月计划，共有 6 个月球车成功登月。上述 3 个登月计划共计有 19 个月球车成功登月，如图 1(本文图件可见图版 IV—XVIII)所示[3]，并在登陆点开展了月壤力学性质、地震、热流、磁场及太阳风等多方面的研究，积累了丰富的科学数据。其中 9 个月球车带回了共计 382 kg 的月岩(壤)样品。

本文原载于《国土资源遥感》，2009，Vol. 21，No. 4，8~13。

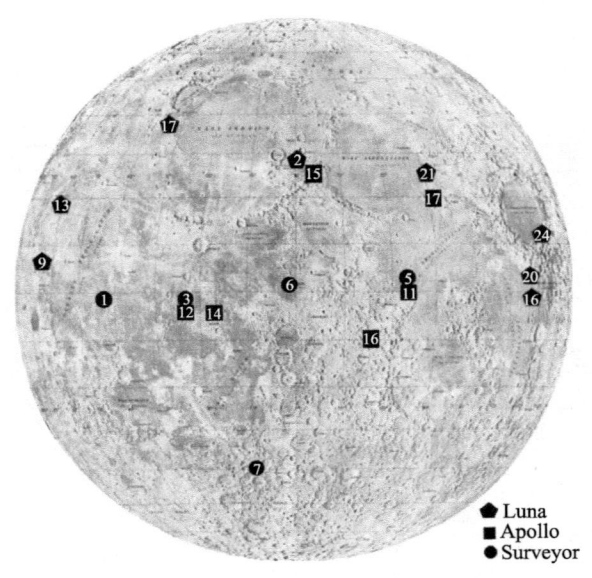

图 1 月球车登月点影像图

3 月球软着陆点的选择

中国月球探测软着陆点的选择应从科学研究目标与工程实现的可行性两方面加以综合分析。科学研究目标要在总结提炼月球探测已取得的成果和存在问题的基础上进行登月点的选择,着重考虑选择科学研究的重点与热点地区,同时考虑 Luna,Surveyor 和 Apollo 登月点的分布特点与规律,力争使获取的科学数据在月球科学研究中与国外已有的数据相互补充,避免出现不必要的重复;工程实现的可行性主要从预选登月点的坡度、坡向及月壤厚度等方面开展对比分析。

3.1 克里普岩——月球软着陆探测的目标

月表岩石主要由高地斜长岩、月海玄武岩及克里普岩(KREEP)3 大类岩石组成。克里普岩是现今月球研究的热点之一,首先发现于 Apollo12 号月岩样品中[4]。该岩石是一类化学成分接近月海玄武岩但更富集 K、P、REE(稀土元素)、Th 与 U 等不相容元素的一类岩石。克里普岩中的 K、P、REE、Th 及 U 等特征元素之间的相对含量大致为常数[5],通过提取其中一种元素在月表的含量分布即可确定克里普岩的分布。利用搭载于 Lunar Prospector 之上的射线仪填绘了月表 Th 的含量分布,并进而确定了克里普岩的分布范围,即主要分布于风暴洋与雨海周围[6,7](图 2)。

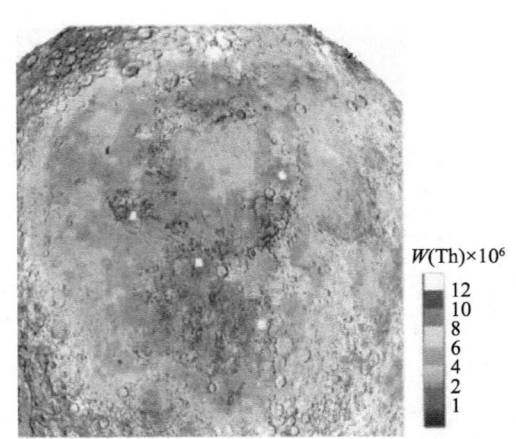

图岩形成于月球"岩浆洋"冷却结晶过程当中,"岩浆洋"是厚约几百 km 的层状熔融外层[5],随着矿物

结晶，矿物根据其与熔体密度的相对大小发生重力分异，橄榄石和辉石等密度较大的矿物下沉形成月幔，斜长石等较轻的富铝硅酸盐矿物上浮形成月壳，K、REE、P、Th 及 U 等不相容元素富集于残留熔体中，最终形成赋存于月壳与月幔之间的过渡层[8,9]。

研究克里普岩的分布和成分特征对于研究月球起源和演化具有重要意义，其中的 Th、U 等放射性元素的含量是研究月球热演化历史的基础性数据[10]。目前，这些基础性数据均来自 Apollo12、14 和 15 等登月点月岩(壤)样品，而这些样品均为富含 Ti 的绳状玻璃碎片[11]，这是由于撞击而覆盖于月表玄武岩之上的溅射物，并非直接采自于克里普岩源区，这在一定程度上影响了分析结果的可靠性，进而影响到月球热演化历史的研究。因此，在克里普岩区开展软着陆探测有助于深入了解克里普岩的化学成分在月表的分布，对研究月球起源与演化具有重要意义。

3.2 撞击坑——月球软着陆探测的窗口

γ 射线仪 Th 含量提取结果表明，风暴洋克里普岩区绝大部分为克里普岩与玄武岩的混合区，真正出露克里普岩的区域很少，绝大部分被一层薄薄的月海洋玄武岩所覆盖[12]。从月球轨道探测器数据难以研究玄武岩覆盖之下克里普岩的物质组成、岩石类型及其在月球的分布和延伸深度。月表撞击坑，尤其是形成年龄较小(受后期改造作用较弱)且撞击深度较大的撞击坑是研究月表玄武岩覆盖以下地质、地球物理特征的窗口。在风暴洋克里普岩分布区选择年龄较小的撞击坑作为软着陆探测点，对于研究克里普岩的特征和成因有重要意义。

从月表地质构造单元讲，Luna、Surveyor 及 Apollo 登月点基本构成了一条贯穿月海玄武岩区与高地斜长岩区的探测剖面[13]，便于研究月海玄武岩与高地斜长岩两大月表组成单元的形成与相互作用。以撞击坑作为窗口进行克里普岩的软着陆探测研究无疑与以往的月球软着陆探测起到相互补充的作用，利于从多层面、多角度对月球开展研究。

4 预选着陆点的对比

Aristarchus 撞击坑 Th 含量分布图(图 3)表明[14]，由撞击坑中央向外，Th 含量逐渐减少，表明风暴洋区克里普岩可能被玄武岩所覆盖，陨石的撞击作用将玄武岩下覆的克里普岩"挖掘"出来，形成了以陨石坑为中心的同心圆状分布模式。结合月海玄武岩区 Th 主要分布于雨海的边缘且呈环状分布和在撞击坑周围分布较高的特征(图 2)，可推测月海仅仅出露部分克里普岩，大部分的克里普岩为玄武岩所覆盖。

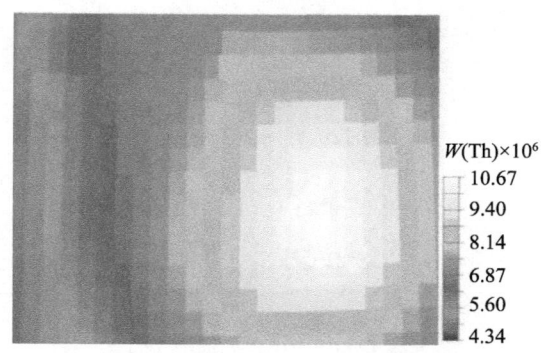

图 3　Aristrarchus 撞击坑 Th 含量的分布[14]

在 Th 元素分布区以 Kepler、Copernicus 和 Aristrarchus 3 个撞击坑作为月球软着陆点的预选点，撞击坑周围均有较明显的溅射辐射纹，表明其形成年龄较小，受后期改造作用较弱，是了解风暴洋玄武岩之下克里普岩的理想窗口(图 4)。

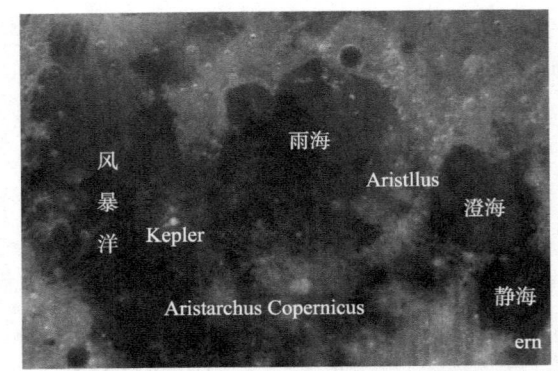

图 4 Kepler、Copernicus 及 Aristarchus 撞击坑位置

综合利用嫦娥一号 CCD 数据、LIDAR 数据以及 Clementine UV/VIS/NIR 数据分别从月形、月貌及物质组成等方面进行分析对比，为预选点的筛选提供参考。利用 Clementine UV/VIS/NIR 数据分别提取了 3 个撞击坑的 TiO_2 与硅酸盐 FeO 的含量分布[15~18]。月表 FeO 分为硅酸盐 FeO 与钛铁矿 FeO 两部分，TiO_2 含量与钛铁矿 FeO 有较高的相关性[18]。为了突出撞击坑物质成分差异与特点，提取了各撞击坑的硅酸盐 FeO 的含量分布。

需要说明的是，本文为了突出撞击坑 CCD 影像的负地形特征，除特殊标识外，撞击坑影像上方为南，下方为北。

4.1 Kepler 撞击坑

Kepler 撞击坑位于风暴洋东缘(图 4)。该区域 TiO_2 以 Kepler 撞击坑为中心呈辐射状分布(图 5)，为陨石坑形成时内部溅射物覆盖于月表所致；硅酸盐 FeO 含量高值区沿陨石坑四壁与中心分布(图 6)。

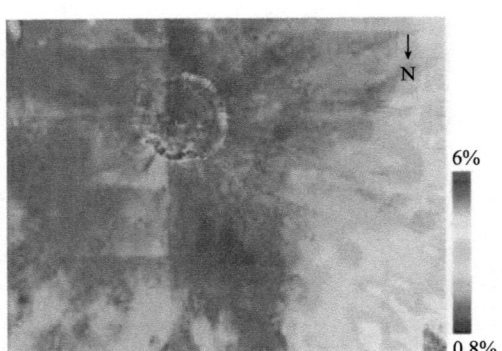

图 5 Kepler 撞击坑区 TiO_2 分布

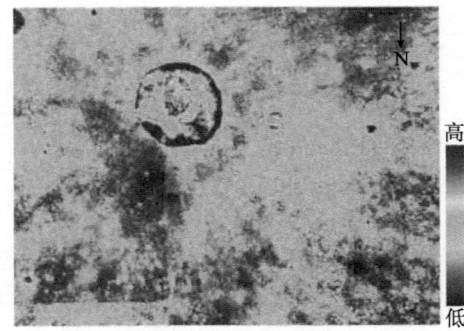

图 6 Kepler 撞击坑区 FeO 分布

Kepler 陨石坑地区的元素分布特征表明，该区域玄武岩明显呈垂向分层结构，高 Fe 低 Ti 的玄武岩之上覆盖了一层较薄的高 Ti 低 Fe 玄武岩。γ 射线探测结果表明，该区域 Th 以 Kepler 陨石坑为中心呈辐射状分布，Kepler 陨石坑为 Th 的富集区(图 2)，表明克里普岩被上伏玄武岩所覆盖，高 Fe 低 Ti 成分的玄武岩可能与克里普岩相对应。

Kepler 陨石坑嫦娥一号 CCD 影像特征与 LiDAR 数据(图 7)表明，该陨石坑北高南低，相差约 610 m，陨石坑深约 1600 m，坑底部可能不存在较平坦宽阔的区域，地形较复杂，软着陆工程实现的难度较大。

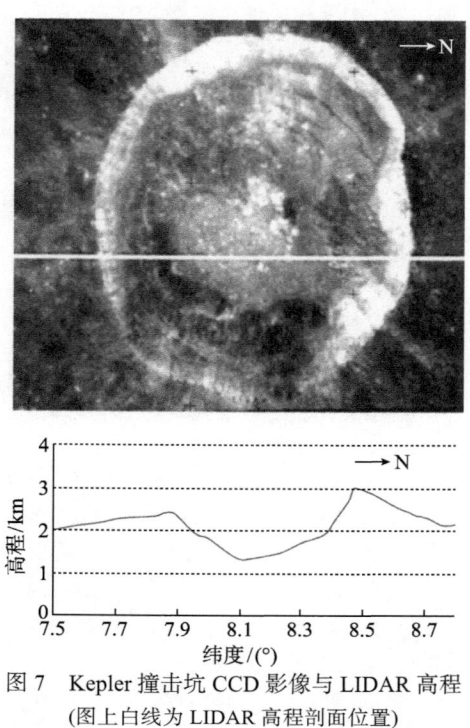

图 7　Kepler 撞击坑 CCD 影像与 LIDAR 高程
(图上白线为 LIDAR 高程剖面位置)

4.2　Aristarchus 撞击坑

Aristarchus 陨石坑位于 Kepler 陨石坑东南方向(图 4)，该区域 TiO_2 以 Aristarchus 撞击坑为中心呈半圆状分布，陨石坑东壁 TiO_2 明显呈层状分布(图 8)，为陨石坑形成时内部溅射物覆盖于月表所致。

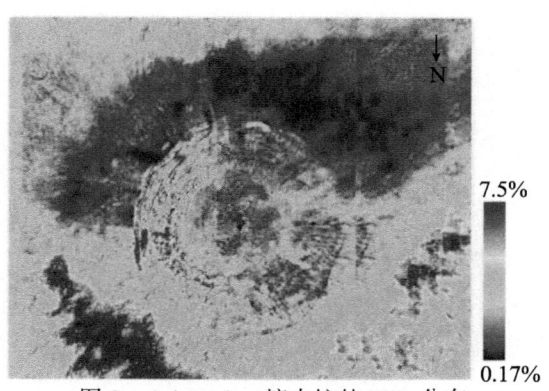

图 8　Aristarchus 撞击坑的 TiO_2 分布

硅酸盐 FeO 含量高值区沿陨石坑四壁分布，明显呈层状(图 9)。TiO_2、FeO 在陨石坑四壁的层状分布特征可能表明玄武岩成分的垂向分层结构。

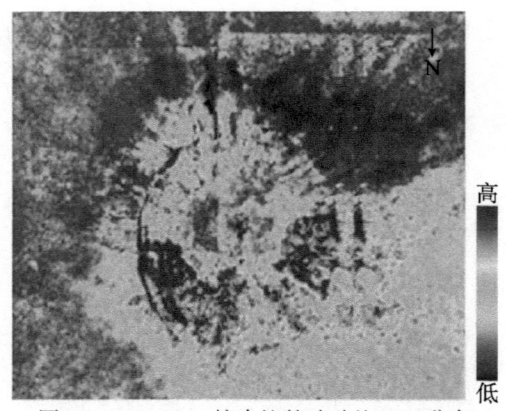

图 9　Aristarchus 撞击坑的硅酸盐 FeO 分布

Aristarchus 撞击坑相对高差约 3 km, 底部相对 Kepler 撞击坑较平坦, 由 CCD 影像特征与 LIDAR 高程数据(图 10)可知, 撞击坑东南部有一面积约为 15~20 km² 的平坦区域, 特别是该区域的南部最为平坦(图 10 中白色曲线内)。

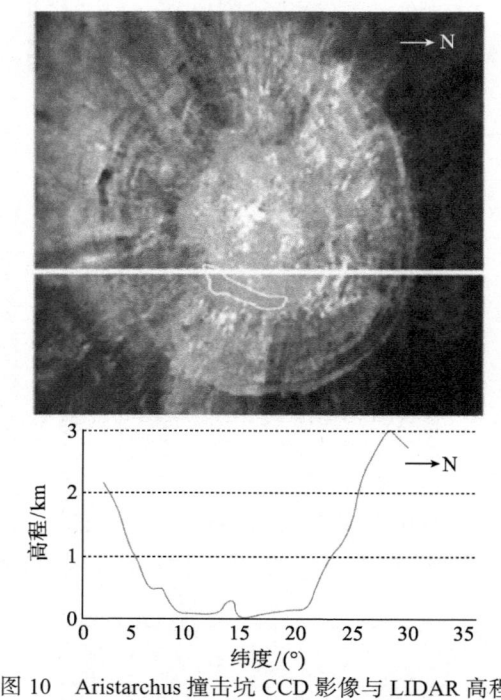

图 10　Aristarchus 撞击坑 CCD 影像与 LIDAR 高程
(白线为 LIDAR 高程剖面位置)

4.3　Copernicus 撞击坑

Copernicus 撞击坑位于 Aristarchus 撞击坑东(图 4)。该区域 TiO_2 以 Copernicus 撞击坑为中心呈半圆状分布, 陨石坑东壁 TiO_2 明显呈层状分布(图 11), 为陨石坑形成时内部溅射物覆盖于月表所致。

硅酸盐 FeO 同样以 Copernicus 撞击坑为中心呈半圆状分布, 撞击坑北壁硅酸盐 FeO 含量较高, 并且明显呈层状(图 12)。

Copernicus 撞击坑相对高差约 3km, 尽管由 CCD 影像特征与 LIDAR 数据(图 13)可推测撞击坑北部存在一平坦区域, 但由 LIDAR 数据可知, Copernicus 撞击坑地形起伏较大, 无大面积的平坦区域, 着陆工程实现的难度较大。

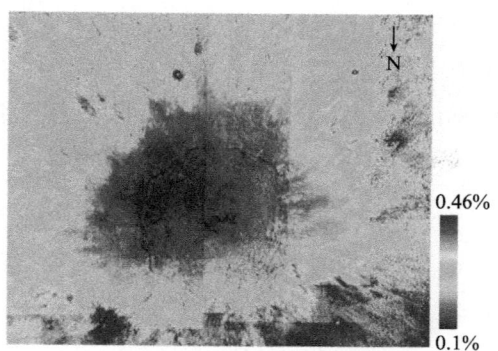

图 11　Copernicus 撞击坑的 TiO_2 分布

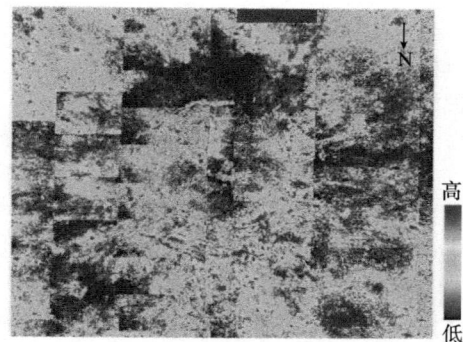

图 12　Copernicus 撞击坑的硅酸盐 FeO 分布

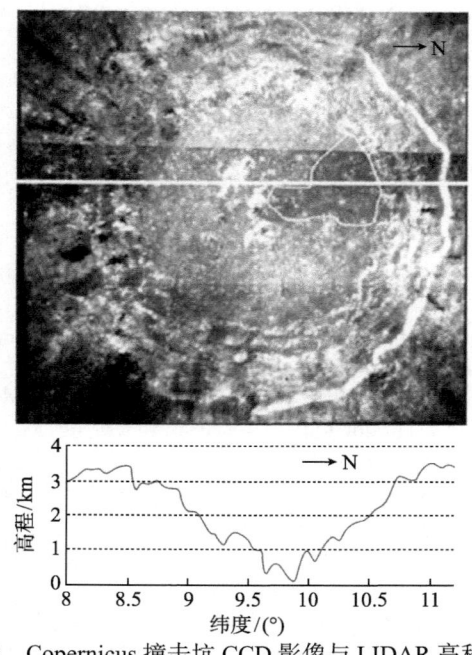

图 13　Copernicus 撞击坑 CCD 影像与 LIDAR 高程剖面
(白线为 LIDAR 高程剖面位置)

5　讨论

基于嫦娥一号 CCD、LIDAR 及 Clementine UV/VIS/NIR 等数据对月球风暴洋克里普岩区 3 个撞击坑从月形月貌及物质成分两大方面进行了初步分析对比,结果表明:3 个撞击坑物质成分上存在较大差异,表明该区域玄武岩成分在垂向上存在较大变化,风暴洋克里普岩区为较薄的一层玄武岩所覆盖,着陆点选择在撞击坑内对于研究克里普岩物质组成、撞击坑结构及玄武岩石成分的垂向变化等科学问题有重要意义。本文的分析仅仅基于有限的月球探测数据，着陆点的选择应在进一步分析 X、γ 射线以及微波、高空间分辨率(米级)等其他月球探测数据的基础上进行更全面的科学分析与论证。

6 结论

(1) 月球风暴洋克里普岩区物质组成在垂向上存在明显变化，克里普岩可能为一层较薄的玄武岩层所覆盖，着陆点选择在撞击坑内部有助于深入了解克里普岩的成分、结构、撞击坑结构及玄武岩石成分的垂向变化等科学问题。

(2) Kepler、Copernicus 撞击坑底部可能无大面积的平坦区域，软着陆工程实现的难度较大。Aristarchus 撞击坑东南部存在面积为 15~20 km^2 的平坦区域，有可能具备作为月球软着陆点的预选点条件，应利用高空间分辨率(米级)的数据对其地形地貌进行进一步分析。

致谢 中国探月工程中心无偿提供了嫦娥一号 CCD 与 LIDAR 数据，在此表示感谢!

参 考 文 献

[1] 欧阳自远.月球探测进展与中国的月球探测[J].地质科技情报，2004(a)，23(4):1—5
[2] 欧阳自远.我国月球探测的总体科学目标与发展战略[J].地球科学进展，2004，19(3):352—358
[3] Clive R N. The Moon 35 Years after Apollo What's Left to Learn? [J]. Chemie Der Erde - Geochemistry, 2009, 69 (1): 3—43
[4] Meyer C, Hubbard N J. High Potassium, High Phosphorous Glass as an important Rock Type in the Apollo12 Soil Samples (Abstract)[J].Meteoritic, 1970, 5: 210—211
[5] Warren P H. The Magma Ocean Concept and Lunar Evolution [J]. Annal Review of Earth and Planetary Science, 1985, 13: 201—217
[6] Lawrence D J, Barraclough B L, Binder A B, et a.l High Resolution Measurements of Absolute Thorium Abundances on the Lunar Surface[J]. Geophysical Research Letters, 1999, 26(17):2681—2684
[7] Haskin L A, Gillis J J, Korotev R L, et al. The Materials of the Lunar Procellaru- KREEP Terrane: A Synthesis of Data from Geormorphological Mapping, Remote Sensing, and Sample Analyses [J]. Journal of Geophysical Research, 2000, 105(E8): 20403—20415
[8] Warren P H, Wasson J T. The Origin of KREEP[J]. Reviews of Geophysics and Space Physics, 1979, 17:73—88
[9] Hubbard N J, Gast P W, Meyer C, et al. Chemical Composition of Lunar Anortho sites and Their Parent Liquids[J]. Earth and Planetary Science Letters, 1971, 13(1):71—75
[10] Solomon S C, Chaiken J. Thermal Expansion and Thermal Stress in the Moon and Terrestrial Planets[A]. Proc Lunar Sc.i Conf 7th[C]. Institute of Lunar and Planetary Sciences, Houston, 1976
[11] Wentworth S J, Mchey D S, Lindstrom D J. Apollo-12 Ropy Glassed Revised[J]. Meteoritics, 1994, 29:323—333
[12] Jolliff B L, Gillis J J, Haskin L A, et al. Major Lunar Crustal Terranes: Surface Expressions and Crust-mentle Origins[J]. Journal of Geophysical Research, 2000, 105(E2):4197—4216
[13] Jolliff B L, Wieczorek M A, Charles K S, et al. New Views of the Moon[R]. Mineralogical Society of America Geochemical Society, 2006
[14] Hagerty J J, Lawrence D J, Hawke B R, et al. Thorium Abunrdances on the Aristarchus Plateau: Insights into the Composition of the Aristarthus Pyroclastic Glass Deposits[J]. Journal of Geophysical Research, 2009, 114(E04002):1—15
[15] Lucey P G, Talor G J, Malaret E. Abundance and Distribution of Iron on the Moon[J]. Science, 1995, 268:1150—1153
[16] Lucey P G, Blewett D T, Hawke B R. Mapping the FeO and TiO_2 Content of the Lunar Surface with Multispectral Imagery[J]. Jourrnal of Geophysical Research, 1998, 103:3679—3699
[17] Lucey P G, Blewett D T, Jolliff B L. Lunar Iron and Titanium Abundance Algorithms Based on Final Processing of Clementine Ultraviolet-visible Images[J]. Jounal of Geophysical Research, 2000, 105:20297—20305
[18] Mouelic S L, Lucey P G, Langevin Y, et al. Calculating Iron Contents of Lunar Highland Materials Surrounding Tycho Crater from Integrated Clementine UV-Visible and Near-infrared data[J]. Journal of Geophysical Research, 2002, 107(E10):4-1—4-9

Advances in the Study of Lunar Opposition Effect

Liu Jian[1,3]   Ouyang Ziyuan[1,2]
Li Chunlai[2]   Zou Yongliao[2]

(1. I Institute of Geochemistry, Chinese Academy of Sciences, Guiyang 550002, China;
2. National Astronomical Observatory, Chinese Academy of Sciences, Beijing 100012, China;
3. Graduate School, Chinese Academy of Sciences, Beijing 100039, China)

Abstracct Photometry is one of the main methods of planetary remote sensing. The opposition effect is a sharp surge in brightness around zero phase angles. Research on opposition effect is an important branch of photometry and also is an important tool in remote sensing of the Moon. In this paper, we reviewed the main laboratory experiments, which depend on simulate samples, lunar soil samples, telescope observation sand spacecraft data, performed by all kinds of work on the lunar opposition effect. And we also reviewed the theoretical development of the lunar opposition effect (i.e., the major causes of the lunar opposition effect): the shadow hiding mechanism causes the lunar opposition effect, which includes the famous models (Hapke model and Lumme & Bowell model) ; then, the coherent back scatter mechanism; and now, the model combining the shadow hiding and coherent back scatter. China has sponsored the Chang'E plan of lunar exploration, and the plan along with the SMART-1 gives a good chance to lunar opposition effect research when the data on the opposition surge at very small phase angles are obtained by the spacecrafts.
Key words   Advance   Lunar opposition effect   Shadow hiding   Coherent back scatter

1 Introduction

The light scattered from the surfaces of solar system objects bears information on the nature of their surfaces, which can improve our knowledge on the origin and evolution of the whole solar system. The opposition effect refers to the narrow peak in the intensity of light scattered from a particulate medium directly back in the direction toward the source, which also means the brightness increases with decreasing solar phase angle (Hapke et al., 1998). It is one of the major challenges in optics of planetary surfaces. Seeliger (1887) discovered firstly a brightness peak in the phase curve of Saturn's rings. And the lunar opposition effect was discovered by Gehrels et al. (1964). He found the brightness of any area on the lunar surface increases by more than 40% between the solar phase angles of 0° and 4°. In order to discover the theoretical mechanism of the lunar opposition effect, numerous laboratory experiments with simulate samples, lunar soil, telescope observation sand spacecraft data were performed on the lunar opposition effect. More and more work showed that both shadow hiding and coherent back scatter contributed to the causes of the lunar opposition effect. Now, the major challenge of it is how to combine the two mechanisms very well.

2 Experiments

2.1 Experiments on simulate samples and lunar soil

More than one hundred years ago, photometric studies of surfaces with complicated structures in laboratories started. Ångström (1885) is a pioneer in laboratory photometry of rough surfaces by visual methods.

Barabashov and Chekirda (1945) carried out the first laboratory measurements with a photo element in a planetology context and it is the beginning of precise photometry of simulators on planetary surfaces. Hapke and Horn (1963) and Diggelen (1964) also made detailed experiments on the opposition effect. Numerous materials such as volcanic ash, rock powder, lichen, rubber, and sponge were used. However, none of the measurements reproduced the lunar opposition effect properly, though one type of lichen (cladonia rangiferina) gave a good match to the lunar phase curve. Oetking (1966) measured samples of smoked magnesium oxide (MgO), which possesses very high albedo, and found that a well-expressed opposition surge with an amplitude of about 25% started at approximately 10°. The effect of coherent back scatter was discovered for very bright surfaces composed of tiny particles.

In order to understand the nature of the narrow and wide opposition surges, Shkuratov et al. (1997) conducted laboratory measurements with a sample formed by MgO smoked deposits at the phase angles (0.2°-3.5°), which suggested two opposition surges ($\alpha<1.5°$ and $\alpha<10°$), both with amplitudes of about 20%. Shkuratov et al. (1999, 2002) carried out simultaneous photometric/polarimetric measurements of laboratory samples to simulate the structure of planetary regoliths. Laboratory research was carried out with a laboratory photometer/polarimeter at phase angles of 0.2°-0.4° and wavelengths of 0.45 and 0.63μm. Their studies confirmed that the degree of polarization for highly reflective dielectric surfaces depends not only on phase angle, but also on surface tilt. Even at exact zero phases the degree of polarization for tilted surfaces can be non-zero. Muinonen et al. (2002b) measured two sets of Sahara sands differring in particle size. The measurements were carried out with the laboratory photometer/polarimeter of Kharkov Astronomical Observatory at the wavelength near 0.65μm. The albedos of the samples, estimated at the phase angle of 2°, are typical for lunar high lands. The results showed dependences of the spike amplitudes and widths on the size of sand particles. It also means that the smaller the particle size, the greater the amplitude. The SMART-1 measurements of the opposition effect could help estimate regional variations in average particle size of regolith.

Lunar regolith is commonly used as a prototype to explain the electromagnetic scattering properties of solar system regolith, so lunar soil was studied at the laboratory using the soil samples sent back by Apollo missions (Hapke et al., 1993). Hapke et al. (1993, 1998) measured the reflectance of Apollo lunar soil samples in linearly and circularly polarized light. The result showed that coherent backscatter was a major contributor to the lunar opposition effect. It was concluded that neither coherent back scatter nor shadow hiding alone can account for the photometric function of the Moon at small phase angles, and the zero-phase peak of lunar regolith was caused by both shadow-hiding and coherent back scatter effects in roughly equal amounts.

2.2 Telescope observations and spacecraft data

Making full use of the telescope observation and spacecraft data is another major research method of the lunar opposition effect. Wildey (1978) produced an image of phase ratio I (2.0°)=I(4.5°), where I (α) is lunar surface brightness at a phase angle. He found the overallmare/high land contrast in this parameter appeared to be very weak. High values of the ratio were observed in regions with moderate albedos. Akimov and Shkuratov (1981) performed similar studies on the phase ratio I (3.2°)=I (14.5°) images of the Moon for two spectral ranges (λ=0.55μm and 0.38μm). Ratio I (3.2°)=I (14.5°) at =0.55μm is much higher for regions with intermediate albedo. As to the ultraviolet range, the ratio increases system atically with albedo. Shkuratov et al. (1994) also conducted studies on lunar phase-ratio images.

The data acquired by the Apollo and Clementine missions on the opposition surge at these small phase angles are an important source of the research. Pohn et al. (1969) and Whitaker (1969) found that brightness ratio I (0°)=I (8°) is basically within the limits of 1.3-1.4. Surface variations in ratio I (0°)=I (5°) were also investigated. Nozette et al. (1994) reported that Clementine discovered a narrow lunar opposition spike with an amplitude of 20% in the last 0.25. However, Shkuratov and Stankevich (1995) and Shkuratov et al. (1997) showed such a spike could not be observed. Each point of the solar disk has its own phase angle, which makes it impossible to observe any details of brightness phase curves because their widths are less than the angular radius of the Sun.

Buratti et al. (1996) analyzed Clementine lunar data. He found the lunar brightness is enhanced by more than 40% between phase angles of 4° and 0°, and the opposition surge is somewhat larger at 0.41μm than at 1.00μm, and the amplitude of the surge depends noticeably on terrain type-it is about 10% greater at lunar highlands. He concluded that shadow hiding was dominant on the sample surface; however, the data cited to support this conclusion could be interpreted in other ways. In particular, photometric differences between the high lands and the maria support coherent backscatter. Helfenstein et al. (1997) analyzed telescopic lunar data and concluded that both shadow hiding and coherent back scatter contributed to the opposition effect of the Moon. However, this conclusion does not have an observational basis as strong as one would like because it rests rather critically on one data point. Hillier (1997) also analyzed the Clementine lunar data. He found that most of the surges could be explained by shadow hiding with a halfwidth of 8°. However, at the brightest region (the high lands at 0.75-1.0μm), a small additional narrow component (halfwidth of <2°) with a total amplitude from 1/5 to 1/4 that of the shadow hiding surge is observed, which may be attributed to the coherent back scatter.

Shkuratov et al. (1999) analyzed the Clementine data from the UVVIS camera. The fit of the calculated curves to the average brightness phase function of the Moon, which was derived from the Clementine data, indicated that the coherent backscatter component is non-zero. The average amplitude of the opposition surge of the Moon in the range of phase angles 0°-1° was approximately 10%. The data also showed a flattening of phase-dependent brightness at angles less than 0.25°, which was caused by the angular size of the solar disk. The lunar brightness phase curves at small phase angles were almost the same at different wave lengths, even though at larger phase angles (5°-50°) the lunar surface would become distinctly redder with increasing phase angle.

The Chang'E program and the SMART-1 (Xu Tao et al., 2005) will give good chances to the lunar opposition effect research when the suited data on the opposition surges at very small phase angles are obtained by the spacecrafts.

3 Theoretical studies

3.1 Shadow hiding mechanism

Seeliger established a traditional way to explain the opposition effect. and the surface particles were considered to hide their own shadows, resulting in increased brightness. Most of the theoretical work on the shadow hiding is based on Seeliger's studies. In a particulate medium the interstices among the particles can be regarded as "tunnels" through which light can penetrate. The mutual shadows cast by particles in the upper regolith were hidden at opposition, which resulted in increased brightness, but the shadows became rapidly visible as the phase angle increased (Irvine, 1966;Hapke, 1986). Because the opposition effect is a sensitive indicator of the surficial compaction state and particle size (Hapke, 1986), observations at small solar phase angles are important to obtain.

The statistical approach to describing particulate surfaces was introduced by some scholars who had made efforts to parameterize the scattering medium in detail. Most of the theoretical work approximates the effect of multiple scattering, and single-particle scattering is usually taken into account in a semi-empirical way using the parameterized phase functions. The two widely used theoretical models, Hapke's law (1981, 1986) and the (H,G) system (Lumme and Bowell, 1981 a, b; Bowell et al., 1989), can give nearly perfect fits to the phase curves, including the opposition effect. These models are very important in the classification of lunar surface units.

Photometric equations were derived from the radiative transfer theory, including corrections for shadowing, to be applied to the planetary surfaces. Hapke's photometric equation (1986) has been widely applied in the interpretation of lunar and planetary photometry, as well as in a great number of laboratory experiments (e.g. Buratti, 1985; Bowell et al., 1989; Verbiscer and Veverka, 1990; Hartman and Domingue, 1998). The following model is the Hapke's photometric model (we only explain the function of the Hapke's photometric model, which describes the opposition effect):

$$r(i,e,g) = \frac{w}{4\pi} \frac{\mu_0}{\mu_0 + \mu} \{[1 + B(g)]p(g) + H(\mu_0)H(\mu) - 1\} \quad (1)$$

$$B(g) = B_0 / [1 + \tan(g/2)/h] \quad (2)$$

where B_0 is the amplitude of the opposition effect; h, the width of the opposition peak, which is related to soil structure; g, the phase angle.

$$B_0 = S(0)/wp(0) \quad (3)$$

where $S(0)$ is the fraction of light scattered from near the surface of a particle at $g=0$; w, the average single scattering albedo; and $p(g)$, the average particle angular scattering function.

The radiative transfer model established by Lumme and Bowell (1981 a, b) has four essential parameters: w is the single scattering albedo which defines the total brightness and the proportion of multiple scattering; g, the asymmetry factor of the single-scattering phase function; p, roughness, the higher value of which was considered to increase the opposition effect; and D, the volume density of the surface material, more porous material would help increase the opposition peak. The H G magnitude system for asteroids, adapted by IAU Commission in 1985, was developed from the Lumme and Bowell scattering model (Bowell et al., 1989). The V-band magnitude $H(\alpha)$ (at phase angle α) of an asteroid including the Moon is defined as

$$H(\alpha) = H - 2.5 \lg[(1-G)\Phi_1(\alpha) + C\Phi_2(\alpha)] \quad (4)$$

where H is the absolute magnitude at $\alpha=0$; G is the slope parameter, selected to be equal to zero for steep phase curves connected with low albedo objects, and unity for shallow phase curves; Φ_1 and Φ_2 are phase functions.

The phase functions Φ_1 and Φ_2 are defined as

$$\Phi_i = \exp\left[-A_i \left(\tan\frac{1}{2}\alpha\right)^{B_i}\right], \quad i=1,2 \quad (5)$$

where $A_1=3.33$, $A_2=1.87$, $B_1=0.63$, and $B_2=1.22$. Φ_1 and Φ_2 are related to surface roughness and porosity and the amount of multiple scattering in the regolith, respectively. The HG system has been applied to all asteroids for which orbits and magnitude observations are available (Bowell et al., 1989, Lagerkvist and Magnusson, 1990). Verbiscer and Veverka (1990) provided the first quantitative translation between the HG system and Hapke's equations.

Both Hapke's and Lumme and Bowell's models were based on the effect of shadow hiding mechanism, which could not predict the sharp opposition peaks observed for moderate and high albedo objects. In fact, these peaks were in contradiction with the shadow hiding mechanism, which predicted that multiple scattering should mask the opposition effect at small phase angles (Helfenstein et al., 1997, Nelson et al., 1998). The ambiguity of the fitting parameters is the major difficulty in the application of both Hapke's and Lumme and Bowell's models in the interpretation of phase curves: a unique set of parameters for a given phase curve can not be determined, even for moderately good quality data (e.g. Mallama et al., 2002). Different parameters have very similar effects on the phasecurves. The lack of measurements and observations at near zero phases, which would be essential in characterizing the opposition surge, especially the narrow components, should be considered (Helfenste in et al., 1997).

3.2 Coherent backscatter mechanism

However, recently the coherent back scatter mechanism has been introduced to explain this phenomenon. Coherent back scatter is a constructive interference effect between two rays traveling in a multiply scattering medium in reversed paths. An intensity peak results in a narrow angular cone around the back scatter direction, caused by the equal phase of the waves.

The coherent back scatter mechanism is firstly related to the lunar opposition effect by Kuga and Ishimaru (1984). The coherent backscatter mechanism was in dependently implied in optics of planetary surfaces in a few works (e.g. Shkuratov, 1988; Muinonen, 1989; Shkuratov and Muinonen, 1991). The coherent backscatter mechanism was suggested as an explanation of the negative linear polarization by Shkuratov (1989) and Muinonen (1990). The effects of polarization have to be taken into account to compute more accurately the amplitude of the intensity peak and explain the sharp negative linear polarization effects, observed for e.g. icy Galilean satellites (Mishchenko and Dlugach, 1992; Rosenbush et al., 1997). The vector computation was used in Mishchenko et al. (2000) to present a vector solution for the polarization opposition effect in the full phase angle ranges.

Muinonen (2002) presented a numerical Monte Carlo algorithm for multiple scattering including vector radiative transfer and coherent backscatter. The parameters which varied in the study are the single scattering albedos and mean-free path parameters, and the wave-length was held constant (Muinonen et al. 2002a).

3.3 Combining shadow hiding and coherent backscatter

The above results showed that both shadow hiding and coherent back scatter contribute to the opposition effect. The radiative transfer model for planetary regoliths must take this dual nature in to account. However, there is no adequate model of the opposition effect which combines shadow hiding and coherent backscatter and accounts for polarization.

Helfenstein et al. (1997) combined the Hapke's photometric model with the Mishchenko's description of coherent back scatter, and fitted the results to the disk-integrated and disk-resolved observations of the Moon. The model contained finally eight parameters, but was reported to have offered improvements as compared to the clasical shadow hiding models.

Hapke (2002) modified his light scattering model to include coherent backscatter. Two additional parameters, the amplitude and width of the coherent backscatter opposition peak, were added, i.e.,

$$r(i,e,g) = \frac{w}{4\pi} \frac{\mu_0}{\mu_0 + \mu}[p(g)B_{SH}(g) + M(\mu_0,\mu)]B_{CB}(g) \qquad (6)$$

where $M(\mu_0,\mu)$ is the new model for multiple scattering. Single scattering is described exactly by $p(g)$, and coherent backscatter $B_{CB}(g)$ has been incorporated. The Ambartsumian-Chandrasekhar H-functions for isotropic scatterers have a more accurate analytic approximation than before.

Shkuratov et al. (1999, 2002) used analytical approximations to interpret more easily the experimental data, e.g. Clementine observations of the lunar surface, and established a new photometric function combining the shadow-hiding and coherent backscatter. The model describes the phase function with three parameters: k describes the shadow hiding effect, d describes the size of the scatterer, and l is the diffusion length of the internally scattered radiation field, which should increase with increasing albedo. The model suggests the photometric function in the form:

$$f(\alpha,b,l) = H(\alpha)D(\alpha,b,l) \qquad (7)$$

where the first factor $H(\alpha)$ accounts for strong dependence of the photometric function on α. The second factor $D(\alpha, b, l)$ presents dependences on the photometric latitude b and longitude l.

All the models including coherent backscatter have succeeded better than the traditional scattering models in describing the sharp brightness increases and explaining the polarization opposition effect as a direct problem, thus contributing to a better understanding of the physics behind the effect.

4 Conclusions

Not only laboratory experiments depending on simulate samples and lunar soil samples, but also telescope observations and spacecraft data were performed by all kinds of work on the lunar opposition effect. At the same time, the first theory of the lunar opposition effect is shadow hiding mechanism, which was first established by Seelinger, and developed very well by others. Many photometric models derived from the radiative transfer theory are established, including Hapke's model and Lumme and Bowell's model. However, many scholars found the coherent back scatter mechanism also contributes to the opposition effect, which is ensured by more and more experimental results. Now the major challenge is to establish the models combining shadow hiding and coherent backscatter. If the established models are used in the inverse problem, .ie., finding unique and realistic parameters from remote sensing data, they have yet to be studied further. Most of the coherent backscatter techniques that include polarization are numerical. Further simplifications are necessary to make them more useful in the interpretation of data other than in comparative sense. The Chang'E program and the SMART-1 will give good chances to solve the above problems involved in lunar opposition effect research if the data on the opposition surge at very small phase angles were obtained very well by the spacecrafts.

References

[1] Akimov L. A. and Shkuratov Yu. G. (1981) Phase-ratio distributions of the lunar surface in two spectral ranges[J]. *Preliminary Studies. In A stronom icheskii Circular*. 1167, 3—6 (in Russian)

[2] Ångström K. (1885) Ueber diffusion der strahlenden Warme von ebenen Flachen [J]. *Ann. Phys. Chem*. 26, 264

[3] Barabashov N. P. and Chekirda A. T. (1945) On light reflection by the surfaces of the Moon and Mars [J]. *Astron. J*. 22, 11—22 (in Russian)

[4] Bowell E., Hapke B., Domingue D., Lumme K., PeltoniemiJ., and Harris A. W. (1989) Application if photometric models to a steroids [M]. *Asteroids II*. p. 524-556. Univ. of Arizona Press, Tucson

[5] Buratti B. (1985) Application of radiative transfer model to bright icy satellites [J]. *Icarus*. 61, 208—217

[6] Buratti B. J., Hillier J. K., and Wang M. (1996) The lunar opposition surge: Observation by Clementine [J]. *Icarus*. 124, 490—499

[7] Diggelenvan J. (1964) The radiance of lunar objects near opposition [J]. *Planet. Space Science*. 13, 271—279

[8] Gehrels T., Coffeen T., and Owings D. (1964) Wavelength dependence of polarization. III. The lunar surface [J]. *Astron. J*. 69, 826

[9] Hapke B. and van Horn H. (1963) Photometric studies of complex surface, with application to the Moon [J]. *J. Geophys. Res*. 68, 4545—4570

[10] Hapke B.(1981) Bidirectional reflectance spectroscopy, 1. Theory [J]. *J. Geophys. Res*. 86, 3039—3054

[11] Hapke B. (1986) Bidirectional reflectance spectroscopy, 4. The extinction coefficient and the opposition effect [J]. *Icarus*. 67, 264—280

[12] Hapke B., Nelson R., and Smith W. (1993) The opposition effect of the Moon: The contribution of coherent backscatter [J]. *Science*. 260, 509—511

[13] HapkeB., Nelson R., and Smith W. (1998) The opposition effect of the Moon: Coherent backscatter and shadow hiding [J]. *Icarus*. 133, 89—97

[14] Hapke B. (2002) Bidirectional reflectance spectroscopy, 5. The coherent back scatter opposition effect and an isotropic scattering [J]. *Icarus*. 157, 523—534

[15] Hartman B. and Domingue D. (1998) Scattering of light by individual particles and the implications for models of planetary surfaces [J]. *Icarus*. 131, 421—448

[16] Helfenstein P., Veverka J., and Hillier J. (1997) The lunar opposition effect: A test of alternative models [J]. *Icarus*. 128, 2-14

[17] Hillier J. K. (1997) Shadow-hiding opposition surge for a two-layer surface [J]. Icarus. 128, 15—27

[18] Irvine W. M. (1966) The shadowing effect in diffuse reflection [J]. *J. Geophys. Res*. 71, 2931—2937

[19] Kuga Y. and Ishimaru A. (1984) Retro reflectance from a dense distribution of spherical particles [J]. *J. Opt. Soc. Am*. A1, 831—835

[20] Lagerkvist C. -I. and Magnusson P. (1990) Analysis of a steroid Light-curves, II. Absolute magnitudes and slope parameters in a generalized HG-System. Astron [J]. *Astrophys*. 86 (sup), 119—165

[21] Lumme K. and Bowell E. (1981a) Radiative transfer in the surfaces of atmosphere less bodies, I. Theory [J]. *Astron. J*. 86, 1694—1704

[22] LummeK. and Bowell E. (1981b) Radiative transfer in the surfaces of atmosphere less bodies, II. Interpretation of phase curves [J]. *Astron. J*. 86, 1705—1721

[23] Mallama A., Wang D., and Howard R. A. (2002) Photometry of mercury from SOHO/LASCO and earth. [J]. *Icarus*. 155, 253—264

[24] Mishchenko M. I. and. Dlugach J. M (1992) Canweak localization of photons explain the opposition effect of Saturn srings? [J]. *Mon. Not. R. Astr. Soc*. 254, 15—18

[25] M ishchenko M. I., Luck J.-M., and Nieuwenhuizen T. M. (2000) Full angular profile of the coherent polarization opposition effect [J]. *J. Opt. Soc. Am*. A5, 888—891

[26] Muinonen K. (1989) Electromagnetic scattering by two interacting dipoles. In *Proceedings of the 1989 URSIE lectromag. Theory Symposium*, Stockholm, p. 428—430

[27] Muinonen K. (1990) Light scattering by inhomogeneous media: backward enhancement and reversal of linear polarization [D]. *Ph. D. Thesis*, Univ. of Helsinki

[28] Muinonen K. (2002) Coherent backscattering by absorbing and scattering media. In *6th Conference on Electromagnetic and Light Scattering by Nonspherical Particles*. p. 223—226. Adelph ,i M D

[29] Muinonen K., Videen G., Zubko E., and Shkuratov Yu. (2002 a) Numerical techniques for backscattering by random media [M]. *Optics of Cosmic Dust*. p. 261—282, Kluwer Academic publishers, Dodrecht

[30] Muinonen K., Shkuratov Yu., Ovcha renko A. et al. (2002b) The SMART-1 AMIE experiment: implication to the lunar opposition effect [J]. *Planetary and Space Science*. 50, 1339—1344

[31] Nelson R. M., Hapke B. W., Smythe W. D., and Horn L. J. (1998) Phase curves of selected particulate materials: the contribution of coherent backscattering to the opposition surge [J]. *Icarus*. 131, 223—230

[32] Nozette S. and 33 Colleagues (1994) The Clementine mission to the Moon: Scientific overview [J]. *Science*. 266, 1835-1839

[33] Oetking P. (1966) Photometric studies of diffusely reflecting surface with application to the brightness of the Moon [J]. *J. Geophys. Research*. 71, 2505—2513

[34] Pohn H. A., Radin H. W., and Wildey R. L. (1969) The Moon's photometric function near zero phase angle from Apollo 8 photography [J]. *Astrophys. J*. 157, 193—195

[35] Rosenbush V. K., Avramchuk V. V., Rosenbush A. E., and Mishchenko M. I. (1997) Polarization properties of the Galileansatellites of Jupiter: observations and preliminary analysis [J] *Astrophys. J*. 487, 402—414

[36] Seeliger H. (1887) Zur Theorie der Beleuchtung der grossen Planeten In sbesondere des Saturn [J]. *Abhandl. Bayer. Akad. Wiss. Math.-Naturw*. 16, 405—516

[37] Shkuratov Yu. (1988) A diffraction mechanism of brightness opposition effect of surface with complicated structure [J]. *Kinemat. Fiz. Nebesnykh Tel*. 4, 33—39 (in Russian)

[38] Shkuratov Yu. G. (1989) A new mechanism for the negative polarization of light scattered by the solid surfaces of cosmic bodies [J]. *Astron*. 23, 176—180. (In Rusian.)

[39] Shkuratov Yu. G. and Muinonen K. (1991) Interpreting asteroid photometry and polarimetry using a model of shadowing and coherent backscatter [M]. *Asteroids, Comets, Meteors*, p. 549—552

[40] Shkuratov Yu., Starukhina L., Kreslavsky M., Opanasenko N. V., Stankevich D. G., and Shevchenko V. G. (1994) Principle of perturbation invariance in photometry of atmosphereless celestial bodies [J]. *Icarus*. 109, 168—190

[41] Shkuratov Yu. G. and Stankevich D. G. (1995) Can lunar opposition spike measured by Clementine exist? In *Lunar Planet. Science. 26th*. p.1295-1296. LPI, Houston

[42] Shkutatov Yu. G., Ovcharenko A. A., Gstankevich D., and Korokh in V. V. (1997) A study of light backscattering from planetary-regolith-type surfaces at phase angles 0.2-3.5 degrees [J]. *Sol. Syst. Res*. 31, 56—63

[43] Shkuratov Yu., Kreslavsky M., Ovcharenko A., Stankevich D., Zubko E., Pieters C., and Arnold G. (1999) Opposition effect from Clementine data and mechanisms of backscatter [J]. *Icarus*. 141, 132—155

[44] Shkuratov Yu., Ovcharenko A., Zubko E., Miloslavskaya O., Muinonen K., Piironen J., Nelson R., Smythe W., Rosenbush V., and Helfenste in P. (2002) The opposition effect and negative polarization of structural analogs for planetary regoliths [J]. *Icarus*. 159, 396—416

[45] Verbiscer A. and Veverka J. (1990) Scattering properties of natural snow and frost: comparison with icy satellite photometry [J]. *Icarus*. 88, 418—428

[46] Whitaker E. A. (1969) An investigation of the lunar heiligenschein. In *Apollo 11 Preliminary Science Report*. NASA SP. 201, 38—39

[47] Wildey R. L. (1978) The Moon in heiligenschein [J]. *Science*. 200, 1265—1267

[48] Xu Tao, Ouyang Ziyuan, Li Chunlai, and Xu Lin. (2005) Advances in lunar exploration detectors [J]. *Chinese Journal of Geochemistry*. 24, 95—100

The Application of the Instantaneous States Reduction to the Orbital Monitoring of Pivotal Arcs of the Chang'E-1 Aatellite

Li Jinling   Guo Li   Qian Zhihan   Ping Jinsong

(Shanghai Astronomical Observatory, Shanghai 200030, China)

Abstracct   In the Chinese lunar exploration project, the Chang'E-1 (CE-1) satellite was jointly monitored by the United S-band range and Doppler and the VLBI technique. A real-time reduction of the tracking data is realized to deduce the time series of the instantaneous state vectors (ISV) (position and velocity vectors) of the CE-1 satellite, and is applied to the orbital monitoring of pivotal arcs. This paper introduces this real-time data reduction method and its application to the orbital monitoring of pivotal arcs of the CE-1 satellite in order to serve as a source of criticism and reference.

Key words   Lunar exploration   Orbital maneuver   Trace monitoring   Instantaneous states   CE-1 satellite

The Chinese lunar exploration project CE-1 was formally set up in January 2004[1,2]. The first Chinese lunar satellite was launched on October 24, 2007. The satellite had flied in an accumulation of 494 days and had been successfully controlled landing on the Moon on March 1, 2009, which gathered precious technical and engineering experiences for the follow-on Chinese lunar explorations. The Chinese United S-band (USB) range and Doppler system and VLBI systems had been jointly tracking this satellite in a long term. We introduce in this paper the orbital monitoring process of the satellite with the real-time reduction method of the instantaneous state vectors (ISV) (position and velocity) during some pivotal arcs such as the lunar capture, the landing on the Moon and other orbital maneuver arcs.

There are similarities and significant differences between the orbital analysis and the ISV reduction of space probes[3-7]. Concerning the fundamental differences, in the orbital analysis a sufficiently long tracking pass is necessary, the observations at various epochs are integrated via the state transfer matrix (STM) and processed simultaneously with the emphasis laying on the precisely modeling of the forces exerting on the probes during the whole pass. While in the ISV reduction, the time series of the positions and velocities at various given epochs are taken as unknowns, the length of the tracking pass is not a crucial prerequisite, the forces exerting on the probe are not necessary to be precisely modeled. With the ISV reduction, the quality of the tracking data could be real-timely monitored and the trace could be measured rapidly and geometrically. Especially during the arcs of orbital maneuver, soft-landing and surface walking, because it is not easy to precisely modeling the forces exerting on the probes, the ISV reduction could be used to the rapid trace measurement[8-11].

1   A brief introduction of the ISV method

It is generally followed in the data analysis of the nowadays observations in such a way that, the unknowns are evaluated with those as accurate as possible initial figures, the theoretical observations and the partials to the unknowns are calculated according to theoretical models, the differences between the observations and theoreticals are expressed as the Taylor expansion of the corrections of unknowns, the corrections are solved from observation equations and the adopted values of unknowns are therefore updated. This process could be formally expressed as follows[12,13]:

o——the observations.

c——the theoreticals calculated from the models with initial (adopted) values of unknowns.

x——the collection of unknowns.

$\frac{\partial c}{\partial x}$ ——the partials to the unknowns deduced from the models.

$o-c$——the difference between the observations and the theoreticals.

dx——the collection of the solved-for parameters, the correction to the adopted initial values of unknowns.

$o-c = \sum \frac{\partial c}{\partial x} dx$ ——observation equation, from which dx is solved.

$x + dx$——the updated value of unknowns via observations.

The ISV reduction is briefly introduced concerning the observation models, the calculation of the theoreticals and partials, the pretreatment of the observations, the constraint of the observation equations and so on as in the following.

1.1 Observation model

The tracking data of the CE-1 satellite include two-way and three-way USB ranges and the corresponding integration Dopplers[14], as well as the VLBI delays and delay rates[4].

1.1.1 USB range ρ

$$\rho = (|r_S(t) - r_1(t_1)| + |r_S(t) - r_2(t_2)|)/2 \tag{1}$$

where t_1, t_2 are respectively the epoch of the uplink signal at emission and the downlink signal at receiving, $r_1(t_1)$ and $r_2(t_2)$ are the position vectors of the observation sites, $r_S(t)$ is the probe position at the transmission epoch t of the signal, ρ is the range at the observation epoch t_2. If the signal is emitted and received at one site then it is a two-way range, while if at two sites it is a three-way range.

1.1.2 USB integration Doppler velocity $\dot{\rho}$

$$\dot{\rho} = (\rho(t_2) - \rho(t_2 - \Delta t))/\Delta t \tag{2}$$

where Δt is the integration time, $\rho(t_2)$ and $\rho(t_2 - \Delta t)$ are the ranges, $\dot{\rho}$ is the integral Doppler velocity at t_2.

1.1.3 VLBI delay τ and delay rate $\dot{\tau}$

$$\tau = \frac{1}{c}(|r_2(t+tr+\tau) - r_S(t)| - |r_1(t+tr) - r_S(t)|) \tag{3}$$

where c is the speed of light in vacuum, tr is the flying time (light time) of the signal wave front from the probe to the reference site, r_1, r_2 and r_s are the position vectors of the reference site, the remote site and the probe, τ is the observation delay at $t + tr$. The delay rate is expressed as

$$\dot{\tau} = \Delta\tau/\Delta t \tag{4}$$

Δt is the integration time. According to the related convention of the CE-1 project, the observation epoch of $\dot{\tau}$ is at the middle of the integration.

Eq. (1) through (4) should be corrected with the delays of the transmission media, instrument, clock and so on. The tracking data of the CE-1 satellite used in this paper have already been corrected by proper means with precisions being in accordance with the requirement of the CE-1 project[5,6].

1.2 The calculation of the theoreticals

As shown by eq. (1) through (4), the theoreticals will be calculated based on the state vectors of the probe and sites at the corresponding epochs. The probe states are unknowns, while the site states are taken as known parameters. The initial values of the probe states at a given epoch could be predicted from the preliminary orbital elements, which are usually conducted in the inertial reference frame. The site states at a given epoch could be calculated based on the corresponding entries in the International Terrestrial Reference Frame (ITRF) and would be corrected by the tidal effects, the plate motion and other geophysical effects and finally transformed into the inertialreference frame.

The probe states are interrelated with the site states by the wave front of a signal. For the probe states at epoch t, the reference epoch of the corresponding site states is well-determined, and vice versa. The difference between the two epochs is the light time of the signal. Regardless whether the probe states or the site states are specified, the reference epoch of the corresponding states at the other end of the link could be iteratively calculated. In the following part of the present subsection, the iteration calculations of the probe states, the site states and the light time are discussed in order to deduce the theoreticals and partials of parameters.

1.2.1 Probe states

Orbital prediction is an important step in the space environment monitoring and in the real-time tracking of probes, and which is usually conducted with analytical methods[15]. To increase as much as possible the prediction precision is the important leverage to improve the solution precision of unknowns as well as the speed of the solution convergence. Taking the earth satellite as an example, the motion equation in the geocentric celestial (inertial) reference frame can be expressed as

$$\ddot{r} = f \qquad (5)$$

where $\ddot{r}$ is the satellite acceleration, f is the acting forces including the attraction of the Earth, the perturbations of the earth figure, the Sun and the Moon as well as other solar bodies, the atmosphere retardation, the solar radiation pressure and so on. If all the acting forces are precisely known, the probe states r and $\dot{r}$ at any epoch could be deduced based on the states r_0 and $\dot{r}_0$ at the reference epoch t_0, that is, eq. (5) requires the following initial conditions,

$$\begin{cases} r_0(t_0) = r_0, \\ \dot{r}_0(t_0) = \dot{r}_0 \end{cases} \qquad (6)$$

Generally speaking, the probe initial states $\dot{r}_0$ and r_0 are not precisely known beforehand, and only the reference values r_0^* and $\dot{r}_0^*$ are available. In the orbital determination analysis, iteration calculations are necessary to improve the precision of the reference values step by step and finally result in the high precision states at the reference epoch. In the ISV reduction, the states r^* and $\dot{r}^*$ at epoch t could be predicted based on the reference states r_0^* and $\dot{r}_0^*$ of the preliminary orbit, the corrections to the predicted values would be solved from observation equations, and finally the precise states would be iteratively reached.

1.2.2 The site states

In the probe ISV reduction, the site states are taken as known and so the precision should be guaranteed in order to decline the influence of the uncertainty on the probe ISV precision. The site states at the observation epoch could be deduced fromthe corresponding entry in ITRF. After various geophysical corrections the states are converted from ITRF into inertial system and which are accompanied with the probe initial states to iteratively calculate the light time of the signal.

1.2.3 Iterative calculation of the light time and the theoreticals

As mentioned in the above text, the reference epochs of the probe and site states are interrelated by the wave front of a signal and the difference between the two epochs is the light time. Taking the observation delay of VLBI as an example, the signal wave front is emitted from the probe and received at the site. The difference between the arrival epochs of the wave front at the two ends of a baseline is just the VLBI observation delay. The following iteration calculation process of the light time is designed during the software development of ISV reduction.

(1) Let tr represent the light time with initial value as 0, that is, $tr0 = 0$.

(2) Predict the probe position $r_s(t)$ at the given epoch t in the inertial frame.

(3) Calculate the reference site position $r_1(t+tr0)$ at $t+tr0$ in the inertial frame.

(4) Update the estimation of the light time $tr = (|r_1(t+tr0)| - |r_s(t)|)/c$. Set $dtr = |tr - tr0|$ and $tr0 = tr$. When dtr is less than a threshold setting, then take tr as the final estimation of the light time and conduct the follow-on calculations. Otherwise, back to step (3), recalculate $r_1(t+tr0)$ and update the estimation of tr.

(5) Let $tr1 = tr$ representing the light time of a signal traveling from the probe to the reference site.

(6) Let τ be the observation delay with initial value as 0, that is $\tau 0 = \tau$,

(7) Calculate the remote site position $r_2(t+tr1+\tau)$ at $t+tr1+\tau$ in the inertial frame.

(8) Update the estimation $\tau = (|r_2(t+tr0)| - |r_s(t)|)/c - tr1$ Let $d\tau = |\tau - \tau_0|$ and $\tau 0 = \tau$ When $d\tau$ is less than a threshold setting, then take τ as the final estimation of the observation delay and continue the follow-on calculations. Otherwise go back to step (7), recalculate $r_2(t+tr1+\tau)$ and update the estimation of τ.

The iteratively deduced τ through the above process is the theoretical VLBI delay. It is easy to accordingly deduce the theoretical delay rate $\dot{\tau}$ with the promise of integration time and so on, as well as to deduce those of the two-way and three-way ranges ρ and integration Dopplers $\dot{\rho}$.

The above iteration calculation process of the light time and theoreticals is characterized by rapid in convergence and high in precision. The validity has been assured by data simulations and data analysis of observations.

1.3 Parameter partials

Just taking the partial of the VLBI delay to the probe coordinate x_s as an example, it is easy to deduce the following from eq. (3),

$$\frac{\partial(c\tau)}{\partial x_s} = \frac{x_1 - x_s}{R_1} - \frac{x_2 - x_s}{R_2} \qquad (7)$$

where R_1 is the distance from the probe to the reference site, x_1 is the coordinate of the reference site and so on.

1.4 The interpolation of observations

Still taking the VLBI delay as an example, the observation of a baseline at an epoch is the measurement of the same wave front at both ends. The observations at a specified epoch from different baselines may not correspond to the same wave front. There could be two modes to deduce the probe ISV at a given epoch. (a) To deduce the probe states at the emission epoch t of a wave front. (b) To deduce the probe states at epoch $t-tr$ (tr represents the light time) corresponding to the observation epoch t of a baseline. In mode (a), the observation of a baseline at the epoch in accordance with the wave front at epoch t may not exist but could be interpolated from the observation time series of the baseline. In mode (b), if taking the reference epochs t of the observations of baseline B as the time nodes, even if the observations are evenly sampled, the reference epochs $t-tr$ of the deduced probe states would still be uneven due to the geometric variation of the probe relative to the baseline. Furthermore, for the baselines besides B, the observations in accordance with the wave front at $t-tr$ may not exist, and so the interpolation of the observation time series is still the need. Therefore, whether to follow mode (a) or (b), the interpolation of the observation series is always necessary.

In the orbit determination, the unknowns are the probe states at the reference epoch. The observations of different baselines at different epochs would correspond to probe states at different epochs, which could be expressed as functions of the states at the reference epoch via the STM, and so the interpolation of the observation series is not necessary. In the ISV reduction, the probe states at all specified epochsare taken as solving-for parameters with the emphasis on the epoch correspondence of the probe states and the observations, rather than to simultaneously process all the observations of the whole tracking pass in the assistance of the STM. The force constraint exerting on the probe during the orbital motion is emphasized in the orbital determination, while in the ISV reduction the probe states are deduced with the emphasis only on the geometric relation without the consideration of the force constraint. The ISV reduction is different from the orbital determination in the data reduction manner and so possesses distinctive qualities. Especially in the data reduction of the tracking pass of orbital maneuver, the ISV reduction is suitable to the project requirements of real-time and rapidness by emphasizing only on the geometry because the precisely modeling of the forces is rather difficult.

There are various interpolation models of observations such as Chebyshev polynomials, Lagrange polynomials, Newton polynomials, cubic spline interpolation (CPI) and so on. The CPI is based on the hypothesis that the interpolation function is smooth in the first derivative, and continuous in the second derivative. The second derivatives at all the epoch nodes of the data series are firstly made once only, then the interpolated function value at any index within the data series could be easily got. Tests of data simulation and reduction of experiment observations show that the CPI method is characterized by rapidness and high precision.

1.5 The constraint of the geocentric distance and rate of a probe

For a lunar satellite, the solid angle of the present Chinese VLBI network is only about 0.5deg, which means that, even with observations from four baselines (three independent delays), the observation equations are still strong in correlation. So the three-dimensional coordinates of the probe could not be precisely determined and the precision of the radial component would be rather lower than that of the tangential component. Accordingly, the VLBI delays could be accompanied with ranges, or be constrained by the geocentric distance of the probe,

$$|r_s| = |r_{s0}|$$

That is to suppose the geocentric distance $|r_s|$ of the probe at epoch t is equal to $|r_{s0}|$, the prediction of the initial orbit. After linearization, one more observation equation is resultant as following.

$$\frac{x_s}{|r_{s0}|}\mathrm{d}x_s + \frac{y_s}{|r_{s0}|}\mathrm{d}y_s + \frac{z_s}{|r_{s0}|}\mathrm{d}z_s = 0 \tag{8}$$

The a *priori* uncertainty of the geocentric distance would be in accordance with that of the initial orbit.

Similarly, in order to improve the conditions of the observation equation and the solution precision of the speed vector, the delay rates could be processed in accompany with Doppler observations, or with the rate constraint of the geocentric distance,

$$|\dot{r}_s| = |\dot{r}_{s0}|$$

That is to suppose $|\dot{r}_s|$, the geocentric distance rate of the probe at epoch t is equal to $|\dot{r}_{s0}|$ the initial orbit prediction. Via linearization we have

$$\frac{\dot{x}_s}{|\dot{r}_{s0}|}\mathrm{d}\dot{x}_s + \frac{\dot{y}_s}{|\dot{r}_{s0}|}\mathrm{d}\dot{y}_s + \frac{\dot{z}_s}{|\dot{r}_{s0}|}\mathrm{d}\dot{z}_s = 0 \tag{9}$$

Again, the a *priori* uncertainty could be determined from the initial orbit.

2 Application examples of the ISV reduction

The Chinese USB and VLBI network were organized by the tracking and control system of the CE-1 satellite respectively on March 17 to 20, 2005, May 29 to June 2, 2006 and January 22 to 24, 2007, to jointly track the Chinese Tance-1 satellite and the Smart-1 satellite of the European Space Agency (ESA). Through the rehearsal of the observation reduction real-timely and afterward, the ISV reduction was improved to some extentconcerning the method and the software development, especially there is significant improvement in the real-time processing of the tracking data and the solution precision of parameters. The CE-1 satellite was successfully launched on October 24, 2007. Since then the Chinese USB and VLBI network had conducted long term tracking, the ISV reduction had successfully been applied to the real-time monitoring of the tracking data quality and the rapid trace measurement of the CE-1 satellite.

2.1 Smart-1 tracking experiment in 2006

In May 29 to June 2, 2006, the Chinese USB and VLBI network jointly tracked Smart-1, an ESA lunar satellite. At that time, the Smart-1 satellite could be continuously tracked for about 12 hours every day for stations within the Chinese territory. Before, after and in the middle of the satellite tracking pass, extragalactic radio sources were interpolated observed in order to correct the clock behavior. The sampling interval of the VLBI delay and rate was 5 s while that of the range and Doppler was 1 s. However, since then due to the large systematic errors in the USB data and which were hard to identify and be corrected in the real-time mode of the ISV reduction, the constraints of the geocentric distance and rate were applied.

Fig. 1 demonstrates the delay residuals of the tracking on June 2 in the real-time ISV reduction. The pre-fit residual is the difference of the delay observation and the initial orbital prediction, as indicated by the black points, while the post-fit residual is the delay residual after the ISV reduction, as indicated by the gray points. It is shown that the pre-fit residuals fluctuate up and down within [−5 m, +5 m], equivalent to about a dozen of nanoseconds, and so the tracking data as well as the predicted theoreticals by the software are good in precision. The post-fit residual is at the level of 1 m, which is obviously less than the pre-fit residuals and which indicates the ISV reduction is successful.

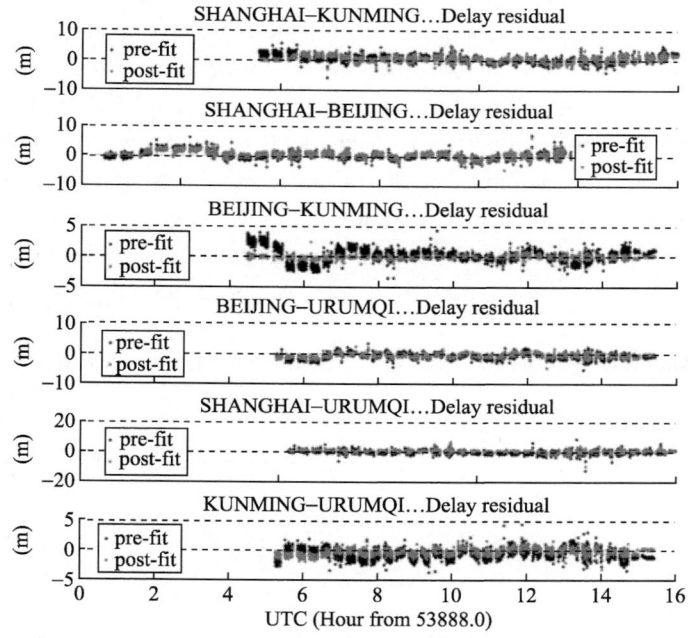

Fig.1 Pre- and post-fit residual of the tracking delays of Smart-1

Figure 2 demonstrates the comparison in the geocentric right ascension and declination between the real-time ISV reduction and the ESA reconstructed orbit, which is high in precision. It is seen that the consistence is good, the standard deviation of the differences is about 0.1 as, which is equivalent to 186 m at the lunar distance. And so the precision of the VLBI observablesas well as the ISV reduction method and software is reliable.

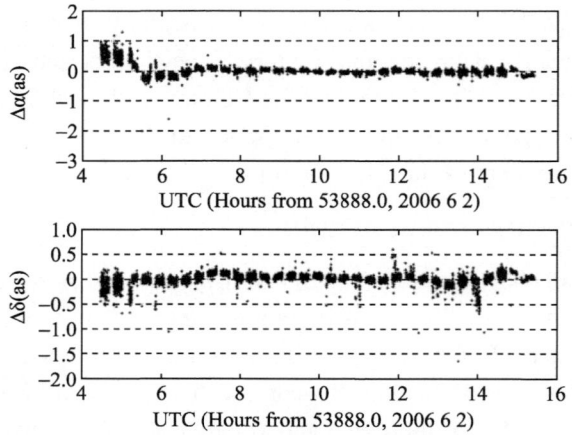

Fig.2　A comparison between the real-time ISV reduction and the ESA reconstructed orbit of Smart-1

2.2　Trace monitoring of the CE-1 satellite at the lunar capture arc

The designed orbit of the CE-1 satellite is in turn the parking, transfer and lunar orbit. During the injection process from the transfer to the lunar orbit, the braking maneuver near the perilune is the key link. The rapid and precise measurement of the probe orbit near the perilune is the critical duty of the orbital monitoring system, and which serves to the identification of properly injection as well as the basis of further orbital maneuvers. During the braking maneuver of the CE-1 satellite near the perilune, there is no sudden change in the satellite position, but in the speed due to the braking. The time series of the satellite positions could be used in the orbital identification, but a certain amount of data accumulation should be necessary. If the ISV is available, the instantaneous orbital elements then could be easily deduced and serve to the rapid orbital identification.

On November 5, 2007, the CE-1 satellite was braking maneuvered near the perilune. The time series of the ISV was real-timely deduced from the VLBI tracking delays and rates as well as the USB range and Doppler. Table 1 shows the corresponding instantaneous orbital elements, from which it is demonstrated that the braking maneuver lasted about 30 min. The evolution of the orbital eccentricity indicates that the lunar satellite orbit gradually changes from a hyperbola, into a parabola, and then into an ellipse, which means a successful capture by the Moon. During the mission period, this successful capture was quickly declared based on the evolution of the orbital eccentricity from ISV reduction only with a delay of about five minutes.

Table 1　Trace monitoring of the CE-1 satellite during the lunar capture arc on November 5, 2007

Epoch(UTC)	Semi-major axis/km	Eccentricity
	Hyperbola	
03:13:01	6618.708	1.29441
03:13:31	6641.511	1.29340
03:14:00	6664.319	1.29238
	Very near to parabola	
03:17:07	3406.991	1.08194
	Ellipse	
03:20:53	2990.028	0.84911
03:20:58	3016.474	0.84918
03:42:01	6205.055	0.68567
03:42:30	6154.900	0.68305
03:43:00	6155.691	0.68311

2.3 Monitoring the maneuver process of the CE-1 satellite in lunar orbit

After the lunar capture of the CE-1 satellite, several orbital maneuvers were conducted in order to let the satellite reject into the designed orbit. The trace evolution of the satellite is monitored via real-time ISV reduction. As an example, Table 2 shows the maneuver monitoring of the CE-1 satellite on November 6, 2007. The maneuver only lasted about 15 min, but the orbital semi-major axis and the eccentricity significantly evolved respectively from 6000 to 2700 km and from 0.68 to 0.28. The orbital

Table 2 Real-time monitoring of the orbital maneuver of the CE-1 satellite on November 6, 2007

Epoch(UTC)	Semi-major axis(km)	Eccentricity	Inclination(deg)	Argument of the ascending node(deg)
03:21:14	6015.141	0.67682	87.524	265.446
03:21:39	5543.681	0.64883	87.522	265.421
03:22:08	4785.971	0.59599	87.496	265.293
03:22:23	4618.108	0.58076	87.500	265.293
03:22:43	4418.871	0.56143	87.503	265.290
03:26:29	3360.283	0.42134	87.555	265.223
03:26:49	3237.410	0.39931	87.559	265.221
03:27:33	3214.633	0.39420	87.561	265.277
03:28:37	3071.672	0.36547	87.573	265.279
03:28:46	3056.361	0.36189	87.581	265.279
03:29:11	2993.481	0.34900	87.578	265.278
03:29:26	2971.901	0.34367	87.583	265.281
03:31:43	2600.635	0.24986	87.607	265.299
03:31:58	2597.407	0.24883	87.605	265.310
03:32:18	2592.536	0.24780	87.603	265.326
03:32:33	2594.346	0.24800	87.602	265.339
03:32:57	2591.983	0.24783	87.594	265.357
03:33:02	2593.461	0.24816	87.593	365.361
03:33:17	2595.940	0.24898	87.589	265.373
03:33:46	2601.796	0.25105	87.578	265.396
03:35:34	2681.237	0.27258	87.535	265.481
03:36:43	2704.682	0.27867	87.531	265.484

inclination and the argument of the ascension node remain relatively stable respectively at about 87 deg and 265 deg, indicating that this maneuver mainly corresponds to the change in the shape of the orbit from applanation to near circular.

2.4 Monitoring the landing trace of the CE-1 satellite on the Moon

The CE-1 satellite had continuously flied for 494 days and was controlled landing on the Moon on March 1, 2009, which accumulated technical and engineering experience for the follow-on Chinese lunar exploration program. Figure 3 shows the landing trace of the CE-1 satellite on the Moon resulted from the ISV reduction of the USB andVLBI tracking data. From the top down-on it is the lunar longitude, latitude and surface height (reference lunar radius as 1738 km) in the lunar primary axis system (LPAS). In the transformation from the geocentric celestial coordinate system to the LPAS, the DE405 ephemeris and its lunar libration parameters are adopted. From Figure 3 it is shown that, around UTC07h45m the lunar longitude and latitude of the trace of CE-1 satellite exhibit significant change in the trend, which is in good consistence with the orbit maneuver period from UTC7h36m to UTC7h56m. Since the points on the trace are deduced independently, the continuity of all the points and the evenness of the curve as a whole indicate that the tracking observation and the ISV reduction are rather successful, which serves strong support to the realization of the trace measurement of the lander in the second stage of the Chinese lunar exploration program as concerns the methodology, software, confidence and so on.

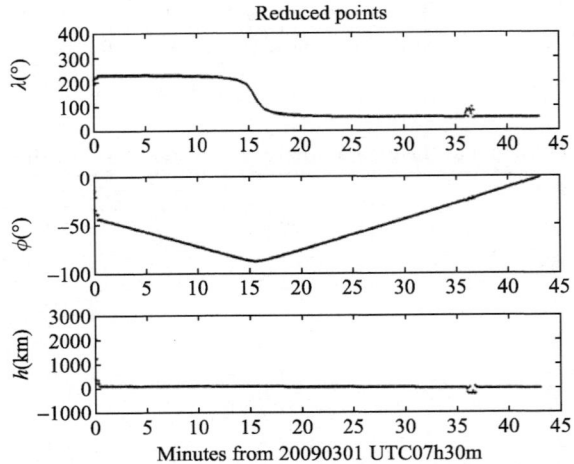

Fig. 3   The landing trace of the CE-1 satellite on the Moon in the lunar primary axis system

The coordinates of the landing point is deduced via further data analysis as shown in Table 3, in which the "Solved" column shows the LPAS coordinates of the CE-1 satellite at the emission epoch of the final signal resultant from the ISV reduction of the tracking data, the uncertainty is the formal error of the solution. The "Smoothed" column shows the smoothed and fitted results of the coordinate series of all the points on the trace, the uncertainty is the standard deviation of the data series.

Table 3   LPAS coordinates of the landing point of CE-1 satellite on the Moon

	Solved	Smoothed
Epoch	20090301UTC8h13m06.514s	
E.long	52.2760deg ±0.0018deg	52.2732deg ±0.0040deg
S.lat	1.6407deg ±0.0031deg	1.6440deg ±0.0091deg
Surface H.	−3.30km ±0.06km	−3.56km ±0.18km

Concerning the landing epoch, it was shown during the correlation processing of the VLBI tracking data, the signal disappeared at the tracking stations at about UTC8h13m7.8s, when the earth-moon distance is about 378336 km and the light time of the signal is about 1.26 s. It is therefore deduced that the landing epoch of the CE-1 satellite on the Moon is at about UTC8h13m6.5s, which is consistent with the epoch shown in Table 3 through ISV reduction.

The uncertainty of the smoothed coordinates in Table 3 corresponds to a tangential linear uncertainty on the lunar surface as 0.30 km, a three-dimensional (3D) uncertainty as 0.35km. Taking consideration of the step length in the ISV reduction being 0.01s, if the uncertainty of the epoch node is three times of the step length, i.e., 0.03s, based on the smoothed series of the coordinates, the corresponding uncertainty of the landing point in the lunar longitude, latitude and surface height is respectively ±0.0001deg, ±0.0017 deg and ±0.00 km. Moreover, in the ISV reduction the integration time of the VLBI delays is 1s, there may exist a bias as about ±0.25 s in the reference epoch of the final data point of every baseline. Based on the smoothed data series the accordingly estimated uncertainties in the three directions are respectively ±0.0006 deg, ±0.014deg and ±0.01km. Taking a synthesized consideration of the coordinate fluctuations of the data series, the step length in the ISV reduction and the integration time in the VLBI correlation processing, the coordinate uncertainty of the landing point on the

Moon in the lunar longitude, latitude and surface height are respectively ±0.0040 deg, ±0.0168 deg and ±0.18 km. The corresponding tangential linear uncertainty is 0.52 km, the 3D uncertainty is 0.55 km. If the uncertainty in the reference epoch of the VLBI delay is about ±0.5 s, the corresponding tangential and 3D uncertainty would be 0.90 and 0.92 km, respectively. Therefore, for a rapidly moving target, it is helpful to improve the positioning precision by reducing the integration time in the VLBI correlation processing. It is deduced from the above analysis that, if in the second stage of the Chinese lunar exploration program the lander will also be jointly tracked by the radio range and Doppler as well as VLBI, even with the present technical specifications it could still be promising to possess a positioning precision of the lander on the Moon as about 1 km.

Table 4 Statistics of the differences of overlapping arcs resultant from different data collections   (unit: m)

Date	USB+VLBI			USB		
	R	T	N	R	T	N
24/25	12.09	36.70	14.38	14.76	149.16	223.49
25/26	3.45	−43.10	−116.50	50.04	−43.90	−63.48
26/27	1.39	−123.35	−19.20	3.9	−196.44	−121.72
27/28	10.83	−10.35	4.02	10.07	250.37	42.97
28/29	1.64	14.87	−0.87	1.28	−446.26	449.87
Mean	5.88	−25.05	−23.63	16.01	−57.41	106.23
STD	5.17	62.47	53.32	19.74	277.54	232.70

3 Discussion

The length of the tracking pass is not a prerequisite for the ISV reduction, which could geometrically realize the monitoring of the quality of tracking data and the rapid trace measurement. The ISV reduction is suitable to the trace measurement during orbital maneuver arcs and which has been assured by experiment observations during the first stage of the Chinese lunar exploration program. In the second stage of theChinese lunar exploration program it is planned to soft-land a lander on the Moon, to release a rover and to patrol and survey the selected lunar district. The ISV reduction is very promising in the landing tracemonitoring of the lander, the determination of coordinates of the landing point, the walking trace measurement of the rover on the Moon and so on.

It is necessary to point out that, the ISV reduction could be conducted only when there are sufficient independent observations of a wave front of signal. The CE-1 satellite was jointly tracked by USB and VLBI technique, which is the basic guarantee of the ISV reduction. The contribution of VLBI technique to the orbital determination of the CE-1 satellite is obvious too, especially for the situation of short tracking arcs. During the on-track tests of the CE-1 satellite from November 8 to 30, 2007, there was generally momentum wheel adjustment once per day. The orbital analysis is made based on one-day's tracking data and the orbit is extrapolated 12 hours, which is compared with the next day's orbital analysis. As examples, Table 4 lists out the coordinate difference in the RTN system (the orbital radial, along-track and normal direction) at the beginning point of five pieces overlapped arcs, from which it is seen that, the standard deviation (STD) of the coordinate difference in three directions from the jointed orbital analysis of the USB and VLBI tracking data is obviously smaller than that only with the USB data. The successful application of the VLBI technique in the first stage of the Chinese lunar exploration program lays a favorable foundation for the follow-on lunar and deep space exploration projects.

Acknowledgements The author expresses their great thanks to the VLBI center of the sub-system of VLBI orbit monitoring of CE-1 satellite for providing the tracking data.

References

[1] Ouyang Z Y. The development strategy and the overall scientific objective of the Chinese lunar exploration. Prog Geosci, 2004, 19(3): 355—357

[2] Ouyang Z Y. The progress of the lunar exploration and the Chinese actions of lunar exploration. Nat J, 2005, 27(4): 187—188

[3] Wang J S, Chen J R, Ma P B, et al. Joint orbit determination for TC-1 satellite with USB and VLBI. J Space TT&C Technol, 2006, 25(1): 31—33

[4] Sovers O J, Fanselow J L. Astrometry and geodesy with radio interferometry: Experiments, models, results. Rev Mod Phys, 1998, 70(4): 1393—1454

[5] Huang Y. A study on the orbit determination of the Chang'E-1 lunar probe. Doctorial Dissertation. Shanghai: Shanghai Astronomical Observatory, Chinese Academy of Sciences, 2006

[6] Guo L. Reduction of the instantaneous state vectors of spacecraft based on VLBI tracking data. Doctorial Dissertation. Shanghai: Shanghai Astronomical Observatory, Chinese Academy of Sciences, 2007

[7] Qiao S B, Li J L, Sun F P. Application analysis of lunar exploration satellite positioning by VLBI technique. Acta Geodaetica et Cartographica Sinica, 2007, 36(3): 262—268

[8] Li J L, Guo L, Zhang B. The Chinese VLBI network and its astrometricrole. In: Jin W J, Platais I, Perryman M A C, eds. Proceedings IAU Symposium No. 248, 2007 Oct 15-19, Shanghai. Cambridge: Cambridge University Press, 2008. 182—185

[9] Guo L, Li J L, Qiao S B, et al. Monitoring the lunar capture of Chang'E-1 satellite by real-time reduction of the instantaneous state vectors. In: Jin W J, Platais I, Perryman M A C, eds. Proceedings IAU Symposium No. 248, 2007 Oct 15-19, Shanghai. Cambridge: Cambridge University Press, 2008. 194—195

[10] Li J L. The contribution of CVN to the CE-1 mission. In: Finkelstein A, Behrend D, eds. Measuring the Future. Saint Petersburg: Nauka, 2008. 193—198

[11] Guo L, Li J L. A discussion on the angular positinoingreduciont of the experiment S7123a. J Space TT&C Technol, 2007, 26(6): 30—35

[12] Press W H, Teukolsky S A, Vetterling W T, et al. Numerical Recipes in Fortran 77: The Art of Scientific Computing. London: Cambridge University Press, 1992

[13] Xu S L. FORTRAN Algorithms Procedures Set. Beijing: Tsinghua University Press, 1995

[14] Li J S. Satellite Precisely Orbit Determination. Beijing: People's Liberation Army Press, 1995

[15] Liu L. Orbital Mechanics of Artificial Earth Satellite. Beijing: Higher Education Press, 1999

A Digital Open-loop Doppler Processing Prototype for Deep-space Navigation

Jian Nianchuan[1]   Shang Kun[1,2]   Zhang Sujun[1,2]   Wang Mingyuan[1,2]
Shi Xian[1,2]   Ping Jingsong[1]   Yan Jianguo[5]   Tang Geshi[3]   Liu Junze[3]
Qiu Shi[1]   Fung Lai-Wo[1]   Zhang Hua[4]   Wang Zhen[4]   Gou Wei[1]

(1. Shanghai Astronomical Observatory of Chinese Academy of Sciences, Shanghai 200030, China;
2. Graduate University of Chinese Academy of Sciences, Beijing 100049, China;
3. Beijing Aerospace Command and Control Center, Beijing 100094, China;
4. Urumqi Astronomical Station of National Astronomical Observatories, CAS, Urumqi 830011, China;
5. State Key Laboratory of Information Engineering in Surveying, Mapping and Remote Sensing, Wuhan University, Wuhan 430070, China)

Abstract   A prototype based on digital radio technology with associated open-loop Doppler signal processing techniques has been developed to measure a spacecraft's line-of-sight velocity. The prototype was tested in China's Chang'E-1 lunar mission relying on S-band telemetry signals transmitted by the satellite, with results showing that the residuals had a RMS value of ~3 mm/s (1σ) using 1-sec integration, which is consistent with the Chinese conventional USB (Unified S-Band) tracking system. Such precision is mainly limited by the short-term stability of the atomic (e.g. rubidium) clock at the uplink ground station. It can also be improved with proper calibration to remove some effects of the transmission media (such as solar plasma, troposphere and ionosphere), and a longer integration time (e.g. down to 0.56 mm/s at 34 seconds) allowed by the spacecraft dynamics. The tracking accuracy can also be increased with differential methods that may effectively remove most of the long-term drifts and some of the short-term uncertainties of the uplink atomic clock, thereby further reducing the residuals to the 1 mm/s level. Our experimental tracking data have been used in orbit determination for Chang'E-1, while other applications (such as the upcoming YH-1 Mars orbiter) based on open-loop Doppler tracking will be initiated in the future. Successful application of the prototype to the Chang'E-1 mission in 2008 is believed to have great significance for China's future deep space exploration.

Key words   Open-loop Doppler   Differential Doppler   Deep space navigation   Chang'E-1

The radio telecommunication system of a deep space probe fulfills two main functions: the first one (uplink) is to transmit commands from a ground station to the probe; the second (downlink) is to send the telemetry signals from the probe to some ground station(s). The radio telecommunication system is used in the two areas, namely, radio science and spacecraft tracking/navigation. Radio science refers to making use of the characteristics (including amplitude, phase and polarization) of the telemetric signals to determine a planet's mass and mass distribution, ionosphere, atmosphere and surface features, and on some occasions conduct relativity tests[1,2]. On the other hand, the radio telecommunication system can be used to get the ranging, velocity and angle information of the spacecraft to support its tracking and navigation. Since 1970 various combinations of tracking data types have been used in different orbital phases of interplanetary spacecrafts, and around that time frame, VLBI (Very

Long Baseline Interferometry) began to become a practical deep space tracking technique. Prior to the 1980s, ranging and Doppler were the main observables, but subsequently differential one-way ranging and differential one-way Doppler observables based on interferometric measurement of side tone phases have prevailed. Since 2000, with high precision and low cost achieved, emphasis has been placed on stability, rapid correction and real-time response that demands better performance for new generations of radio and optical tracking systems[3].

According to the geometry of uplink and downlink, radiometric tracking can be basically divided into one-way, two-way and three-way models. On the other hand, according to the implementation of the radio links, these tracking models can be classified as either closed-loop or open-loop. The one-way tracking model is in open-loop mode, in which the tracking signals are generated by the spacecraft-equipped USO (ultra stable oscillator). But the two-way tracking model is closed-loop, in which the signals are generated by an atomic clock on an uplink ground station, and then re-transmitted after frequency multiplication by a transponder on the spacecraft. The downlink signals, coherent with the original uplink signals, will be received later by the same ground station where the two signals are cross-correlated in real time. The three-way tracking model is the same as the two-way model, with the only exception that the transmitting and receiving stations are not the same, and is therefore in open-loop mode. Figure 1 reveals differences between the various tracking models. The International Telecommunication Union (ITU) has defined the radio frequency bands (listed in Table 1) for deep-space navigation[4]. Table 2 gives the ratios (multiplying factors) of downlink and uplink frequencies used by NASA's Deep Space Network (DSN), and recommended by the Consultative Committee for Space Data Systems (CCSDS).

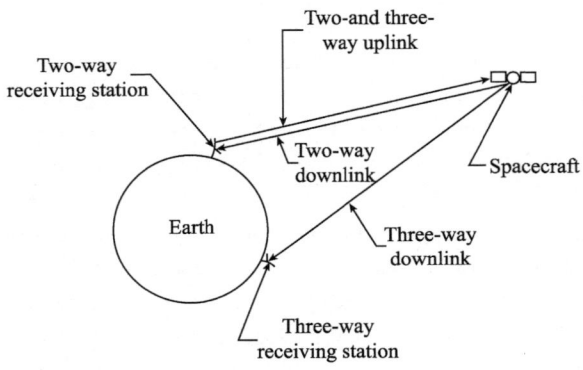

Fig.1 Two-way and three-way tracking modes[5]

Table 1 Uplink and downlink frequencies for deep-space communication

Frequency band	Uplink frequency(MHz)	Downlink frequency(MHz)
S	2110—2120	2290—2300
X	7145—7190	8400—8450
Ka	34200—34700	31800—32300

Table 2 Frequency multiplying ratios

Uplink band	Downlink band	Ratio (downlink/uplink)
S	S	240/221
S	X	880/221
S	Ka	3344/221
X	S	240/749
X	X	880/749
X	Ka	3344/749

Among these basic tracking modes, the two-way tracking is a closed-loop model where the receiving and transmitting stations are the same. But one-way and three-way tracking are open-loop models. The main difference between closed-loop and open-loop tracking models is that the transmitted and received signals are coherent in the closed-loop mode, whereas in the open-loop model the signals are non-coherent. The open-loop measuring process is more difficult than the closed-loop process because of the instability of the clock that generates the tracking signals and the systematic errors of the non-coherent signals. Earlier deep-space navigation missions were dominated by closed-loop tracking, such as Apollo, Viking and Voyager. The closed-loop tracking model has been perfected as the current USB (Unified S-Band) system. In the late 1980s, the stability of on-board ultra-stable oscillators (USO) has been greatly enhanced, making open-loop tracking possible. Open-loop tracking models include one-way or multi-way Doppler, and differential one-way or multi-way Doppler/ranging[3].

Compared with traditional closed-loop measurements, the open-loop measuring mode has the following advantages. First of all, since no uplink station is needed in one-way tracking, a lot of resources can be saved and the model can be conveniently used in very-remote deep space missions (such as Voyager). Secondly, in those very remote missions, radio signal propagation may take an extremely long time (sometimes up to several hours) —a situation that is really bad for conventional USB tracking whereas the three-way tracking model would be more practical. Third, in some special orbital phases such as transitioning between planets, the one-way VLBI/DOR tracking model has some obvious advantages. In the last 30 years, the NASA/DSN has used the open-loop tracking models in dozens of deep-space exploration missions, through which the models have matured and have become particularly important today[6].

The differential tacking types include one-way and multi-way (generally 3- or 4-way) Doppler/ranging. The precision of differential observables is higher than that of the direct observables because the instability of the uplink clock and some of the path effects can be removed by differencing the two individual Doppler measurements[7, 8]. On the other hand, most of the line-of-sight information about the spacecraft motion has been eliminated after differencing. Hence differential observables *alone* should not be used for orbit determination, but must be combined with other non-differential observables in order to recover some of the missing information about the spacecraft dynamics[9].

On October 24, 2007, the first Chinese lunar exploration satellite Chang'E-1 was launched on the Xichang launching site, and on December 11, 2007 Chang'E-1 began to orbit the moon and send back a large amount of scientific data. The tracking model of Chang'E-1 mainly depends on China's USB network which can support closed-loop two-way Doppler and ranging data for orbit determination. In addition to USB tracking, the Chinese VLBI network (CVN) also uses angle tracking data for orbit determination. In the USB tracking model, an up-link station transmits S-band signals to the satellite, which are then re-transmitted by a transponder on the satellite. Later the transmitted and received signals are brought together and processed, so that the embedded Doppler and ranging information can be extracted for tracking in a typical closed-loop model.

Following Chang'E-1 will be the Sino-Russian joint Mars mission to be carried out in September or October 2009. The Russian Photos-Grunt and the Chinese YH-1 satellites will be launched by a Russian carrier rocket, which is expected to arrive in August or September, 2010 at Mars, about 2 AU (astronomical units) from the earth. As China has no powerful uplink station for remote telecommunication, tracking the YH-1 must rely on the open-loop one-way model and the CVN will be used for such tracking activities. In the last couple of years, Chang'E-1 has provided a very good test platform for developing China's primitive open-loop tracking technique. During the mission, three-way Doppler and differential three-way Doppler demonstrations have been successfully completed. The Doppler data were collected from three CVN stations (namely the Nanshan station in Shanghai, Kunming station in Yunnan province, and Nanshan station in Urumqi) and then sent to Shanghai through special high-speed Internet links for post-processing. The three-way Doppler and the corresponding differential Doppler results have been fully verified for future open-loop tracking of the YH-1 satellite between 2010-2011.

1 Principles of open-loop Doppler measurement

1.1 Open-loop Doppler formulas

Doppler effects of electromagnetic waves arise from the relative velocities and different time scales between the transmitter and receiver. If we ignore the different time scales in a gravitational field, the first-order approximation of Doppler can be expressed in the framework of the special theory of relativity[5] as follows:

$$f_R^{1-way} \approx \left(1 - \frac{v}{c}\right) f_T \quad (1)$$

In this formula, f_T is the transmitted frequency and v is the velocity of the transmitter relative to the receiver. When the distance between the transmitter and receiver is increased, v is positive.

In the case of three-way Doppler, the signal propagation can be divided into two processes: uplink and downlink. For the uplink process, the relationship between the transmitted frequency f_T, the satellite received frequency f_S, the relative velocity of the uplink station and satellite v_1 and the speed of light c is given by[5]

$$f_s \approx \left(1 - \frac{v_1}{c}\right) f_T \quad (2)$$

For the downlink process, the relationship between the satellite received frequency f_S, the frequency multiplying faction M for re-transmission, the received frequency f_R at the downlink station, and the relative velocity of downlink station and satellite v_2 is given by

$$f_R \approx \left(1 - \frac{v_2}{c}\right) f_s \cdot M \quad (3)$$

So the three-way Doppler frequency can be expressed as

$$f_R \approx \left(1 - \frac{v_1}{c}\right)\left(1 - \frac{v_2}{c}\right) f_T \cdot M \approx \left(1 - \frac{v_1 + v_2}{c}\right) f_T \cdot M \quad (4)$$

Here second- and higher-order small quantities have been ignored.

For convenience, the three-way Doppler observable is defined as the average of v_1 and v_2.

$$f_R^{3-way} \approx \left(1 - \frac{2v_{3w}}{c}\right) f_T \cdot M \quad (5)$$

$$v_{3w} \approx -\frac{1}{2} \cdot \frac{f_R^{3-way} - f_T \cdot M}{f_T \cdot M} \cdot c \quad (6)$$

Equation (6) is the first-order formula of three-way Doppler which depends on the transmitted frequency, received frequency, re-transmission frequency ratio and the velocity of light.

1.2 Differential Doppler formulas

If the two receiving stations record the downlink signals *simultaneously*, differential one-way Doppler can be formulated. The differential one-way Doppler can be expressed as the difference of frequencies recorded at the two stations.

$$D\_f_R^{1-way} = f_R^{1-way,S2}(t_R) - f_R^{1-way,S1}(t_R)$$
$$= \left(1 - \frac{v^{S2}}{c}\right) f_T(t_{T2}) - \left(1 - \frac{v^{S1}}{c}\right) f_T(t_{T1}) \quad (7)$$

The corresponding formula of differential three-way Doppler then becomes

$$D\_f_R^{3-way} = f_R^{3-way,S2}(t_R) - f_R^{3-way,S1}(t_R)$$

$$= \left(1 - \frac{2v_{3w}^{S2}}{c}\right) f_T(t_{T2})M - \left(1 - \frac{2v_{3w}^{S1}}{c}\right) f_T(t_{T1})M \tag{8}$$

In our demonstration the main measurement errors are the frequency deviations of the USO (on the satellite) or the atomic clock (at the uplink station). The frequency source involves mainly three indices, namely, frequency accuracy, frequency stability and frequency drift rate. Frequency accuracy is the *systematic* deviation between the measurements and the nominal value. Frequency stability refers to the frequency dispersion due to the environment. Frequency drift rate is related to device aging. The Doppler systematic deviations can be removed by differencing.

The simple frequency evolution model is given by

$$f_T = f_{T_0} + Bt + A \cdot \text{rand}(t) \tag{9}$$

where f_{T_0} is the nominal frequency, B is the linear evolution factor, and A is the noise intensity expressed as the Allen variance.

In the one-way Doppler measurement, the accuracy of the observable depends on the USO on the satellite. The relationship between the systematic measurement error and USO accuracy is given by

$$\Delta f_R^{1-\text{way}} = \Delta f_T \left(1 - \frac{v}{c}\right) \approx \Delta f_T \tag{10}$$

In the three-way Doppler measurement, the accuracy of the observable depends on the atomic clock at the uplink station. The relationship between the systematic measurement errors and the clock accuracy is given by

$$\Delta f_R^{3-\text{way}} = \Delta f_T M \left(1 - \frac{2v_{3w}}{c}\right) \approx \Delta f_T M \tag{11}$$

The one-way differential Doppler measurement errors are given by

$$\Delta D\_f_{R_{1,2}}^{1-\text{way}} \approx \Delta f_T(t_{T2}) - \Delta f_T(t_{T1}) \tag{12}$$

in which t_{T1} and t_{T2} are related to two different transmission moments. Finally, the three-way differential Doppler measurement error is given by

$$\Delta D\_f_{R_{1,2}}^{3-\text{way}} \approx [\Delta f_T(t_{T2}) - \Delta f_T(t_{T1})] \cdot M \tag{13}$$

Through the above analysis we can draw the following conclusions: (1) The systematic accuracy of one-way Doppler is equal to the accuracy of the on-board USO; (2) The systematic accuracy of the three-way Doppler is equal to M times of the ground station clock accuracy; (3) The differential Doppler accuracy depends on the frequency change between the two transmission moments t_{T1} and t_{T2}, so the non-differential measurement accuracy depends on the frequency source accuracy whereas the differential measurement accuracy depends on the short-term stability of the source. The simulation results show that the time difference of the two transmission moments is about 0.01 second and so the frequency change is very small (probably under 1 mHz). On the other hand, the accuracy of the source includes long-term drifts that can reach several Hz. Thus the accuracy of the differential measurements is much higher than the non-differential case.

2 Open-loop Doppler experiments on Chang'E-1

2.1 Introducing the Chang'E-1 beacon

The Chang'E-1 satellite transmits both S- and X-band signals. The X-band signals, which are white noise and cannot be used to extract Doppler, are intended for VLBI tracking only. The S-band has two channels near 2210 MHz, one of which is for one-way tracking while the other is reserved for re-transmitting the uplink signals. Our experimental results show that the stability of the one-way signals is not too good, with a dispersion of about 1 Hz. The corresponding stability of the USO on the satellite is about 4.5×10^{-10} and so the velocity precision is about 0.1 m/s. Hence one-way Doppler cannot be used for orbit determination because of its poor precision. The uplink stations of Chang'E-1 are Qingdao and Kashi stations, both of which have relatively stable rubidium clocks. Three-way Doppler data analysis indicates that the short-term stability of these clocks is about 1.31×10^{-11} so that the precision of three-way Doppler is 20 mHz and the corresponding precision of velocity is 3 mm/s, which can be used only for common orbit determination. The following discussion is based on the three-way Doppler and differential three-way Doppler models.

2.2 Introducing the data recording devices and data flow

Experiments are supported by the CVN network and three CVN stations mentioned above are equipped with the same data recording device. For trial we use two different recording devices which are NUDAQ (PCI-9812/9810) and K5/vssp32 VLBI data sampling card provided respectively by ADLINK and NICT (National Institute of Information and Communications Technology) of Japan. NUDAQ is a 32-bit bus-based high-performance data acquisition card that can sample data at 20 Mbps and record the data simultaneously on hard disks. The ADLINK card has been further developed, including an outer clock trigger, an outer clock pre-trigger, an outer frequency standard module[10], as well as long time recording function.

The K5/vssp32 designed for VLBI data recording has two types of working models. The one-channel module has a sampling rate of up to 64 Mbps, while the four-channel model is up to 256 Mbps. There are four quantization modes: 1 bit, 2 bits, 4 bits and 8 bits.

The radio frequency signals transmitted by the spacecraft are received by an antenna at a ground station where they are subsequently down-converted to intermediate frequency. The intermediate frequency signals are then down-converted to baseband signals by a base-band converter (BBC). In our experiment we properly set the local oscillator frequency and let the carrier signal of Chang'E-1 be located at about 500 kHz in the baseband. The whole recording process can be controlled by a remote computer. After recording, post-progressing is performed to extract the Doppler information from the recorded data as shown in Figure 2.

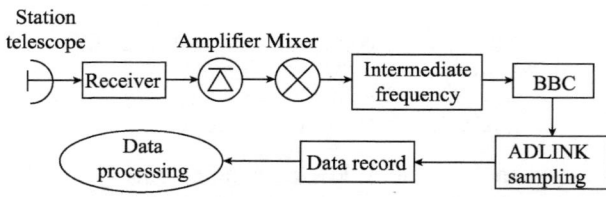

Fig.2　The flow chart for open-loop Doppler data processing

2.3　Algorithms for open-loop Doppler extraction

The signals transmitted by the Chang'E-1 satellite are different than those from natural radio sources in that they have higher energy levels and SNR (signal-to-noise ratios). We have tried two methods, which are identical in the same integration time, to extract open-loop Doppler information from the recorded data.

(1) After being down-converted and filtered, the original data are smoothed and compressed. Then the Doppler values can be extracted from the baseband signals using phase counting algorithms.

(2) The Doppler values can also be extracted from the baseband using polynomial fitting algorithms that use polynomials to fit the data phase.

Two algorithms are tried to compute the open-loop differential Doppler value.

(1) By the time-domain method (also known as "spacecraft narrowband interferometry" or INS), the original signals received by the two stations are mixed and the differential Doppler is extracted.

(2) The frequency-domain method (also known as "differenced Doppler") is to *separately* extract the open-loop three-way Doppler from the original data at each station, and then difference them to form the differential three-way Doppler.

In theory, observables based on these two methods are *equivalent*, but in practice they may differ a lot as they have rather different accuracies that may depend on a number of factors. Our test results indicate that the two algorithms agree with each other quite well. Since the second method takes less computing resources and does not require bringing the voluminous data together for cross-correlation processing, we often use the second method in practical applications if we can tolerate some loss of precision.

Using eqs. (1)–(8) that relate Doppler frequencies and changes in relative velocity, we can compute the Chang'E-1 satellite's Doppler velocities and differential Doppler velocities for orbit determination. In practical applications the integration time is usually 1 second for the three-way Doppler and the differential Doppler, but other integration times may be tried for better results as discussed below.

3 Open-loop Doppler measurements for the Chang'E-1 mission

3.1 Open-loop three-way Doppler experiment

The open-loop three-way Doppler observation of Chang'E-1 began in May, 2008. The sampling equipments were set up at Sheshan station (Shanghai) and at the same time the signal processing algorithm was being developed. The same equipments were set up at Nanshan station (Urumqi) and Kunming station (Yunnan) in the experiment. The whole system was tested many times during the Chang'E-1 mission. This paper takes the observation of Chang'E-1 on December 18, 2008 as an example to illustrate the data processing task.

The three-way Doppler observation carried out on December 18, 2008 at Sheshan station is the longest observation among the several experiments we performed. The Chang'E-1 perilune is 17 km and the apolune is about 100 km. The uplink stations are Kashi and Qingdao stations. The effective open-loop three-way Doppler observation lasted for about 9 hours, during which the uplink site is moved from Kashi to Qingdao. At the same time the Chinese USB network provides the closed-loop Doppler tracking data that can be used to validate the open-loop three-way Doppler data.

Figure 3 gives the open-loop and USB closed-loop Doppler observation results covering the period from UTC 18:00, December 18 to UTC 24:00, December 19. Some outliers have been eliminated and the discontinuous data in the figure is due to the change of uplink station. Figure 3(a) shows the three-way Doppler result while Figure 3(b) stands for the USB result. The effective three-way data covers about 9 hours as opposed to 13 hours of USB data.

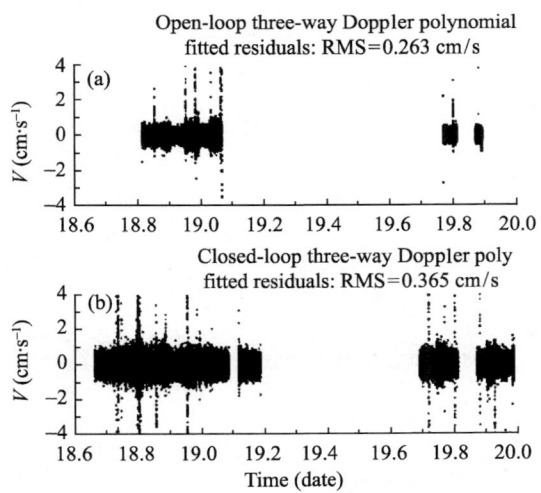

Fig.3 Fitted Doppler residuals
(a) The open-loop three-way; (b) the closed-loop two-way (USB)

A direct polynomial segment fitting is performed on the three-way Doppler data as mentioned above. The RMS (Root Mean Square) of the fitting residuals is used to represent the Doppler precision. The polynomial segment fitting residuals are shown in Figure 3, most of which vary within ±1 cm but some are outside the ±2 cm range due to discontinuities in the data recording. The statistics of fitted residuals without outliers[11,12] show that the precision of the open-loop three-way Doppler is about 2.63 mm/s and that of the closed-loop USB is about 3.65 mm/s. Hence the precision of the open-loop three-way Doppler is slightly higher than that of the closed-loop two-way Doppler (USB), both using 1-sec integration time.

The precision of the open-loop three-way Doppler can be evaluated for orbit determination with the GEODYN II software package provided by NASA/ GSFC to examine the data residuals. We use the publicized precise orbit as the initial orbit and use the three-way Doppler data to modify the initial orbit. This method has the advantage to remove systematic errors so that the random errors can be revealed. Figure 4 gives the residuals of Doppler that can reflect the seriousness of random errors. Figure 4(a) shows the residuals of three-way Doppler assuming the 100×100 lunar gravity model. Figure 4(b) shows the residuals of the two-way Doppler (USB). Both sub-figures show that the residuals are not normally distributed and there are obvious "structures" in the residuals. The long-term trend of the structure in Figure 5 is related to the periodic motion of Chang'E-1, which is about 2 hours per cycle. The short-term features of the structure are perhaps contributed by the high orders of the lunar gravity field[13,14]. The three-way Doppler residuals and the two-way Doppler residuals show obvious correlation, indicating that the two tracking models have similar precision levels. In our experiment the open-loop Doppler data can be used only to evaluate the observation precision but cannot be used to evaluate the orbit determination precision because of deficient tracking data. Application of the open-loop tracking data to orbit determination will be the focus of future work.

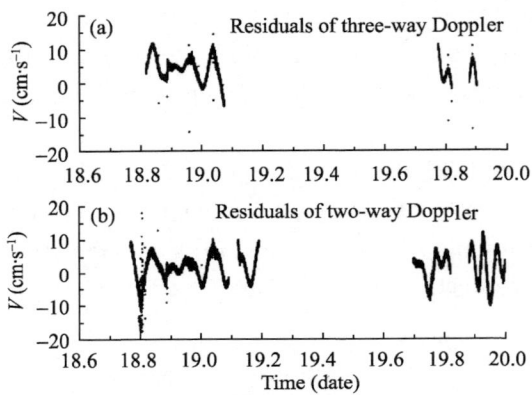

Fig.4 Doppler residuals. (a) The open-loop three-way; (b) the closed- loop

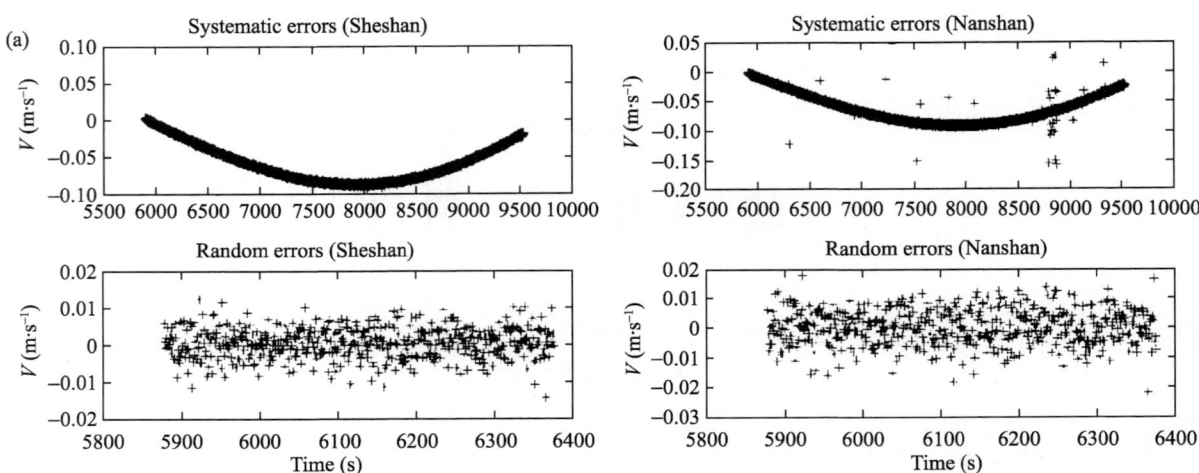

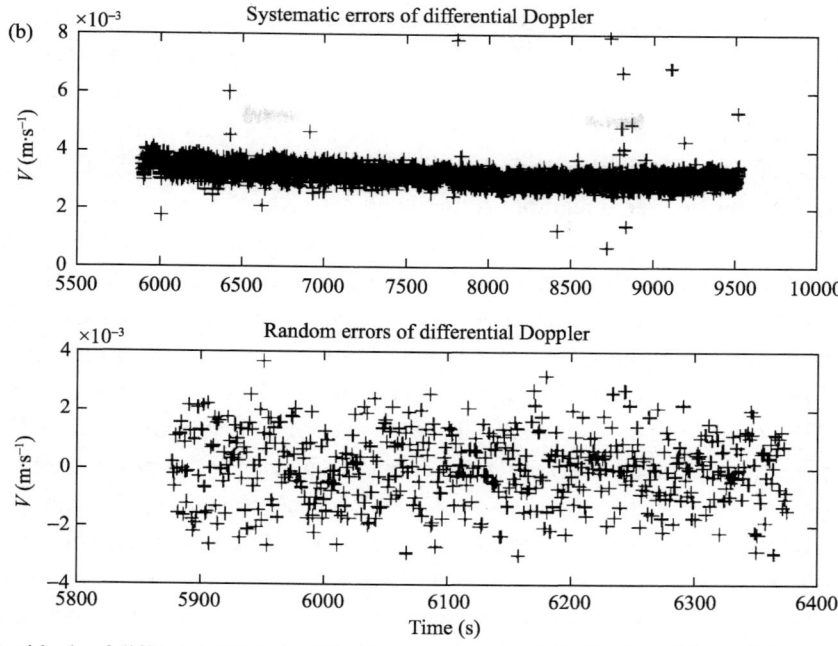

Fig.5 Residuals of differential Doppler (the first arc of August 29). (a) Non-differential; (b) differential

3.2 Differential open-loop three-way Doppler

On August 29, 2008 an open-loop three-way Doppler experiment was carried out. The original Doppler data of two arcs were recorded at the Sheshan and Nanshan stations from which we obtained the differential three-way Doppler data for the first time. We developed some simple software to compute the *theoretical* differential three-way Doppler values because GEODYN II does not support this observation data type. Table 3 gives the statistics of the systematic and random errors. Figure 8 shows the differential results of the first arc on August 29.

Table 3 Statistics of open-loop Doppler residuals for Chang'E-1

Different three-way Doppler residuals		Systematic errors (mean)	Random errors(σ)
August 29 First arc	Sheshan	4.3 cm/s	3.3 mm/s
	Nanshan	4.5 cm/s	3.5 mm/s
	Differential	4.0 mm/s	1.0 mm/s
August 29 Second arc	Sheshan	5.2 cm/s	3.2 mm/s
	Nanshan	5.4 cm/s	4.0 mm/s
	Differential	3.9 mm/s	1.2 mm/s

4 Discussion and conclusion

From May 2008 through December 2008 we performed twenty open-loop Doppler tracking experiments for Chang'E-1, with results indicating that the precision (~3 mm/s) of the non-differential three-way Doppler is in agreement with the short-term stability (about 10^{-11}) of the rubidium clock at the uplink station. Such performance is expected to be further improved when the ground-station uplink clock is replaced in the future by a more stable atomic clock (such as a hydrogen clock) with short-term stability as good as 10^{-15}-10^{-16}, But the degree of improvement is ultimately limited by other factors such as noise levels of the receiving stations and radio signal transmission media (viz. solar plasma, troposphere and ionosphere). Media effects are among the most important error sources of the Doppler observable and so must be properly corrected (see sec. 13.3.2 of ref. [16]) for better performance in both tracking and radio science applications.

Our experimentation shows that longer integration times can help to increase the measurement accuracy by averaging out some of the noise in the signals. The residuals error RMS (Sheshan station) drops from 3.2 mm/s to as little as 0.56 mm/s if the integration time is increased from 1 second to 34 seconds, whereas the error of Nanshan station, due to differences in receiver electronics, drops from 4.0 mm/s to only 0.8 mm/s. These results are comparable to the best S-band Doppler performance reported by NASA/JPL[3] in the 1980s. However, we must be aware of the fact that longer integration times will not always result in better Doppler measurements because as the integration time increases to a certain point, other sources of error may appear. In our tests, the error RMS increases if the integration time is longer than 34 seconds.

In applications (such as spacecraft tracking in the cruising stage) in which the spacecraft dynamics are much weaker, very long integration times can be invoked to get highly accurate Doppler measurements. For example, for the radio science project in the Cassini mission, a whopping 300-sec integration time was used to get an extremely low error RMS of about 0.0022 mm/s[15]! To achieve such an impressive result takes careful calibration based on a multi-frequency link strategy working in the X and Ka bands to minimize the effects of the radio signal transmission media in some highly demanding radio science experiments, such as verification of general relativity theories and PPN parameter testing. These measures are also the key to high-performance spacecraft tracking for future Chinese deep-space exploration.

Our data have also demonstrated the effective application of differential techniques to remove most of the long-term drifts and some of the random errors of the ground station clock. Higher precisions of ~1 mm/s and 0.8 mm/s have been obtained using 1-sec and 34-sec integration times respectively. It must be emphasized, however, that the differential three-way Doppler observable must be combined with some other non-differential observables for high-precision orbit determination, since the differencing has undesirably eliminated a great deal of information about the spacecraft dynamics.

"Traditional" VLBI (range and range rate measurements on random "white noise" signals from spacecraft) and differential one-way Doppler measurements based on open-loop signal recording will be the main data processing modes for China's upcoming YH-1 Mars orbiter mission. Three carriers with test tones in the X-band will be used for spacecraft tracking and orbit determination. Stability of the on-board USO is about 10^{-12} which is expected to provide higher precision than Chang'E-1. In radio science research, open-loop Doppler can be used for planetary gravity field inversion, planetary occultation, planetary ionosphere and magnetic field studies[2].

Compared with other key observables (such as differential one-way ranging DOR), differential one-way Doppler has two advantages: lower sensitivity to receiver delay calibration, and no need for an extremely stable USO. NASA's Jet Propulsion Laboratory has shown that when the interferometric technique (i.e. the so-called "time-domain method" described in sec. 2.3 above) is used with support of a top-quality signal processing algorithm and a high-precision orbit determination program, an USO with 10^{-12} short-term stability is adequate to provide km-level accuracy, while a tracking system with a 10^{-13} USO can perform as well as a two-way Doppler system[17].

Further research will be aimed at accurately estimating the frequency/phase information from low-intensity signals received from spacecrafts at Martian distances. We will improve the Doppler extracting algorithms for low-SNR data, enhance the efficiency of the processing software, and increase the stability of USO for future Mars missions.

Acknowledgements The authors would like to thank the Chinese USB network and CVN network for providing the tracking data. We also thank NASA/GSFC for providing the GEODYN II POD software package.

References

[1] Armstrong J W. Low-frequency gravitational wave searches using spacecraft Doppler tracking. Living Rev Relativ, 2006, 9: 1—10

[2] Asmar S W, Renzetti N A. The Deep Space Network as an instrument for radio science research. NASA contractor report, 1993. 80—93

[3] Thornton C L, Border J S. Radiometric Tracking Techniques for Deep-Space Navigation, (JPL DESCANSO Book Series, Vol. 1). New York: John Wiley & Sons, Inc., 2003
[4] Ruggier C J. Frequency and Channel Assignments, 5-7, DSMS Telecommunications Link Design Handbook. Pasadena:, 2000
[5] Yuen J H. Deep Space Telecommunications Systems Engineering. New York: Plenum Press, 1983. 123—178
[6] Ellis J. Deep space navigation with noncoherent tracking data. TDA Progress Report. 1983, 42—74: 12
[7] O'Reilly B D, Chao C C. An evaluation of QVLBI OD analysis of Pioneer 10 encounter data in the presence of unmodeled satellite accelerations. DSN Progress Report, 1974, 42—22: 12
[8] Rourke K H, Ondrasik V J. Application of differenced tracking data types to the zero declination and process noise problems. DSN Progress Report, 1971, 32—1526: 12
[9] Bhaskaran S. The application of noncoherent Doppler data types for deep space navigation. TDA Progress Report, 1995, N95-32221: 54—65
[10] Yang X N, Lou C Y, Xu J L. The Principle And Application Software Defined Radio. Beijing: Electronics Industry Press, 2001. 21—30
[11] Shi L, Zhao S J. Digital Signal Process. Beijing: Science Press, 2007
[12] Ding M Y, Gao X Q. Digital Signal Process. Xi'an: Xi'an Electronic Science and Technology University Press, 2001
[13] Ping J S, Frank W, Yusuke K, et al. High frequency components in LP Doppler data. J Planet Geodesy, 2001, 36: 15—22
[14] Ping J S, Yusuke K, Frank W, et al. A method of getting velocity information from S/C VLBI observation. J Beijing Normal Univ (Natural Science), 2000, 36: 769—774
[15] Bertotti B, Iess L, Tortora P. A test of general relativity using radio links with the cassini spacecraft. Nature, 2003, 425: 374—376[doi]
[16] Moyer T D. Formulation for Observed and Computed Values of Deep Space Network Data Types for Navigation (JPL DESCANSO Book Series, Vol. 2). New York: John Wiley & Sons, Inc., 2003
[17] Highsmith D E. The Effect of USO Stability on One-Way Doppler Navigation of the Mars Reconnaissance Orbiter. Jet Propulsion Laboratory, Pasadena, California

Chang'E-1 Orbiter Discovers a Lunar Nearside Volcano: YUTU Mountain

Ping Jinsong[1,2] Huang Qian[1,2] Su Xiaoli[1] Tang Geshi[2] Shu Rong[3]
Xiao Long[4] Huang Jun[4]

(1. Shanghai Astronomical Observatory, Chinese Academy of Sciences, Shanghai 200030, China;
2. Flight Dynamics Laboratory of BACC, Beijing 100094, China;
3. Shanghai Institute of Technical Physics, Chinese Academy of Sciences, Shanghai 200083, China;
4. Institute of Geophysics and Geomatics, China University of Geosciences, Wuhan 430074, China)

Abstract In the day time of the Moon surface, the strong illumination from high altitude and high albedo rate radical craters will introduce the illumination effect on observing the nearby low altitude, low albedo rate and shallow small slop rate area seriously, and even can "hide" the later area from the light. Based on the lunar global topography model obtained by Chang'E-1 mission, and by comparing with the lunar gravity model, a volcano named "YUTU Mountain" has been identified. It is a volcano with diameter of ~300 km and height of ~2 km located at (14°N, 308°E) in Oceanus Procellarum. Besides, the DEM of another volcano named "GUISHU Mountain" in the same area has been improved. This new discovery will benefit the study of lunar magmatism and volcanism evolution in the nearside of the Moon.
Keywords Chang'E-1 Lunar topography Volcano YUTU Mountain

Since the first detection of lunar global topography by the Clementine mission[1,2], a long period of silence had elapsed before the two new laser altimeters aboard the Japanese lunar explorer Selenological and Engineering Explorer (SELENE//KAGUYA) and the Chinese lunar orbiter Chang'E-1 (CE-1) independently obtained the highly similar lunar global topographical maps that help improve basic parameters of the lunar form and shape with extremely high resolution and precision. These ad-vanced instruments were named STM359_grid-02[3] and CLTM-s01[4], respectively. CE-1 also obtained lunar global image with a resolution of 120 m, which will help improve the studies of lunar craters[5].

Through frequency-domain (coherence and admit-tance) and spatial-domain analysis, our new maps are compared with the existing lunar global gravity models such as LP150Q[6] and SGM90d[7] to study the dicho-tomy and the subsurface structure of the Moon in detail. As shown in Figure 1, the lunar topography and gravity are found to be strongly coherent at the middle wave-lengths (100 – 300 km). Although this is not revealed in the figure mainly because of poor resolution of the grav-ity model at higher frequencies, strong coherence should be expected at shorter wavelengths. Some researcher also used the method of seismic tomography to study the inner structure of the Moon[8].

Comparing the CLTM-s01[4] grid map (0.0625°×0.0625°) with ULCN2005[9] and the Clementine image data (http://www.lpi.usra.edu/resource/mapcatalog/), we notice some new features at the middle and shorter wave-lengths[10], from which we have identified a large volcano highland. This finding is also supported by the STM359_grid-02 model[3].

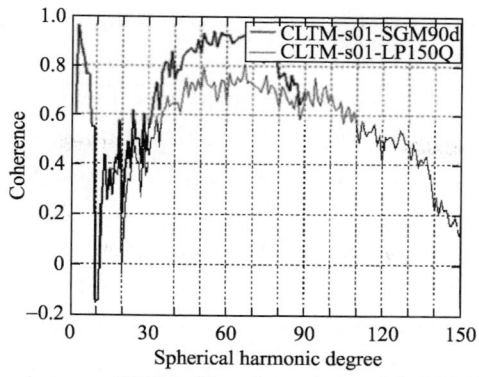

Fig.1 Coherence between CLTM-s01 and gravity models LP150Q and SGM90D

In the area of (0°N~30°N, 300°E~330°E) in the Oceanus Procellarum, two highlands are clearly shown in the western region (Figure 2, Plate XIX). The one in the north (25°N, 310°E) has a diameter of ~250 km and an altitude of ~3 km, which can be identified in the camera images from most missions. The radial impact craters, the Aristarchus and Herodotus, as well as the Vallia Schroteri are located in this area. The one in the south, inside the solid black circle of Figure 2, has a diameter of ~300 km and a height of ~2 km from the bottom of the basin, with its center located at (14°N, 308°E). Before they are officially recognized, we call the northern and southern highlands GUISHU and YUTU Mountains, named respectively after the cherry bay and the jade rabbit from an ancient Chinese legend. Both of them can also be clearly seen in STM359_grid-02.

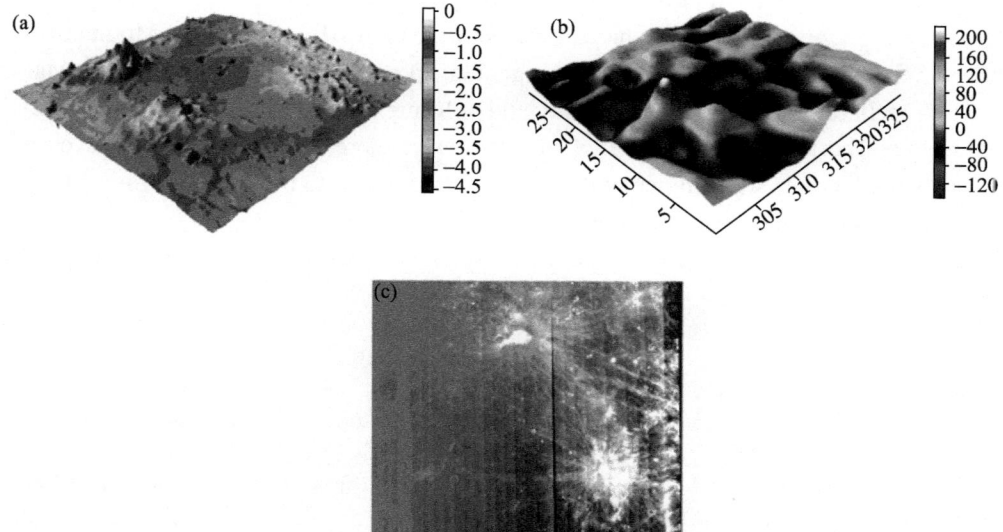

Fig.2 Regional lunar topography, gravity anomaly, and images of GUISHU and YUTU provided by the CLTM-s01 (a), LP150Q (b) and CE-1 camera data (c)

In the early gravity data of the LP150Q model, evidence of positive gravity anomaly of ~250 mgal at the YUTU Mountain has been noticed (see Figure 2). How- ever, its high altitude could not be seen either in the Clementine mission or in the CE-1 camera image. The Clementine laser data even missed it altogether. In the DEM of ULCN2005, the YUTU Mountain area shows a weak apophysis, or a flat gradient dome similar to the relief of a lunar mare. The LPI lunar map shows complicated relief characteristics (such as the radial line, the mountain, the crater, the valley and the rims) in the GUISHU Mountain area. However, in the YUTU Moun-tain vicinity that looks like a very simple geological area, only the Crater Marius and Rima Marius have been identified and named. Before the CE-1 and SELENE missions, we even guessed that there should be some hidden mass under the surface of this region.

We attempt here to provide a reason why the YUTU Mountain has always been hiding away from optical detectors. The slope rates for the YUTU on its west and east sides are only 1% and 2% respectively. In earlier archival images and the image obtained by the CE-1 camera (see Figure 2), it is found that, due to the special illumination effects in the day time[11], the strong illumination from two big nearby radial compact craters, the Aristarchus and the Kepler, is hiding the whole area of the YUTU Mountain with very low albedos and also part of the GUISHU Mountain, especially the sides with gentle slopes. Neither polar orbital missions nor low-inclination missions can image the YUTU area as a result of severe limitations due to the strong illumination effects. Such effects may also come from the famous Copernicus Crater lying in the south-eastern direction some distance away. For this reason, even powerful optical telescopes on the Earth could not identify the YUTU Mountain in the long human history.

The two mountains are located in the igneous area of Oceanus Procellarum. Early studies indicated that the domes in various scales such as the YUTU and GUISHU Mountains in the lunar mares were believed to be volcanic shields[12,13]. At the south-west bottom of the YUTU Mountain, a relief that looks like flow fronts appears beside the Reiner Gamma near the impact crater Reiner and places surrounding the Mountain. As revealed by Figure 2, terrain and gravity have highly positive correlation at the YUTU Mountain; in contrast, highly negative correlation appears at the GUISHU Mountain and at locations such as (8°N, 308°E), (2°N, 317°E) and their neighborhoods. Either four mantle plumes may be implied under them, or the YUTU Mountain and the area surrounding (2°N, 317°E) may be standing on two separated channels that could have originated from the same plume. The strong positive and negative correlations between the terrain and gravity data may reveal the existence of some very complicated, alternating geological features lying in this area. The relationship between the topography and gravity in this area strongly supports the original idea of a shielded volcano.

The similar relief, gravity and optical characteristics of the GUISHU and YUTU Mountains imply that they may have similar origins and evolution histories. Their geological eras are probably before the Mare Imbrium but after the Oceanus Procellarum. The low crater density and volcanic relief of the YUTU may also imply a younger Lower-Imbrium Series. The topography and relief of this area is more complicated than our current knowledge could explain. The SELENE data with higher spatial resolution and data to be obtained from future missions may benefit further detailed investigation of this area.

Acknowledgements The CE-1 data used in this research were supported by China Lunar Ex- ploration Engineering Center and by the CE-1 application system.

References

[1] Zuber M T, Smith D E, Lemoine F G, et al. The shape and internal structure of the moon from the clementine mission. Science, 1994, 266: 1839—1843[doi]

[2] Smith D E, Zuber M T, Neumann G A, et al. Topography of the moon from the Clementine lidar. J Geophys Res, 1997, 102:1591—611[doi]

[3] Araki H, Tazawa S, Noda H, et al. Lunar global shape and polar topography derived from Kaguya-LALT laser altimetry. Science, 2009, 323: 897—900[doi]

[4] Ping J S, Huang Q, Yan J, et al. Lunar topographic model CLTM-s01 from Chang'E-1 laser altimeter. Sci China Ser G-Phys Mech Astron, 2009, 52: 1105—1114[doi]

[5] Yue Z, Liu J, Wu G. Automated detection of lunar craters based on object-oriented approach. Chinese Sci Bull, 2008, 53: 3699—3704[doi]

[6] Konopliv A S, Asmar S W, Yuan D N. Recent gravity models as a result of the lunar prospector mission. Icarus, 2001, 150: 1—18[doi]

[7] Namiki N, Iwata T, Matsumoto K, et al. Farside gravity field of the Moon from four-way Doppler measurements of SELENE (Kaguya). Science, 2009, 323: 900—905[doi]

[8] Zhao D, Lei J, Liu L. Seismic tomography of the Moon. Chinese Sci Bull, 2008, 53: 3897—3907[doi]

[9] Archinal B A, Rosiek M R, Kirk R L, et al. The unified lunar con- trol network 2005. U.S. Geological Survey Open-File Report, 2006, Version 1.0, 2006—1367

[10] Huang Q, Ping J S, Su X, et al. New features of the Moon revealed and identified by CLTM-s01. Sci China Ser G-Phys Mech

Astron, in press
[11] Wu S S C. The effect of illumination on the precision of photogram- metric measurements using Apollo metric camera photographs. In: American Society of Photogrammetry and American Congress on Surveying and Mapping, Fall Convention, Phoenix, Ariz., 1975. Pro- ceedings (A76-38501 19-43) Falls Church, Va., American Society of Photogrammetry, 1976. 99—118
[12] Macauley J F. The nature of the lunar surface determined by sys- tematic geological mapping. In: Runcorn S K, ed. Mantles of the Earth and Terrestrial Planets. London: Interscience, 1968. 431—460
[13] Elston W E. Evidence for lunar volcano-tectonic features. J Geophys Res, 1971, 76: 5690—5702[doi]

Space Operation System for Chang'E Program and its Capability Evaluation

Yu Zhijian   Lu Lichang   Liu Yungchun   Dong Guangliang

(Beijing Institute of Tracking and Telecommunications Technology, P.O. Box 5131, Beijing 100094, China)

Abstract Space operation for China's first lunar exploration program, Chang'E will be provided by the S-band aerospace Telemetry, Tracking and Command (TT&C) network designed for China's manned space program. This is undoubtedly a great challenge to the ground TT&C system. The largest antennas of China's S-band aerospace TT&C network has an aperture of only 12 m. A series of technical measures have been taken into the designing of the spacecraft-ground TT&C system to ensure that such antennas can communicate with Chang'E-1 lunar probe 400,000 km away. These include installation of high-gain directional antennae and medium-gain omni-directional antennae for the probe, adding channel encoding to the downlink channel, using both high and low data rates for information transmission and upgrade and design of ground equipment terminals. Among them, the omni-directional antenna will operate in the earth-ground transfer orbit phase and the directional antenna will operate in the lunar orbit phase. These measures satisfy the spacecraft-ground link and program design requirements.

To provide accurate navigation for the probe during its Earth–Moon flight and initial lunar orbiting flight, China's VLBI system designed for astronomical observations, will also be used besides the ranging and range rate measurement capabilities of the S-band TT&C network. The purpose is to provide 100 m accuracy in position determination during lunar orbit. This paper describes the system design, technical challenges, solutions and capability evaluation of space operation for Chang'E-1.

Keywords Chang'E program   Lunar exploration   Space operations   TT&C design

1 Basic configuration of the space operation system for Chang'E program

China will implement its lunar exploration program in the next few years. The program encompasses five major systems: launch vehicle, spacecraft, launch site, application system and TT&C system. Among these, the main tasks of the TT&C system will be tracking of the spacecraft through ground operation, data reception and space operation control. Large-aperture deep space TT&C facilities are usually built to provide TT&C support for lunar exploration programs in the world. However, such facilities require heavy investment. Due to the constraint in program cost, space operation for China's first lunar exploration program will be provided by the aerospace TT&C network designed for China's manned space program. The TT&C network consists of a flight control center and a number of S-band TT&C stations and has already successfully provided reliable TT&C support service for many domestic and foreign spacecrafts and for five spaceships in flight experiments of China's manned space program. All technical specifications of the network have been proven to meet the design requirements.

To provide tracking and instrumentation for Chang'E-1 lunar probe at a long range, a series of advance technologies have been incorporated in the TT&C system, so that the 12 m S-band stations can cooperate with onboard equipment and support the mission at the available signal level.

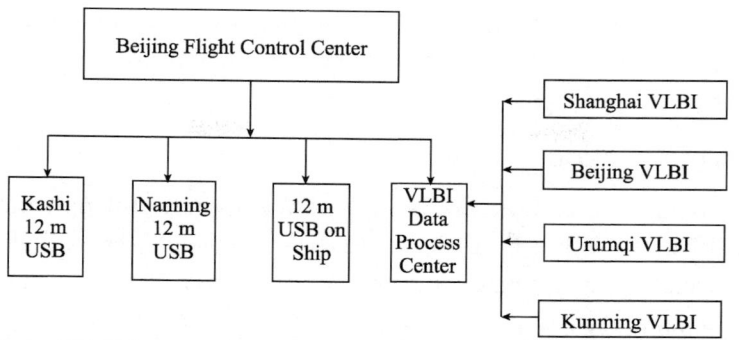

Fig.1 The Telemetry, Tracking and Command System for Chang'E program

In addition to the use of the ranging and range rate capabilities of the S-band TT&C network, the VLBI system made for astronomical observations in China will be used to provide initial accurate navigation for the Earth–Moon flight and moon orbiting phase of Chang'E-1 probe. Use of the VLBI system is also aimed at achieving an accuracy of 100 m for orbit determination during the lunar orbiting phase as required for accomplishing scientific objectives.

China completed its international standard S-band aerospace TT&C network in 1998 to provide support for its manned space missions. The network consists of a network control center and three S-band TT&C stations, having the largest antenna aperture of 12 m. Besides, international cooperation is also sought with some foreign S-band TT&C stations to provide support for this mission.

The 12 m S-band TT&C systems are located in Kashi, Nanning and Yuanwang instrumentation ships. China's VLBI measurement system has four observation stations located in Shanghai, Beijing, Urumqi and Kunming. Their observation data can be transmitted in real time to the data processing center in Shanghai for processing, and the processing result can be sent to Beijing Flight Control Center, which is responsible for command, control and data processing for the Chang'E program.

2 Design of the space-ground TT&C link

2.1 Computation of the TT&C link margin

The main technical specifications of the 12 m USB (Unified S-Band) stations are as follows:
- System uplink EIRP: 71 dBW
- Ground station system G/T: 22.5 dB/K
- Measurement accuracy:

Angular: $S/\Phi|_C = 43$ dBHz; $\sigma_{AE} = 0.01°$, $\Delta_{AE} = 0.02°$
Ranging: $S/\Phi|_{\text{Major tone}} = 35$ dBHz; $\sigma_R = 13$ m, $\Delta_R = 8$ m
Range rate: $S/\Phi|_C = 43$ dBHz; $\sigma = 0.05$ m/s

- TM bit rate 0.1 to ~64 kbps, modulation PCM-PSK/PCM-DPSK
- TC bit rate 0.1 to ~8 kbps modulation PCM-PSK
- Data signal bit rate 2Mbps, modulation PCM/QPSK

G/T of China's 12 m antenna systems is 22.5 dB/K and the EIRP is 71 dBW.
The equipment onboard Chang'E-1 probe has a G/T of −30 dB/K, EIRP 13 dBW
- The effective space-ground range is 400000 km

The omni-directional antenna operates in the high-gain zone (antenna gain: 0 dBi; EIRP: 13 dBW). Besides, atmospheric attenuation, polarization loss, space noise, etc. have to be taken into consideration. Telecommand and telemetry bit rate for lunar spacecraft is usually low for a lunar exploration mission. Bit rate of telemetry usually does not exceed 1 kbps; we use this value for telemetry. Bit rate for the telecommand is 250 bps. Based on preliminary analysis, the S-band TT&C stations should be able to meet the requirements of the lunar orbiter program after appropriate adaptive modifications. Directional antenna operation mode is used for downlinking. Because the EIRP of the directional antenna is 5 dB greater than the high-gain zone of the omni-direction antenna, 5 dB margin is added to the downlink signal.

2.2 Results of computation

Our calculations show that the requirement on G/T of the spacecraft-borne equipment is not high, and mission requirements will be met with −36 dB/K. However, the downlink has a high requirement of over 13 dBW on EIRP of the spacecraft-borne antenna. Therefore, the spacecraft-borne transmitter should have a power of 20 W. As a result, a high requirement is imposed on the spacecraft-borne antenna pattern. Besides, near-omni-directional earth coverage is required from the spacecraft-borne antenna because the lunar probe will undergo significant changes of attitude during its flight and, at the same time, TT&C of the spacecraft should still be possible when the attitude of the spacecraft is not stable. It is a challenging task for an omni-directional antenna to attain such high performance.

Multiple operation modes can be used by a unified carrier S-band (USB) system, and these include:
Mode 1: Telemetry only
Mode 2: Telecommand, telemetry and ranging
Mode 3: Ranging acquisition and telemetry
Mode 4: Telecommand and telemetry
The Table 1 gives margins of system level for different operation modes.

Table 1 Calculation of margins of the TT&C channels (unit: dB)

	Mode 1 Telemetry only	Mode 2 Telecommand, telemetry and ranging	Mode 3 Ranging acquisition and telemetry	Mode 4 Telecommand and telemetry
TC	–	14	–	17
TM	8	6	6	8
Major tone	–	5	5	–
Range rate	8	7	7	8

2.3 Technical challenges

It is a great challenge to use the existing S-band TT&C network to provide support for the lunar spacecraft. To meet this challenge, we have to take measures in the designing of the spacecraft-borne and ground TT&C systems to increase their technical capabilities. These measures include:
• Layout of the transponders and antennas should be optimized to reduce loss of feedlines;
• Allocation of signal energy can be optimized through an appropriate adjustment of uplink and downlink modulation.
• A range ambiguity-resolution technology combining software and hardware will be used.

Besides, Low Density Parity Check (LDPC) encoding technology can be used for reception of telemetry signals in designing the ground system.

LDPC code was proposed by R G Gallager in 1963, but it was not possible to analyze and verify its performance through simulation because of the low level of development of computer technology and hardware. LDPC became a hot subject of research in 1995 when the Turbo code was put forth and received wide attention. Following that, M Luby *et al* made improvements on LDPC code and formed irregular LDPC code, that is, they changed the columns and rows of the original LDPC code parity rectangular matrix from single order to multiple orders so that LDPC acquires performance approaching Shannon limit. Its technical advantages are:
• Low decoding threshold: its encoding gain is far greater than convolution code. At an ideal condition of BER of 10^{-5}, LDPC code has 2.8 dB more performance gain over (2, 1, 7) convolution code.

- Having strong anti-burst error capability.
- Low decoding complexity.

Its delay is similar to that of convolution encoding/ Viterbi decoding.

We have got the following conclusion based on testing on hardware:

At BER of 10^{-5}:

- $E_b/N_0 = 9.7$ dB in case of non-encoding.
- When the demodulator outputs 3-bit soft decision information, $E_b/N_0 = 3.1$dB when LDPC code is used for coding and decoding, and $E_b/N_0 = 4.9$ dB when convolution coding/ Viterbi decoding is used. LDPC coding and decoding is 1.8 dB higher than convolution coding/ Viterbi decoding.
- Computer-aided instrumentation method assisted by channel information is used. When the demodulator outputs 3-bit soft decision information, $E_b/N_0 = 2.9$ dB when LDPC code is used for coding and decoding. When the demodulator outputs 6-bit soft decision information, $E_b/N_0 = 2.2$ dB when LDPC code is used for coding and decoding. LDPC coding and decoding is 2.7 dB higher than convolution coding/Viterbi decoding

3 TT&C operation modes

3.1 Operation mode of the probe-borne equipment

As in operation mode of the spacecraft-borne equipments, the spacecraft should use omni-directional antenna for reception as well as for transmission because of significant changes in the attitude of the spacecraft during its flight.

Besides, spacecraft-borne directional antenna is a key to deep space communication technology, high-data rate science data takes wide channel bandwidth, and it is no longer possible to receive the data with a non-directional antenna. A directional antenna can also be used for tracking and control of the spacecraft and can serve as backup. In an abnormal case when attitude of the spacecraft is not stable or the directional antenna fails, the ground will have to rely on the omni-directional antenna to provide emergency support.

3.2 Operation mode of the ground TT&C network

Beijing Flight Control Center organizes implementation of the TT&C missions, carries out remote monitoring of the S-band stations including setting of operational parameters and monitoring of the operation status. The center will also provide accurate spacecraft orbit data and satellite telemetry parameter reports for operation and control of the spacecraft. Instrumentation data of the VLBI facilities are sent through data links to the flight control center and incorporated in orbit analysis and computation. VLBI instrumentation data are relatively independent of the S-band network and are only used for comprehensive data processing.

4 Orbit measurement and orbit determination

Deep space radio tracking relied solely on Doppler and range systems in the early deep space explorations. Though tracking of vehicles at lunar distances could be accomplished through a variety of radio and optical techniques, the radio communication provides the main tracking data. The navigation of Chang'E-1 will also rely on the range and Doppler data. However, two ground stations cannot take range and range-rate data simultaneously due to their limited capability. In addition, the radiometric tracking using S-band is done by single-frequency downlink. It is not possible to reduce the effects of the ionosphere and solar plasma. The orbit determination will mostly be based upon separated data observed by a single station.

To obtain preliminary estimates of such radio navigation capabilities, we have carried out simulation with the batch filtering technique of conventional radio data from stations considering the effects of error sources:

- Data coverage
- Length of data arc

- Constant station location errors
- Constant and stochastic spacecraft accelerations
- *A priori* knowledge of the lunar ephemeris

Simulation results indicate that navigation accuracy is associated with the geometry of the ground stations and the length of the tracking arc. Therefore the orbit determination capabilities could be enhanced with the availability of radio data tracked by stations located on sea.

In order to ensure the accuracy of the flight path of Chang'E-1 and provide navigation during its initial flight period around the Moon, the Chinese astronomical observation VLBI system is used besides USB, to measure the range and Doppler to accurately determine orbit. In the VLBI system, four antennas provide two approximately orthogonal baselines for satellite positioning. The evaluation of orbit determination capabilities of these two combined systems was also performed. The positional accuracy of Chang'E in its lunar orbit is expected to be better than 100 m.

5   Evaluation of the comprehensive performance of the system

When the omni-direction antenna is used onboard the spacecraft, the 12 m S-band ground TT&C stations will be able to provide support for the spacecraft if the high-gain zone (EIRP=13 dBW) of the probe-borne direction omni-antenna covers the Earth. An omni-direction antenna has different degrees of low-gain zones (EIRP=8 dBW). Based on link calculations, the ground system will have a small level margin in the low-gain zone earth coverage arcs. However, it can still provide TT&C support for the spacecraft after appropriate measures are taken.

When the directional antenna is used onboard the spacecraft, the 12 m S-band ground stations will be able to provide the TT&C support required for the spacecraft orbiting around the Moon and provide accurate measurement of the spacecraft range.

Using the ranging and range rate measurement information of the S-band TT&C network and the VLBI system used for astronomical observations in China, the 100 m accuracy requirement of science targets for orbit determination during the lunar orbit phase could be met. Based on the above analysis, China will be able to use its own aerospace network and facilities to provide operation and control support for the lunar exploration spacecraft. The use of existing deep space network facilities in the world to improve the reliability and support capabilities for the Chang'E program is also taken into consideration in the system design. Possible use of ESA's deep space stations in Spain and Perth is under consideration.

Preliminary Evaluation of Radio Data Orbit Determination Capabilities of China's First Lunar Orbiter

Yingchun Liu[1,2]　Zhijian Yu[1]　Guangliang Dong[1]

(1. Beijing Institute of Technology of Tracking and Telecommunication, P. O. Box 5131-6, Beijing, P. R. China, 100094;
2. Centre for Space Science and Applied Research, Chinese Academy of Science, P. O. Box 8701, Beijing, P. R. China, 10008)

Abstract　Orbit determination accuracies with radio tracking of China's first lunar orbiter (Chang'E-1) have been investigated. The challenge is that China has no global deep space network. Furthermore, two stations couldn't take range and range-rate data simultaneously due to the platform's capability. The orbit determination will mostly be based upon a short arc of data observed by a single station and no differenced radio data could be employed. To obtain preliminary estimates of such radio navigation capabilities, simulations has been carried out with batch filtering of conventional radio data from separated stations. Analysis results indicate that navigation accuracy is associated with the geometry of the ground stations. Therefore the orbit determination capabilities could be enhanced with the availability of overseas stations' radio data.

1　Introduction

China is planning a three-phase Moon program: orbiting, landing, and returning from the Moon. In the first phase, a lunar orbiter, Chang'E-1, named after a famous Chinese fairy, will be send to the lunar polar orbit, circling the moon and mapping its surface. In the mission design, it is necessary to evaluate the navigational capabilities associated with the ground tracking system.

Deep space radio tracking relied solely upon Doppler and range systems in the early deep space explorations[1]. Though tracking of vehicles at lunar distances could be accomplished through a variety of radio and optical techniques, the radio data are still the main type of tracking data. The Lunar Prospector mission succeed in trajectory operation and updated the lunar gravity model mainly with the range and Doppler data provided by the Deep Space Network (DSN) located in California, Australia, and Spain[2].

The navigation of Chang'E-I will also mainly rely on the range and Doppler data. However, there are still some challenges for China to perform the trajectory operation for its first lunar orbiter.

China has no deep space network. As a result of the project cost limit, the space operation of China's first lunar exploration project will be undertaken by the Unified S-Band (USB) system, which is mainly designed for the manned space flight[3]. Though the system has provided the wide coverage and reliable TT&C and Communication services for Shenzhou-5 manned spaceship, it has only three 12-meter-antennas which receive the signal transmitted from 400,000 kilometers far away. It could not provide the 100% coverage for the

Earth-to-Moon transfer trajectory.

And two stations couldn't take rang and Doppler data simultaneously due to the platform's capability. Therefore, the orbit determination will mostly be based upon the data observed by one station and no differenced radio data could be employed.

In additional, the radiometric tracking at S-band is single-frequency downlink. It is not available to use the differenced data to reduce the effects of the ionosphere and solar plasma.

Under the above constraint conditions, the preliminary evaluation of radio data orbit determination capabilities of Chang'E-1 has been performed. The strategy of orbit determination for various phases are not similar. The precise orbit determination for lunar polar orbiter has been discussed formerly[4]. In this paper, the orbit determination capabilities of the trans-lunar trajectory are reported.

2 Erath-to-Moon trajectory of Chang'E-1

A view of orbit design for Chang'E in on-orbit configuration is shown in Figure 1.

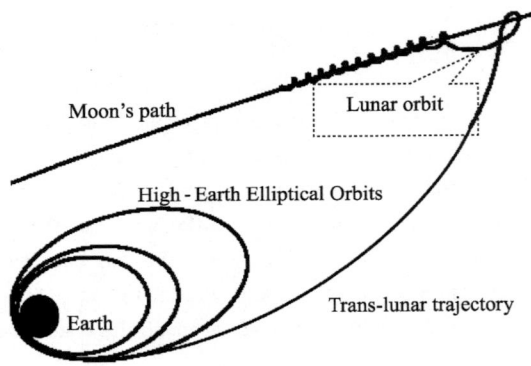

Fig.1 Earth-to-Moon trajectory of Chang'E-1

LM-3 rocket will boost Chang'E-1 into ultra-high elliptical orbit with a 31-degree inclination and a 200-km perigee. Several impulse thrusts (Δv_i) will be executed until the spacecraft could be injected into its nearly five-days trans-lunar trajectory[5, 6]. During the cruise phase, two or three midcourse corrections are necessary[7]. At the nearest approach to the Moon, a serial of Hohmann transfers will be executed to reduce the velocity and put the spacecraft into the lunar circular polar orbit. Chang'E is hoped to work in this orbit more than one year. A trans-lunar trajectory is given by paper[7]:

Insert epoch: 02:50:00 2004-08-27
End epoch: 02:50:00 2004-09-01
Earth-to-Moon Transfer time: 120 hours
Velocity at the perigee: 10.5946 km/s
RA of ascend Node: 7.9996°
Argument of perigee: 180.2144°

The trans-lunar trajectory was divided into three arcs by two mid-course corrections, signed as TLI1, TLI2, and TLI3. The first mid-course correction was planed in the 24 hours after the orbit insert, and the second one was planed 24 hours before the satellite captured by the moon.

3 Ground tracking system

Figure 2 illustrated the ground tracking system for Chang'E-1. The Chinese ground network, including two stations in the mainland and one vessel located in the Pacific, is illustrated. Kashi is located the Northwest China.

Nanning is located in the Southeast China. A vessel, called Yuanwang, will be located in the Pacific. Besides these stations, three overseas ground stations are simulated to tracking the lunar orbiter, which are Villafranca, New norcia, and CEE of Chile. The 9-m antenna of CEE is only available in the early trans-lunar trajectory, TLI1.

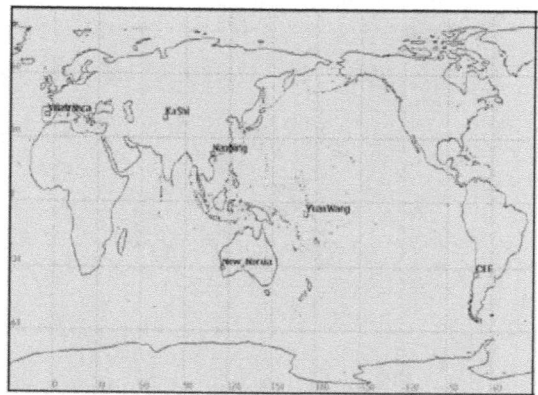

Fig. 2 Ground network

The access time between the Chinese ground station system and the global system are illustrated in Fig. 3.

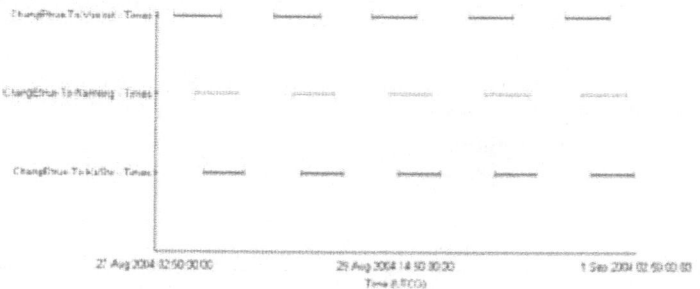

Fig. 3 (a) Access time (of the Chinese ground network)

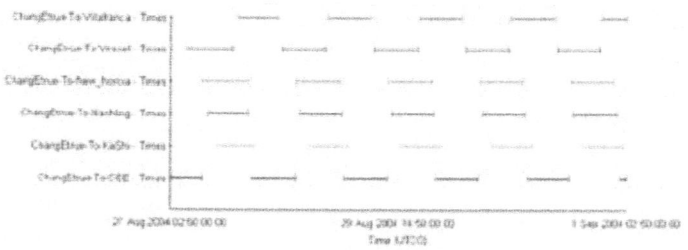

Fig. 3 (b) Access time (of the global ground network)

4 Methodologies

In the difference phases of Chang'E-I, the motion of the spacecraft could be represented in the different equations[8, 9]. For the trans-lunar phase, the equation of the satellite motion is represented as an initial problem of the differential equation in ECI, i.e.

$$\begin{cases} \ddot{\vec{r}} = \vec{F}_e(\vec{r}) + \vec{F}_m(\vec{r}) + \vec{F}_\varepsilon(\vec{r},t) \\ t_0 : \vec{r}(t) = \vec{r}_0, \dot{\vec{r}}(t) = \dot{\vec{r}}_0 \end{cases} \quad (1)$$

where $\vec{r}, \dot{\vec{r}}$, and $\ddot{\vec{r}}$, respectively, represent the position vector, the velocity vector and the acceleration vector of the satellite in the Mean of J2000.0 coordinate system. $\vec{F}_e(\vec{r})$ is the earth central gravity, $\vec{F}_m(\vec{r})$ is the lunar central gravity, $\vec{F}_\varepsilon$ includes the accelerations due to the perturbations.

Supposed the tracking data are the range and Doppler data. The relationship between the observation vector Y and the state vector X is represented as

$$Y = \begin{pmatrix} \rho \\ \dot{\rho} \end{pmatrix} \tag{2}$$

where

$$\rho(\vec{r}, \vec{R}) = (\vec{r} - \vec{R}) \tag{3}$$

and

$$\dot{\rho} = (\vec{r} - \vec{R}) \cdot (\dot{\vec{r}} - \dot{\vec{R}}) \tag{4}$$

5 Simualation and result

To obtain preliminary estimates of such radio navigation capabilities, analyses were performed based upon the two types of radio data. The simulation has been carried out with the batch filtering of conventional radio data from stations considering the effects of error sources. The stations are supposed tracking in sequence.

In addition to the measurement bias and noise, the solutions is also sensitive to the following error sources[10]:

(1) Data coverage
(2) Length of data arc
(3) Constant station location errors
(4) Constant and stochastic spacecraft accelerations
(5) A priori knowledge of the lunar ephemeris

In Table 1, the errors sources considered are listed.

Table 1　Errors sources

Error sources			Magnitude	Estimated
Data error	range	bias	10m	yes
		noise	10m	
	Doppler	bias		
		noise	5cm/s	
Station location error	only for Vessle	constant	10m	yes
Spacecraft acceleration	along track	constant	10×10^{-12} m/s^2	no
Lunar ephemeris error			Not considered	

Firstly, the effect of data coverage and length of the data arc are analyzed. The orbit solution errors in position for TLI1, TLI2 and TLI3 are showed in Figure 4, Figure 5 and Figure 6 respectively, where a is the solution of the data tracked by the Chinese ground network and b is the solution of the data tracked by the global ground network. The tracking geometry of the global ground network is better than the Chinese ground network. And the length of data arc is extended. For all the three phases of the lunar trajectory, the solutions of the global tracking system are obviously better than those of the Chinese ground system. The accuracies are doubled or even improved more than one order.

Secondly, the orbit determination capabilities of the two types of tracking data are analyzed. The solution error of TLI1 is showed in Figure 7, where a is the solution of range data only and b is the solutions of Doppler data only. It could be concluded that the accuracy of Doppler data of the Chinese facilities are much worse than that of the range data.

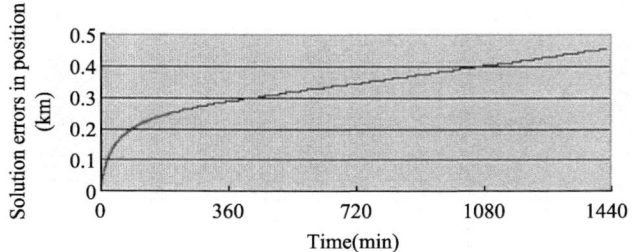

Fig.4 (a)　Solution errors in position of TLI1
by the Chinese ground network

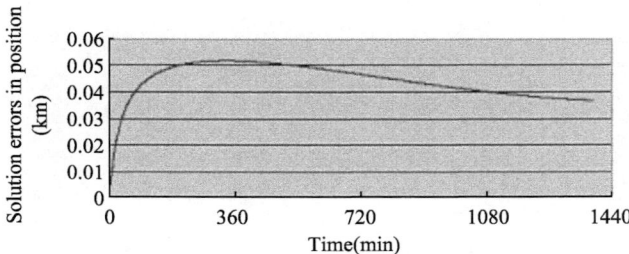

Fig.4 (b)　Solution errors in position of TLI1
by the global ground network

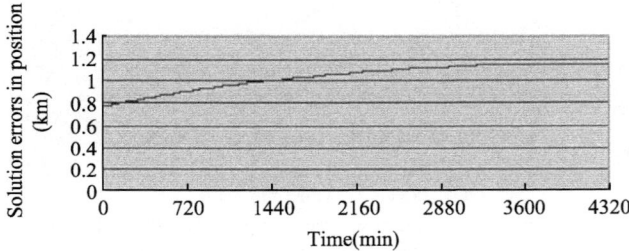

Fig.5 (a)　Solution errors in position of TLI2
by the Chinese ground network

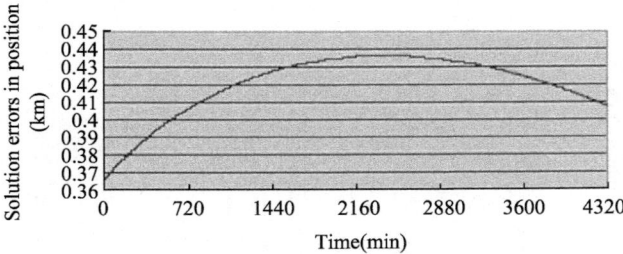

Fig.5 (b)　Solution errors in position of TLI2
by the global ground network

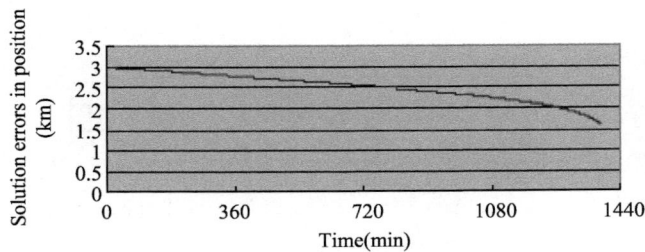

Fig.6 (a)　Solution errors in position of TLI3
by the Chinese ground network

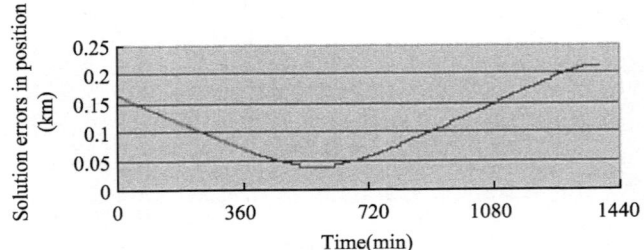

Fig.6 (b)　Solution errors in position of TLI3
by the Chinese ground network

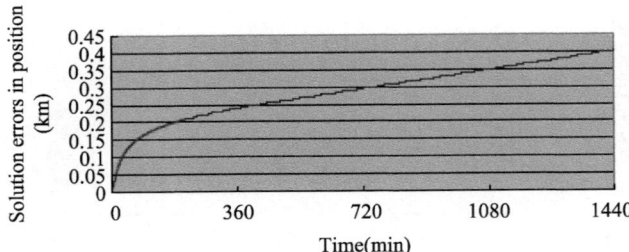

Fig.7 (a)　Solution errors in position of TLI1
range data of the Chinese network only

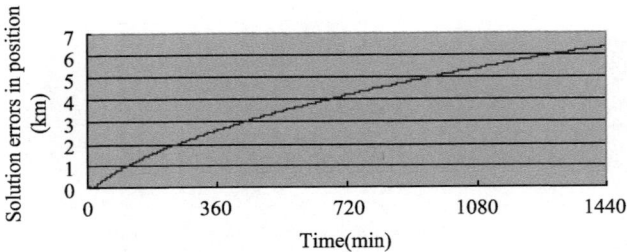

Fig.7 (b)　Solution errors in position of TLI1
Doppler data of the Chinese network only

　　The effects of constant unmodeled spacecraft acceleration and the station location error are also analyzed. Figure 8(a) and 8(b) are the solution errors in position. As the station location error couldn't be solved entirety, its effect is obvious.

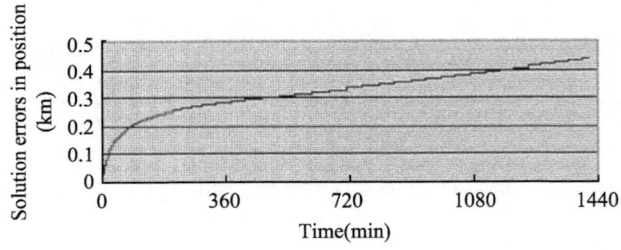

Fig.8 (a)　Solution errors in position of TLI1 with unmolded acceration

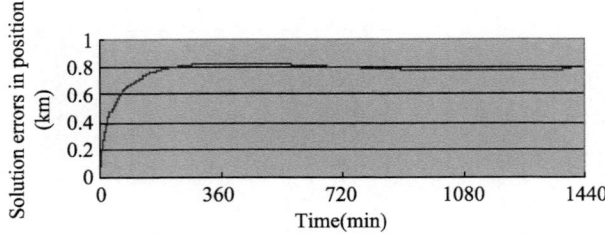

Fig.8 (b)　Solution errors in position of TLI1 with unmolded acceration

6 Conclusions

Simulation results indicate that navigation accuracy is associated with the geometry of the ground stations and the length of the tracking arc. Therefore the orbit determination capabilities could be enhanced with the availability of radio data tracked by oversea stations. It has been within the system design consideration using the overseas deep space facility to enhance the reliability and the support ability of the space operation for the Chang'E project[11].

In order to guarantee the flight path of the Chang'E and provide the precise navigation during its initial flight period around the moon, the Chinese astronomical observation VLBI system is also introduced besides using USB to measure the range and Doppler to achieve the proposed precise orbit required by the scientific instruments[11]. In Chinese VLBI system, four antennas provide two approximately orthogonal baselines for satellite positioning. The Evaluation of orbit determination capabilities of these two systems combined will be reported later.

References

[1] C. L. Thornton and J. S. Border, *Radiometric Tracking Techniques for Deep-Space Navigation*, John Wiley & Sons, USA, 2003.
[2] David Lozier, et al. Lunar Prospector Mission Design and Trajectory Support, *AAS* 98—323, 1998
[3] Zhijian Yu, TT&C and Communication System of Chinese Manned Space Flight, *Chinese Journal of Astronautics*, Vol.25, No.1, pp247—250, 2004.
[4] Liu Lin, Liu Yingchun, Precise Orbit Determination for Lunar Satellite, *Acta Astronautics*, Vol 51, No.1—9
[5] Weilian Yang, Research on the trans-lunar trajectory of lunar polar orbiter, *Spacecraft Engineer*, Vol. 12, No.3, pp19—33, 1997.
[6] *Deep space Exploration*, Vol. 1, 2004.
[7] Wenyan Zhou, Weilian Yang, Mid-correction of trans-lunar trajectory of lunar explorer, *Chinese Journal of Astronautics*, Vol.25, No.1, pp89—93, 2004.
[8] Liu Lin, Wang Xin, on the orbit dynamics of lunar satellite, *Progress in Astronomy*, Vol. 21, No. 4, pp281—238, 2003.
[9] C. Yee, et al., Orbit Determination Support for the Clementine Mission during Lunar Orbit Phase, *AAS Paper* 95—388 • 1995
[10] V.J.Ondrasik, C.E. Hildebrand, and G. A. Ransford, Preliminary Evaluation of Radio Data Orbit Determination Capablities for the Saturn Portion of a Jupiter-Saturn-Pluto 1977 Mission, *JPL Technical Report 32-1526*, Vol. X., pp. 59—75, 1971
[11] Zhijian Yu, Lichang Lu, Guangliang Dong, Yingchun Liu, Chang E project space operation system design and ability Evaluation, submitted to *International Conference on Exploration and Utilization of the Moon (ICEUM-6)*, 2004, Udaipur, India.

LUT: A Lunar-based Ultraviolet Telescope

Cao Li[1]   Ruan Ping[2]   Cai Hongbo[1]   Deng Jinsong[1]   Hu Jingyao[1]
Jiang Xiaojun[1]   Liu Zhaohui[2]   Qiu Yulei[1]   Wang Jing[1]
Wang Shen[1]   Yang Jianfeng[2]   Zhao Fei[1]   Wei Jianyan[1]

(1. National Astronomical Observatories, Chinese Academy of Sciences, Beijing 100012, China;
2. Xi'an Institute of Optics and Precision Mechanics, Chinese Academy of Sciences,
Xi'an 710119, China)

Abstract   The Lunar-based Ultraviolet Telescope (LUT) is a funded lunar-based ultraviolet telescope dedicated to continuously monitoring variable stars for as long as dozens of days and performing low Galactic latitude sky surveys. The slow and smooth spin of the Moon makes its step by step pointing strategy possible. A flat mirror mounted on a gimbal mount is configured to enlarge the sky coverage of the LUT. A Ritchey-Chrétien telescope with a Nasmyth focus configuration is adopted to reduce the total length of the system. A UV enhanced back illuminated AIMO CCD 47-20 chip together with the low noise electric design will minimize the instrumental influence on the system. The preliminary proposal for astrometric calibration and photometric calibration are also presented.
Keywords   Ultraviolet   Space-based ultraviolet telescope   Photometric instrumentation   Lunar probe

1   Introduction

As the unique natural satellite of the Earth, the Moon is an ideal site to carry out astronomical observations. Lunar-based observations have several advantages over both ground-based and space-based ones:

(1) It was known for centuries that the Moon has an extremely tenuous atmosphere. With the breakthrough made by the Apollo mission, the surface concentration of the atmosphere was measured to be about $\sim 10^{4\sim5}$ molecules cm^{-3} at night, and a much higher value $\sim 10^{7\sim8}$ molecules cm^{-3} during daytime[1,2]. The extremely tenuous atmosphere implies that the observational effects due to atmospheric radiation, wave-front distortion, and light absorption are negligible on the Moon.

(2) Unlike space-based observations, the Moon provides a large stable platform for maintaining astronomical instruments in permanently stable configurations. The rotation angular velocity of the Moon is about 0.55″ s^{-1}, which is about 27 times slower than the Earth. The slow rotation allows long term monitoring without interruption for as long as about 10 days.

(3) The temperature in the permanent shadow regions (PSRs) at both poles of the Moon could be as low as 30 K [3]. PSRs are therefore rare ideal conditions for infrared observations.

(4) Observations at very low frequencies (<10 MHz, VLF) are feasible on the surface of the Moon, but for the Earth, this kind of observation is impossible, since the VLF electromagnetic wave is highly scattered by Earth's ionosphere.

The Moon is therefore an ideal location for Ultraviolet astronomy by taking into account all of these advantages. In fact, the Apollo-16 mission performed the first, and thus far only, far-ultraviolet observations on the lunar surface in 1976 [4~6]. The observations were carried out with a 3-inch Schmidt telescope equipped with a far-ultraviolet camera/spectrograph operating in the wavelength range from 100~160 nm. The field-of-view was 20 degrees, and limiting magnitude was 11^m. In total, 178 images of objects including Earth, star clusters and the LMC were taken by the telescope, and delivered to Earth by the astronauts. The telescope was placed in the shadow of the Lunar Module to avoid heating by the Sun. These observations, however, contributed little to astronomy due to the low level of technology used at that time.

As the forerunners of future exploration on the Moon, we here propose the concept of a small automatic Lunar-based Ultraviolet Telescope (LUT) with modern technology. This exploration can not only provide important scientific results that cannot be achieved on the ground in many astronomical fields (e.g., stellar evolution, compact stars and black holes), but also develop significant experience for lunar-based astronomical observation (e.g., gimbals, CCD detectors and onboard data processing).

2 Scientific objective

LUT is a small lunar-based astronomical telescope working in the near-ultraviolet band, which is not available on the ground because of the heavy (almost 100%) absorption by the Earth's dense atmosphere.

The main scientific goal of LUT is to continuously monitor variable stars and galaxies in the near-ultraviolet band without interruption for as long as dozens of days, taking advantage of both the very slow rotation and extremely tenuous atmosphere of the Moon. To our current knowledge, the temperatures of many variable objects are as high as 10^4 K, which results in the fact that the brightness variation is usually much stronger in the near-ultraviolet band than in other bands. The objects with large brightness variation in the near-ultraviolet band are binaries with compact objects including cataclysmic variable stars, large/small mass binaries and novae, quasars and Blazars, dMe stars, and Lyr RR stars. Table 1 lists the typical NUV-to-optical colors, variation mechanisms, amplitudes and timescales for these objects.

Table 1 Main scientific objectives of LUT

Objects	NUV-r'	Variation mechanisms	Amplitude	Timescale
Binaries with compact stars	0–2	flare	>1^m	Burst may last as long as about 10 days. Brightness reaches its maximum in several dozen seconds.
		sparking	~0.1–1^m	several seconds to several minutes
Quasars and blazers	0–2	non-relativistic	0.1–1^m	from about 1 day to 100 days
		relativistic	0.1–1^m	from less than about 1 hour to 1 day
Solar-like chromosphere active stars & M Dwarf Flare Stars	>3	flare	>1^m	Burst can last as long as about 1 hour. Brightness reaches maximum in several dozen seconds.
		rotation of on-spot	~0.1^m	about 1 day
Lyr RR stars	3	short term	~0.1^m	from 0.5 to 1 day

Analyzing results of the light curves (e.g., amplitude, timescale, period, power spectrum) allows us to determine the variation of the temperature, radius, and accretion rate of these compact objects, which is valuable for verifying the current stellar atmosphere model and for calculation of the transparency of the stellar atmosphere, for estimating the metal abundance of stars, and for calibrating the metal abundance as a function of the V-band absolute magnitude for Lyr RR stars. The light curve analysis is also a powerful tool to investigate the origin of the instability occurring in the central part of some objects.

In addition to the monitoring observations, LUT will also perform a series of low Galactic latitude sky surveys. GALEX (Galaxy Evolution Explorer) is a NASA small telescope working in both near- and far-ultraviolet bands [7]. Since its launch on April 28th, 2003, its pointing observations have covered about 26,000 square degrees of sky area. However, the instruments onboard GALEX are very sensitive since GALEX is dedicated to observing distant faint galaxies, which results in the fact that the sky areas (usually with low galactic latitude) that are dominated by bright objects (NUV <10 mag) are always avoided in its observations. In fact, there are a group of proposed LUT objects with strong NUV emission near the galactic plane, including e.g., OB associations, massive star formation regions and planetary nebulae. Surveys at low galactic latitude may present the distribution in the Galaxy for massive stars and cataclysmic variable stars. The combination of the survey data and models could be used to estimate the dust extinction/gas absorption for each of these line-of-sight observations and the distribution of dust/gas in the Galaxy. The distribution is in fact essential for our understanding of the formation and evolution of our own Galaxy.

3 Instrument

3.1 Constraints

There are two critical constraints on the instrumental design of LUT: One is that LUT is located inside a cabin. The other is that LUT has to observe during lunar daytime.

In order to enlarge the observing sky area for an observation inside a cabin, a flat mirror with a gimbal mount has been configured. The azimuth axis of the mount is along the optical axis of the primary mirror, and the altitude axis is perpendicular to the plane of the light path. The pointing flat mirror provides over 1200 square degrees of visible sky area for LUT.

The stray light from the Sun is the main problem introduced by the daytime observation. Choosing an ultraviolet pass band partly accounts for avoiding the optical band in which the Sun radiates most of its energy. In order to greatly reduce the influence of the Sun's light, many strategies have been taken, e.g. a solar blinded filter design has been adopted, the flat mirror will be set almost pointing in the direction away from the Sun, etc.

3.2 System profile

The system profile of LUT is shown in Fig. 1. The lights from celestial objects enter the payload cabin and reach the pointing flat mirror, which is set on a two dimensional gimbal mount. Then the lights are reflected to the optics of the telescope and finally focused on the CCD detector in the focal plane. The image data and status information of the CCD, the CCD's Peltier cooler and the mount are sent to the payload's electric cabinet. The LUT software runs in the computer unit inside the payload's electric cabinet. It has six functions, a) resolving the observation plans, b) executing the observation plans, c) astronomical data processing, d) controlling the gimbal mount and collecting its status in- formation, e) controlling the CCD detector and collecting its status information, and f) controlling the CCD's Peltier cooler and collecting its status information.

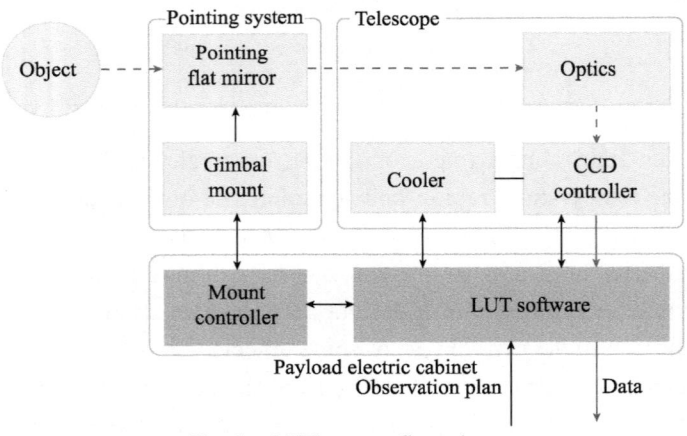

Fig. 1 LUT system flow chart

3.3 Telescope

The optical system of the LUT is a 150-mm-diameter aperture and a F/3.75 Ritchey-Chrétien telescope working at the Nasmyth focus. The Nasmyth focus is chosen to reduce the entire length of the system. A pointing flat mirror is configured to point to the preferred sky area and a field corrector lens is used to correct the field curvature. The optical design is briefly illustrated in Fig. 2.

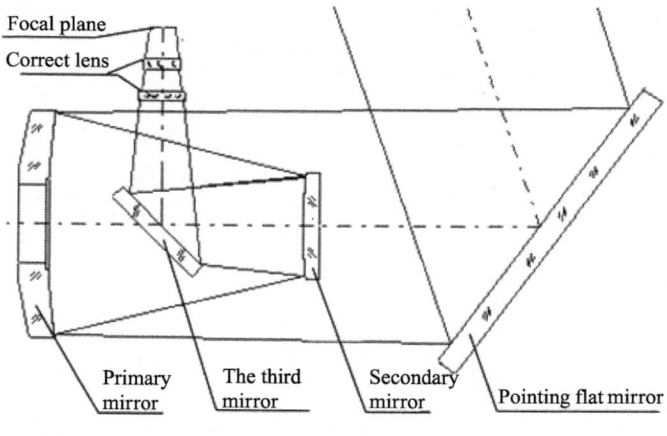

Fig. 2 LUT light path

The quantum efficiency of the CCD is quite high in the optical band (as shown in Fig. 4). A filter shall be solar blinded to reduce the red leak and block the optical band stray light introduced by the Sun by the coatings on the field corrector. The transmission curve of the filter is shown in Fig. 3.

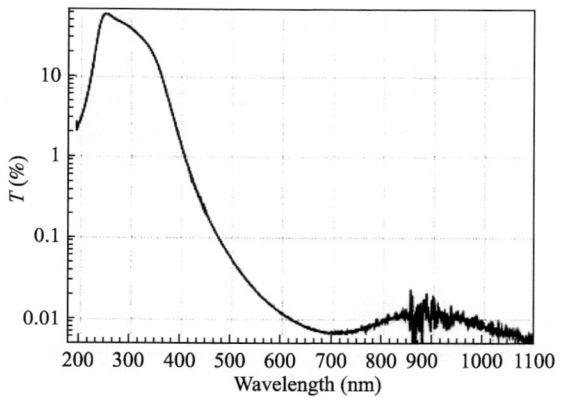

Fig. 3 The transmission curve of the filter of LUT

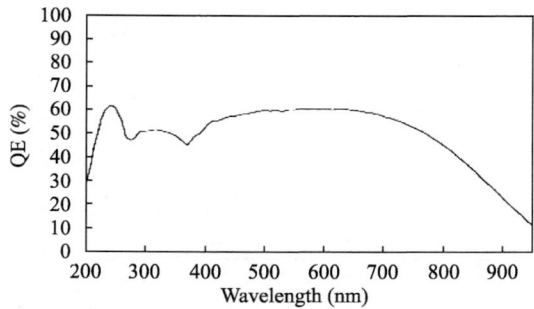

Fig. 4 Typical QE at −40°C for the UV enhanced processed AIMO device without a window

3.4 Detector

AUV enhanced back illuminated AIMO CCD 47-20 chip in the Peltier package with a window is selected as the focal plane detector. It is a frame transfer device, which is perfect for a space application since it does not require a mechanical shutter, which may increase the dependability of the whole system.

3.5 Summary

The primary parameters of the LUT system are presented in Table 2.

Table 2 LUT parameters

Telescope optics		Detector	
Diameter (mm)	150	Readout noise (e^-/pixel)	8
Focal length (mm)	562.5	Dark current (e^-/pixel/s) (40°C)	< 0.1
Pass band (nm)	1,245–345	Active pixels	1024×1024
		pixel size (μm)	13×13

4 Observation strategy

The slow and smooth rotation of the Moon greatly benefits LUT. It makes the long term continuous monitoring of a variable star possible and makes the requirements for the tracking capabilities lower than that on Earth. A topocentric telescope has to constantly adjust attitude to compensate for the diurnal apparent motion of the stars. However, the apparent motion of the stars on the Moon can be very slow. It is about 27.3 times slower than that observed from the Earth. If we set the exposure time to be relatively short, there is no need to track the object constantly. For the same reason, the interval between subsequent attitude repositionings can be several dozen minutes as long as the objects of interest still remain in the central region of the image.

Table 3 shows the apparent speed of stars on the Moon, the exposure time, and repositioning interval for different lunar declinations in the selenocentric equatorial coordinate system accordingly. The exposure time is equal to the time which it takes for a star to travel two pixels' distance on the focal plane due to apparent motion. The repositioning interval is equal to the time which it takes for a star to travel 100 pixels' distance on the focal plane.

Table 3 The observation strategy: Exposure time and repositioning interval

|Dec| /(°)	Apparent motion speed /("/s)	Exposure time /s	Repositioning interval /m
0	0.55	17	14
10	0.54	18	15
30	0.48	20	17
40	0.42	23	19
50	0.35	27	23
60	0.27	35	29
70	0.19	51	42
80	0.10	100	83
85	0.05	199	166

The observation flow chart can be found in Fig. 5. For the case of variable stars (i.e., Pointing observation), the observation will continue until the object moves out of the available sky region of the telescope or the amount of accumulated data is enough. As for the low galactic latitude sky surveys, there is no need to frequently readjust the pointing flat mirror's attitude. The apparent motion of celestial objects will make the telescope scan a long slit sky area even if the attitude of the telescope remains constant.

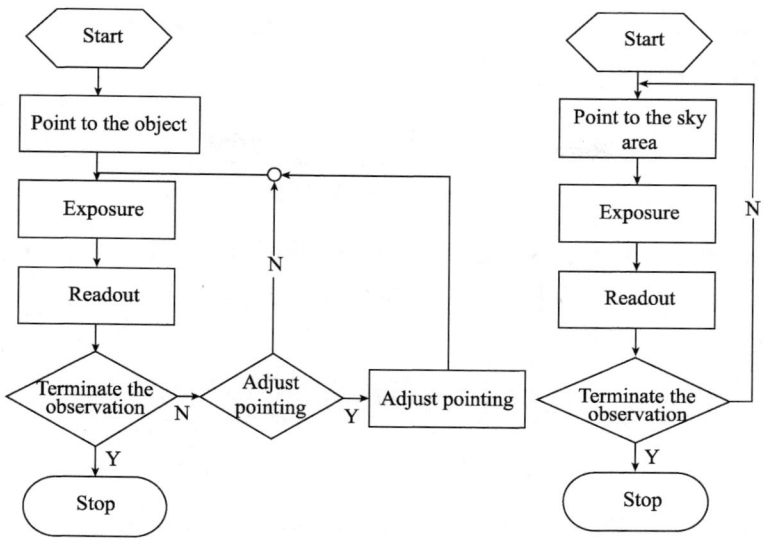

Fig. 5 Observation flow chart (left: Pointing observation; right: Survey Observation)

5 Calibration

5.1 Astrometric calibration

The Lunar Lander on which LUT will be mounted is an unmanned explorer. The conditions make setting up the telescope mount on the Moon impossible. Due to the uncertainty of the landing position and 3-axes of attitude[8], great efforts have to be devoted to the attitude calibration.

In order to observe a certain celestial object, the mean equatorial coordinate at J2000 in the heliocentric coordinate system should first be converted to the observational horizon coordinate in the Lunar topocentric coordinate system [9,10]. The observational coordinate of the object shall subsequently be converted to the attitude of the mount, i.e. the azimuth angle and the altitude angle. The above step needs to consider the parameters such as landing position, 3-axis attitude of the Lunar Lander and the relative position of the LUT on the Lunar Lander. All of these parameters shall be determined by the attitude calibration process through observations after the LUT lands on the Moon. A ground verification system has been established to study the attitude calibration process and associated method.

Another key point related to the astrometric calibration is to establish a guide star catalog for the LUT system. There are two approaches to carry out this task. One can use the existing GALEX NUV catalog that has a similar pass band as LUT. Another way is to generate an astrometry catalog from the Hipparcos catalog by establishing the magnitude transform relationship between the LUT magnitude and Tycho B and V magnitudes for each spectral type.

5.2 Photometric calibration

Since the LUT does not follow any standard photometric system, an LUT photometric system must be established. A photometric system is defined by a list of standard magnitudes and colors measured at specific bandpasses for a set of stars that are well distributed around the sky [11]. There is only one bandpass in LUT. The definition of the LUT photometric system is made up of two steps: ①to define the LUT magnitude; ②to define the LUT magnitudes for a set of standard stars.

The definition of magnitude for the LUT system uses the AB_v system that was first defined by Oke & Gunn [12] and gets widely used from its application in the Sloan Digital Sky Survey:

$$m_{LOT} = -2.5 \log \tilde{F}_v - 48.60 \qquad (1)$$

where $\tilde{F}_v$ is the appropriately averaged monochromatic flux (in units of ergs s^{-1} cm^{-2} Hz^{-1}) at the effective wave-length of a bandpass (following the description in [13]). We adopt the definition of $\tilde{F}_v$ as taking the system response average of the flux as in [14],

$$\tilde{F}_v = \frac{\int_{\lambda_1}^{\lambda_2} f_v(\lambda) \cdot R(\lambda) \cdot d\lambda}{\int_{\lambda_1}^{\lambda_2} R(\lambda) \cdot d\lambda} \qquad (2)$$

where $R(\lambda)$ is the system's response, f_v is the spectral energy distribution of an object, and λ_1 and λ_2 are the lower and upper wavelength limits of the LUT bandpass respectively. For telescope systems with a photoelectric detector such as a CCD, $\tilde{F}_v$ is usually written as

$$\tilde{F}_v = \frac{\int_{\lambda_1}^{\lambda_2} f_v(\lambda) \cdot R(\lambda) \cdot d(\ln v)}{\int_{\lambda_1}^{\lambda_2} R(\lambda) \cdot d(\ln v)} = \frac{\int_{\lambda_1}^{\lambda_2} \frac{1}{hv} f_v(\lambda) \cdot R(\lambda) \cdot dv}{\int_{\lambda_1}^{\lambda_2} \frac{1}{hv} R(\lambda) \cdot dv} \qquad (3)$$

The system response ($R(\lambda)$), which includes the filter transmission and the quantum efficiency of the CCD, will be determined through ground tests.

The primary standard stars that define the LUT photometric zero point should be chosen from standard spectrophotometric stars with available accurate absolute specific flux densities. There are many constraints for choosing LUT standard stars, e.g. the standard stars should be within the sky coverage of LUT, the brightness of the stars should fit the dynamic range of the LUT, and the absolute spectrophotometric data in the ultraviolet band should be available. There are many sources for spectrophotometric standard stars [15,16].

This work was supported by the Ministry of Science and Technology of China and the National Natural Science Foundation of China (Grant Nos. 10803008, 10978020 and 10878019).

References

[1] Heiken G H, Vaniman D T, French B M. Lunar Sourcebook—A User's Guide to the Moon. 1st ed. New York: Cambridge University Press, 1991

[2] Sridharan R, Ahmed S M, Pratim Das T, et al. 'Direct' evidence for water H_2O in the sunlit lunar ambience from CHACE on MIP of Chandrayaan I. Planet Space Sci, 2010, 58: 947—950

[3] Stern S A. The lunar atmosphere: History, status, current problems, and context. Rev Geophys, 1999, 37: 453—492

[4] Page T L, Carruthers G R. Apollo 16 far-ultraviolet imagery and spectra of the Large Magellanic Cloud. Space Res XVII, 1977

[5] Carruthers G R, Page T. Apollo-16 far-ultraviolet spectra in the Large Magellanic Cloud. Astrophys J, 1977, 211: 728—736

[6] Carruthers G R, Page T. Apollo 16 Far-ultraviolet camera/spectrograph: Earth observations. Science, 1972, 177: 788—791

[7] Martin D C, Fanson J, Schiminovich D, et al. The galaxy evolution explorer: A space ultraviolet survey mission. Astrophys J. 2005, 619: L1-L6

[8] Peng J, Liu Z, Zhang H. Conceptual design of a lunar lander. Spacecraft Eng, 2008, 17 (1): 6

[9] Kolaczek B. Selenocentric and Lunar Topocentric Coordinates of Different Spherical Systems. SAO Special Report, 1968, 286

[10] Kolaczek B. Selenocentric and lunar topocentric spherical coordinates on the base of the general formulas of spherical coordinate transformation. Astron J, 1968, 73: 20

[11] Bessell M S. Standard photometric systems. Annu Rev Astron Astrophys, 2005, 43: 293—336

[12] Oke J B, Gunn J E. Secondary standard stars for absolute spectrophotometry. Astrophys J, 1983, 266: 713—717

[13] Fukugita M, Ichikawa T, Gunn J E, et al. The sloan digital sky survey photometric system. Astron J, 1996, 111: 1748

[14] Yan H, Burstein D, Fan X, et al. Calibration of the BATC survey: Methodology and accuracy. Pub Astron Soc Pacific, 2000, 112: 691—702

[15] Bohlin R C, Dickinson M E, Calzetti D. Spectrophotometric standards from the far-ultraviolet to the near-infrared: STIS and NICMOS fluxes. Astron J, 2001, 122: 2118—2128

[16] Stritzinger M, Suntzeff N B, Hamuy M, et al. An atlas of spectrophotometric landolt standard stars. Pub Astron Soc Pacific, 2005, 117: 810—822

Measurements of Electronic Properties of the Miyun 50 m Radio Telescope

Zhang Xi-Zhen   Zhu Xinying   Kong Deqing   Zheng Lei   Yao Cheng
Zhang Hongbo   Suyan   Piao Tingyi

(National Astronomical Observatories, Chinese Academy of Sciences, Beijing 100012, China)

Abstract   Measurement results of some properties of the Miyun 50 m radio telescope (MRT50) of the National Astronomical Observatories, such as pointing calibration, antenna beams, system noise temperature, gain and gain variations with elevation are introduced. By using a new de-convolution technique developed by our group, the broadening effect on measured beams caused by the width of an extended radio source has been re- moved so that we obtained higher accuracy on the measurements of MRT50 beams.

Key words   Instrument   telescope — antenna — method

1 Introduction

In 1999, after the Miyun Synthesis Radio Telescope (MSRT) finished its main scientific goal: the meter- wave radio source survey (Zhang et al. 1997), Miyun radio group proposed to build a 45 m single dish telescope (Zhang et al. 2000) for radio astronomy research at low frequency bands. Afterwards, it was suggested that the project construct a more multi-function 50 m radio telescope for both radio astronomical observations and space communications in 2002 (Wang 2002). The MRT50 was put into test operation in 2005 and passed its final reception test in 2006. Section 2 gives a short introduction to the MRT50.

In general, some necessary tests, such as pointing calibration, beam-size/efficiency determination, gain measurements, and so on, have to be done before the telescope can be used correctly. First of all, pointing calibration was carried out during the test period of MRT50 in 2005. The pointing accuracy reached about $30''$ after the first calibration. This made the observation of SMART-1 successful in the summer of 2005. The second round of pointing calibration was continued in 2006 (after Meeks et al.1968; Yuan et al. 1986). We will introduce the pointing calibration method and result in Section 3.

To determine antenna gain with a cosmic source, it is expected that the source is small in angular size and strong in intensity. If a large angular source is used, the measured beam width has to be corrected (Yuan et al. 1986). Because of the sensitivity of MRT50, the radio source Cyg A was adopted to determine the beams. To reduce the effect of Cyg A structure, a new de-convolution method, developed by our group, was used to remove the broadening effect. The de-convolution method, results of antenna gain and gain variations with antenna elevation will be introduced in Section 4. Related observations were performed in 2007. In Section 5, we will give the results of equivalent noise temperature at the input-port of LNA and its variations with antenna elevation.

2 Miyun 50 m Radio telescope

The MRT50 is located at the Miyun Station of the National Astronomical Observatories, with longitude 116.976 E, latitude 40.558 N, and elevation 155 m (see Fig.1, Plate XIX).

Fig.1　The Miyun 50 m Radio Telescope

The Miyun Radio Telescope (MRT50), which is 50 meters in diameter, is a wheel-on-track system with an Azimuth-Elevation frame structure. It consists of a parabolic reflector, which is made from a solid filled panel in the inner 30 m region and wire mesh in the outer 30 m region, and some multi-function receiver back-ends. The antenna has primary feeds working at UHF1/UHF2, L, S/X and Ku bands. The feed selected can be rotated to the right position automatically to change the observing frequency and moved along the reflector axis. Both the rotation and movement of the selected feed can compensate the variation of the reflector focus caused by gravitation. Tables 1 and 2 list the mechanical and electronic properties of the antenna respectively (measured by the antenna maker, the 54th institute of CETC, in 2006).

Table 1　Main Mechanical Specifications of the Miyun 50 m Antenna

Items	Azimuth	Elevation
Movable range	±270°	+5°~+90°
Maximum slewing speed (deg s^{-1})	1.0	0.5
Slewing speed while accurately controlling (deg s^{-1})	0.003°~0.3°	0.0015°~0.15°
Maximum acceleration (deg s^{-2})	0.5	0.3
Acceleration while accurately controlling (deg s^{-2})	0.015	0.015
Surface rms (<30m)	0.75 mm	
Surface rms (30–50m)	1.06 mm	
Optics	Prime focus, F/D=0.35	
Maximum wind speed while accurately controlling	17 m s^{-1}	

By making use of the UHF1/UHF2 and S/X frequency groups, a design of interplanetary scintillation (IPS) monitoring system has been performed recently (Zhang 2007). An L-band pulsar receiver is now under construction for the MRT50 (Jin et al. 2006). The MRT50 is one of the key-sites for downloading scientific data from the CE-1 lunar satellite and taking part in measurements of its orbit.

3　Pointing calibration of the MRT50

In general, a new large aperture radio telescope ($D>100\lambda$) cannot point at a radio source using the coder only because there are a lot of system errors such as axial error, basic plane error, and so on.

Table 2 Main Electronic Specifications of the Miyun 50m Antenna

Tracking mode	Programming, Automatic
Pointing accuracy	19″ (r.m.s)
Frequency/band width (MHz)	327/30,611/60,1650/330,2300/460, 8400/1600,12100/800
VSWR	≤1.3 (single feed), ≤1.5 (dual-frequency feed)
First side-lobe(dB)	≤−18
Cross couple between two polarization channel (dB)	≤−20
Cross couple between different frequency channel (dB)	<−20
Antenna temperature at feed output point	< 17K for all bands
Antenna efficiency	57.3%@345 (MHz) 57.0%@611 (MHz) 60.2%@1537 (MHz) 59.9%@2500 (MHz) 68.4%@7600 (MHz)(43m diameter) 62.9%@12250 (MHz)(33m diameter)

Table 3 Information of the Pointing Calibration Sources for MRT50

Source name	R.A.(2000)	Dec.(2000)	Type/size(″)	Flux(Jy)X band
3C 48	01 37 41.27	33 09 35.7	QSO /<1.5	3.5
3C 84	03 19 48.16	41 30 42.1	Gal./<1.0	40.0
3C 123	04 37 04.17	29 40 15.1	Gal./23	12.0
3C 144	05 34 32.	22 00 58.	SNR /180 × 240	560
3C 147	05 42 36.14	49 51 07.2	QSO /<1.0	6.0
3C 161	06 27 10.10	−05 53 04.8	Gal./3	4.5
3C 218	09 18 05.70	−12 05 44.0	Gal./< 47 × 47	9.0
3C 273B	12 29 06.70	02 03 08.60	QSO/<20	40.0
3C 274	12 30 49.42	12 23 28.0	Gal./150 × 250	55.0
3C 279	12 56 11.17	−05 47 21.5	QSO/<2.0	13.0
3C 286	13 31 08.29	30 30 33.0	QSO/<2.0	6.0
3C 295	14 11 20.65	52 12 09.1	Gal./5 × 1	4.5
3C 348	16 51 08.20	04 59 33.0	Gal./170.	9.0
3C 353	17 20 28.20	−0058 48.0	Gal./240 × 240	18.0
3C 380	18 29 31.72	48 44 47.0	QSO/<1.0	5.0
3C 405	19 59 28.4	40 44 02.	Gal./170 × 45	190
3C 461	23 23 24.8	58 48 59.0	SNR/240 × 240	610
DR 21	05 55 30.81	39 48 49.2	QSO/<20	20

So, carrying out system pointing calibration is the first step to use the telescope. There are a number of pioneers who introduced the principle and method to calibrate pointing of a telescope (Meeks et al.1968; Yuan et al. 1986; Himwich 1993; Iesber 1967; Ulich 1981, 1982). Instead of the fitting method of least squares, which is widely adopted in model fitting of pointing calibration, Kong (2008) introduced a new fitting method, which is named the Generalized Extended Interpolation Correction Method. Zhang (2007) also proposed another model fitting method, named the Least Squares Support Vector Machines (LSSVM). The beam-width of MRT50 at X band is about 200″, which is the highest frequency available so far. The expected pointing calibration accuracy is about 20″.

Sources for calibration observation should have small angular diameter, strong intensity and uniform distribution in the sky. Table 3 lists the sources we used. Some sources with small angular size but are somewhat weak were excluded from this calibration because of the MRT50 sensitivity, whereas some strong and extended sources, for example Cyg A, Cas A, Tau A and Vir A, were used in the first round of observations. Each source was observed at different positions in the sky. In total, 152 positions, which have an almost uniform distribution in the sky, were observed in 2006. Fig. 2 gives the distribution of the 152 positions in the sky.

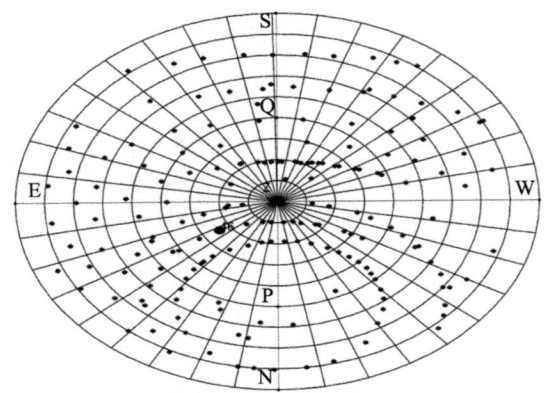

Fig.2 The sky coverage of the pointing calibration observations

The observations were carried out using two dimensional scans. The measurement values at each position are defined as $\Delta A = A_{obs} - A_{cal}$, and $\Delta E = E_{obs} - E_{cal}$. Here A is the antenna azimuth and E is its elevation. To increase the measurement accuracy of ΔA and ΔE, we use a pair of scan-curve median points (Half Maximum Width) to measure A_{obs}, and E_{obs}. This method can reduce the errors because of the different grades near the peak and median points.

The pointing calibration models for MRT50 were built as Equations (1) and (2) (refer to Meeks et al. 1968) from the 152 observations.

$$\Delta A = a_1 \cdot x_1 + a_2 \cdot x_2 + a_3 \cdot x_3 + a_4 \cdot x_4 + a_5 \cdot x_5 \quad (1)$$
$$\Delta E = b_1 \cdot y_1 + b_2 \cdot y_2 + b_3 \cdot y_3 + b_4 \cdot y_4 + b_5 \cdot y_5 \quad (2)$$

where $a_1 = \tan(E) \times \cos(A)$, $a_2 = \tan(E) \times \sin(A)$, $a_3 = \tan(E)$, $a_4 = -\sec(E)$, $a_5 = 1$; $x_1 = -\phi \times \cos(\kappa)$, $x_2 = -\phi \times \sin(\kappa)$, $x_3 = \varepsilon$, $x_4 = \delta$, $x_5 = $KA; $b_1 = -\sin(A)$, $b_2 = \cos(A)$, $b_3 = \cos(E)$, $b_4 = 1/\tan(E)$, $b_5 = 1$; $y_1 = -\phi \times \cos(\kappa)$, $y_2 = -\phi \times \sin(\kappa)$, $y_3 = \beta$, $y_4 = $ato, $y_5 = $KH.

The parameters used in (1)/(2) are defined as the followings. They are: ϕ –azimuth-axis tilt, $(90° - \kappa)$ – the azimuth toward which the azimuth axis is tilted, $(90° - \varepsilon)$ – the elongation between azimuth-axis and the elevation-axis, δ– the collimation error (the axis of the antenna beam is not exactly perpendicular to the elevation axis), KA – constant (the azimuth-axis coding zero-error), β – gravitational deflection error, 'ato' – the residue after calibration of the atmospheric model, KH – constant (the elevation-axis coding zero-error). The parameter values obtained are given in Table 4. Fig. 3(Plate XX) and 4 show the distribution of MRT50 pointing residual errors, and the models applied respectively. Our results show that the pointing calibration accuracy (13.5″ in azimuth, 17.5″ in elevation) reached 1/10 beam-width.

Table 4 Parameter Values Fitted from the 152 Observational Data
(unit is degree for all parameters)

KA	ϕ	κ	ε	δ	KH	ato	β
−0.37358	0.01081	−16.75	−0.00308	0.06442	0.028468	0.00208	−0.03919

For further pointing calibration, using the Ku band may be a good choice. The beam at Ku band of the MRT50 is about 150″ which makes the reachable accuracy of about 15″. In addition, the MRT50 antenna maker, the 54th institute of CETC, measured an epsilon value using a mechanical method. The value they found was 12″

while our value is about 11″.

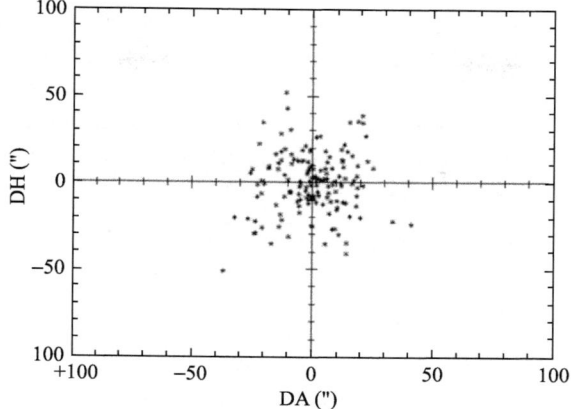

Fig.3　Distribution of the pointing calibration residues

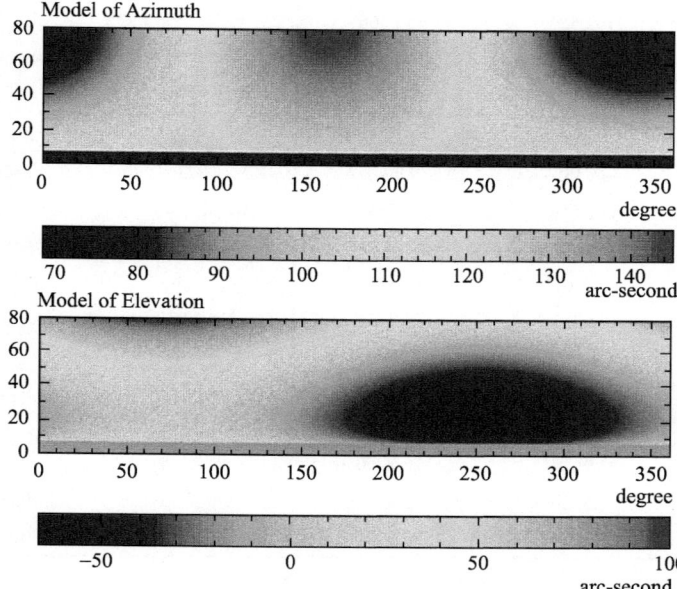

Fig.4　The azimuth model (top) and the elevation model (bottom) used for the pointing correction of MRT50. The constant term of the model is not displayed in this figure

4　Measurements of the main beam and gain of the MRT50

Ideal radio sources for beam measurement of a radio telescope should be small and strong. However these kinds of sources cannot be used to measure a side-lobe as low as −20 dB for MRT50 because of its sensitivity limitation. The observed beam will be broadened if an extended source is used to measure the beam. Cyg A is a strong radio source with two-point components. It is possible to deal with Cyg A as two small angular sources because of its special structure.

We developed a program to fit the observed scan curve with a calculated curve which convolutes an adjustable beam with a two-point model of Cyg A. The model of Cyg A is defined as the following: two point sources, each is 30″ in diameter, and the angular spacing between the two points varies with the azimuth and elevation (see Table 5).When the residue between the observed curve and calculated one is small enough, the adjustable beam should be the beam which has no broadening effect on the two-point structure. Fig. 5 gives examples at S/X bands. It shows that the residue is smaller than 1% and the fit between the observed curve and the calculated one is quite good. The fit suggests that the beam shape of the MRT50 at X band can be well represented by a Gaussian function. The feed of the S/X band is a compound one. The X band feed is in the center part surrounded by the S band feed. Hence, the beam shape of S band cannot be represented very well by a Gaussian function. It is represented by a modulated Gaussian function. In this case, the residue is also small enough.

Table 5 Distances between the two-components of the Cyg A in the western and eastern parts of the sky ($da = (a_1 - a_2) \times \cos(h)$, $dh = h_1 - h_2$), where a is the azimuth and h is the elevation in degrees

West	$a_1(°)$	$a_2(°)$	$h_1(°)$	$h_2(°)$	$da(')$	$dh(')$
	275.0571	274.9711	80.9007	80.9317	0.8160	−1.8640
	275.5492	275.4730	79.3877	79.4185	0.8424	−1.8521
	276.0838	276.0150	77.8760	77.9067	0.8676	−1.8402
	276.6475	276.5844	76.3659	76.3964	0.8913	−1.8292
	—	—	—	—	—	—
	311.9667	311.9357	14.2684	14.2840	1.8047	−0.9369
	313.0168	312.9857	13.1479	13.1630	1.8223	−0.9036
	314.0834	314.0520	12.0466	12.0611	1.8409	−0.8692
	315.1666	315.1351	10.9652	10.9791	1.8569	−0.8340
	316.2668	316.2351	9.9042	9.9175	1.8705	−0.7979
East	$a_1(°)$	$a_2(°)$	$h_1(°)$	$h_2(°)$	$da(')$	$dh(')$
	43.7332	43.7185	9.9042	9.8736	0.8699	1.8388
	44.8334	44.8192	10.9652	10.9342	0.8348	1.8556
	45.9166	45.9030	12.0466	12.0154	0.7993	1.8712
	46.9831	46.9701	13.1479	13.1165	0.7640	1.8856
	48.0333	48.0207	14.2684	14.2367	0.7289	1.8994
	—	—	—	—	—	—
	82.7677	82.8011	74.8575	74.8247	−0.5226	1.9666
	83.3525	83.3913	76.3659	76.3332	−0.5476	1.9597
	83.9162	83.9617	77.8760	77.8435	−0.5733	1.9524
	84.4508	84.5050	79.3877	79.3553	−0.5995	1.9441
	84.9429	85.0089	80.9007	80.8684	−0.6268	1.9359

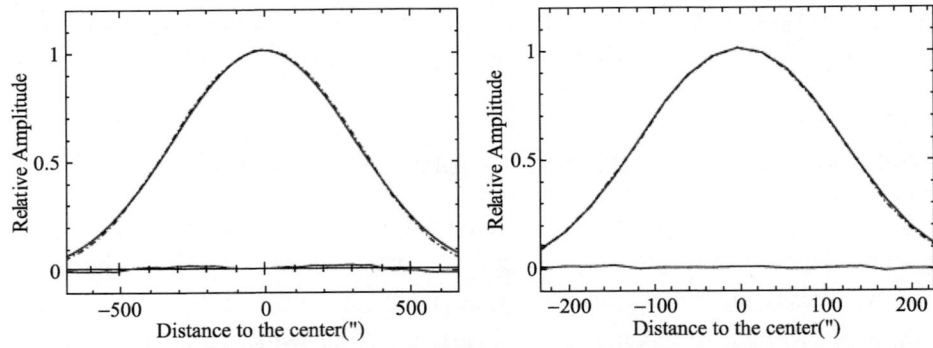

Fig.5 An example of the de-convolution fit at S band (left) and X band (right), continuous line (up) is the observed curve and the dashed line (up) is the calculated curve, and the lower line is the residue

4.1 Two Dimensional Main Beam

By using the method described above, we determined the two dimensional MRT50 beams at S/X bands. Cyg A was used as the emitting source. Scans were carried out in the azimuth direction with the intervals of 120″ at S band and 60″ at X band in the elevation direction.

Fig. 6 (Plate XX) shows the measured S and X band beams, which are both symmetric. The beam-width of 3 dB and 10 dB at S band is $11.816'(A) \times 11.779'(E)$ and $19.945'(A) \times 19.912'(E)$ at elevation 30° respectively, while at X band it is about $3.527'(A) \times 3.187'(E)$ and $6.430'(A) \times 5.810'(E)$, respectively. The corresponding S/X band gains are 58.8 dB and 69.4 dB, respectively. Our measurements were carried out at 2300MHz (S band) and 8400MHz (X band). The MRT50 antenna maker, the CTI 54th institute, measured one-dimensional scans using a satellite signal as the emitting source in the azimuth and altitude planes at 2500MHz and 7600MHz (not our S/X working frequency) respectively. The MRT50 gains at 2500/7600MHz were 59.85 dB and 68.37 dB respectively. After changing our gain values to the 2500/7600MHz, we got 59.52 dB and 68.21 dB. We found that our gain value at X band agrees with their result well while the gains at S band have a 0.33 dB difference. According to the measured two-dimensional beams, the main beam efficiencies, effective receiving areas, aperture efficiencies, and other parameters at the S/X bands were calculated. Table 6 lists the values. The meanings of the parameters in Table 6 should be clear without any explanation except for the η_{ANT} parameter. It is the efficiency of the antenna related to the loss of the reflector shape-errors and the feed (Stutzman 1998). Although the illuminating intensity at the reflector edge (50 m) is about –10 dB and –16 dB for S/X bands respectively, we take the whole reflector area to calculate the aperture efficiencies for both S and X bands (see Table 6). On the other hand, the effective receiving areas we measured come from the radiating property of the whole reflector (50 m), so the whole reflector area should be taken to calculate the aperture efficiencies. This is different from what the 54th institute calculated. The illuminating diameter they used at X band is 43m so that they achieved a higher aperture efficiency number at X band.

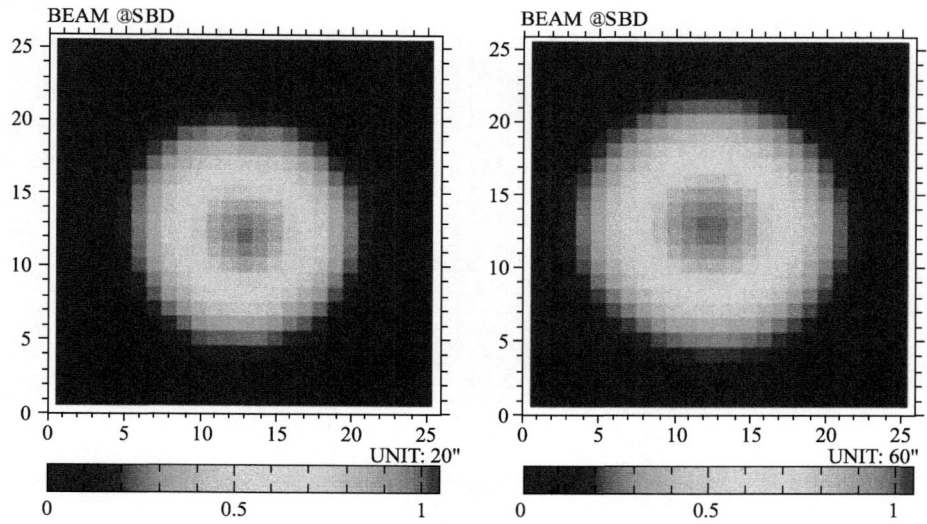

Fig.6 Main beam of MRT50 at X band (left) and S band (right)

Table 6 Calculated Parameter Values of the MRT50 Antenna

Term	S band (2300 MHz)	X band (8400 MHz)
η_{ANT}	92%	88%
Ω_M	1.3341×10^{-5}	1.07747×10^{-6}
Ω_A	1.52685×10^{-5}	1.27086×10^{-6}
D_0	59.154 dB	69.951 dB
η_{MB}	87.4%	84.8%
A_E	1107 m^2	1020 m^2
η_{AP}	56.4%	52%

Explanation can be found in Section 4.1

4.2 Gain and gain variation with elevation of MRT50

The gain of an antenna may vary with its elevation because gravitation may cause its reflector's shape to change. To measure this property, we measured the gains from 10°–80° elevation of MRT50. Two scans in the azimuth and elevation planes were recorded with an interval of 5° from 10°–80°. These two scan data were reduced by using the method mentioned above to get a new beam curve, named as C-beam, from which the effect of Cyg A broadening has been removed. Then, the 3 dB and 10 dB beam-width of the C-beam in the two planes were measured. The antenna gain at this elevation position was then calculated by Equation

$$G(\text{dB}) = 10\lg\left[\frac{1}{2}\left(\frac{31000}{\theta_{3\text{dBAz}} \times \theta_{3\text{dBEl}}} + \frac{91000}{\theta_{10\text{dBAz}} \times \theta_{10\text{dBEl}}}\right)\right] - \Delta G_F - \Delta G_\sigma \tag{3}$$

Table 7 and Fig. 7 show the MRT50 gain variation with elevation at the S/X bands. There is a maximum in gain around elevation 25° at the S/X bands. This may be related to the MRT50 assembly procedure. The MRT50 maker made a reflector-shape adjustment at about elevation 30°. The gravitational change of the reflector-shape at X band may be smaller than that of the S band because the effective illuminating diameter of the MRT50 at S band is larger than that at X band. On the other hand, the outer part of the reflector may cause more gravitational shape-variation than the inner part does. This may be responsible for the obvious turn over in gain variation of S band in the low elevation region in the right panel of Fig. 7.

Table 7 Measured Gains of MRT50 at S/X Bands

Gain at X band		Gain at S band	
Elevation (°)	Gain (dB)	Elevation (°)	Gain (dB)
11	69.48	10	58.85
15	69.46	15	58.78
20	69.61	20	58.77
25	69.53	25	59.11
30	69.41	30	58.80
35	69.43	35	58.85
40	69.33	40	58.95
45	69.35	45	59.08
50	69.31	50	58.78
55	69.25	55	58.70
60	69.22	60	58.62
65	69.21	65	58.81
70	69.17	70	58.57
75	69.18	74	58.56
80	69.20		

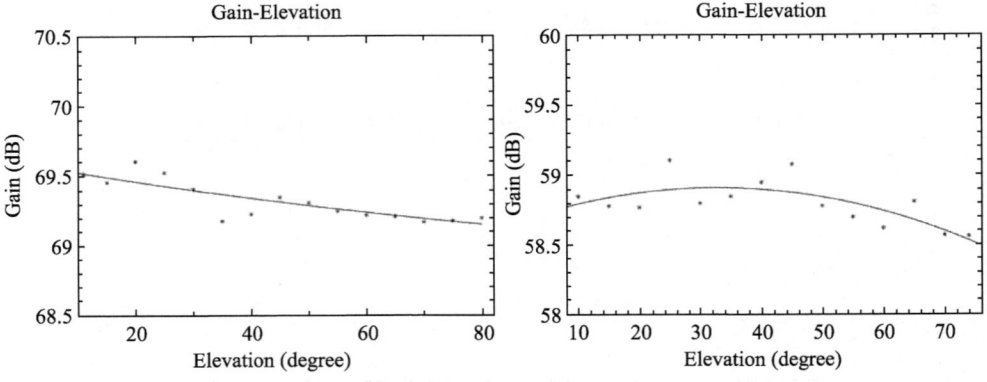

Fig.7 Gain variations at X band (left) and S band (right)

The method we used to measure the gain variation is not to determine the absolute amplitude of the gain at different elevations. We measure the beams in the two perpendicular planes at a series of elevations only so that the effect from atmospheric attenuation on our measurements could be ignored. In addition, the observations were carried out during the dry and clear days, Sep. –Oct. 2007, the driest season in Beijing, so that atmospheric attenuation at X band may not be so serious. All of these may explain why the gain variations with elevation at X band and S band may not have a large difference.

5 Equivalent noise temperature and its variation

The system temperature of a radio telescope comes from receiver noise, background sky emission, ground emission, atmospheric emission and the antenna structure itself. Except for the receiver noise, the others vary with elevation of the antenna. This is also an important property to measure. The method we used is the same as that described in papers (Huang et al. 1988; Liu et al. 2002). In this paper, we take the LNA input-port as the reference point. A noise signal is injected at this point as the calibrator. All received noise power (within the same bandwidth) has been scaled to the temperature at this point. The measured results of system temperature variations are shown in Figure 8.

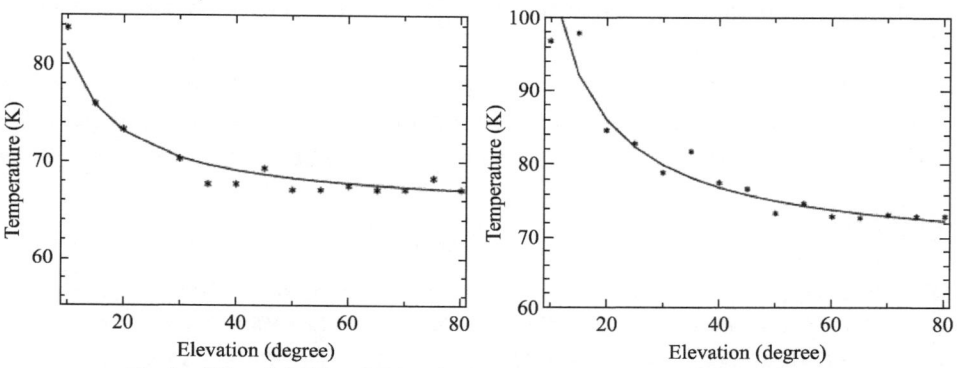

Fig.8 X band (left) and S band (right) system temperatures of MRT50

System temperature, the total noise temperature at the reference point, can be expressed as Equation (4) (Huang et al. 1984).

$$T_{sys} = T_r + (1 - \eta)T_0 + \eta T_{sky} \tag{4}$$

where T_0 is the ambient temperature, T_{sky} is the equivalent temperature contributed by the sky background emission, ground emission, atmospheric emission, and antenna structure at the feed output, T_r is the equivalent temperature of the receiver at the LNA input port, and η is the transmission efficiency between the feed output-port and the input-port of the LNA.

The Equation (4) can be re-written as $T_{sys} = T_{const.} + T_{var.}$, where $T_{const.}$ can be represented by the tangent-lines in the bottoms of Figures 10 and 11. The $T_r + (1-\eta)T_0$ and the T_{sky} toward the zenith give the contribution to the $T_{const.}$ from which we can estimate the transmission efficiency (η) between the feed output port and the input port of the LNA. Here, we give an example. For S band, the $T_{const.}$ is about 72K, T_r is about 20K and T_0 is 291K. The T_{sky} for the zenith direction can be estimated as follows (Anderson et al. 1991)

$$T_{sky} = T_{sky-bg}(\sim 2.7K) + T_{galac}(\sim 1K) + T_{atmos}(\sim 4K) + T_{spillover}(4\sim 5K)$$
$$+ T_{scatter-support}(3-4K) + T_{leakage-mesh}(3-4K) \tag{5}$$

Then we get $\eta_S = 0.86\pm 0.02$.

For X band, two main things are different. One is that the reflector for X band is the inner part of the whole reflector. Another is that the ratios of mesh surface to solid panel surface for X/S bands are 1.05 and 1.78 respectively. So both $T_{spillover}$ and $T_{leakage-mesh}$ at X band should be smaller than that at S band, whereas the T_{atmos}

at X band is slightly larger than its value at S band. Taking these factors into account, we found $\eta_X = 0.90\pm0.02$.

6 Summary

The main points of this paper are listed as follows.

(1) Pointing accuracy of the MRT50 (rms) is 13.5″. in the azimuth direction and 17.5″ in the elevation directions, respectively.

(2) Measured the two-dimensional main beams at S/X bands. The main beam efficiency is 87.4% and 84.8% at S/X bands respectively. The beams show a symmetric shape.

(3) Measured gains and its variation with elevation at S/X bands. No large gain variations with elevation from 10°– 80° elevation were found, although the trends of S/X band show some differences. The tune-over of S band gain variation in the low elevation region may be caused mainly by the outer part of the reflector.

(4) Measured system temperature and its variation with elevation at S/X bands. The system temperature varies from 62–76K and 72–100K at X/S bands for elevation 80°–10° and gets the estimated transmission efficiencies between the feed output port and the input port of the LNA at S/X bands, which is $\eta_S = 0.86\pm0.02$ and $\eta_X = 0.90\pm0.02$ respectively.

(5) Developed a new program to remove the broadening effect on beam measurement when using an extended source as the emitting source.

Acknowledgements Authors thank Prof. Liang Shiguang for the long-term cooperation in radio astronomy and especially for the preparation of the new radiometer and thank Mr. Cao Xiandong for his helps with the sampling program development. We are also thankful to the MRT50 operators for their assistance during all the observations.

References

[1] Anderson, M. D., Routedge, D. V., Vaneldik, J. F., & Landecker, T. L. 1991, Radio Science, 26, 353
[2] Himwich, W. E. 1993, Pointing Model Derivation, Operation Manual of VLBI Mark IV Field System
[3] Huang, X. Y., & Gao, Y. H. 1991, Annals of Shanghai Observatory Academia Sinica, 12, 127
[4] Huang, Y. R., Xu, A. A., Tang, Y. H., et al. 1984, Observational Astrophysics, 483 (in Chinese)
[5] Isber, A. M. 1967, Microwaves, August, 40
[6] Jin, C., Cao, Y., Chen, H., et al. 2006, ChJAA(Chin. J. Astron. Astrophys.), 6S2, 319
[7] Johannsen, K. G., & Titus, L. 1986, IEEE ON IM, IM-35, 344
[8] Kong, D. Q., Shi, H. L., Zhang, X. Z., & Zhang, H. B. 2008, Journal of Xidian University, 35, 157
[9] Liu, X. 2002, Technique Report of the Urumiq Astronomical Observatory, 134
[10] Meeks, M. L., Ball, J. A., & Hull, A. B. 1968, IEEE Trans. ON AP, AP-16, 746
[11] Stutzman, W. L. 1998, IEEE Trans. On AP, AP-40, 7
[12] Ulich, B. L. 1981, International Journal of Infrared and Millimeter Waves, 2, 293
[13] Ulich, B. L. 1982, SPIE, 332, 33
[14] Wang, S. G., Zhu, Z. H., Zhou Z. L., & Zhang, Y. Z. 2002, IJMPD, 11, 1061
[15] Yuan, H. R., Peng, Y. L., & Xue, Y. Z. 1986, Antenna Parameter Measurement Using Radio Astronomical
[16] Technique, Electronic and Industry Press, 147 (in Chinese)
[17] Zhang, X. Z., et al. 2000, in Proceedings of Kanasi Pulsar Observation and Research Meeting, ed. Zhang J., 139
[18] Zhang, J. Y. 2007, PhD Thesis, 2007, 101
[19] Zhang, X. Z., Zheng, Y. J., Chen, H. S., et al. 1997, A&AS, 121, 59
[20] Zhang, X. Z. 2007, ChJAA(Chin. J. Astron. Astrophys.), 7, 712

Design and Implementation of Space Dust Database

Zuo Wei[1,2,3]   Li Chunlai[2]   Xu Lin[1,2,3]   Liu Jian[1,2,3]   Liu Jianjun[2]

(1. Institute of Geochemistry, Chinese Academy of Sciences, Guiyang 550002, China;
2. National Astronomical Observatories, Chinese Academy of Sciences, Beijing 100012, China;
3. Graduate School of Chinese Academy of Sciences, Beijing 100039, China)

Abstract   Space debris is very dangerous to the security of on-the-orbit spacecrafts, and it is increasing in number at high speed with the expansion of human space exploration. Space debris has become a serious space pollutant noticed by many astronomers. The increase of space dust sources and the development of research on space dust urgently need space dust data sharing and exchanging. It is necessary for us to establish the Space Dust Database to realize the sharing and canonical management of the data. The Space Dust Database (SDD) management system, based on the 3-layer B/S computer mode, was designed and implemented in this paper. The system's features include significantly improved runtime efficiency, good scalability and maintainability. The Space Dust Database can provide some scientific bases for the study of the chemical constituents, mineral composition, origin and sources of space dust, but also provide excellent data services and decision-making support for the protection of space and model construction of space dust.

Key words   Space dust   Database   B/S computer mode   ASP/ADO

1 Introduction

Since the first artificial satellite was successfully launched in October, 1957, people have conducted a series of space activities. Through 45 years of outer space exploration in the past, there is no doubt that human's knowledge and civilization has been promoted to a great extent, but a great amount of space debris has been instantaneously discarded in space. According to the statistical data available, there have been more than 4000 times of space launching over the past years, the objects that have been launched to space are more than 26000 in number, one third of which still remain in space, moving on their orbits. At present, there have been more than 8000 large space objects that can be observed. Their orbits can be determined at ground, but just six percent of the space objects are space shuttles that can work well, and the rest are space debris. The amount of small space debris is several times greater than that of large space debris. Space debris, large than one meter in diameter, numbers over 110000; one millimeter in diameter, over 35000000 (Li Chunlai et al., 2002; Ouyang Ziyuan, 1988). It is very terrible that the amount of space debris is still increasing at a high speed with the expansion of space resource exploration and application, and meanwhile the space environment that surrounds the Earth is becoming increasingly polluted by space debris.

Space debris is very dangerous to the security of spacecrafts. Debris collision is one of the main mechanisms leading to spacecraft destruction, and is also a main factor affecting the life of spacecraft. Sometimes, space debris even threatens the life of astronauts, as well as astronomic observation and research. Owing to the growing awareness of space debris risk, most of the countries all over the world are commonly acquainted with the importance to take measures to control the increase of space debris. Simultaneously, they also have invested large manpower and funds to study and develop new technologies to shield spacecrafts from orbital debris.

In recent years, methodology and technology have been developing rapidly in reducing space debris and alleviating the damage of space debris. Chinese scientists have also accumulated lots of primary data and information on space debris. By analyzing these data, we not only could understand the mode of occurrence of space debris which can be applied to evaluating the risk and harm of space debris, estimating the detected ratio of space debris and improving the forecast of how to avoid flying airplanes and evaluating the analytical effects of methods to eliminate space debris, but also could acquire significantly scientific evidence for better design of spacecraft in choice of shielding material and structure so as to shield on-the-orbit spacecrafts from the attack of space debris. Computer science and information technology have greatly improved human's capability of acquiring, processing and applying the information. With the development of shielding process against space debris and the increasing availability of space debris data and information, it is imperative that we should establish the Space Dust Database System(SDDS) that can support data and information sharing and exchanging by means of the advanced information management technology. And it is also possible for us to realize a goal that information can be managed effectively and used widely. Therefore, we intend to establish an open and highly safe SDDS that is of good scalability and maintainability and can be easily, expediently visited by guests. The data of SDDS will be organized based on the criteria of metadata, and then incorporated into the SDDS. We can query the information via internet. In this paper, combining with part of the work on "space database and chemical types", emphasis is given on the design and implementation of SDDS. In the SDDS design the distributed database technology is employed, and the SDDS is constructed under DBMS-SQL 2000.

2 The design of the Space Dust Database

2.1 The objective and significance of cosmic dust database

Establishing a cosmic dust database towards protection and model construction is very helpful for users to know the chemical composition, mineral composition, origin and material source of cosmic dust. And based on the cosmic dust database, we can also estimate the proportions of all chemical species in cosmic dust by means of statistical analysis, study the source of cosmic dust and the factors leading to spatial pollution, and thus we can put forward some advice or effective measures to reduce spatial debris and provide some scientific bases for the protection of cosmic space and the model construction of cosmic dust. In order to realize the above functions, the system structure is recommended to have the advantages of credibility, flexibility, ease expansion and maintenance. In addition, the operation of client should be as simple as possible, with high practicability. At the same time, the system should provide convenient searching means and maintenance function, and possesses safety precautions of network accessing.

2.2 System structure

On the basis of the objective of system design, we decide to adopt the three-layer browse/server computer model as the integrated development model of the system, and this model has integrated web technology and database technology, as shown in Fig. 1.

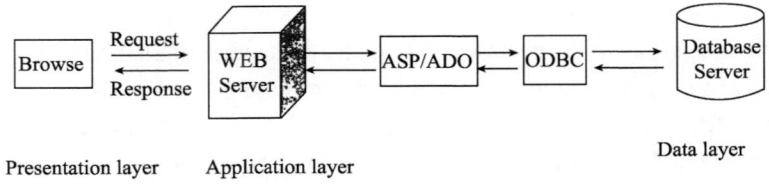

Fig. 1 Sketch showing the 3-layer B/S structure

The 3-layer B/S computer model is an advanced development model which is coordinated with the applied program. According to this scheme, the system is divided into 3 layers in terms of their service functions, all the three layers constitute an applied program. This 3-layer service program includes: 1) client server, called the browser of client (such as IE, Netscape, etc.), is used to receive user's input information and to display what is returned from the Web server; 2) applied server, namely web server programmer (such as IIS, Apache, etc.), is used to receive the client information from the browser, send query request to the database server, and return the query results to the browser; and 3) data server, namely the database server (such as SQL Server, Oracle and so on), is used to finish data operation involving definition, query, update and so on, and protect data security and integrity. The 3-layer B/S computer model is characterized by concentrative management, i.e., all software programs, databases and other modules are concentrated at the server terminal, and users can easily acquire reliable and all-around information through the installed browser at the client terminal.

2.3 Configuration of the software and hardware

According to the demands of design, our local area network is equipped with 2 servers and five personal computers at present. Network trunk line is connected with the center server of the network, the server of user management, the server of catalog management and other large-capacity data storage devices through a 100 M network line. Branch line is connected with the local area network information browser and query server through a 10 M common network line by the data exchange machine, such as WWW hyper text browser, FTP file transport, E-mail mailbox, TELNET long-distance accessing, etc. They are medium or small data capacity network facilities, laterally connected with each client terminal and relevant equipment through a common network line. The operating system of the main central server is Windows 2000 Advanced Server, with RDBMS-SQL server 2000 and Visual studio 7.0 to develop the distributing database applied system installed. The client terminal just needs to install the windows operating system.

2.4 The structure of the cosmic dust database

The cosmic dust database consists mainly of four sub-databases, each of which makes up the tables of the large relational database SQL Server 2000. The name and content of each sub-database are listed as follows:

(1) User Table: This table includes all the information about users, such as user's name, login name, password, contacting manner, the permission status of each user, etc.

(2) Classification of the cosmic dust table: This table includes the classification information about the collected cosmic dust, and provides a convenient and rapid approach to acquiring the information about cosmic dust. Through the classification index, users can look up the detailed information on cosmic dust they want to know. The content includes mainly index ID, classified name and arranged sort, etc.

(3) Information on the cosmic dust table: This table includes the detailed observational and analytical data of each collected cosmic dust. The relations are set up for the classification table of cosmic dust with a field of this table, and then the classification and storage of cosmic dust can be realized. The content of this table includes cosmic dust's name, size, morphology, color, type, SEM photo, etc.

(4) Archives of the cosmic dust table: This table includes the general information on cosmic dust. It is free of any restriction on the type and format of information. The main fields of this table are title, abstract, text, etc. Inputting the keyword, users can obtain all the relevant cosmic dust information matched with any of the title, abstract and text.

3 The implementation database functions

3.1 Security of the database

Security of the database is mainly dependent on the following three aspects:

(1) Security of the database server: Safety precautions such as login security setting, users' role management, perfect data backup and recovery are taken to ensure safe operation of the database server.

(2) Security of the network: We will adopt several key technologies of network safety precautions such as fireworks, encrypt, network supervision and deputized server deployment, scanning system at scheduled time, repairing security hole in time, running the inspection system of network attack, inspecting and checking system incessantly to ensure the safe operation of the network system and network protocol server, the availability of network layer and the security & validity of data transmission.

(3) Security of the applied program: In the ASP program, we should adopt users' ID, set visiting overtime, set up visiting log and other corresponding component technologies to improve the stability and security of the applied program.

3.2 Management of the database

The system of the database consists of three management modules as listed below:

(1) The module of collecting users' information: There are three kinds of users in this system, which are administrators, registered users and common users, respectively. The administrators enjoy the highest management permission, and can realize the management of users, the maintenance of the system and the updating of the data through manipulating the system management module. The registered users are divided into different groups. Each group possesses the permission of different system modules so that the users can edit and browse the information that they manage. The common users just possess such a permission as to browse and search for information about cosmic dust. Through the login page, the system is able to identify which kind of browser is being used by the users, and then the system automatically allots the permission to the browser based on users' login name and password. This procedure is shown in Fig. 2.

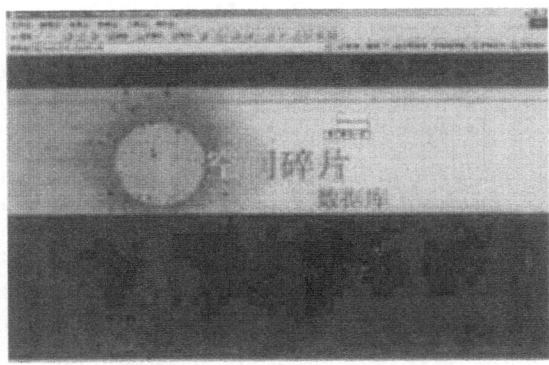

Fig. 2 The login page of the cosmic dust database

(2) Integrated information management module: It includes the functions of information processing and information browsing. Information processing includes the functions of cosmic dust classification, edition and updating of information and input of new data of cosmic dust.

(3) System management module: It not only takes care of the system user management, permission setting and allotting, but also controls the daily maintenance, timing management and data updating and backup of the whole database system. \; ASP and ADO components technologies are used to develop the system of cosmic dust database and to realize data management. ASP/ADO is a perfect settlement project of the network database system, which is developed by Microsoft. Through ActiveX Scripting, it can be easily connected with any database that is compatible to ODBC (Microsoft corporation, 1998-2004).

Once users succeed in login, they can carry out corresponding data operation-query and update (insert, delete and modify) according to the permission endowed by the system. Data updating operators can accomplish these operations just by using a simple applied interface at the client terminal, and these operators are not required to master any complex technology about relational database and SQL language, while integrality checkout of the data and function constraints are hidden at the server terminal.

3.3 Database query

The query subsystem can realize the search function of cosmic dust database under internet environment. Using ASP/ADO, ActiveX and other components, we have developed the applied internet program. The query system can provide various and agile query methods such as exact and blur query.

4 Conclusions

Based on the developed cosmic dust database of 3-layer B/S computer mode, the distributed share service of cosmic dust data is realized successfully. At present, the development of the system has been almost completed. As a case, we have recorded 468 data items for cosmic dust particles collected from the lunar surfaces (L2021 and L2036) (Cosmic Dust Preliminary Examination Team, 2002). This has proved that the system has perfect functions, and realized the objective of project design. The cosmic dust database can provide excellent data service and decision-making support for the study of the chemical composition, mineral composition, origin and sources of cosmic dust and for the scientific chemical classification of cosmic dust. In order to meet users' increasing needs for information, we should keep on improving the functions of the database and adding more information so as to make the database system become more reliable and more practical.

References

[1] Cosmic Dust Preliminary Examination Team, Cosmic Dust Catalog Volume 15 Contents, http://curator.jsc.nasa.gov/dust/cdcat15/Introduction.html, 2002

[2] Li Chunlai, Ouyang Ziyuan, and Du Hen, 2002, Space debris and space environment [J]: Quaternary Sciences, v. 22, n. 6, p. 540—551 (in Chinese with English abstract)

[3] Microsoft corporation, ADO Programmer's Guide, http://msdn.microsoft.com/library/default.asp?url=/nhp/default .asp?contentid=28001860, 1998—2004

[4] Ouyang Ziyuan, 1988, Cosmochemistry [M]: Beijing, Science Press, p. 273—289 (in Chinese)

Introduction to the Payloads and the Initial Observation Results of Chang'E-1

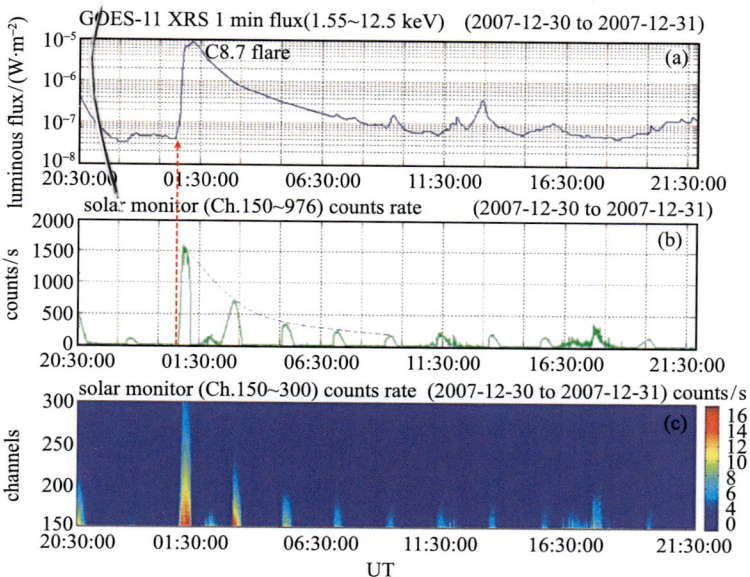

Fig.10   Counts rate (1.5~10 keV) and instantaneous spectra (1.5~3 keV) from solar X-ray monitor (b) and (c). Data from GOES are shown (in units of W/m^2) for comparison (a)

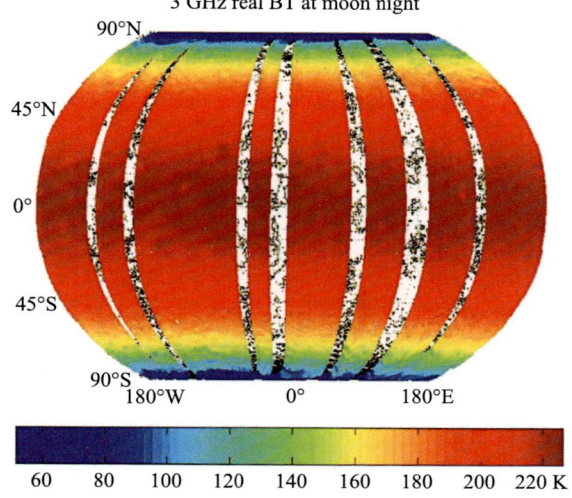

Fig.14   lunar night Brightness Temperature (BT) at 3 GHz from Dec. 4 to Dec. 30 of 2007

图版 II

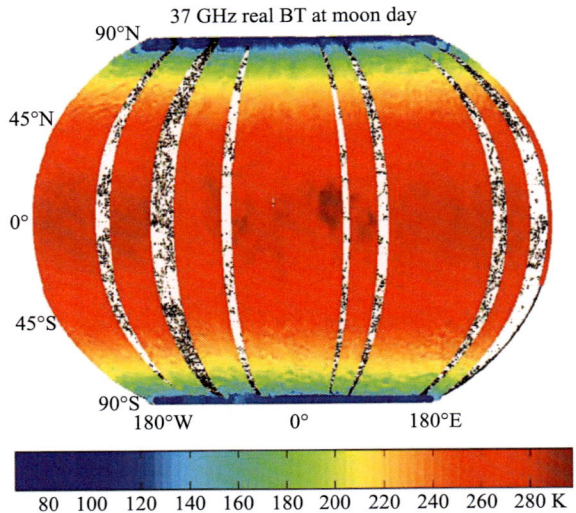

Fig.15　lunar day Brightness Temperature (BT) at 37 GHz from Dec. 4 to Dec. 30 of 2007

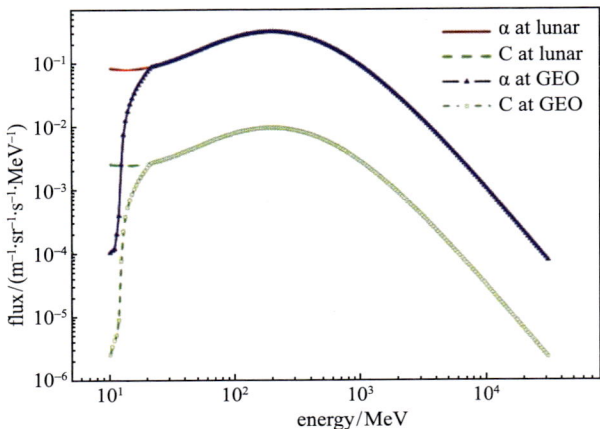

Fig.20　Differential fluxes of cosmic Carbon and a-particles at GEO and lunar orbit

嫦娥一号卫星 CCD 立体相机的设计与在轨运行

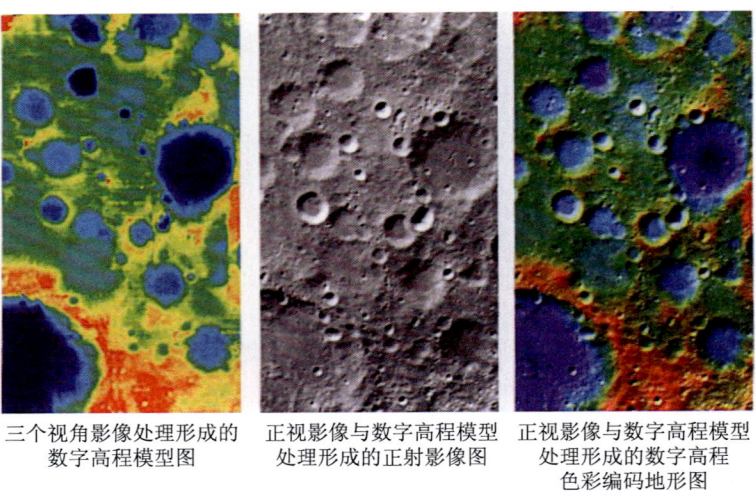

三个视角影像处理形成的　　正视影像与数字高程模型　　正视影像与数字高程模型
数字高程模型图　　　　　　处理形成的正射影像图　　　处理形成的数字高程
　　　　　　　　　　　　　　　　　　　　　　　　　　色彩编码地形图

图 9　月面图像局域区域形貌图

图版 III

图 11　由激光高度计数据绘制的全月球 DEM 图

嫦娥一号 IIM 数据应用处理流程分析

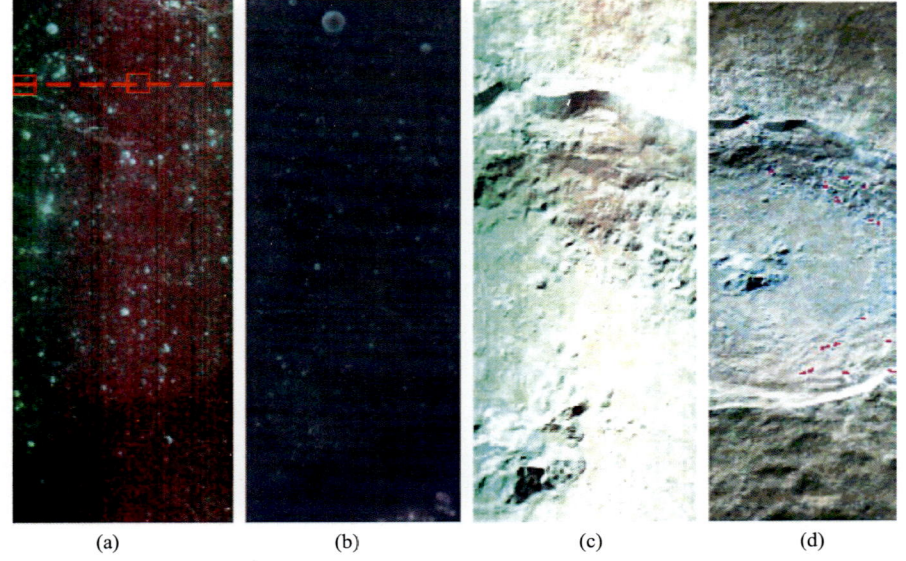

图 1　MS2(a)和 Tycho 撞击坑(c)IIM 真彩色合成图像及相应的克莱门汀图像(b)和(d)

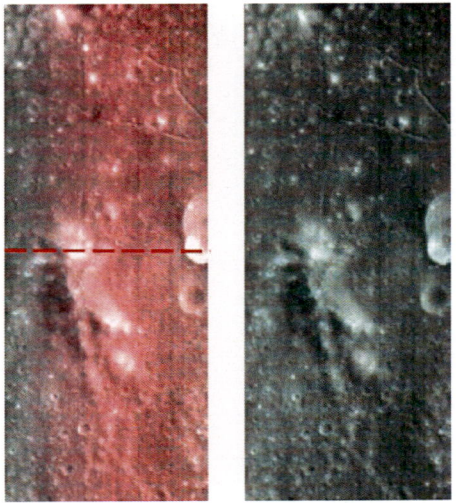

图 6　Arzachel 撞击坑校正前(左)、后(右)彩色合成图像

图版 IV

图 8 Aristarchus 高地 B_{757}/B_{644}(R)、B_{757}/B_{865}(G) 和 B_{644}/B_{757}(B) 校正前的比值(左)和校正后的比值(右)合成图像(2 897、2 898 和 2 899 轨镶嵌)

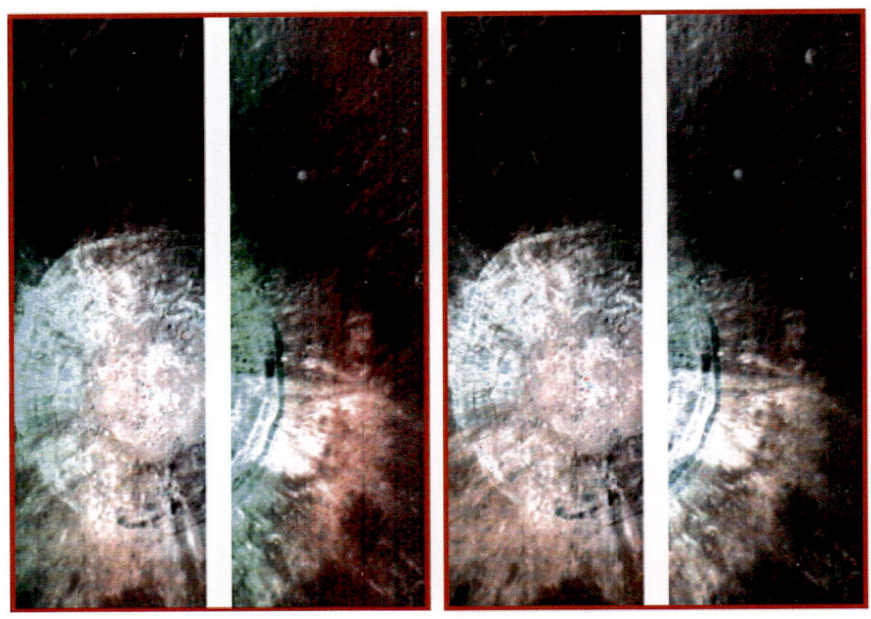

图 9 Aristarchus 撞击坑校正前(左)和校正后(右) B_{918}(R)、B_{757}(G) 和 B_{618}(B) 合成图像

嫦娥一号IIM数据应用处理流程分析　　　　　　　　　　　　　　　　　　　　　　　　　　　　　　　　　　图版 V

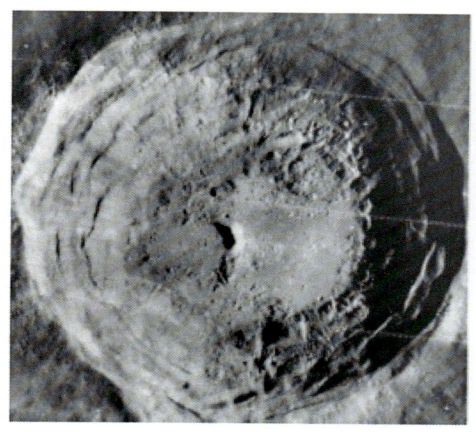
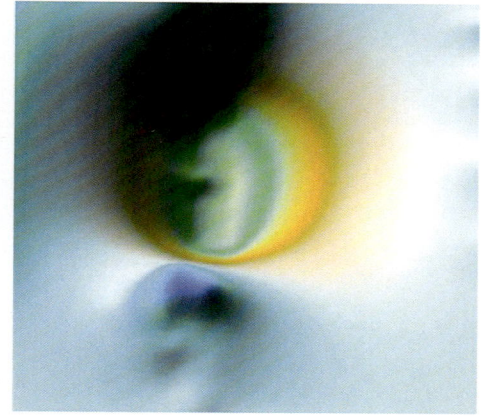

图10　Aristarchus 撞击坑 Lunar Orbiter 高清晰照片(左)和 DEM(右)

嫦娥一号IIM数据绝对定标与初步应用

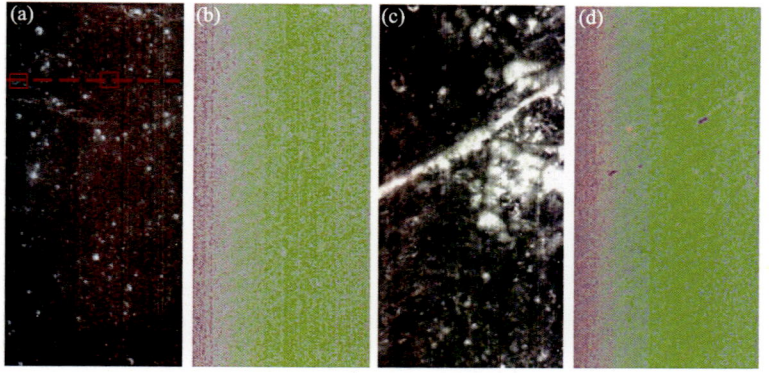

图3　MS2(a)和 Petavius 撞击坑(c)真彩色合成影像以及两个地区相应的 891/918 nm(R)，918/891 nm(G)，757/918 nm(B)比值合成影像(b)和(d)

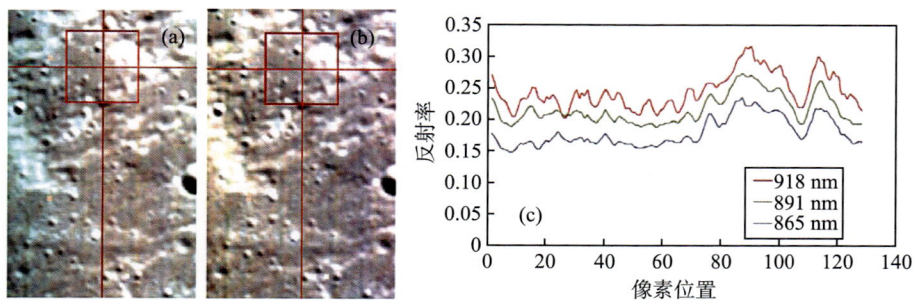

图7　高纬度、斜长岩地区校正前(a)后(b)影像和校正后的横向光谱剖面(c)

图版 VI

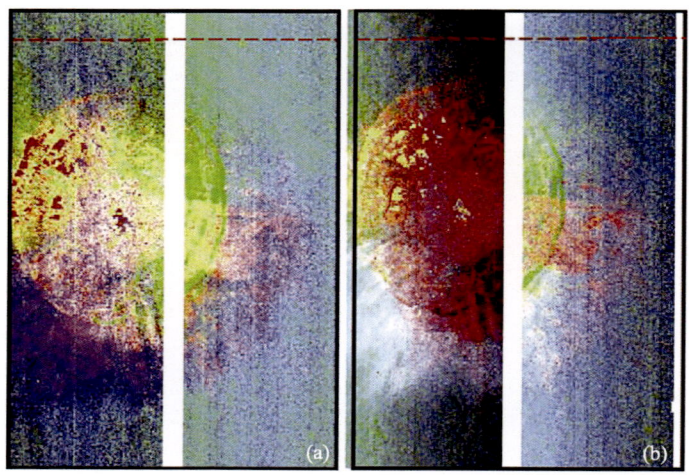

图 9　Aristarchus 撞击坑校正前(a)和校正后(b)PC1(R)，PC2(G)，PC3(B)合成图

嫦娥一号 IIM 数据定标的改进方法

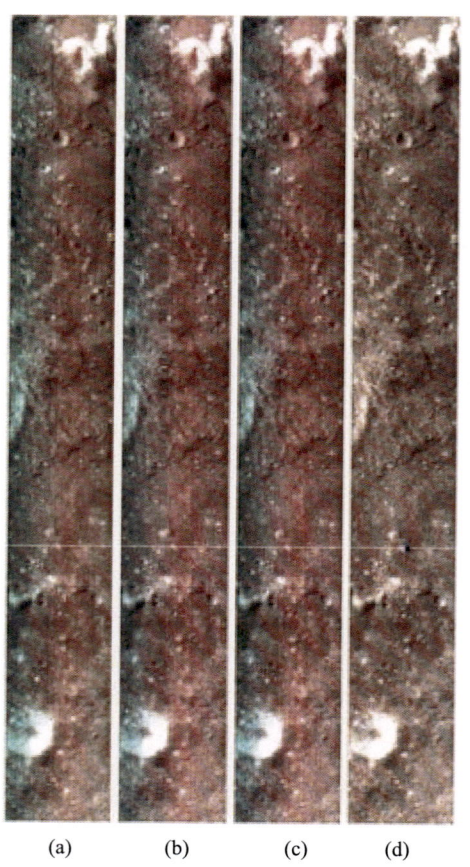

图 1　原始2225轨(a)、坏线修复(b)、坏点修复(c)以及条纹去除后(d)的RGB
(R：918 nm，G：757 nm，B：658 nm)彩色合成影像

(a)~(c)中均可见明显的色调不均匀，左侧偏蓝，中央呈深红，而右侧呈暗红，表明从左到右的响应不一致。(d)是条纹去除后的 RGB 合成影像，色调均一，表明消除了平场效应。(d)中蓝色方框所示范围为定标区域。(a)~(d)中白色横线为图 2 的切面位置

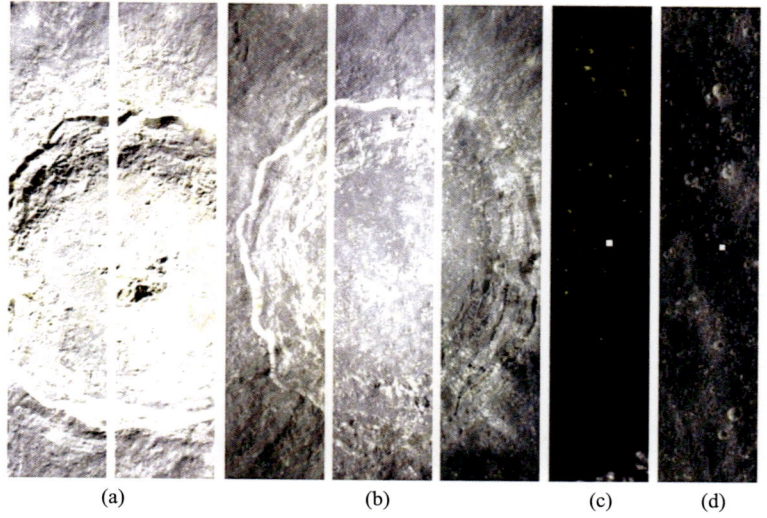

图 11 Tycho 撞击坑(a)、Copernicus 撞击坑(b)、Mare Serenitatis 月海(c)和 Apollo 14 着陆点(d)的 RGB 合成影像图(R 为 658 nm，G 为 757nm，B 为 918nm)

A Preliminary Experience in the Use of Chang'E-1 IIM Data

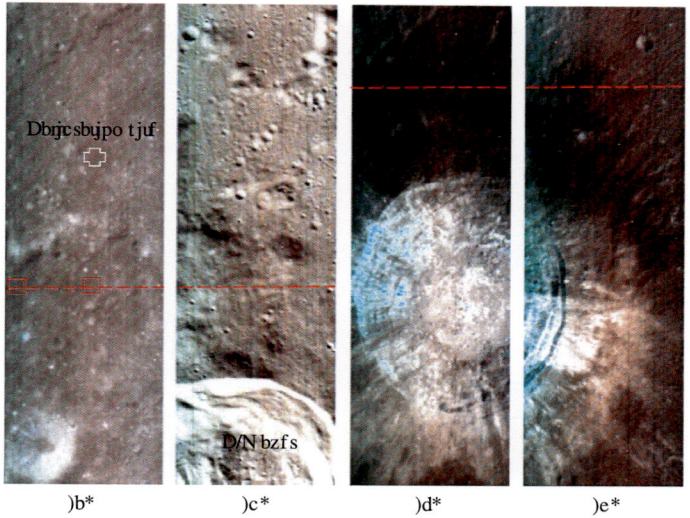

Fig.2 Composite image of 918 nm (red), 757 nm (green) and 618 nm (blue) of IIM data for calibration site (a), the crater C. Mayer (b) and Aristarchus area (c and d). For each plot the composition of the surface from left to right crossed by the dashed line is approximately the same

图版 VIII

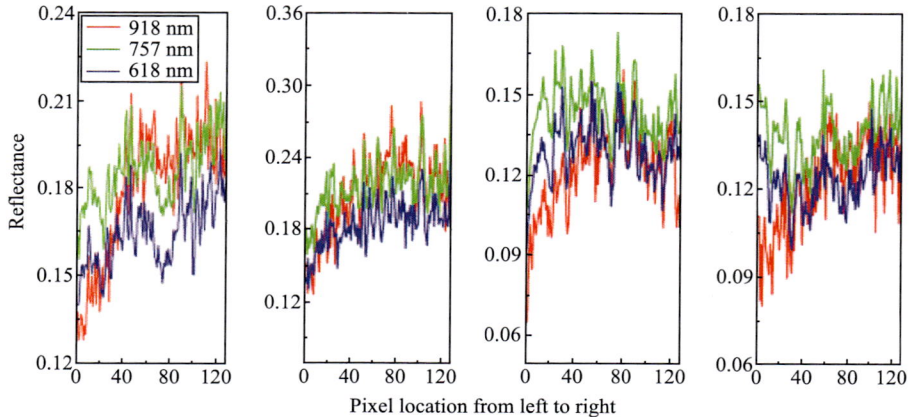

Fig.3　Reflectance profile of the dashed lines shown in Fig.2 parts A–D, for 918, 757 and 618 nm

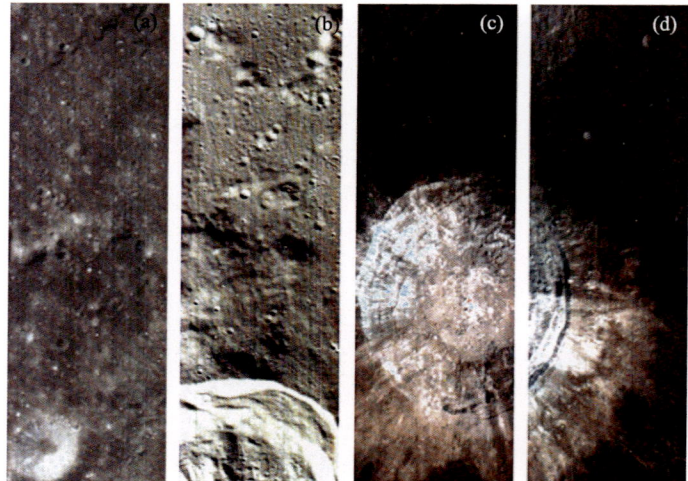

Fig.5　Composite image of the same area as Fig.2 but with the corrected data

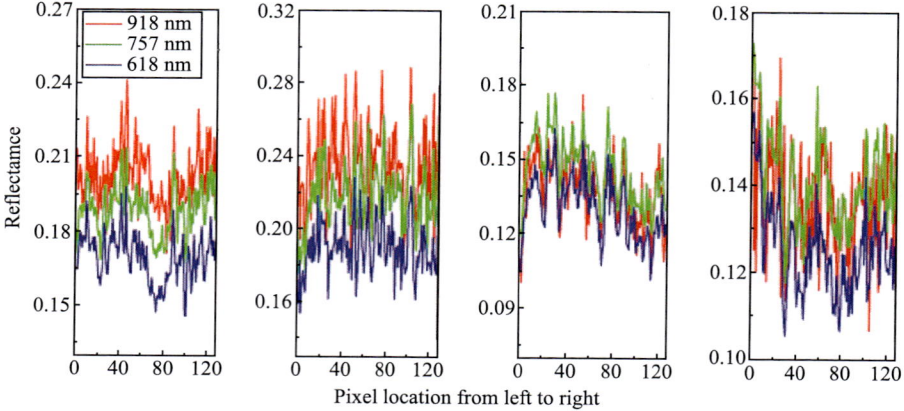

Fig.6　Reflectance profile same as Fig.3 but with the corrected data

A Preliminary Experience in the Use of Chang'E-1 IIM Data

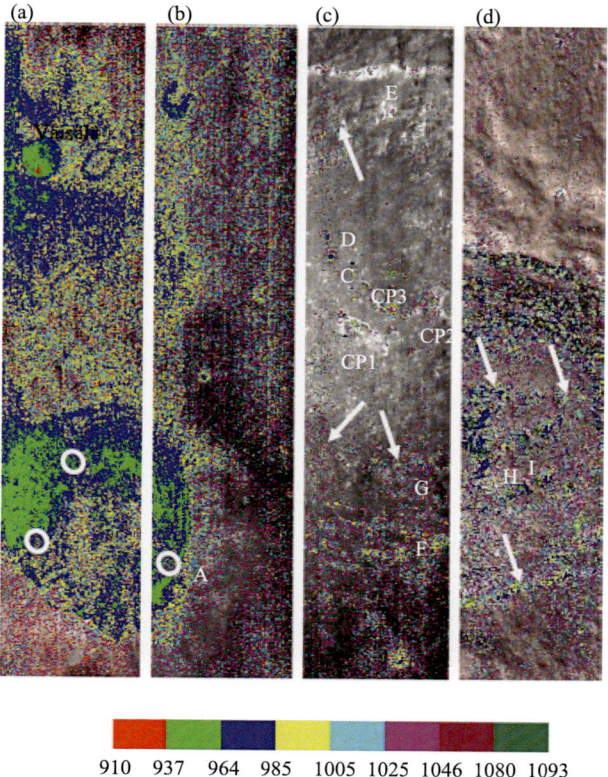

Fig.9　Map of the absorption band center for Aristarchus (a and b), Copernicus (c) and Zucchius (d). The ellipses denote the location of the telescopic spectra

Absolute Calibration of the Chang'E-1 IIM Camera and its Preliminary Application

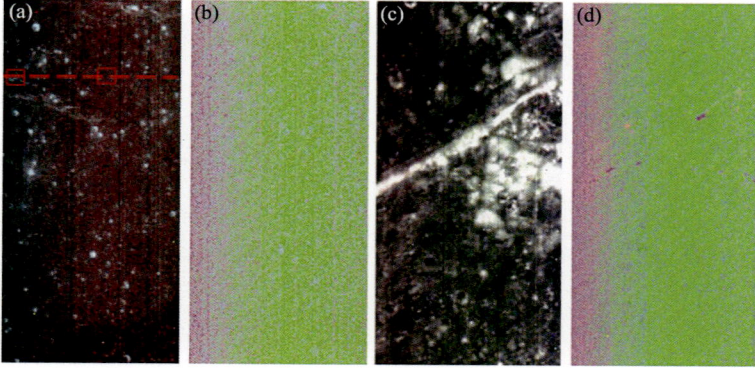

Fig. 3　True color image of IIM data for MS2 (a) and Petavius (c) area, and the ratio composite image of 891/918 nm (R), 918/891 nm (G), 757/918 nm(B) of the two areas ((b) and (d))

图版 X   Absolute Calibration of the Chang'E-1 IIM Camera and its Preliminary Application

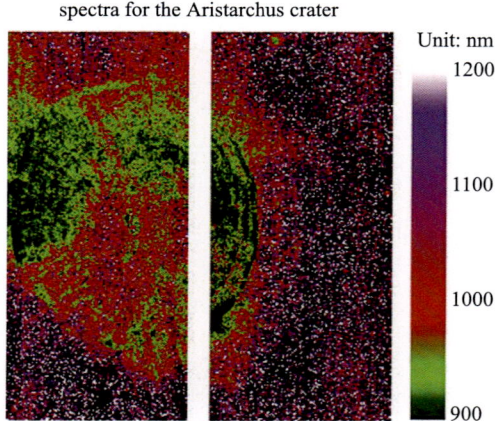

Fig. 11 Map of the absorption center of the Aristarchus crater

嫦娥一号卫星 X 射线谱仪的性能模拟

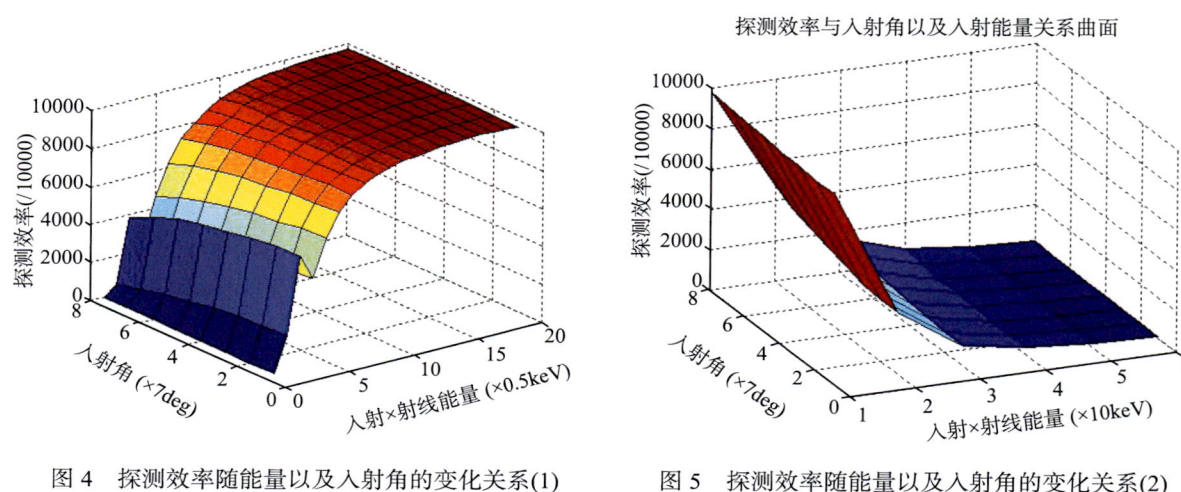

图 4 探测效率随能量以及入射角的变化关系(1)   图 5 探测效率随能量以及入射角的变化关系(2)

Time Series Data Correction for the Chang'E-1 Gamma-ray Spectrometer

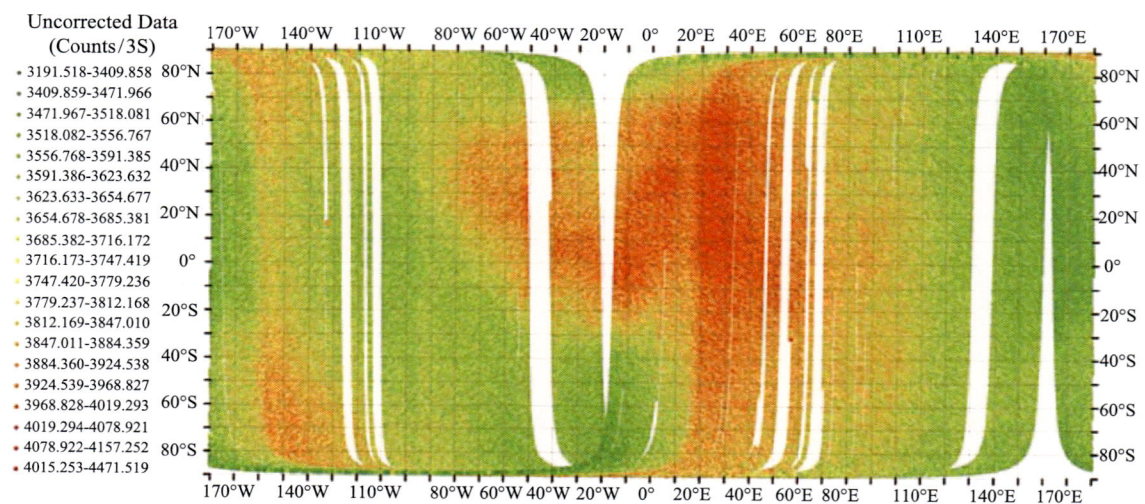

Time Series Data Correction for the Chang'E-1 Gamma-ray Spectrometer

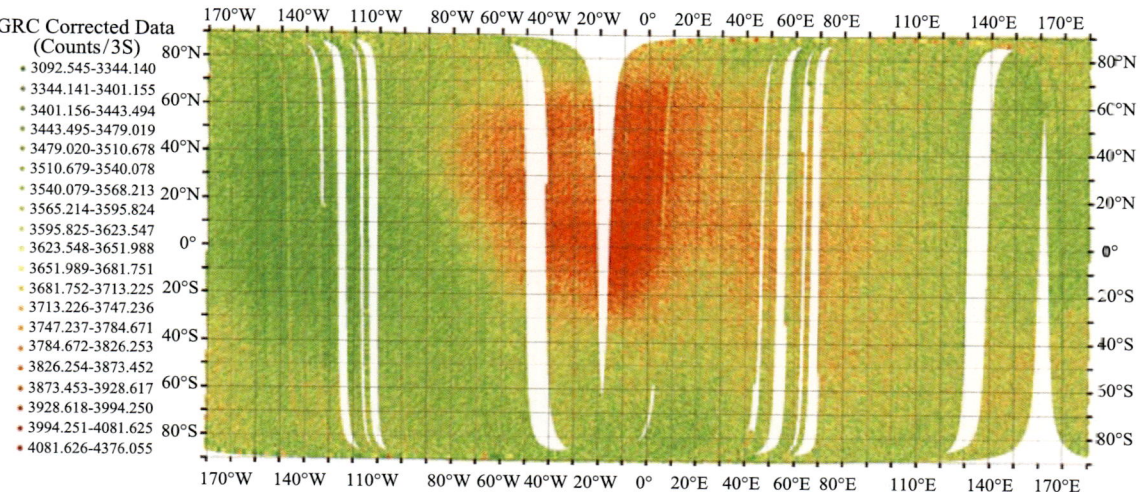

Fig.12 *Top panel*: the uncorrected count distribution; *bottom panel*: the GCR corrected count distribution

月表温度剖面对于"嫦娥一号"卫星微波探测仪探测亮温影响的模拟研究

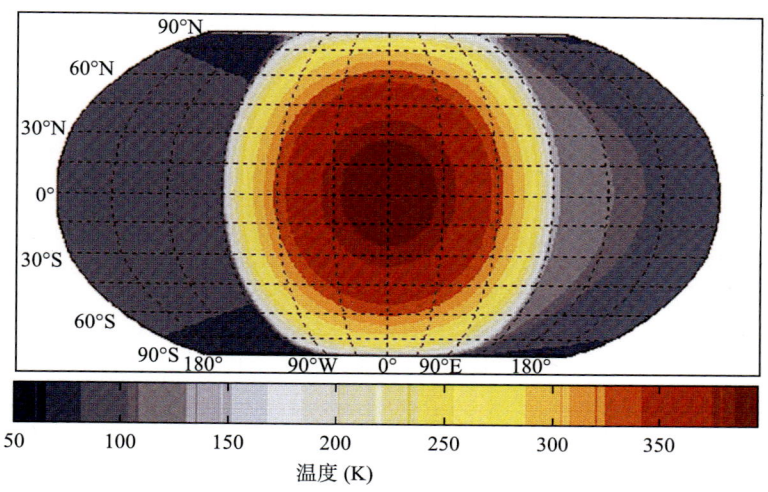

图 3　月球赤道正对地球中心点正午时刻全月的温度分布模拟结果

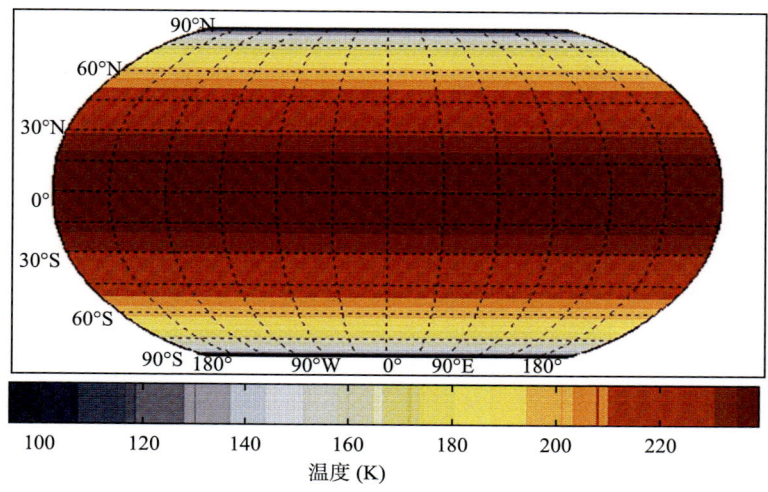

图 6　全月 6 m 深度的温度分布

图版 XII

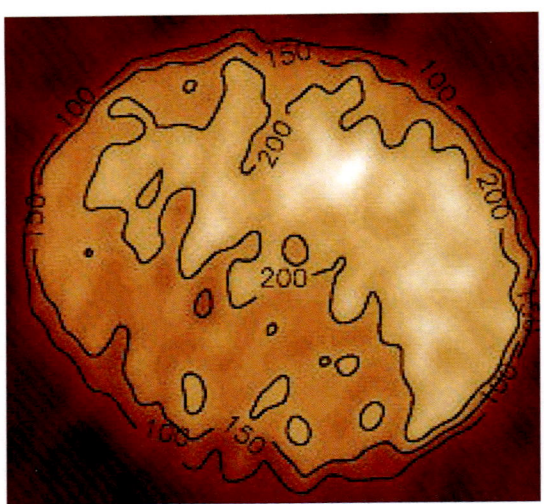

Fig.3 Microwave brightness temperature image of the Moon at 5.7 GHz

The Analysis of Affections to the Cold Space Calibration Source of Chang'E-1 Payload Microwave Detector

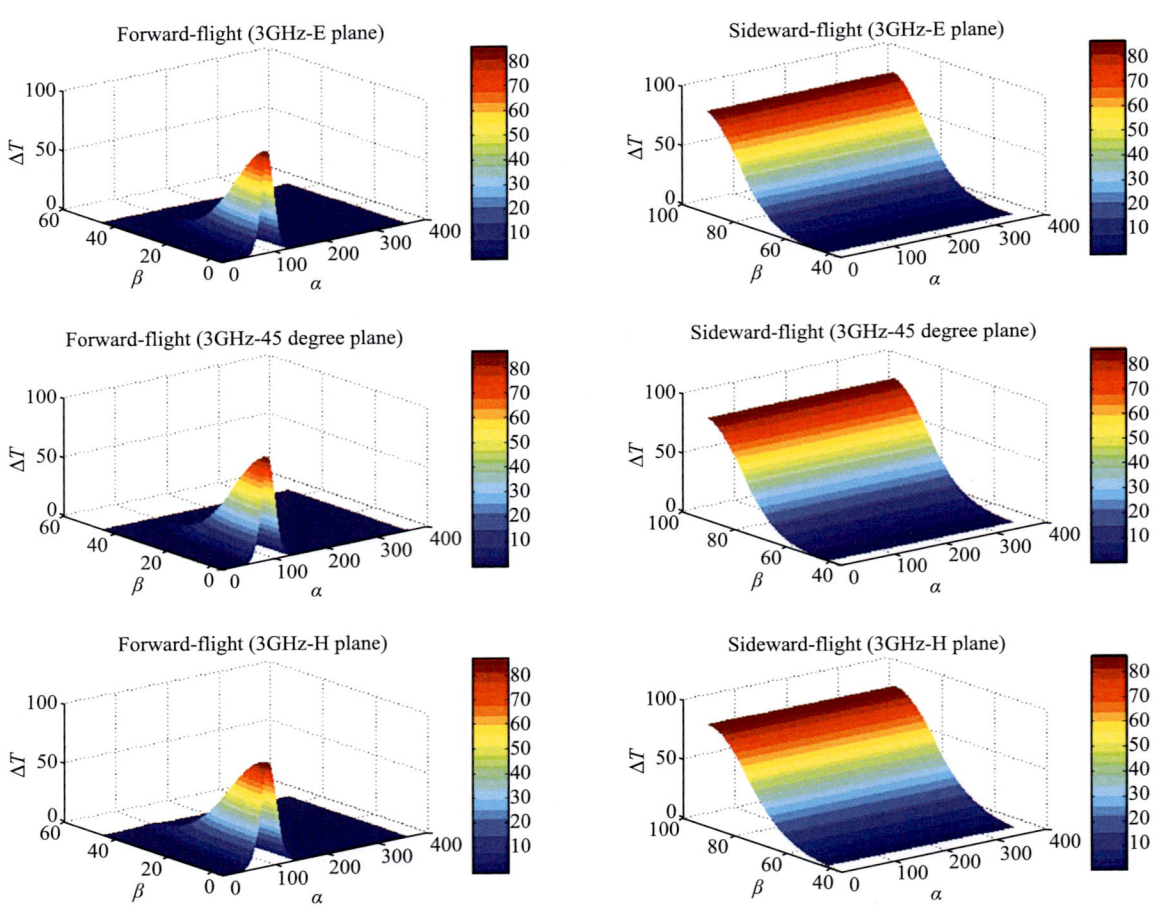

Fig.5 The relationship between α, β and ΔT in forward-flight (3GHz)

The Analysis of Affections to the Cold Space Calibration Source of Chang'E-1 Payload Microwave Detector

图版 XII

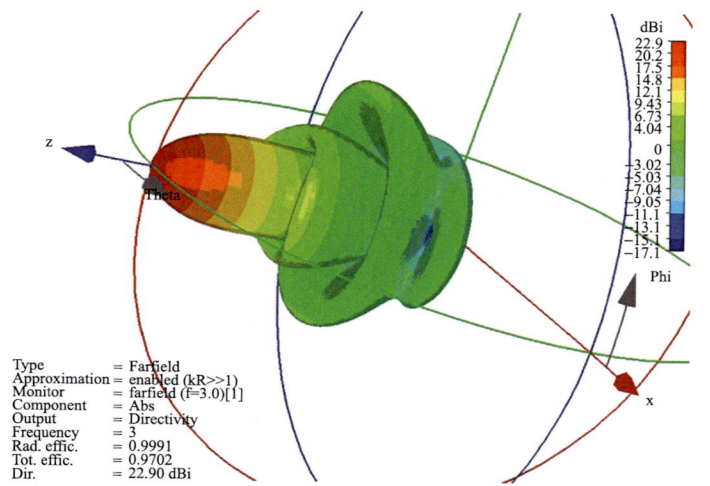

Fig.6　The 3D emulational model of channel 1 (3 GHz)'s antenna radiation pattern

嫦娥一号绕月卫星对月球重力场模型的优化

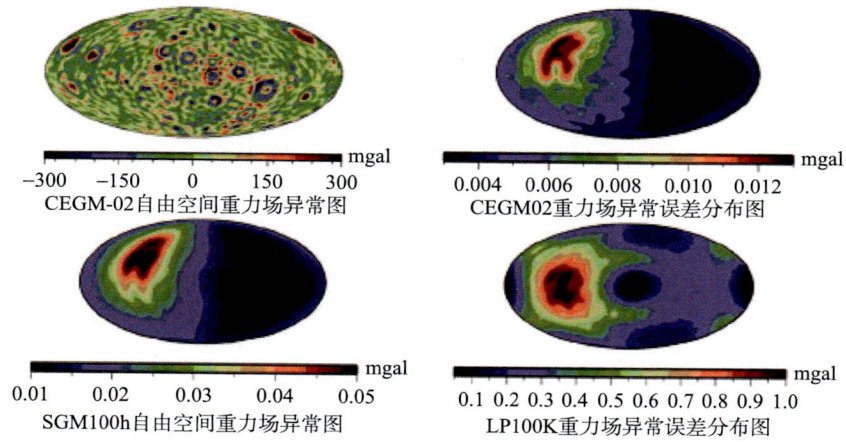

图 3　CEGM02 模型重力场异常以及各模型截断至 10 阶次的重力场异常误差分布图
重力场异常图对称中心为(180°E, 0°N)；误差图的对称中心为(270°E, 0°N)，左边为月球背面、右边为月球正面

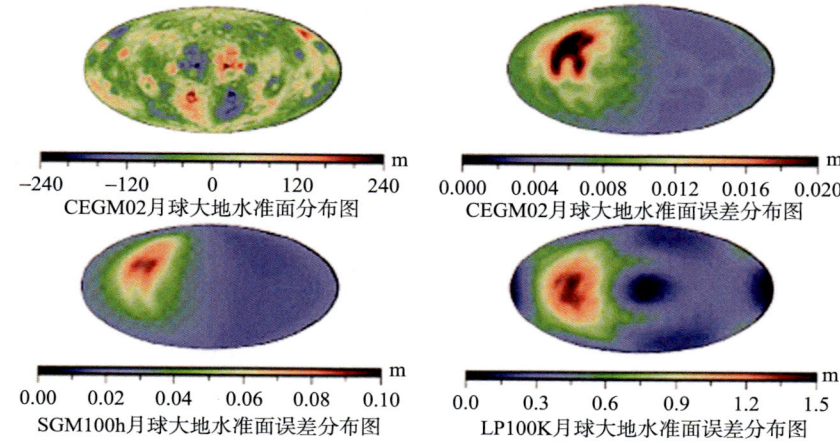

图 4　CEGM02 模型月球大地水准面以及各模型截断至 10 阶次的月球大地水准面误差分布图
月球大地水准面分布图对称中心为(180°E, 0°N)；误差图的对称中心为(270°E, 0°N)，左边为月球背面、右边为月球正面

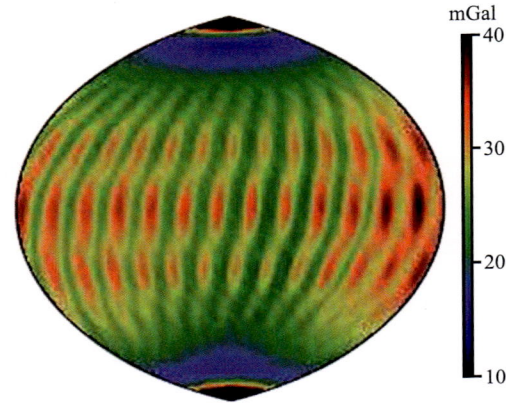

图 3　77°倾角卫星解算重力场模型对应的月球正面重力异常误差分布

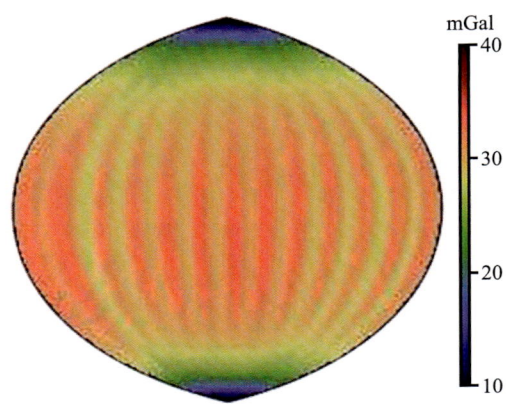

图 4　极轨卫星解算重力场模型对应的月球正面重力异常误差分布

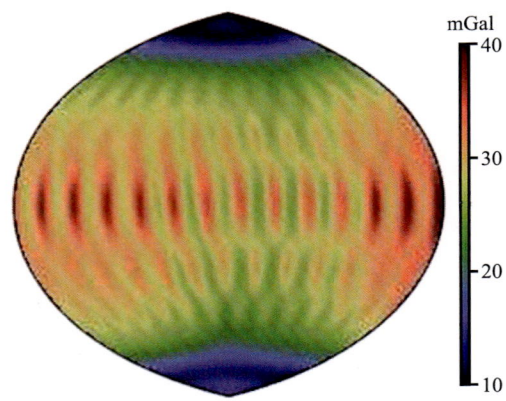

图 5　极轨卫星与77°倾角卫星联合解算重力场模型对应的月球正面重力异常误差分布

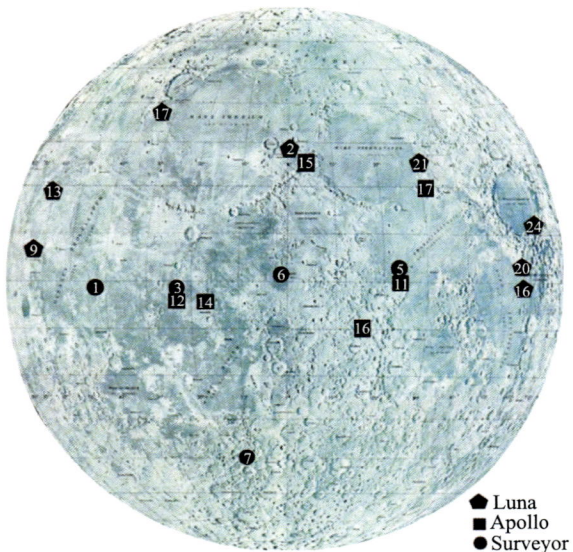

图 1　月球车登月点影像图

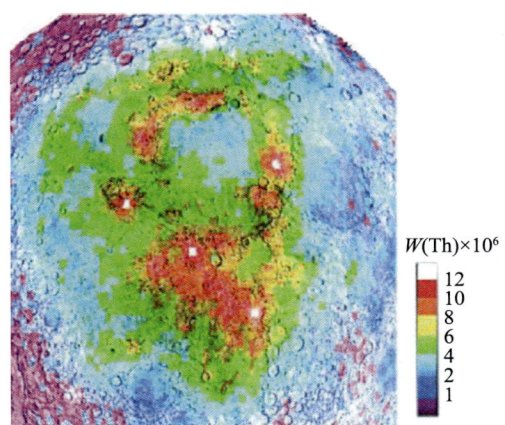

图 2　月球风暴洋 Th 的含量分布[7]

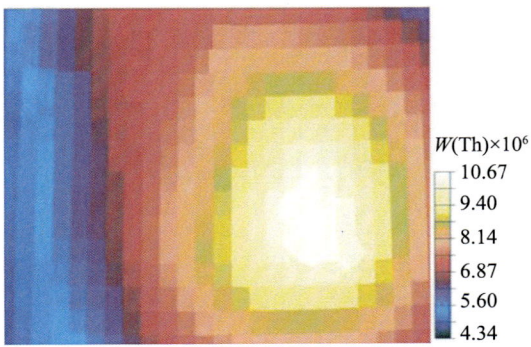

图 3　Aristrarchus 撞击坑 Th 含量的分布[14]

图版 XVI

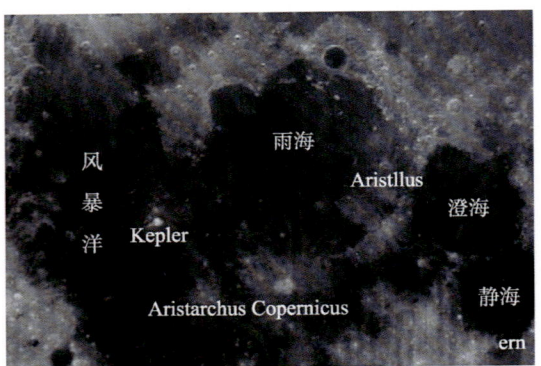

图 4　Kepler、Copernicus 及 Aristarchus 撞击坑位置

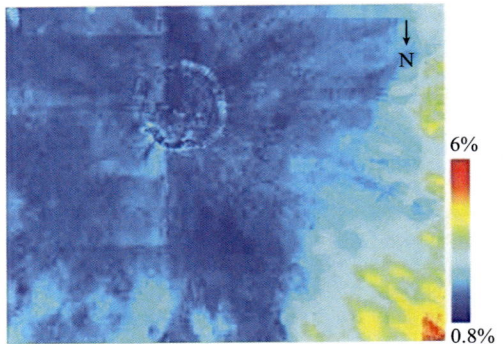

图 5　Kepler 撞击坑区 TiO$_2$ 分布

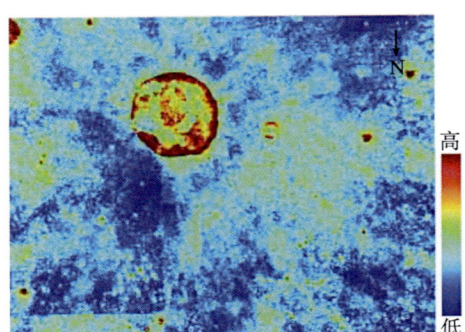

图 6　Kepler 撞击坑区 FeO 分布

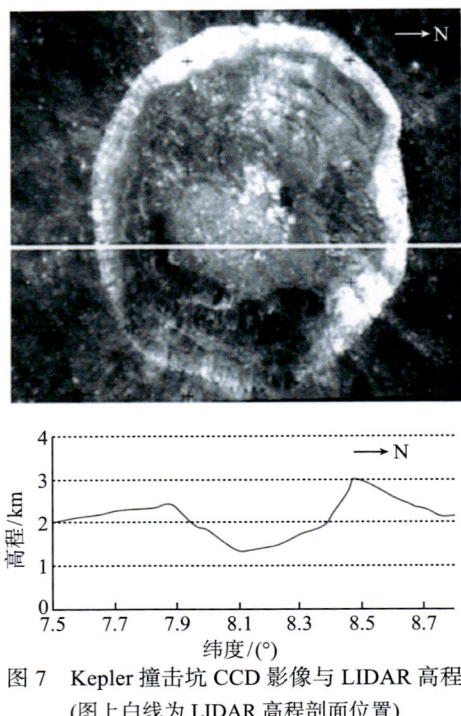

图 7　Kepler 撞击坑 CCD 影像与 LIDAR 高程
(图上白线为 LIDAR 高程剖面位置)

图版 XVII

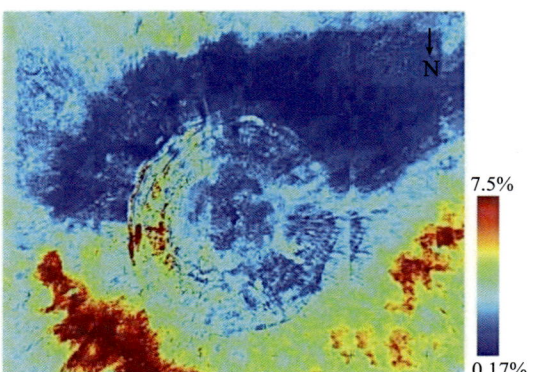

图 8 Aristarchus 撞击坑的 TiO_2 分布

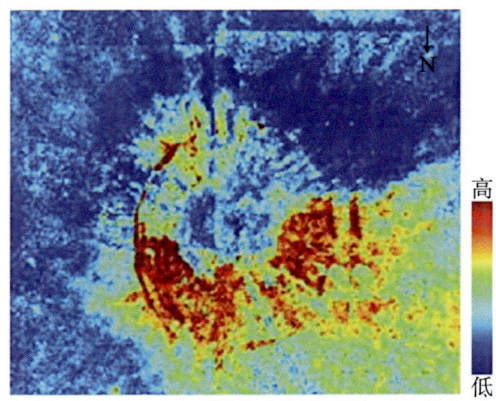

图 9 Aristarchus 撞击坑的硅酸盐 FeO 分布

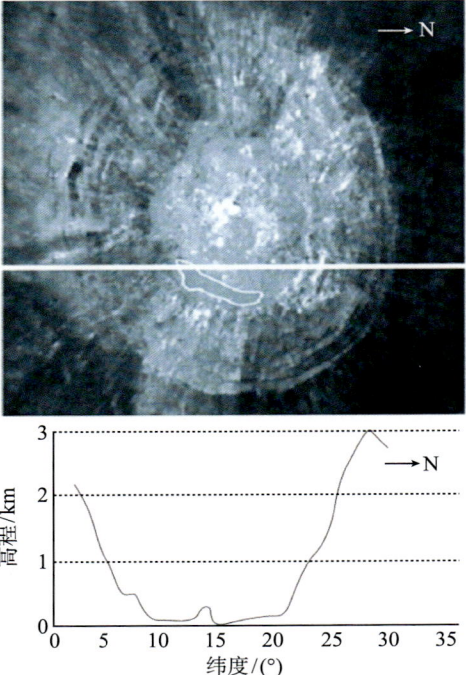

图 10 Aristarchus 撞击坑 CCD 影像与 LIDAR 高程
(白线为 LIDAR 高程剖面位置)

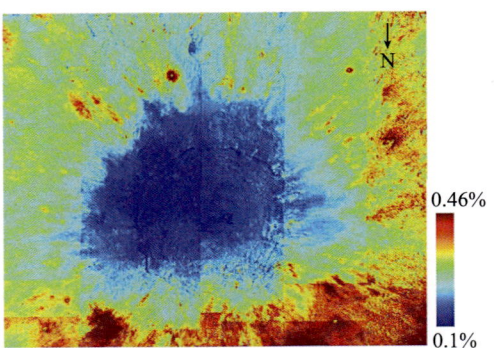

图 11 Copernicus 撞击坑的 TiO_2 分布

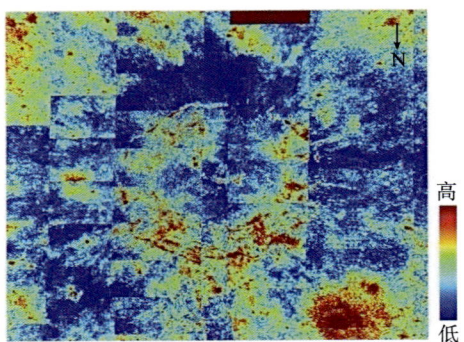

图 12 Copernicus 撞击坑的硅酸盐 FeO 分布

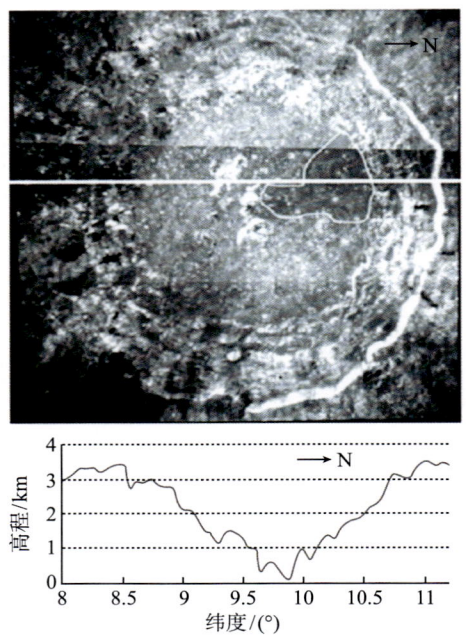

图 13 Copernicus 撞击坑 CCD 影像与 LIDAR 高程剖面
(白线为 LIDAR 高程剖面位置)